FLOWERING PLANT INDEX

OF

ILLUSTRATION AND INFORMATION

THE GARDEN CENTER

OF GREATER CLEVELAND

Compiled by

Richard T. Isaacson

Volume 1

A—J

G. K. HALL & CO., 70 LINCOLN STREET, BOSTON, MASSACHUSETTS
1979

This publication is printed on permanent/durable acid-free paper.

ISBN 0-8161-0301-1

INTRODUCTION

The *Flowering Plant Index of Illustration and Information* is an index of colored flowering plant illustrations. This 55,000 entry reference work, developed through sponsorship of The Garden Center of Greater Cleveland, gives the user quick access to illustrations and information on many flowering plants of the world. Works indexed are popular post-1935 titles that are in the collections of many libraries. To be included in the *Index,* a title's illustrations must be in color (either photograph or artist's representation), the botanical name must be indicated, and the illustrations must be of flowering plants. Each entry gives access to plant illustrations and will also lead the user to such plant information as plant use, history, cultural information, plant lore, native habitat and other data.

The *Flowering Plant Index* was designed for ease of use by professional plant researchers or for those with little formal botanical knowledge. To find indexed information on any plant it is necessary only to know the common or botanical name of the plant. Common names are cross-referenced to their botanical name. Each entry under botanical name gives citation of indexed work, indication whether flower, fruit or habit is pictured, and the page/plate number on which it is found. Fuller citations of indexed works can be found in the annotated listing of indexed works following this introduction.

Uniformity in the indication of whether flower (fl.), fruit (fr.), or general habit (hab.) is pictured has been attempted. The term habit is used to indicate any nonflowering plant part that can be useful for identification, whether bark, foliage characteristics or general growth habit. Thus habit can indicate an illustration of a foliage houseplant, the general growth habit of a conifer, or a detail of the foliage of any flowering plant.

Arrangement is alphabetical by taxa with common names interfiled. One exception to recognized filing rules is that common names are filed according to their smallest correct spelling subdivision. Thus, bellflower is filed as bell flower and bellwort as bell wort.

English common names are cross-referenced to their botanical name. As botanical names are given fully only to their species level, cultivars are indicated by the abbreviation, "var." and hybrid entries are indicated by the term "hybrid". Bigeneric hybrids are indicated by an 'x' in front of the genus, and a ' + ' is used for graft hybrids. All hybrid species as accepted by *Hortus Third* (indicated by an 'x' in front of the species) have been retained in this *Index.* Thus Cornus florida 'Fastigiata' is indexed Cornus florida var., Hemerocallis 'Moon Wind' as Hemerocallis hybrid, and Magnolia x soulangiana 'Alba' as Magnolia x soulangiana var. All genus listings have been checked against standard reference works. With very few exceptions, if the botanical name listed was once considered correct no effort has been made to update it to currently accepted nomenclature. Botanical nomenclature and spellings are accepted as being correct unless inconsistencies are found. Thus, for instance, variance in spellings have been resolved. It is estimated this has meant the correction mainly in spelling of 25% of the indexed botanical name entries.

Titles indexed are those that have been found to include acceptable illustrations. Works indexed are those published since 1935, as pre-1935 works have been indexed in the monumental *Index Londinensis.* Titles purposely excluded include such titles as *Tropica* (over 7,000 tropical plants in color) and H.D. Rickett's multivolume series on the wildflowers of the U.S. There are, also, many excellent regional U.S. wildflower guides that were ignored as they duplicate other works already indexed. Also ignored were orchid periodicals which have generally been indexed elsewhere.

Introduction

The *Flowering Plant Index* has been developed through the support of The Garden Center of Greater Cleveland, a non-profit educational institution. As this *Index* is an ongoing project, further supplements are planned. The Garden Center, the oldest civic garden center in the country, houses the Eleanor Squire Library, a horticultural library with about 10,000 volumes. The index was begun under the leadership of Howard W. Swift in the 1960's. It has been greatly assisted through the volunteer efforts of the late Ralph S. Schmitt, and from Mrs. Samuel H. Lamport and numerous other volunteers. Many libraries and librarians of the Council on Botanical and Horticultural Libraries (CBHL) generously supported this project. The compiler would like to thank Mr. C. W. Eliot Paine, Director, the staff and Executive Committee of The Garden Center and the many volunteers who assisted in this project.

RICHARD T. ISAACSON

July, 1979
Cleveland

The following is an annotated listing of the indexed
titles prepared by the compiler. Artist's representations or photographs are indicated. All works are
completely indexed with the exception of a few periodicals.
In these selectively indexed periodicals most hybrids,
illustrations of very inferior quality, and those with
unnamed plants are omitted. Provenance is indicated if
different from title or publication origin.

Addisonia, see New York Botanical Garden.

Alaska-Yukon Wild Flowers Guide. Anchorage,
Alaska Northwest Publ., 1974. Photographs.
Small paperback with numerous illustrations.

Alpine Flowers of the Kosciusko State Park.
Sydney, Australia, K.G. Murray, 1962. Photographs. 16 plates.

Alpine Garden Society. Quarterly Bulletin.
vol 1-39, 1930-71. England. Photographs &
Artist's Representations. Small sized and of
uneven quality.

American Begonia Society. Begonian. vol 39-45,
1972-78. Few color illustrations. Selectively
indexed.

American Gloxinian & Gesneriad Society. The
Gloxinian. vol 1-28, 1951-78. Mainly photo-
graphs. Illustrations increase in mid-1970's.
Selectively indexed.

American Horticultural Society.
National Horticultural Magazine. vol 1-38,
1922-59.
American Horticultural Magazine. vol 39-50,
1960-71.
American Horticulturist. vol 51-57, 1972-78.
Mainly photographs. Few illustrations in
color prior to 1960's. Selectively indexed.

American Rhododendron Society. Quarterly Bulletin.
vol 1-32, 1947-78. Photographs. Illustrations
increase in the 1960's. Selectively indexed.

Ary, S. & M. Gregory. The Oxford Book of Wildflowers. London, Oxford U. Press, 1960.
Artist's Representations. States "In the British
Isles." "550 species all illustrated in full
colour".

Backeberg, Curt. Cactus Lexicon. Dorset, England,
Blandford Press, 1977. Germany. Photographs.

Baker, H.A. & E.G.H. Oliver. Ericas in Southern
Africa. Cape Town, Purnell, 1967. Artist's
Representations. "167 species are illustrated
in full colour".

Bally, Peter R.O. The Genus Mondenium. Berne,
Benteli, 1961. Artist's Representations. Few
in color.

Barneby, T.P. European Alpine Flowers in Color.
New York, Nelson, 1967. England. Photographs.
576 plants in color. Usually 6 per page.

Bartels, Andreas. Das Grosse Buch der Gartengehoze.
Stuttgart, Eugen Ulmer, 1973. Photographs.
"120 Farbfotos".

Bartlett, Mary. Gentians. London, Blandford
Press, 1975. Photographs. "21 species in full
colour".

Batson, Wade T. Wild Flowers in South Carolina.
Columbia, S.C., U. of South Carolina Press,
1964. Photographs. "Two hundred illustrations
in color". Usually 2 per page.

Beck, Christabel. Fritillaries. London, Faber &
Faber, 1953. Artist's Representations. 4
color plates.

Bedford, Roger B. A Guide to Native Australian
Orchids. Sydney, Angus & Robertson, 1969.
Artist's Representations. 7 color plates.

Bianchini, Francesco & F. Corbetta. The Complete
Book of Fruits & Vegetables. New York, Crown,
1975. Italian. Artist's Representations "400
plants".
_____. Health Plants of the World. New York,
Newsweek Books, 1977. Italian. Artist's Representations. 82 plates with 2 or 3 plants per
plate.

Blombery, Alex. What Wildflower is That? New York,
Paul Hamlyn, 1972. Australian. Photographs.
"750 illustrations".

Blunt, Wilfrid. Flora Superba. London, Tryon
Gallery, 1971. Artist's Representations.
16 plates in large folio.

Boom, B.K. & H. Klein. The Glory of the Tree.
Garden City, N.Y., Doubleday, 1966. Netherlands.
Photographs. "195 colour illustrations".

Bramwell, David & Zoe I. Bramwell. Wild Flowers
of the Canary Islands. London, Stanley Thornes,
1974. Photographs. 205 species illustrated.

Bromeliad Society.
Bulletin. vol 1-20, 1951-70.
Journal. vol 21-28, 1971-78.
U.S. Photographs. Selectively indexed.

Brooke, Jocelyn. The Wild Orchids of Britain.
London, Bodley Head, 1950. Artist's Representations. 40 plates.

Brown, Claire A. Wildflowers of Louisiana and
Adjoining States. Baton Rouge, Louisiana
State U. Press, 1972. Photographs. Over 200
pages with usually 2 plants per page.

Bruggeman, L. Tropical Plants & Their Culture.
London, Thames & Hudson, 1957, Netherlands.
Artist's Representations. "292 illustrations
in colour".

Burgis, D.S. and J.R. Oresigo. Florida Weeds. Part One. A Supplement to Weeds of the Southern United States. Gainesville, Fla., U. of Florida, 1969. Circular 331. Photographs. Small pamphlet. 27 plates.

Cactus & Succulent Society of America. Journal. vol 1-50, 1930-78. Mainly photographs. Selectively indexed.

Cady, Leo & E.R. Rotherham. Australian Native Orchids in Colour. Rutland, Vt., Charles E. Tuttle, 1970. Australia. Photographs. 100 plates with 2 plants per plate.

Campbell, Carlos C. Great Smoky Mountain Wildflowers. Third edition. Knoxville, U. of Kentucky Press, 1962, 1970. Photographs. Paperback.

Chickering, Carol Rogers. Flowers of Guatemala. Norman, Okla., U. of Oklahoma Press, 1973. Artist's Representations. "Fifty full color portraits".

Clark, Alice M. Begonia Portraits. San Marcos, Ca., author, 1977. Artist's Representations. "41 color plates".

Clark, Lewis J. Wildflowers of British Columbia. Sidney, B.C., Gray's Publ., 1973. Photographs. 573 plates.

Cochrane, G.R., et al. Flowers & Plants of Victoria. Sydney, A.H. & A.W. Reed, 1968. Photographs. "543 colour illustrations".

Codd, L.E.W. Trees & Shrubs of the Kruger National Park. Pretoria, Government Printing/Dept. of Agra., Division of Botany & Plant Pathology, 1951. (Botany Survey Memoir No.26). Artist's Representations. Few in color.

Color Treasury of Herbs & Other Medicinal Plants. New York, Crescent, 1972. Italian. Photographs.

Courtenay, Booth & James Hill Zimmerman. Wildflower & Weeds. New York, Van Nostrand Reinhold, n.d. Photographs. "650 color".

Coyle, Jeannette & Norman C. Roberts. A Field Guide to the Common & Interesting Plants of Baja California. LaJolla, Ca., Ca. Natural History Publ., 1975. Photographs. "189 color".

Crockett, James Underwood, ed. Annuals. New York, Time-Life Books, 1971.

_____. Evergreens. New York, Time-Life Books, 1971.

_____. Lawns & Ground Covers. New York, Time-Life Books, 1971. Mostly Artist's Representations. Small sized drawings.

Curtis, Winifred. Endemic Flora of Tasmania. London, Ariel Press, 1967-78. 6 vols. Artist's Representations. Large folio.

Curtis's Botanical Magazine. London, Royal Hort. Soc. vol 159-177, 1936-1972. London, Royal Botanic Gardens, Kew. vol 178-181, 1972-77. Artist's Representations. 1 plant per plate.

Dean, Blanche E. Wildflowers of Alabama and Adjoining States. University, Ala., U. of Alabama Press, 1973. Photographs. 100 plates with 4 plants per plate.

Duncan, Wilbur H. Wildflowers of the Southeastern United States. Athens, U. of Georgia Press, 1975. Photographs. "500 color".

Dunsterville, G.C.K. Introduction to the World of Orchids. Garden City, N.Y., Doubleday, 1964.

Photographs. "full-color photographs of forty-one species of Venezuelan orchids".

Ebel, Friedrich & Otfried Birnbaum. The Strange and Beautiful World of Orchids. New York, Van Nostrand Reinhold, 1972. German. Photographs. "70 orchids".

Elbert, Virginie F. & George A. Elbert. The Miracle Houseplants. New York, Crown Publ., 1976. Photographs. 29 gesneriads in color.

Eliovson, Sima. Namaqualand in Flower. Johannesburg, Macmillan, 1972. Photographs. 56 plates.

Encke, Fritz. Zimmerpflanzen. Stuttgart, Eugen Elmer, 1952, 1975. Photographs. "104 Farbfotos".

Everett, T.H. Living Trees of the World. New York, Doubleday, n.d. Photographs. "67 in full color".

Everett, T.H., ed. New Illustrated Encyclopedia of Gardening. New York, Greystone Press, 1960. 14 vols. Photographs. Variable quality color.

Feinbrun-Dothan, Naomi. Wild Plants in the Land of Israel. Tel-Aviv, Hakibbutz Hameuchad, 1960. Artist's Representations. 52 plates in color.

Felsko, Elsa. A Book of Wildflowers. New York, Thomas Yoseloff, 1956 & 1959. 2 vols. Germany. Artist's Representations. 160 & 140 European wildflowers.

Ferguson, Mary & Richard M. Saunders. Wildflowers. New York, Van Nostrand Reinhold 1976. Canada. Photographs. "150 full color".

Flemer, William et al. Shade & Ornamental Trees in Color. New York, Grossett & Dunlap, 1965. Photographs. "100 photographs in full color".

Fleming, Glenn. Wild Flowers of Florida. Miami, Banyon Books, 1976. Photographs. "156 photographs in full color".

Flowering Plants of Africa. vol 25-44, 1945-77 (continuation of Flowering Plants of South Africa).

Flowering Plants of South Africa. Johannesburg, Speciality Press of South Africa. vol 16-24, 1936-44. (continued as Flowering Plants of Africa). Artist's Representations. Usually 1 drawing per page.

Fowlie, J.A. The Brazilian Bifoliate Cattleyas and Their Color Varieties. Pomona, Ca., author, 1977.

_____. The Genus Lycaste. Pomona, Ca., author, 1970. Artist's Representations. Folios.

Fries, Mary A. Wildflowers of Mount Ranier and the Cascades. Seattle, Washington, The Mount Ranier Natural History Assn., 1970. Photographs. Paperback.

Furrer, D. Blumen am Wege. Zurich, Silva-Bilderdienst, 1950. Artist's Representations. 60 plates.

Gault, S. Millar. The Color Dictionary of Shrubs. New York, Crown Publ., 1976. England. Photographs. "over 500 plates".

Gossner, Gabriele. Vielfalt der Rosen. Munich, author, 1976. Artist's Representations. "17 Aquarelle".

Goulimis, Constantine N. & Niki A. Goulandris. Wild Flowers of Greece. Kifissia, Greece, Goulandris Botanical Museum, 1968. Artist's Representation. "selection of 110 species". Folio.

Grove, Alfred. Supplement to Elwes' Monograph of the Genus Lilium. London, Taylor & Francis, 1933-40. Artist's Representations. 30 large folio plates.

Grubb, Roy. Selected Orchidaceous Plants. Caterham, Surrey, England, author, 1961 & 1970. 2 vols. Artist's Representations. 39 & 14 color plates.

Hall, Clarence E. Flowers of the Islands in the Sun. New York, A.S. Barnes, n.d. Artist's Representations. 32 plates.

Hall, A. Daniel. The Genus Tulipa. London, Royal Horticultural Society, 1940. Artist's Representations. 40 color plates.

Hansen, Richard & Friedrich Stahl. Baume und Straucher im Garten. Stuttgart, Eugen Ulmer, 1976. Photographs. "64 Farbfotos".

Hara, Hiroshi, comp. Photo-Album of Plants of Eastern Himalaya. Tokyo, Inoue Book, 1968. Photographs.

Hargreaves, Dorothy & Bob Hargreaves. Tropical Blossoms of the Caribbean. Portland, Oregon, Hargreaves International, n.d. Photographs. "over 100 full color". Paperback.

Harrison, Richmond E. Climbers & Trailers. Rutland, Vt., Charles E. Tuttle, 1973. Australian. Photographs. "260 colour plates".

_____. Trees & Shrubs. Rutland, Vt., Charles E. Tuttle, 1965. Australian. Photographs. "582 plates in natural colour".

Harvey, Norman. New Zealand Botanical Paintings. Christchurch, N.Z., Whitcombe & Tombs, 1969. Artist's Representations. 40 plates.

Hay, Roy. The Color Dictionary of Flowers and Plants for Home and Garden. New York, Crown Publ., 1969. England. Photographs. "2048 illustrations in full color".

_____. The Dictionary of House Plants. New York, McGraw-Hill, 1974. England. Photographs. 506 illustrations.

Heller, Christine. Wild Flowers of Alaska. Portland, Oregon, Graphic Arts Center, 1966. Photographs. 307 plates. Paperback.

Hellyer, A.G.L. Shrubs in Colour. Garden City, N.Y., Doubleday, 1966. Artist's Representations. 264 in color.

Herklots, Geoffrey. Flowering Tropical Climbers. New York, Science History Publ./Neale Watson Academic Publ., 1976. Artist's Representations.

Herre, H. The Genera of the Mesembryanthemaceae. Rotterdam, A.A. Balkema, 1973. Artist's Representations. "124 colour plates".

Hersey, Jean. The Woman's Day Book of House Plants. New York, Simon & Schuster, 1965. Artist's Representations. 107 pages. Small illustrations.

Hilliard, O.M. & B.L. Burtt. Streptocarpus: An African Plant Study. Pietermaritzburg, U. of Natal Press, 1971. Photographs. 16 plates.

Holliday, Ivan. Eucalypts. London, Robert Hale, 1973. Australia. Photographs. 31 pages.

Hu, Shiu-Ying. The Genera of Orchidaceae in Hong Kong. Hong Kong, Chinese U. Press, 1977. Mainly Photographs. 5 pages color plates.

Huxley, Anthony, ed. Deciduous Garden Trees & Shrubs. New York, Macmillan Publ., 1973. Denmark. Artist's Representations. 302 plants.

_____. Evergreen Garden Trees & Shrubs. New York, Macmillan Publ., 1973. Denmark. Artist's Representations. 200 plants.

_____. Garden Annuals and Bulbs. New York, Macmillan Publ., 1971. Denmark. Artist's Representations. 322 plants.

_____. Garden Perennials & Water Plants. New York, Macmillan Publ., 1970. Denmark. Artist's Representations. 322 plants.

Hvass, Else. Plants That Feed & Serve Us. New York, Hippocrene Books, 1973. Artist's Representations. 293 plants.

Jennings, O.E. Wildflowers of Western Pennsylvania and the Upper Ohio Basin. Pittsburgh, U. of Pittsburgh Press, 1953. 2 vols. Artist's Representations. 200 quarto plates.

Johnstone, G.H. Asiatic Magnolias in Cultivation. London, Royal Horticultural Society, 1955. Artist's Representations. 14 plates.

Jones, Paul. Flora Magnifica. London, Tryon Gallery, 1976. Artist's Representations. 16 folio plates.

Kamemoto, Haruyuki. Beautiful Thai Orchid Species. Bangkok, Orchid Soc. of Thailand, 1975. Artist's Representations. over 50 pages of color plates.

Kariyone, Tatsuo. Atlas of Medicinal Plants. Osaka, Takedu Chemical Industries, 1971. Artist's Representations. 150 plates.

Kiaer, Eigil. Indoor Plants in Color. London, Blandford Press, 1955. Denmark. Artist's Representations. 372 plants.

Kimber, Sheila. A Handbook of Orchids. Kent, England, author, 1964. Artist's Representations. 22 plates.

Kimura, Koiti. Japanese Medicinal Plants. Tokyo, Hirokawa Publ., 1958-60. 2 vols. Photographs. 240 plates.

Kitamura, Siro. Colored Illustrations of the Trees & Shrubs of Japan. Osaka, Hoikusha, 1958. Photographs. 451 plates. Small sized illustrations.

Klaber, Doretta. Violets of the United States. South Brunswick, N.J., A.S. Barnes, 1976. Artist's Representations. "108 full-color plates".

Kleijn, H. & P. Vermeulen. The Beauty of the Wild Plant. New York, Taplinger, 1966. Netherlands. Photographs. "190 colour".

Klimas, John E. & James A. Cunningham. Wildflowers of Eastern America. New York, Knopf, 1974. Photographs. "304 color".

Klute, Jeannette. Woodland Portraits. Boston, Little, Brown, 1954. Photographs. 50 plates, not all of flowering plants.

Kohlhaupt, Paula. Fleurs des Alpages. Paris, Hatier, 1966. 2 vols. Germany. Photographs. 242 plates.

Kramer, Jack. Orchids; Flowers of Romance and Mystery. New York, Harry N. Abrams, 1975. Photographs. "92 plates in full color".

Kromdijk, G. 200 House Plants in Color. New York, Herder & Herder, 1971. Netherlands. Photographs.

Kupper, Walter. Cacti. New York, Thomas Nelson, 1960. Germany. Artist's Representations. 1960 edition has corrected names.

_____. Orchids. New York, Nelson, 1961. Germany. Artist's Representations. 60 plates.

Lamb, Edgar & Brian Lamb. Colorful Cacti of the American Deserts. New York, Macmillan, 1974. England. Photographs. "140 photographs".

_____. Pocket Encyclopedia of Cacti & Succulents in Color. New York, Macmillan, 1970. England. Photographs. "326 photographs".

_____. Popular Exotic Cacti in Color. New York, Collier Books/Macmillan, 1965. England. Photographs.

_____. Stapeliads in Cultivation. London, Blandford Press, 1957. Photographs. "27 coloured".

Lemmon, Robert S. Wildflowers of North America in Full Color. Garden City, N.Y., Hanover House, 1961. Photographs. 435 photographs.

Lewis, G. Joyce & A. Amelia Obermeyer. Gladiolus; A Revision of the South African Species. Cape Town, Purnell, 1972. Artist's Representations. "32 illustrations in full colour".

Lind, E.M. Some Common Flowering Plants of Uganda. London, Oxford U. Press, 1962. Artist's Representations. 16 plates.

Lindman, C.A.M. Nordens Flora. Stockholm, Wahlstrom & Widstrand, 1964, 1974. 10 vols. Artist's Representations. 663 full page plates.

Loewenfeld, Claire & Philippa Beck. The Complete Book of Herbs & Spices. New York, Putnam, 1974. England, Artist's Representations. "56 in full color". Paperback edition does not have colored plates.

Luer, Carlyle A. The Native Orchids of Florida. New York, New York Botanical Garden, 1972. Photographs. 96 plates.

_____. The Native Orchids of the United States and Canada Excluding Florida. New York, New York Botanical Garden, 1975. Photographs. 84 plates.

Macoby, Stirling. What Flower is That? New York, Crown Publ., 1971. Australian. Photographs. 1000 plates.

Martin, W. Keble. Concise British Flora in Colour. London, Ebury Press/Michael Joseph, 1965. England. Artist's Representations. "1486 species illustrated".

Masefield, G.B. The Oxford Book of Food Plants. London, Oxford U. Press, 1969. Artist's Representations. 420 plant varieties.

Massachusetts Horticultural Society. Horticulture. vol 31-56, 1953-78. Selectively indexed.

Mathew, Brian. Dwarf Bulbs. New York, Arco Publ., 1973. England. Photographs. 49 in color.

_____. The Larger Bulbs. London, B.T. Batsford/Royal Horticultural Society, 1978. Photographs. 9 in color.

Mathias, Mildred E., ed. Color for the Landscape. Los Angeles, California Arboretum Foundation, 1973. Photographs. About 200 illustrations.

Mee, Margaret. Flowers of the Brazilian Forests. London, Tryon Gallery, 1969. Artist's Representations. 31 large folio plates.

Megaw, Elektra. Wild Flowers of Cyprus. London, Phillimore, 1973. Artist's Representations. 40 plates.

Meikle, R.D. British Trees & Shrubs. London, Eyre & Spottiswoode, 1958. Artist's Representations. 15 color.

_____. Garden Flowers. London, Eyre & Spottiswoode, 1963. Artist's Representations. 15 color.

Menninger, Edwin A. Flowering Vines of the World. New York, Hearthside Press, 1970. Photographs. 200 color. Small sized illustrations.

Miles, Bebe. Bulbs for the Home Gardener. New York, Grosset & Dunlap, 1976. Photographs & Artist's Representations. Numerous illustrations.

Milne, Lorus & Margery Milne. Living Plants of the World. New York, Random House, 1967. Photographs. "176 in color".

Misono, Isamu. Begonias. Japan, Bujka Shuppansha, 1974? Photographs. 298 illustrations. (Translation of text published by American Begonia Society in 1978.)

Morcombe, M.K. Australia's Western Wildflowers. Perth, Landfall Press, 1968. Photographs. "130 color".

Moriarty, Audrey. Wild Flowers of Malawi. Cape Town, Purnell, 1975. Artist's Representations. "360 paintings".

Morley, Brian D. Wild Flowers of the World. New York, G.P. Putnam, 1970. England. Artist's Representations. "1000 of the world's exotic & colourful plants".

Moyle, John B. & Evelyn W. Moyle. Northland Wild Flowers. Minneapolis, U. of Minnesota Press, 1977. Photographs. "308 color photographs".

Mullins, Barbara & Douglas Baglin. Australian Wildflowers in Colour. Sydney, A.H. & A.W. Reed, 1969. Photographs. 109 plates.

Munoz Pizarro, Carlos. Flores Silvestres de Chile. Santiago, Ediciones de la Universidad de Chile, 1966. Artist's Representations. 50 plates.

Munz, Philip A. California Desert Wildflowers. Berkeley, U. of California Press, 1962. Photographs. 96 color. Paperback.

_____. California Mountain Wildflowers. Berkeley, U. of California Press, 1972. Photographs. 96 color. Paperback.

_____. California Spring Wildflowers. Berkeley, U. of California Press, 1972. Photographs. 96 color. Paperback.

Neufeld, J.B. Wild Flowers of the Prairies. Saskatoon, Saskatchewan, Modern Press/Prairie Book Service, 1968. Photographs. 28 photographs. Paperback.

New York Botanical Garden. Addisonia: Colored Illustrations and Popular Descriptions of Plants. vol 19-25, 1935-64. Artist's Representations. Full page.

Nicolaisen, A. The Pocket Encyclopedia of Indoor Plants in Colour. London, Macmillan, 1969. Denmark. Artist's Representations. 150 in color.

Oakman, Harry. Colourful Trees for Landscapes and Gardens. London, Angus & Robertson, 1967. Australian. Photographs. Numerous illustrations.

O'Gorman, Helen. <u>Mexican Flowering Trees & Plants</u>. Mexico City, Ammex Associados, 1961. Artist's Representations. About 100 drawings.

Orr, Robert T. & Margaret C. Orr. <u>Wildflowers of Western America</u>. New York, Knopf, 1974. Photographs. 300 plates.

Ospina, M. <u>Orquideas de las Americas</u>. Bogota, Colombia, Litografia Arco, 1974. Photographs. 194 figures.

Pabst, G.F.J. & F. Dungs. <u>Orchidaceae Brasilienses</u>. Hildesheim, Kurt Schmerson, 1975-77. 2 vols. Artist's Representations.

Pacific Horticultural Foundation. <u>California Horticultural Journal</u>. vol 31-36, 1970-75. <u>Pacific Horticulture</u>. vol 37-39, 1976-78. Selectively indexed.

Pacific Tropical Botanical Garden. <u>Bulletin</u>. vol 1-8, 1971-78. Cover illustrations only.

Padilla, Victoria. <u>Bromeliads</u>. New York, Crown Publ., 1973. Photographs. 8 plates.

Padilla, Victoria, ed. <u>Bromeliads in Color and Their Culture</u>. Bromeliad Society, 1966. Photographs.

Palgrave, K.C. & Olive H.C. Palgrave. <u>Trees of Central Africa.</u> Salisbury, Southern Rhodesia, National Publications Trust, 1957. Artist's Representations.

Palmer, Eve. <u>Trees of South Africa</u>. Cape Town, Balkema, 1961. Artist's Representations. 30 plates.

Parham, J.W. <u>Plants of the Fiji Islands</u>. Revised edition. Suva, Government Printer, 1972. Artist's Representations. 4 plates in color.

Payne, Helen E. <u>Plant Jewels of the High Country</u>. Medford, Oregon, Pine Cone Publ., 1972. Photographs. 111 color plates.

Pennsylvania Horticultural Society. <u>The Green Scene</u>. vol 1-7, 1973-78. Selectively indexed.

Perrot, Emile & Rene Paris. <u>Les Plantes Medicinale</u>. Paris, Presses Universitaires, 1971. 2 vols. Artist's Representations. Numerous plates.

Perry, Frances & Leslie Greenwood. <u>Flowers of the World</u>. New York, Crown Publ., 1972. England. Artist's Representations. "828 illustrations in color".

Pertchik, Bernard & Harriet Pertchik. <u>Flowering Trees of the Caribbean</u>. New York, Rinehart, 1951. Artist's Representations. Full page plates.

Piers, Frank. <u>Orchids of East Africa</u>. 2nd edition. Lehre, J. Cramer, 1968. Photographs. 2 plates. Paperback.

Polunin, Oleg. <u>Flowers of Europe</u>. New York, Oxford U. Press, 1969. England. Photographs. "Over 1,000 plants are illustrated by colour".

Polunin, Oleg & Anthony Huxley. <u>Flowers of the Mediterranean</u>. London, Chatto & Windus, 1965. Photographs. 311 in color.

Polunin, Oleg & Barbara Everard. <u>Trees and Bushes of Europe.</u> New York, Oxford U. Press, 1976. England. Photographs & Artist's Representations. "Over 1,000 illustrations in colour".

Pond, Barbara. <u>Sampler of Wayside Herbs</u>. Riverside, Conn., Chatham Press, 1974. Artist's Representations. 32 color plates.

Porsild, A.E. <u>Rocky Mountain Wild Flowers</u>. Ottawa, National Museum of Natural Sciences/National Museums of Canada, 1974. Natural History Series, No. 2. Artist's Representations. 250 in color.

Rauh, Werner. <u>Bromeliads for Home, Garden and Greenhouse</u>. Dorset, England, Blandford Press, 1979. Germany. Photographs. "134 coloured plates".

Raven, John. <u>Mountain Flowers</u>. New York, Macmillan, 1956. 16 plates.

Riefel, Carlos von. <u>A Folio of Fruit</u>. London, Ariel Press, 1957. Austrian. Artist's Representations. 12 folio plates.

Robert, Paul A. <u>Alpine Flowers</u>. New York, Oxford U. Press, 1945. England. Artist's Representations. 18 plates.

Roberts, June Carver. <u>Born in the Spring</u>. Athens, Ohio, Ohio U. Press, 1976. Artist's Representations. 46 plates in color.

Rosengarten, Frederick, Jr. <u>The Book of Spices</u>. Wynnewood, Pa., Livingston Publ., 1969. Artist's Representations. "73 in color".

Roxburgh, William. <u>Icones Roxburghianae; or, Drawings of Indian Plants</u>. Calcutta, Botanical Survey of India, 1964. Artist's Representations. 24 plates.

Royal Horticultural Society. <u>Journal</u>. vol 59-100, 1935-75. <u>The Garden</u>. vol 100-103, 1975-78. Selectively indexed.

Schnell, Donald E. <u>Carnivorous Plants of the United States and Canada</u>. Winston-Salem, N.C., Blair, 1976. Photographs. Numerous illustrations.

Shaw, Richard J. <u>Field Guide to the Vascular Plants of Grand Teton National Park and Teton County, Wyoming</u>. Logan, Utah, Utah State U. Press, 1976. Photographs. 11 plates.

Smith, Lyman B. & Margaret Mee. <u>The Bromeliads</u>. South Brunswick, N.J., A.S. Barnes, 1969. Artist's Representations. 32 plates.

Stern, F.C. <u>Snowdrops and Snowflakes</u>. London, Royal Horticultural Society, 1956. Artist's Representations. 15 plates.

_____. <u>A Study of the Genus Paeonia</u>. London, Royal Horticultural Society, 1946. Artist's Representations. 15 plates.

Stewart, Joyce & Bob Campbell. <u>Orchids of Tropical Africa</u>. Cranbury, N.J., A.S. Barnes, 1970. Photographs. 45 plates.

Stodola, Jiri. <u>Encyclopedia of Water Plants</u>. Jersey City, N.J., T.F.H. Publ., 1967. Czechoslovakia. Artist's Representations. 200 figures.

Subik, Rudolf. <u>Decorative Cacti</u>. New York, Hamblyn, 1972. Czechoslovakia. Artist's Representations. 56 plates.

Szczawinski, Adam F. & Nancy J. Turner. <u>Edible Garden Weeds of Canada</u>. Ottawa, National Museums of Canada, 1978. Photographs. Paperback.

Takatori, Jisuke. <u>Medicinal Plants of Japan</u>. Tokyo, Hirokawa Publ., 1966. Artist's Representations. "80 coloured plants".

Tarver, David P., et al. <u>Aquatic and Wetland Plants of Florida</u>. Tallahassee, Florida Dept. of Natural Resources, 1978. Photographs. Pamphlet.

Taylor, A.W. <u>Wild Flowers of Spain and Portugal</u>. London, Chatto & Windus, 1972. Photographs.

Taylor, Ronald J. & George W. Douglas. <u>Mountain Wild Flowers of the Pacific Northwest</u>. Portland, Oregon, Binford & Mort, 1975. Photographs. About 150 plates.

Tosco, Uberto. <u>The World of Mountain Flowers</u>. New York, Crown Publ./Bounty Books, 1973. Italy. Photographs. 200 photographs.

Treseder, Neil G. <u>Magnolias</u>. Boston, Faber & Faber, 1978. England. Photographs. 48 color illustrations.

Tsukamoto, Yotaro. <u>Colored Illustrations of Garden Flowers</u>. Japan, Osaka Hoikusha, 1955. vol 9-10. Photographs. 236 + 217 small sized illustrations.

Turner, Nancy J. & Adam F. Szczawinski. <u>Wild Coffee and Tea Substitutes of Canada</u>. Ottawa, National Museums of Canada, 1978. Photographs. Paperback.

Turrill, W.B. <u>A Supplement to Elwes' Monograph of the Genus Lilium</u>. London, Royal Horticultural Society, 1960, 62. vol 8-9. Artist's Representations. 10 large folio plates.

Urquhart, Beryl Leslie, ed. <u>The Rhododendron</u>. London, Leslie Urquhart Press, 1958, 62. 2 vols. Artist's Representations. 36 folio plates.

van der Walt, J.J. <u>Pelargoniums of Southern Africa</u>. Cape Town, Purnell, 1977. Artist's Representations. 50 plates.

Van Laren, A. <u>Cactus</u>. Los Angeles, Abbey San Encino Press, 1935. Netherlands. Artist's Representations. 114 figures.

Vedel, H.J. Lange. <u>Arbres et Arbustes de nos Forets et de nos Jardins</u>. Paris, Fernand Nathan, n.d. Denmark. Artist's Representations. 120 plants.

Vilmorin, Roger de. <u>Plantes Alpines dans les Jardins</u>. Paris, Flammarion, 1956. Photographs. 80 illustrations.

Walcott, Mary Vaux. <u>North American Wild Flowers</u>. Washington, D.C., Smithsonian Institution, 1925-29. 5 vols. Artist's Representations. Usually in looseleaf form.

Walden, Beryl M. <u>Wild Flowers of Hong Kong</u>. Hong Kong, Chung Chi College/Arnold Arboretum, 1977. Artist's Representations. 255 illustrations.

Walther, Eric. <u>Echeveria</u>. San Francisco, California Academy of Sciences, 1972. Photographs & Artist's Representations. 16 plates.

Webster, Mary. <u>Flora of Maray, Nairn & East Inverness</u>. Aberdeen, Aberdeen U. Press, 1978. Photographs. 20 plates.

<u>Weeds of the Southern United States</u>. Athens, Ga., Cooperative Extension Service, U. of Georgia, College of Agr., n.d. Photographs. Pamphlet.

Welsh, Stanley L. & Bill Ratcliffe. <u>Flowers of the Canyon Country</u>. Provo, Utah, Brigham Young U. Press, 1971. Photographs. 50 pages of plates. Paperback.

_____. <u>Flowers of the Mountain Country</u>. Provo, Utah, Brigham Young U. Press, 1971. Photographs. 81 pages of plates.

Wendelbo, Per. <u>Tulips and Irises of Iran and Their Relatives</u>. Tehran, Botanical Institute of Iran/Ariamehr Botanical Garden, 1977. Photographs. 84 illustrations.

Weninger, Del. <u>Cacti of the Southwest</u>. Austin, U. of Texas Press, 1971. Photographs. 64 plates.

Wharton, Mary & Roger W. Barbour. <u>A Guide to the Wildflowers and Ferns of Kentucky</u>. Lexington, U. Press of Kentucky, 1971. Photographs. 500 illustrations.

_____. <u>Trees and Shrubs of Kentucky</u>. Lexington, U. Press of Kentucky, 1973. Photographs. 260 illustrations.

Williamson, Graham. <u>The Orchids of South Central Africa</u>. New York, David McKay, 1978. England. Photographs. 185 plates.

Wilson, Robert G. & Catherine Wilson. <u>Bromeliads in Cultivation</u>. Coconut Grove, Fla., Hurricane House Publ., 1963. vol 1. Photographs. 126 illustrations.

Wit, H.C.D. de. <u>Plants of the World: The Higher Plants</u>. New York, E.P. Dutton, 1968. 2 vols. Netherlands. Photographs. 187 & 175 in color.

Wright, Norman Pelham. <u>Orquideas de Mexico</u>. Mexico, Editorial Fournier, 1958. Photographs. 40 plates.

Aaron's Beard

see

Hypericum calycinum

Aaron's beard

see

Hypericum chinense

Aaron's beard

see

Saxifraga stolonifera

Aaron's Rod

see

Verbascum thapsus

Aaronsohnia faktorovskyi

Feinbrun-Dothan, Naomi
Wild Plants in the Land
of Israel

fl. hab.   p 152

Abacus Plant, Hairy Fruited

see

Glochidion eriocarpum

Abelia chinensis

Curtis's Botanical Magazine
Vol 168   1951

fl.   168

Abelia floribunda

Harrison, R. E.
Trees and Shrubs

fl.   P 17

Abelia floribunda

Royal Hort. Soc.
Journal of the Royal Hort. Soc.
    vol 92, No. 11   1967

fl.   p. 478   Pl. 266

Abelia, glossy

see

Abelia grandiflora

Abelia x grandiflora

Crockett, James Underwood
Evergreens

hab. fl.   p. 111

Abelia x grandiflora

Everett, T.H., ed.
New Illustrated Encyclopedia of
    Gardening vol. 1

fl.   p. 6

Abelia x grandiflora

Hay, Roy
The Color Dictionary of Flowers
& Plants

fl.   p. 179   Pl. 1425

Abelia x grandiflora

Hellyer, A.G.L.
Shrubs in Colour

fl.   p. 10

Abelia x grandiflora

Macoby, Stirling
What Flower is That

fl.   p. 26

Abelia x grandiflora

Massachusetts Hort. Soc.
Horticulture
    vol. 32, No. 8   1954

hab., fl.   backcover

Abelia x grandiflora

Menninger, Edwin A.
Flowering Vines of the World

fl.   Pl. 45

Abelia x grandiflora

Tsukamoto, Yotaro   Vol. 10
Coloured Illustrations of Garden
    Flowers

fl.   opp. p. 41   Pl. 130

Abelia x grandiflora var.

Harrison, R. E.
Trees and Shrubs

fl.   p. 17

Abelia, Himalayan

see

Abelia triflora

Abelia schumannii

Macoby, Stirling
What Flower is That

fl.   P 26

Abelia serrata var.

Kitamura, Siro
Coloured Illustrations of Trees and
    Shrubs of Japan

fl.   p. 237

Abelia serrata var.

Kitamura, Siro
Coloured Illustrations of Trees &
    Shrubs of Japan

fl.   444

Abelia spathulata

Kitamura, Siro
Coloured Illustrations of Trees &
    Shrubs of Japan

fl.   445

Abelia sp.

Crockett, James Underwood
Evergreens

fl.   p. 69

Abelia triflora

Morley, Brian D.
Wild Flowers of the World

fl.   Pl. 94

Abeliophyllum distichum

Bartels, Andreas
Das Grosse Buch der Gartengeholze

fl.   p 106

Abeliophyllum distichum

Curtis's Botanical Magazine
Vol 165   1948

fl.   10

Abeliophyllum distichum

Gault, S. Millar
The Color Dictionary of Shrubs

fl.   Pl. 1

Abeliophyllum distichum

Hay, Roy
The Color Dictionary of Flowers &
    Plants

fl.   p. 179   Pl. 1426

Abeliophyllum distichum

Perry, Frances
Flowers of the World

fl.   p 199

Abeliophyllum distichum

Royal Hort. Soc.
The Garden
    Vol. 101, No. 9   1976

fl.   p. 489

Abelmoschus manihot

Kimura, Koiti
Japanese Medicinal Plants, Vol. I

fl.   p. 58

Abies alba

Boom, B. K.
The glory of the tree

hab.   opp. p. 17   Pl. 19

Abies alba

Edlin, Herbert
The Illustrated Encyclopedia
   of Trees

hab.   p 102

Abies alba

Hay, Roy
The Color Dictionary of Flowers &
   Plants

fr.   p. 251   Pl. 2001

Abies alba

Huxley, Anthony
Evergreen Garden Trees and Shrubs

fr. hab.   Pl. 1, 1a

Abies alba

Polunin, Oleg
Trees and Bushes of Europe

fr. hab.   p 2

Abies alba

Royal Hort. Soc.
The Garden
   vol 101, No. 9 1976

hab.   p. 466

Abies alba

Vedel, H.
Arbres et Arbustes

fr. hab.   p. 18, 36

Abies amabilis

Curtis's Botanical Magazine
Vol 171   1956-57

           306

Abies amabilis

Edlin, Herbert
The Illustrated Encyclopedia
   of Trees

fl. hab.   p 101

Abies balsamea

Huxley, Anthony
Evergreen Garden Trees and Shrubs

fr. hab.   pl. 2,

Abies cephalonica

Edlin, Herbert
The Illustrated Encyclopedia
   of Trees

fr. hab.   p 103

Abies cephalonica

Polunin, Oleg
Flowers of Europe

fr.   Pl. 1 #1

Abies cephalonica

Polunin, Oleg
Flowers of the Mediterranean

fr.   Pl. 5

Abies cephalonica

Polunin, Oleg
Trees and Bushes of Europe

fr. hab.   p 2

Abies concolor

Boom, B. K.
The glory of the tree

hab.   opp. p. 17   pl. 18

Abies concolor

Coyle, Jeanette
A Field Guide to the Common and
   Interesting Plants of Baja
   California

fr.   p. 39

Abies concolor

Crockett, James Underwood
Evergreens

hab. fr.   p. 90

Abies concolor

Edlin, Herbert
The Illustrated Encyclopedia
   of Trees

fr.   p 101

Abies concolor

Hansen, Richard
Baume und Straucher im Garten

hab.   p. 207

Abies concolor

Huxley, Anthony
Evergreen Garden Trees and Shrubs

fr. hab.   Pl. 3, 3a

Abies concolor var.

Royal Hort. Soc.
Journal of the Royal Hort. Soc.
   vol 98, No. 10 1973

hab.   p. 442   Pl. 225

Abies delavayi var.

Royal Hort. Soc.
Journal of the Royal Hort. Soc.
   vol 98, No. 3 1973

fr.   p. 122   Pl. 64

Abies forrestii

Hay, Roy
The Color Dictionary of Flowers &
   Plants

fr.   p. 251   Pl. 2002

Abies grandis

Edlin, Herbert
The Illustrated Encyclopedia
   of Trees

fr. hab.   p 100

Abies grandis

Huxley, Anthony
Evergreen Garden Trees and Shrubs

hab.   Pl. 4

Abies grandis

Polunin, Oleg
Trees and Bushes of Europe

hab.   p 4

Abies grandis

Vedel, H.
Arbres et Arbustes

fr. hab.   p. 21, 37

Abies koreana

Boom, B. K.
The glory of the tree

fr., hab opp. p. 17   Pl. 20-21

Abies koreana

Curtis's Botanical Magazine
Vol 165   1948

fl. fr.   40

Abies koreana

Edlin, Herbert
The Illustrated Encyclopedia
   of Trees

fr. hab.   p 102-3

Abies koreana

Hansen, Richard
Baume und Straucher im Garten

hab.   p. 189

Abies koreana

Wit, H. C. D. de
Plants of the World;
   The Higher Plants, Vol I

fr.   p 54, Pl. 13

Abies lasiocarpa

Curtis's Botanical Magazine
Vol 162   1939

fl. fr.   No. 9600

Abies lasiocarpa

Porsild, A. E.
Rocky Mountain Wild Flowers

fr.   p. 31

Abies lasiocarpa

Walcott, Mary Vuax
North American Wild Flowers

fr.   Pl. 18

Abies magnifica

Everett, Thomas H.
Living Trees of the World

hab.   p. 19

Abies magnifica

Royal Hort. Soc.
Journal of the Royal Hort. Soc.
    vol 98, No. 10   1973

hab.   p. 442   Pl. 224

Abies mariesii

Curtis's Botanical Magazine
Vol 166   1949

fl. fr.   45

Abies mariesii

Kitamura, Siro
Coloured Illustrations of Trees &
    Shrubs of Japan

fr.   14

Abies nobilis

Vedel, H.
Arbres et Arbustes

fl., hab.   p. 20, 36

Abies nordmanniana

Boom, B. K.
The glory of the tree

fr.   opp. p. 35   pl. 25

Abies nordmanniana

Edlin, Herbert
The Illustrated Encyclopedia
    of Trees

fr. hab.   p 102

Abies nordmanniana

Huxley, Anthony
Evergreen Garden Trees and Shrubs

hab.   Pl. 5

Abies nordmanniana

Polunin, Oleg
Trees and Bushes of Europe

fr. hab.   p 3

Abies nordmanniana

Vedel, H.
Arbres et Arbustes

fl., fr., hab.   p. 19, 36

Abies pectinata

Perrot, Emile
Les Plantes Medicinales

fl. fr. hab.   p 143

Abies pinsapo

Edlin, Herbert
The Illustrated Encyclopedia
    of Trees

hab.   p 103

Abies pinsapo

Huxley, Anthony
Evergreen Garden Trees and Shrubs

hab.   Pl. 7

Abies pinsapo

Polunin, Oleg
Trees and Bushes of Europe

hab.   p 3

Abies pinsapo var.

Curtis's Botanical Magazine
    vol. 170   1954-55

fl. fr.   p. 242

Abies procera

Boom, B. K.
The glory of the tree

fr.   opp. p. 24   pl. 22

Abies procera

Edlin, Herbert
The Illustrated Encyclopedia of Trees

fr., hab.   p. 100

Abies procera

Polunin, Oleg
Trees and Bushes of Europe

fr.   p 4

Abies procera var.

Bartels, Andreas
Das Grosse Buch der Gartengeholze

fr.   p 244

Abies procera var.

Harrison, R. E.
Trees and Shrubs

hab.   Pl. 563

Abies procera var.

Huxley, Anthony
Evergreen Garden Trees and Shrubs

hab.   Pl. 6

Abies sachalinensis

Kitamura, Siro
Coloured Illustrations of Trees &
    Shrubs of Japan

fr.   12

Abies spectabilis

Hara, Hiroshi, comp.
Photo-Album of Plants of
    Eastern Himalaya

hab.   Pl. 1

Abies veitchii

Boom, B. K.
The glory of the tree

fr.   opp. p.24   Pl. 23

Abies veitchii

Edlin, Herbert
The Illustrated Encyclopedia
    of Trees

fr.   p 103

Abronia fragrans

Welsh, Stanley L.
Flowers of the Canyon Country

fl. hab.   p 6

Abronia latifolia

Clark, Lewis J.
Wild Flowers of British Columbia

fl. hab.   p 126

Abronia latifolia

Munz, Philip A.
California Spring Wildflowers

fl.   p 28, Pl. 5

Abronia latifolia

Orr, Robert T.
Wildflowers of Western America

fl.   Pl. 126

Abronia maritima

Coyle, Jeanette
A Field Guide to the Common and
    Interesting Palnts of Baja California

fl.   p 73

Abronia umbellata

Crockett, James Underwood
Annuals

fl.   P 90

Flowering Plant Index of Illustration and Information

Abronia umbellata
Massachusetts Hort. Soc.
Horticulture
vol 41, No. 2   1963
fl.   p. 86

Abronia umbellata
Orr, Robert T.
Wildflowers of Western America
fl.   Pl. 220

Abronia villosa
Lemmon, Robert S.
Wildflowers of North America in
Full Color
fl.   p. 67   Pl. 107

Abronia villosa
Menninger, Edwin A.
Flowering Vines of the World
fl.   Pl. 132

Abronia villosa
Orr, Robert T.
Wildflowers of Western America
fl.   Pl. 259

Abrotanella forsterioides
Curtis, Winifred
The Endemic Flora of Tasmania   Vol. 2
fl. fr.   Pl. 76

Abrotanella scapigera
Curtis, Winifred
The Endemic Flora of Tasmania Vol. 2
fl.   Pl. 75

Abrus precatorius
Burgis, D.S.
Florida Weeds
fl., fr.   p. 8

Abrus precatorius
Fleming, Glenn
Wild Flowers of Florida
fr.   p 76

Abrus precatorius
Mullins, Barbara
Australian Wildflowers in Colour
fr.   p 21, Pl. 10

Abrus precatorius
Wit, H. C. D. de
Plants of the World;
The Higher Plants, Vol I
fl. fr.   p 286, Pl. 183

Absinthe
see
Artemisia absinthium

Abutilon auritum
Blombery, Alec M.
What Wildflower is That
fl.   p. 30   Pl. 47

Abutilon avicennae
Kimura, Koiti
Japanese Medicinal Plants, Vol II
fl. fr.   p 199

Abutilon, Brazilian
see
Abutilon megapotamicum

Abutilon hybrid
Encke, Fritz
Zimmerpflanzen
fl.   p. 48

Abutilon hybrid
Kromdijk, G.
200 House Plants in Colour
fl.   pl. 1

Abutilon hybrid
Macoby, Stirling
What Flower is That
fl.   p. 26

Abutilon hybrid
Pacific Hort. Foundation
Pacific Horticulture
vol 37, No. 3   1976
fl.   p. 41

Abutilon hybrid
Tsukamoto, Yotaro
Coloured Illustrations of Garden
Flowers Vol. 9
hab opp. p. 28

Abutilon hybridum
Crockett, James Underwood
Annuals
fl.   p. 90

Abutilon hybridum
Hersey, Jdan
Woman's Day Book of House Plants
fl.   p. 63

Abutilon hybridum
Kiaer, Eigil
Indoor Plants in Colour
fl.   p. 5

Abutilon hybridum var.
Harrison, R. E.
Trees and Shrubs
fl.   p. 17

Abutilon hybridum var.
Perry, Frances
Flowers of the World
fl.   p. 184

Abutilon insigne
Morley, Brian D.
Wild Flowers of the World
fl.   Pl. 177

Abutilon integerrimum
Pertchik, Bernard
Flowering Trees of the
Caribbean
fl.   p 27

Abutilon megapotamicum
Hay, Roy
The Color Dictionary of Flowers &
Plants
fl.   p. 179   Pl. 1427

Abutilon megapotamicum
Hay, Roy
The Dictionary of House Plants
fl.   Pl. 1

Abutilon megapotamicum
Hellyer, A.G.L.
Shrubs in Colour
fl.   p. 10

Abutilon megapotamicum
Kiaer, Eigil
Indoor Plants in Colour
fl.   p. 5

Abutilon megapotamicum
Morley, Brian D.
Wild Flowers of the World
fl.   pl. 177

Abutilon megapotamicum
Nicolaisen, Age
Pocket Encyclopedia of Indoor Plants
fl.   Pl. 80

Abutilon megapotamicum

Perry, Frances
Flowers of the World

fl.   titlepage, p. 184

Abutilon megapotamicum

Royal Hort. Soc.
Journal of Royal Hort. Soc.
vol 84, No. 3 1959

fl.   p. 122   Pl. 37

Abutilon megapotamicum var.

Harrison, R.E.
Trees and Shrubs

fl.   p. 17

Abutilon, Mexican

see

Abutilon megapotamicum

Abutilon x milleri

Macoby, Stirling
What Flower is That

fl.   p. 26

Abutilon x milleri var.

Royal Hort. Soc.
Journal of the Royal Hort. Soc.
vol 99, No. 10   1974

fl., hab.   p. 442   Pl. 207

Abutilon ochsenii

Curtis's Botanical Magazine
1964-65   **vol 175**

fl   Pl. 445

Abutilon striatum

Macoby, Stirling
What Flower is That

fl.   P 27

Abutilon striatum var.

Hay, Roy
The Dictionary of House Plants

fl.   Pl. 2

Abutilon striatum var.

Kiaer, Eigil
Indoor Plants in Colour

fl.   p. 5

Abutilon striatum var.

Nicolaisen, Age.
Pocket Encyclopedia of Indoor
     Plants

fl.   Pl. 81

Abutilon x suntense

Curtis's Botanical Magazine
Vol. 180   1974

fl.   Pl. 679

Abutilon x suntense

Gault, S. Millar
The Color Dictionary of Shrubs

fl.   Pl. 2

Abutilon x suntense

Pacific Hort. Foundation
Pacific Horticulture
vol 37, No. 3   1976

fl.   p. 41

Abutilon x suntense

Royal Hort. Soc.
Journal of the Royal Hort. Soc.
vol 96, No. 6   1971

fl.   p. 258   Pl. 122

Abutilon x suntense

Royal Hort. Soc.
The Garden
vol 103, No. 12   1978

fl.   p. 496

Abutilon theophrasti

Courtenay, Booth
Wildflowers & Weeds

fl.   p 45

Abutilon theophrasti

Klimas, John E.
Wildflowers of Eastern America

fl.   Pl. 165

Abutilon virgatum

Wit, H. C. D. de
Plants of the World;
    The Higher Plants, Vol I

fl.   p 220, Pl. 133

Abutilon vitifolium

Hay, Roy
The Color Dictionary of Flowers & Plants

fl.   p 179, Pl. 1428

Abutilon vitifolium

Royal Hort. Soc.
The Garden
vol 103, No. 6   1978

fl.   p. 248

Abutilon vitifolium var.

Gault, S. Millar
The Color Dictionary of Shrubs

fl.   Pl. 3

Acacallis cyanea

Kupper, Walter
Orchids

fl.   p 39

Acacallis cyanea

Ospina, Mariano
Orquideas de las americas

fl. hab.   Pl. 97

Acacallis cyanea

Pabst, G. F. J.
Orchidaceae Brasilienses,
    Vol 2

Acacia adunca

Curtis's Botanical Magazine
    Vol 173   1960

fl.   Pl. 372

Acacia alata

Blombery, Alec M.
What Wildflower is That

fl.   p 31, Pl. 48

Acacia alata

Curtis's Botanical Magazine
    Vol 172   1958-59

fl., fr.   Pl. 311

Acacia albida

Edlin, Herbert'
The Illustrated Encyclopedia
    of Trees

hab.   p 229

Acacia albida

Palmer, Eve.
Trees of South Africa

hab.   p. 160   Pl. V

Acacia alpina

Cochrane, G. R.
Flowers and Plants of Victoria

fl.   Pl. 527

Acacia aneura

Blombery, Alec M.
What Wildflower is That

hab.   p. 26   Pl. 36

Acacia armata

Cochrane, G. R.
Flowers and Plants of Victoria

fl.   Pl. 335

Acacia armata

Hay, Roy
The Dictionary of House Plants

fl.   Pl. 3

Acacia ataxacantha

Flowering Plants of Africa
        vol 42   1972-73

fl., fr.   Pl. 1652

Acacia auriculiformis

Bruggeman, L.
Tropical Plants

fl.   Pl. 200

Acacia axillaris

Curtis, Winifred
The Endemic Flora of Tasmania
        Vol. 2

fl. fr.   Pl. 43

Acacia baileyana

Blombery, Alec M.
What Wildflower is That

fl.   p 31, Pl. 49

Acacia baileyana

Crockett, James Underwood
Evergreens

fl. hab.   p. 111

Acacia baileyana

Everett, Thomas H.
Living Trees of the World

hab., fl.   p. 177

Acacia baileyana

Flemer, William
Shade and Ornamental trees in Color

hab.   p. 81

Acacia baileyana

Harrison, R. E.
Trees and Shrubs

fl.   P 18

Acacia baileyana

Macoby, Stirling
What Flower is That

fl.   P 27

Acacia baileyana

Mathias, Mildred E.
Color for the Landscape

fl.   hab.   p 36

Acacia baileyana

Mullins, Barbara
Australian Wildflowers in Colour

fl.   p. 13   Pl. 2

Acacia baileyana

Oakman, Harry
Colorful Trees

Hab. Fl   P 13

Acacia baileyana

Royal Hort. Soc.
Journal of the Royal Hort. Soc.
        vol 93, No. 8   1968

fl.   p. 336   Pl. 168

Acacia baueri

Blombery, Alec M.
What Wildflower is That

fl.   p 31, Pl. 50

Acacia, black-wooded

see

Acacia melanoxylon

Acacia blakelyi

Morcombe, M. K.
Australia's Western Wildflowers

fl.   p 92

Acacia boormanii

Cochrane, G. R.
Flowers and Plants of Victoria

fl.   Pl. 463

Acacia boormanii

Curtis's Botanical Magazine
Vol 177   1969-70

fl.   569

Acacia botrycephala

Blombery, Alec M.
What Wildflower is That

fl.   p. 31   Pl. 51

Acacia buxifolia

Blombery, Alec M.
What Wildflower is That

fl.   p. 32   Pl. 52

Acacia caffra

Flowering Plants of Africa
        vol 40   1969-70

fl., fr.   Pl. 1586

Acacia caffra

Palmer, Eve
Trees of South Africa

fl.   p. 160   Pl. IV

Acacia, Catclaw

see

Acacia greggii

Acacia celastrifolia

Morcombe, M. K.
Australia's Western Wildflowers

fl.   p 16

Acacia confusa

Walden, Beryl M.
Wild Flowers of Hong Kong

fl. fr.   Pl. 41(108)

Acacia constricta

Lemmon, Robert S.
Wildflowers of North America in
        Full Color

fl.   pg. 72   pl. 116

Acacia cultriformis

Boom, B. K.
The glory of the tree

fl.   p.84   Pl. 137

Acacia cultriformis

Curtis's Botanical Magazine
        Vol. 172   1958-59

fl.   Pl. 322

Acacia cultriformis

Morley, Brian D.
Wild Flowers of the World

fl. fr.   Pl. 129

Acacia cultriformis

Tsukamoto, Yotaro
Coloured Illustrations of Garden
        Flowers Vol. 10

fl.   opp. p. 41   Pl. 131

Acacia cyanophylla

Huxley, Anthony
Evergreen Garden Trees and Shrubs

fl. hab.   Pl. 177

Acacia cyanophylla

Mullins, Barbara
Australian Wildflowers in Colour

fl.   p. 11   Pl. 1

Acacia cyanophylla

Nicolaisen, Age
Pocket Encyclopedia of Indoor Plants

fl.   Pl. 98

Acacia cyanophylla

Polunin, Oleg
Flowers of the Mediterranean

fl.   Pl. 50

Acacia cyanophylla

Polunin, Oleg
Trees and Bushes of Europe

fl.   p 103

Acacia cyclops

Polunin, Oleg
Trees and Bushes of Europe

fr.   p 104

Acacia dealbata

Boom, B. K.
The glory of the tree

fr.   p. 84   Pl. 136

Acacia dealbata

Edlin, Herbert
The Illustrated Encyclopedia
of Trees

fl.   p 242

Acacia dealbata

Hay, Roy
The Dictionary of House Plants

fl.   Pl. 4

Acacia dealbata

Oakman, Harry
Colorful Trees

Fl.   P 23

Acacia dealbata

Polunin, Oleg
Trees and Bushes of Europe

fl. fr. hab.   p 102

Acacia dealbata var.

Nicolaisen, Age
Pocket Encyclopedia of Indoor
Plants

fl.   Pl. 99

Acacia decurrens

Blombery, Alec M.
What Wildflower is That

fl.   p. 32   Pl. 53

Acacia decurrens

Oakman, Harry
Colorful Trees

fl.   P 24

Acacia decurrens var.

Hay, Roy
Color Dictionary of Flowers
& Plants

fl.   p 179, Pl. 1429

Acacia diffusa

Cochrane, G. R.
Flowers and Plants of Victoria

fl.   Pl. 322

Acacia dodonaeifolia

Harrison, R. E.
Trees and Shrubs

fl.   P 18

Acacia drepanolobium

Edlin, Herbert
The Illustrated Encyclopedia
of Trees

hab.   p 229

Acacia drepanolobium

Milne, Lorus
Living Plants of the World

hab.   p 108

Acacia drummondii

Blombery, Alec M.
What Wildflower is That

fl.   p 32, Pl. 54

Acacia drummondii

Morcombe, M. K.
Australia's Western Wildflowers

fl.   p 18

Acacia echinula

Mullins, Barbara
Australian Wildflowers in Colour

fl.   P. 13   Pl. 3

Acacia elata

Blombery, Alec M.
What Wildflower is That

fl.   p 33, Pl. 55

Acacia elata

Morcombe, M. K.
Australia's Western Wildflowers

fl.   p 84

Acacia elongata

Blombery, Alec M.
What Wildflower is That

fl.   p 33, Pl. 56

Acacia erioloba

Flowering Plants of Africa
vol 44   1977

fl.   Pl. 1749

Acacia, false

see

Robinia pseudoacacia

Acacia farnesiana

Brown, Clair A.
Wildflowers of Louisiana and
Adjoining States

fl. fr.   p 72

Acacia farnesiana

Bruggeman, L.
Tropical Plants

fl. fr.   Pl. 216

Acacia farnesiana

O'Gorman, Helen
Mexican Flowering Trees and Plants

fl.   p 11

Acacia farnesiana

Perrot, Emile
Les Plantes Medicinales

fl. fr. hab.   p 52

Acacia farnesiana

Polunin, Oleg
Trees and Bushes of Europe

fl. hab.   p 102

Acacia frimbriata

Oakman, Harry
Colorful Trees

fl.,fr.,hab.   p. 15

Acacia galpinii

Palgrave, K. C.
Trees of Central Africa

fl., fr.   p. 240

Acacia giraffae

Palmer, Eve.
Trees of South Africa

hab.   p. 160   Pl. VI

Acacia glandulicarpa

Mathias, Mildred E.
Color for the Landscape

fl. hab.   p 37

Acacia greggii

Coyle, Jeanette
A Field Guide to the Common and
Interesting Plants of Baja California

fl.   p 87

Acacia haemotoxylon

Palmer, Eve
Trees of South Africa

hab.   p. 160   Pl. VII

Acacia hakeoides var.

Cochrane, G. R.
Flowers and Plants of Victoria

fl.   Pl. 314

Acacia howittii

Cochrane, G. R.
Flowers and Plants of Victoria

fl.   Pl. 410

Acacia howittii

Curtis's Botanical Magazine
Vol 171   1956-57

fl.   271

Acacia juniperina

Perry, Frances
Flowers of the World

fl.   p 157

Acacia karroo

Eliovson, Sima
Namaqualand in Flower

fl. hab.   Pl. 45, 2/3

Acacia karroo

Everett, Thomas H.
Living Trees of the World

fl.   p. 177

Acacia karroo

Flowering Plants of Africa
vol 31   1956

fl.   Pl. 1220

Acacia karroo

Palgrave, K. C.
Trees of Central Africa

fl., fr.   p. 244

Acacia karroo

Polunin, Oleg
Trees and Bushes of Europe

fl. fr. hab.   p. 101, 102

Acacia lanigera

Cochrane, G. R.
Flowers and Plants of Victoria

fl. hab.   Pl. 371

Acacia ligulata

Cochrane, G. R.
Flowers and Plants of Victoria

fl.   Pl. 182

Acacia lineata

Blombery, Alec M.
What Wildflower is That

fl.   p 33, Pl. 57

Acacia longifolia

Blombery, Alec M.
What Wildflower is That

fl.   p 33, Pl. 58

Acacia longifolia

Harrison, R. E.
Trees and Shrubs

fl.   P. 18

Acacia longifolia

Hay, Roy
The Dictionary of House Plants

fl.   Pl. 5

Acacia longifolia

Macoby, Stirling
What Flower is That

fl.   P 27

Acacia longifolia

Mullins, Barbara
Australian Wildflowers in Colour

fl.   P. 13   Pl. 4

Acacia longifolia

Oakman, Harry
Colorful Trees

fl., hab.   p. 14

Acacia longifolia

Polunin, Oleg
Flowers of Europe

fl.   Pl. 49 #488

Acacia longifolia

Polunin, Oleg
Trees and Bushes of Europe

fl. fr. hab.   p 103

Acacia longifolia

Taylor, A. W.
Wild Flowers of Spain and Portugal

fl.   p. 51

Acacia macrothyrsa

Palgrave, K. C.
Trees of Central Africa

fl., fr.   p. 248

Acacia melanoxylon

Blombery, Alec M.
What Wildflower is That

fl.   p 34, Pl. 59

Acacia melanoxylon

Cochrane, G. R.
Flowers and Plants of Victoria

fl.   Pl. 240

Acacia melanoxylon

Oakman, Harry
Colorful Trees

Fl.   P 25

Acacia melanoxylon

Polunin, Oleg
Trees and Bushes of Europe

hab.   p 103

Acacia mellei

Flowering Plants of South Africa
vol XXII   1942

fl.   Pl. 860

Acacia mitchellii

Cochrane, G. R.
Flowers and Plants of Victoria

fl.   Pl. 22

Acacia, Mountain

see

Brachystegia tamarindoides

Acacia mucronata

Cochrane, G. R.
Flowers and Plants of Victoria

fl.   Pl. 406

Acacia myrtifolia

Blombery, Alec M.
What Wildflower is That

fl.   p 34, Pl. 60

Acacia nigrescens

Palgrave, K. C.
Trees of Central Africa

fl., fr.   p. 252

Acacia nigrescens

Palmer, Eve
Trees of South Africa

hab.   p. 192   Pl. VIII

Acacia nilotica var.

Flowering Plants of Africa
vol 41   1970-71

fl.   Pl. 1636

Acacia oxycedrus

Blombery, Alec M.
What Wildflower is That

fl.   p 34, Pl. 61

Acacia oxycedrus

Cochrane, G. R.
Flowers and Plants of Victoria

fl.   Pl. 53

Acacia pataczekii

Curtis, Winifred
The Endemic Flora of Tasmania
Vol VI

fl. fr.   Pl. 245

Acacia peninsularis

Coyle, Jeanette
A Field Guide to the Common and
Interesting Plants of Baja California

fl.   p 87

Acacia podalyriifolia

Blombery, Alec M.
What Wildflower is That

fl.   p 34, Pl. 62

Acacia podalyriifolia

Curtis's Botanical Magazine
Vol 163   1940 - 42

fl.   No. 9604

Acacia podalyriifolia

Macoby, Stirling
What Flower is That

fl.   P. 27

Acacia podalyriifolia

Morley, Brian D.
Wild Flowers of the World

fl.   Pl. 130

Acacia podalyriifolia

Perry, Frances
Flowers of the World

fl.   p 157

Acacia polyacantha

Palgrave, K. C.
Trees of Central Africa

fl., fr.   p. 237

Acacia pravissima

Royal Hort. Soc.
The Garden
      vol 102, No. 12   1977

fl.   p. 510

Acacia princefolia

Blombery, Alec M.
What Wildflower is That

fl.   p 35, Pl. 63

Acacia prominens

Blombery, Alec M.
What Wildflower is That

fl.   p 35, Pl. 64

Acacia pulchella

Morcombe, M. K.
Australia's Western Wildflowers

fl.   p 105

Acacia pycnantha

Blombery, Alec M.
What Wildflower is That

fl.   p 35, Pl. 65

Acacia pycnantha

Cochrane, G. R.
Flowers and Plants of Victoria

fl.   Pl. 308

Acacia pycnantha

Macoby, Stirling
What Flower is That

fl.   P 27

Acacia pycnantha

Mullins, Barbara
Australian Wildflowers in Colour

fl.   P. 15   Pl. 5

Acacia pycnantha

Polunin, Oleg
Trees and Bushes of Europe

fl. fr.   p 101, 104

Acacia raddiana var.

Edlin, Herbert
The Illustrated Encyclopedia
  of Trees

hab.   p 228

Acacia retinodes

Polunin, Oleg
Trees and Bushes of Europe

fl. fr. hab.   p 104

Acacia riceana

Curtis, Winifred
The Endemic Flora of Tasmania   Vol. 2

fl. fr.   Pl. 42

Acacia riceana

Harrison, R.E.
Trees and Shrubs

fl.   P 18

Acacia riceana

Royal Hort. Soc.
The Garden
      vol 102, No. 4   1977

hab., fr.   p. 168

Acacia rigens

Cochrane, G. R.
Flowers and Plants of Victoria

fl.   Pl. 126

Acacia robusta

Flowering Plants of South Africa
      Vol. XXII

fl. fr.   Pl. 851

Acacia robusta

Palmer, Eve
Trees of South Africa

fl.   p. 192   Pl. IX

Acacia, Rose

see

Robinia hispida

Acacia rossei

Blombery, Alec M.
What Wildflower is That

fl.   p 36, Pl. 66

Acacia saligna

Encke, Fritz
Zimmerpflanzen

fl   p. 48

Acacia senegal

Hvass, Elsie
Plants That Feed and Serve Us

fl.   p 108, Pl. 231

Acacia sieberiana var.

Flowering Plants of Africa
Vol. 42

fl.   Pl. 1653

Acacia sieberiana var.

Palgrave, K. C.
Trees of Central Africa

fl., fr.   p. 256

Acacia sieberiana var.

Palmer, Eve
Trees of South Africa

hab.   p. 192   Pl. X

Acacia sophorae

Cochrane, G. R.
Flowers and Plants of Victoria

fl.   Pl. 283

Acacia sp.

Everett, T.H., ed.
New Illustrated Encyclopedia of
    Gardening vol. 1

fl.   p. 6

Acacia spinescens

Cochrane, G. R.
Flowers and Plants of Victoria

fl.   Pl. 85

Acacia stricta

Cochrane, G. R.
Flowers and Plants of Victoria

fl.   Pl. 353

Acacia, Sweet

    see

Acacia farnesiana

Acacia tortilis

Everett, Thomas H.
Living Trees of the World

hab.   p. 191 top

Acacia tortilis

Palmer, Eve
Trees of South Africa

hab.   p. 192   Pl. XI

Acacia triptera

Cochrane, G. R.
Flowers and Plants of Victoria

fl.   Pl. 198

Acacia ulicifolia

Blombery, Alec M.
What Wildflower is That

fl.   p 36, Pl. 67

Acacia ulicifolia

Cochrane, G. R.
Flowers and Plants of Victoria

fl.   Pl. 351

Acacia verniciflua

Cochrane, G. R.
Flowers and Plants of Victoria

fl.   Pl. 328

Acacia verticillata

Cochrane, G. R.
Flowers and Plants of Victoria

fl.   Pl. 390

Acacia xanthophloea

Flowering Plants of Africa
vol 41   1970-71

fl.   Pl. 1637

Acacia xiphophylla

Morcombe, M. K.
Australia's Western Wildflowers

fl.   p 30

Acaena anserinifolia

Cochrane, G. R.
Flowers and Plants of Victoria

fl.   Pl. 361

Acaena montana

Curtis, Winifred
The Endemic Flora of Tasmania
                        vol 3
fl., fr.   p 183

Acalypha hamiltoniana

Bruggeman, L.
Tropical Plants

hab.   Pl. 279

Acalypha hispida

Hall, Clarence E.
Flowers of the Islands in the Sun

fl.   p. 39   Pl. 6

Acalypha hispida

Hargreaves, Dorothy
Tropical Blossoms of the Caribbean

fl.,hab.   p. 11

Acalypha hispida

Hay, Roy
The Color Dictionary of Flowers & Plants

fl.   p 51, Pl. 401

Acalypha hispida

Hay, Roy
The Dictionary of House Plants

fl.   Pl. 6

Acalypha hispida

Hersey, Jean
Woman's Day Book of House Plants

fl.   p 47

Acalypha hispida

Kiaer, Eigil
Indoor Plants in Colour

fl.   p. 6

Acalypha hispida

Macoby, Stirling
What Flower is That

fl.   P 28

Acalypha hispida

Perry, Frances
Flowers of the World

fl,   p 117

Acalypha rhomboidea

Courtenay, Booth
Wildflowers & Weeds

fl.   p. 67

Acalypha sanderi

Massachusetts Hort. Soc.
Horticulture
    vol 38, No. 12   1960

fl.   p. 625

Acalypha stuhlmannii

Moriarty, Audrey
Wild Flowers of Malawi

fl.   Pl. 51; 4a, 4b

Acalypha virginica

Weeds of the Southern United States

hab.   p. 21

Acalypha wilkesiana

Bruggeman, L.
Tropical Plants

fl.   Pl. 281

Acalypha wilkesiana

Hersey, Jean
Woman's Day Book of House Plants

fl.   p 49

Acalypha wilkesiana

Macoby, Stirling
What Flower is That

hab   P 28

Acalypha wilkesiana var.

Kiaer, Eigil
Indoor Plants in Colour

fl.   p. 6

Acampe multiflora

Walden, Beryl M.
Wild Flowers of Hong Kong

fl. hab.   Pl. 60 (176)

Acampe pachyglossa

Flowering Plants of Africa
    vol 30   1954-55

fl., fr., hab.   Pl. 1175

Acampe pachyglossa

Williamson, Graham
The Orchids of South Central Africa

fl.   Pl. 164

Acampe pachyglossa var.

Stewart, Joyce
Orchids of Tropical Africa

fl.   Pl. 1

Acamptopappus shockleyi

Munz, Philip A.
California Desert Wildflowers

fl.   p. 52   Pl. 76

Acanthephippium mantinianum

Kramer, Jack
Orchids; Flowers of Romance
    and Mystery

fl.   p. 153

Acanthephippium sinense

Walden, Beryl M.
Wild Flowers of Hong Kong

fl. hab.   Pl. 44

Acanthephippium vitiense

Parham, J.W.
Plants of the Fiji Islands

fl., hab.   frontispiece

Acanthocalycium veriiflorum

Backeberg, Curt
Cactus Lexicon

fl.   p 547, Pl. 3

Acanthocalycium violaceum

Lamb, Edgar
The Pocket Encyclopedia of Cacti
    and Succulents in Color

fl.   Pl. 1

Acanthocereus pentagonus

Weniger, Del
Cacti of the Southwest

fl.   P 18

Acantholimon sp.

Royal Hort. Soc.
Journal of the Royal Hort. Soc.
    vol 88, No. 5   1963

fl., hab.   p. 210   Pl. 73

Acantholimon venustum

Morley, Brian D.
Wild Flowers of the World

fl.,hab.   Pl. 48

Acantholimon venustum

Perry, Frances
Flowers of the World

fl.   p 232

Acanthopale tetrasperma

Walden, Beryl M.
Wild Flowers of Hong Kong

fl.   Pl. 73 (210)

Acanthopanax divaricatum

Kitamura, Siro
Coloured Illustrations of Trees &
    Shrubs of Japan

fl. fr.   368

Acanthopanax sciadophylloides

Kitamura, Siro
Coloured Illustrations of Trees &
    Shrubs of Japan

hab.   369

Acanthopanax sieboldianus

Kitamura, Siro
Coloured Illustrations of Trees &
    Shrubs of Japan

fl.   367

Acanthopanax trifoliatus

Kiaer, Eigil
Indoor Plants in Colour

hab.   p. 7

Acanthorhipsalis monacantha

Backeberg, Curt
Cactus Lexicon

fl.   P. 550

Acanthospermum hispidum

Weeds of the Southern United States

hab., fl.   p. 8

Acanthostachys strobilacea

Massachusetts Hort. Soc.
Horticulture
    vol 36, No. 11   1958

hab.   p. 577

Acanthostachys strobilacea

Massachusetts Hort. Soc.
Horticulture
    vol 43, No. 6   1965

hab.   p. 25

Acanthostachys strobilacea

Padilla, Victoria
Bromeliads

fl.   p. 63   Pl. 1

Acanthostachys strobilaceae

Padilla, Victoria, ed.
Bromeliads in Color and
    Their Culture

fl.   p 79

Acanthostachys strobilacea

Wilson, Robert Gardner
Bromeliads in Cultivation

fl. hab.   Pl. 2, 3

Acanthus longifolius

Royal Hort. Soc.
Journal of the Royal Hort. Soc.
    vol 98, No. 8   1973

fl.   p. 354   Pl. 184

Acanthus mollis

Macoby, Stirling
What Flower is That

fl.   P 28

Acanthus mollis

Polunin, Oleg
Flowers of Europe

fl.   Pl. 130 #1226

Acanthus montanus

Wit, H. C. D. de
Plants of the World;
   The Higher Plants, Vol II

fl.   Pl. 100, 101

Acanthus pubescens

Flowering Plants of Africa
   vol 32   1957-58

fl.   Pl. 1268

Acanthus spinosus

Hay, Roy
The Color Dictionary of Flowers &
   Plants

fl.   p. 119   Pl. 945

Acanthus spinosus

Huxley, Anthony
Garden Perennials and Water Plants

fl.   Pl. 23

Acanthus spinosus

Morley, Brian D.
Wild Flowers of the World

fl.   Pl. 21F

Acanthus spinosus

Perry, Frances
Flowers of the World

fl.   p 12

Acanthus spinosus

Polunin, Oleg
Flowers of Europe

fl.   Pl. 130 #1267

Acer aidzuense

Kitamura, Siro
Coloured Illustrations of Trees &
   Shrubs of Japan

fr.   309

Acer argutum

Royal Hort. Soc.
Journal of the Royal Hort. Soc.
   vol 92, No. 10   1967

hab.   p. 432   Pl. 237

Acer argutum

Royal Hort. Soc.
Journal of the Royal Hort. Soc.
   vol 99, No. 9   1974

hab.   p. 394   Pl. 190

Acer campbellii

Royal Hort. Soc.
The Garden
   vol 101, No. 12   1976

hab.   p. 592

Acer campestre

Edlin, Herbert
The Illustrated Encyclopedia
   of Trees

fr. hab.   p 189

Acer campestre

Hay, Roy
The Color Dictionary of Flowers & Plants

hab.   p 179, Pl. 1430

Acer campestre

Huxley, Anthony
Deciduous Garden Trees and Shrubs

fr.   Pl. 1

Acer campestre

Lindman, C. A. M.
Nordens Flora, Vol 6

fl. fr.   Pl. 381

Acer campestre

Martin, W. Keble
The Concise British Flora in Colour

fr. hab.   Pl. 20

Acer campestre

Perrot, Emile
Les Plantes Medicinales

fl. fr. hab.   p 92

Acer campestre

Polunin, Oleg
Trees and Bushes of Europe

fl. fr. hab.   p 123

Acer campestre

Vedel, H.
Arbres et Arbustes

fl. fr. hab.   p 94, Pl. 97

Acer cappadocicum

Boom, B. K.
The glory of the tree

hab.   p. 96   Pl. 160

Acer cappadocicum

Hansen, Richard
Baume und Straucher im Garten

hab.   p. 82

Acer cappadocicum

Royal Hort. Soc.
Journal of the Royal Hort. Soc.
   vol 89, No. 2   1964

hab.   p. 84   Pl. 25

Acer carolinianum

Walcott, Mary Vaux
North American Wild Flowers

fr.   Pl. 138

Acer carpinifolium

Kitamura, Siro
Coloured Illustrations of Trees &
   Shrubs of Japan

fr.   312

Acer circinatum

American Hort. Soc.
American Horticulturist
   vol 54, No. 5   1975

hab.   cover

Acer crataegifolium

Kitamura, Siro
Coloured Illustrations of Trees &
   Shrubs of Japan

fr.   313

Acer davidii

Bartels, Andreas
Das Grosse buch der Gartengeholze

hab.   p 106

Acer diabolicum

Kitamura, Siro
Coloured Illustrations of Trees &
   Shrubs of Japan

fl.   315

Acer diabolicum

Royal Hort. Soc.
Journal of the Royal Hort. Soc.
   vol 99, No. 9   1974

hab.   p. 394   Pl. 191

Acer distylum

Royal Hort. Soc.
Journal of the Royal Hort. Soc.
   vol 99, No. 9   1974

hab.   p. 394   Pl. 192

Acer drummondii

Brown, Clair A.
Wildflowers of Louisiana and
   Adjoining States

fr.   p 106

Acer forrestii

Royal Hort. Soc.
Journal of the Royal Hort. Soc.
   vol 92, No. 10   1967

hab.   p. 432   Pl. 236

Acer ginnala

Bartels, Andreas
Das Grosse Buch der Gartengeholze

fr.   p 106

Acer granatense

Polunin, Oleg
Trees and Bushes of Europe

fl. hab.   p 125

Acer grandidentatum

American Hort. Soc.
American Horticulturist
        vol 54, No. 5 1975

hab.   p. 19

Acer griseum

American Hort. Soc.
American Horticulturist
        vol 55, No. 5   1976

hab.   p. 26-27

Acer griseum

Boom, B. K.
The glory of the tree

hab.   p. 96   Pl. 162

Acer griseum

Hay, Roy
The Color Dictionary of Flowers &
        Plants

hab.   P. 179   Pl. 1431

Acer griseum

Royal Hort. Soc.
Journal of the Royal Hort. Soc.
        vol 92, No. 3 1967

hab.   p. 116   Pl. 43-44

Acer griseum

Royal Hort. Soc.
Journal of the Royal Hort. Soc.
        vol 94, No. 12   1969

hab.   p. 514   Pl. 284

Acer heldreichii

Polunin, Oleg
Trees and Bushes of Europe

fr. hab.   p 124

Acer hersii

Royal Hort. Soc.
Journal of the Royal Hort. Soc.
        vol 92, No. 10   1967

hab.   p. 432   Pl. 235

Acer hyrcanum

Polunin, Oleg
Trees and Bushes of Europe

fr. hab.   p 125

Acer japonicum

Huxley, Anthony
Deciduous Garden Trees and Shrubs

fl.   Pl. 2

Acer japonicum

Kitamura, Siro
Coloured Illustrations of Trees &
        Shrubs of Japan

fr.   306

Acer japonicum var.

Huxley, Anthony
Deciduous Garden Trees & Shrubs

fr.   Pl. 3a

Acer macrophyllum

Edlin, Herbert
The Illustrated Encyclopedia
        of Trees

fr. hab.   p 189

Acer macrophyllum

Pacific Hort. Foundation
Pacific Horticulture
        vol 37, No. 4   1976

hab.   p. 47

Acer macrophyllum

Royal Hort. Soc.
Journal of the Royal Hort. Soc.
        vol 93, No. 5   1968

hab.   p. 204   Pl. 102

Acer micranthum

Royal Hort. Soc.
Journal of the Royal Hort. Soc.
        vol. 100, No. 2   1975

hab.   p. 78   Pl. 30

Acer mono

Royal Hort. Soc.
Journal of the Royal Hort. Soc.
        vol 99, No. 9   1974

hab.   p. 394   Pl. 193

Acer mono var.

Kitamura, Siro
Coloured Illustrations of Trees & Shrubs
        of Japan

fr.   310, 311

Acer monspessulanum

Polunin, Oleg
Trees and Bushes of Europe

fl. fr. hab.   p 126

Acer negundo

Oakman, Harry
Colorful Trees

Hab. fr.   P 123

Acer negundo

Polunin, Oleg
Trees and Bushes of Europe

fl. fr. hab.   p 127

Acer negundo

Wharton, Mary E.
Trees & Shrubs of Kentucky

fl.   p 110, Pl. 4.9b

Acer negundo

Wit, H. C. D. de
Plants of the World;
        The Higher Plants, Vol II

fl.   p 58, Pl. 24

Acer negundo var.

Edlin, Herbert
The Illustrated Encyclopedia
        of Trees

fr. hab.   p 189

Acer negundo var.

Hansen, Richard
Baume und Straucher im Garten

hab.   p. 82

Acer negundo var.

Harrison, R. E.
Trees and Shrubs

hab.   Pl. 20

Acer negundo var.

Huxley, Anthony
Deciduous Garden Trees and Shrubs

hab.   Pl. 4

Acer negundo var.

Macoby, Stirling
What Flower is That

hab.   p. 28

Acer nikoense

American Hort. Soc.
American Horticulturist
        vol 56, No. 6 1977

hab.   p. 42

Acer nikoense

Curtis's Botanical Magazine
        Vol 173   1960

fl., fr.   Pl. 387

Acer nikoense

Hay, Roy
The Color Dictionary of Flowers & Plants

hab.   p 180, Pl. 1433

Acer nikoense

Kitamura, Siro
Coloured Illustrations of Trees &
   Shrubs of Japan

fr.   317

Acer obtusatum

Polunin, Oleg
Trees and Bushes of Europe

fl. fr.   p 125

Acer opalus

Polunin, Oleg
Trees and Bushes of Europe

fl. fr. hab. p 122, 125

Acer palmatum

Harrison, R. E.
Trees and Shrubs

hab.   P 19

Acer palmatum

Huxley, Anthony
Deciduous Garden Trees and Shrubs

hab.   Pl. 5

Acer palmatum

Macoby, Stirling
What Flower is That

hab P 29

Acer palmatum

Oakman, Harry
Colorful Trees

Hab.   P 126

Acer palmatum

Pacific Hort. Foundation
Pacific Horticulture
   vol 37, No. 4   1976

hab.   p. 46

Acer palmatum var.

American Hort. Soc.
American Horticulturist
   vol 52, No. 2 1973

hab.   p. 10-11, 14

Acer palmatum var.

American Hort. Soc.
American Horticulturist
   vol 53, No. 3 1974

hab.   p. 29

Acer palmatum var.

Bartels, Andreas
Das Grosse Buch der Gartengeholze

hab.   p 107

Acer palmatum var.

Edlin, Herbert
The Illustrated Encyclopedia
   of Trees

hab.   p 189

Acer palmatum var.

Everett, T.H., ed.
New Illustrated Encyclopedia of
Gardening vol. 1

hab.   p. 7

Acer palmatum var.

Flemer, William
Shade and Ornamental Trees in
   Color

hab.   p. 59

Acer palmatum var.

Gault, S. Millar
The Color Dictionary of Shrubs

hab.   Pl. 4, 5, 6

Acer palmatum var.

Hansen, Richard
Baume und Straucher im Garten

hab.   P. 117

Acer palmatum var.

Harrison, R. E.
Trees and Shrubs

hab.   Pl. 19

Acer palmatum var.

Hay, Roy
The Color Dictionary of Flowers & Plants

hab.   p 180, Pl. 1434, 1435, 1436

Acer palmatum var.

Hellyer, A.G.L.
Shrubs in Colour

hab.   p..10

Acer palmatum var.

Huxley, Anthony
Deciduous Garden Trees & Shrubs

hab.   Pl. 6

Acer palmatum var.

Kitamura, Siro
Coloured Illustrations of Trees &
   Shrubs of Japan

fl. hab.   307, 308

Acer palmatum var.

Macoby, Stirling
What Flower is That

hab.   P. 29

Acer penctinatum

Royal Hort. Soc.
The Garden
   vol 101, No. 12   1976

hab.   p. 593

Acer pensylvanicum

American Hort. Soc.
American Horticulturist
   vol 55, No. 5   1976

hab.   p. 26

Acer pensylvanicum

Bartels, Andreas
Das Grosse Buch der Gartengeholze

hab.   p 107

Acer pensylvanicum

Campbell, Carlos C.
Great Smoky Mountain Wildflowers

fl.   p 23

Acer pensylvanicum

Edlin, Herbert
The Illustrated Encyclopedia
   of Trees

fr. hab.   p 189

Acer pensylvanicum

Hay, Roy
The Color Dictionary of Flowers & Plants

hab.   p 180, Pl. 1437

Acer pensylvanicum

Royal Hort. Soc.
Journal of the Royal Hort. Soc.
   Vol. 94, No. 12   1969

hab.   p. 514   Pl. 286

Acer platanoides

Boom, B. K.
The glory of the tree

fl.   p. 96   Pl. 163

Acer platanoides

Edlin, Herbert
The Illustrated Encyclopedia
   of Trees

fl. fr. hab.   p 189, 191

Acer platanoides

Flemer, William III
Shade & Ornamental Trees in Color

hab. p. 60

Acer platanoides

Hay, Roy
The Color Dictionary of Flowers
   & Plants

fl.   p. 180   Pl. 1438

Acer platanoides

Huxley, Anthony
Deciduous Garden Trees and Shrubs

fr.   Pl. 7a

Acer platanoides

Hvass, Elsie
Plants That Feed and Serve Us

fr.   p 124, Pl. 272

Acer platanoides

Lindman, C. A. M.
Nordens Flora, Vol 6

fl. fr.   Pl. 380

Acer platanoides

Morley, Brian D.
Wild Flowers of the World

fl. fr.   Pl. 19B

Acer platanoides

Perrot, Emile
Les Plantes Medicinales

fl. fr. hab.   p 92

Acer platanoides

Polunin, Oleg
Flowers of Europe

Acer platanoides

Polunin, Oleg
Trees and Bushes of Europe

fl. fr. hab.   p 122, 123

Acer platanoides

Vedel, H.
Arbres et Arbustes

fl. fr. hab.   p 92, Pl. 95

Acer platanoides var.

Boom, B.K.
The Glory of the Tree

fl.   p. 93   Pl. 158

Acer platanoides var.

Flemer, William III
Shade & Ornamental Trees in Color

hab.   p. 60, 61

Acer pseudo-platanus

Bartels, Andreas
Das Grosse Buch der Gartengeholze

fl.   p 106

Acer pseudoplatanus

Boom, B. K.
The glory of the tree

fl.   p. 96   Pl. 164

Acer pseudoplatanus

Edlin, Herbert
The Illustrated Encyclopedia
   of Trees

hab.   p 188, 191

Acer pseudoplatanus

Huxley, Anthony
Deciduous Garden Trees and Shrubs

fr.   Pl. 8a

Acer pseudoplatanus

Morley, Brian D.
Wild Flowers of the World

fl. fr.   Pl. 19C

Acer pseudoplatanus

Perrot, Emile
Les Plantes Medicinales

fl. fr. hab.   p. 92

Acer pseudoplatanus

Polunin, Oleg
Flowers of Europe

fl.   Pl. 69 #708

Acer pseudoplatanus

Polunin, Oleg
Trees and Bushes of Europe

fl. fr. hab.   p 124

Acer pseudoplatanus

Vedel, H.
Arbres et Arbustes

fl. fr. hab.   p 93, Pl. 96

Acer pseudoplatanus var.

Harrison, R.E.
Trees and Shrubs

hab.   pl. 20

Acer pseudoplatanus var.

Hay, Roy
The Color Dictionary of Flowers & Plants

hab.   p 180, Pl. 1439

Acer pycnanthum

Kitamura, Siro
Coloured Illustrations of Trees &
   Shrubs of Japan

fl. hab.   316

Acer rubrum

American Hort. Soc.
American Horticulturist
   vol 54, No. 5 1975

hab.   p. 18

Acer rubrum

American Hort. Soc.
American Horticulturist
   vol 55, No. 5   1976

hab.   p. 26

Acer rubrum

American Hort. Soc.
American Horticulturist
   vol 55, No. 6 1976

fr.   p. 30

Acer rubrum

Boom, B. K.
The glory of the tree

hab.   p. 93   Pl. 156

Acer rubrum

Edlin, Herbert
The Illustrated Encyclopedia
   of Trees

fr. hab.   p 189

Acer rubrum

Flemer, William III
Shade & Ornamental Trees in Color

hab.   p. 62

Acer rubrum

Hansen, Richard
Baume und Straucher im Garten

hab.   p. 82

Acer rubrum

Massachusetts Hort. Soc.
Horticulture
   vol 44, No. 10 1966

hab.   cover, backcover

Acer rubrum

Oakman, Harry
Colorful Trees

hab   p. 124

Acer rubrum

Walcott, Mary Vaux
North American Wild Flowers

fl.   Pl. 137

Acer rubrum

Wharton, Mary E.
Trees & Shrubs of Kentucky

fl.   p 98, Pl. 2.34
fr.   p 117, Pl. 1.5
hab.   p 27

Acer rubrum

Wit, H. C. D. de
Plants of the World;
The Higher Plants, Vol. II

hab.   p. 55   Pl. 18

Acer rubrum var.

Hay, Roy
The Color Dictionary of Flowers & Plants

hab.   p 180, Pl. 1440

Acer rufinerve

Kitamura, Siro
Coloured Illustrations of Trees &
Shrubs of Japan

fr.   314

Acer saccharinum

Boom, B. K.
The glory of the tree

hab.   p. 96   Pl. 159

Acer saccharinum

Edlin, Herbert
The Illustrated Encyclopedia of Trees

hab., fr.   p. 189

Acer saccharinum

Everett, T.H., ed.
New Illustrated Encyclopedia of
Gardening vol. 1

hab.   p. 7

Acer saccharinum

Flemer, William III
Shade & Ornamental Trees in Color

hab.   p. 62

Acer saccharinum

Oakman, Harry
Colorful Trees

hab P 125

Acer saccharum

Edlin, Herbert
The Illustrated Encyclopedia
of Trees

hab.   p. 188, 191

Acer saccharum

Everett, Thomas H.
Living Trees of the World

hab.   p. 192

Acer saccharum

Everett, T.H., ed.
New Illustrated Encyclopedia of
Gardening vol. 12

hab.   opp. flyleaf

Acer saccharum

Flemer, William III
Shade & Ornamental Trees in Color

hab.   p. 63

Acer saccharum

Hvass, Elsie
Plants That Feed and Serve Us

fl. fr.   p. 14   Pl. 20

Acer Saccharum

Masefield, G. B.
The Oxford Book of Food Plants

fr.   Pl. 17, 4

Acer saccharum

Milne, Lorus
Living Plants of the World

hab.   p 128

Acer saccharum

Wharton, Mary E.
Trees & Shrubs of Kentucky

fl.   p 110, Pl. 4.9a
hab.   p 25

Acer sempervirens

Polunin, Oleg
Trees and Bushes of Europe

fr. hab.   p 126

Acer sikkimense

Royal Hort. Soc.
The Garden
vol 101, No. 12   1976

hab.   p. 592

Acer spicatum

Campbell, Carlos C.
Great Smoky Mountain Wildflowers

fl.   p 23

Acer spicatum

Wharton, Mary E.
Trees & Shrubs of Kentucky

fl.   p 74, Pl. 2.11

Acer tataricum

Polunin, Oleg
Trees and Bushes of Europe

fr. hab.   p 124

Acer tetramerum

Royal Hort. Soc.
Journal of the Royal Hort. Soc.
vol 92, No. 10   1967

hab.   p. 432   Pl. 238

Acer tonkinense

Royal Hort. Soc.
The Garden
vol 100, No. 10   1975

hab.   p. 497

Aceras anthropophorum

Ary, S.
The Oxford Book of Wildflowers

fl.   p 44, Pl. 2

Aceras anthropophorum

Barneby, T. P.
European Alpine Flowers in Colour

fl. hab.   Pl. 10, 1

Aceras anthropophorum

Brooke, Jocelyn
The Wild Orchids of Britain

fl.   Pl. 20

Aceras anthropophorum

Goulimis, Constantine N.
Wild Flowers of Greece

fl. hab.   p. 189

Aceras anthropophorum

Morley, Brian D.
Wild Flowers of the World

fl.   Pl. 26

Aceras anthropophorum

Perrot, Emile
Les Plantes Medicinales

fl.   p. 165

Aceras anthropophorum

Polunin, Oleg
Flowers of Europe

fl.   Pl. 189 #1905

Aceras anthropophorum

Polunin, Oleg
Flowers of the Mediterranean

fl.   Pl. 306

Achania malvaviscus

Massachusetts Hort. Soc.
Horticulture
vol 55, No. 1   1977

fl.   p. 36

Achillea atrata

Barneby, T. P.
European Alpine Flowers in Colour

fl. hab.   Pl. 88, 3

Achillea atrata

Kleijn, H.
Beauty of the Wild Plant

fl.   p. 96   Pl. 134

Achillea atrata

Kohlhaupt, Paula
Fleurs des Alpages Vol 1

fl.   P 110

Achillea borealis

Alaska-Yukon Wild Flowers Guide

fl. hab.   p 50

Achillea borealis

Heller, Christine
Wild Flowers of Alaska

fl.   Pl. 210

Achillea borealis var.

Munz, Philip A.
California Spring Wildflowers

fl.   p. 86   Pl. 84

Achillea clavenae

Barneby, T. P.
European Alpine Flowers in Colour

fl. hab.   Pl. 88, 5

Achillea clavenae

Kohlhaupt, Paula
Fleurs des Alpages Vol 2

Fl.   P 111

Achillea clypeolata

Hay, Roy
Color Dictionary of Flowers &
Plants

fl.   p. 119   Pl. 946

Achillea condensata

Bruggeman, L.
Tropical Plants

fl.   Pl. 64

Achillea filipendulina

Macoby, Stirling
What Flower is that

fl.   P 29

Achillea filipendulina

Massachusetts Hort. Soc.
Horticulture
    vol 44, No. 7   1966

fl.   p. 24-25

Achillea filipendulina

Pennsylvania Hort. Soc.
The Green Scene
    vol 6, No. 6   1978

fl.   backcover

Achillea filipendulina var.

Hay, Roy
Color Dictionary of Flowers & Plants

fl.   p. 119

Achillea filipendulina var.

Huxley, Anthony
Garden Perennials & Water Plants

fl.   Pl. 20

Achillea filipendulina var.

Perry, Frances
Flowers of the World

fl.   p. 85

Achillea hybrid

Hay, Roy
Color Dictionary of Flowers &
Plants

fl.   p. 119   Pl. 948-50

Achillea lanulosa

Walcott, Mary Vaux
North American Wild Flowers

fl. hab.   Pl. 151

Achillea macrophylla

Barneby, T. P.
European Alpine Flowers in Colour

fl.   Pl. 88, 6

Achillea millefolium

American Hort. Soc.
American Horticulturist
    vol 55, No. 4   1976

fl.   p. 6

Achillea millefolium

Ary, S.
The Oxford Book of Wildflowers

fl.   p 98, Pl. 1

Achillea millefolium

Bianchini, Francesco
Health Plants of the World

fl. hab.   p 168

Achillea millefolium

Brown, Clair A.
Wildflowers of Louisiana and
    Adjoining States

fl. hab.   p 183

Achillea millefolium

Bruggeman, L.
Tropical Plants

fl.   Pl. 65

Achillea millefolium

Campbell, Carlos C.
Great Smoky Mountain Wildflowers

fl.   p 89

Achillea millefolium

Clark, Lewis J.
Wild Flowers of British Columbia

fl.   p 510

Achillea millefolium

Color Treasury of Herbs & Other
    Medicinal Plants

fl.   p 59   Pl. 95

Achillea millefolium

Courtenay, Booth
Wildflowers & Weeds

fl.   p 110

Achillea millefolium

Crockett, James Underwood
Lawns & Ground Covers

fl   p. 73

Achillea millefolium

Dean, Blanche E.
Wildflowers of Alabama and
    Adjoining States

fl.   p. 201

Achillea millefolium

Felsko, Elsa
A Book of Wildflowers

fl.   P 114

Achillea millefolium

Furrer, D.
Blumen am Wege

fl. hab.   p 109

Achillea millefolium

Klimas, John E.
Wildflowers of Eastern America

fl.   Pl. 44

Achillea millefolium

Lemmon, Robert S.
Wildflowers of North America in
    Full Color

fl.   pg. 212   pl. 336

Achillea millefolium

Lindman, C. A. M.
Nordens Flora, Vol 9

fl.   Pl. 601

Achillea millefolium

Martin, W. Keble
The Concise British Flora in Colour

fl. hab.   Pl. 45

Achillea millefolium

Moyle, John B.
Northland Wild Flowers

fl. hab.   p 164, Pl. 197

Achillea millefolium

Perrot, Emile
Les Plantes Medicinales

fl. fr. hab.   p 150

Achillea millefolium

Pond, Barbara
A Sampler of Wayside Herbs

fl.   Pl. XXII

Achillea millefolium

Taylor, Ronald J.
Mountain Wild Flowers

fl. hab.   p 152

Achillea millefolium

Tosco, Uberto
The World of Mountain Flowers

fl.   p 20

Achillea millefolium

Weeds of the Southern United States

fl.   p. 8

Achillea millefolium

Welsh, Stanley L.
Flowers of the Mountain Country

fl.   p. 3

Achillea millefolium

Wharton, Mary E.
A Guide to the Wildflowers & Ferns
   of Kentucky

fl.   p 262   Pl. 3.39

Achillea millefolium var.

Hay, Roy
The Color Dictionary of Flowers &
   Plants

fl.   p. 119   Pl. 949

Achillea millefolium var.

Huxley, Anthony
Garden Perennials & Water Plants

fl.   Pl. 21

Achillea millefolium var.

Massachusetts Hort. Soc.
Horticulture
   vol 41, No. 6 1963

fl.   cover

Achillea moschata

Barneby, T. P.
European Alpine Flowers in Colour

fl. hab.   Pl. 88, 2

Achillea nana

Barneby, T. P.
European Alpine Flowers in Colour

fl.   Pl. 88, 4

Achillea nana

Color Treasury of Herbs & Other
   Medicinal Plants

fl. p 58 Pl. 92

Achillea nana

Kohlhaupt, Paula
Fleurs des Alpages Vol 1

fl. P 109

Achillea nana

Polunin, Oleg
Flowers of Europe

fl.   Pl. 147 #1419

Achillea nigrescens

Porsild, A. E.
Rocky Mountain Wild Flowers

fl. hab.   p 383

Achillea oxyloba

Kohlhaupt, Paula
Fleurs des Alpages Vol 2

Fl. P 110

Achillea ptarmica

Ary, S.
The Oxford Book of Wildflowers

fl.   p 98, Pl. 4

Achillea ptarmica

Lindman, C. A. M.
Nordens Flora, Vol 9

fl.   Pl. 600

Achillea ptarmica

Martin, W. Keble
The Concise British Flora in Colour

fl. hab.   Pl. 45

Achillea ptarmica var.

Hay, Roy
The Color Dictionary of Flowers &
   Plants

fl.   p. 119   Pl. 951

Achillea ptarmica var.

Huxley, Anthony
Garden Perennials & Water Plants

fl.   Pl. 22

Achillea sibirica

Heller, Christine
Wild Flowers of Alaska

fl.   Pl. 211

Achillea tomentosa

Barneby, T. P.
European Alpine Flowers in Colour

fl. hab.   Pl. 88, 1

Achillea tomentosa

Crockett, James Underwood
Lawns & Ground Covers

fl. p. 120

Achillea tomentosa

Polunin, Oleg
Flowers of Europe

fl.   Pl. 147 #1421

Achimenes grandiflora

Tsukamoto, Yotaro
Coloured Illustrations of Garden
   Flowers   Vol. 9

fl.   opp. p. 1

Achimenes heterophylla var.

Hay, Roy
The Dictionary of House Plants

fl.   Pl. 7

Achimenes hybrid

Encke, Fritz
Zimmerpflanzen

fl.   p. 48

Achimenes hybrid

Kromdijk, G.
200 House Plants in Colour

fl.   Pl. 4

Achimenes hybrid

Miles, Bebe
Bulbs for the Home Gardener

fl.   p 150

Achimenes longiflora

American Gloxinia Society
The Gloxinian
     vol 6, No. 4-5   1957

fl.   cover

Achimenes longiflora

Chickering, Carol Rogers
Flowers of Guatemala

fl.   p. 31

Achimenes longiflora

Kiaer, Eigil
Indoor Plants in Colour

fl.   p. 7

Achimenes longiflora

Macoby, Stirling
What Flower is That

fl.   p. 29

Achimenes longiflora

Nicolaisen, Age
Pocket Encyclopedia of Indoor Plants

fl.   Pl. 119

Achlys triphylla

Clark, Lewis J.
Wild Flowers of British Columbia

fl.   p 174

Achlys triphylla

Fries, Mary A.
Wildflowers of Mount Ranier and
the Cascades

fl.   P 25

Achlys triphylla

Lemmon, Robert S.
Wildflowers of North America in
Full Color

fl.   pg. 253   pl. 398

Achlys triphylla

Orr, Robert T.
Wildflowers of Western America

fl.   Pl. 34

Achlys triphylla

Taylor, Ronald J.
Mountain Wild Flowers

fl. hab.   p 69

Achras sapota

Hvass, Elsie
Plants That Feed and Serve Us

fl. fr.   p 107, Pl. 230

Achras sapota

Masefield, G. B.
The Oxford Book of Food Plants

fl. fr.   Pl. 99, 3

Achyranthes aspera

Moriarty, Audrey
Wild Flowers of Malawi

fl.   Pl. 44; 4

Achyranthes fauriei

Kariyone, Tatsuo
Atlas of Medicinal Plants

fl., hab.   Pl. 134

Achyranthes indica

Burgis, D.S.
Florida Weeds

hab., fr.   p. 3

Achyranthes japonica

Kimura, Koiti
Japanese Medicinal Plants, Vol I

fl.   p 24

Achyranthes japonica

Takatori, Jisuke
Color Atlas of Medicinal Plants of Japan

fl.   Fig. 57

Acianthus caudatus

Blombery, Alec M.
What Wildflower is That

fl.   p 36, Pl. 68

Acianthus caudatus

Cady, Leo
Australian Native Orchids in Colour

fl.   Pl. 63

Acianthus exsertus

Cady, Leo
Australian Native Orchids in Colour

fl.   Pl. 64

Acianthus reniformus

Blombery, Alec M.
What Wildflower is That

fl.   p 36, Pl. 69

Acidanthera bicolor

Everett, T.H., ed.
New Illustrated Encyclopedia of
Gardening vol. 1

fl.   p. 22

Acidanthera bicolor

Huxley, Anthony
Garden Annuals and Bulbs

hab. fl.   Pl. 145, 195

Acidanthera bicolor

Wit, H. C. D. de
Plants of the World;
The Higher Plants, Vol II

fl.   Pl. 140

Acidanthera bicolor var.

Hay, Roy
The Color Dictionary of Flowers & Plants

fl.   p. 83   Pl. 657

Acidanthera bicolor var

Tsukamoto, Yotaro
Coloured Illustrations of Garden
Flowers   Vol 9

fl.   opp. p. 1

Acidanthera murielae

Miles, Bebe
Bulbs for the Home Gardener

fl.   p. 151

Acineta erythroxantha

Ospina, Mariano
Orquideas de las americas

fl.   Pl. 151

Acineta humboldtii

Kramer, Jack
Orchids; Flowers of Romance
and Mystery

fl.   p 175

Acineta superba

Dunsterville, G. C. K.
Introduction to the World
of Orchids

fl.   Pl. 1

Acinos alpinus

Polunin, Oleg
Flowers of Europe

fl.   Pl. 114 #1157

Acinos arvensis

Ary, S.
The Oxford Book of Wildflowers

fl.   p 142, Pl. 2

Acinos arvensis

Martin, W. Keble
The Concise British Flora in Colour

fl. hab.   Pl. 68

Aciphylla glacialis

Alpine Flowers of the Kosciusko
    State Park

fl.   Pl. 9

Aciphylla glacialis

Cochrane, G. R.
Flowers and Plants of Victoria

fl. hab.   Pl. 532, 535

Aciphylla procumbens

Curtis, Winifred
The Endemic Flora of Tasmania Vol 3

fl. fr.   p 169

Aciphylla spedeni

Royal Hort. Soc.
The Garden
    vol 103, No. 11   1978

hab., fl.   p. 441

Acmena smithii

Blombery, Alec M.
What Wildflower is That

fr.   p 37, Pl. 70

Acmena smithii

Cochrane, G. R.
Flowers and Plants of Victoria

fr.   Pl. 462

Acokanthera spectabilis

Macoby, Stirling
What Flower is That

fl.   P 30

Acokanthera venenata

Addisonia   Vol 24   1960-64

fl., fr.   opp. p. 23   Pl. 780

Aconite

    see

Aconitum

Aconite, winter

    see

Eranthis hyemalis

Aconitum anglicum

Martin, W. Keble
The Concise British Flora in Colour

fl. fr. hab.   Pl. 4

Aconitum anthora

Barneby, T. P.
European Alpine Flowers in Colour

fl.   Pl. 23, 1

Aconitum x cammarum var.

Perry, Frances
Flowers of the World

fl.   p. 251

Aconitum Carmichaeli

Kimura, Koiti
Japanese Medicinal Plants, Vol. II

fl.   p. 159

Aconitum columbianum

Clark, Lewis J.
Wild Flowers of British Columbia

fl.   p 150

Aconitum columbianum

Lemmon, Robert S.
Wildflowers of North America in
    Full Color

fl.   pg. 126   pl. 201

Aconitum columbianum

Munz, Philip A.
California Mountain Wildflowers

fl.   p 48, Pl. 64

Aconitum columbianum

Shaw, Richard J.
Field Guide to the Vascular Plants of
    Grand Teton National Park

fl.   Pl. 10

Aconitum columbianum

Welsh, Stanley L.
Flowers of the Mountain Country

fl.   p. 66, 70

Aconitum delphinifolium

Heller, Christine
Wild Flowers of Alaska

fl.   Pl. 228

Aconitum delphinifolium var.

Heller, Christine
Wild Flowers of Alaska

fl.   Pl. 229

Aconitum fauriei

Kariyone, Tatsuo
Atlas of Medicinal Plants

fl., hab.   Pl. 124

Aconitum fletcheranum

Royal Hort. Soc.
Journal of the Royal Hort. Soc.
    vol 77, No. 7   1952

fl.   p. 248   Pl. 108

Aconitum henryi var.

Perry, Frances
Flowers of the World

fl.   p. 251

Aconitum japonicum

Takatori, Jisuke
Color Atlas of Medicinal Plants of Japan

Fl.
fr.   Fig. 51

Aconitum lycoctonum

Barneby, T. P.
European Alpine Flowers in Colour

fl.   Pl. 23, 2

Aconitum lycoctonum

Felsko, Elsa
A Book of Wild Flowers 2nd Series

fl.   p. 16

Aconitum nagarum

Curtis's Botanical Magazine
    Vol 163   1940 - 42

fl.   No. 9613

Aconitum napellus

Barneby, T. P.
European Alpine Flowers in Colour

fl.   Pl. 22, 5

Aconitum napellus

Bianchini, Francesco
Health Plants of the World

hab.   p. 83

Aconitum napellus

Color Treasury of Herbs & Other
    Medicinal Plants

fl.   p 28   Pl. 24

Aconitum napellus

Felsko, Elsa
A Book of Wildflowers

fl.   P 67

Aconitum napellus

Huxley, Anthony
Garden Perennials and Water Plants

hab.   Pl. 1

Aconitum napellus

Kohlhaupt, Paula
Fleurs des Alpages Vol. 1

fl.   P. 16

Aconitum napellus

Loewenfeld, Claire
The Complete Book of Herbs and Spices

fl. hab.   p 112

Aconitum napellus

Perrot, Emile
Les Plantes Medicinales

fl. fr. hab.   p. 2

Aconitum napellus

Polunin, Oleg
Flowers of Europe

fl.   Pl. 19 #210

Aconitum napellus

Robert, Paul A.
Alpine Flowers

fl. Pl. 5

Aconitum napellus

Tosco, Uberto
The World of Mountain Flowers

fl.   p 55

Aconitum napellus var.

Hay, Roy
Color Dictionary of Flowers &
    Plants

fl.   p. 119 Pl. 952

Aconitum napellus var

Huxley, Anthony
Garden Perennials & Water Plants

fl   Pl. 24

Aconitum noveboracense

Addisonia
   Vol 20   1937-38

fl.   p. 19 Pl. 650

Aconitum paniculatum

Barneby, T. P.
European Alpine Flowers in Colour

fl.   Pl. 22, 6

Aconitum septentrionale

Lindman, C. A. M.
Nordens Flora, Vol 4

fl. fr.   Pl. 224

Aconitum uncinatum

Campbell, Carlos C.
Great Smoky Mountain Wildflowers

fl.   p 97

Aconitum uncinatum

Courtenay, Booth
Wildflowers & Weeds

fl.   p 31

Aconitum uncinatum

Dean, Blanche E.
Wildflowers of Alabama and
    Adjoining States

fl.   p 59

Aconitum uncinatum

Klimas, John E.
Wildflowers of Eastern America

fl.   Pl. 287

Aconitum variegatum

Bianchini, Francesco
Health Plants of the World

fl. hab.   p. 83

Aconitum vulparia

Kleijn, H.
Beauty of the Wild Plant

fl.   p. 85   Pl. 114

Aconitum vulparia

Kohlhaupt, Paula
Fleurs des Alpages Vol 1

fl. P 15

Aconitum vulparia

Polunin, Oleg
Flowers of Europe

fl.   Pl. 19 #208

Acorus americanus

Wharton, Mary E.
A Guide to the Wildflowers & Ferns
   of Kentucky

fl. p 84 Pl. 2.2

Acorus calamus

Bianchini, Francesco
Health Plants of the World

fl. hab.   p 45

Acorus calamus

Courtenay, Booth
Wildflowers and weeds

fl.   p. 3

Acorus calamus

Huxley, Anthony
Garden Perennials and Water Plants

fl. fr.   Pl. 283

Acorus calamus

Jennings, O. E.
Wild Flowers of Western Pennsylvania
    Vol. II

fl. fr. hab.   Pl. 4

Acorus calamus

Klimas, John E.
Wildflowers of Eastern America

fl.   Pl. 91

Acorus calamus

Lindman, C. A. M.
Nordens Flora. Vol 1

fl. fr. hab.   Pl. 31

Acorus calamus

Martin, W. Keble
The Concise British Flora in Colour

fl. hab.   Pl. 88

Acorus calamus

Moyle, John B.
Northland Wild Flowers

fl. hab.   p 203, Pl. 265

Acorus calamus

Perrot, Emile
Les Plantes Medicinales

fl. fr. hab.   p 3

Acorus calamus

Polunin, Oleg
Flowers of Europe

fl.   Pl. 183 #1816

Acorus calamus

Stodola, Jiri
Encyclopedia of Water Plants

fl.   p. 342

Acorus calamus var.

Kariyone, Tatsuo
Atlas of Medicinal Plants

fr. hab.   Pl. 6

Acorus calamus var.

Stodola, Jiri
Encylopedia of Water Plants

hab.   p. 78, 82

Acorus gramineus

Stodola, Jiri
Encyclopedia of Water Plants

hab.   P. 79

Acorus gramineus

Takatori, Jisuke
Color Atlas of Medicinal Plants of Japan

fl.   Fig. 77A

Acorus gramineus var.

Kiaer, Eigil
Indoor Plants in Colour

hab.   p. 8

Acradenia frankliniae

Curtis, Winifred
The Endemic Flora of Tasmania, Vol. 5

fl. fr.   Pl. 167

Acridocarpus natalitius

Morley, Brian D.
Wild Flowers of the World

fl.   Pl. 76

Acridocarpus smeathmannii

Wit, H. C. D. de
Plants of the World;
  The Higher Plants, Vol II

fl. fr.   p 63, Pl. 39

Acrocephalus callianthus

Flowering Plants of South Africa
  vol XXII   1942

fl.   Pl. 847

Acrocephalus callianthus

Moriarty, Audrey
Wild Flowers of Malawi

fl.   Pl. 53; 5

Acroclinium roseum

Macoby, Stirling
What Flower is That

fl.   P 30

Acrodon bellidiflorus

Herre, H.
The Genera of the Mesembryanthemaceae

fl. fr.   p. 63

Acronychia pedunculata

Walden, Beryl M.
Wild Flowers of Hong Kong

fl.   Pl. 46(129)

Acrotriche prostrata

Cochrane, G. R.
Flowers and Plants of Victoria

fl.   Pl. 427

Actaea alba

Courtenay, Booth
Wildflowers & Weeds

fl.   p 30

Actaea alba

Jennings, O. E.
Wild Flowers of Western Pennsylvania
  Vol. II

fr.   Pl. 58

Actaea alba

Lemmon, Robert S.
Wildflowers of North America

fl. fr.   p. 252   Pl. 396

Actaea alba

Massachusetts Hort. Soc.
Horticulture
  vol 39, No. 9 1961

fr.   inside cover

Actaea arguta

Heller, Christine
Wild Flowers of Alaska

fl., fr.   Pl. 135-136

Actaea arguta

Porsild, A. E.
Rocky Mountain Wild Flowers

fr.   p 169

Actaea arguta

Walcott, Mary Vaux
North American Wild Flowers

fr. hab.   Pl. 73

Actaea pachypoda

American Hort. Soc.
American Horticulturist
  vol 55, No. 6 1976

fr.   p. 14

Actaea pachypoda

Campbell, Carlos C.
Great Smoky Mountain Wildflowers

fl. fr.   p 105

Actaea pachypoda

Dean, Blanche E.
Wildflowers of Alabama and
  Adjoining States

fl.   p 61

Actaea pachypoda

Duncan, Wilbur H.
Wildflowers of the Southeastern
  United States

fr.   p 41

Actaea pachypoda

Klimas, John E.
Wildflowers of Eastern America

fl.   Pl. 81

Actaea rubra

Clark, Lewis J.
Wild Flowers of British Columbia

fr.   p 150

Actaea rubra

Moyle, John B.
Northland Wild Flowers

fl. fr. hab.   p 55, Pl. 17, 17a

Actaea rubra

Shaw, Richard J.
Field Guide to the Vascular Plants of
  Grand Teton National Park

fr.   Pl. 9

Actaea rubra var.

Munz, Philip A.
California Mountain Wildflowers

fl.   p. 40   Pl. 41

Actaea spicata

Ary, S.
The Oxford Book of Wildflowers

fl. fr.   p 66, Pl. 3

Actaea spicata

Barneby, T. P.
European Alpine Flowers in Colour

fr.   Pl. 21, 3

Actaea spicata

Felsko, Elsa
A Book of Wild Flowers 2nd Series

fl. fr.   p 77

Actaea spicata

Lindman, C. A. M.
Nordens Flora, Vol 4

fl. fr.   Pl. 220

Actaea spicata

Martin, W. Keble
The Concise British Flora in Colour

fl. hab.   Pl. 4

Actinidia chinensis

Bartels, Andreas
Das Grosse Buch der Gartengeholze

fl.   p 110

Actinidia chinensis

Harrison, Richmond E.
Climbers and Trailers

fl. fr. p 17 Pl. 1

Actinidia chinensis

Macoby, Stirling
What Flower is That

fr.   P 31

Actinidia chinensis

Mathias, Mildred E.
Color for the Landscape

fl.   p 97

Actinidia kolomikta

American Hort. Soc.
American Horticulturist
    vol 52, No. 1   1973

hab.   p. 24

Actinidia kolomikta

Harrison, Richmond E.
Climbers and Trailers

hab. p 17 Pl. 2

Actinidia kolomikta

Hay, Roy
The Color Dictionary of Flowers & Plants

hab.   p 245, Pl. 1957

Actinidia kolomikta

Hellyer, A.G.L.
Shrubs in Colour

hab.   Pg. 11

Actinidia kolomikta

Huxley, Anthony
Deciduous Garden trees & Shrubs

Actinidia kolomikta

Kitamura, Siro
Coloured Illustrations of Trees &
    Shrubs of Japan

hab.   340

Actinidia kolomikta

Royal Hort. Soc.
The Garden
    vol 101, No. 1   1976

fl., hab.   p. 51

Actinidia latifolia

Walden, Beryl M.
Wild Flowers of Hong Kong

fl.   Pl. 47 (133)

Actinidia polygama

Kitamura, Siro
Coloured Illustrations of Trees &
    Shrubs of Japan

fl.   339

Actinidia sinensis

Bianchini, F.
The Complete Book of Fruits & Vegetables

fr.   p 175

Actinobole uliginosum

Cochrane, G. R.
Flowers and Plants of Victoria

fl. hab.   Pl. 127

Actinodaphne longifolia

Kitamura, Siro
Coloured Illustrations of Trees &
    Shrubs of Japan

fr.   190, 191

Actinodium cunninghamii

Blombery, Alec M.
What Wildflower is That

fl.   p. 38   Pl. 72

Actinodium cunninghamii

Harrison, R. E.
Trees and Shrubs

fl.   P 21

Actinodium cunninghamii

Macoby, Stirling
What Flower is That

fl.   P 31

Actinomeris alternifolia

Jennings, O. E.
Wild Flowers of Western Pennsylvania
    Vol. II

fl. hab.   Pl. 187

Actinomeris alternifolia

Wharton, Mary E.
A Guide to the Wildflowers & Ferns
    of Kentucky

fl. p 254   Pl. 3.21

Actinotus helianthi

Blombery, Alec M.
What Wildflower is That

fl.   p. 37   Pl. 71

Actinotus helianthi

Blunt, Wilfrid
Flora Superba

fl. hab.   Pl. X

Actinotus helianthi

Macoby, Stirling
What Flower is That

fl.   P 31

Actinotus helianthi

Morley, Brian D.
Wild Flowers of the World

fl.   Pl. 133

Actinotus helianthi

Mullins, Barbara
Australian Wildflowers in Colour

fl.   P. 87   Pl. 84

Actinotus moorei

Curtis, Winifred
The Endemic Flora of Tasmania
vol 3
fl.   p 169

Ada aruantiaca

Ospina, Mariano
Orquideas de las americas

fl.   Pl. 166

Adam and Eve

see

Aplectrum hyemale

Adam and Eve

see

Rhoeo discolor

Adam's needle

see

Yucca filamentosa

Adam's Tree

see

Fouquieria diguetii

Adansonia digitata

Edlin, Herbert
The Illustrated Encyclopedia
of Trees

hab.   p 231

Adansonia digitata

Everett, Thomas H.
Living Trees of the World

hab. fr. fl.   p. 227

Adansonia digitata

Morley, Brian D.
Wild Flowers of the World

fl.,fr.   Pl. 76

Adansonia digitata

Palgrave, K. C.
Trees of Central Africa

fr., fl.   p. 52-53   Pl.1-2

Adansonia digitata

Wit, H. C. D. de
Plants of the World;
The Higher Plants, Vol I

hab.   p 217, Pl. 127

Adansonia madagascariensis

Cactus & Succulent Society of America
Journal
    vol 49, No. 4   1977

hab.   p. 160

Adder's Mouth

    see

Malaxis

Adder's Mouth, Bog

    see

Malaxis paludosa

Adders-Mouth, Green

    see

Malaxis unifolia

Adder's Mouth, Little

    see

Malaxis paludosa

Adder's Tongue

    see

Erythronium revolutum

Adenandra fragrans

Harrison, R. E.
Trees and Shrubs

fl.   P 21

Adenandra uniflora

Harrison, R. E.
Trees and Shrubs

fl.   P 21

Adenanthera pavonina

Bruggeman, L.
Tropical Plants

fl.   pl. 189

Adenanthos barbigera

Blombery, Alec M.
What Wildflower is That

fl.   p 38, Pl. 73

Adenanthos obovata

Blombery, Alec M.
What Wildflower is That

fl.   p 39, Pl. 74

Adenanthos obovata

Morcombe, M. K.
Australia's Western Wildflowers

fl.   p 40

Adenia glauca

Cactus & Succulent Soc. of America
Journal
    vol 45, No. 2   1973

fr.   p. 74

Adenia glauca

Flowering Plants of South Africa
    Vol. XX

fl. fr.   Pl.770

Adenium boehmianum

Flowering Plants of South Africa
    Vol. XVIII

fl.   Pl. 701

Adenium coetaneum

Curtis's Botanical Magazine
Vol 171   1956-57

fl.   277

Adenium coetaneum

Flowering Plants of South Africa
    Vol. XIX

fl.   Pl. 753

Adenium multiflorum

Cobb, L.E.W.
Trees & Shrubs of the Kruger
    National Park

fl.   p. 130   Pl. V

Adenium multiflorum

Palgrave, K. C.
Trees of Central Africa

fl., fr.   p. 15

Adenium multiflorum

Royal Hort. Soc.
Journal of the Royal Hort. Soc.
    vol 87, No. 1   1962

fl., hab.   p. 54   Pl. 16

Adenium obesum

Morley, Brian D.
Wild Flowers of the World

fl.   Pl. 61

Adenium obesum var.

Cactus & Succulent Society of America
Journal
    vol 50, No. 2   1978

fl.   cover

Adenium swazicum

Flowering Plants of South Africa
    Vol. XVII

fl.   Pl. 664

Adenocarpus bacquei

Royal Hort. Soc.
Journal of the Royal Hort. Soc.
    vol 90, No. 9   1965

fl.   p. 382   Pl. 171

Adenocarpus complicatus

Polunin, Oleg
Flowers of Europe

fl.   Pl. 55 #518

Adenocarpus complicatus

Polunin, Oleg
Trees and Bushes of Europe

fl. fr.   p 111

Adenocarpus decorticans

Curtis's Botanical Magazine
Vol 166   1949

fl.   48

Adenocarpus decorticans

Royal Hort. Soc.
Journal of the Royal Hort. Soc.
    vol 89, No. 1   1964

fl.   p. 22   Pl. 11

Adenocarpus foliosus

Bramwell, David
Wild Flowers of the Canary Islands

fl.   Pl. 165, 166

Adenocarpus viscosus

Bramwell, David
Wild Flowers of the Canary Islands

fl.   Pl. 164

Adenocaulon bicolor

Clark, Lewis J.
Wild Flowers of British Columbia

fl.   p 518

Adenocaulon bicolor

Taylor, Ronald J.
Mountain Wild Flowers

fl. hab.   p 63

Adenochilus nortonii

Cady, Leo
Australian Native Orchids in Colour

fl. hab.   Pl. 42

Adenophora coelestis

Curtis's Botanical Magazine
    Vol 163   1940 - 42

fl.   No. 9626

Adenophora morrisonensis

Curtis's Botanical Magazine
    Vol 159   1936

fl.   No. 9444

Adenophora triphylla var.

Kimura, Koiti
Japanese Medicinal Plants, Vol II

fl.   p 233

Adenosma glutinosa

Walden, Beryl M.
Wild Flowers of Hong Kong

fl.   Pl. 41(117)

Adenostoma fasciculatum

Coyle, Jeanette
A Field Guide to the Common and
  Interesting Plants of Baja California

fl. hab.   p 81

Adenostoma fasciculatum

Munz, Philip A.
California Spring Wildflowers

fl.   p 87, Pl. 86

Adenostoma sparsifolium

Coyle, Jeanette
A Field Guide to the Common and
  Interesting Plants of Baja California

hab.   p 81

Adenostyles alliariae

Barneby, T. P.
European Alpine Flowers in Colour

fl.   Pl. 86, 4

Adenostyles alliariae

Kleijn, H.
Beauty of the Wild Plant

fl.   opp. p. 88   Pl. 119

Adenostyles alliariae

Kohlhaupt, Paula
Fleurs des Alpages Vol 2

Fl.   P 109

Adenostyles alliariae

Polunin, Oleg
Flowers of Europe

fl.   Pl. 150 #1445

Adenostyles glabra

Barneby, T. P.
European Alpine Flowers in Colour

fl.   Pl. 86, 3

Adenostyles leucophylla

Barneby, T. P.
European Alpine Flowers in Colour

fl.   Pl. 86, 5

Adhatoda cydoniaefolia

Menninger, Edwin A.
Flowering Vines of the World

fl.   Pl. 1

Adina microcephala

Palgrave, K. C.
Trees of Central Africa

fl., fr.   p. 383

Adlumia fungosa

Courtenay, Booth
Wildflowers & Weeds

fl.   p 22

Adlumia fungosa

Jennings, O. E.
Wild Flowers of Western Pennsylvania
    Vol. II

fl.   Pl. 69

Adlumia fungosa

Pennsylvania Hort. Soc.
The Green Scene
    vol 5, No. 2   1976

fl.   p. 33

Adonis aestivalis

Crockett, James Underwood
Annuals

fl.   P 90

Adonis aestivalis

Felsko, Elsa
A Book of Wild Flowers 2nd Ser.

fl. fr.   p 78

Adonis aestivalis

Wit, H. C. D. de
Plants of the World;
    The Higher Plants, Vol I

fl.   p 98, Pl. 41

Adonis amurensis

Hay, Roy
The Color Dictionaey of Flowers & Plants

fl.   p 1, Pl. 1

Adonis amurensis

Huxley, Anthony
Garden Perennials and Water Plants

fl.   Pl. 25

Adonis amurensis

Kariyone, Tatsuo
Atlas of Medicinal Plants

fl., hab.   Pl. 125

Adonis amurensis

Kimura, Koiti
Japanese Medicinal Plants, Vol I

fl. hab.   p 28

Adonis amurensis

Royal Hort. Soc.
Journal of the Royal Hort. Soc.
    vol 97, No. 4   1972

fl.   p. 170   Pl. 75

Adonis amurensis

Royal Hort. Soc.
The Garden
    vol 101, No. 3   1976

fl.   p. 149

Adonis amurensis var.

Royal Hort. Soc.
The Garden
    vol 101, No. 3   1976

fl.   p. 149

Adonis annua

Ary, S.
The Oxford Book of Wildflowers

fl.   p 104, Pl. 3

Adonis annua

Martin, W. Keble
The Concise British Flora in Colour

fl. hab.   Pl. 1

Adonis annua

Morley, Brian D.
Wild Flowers of the World

fl.,fr.   Pl. 27

Adonis annua

Polunin, Oleg
Flowers of Europe

fl.   Pl. 24 #230

Adonis annua

Polunin, Oleg
Flowers of the Mediterranean

fl.   Pl. 32

Adonis brevistyla

Curtis's Botanical Magazine
Vol 165   1948

fl.   24

Adonis brevistyla

Hara, Hiroshi, comp.
Photo-Album of Plants of
Eastern Himalaya

fl.   Pl. 170

Adonis chrysocyathus

Massachusetts Hort. Soc.
Horticulture
   vol 51, No. 1   1973

fl.   p. 37

Adonis chrysocyathus

Morley, Brian D.
Wild Flowers of the World

fl.   Pl. 91

Adonis chrysocyathus

Royal Hort. Soc.
Journal of the Royal Hort. Soc.
   vol 93, No. 6   1968

fl.   p. 246   Pl. 129

Adonis chrysocyathus

Royal Hort. Soc.
The Garden
   vol 102, No. 11   1977

fl.   p. 454

Adonis flammeus

Barneby,. T. P.
European Alpine Flowers in Colour

fl.   Pl. 27, 3

Adonis hybrid

Royal Hort. Soc.
Journal of the Royal Hort. Soc.
   vol 89, No. 3   1964

fl.   p. 114   Pl. 42-43,
                        48-51

Adonis pyrenaica

Alpine Garden Society
Bulletin
   vol. 39, No. 1   1971

fl., hab.   p. 36

Adonis, Spring

   see

Adonis vernalis

Adonis, summer

   see

Adonis aestivalis

Adonis vernalis

Barneby, T. P.
European Alpine Flowers in Colour

fl. hab.   Pl. 27, 2

Adonis vernalis

Bianchini, Francesco
Health Plants of the World

fl. hab.   p 79

Adonis vernalis

Felsko, Elsa
A Book of Wildflowers

fl.   P 8

Adonis vernalis

Lindman, C. A. M.
Nordens Flora, Vol 4

fl. fr. hab.   Pl. 243

Adonis vernalis

Massachusetts Hort. Soc.
Horticulture
   vol 42, No. 3   1964

fl.   p. 35

Adonis vernalis

Perrot, Emile
Les Plantes Medicinales

fl. fr. hab.   p 4

Adonis vernalis

Polunin, Oleg
Flowers of Europe

fl.   Pl. 24 #231

Adonis vernalis

Royal Hort. Soc.
Journal of the Royal Hort. Soc.
   vol 89, No. 7   1964

fl.   p. 290   Pl. 118

Adonis vernalis

Royal Hort Soc.
The Garden
   vol 101, No. 3   1976

fl. hab. p. 149

Adonis vernalis

Vilmorin, Roger de
Plantes Alpines dans les Jardins

fl. hab.   Pl. XIX

Adoxa moschatellina

Ary, S.
The Oxford Book of Wildflowers

fl.   p 48, Pl. 5

Adoxa moschatellina

Felsko, Elsa
A Book of Wild Flowers   2nd Ser.

fl. hab.   p. 111

Adoxa moschatellina

Heller, Christine
Wild Flowers of Alaska

hab., fl.   Pl. 300

Adoxa moschatellina

Lindman, C. A. M.
Nordens Flora, Vol 8

fl. hab.   Pl. 553

Adoxa moschatellina

Martin, W. Keble
The Concise British Flora in Colour

fl. fr. hab.   Pl. 41

Adoxa moschatellina

Morley, Brian D.
Wild Flowers of the World

fl.   Pl. 16A

Adoxa moschatellina

Polunin, Oleg
Flowers of Europe

fl. Pl. 137 #1308

Adriana quadripartita

Cochrane, G. R.
Flowers and Plants of Victoria

fl.   Pl. 289

Adromischus cristatus

Hay, Roy
The Dictionary of House Plants

hab.   Pl. 10

Adromischus cristatus

Subik, Rudolf
Decorative Cacti

fl. hab. p 93

Adromischus pulchellus

Lamb, Edgar
Popular Exotic Cacti in Color

hab.   p 19

Adromischus rotundifolius

Hay, Roy
The Dictionary of House Plants

hab.   Pl. 11

Adromischus saxicola

Lamb, Edgar
The Pocket Encyclopedia of Cacti
 and Succulents in Color

hab. Pl. 214

Adromischus trigynus

Lamb, Edgar
The Pocket Encyclopedia of Cacti
 and Succulents in Color

hab. Pl. 215

Aechmea allenii

Bromeliad Society
Journal
 vol 24, No. 6 1974

fl. p. 236

Aechmea allenii

Bromeliad Society
Journal
 vol 27, No. 4 1977

fl. p. 169

Aechmea angustifolia

Wilson, Robert Gardner
Bromeliads in Cultivation

fl.   Pl. 4

Aechmea apocalyptica var.

Bromeliad Society
Journal
 vol 27, No. 4 1977

fl., hab. p. 169

Aechmea araneosa

Bromeliad Society
Journal
 vol 28, No. 6 1978

fl. p. 272

Aechmea blumenavii

Wilson, Robert Gardner
Bromeliads in Cultivation

fl.   Pl. 6

Aechmea bracteata

Bruggeman, L.
Tropical Plants

fl.   Pl. 66

Aechmea bracteata

Rauh, Werner
Bromeliads for Home, Garden and
 Greenhouse

fr.   Pl. 64

Aechmea bracteata

Wilson, Robert Gardner
Bromeliads in Cultivation

fl. hab. Pl. 5

Aechmea bromeliifolia

Padilla, Victoria
Bromeliads

fl. p. 63 Pl. 1

Aechmea bromeliifolia

Wilson, Robert Gardner
Bromeliads in Cultivation

fl.   Pl. 7

Aechmea calyculata

Wilson, Robert Gardner
Bromeliads in Cultivation

fl. hab. Pl. 8

Aechmea caudata

Smith, Lyman B.
The Bromeliads

fl. hab. Pl. 25

Aechmea caudata var.

Macoby, Stirling
What Flower is That

fl. P 32

Aechmea caudata var.

Rauh, Werner
Bromeliads for Home, Garden and
 Greenhouse

fl.   Pl. 65

Aechmea caudata var.

Wilson, Robert Gardner
Bromeliads in Cultivation

hab. Pl. 9, 10

Aechmea chantinii

Bromeliad Society
Bulletin
 vol 12, No. 1 1962

fl., hab. p. 10

Aechmea chantinii

Bromeliad Society
Journal
 vol 25, No. 2 1975

fl. p. 63

Aechmea chantinii

Hay, Roy
The Dictionary of House Plants

fl.   Pl. 12

Aechmea chantinii

Padilla, Victoria
Bromeliads

fl. p. 63 Pl. 1

Aechmea chantinii

Padilla, Victoria, ed.
Bromeliads in Color and
 Their Culture

fl.   p 94

Aechmea chantinii

Rauh, Werner
Bromeliads for Home, Garden and
 Greenhouse

fl.   Pl. 66

Aechmea chantinii

Royal Hort. Soc.
Journal of the Royal Hort. Soc.
 Vol. 91, No. 8 1966

fl. p. 338 Pl. 176

Aechmea chantinii

Wilson, Robert Gardner
Bromeliads in Cultivation

fl. hab. Pl. 15

Aechmea chantinii var.

Bromeliad Society
Journal
 vol 22, No. 4 1972

fl., hab. p. 95

Aechmea chantinii var.

Wilson, Robert Gardner
Bromeliads in Cultivation

fl.   Pl. 17

Aechmea coelestis

Padilla, Victoria
Bromeliads

fl.   p. 63   Pl. 1

Aechmea coelestis var.

Wilson, Robert Gardner
Bromeliads in Cultivation

hab.   Pl. 12

Aechmea comata

Rauh, Werner
Bromeliads for Home, Garden and
    Greenhouse

fl.   Pl. 67

Aechmea corymbosa

Bromeliad Society
Journal
    vol 28, No. 2   1978

fl.   p. 76

Aechmea corymbosa var.

Bromeliad Society
Journal
    vol 27, No. 4   1977

fl.   p. 168

Aechmea cylindrata

Bromeliad Society
Journal
    vol 22, No. 3   1972

fl.   p. 70

Aechmea cylindrata

Rauh, Werner
Bromeliads for Home, Garden and
    Greenhouse

fl.   Pl. 96

Aechmea cylindrata

Smith, Lyman B.
The Bromeliads

fl. hab.   Pl. 26

Aechmea dactylina

Wilson, Robert Gardner
Bromeliads in Cultivation

fl.   Pl. 11

Aechmea dealbata

Rauh, Werner
Bromeliads for Home, Garden and
    Greenhouse

fl.   Pl. 69

Aechmea dichlamydea var.

Bromeliad Society
Bulletin
    vol 18, No. 5   1968

fl.   cover (p. 97)

Aechmea dichlamydea var

Padilla, Victoria
Bromeliads

fl.   p. 63   Pl. 1

Aechmea distichantha

American Hort. Soc.
American Horticulturist
    vol 57, No. 6   1978

fl.   p. 27

Aechmea distichantha

Padilla, Victoria, ed.
Bromeliads in Color and
    Their Culture

fl.   p 30

Aechmea distichantha var.

Bromeliad Society
Journal
    vol 27, No. 4   1977

fl.   p. 168

Aechmea distichantha var.

Rauh, Werner
Bromeliads for Home, Garden and
    Greenhouse

fl.   Pl. 71

Aechmea distichantha var.

Wilson, Robert Gardner
Bromeliads in Cultivation

fl.   Pl. 13

Aechmea fasciata

Bromeliad Society
Bulletin
    vol 12, No. 1   1962

fl., hab.   p. 11

Aechmea fasciata

Bromeliad Society
Journal
    vol 27, No. 1   1977

fl.   p. 25

Aechmea fasciata

Encke, Fritz
Zimmerpflanzen

fl.   p. 57

Aechmea fasciata

Hay, Roy
The Dictionary of House Plants

fl.   Pl. 13

Aechmea fasciata

Massachusetts Hort. Soc.
Horticulture
    vol 36, No. 11   1958

fl.   p. 577

Aechmea fasciata

Massachusetts Hort. Soc.
Horticulture
    vol 37, No. 11   1959

fl., hab.   p. 585

Aechmea fasciata

Massachusetts Hort. Soc.
Horticulture
    vol 43, No. 6   1965

fl., hab.   p. 25

Aechmea fasciata

Massachusetts Hort. Soc.
Horticulture
    vol 45, No. 8   1967

fl., hab.   p. 25

Aechmea fasciata

Massachusetts Hort. Soc.
Horticulture
    vol 56, No. 12   1978

fl.   p. 23

Aechmea fasciata

Mee, Margaret
Flowers of the Brazilian Forests

fl., hab.   Pl. 26

Aechmea fasciata

Nicolaisen, Age
Pocket Encyclopedia of Indoor
    Plants

Aechmea fasciata

Padilla, Victoria
Bromeliads

fl.   p. 63   Pl. 1

Aechmea fasciata

Padilla, Victoria, ed.
Bromeliads in Color and
    Their Culture

fl.   p 15, 110

Aechmea fasciata

Perry, Frances
Flowers of the World

fl., hab.   p. 52-53

Aechmea fasciata

Rauh, Werner
Bromeliads for Home, Garden and
    Greenhouse

fl.   Pl. 68

Aechmea fasciata

Wilson, Robert Gardner
Bromeliads in Cultivation

fl. hab.   Pl. 18

Aechmea fasciata var.

Bromeliad Society
Bulletin
   vol 20, No. 3 1970

fl., hab. cover (p. 49)

Aechmea fasciata var.

Padilla, Victoria
Bromeliads

fl., hab. p. 63   Pl. 1

Aechmea fasciata var.

Perry, Frances
Flowers of the World

fl. p. 53.

Aechmea fasciata var.

Wilson, Robert Gardner
Bromeliads in Cultivation

fl. hab. Pl. 19, 20, 22, 126

Aechmea fendleri

Bromeliad Society
Journal
   vol 27, No. 4 1977

fl. p. 168

Aechmea fernandae

Mee, Margaret
Flowers of the Brazilian Forests

fl., hab. Pl. 27

Aechmea ferruginea

Rauh, Werner
Bromeliads for Home, Garden and
   Greenhouse

fl. Pl. 77

Aechmea fosteriana

Smith, Lyman B.
The Bromeliads

fl. hab. Pl. 24

Aechmea fosteriana

Wilson, Robert Gardner
Bromeliads in Cultivation

fl. hab. Pl. 21, 23

Aechmea fulgens

Bromeliad Society
Journal
   vol 25, No. 1 1975

fl., hab. cover (p. 1)

Aechmea fulgens

Everett, T.H., ed.
New Illustrated Encyclopedia of
   Gardening vol. 1

fl. p. 22

Aechmea fulgens

Hay, Roy
The Color Dictionary of Flowers & Plants

fl. p 51, Pl. 402

Aechmea fulgens

Massachusetts Hort. Soc.
Horticulture
   vol 36, No. 11 1958

fl. p. 577

Aechmea fulgens

Perry, Frances
Flowers of the World

fl. p 53

Aechmea fulgens var.

Addisonia
   Vol 21 1939-42

fl. . p. 17 Pl. 681

Aechmea fulgens var.

Bromeliad Society
Journal
   vol 22, No. 3 1972

fl. p. 72

Aechmea fulgens var.

Massachusetts Hort. Soc.
Horticulture
   vol 43, No. 6 1965

fl., hab. p. 25

Aechmea fulgens var.

Rauh, Werner
Bromeliads for Home, Garden and
   Greenhouse

fl. Pl. 72

Aechmea fulgens var.

Wilson, Robert Gardner
Bromeliads in Cultivation

fl., hab. Pl. 14, 16, 122

Aechmea germinyana

Wilson, Robert Gardner
Bromeliads in Cultivation

fl. Pl. 24

Aechmea hybrid

Hersey, Jean
Woman's Day Book of House Plants

fl. p. 78

Aechmea hybrid

Macoby, Stirling
What Flower is That

fl. p. 32

Aechmea hybrid.

Padilla, Victoria, ed.
Bromeliads in Color
   and Their Culture

fl. p 15

Aechmea hybrid

Wilson, Robert Gardner
Bromeliads in Cultivation

fl. hab. Pl. 52 - 60

Aechmea kerteszia

Rauh, Werner
Bromeliads for Home, Garden and
   Greenhouse

fl. Pl. 70

Aechmea lamarchei

Bromeliad Society
Bulletin
   vol 16, No. 2 1966

fl. p. 48

Aechmea luddemanniana

Padilla, Victoria
Bromeliads

fl. p. 63   Pl. 1

Aechmea luddemanniana

Wilson, Robert Gardner
Bromeliads in Cultivation

fl. Pl. 25

Aechmea luddemanniana var.

Bromeliad Society
Journal
   vol 21, No. 6 1971

hab. p. 123

Aechmea luddemanniana var.

Bromeliad Society
Journal
   vol 26, No. 2 1976

fl. p. 67

Aechmea magdalenae var.

Bromeliad Society
Bulletin
   vol 16, No. 2 1966

fl., hab. cover (p. 25)

Aechmea mariae-reginae

Bromeliad Society
Bulletin
   vol6, No. 6 1956

fl. cover (p. 81)

Aechmea mariae-reginae

Wilson, Robert Gardner
Bromeliads in Cultivation

fl. hab. Pl. 26

Aechmea mariae-reginae

Rauh, Werner
Bromeliads for Home, Garden and
    Greenhouse

fl.   Pl. 4

Aechmea marie var.

Bromeliad Society
Bulletin
    vol 20, No. 5 1970

fl.   p. 120

Aechmea marmorata

Wilson, Robert Gardner
Bromeliads in Cultivation

fl., hab.   Pl. 28, 125

Aechmea melinonii

Bromeliad Society
Bulletin
    vol 19, No. 1 1969

fl.   p. 24

Aechmea mexicana

Bromeliad Society
Bulletin
    vol 15, No. 2 1965

hab.   p. 44

Aechmea mexicana

Bromeliad Society
Journal
    vol 26, No. 2 1976

fl., hab.   p. 67

Aechmea mexicana

Wilson, Robert Gardner
Bromeliads in Cultivation

fl, hab.   Pl. 31

Aechmea miniata var.

Padilla, Victoria
Bromeliads

fl., hab.   p. 64   Pl. 2

Aechmea miniata var.

Padilla, Victoria, ed.
Bromeliads in Color and
    Their Culture

fl.   p 15

Aechmea miniata var.

Rauh, Werner
Bromeliads for Home, Garden and
    Greenhouse

fl.   Pl. 73

Aechmea miniata var.

Wilson, Robert Gardner
Bromeliads in Cultivation

fl.   Pl. 27

Aechmea mooreana

Bromeliad Society
Bulletin
    vol 15, No. 6 1965

fl.   p. 140

Aechmea mulfordii

Bromeliad Society
Journal
    vol 25, No. 2 1975

fl.   p. 58

Aechmea mulfordii

Bromeliad Society
Journal
    vol 26, No. 2 1976

fl.   cover (p. 45)

Aechmea nidularioides

Bromeliad Society
Bulletin
    vol 14, No. 3 1964

hab., fl.   p. 56

Aechmea nidularioides

Bromeliad Society
Bulletin
    vol 20, No. 6 1970

fl.   p. 136

Aechmea nidularioides

Bromeliad Society
Journal
    vol 27, No. 4 1977

fl., hab.   p. 169

Aechmea nidularioides

Wilson, Robert Gardner
Bromeliads in Cultivation

fl.   Pl. 29

Aechmea nudicaulis

Bromeliad Society
Bulletin
    vol 15, No. 6 1965

fl.   cover (p. 117)

Aechmea nudicaulis

Bromeliad Society
Journal
    vol 24, No. 3 1974

fl.   p. 95

Aechmea nudicaulis

Bromeliad Society
Journal
    vol 26, No. 2 1976

fl.   p. 67

Aechmea nudicaulis

Padilla, Victoria
Bromeliads

fl.   p. 64, 70   Pl. 2, 8

Aechmea nudicaulis var.

Rauh, Werner
Bromeliads for Home, Garden and
    Greenhouse

fl.   Pl. 74

Aechmea nudicaulis var.

Wilson, Robert Gardner
Bromeliads in Cultivation

fl. hab.   Pl. 30, 34

Aechmea orlandiana

Bromeliad Society
Journal
    vol 21, No. 5 1971

fl.   cover (p. 97)

Aechmea orlandiana

Massachusetts Hort. Soc.
Horticulture
    vol 36, No. 11 1958

hab.   p. 577

Aechmea orlandiana

Massachusetts Hort. Soc.
Horticulture
    vol 43, No. 6 1965

fl., hab.   p. 25

Aechmea orlandiana

Padilla, Victoria
Bromeliads

fl.   p. 64   Pl. 2

Aechmea orlandiana

Rauh, Werner
Bromeliads for Home, Garden and
    Greenhouse

fl.   Pl. 75

Aechmea orlandiana

Smith, Lyman B.
The Bromeliads

fl. hab.   Pl. 23

Aechmea orlandiana

Wilson, Robert Gardner
Bromeliads in Cultivation

fl. hab.   Pl. 32

Aechmea orlandiana var.

Bromeliad Society
Bulletin
    vol 20, No. 4 1970

hab.   p. 96

Aechmea orlandiana var.

Bromeliad Society
Journal
    vol 22, No. 4 1972

fl., hab.   p. 102

Aechmea orlandiana var.

Bromeliad Society
Journal
    vol 26, No. 6   1976

fl., hab.   p. 268

Aechmea ornata

Bromeliad Society
Journal
    vol 24, No. 2   1974

fl.   p. 80

Aechmea ornata

Rauh, Werner
Bromeliads for Home, Garden and
    Greenhouse

fl.   Pl. 76

Aechmea ornata var.

Wilson, Robert Gardner
Bromeliads in Cultivation

hab.   Pl. 36

Aechmea pectinata

Bromeliad Society
Journal
    vol 21, No. 6   1971

hab., fl.   p. 148

Aechmea penduliflora

Wilson, Robert Gardner
Bromeliads in Cultivation

fl.   Pl. 37

Aechmea pimenti-velosoi

Wilson, Robert Gardner
Bromeliads in Cultivation

fl.   Pl. 38

Aechmea pineliana

Rauh, Werner
Bromeliads for Home, Garden and
    Greenhouse

fl.   Pl. 78

Aechmea pineliana

Wilson, Robert Gardner
Bromeliads in Cultivation

fl. hab.   Pl. 39

Aechmea pineliana var.

Bromeliad Society
Journal
    vol 22, No. 4   1972

hab., fl.   p. 104

Aechmea pittieri

Wilson, Robert Gardner
Bromeliads in Cultivation

fl. hab.   Pl. 33

Aechmea poetaei

Bromeliad Society
Bulletin
    vol 14, No. 3   1963

hab., fl.   p. 58

Aechmea politii

Bromeliad Society
Journal
    vol 28, No. 3   1978

fl.   cover (p. 97)

Aechmea pubescens

Wilson, Robert Gardner
Bromeliads in Cultivation

fl.   Pl. 40

Aechmea purpureo-rosea

Wilson, Robert Gardner
Bromeliads in Cultivation

fl. hab.   Pl. 35

Aechmea racinae

Bromeliad Society
Bulletin
    vol 13, No. 6   1963

fl.   p. 123

Aechmea racinae

Padilla, Victoria
Bromeliads

fl.   p. 64   Pl. 2

Aechmea racinae

Padilla, Victoria, ed.
Bromeliads in Color and
    Their Culture

fl.   p 30

Aechmea racinae

Wilson, Robert Gardner
Bromeliads in Cultivation

fl., hab.   Pl. 41, 120

Aechmea ramosa

Wilson, Robert Gardner
Bromeliads in Cultivation

fl.   Pl. 42

Aechmea ramosa var.

Bromeliad Society
Bulletin
    vol 20, No. 3   1970

fl., hab.   p. 72

Aechmea recurvata

Bromeliad Society
Journal
    vol 26, No. 2   1976

fl., hab.   p. 66

Aechmea recurvata var.

Bromeliad Society
Bulletin
    vol 15, No. 3   1965

fl., hab.   cover (p. 45)

Aechmea recurvata var.

Bromeliad Society
Bulletin
    vol 16, No. 5   1966

fl., hab.   p. 120

Aechmea recurvata var.

Bromeliad Society
Bulletin
    vol 18, No. 1   1968

fl.   cover (p. 1), 24

Aechmea recurvata var.

Bromeliad Society
Bulletin
    vol 20, No. 6   1970

fl.   p. 136

Aechmea recurvata var.

Padilla, Victoria
Bromeliads

fl.   p. 64   Pl. 2

Aechmea recurvata var.

Rauh, Werner
Bromeliads for Home, Garden and
    Greenhouse

fl.   Pl. 80, 81

Aechmea recurvata var.

Wilson, Robert Gardner
Bromeliads in Cultivation

hab.   Pl. 44, 46

Aechmea selloana

Addisonia
    Vol 24   1960-64

fl., fr.   p. 61   Pl. 799

Aechmea servitensis

Bromeliad Society
Journal
    vol 28, No. 2   1978

fl.   p. 76

Aechmea setigera

Wilson, Robert Gardner
Bromeliads in Cultivation

fl. hab.   Pl. 45

Aechmea tessmannii

Bromeliad Society
Journal
    vol 26, No. 2   1976

fl.   p. 66

Aechmea tessmannii

Bromeliad Society
Journal
        vol. 27, No. 1   1977

fl.   cover

Aechmea teesmannii

Royal Hort. Soc.
Journal of the Royal Hort. Soc.
    Vol. 91, No. 8   1966

hab.   p. 338   Pl. 174

Aechmea tillandsioides

Bromeliad Society
Journal
    vol 24, No. 3   1974

fl., hab.   p. 116

Aechmea tillandsioides

Bromeliad Society
Journal
    vol 25, No. 5   1975

fl.   p. 200

Aechmea tillandsioides

Padilla, Victoria, ed.
Bromeliads in Color and
    Their Culture

fl.   p. 30

Aechmea tillandsioides

Perry, Frances
Flowers of the World

fl.   p 52

Aechmea tillandsioides var.

Bromeliad Society
Bulletin
    vol 16, No. 1   1966

fl.   p. 24

Aechmea tillandsioides var.

Bromeliad Society
Journal
    vol 24, No. 1   1974

fl.   p. 10

Aechmea tillandsioides var.

Wilson, Robert Gardner
Bromeliads in Cultivation

fl.   Pl. 48

Aechmea triangularis

Bromeliad Society
Bulletin
    vol 14, No. 6   1964

fl.   cover(p. 105)

Aechmea triangularis

Padilla, Victoria
Bromeliads

fl.   p. 64   Pl. 2

Aechmea triangularis

Wilson, Robert Gardner
Bromeliads in Cultivation

hab.   Pl. 43

Aechmea vanhoutteana

Smith, Lyman B.
The Bromeliads

fl.   Pl. 27

Aechmea veitchii

Rauh, Werner
Bromeliads for Home, Garden and
    Greenhouse

fl.   Pl. 79

Aechmea veitchii

Wilson, Robert Gardner
Bromeliads in Cultivation

fl.   Pl. 49

Aechmea victoriana

Bromeliad Society
Journal
    vol 27, No. 1   1977

fl.   p. 24

Aechmea victoriana

Wilson, Robert Gardner
Bromeliads in Cultivation

fl.   Pl. 47

Aechmea weilbachii

Massachusetts Hort. Soc.
Horticulture
    vol 36, No. 11   1958

fl.   p. 577

Aechmea weilbachii

Massachusetts Hort. Soc.
Horticulture
    vol 43, No. 6   1965

hab., fl.   p. 25

Aechmea weilbachii

Massachusetts Hort. Soc.
Horticulture
    vol 45, No. 8   1967

fl., hab.   p. 25

Aechmea weilbachii

Rauh, Werner
Bromeliads for Home, Garden and
    Greenhouse

fl.   Pl. 83

Aechmea weilbachii

Wilson, Robert Gardner
Bromeliads in Cultivation

fl.   Pl. 50

Aechmea zebrina

Bromeliad Society
Journal
    vol 27, No. 3   1977

fl., hab.   p. 130

Aechmea zebrina

Wilson, Robert Gardner
Bromeliads in Cultivation

fl.   Pl. 51

Aegilops ovata

Polunin, Oleg
Flowers of Europe

fr.   Pl. 178 #1730

Aeginetia indica

Walden, Beryl M.
Wild Flowers of Hong Kong

fl.   Pl. 55 (165)

Aegopodium podagraria

Ary, S.
The Oxford Book of Wildflowers

fl.   p 90, Pl. 2

Aegopodium podagraria

Lindman, C. A. M.
Nordens Flora, Vol 7

fl, fr.   Pl. 426

Aegopodium podagraria

Martin, W. Keble
The Concise British Flora in Colour

fl. hab.   Pl. 38

Aegopodium podagraria var.

Crockett, James Underwood
Lawns & Ground Covers

fl.   p. 120

Aeolanthus njassae

Moriarty, Audrey
Wild Flowers of Malawi

fl.   Pl. 54; 3

Aeonium arboreum

Everett, T.H., ed.
New Illustrated Encyclopedia of
    Gardening vol. 1

fl.   p. 71

Aeonium arboreum

Hay, Roy
The Dictionary of House Plants

hab.   Pl. 14

Aeonium arboreum

Kiaer, Eigil
Indoor Plants in Colour

hab.   p. 9

Aeonium arboreum

Lamb, Edgar
The Pocket Encyclopedia of Cacti
    and Succulents in Color

fl. hab.   Pl. 196, 197

Aeonium arboreum

Lamb, Edgar
Popular Exotic Cacti in Color

fl. hab.   p 21

Aeonium arboreum

Polunin, Oleg
Flowers of Europe

fl.   Pl. 39 #386

Aeonium arboreum var.

Kiaer, Eigil
Indoor Plants in Colour

hab.,p. 9

Aeonium canariense var.

Mathias, Mildred E.
Color for the Landscape

fl. hab.   p. 157, 158

Aeonium cuneatum

Bramwell, David
Wild Flowers of the Canary Islands

fl.   Pl. 147

Aeonium decorum

Addisonia
    Vol 24   1960-64

fl.   p. 41   Pl. 789

Aeonium holochrysum

Bramwell, David
Wild Flowers of the Canary Islands

fl., hab.   Pl. 150

Aeonium lancerottense

Bramwell, David
Wild Flowers of the Canary Islands

hab.   Pl. 155

Aeonium manriqueorum

Bramwell, David
Wild Flowers of the Canary Islands

fl., hab.   Pl. 152

Aeonium nobile

Bramwell, David
Wild Flowers of the Canary Islands

fl.   Pl. 159

Aeonium palmense

Lamb, Edgar
The Pocket Encyclopedia of Cacti
    and Succulents in Color

hab.   Pl. 314

Aeonium rubrolineatum

Lamb, Edgar
The Pocket Encyclopedia of Cacti
    and Succulents in Color

fl. hab.   Pl. 198

Aeonium sedifolium

Bramwell, David
Wild Flowers of the Canary Islands

fl., hab.   Pl. 153

Aeonium simsii

Bramwell, David
Wild Flowers of the Canary Islands

fl.   Pl. 158

Aeonium smithii

Bramwell, David
Wild Flowers of the Canary Islands

fl.   Pl. 160

Aeonium spathulatum

Bramwell, David
Wild Flowers of the Canary Islands

hab.   Pl. 157

Aeonium subplanum

Bramwell, David
Wild Flowers of the Canary Islands

hab.   Pl. 148

Aeonium tabuliforme

Bramwell, David
Wild Flowers of the Canary Islands

hab.   Pl. 151

Aeonium tabuliforme

Hay, Roy
The Dictionary of House Plants

hab.   Pl. 15

Aeonium undulatum

Bramwell, David
Wild Flowers of the Canary Islands

fl., hab.   Pl. 149

Aeonium urbicum

Bramwell, David
Wild Flowers of the Canary Islands

hab.   Pl. 154

Aeonium valverdense

Bramwell, David
Wild Flowers of the Canary Islands

fl., hab.   Pl. 156

Aerangis citrata

Grubb, Roy
Selected Orchidaceous Plants
    Vol I

fl. hab.   p 123

Aerangis citrata

Stewart, Joyce
Orchids of Tropical Africa

fl.   Pl. 2

Aerangis collum-cygni

Williamson, Graham
The Orchids of South Central Africa

fl.   Pl. 176

Aerangis coriacea

Curtis's Botanical Magazine
    Vol 171   1956-57

fl.   297

Aerangis friesiorum

Stewart, Joyce
Orchids of Tropical Africa

fl. hab.   Pl. 3

Aerangis kotschyana

Williamson, Graham
The Orchids of South Central Africa

fl.   Pl. 175

Aerangis mystacidii

Williamson, Graham
The Orchids of South Central Africa

fl.   Pl. 178

Aerangis rhodosticta

Curtis's Botanical Magazine
    Vol 171   1956-57

fl.   276

Aerangis rhodosticta

Stewart, Joyce
Orchids of Tropical Africa

fl.   Pl. 4

Aerangis rhodosticta var.
Piers, Frank
Orchids of East Africa
fl.   Pl. IIh

Aerangis sp.
Piers, Frank
Orchids of East Africa
fl., hab.   Pl. IIk

Aerangis sp.
Williamson, Graham
The Orchids of South Central Africa
fl.   Pl. 177

Aerangis thomsonii
Grubb, Roy
Selected Orchidaceous Plants
Vol I
fl. hab.   p 79

Aeranthes arachnites
Kramer, Jack
Orchids; Flowers of Romance
  and Mystery
fl.   p 279

Aeranthes caudata
Curtis's Botanical Magazine
  Vol. 180   1974
fl. hab.   Pl. 685

Aeranthes grandiflora
Stewart, Joyce
Orchids of Tropical Africa
fl.   Pl. 5

Aerides crassifolia
Kamemoto, Haruyuki
Beautiful Thai Orchid Species
fl.   p 120

Aerides falcata
Kamemoto, Haruyuki
Beautiful Thai Orchid Species
fl.   p 118

Aerides falcata var.
Kamemoto, Haruyuki
Beautiful Thai Orchid Species
fl.   p 118

Aerides falcata var.
Kupper, Walter
Orchideen
fl.   p. 115

Aerides flabellata
Kamemoto, Haruyuki
Beautiful Thai Orchid Species
fl.   p 119

Aerides houlletianum
Curtis's Botanical Magazine
Vol 170   1954-55
fl.   260

Aerides houlletianum
Kupper, Walter
Orchids
fl. hab.   p 115

Aerides mitrata
Kamemoto, Haruyuki
Beautiful Thai Orchid Species
fl.   p 120

Aerides multiflora
Kamemoto, Haruyuki
Beautiful Thai Orchid Species
fl.   p 119

Aerides odorata
Kamemoto, Haruyuki
Beautiful Thai Orchid Species
fl.   p 117

Aerva leucura
Moriarty, Audrey
Wild Flowers of Malawi
fl.   Pl. 44; 6

Aeschynanthus chinensis
Walden, Beryl M.
Wild Flowers of Hong Kong
fl.   Pl. 1(9), 3(9)

Aeschynanthus ellipticus
American Gloxinia & Gesneriad Society
The Gloxinian
     vol 19, No. 4   1970
fl.   cover

Aeschynanthus ellipticus
American Gloxinia & Gesneriad Society
The Gloxinian
     vol 27, No. 1   1977
fl.   p. 21

Aeschynanthus lobbianus
Encke, Fritz
Zimmerpflanzen
fl.   p. 48

Aeschynanthus lobbianus
Hay, Roy
The Dictionary of House Plants
fl.   Pl. 16

Aeschynanthus lobbianus
Menninger, Edwin A.
Flowering Vines of the World
fl.   Pl. 84

Aeschynanthus longiflorus
Morley, Brian D.
Wild Flowers of the World
fl. fr.   Pl. 122

Aeschynanthus mimetes
Curtis's Botanical Magazine
Vol 162   1939
fl.   No. 9595

Aeschynanthus obconicus
Elbert, Virginie F.
The Miracle Houseplants
fl.   p. 116

Aeschynanthus obconicus
Massachusetts Hort. Soc.
Horticulture
    vol 51, No. 6   1973
fl.   p. 23

Aeschynanthus pulcher
American Gloxinia & Gesneriad Society
The Gloxinian
    vol 27, No. 1   1977
fl.   p. 21

Aeschynanthus pulcher
Massachusetts Hort. Soc.
Horticulture
    vol 34, No. 1   1956
fl.   p. 20

Aeschynanthus pulcher
Massachusetts Hort. Soc.
Horticulture
    vol 35, No. 10   1957
fl.   p. 500

Aeschynanthus pulcher
Menninger, Edwin A.
Flowering Vines of the World
fl.   Pl. 80

Aeschynanthus pullobia
American Gloxinia & Gesneriad Society
The Gloxinian
    vol 27, No. 1   1977
fl.   cover

Aeschynanthus radicans

American Gloxinia & Gesneriad Society
The Gloxinian
vol 27, No. 1   1977

fl.   p. 20

Aeschynanthus speciosus

Perry, Frances
Flowers of the World

fl.   p 128

Aeschynanthus speciosus

Wit, H. C. D. de
Plants of the World;
The Higher Plants, Vol II

fl.   Pl. 96

Aeschynanthus x splendidus

American Gloxinia & Gesneriad Soc.
The Gloxinian
vol 27, No. 1   1977

fl.   p. 20

Aeschynanthus x splendidus

Menninger, Edwin A.
Flowering Vines of the World

fl.   Pl. 80

Aeschynanthus tricolor

Morley, Brian D.
Wild Flowers of the World

fl.   Pl. 122

Aeschynomene abyssinica

Moriarty, Audrey
Wild Flowers of Malawi

fl. fr.   Pl. 68; 2

Aeschynomene schimperi

Moriarty, Audrey
Wild Flowers of Malawi

fl. fr.   Pl. 68; 1

Aesculus californica

Mathias, Mildred E.
Color for the Landscape

fl. hab.   p 192

Aesculus californica

Munz, Philip A.
California Spring Wildflowers

fl.   p 34, Pl. 24

Aesculus x carnea

Boom, B. K.
The glory of the tree

fl.   p. 97   Pl. 168

Aesculus x carnea

Edlin, Herbert
The Illustrated Encyclopedia
of Trees

fl. hab.   p. 193

Aesculus x carnea

Hay, Roy
The Color Dictionary of Flowers
& Plants

fl.   p. 181   Pl. 1441

Aesculus x carnea

Huxley, Anthony
Deciduous Garden Trees and Shrubs

fl.   Pl. 10

Aesculus x carnea

Perry, Frances
Flowers of the World

fl.   p. 138

Aesculus x carnea

Polunin, Oleg
Flowers of Europe

fl.   Pl. 69 #713

Aesculus x carnea

Polunin, Oleg
Trees and Bushes of Europe

fl.   p. 128

Aesculus x carnea var.

Flemer, William III
Shade & Ornamental Trees in Color

hab.   p. 87

Aesculus x carnea var.

Harrison, R. E.
Trees & Shrubs

fl.   p. 21

Aesculus x carnea var.

Pacific Hort. Foundation
Pacific Horticultue
vol. 38, No. 1   1977

fl.   p. 46

Aesculus flava

Edlin, Herbert
The Illustrated Encyclopedia
of Trees

fr. hab.   p 193

Aesculus glabra

Wharton, Mary E.
Trees & Shrubs of Kentucky

fl.   p 59, Pl. 1.26a

Aesculus hippocastanum

Ary, S.
The Oxford Book of Wildflowers

fl. fr.   p 188, Pl. 5

Aesculus hippocastanum

Bartels, Andreas
Das Grosse Buch der Gartengeholze

hab.   p 110

Aesculus hippocastanum

Bianchini, Francesco
Health Plants of the World

fr.   p 175

Aesculus hippocastanum

Boom, B. K.
The glory of the tree

hab. fl. fr.   p.96   Pl. 165,
                                    167

Aesculus hippocastanum

Edlin, Herbert
The Illustrated Encyclopedia
of Trees

fl. fr. hab.   p 192

Aesculus hippocastanum

Huxley, Anthony
Deciduous Garden Trees and Shrubs

fl. fr.   Pl. 9a, 9b

Aesculus hippocastanum

Meikle, R. D.
British Trees and Shrubs

fl. fr.   Pl. 3

Aesculus hippocastanum

Morley, Brian D.
Wild Flowers of the World

fl. fr.   Pl. 19A

Aesculus hippocastanum

Oakman, Harry
Colorful Trees

Fl.   P 39

Aesculus hippocastanum

Perrot, Emile
Les Plantes Medicinales

fl. fr. hab.   p 140

Aesculus hippocastanum

Polunin, Oleg
Flowers of Europe

fl.   Pl. 69 #713

Aesculus hippocastanum

Polunin, Oleg
Trees and Bushes of Europe

fl. fr. hab.   p 128

Aesculus hippocastanum

Vedel, H.
Arbres et Arbustes

fl. fr. hab.   p 95, Pl. 98

Aesculus indica

Edlin, Herbert
The Illustrated Encyclopedia
of Trees

fl. hab.   p 193

Aesculus indica var.

Royal Hort. Soc.
Journal of the Royal Hort. Soc.
vol 93, No. 4   1968

fl., hab.   p. 164   Pl. 72

Aesculus x mutabilis var.

Cault, S. Millar
The Color Dictionary of Shrubs

fl.   Pl. 7

Aesculus octandra

Boom, B. K.
The glory of the tree

fl.fr.   p. 97   pl. 170-71

Aesculus octandra

Campbell, Carlos C.
Great Smoky Mountain Wildflowers

fl.   p 31

Aesculus octandra

Hay, Roy
The Color Dictionary of Flowers & Plants

fl.   p 181, Pl. 1442

Aesculus octandra

Wharton, Mary E.
Trees & Shrubs of Kentucky

fl.   p 60, Pl. 1.26b

Aesculus parryi

Coyle, Jeanette
A Field Guide to the Common and
Interesting Plants of Baja California

hab.   p 119

Aesculus parviflora

Bartels, Andreas
Das Grosse Buch der
Gartengeholze

fl.   p 110

Aesculus parviflora

Dean, Blanche E.
Wildflowers of Alabama and
Adjoining States

fl.   p 101

Aesculus parviflora

Everett, T.H., ed.
New Illustrated Encyclopedia of
Gardening vol. 1

fl.   p. 70

Aesculus parviflora

Hellyer, A.G.L.
Shrubs in Colour

fl.   p. 11

Aesculus parviflora var.

American Hort. Soc.
American Horticulturist
vol 51., No. 3 1972

fl., hab.   p. 11

Aesculus pavia

Batson, Wade T.
Wild Flowers in South Carolina

fl.   p 73

Aesculus pavia

Brown, Clair A.
Wildflowers of Louisiana and
Adjoining States

fl. hab.   p 107

Aesculus pavia

Walcott, Mary Vaux
North American Wild Flowers

fl. hab.   Pl. 47

Aesculus pavia

Wharton, Mary E.
Trees & Shrubs of Kentucky

fl.   p 60, Pl. 1.26c

Aesculus pavia var.

Boom, B.K.
The Glory of the Tree.

fl.   p. 97   Pl. 169

Aesculus turbinata

Kitamura, Siro
Coloured Illustrations of Trees &
Shrubs of Japan

fl.   318

Aethephyllum pinnatifidum

Herre, H.
The Genera of the Mesembryanthemaceae

fl. fr.   p 65

Aethionema grandiflora

Royal Hort. Soc.
Journal of the Royal Hort. Soc.
vol 90, No. 2   1965

fl.   p. 84   Pl. 29

Aethionema grandiflora

Vilmorin, Roger de
Plantes Alpines dans les Jardins

fl. hab.   Pl. III

Aethionema hybrid

Hay, Roy
The Color Dictionary of Flowers
& Plants

fl.   p. 1 Pl. 2

Aethionema saxatile

Polunin, Oleg
Flowers of Europe

fl.   Pl. 36 #346

Aethionema saxatile var.

Polunin, Oleg
Flowers of the Mediterranean

fl.   Pl. 43

Aethusa cynapium

Ary, S.
The Oxford Book of Wildflowers

fl.   p 86, Pl. 4

Aethusa cynapium

Klimas, John E.
Wildflowers of Eastern America

fl.   Pl. 85

Aethusa cynapium

Lindman, C. A. M.
Nordens Flora, Vol 7

fl. fr. hab.   Pl. 429

Aethusa cynapium

Martin, W. Keble
The Concise British Flora in Colour

fl. fr. hab.   Pl. 39

Aethusa cynapium

Perrot, Emile
Les Plantes Medicinales

fl. fr. hab.   p 68

Afgekia sericea

Menninger, Edwin A.
Flowering Vines of the World

fl.   Pl. 94

Aframomum angustifolium

Moriarty, Audrey
Wild Flowers of Malawi

fl.   Pl. 18; 6

Aframomum sceptrum

Morley, Brian D.
Wild Flowers of the World

fl.,hab.   Pl. 68

Aframomum sulcatum

Wit, H. C. D. de
Plants of the World;
The Higher Plants, Vol II

fr.   Pl. 147

Afrormosia angolensis

Palgrave, K. C.
Trees of Central Africa

fl., fr.   p. 311

Afternoon Flower

see

Lampranthus blandus

Afzelia cuanzensis

Palgrave, K. C.
Trees of Central Africa

fl., fr.   p. 63

Afzelia cuanzensis

Palmer, Eve
Trees of South Africa

fr.   p. 144   Pl. 1f

Afzelia quanzensis

Morley, Brian D.
Wild Flowers of the World

fl.,fr.   Pl. 54

Agalinis fasciculata

Brown, Clair A.
Wildflowers of Louisiana and
Adjoining States

fl.   p 166

Agalinis fasciculata

Duncan, Wilbur H.
Wildflowers of the Southeastern
United States

fl.   p. 175

Agalinis fasciculata

Fleming, Glenn
Wild Flowers of Florida

fl. hab.   p 85

Agalinis purpurea

Dean, Blanche E.
Wildflowers of Alabama and
Adjoining States

fl.   p 167

Agalinis purpurea

Duncan, Wilbur H.
Wildflowers of the Southeastern
United States

fl.   p. 177

Agalinis tenuifolia

Duncan, Wilbur H.
Wildflowers of the Southeastern
United States

fl. fr.   p. 177

Agalmyla parasitica

American Gloxinia & Gesneriad Society
The Gloxinian
vol 27, No. 6   1977

fl.   p. 21

Agalmyla parasitica

Elbert, Virginie F.
The Miracle Houseplants

fl.   p 116

Aganisia pulchella

American Hort. Soc.
American Horticulturist
vol 51, No. 4   1972

fl.   p. 28

Agapanthus africanus

Bruggeman, L.
Tropical Plants

fl.   Pl. 67

Agapanthus africanus

Hay, Roy
The Dictionary of House Plants

fl.   Pl. 17

Agapanthus africanus

Kiaer, Eigil
Indoor Plants in Colour

fl.   p. 9

Agapanthus africanus

Macoby, Stirling
What Flower is That

fl.   P 32

Agapanthus africanus

Mathias, Mildred E.
Color for the Landscape

fl. hab.   p 138

Agapanthus africanus

Tsukamoto, Yotaro
Coloured Illustrations of Garden
Flowers   Vol. 9

fl.   p. 28

Agapanthus campanulatus

Hay, Roy
The Dictionary of House Plants

fl.   Pl. 18

Agapanthus campanulatus var.

Flowering Plants of Africa
vol 37   1965-66

fl.   Pl. 1478

Agapanthus caulescens

Royal Hort. Soc.
Journal of the Royal Hort. Soc.
vol 92, No. 8   1967

fl.   p. 344   Pl. 176

Agapanthus caulescens var.

Curtis's Botanical Magazine
Vol 179   1972

fl.   Pl. 632

Agapanthus hybrid

Hay, Roy
The Color Dictionary of Flowers & Plants

fl.   p 120, Pl. 953

Agapanthus hybrid

Perry, Frances
Flowers of the World

fl.   p. 20

Agapanthus hybrid

Royal Hort. Soc.
The Garden
vol 103, No. 8   1978

fl.   p. 318

Agapanthus inapertus

Curtis's Botanical Magazine
Vol 163   1940-42

fl.   No. 9621

Agapanthus inapertus var.

Flowering Plants of Africa
Vol. 37   1965-66

fl.   Pl. 1479, 80

Agapanthus orientalis

Nicolaisen, Age
Pocket Encyclopedia of Indoor
Plants

fl.   Pl. 18

Agapanthus patens

Curtis's Botanical Magazine
Vol. 173   1960

fl.   Pl. 380

Agapanthus, pink

see

Nerine bowdenii

Agapanthus praecox var.

Flowering Plants of Africa
vol 37   1965-66

fl.   Pl. 1476, 77

Agapanthus walshiii

Flowering Plants of Africa
Vol. 42   1972-73

fl., fr.   Pl. 1675

Agapetes hosseana

Royal Hort. Soc.
The Garden
vol 102, No. 2   1977

fl.   p. 66

Agapetes macrantha

Perry, Frances
Flowers of the World

fl.   p 112

Agapetes odontocera

Morley, Brian D.
Wild Flowers of the World

fl.   Pl. 101

Agapetes serpens

Perry, Frances
Flowers of the World

fl.   p 112

Agastache foeniculum

Moyle, John B.
Northland Wild Flowers

fl.   p 137, Pl. 152

Agastache mexicana

Curtis's Botanical Magazine
Vol 164   1943-48

fl.   No. 9685

Agastache mexicana

Macoby, Stirling
What Flower is That

fl.   P 33

Agastache mexicana

Meikle, R. D.
Garden Flowers

fl.   p. 353   Pl. 11

Agastache nepetoides

Wharton, Mary E.
A Guide to the Wildflowers & Ferns
of Kentucky

fl.   p 208   Pl. 2.35

Agastache rugosa

Kimura, Koiti
Japanese Medicinal Plants, Vol II

fl.   p 219

Agastache scrophulariaefolia

Courtenay, Booth
Wildflowers & Weeds

fl.   p 95

Agastachys odorata

Curtis, Winifred
The Endemic Flora of Tasmania
Vol 3

fl.   p 189

Agathis alba

Bruggeman, L.
Tropical Plants

fr.   Pl. 185

Agathis australis

Harvey, Norman B.
New Zealand Botanical Paintings

fr., hab.   p. 16-18   Pl. 5,6

Agathis damara

Hvass, Elsie
Plants That Feed and Serve Us

fl.   p 109, Pl. 233

Agathisanthemum globosum

Moriarty, Audrey
Wild Flowers of Malawi

fl.   Pl. 48; 2

Agathosma cerefolium

Harrison, R. E.
Trees and Shrubs

fl.   P 22

Agave americana

Kromdijk, G.
200 House Plants in Colour

hab.   Pl. 8

Agave americana

Lamb, Edgar
The Pocket Encyclopedia of Cacti
and Succulents in Color

hab.   Pl. 201

Agave americana

Nicolaisen, Age
Pocket Encyclopedia of Indoor
Plants

hab.   Pl. 33

Agave americana

Polunin, Oleg
Flowers of Europe

fl.   Pl. 171 #1660

Agave americana

Polunin, Oleg
Flowers of the Mediterranean

fl.   Pl. 255

Agave americana

Polunin, Oleg
Trees and Bushes of Europe

hab.   p 181

Agave americana

Wit, H. C. D. de
Plants of the World;
The Higher Plants, Vol II

fl. hab.   Pl. 129

Agave americana var

Bruggeman, L.
Tropical Plants

hab.   Pl. 68

Agave americana var.

Hay, Roy
The Dictionary of House Plants

hab.   Pl. 19, 20, 21

Agave americana var.

Kiaer, Eigil
Indoor Plants in Colour

hab.   p. 10

Agave americana var.

Lamb, Edgar
The Pocket Encyclopedia of Cacti
and Succulents in Color

hab.   Pl. 200

Agave americana var.

Lamb, Edgar
Popular Exotic Cacti in Color

hab.   p 23

Agave americana var.

Macoby, Stirling
What Flower is That

hab.   p. 33

Agave americana var.

Perry, Frances
Flowers of the World

hab. P. 15

Agave angustifolia var.

Massachusetts Hort. Soc.
Horticulture
vol 53, No. 11 1975

hab.   cover, backcover

Agave attenuata

Macoby, Stirling
What Flower is That

fl.   P 33

Agave attenuata

Morley, Brian D.
Wild Flowers of the World

fl.   Pl. 168

Agave attenuata

Royal Hort. Soc.
Journal of the Royal Hort. Soc.
vol 91, No. 10 1966

fl.   p. 426   Pl. 215

Agave deserti

Coyle, Jeanette
A Field Guide to the Common and
Interesting Plants of Baja California

fl. hab.   p 59

Agave deserti

Lemmon, Robert S.
Wildflowers of North America in
Full Color

fl., hab. p. 65   Pl. 105

Agave deserti

Munz, Philip A.
California Desert Wildflowers

fl.   p 29, Pl. 8, 9

Agave filifera

Hay, Roy
The Dictionary of House Plants

hab.   Pl. 22

Agave filifera

Lamb, Edgar
The Pocket Encyclopedia of Cacti
and Succulents in Color

hab.   Pl. 199

Agave parrasana

Lamb, Edgar
The Pocket Encyclopedia of Cacti
and Succulents in Color

hab.   Pl. 202

Agave parviflora

Lamb, Edgar
The Pocket Encyclopedia of Cacti
and Succulents in Color

hab.   Pl. 203

Agave shawii

Coyle, Jeanette
A Field Guide to the Common and
Interesting Plants of Baja California

fl. hab.   p 59

Agave shawii

Mathias, Mildred E.
Color for the Landscape

fl. hab.   p 161

Agave shawii

Orr, Robert T.
Wildflowers of Western America

fl.   Pl. 104

Agave sisalana

Hvass, Elsie
Plants That Feed and Serve Us.

fl.   p. 103   Pl. 221

Agave utahensis var.

Lamb, Edgar
The Pocket Encyclopedia of Cacti
and Succulents in Color

hab.   Pl. 204

Agave victoriae-reginae

Cactus & Succulent Soc. of America
Journal
vol 45, No. 2 1973

hab.   cover (p. 49)

Agave victoriae-reginae

Hay, Roy
The Dictionary of House Plants

hab.   Pl. 23

Agave victoriae-reginae

Kiaer, Eigil
Indoor Plants in Colour

hab.   p. 10

Agave victoriae-reginae

Lamb, Edgar
The Pocket Encyclopedia of Cacti
and Succulents in Color

hab.   Pl. 205

Agave victoriae-reginae

Subik, Rudolf
Decorative Cacti

hab.   p 95

Agave virginica

Brown, Clair A.
Wildflowers of Louisiana and
Adjoining States

fl. fr.   p 24

Agave virginica

Dean, Blanche E.
Wildflowers of Alabama and
Adjoining States

fl.   p 29

Agave virginica

Wharton, Mary E.
A Guide to the Wildflowers & Ferns
of Kentucky

fl.   p 57 Pl. 1.21

Ageratum conyzoides

Walden, Beryl M.
Wild Flowers of Hong Kong

fl.   Pl. 36(97)

Ageratum corymbosum

O'Gorman, Helen
Mexican Flowering Trees and Plants

fl. hab.   p 45

Ageratum houstonianum

Crockett, James Underwood
Annuals

fl.   P 91

Ageratum houstonianum

Hay, Roy
The Color Dictionary of Flowers & Plants

fl.   p 29, Pl. 225

Ageratum houstonianum

Macoby, Stirling
What Flower is That

fl.   P 33

Ageratum houstonianum

Tsukamoto, Yotaro
Coloured Illustrations of Garden
Flowers   Vol. 10

fl. Illus. 1   p. 1

Ageratum houstonianum var

Huxley, Anthony
Garden Annuals & Bulbs

fl.   Pl. 2b

Ageratum houstonianum var.

Kiaer, Eigil
Indoor Plants in Colour

fl.   p. 10

Ageratum, Wild

see

Eupatorium coelestinum

Aglaonema commutatum

Encke, Fritz
Zimmerpflanzen

fr., hab.   p. 52

Aglaonema commutatum

Everett, T.H., ed.
New Illustrated Encyclopedia of
    Gardening vol. 1

fr.   p. 71

Aglaonema commutatum

Hay, Roy
The Color Dictionary of Flowers & Plants

fl.   p 51, Pl. 403

Aglaonema commutatum

Hay, Roy
The Dictionary of House Plants

hab.   Pl. 24

Aglaonema commutatum

Hersey, Jean
Woman's Day Book of House Plants

hab.   p. 47

Aglaonema commutatum

Kiaer, Eigil
Indoor Plants in Colour

hab.   p. 11

Aglaonema crispum var.

Hay, Roy
The Dictionary of House Plants

hab.   Pl. 25

Aglaonema hybrid

Hay, Roy
The Color Dictionary of Flowers &
    Plants

hab.   p. 51   Pl. 404

Aglaonema hybrid

Kromdijk, G.
200 House Plants in Colour

hab.   Pl. 9

Aglaonema modestum

Addisonia
    Vol 19   1935-36

fl.   p. 5   Pl. 611

Aglaonema modestum

Kiaer, Eigil
Indoor Plants in Colour

hab.   p. 11

Aglaonema oblongifolium var.

Nicolaisen, Age.
Pocket Encyclopedia of Indoor Plants

hab.   Pl. 7

Aglaonema pseudo-bracteata

Macoby, Stirling
What Flower is That

hab. P 34

Aglaonema pseudo-bracteata

Nicholaisen, Age
Pocket Encyclopedia of Indoor Plants

hab.   Pl. 7

Aglaonema roebillinii

Nicolaisen, Age
Pocket Encyclopedia of Indoor
    Plants

hab.   Pl. 7

Aglaonema treubii

Hay, Roy
The Color Dictionary of Flowers & Plants

hab.   p 51, Pl. 405

Aglaonema treubii

Hay, Roy
The Dictionary of House Plants

hab.   Pl. 26

Aglaonema treubii

Hersey, Jean
Woman's Day Book of House Plants

fr.   p. 47

Agonis flexuosa

Blombery, Alec M.
What Wildflower is That

fl.   p 40, Pl. 76

Agonis flexuosa

Macoby, Stirling
What Flower is That

fl.   p. 34

Agonis flexuosa var.

Harrison, R. E.
Trees and Shrubs

hab.   p. 22

Agonis obtusissima

Blombery, Alec M.
What Wildflower is That

fl.   p 40, Pl. 77

Agoseris aurantiaca

Orr, Robert T.
Wildflowers of Western America

fl.   Pl. 83

Agoseris aurantiaca

Porsild, A. E.
Rocky Mountain Wild Flowers

fl. hab.   p 383

Agoseris aurantiaca

Taylor, Ronald J.
Mountain Wild Flowers

fl. hab.   p 77

Agoseris glauca

Moyle, John B.
Northland Wild Flowers

fl. hab.   p 196, Pl. 253

Agoseris glauca

Orr, Robert T.
Wildflowers of Western America

fl.   Pl. 82

Agoseris glauca

Welsh, Stanley L.
Flowers of the Canyon Country

fl. hab.   p 42

Agoseris glauca

Welsh, Stanley L.
Flowers of the Mountain Country

fl.   p. 48

Agoseris gracilens

Walcott, Mary Vaux
North American Wild Flowers

fr. hab.   Pl. 89

Agoseris graminifolia

Walcott, Mary Vaux
North American Wild Flowers

fl. fr. hab.   Pl. 88

Agoseris, Crassleaf

see

Agoseris graminifolia

Agoseris, Palo

see

Agoseris glauca

Agoseris, Slender

see

Agoseris gracilens

Agoseris villosa

Walcott, Mary Vaux
North American Wild Flowers

fl. hab.   Pl. 195

Agrimonia eupatoria

Ary, S.
The Oxford Book of Wildflowers

fl.   p 16, Pl. 6

Agrimonia eupatoria

Bianchini, Francesco
Health Plants of the World

fl.   p 171

Agrimonia eupatoria

Felsko, Elsa
A Book of Wild Flowers 2nd Ser.

fl.   p 17

Agrimonia eupatoria

Kleijn, H.
Beauty of the Wild Plant

fl.   p. 68   Pl. 84

Agrimonia eupatoria

Lindman, C. A. M.
Nordens Flora, Vol 5

fl. fr.   Pl. 323

Agrimonia eupatoria

Martin, W. Keble
The Concise British Flora in Colour

fl. hab.   Pl. 26

Agrimonia eupatoria

Perrot, Emile
Les Plantes Medicinales

fl. fr. hab.   p 30

Agrimonia gryp, sepala

Courtenay, Booth
Wildflowers & Weeds

fl.   p 41

Agrimonia parviflora

Wharton, Mary E.
A Guide to the Wildflowers & Ferns
of Kentucky

fl. p 149   Pl. 2.1

Agrimonia pubescens

Klimas, John E.
Wildflowers of Eastern America

fl.   Pl. 161

Agrimony

see

Agrimonia

Agrimony, Hemp

see

Eupatorium cannabinum

Agrimony, hemp

See Eupatorium purpureum

Agropyron junceiforme

Polunin, Oleg
Flowers of Europe

fl.   Pl. 178 #1727

Agropyron repens

Bianchini, Francesco
Health Plants of the World

fl. hab.   p 135

Agropyron repens

Crockett, James Underwood
Lawns & Ground Covers

fl.   p. 73

Agropyron repens

Perrot, Emile
Les Plantes Medicinales

fl. fr. hab.   p. 66

Agropyron repens

Polunin, Oleg
Flowers of Europe

fr. Pl. 178 #1728

Agropyron repens

Pond, Barbara
A Sampler of Wayside Herbs

fl.   Pl. XXVII

Agrostemma githago

Ary, S.
The Oxford Book of Wildflowers

fl.   p 106, Pl. 2

Agrostemma githago

Courtenay, Booth
Wildflowers & Weeds

fl.   p 25

Agrostemma githago

Crockett, James Underwood
Annuals

fl.   P 91

Agrostemma githago

Duncan, Wilbur H.
Wildflowers of the Southeastern
United States

fl.   p 35

Agrostemma githago

Felsko, Elsa
A Book of Wildflowers

fl.   P. 105

Agrostemma githago

Kleijn, H.
Beauty of the Wild Plant

fl.   pg. 39   pl. 43

Agrostemma githago

Lindman, C. A. M.
Nordens Flora, Vol 3

fl. fr. hab.   Pl. 206

Agrostemma githago

Martin, W. Keble
The Concise British Flora in Colour

fl. hab.   Pl. 14

Agrostemma githago

Polunin, Oleg
Flowers of Europe

fl.   Pl. 16 #163

Agrostemma githago

Weeds of the Southern United States

fl., hab.   p. 6

Agrostemma githago

Wharton, Mary E.
A Guide to the Wildflowers & Ferns
   of Kentucky

fl.   p. 124   Pl. 1.59

Agrostemma githago var.

Hay, Roy
The Color Dictionary of Flowers &
   Plants

fl.   p. 29   Pl. 226

Agrostemma githago var.

Huxley, Anthony
Garden Annuals & Bulbs

fl.   Pl. 1

Agrostis alba

Crockett, James Underwood
Lawns & Ground Covers

fr., hab.   p. 114

Agrostis humilis

Porsild, A. E.
Rocky Mountain Wild Flowers

fl. hab.   p 47

Agrostis nebulosa

Crockett, James Underwood
Annuals

fr, hab P 91

Agrostis tenuis

Crockett, James Underwood
Lawns & Ground Covers

fr., hab.   p. 78, 114

Agrostis tenuis

Lindman, C. A. M.
Nordens Flora, Vol 2

fr. hab.   Pl. 99

Agrostocrinum scabrum

Blombery, Alec M.
What Wildflower is That

fl.   p 40, Pl. 78

Aichryson bollei

Bramwell, David
Wild Flowers of the Canary Islands

fl., hab.   Pl. 145

Aichryson palmense

Bramwell, David
Wild Flowers of the Canary Islands

hab.   Pl. 146

Aichryson parlatorei

Bramwell, David
Wild Flowers of the Canary Islands

fl.   Pl. 144

Ailanthus altissima

Boom, B. K.
The glory of the tree

hab., fr.   p. 89   Pl. 149,
                                        151

Ailanthus altissima

Edlin, Herbert
The Illustrated Encyclopedia
   of Trees

hab.   p 184

Ailanthus altissima

Huxley, Anthony
Deciduous Garden Trees and Shrubs

fl. fr.   Pl. 11

Ailanthus altissima

Kitamura, Siro
Coloured Illustrations of Trees &
   Shrubs of Japan

fl.   267

Ailanthus altissima

Macoby, Stirling
What Flower is That

fl P 34

Ailanthus altissima

Polunin, Oleg
Flowers of Europe

fl.   Pl. 68 #694

Ailanthus altissima

Polunin, Oleg
Trees and Bushes of Europe

fl. fr. hab.   p 117

Ailanthus altissima

Wharton, Mary E.
Trees & Shrubs of Kentucky

fl.   p 87, Pl. 2.25
fr.   p 117, Pl. 1.4

Ainsliaea fragrans

Walden, Beryl M.
Wild Flowers of Hong Kong

fl. hab.   Pl. 68 (203)

Aipyanthus echioides

Morley, Brian D.
Wild Flowers of the World

fl.,fr.,hab.   Pl. 49

Air Plant

   see

Bryophyllum pinnatum

   Air Plant

      see

   forms of Bromeliaceae

Aira caryophyllea

Lindman, C. A. M.
Nordens Flora, Vol 2

fr. hab.   Pl. 93B

Aira praecox

Lindman, C. A. M.
Nordens Flora, Vol 2

fr. hab.   Pl. 93A

   Aivrons

      see

   Rubus chamaemorus

Ajuga australis

Blombery, Alec M.
What Wildflower is That

fl.   p 41, Pl. 79

Ajuga australis

Cochrane, G. R.
Flowers and Plants of Victoria

fl.   Pl. 149

Ajuga chamaepitys

Ary, S.
The Oxford Book of Wildflowers

fl.   p 28, Pl. 5

Ajuga chamaepitys

Bianchini, Francesco
Health Plants of the World

fl. hab.   p 149

Ajuga chamaepitys

Martin, W. Keble
The Concise British Flora in Colour

fl. hab.   Pl. 70

Ajuga chamaepitys

Polunin, Oleg
Flowers of Europe

hab.   Pl. 106 #1094

Ajuga chamaepitys

Polunin, Oleg
Flowers of the Mediterranean

fl.   Pl. 159

Ajuga genevensis

Courtenay, Booth
Wildflowers & Weeds

fl.   p 95

Ajuga genevensis

Martin, W. Keble
The Concise British Flora in Colour

fl. hab.   Pl. 70

Ajuga genevensis

Morley, Brian D.
Wild Flowers of the World

fl.   Pl. 20C

Ajuga genevensis

Polunin, Oleg
Flowers of Europe

fl.   Pl. 105 #1090

Ajuga pyramidalis

Barneby, T. P.
European Alpine Flowers in Colour

fl. hab.   Pl. 67, 6

Ajuga pyramidalis

Hay, Roy
The Color Dictionary of Flowers & Plants

fl.   p 120, Pl. 954

Ajuga pyramidalis

Lindman, C. A. M.
Nordens Flora, Vol 8

fl. hab.   Pl. 496

Ajuga pyramidalis

Martin, W. Keble
The Concise British Flora in Colour

fl. hab.   Pl. 70

Ajuga pyramidalis

Polunin, Oleg
Flowers of Europe

fl.   Pl. 106 #1091

Ajuga pyramidalis

Tosco, Uberto
The World of Mountain Flowers

fl. hab.   p 21

Ajuga reptans

Ary, S.
The Oxford Book of Wildflowers

fl.   p. 144   Pl. 1

Ajuga reptans

Barneby, T. P.
European Alpine Flowers in Colour

fl.   Pl. 68, 1

Ajuga reptans

Crockett, James Underwood
Lawns & Ground Covers

fl.   p. 120

Ajuga reptans

Felsko, Elsa
A Book of Wildflowers

fl.   P 48

Ajuga reptans

Furrer, D.
Blumen am Wege

fl. hab.   p 85

Ajuga reptans

Lindman, C. A. M.
Nordens Flora, Vol 8

fl. hab.   Pl. 495

Ajuga reptans

Macoby, Stirling
What Flower is That

fl.   P 35

Ajuga reptans

Martin, W. Keble
The Concise British Flora in Colour

fl. hab.   Pl. 70

Ajuga reptans

Mathias, Mildred E.
Color for the Landscape

fl. hab.   p 136

Ajuga reptans

Perrot, Emile
Les Plantes Medicinales

fl. hab.   p 28

Ajuga reptans

Polunin, Oleg
Flowers of Europe

fl.   Pl. 106 #1089

Ajuga reptans var.

Hay, Roy
The Color Dictionary of Flowers & Plants

fl.   p. 120   Pl. 955

Ajuga reptans var.

Huxley, Anthony
Garden Perennials & Water Plants

fl.   Pl. 26

Ajuga reptans var.

Perry, Frances
Flowers of the World

fl., hab.   p. 152

Akebia, Five-Leaf

see

Akebia quinata

Akebia quinata

Crockett, James Underwood
Lawns & Ground Covers

fl.   p. 121

Akebia quinata

Harrison, R.E.
Climbers & trailers

fl.   p. 17 Pl. 3

Akebia quinata

Hellyer, A.G.L.
Shrubs in Colour

fl.   p. 11

Akebia quinata

Kariyone, Tatsuo
Atlas of Medicinal Plants

fl.   Pl. 121

Akebia quinata

Kimura, Koiti
Japanese Medicinal Plants, Vol II

fl.   p 161

Akebia quinata

Kitamura, Siro
Coloured Illustrations of Trees & Shrubs of Japan

fl.   153

Akebia quinata

Macoby, Stirling
What Flower is That

fl.   P 35

Akebia quinata

Morley, Brian D.
Wild Flowers of the World

fl.   Pl. 90

Akebia quinata

Perry, Frances
Flowers of the World

fl.   p 156

Akebia trifoliata

Kitamura, Siro
Coloured Illustrations of Trees &
    Shrubs of Japan

fl.   155

Akee

See Blighia sapida

Alamania punicea

Ospina, Mariano
Orquideas de las americas

fl.   Pl. 48

Alangium handelii

Walden, Beryl M.
Wild Flowers of Hong Kong

fl.   Pl. 31(82)

Alangium platanifolium var.

Kitamura, Siro
Coloured Illustrations of Trees &
    Shrubs of Japan

fl.   362

Alberta magna

Flowering Plants of Africa
    vol 29   1952-53

fl.   Pl. 1125

Alberta magna

Harrison, R. E.
Trees and Shrubs

fl.   P. 22

Alberta magna

Palmer, Eve
Trees of South Africa

fl.   p. 304   Pl. XXIX

Albizia amara var.

Palgrave, K. C.
Trees of Central Africa

fl., fr.   p. 259

Albizia antunesiana

Palgrave, K. C.
Trees of Central Africa

fl., fr.   p. 263

Albizia distachya

Kiaer, Eigil
Indoor Plants in Colour

fl.   p. 11

Albizia gummifera

Palgrave, K. C.
Trees of Central Africa

fl., fr.   p. 267

Albizia julibrissin

Everett, T.H., ed.
New Illustrated Encyclopedia of
    Gardening vol. 1

fl., hab.   p. 23

Albizia julibrissin

Flemer, William III
Shade and Ornamental Trees in Color

fl., hab.   p. 76, 100

Albizia julibrissin

Kitamura, Siro
Coloured Illustrations of Trees &
    Shrubs of Japan

fl fr.   241, 242

Albizia julibrissin

Massachusetts Hort. Soc.
Horticulture
    vol. 37, No. 6   1959

fl.   backcover

Albizia julibrissin

Mathias, Mildred E.
Color for the Landscape

fl. hab.   p. 7

Albizia julibrissin

Polunin, Oleg
Flowers of Europe

fl.   Pl. 50 #493

Albizia julibrissin

Polunin, Oleg
Trees and Bushes of Europe

fl. hab.   p 105

Albizia julibrissin var.

Massachusetts Hort. Soc.
Horticulture
    vol. 42, No. 7   1964

fl.   inside backcover

Albizia julibrissin var.

Oakman, Harry
Colorful Trees

Hab., fr.   p. 89

Albizia lebbeck

Everett, Thomas H.
Living Trees of the World

fr.   p. 177

Albizia lebbedoides

Oakman, Harry
Colorful Trees

Hab., fr.   p. 77

Albizia lophantha

Blombery, Alec M.
What Wildflower is That

fl.   p. 42   Pl. 80

Albizia lophantha

Curtis's Botanical Magazine
Vol 171   1956-57

fl.   304

Albizia lophantha

Macoby, Stirling
What Flower is That

fl.   P 35

Albizia lophantha

Royal Hort. Soc.
Journal of the Royal Hort. Soc.
    Vol. 93, No. 4   1968

fl.   p. 164   Pl. 82

Albizia, Plume

see

Albizia distachya

Albizia, plume

see

Albizia lophantha

Albizia rhodesica

Palgrave, K. C.
Trees of Central Africa

fl., fr.   p. 271

Albuca aperta

Flowering Plants of Africa
    vol 33   1959

fl., hab.   Pl. 1314

Albuca bainesii

Flowering Plants of South Africa
vol XX   1940

fl., hab.   Pl. 773

Albuca namaquensis

Flowering Plants of Africa
vol 32   1957-58

fl., hab.   Pl. 1255

Albuca nyikensis

Moriarty, Audrey
Wild Flowers of Malawi

fl.   Pl. 11; 2

Albuca patersoniae

Flowering Plants of Africa
vol XXVI   1947

fl., hab.   Pl. 1022

Albuca sp.

Moriarty, Audrey
Wild Flowers of Malawi

fl.   Pl. 11; 3

Albuca spiralis

Eliovson, Sima
Namaqualand in Flower

fl.   Pl. 25, 2

Albuca transvaalensis

Flowering Plants of Africa
vol XXVI   1947

fl., fr., hab.   Pl. 1009

Alcea pallida

Polunin, Oleg
Flowers of Europe

fl.   Pl. 72 #748

Alcea rosea

Perry, Frances
Flowers of the World

fl.   p 182

Alchemilla alpina

Martin, W. Keble
The Concise British Flora in Colour

fl. hab.   Pl. 26

Alchemilla conjuncta

Barneby, T. P.
European Alpine Flowers in Colour

fl. hab.   Pl. 38, 2

Alchemilla conjuncta

Martin, W. Keble
The Concise British Flora in Colour

fl. hab.   Pl. 26

Alchemilla conjuncta

Morley, Brian D.
Wild Flowers of the World

fl. fr.   Pl. 9E

Alchemilla filicaulis var.

Martin, W. Keble
The Concise British Flora in Colour

fl. hab.   Pl. 26

Alchemilla glabra

Martin, W. Keble
The Concise British Flora in Colour

fl. hab.   Pl. 26

Alchemilla hoppeana

Kohlhaupt, Paula
Fleurs des Alpages Vol 2

Fl.   P 55

Alchemilla mollis

Hay, Roy
The Color Dictionary of Flowers & Plants

fl.   p 120, Pl. 956

Alchemilla mollis

Royal Hort. Soc.
The Garden
vol 100, No. 9   1975

fl., hab.   p. 414

Alchemilla pentaphylla

Barneby, T. P.
European Alpine Flowers in Colour

fl. hab.   Pl. 38, 1

Alchemilla vulgaris

Ary, S.
The Oxford Book of Wildflowers

fl.   p 48, Pl. 1

Alchemilla vulgaris

Felsko, Elsa
A Book of Wild Flowers
2nd Ser.
fl. hab.   p 120

Alchemilla vulgaris

Kohlhaupt, Paula
Fleurs des Alpages   Vol 2

Fl.   P 56

Alchemilla vulgaris

Lindman, C. A. M.
Nordens Flora, Vol 5

fl. hab.   Pl. 317

Alchemilla vulgaris

Loewenfeld, Claire
The Complete Book of Herbs and Spices

fl. hab.   p 129

Alchemilla vulgaris

Perrot, Emile
Les Plantes Medicinales

fl. fr. hab.   p. 7

Alchornea cordifolia

Wit, H. C. D. de
Plants of the World;
The Higher Plants, Vol I

fr.   p 222, Pl. 140

Alder

see

Alnus

Alder, Black

See

Alnus glutinosa

Alder, Cherry

see

Eugenia luehmanni

Alder, Common

see

Alnus glutinosa

Alder, European

See

Alnus viridis var.

Alder, Grey

see

Alnus incana

Alder, Hazel

see

Alnus rugosa

Alder, Italian

see

Alnus cordata

Alder, Mountain

see

Alnus crispa

Alder, Oregon

see

Alnus rubra

Alder, Red

see

Alnus rubra

Alder, Western Green

see

Alnus sinuata

Alder, White

see

Clethra alnifolia

Aldrovanda vesiculosa

Stodola, Jiri
Encyclopedia of Water Plants

hab.   P. 63

Alecost

see

Chrysanthemum balsamita

Alectra sessiflora

Moriarty, Audrey
Wild Flowers of Malawi

fl.   Pl. 40; 1

Alectryon connatus

Wit, H. C. D. de
Plants of the World;
    The Higher Plants, Vol II

fl. fr.   p 58, Pl. 25
                p 59, Pl. 29

Alectryon excelsus

Curtis's Botanical Magazine
Vol 177   1969-70

fl. fr.   542

Alepidea longifolia

Moriarty, Audrey
Wild Flowers of Malawi

fl.   Pl. 71; 2

Alerce

see

Tetraclinis articulata

Aletris aurea

Brown, Clair A.
Wildflowers of Louisiana and
    Adjoining States

fl.   p 14

Aletris farinosa

Courtenay, Booth
Wildflowers and weeds

fl.   p. 7

Aletris farinosa

Dean, Blanche E.
Wildflowers of Alabama and
    Adjoining States

fl.   p 19

Aletris farinosa

Duncan, Wilbur H.
Wildflowers of the Southeastern
    United States

fl.   p 261

Aletris farinosa

Klimas, John E.
Wildflowers of Eastern America

fl.   Pl. 29

Aletris farinosa

Wharton, Mary E.
A Guide to the Wildflowers & Ferns
    of Kentucky

fl.   p 55 Pl. 1.17

Aletris lutea

Brown, Clair A.
Wildflowers of Louisiana and
    Adjoining States

fl.   p 15

Aletris lutea

Duncan, Wilbur H.
Wildflowers of the Southeastern
    United States

fl.   p 261

Aletris, White-Flowered American

See

Aletris farinosa

Aleurites fordii

American Hort. Soc.
American Horticulturist
    vol 55, No. 6   1976

hab.   p. 7

Aleurites fordii

Everett, T.H., ed.
New Illustrated Encyclopedia of
    Gardening vol. 1

fl.   p. 55

Aleurites fordii

Kitamura, Siro
Coloured Illustrations of Trees &
    Shrubs of Japan

fr.   275

Alexander, Golden

see

Alexander, Golden

see

Zizia aptera

Alexander, Golden

see

Zizia aurea

Alexandergrass

see

Brachiaria plantaginea

Alexanders

see

Smyrnium olusatrum

Alfalfa

see

Medicago sativa

Algodon

see

Gossypium davidsonii

Algodoncillo of the Mountain

see

Abutilon integerrimum

Alisma gramineum

Stodola, Jiri
Encyclopedia of Water Plants

fl.   P. 175

Alisma lanceolatum

Martin, W. Keble
The Concise British Flora in Colour

fl. fr. hab.   Pl. 79

Alisma orientale

Kimura, Koiti
Japanese Medicinal Plants, Vol II

fl.   p 127

Alisma plantago-aquatica

Ary, S.
The Oxford Book of Wildflowers

fl.   p 108, Pl. 2

Alisma plantago-aquatica

Clark, Lewis J.
Wild Flowers of British Columbia

fl. fr.   p 6

Alisma plantago-aquatica

Courtenay, Booth
Wildflowers and weeds

fl. p. 1

Alisma plantago-aquatica

Huxley, Anthony
Garden Perennials and Water Plants

fl.   Pl. 282

Alisma plantago-aquatica

Lindman, C. A. M.
Nordens, Flora   Vol. 1

fl.   Pl. 46

Alisma plantago-aquatica

Martin, W. Keble
The Concise British Flora in Colour

fl. hab.   Pl. 79

Alisma plantago-aquatica

Stodola, Jiri
Encyclopedia of Water Plants

fl.   P. 175

Alisma plantago-aquatica

Takatori, Jisuke
Color Atlas of Medicinal Plants of Japan

fl.   Fig. 79A

Alisma plantago-aquatica var.

Kariyone, Tatsuo
Atlas of Medicinal Plants

fl., hab.   Pl. 25

Alisma subcordatum

Wharton, Mary E.
A Guide to the Wildflowers & Ferns
of Kentucky

fl.   p 82 Pl. 1.13

Alisma triviale

Klimas, John E.
Wildflowers of Eastern America

fl.   Pl. 50

Alkanet

see

Alkanna tinctoria

Alkanet

see

Anchusa azurea

Alkanet

See Pentaglottis sempervirens

Alkanet, Common

see

Anchusa officinalis

Alkanet, Italian

see

Anchusa italica

Alkanet, wood

See Anchusa myosotidiflora

Alkanna incana

Royal Hort. Soc.
Journal of Royal Hort. Soc.
  vol 82, No. 4   1957

fl.   p. 162   Pl. 57

Alkanna orientalis

Polunin, Oleg
Flowers of the Mediterranean

fl.   Pl. 152

Alkanna tinctoria

Morley, Brian D.
Wild Flowers of the World

fl.   Pl. 37

Alkanna tinctoria

Perrot, Emile
Les Plantes Medicinales

fl. fr. hab.   p 164

Alkanna tinctoria

Polunin, Oleg
Flowers of Europe

fl.   Pl. 103 #1061

Alkekengi

see

Physalis pubescens

All-Good

see

Chenopodium bonus-henricus

Allheal, Clown's

see

Stachys palustris

All-Seed

see

Chenopodium polyspermum

All-seed

see

Radiola linoides

Allspice

see

Pimenta dioica

Allspice

see

Pimenta officinalis

Allspice, Carolina

see

Calycanthus fertilis

Allspice, Carolina

see

Calycanthus floridus

Allamanda cathartica

Bruggeman, L.
Tropical Plants
fl.   pl. 1

Allamanda cathartica

Encke, Frits
Zimmerpflanzen

fl.   p. 52

Allamanda cathartica

Everett, T.H., ed.
New Illustrated Encylopedia of
   Gardening Vol 13

fl.   opp. p. 2407

Allamanda cathartica

Hall, Clarence E.
Flowers of the Islands in the Sun

fl.   p. 55   Pl. 10

Allamanda cathartica

Harrison, Richomnd E.
Climbers and Trailers

fl. p 18   Pl. 5

Allamanda cathartica

Hay, Roy
The Color Dictionary of Flowers & Plants

fl.   p 51, Pl. 406

Allamanda cathartica

Kromdijk, G.
200 House Plants in Colour

fl.   Pl. 10

Allamanda cathartica

Milne, Lorus
Living Plants of the World

fl.   p 201

Allamanda cathartica var.

Blunt, Wilfrid
Flora Superba

fl.   Pl. III

Allamanda cathartica var.

Hargreaves, Dorothy
Tropical Blossoms of the Caribbean

fl.   p. 3

Allamanda cathartica var.

Hay, Roy
The Dictionary of House Plants

fl.   Pl. 27

Allamanda cathartica var.

Kiaer, Eigil
Indoor Plants in Colour

fl.   p. 12

Allamanda cathartica var.

Perry, Francis
Flowers of the World

fl.   p. 29

Allamanda neriifolia

Harrison, Richomnd E.
Climbers and Trailers

fl.   p 18   Pl. 6

Allamanda neriifolia

Hay, Roy
The Dictionary of House Plants

fl.   Pl. 28

Allamanda neriifolia

Macoby, Stirling
What Flower is That

fl.   P 36

Allamanda neriifolia

Nicolaisen, Age
Pocket Encyclopedia of Indoor
   Plants

fl.   Pl. 134

Allamanda neriifolia

Wit, H. C. D. de
Plants of the World;
The Higher Plants, Vol II

fl.   p 102, Pl. 55

Allamanda violacea

Harrison, Richmond E.
Climbers and Trailers

fl.   p 19 Pl. 7

Allamanda violacea

Macoby, Stirling
What Flower is That

fl.   P 36

Allamanda violacea

Morley, Brian D.
Wild Flowers of the World

fl.   Pl. 180

Allamanda, yellow

See

Allamanda cathartica var.

Allegheny Vine

see

Adlumia fungosa

Alliaria officinalis

Klimas, John E.
Wildflowers of Eastern America

fl.   Pl. 15

Alliaria officinalis

Lindman, C. A. M.
Nordens Flora, Vol 4

fl. fr. hab.   Pl. 275

Alliaria officinalis

Wharton, Mary E.
A Guide to the Wildflowers & Ferns
   of Kentucky

fl. p 149   Pl. 2.3

Alliaria petiolata

Ary, S.
The Oxford Book of Wildflowers

fl.   p 68, Pl. 2

Alliaria petiolata

Felsko, Elsa
A Book of Wild Flowers 2nd Ser.

fl. hab.   p 113

Alliaria petiolata

Martin, W. Keble
The Concise British Flora in Colour

fl. hab.   Pl. 8

Alliaria petiolata

Polunin, Oleg
Flowers of Europe

fl.   Pl. 31 #289

Alligator Weed

see

Alternanthera philoxeroides

Allionia incarnata

Lemmon, Robert S.
Wildflowers of North America in
   Full Color

fl.   pg. 67   Pl. 108

Allionia incarnata

Welsh, Stanley L.
Flowers of the Canyon Country

fl.   p 13

Allium acuminatum

Clark, Lewis J.
Wild Flowers of British Columbia

fl.   p 7

Allium acuminatum

Curtis's Botanical Magazine
Vol 169   1952-53

fl.   213

Allium acuminatum

Welsh, Stanley L.
Flowers of the Canyon Country

fl.   p 19

Allium aflatunense

American Hort. Soc.
American Horticulturist
    vol 57, No. 5   1978

fl.   p. 22

Allium aflatunense

Hay, Roy
The Color Dictionary of Flowers & Plants

fl.   p 83, Pl. 658

Allium akaka

Mathew, Brian
Dwarf Bulbs

fl.   p. 16   Pl. 4

Allium akaka

Royal Hort. Soc.
Journal of the Royal Hort. Soc.
    vol 95, No. 4   1970

fl.   p. 166   Pl. 103

Allium akaka var.

Curtis's Botanical Magazine
Vol 160   1937

fl.   No. 9506

Allium albopilosum

Huxley, Anthony
Garden annuals and bulbs

fl.   Pl. 192

Allium albopilosum

Massachusetts Hort. Soc.
Horticulture
    vol 55, No. 3   1977

fl.   p. 36-37

Allium ampeloprasum

Polunin, Oleg
Flowers of Europe

fl.   Pl. 166 #1615

Allium ampeloprasum var.

Bianchini, F.
The Complete Book of Fruits & Vegetables

hab.   p 87

Allium ampeloprasum var.

Masefield, G. B.
The Oxford Book of Food Plants

fl.   Pl. 169, 2

Allium arenicola

Brown, Clair A.
Wildflowers of Louisiana and
    Adjoining States

fl.   p 16

Allium ascalonicum

Bianchini, F.
The Complete Book of Fruits & Vegetables

hab.   p 85

Allium atroviolaceum

Wendelbo, Per
Tulips and Irises of Iran and
    Their Relatives

fl.   p 19, Pl. 11

Allium beesianum

Hay, Roy
The Color Dictionary of Flowers & Plants

fl.   p 83, Pl. 659

Allium bivalve

Dean, Blanche E.
Wildflowers of Alabama and
    Adjoining States

fl.   p 27

Allium bodeanum

Wendelbo, Per
Tulips and Irises of Iran and
    Their Relatives

fl.   p 23, Pl. 15

Allium brandegei

Welsh, Stanley L.
Flowers of the Mountain Country

fl.   p. 6

Allium bulgaricum

Curtis's Botanical Magazine
Vol 170   1954-55

fl.   257

Allium bungei

Wendelbo, Per
Tulips and Irises of Iran and
    Their Relatives

fl.   p 19, Pl. 10

Allium caeruleum

Miles, Bebe
Bulbs for the Home Gardener

fl.   p 50

Allium calocephalum

Royal Hort. Soc.
Journal of the Royal Hort. Soc.
    vol 95, No. 4   1970

fl.   p. 166   Pl. 104

Allium canadense

Brown, Clair A.
Wildflowers of Louisiana and
    Adjoining States

fl.   p 15

Allium canadense

Courtenay, Booth
Wildflowers & Weeds

fl.   p 8

Allium canadense

Duncan, Wilbur H.
Wildflowers of the Southeastern
    United States

fl.   p 249

Allium canadense

Klimas, John E.
Wildflowers of Eastern America

fl.   Pl. 196

Allium canadense

Moyle, John B.
Northland Wild Flowers

fl.   p 206, Pl. 270

Allium canadense

Wharton, Mary E.
A Guide to the Wildflowers & Ferns
    of Kentucky

fl.   p 58 Pl. 1.23

Allium carinatum

Polunin, Oleg
Flowers of Europe

fl.   Pl. 165 #1607

Allium caspium

Mathew, Brian
Dwarf Bulbs

fl.   p. 16   Pl. 5

Allium caspium

Royal Hort. Soc.
Journal of the Royal Hort. Soc.
    vol 95, No. 4   1970

fl.   p. 166   Pl. 106

Allium cepa

Bianchini, F.
The Complete Book of Fruits & Vegetables

hab.   p 83

Allium cepa

Bianchini, Francesco
Health Plants of the World

hab.   p 153

Allium cepa

Color Treasury of Herbs & Other
    Medicinal Plants

fr.   p. 23   Pl. 15

Allium cepa

Hvass, Elsie
Plants That Feed and Serve Us

fl.   p 36, Pl. 67

Allium cernuum

Addisonia, Vol. 24   1960-64

fl.   opp. p. 17   Pl. 777

Allium cernuum

Clark, Lewis J.
Wild Flowers of British Columbia

fl.   p. 6

Allium cernuum

Courtenay, Booth
Wildflowers & Weeds

fl.   p 8

Allium cernuum

Hay, Roy
The Color Dictionary of Flowers & Plants

fl.   p 83, Pl. 660

Allium cernuum

Porsild, A. E.
Rocky Mountain Wild Flowers

fl. hab.   p 95

Allium cernuum

Walcott, Mary Vaux
North American Wild Flowers
    vol. 4

fl.   Pl. 304

Allium cernuum

Wharton, Mary E.
A Guide to the Wildflowers & Ferns
    of Kentucky

fl.   p 58   Pl. 1.22

Allium christophii

Hay, Roy
The Color Dictionary of Flowers & Plants

fl.   p 83, Pl. 661

Allium christophii

Pacific Hort. Foundation
Pacific Horticulture
    vol 39, No. 4   1978-79

fr.   p. 40

Allium christophii

Wendelbo, Per
Tulips and Irises of Iran and
    Their Relatives

fl.   p. 21   Pl. 14

Allium crenulatum

Clark, Lewis J.
Wild Flowers of British Columbia

fl.   p 23

Allium cuthbertii

Dean, Blanche E.
Wildflowers of Alabama and
    Adjoining States

fl.   p 25

Allium cuthbertii

Duncan, Wilbur H.
Wildflowers of the Southeastern
    United States

fl.   p 251

Allium cyaneum

Curtis's Botanical Magazine
    Vol 160   1937

fl. fr. hab.   No. 9483

Allium cyaneum

Perry, Frances
Flowers of the World

fl.   p 19

Allium cyathophorum var.

Curtis's Botanical Magazine
    Vol. 170   1954-55

fl.   252

Allium derderianum

Wendelbo, Per
Tulips and Irises of Iran and
    Their Relatives

fl.   p 21, Pl. 13

Allium egelii

Royal Hort. Soc.
Journal of the Royal Hort. Soc.
    vol 93, No. 2   1968

fl.   p. 94   Pl. 35

Allium falcifolium

Orr, Robert T.
Wildflowers of Western America

fl.   Pl. 281

Allium fimbriatum

Lemmon, Robert S.
Wildflowers of North America in
    Full Color

fl.   Pg. 64   Pl. 104

Allium fistulosum

Loewenfeld, Claire
The Complete Book of Herbs and Spices

hab.   p. 113

Allium fistulosum

Masefield, G. B.
The Oxford Book of Food Plants

hab.   Pl. 167, 6

Allium flavum

Hay, Roy
The Color Dictionary of Flowers &
    Plants

fl.   p. 83   Pl. 662

Allium flavum

Polunin, Oleg
Flowers of Europe

fl.   Pl. 165 #1606

Allium geyeri var.

Clark, Lewis J.
Wild Flowers of British Columbia

fl.   p 22

Allium giganteum

American Hort. Soc.
American Horticulturist
    vol 56, No. 6 1977

fl.   p. 7

Allium giganteum

Macoby, Stirling
What Flower is That

fl.   P 36

Allium giganteum

Massachusetts Hort. Soc.
Horticulture
    vol 49, No. 1   1971

fl.   p. 26

Allium giganteum

Miles, Bebe
Bulbs for the Home Gardener

fl.   p 51

Allium giganteum

Wendelbo, Per
Tulips and Irises of Iran and
    Their Relatives

fl.   p 23, Pl. 16

Allium helicophyllum

Wendelbo, Per
Tulips and Irises of Iran and
    Their Relatives

fl. hab.   p 23, Pl. 17

Allium insubricum

Barneby, T. P.
European Alpine Flowers in Colour

fl.   Pl. 3, 6

Allium insubricum

Kohlhaupt, Paula
Fleurs des Alpages Vol 2

fl.   P 10

Allium karataviense

American Hort. Soc.
American Horticulturist
    vol 56, No. 6   1977

fl.   p. 7

Allium karataviense

Hay, Roy
The Color Dictionary of Flowers & Plants

fl.   p 83, Pl. 663

Allium karataviense

Miles, Bebe
Bulbs for the Home Gardener

fl., fr.   p. 50, 199

Allium karataviense

Royal Hort. Soc.
Journal of the Royal Hort. Soc.
    vol 95, No. 4   1970

fl., hab.   cover

Allium lacunosum

Orr, Robert T.
Wildflowers of Western America

fl.   Pl. 59

Allium mirum

Mathew, Brian
Dwarf Bulbs

fl.   p. 16   Pl. 3

Allium mirum

Royal Hort. Soc.
Journal of the Royal Hort. Soc.
    vol 93, No. 2   1968

fl.   p. 54   Pl. 31

Allium mirum

Royal Hort. Soc.
Journal of the Royal Hort. Soc.
    vol 95, No. 4   1970

fl.   p. 166   Pl. 102

Allium moly

American Hort. Soc.
American Horticulturist
    vol 56, No. 6   1977

fl.   p. 7

Allium moly

Everett, T. H., ed.
New Illustrated Encyclopedia of
    Gardening   Vol. 10

fl.   p. 1847

Allium moly

Hay, Roy
The Color Dictionary of Flowers & Plants

fl.   p 83, Pl. 664

Allium moly

Huxley, Anthony
Garden Annuals and Bulbs

hab., fl   Pl. 143, 194

Allium moly

Massachusetts Hort. Soc.
Horticulture
    vol 51, No. 9   1973

fl.   p. 33

Allium moly

Miles, Bebe
Bulbs for the Home Gardener

fl.   p 50

Allium moly

Morley, Brian D.
Wild Flowers of the World

fl., hab.   Pl. 24

Allium moly

Perry, Frances
Flowers of the World

fl.   p 19

Allium moly

Royal Hort. Soc.
Journal of the Royal Hort. Soc.
    vol 95, No. 4   1970

fl.   p. 166   Pl. 98

Allium moly

Tsukamoto, Yotaro
Coloured Illustrations of Garden
    Flowers   Vol. 9

fl.   p. 1

Allium moly

Vilmorin, Roger de
Plantes Alpines dans les Jardins

fl. hab.   Pl. XXV

Allium montanum

Royal Hort. Soc.
The Garden
    vol 101, No. 3   1976

fl.   p. 125

Allium narcissiflorum

Hay, Roy
The Color Dictionary of Flowers & Plants

fl.   p 84, Pl. 665

Allium narcissiflorum

Mathew, Brian
Dwarf Bulbs

fl.   p. 16   Pl. 2

Allium narcissiflorum

Perry, Frances
Flowers of the World

fl.   p 19

Allium narcissiflorum

Royal Hort. Soc.
Journal of the Royal Hort. Soc.
    vol 95, No. 4   1970

fl.   p. 166   Pl. 105

Allium narcissiflorum

Royal Hort. Soc.
The Garden
    vol 101, No. 4   1976

fl.   p. 187

Allium narcissiflorum var.

Royal Hort. Soc.
Journal of the Royal Hort. Soc.
    vol 91, No. 7   1966

fl.   p. 294   Pl. 157

Allium neapolitanum

American Hort. Soc.
American Horticulturist
    vol 56, No. 6   1977

fl.   p. 7

Allium neapolitanum

Miles, Bebe
Bulbs for the Home Gardener

fl.   p 50

Allium neapolitanum

Polunin, Oleg
Flowers of Europe

fl.   Pl. 165 #1609

Allium neapolitanum

Polunin, Oleg
Flowers of the Mediterranean

fl.   Pl. 250

Allium noeanum

Wendelbo, Per
Tulips and Irises of Iran and
    Their Relatives

fl. hab.   p 19, Pl. 12

Allium oleraceum

Lindman, C. A. M.
Nordens Flora, Vol 1

fl. fr.   Pl. 54A

Allium oleraceum

Martin, W. Keble
The Concise British Flora in Colour

fl. hab.   Pl. 85

Allium oreophilum var.

Hay, Roy
The Color Dictionary of Flowers
    & Plants

fl.   p. 84   Pl. 666

Allium ostrowskianum

Huxley, Anthony
Garden Annuals and Bulbs

hab., fl.   Pl. 144, 193

Allium ostrowskianum

Miles, Bebe
Bulbs for the Home Gardener

fl.   p 50

Allium praecox

Munz, Philip A.
California Spring Wildflowers

fl.   p. 80   Pl. 64

Allium pulchellum

Barneby, T. P.
European Alpine Flowers in Colour

fl.   Pl. 3, 5

Allium pulchellum

Hay, Roy
The Color Dictionary of Flowers & Plants

fl.   p 84, Pl. 667

Allium pulchellum

Royal Hort. Soc.
Journal of the Royal Hort. Soc.
    Vol. 95, No. 4   1970

fl.   p. 166   Pl. 100

Allium pulchellum var.

Royal Hort. Soc.
Journal of the Royal Hort. Soc.
    vol 96, No. 4   1970

fl.   p. 166   Pl. 101

Allium regelii

Mathew, Brian
Dwarf Bulbs

fl.   p. 16   Pl. 6

Allium regelii

Royal Hort. Soc.
Journal of the Royal Hort. Soc.
    vol 93, No. 1   1968

fl.   p. 24   Pl. 9

Allium rosenbachianum

Mathew, Brian
The Larger Bulbs

fl.   p. 64

Allium rosenbachianum

Royal Hort. Soc.
Journal of the Royal Hort. Soc.
    vol 96, No. 9   1971

hab., fl.   p. 404   Pl. 175

Allium roseum

Polunin, Oleg
Flowers of Europe

fl.   Pl. 166 #1611

Allium roseum

Polunin, Oleg
Flowers of the Mediterranean

fl.   Pl. 251

Allium roseum

Taylor, A. W.
Wildflowers of Spain and Portugal

fl.   p. 15

Allium sativum

Bianchini, F.
The Complete Book of Fruits & Vegetables

hab.   p 85

Allium sativum

Bianchini, Francesco
Health Plants of the World

hab.   p 33

Allium sativum

Color Treasury of Herbs & Other
    Medicinal Plants

fr.   p 23   Pl. 14

Allium sativum

Hvass, Elsie
Plants That Feed and Serve Us

fl.   p 36, Pl. 68

Allium sativum

Kariyone, Tatsuo
Atlas of Medicinal Plants

fr., hab.   Pl. 14

Allium sativum

Kimura, Koiti
Japanese Medicinal Plants, Vol II

fr.   p 133

Allium sativum

Masefield, G. B.
The Oxford Book of Food Plants

fl.   Pl. 169, 3

Allium sativum

Perrot, Emile
Les Plantes Medicinales

fl. fr. hab.   p 210

Allium sativum

Rosengarten, Frederic, Jr.
The Book of Spices

fr. p 322

Allium scabriscapum

Wendelbo, Per
Tulips and Irises of Iran and
    Their Relatives

fl.   p 19, Pl. 9

Allium schoenoprasum

Alaska-Yukon Wild Flowers Guide

fl.   p 8

Allium schoenoprasum

Hay, Roy
The Color Dictionary of Flowers & Plants

fl.   p 84, Pl. 668

Allium schoenoprasum

Heller, Christine
Wild Flowers of Alaska

fl.   Pl. 224

Allium schoenoprasum

Hvass, Elsie
Plants That Feed and Serve Us

fl.   p 37, Pl. 70

Allium schoenoprasum

Kohlhaupt, Paula
Fleurs des Alpages Vol 1

fl. P 25

Allium schoenoprasum

Lindman, C. A. M.
Nordens Flora, Vol. 1

fl. fr. hab.   Pl. 53

Allium schoenoprasum

Loewenfeld, Claire
The Complete Book of Herbs and Spices

fl. hab.   p 113

Allium schoenoprasum

Martin, W. Keble
The Concise British Flora in Colour

fl. hab.   Pl. 85

Allium schoenoprasum

Masefield, G. B.
The Oxford Book of Food Plants

fl.   Pl. 167, 5

Allium schoenoprasum

Miles, Bebe
Bulbs for the Home Gardener

fl.   p 51

Allium schoenoprasum

Polunin, Oleg
Flowers of Europe

fl.   Pl. 165 #1603

Allium schoenoprasum

Royal Hort. Soc.
Journal of the Royal Hort. Soc.
vol 95, No. 4   1970

fl.   p. 166   Pl. 97

Allium schoenoprasum var.

Martin, W. Keble
The Concise British Flora in Colour

fl. hab.   Pl. 85

Allium scorodoprasum

Martin, W. Keble
The Concise British Flora in Colour

fl. fr. hab.   Pl. 84

Allium sibiricum

Barneby, T. P.
European Alpine Flowers in Colour

fl.   Pl. 3, 4

Allium sibiricum

Walcott, Mary Vaux
North American Wild Flowers
vol. 5

fl.   Pl. 383

Allium siculum

Hay, Roy
The Color Dictionary of Flowers & Plants

fl.   p 84, Pl. 669

Allium sphaerocephalon

Polunin, Oleg
Flowers of Europe

fl.   Pl. 165 #1603

Allium sphaerocephalon

Royal Hort. Soc.
Journal of the Royal Hort. Soc.
vol 95, No. 4   1970

fl.   p. 166   Pl. 99

Allium sphaerocephalon

Royal Hort. Soc.
The Garden
vol 103, No. 12   1978

fl.   p. 474

Allium stellatum

Addisonia
Vol 20   1937-38

Allium stellatum

Klimas, John E.
Wildflowers of Eastern America

fl.   Pl. 221

Allium stellatum

Moyle, John B.
Northland Wild Flowers

fl.   p 207, Pl. 271

Allium stocksianum

Royal Hort. Soc.
Journal of the Royal Hort. Soc.
vol. 98, No. 7   1973

fl.   p. 306   Pl. 154

Allium subhirsutum

Polunin, Oleg
Flowers of the Mediterranean

fl.   Pl. 249

Allium tricoccum

Campbell, Carlos C.
Great Smoky Mountain Wildflowers

fl. hab.   p 61

Allium tricoccum

Courtenay, Booth
Wildflowers & Weeds

fl.   p 8

Allium tricoccum

Klimas, John E.
Wildflowers of Eastern America

fl.   Pl. 65

Allium tricoccum

Wharton, Mary E.
A Guide to the Wildflowers & Ferns
of Kentucky

fl. p 90 Pl. 3.1

Allium triquetrum

Martin, W. Keble
The Concise British Flora in Colour

fl. hab.   Pl. 85

Allium triquetrum

Polunin, Oleg
Flowers of Europe

fl.   Pl. 165 #1610

Allium triquetrum

Polunin, Oleg
Flowers of the Mediterranean

fl.   Pl. 248

Allium tuberosum

American Hort. Soc.
American Horticulturist
vol 53, No. 1   1974

fl., hab.   p. 33

Allium tuberosum

Curtis's Botanical Magazine
Vol 173   1960

fl.   Pl. 386

Allium tuberosum

Kariyone, Tatsuo
Atlas of Medicinal Plants

fl., hab.   Pl. 15

Allium tuberosum

Kimura, Koiti
Japanese Medicinal Plants, Vol I

fl.   p 8

Allium unifolium

Royal Hort. Soc.
Journal of the Royal Hort. Soc.
vol 95, No. 4   1970

fl.   p. 166   Pl. 107

Allium ursinum

Ary, S.
The Oxford Book of Wildflowers

fl.   p 102, Pl. 3

Allium ursinum

Felsko, Elsa
A Book of Wild Flowers 2nd Ser.

fl.   p 114

Allium ursinum

Kleijn, H.
Beauty of the Wild Plant

fl.   p. 44   Pl. 48

Allium ursinum

Lindman, C. A. M.
Nordens Flora, Vol 1

fl. fr.   Pl. 52

Allium ursinum

Martin, W. Keble
The Concise British Flora in Colour

fl. hab.   Pl. 85

Allium ursinum

Polunin, Oleg
Flowers of Europe

fl.   Pl. 166 #1616

Allium validum

Munz, Philip A.
California Mountain Wildflowers

fl.   p. 28   Pl. 6

Allium victorialis

Barneby, T. P.
European Alpine Flowers in Colour

fl.   Pl. 3, 3

Allium victorialis

Kohlhaupt, Paula
Fleurs des Alpages Vol 2

fl.   P 9

Allium vineale

Ary, S.
The Oxford Book of Wildflowers

fl.   p 162, Pl. 5

Allium vineale

Crockett, James Underwood
Lawns & Ground Covers

fl.   p. 71

Allium vineale

Klimas, John E.
Wildflowers of Eastern America

fl.   Pl. 195

Allium vineale

Lindman, C. A. M.
Nordens Flora, Vol 1

fl. hab.   Pl. 54B

Allium vineale

Martin, W. Keble
The Concise British Flora in Colour

fl. fr.   Pl. 84

Allium vineale

Weeds of the Southern United States

hab., fr.   p. 31

Allotropa virgata

Clark, Lewis J.
Wild Flowers of British Columbia

fl.   p 374

Allotropa virgata

Orr, Robert T.
Wildflowers of Western America

fl.   Pl. 160

Allotropa virgata

Taylor, Ronald J.
Mountain Wild Flowers

fl. hab.   p 44

Alluaudia dumosa

Cactus & Succulent Society of America
Journal
vol 48, No. 2   1976

fl., hab. p. 76

Alluaudia montagnacii

Cactus & Succulent Society of America
Journal
vol 48, No. 2   1976

hab.   p. 76

Alluaudiopsis fiherenensis

Cactus & Succulent Society of America
Journal
vol 48, No. 2   1976

fl.   p. 76

Alluaudiopsis marnierana

Cactus & Succulent Society of America
Journal
vol 50, No. 2   1978

fl.   p. 86

Almond

see

Prunus amygdalus

Almond

see

Prunus communis

Almond

see

Prunus dulcis

Almond, dwarf flowering

see

Prunus glandulosa var.

Almond, Dwarf Russian

see

Prunus tenella

Almond, Indian

see

Terminalia catappa

Almond, Tropical

see

Terminalia catappa

Alnus cordata

Boom, B. K.
The glory of the tree

fr.   p. 44   Pl. 64

Alnus cordata

Edlin, Herbert
The Illustrated Encyclopedia
of Trees

fl. fr.   p 142-3

Alnus cordata

Hay, Roy
The Color Dictionary of Flowers & Plants

fr.   p 181, Pl. 1443

Alnus cordata

Polunin, Oleg
Trees and Bushes of Europe

fr. hab.   p 46

Alnus crispa var.

Porsild, A. E.
Rocky Mountain Wild Flowers

fl. fr.   p 137

Alnus firma var.

Kitamura, Siro
Coloured Illustrations of Trees &
    Shrubs of Japan

fl.   110

Alnus glutinosa

Boom, B. K.
The glory of the tree

hab.   p. 44   Pl. 62

Alnus glutinosa

Edlin, Herbert
The Illustrated Encyclopedia
    of Trees

fr. hab.   p 142

Alnus glutinosa

Huxley, Anthony
Deciduous Garden Trees and Shrubs

fr.   Pl. 12a

Alnus glutinosa

Hvass, Elsie
Plants That Feed and Serve Us

fr.   p 123, Pl. 270

Alnus glutinosa

Lindman, C. A. M.
Nordens Flora, Vol 3

fr.   Pl. 168

Alnus glutinosa

Martin, W. Keble
The Concise British Flora in Colour

fl. fr. hab.   Pl. 77

Alnus glutinosa

Perrot, Emile
Les Plantes Medicinales

fl. fr. hab.   p 23

Alnus glutinosa

Polunin, Oleg
Flowers of Europe

fl. fr.   Pl. 4 #40

Alnus glutinosa

Polunin, Oleg
Trees and Bushes of Europe

fl. fr. hab.   p 44, 45

Alnus glutinosa

Vedel, H.
Arbres et Arbustes

fl. hab.   p 52, Pl. 45

Alnus glutinosa

Wit, H, C. D. de
Plants of the World;
    The Higher Plants, Vol I

fl. fr.   p 108, Pl. 63

Alnus glutinosa var.

Edlin, Herbert
The Illustrated Encyclopedia
    of Trees

hab.   p 142

Alnus hirsuta var.

Kitamura, Siro
Coloured Illustrations of Trees &
    Shrubs of Japan

fr.   104

Alnus incana

Edlin, Herbert
The Illustrated Encyclopedia
    of Trees

hab.   p 143

Alnus incana

Huxley, Anthony
Deciduous Garden Trees and Shrubs

fr.   Pl. 13a

Alnus incana

Lindman, C. A. M.
Nordens Flora, Vol 3

fr.   Pl. 169

Alnus incana

Polunin, Oleg
Flowers of Europe

fl.   Pl. 4 #41

Alnus incana

Polunin, Oleg
Trees and Bushes of Europe

fl. fr. hab.   p 45

Alnus incana

Vedel, H.
Arbres et Arbustes

fl. hab.   p 53, Pl. 46

Alnus incana var.

Royal Hort. Soc.
Journal of the Royal Hort. Soc.
    vol 96, No. 7   1971

fl.   p. 304   Pl. 140

Alnus inokumae

Bartels, Andreas
Das Grosse Buch der Gartengeholze

fl.   p 111

Alnus japonica

Kitamura, Siro
Coloured Illustrations of Trees &
    Shrubs of Japan

fr.   111

Alnus maximowiczii

Bartels, Andreas
Das Grosse Buch der Gartengeholze

fl.   p 111

Alnus maximowiczii

Kitamura, Siro
Coloured Illustrations of Trees &
    Shrubs of Japan

fr.   105

Alnus pendula

Kitamura, Siro
Coloured Illustrations of Trees &
    Shrubs of Japan

fl. fr.   106, 107

Alnus rubra

Edlin, Herbert
The Illustrated Encyclopedia
    of Trees

fr. hab.   p 143

Alnus rugosa

Polunin, Oleg
Trees and Bushes of Europe

fl. fr. hab.   p 46

Alnus rugosa

Walcott, Mary Vuax
North American Wild Flowers

fr.   Pl. 16

Alnus serrulata

Wharton, Mary E.
Trees & Shrubs of Kentucky

fl.   p 102, Pl. 3.5

Alnus sieboldiana

Kitamura, Siro
Coloured Illustrations of Trees &
    Shrubs of Japan

fl. fr.   108, 109

Alnus sinuata

Royal Hort. Soc.
Journal of the Royal Hort. Soc.
vol 94, No. 2   1969

fl.   p. 86   Pl. 33

Alnus sinuata

Walcott, Mary Vaux
North American Wild Flowers

fl. hab.   Pl. 185

Alnus viridis

Polunin, Oleg
Trees and Bushes of Europe

fl. fr. hab.   p 46

Alnus viridis var.

Curtis's Botanical Magazine
Vol 173   1960

fl., fr.   Pl. 382

Alocasia, Giant

See

Alocasia macrorrhiza

Alocasia macrorrhiza

Hargreaves, Dorothy
Tropical Blossoms of the Caribbean

fl.   p. 4

Alocasia macrorrhiza

Kiaer, Eigil
Indoor Plants in Colour

hab.   p. 12

Alocasia macrorrhiza

Macoby, Stirling
What Flower is That

fl.   p. 36

Alocasia macrorrhiza var.

Macoby, Stirling
What Flower is That

hab.   p. 36

Alocasia sanderiana

Massachusetts Hort. Soc.
Horticulture
vol 32, No. 2   1954

hab.   p. 77

Alocasia thibautiana

Massachusetts Hort. Soc.
Horticulture
vol 32, No. 2   1954

hab.   p. 77

Aloe affinis

Flowering Plants of South Africa
vol XIX   1939

fl.   Pl. 759

Aloe africana

American Hort. Soc.
American Horticulturist
vol 56, No. 4   1977

fl., hab.   p. 14

Aloe africana

Mathias, Mildred E.
Color for the Landscape

fl. hab.   p 163

Aloe albiflora

Curtis's Botanical Magazine
Vol. 180   1974

fl. hab.   Pl. 674

Aloe albiflora

Flowering Plants of Africa
vol 37   1965-66

fl., hab.   Pl. 1466

Aloe angelica

Cactus & Succulent Society of America
Journal
vol 47, No. 4   1975

hab.   cover

Aloe angustifolia

Flowering Plants of South Africa
vol XVIII   1938

fl.   Pl. 708

Aloe arborescens

Hersey, Jean
Woman's Day Book of House Plants

hab.   p. 28

Aloe arborescens

Kiaer, Eigil
Indoor Plants in Colour

fl.   p. 13

Aloe arborescens

Kimura, Koiti
Japanese Medicinal Plants, Vol I

fl.   p 8

Aloe arborescens

Kromdijk, G.
200 House Plants in Colour

hab.   Pl. 11

Aloe arborescens

Macoby, Stirling
What Flower is That

fl. hab.   P 37

Aloe arborescens

Royal Hort. Soc.
The Garden
vol 103, No. 4   1978

fl.   cover

Aloe arborescens

Royal Hort. Soc.
The Garden
vol 103, No. 12   1978

fl.   p. 496

Aloe arborescens

Takatori, Jisuke
Color Atlas of Medicinal Plants of Japan

fl.   Fig. 72A

Aloe arborescens var.

Cactus & Succulent Society of America
Journal
vol 2, No. 11   1931

fl., hab.   cover (p. 465)

Aloe archeri

Cactus & Succulent Society of America
Journal
vol 49, No. 2   1977

hab.   p. 74

Aloe arenicola

Eliovson, Sima
Namaqualand in Flower

fl.   Pl. 49, 3

Aloe arenicola

Flowering Plants of Africa
vol 37   1965-66

fl.   Pl. 1467

Aloe asperifolia

Flowering Plants of Africa
vol 44   1977

fl.   Pl. 1753

Aloe ballii

Flowering Plants of Africa
vol 40   1969-70

fl.   Pl. 1589

Aloe barkeri

Flowering Plants of Africa
vol 36   1963-64

fl., hab.   Pl. 1401

Aloe barteri

Wit, H. C. D. de
Plants of the World;
The Higher Plants, Vol II

fl. hab.   Pl. 122

Aloe bellatula

Flowering Plants of Africa   Vol. 36
        1963-64
fl.   Pl. 1402

Aloe boylei

Flowering Plants of South Africa
    vol XVI   1936

fl.   Pl. 634

Aloe branddraaiensis

Flowering Plants of South Africa
    vol XX   1940

fl.   Pl. 761

Aloe brevifolia

Flowering Plants of South Africa
    vol XVI   1936

fl., hab.   Pl. 604

Aloe broomii

Flowering Plants of South Africa
    vol XVI   1936

fl., hab.   Pl. 605

Aloe buchananii

Flowering Plants of South Africa
    vol XX   1940

fl.   Pl. 763

Aloe buchananii

Moriarty, Audrey
Wild Flowers of Malawi

fl.   Pl. 14; 3

Aloe bulbicaulis

Flowering Plants of South Africa
    vol XVI   1936

fl.   Pl. 630

Aloe, candelabra

    see

Aloe arborescens

Aloe candelabrum

Flowering Plants of South Africa
    vol XXIV   1944

fl., fr.   Pl. 945

Aloe capitata

Flowering Plants of Africa
    vol 40   1969-70

fl.   Pl. 1595

Aloe chabaudii var.

Flowering Plants of South Africa
    vol XVIII   1938

fl.   Pl. 698 - 99

Aloe chimanimaniensis

Flowering Plants of South Africa
    vol XVI   1936

fl.   Pl. 639

Aloe chrysostachys

Cactus & Succulent Society of America
Journal
    vol 48, No. 6   1976

fl., hab.   p. 278

Aloe ciliaris

Addisonia
    Vol. 24   1960-64

fl.   opp. p. 33   Pl. 785

Aloe ciliaris

Lamb, Edgar
The Pocket Encyclopedia of Cacti
    and Succulents in Color

fl. hab.   Pl. 206

Aloe ciliaris

Mathias, Mildred E.
Color for the Landscape

fl. hab.   p 162

Aloe ciliaris

Menninger, Edwin A.
Flowering Vines of the World

fl.   Pl. 116

Aloe claviflora

Eliovson, Sima
Namaqualand in Flower

fl.   Pl. 49, 1

Aloe, Climbing

    see

Aloe ciliaris

Aloe, Climbing

    see

Aloe tenuior

Aloe concinna

Subik, Rudolf
Decorative Cacti

fl. hab.   p 97

Aloe, cushion

    see

Haworthia margaritifera

Aloe davyana var.

Flowering Plants of South Africa
    vol XIX   1939

Fl.   Pl. 732

Aloe debrana

Flowering Plants of Africa
    vol XXVI   1947

fl.   Pl. 1016

Aloe deltoideodonta

Lamb, Edgar
The Pocket Encyclopedia of Cacti
    and Succulents in Color

hab.   Pl. 207

Aloe deserti

Flowering Plants of Africa
    vol 29   1952-53

fl.   Pl. 1151

Aloe dewetii

Flowering Plants of South Africa
    vol XVIII   1938

fl.   Pl. 692

Aloe dewinteri

Flowering Plants of Africa
    vol 44   1977

fl.   Pl. 1752

Aloe dhalensis

Flowering Plants of Africa
    vol 39   1968-69

fl.   Pl. 1522

Aloe dichotoma

Edlin, Herbert
The Illustrated Encyclopedia
    of Trees

hab.   p 231

Aloe dichotoma

Eliovson, Sima
Namaqualand in Flower

hab.   Pl. 50, 51,1

Aloe dichotoma

Flowering Plants of South Africa
vol XVIII   1938

fl.   Pl. 709

Aloe dichotoma

Lamb, Edgar
The Pocket Encyclopedia of Cacti
   and Succulents in Color

fl.   hab.   Pl. 316

Aloe dichotoma

Mathias, Mildred E.
Color for the Landscape

hab.   p 163

Aloe dinteri

Flowering Plants of South Africa
vol XVI   1936

fl.   Pl. 637

Aloe distans

Kiaer, Eigil
Indoor Plants in Colour

hab.   p. 13

Aloe doei

Cactus & Succulent Society of America
Journal
   vol 42, No. 6   1970

fl.   p. 262

Aloe doei var.

Cactus & Succulent Society of America
Journal
   vol 42, No. 6   1970

fl.   p. 262

Aloe ecklonis

Flowering Plants of South Africa
vol XVI   1936

fl.   Pl. 609

Aloe erensii

Flowering Plants of South Africa
vol XX   1940

fl.   Pl. 797

Aloe eylesii

Flowering Plants of South Africa
vol XVI   1936

fl.   Pl. 638

Aloe, False

see

Agave virginica

Aloe ferox

American Hort. Soc.
American Horticulturist
   vol 53, No. 3 1974

fl.   p. 34

Aloe ferox

American Hort. Soc.
American Horticulturist
   vol 56, No. 4 1977

fl., hab   p. 15

Aloe ferox

Kariyone, Tatsuo
Atlas of Medicinal Plants

fl., hab.   Pl. 16

Aloe ferox

Perrot, Emile
Les Plantes Medicinales

fl, fr, hab.   p 9

Aloe ferox

Royal Hort. Soc.
Journal of the Royal Hort. Soc.
   vol 90, No. 6   1965

fl., hab.   p. 250   Pl. 103

Aloe fibrosa

Cactus & Succulent Society of America
Journal
   vol 48, No. 6   1976

fl., hab.   p. 275

Aloe fosteri

Flowering Plants of South Africa
vol XVI   1936

fl.   Pl. 612

Aloe framesii

Flowering Plants of South Africa
vol XIX   1939

fl.   Pl. 731

Aloe gariepensis

Flowering Plants of Africa
vol 42   1972-73

fl.   Pl. 1654

Aloe globuligemma

Lamb, Edgar
The Pocket Encyclopedia of Cacti
   and Succulents in Color

fl.   hab.   Pl. 315

Aloe greatheadii

Morley, Brian D.
Wild Flowers of the World

fl.   Pl. 68

Aloe hlangapies

Flowering Plants of South Africa
vol XVIII   1938

fl.   Pl. 710

Aloe humilis

Lamb, Edgar
The Pocket Encyclopedia of Cacti
   and Succulents in Color

fl. hab.   Pl. 209, 210

Aloe hybrid

Massachusetts Hort. Soc.
Horticulture
   vol 41, No. 8   1963

fl., hab.   backcover

Aloe inermis

Flowering Plants of Africa
vol 38   1967

fl.   Pl. 1516

Aloe integra

Flowering Plants of South Africa
vol XVI   1936

fl.   Pl. 607

Aloe inyangensis

Flowering Plants of South Africa
vol XVI   1936

fl.   Pl. 640

Aloe isaloensis

Flowering Plants of Africa
vol 36   1963-64

fl.   Pl. 1419

Aloe jucunda

Flowering Plants of Africa
vol 35   1962

fl., hab.   Pl. 1390

Aloe karasbergensis

Flowering Plants of South Africa
vol XVIII   1938

fl.   Pl. 720

Aloe kniphofioides

Flowering Plants of Africa
vol 28   1950-51

fl., hab.   Pl. 1120

Aloe krapohliana

Eliovson, Sima
Namaqualand in Flower

fl.   Pl. 49, 2

Aloe krausnii

Flowering Plants of South Africa
vol XVI   1936

fl.   Pl. 635

Aloe latifolia

Kiaer, Eigil
Indoor Plants in Colour

hab.   p. 13

Aloe lettyae

Flowering Plants of South Africa
vol XX   1940

fl.   Pl. 764

Aloe linearifolia

Flowering Plants of South Africa
vol XXII   1942

fl.   Pl. 849

Aloe lutescens

Flowering Plants of South Africa
vol XVIII   1938

fl.   Pl. 707

Aloe mcloughlinii

Flowering Plants of Africa
vol 28   1950-51

fl.   Pl. 1112

Aloe magnidentata

Flowering Plants of Africa
vol XXVI   1947

fl.   Pl. 1015

Aloe marlothii

American Hort. Soc.
American Horticulturist
vol 53, No. 3 1974

fl., hab.   p. 35

Aloe marlothii

Cactus & Succulent Society of America
Journal
vol 23, No. 6   1951

fl., hab.   p. 177

Aloe marlothii

Macoby, Stirling
What Flower is That

fr.   P. 37

Aloe marsabitensis

Flowering Plants of South Africa
vol XX   1940

fl.   Pl. 798

Aloe mawii

Moriarty, Audrey
Wild Flowers of Malawi

fl.   Pl. 14; 1

Aloe melsetterensis

Flowering Plants of South Africa
vol XVIII   1938

fl.   Pl. 697

Aloe mketiensis

Flowering Plants of South Africa
vol XX   1940

fl.   Pl. 785

Aloe monotropa

Flowering Plants of Africa
vol 34   1960-61

fl.   Pl. 1342

Aloe morogoroensis

Flowering Plants of Africa
vol XXV   1945-46

fl.   Pl. 991

Aloe munchii

Flowering Plants of Africa
vol 28   1950-51

fl.   Pl. 1091

Aloe mutabilis

Flowering Plants of South Africa
vol XVI   1936

fl.   Pl. 611

Aloe mutans

Flowering Plants of South Africa
vol XVI   1936

fl.   Pl. 602

Aloe mzimbana

Flowering Plants of South Africa
vol XXI   1941

fl.   Pl. 838

Aloe mzimbana

Moriarty, Audrey
Wild Flowers of Malawi

fl.   Pl. 14; 2

Aloe namibensis

Flowering Plants of Africa
vol 44   1977

fl.   Pl. 1730

Aloe nubigena

Flowering Plants of South Africa
vol XVI   1936

fl., hab.   Pl. 628

Aloe nuttii

Flowering Plants of South Africa
vol XX   1940

fl.   Pl. 762

Aloe nyeriensis

Flowering Plants of Africa
vol 29   1952-53

fl.   Pl. 1126

Aloe ortholopha

Flowering Plants of South Africa
vol XXIII   1943

fl.   Pl. 882

Aloe parvula

Flowering Plants of Africa
vol 31   1956

fl., hab.   Pl. 1234

Aloe peckii

Flowering Plants of Africa
vol 31   1956

fl.   Pl. 1214

Aloe peglerae

Lamb, Edgar
The Pocket Encyclopedia of Cacti
and Succulents in Color

hab.   Pl. 212

Aloe personii

Flowering Plants of Africa
Vol. 40 1969-70

fl.   Pl. 1594

Aloe pluridens

Flowering Plants of South Africa
vol XVI   1936

fl.   Pl. 610

Aloe pole-evansii

Flowering Plants of South Africa
vol XX   1940

fl.   Pl. 782

Aloe polyphylla

Royal Hort. Soc.
Journal of the Royal Hort. Soc.
vol 90, No. 6   1965

fl.   p. 250   Pl. 116

Aloe pongolensis

Flowering Plants of South Africa
vol XVI   1936

fl.   Pl. 603

Aloe prinslooi

Flowering Plants of Africa
vol 37   1965-66

fl., fr.   Pl. 1453

Aloe rauhii

Flowering Plants of Africa
vol 38   1967

fl.   Pl. 1517

Aloe recurvifolia

Flowering Plants of South Africa
vol XVI   1936

fl., hab.   Pl. 601

Aloe reitzii

Flowering Plants of South Africa
vol XXIII   1943

fl., hab.   Pl. 911

Aloe rubrolutea

Flowering Plants of South Africa
vol XXI   1941

fl.   Pl. 802

Aloe rubroviolacea

Flowering Plants of Africa
vol 41   1970-71

fl., hab.   Pl. 1610

Aloe scobinifolia

Flowering Plants of Africa
vol 35   1962

fl.   Pl. 1382

Aloe secundiflora

Flowering Plants of South Africa
vol 34   1960-61

fl.   Pl. 1341

Aloe sladeniana

Flowering Plants of Africa
vol 29   1952-53

fl.   Pl. 1122

Aloe Socotrina

Hvass, Elsie
Plants That Feed and Serve Us

fl.   p 96, Pl. 205

Aloe solaiana

Flowering Plants of South Africa
vol XX   1940

fl., fr.   Pl. 781

Aloe soutpansbergensis

Cactus & Succulent Society of America
Journal
vol 41, No. 4   1969

fl., hab.   p. 165

Aloe soutpansbergensis

Flowering Plants of Africa
vol 35   1962

fl.   Pl. 1391

Aloe sp.

Color Treasury of Herbs & Other Medicinal
Plants

fr.   p 22   Pl. 13

Aloe speciosa

Flowering Plants of South Africa
vol XVI   1936

fl., hab.   Pl. 606

Aloe speciosa

Macoby, Stirling
What Flower is That

fl.   P 37

Aloe speciosa

Royal Hort. Soc.
Journal of the Royal Hort. Soc.
vol 93, No. 8   1968

fl.   p. 336   Pl. 172

Aloe squarrosa

Flowering Plants of Africa
vol 41   1970-71

fl., fr.   Pl. 1611

Aloe striata

Lamb, Edgar
The Pocket Encyclopedia of Cacti
and Succulents in Color

fl. hab.   Pl. 211, 213

Aloe striata

Massachusetts Hort. Soc.
Horticulture
vol 40, No. 1   1962

hab., fl.   p. 38

Aloe striata var.

Mathias, Mildred E.
Color for the Landscape

fl., hab.   p. 156

Aloe striatula var.

Flowering Plants of South Africa
vol XVI   1936

fl.   Pl. 633

Aloe suprafoliata

Flowering Plants of South Africa
vol XIX   1939

fl.   Pl. 733

Aloe suzannae

Cactus & Succulent Society of America
Journal
vol 47, No. 2   1975

hab.   cover

Aloe tenuior

Flowering Plants of Africa
vol 34   1960-61

fl., fr.   Pl. 1352

Aloe tenuior

Harrison, Richmond E.
Climbers and Trailers

fl.   p 18   Pl. 4

Aloe thompsoniae

Flowering Plants of Africa
vol XXV   1945-46

fl., hab.   Pl. 980

Aloe thraskii

Flowering Plants of South Africa
vol XXIV   1944

fl.   Pl. 923

Aloe thraskii

Morley, Brian D.
Wild Flowers of the World

fl.   Pl. 85

Aloe tidmarshi

Flowering Plants of South Africa
vol XXIII   1943

fl.   Pl. 910

Aloe, Tongue Common

see

Aloe lingua

Aloe tororoana

Flowering Plants of Africa
vol 29   1952-53

fl.   Pl. 1144

Aloe torrei

Flowering Plants of Africa
vol XXV   1945-46

fl.   Pl. 987

Aloe transvaalensis

Flowering Plants of South Africa
vol XVI   1936

fl.   Pl. 636

Aloe, Tree

see

Aloe arborescens

Aloe trichosantha

Flowering Plants of Africa
vol XXVI   1947

fl.   Pl. 1014

Aloe vanbalenii

Flowering Plants of South Africa
vol XVI   1936

fl.   Pl. 608

Aloe variegata

Cactus & Succulent Society of America
Journal
vol 26, No. 1   1954

fl., hab.   p. 19

Aloe variegata

Hay, Roy
The Color Dictionary of Flowers &
Plants

hab.   p. 51   Pl. 407

Aloe variegata

Hay, Roy
The Dictionary of House Plants

hab.   Pl. 29

Aloe variegata

Kromdijk, G.
200 House Plants in Colour

hab.   Pl. 12

Aloe variegata

Nicolaisen, Age
Pocket Encyclopedia of Indoor
Plants

fl.   Pl. 19

Aloe variegata

Subik, Rudolf
Decorative Cacti

fl. hab.   p 99

Aloe vera

Hersey, Jean
Woman's Day Book of House Plants

hab.   p 39

Aloe vera

Perrot, Emile
Les Plantes Medicinales

fl. fr. hab.   p 9

Aloe verdoorniae

Flowering Plants of South Africa
vol XXII   1942

fl.   Pl. 879

Aloe viridiflora

Flowering Plants of Africa
vol 40   1969-70

fl.   Pl. 1598

Aloe vryheidensis

Flowering Plants of South Africa
vol XXI   1941

fl.   Pl. 805

Aloe, warty

see

Gasteria verrucosa

Aloe woolliana

Lamb, Edgar
The Pocket Encyclopedia of Cacti
and Succulents in Color

fl.   Pl. 208

Aloe woolliana var.

Lamb, Edgar
Popular Exotic Cacti in Color

fl. hab.   p. 25

Aloinopsis malherbei

Flowering Plants of Africa
vol XXVI   1947

fl., hab.   Pl. 1035

Aloinopsis rosulata

Lamb, Edgar
Popular Exotic Cacti in Color

fl. hab.   p 27

Aloinopsis spathulata

Herre, H.
The Genera of the Mesembryanthemaceae

fl. fr.   p 67

Alona rostrata

Munoz Pizarro, Carlos
Flores Silvestres de Chile

fl.   Pl. 7

Alonsoa incisifolia

Munoz Pizarro, Carlos
Flores Silvestres de Chile

fl.   Pl. 15

Alonsoa warscewiczii

Crockett, James Underwood
Annuals

Fl.   P 92

Alonsoa warscewiczii

Hay, Roy
The Color Dictionary of Flowers & Plants

fl.   p 29, Pl. 227

Alonsoa warscewiczii

Hay, Roy
The Dictionary of House Plants

fl.   Pl. 30

Alopecurus pratensis

Hvass, Else
Plants That Feed and Serve Us

fr.   p. 62   Pl. 121

Alopecurus pratensis

Lindman, C. A. M.
Nordens Flora, Vol 2

fr. hab.   Pl. 100

Alopecurus pratensis

Polunin, Oleg
Flowers of Europe

fl.   Pl. 181 #1780

Alophia drummondii

Brown, Clair A.
Wildflowers of Louisiana and
Adjoining States

fl. hab.   p 29

Alphitonia excelsa

Blombery, Alec M.
What Wildflower is That

fl.   p 42, Pl. 81

Alpinia chinensis

Walden, Beryl M.
Wild Flowers of Hong Kong

fl.   Pl. 39(106)

Alpinia mutica var.

Perry, Frances
Flowers of the World

fl.   p. 311

Alpinia nutans

Hargreaves, Dorothy
Tropical Blossoms of the Caribbean

fl.   p. 14

Alpinia purpurata

American Hort. Soc.
American Horticulturist
     vol 53, No. 4   1974

fl.   p. 21

Alpinia purpurata

Hargreaves, Dorothy
Tropical Blossoms of the Caribbean

fl.   p. 14

Alpinia purpurata

Macoby, Stirling
What Flower is That

fl.   P 37

Alpinia sanderae

American Hort. Soc.
American Horticulturist
     vol 53, No. 4   1974

hab.   p. 23

Alpinia speciosa

Macoby, Stirling
What Flower is That

fl.   P 37

Alpinia speciosa

Perry, Frances
Flowers of the World

fl.   p 309

Alpinia speciosa

Walden, Beryl M.
Wild Flowers of Hong Kong

fl.   Pl. 14(41)

Alpinia speciosa

Wit, H. C. D. de
Plants of the World;
    The Higher Plants, Vol II

fl.   Pl. 148

Alpinia tricolor

Tsukamoto, Yotaro
Coloured Illustrations of
    Garden Vlowers Vol. 9

fl.   p. 28

Alpinia vitellina

Bruggeman, L.
Tropical Plants

fl.   Pl. 69

Alpinia serumbet

American Hort. Soc.
American Horticulturist
    vol 53, No. 4   1974

fl.   p. 22

Alsike

see

Trifolium hybridum

Alsine laricifolia

Tosco, Uberto
The World of Mountain Flowers

fl. hab.   p 70

Alstroemeria aurantiaca

Hay, Roy
The Color Dictionary of Flowers &
    Plants

fl.   p. 120   Pl. 957

Alstroemeria aurantiaca

Macoby, Stirling
What Flower is That

fl.   P 38

Alstroemeria aurantiaca var.

Everett, T.H., ed.
New Illustrated Encyclopedia of
    Gardening vol. 1

fl.   p. 22

Alstroemeria aurantiaca var.

Perry, Frances
Flowers of the World

fl.   p. 21

Alstroemeria campaniflora

Curtis's Botanical Magazine
Vol 164   1943-48

fl.   No. 9664

Alstroemeria, climbing

see

Bomarea shuttleworthii

Alstroemeria ligtu var.

Hay, Roy
The Color Dictionary of Flowers & Plants

fl.   p 120, Pl. 958

Alstroemeria ligtu var.

Munoz Pizarro, Carlos
Flores Silvestres de Chile

fl.   Pl. 30

Alstroemeria ligtu var.

Perry, Frances
Flowers of the World

fl.   p. 21

Alstroemeria pulchella

Macoby, Stirling
What Flower is That

Alstroemeria pulchella

Perry, Frances
Flowers of the World

fl.   p. 21

Alstroemeria sierrae

Munoz Pizarro, Carlos
Flores Silvestres de Chile

fl.   Pl. 8

Alstroemeria violacea

Curtis's Botanical Magazine
Vol 165   1948

fl.   42

Altensteinia fimbriata

Ospina, Mariano
Orquideas de las americas

fl.   Pl. 25

Alternanthera amoena

Crockett, James Underwood
Annuals

hab.   P 92

Alternanthera amoena

Macoby, Stirling
What Flower is That

hab. P 38

Alternanthera philoxeroides

Stodola, Jiri
Encyclopedia of Water Plants

fl., hab. P. 278

Alternanthera philoxeroides

Tarver, David P.
Aquatic and Wetland Plants
    of Florida

fl. hab.   p 14

Althaea hirsuta

Martin, W. Keble
The Concise British Flora in Colour

fl. fr. hab.   Pl. 18

Althaea officinalis

American Hort. Soc.
American Horticulturist
    vol 55, No. 4   1976

fl.   p. 7

Althaea officinalis

Bianchini, Francesco
Health Plants of the World

fl. hab.   p 87

Althaea officinalis

Kleijn, H.
Beauty of the Wild Plant

fl.   p. 76   Pl. 97

Althaea officinalis

Martin, W. Keble
The Concise British Flora in Colour

fl. fr. hab.   Pl. 18

Althaea officinalis

Milne, Lorus
Living Plants of the World

fl.   p 142

Althaea officinales

Perrot, Emile
Les Plantes Medicinales

fl. fr. hab.   p. 114

Althaea officinalis

Polunin, Oleg
Flowers of Europe

fl.   Pl. 72 #747

Althaea rosea

Bianchini, Francesco
Health Plants of the World

fl.   p 87

Althaea rosea

Color Treasury of Herbs & Other
    Medicinal Plants

fl.   p 36   Pl. 44

Althaea rosea

Crockett, James Underwood
Annuals

fl.   P 92

Althaea rosea

Everett, T.H., ed.
New Illustrated Encyclopedia of
    Gardening vol. 1

fl.   p. 54

Althaea rosea

Hay, Roy
The Color Dictionary of Flowers &
    Plants

fl.   p. 29   Pl. 228

Althaea rosea

Macoby, Stirling
What Flower is That

fl.   p. 38

Althaea rosea

Perrot, Emile
Les Plantes Medicinales

fl. fr. hab.   p 206

Althaea rosea

Wit, H. C. D. de
Plants of the World;
    The Higher Plants, Vol I

fl.   p 220, Pl. 135

Althaea rosea var.

Hay, Roy
The Color Dictionary of Flowers &
    Plants

fl.   p. 29   Pl. 229

Althaea rosea var.

Huxley, Anthony
Garden Perennials and Water Plants

Althaea, Shrub

see

Hibiscus syriacus

Alum-root

see

Heuchera

Aluminum plant

see

Pilea cadierei

Alyogyne hakeifolia

Blombery, Alec M.
What Wildflower is That

fl.   p. 42   Pl. 82

Alyssum alpestre

Kohlhaupt, Paula
Fleurs des Alpages Vol 2

Fl.   P 74

Alyssum, golden

see

Alyssum saxatile

Alyssum, hoary

see

Berteroa incana

Alyssum maritimum var.

Hay, Roy
The Color Dictionary of Flowers &
    Plants

fl.   p. 29   Pl. 230, 231

Alyssum procumbens var.

Huxley, Anthony
Garden Annuals and Bulbs

fl.   Pl. 5

Alyssum saxatile

Everett, T.H., ed
New Illustrated Encyclopedia of
    Gardening vol. 1

fl.   p. 55

Alyssum saxatile

Everett, T.H., ed.
New Illustrated Encyclopedia of
    Gardening Vol 10

fl.   p. 1830

Alyssum saxatile

Felsko, Elsa
A Book of Wildflowers

Fl.   P 9

Alyssum saxatile

Huxley, Anthony
Garden Perennials and Water

hab.   Pl. 2

Alyssum saxatile

Macoby, Stirling
What Flower is That

fl.   P 39

Alyssum saxatile

Massachusetts Hort. Soc.
Horticulture
    vol 49, No. 11 1971

fl.   p. 39

Alyssum saxatile

Polunin, Oleg
Flowers of the Mediterranean

fl.   Pl. 40

Alyssum saxatile

Vilmorin, Roger de
Plantes Alpines dans les Jardins

fl. hab.   Pl. XIV, XXXIII

Alyssum saxatile var.

Hay, Roy
The Color Dictionary of Flowers &
Plants

fl.   p. 1   Pl. 3

Alyssum saxatile var.

Huxley, Anthony
Garden Perennials and Water Plants

fl.   Pl. 28

Alyssum saxatile var.

Massachusetts Hort. Soc.
Horticulture
     vol 46, No. 12   1968

fl.   p. 22

Alyssum, Sweet

     see

Lobularia maritima

Alyxia buxifolia

Cochrane, C. R.
Flowers and Plants of Victoria

fl. fr.   Pl. 276

Alyxia pubescens

Curtis's Botanical Magazine
Vol 171   1956-57

fl. fr.   266

Amana edulis

Hall, Daniel
The Genus Tulipa

fl., hab.   Pl. XL

Amaranth, Globe

     see

Gomphrena globosa

Amaranth, Spiny

     see

Amaranthus spinosus

Amaranthus atropurpureus

Bruggeman, L.
Tropical Plants

fl.   Pl. 28

Amaranthus caudatus

Crockett, James Underwood
Annuals

fl.   P 93

Amaranthus caudatus

Hay, Roy
The Color Dictionary of Flowers & Plants

fl.   p 29, Pl. 232

Amaranthus caudatus

Huxley, Anthony
Garden Annuals and Bulbs

fl.   Pl. 3

Amaranthus caudatus

Perry, Frances
Flowers of the World

fl.   p 22

Amaranthus caudatus

Tsukamoto, Yotaro
Coloured Illustrations of Garden
     Flowers   Vol. 10

fl. Illus. 4   p. 2

Amaranthus gangeticus var.

Crockett, James Underwood
Annuals

fl.   p. 93

Amaranthus graecizans

Courtenay, Booth
Wildflowers & Weeds

fl.   p 69

Amaranthus hybridus var.

Moriarty, Audrey
Wild Flowers of Malawi

fl.   Pl. 44; 5

Amaranthus hypochondriacus

Huxley, Anthony
Garden Annuals and Bulbs

fl.   Pl. 4

Amaranthus hypochondriacus

Macoby, Stirling
What Flower is That

hab.   p. 39

Amaranthus retroflexus

Clark, Lewis J.
Wild Flowers of British Columbia

fl.   p 123

Amaranthus retroflexus

Courtenay, Booth
Wildflowers & Weeds

fl.   p 69

Amaranthus retroflexus

Polunin, Oleg
Flowers of Europe

fl.   Pl. 10 #117

Amaranthus retroflexus

Weeds of the Southern United States

fr. hab.   p. 3

Amaranthus salicifolius

Bruggeman, L.
Tropical Plants

fl.   Pl. 29

Amaranthus sp.

Masefield, G. B.
The Oxford Book of Food Plants

fl.   Pl. 161, 6

Amaranthus spinosus

Weeds of the Southern United States

fr. hab.   p. 3

Amaranthus tricolor

Everett, T.H., ed.
New Illustrated Encyclopedia of
     Gardening vol. 1

fl.   p. 55

Amaranthus tricolor

Everett, T.H., ed.
New Illustrated Encyclopedia of
     Gardening vol. 13

hab.   p. 2406

Amaranthus tricolor

Hay, Roy
The Color Dictionary of Flowers &
Plants

hab.   p. 30   Pl. 233

Amaranthus tricolor

Macoby, Stirling
What Flower is That

hab. P 39

Amaranthus tricolor var.

Tsukamoto, Yotaro
Coloured Illustrations of Garden
    Flowers   Vol. 10

fl.   Illus. 3   p. 1

x Amarcrinum howardii

Tsukamoto, Yotaro
Coloured Illustrations of Garden
    Flowers   Vol. 9

fl.   p. 2

x Amarcrinum hybrid

Miles, Bebe
Bulbs for the Home Gardener

fl.   p 171

x Amarine tubergenii var.

Royal Hort. Soc.
Journal of the Royal Hort. Soc.
    vol 99, No. 11   1974

fl.   p. 490   Pl. 220

Amaryllis

    see

Hippeastrum

Amaryllis belladonna

Everett, T.H., ed.
New Illustrated Encyclopedia of
    Gardening vol. 1

fl.   p. 70

Amaryllis belladonna

Flowering Plants of Africa
    vol 30   1954-55

fl.   Pl. 1200

Amaryllis belladonna

Hay, Roy
The Color Dictionary of Flowers & Plants

fl.   p 84, Pl. 670

Amaryllis belladonna

Hay, Roy
The Dictionary of House Plants

fl.   Pl. 31

Amaryllis belladonna

Macoby, Stirling
What Flower is That

fl.   P 39

Amaryllis belladonna

Massachusetts Hort. Soc.
Horticulture
    vol 51, No. 8   1973

fl.   p. 23

Amaryllis bella-donna

Morley, Brian D.
Wild Flowers of the World

fl.   Pl. 86

Amaryllis belladonna

Perry, Frances
Flowers of the World

fl.   p 24

Amaryllis belladonna

Royal Hort. Soc.
Journal of the Royal Hort. Soc.
    vol 75, No. 1   1950

fl.   p. 26   Pl. 2

Amaryllis belladonna

Wit, H. C. D. de
Plants of the World;
    The Higher Plants, Vol II

fl.   Pl. 134

Amaryllis belladonna var.

Royal Hort. Soc.
Journal of the Royal Hort. Soc.
    vol 76, No. 11   1951

fl.   p. 402   Pl. 178, 185-86

Amaryllis, blue

    see

Griffinia hyacinthina

Amaryllis, Blue

    see

Worsleya procera

Amaryllis evansiae

American Hort. Soc.
National Horticultural Magazine
    vol 36, No. 2   1957

fl.   cover

Amblyopetalum coccineum

Curtis's Botanical Magazine
Vol 174   1962-63

fl.   Pl. 431

Ambrosia artemisiifolia

Courtenay, Booth
Wildflowers & Weeds

fl.   p 113

Ambrosia artemisiifolia

Weeds of the Southern United States

fl.   p. 8

Ambrosia trifida

Courtenay, Booth
Wildflowers & Weeds

fl.   p 113

Ambrosia trifida

Klimas, John E.
Wildflowers of Eastern America

fl.   Pl. 102

Ambrosia trifida

Lindman, C. A. M.
Nordens Flora, Vol 9

fl. fr.   Pl. 596

Ambrosia trifida

Weeds of the Southern United States

fr.   p. 9

Amelanchier alnifolia

Clark, Lewis J.
Wild Flowers of British Columbia

fl.   p 231

Amelanchier alnifolia

Heller, Christine
Wild Flowers of Alaska

fl., fr.   Pl. 143-144

Amelanchier alnifolia

Porsild, A.E.
Rocky Mountain Wild Flowers

fr.   p 231

Amelanchier alnifolia

Shaw, Richard J.
Field Guide to the Vascular Plants of
    Grand Teton National Park

fr.   Pl. 6

Amelanchier alnifolia

Walcott, Mary Vaux
North American Wild Flowers

fr. hab.   Pl. 117

Amelanchier alnifolia var.

Royal Hort. Soc.
Journal of the Royal Hort. Soc.
    vol 95, No. 2   1970

hab., fr.   p. 86   Pl. 83-85

Amelanchier arborea

Batson, Wade T.
Wild Flowers in South Carolina

fl.   p. 56

Amelanchier arborea

Brown, Clair A.
Wildflowers of Louisiana and
    Adjoining States

fl.   p 66

Amelanchier arborea

Jennings, O. E.
Wild Flowers of Western Pennsylvania
                        Vol II
fl.   Pl. 47

Amelanchier arborea

Wharton, Mary E.
Trees & Shrubs of Kentucky

fl.   p 44, Pl. 1.7
fr.   p 134, Pl. 2.23

Amelanchier asiatica

Kitamura, Siro
Coloured Illustrations of Trees &
    Shrubs of Japan

fl.   235

Amelanchier canadensis

Harrison, R. E.
Trees andShrubs

fl. hab.   P 23

Amelanchier canadensis

Hay, Roy
The Color Dictionary of Flowers & Plants

fl. hab.   p 181, Pl. 1444, 1445

Amelanchier canadensis

Massachusetts Hort. Soc.
Horticulture
    vol 53, No. 7   1975

fr.   p. 37

Amelanchier canadensis

Morley, Brian D.
Wild Flowers of the World

fl. fr.   Pl. 152

Amelanchier confusa

Gault, S. Millar
The Color Dictionary of Shrubs

hab.   Pl. 8

Amelanchier florida var.

Curtis's Botanical Magazine
    Vol 160   1937

fl. fr.   No. 9496

Amelanchier x grandiflora

Massachusetts Hort. Soc.
Horticulture
    vol. 34, No. 6   1956

fl.   p. 346

Amelanchier x grandiflora

Polunin, Oleg
Trees and Bushes of Europe

fl.,fr.,hab.   p. 85

Amelanchier laevis

Campbell, Carlos C.
Great Smoky Mountain Wildflowers

fl.   p 73

Amelanchier laevis

Huxley, Anthony
Deciduous Garden Trees and Shrubs

fl.   Pl. 14

Amelanchier lamarckii

Hansen, Richard
Baume und Straucher im Garten

fl.   p. 99

Amelanchier ovalis

Barneby, T. P.
European Alpine Flowers in Colour

fl.   Pl. 36; 5

Amelanchier ovalis

Kohlhaupt, Paula
Fleurs des Alpages Vol 2

Fl.   P 58

Amelanchier ovalis

Polunin, Oleg
Flowers of Europe

fl.   Pl. 47 #471

Amelanchier ovalis

Polunin, Oleg
Trees and Bushes of Europe

fl. fr. hab.   p 84, 85

Amelanchier spicata

Vedel, H.
Arbres et Arbustes

fl. fr. hab.   p. 68   Pl. 64

Amelanchier utahensis

Welsh, Stanley L.
Flowers of the Mountain Country

fl.   p. 4

Amerorchis rotundifolia

Alaska-Yukon Wild Flowers Guide

fl.   p 17

Amerorchis rotundifolia

Luer, Carlyle A.
The Native Orchids of the
    United States and Canada

fl. hab.   Pl. 37

Amethystanthus japonicus

Kimura, Koiti
Japanese Medicinal Plants, Vol I

fl.   p 78

Amherstia nobilis

Blunt, Wilfrid
Flora Superba

fl.   Pl. XIII

Amherstia nobilis

Bruggeman, L.
Tropical Plants

fl.   Pl. 201

Amherstia nobilis

Hargreaves, Dorothy
Tropical Blossoms of the Caribbean

fl.   p. 39

Amherstia nobilis

Morley, Brian D.
Wild Flowers of the World

fl.fr.   Plate 113

Amherstia nobilis

Pertchik, Bernard
Flowering Trees of the
    Caribbean

fl.   p 63

Amianthium muscaetoxicum

Addisonia
Vol 20   1937-38

fl., fr.   p. 55   Pl. 668

Amianthium muscaetoxicum

Batson, Wade T.
Wild Flowers in South Carolina

fl. hab.   p 30

Amianthium muscaetoxicum

Campbell, Carlos C.
Great Smoky Mountain Wildflowers

fl.   p 87

Amianthium muscaetoxicum

Dean, Blanche E.
Wildflowers of Alabama and
    Adjoining States

fl.   p 11

Amianthium muscaetoxicum

Klimas, John E.
Wildflowers of Eastern America

fl.   Pl. 32

Ammi majus

Perrot, Emile
Les Plantes Medicinales

fr.   p 10

Ammi visnaga

Perrot, Emile
Les Plantes Medicinales

fl. fr. hab.   p 10

Ammobium alatum

Blombery, Alec M.
What Wildflower is That

fl.   p 42, Pl. 83

Ammobium alatum

Crockett, James Underwood
Annuals

fl.   P 93

Ammobium alatum

Morley, Brian D.
Wild Flowers of the World

fl. fr.   pl. 140

Ammobium alatum var.

Hay, Roy
The Color Dictionary of Flowers &
     Plants

fl.   p. 30   Pl. 234

Ammocharis coranica

Flowering Plants of South Africa
     vol XVIII   1938

fl.   Pl. 712

Ammophila arenaria

Lindman, C. A. M.
Nordens Flora, Vol 2

fr. hab.   Pl. 96

Ammophila arenaria

Polunin, Oleg
Flowers of Europe

fr.   Pl. 181 #1775

Amoebophyllum angustum

Herre, H.
The Genera of the Mesembryanthemaceae

fl. fr.   p 69

Amoreuxia palmatifida

Chickering, Carol Rogers
Flowers of Guatemala

fl. fr.   p 33

Amorpha canescens

Bartels, Andreas
Das Grosse Buch der Gartengeholze

fl.   p 113

Amorpha canescens

Courtenay, Booth
Wildflowers & Weeds

fl.   p 51

Amorpha canescens

Moyle, John B.
Northland Wild Flowers

fl. hab.   p 77, Pl. 51

Amorpha fruticosa

Batson, Wade T.
Wild Flowers in South Carolina

fl. hab.   p 64

Amorpha fruticosa

Brown, Clair A.
Wildflowers of Louisiana and
     Adjoining States

fl. hab.   p 72

Amorpha fruticosa

Curtis's Botanical Magazine
Vol 178   1970-72

fl.   604

Amorpha fruticosa

Dean, Blanche E.
Wildflowers of Alabama and
     Adjoining States

fl.   p 87

Amorpha fruticosa

Polunin, Oleg
Trees and Bushes of Europe

fl. fr.   p 113

Amorpha fruticosa

Wharton, Mary E.
Trees & Shrubs of Kentucky

fl.   p 73, Pl. 2.8

Amorphophallus abyssinicus

Flowering Plants of Africa
     vol 32   1957-58

fl.   Pl. 1251

Amorphophallus abyssinicus

Morley, Brian D.
Wild Flowers of the World

fl.,hab.   Pl. 56

Amorphophallus kiusianus

Kimura, Koiti
Japanese Medicinal Plants, Vol II

fr.   p 131

Amorphophallus variabilis

Walden, Beryl M.
Wild Flowers of Hong Kong

fl. fr. hab.   Pl. 28(77)

Ampelamus albidus

Weeds of the Southern United States

hab., fr.   p. 4

Ampelodesma mauritanica

Polunin, Oleg
Flowers of Europe

fr.   Pl. 182 #1741

Ampelopsis aconitifolia var.

Addisonia
Vol 22   1943-46

fl., fr.   p. 21   Pl 715

Ampelopsis arborea

Weeds of the Southern United States

fr., hab.   p. 42

Ampelopsis, Basket

see

Ampelopsis brevipedunculata var.

Ampelopsis brevipedunculata

Harrison, Richmond E.
Climbers and Trailers

fr.   p 19   Pl. 8

Ampelopsis brevipedunculata var.

Kiaer, Eigil
Indoor Plants in Colour

hab.   p. 15

Ampelopsis brevipedunculata var.

Kromdijk, G.
200 House Plants in Colour

hab.   Pl. 13

Ampelopsis brevipedunculata var.

Macoby, Stirling
What Flower is That

hab.   p. 40

Ampelopsis brevipedunculata var.

Massachusetts Hort. Soc.
Horticulture
   vol 49, No. 8   1971

fr.   p. 32

Ampelopsis brevipedunculata var.

Perry, Frances
Flowers of the World

fr.   p. 307

Amphianthus pusillus

Duncan, Wilbur H.
Wildflowers of the Southeastern
   United States

fl.   p 173

Amphibolia gydouwensis

Herre, H.
The Genera of the Mesembryanthemaceae

fl. fr.   p 71

Amphicarpa africana

Moriarty, Audrey
Wild Flowers of Malawi

fl.   Pl. 65; 4

Amphicarpa bracteata

Courtenay, Booth
Wildflowers & Weeds

fl.   p 52

Amphicarpa bracteata

Moyle, John B.
Northland Wild Flowers

fl. hab.   p 84, Pl. 63

Amphicarpa bracteata

Wharton, Mary E.
A Guide to the Wildflowers & Ferns
   of Kentucky

fl. p 181   Pl. 1.5

Amsinckia douglasiana

Munz, Philip A.
California Spring Wildflowers

fl.   p 77, Pl. 56

Amsinckia intermedia

Clark, Lewis J.
Wild Flowers of British Columbia

fl.   p 446

Amsinckia intermedia

Orr, Robert T.
Wildflowers of Western America

fl.   Pl. 124

Amsonia ciliata

Batson, Wade T.
Wild Flowers in South Carolina

fl. hab.   p 94

Amsonia rigida

Addisonia
Vol 20   1937-38

fl., fr.   p. 25   Pl. 653

Amsonia salicifolia

Hay, Roy
The Color Dictionary of Flowers & Plants

fl.   p 120, Pl. 959

Amsonia tabernaemontana

Brown, Clair A.
Wildflowers of Louisiana and
   Adjoining States

fl. hab.   p 142

Amsonia tabernaemontana

Dean, Blanche E.
Wildflowers of Alabama and
   Adjoining States

fl.   p 137

Amsonia tabernaemontana

Duncan, Wilbur H.
Wildflowers of the Southeastern
   United States

fl.   p 127

Amsonia tabernaemontana

Wharton, Mary E.
A Guide to the Wildflowers & Ferns
   of Kentucky

fl. p 145   Pl. 1.102

Amyema congener

Blombery, Alec M.
What Wildflower is That

fl.   p 43, Pl. 84

Amyema fasciculatum

Morley, Brian D.
Wild Flowers of the World

fl.   Plate 118

Amyema pendula

Cochrane, G. R.
Flowers and Plants of Victoria

fl.   Pl. 366

Amygdalus communis

Linnell, T.
Plantes Utiles du Monde entier

fl. fr.   p. 19   Pl. 28

Amygdalus communis

Perrot, Emile
Les Plantes Medicinales

fl. fr. hab.   p 174

Amygdalus orientalis

Morley, Brian D.
Wild Flowers of the World

fl.,fr.   Pl. 45

Ana Tree

   see

Acacia albida

Anabasis articulata

Feinbrun-Dothan, Naomi
Wild Plants in the Land
   of Israel

fl. fr.   p 142

Anacampseros rufescens

Lamb, Edgar
The Pocket Encyclopedia of Cacti
   and Succulents in Color

fl. hab.   Pl. 216

Anacamptis pyramidalis

Ary, S.
The Oxford Book of Wildflowers

fl.   p 158, Pl. 4

Anacamptis pyramidalis

Barneby, T. P.
European Alpine Flowers in Colour

fl.   Pl. 9, 6

Anacamptis pyramidalis

Brooke, Jocelyn
The Wild Orchids of Britain

fl., hab.   Pl. 21

Anacamptis pyramidalis

Martin, W. Keble
The Concise British Flora in Colour

fl.   Pl. 81

Anacamptis pyramidalis

Morley, Brian D.
Wild Flowers of the World

fl.   Pl. 26

Anacamptis pyramidalis

Polunin, Oleg
Flowers of Europe

fl.   Pl. 190 #1908

Anacamptis pyramidalis

Polunin, Oleg
Flowers of the Mediterranean

fl.   Pl. 307

Anacardium occidentale

Bianchini, F.
The Complete Book of Fruits & Vegetables

fr.   p 199

Anacardium occidentale

Hargreaves, Dorothy
Tropical Blossoms of the Caribbean

fl.,fr.   p. 62

Anacardium occidentale

Masefield, G. B.
The Oxford Book of Food Plants

fr.   Pl. 31, 2

Anacardium occidentale

Milne, Lorus
Living Plants of the World

fr.   p 127

Anacyclus depressus

Hay, Roy
The Color Dictionary of Flowers & Plants

fl.   p 1, Pl. 4

Anacyclus pyrethrum

Perrot, Emile
Les Plantes Medicinales

fl. fr. hab.   p 192

Anagallis arvensis

Ary, S.
The Oxford Book of Wildflowers

fl.   p 104, Pl. 1

Anagallis arvensis

Clark, Lewis J.
Wild Flowers of British Columbia

fl.   p 414

Anagallis arvensis

Crockett, James Underwood
Annuals

fl.   P 94

Anagallis arvensis

Dean, Blanche E.
Wildflowers of Alabama and
   Adjoining States

fl.   p 129

Anagallis arvensis

Felsko, Elsa
A Book of Wildflowers

fl.fr.   P 96

Anagallis arvensis

Furrer, D.
Blumen am Wege

fl. hab.   p 69

Anagallis arvensis

Kleijn, H.
Beauty of the Wild Plant

fl.   p. 39   Pl. 46

Anagallis arvensis

Lemmon, Robert S.
Wildflowers of North America in
   Full Color

fl.   p. 37   Pl. 62

Anagallis arvensis

Lindman, C. A. M.
Nordens Flora, Vol 7

fl. fr. hab.   Pl. 469

Anagallis arvensis

Martin, W. Keble
The Concise British Flora in Colour

fl. hab.   Pl. 58

Anagallis arvensis

Perry, Frances
Flowers of the World

fl.   p 243

Anagallis arvensis

Polunin, Oleg
Flowers of the Mediterranean

fl.   Pl. 128

Anagallis arvensis

Pond, Barbara
A Sampler of Wayside Herbs

fl.   Pl. VI

Anagallis arvensis

Wharton, Mary E.
A Guide to the Wildflowers & Ferns
   of Kentucky

fl.   p 177   Pl. 2.63

Anagallis arvensis var.

Martin, W. Keble
The Concise British Flora in Colour

fl. hab.   Pl. 58

Anagallis arvensis var.

Morley, Brian D.
Wild Flowers of the World

fl.,hab.   Pl. 33

Anagallis coerulea

Feinbrun-Dothan, Naomi
Wild Plants in the Land
   of Israel

fl. hab.   p 112

Anagallis linifolia

Crockett, James Underwood
Annuals

fl.   P 94

Anagallis linifolia

Polunin, Oleg
Flowers of Europe

fl.   Pl. 93 #967

Anagallis linifolia

Royal Hort. Soc.
Journal of the Royal Hort. Soc.
   vol 89, No. 1   1964

fl.   p. 22   Pl. 9

Anagallis linifolia var.

Hay, Roy
The Color Dictionary of Flowers &
   Plants

fl.   p. 1   Pl. 5, 6

Anagallis minima

Martin, W. Keble
The Concise British Flora in Colour

hab.   Pl. 58

Anagallis monellii var.

Morley, Brian D.
Wild Flowers of the World

fl.,hab.   Pl. 33

Anagallis tenella

Ary, S.
The Oxford Book of Wildflowers

fl.   p 124, Pl. 7

Anagallis tenella

Martin, W. Keble
The Concise British Flora in Colour

fl. hab.   Pl. 58

Anagyris foetida

Polunin, Oleg
Flowers of the Mediterranean

fr.   Pl. 49

Anagyris foetida

Polunin, Oleg
Trees and Bushes of Europe

fl.   p 106

Ananas bracteatus var.

Bromeliad Society
Bulletin
    vol 20, No. 1 1970

hab.   p. 24

Ananas bracteatus var.

Bromeliad Society
Journal
    vol 24, No. 1 1974

fl. p. 10

Ananas bracteatus var.

Hay, Roy
The Dictionary of House Plants

fr.   Pl. 32

Ananas bracteatus var.

Macoby, Stirling
What Flower is That

hab.   p. 40

Ananas bracteatus var.

Wilson, Robert Gardner
Bromeliads in Cultivation

fl. hab.   Pl. 61

Ananas comosus

Bianchini, F.
The Complete Book of Fruits & Vegetables

fl.   p 171

Ananas comosus

Kromdijk, G.
200 House Plants in Colour

hab.   Pl. 14

Ananas comosus

Masefield, G. B.
The Oxford Book of Food Plants

fl. fr.   Pl. 97, 1

Ananas comosus var.

Bromeliad Society
Bulletin
    vol 15, No. 1 1965

fl., hab.   p. 20

Ananas comosus var.

Padilla, Victoria
Bromeliads

fr., hab.   p. 64   Pl. 2

Ananas comosus var.

Padilla, Victoria, ed.
Bromeliads in Color and
    Their Culture

fr.   frontis.

Ananas comosus var.

Perry, Frances
Flowers of the World

fl.   p. 55

Ananas comosus var.

Rauh, Werner
Bromeliads for Home, Garden and
    Greenhouse

fl.   Pl. 84

Ananas comosus var.

Royal Hort. Soc.
Journal of the Royal Hort. Soc.
    vol 87, No. 8   1962

fl., hab.   p. 358   Pl. 109

Ananas comosus var.

Royal Hort. Soc.
The Garden
    vol 100, No. 11   1975

fr., hab.   p. 533

Ananas comosus var.

Wilson, Robert Gardner
Bromeliads in Cultivation

fr. hab.   Pl. 63, 66

Ananas erictifolius

Wilson, Robert Gardner
Bromeliads in Cultivation

fr. hab.   Pl. 69

Ananas nanas

Wilson, Robert Gardner
Bromeliads in Cultivation

fr. hab.   Pl. 69

Ananas sagenaria

Wit, H. C. D. de
Plants of the World;
    The Higher Plants, Vol II

fr.   Pl. 137

Ananas sativus

Hvass, Elsie
Plants That Feed and Serve Us

fr.   p 54, Pl. 100

Anapalina caffra

Flowering Plants of Africa
    vol 35   1962

fl., hab.   Pl. 1400

Anaphalis margaritacea

Clark, Lewis J.
Wild Flowers of British Columbia

fl.   p 518

Anaphalis margaritacea

Courtenay, Booth
Wildflowers & Weeds

fl.   p. 126

Anaphalis margaritacea

Huxley, Anthony
Garden Perennials and Water Plants

fl.   Pl. 29

Anaphalis margaritacea

Munz, Philip A.
California Mountain Wildflowers

fl.   p 47, Pl. 61

Anaphalis margaritacea

Taylor, Ronald J.
Mountain Wild Flowers

fl. hab.   p 43

Anaphalis margaritacea

Walcott, Mary Vaux
North American Wild Flowers
    vol. 4

fl.   Pl. 289

Anaphalis margaritacea

Welsh, Stanley L.
Flowers of the Mountain Country

fl.   p. 6

Anaphalis nubigena

Hay, Roy
The Color Dictionary of Flowers & Plants

fl.   p 120, Pl. 960

Anaphalis triplinervis

Hay, Roy
The Color Dictionary of Flowers & Plants

fl.   p 121, Pl. 961

Anaphalis triplinervis

Tosco, Uberto
The World of Mountain Flowers

fl.   p 91

Anaphalis yedoensis

Hay, Roy
The Color Dictionary of Flowers & Plants

fl.   p. 121   Pl. 962

Anarrhinum bellidifolium

Polunin, Oleg
Flowers of Europe

fl.   Pl. 123 #1211

Anastatica hierochuntica

Morley, Brian D.
Wild Flowers of the World

fl.,fr.,hab.   Pl. 46

Anchusa angustissima

Curtis's Botanical Magazine 1962-63
        Vol. 174

fl.   Pl. 411

Anchusa arvensis

Lindman, C. A. M.
Nordens Flora, Vol 7

fl. fr. hab.   Pl. 487

Anchusa azurea

Polunin, Oleg
Flowers of Europe

fl.   Pl. 102 #1056

Anchusa azurea

Polunin, Oleg
Flowers of the Mediterranean

fl.   Pl. 144

Anchusa azurea

Tsukamoto, Yotaro
Coloured Illustrations of Garden
        Flowers Vol. 10

fl. Illus. 7   p. 3

Anchusa azurea var.

Hay, Roy
The Color Dictionary of Flowers
        & Plants

fl.   p 121   Pl. 963

Anchusa azurea var.

Huxley, Anthony
Garden Perennials and Water Plants

fl.   Pl 30

Anchusa azurea var.

Perry, Frances
Flowers of the World

fl.   p. 49

Anchusa, capensis

Crockett, James Underwood
Annuals

fl.   p. 94.

Anchusa capensis

Tsukamoto, Yotaro
Coloured Illustrations of Garden
        Flowers Vol. 10

fl. Illus. 8   p. 3

Anchusa capensis var.

Crockett, James Underwood
Annuals

fl.   p 94

Anchusa capensis var.

Hay, Roy
The Color Dictionary of Flowers &
        Plants

fl.   p. 30   Pl. 235

Anchusa capensis var.

Hay, Roy
The Dictionary of House Plants

fl.   Pl. 33

Anchusa hybrida

Polunin, Oleg
Flowers of the Mediterranean

fl.   Pl. 143

Anchusa italica

Barneby, T. P.
European Alpine Flowers in Colour

fl.   Pl. 66, 1

Anchusa myosotidiflora

Huxley, Anthony
Garden Perennials and Water Plants

fl.   Pl. 31

Anchusa officinalis

Felsko, Elsa
A Book of Wildflowers

fl.   P 57

Anchusa officinalis

Lindman, C. A. M.
Nordens Flora, Vol 7

fl. fr.   Pl. 486

Anchusa officinalis

Polunin, Oleg
Flowers of Europe

fl.   Pl. 103 #1055

Ancistrocactus scheeri

Lamb, Edgar
Colorful Cacti of the American Deserts

hab.   Pl. 5

Ancistrochilus rothschildianus

Piers, Frank
Orchids of East Africa

fl   Pl. Id

Ancistrochilus rothschildianus

Stewart, Joyce
Orchids of Tropical Africa

fl.   Pl. 6

Ancistrochilus thomsonianus

Kupper, Walter
Orchids

fl. hab.   p 49

Andersonia caerulea

Blombery, Alec M.
What Wildflower is That

fl.   p 44, Pl. 85

Andersonia echinocephala

Blombery, Alec M.
What Wildflower is That

fl.   p 44, Pl. 86

Andrachne phyllanthoides

Dean, Blanche E.
Wildflowers of Alabama and
        Adjoining States

fl. fr.   p 97

Androcymbium ciliolatum

Eliovson, Sima
Namaqualand in Flower

fl.   Pl. 33, 2

Androcymbium melanthioides

Moriarty, Audrey
Wild Flowers of Malawi

fl.   Pl. 11; 4

Androcymbium roseum

Flowering Plants of Africa
        vol 31   1956

fl., hab.   Pl. 1225

Androlepis skinneri

Bromeliad Society
Journal
        vol 26, No. 3   1976

fl.   p. 113

Androlepis skinneri

Wilson, Robert Gardner
Bromeliads in Cultivation

hab.   Pl. 67

Andromeda

    see

Pieris japonica

Andromeda glaucophylla

Courtenay, Booth
Wildflowers & Weeds

fl.   p 77

Andromeda glaucophylla

Ferguson, Mary
Wildflowers

fl.   p. 67

Andromeda glaucophylla

Moyle, John B.
Northland Wild Flowers

fl. hab.   p 116, Pl. 116

Andromeda, Japanese

    see

Pieris japonica

Andromeda, Marsh

    see

Andromeda polifolia

Andromeda, mountain

    see

Pieris floribunda

Andromeda polifolia

Alaska-Yukon Wild Flower Guide

fl.   p 117

Andromeda polifolia

Ary, S.
The Oxford Book of Wildflowers

fl.   p 118, Pl. 4

Andromeda polifolia

Felsko, Elsa
A Book of Wildflowers

fl.   P 86

Andromeda polifolia

Heller, Christine
Wild Flowers of Alaska

fl.   Pl. 108

Andromeda polifolia

Hellyer, A.G.L.
Shrubs in Colour

fl.   p. 14

Andromeda polifolia

Kleijn, H.
Beauty of the Wild Plant

fl.   p. 57   Pl. 71

Andromeda polifolia

Lindman, C. A. M.
Nordens Flora, Vol 7

fl. fr.   Pl. 446

Andromeda polifolia

Martin, W. Keble
The Concise British Flora in Colour

fl. fr. hab.   Pl. 55

Andromeda polifolia

Polunin, Oleg
Flowers of Europe

fl.   Pl. 88 #923

Andromeda, Privet

    see

Lyonia ligustrina

Andropogon amplectens

Flowering Plants of South Africa
vol XXIV   1944

fr., hab.   Pl. 922

Andropogon capillipes

Duncan, Wilbur H.
Wildflowers of the Southeastern
United States

fl.   p 233

Andropogon ternarius

Batson, Wade T.
Wild Flowers in South Carolina

fl.   p 25

Andropogon virginicus

Batson, Wade T.
Wild Flowers in South Carolina

hab.   p 25

Andropogon virginicus

Weeds of the Southern United States

fr., hab.   p. 23

Andropogon virginicus var.

Batson, Wade T.
Wild Flowers in South Carolina

fl.   p 25

Androsace alpina

Barneby, T.P.
European Alpine Flowers in Colour

fl. hab.   Pl. 61, 3

Androsace alpina

Kleijn, H.
Beauty of the Wild Plant

fl.   p. 93   Pl. 128

Androsace alpina

Polunin, Oleg
Flowers of Europe

fl.   Pl. 91 #949

Androsace carinata

Walcott, Mary Vaux
North American Wild Flowers

fl. hab.   Pl. 107

Androsace carnea

Barneby, T. P.
European Alpine Flowers in Colour

fl. hab.   Pl. 61, 1

Androsace carnea

Kohlhaupt, Paula
Fleurs des Alpages Vol 2

Fl.   P 42

Androsace carnea

Polunin, Oleg
Flowers of Europe

fl.   Pl. 90 #953

Androsace chamaejasme

Alaska-Yukon Wild Flower Guide

fl. hab.   p 126

Androsace chamaejasme

Barneby, T. P.
European Alpine Flowers in Colour

fl. hab.   Pl. 60, 5

Androsace chamaejasme

Kohlhaupt, Paula
Fleurs des Alpages Vol 1

fl. P 37

Androsace Chamaejasme

Porsild, A. E.
Rocky Mountain Wild Flowers

fl. hab.   p 325

Androsace chamaejasme var.

Heller, Christine
Wild Flowers of Alaska

fl.   Pl. 189

Androsace helvetica

Hay, Roy
The Color Dictionary of Flowers & Plants

fl.   p 1, Pl. 7

Androsace helvetica

Kohlhaupt, Paula
Fleurs des Alpages Vol 2

Fl. P 43

Androsace helvetica

Tosco, Uberto
The World of Mountain Flowers

fl. hab.   p 64

Androsace imbricata

Alpine Garden Society
Bulletin
    vol. 6, No. 4   1938

fl., hab.   p. 341

Androsace imbricata

Royal Hort. Soc.
Journal of the Royal Hort. Soc.
    vol 85, No. 6   1960

fl.   p. 266   Pl. 78

Androsace jacquemontii

Hay, Roy
The Color Dictionary of Flowers & Plants

fl.   p 1, Pl. 8

Androsace lactea

Barneby, T. P.
European Alpine Flowers in Colour

fl. hab.   Pl. 61, 2

Androsace lanuginosa

Hay, Roy
The Color Dictionary of Flowers & Plants

fl.   p 2, Pl. 9

Androsace obtusifolia

Barneby, T. P.
European Alpine Flowers in Colour

fl. hab.   Pl. 60, 6

Androsace obtusifolia

Kohlhaupt, Paula
Fleurs des Alpages   Vol 2

Fl.   P 41

Androsace occidentalis

Courtenay, Booth
Wildflowers & Weeds

fl.   p 84

Androsace ochotensis

Heller, Christine
Wild Flowers of Alaska

fl.   Pl. 76

Androsace, Pygmy

see

Androsace subumbellata

Androsace pyrenaica

Hay, Roy
The Color Dictionary of Flowers & Plants

fl.   p 2, Pl. 10

Androsace sarmentosa

Huxley, Anthony
Garden Perennials and Water Plants

fl.   Pl. 32

Androsace sarmentosa var.

Vilmorin, Roger de
Plantes Alpines dans les Jardins

fl. hab.   Pl. XXXII

Androsace septentrionalis

Lindman, C. A. M.
Nordens Flora, Vol 7

fl, fr, hab.   Pl. 461

Androsace septentrionalis

Porsild, A. E.
Rocky Mountain Wild Flowers

fl. hab.   p. 325

Androsace strigillosa

Morley, Brian D.
Wild Flowers of the World

fl.   Pl. 98

Androsace subumbellata

Walcott, Mary Vaux
North American Wild Flowers

fl. hab.   Pl. 184

Androsace triflora

Morley, Brian D.
Wild Flowers of the World

fl. fr.   Pl. 48

Androsace villosa var.

Hay, Roy
The Color Dictionary of Flowers &
    Plants

fl.   p. 2   Pl. 11

Andryala cheiranthifolia

Bramwell, David
Wild Flowers of the Canary Islands

fl.   Pl. 319

Andryala integrifolia

Polunin, Oleg
Flowers of Europe

fl.   Pl. 159 #1533

Andryala integrifolia

Polunin, Oleg
Flowers of the Mediterranean

fl.   Pl. 217

Aneilema aequinoctiale

Flowering Plants of Africa
    vol. 33   1959

fl.   Pl. 1302

Aneilema aequinoctiale

Moriarty, Audrey
Wild Flowers of Malawi

fl.   Pl. 20; 5

Aneilema hockii

Moriarty, Audrey
Wild Flowers of Malawi

fl.   Pl. 20; 4

Aneilema nudiflorum

Walden, Beryl M.
Wild Flowers of Hong Kong

fl.   Pl. 61 (188)

Aneilema welwitschii

Moriarty, Audrey
Wild Flowers of Malawi

fl.   Pl. 20; 3

Anemarrhena asphodeloides

Kariyone, Tatsuo
Atlas of Medicinal Plants

fl., hab.   Pl. 17

Anemarrhena asphodeloides

Kimura, Koiti
Japanese Medicinal Plants, Vol II

fl.   p 135

Anemarrhena asphodeloides

Takatori, Jisuke
Color Atlas of Medicinal Plants of Japan

fl.   Fig. 71

Anemone, Alaska Blue

see

Anemone multiceps

Anemone, Alpen

see

Pulsatilla alpina

Anemone apennina

Polunin, Oleg
Flowers of Europe

fl.   Pl. 22 #214

Anemone apennina

Royal Hort. Soc.
Journal of the Royal Hort. Soc.
vol 93, No. 4   1968

fl.   cover

Anemone apennina var.

Royal Hort. Soc.
The Garden
vol. 101, No. 7   1976

fl.   p. 351

Anemone baldensis

Barneby, T. P.
European Alpine Flowers in Colour

fl. hab.   Pl. 24, 2

Anemone baldensis

Kohlhaupt, Paula
Fleurs des Alpages   Vol. 2

fl.   P. 2

Anemone biflora

Curtis's Botanical Magazine
Vol 161   1938

fl. hab.   No. 9551

Anemone biflora

Mathew, Brian
Dwarf Bulbs

fl.   p. 40   Pl. 12

Anemone biflora

Royal Hort. Soc.
Journal of the Royal Hort. Soc.
vol 88, No. 5   1963

fl.   p. 210   Pl. 72

Anemone biflora var.

Royal Hort. Soc.
Journal of the Royal Hort. Soc.
vol 90, No. 1   1965

fl.   p. 22   Pl. 9

Anemone blanda

American Hort. Soc.
American Horticulturist
vol 56, No. 2   1977

fl.   p. 22

Anemone blanda

Curtis's Botanical Magazine
Vol 178   1970-72

fl.   598

Anemone blanda

Huxley, Anthony
Garden Annuals and Bulbs

hab.   Pl. 146

Anemone blanda

Massachusetts Hort. Soc.
Horticulture
vol 43, No. 3   1965

fl.   p. 29

Anemone blanda

Massachusetts Hort. Soc.
Horticulture
vol 47, No. 9   1969

fl.   p. 32

Anemone blanda

Megaw, Elektra
Wild Flowers of Cyprus

fl., hab.   p. 7   Pl. 3

Anemone blanda

Miles, Bebe
Bulbs for the Home Gardener

fl.   p 52

Anemone blanda

Morley, Brian D.
Wild Flowers of the World

fl. fr.   Pl. 8D

Anemone blanda

Pennsylvania Hort. Soc.
The Green Scene
vol 6, No. 1   1977

fl.   p. 15

Anemone blanda

Polunin, Oleg
Flowers of Europe

fl.   Pl. 21 #214

Anemone blanda

Polunin, Oleg
Flowers of the Mediterranean

fl.   Pl. 21

Anemone blanda

Royal Hort. Soc.
Journal of the Royal Hort. Soc.
vol 100, No. 3   1975

fl.   cover

Anemone blanda var.

Hay, Roy
The Color Dictionary of Flowers &
Plants

fl.   p. 84   Pl. 671

Anemone blanda var

Huxley, Anthony
Garden Annuals and Bulbs

fl.   197 a & b

Anemone blanda var.

Mathew, Brian
Dwarf Bulbs

fl.   p. 40   Pl. 11

Anemone blanda var.

Perry, Frances
Flowers of the World

fl.   p. 252

Anemone blanda var.

Royal Hort. Soc.
Journal of the Royal Hort. Soc.
vol 91, No. 6   1966

fl.   p. 250   Pl. 128

Anemone bucharica

Curtis's Botanical Magazine
Vol 179   1972

fl.   622

Anemone canadensis

Courtenay, Booth
Wildflowers & Weeds

fl.   p 34

Anemone canadensis

Ferguson, Mary
Wildflowers

fl.   p 46

Anemone canadensis

Jennings, O. E.
Wild Flowers of Western Pennsylvania
        Vol. II

fl.   Pl. 63

Anemone canadensis

Lemmon, Robert S.
Wildflowers of North America in
        Full Color

fl.   pg. 248   pl. 390

Anemone canadensis

Moyle, John B.
Northland Wild Flowers

fl. hab.   p 47, Pl. 4

Anemone canadensis

Neufeld, J.B.
Wild Flowers of the Prairies

fl.   p. 27

Anemone capensis

Flowering Plants of Africa
        vol. 40   1969-70

fl.   Pl. 1569

Anemone caroliniana

Brown, Clair A.
Wildflowers of Louisiana and Adjoining
        States

fl.   p 50

Anemone caroliniana

Dean, Blanche E.
Wildflowers of Alabama and
        Adjoining States

fl.   p 67

Anemone coronaria

Alpine Garden Society
Bulletin
        vol. 9, No. 2   1941

fl.   p. 120

Anemone coronaria

Huxley, Anthony
Garden Annuals and Bulbs

hab. fl.   Pl. 147, 198

Anemone coronaria

Kleijn, H.
Beauty of the Wild Plant

fl.   p. 113   pl. 164

Anemone coronaria

Macoby, Stirling
What Flower is Yhsy

fl.   P. 40

Anemone coronaria

Massachusetts Hort. Soc.
Horticulture
        vol 54, No. 12   1976

fl.   p. 31

Anemone coronaria

Megaw, Elektra
Wild Flowers of Cyprus

fl., hab.   p. 7   Pl. 2

Anemone coronaria

Milne, Lorus
Living Plants of the World

fl.   p 53

Anemone coronaria

Morley, Brian D.
Wild Flowers of the World

fl.,hab.   Pl. 27

Anemone coronaria

Polunin, Oleg
Flowers of Europe

fl.   Pl. 21 #216

Anemone coronaria

Polunin, Oleg
Flowers of the Mediterranean

fl.   Pl. 24, 26, 27

Anemone coronaria

Royal Hort. Soc.
Journal of Royal Hort. Soc.
        vol 82, No. 4   1957

fl.   p. 162   Pl. 44

Anemone coronaria

Tsukamoto, Yotaro
Coloured Illustrations of Garden
        Flowers   Vol. 9

fl.   p. 3

Anemone coronaria var.

Everett, T.H., ed.
New Illustrated Encyclopedia of
        Gardening vol. 1

fl.   p. 22

Anemone coronaria var.

Feinbrun-Dothan, Naomi
Wild Plants in the Land of Israel

fl.,fr.,hab.   p. 90, 95

Anemone coronaria var.

Hay, Roy
The Color Dictionary of Flowers &
        Plants

fl.   p. 84   Pl. 672

Anemone coronaria var.

Massachusetts Hort. Soc.
Horticulture
        vol 51, No. 3   1973

fl.   cover, backcover

Anemone coronaria var.

Miles, Bebe
Bulbs for the Home Gardener

fl.   p 151

Anemone crassifolia

Curtis, Winifred
The Endemic Flora of Tasmania   Vol. 3

fl.   p. 167

Anemone cylindrica

Courtenay, Booth
Wildflowers & Weeds

fl.   p 34

Anemone cylindrica

Moyle, John B.
Northland Wild Flowers

fl. fr. hab.   p 48, Pl. 5

Anemone deltoidea

Walcott, Mary Vaux
North American Wild Flowers
        vol. 4

fl.   Pl. 306

Anemone, Double-flowered

see

Anemonella thalictroides var.

Anemone drummondii

Clark, Lewis J.
Wild Flowers of British Columbia

fl.   p 147

Anemone drummondii

Orr, Robert T.
Wildflowers of Western America

fl.   Pl. 28

Anemone Drummondii

Porsild, A. E.
Rocky Mountain Wild Flowers

hab.   p. 171

Anemone drummondii

Royal Hort. Soc.
Journal of the Royal Hort. Soc.
vol 89, No. 4   1964

fl.   p. 158   Pl. 61

Anemone drummondii

Taylor, Ronald J.
Mountain Wild Flowers

fl. hab.   p 79

Anemone, False Rue

see

Isopyrum biternatum

Anemone fanninii

Flowering Plants of Africa
vol 37   1965-66

fl.   Pl. 1441

Anemone, field

see

Anemone pratensis

Anemone, Florists'

see

Anemone coronoria

Anenone, Forest

see

Anemone deltoidea

Anemone, Globe

see

Anemone globosa

Anemone globosa

Walcott, Mary Vaux
North American Wild Flowers
vol. 4

fl.   Pl. 291

Anemone, Great

see

Anemone sylvestris

Anemone halleri

Addisonia   Vol. 23   1954-59

fl., fr.   p. 51   Pl. 762

Anemone hepatica

Bianchini, Francesco
Health Plants of the World

fl. hab.   p 115

Anemone hepatica

Felsko, Elsa
A Book of Wildflowers

fl.   P 43

Anemone hepatica

Furrer, D.
Blumen am Wege

fl. hab.   p 23

Anemone hepatica

Lindman, C. A. M.
Nordens Flora, Vol 4

fl. fr. hab.   Pl. 225

Anemone hortensis

Polunin, Oleg
Flowers of Europe

fl.   Pl. 21 #216

Anemone hortensis

Polunin, Oleg
Flowers of the Mediterranean

fl.   Pl. 23

Anemone hortensis var.

Polunin, Oleg
Flowers of the Mediterranean

fl.   Pl. 22, 25

Anemone hupehensis

Hay, Roy
The Color Dictionary of Flowers & Plants

fl.   p 121, Pl. 964

Anemone hupehensis

Macoby, Stirling
What Flower is That

fl.   P 40

Anemone x hybrida

Huxley, Anthony
Garden Perennials and Water Plants

hab.   Pl. 3

Anemone x hybrida

Perry, Frances
Flowers of the World

fl.   p. 252

Anemone x hybrida var.

Huxley, Anthony
Garden Perennials and Water Plants

fl.   Pl. 33b, a

Anemone, Japanese

see

Anemone hupehensis

Anemone japonica

Massachusetts Hort. Soc.
Horticulture
vol 51, No. 9   1973

fl.   p. 35

Anemone lancifolia

Batson, Wade T.
Wild Flowers in South Carolina

fl.   p 45

Anemone lancifolia

Campbell, Carlos C.
Great Smoky Mountain Wildflowers

fl.   p 51

Anemone lancifolia

Duncan, Wilbur H.
Wildflowers of the Southeastern
United States

fl.   p 43

Anemone x lesseri

Hay, Roy
The Color Dictionary of Flowers &
Plants

fl.   p. 121   Pl. 965

Anemone lyallii

Clark, Lewis J.
Wild Flowers of British Columbia

fl.   p 151

Anemone, Mountain

see

Anemone apennina

Anemone multiceps

Heller, Christine
Wild Flowers of Alaska

fl.   Pl. 226

Anemone multifida

Alaska-Yukon Wild Flower Guide

fl.   p. 44

Anemone multifida

Clark, Lewis J.
Wild Flowers of British Columbia

fl.   p 147

Anemone multifida

Heller, Christine
Wild Flowers of Alaska

fl.   Pl. 65

Anemone narcissiflora

Alaska-Yukon Wild Flowers Guide

fl.   p 43

Anemone narcissiflora

Barneby, T. P.
European Alpine Flowers in Colour

fl. hab.   Pl. 23, 6

Anemone narcissiflora

Hay, Roy
The Color Dictionary of Flowers & Plants

fl.   p 121, Pl. 966

Anemone narcissiflora

Heller, Christine
Wild Flowers of Alaska

fl.   Pl. 137

Anemone narcissiflora

Kohlhaupt, Paula
Fleurs des Alpages   Vol. 1

fl.   P. 6

Anemone narcissiflora

Polunin, Oleg
Flowers of Europe

fl.   Pl. 22 #218

Anemone narcissiflora

Robert, Paul A.
Alpine Flowers

fl., hab.   Pl. 13

Anemone narcissiflora

Tosco, Uberto
The World of Mountain Flowers

fl.   p 111

Anemone nemorosa

Ary, S.
The Oxford Book of Wildflowers

fl.   p 66, Pl. 1

Anemone nemorosa

Felsko, Elsa
A Book of Wildflowers

fl.   P 126

Anemone nemorosa

Furrer, D.
Blumen am Wege

fl. hab.   p 25

Anemone nemorosa

Lindman, C. A. M.
Nordens Flora, Vol 4

fl. fr. hab.   Pl. 226

Anemone nemorosa

Martin, W. Keble
The Concise British Flora in Colour

fl. hab.   Pl. 1

Anemone nemorosa

Polunin, Oleg
Flowers of Europe

fl.   Pl. 22 #214

Anemone nemorosa

Tosco, Uberto
The World of Mountain Flowers

fl. hab.   p 26, 111

Anemone nemorosa var.

Hay, Roy
The Color Dictionary of Flowers & Plants

fl.   p. 85   Pl. 673

Anemone, Northern

see

Anemone parviflora

Anemone obtusiloba

Royal Hort. Soc.
Journal of the Royal Hort. Soc.
   vol 93, No. 6   1968

fl.   p. 246   Pl. 127

Anemone occidentalis

Clark, Lewis J.
Wild Flowers of British Columbia

fl. hab.   p 158, 159

Anemone occidentalis

Ferguson, Mary
Wildflowers

fl.   p 83

Anemone occidentalis

Fries, Mary A.
Wildflowers of Mount Ranier and
the Cascades

fl.   P 100 & 101

Anemone occidentalis

Orr, Robert T.
Wildflowers of Western America

fl.   Pl. 2

Anemone occidentalis

Taylor, Ronald J.
Mountain Wild Flowers

fl. fr. hab.   p 78

Anemone palmata

Polunin, Oleg
Flowers of Europe

fl.   Pl. 21 #217

Anemone parviflora

Alaska-Yukon Wild Flowers Guide

fl.   p 42

Anemone parviflora

Heller, Christine
Wild Flowers of Alaska

fl.   Pl. 138

Anemone parviflora

Walcott, Mary Vaux
North American Wild Flowers
vol. 5

fl.   Pl. 371

Anemone patens

Courtenay, Booth
Wildflowers & Weeds

fl.   p 35

Anemone patens

Ferguson, Mary
Wildflowers

fl.   p 116

Anemone patens

Lemmon, Robert S.
Wildflowers of North America in
   Full Color

fl.   pg. 177   pl. 278

Anemone patens

Moyle, John B.
Northland Wild Flowers

fl. hab.   p 46, Pl. 2

Anemone patens

Orr, Robert T.
Wildflowers of Western America

fl.   Pl. 268/269

Anemone patens

Welsh, Stanley L.
Flowers of the Mountain Country

fl.   p. 71

Anemone patens var.

Clark, Lewis J.
Wild Flowers of British Columbia

fl.   p 151

Anemone patens var.

Heller, Christine
Wild Flowers of Alaska

fl.   Pl. 227

Anemone pavonina

Hay, Roy
The Color Dictionary of Flowers &
    Plants

fl.   p. 85   Pl. 674

Anemone pavonina

Polunin, Oleg
Flowers of Europe

fl.   Pl. 21 #216

Anemone pavonina var.

Alpine Garden Society
Bulletin
    vol. 45, No. 2   1977

fl.   p. 115

Anemone pavonina var.

Goulimis, Constantine N.
Wild Flowers of Greece

fl.   p. 17, 19

Anemone, Plume

    see

Pulsatilla occidentalis

Anemone, poppy

    see

Anemone coronaria

Anemone potentilloides

Alpine Garden Society
Bulletin
    vol. 8, No. 4   1940

fl., hab.   p. 299

Anemone pratensis

Felsko, Elsa
A Book of Wildflowers

fl.   P 61

Anemone pratensis

Lindman, C. A. M.
Nordens Flora, Vol 4

fl. fr. hab.   Pl. 229

Anemone pratensis

Milne, Lorus
Living Plants of the World

fl.   p 54

Anemone pulsatilla

Ary, S.
The Oxford Book of Wildflowers

fl.   p 138, Pl. 2

Anemone pulsatilla

Color Treasury of Herbs & Other
    Medicinal Plants

fl.   p 26   Pl. 22

Anemone pulsatilla

Felsko, Elsa
A Book of Wild Flowers 2nd Ser.

fl. fr. hab.   p. 38

Anemone pulsatilla

Lindman, C. A. M.
Nordens Flora, Vol 4

fl. fr. hab.   Pl. 228

Anemone pulsatilla

Massachusetts Hort. Soc.
Horticulture
    vol 42, No. 3   1964

fl.   p. 35

Anemone pulsatilla.

Massachusetts Hort. Soc.
Horticulture
    vol 46, No. 2   1968

fl.   cover, backcover

Anemone pulsatilla

Massachusetts Hort. Soc.
Horticulture
    vol 47, No. 8   1969

fl.   p. 19

Anemone pulsatilla

Miles, Bebe
Bulbs for the Home Gardener

fl.   p. 6

Anemone pulsatilla

Neufeld, J.B.
Wild Flowers of the Prairies

fl.   p. 7

Anemone pulsatilla

Pacific Hort. Foundation
Pacific Horticulture
    vol 39, No. 4   1978-79

fr.   p. 40

Anemone pulsatilla

Perrot, Emile
Les Plantes Medicinales

fl. fr. hab.   p 11

Anemone pulsatilla

Wit, H. C. D. de
Plants of the World:
    The Higher Plants, Vol I

fl.   P 97, Pl. 38

Anemone quinquefolia

Courtenay, Booth
Wildflowers & Weeds

fl.   p 34

Anemone quinquefolia

Dean, Blanche E.
Wildflowers of Alabama and
    Adjoining States

fl.   p 65

Anemone quinquefolia

Jennings, O. E.
Wild Flowers of Western Pennsylvania
    Vol. II

fl. hab.   Pl. 112

Anemone quinquefolia

Lemmon, Robert S.
Wildflowers of North America in
    Full Color

fl.   pg. 248   pl. 389

Anemone quinquefolia

Moyle, John B.
Northland Wild Flowers

fl. hab.   p 47, Pl. 3

Anemone quinquefolia

Wharton, Mary E.
A Guide to the Wildflowers & Ferns
    of Kentucky

fl.   p 110 Pl. 1.32

Anemone ranunculoides

Barneby, T. P.
European Alpine Flowers in Colour

fl.   Pl. 24, 1

Anemone ranunculoides

Bianchini, Francesco
Health Plants of the World

fl.   p. 115

Anemone ranunculoides

Felsko, Elsa
A Book of Wild Flowers 2nd Ser.

fl. hab.   p. 2

Anemone ranunculoides

Huxley, Anthony
Garden Perennials and Water Plants

fl.   Pl. 35

Anemone richardsonii

Heller, Christine
Wild Flowers of Alaska

fl.   Pl. 17

Anemone, Rue

see

Anemonella thalictroides

Anemone rupicola

Curtis's Botanical Magazine
   Vol 160   1937

fl.   No. 9476

Anemone somaliensis

Flowering Plants of Africa
   vol 43   1974-76

fl., hab.   Pl. 1696

Anemone sp.

Royal Hort. Soc.
Journal of the Royal Hort. Soc.
   vol 76, No. 4   1951

fl.   p. 126   Pl. 60

Anemone stellata var.

Alpine Garden Society
Bulletin
   vol. 45, No. 2 1977

fl.   p. 115

Anemone sumatrana

Royal Hort. Soc.
The Garden
   vol 102, No. 2 1977

fl.   p. 63

Anemone sylvestris

Felsko, Elsa
A Book of Wild Flowers 2nd Ser.

fl.   p. 115

Anemone sylvestris

Huxley, Anthony
Garden Perennials and Water Plants

fl.   Pl. 36

Anemone sylvestris

Lindman, C. A. M.
Nordens Flora, Vol. 4

fl. fr.   Pl. 227

Anemone sylvestris

Wit, H. C. D. de
Plants of the World;
   The Higher Plants, Vol. I

fl.   p. 62   Pl. 33

Anemone, Tall

see

Anemone virginiana

Anemone tetrasepala

Royal Hort. Soc.
The Garden
   vol 102, No. 11   1977

fl.   p. 452

Anemone thalictroides

Lemmon, Robert S.
Wildflowers of North America in
   Full Color

fl.   pg. 249   pl. 391

Anemone, Tree

see

Carpenteria californica

Anemone trifolia

Barneby, T. P.
European Alpine Flowers in Colour

fl. hab.   Pl. 23, 5

Anemone trullifolia

Hara, Hiroshi, comp.
Photo-Album of Plants of
   Eastern Himalaya

fl.   Pl. 223

Anemone tschernjaewi

Mathew, Brian
Dwarf Bulbs

fl.   p. 40   Pl. 13

Anemone tschernjaewi

Royal Hort. Soc.
Journal of the Royal Hort. Soc.
   vol 90, No. 11   1965

fl.   p. 470   Pl. 213

Anemone vernalis

Alpine Garden Society
Bulletin
   vol. 9, No. 3   1941

fl., hab.   p. 210

Anemone vernalis

Kleijn, H.
Beauty of the Wild Plant

fl.   p. 101   Pl. 148

Anemone virginiana

Wharton, Mary E.
A Guide to the Wildflowers & Ferns
   of Kentucky

fl. fr.   P. 110   Pl. 1.33

Anemone, western

see

Anemone occidentalis

Anemone, wood

see

Anemone nemorosa

Anemone, Wood

see

Anemone quinquefolia

Anemone, Yellow

see

Anemone ranunculoides

Anemone, Yellow

see

Anemone richardsonii

Anemonella

see

Syndesmon thalictroides

Anemonella thalictroides

Batson, Wade T.
Wild Flowers in South Carolina

fl.   p 45

Anemonella thalictroides

Campbell, Carlos C.
Great Smoky Mountain Wildflowers

fl.   back cover

Anemonella thalictroides

Courtenay, Booth
Wildflowers & Weeds

fl.   p 33

Anemonella thalictroides

Duncan, Wilbur H.
Wildflowers of the Southeastern
     United States

fl. hab.   p 43

Anemonella thalictroides

Jennings, O. E.
Wild Flowers of Western Pennsylvania
               Vol. II

fl.   Pl. 98

Anemonella thalictroides

Klimas, John E.
Wildflowers of Eastern America

fl.   Pl. 23

Anemonella thalictroides

Miles, Bebe
Bulbs for the Home Gardener

fl.   p 53

Anemonella thalictroides

Moyle, John B.
Northland Wild Flowers

fl. hab.   p. 49   Pl. 6

Anemonella thalictroides

Wharton, Mary E.
A Guide to the Wildflowers & Ferns
     of Kentucky

fl.   p 109 Pl. 1.30

Anemonella thalictroides var.

Lemmon, Robert S.
Wildflowers of North America in
          Full Color

fl.   pg. 249   pl. 392

Anethum graveleons

Hvass, Elsie
Plants That Feed and Serve Us

fl. fr.   p 66, Pl. 134

Anethum graveolens

Loewenfeld, Claire
The Complete Book of Herbs and Spices

fr., hab.   p.128

Anethum graveolens

Masefield, G. B.
The Oxford Book of Food Plants

fl. fr.   Pls. 139, 4; 147, 2

Anethum graveolens

Perrot, Emile
Les Plantes Medicinales

fl. fr. hab.   p 13

Anethum graveolens

Rosengarten, Frederic, Jr.
The Book of Spices

fl.   p 232

Angelica acutiloba

Kariyone, Tatsuo
Atlas of Medicinal Plants

fl., hab.   Pl. 68

Angelica acutiloba

Kimura, Koiti
Japanese Medicinal Plants, Vol I

fl.   p 66

Angelica archangelica

Bianchini, Francesco
Health Plants of the World

fr.   p 41

Angelica archangelica

Milne, Lorus
Living Plants of the World

fl. hab.   p 176

Angelica archangelica

Polunin, Oleg
Flowers of Europe

fl.   Pl. 85 #895

Angelica atropurpurea

Courtenay, Booth
Wildflowers & Weeds

fl.   p 58

Angelica dawsonii

Clark, Lewis J.
Wild Flowers of British Columbia

fl.   p 350

Angelica, Eurasian

see

Angelica archangelica

Angelica, Filmy

see

Angelica triquinata

Angelica, Hairy

see

Angelica venenosa

Angelica pubescens

Kimura, Koiti
Japanese Medicinal Plants, Vol II

fl.   p 207

Angelica sylvestris

Ary, S.
The Oxford Book of Wildflowers

fl.   p 86, Pl. 6

Angelica sylvestris

Lindman, C. A. M.
Nordens Flora, Vol. 7

fl. fr.   Pl. 431

Angelica sylvestris

Martin, W. Keble
The Concise British Flora in Colour

fl. fr. hab.   Pl. 40

Angelica Tree, Japanese

see

Aralia elata

Angelica triquinata

Campbell, Carlos C.
Great Smoky Mountain Wildflowers

fl.   p 93

Angelica venenosa

Duncan, Wilbur H.
Wildflowers of the Southeastern
     United States

Angelica venenosa

Wharton, Mary E.
A Guide to the Wildflowers & Ferns
     of Kentucky

fl.   p 160   Pl. 2.26

Angels Fishing Rod

see

Dierama pulcherrima

Angel's tears

See Datura candida

Angel's tears

see

Datura sanguinea

Angel's Tears

see

Helxine soleirolli

Angel's Tears

see

Narcissus triandrus

Angel's trumpet

see

Datura sanguinea

Angel's Trumpet

see

Datura suaveolens

Angel's Trumpet Tree

see

Datura candida

Angianthus eriocephalus

Curtis, Winifred
The Endemic Flora of Tasmania Vol. 3

fl.   p. 197

Angianthus preissianus

Cochrane, G. R.
Flowers and Plants of Victoria

fl.   Pl. 197

Angkalanthus transvaalensis

Flowering Plants of Africa
      vol 31   1956

fl.   Pl. 1227

Angle-Pod

see

Gonolobus carolinensis

Angle-pod

see

Gonolobus shortii

Angophora bakeri

Blombery, Alec M.
What Wildflower is That

fl.   p 45, Pl. 87

Angophora cordifolia

Blombery, Alec M.
What Wildflower is That

fl.   p 46, Pl. 88

Angophora cordifolia

Macoby, Stirling
What Flower is That

fl.   P 41

Angraecopsis gracillima

Stewart, Joyce
Orchids of Tropical Africa

fl. hab.   Pl. 7

Angraecopsis tenerrima

Piers, Frank
Orchids of East Africa

fl.   Pl. IIj

Angraecum bilobum var.

Pennsylvania Hort. Soc.
The Green Scene
      vol 5, No. 2   1976

fl.   cover

Angraecum conchiferum

Flowering Plants of South Africa
      vol XXII   1942

fl., hab.   Pl. 852

Angraecum eichleranum

Ebel, Friedrich
The Strange and Beautiful
World of Orchids

fl.   p 109

Angraecum germinyanum

Stewart, Joyce
Orchids of Tropical Africa

fl. hab.   Pl. 8a, 8b

Angraecum infundibulare

Morley, Brian D.
Wild Flowers of the World

fl.   Pl. 69

Angraecum infundibulare

Stewart, Joyce
Orchids of Tropical Africa

fl. hab.   Pl. 9

Angraecum leonis

Stewart, Joyce
Orchids of Tropical Africa

fl.   Pl. 10

Angraecum magdalenae

Curtis's Botanical Magazine
Vol 178   1970-72

fl.   591

Angraecum montanum

Stewart, Joyce
Orchids of Tropical Africa

fl. hab.   Pl. 11

Angraecum scottianum

Stewart, Joyce
Orchids of Tropical Africa

fl. hab.   Pl. 12

Angraecum sesquipedale

Ebel, Friedrich
The Strange and Beautiful
World of Orchids

fl.   p 53

Angraecum sesquipedale

Kupper, Walter
Orchids

fl. hab.   p 127

Angraecum sesquipedale

Perry, Frances
Flowers of the World

fl.   p 208

Angraecum sesquipedale

Stewart, Joyce
Orchids of Tropical Africa

fl.   Pl. 13

Angraecum stolzii

Williamson, Graham
The Orchids of South Central Africa

fl.   Pl. 165

Anguillaria dioica

Blombery, Alec M.
What Wildflower Is That

fl.   p 47, Pl. 89, 90

Anguillaria dioica

Cady, Leo
Australian Native Orchids in Colour

fl.   Pl. 1

Anguillaria dioica

Cochrane, C. R.
Flowers and Plants of Victoria

fl.   Pl. 321

Anguloa cliftonii

Kramer, Jack
Orchids; Flowers of Romance
   and Mystery

fl.   p 214

Anguloa cliftonii

Ospina, Mariano
Orquideas de las americas

fl.   Pl. 98

Anguloa clowesii

Kramer, Jack
Orchids; Flowers of Romance
   and Mystery

fl.   p 215

Anguloa ruckeri

Kramer, Jack
Orchids; Flowers of Romance
   and Mystery

fl.   p 216

Anguloa ruckeri

Kupper, Walter
Orchids

fl.   p 81

Ania hongkongensis

Walden, Beryl M.
Wild Flowers of Hong Kong

fl. hab.   Pl. 26(69)

Ania hookeriana

Curtis's Botanical Magazine
   Vol 161   1938

fl.   No. 9553

Anigozanthos bicolor

Blombery, Alec M.
What Wildflower is That

fl.   p. 47   Pl. 91

Anigozanthos flavidus

Blombery, Alec M.
What Wildflower is That

fl.   p. 47   Pl. 92

Anigozanthos flavidus

Morley, Brian D.
Wild Flowers of the World

fl.   Pl. 144

Anigozanthos flavidus var.

Perry, Frances
Flowers of the World

fl.   p. 135

Anigozanthos humilus

Blombery, Alec M.
What Wildflower is That

fl.   p. 47   Pl. 93

Anigozanthos manglesii

Blombery, Alec M.
What Wildflower is That

fl.   p. 48   Pl. 94

Anigozanthos manglesii

Macoby, Stirling
What Flower is That

fl. P 41

Anigozanthos manglesii

Mullins, Barbara
Australian Wildflowers in Color

fl.   P. 107   Pl. 106

Anigozanthos manglesii

Royal Hort. Soc.
Journal of the Royal Hort. Soc.
   Vol. 95, No. 7   1970

fl., hab.   p. 304   Pl. 162

Anigozanthos preissii

Blombery, Alec M.
What Wildflower is That

fl.   P. 48   Pl. 95

Anigozanthos preissii

Royal Hort. Soc.
Journal of the Royal Hort. Soc.
   Vol. 93, No. 11   1968

fl.   p. 470   Pl. 258

Anigozanthos pulcherrima

Blombery, Alec M.
What Wildflower is That

fl.   p. 48   Pl. 96

Anigozanthos pulcherrima

Morcombe, M. K.
Australia's Western Wildflowers

fl.   p. 7

Angiozanthos rufa

Blombery, Alec M.
What Wildflower is That

fl.   p. 48   Pl. 97

Anigozanthos rufa

Morcombe, M. K.
Australia's Western Wildflowers

fl.   p. 70

Angiozanthos viridis

Blombery, Alec M.
What Wildflower is That

fl.   p. 48   Pl. 98

Anise

see

Pimpinella anisum

Anise, Chinese

see

Illicium anisatum

Anise Root

see

Osmorhiza longistylis

Anise, Star

see

Illicum floridanum

Anise, Star

see

Illicium verum

Anise, Star Dunn's

see

Illicium dunnianum

Anise, Sweet

see

Osmorhiza longistylis

Anise Tree

see

Illicium

Aniseed

see

Pimpinella anisum

Anisocalyx vaginatus

Herre, H.
The Genera of the mesembryanthemaceae

fl.   fr.   p 73

Anisocoma acaulis

Munz, Philip A.
California Desert Wildflowers

fl.   p 58, Pl. 95

Anisodontea triloba

Eliovson, Sima
Namaqualand in Flower

fl.   Pl. 23, 2

Anisopappus lastii var.

Moriarty, Audrey
Wild Flowers of Malawi

fl.   Pl. 78; 5

Anisostichus capreolata

Dean, Blanche E.
Wildflowers of Alabama and
    Adjoining States

fl.   p 167

Anisostichus capreolata

Walcott, Mary Vaux
North American Wild Flowers
    vol. 4

fl.   Pl. 261

Annatto Tree

see

Bixa orellana

Annona cherimola

Bianchini, F.
The Complete Book of Fruits & Vegetables

fr.   p 177

Annona cherimola

Macoby, Stirling
What Flower is That

fr.   P. 42

Annona cherimola

Masefield, G. B.
The Oxford Book of Food Plants

fl. fr.   Pl. 97, 4

Annona glabra

Luer, Carlyle A.
The Native Orchids of Florida

hab.   Pl. 2; 4

Annona muricata

Masefield, G. B.
The Oxford Book of Food Plants

fr.   Pl. 97, 3

Annona muricata

Wit, H. C. D. de
Plants of the World;
    The Higher Plants, Vol I

fr.   P 59, Pl. 27

Annona sqamosa

Masefield, G. B.
The Oxford Book of Food Plants

fr.   Pl. 97, 3

Anoda cristata

Curtis's Botanical Magazine
Vol 171   1956-57

fl.   288

Anoda cristata

O'Gorman, Helen
Mexican Flowering Trees and Plants

fl. fr. hab.   p 125

Anodopetalum biglandulosum

Curtis, Winifred
The Endemic Flora of Tasmania   Vol. 1

fl. fr.   Pl. 19

Anoectochilus albolineatus

Morley, Brian D.
Wild Flowers of the World

fl.   Pl. 126

Anoectochilus brevilabris

Grubb, Roy
Selected Orchidaceous Plants
    Vol. I

hab.   p. 143

Anoectochilus brevilabris

Grubb, Roy
Selected Orchidaceous Plants
    Vol. II

hab.   frontispiece

Anoectochilus lineatus

Grubb, Roy
Selected Orchidaceous Plants
    Vol II

hab.   p 45

Anoectochilus roxburghii

Curtis's Botanical Magazine
Vol 161   1938

fl.   No. 9529

Anoectochilus sandvicensis

Luer, Carlyle A.
The Native Orchids of the
    United States and Canada

fl. hab.   Pl. 32

Anoectochilus sp.

Perry, Frances
Flowers of the World

hab.   p. 213

Anoectochilus yungianus

Walden, Beryl M.
Wild Flowers of Hong Kong

fl. hab. frontispiece   Pl. 77 (237)

Anomatheca grandiflora

Moriarty, Audrey
Wild Flowers of Malawi

fl.   Pl. 7; 1

Anopterus glandulosus

Curtis, Winifred
The Endemic Flora of Tasmania   Vol. 2

fl. fr.   Pl. 44

Ansellia africana

Ebel, Friedrich
The Strange and Beautiful
    World of Orchids

fl. hab.   p 35

Ansellia africana

Kramer, Jack
Orchids; Flowers of Romance
    and Mystery

fl.   p 142

Ansellia gigantea var.

Flowering Plants of South Africa
    vol XVIII   1938

fl., hab.   Pl. 703

Ansellia gigantea var.

Morley, Brian D.
Wild Flowers of the World

fl.,hab.   Pl. 69

Ansellia gigantea var.

Piers, Frank
Orchids of East Africa

fl.   Pl. Ic

Ansellia gigantea var.

Stewart, Joyce
Orchids of Tropical Africa

fl.   Pl. 14

Ansellia gigantea var.

Williamson, Graham
The Orchids of South Central Africa

fl.   Pl. 106-107-108-
                                     109

Ansellia nilotica var.

Piers, Frank
Orchids of East Africa

fl.   Pl. Ia

Antegibbaeum fissoides

Herre, H.
The Genera of the Mesembryanthemaceae

fl. fr.   p. 75

Antelope Bush

        see

Purshia tridentata

Antelope Horn

        see

Asclepiodora viridis

Antennaria alpina

Lindman, C. A. M.
Nordens Flora, Vol 9

fl. hab.   Pl. 587

Antennaria carpatica

Barneby, T. P.
European Alpine Flowers in Colour

fl.   Pl. 89, 2

Antennaria carpatica

Lindman, C. A. M.
Nordens Flora, Vol. 9

fl. hab.   Pl. 588

Antennaria dioica

Ary, S.
The Oxford Book of Wildflowers

fl.   p 126, Pl. 6

Antennaria dioica

Barneby, T. P.
European Alpine Flowers in Colour

fl. hab.   Pl. 89, 1

Antennaria dioica

Lindman, C. A. M.
Nordens Flora, Vol. 9

fl. hab.   Pl. 586

Antennaria dioica

Martin, W. Keble
The Concise British Flora in Colour

fl. hab.   Pl. 44

Antennaria dioica

Perrot, Emile
Les Plantes Medicinales

fl. fr. hab.   p 179

Antennaria dioica

Polunin, Oleg
Flowers of Europe

fl.   Pl. 143 #1378

Antennaria bowellii

Walcott, Mary Vaux
North American Wild Flowers

fr. hab.   Pl. 104

Antennaria lanata

Porsild, A. E.
Rocky Mountain Wild Flowers

fl. hab.   p 387

Antennaria lanata

Taylor, Ronald J.
Mountain Wild Flowers

fl. hab.   p 133

Antennaria luzuloides

Walcott, Mary Vaux
North American Wild Flowers

fl. hab.   Pl. 171

Antennaria media

Porsild, A. E.
Rocky Mountain Wild Flowers

fl. hab.   p 389

Antennaria neglecta

Courtenay, Booth
Wildflowers & Weeds

fl.   p 126

Antennaria neglecta

Klimas, John E.
Wildflowers of Eastern America

fl.   Pl. 3

Antennaria neglecta

Lemmon, Robert S.
Wildflowers of North America in
        Full Color

fl.   pg. 48   pl. 81

Antennaria parvifolia

Welsh, Stanley L.
Flowers of the Canyon Country

fl. hab.   p 4

Antennaria plantaginifolia

Brown, Clair A.
Wildflowers of Louisiana and
        Adjoining States

fl.   p 185

Antennaria plantaginifolia

Duncan, Wilbur H.
Wildflowers of the Southeastern
        United States

fl.   p 207

Antennaria plantaginifolia

Moyle, John B.
Northland Wild Flowers

fl. hab.   p 183, Pl. 230

Antennaria racemosa

Porsild, A. E.
Rocky Mountain Wild Flowers

fl. hab.   p 391

Antennaria rosea

Clark, Lewis J.
Wild Flowers of British Columbia

fl.   p 518

Antennaria rosea

Porsild, A. E.
Rocky Mountain Wild Flowers

fl. hab.   p 393

Antennaria rosea

Taylor, Ronald J.
Mountain Wild Flowers

fl. hab.   p 133

Antennaria rosea

Walcott, Mary Vaux
North American Wild Flowers
        vol. 4

fl., hab.   Pl. 286

Antennaria solitaria

Wharton, Mary E.
A Guide to the Wildflowers & Ferns
        of Kentucky

fl.   p 233   Pl. 1.1

Antennaria umbrinella

Clark, Lewis J.
Wild Flowers of British Columbia

fl.   p 511

84

Anthemis arvensis

Ary, S.
The Oxford Book of Wildflowers

fl.   p 98, Pl. 6

Anthemis arvensis

Clark, Lewis J.
Wild Flowers of British Columbia

fl.   p 519

Anthemis arvensis

Furrer, D.
Blumen am Wege

fl. hab.   p 123

Anthemis arvensis

Kleijn, H.
Beauty of the Wild Plant

fl.   p. 33   Pl. 36

Anthemis arvensis

Martin, W. Keble
The Concise British Flora in Colour

fl. hab.   Pl. 45

Anthemis chia

Polunin, Oleg
Flowers of the Mediterranean

fl.   Pl. 203

Anthemis cotula

Brown, Clair A.
Wildflowers of Louisiana and
   Adjoining States

fl. hab.   p 184

Anthemis cotula

Courtenay, Booth
Wildflowers & Weeds

fl.   p. 110

Anthemis cotula

Martin, W. Keble
The Concise British Flora in Colour

fl. hab.   Pl. 45

Anthemis cotula

Moyle, John B.
Northland Wild Flowers

fl.   p 166, Pl. 200

Anthemis cotula

Weeds of the Southern United States

fl.   p. 9

Anthemis cretica

Goulimis, Constantine N.
Wild Flowers of Greece

fl. hab. p 111

Anthemis cupaniana

Hay, Roy
The Color Dictionary of Flowers & Plants

fl.   p 2, Pl. 12

Anthemis nobilis

Bianchini, Francesco
Health Plants of the World

fl.   p 157

Anthemis nobilis

Crockett, James Underwood
Lawns & Ground Covers

fl.   p. 121

Anthemis nobilis

Hvass, Elsie
Plants That Feed and Serve Us

fl. p 85, Pl. 178

Anthemis nobilis

Loewenfeld, Claire
The Complete Book of Herbs and Spices

fl. hab. p. 113

Anthemis nobilis

Perrot, Emile
Les Plantes Medicinales

fl. fr. hab. p 46

Anthemis pusilla

Goulimis, Constantine N.
Wild Flowers of Greece

fl., hab. p. 111

Anthemis sancti-johannis

Hay, Roy
The Color Dictionary of Flowers & Plants

fl.   p 121, Pl. 967

Anthemis tinctoria

Bianchini, Francesco
Health Plants of the World

fl.   p 157

Anthemis tinctoria

Felsko, Elsa
A Book of Wild Flowers 2nd Ser.

fl. hab.   p. 31

Anthemis tinctoria

Lindman, C. A. M.
Nordens Flora, Vol 9

fl.   Pl. 599

Anthemis tinctoria

Macoby, Stirling
What Flower is That

fl.   P. 42

Anthemis tinctoria

Martin, W. Keble
The Concise British Flora in Colour

fl. hab.   Pl. 45

Anthemis tinctoria

Milne, Lorus
Living Plants of the World

fl.   p 237

Anthemis tinctoria

Polunin, Oleg
Flowers of Europe

fl.   Pl. 147 #1409

Anthemis tinctoria var.

Hay, Roy
The Color Dictionary of Flowers & Plants

fl. p. 121, 122 Pl. 968, 969,
         970

Anthemis tricolor

Morley, Brian D.
Wild Flowers of the World

fl.,hab.   Pl. 40

Anthericum, Branched

see

Anthericum ramosum

Anthericum chandleri

Addisonia, Vol. 21 1939-42

fl. fr. opp. p. 49 pl. 697

Anthericum liliago

Barneby, T. P.
European Alpine Flowers in Colour

fl.   Pl. 1, 5

Anthericum liliago

Felsko, Elsa
A Book of Wild Flowers 2nd Ser.

fl.   p. 121

Anthericum liliago

Hay, Roy
The Color Dictionary of Flowers & Plants

fl.   p 122, Pl. 971

Anthericum liliago

Kleijn, H.
Beauty of the Wild Plant

fl.   p. 117   Pl. 171

Anthericum liliago

Polunin, Oleg
Flowers of Europe

fl.   Pl. 164 #1596

Anthericum ramosum

Barneby, T. P.
European Alpine Flowers in Colour

fl.   Pl. 1, 6

Anthericum ramosum

Lindman, C. A. M.
Nordens Flora, Vol 1

fl. fr. hab.   Pl. 51

Anthericum salteri

Flowering Plants of South Africa
        vol XVIII   1938
fl., hab.   Pl. 687

Anthericum, walking

            see

Chlorophytum elatum var.

Antherotoma naudinii

Moriarty, Audrey
Wild Flowers of Malawi

fl.   Pl. 72; 5

Anthers-in-a-Boat

            see

Androcymbium ciliolatum

Anthocercis frondosa

Cochrane, G. R.
Flowers and Plants of Victoria

fl.   Pl. 83

Anthocercis littorea

Blombery, Alec M.
What Wildflower is That

fl.   p. 49   Pl. 99

Anthocercis myosotidea

Cochrane, G. R.
Flowers and Plants of Victoria

fl.   Pl. 173

Anthocercis tasmanica

Curtis, Winifred
The Endemic Flora of Tasmania, Vol. 5

fl. fr.   Pl. 199

Anthocercis viscosa

Blombery, Alec M.
What Wildflower is That

fl.   p. 49   Pl. 100

Antholyza ringens

Morley, Brian D.
Wild Flowers of the World

fl.,hab.   Pl. 87

Anthoxanthum odoratum

Lindman, C. A. M.
Nordens Flora, Vol 2

fr. hab.   Pl. 104

Anthriscus caucalis

Martin, W. Keble
The Concise British Flora in Colour

fr. hab.   Pl. 38

Anthriscus cerefolium

Hvass, Elsie
Plants That Feed and Serve Us

fl. fr.   p 68, Pl. 139

Anthriscus cerefolium

Masefield, G. B.
The Oxford Book of Food Plants

fl. fr.   Pl. 147, 3

Anthriscus cerefolium

Rosengarten, Frederic, Jr.
The Book of Spices

fl.   p. 180

Anthriscus sylvestris

Ary, S.
The Oxford Book of Wildflowers

fl.   p 86, Pl. 5

Anthriscus sylvestris

Felsko, Elsa
A Book of Wild Flowers 2nd Ser.

fl. hab.   p. 116

Anthriscus sylvestris

Lindman, C. A. M.
Nordens Flora, Vol. 7

fl. fr.   Pl. 417

Anthriscus sylvestris

Martin, W. Keble
The Concise British Flora in Colour

fl. fr. hab.   Pl. 38

Anthurium andraeanum

American Hort. Soc.
American Horticulturist
vol 54, No. 4 1975

fl.   p. 25

Anthurium andraeanum

Hargreaves, Dorothy
Tropical Blossoms of the Caribbean

fl.   p. 6

Anthurium andraeanum

Hay, Roy
The Color Dictionary of Flowers & Plants

fl.   p. 51   Pl. 408

Anthurium andraeanum

Hay, Roy
The Dictionary of House Plants

fl.   Pl. 34

Anthurium andraeanum

Hersey, Jean
Woman's Day Book of House Plants

fl.   p. 29

Anthurium andraeanum

Kiaer, Eigil
Indoor Plants in Colour

fl.   p. 15

Anthurium andraeanum

Kromdijk, G.
200 House Plants in Colour

fl.   Pl. 15

Anthurium andraeanum

Menninger, Edwin A.
Flowering Vines of the World

fl.   Pl. 37

Anthurium andraeanum

Perry, Frances
Flowers of the World

fl.   p. 35

Anthurium andraeanum

Tsukamoto, Yotaro
Coloured Illustrations of Garden
    Flowers Vol. 9

fl.   p. 29

Anthurium andraeanum var.

Macoby, Stirling
What Flower is That

fl.   P. 42

Anthurium crystallinum

Bruggeman, L.
Tropical Plants

hab.   Pl. 70

Anthurium crystallinum

Hersey, Jean
Woman's Day Book of House Plants

hab.   p. 30

Anthurium crystallinum

Kiaer, Eigil
Indoor Plants in Colour

hab.   p. 14

Anthurium crystallinum

Kromdijk, G.
200 House Plants in Colour

hab.   Pl. 16

Anthurium crystallinum

Massachusetts Hort. Soc.
Horticulture
    vol 35, No. 1   1957

hab.   p. 37 (see errata p. 108)

Anthurium crystallinum

Nicolaisen, Age
Pocket Encyclopedia of Indoor Plants

hab.   Pl. 9

Anthurium crystallinum

Royal Hort. Soc.
The Garden
    vol 103, No. 1   1978

hab.   p. 24

Anthurium x ferrierense

Bruggeman, L.
Tropical Plants

fl.   Pl. 71

Anthurium hybrid

Massachusetts Hort. Soc.
Horticulture
    vol 40, No. 12   1962

fl.   inside cover

Anthurium hybridum

Tsukamoto, Yotaro
Coloured Illustrations of Garden
    Flowers   Vol. 9

fl.   p.29

Anthurium scherzeranum

Encke, Fritz
Zimmerpflanzen

fl.   p. 52

Anthurium Scherzeranum

Everett, T.H., ed.
New Illustrated Encyclopedia of
    Gardening vol. 1

fl.   p. 71

Anthurium scherzeranum

Hay, Roy
The Dictionary of House Plants

fl.   Pl. 35

Anthurium scherzeranum

Hersey, Jean
Woman's Day Book of House Plants

fl.   p. 30

Anthurium scherzeranum

Kiaer, Eigil
Indoor Plants in Colour

fl.   p. 14

Anthurium scherzeranum

Kromdijk, G.
200 House Plants in Colour

fl.   Pl. 17

Anthurium scherzeranum

Macoby, Stirling
What Flower is That

fl.   P. 42

Anthurium scherzeranum

Massachusetts Hort. Soc.
Horticulture
    Vol. 42, No. 9   1964

fl.   p. 23

Anthurium scherzeranum

Milne, Lorus
Living Plants of the World

fl.   p. 291

Anthurium scherzeranum

Nicolaisen, Age
Pocket Encyclopedia of Indoor Plants

fl.   Pl. 8

Anthurium scherzeranum

Perry, Frances
Flowers of the World

fl.   p 34

Anthurium scherzeranum

Tsukamoto, Yotaro
Coloured Illustrations of Garden
    Flowers Vol. 9

fl.   p. 29

Anthurium sp.

American Hort. Soc.
American Horticulturist
    vol 57, No. 6   1978

fl.   p. 5

Anthyllis montana

Barneby, T. P.
European Alpine Flowers in Colour

fl.   Pl. 44, 5

Anthyllis montana

Curtis's Botanical Magazine
    vol 172   1958-59

fl.   Pl. 333

Anthyllis montana

Felsko, Elsa
A Book of Wild Flowers 2nd Ser.

fl. hab.   p. 79

Anthyllis montana

Polunin, Oleg
Flowers of Europe

fl.   Pl. 61 #621

Anthyllis tetraphylla

Polunin, Oleg
Flowers of Europe

fl., fr.   Pl. 61 #623

Anthyllis vulneraria

Ary, S.
The Oxford Book of Wildflowers

fl.   p 22, Pl. 1

Anthyllis vulneraria

Felsko, Elsa
A Book of Wildflowers

fl.   P. 12

Anthyllis vulneraria

Lindman, C. A. M.
Nordens Flora, Vol 5

fl. fr. hab.   Pl. 343

Anthyllis vulneraria

Martin, W. Keble
The Concise British Flora in Colour

fl. fr. hab.   Pl. 23

Anthyllis vulneraria

Perrot, Emile
Les Plantes Medicinales

fl. hab.   p 245

Anthyllis vulneraria

Polunin, Oleg
Flowers of Europe

fl.   Pl. 61 #622

Anthyllis vulneraria

Tosco, Uberto
The World of Mountain Flowers

fl.   p 49

Anthyllis vulneraria var.

Barneby, T. P.
European Alpine Flowers in Colour

fl.   Pl. 44, 4

Anthyllis vulneraria var.

Kohlhaupt, Paula
Fleurs des Alpages   Vol. 2

fl.   P. 59

Anthyllis vulneraria var.

Martin, W. Keble
The Concise British Flora in Colour

fl. fr. hab.   Pl. 23

Anthyllis vulneraria var.

Polunin, Oleg
Flowers of the Mediterranean

fl.   Pl. 67

Antigonon leptopus

American Hort. Soc.
American Horticulturist
   vol 52, No. 4   1973

fl.   p. 11

Antigonon leptopus

Bruggeman, L.
Tropical Plants

fl.   Pl. 2

Antigonon leptopus

Coyle, Jeanette
A Field Guide to the Common and
   Interesting Plants of Baja California

fl.   p 71

Antigonon leptopus

Everett, T.H., ed.
New Illustrated Encyclopedia of
   Gardening   vol. 1

fl.   p. 55

Antigonon leptopus

Hall, Clarence E.
Flowers of the Islands in the Sun

fl.   p. 47   Pl. 8

Antigonon leptopus

Hargreaves, Dorothy
Tropical Blossoms of the Caribbean

fl.   p. 24

Antigonon leptopus

Harrison, Richmond E.
Climbers and Trailers

fl.   p 19   Pl. 9

Antigonon leptopus

Macoby, Stirling
What Flower is That

fl.   p. 43

Antigonon leptopus

Mathias, Mildred E.
Color for the Landscape

fl. hab.   p 98

Antigonon leptopus

Menninger, Edwin A.
Flowering Vines of the World

fl.   Pl. 148

Antigonon leptopus

O'Gorman, Helen
Mexican Flowering Trees and Plants

fl. hab.   p 85

Antigonon leptopus

Perry, Frances
Flowers of the World

fl.   p 236

Antigonon leptopus var.

Menninger, Edwin A.
Flowering Vines of the World

fl.   Pl. 150

Antirrhinum asarina

Vilmorin, Roger de
Plantes Alpines dans les Jardins

fl. hab.   Pl. IX

Antirrhinum hybrid

Perry, Frances
Flowers of the World

fl.   p. 275

Antirrhinum latifolium

Kleijn, H.
Beauty of the Wild Plant

fl.   p. 116   Pl. 169

Antirrhinum latifolium

Polunin, Oleg
Flowers of Europe

fl.   Pl. 121 #1198

Antirrhinum latifolium

Polunin, Oleg
Flowers of the Mediterranean

fl.   Pl. 171

Antirrhinum majus

Crockett, James Underwood
Annuals

fl.   P 95

Antirrhinum majus

Macoby, Stirling
What Flower is That

fl.   P 43

Antirrhinum majus

Tsukamoto, Yotaro
Coloured Illustrations of Garden
   Flowers   Vol. 10

fl.   p.3-4   Pl. 9,10,11

Antirrhinum majus var.

Curtis's Botanical Magazine 1960
          Vol. 173
fl.   Pl. 393

Antirrhinum majus var.

Hay, Roy
The Color Dictionary of Flowers &
   Plants

fl.   p. 30   Pl. 236,237

Anthirrhinum majus var.

Huxley, Anthony
Garden Annuals and Bulbs

fl.   Pl. 6a,b,c 7a,b,c

Antirrhinum orontium

Polunin, Oleg
Flowers of Europe

fl.   Pl. 122 #1199

Antirrhinum orontium

Polunin, Oleg
Flowers of the Mediterranean

fl.   Pl. 172

Antirrhinum orontium

Wit, H. C. D. de
Plants of the World;
  The Higher Plants, Vol II

fl.   Pl. 87

Antirrhinum vulgaris

de Wit, H. C. D.
Plants of the World; The Higher Plants Vol II

fl.   p 166, Pl. 87

Anu

  See Tropaeolum tuberosum

Anubias congensis

Stodola, Jiri
Encyclopedia of Water Plants

fl.   P. 118

Anubias lanceolata

Stodola, Jiri
Encyclopedia of Water Plants

fl.   P. 119

Aotus ericoides

Blombery, Alec M.
What Wildflower is That

fl.   P. 50   Pl. 101

Aotus ericoides

Cochrane, G. R.
Flowers and Plants of Victoria

fl.   Pl. 55

Aotus ericoides

Mullins, Barbara
Australian Wildflowers in Colour

fl.   P. 17   Pl. 8

Aotus villosa

Blombery, Alec M.
What Wildflower is That

fl.   p. 50   Pl. 102

Apache Plume

  see

Fallugia paradoxa

Apalochlamys spectabilis

Cochrane, G. R.
Flowers and Plants of Victoria

fl.   Pl. 50

Apatesia maughanii

Herre, H.
The Genera of the Mesembryanthemaceae

fl. fr.   p 77

Aphananthe aspera

Kitamura, Siro
Coloured Illustrations of Trees &
  Shrubs of Japan

fr.   137

Aphanes arvensis

Martin, W. Keble
The Concise British Flora in Colour

hab.   Pl. 26

Aphanopetalum resinosum

Blombery, Alec M.
What Wildflower is That

fl.   p. 50   Pl. 103

Aphanostephus skirrhobasis

Brown, Clair A.
Wildflowers of Louisiana and
  Adjoining States

fl. hab.   p. 185

Aphanostephus skirrhobasis

Crockett, James Underwood
Annuals

fl.   P 95

Aphanostephus skirrhobasis

Dean, Blanche E.
Wildflowers of Alabama and
  Adjoining States

fl.   p. 189

Aphelandra aurantiaca var

Addisonia, Vol. 23   1954-59

fl.   p. 17   Pl. 745

Aphelandra schiediana

Chickering, Carol Rogers
Flowers of Guatemala

fl.   p 35

Aphelandra squarrosa

Hay, Roy
The Dictionary of House Plants

fl.   Pl. 36

Aphelandra squarrosa

Macoby, Stirling
What Flower is That

fl.   P 43

Aphelandra squarrosa

Nicolaisen, Age
Pocket Encyclopedia of Indoor Plants

fl., hab.   Pl. 125

Aphelandra squarrosa var.

Encke, Fritz
Zimmerpflanzen

fl.   p. 52

Aphelandra squarrosa var.

Kiaer, Eigil
Indoor Plants in Colour

fl.   p. 16

Aphelandra squarrosa var

Kromdijk, G.
200 House Plants in Colour

fl.   Pl. 18

Aphelandra squarrosa var.

Massachusetts Hort. Soc.
Horticulture
  vol 38, No. 12   1960

fl., hab.   p. 603

Aphelandra squarrosa var.

Perry, Frances
Flowers of the World

fl.   p. 13

Aphelandra squarrosa var.

Royal Hort. Soc.
Journal of the Royal Hort. Soc.
  vol 90, No. 5   1965

fl., hab.   p. 206   Pl. 88

Aphyllanthes monspeliensis

Barneby, T. P.
European Alpine Flowers in Colour

fl.   Pl. 5, 5

Aphyllanthes monspeliensis

Polunin, Oleg
Flowers of Europe

fl.   Pl. 162 #1585

Aphyllanthes monspeliensis

Polunin, Oleg
Flowers of the Mediterranean

fl.   Pl. 252

Apios americana

Brown, Clair A.
Wildflowers of Louisiana and
    Adjoining States

fl. hab.   p 73

Apios americana

Courtenay, Booth
Wildflowers & Weeds

fl.   p 52

Apios americana

Dean, Blanche E.
Wildflowers of Alabama and
    Adjoining States

fl.   p 89

Apios americana

Duncan, Wilbur H.
Wildflowers of the Southeastern
    United States

fl.   p 85

Apios americana

Jennings, O. E.
Wild Flowers of Western Pennsylvania
    Vol. II

fl.   Pl. 92

Apios americana

Klimas, John E.
Wildflowers of Eastern America

fl.   Pl. 257

Apios americana

Lemmon, Robert S.
Wildflowers of North America in
    Full Color

fl.   pg. 187   pl. 294

Apios americana

Morley, Brian D.
Wild Flowers of the World

fl.   Pl. 153

Apios americana

Moyle, John B.
Northland Wild Flowers

fl. hab.   p 83, Pl. 62

Apios americana

Wharton, Mary E.
A Guide to the Wildflowers & Ferns
    of Kentucky

fl. p 186   Pl. 1.15

Apium graveolens

Ary, S.
The Oxford Book of Wildflowers

fl.   p 46, Pl. 1

Apium graveolens

Bianchini, F.
The Complete Book of Fruits & Vegetables

hab.   p 99

Apium graveolens

Loewenfeld, Claire
The Complete Book of Herbs and Spices

fl. fr.   p 208

Apium graveolens

Martin, W. Keble
The Concise British Flora in Colour

fl. fr. hab.   Pl. 37

Apium graveolens

Masefield, G. B.
The Oxford Book of Food Plants

fl.   Pl. 149, 1

Apium graveolens

Perrot, Emile
Les Plantes Medicinales

fl. fr. hab.   p 137

Apium graveolens

Rosengarten, Frederic, Jr.
The Book of Spices

fl. fr.   p 174

Apium graveolens var.

Bianchini, F.
The Complete Book of Fruits & Vegetables

hab.   p 99

Apium graveolens var.

Hvass, Elsie
Plants That Feed and Serve Us

hab.   p. 22   Pl. 35

Apium graveolens var.

Masefield, G. B.
The Oxford Book of Food Plants

hab.   Pl. 175, 3

Apium inundatum

Ary, S.
The Oxford Book of Wildflowers

fl.   p 88, Pl. 8

Apium inundatum

Lindman, C. A. M.
Nordens Flora, Vol 7

fl. fr. hab.   Pl. 422

Apium inundatum

Martin, W. Keble
The Concise British Flora in Colour

fl. hab.   Pl. 37

Apium nodiflorum

Ary, S.
The Oxford Book of Wildflowers

fl.   p 88, Pl. 1

Apium nodiflorum

Martin, W. Keble
The Concise British Flora in Colour

fl. fr. hab.   Pl. 37

Apium nodiflorum

Polunin, Oleg
Flowers of Europe

fl.   Pl. 85 #886

Apium repens

Martin, W. Keble
The Concise British Flora in Colour

fl. hab.   Pl. 37

Aplectrum hyemale

Courtenay, Booth
Wildflowers & Weeds

fl.   p 15

Aplectrum hyemale

Dean, Blanche E.
Wildflowers of Alabama and
    Adjoining States

fl.   p 41

Aplectrum hyemale

Luer, Carlyle A.
The Native Orchids of Florida

fl. hab.   Pl. 1; 5

Aplectrum hyemale

Luer, Carlyle A.
The Native Orchids of the
    United States and Canada

fl. fr. hab.   Pl. 88

Aplectrum hyemale

Wharton, Mary E.
A Guide to the Wildflowers & Ferns
    of Kentucky

fl. hab. p 91 Pl. 3.3

Apocynum androsaemifolium

Clark, Lewis J.
Wild Flowers of British Columbia

fl.   p 391

Apocynum androsaemifolium

Courtenay, Booth
Wildflowers & Weeds

fl.   p 88

Apocynum androsaemifolium

Duncan, Wildflowers of the Southeastern
    United States

fl.   p 127

Apocynum androsaemifolium

Klimas, John E.
Wildflowers of Eastern America

fl.   Pl. 63

Apocynum androsaemifolium

Moyle, John B.
Northland Wild Flowers

fl. hab.   p 124, Pl. 130

Apocynum androsaemifolium

Orr, Robert T.
Wildflowers of Western America

fl.   Pl. 171

Apocynum androsaemifolium

Wharton, Mary E.
A Guide to the Wildflowers & Ferns
    of Kentucky

fl.   p 175   Pl. 2.58

Apocynum cannabinum

Moyle, John B.
Northland Wild Flowers

fl. hab.   p. 125   Pl. 131

Apocynum cannabinum

Weeds of the Southern United States

hab., fr.   p. 4

Apocynum cannabinum

Wharton, Mary E.
A Guide to the Wildflowers & Ferns
    of Kentucky

fl.   p 175   Pl. 2.57

Apocynum sibiricum

Courtenay, Booth
Wildflowers & Weeds

fl.   p 88

Apodytes dimidiata

Flowering Plants of Africa
    vol 43   1974-76

fl., fr.   Pl. 1695

Apollonias barbusana

Bramwell, David
Wild Flowers of the Canary Islands

fr.   Pl. 132

Aponogeton bernierianus

Stodola, Jiri
Encyclopedia of Water Plants

fl.   P. 162

Aponogeton crispus

Stodola, Jiri
Encyclopedia of Water Plants

fl.   P. 163

Aponogeton distachyus

Flowering Plants of Africa
    vol 41   1970-71

fl., hab.   Pl. 1618

Aponogeton distachyus

Hay, Roy
The Color Dictionary of Flowers & Plants

fl.   p 122, Pl. 972

Aponogeton distachyus

Morley, Brian D.
Wild Flowers of the World

fl., hab.   Pl. 84

Aponogeton distachyus

Perry, Frances
Flowers of the World

fl.   p. 31

Aponogeton distachyus

Stodola, Jiri
Encyclopedia of Water Plants

fl.   P. 214

Aponogeton elongatus

Stodola, Jiri
Encyclopedia of Water Plants

fl.   P. 166

Aponogeton fenestralis

Perry, Frances
Flowers of the World

fl.   p 31

Aponogeton fenestralis

Stodola, Jiri
Encyclopedia of Water Plants

fl.   P. 167

Aponogeton henkelianus

Stodola, Jiri
Encyclopedia of Water Plants

hab.   P. 170

Aponogeton juceus var.

Flowering Plants of Africa
    vol 37   1965-66

fl., hab.   Pl. 1449

Aponogeton natans

Wit, H. C. D. de
Plants of the World;
    The Higher Plants, Vol II

fl.   Pl. 117

Aponogeton ulvaceus

Stodola, Jiri
Encyclopedia of Water Plants

fl.   P. 171

Aponogeton undulatus

Stodola, Jiri
Encyclopedia of Water Plants

fl.   P. 174

Aporocactus flagelliformis

Hay, Roy
The Dictionary of House Plants

fl.   Pl. 37

Aporocactus flagelliformis

Hersey, Jean
Woman's Day Book of House Plants

fl., hab.   p 44

Aporocactus flagelliformis

Kiaer, Eigil
Indoor Plants in Colour

fl.   p. 17

Aporocactus flagelliformis

Kupper, Walter
Cacti

fl.   p. 109   Pl. 51

Aporocactus flagelliformis

Lamb, Edgar
Popular Exotic Cacti in Color

fl. hab.   p 29

Aporocactus flagelliformis

O'Gorman, Helen
Mexican Flowering Trees and Plants

fl. hab.   p 201

Aporocactus flagelliformis
Van Laren, A. J.
Cactus

fl.   P 26   Fig. H.

Aposeris foetida
Barneby, T. P.
European Alpine Flowers in Colour

fl. hab.   Pl. 94, 4

Apostle plant

see

Neomarica northiana

Apple

see

Malus

Apple, Adam's

see

Ervatamia coronaria

Apple, Adams

see

Musa paradisiaca

Apple, Argyle

see

Eucalyptus cinerea

Apple, Baked

see

Rubus Chamaemorus

Apple, Balsam

see

Echinocystis lobata

Apple, Balsam Wild

see

Momordica charantia

Apple, Bechlel's crab

see

Malus ioensis var.

Apple-Berry

see

Billardiera scandens

Apple Berry, Sweet

see

Billardiera cymosa

Apple, Bitter

see

Citrullus colocynthis

Appleblossom

see

Cassia javanica

Apple Blossom

see

Weigela

Apple, Cane

see

Arbutus unedo

Apple, common

see

Malus pumila

Apple, Crab Bechtel

see

Malus ioensis var.

Apple, Crab Carmine

see

Malus atrosanguinea

Apple, Crab European

see

Malus sylvestris

Apple, Crab Japanese

see

Malus floribunda

Apple, Crab Sargent

see

Malus sargentii

Apple, Crab Showy

see

Malus floribunda

Apple, Crab Siberian

see

Malus baccata

Apple, custard

see

Annona cherimolia

Apple, Dwarf

see

Angophora cordifolia

Apple, Eve's

see

Tabernaemontana dichotoma

Apple, floral

see

Malus ioensis var.

Applegum, dwarf

see

Angophora cordifolia

Apple, Indian

see

Monotropa brittonii

Apple, Kangaroo

see

Solanum aviculare

Apple, Kangaroo

see

Solanum laciniatum

Apple, Malay

see

Eugenia malaccensis

Apple, May

see

Podophyllum emodi

Apple, Monkey

see

Strychnos spinosa var.

Apple-of-Peru

see

Nicandra physalodes

Apple of Peru

see

Nicandra violacea

Apple, Pond

see

Annona glabra

Apple, rose

see

Eugenia australis

Apple, Rose

see

Syzygium jambos

Apple, Sand

see

Parinari mobola

Apple, Scrub

see

Angophora cordifolia

Apple Shrub

see

Calycanthus floridus

Apple, Soda

see

Solanum ciliatum

Apple, Thorn

see

Crataegus monogyna

Apple, Thorn

see

Datura stramonium

Apricot

see

Prunus armeniaca

Apricot, Desert

see

Prunus fremontii

Apricot, flowering

see

Prunus mume

Apricot, Japanese

see

Prunus mume

April Fool

see

Haemanthus coccineus

Aptenia cordifolia

Herre, H.
The Genera of the Mesembryanthemaceae

fl. fr.   p 79

Aptosimum depressum

Eliovson, Sima
Namaqualand in Flower

fl.   Pl. 21

Aptosimum spinescens

Eliovson, Sima
Namaqualand in Flower

fl.   Pl. 23, 2

Aquilegia alpina

Barneby, T. P.
European Alpine Flowers in Colour

fl.   Pl. 22, 2

Aquilegia alpina

Kohlhaupt, Paula
Fleurs des Alpages Vol. 1

fl.   P. 13

Aquilegia alpina

Polunin, Oleg
Flowers of Europe

fl.   Pl. 27 #254

Aquilegia atrata

Tosco, Uberto
The World of Mountain Flowers

fl.   p 100

Aquilegia aurea

Royal Hort. Soc.
Journal of the Royal Hort. Soc.
  vol 87, No. 1   1962

fl.   p. 22   Pl. 5

Aquilegia brevistyla

Heller, Christine
Wild Flowers of Alaska

fl.   Pl. 241

Aquilegia brevistyla

Walcott, Mary Vaux
North American Wild Flowers
  vol. 4

fl.   Pl. 292

Aquilegia caerulea

Orr, Robert T.
Wildflowers of Western America

fl.   Pl. 232

Aquilegia caerulea

Welsh, Stanley L.
Flowers of the Canyon Country

fl.   p 11

Aquilegia caerulea

Welsh, Stanley L.
Flowers of the Mountain Country

fl.   p. 72

Aquilegia caerulea var.

Lemmon, Robert S.
Wildflowers of North America
  in Full Color

fl.   pg. 127   Pl. 202

Aquilegia canadensis

Campbell, Carlos C.
Great Smoky Mountain Wildflowers

fl.   p 25

Aquilegia canadensis

Courtenay, Booth
Wildflowers & Weeds

fl.   p 31

Aquilegia canadensis

Dean, Blanche E.
Wildflowers of Alabama and
    Adjoining States

fl.   p 57

Aquilegia canadensis

Duncan, Wilbur H.
Wildflowers of the Southeastern
    United States

fl.   p 43

Aquilegia canadensis

Ferguson, Mary
Wildflowers

fl.   p 106

Aquilegia canadensis

Jennings, O. E.
Wild Flowers of Western Pennsylvania
    Vol. II

fl. hab.   Pl. 61

Aquilegia canadensis

Klimas, John E.
Wildflowers of Eastern America

fl.   Pl. 191

Aquilegia canadensis

Lemmon, Robert S.
Wildflowers of North America
    in Full Color

fl.   pg. 251   Pl. 395

Aquilegia canadensis

Moyle, John B.
Northland Wild Flowers

fl.   p 53, Pl. 14

Aquilegia canadensis

Roberts, June Carver
Born in the Spring

fl. fr. hab.   p 85

Aquilegia canadensis

Walcott, Mary Vaux
North American Wild Flowers

fl. hab.   Pl. 141

Aquilegia canadensis

Wharton, Mary E.
A Guide to the Wildflowers & Ferns
    of Kentucky

fl.   p 121   Pl. 1.53

Aquilegia chrysantha

Milne, Lorus
Living Plants of the World

fl.   p 49

Aquilegia chrysantha

Orr, Robert T.
Wildflowers of Western America

fl.   Pl. 78

Aquilegia chrysantha

Vilmorin, Roger de
Plantes Alpines dans les Jardins

fl. hab.   Pl. XLI

Aquilegia chrysantha var.

Huxley, Anthony
Garden Perennials and Water Plants

fl.   Pl. 39

Aquilegia einseleana

Barneby, T. P.
European Alpine Flowers in Colour

fl.   Pl. 22, 3

Aquilegia elegantula

Massachusetts Hort. Soc.
Horticulture
    vol 53, No. 2   1975

fl.   cover, backcover

Aquilegia eximia

Munz, Philip A.
California Spring Wildflowers

fl.   p 42, Pl. 47

Aquilegia flabellata

Tsukamoto, Yotaro
Coloured Illustrations of Garden
    Flowers   Vol. 9

fl.   p. 30

Aquilegia flavescens

Clark, Lewis J.
Wild Flowers of British Columbia

fl.   p 162

Aquilegia flavescens

Porsild, A. E.
Rocky Mountain Wild Flowers

fl. hab.   p 173

Aquilegia flavescens

Walcott, Mary Vaux
North American Wild Flowers

fl. hab.   Pl. 201

Aquilegia flavescens

Welsh, Stanley L.
Flowers of the Canyon Country

fl.   p 30

Aquilegia formosa

Alaska-Yukon Wild Flowers Guide

fl.   p 38, 165

Aquilegia formosa

Clark, Lewis J.
Wild Flowers of British Columbia

fl. hab.   p 162

Aquilegia formosa

Fries, Mary A.
Wildflowers of Mount Ranier and
the Cascades

fl.   P 125

Aquilegia formosa

Heller, Christine
Wild Flowers of Alaska

fl.   Pl. 64

Aquilegia formosa

Orr, Robert T.
Wildflowers of Western America

fl.   Pl. 150

Aquilegia formosa

Taylor, Ronald J.
Mountain Wild Flowers

fl.   p 91

Aquilegia glandulosa var.

Royal Hort. Soc.
Journal of the Royal Hort. Soc.
    vol 93, No. 6   1968

fl.   p. 246   Pl. 125

Aquilegia hybrid

Huxley, Anthony
Garden Perennials and Water Plants

fl.   Pl. 37

Aquilegia hybrid

Perry, Frances
Flowers of the World

fl.   p. 253

Aquilegia x hybrida

Hay, Roy
The Color Dictionary of Flowers
& Plants

fl.   p. 122   Pl. 973,
                                  974

Aquilegia x hybrida

Tsukamoto, Yotaro
Coloured Illustrations of
  Garden Flowers   Vol. 9

fl.   p. 30

Aquilegia micrantha

Welsh, Stanley L.
Flowers of the Canyon Country

fl.   p 46

Aquilegia nivalis

Alpine Garden Society
Bulletin
  vol. 38, No. 3   1970

fl., hab.   p. 247

Aquilegia nivalis

Royal Hort. Soc.
The Garden
  vol 102, No. 11   1977

fl.   p. 452

Aquilegia pyrenaica

Curtis's Botanical Magazine   Vol. 174
                  1962-63

fl.   Pl. 435

Aquilegia vulgaris

Barneby, T.P.
European Alpine Flowers in Colour

fl.   Pl. 22, 1

Aquilegia vulgaris

Bianchini, Francesco
Health Plants of the World

fl.   p 171

Aquilegia vulgaris

Felsko, Elsa
A Book of Wildflowers

fl.   P 62

Aquilegia vulgaris

Kleijn, H.
Beauty of the Wild Plant

fl.   p. 52   Pl. 64

Aquilegia vulgaris

Lindman, C. A. M.
Nordens Flora, Vol 4

fl. fr.   Pl. 218

Aquilegia vulgaris

Macoby, Stirling
What Flower is That

fl.   P 44

Aquilegia vulgaris

Martin, W. Keble
The Concise British Flora in Colour

fl. hab.   Pl. 4

Aquilegia vulgaris

Morley, Brian D.
Wild Flowers of the World

fl.   Pl. 8F

Aquilegia vulgaris

Polunin, Oleg
Flowers of Europe

fl.   Pl. 27 #253

Aquilegia vulgaris

Tosco, Uberto
The World of Mountain Flowers

fl.   p 66

Arabidopsis thaliana

Ary, S.
The Oxford Book of Wildflowers

fl. fr.   p 70, Pl. 4

Arabidopsis thaliana

Lindman, C. A. M.
Nordens Flora, Vol 4

fl. fr. hab.   Pl. 272B

Arabidopsis thaliana

Martin, W. Keble
The Concise British Flora in Colour

fl. fr. hab.   Pl. 8

Arabis albida

Crockett, James Underwood
Lawns & Ground Covers

fl.   p. 121

Arabis albida var.

Hay, Roy
The Color Dictionary of Flowers &
Plants

fl.   p. 2   Pl. 13

Arabis albida var.

Macoby, Stirling
What Flower is That

fl.   P. 44

Arabis alpina

Barneby, T. P.
European Alpine Flowers in Colour

fl. hab.   Pl. 31, 1

Arabis alpina

Lindman, C. A. M.
Nordens Flora, Vol 4

fl. fr. hab.   Pl. 271

Arabis alpina

Martin, W. Keble
The Concise British Flora in Colour

fl. fr. hab.   Pl. 7

Arabis alpina

Polunin, Oleg
Flowers of Europe

fl.   Pl. 34 #318

Arabis bellidifolia

Barneby, T. P.
European Alpine Flowers in Colour

fl. hab.   Pl. 31, 2

Arabis canadensis

Courtenay, Booth
Wildflowers & Weeds

fl.   p 18

Arabis canadensis

Klimas, John E.
Wildflowers of Eastern America

fl.   Pl. 18

Arabis caucasica var.

Huxley, Anthony
Garden Perennials and Water Plants

fl.   Pl. 41

Arabis cypria

Curtis's Botanical Magazine Vol. 175
                  1964-65

fl.   Pl. 464

Arabis cypria

Megaw, Elektra
Wild Flowers of Cyprus

fl., hab   frontispiece

Arabis hirsuta

Ary, S.
The Oxford Book of Wildflowers

fl. fr.   p 70, Pl. 2

Arabis hirsuta
Martin, W. Keble
The Concise British Flora in Colour
fl. fr. hab.   Pl. 7

Arabis laevigata
Wharton, Mary E.
A Guide to the Wildflowers & Ferns
   of Kentucky
fl.   p 150   Pl. 2.6

Arabis Lyallii
Porsild, A.E.
Rocky Mountain Wild Flowers
fl. fr. hab.   p 199

Arabis lyallii
Taylor, Ronald J.
Mountain Wild Flowers
fl. hab.   p 134

Arabis lyrata
Courtenay, Booth
Wildflowers & Weeds
fl.   p 18

Arabis pumila
Barneby, T. P.
European Alpine Flowers in Colour
fl. hab.   Pl. 31, 3

Arabis stricta
Martin, W. Keble
The Concise British Flora in Colour
fl. fr. hab.   Pl. 7

Arabis verna
Polunin, Oleg
Flowers of Europe
fl.   Pl. 35 #319

Arachis hypogaea
Bianchini, F.
The Complete Book of Fruits & Vegetables
fr.   p. 221,235

Arachis hypogaea
Hvass, Elsie
Plants That Feed and Serve Us
fl. fr.   p 16, Pl. 24

Arachis hypogaea
Masefield, G. B.
The Oxford Book of Food Plants
fl. fr.   Pl. 23, 3

Arachis hypogaea
Perrot, Emile
Les Plantes Medicinales
fl. fr. hab.   p 14

Arachnis clarkei
Kimber, Sheila
A Handbook of Orchids
fl., hab.   p. 10

Arachnis flos-aeris
Macoby, Stirling
What Flower is That
fl.   P. 44

Arachnis flos-aeris
Morley, Brian D.
Wild Flowers of the World
fl.   Pl. 127

Arachus bypogea
Linnell, T.
Plantes Utiles du Monde entier
fl. fr.   p. 16   Pl. 24

Araeococcus flagellifolius
Bromeliad Society
Journal
vol 21, No. 5 1971
fl.   p. 120

Araeococcus flagellifolius
Padilla, Victoria
Bromeliads
hab., fl.   p. 65   Pl. 3

Araeococcus flagellifolius
Wilson, Robert Gardner
Bromeliads in Cultivation
hab.   Pl. 68

Aralia
see
Tetrapanax papyriferus

Aralia cordata
Kimura, Koiti
Japanese Medicinal Plants, Vol I
fl.   p 62

Aralia elata
Everett, T.H., ed.
New Illustrated Encyclopedia of
   Gardening vol. 1
fl., hab.   p. 23

Aralia elata
Huxley, Anthony
Deciduous Garden Trees and Shrubs
fl.   Pl. 15

Aralia elata
Kimura, Koiti
Japanese Medicinal Plants, Vol I
fl.   p 64

Aralia elata
Kitamura, Siro
Coloured Illustrations of Trees and
   Shrubs of Japan
fl.   opp. p. 169

Aralia elata var.
Gault, S. Millar
The Color Dictionary of Shrubs
hab.   Pl. 9

Aralia elata var.
Hay, Roy
The Color Dictionary of Flowers &
   Plants
hab.   p. 181   Pl. 144

Aralia elata var.
Royal Hort. Soc.
Journal of the Royal Hort. Soc.
   vol 98, No. 2 1973
hab.   p. 56   Pl. 35

Aralia, False
see
Dizygotheca elegantissima

Aralia, Finger
see
Dizygotheca elegantissima

Aralia hispida
Courtenay, Booth
Wildflowers & Weeds
fl.   p 60

Aralia, Japanese
see
Fatsia japonica var.

Aralia nudicaulis
Clark, Lewis J.
Wild Flowers of British Columbia
fl.   p 338

Aralia nudicaulis

Courtenay, Booth
Wildflowers & Weeds

fl.   p 60

Aralia nudicaulis

Lemmon, Robert S.
Wildflowers of North America in
    Full Color

fl.   p. 138   Pl. 219

Aralia nudicaulis

Moyle, John B.
Northland Wild Flowers

fl. hab.   p 99, Pl. 89

Aralia racemosa

Courtenay, Booth
Wildflowers & Weeds

fl.   p 60

Aralia racemosa

Moyle, John B.
Northland Wild Flowers

fl. hab.   p 100, Pl. 90

Aralia racemosa

Pennsylvania Hort. Soc.
The Green Scene
    vol 5, No. 5   1977

fl.   p. 23

Aralia racemosa

Wharton, Mary E.
A Guide to the Wildflowers & Ferns
    of Kentucky

fl.   fr.   p 163   Pl. 2.33

Aralia spinosa

Batson, Wade T.
Wild Flowers in South Carolina

hab.   p. 79

Aralia spinosa

Brown, Clair A.
Wildflowers of Louisiana and
    Adjoining States

fl. hab.   p 122

Aralia spinosa

Campbell, Carlos C.
Great Smoky Mountain Wildflowers

fl.   p 105

Aralia spinosa

Wharton, Mary E.
Trees & Shrubs of Kentucky

fl. hab.   p 69, Pl. 2.2
fr.   p 133, Pl. 2.22

Araucaria angustifolia

Royal Hort. Soc.
The Garden
    vol 103, No. 12   1978

hab.   p. 481

Araucaria araucana

Boom, B. K.
The glory of the tree

hab.   opp. p. 16   Pl. 15

Araucaria araucana

Edlin, Herbert
The Illustrated Encyclopedia
    of Trees

hab.   p 246

Araucaria araucana

Huxley, Anthony
Evergreen Garden Trees and Shrubs

hab.   Pl. 8

Araucaria bidwillii

Blombery, Alec M.
What Wildflower is That

hab.   p. 51   Pl. 104

Araucaria cunninghamii

Bruggeman, L.
Tropical Plants

hab.   Pl. 186

Araucaria excelsa

Hay, Roy
The Dictionary of House Plants

hab.   Pl. 38

Araucaria excelsa

Hersey, Jean
Woman's Day Book of House Plants

hab.   p. 80

Araucaria excelsa

Kiaer, Eigil
Indoor Plants in Colour

hab.   p. 17

Araucaria excelsa

Massachusetts Hort. Soc.
Horticulture
    vol 52, No. 12   1974

hab.   p. 27

Araucaria excelsa

Nicolaisen, Age
Pocket Encyclopedia of Indoor Plants

hab.   Pl. 3

Araucaria heterophylla

Crockett, James Underwood
Evergreens

hab.   p. 90

Araucaria heterophylla

Edlin, Herbert
The Illustrated Encyclopedia
    of Trees

hab.   p 246

Araucaria heterophylla

Kromdijk, G.
200 House Plants in Colour

hab.   Pl. 19

Araujia sericifera

Harrison, Richmond E.
Climbers and Trailers

fl. fr.   p. 20   Pl. 10

Araujia sericifera

Polunin, Oleg
Trees and Bushes of Europe

fr.   p 168

Araujia sericifera

Royal Hort. Soc.
The Garden
    Vol. 101, No. 2   1976

fl.   p. 104

Arbor vine, Spanish

See Ipomoea tuberosa

Arborvitae

see

Thuja

Arborvitae, American

see

Thuja occidentalis

Arborvitae, Douglas

see

Thuja occidentalis var.

Arborvitaea, Giant

see

Thuja plicata

Arborvitae, Hiba False

see

Thujopsis dolabrata

Arborvitae, Japanese

see

Thuja standishii

Arbutus andrachne

Boom, B. K.
The glory of the tree

fl.   p. 104   pl. 178

Arbutus andrachne

Feinbrun-Dothan, Naomi
Wild Plants in the Land
 of Israel

fl. fr.   p 44

Arbutus andrachne

Megaw, Elektra
Wild Flowers of Cyprus

fr.   p. 9   Pl. 11

Arbutus andrachne

Polunin, Oleg
Flowers of the Mediterranean

fl.   Pl. 120

Arbutus andrachne

Polunin, Oleg
Trees and Bushes of Europe

fl. fr. hab.   p 158, 159

Arbutus andrachne

Royal Hort. Soc.
Journal of the Royal Hort. Soc.
 vol 100, No. 3   1975

hab.   p. 116   Pl. 32

Arbutus andrachnoides

Hay, Roy
The Color Dictionary of Flowers &
 Plants

fl., hab.   p. 181   Pl. 1447, 1448

Arbutus menziesii

Boom, B. K.
The glory of the tree

fl.   p. 104   Pl. 177

Arbutus menziesii

Curtis's Botanical Magazine
Vol 171   1956-57

fl. fr.   275

Arbutus menziesii

Hay, Roy
The Color Dictionary of Flowers & Plants

fr.   p 182, Pl. 1449

Arbutus menziesii

Massachusetts Hort. Soc.
Horticulture
 vol 50, No. 7   1972

fl.   p. 37

Arbutus menziesii

Munz, Philip A.
California Spring Wildflowers

fl.   p 29, Pl. 8

Arbutus, Trailing

see

Epigaea repens

Arbutus unedo

Bianchini, F.
The Complete Book of Fruits & Vegetables

fr.   p 165

Arbutus unedo

Bianchini, Francesco
Health Plants of the World

fr. hab.   p 65

Arbutus unedo

Boom, B. K.
The glory of the tree

hab., fr.   p. 104   Pl. 175-76

Arbutus unedo

Edlin, Herbert
The Illustrated Encyclopedia
 of Trees

fl. fr. hab.   p 200-1

Arbutus unedo

Harrison, R. E.
Trees and Shrubs

fr.   P. 24

Arbutus unedo

Hay, Roy
The Color Dictionary of Flowers & Plants

fr.   p 182, Pl. 1450

Arbutus unedo

Macoby, Stirling
What Flower is That

fr.   P 45

Arbutus unedo

Martin, W. Keble
The Concise British Flora in Colour

fl. fr. hab.   Pl. 55

Arbutus unedo

Masefield, G. B.
The Oxford Book of Food Plants

fl. fr.   Pl. 83, 5

Arbutus unedo

Massachusetts Hort. Soc.
Horticulture
 vol 52, No. 11   1974

fl., fr.   cover, backcover

Arbutus unedo

Oakman, Harry
Colorful Trees

fr.   P 132

Arbutus unedo

Perrot, Emile
Les Plantes Medicinales

fl. fr. hab.   p 15

Arbutus unedo

Perry, Frances
Flowers of the World

fl. fr.   p. 108

Arbutus unedo

Polunin, Oleg
Flowers of Europe

fl.   Pl. 90 #924

Arbutus unedo

Polunin, Oleg
Flowers of the Mediterranean

fl. fr.   Pl. 119

Arbutus unedo

Polunin, Oleg
Trees and Bushes of Europe

fl. fr. hab.   p 158

Arbutus unedo

Royal Hort. Soc.
Journal of the Royal Hort. Soc.
 vol 90, No. 5   1965

fr., hab.   p. 206   Pl. 101-02

Arbutus unedo

Wit, H. C. D. de
Plants of the World:
 The Higher Plants, Vol I

fl. fr.   p 172, Pl. 98
          p 211, Pl. 113

Arbutus unedo var.

Crockett, James Underwood
Evergreens

hab. fl. fr.   P. 112

Arbutus unedo var.

Curtis's Botanical Magazine
        Vol. 169   1952-53

fl. fr.   203

Arbutus unedo var.

Hay, Roy
The Color Dictionary of Flowers &
        Plants

fl.   p. 182   Pl. 1451

Arbutus unedo var.

Royal Hort. Soc.
Journal of the Royal Hort. Soc.
        vol 79, No. 1   1954

fl., fr.   p. 24   Pl. 1?

Arceuthobium americanum

Porsild, A. E.
Rocky Mountain Wild Flowers

fr.   p. 141

Archangel, Yellow

        see

Galeobdolon luteum

Archangelica officinales

Perrot, Emile
Les Plantes Medicinales

fl. fr. hab.   p 12

Archeria comberi

Curtis, Winifred
The Endemic Flora of Tasmania   Vol. 3

fl. fr.   p. 161

Archeria eriocarpa

Curtis, Winifred
The Endemic Flora of Tasmania, Part IV

fl. fr. hab.   Pl. 155

Archeria hirtella

Curtis, Winifred
The Endemic Flora of Tasmania, Part IV

fl. hab.   Pl. 154

Archeria serpyllifolia

Curtis, Winifred
The Endemic Flora of Tasmania   Vol. 3

fl. fr.   p. 161

Archontophoenix alexandrae

Massachusetts Hort. Soc.
Horticulture
        vol 32, No. 2   1954

fr., hab.   cover

Archontophoenix cunninghamiana

Blomberg, Alec M.
What Wildflower is That

fl. hab.   p 51, Pl. 105

Arctium lappa

Ary, S.
The Oxford Book of Wildflowers

fl.   p 154, Pl. 6

Arctium lappa

Bianchini, Francesco
Health Plants of the World

fl.   p 161

Arctium Lappa

Kimura, Koiti
Japanese Medicinal Plants, Vol I

fl. fr.   p116

Arctium lappa

Martin, W. Keble
The Concise British Flora in Colour

fl.   Pl. 47

Arctium lappa

Pond, Barbara
A Sampler of Wayside Herbs

fl.   Pl. XXIX

Arctium minus

Clark, Lewis J.
Wild Flowers of British Columbia

fl.   p 518

Arctium minus

Courtenay, Booth
Wildflowers & Weeds

fl.   p 107

Arctium minus

Klimas, John E.
Wildflowers of Eastern America

fl.   Pl. 271

Arctium minus

Martin, W. Keble
The Concise British Flora in Colour

fl. hab.   Pl. 47

Arctium minus

Szczawinski, Adam F.
Edible Garden Weeds of Canada

fl., fr.   p. 39

Arctium pubens

Martin, W. Keble
The Concise British Flora in Colour

fl. hab.   Pl. 47

Arctium tomentosum

Lemmon, Robert S.
Wildflowers of North America
        in Full Color

fl.   p 214, Pl. 339

Arctium tomentosum

Lindman, C. A. M.
Nordens Flora, Vol 9

fl.   Pl. 623

Arctium tomentosum

Polunin, Oleg
Flowers of Europe

fl.   Pl. 152 #1472

Arctophila fulva var.

Lindman, C. A. M.
Nordens Flora, Vol 2

fr.   Pl. 88

Arctostaphylos alpina

Heller, Christine
Wild Flowers of Alaska

fr.   Pl. 206

Arctostaphylos alpina

Lindman, C. A. M.
Nordens Flora, Vol 7

fl. fr. hab.   Pl. 449

Arctostaphylos alpina

Polunin, Oleg
Flowers of Europe

fl.   Pl. 88 #925

Arctostaphylos alpina

Porsild, A. E.
Rocky Mountain Wild Flowers

fr.   p 305

Arctostaphylos alpina

Webster, Mary
Flora of Moray, Nairn & East
        Inverness

fr.   p. 228   Pl. 13

Arctostaphylos alpina var.

Clark, Lewis J.
Wild Flowers of British Columbia

fr.   p 375

Arctostaphylos andersonii

Curtis's Botanical Magazine
Vol 171   1956-57

fl.   280

Arctostaphylos canescens

American Hort. Soc.
American Horticulturist
vol 52, No. 3   1973

fl.   p. 27

Arctostaphylos columbiana

Clark, Lewis J.
Wild Flowers of British Columbia

fl.   p 378

Arctostaphylos cushingiana var.

American Hort. Soc.
American Horticulturist
vol 52, No. 3   1973

fl.   p. 31

Arctostaphylos franciscana

American Hort. Soc.
Amiercan Horticulturist
vol 52, No. 3   1973

hab.   p. 28

Arctostaphylos glauca

American Hort. Soc.
American Horticulturist
vol 52, No. 3   1973

fr.   p. 32

Arctostaphylos insularis

Mathias, Mildred E.
Color for the Landscape

fl. hab.   p 172, 173

Arctostaphylos manzanita

Orr, Robert T.
Wildflowers of Western America

fl.   Pl. 46

Arctostaphylos manzanita

Pacific Hort. Foundation
Pacific Horticulture
vol 38, No. 2   1977

hab.   cover

Arctostaphylos media

Clark, Lewis J.
Wild Flowers of British Columbia

fl.   p 386

Arctostaphylos patula

American Hort. Soc.
American Horticulturist
vol 52, No. 3 1973

fr.   p. 27

Arctostaphylos patula

Welsh, Stanley L.
Flowers of the Mountain Country

fl.   p. 25

Arctostaphylos pringlei

Coyle, Jeanette
A Field Guide to the Common and Interesting
 Plants of Baja California

fr.   p 149

Arctostaphylos pringlei var.

Mathias, Mildred E.
Color for the Landscape

fl. hab.   p 173

Arctostaphylos rubra

Porsild, A. E.
Rocky Mountain Wild Flowers

fr.   p 305

Arctostaphylos stanfordiana

Munz, Philip A.
California Spring Wildflowers

fl.   p 76, Pl. 52

Arctostaphylos uva-ursi

American Hort. Soc.
American Horticulturist
 vol. 52, No. 3   1973

fr.   p. 31

Arctostaphylos uva-ursi

Ary, S.
The Oxford Book of Wildflowers

fl. fr.   p 120, Pl. 1

Arctostaphylos uva-ursi

Barneby, T. P.
European Alpine Flowers in Colour

fl. fr.   Pl. 57, 1

Arctostaphylos uva-ursi

Bartels, Andreas
Das Grosse Buch der Gartengeholze

fl.   p 113

Arctostaphylos uva-ursi

Bianchini, Francesco
Health Plants of the World

fr.   p 137

Arctostaphylos uva-ursi

Clark, Lewis J.
Wild Flowers of British Columbia

fl. fr.   p 386

Arctostaphylos uvaursi

Courtenay, Booth
Wildflowers & Weeds

fl.   p. 77

Arctostaphylos uva-ursi

Crockett, James Underwood
Evergreens

hab. fr.   p. 112

Arctostaphylos uva-ursi

Crockett, James Underwood
Lawns & Ground Covers

fl.   p. 122

Arctostaphylos uva-ursi

Ferguson, Mary
Wildflowers

fl.   p 71

Arctostaphylos uva-ursi

Gault, S. Millar
The Color Dictionary of Shrubs

fl.   Pl. 10

Arctostaphylos uva-ursi

Heller, Christine
Wild Flowers of Alaska

fl. fr.   Pl. 109

Arctostaphylos uva-ursi

Hvass, Elsie
Plants That Feed and Serve Us

fl. fr.   p. 88   Pl. 186

Arctostaphylos uva-ursi

Kohlhaupt, Paula
Fleurs des Alpages Vol. 1

fr.   P. 87-88

Arctostaphylos uva-ursi

Lindman, C. A. M.
Nordens Flora, Vol 7

fl. fr.   Pl. 448

Arctostaphylos uva-ursi

Martin, W. Keble
The Concise British Flora in Colour

fl. fr. hab.   Pl. 55

Arctostaphylos uva-ursi

Perrot, Emile
Les Plantes Medicinales

fl. fr. hab.   p 43

Arctostaphylos uva-ursi

Polunin, Oleg
Flowers of Europe

fl.   Pl. 88 #925

Arctostaphylos uva-ursi

Walcott, Mary Vaux
North American Wild Flowers

fl. fr. hab.   Pl. 111, 112

Arctostaphylos uva-ursi

Webster, Mary
Flora of Moray, Nairn & East Inverness

fl., fr.   p. 228   Pl. 13

Arctotheca calendula

Eliovson, Sima
Namaqualand in Flower

fl.   Pl. 20, 2

Arctotheca calendula

Mathias, Mildred E.
Color for the Landscape

fl. hab.   p 139

Arctotis breviscapa

Hay, Roy
The Color Dictionary of Flowers & Plants

fl.   p 30, Pl. 238

Arctotis breviscapa

Hay, Roy
The Dictionary of House Plants

fl.   Pl. 39

Arctotis canescens

Eliovson, Sima
Namaqualand in Flower

fl.   Pl. 18/19

Arctotis cuprea

Flowering Plants of Africa
vol 34   1960-61

fl., fr.   Pl. 1353

Arctotis fastuosa

Eliovson, Sima
Namaqualand in Flower

fl.   Pl. 5

Arctotis grandis

Crockett, James Underwood
Annuals

Fl.   P 95

Arctotis hybrid

Hay, Roy
The Color Dictionary of Flowers & Plants

fl.   p 30, Pl. 239

Arctotis hybrid

Macoby, Stirling
What Flower is That

fl.   P 45

Arctotis x hybrida

Hay, Roy
The Dictionary of House Plants

fl.   Pl. 40

Arctotis stoechadifolia

Tsukamoto, Yotaro
Coloured Illustrations of Garden
Flowers Vol. 10

fl. Illus. 12   p. 4

Arctotis venusta

Morley, Brian D.
Wild Flowers of the World

fl.   Pl. 83

Arctous alpina

Martin, W. Keble
The Concise British Flora in Colour

fl. hab.   Pl. 55

Arctous alpina

Morley, Brian D.
Wild Flowers of the World

fl.   Pl. 6D

Arctous alpina

Walcott, Mary Vaux
North American Wild Flowers
vol. 5

fr.   Pl. 355

Ardisia crenata

Encke, Fritz
Zimmerpflanzen

fr.   p. 53

Ardisia crenata

Kitamura, Siro
Coloured Illustions of Trees &
Shrubs of Japan

fr.   400

Ardisia crenata

Macoby, Stirling
What Flower is That

Ardisia crenata

Walden, Beryl M.
Wild Flowers of Hong Kong

fl.   Pl. 39(121)

Ardisia crispa

Everett, T.H., ed.
New Illustrated Encyclopedia of
Gardening vol. 1

fr.   p. 71

Ardisia crispa

Harrison, R. E.
Trees and Shrubs

fl., fr.   p. 23

Ardisia crispa

Hay, Roy
The Color Dictionary of Flowers & Plants

fr.   P 52, Pl. 409

Ardisia crispa

Hay, Roy
The Dictionary of House Plants

fr.   Pl. 41

Ardisia crispa

Kromdijk, G.
200 Houst Plants in Colour

fr.   Pl. 20

Ardisia crispa

Massachusetts Hort. Soc.
Horticulture
vol 39, No. 1   1961

fr., hab.   inside cover

Ardisia crispa

Perry, Frances
Flowers of the World

fl. fr. p 190

Ardisia crispa var.

Harrison, R. E.
Trees and Shrubs

fl.   P. 23

Ardisia primulaefolia

Walden, Beryl M.
Wild Flowers of Hong Kong

fl. hab.   Pl. 47 (138), 51 (138)

Areca catechu

Edlin, Herbert
The Illustrated Encyclopedia
    of Trees

fr.   p 236

Areca catechu

Hvass, Elsie
Plants That Feed and Serve Us

fr.   p 82, Pl. 169

Areca catechu

Perrot, Emile
Les Plantes Medicinales

fl. fr. hab.   p 16

Aregelia sp.

Kromdijk, G.
200 House Plants in Colour

hab.   Pl. 21

Arenaria balearica

Pacific Hort. Foundation
Pacific Horticulture
    vol 37, No. 3   1976

fl.   p. 48

Arenaria balearica

Vilmorin, Roger de
Plantes Alpines dans les Jardins

fl. hab.   Pl. XXXVII

Arenaria biflora

Barneby, T. P.
European Alpine Flowers in Colour

fl. hab.   Pl. 20, 4

Arenaria biflora

Kohlhaupt, Paula
Fleurs des Alpages Vol. 2

fl.   P. 34

Arenaria capillaris var.

Porsild, A. E.
Rocky Mountain Wild Flowers

fl. hab.   p 157

Arenaria caroliniana

Batson, Wade T.
Wild Flowers in South Carolina

fl. hab.   p 43

Arenaria caroliniana

Duncan, Wilbur H.
Wildflowers of the Southeastern
    United States

fl. hab.   p 35

Arenaria ciliata

Barneby, T. P.
European Alpine Flowers in Colour

fl. hab.   Pl. 20, 5

Arenaria ciliata

Kohlhaupt, Paula
Fleurs des Alpages Vol. 2

fl.   P. 33

Arenaria ciliata var.

Martin, W. Keble
The Concise British Flora in Colour

fl. hab.   Pl. 15

Arenaria gothica

Lindman, C. A. M.
Nordens Flora, Vol 3

fl. fr. hab.   Pl. 199

Arenaria grandiflora

Barneby, T. P.
European Alpine Flowers in Colour

fl. hab.   Pl. 20, 6

Arenaria groenlandica

Klimas, John E.
Wildflowers of Eastern America

fl.   Pl. 71

Arenaria lateriflora

Courtenay, Booth
Wildflowers & Weeds

fl.   p 22

Arenaria lateriflora

Moyle, John B.
Northland Wild Flowers

fl. hab.   p 110, Pl. 107

Arenaria leptoclados

Martin, W. Keble
The Concise British Flora in Colour

fl. hab.   Pl. 15

Arenaria macrophylla

Clark, Lewis J.
Wild Flowers of British Columbia

fl.   p 127

Arenaria montana

Polunin, Oleg
Flowers of Europe

fl.   Pl. 12 #126

Arenaria norica

Alpine Garden Society
Bulletin
    vol. 2, No. 6   1939

fl., hab.   p. 234

Arenaria norvegica var.

Martin, W. Keble
The Concise British Flora in Colour

fl. hab.   Pl. 15

Arenaria patula

Wharton, Mary E.
A Guide to the Wildflowers & Ferns
    of Kentucky

fl.   p 171   Pl. 2.49

Arenaria physodes

Heller, Christine
Wild Flowers of Alaska

fl.   Pl. 160

Arenaria procera

Morley, Brian D.
Wild Flowers of the World

fl.   Pl. 13F

Arenaria pulvinata

Vilmorin, Roger de
Plantes Alpines dans les Jardins

hab.   Pl. XIII

Arenaria rubella

Clark, Lewis J.
Wild Flowers of British Columbia

fl.   p 127

Arenaria serpyllifolia

Ary, S.
The Oxford Book of Wildflowers

fl.   p 76, Pl. 5

Arenaria serpyllifolia

Lindman, C. A. M.
Nordens Flora, Vol 3

fl. fr. hab.   Pl. 200A

Arenaria serpyllifolia

Martin, W. Keble
The Concise British Flora in Colour

fl. hab.   Pl. 15

Arenaria stricta

Courtenay, Booth
Wildflowers & Weeds

fl.   p 22

Arenaria verna

Alpine Garden Society
Bulletin
vol. 2, No. 6   1939

fl., hab.   p. 234

Arenga pinnata

Edlin, Herbert
The Illustrated Encyclopedia
of Trees

hab.   p 235

Arenga pinnata

Hvass, Elsie
Plants That Feed and Serve Us

fl. fr.   p 14, Pl. 19

Arenga saccharifera

Masefield, G. B.
The Oxford Book of Food Plants

fl.   Pl. 17, 2
185, 2

Arenifera pillansii

Herre, H.
The Genera of the Mesembryanthemaceae

fl.   fr.   p 81

Arethusa bulbosa

Courtenay, Booth
Wildflowers & Weeds

fl.   p 14

Arethusa bulbosa

Ferguson, Mary
Wildflowers

fl.   p 44

Arethusa bulbosa

Klimas, John E.
Wildflowers of Eastern America

fl.   Pl. 197

Arethusa bulbosa

Klute, Jeannette
Woodland Portraits

fl.   Pl. 22

Arethusa bulbosa

Luer, Carlyle A.
The Native Orchids of the
United States and Canada

fl.   Pl. 72
Pl. 73

Arethusa bulbosa

Moyle, John B.
Northland Wild Flowers

fl.   hab.   p 223, Pl. 296

Arethusa bulbosa

Ospina, Mariano
Orquideas de las americas

fl.   Pl. 37

Arethusa bulbosa

Walcott, Mary Vaux
North American Wild Flowers

fl. hab.   Pl. 57

Aretia alpina

Kohlhaupt, Paula
Fleurs des Alpages Vol. 1

fl.   P. 38

Argemone albiflora

Brown, Clair A.
Wildflowers of Louisiana and
Adjoining States

fl. hab.   p 58

Argemone albiflora

Dean, Blanche E.
Wildflowers of Alabama and
Adjoining States

fl.   p 71

Argemone corymbosa

Munz, Philip A.
California Desert Wildflowers

fl.   p 32, Pl. 17

Argemone corymbosa

Welsh, Stanley L.
Flowers of the Canyon Country

fl.   p 9

Argemone grandiflora

Crockett, James Underwood
Annuals

fl.   P 96

Argemone intermedia

Lemmon, Robert S.
Wildflowers of North America in
Full Color

fl.   pg. 178   Pl. 280

Argemone mexicana

Bruggeman, L.
Tropical Plants

fl.   Pl. 30

Argemone mexicana

Burgis, D.S.
Florida Weeds

fl., hab.   p. 9

Argemone mexicana

Hay, Roy
The Color Dictionary of Flowers & Plants

fl.   p 30, Pl. 240

Argemone mexicana

Lemmon, Robert S.
Wildflowers of North America in
Full Color

fl.   pg. 21   Pl. 34

Argemone mexicana

Macoby, Stirling
What Flower is That

fl.   P 46

Argemone mexicana

Moriarty, Audrey
Wild Flowers of Malawi

fl.   Pl. 61; 1

Argemone mexicana

Morley, Brian D.
Wild Flowers of the World

fl. fr.   Pl. 154

Argemone mexicana

O'Gorman, Helen
Mexican Flowering Trees and Plants

fl. hab.   p 127

Argemone mexicana

Perry, Frances
Flowers of the World

fl.   p 224

Argemone mexicana

Polunin, Oleg
Flowers of Europe

fl.   Pl. 29 #270

Argemone mexicana

Royal Hort. Soc.
The Garden
vol 101, No. 1   1976

fl., hab.   p. 49

Argemone mexicana

Tsukamoto, Yotaro
Coloured Illustrations of Garden
Flowers Vol. 10

fl. Illus. 6   p. 2

Argemone munita

Munz, Philip A.
California Spring Wildflowers

fl.   p 40, Pl. 40

Argemone platyceras

Lemmon, Robert S.
Wildflowers of North America in
Full Color

fl.   pg. 68   Pl. 109

Argemone platyceras

Orr, Robert T.
Wildflowers of Western America

fl.   Pl. 52

Argemone platyceras

Royal Hort. Soc.
Journal of the Royal Hort. Soc.
vol 93, No. 7   1968

fl.   p. 290   Pl. 150

Argemone platyceras var.

Addisonia
Vol. 20   1937-38

fl., fr.   p. 49   Pl. 665

Argylia radiata var.

Munoz Pizarro, Carlos
Flores Silvestres de Chile

fl., hab.   Pl. 4

Argyranthemum broussonetii

Bramwell, David
Wild Flowers of the Canary Islands

fl.   Pl. 280

Argyranthemum escarrei

Bramwell, David
Wild Flowers of the Canary Islands

fl.   Pl. 283

Argyranthemum foeniculaceum

Bramwell, David
Wild Flowers of the Canary Islands

fl.   Pl. 281

Argyranthemum frutescens

Bramwell, David
Wild Flowers of the Canary Islands

fl.   Pl. 288 , 289

Argyranthemum gracile

Bramwell, David
Wild Flowers of the Canary Islands

fl.   Pl. 279

Argyranthemum haouarytheum

Bramwell, David
Wild Flowers of the Canary Islands

fl.   Pl. 282

Argyranthemum lidii

Bramwell, David
Wild Flowers of the Canary Islands

fl.   Pl. 286

Argyranthemum ochroleucum

Bramwell, David
Wild Flowers of the Canary Islands

fl.   Pl. 285

Argyranthemum teneriffae

Bramwell, David
Wild Flowers of the Canary Islands

fl.   Pl. 284

Argyranthemum winteri

Bramwell, David
Wild Flowers of the Canary Islands

fl.   Pl. 287

Argyroderma ovale

Lamb, Edgar
Popular Exotic Cacti in Color

fl. hab.   p 31

Argyroderma planum

Flowering Plants of Africa
vol 27   1948-49

fl., hab.   Pl. 1057

Argyroderma sp.

Herre, H.
The Genera of the Mesembryanthemaceae

fl. fr.   p. 83

Argyroderma testiculare

Nicolaisen, Age
Pocket Encyclopedia of Indoor Plants

fl.   Pl. 56

Argyrolobium tomentosum

Flowering Plants of Africa
vol 41   1970-71

fl.   Pl. 1602

Argyroxiphium macrocephalum

Milne, Lorus
Living Plants of the World

fl. hab.   p 237

Argyroxiphium sandwicense

Morley, Brian D.
Wild Flowers of the World

fl.   Pl. 149

Argyroxiphium sandwicense

Pacific Tropical Botanical Garden
Bulletin
vol. 3, No. 4   1973

fl., hab.   cover

Aridaria noctiflora

Herre, H.
The Genera of the Mesembryanthemaceae

fl. fr.   p 85

Ariocarpus fissuratus

Lamb, Edgar
Colorful Cacti of the American Deserts

fl. hab.   Pl. 37, 38

Ariocarpus fissuratus

Lamb, Edgar
The Pocket Encyclopedia of Cacti
and Succulents in Color

fl. hab.   Pl. 3

Ariocarpus fissuratus

Van Laren, A. J.
Cactus

fl.   P 42   Fig. 38

Ariocarpus fissuratus

Weniger, Del
Cacti of the Southwest

fl.   P 26

Ariocarpus kotschoubeyanus

Lamb, Edgar
Popular Exotic Cacti in Color

fl. hab.   p 33

Ariocarpus kotschoubeyanus

Van Laren, A. J.
Cactus

fl.   P 42   Fig. 39

Ariocarpus retusus

Van Laren, A. J.
Cactus

hab   P 42   Fig. 37

Ariocarpus scapharostrus

Cactus & Succulent Society of America
Journal
vol 46, No. 4   1974

fl., hab.   p. 173

Ariocarpus trigonus

Cactus & Succulent Society of America
Journal
vol 46, No. 4   1974

fl.   p. 172

Ariocarpus trigonus
Cactus & Succulent Society of America
Journal
vol 49, No. 3   1977

hab.   p. 122

Ariocarpus trigonus
Kupper, Walter
Cacti

hab. p. 103   Pl. 48

Ariopsis peltata
Hara, Hiroshi, comp.
Photo-Album of Plants of
Eastern Himalaya

fl.   Pl. 24

Arisaema acuminatum
Fleming, Glenn
Wild Flowers of Florida

fl.   p 39

Arisaema atrorubens
Roberts, June Carver
Born in the Spring

fl. fr. hab.   p 87

Arisaema atrorubens
Wharton, Mary E.
A Guide to the Wildflowers & Ferns
of Kentucky

fl. fr. p 74 Pl. 1.1

Arisaema atrorubens var.
Jennings, O. E.
Wild Flowers of Western Pennsylvania
Vol. II

fl. fr.   Pl. 8

Arisaema candidissimum
Curtis's Botanical Magazine
Vol 161   1938

fl.   No. 9549

Arisaema candidissimum
Mathew, Brian
The Larger Bulbs

fl.   p. 64

Arisaema candidissimum
Morley, Brian D.
Wild Flowers of the World

fl.   Pl. 108

Arisaema candidissimum
Perry, Frances
Flowers of the World

fl.   p 32

Arisaema candidissimum
Royal Hort. Soc.
Journal of the Royal Hort. Soc.
vol 93, No. 6   1968

fl.   p. 246   Pl. 142

Arisaema candidissimum
Royal Hort. Soc.
Journal of the Royal Hort. Soc.
vol 95, No. 6   1970

fl.   p. 258   Pl. 152

Arisaema dracontium
Courtenay, Booth
Wildflowers and weeds

fl. p. 3

Arisaema dracontium
Jennings, O. E.
Wild Flowers of Western Pennsylvania
Vol. II

fl. fr. hab.   Pl. 7

Arisaema dracontium
Klimas, John E.
Wildflowers of Eastern America

fl.   Pl. 87

Arisaema dracontium
Walcott, Mary Vaux
North American Wild Flowers

fl. hab.   Pl. 22

Arisaema dracontium
Wharton, Mary E.
A Guide to the Wildflowers & Ferns
of Kentucky

fl. p 75 Pl. 1.2

Arisaema exappendiculatum
Hara, Hiroshi, comp.
Photo-Album of Plants of
Eastern Himalaya

fl.   Pl. 27

Arisaema fimbriatum
Morley, Brian D.
Wild Flowers of the World

fl.   Plate 125

Arisaema japonicum
Royal Hort. Soc.
Journal of the Royal Hort. Soc.
vol 93, No. 6   1968

fl.   p. 246   Pl. 145

Arisaema pradhanii
Curtis's Botanical Magazine
Vol 159   1936

hab.   No. 9425

Arisaema schimperianum
Curtis's Botanical Magazine 1964-65
Vol. 175

fl.   Pl. 474

Arisaema sikokianum
Curtis's Botanical Magazine
Vol. 162   1939

fl.   No. 9589

Arisaema sikokianum
Morley, Brian D.
Wild Flowers of the World

fl.   Pl. 108

Arisaema sikokianum
Royal Hort. Soc.
Journal of the Royal Hort. Soc.
vol 93, No. 6   1968

fl.   p. 246   Pl. 144

Arisaema sikokianum
Royal Hort. Soc.
Journal of the Royal Hort. Soc.
Vol. 94, No. 3   1969

fl.   p. 118   Pl. 63

Arisaema speciosum
Royal Hort. Soc.
Journal of the Royal Hort. Soc.
vol 97 No. 1 1972

fl.   p. 22   Pl. 21

Arisaema stewardsonii
Addisonia, Vol. 23, 1954-59

fl., fr.   p. 49   Pl. 761

Arisaema stewardsonii
Jennings, O. E.
Wild Flowers of Western Pennsylvania
Vol. II

fl. fr.   Pl. 9

Arisaema triphyllum
American Hort. Soc.
American Horticulturist
vol 55, No. 6   1976

fl.   p. 39

Arisaema triphyllum
Batson, Wade T.
Wild Flowers in South Carolina

fl. hab.   p. 27

Arisaema triphyllum
Brown, Clair A.
Wildflowers of Louisiana and
Adjoining States

fl. hab.   p. 8

Arisaema triphyllum

Campbell, Carlos C.
Great Smoky Mountain Wildflowers

fl.   p 77

Arisaema triphyllum

Courtenay, Booth
Wildflowers and weeds

fl. p. 3

Arisaema triphyllum

Dean, Blanche E.
Wildflowers of Alabama and
    Adjoining States

fl.   p 5

Arisaema triphyllum

Duncan, Wilbur H.
Wildflowers of the Southeastern
    United States

fl.   p 239

Arisaema triphyllum

Ferguson, Mary
Wildflowers

fl.   p 55

Arisaema triphyllum

Klute, Jeannette
Woodland Portraits

fl.   Pl. 10

Arisaema triphyllum

Lemmon, Robert S.
Wildflowers of North America in
    Full Color

fl., fr.   p. 220   Pl. 346-47

Arisaema triphyllum

Massachusetts Hort. Soc.
Horticulture
    vol 41, No. 3   1963

fl.   inside backcover

Arisaema triphyllum

Miles, Bebe
Bulbs for the Home Gardener

fl.,fr.   p. 54, 138

Arisaema triphyllum

Moyle, John B.
Northland Wild Flowers

fl. fr. hab.   p 201, Pl. 262, 262a

Arisaema triphyllum

Royal Hort. Soc.
Journal of the Royal Hort. Soc.
    vol 93, No. 6   1968

fl.   p. 246   Pl. 143

Arisaema triphyllum

Walcott, Mary Vaux
North American Wild Flowers
    vol. 5

fl.   Pl. 331

Arisaema Wallichianum

Hara, Hiroshi, comp.
Photo-Album of Plants of
    Eastern Himalaya

fl. hab.   Pl. 25

Arisaema wallichianum

Royal Hort. Soc.
The Garden
    vol 100, No. 8   1975

fl., hab.   p. 358

Arisarum proboscideum

Perry, Frances
Flowers of the World

fl.   p 33

Arisarum vulgare

Megaw, Elektra
Wild Flowers of Cyprus

fl., hab.   Pl. 39

Arisarum vulgare

Polunin, Oleg
Flowers of Europe

fl.   Pl. 183 #1821

Arisarum vulgare

Polunin, Oleg
Flowers of the Mediterranean

fl.   Pl. 221

Arisarum vulgare

Taylor, A. W.
Wildflowers of Spain and Portugal

fl.   p. 46

Aristea alata var.

Moriarty, Audrey
Wild Flowers of Malawi

fl.   Pl. 6; 2

Aristea coerulea

Flowering Plants of Africa
    vol 28   1950-51

fl.   Pl. 1083

Aristea ensifolia

Morley, Brian D.
Wild Flowers of the World

fl.   Pl. 87

Aristea lugens

Flowering Plants of Africa
    vol 43   1974-76

fl.   Pl. 1708

Aristea macrocarpa

Flowering Plants of Africa
    vol 28   1950-51

fl., fr.   Pl. 1084

Aristea thyrsiflora

Royal Hort. Soc.
Journal of the Royal Hort. Soc.
    vol 93, No. 11   1968

fl.   p. 470   Pl. 259

Aristolochia baetica hab.

Taylor, A. W.
Wild Flowers of Spain and Portugal

fl.   p. 35

Aristolochia californica

Munz, Philip A.
California Spring Wildflowers

fl.   p 79, Pl. 61

Aristolochia californica

Orr, Robert T.
Wildflowers of Western America

fl.   Pl. 193

Aristolochia clematitis

Bianchini, Francesco
Health Plants of the World

fl. hab.   p 73

Aristolochia clematitis

Felsko, Elsa
A Book of Wildflowers

fl.   P 13

Aristolochia clematitis

Kleijn, H.
Beauty of the Wild Plant

fl.   p. 52   Pl. 62

Aristolochia clematitis

Lindman, C. A. M.
Nordens Flora, Vol 4

fl. fr.   Pl. 245

Aristolochia clematitis

Martin, W. Keble
The Concise British Flora in Colour

fl. hab.   Pl. 76

Aristolochia clematitis

Perrot, Emile
Les Plantes Medicinales

fl. fr. hab.   p. 17

Aristolochia clematitis

Polunin, Oleg
Flowers of Europe

fl.   Pl. 7 #76

Aristolochia durior

Campbell, Carlos C.
Great Smoky Mountain Wildflowers

fl.   p 21

Aristolochia durior

Hellyer, A.C.L.
Shrubs in Colour

fl.   pg. 14

Aristolochia durior

Huxley, Anthony
Deciduous Garden Trees and Shrubs

fl.   Pl. 281

Aristolochia durior

Klimas, John E.
Wildflowers of Eastern America

fl.   Pl. 300

Aristolochia durior

Wharton, Mary E.
Trees & Shrubs of Kentucky

fl. hab.   p. 68   Pl. 1.33

Aristolochia elegans

Hay, Roy
The Color Dictionary of Flowers & Plants

fl.   p 52, Pl. 410

Aristolochia elegans

Hay, Roy
The Dictionary of House Plants

fl.   Pl. 42

Aristolochia elegans

Kiaer, Eigil
Indoor Plants in Colour

fl.   p. 16

Aristolochia elegans

Mathias, Mildred E.
Color for the Landscape

fl. hab.   p 99

Aristolochia elegans

Perry, Frances
Flowers of the World

fl.   p 37

Aristolochia fimbriata

Addisonia, Vol. 24, 1960-64

fl., fr.   p. 51   Pl. 794

Aristolochia gigantea

Harrison, Richmond E.
Climbers and Trailers

fl.   p 20   Pl. 11

Aristolochia gigantea

Morley, Brian D.
Wild Flowers of the World

fl.   Pl. 170

Aristolochia grandiflora

Macoby, Stirling
What Flower is That

fl.   P. 46

Aristolochia grandiflora

O'Gorman, Helen
Mexican Flowering Trees and Plants

fl.   p 87

Aristolochia grandiflora

Wit, H. C. D. de
Plants of the World;
   The Higher Plants, Vol I

fl.   p 64, Pl. 36

Aristolochia Griffithii

Hara, Hiroshi, comp.
Photo-Album of Plants of
   Eastern Himalaya

fl.   Pl. 74

Aristolochia hirta

Goulimis, Constantine N.
Wild Flowers of Greece

fl.   p 3

Aristolochia kewensis

Bruggeman, L.
Tropical Plants

fl.   pl. 4

Aristolochia lindneri

Curtis's Botanical Magazine
Vol 162   1939

fl.   No. 9596

Aristolochia Longa

Goulimis, Constantine N.
Wild Flowers of Greece

fl., hab.   p. 5

Aristolochia macrophylla

Dean, Blanche E.
Wildflowers of Alabama and
   Adjoining States

fl.   p 49

Aristolochia Nakaoi

Hara, Hiroshi, comp.
Photo-Album of Plants of
   Eastern Himalaya

fl.   Pl. 73

Aristolochia ringens

Wit, H. C. D. de
Plants of the World;
   The Higher Plants, Vol I

fl.   p 64, Pl. 37

Aristolochia rotunda

Goulinis, Constantine N.
Wild Flowers of Greece

fl.   p 5

Aristolochia rotunda

Morley, Brian D.
Wild Flowers of the World

fl.,hab.   Pl. 31

Aristolochia rotunda

Polunin, Oleg
Flowers of Europe

fl.   Pl. 8 #77

Aristolochia rotunda

Polunin, Oleg
Flowers of the Mediterranean

fl.   Pl. 9

Aristolochia saccata

Morley, Brian D.
Wild Flowers of the World

fl.   Pl. 110

Aristolochia saccata

Perry, Frances
Flowers of the World

fl.   p 36

Aristolochia salpinx

Perry, Frances
Flowers of the World

fl.   p 37

Aristolochia sempervirens

Addisonia, Vol. 22, 1943-47

fl.   p. 57   Pl. 733

Aristolochia sempervirens

Polunin, Oleg
Flowers of the Mediterranean

fl.   Pl. 8

Aristolochia serpentaria

Dean, Blanche E.
Wildflowers of Alabama and
   Adjoining States

fl.   p 47

Aristolochia serpentaria

Jennings, O. E.
Wild Flowers of Western Pennsylvania
   Vol. II

fl. hab.   Pl. 48

Aristolochia sp.

Royal Hort. Soc.
Journal of the Royal Hort. Soc.
   vol 91, No. 8   1966

fl.   p. 338   Pl. 188

Aristolochia trilobata

Kiaer, Eigil
Indoor Plants in Colour

fl.   p. 16

Aristotelia peduncularis

Curtis, Winifred
The Endemic Flora of Tasmania
   Vol. 1

fl. fr.   Pl. 12

Armatocereus matucanensis

Backeberg, Curt
Cactus Lexicon

fl.   p 554

Armeria alpina

Barneby, T. P.
European Alpine Flowers in Colour

fl.   Pl. 61, 6

Armeria alpina

Kohlhaupt, Paula
Fleurs des Alpages Vol. 1

fl.   P. 69

Armeria alpina

Tosco, Uberto
The World of Mountain Flowers

fl. hab.   p 65

Armeria caespitosa var.

Hay, Roy
The Color Citionary of Flowers &
   Plants

fl.   p. 2   Pl. 14

Armeria fasciculata

Polunin, Oleg
Flowers of Europe

fl. Pl. 94 #976

Armeria maritima

Ary, S.
The Oxford Book of Wildflowers

fl.   p 126, Pl. 5

Armeria maritima

Clark, Lewis J.
Wild Flowers of British Columbia

fl.   p 434

Armeria maritima

Crockett, James Underwood
Lawns & Ground Covers

fl.   p. 122

Armeria maritima

Kleijn, H.
Beauty of the Wild Plant

fl.   p. 81   Pl. 108

Armeria maritima

Lindman, C. A. M.
Nordens Flora, Vol 7

fl. hab.   Pl. 470

Armeria maritima

Macoby, Stirling
What Flower is That

fl.   P 46

Armeria maritima

Perry, Frances
Flowers of the World

fl.   p 232

Armeria maritima

Polunin, Oleg
Flowers of Europe

fl.   Pl. 94 #975

Armeria maritima var.

Hay, Roy
The Color Dictionary of Flowers &
   Plants

fl.   p. 2   Pl. 15

Armeria maritima var.

Huxley, Anthony
Garden Perennials and Water Plants

fl.   Pl. 43

Armeria maritima var.

Lemmon, Robert S.
Wildflowers of North America in
   Full Color

fl.   pg. 37   Pl. 63

Armeria maritima var.

Martin, W. Keble
The Concise British Flora in Colour

fl. hab.   Pl. 56

Armeria vulgaris

Felsko, Elsa
A Book of Wildflowers

fl.   P 85

Armoracia lapathifolia

Pond, Barbara
A Sampler of Wayside Herbs

fl.   Pl. III

Armoracia rusticana

Courtenay, Booth
Wildflowers & Weeds

fl.   p 19

Armoracia rusticana

Masefield, G. B.
The Oxford Book of Food Plants

hab.   Pl. 135, 4

Arnebia benthamii

Morley, Brian D.
Wild Flowers of the World

fl. Plate 105

Arnebia echioides

Hay, Roy
The Color Dictionary of Flowers & Plants

fl.   p 122, Pl. 975

Arnebia echioides

Perry, Frances
Flowers of the World

fl. p 51

Arnebia echioides

Royal Hort. Soc.
Journal of the Royal Hort. Soc.
   vol 87, No. 1   1962

fl.   p. 22   Pl. 9

Arnica acaulis
Batson, Wade T.
Wild Flowers in South Carolina
fl.   p 124

Arnica alpina
Alaska-Yukon Wild Flowers Guide
fl. hab.   p 50, 179

Arnica alpina
Clark, Lewis J.
Wild Flowers of British Columbia
fl.   p 519

Arnica alpina
Heller, Christine
Wild Flowers of Alaska
fl.   Pl. 57

Arnica alpina
Lindman, C. A. M.
Nordens Flora, Vol 9
fl. hab.   Pl. 618

Arnica, Broadleaf
see
Arnica latifolia

Arnica cordifolia
Clark, Lewis J.
Wild Flowers of British Columbia
fl.   p 522

Arnica cordifolia
Lemmon, Robert S.
Wildflowers of North America in Full Color
fl.   pg. 160   pl. 256

Arnica cordifolia
Orr, Robert T.
Wildflowers of Western America
fl.   Pl. 81

Arnica cordifolia
Porsild, A. E.
Rocky Mountain Wild Flowers
fl.   p 395

Arnica frigida
Alaska-Yukon Wild Flower Guide
fl. hab.   o 176

Arnica fulgens
Clark, Lewis J.
Wild Flowers of British Columbia
fl.   p 519

Arnica, Heartleaf
see
Arnica cordifolia

Arnica, Lake Louise
see
Arnica louiseana

Arnica latifolia
Alaska-Yukon Wild Flower Guide
fl.   p 178

Arnica latifolia
Clark, Lewis J.
Wild Flowers of British Columbia
fl.   p 523

Arnica latifolia
Heller, Christine
Wild Flowers of Alaska
fl.   Pl. 59

Arnica latifolia
Taylor, Ronald J.
Mountain Wild Flowers
fl. hab.   p 80

Arnica lessingii
Heller, Christine
Wild Flowers of Alaska
fl.   Pl. 58

Arnica lonchophylla
Porsild, A. E.
Rocky Mountain Wild Flowers
fl. hab.   p 397

Arnica louiseana
Walcott, Mary Vaux
North American Wild Flowers
fl. hab.   Pl. 10

Arnica louiseana var.
Porsild, A. E.
Rocky Mountain Wild Flowers
fl. hab.   p 399

Arnica mollis
Clark, Lewis J.
Wild Flowers of British Columbia
fl. hab.   p 523

Arnica mollis
Porsild, A. E.
Rocky Mountain Wild Flowers
fl. hab.   p 397

Arnica montana
Barneby, T. P.
European Alpine Flowers in Colour
fl. hab.   Pl. 95, 2

Arnica montana
Bianchini, Francesco
Health Plants of the World
fl.   p 169

Arnica montana
Color Treasury of Herbs & Other Medicinal Plants
fl.   p 55   Pl. 86

Arnica montana
Felsko, Elsa
A Book of Wild Flowers 2nd Ser.
fl.   p. 18

Arnica montana
Kleijn, H.
Beauty of the Wild Plant
fl.   p. 100   Pl. 145

Arnica montana
Kohlhaupt, Paula
Fleurs des Alpages Vol. 1
fl.   P. 122

Arnica montana
Lindman, C. A. M.
Nordens Flora, Vol 9
fl.   Pl. 617

Arnica montana
Perrot, Emile
Les Plantes Medicinales
fl. fr. hab.   p 19

Arnica montana
Polunin, Oleg
Flowers of Europe
fl.   Pl. 151 #1446

Arnica montana

Robert, Paul A
Alpine Flowers

fl.   Pl. 11

Arnica nevadensis

Taylor, Ronald J.
Mountain Wild Flowers

fl. hab.   p 80

Arnica tomentosa

Walcott, Mary Vaux
North American Wild Flowers
        vol. 5

fl.   Pl. 348

Arnica unalaschcensis

Alaska-Yukon Wild Flower Guide

fl. hab.   p 177

Arnica, Woolly

    see

Arnica tomentosa

Arnoseris minima

Lindman, C. A. M.
Nordens Flora, Vol 10

fl. hab.   Pl. 637

Arnoseris minima

Martin, W. Keble
The Concise British Flora in Colour

fl. hab.   Pl. 50

Aroid, Snake

    see

Amorphophallus variabilis

Aronia arbutifolia

Batson, Wade T.
Wild Flowers in South Carolina

fl.   P 55

Aronia arbutifolia

Walcott, Mary Vaux
North American Wild Flowers

fl. fr.   Pl. 31, 31a

Aronia melanocarpa

Royal Hort. Soc.
Journal of the Royal Hort. Soc.
        vol 98, No. 11   1973

hab.   cover

Aronia melanocarpa

Wharton, Mary E.
Trees & Shrubs of Kentucky

fl.   p. 82   Pl. 2.20
fr.   p. 134   Pl. 2.24

Arpophyllum giganteum

Ospina, Mariano
Orquideas de las americas

fl.   Pl. 43

Arrabidaea magnifica

Bruggeman, L.
Tropical Plants

fl.   Pl. 3

Arrhenatherum elatius

Lindman, C. A. M.
Nordens Flora, Vol 2

fr. hab.   Pl. 90

Arrhenatherum pubescens

Lindman, C. A. M.
Nordens Flora, Vol 2

fr. hab.   Pl. 91

Arrojadoa aureispina

Backeberg, Curt
Cactus Lexicon

fl.   p 797

Arrojadoa canudosensis

Cactus & Succulent Soc. of America
Journal
        vol 44, No. 3   1972

fl.   p. 111

Arrojadoa dinae

Backeberg, Curt
Cactus Lexicon

fr. hab.   p 798

Arrojadoa eriocaulis

Backeberg, Curt
Cactus Lexicon

fl.   p 798

Arrojadoa penicillata

Backeberg, Curt
Cactus Lexicon

fl.   p 556

Arrojadoa rhodantha

Kupper, Walter
Cacti

fl., hab.   p. 69   Pl. 31

Arrow-Grass

    see

Triglochin

Arrow-head

    see

Sagittaria

Arrowhead, Arum

    see

Sagittaria cuneata

Arrow-head, broad-leaved

    see

Sagittaria latifolia

Arrowhead, Grass-leaf

    see

Sagittaria graminea

Arrowroot

    see

Maranta arundinacea

Arrow-Wood

    see

Viburnum dentatum

Arrowwood, Ash's

    see

Viburnum ashei

Artemisia abrotanum

Hellyer, A.G.L.
Shrubs in Colour

hab.   pg. 14

Artemisia abrotanum

Masefield, G. B.
The Oxford Book of Food Plants

fl.   Pl. 145, 2

Artemisia abrotanum

Perrot, Emile
Les Plantes Medicinales

fl. hab.   p 104

Artemisia absinthium
Ary, S.
The Oxford Book of Wildflowers
fl.   p 32, Pl. 3

Artemisia absinthium
Bianchini, Francesco
Health Plants of the World
fl. hab.   p 55

Artemisia absinthium
Color Treasury of Herbs & Other
Medicinal Plants
fl.   p 57   Pl. 89

Artemisia absinthium
Hvass, Elsie
Plants That Feed and Serve Us
fl.   p 85, Pl. 180

Artemisia absinthium
Lindman, C. A. M.
Nordens Flora, Vol 9
fl.   Pl. 612

Artemisia absinthium
Martin, W. Keble
The Concise British Flora in Colour
fl. hab.   Pl. 46

Artemisia absinthium
Masefield, G. B.
The Oxford Book of Food Plants
fl.   Pl. 137, 4

Artemisia absinthium
Massachusetts Hort. Soc.
Horticulture
    vol. 54, No. 3   1976
fl.   p. 55

Artemisia absinthium
Moyle, John B.
Northland Wild Flowers
fl. hab.   p 168, Pl. 203

Artemisia absinthium
Perrot, Emile
Les Plantes Medicinales
fl. fr. hab.   p 1

Artemisia absinthium
Polunin, Oleg
Flowers of Europe
fl.   Pl. 149 #1436

Artemisia absinthium var.
Hay, Roy
The Color Dictionary of Flowers &
Plants
hab.   p. 122   Pl. 976

Artemisia albula
Everett, T.H., ed.
New Illustrated Encyclopedia of
    Gardening vol. 1
hab.   p. 70

Artemisia arborescens
Gault, S. Millar
The Color Dictionary of Shrubs
hab.   Pl. 11

Artemisia arborescens
Hay, Roy
The Color Dictionary of Flowers & Plants
hab.   p 182, Pl. 1452

Artemisia arborescens var.
Royal Hort. Soc.
Journal of the Royal Hort. Soc.
vol 89, No. 11   1964
hab.   p. 462   Pl. 179 -80

Artemisia arctica
Heller, Christine
Wild Flowers of Alaska
fl.   Pl. 307

Artemisia borealis
Morley, Brian D.
Wild Flowers of the World
fl.   Pl. 6F

Artemisia campestris
Courtenay, Booth
Wildflowers & Weeds
fl.   p 113

Artemisia campestris
Lindman, C. A. M.
Nordens Flora, Vol 9
fl. hab.   Pl. 609B

Artemisia campestris
Martin, W. Keble
The Concise British Flora in Colour
fl. hab.   Pl. 46

Artemisia campestris var.
Alaska-Yukon Wild Flower Guide
fl.   p 175

Artemisia capillaris
Kimura, Koiti
Japanese Medicinal Plants, Vol I
fl.   p 98

Artemisia capillaris
Takatori, Jisuke
Color Atlas of Medicinal Plants of Japan
fl.   Fig. 2A

Artemisia cina
Hvass, Elsie
Plants That Feed and Serve Us
fl.   p 85, Pl. 179

Artemisia discolor
Walcott, Mary Vaux
North American Wild Flowers
    vol. 4
fl.   Pl. 288

Artemisia dracunculus
Masefield, G. B.
The Oxford Book of Food Plants
fl.   Pl. 145, 1

Artemisia dracunculus
Perrot, Emile
Les Plantes Medicinales
fl. fr. hab.   p. 80

Artemisia dracunculus
Rosengarten, Frederic, Jr.
The Book of Spices
fl.   p 428

Artemisia filifolia
Welsh, Stanley L
Flowers of the Canyon Country
hab.   p 40

Artemisia frigida
Clark, Lewis J.
Wild Flowers of British Columbia
fl.   p 530

Artemisia glacialis
Barneby, T. P.
European Alpine Flowers in Colour
fl. hab.   Pl. 89, 6

Artemisia glacialis
Kohlhaupt, Paula
Fleurs des Alpages Vol. 1
fl.   P. 111

Artemisia glacialis

Perrot, Emile
Les Plantes Medicinales

fl. hab.   p 104

Artemisia glacialis

Tosco, Uberto
The World of Mountain Flowers

fl.   p 78

Artemisia glacialis var.

Tosco, Uberto
The World of Mountain Flowers

hab.   p 78

Artemisia globularia

Alaska-Yukon Wild Flower Guide

fl. hab.   p 173

Artemisia kurramensis

Kariyone, Tatsuo
Atlas of Medicinal Plants

fl.   Pl. 26

Artemisia kurramensis

Takatori, Jisuke
Color Atlas of Medicinal Plants of Japan

fl.   Fig. 2B

Artemisia laciniata

Lindman, C. A. M.
Nordens Flora, Vol 9

fl. hab.   Pl. 610

Artemisia lactiflora

Bruggeman, L.
Tropical Plants

fl.   Pl. 72

Artemisia lactiflora

Hay, Roy
The Color Dictionary of Flowers & Plants

fl.   p 123, Pl. 977

Artemisia lactiflora

Huxley, Anthony
Garden Perennials and Water Plants

fl.   Pl. 46

Artemisia laxa

Barneby, T. P.
European Alpine Flowers in Colour

fl. hab.   Pl. 89, 5

Artemisia ludoviciana

Courtenay, Booth
Wildflowers & Weeds

fl.   p 113

Artemisia ludoviciana

Moyle, John B.
Northland Wild Flowers

fl. hab.   p 167, Pl. 202

Artemisia maritima

Ary, S.
The Oxford Book of Wildflowers

fl.   p 60, Pl. 6

Artemisia maritima

Lindman, C. A. M.
Nordens Flora, Vol 9

fl. hab.   Pl. 607

Artemisia maritima

Martin, W. Keble
The Concise British Flora in Colour

fl. hab.   Pl. 46

Artemisia maritima

Perrot, Emile
Les Plantes Medicinales

fl. fr. hab.   p 1

Artemisia maritima var.

Hay, Roy
The Color Dictionary of Flowers & Plants

hab.   p. 123   Pl. 978

Artemisia Michauxiana

Porsild, A. E.
Rocky Mountain Wild Flowers

fl. hab.   p 401

Artemisia monosperma

Feinbrun-Dothan, Naomi
Wild Plants in the Land of Israel

fl.   p 132

Artemisia montana

Kimura, Koiti
Japanese Medicinal Plants, Vol II

fl.   p 235

Artemisia mutellina

Color Treasury of Herbs & Other Medicinal Plants

hab. p 58   Pl. 91

Artemisia mutellina

Kohlhaupt, Paula
Fleurs des Alpages Vol. 2

fl.   P. 112

Artemisia mutellina

Perrot, Emile
Les Plantes Medicinales

fl. fr. hab.   p. 104

Artemisia norvegica

Lindman, C. A. M.
Nordens Flora, Vol 9

fl. hab.   Pl. 611

Artemisia rupestris

Lindman, C. A. M.
Nordens Flora, Vol 9

fl. hab.   Pl. 609A

Artemisia schmidtiana var.

Crockett, James Underwood
Lawns & Ground Covers

hab.   p. 122

Artemisia Schmidtiana var.

Everett, T.H., ed.
New Illustrated Encyclopedia of Gardening vol. 1

hab.   p. 70

Artemisia senjavinensis

Alaska-Yukon Wild Flower Guide

fl. hab.   p 174

Artemisia senjavinensis

Heller, Christine
Wild Flowers of Alaska

fl.   Pl. 60

Artemisia, Silver King

see

Artemisia albula

Artemisia spicata

Perrot, Emile
Les Plantes Medicinales

hab.   p 104

Artemisia suksdorfii

Clark, Lewis J.
Wild Flowers of British Columbia

fl.   p 530

Artemisia tridentata

Clark, Lewis J.
Wild Flowers of British Columbia

fl.   p 530

Artemisia tridentata

Coyle, Jeanette
A Field Guide to the Common and
   Interesting Plants of Baja California

hab.   p 169

Artemisia villarsii

Perrot, Emile
Les Plantes Medicinales

fl. hab.   p. 104

Artemisia vulgaris

Ary, S.
The Oxford Book of Wildflowers

fl.   p 60, Pl. 5

Artemisia vulgaris

Felsko, Elsa
A Book of Wildflowers

fl.   P 39

Artemisia vulgaris

Lindman, C. A. M.
Nordens Flora, Vol 9

fl.   Pl. 608

Artemisia vulgaris

Martin, W. Keble
The Concise British Flora in Colour

fl. hab.   Pl. 46

Artemisia vulgaris

Perrot, Emile
Les Plantes Medicinales

fl. fr. hab.   p 18

Artemisia vulgaris

Pond, Barbara
A Sampler of Wayside Herbs

fl.   Pl. XXVI

Arthrocnemum fruticosum

Polunin, Oleg
Flowers of Europe

hab.   Pl. 10 #113

Arthropodium cirrhatum

Harvey, Norman B.
New Zealand Botanical Paintings

fl.   p. 68   Pl. 31

Arthropodium cirrhatum

Menninger, Edwin A.
Flowering Vines of the World

fl.   Pl. 113

Arthropodium minus

Blombery, Alec M.
What Wildflower is That

fl.   p. 51   Pl. 106

Artichoke

see

Cynara scolymus

Artichoke, Globe

see

Cynara cardunculus

Artichoke, globe

see

Cynara scolymus

Artichoke, Jerusalem

see

Helianthus tuberosus

Artichoke, Wild

see

Cynara cardunculus var.

Artillery Plant

see

Pilea microphylla

Artillery plant

see

Pilea nummularifolia

Artocarpus altilis

Milne, Lorus
Living Plants of the World

fr.   p 59

Artocarpus altilis

Pacific Tropical Botanical Garden
Bulletin
   vol 6, No. 1   1976

fr.   cover

Artocarpus communis

Masefield, G. B.
The Oxford Book of Food Plants

fl. fr.   Pl. 115, 2

Artocarpus communis

Wit, H. C. D. de
Plants of the World;
   The Higher Plants, Vol I

fr.   p 104, Pl. 53

Artocarpus heterophyllus

Edlin, Herbert
The Illustrated Encyclopedia
   of Trees

fr.   p 220

Artocarpus incisus

Hvass, Elsie
Plants That Feed and Serve Us

fl. fr.   p 12, Pl. 15

Arum, Arrow

see

Peltandra virginica

Arum, bog

see

Calla palustris

Arum creticum

Curtis's Botanical Magazine
Vol 167   1950

fl.   101

Arum creticum

Hay, Roy
The Color Dictionary of Flowers & Plants

fl.   p 52, Pl. 411

Arum creticum

Hay, Roy
The Dictionary of House Plants

fl.   Pl. 43

Arum creticum

Royal Hort. Soc.
Journal of the Royal Hort. Soc.
   vol 97, No. 11   1972

fl.   p. 484   Pl. 238

Arum dioscoridis

Megaw, Elektra
Wild Flowers of Cyprus

fl.   Pl. 38

Arum dioscoridis

Polunin, Oleg
Flowers of the Mediterranean

fl.   Pl. 224

Arum, Dragon

see

Dracunculus vulgaris

Arum italicum

American Hort. Soc.
American Horticulturist
    vol 54, No. 1   1975

fl., fr., hab.   p. 15

Arum italicum

Macoby, Stirling
What Flower is That

fl.   P 47

Arum italicum

Martin, W. Keble
The Concise British Flora in Colour

fl.   Pl. 88

Arum italicum

Polunin, Oleg
Flowers of Europe

fl.   Pl. 183 #1818

Arum italicum

Polunin, Oleg
Flowers of the Mediterranean

fl.   Pl. 220

Arum italicum var.

Hay, Roy
The Color Dictionary of Flowers
   & Plants

fr.   p. 85   Pl. 676

Arum italicum var.

Royal Hort. Soc.
Journal of the Royal Hort. Soc.
    vol 99, No. 2   1974

hab., fr.   p. 72   Pl. 35,36

Arum italicum var.

Royal Hort. Soc.
The Garden
   vol 100, No. 12   1975

fr.   p. 608

Arum, ivy

see

Scindapsus aureus

Arum, lizard

see

Sauromatum guttatum

Arum maculatum

Ary, S.
The Oxford Book of Wildflowers

fl. fr.   p 62, Pl. 3

Arum maculatum

Color Treasury of Herbs & Other
   Medicinal Plants

fr.   p 21   Pl. 10

Arum maculatum

Felsko, Elsa
A Book of Wildflowers

fl.   P 131

Arum maculatum

Kleijn, H.
Beauty of the Wild Plant

fl.   p. 44   pl. 47, 47a

Arum maculatum

Lindman, C. A. M.
Nordens Flora, Vol 1

fr. hab.   Pl. 33

Arum maculatum

Martin, W. Keble
The Concise British Flora in Colour

fl. hab.   Pl. 88

Arum maculatum

Perrot, Emile
Les Plantes Medicinales

fl. fr. hab.   p 3

Arum palaestinum

Macoby, Stirling
What Flower is That

fl.   P 46

Arum, Pink

see

Zantedeschia rehmannii

Arum, Virginia

see

Peltandra virginica

Arum, Water

see

Calla palustris

Arum, White

see

Peltandra sagittaefolia

Aruncus dioicus

Campbell, Carlos C.
Great Smoky Mountain Wildflowers

fl.   p 79

Aruncus dioicus

Courtenay, Booth
Wildflowers & Weeds

fl.   p 38

Aruncus dioicus

Dean, Blanche E.
Wildflowers of Alabama and
    Adjoining States

fl.   p 83

Aruncus dioicus

Duncan, Wilbur H.
Wildflowers of the Southeastern
    United States

fl.   p 63

Aruncus dioicus

Klimas, John E.
Wildflowers of Eastern America

fl.   Pl. 48

Aruncus dioicus

Polunin, Oleg
Flowers of Europe

fl.   Pl. 44 #420

Aruncus dioicus

Wharton, Mary E.
A Guide to the Wildflowers & Ferns
    of Kentucky

fl.   p 165   Pl. 2.37

Aruncus sylvester

Alaska-Yukon Wild Flowers Guide

fl.   p 71

Aruncus sylvester

Barneby, T. P.
European Alpine Flowers in Colour

fl.   Pl. 36; 4

Aruncus sylvester

Clark, Lewis J.
Wild Flowers of British Columbia

fl.   p 227

Aruncus sylvester

Hay, Roy
The Color Dictionary of Flowers & Plants

fl.   p 123, Pl. 979

Aruncus sylvester

Heller, Christine
Wild Flowers of Alaska

fl.   Pl. 154

Aruncus sylvester

Huxley, Anthony
Garden Perennials and Water Plants

fl.   Pl. 47

Aruncus sylvester

Kleijn, H.
The Beauty of the Wild Plant

fl.   p. 88   Pl. 120

Aruncus sylvester

Taylor, Ronald J.
Mountain Wild Flowers

fl. hab.   p 35

Aruncus sylvester var.

Kleijn, H.
Beauty of the Wild Plant

fl.   p. 88   Pl. 120

Arundina bambusifolia

Luer, Carlyle A.
The Native Orchids of the United
    States and Canada

fl. hab.   Pl. 78; 2-3

Arundina chinensis

Hu, Shiu-ying
The Genera of Orchidaceae in
    Hong Kong

fl.   p 122

Arundina chinensis

Walden, Beryl M.
Wild Flowers of Hong Kong

fl.   Pl. 57 (152)

Arundina graminifolia

Kamemoto, Haruyuki
Beautiful Thai Orchid Species

fl.   p 121

Arundina speciosa

Bruggeman, L.
Tropical Plants

fl.   Pl. 73

Arundinaria japonica

Hellyer, A.G.L.
Shrubs in Colour

hab.   p. 14

Arundinaria japonica

Huxley, Anthony
Evergreen Garden Trees and Shrubs

hab.   Pl. 164

Arundinaria japonica

Polunin, Oleg
Trees and Bushes of Europe

hab.   p 182

Arundinaria murieliae

Huxley, Anthony
Evergreen Garden Trees and Shrubs

hab.   Pl. 165

Arundinaria nitida

Hellyer, A. G. L.
Shrubs in Colour

hab.   p. 14

Arundinaria nitida

Huxley, Anthony
Evergreen Garden Trees and Shrubs

hab.   Pl. 166

Arundinaria variegata

Gault, S. Millar
The Color Dictionary of Shrbus

hab.   Pl. 12

Arundo donax

Batson, Wade T.
Wild Flowers in South Carolina

fl. hab.   p 24

Arundo donax

Coyle, Jeanette
A Field Guide to the Common and
    Interesting Plants of Baja California

hab.   p 47

Arundo donax

Polunin, Oleg
Trees and Bushes of Europe

fl. fr. hab.   p 182

Arundo donax var.

Macoby, Stirling
What Flower is That

hab.   p. 47

Asarabacca

see

Asarum europaeum

Asarabaca, Canadian

see

Asarum candense

Asarina procumbens

Polunin, Oleg
Flowers of Europe

fl.   Pl. 121 #1200

Asarina scandens

Perry, Frances
Flowers of the World

fl.   p 277

Asarum acuminatum

Campbell, Carlos C.
Great Smoky Mountain Wildflowers

fl. hab.   p 17

Asarum arifolium

Campbell, Carlos C.
Great Smoky Mountain Wildflowers

fl. hab.   p 51

Asarum arifolium

Royal Hort. Soc.
The Garden
    vol. 101, No. 6   1976

hab.   p. 335

Asarum arifolium

Wharton, Mary E.
A Guide to the Wildflowers & Ferns
    of Kentucky

fl.   p 275   Pl. 2.2

Asarum canadense

Courtenay, Booth
Wildflowers & Weeds

fl.   p 79

Asarum canadense

Dean, Blanche E.
Wildflowers of Alabama and
    Adjoining States

fl. hab.   p 49

Asarum canadense

Duncan, Wilbur H.
Wildflowers of the Southeastern
    United States

fl. hab.   p 25

Asarum canadense

Ferguson, Mary
Wildflowers

fl. hab.   p 145

Asarum canadense

Klimas, John E.
Wildflowers of Eastern America

fl.   Pl. 297

Asarum canadense

Klute, Jeannette
Woodland Portraits

fl. hab.   Pl. 11

Asarum canadense

Lemmon, Robert S.
Wildflowers of North America in
    Full Color

fl.   p. 246   pl. 386

Asarum canadense

Moyle, John B.
Northland Wild Flowers

hab.   p 106, Pl. 100

Asarum canadense

Roberts, June Carver
Born in the Spring

fl. hab.   p 57

Asarum canadense

Walcott, Mary Vaux
North American Wild Flowers

fl. hab.   Pl. 127

Asarum canadense

Wharton, Mary E.
A Guide to the Wildflowers & Ferns
    of Kentucky

fl. p 275 Pl. 2.1

Asarum canadense var.

Jennings, O. E.
Wild Flowers of Western Pennsylvania
    Vol. II

fl.   Pl. 47

Asarum caudatum

Clark, Lewis J.
Wild Flowers of British Columbia

fl.   p 111

Asarum caudatum

Crockett, James Underwood
Lawns & Ground Covers

fl.   p. 123

Asarum caudatum

Fries, Mary A.
Wildflowers of Mount Ranier and
the Cascades

fl.   P 33

Asarum caudatum

Morley, Brian D.
Wild Flowers of the World

fl.   Pl. 150

Asarum caudatum

Orr, Robert T.
Wildflowers of Western America

fl.   Pl. 287

Asarum caudatum

Perry, Frances
Flowers of the World

fl.   p 36

Asarum caudatum

Taylor, Ronald J.
Mountain Wild Flowers

fl. hab.   p 34

Asarum europaeum

Bianchini, Francesco
Health Plants of the World

hab.   p 51

Asarum europaeum

Felsko, Elsa
A Book of Wild Flowers 2nd Ser.

fl. hab.   p. 73

Asarum europaeum

Huxley, Anthony
Garden Perennials and Water Plants

hab.   Pl. 45

Asarum eupopaeum

Martin, W. Keble
The Concise British Flora in Colour

fl. hab.   Pl. 76

Asarum europaeum

Perrot, Emile
Les Plantes Medicinales

fl. fr. hab.   p. 17

Asarum europaeum

Polunin, Oleg
Flowers of Europe

fl.   Pl. 7 #74

Asarum shuttleworthii

American Hort. Soc.
American Horticulturist
    vol 55, No. 6   1976

hab.   p. 14

Asarum sieboldii

Kariyone, Tatsuo
Atlas of Medicinal Plants

fl., hab.   Pl. 118

Asarum virginica

Royal Hort. Soc.
The Garden
    vol 101, No. 6   1976

hab.   p. 335

Asclepias amplexicaulis

Batson, Wade T.
Wild Flowers in South Carolina

fl. hab.   p 95

Asclepias amplexicaulis

Courtenay, Booth
Wildflowers & Weeds

fl.   p 28

Asclepias amplexicaulis

Dean, Blanche E.
Wildflowers of Alabama and
    Adjoining States

fl.   p 139

Asclepias amplexicaulis

Duncan, Wilbur H.
Wildflowers of the Southeastern
    United States

fl.   p 129

Asclepias asperula

Welsh, Stanley L.
Flowers of the Canyon Country

fl.   p 45

Asclepias connivens

Duncan, Wilbur H.
Wildflowers of the Southeastern
    United States

fl.   p 133

Asclepias cordifolia

Orr, Robert T.
Wildflowers of Western America

fl.   Pl. 250

Asclepias curassavica

Bruggeman, L.
Tropical Plants

fl.   Pl. 74

Asclepias curassavica

Crockett, James Underwood
Annuals

fl.   P. 96

Asclepias curassavica

O'Gorman, Helen
Mexican Flowering Trees and Plants

fl. hab.   p 129

Asclepias curassavica

Tsukamoto, Yotaro
Coloured Illustrations of
    Garden Flowers   Vol. 9

fl.   p. 31

Asclepias curassavica

Walden, Beryl M.
Wild Flowers of Hong Kong

fl.   Pl. 2(5)

Asclepias curassavica

Wit, H. C. D. de
Plants of the World;
    The Higher Plants, Vol II

fl.   p 97, Pl. 42

Asclepias curassavica var.

Hay, Roy
The Dictionary of House Plants

fl.   Pl. 44

Asclepias engelmanniana

Orr, Robert T.
Wildflowers of Western America

fl.   Pl. 57

Asclepias exaltata

Courtenay, Booth
Wildflowers & Weeds

fl.   p 27

Asclepias exaltata

Duncan, Wilbur H.
Wildflowers of the Southeastern
    United States

fl.   p 131

Asclepias exaltata

Klimas, John E.
Wildflowers of Eastern America

fl.   Pl. 66

Asclepias exaltata

Moyle, John B.
Northland Wild Flowers

fl. hab.   p 128, Pl. 137

Asclepias feayi

Fleming, Glenn
Wild Flowers of Florida

fl.   p 34

Asclepias fruticosa

Macoby, Stirling
What Flower is That

fr.   P 47

Asclepias hirtella

Courtenay, Booth
Wildflowers & Weeds

fl.   p 27

Asclepias humistrata

Dean, Blanche E.
Wildflowers of Alabama and
    Adjoining States

fl.   p 139

Asclepias humistrata

Duncan, Wilbur H.
Wildflowers of the Southeastern
    United States

fl.   p 135

Asclepias humistrata

Lemmon, Robert S.
Wildflowers of North America in
    Full Color

fl.   pg. 41   pl. 68

Asclepias incarnata

Brown, Clair A.
Wildflowers of Louisiana and
    Adjoining States

fl. hab.   p 142

Asclepias incarnata

Courtenay, Booth
Wildflowers & Weeds

Fl.   p 28

Asclepias incarnata

Duncan, Wilbur H.
Wildflowers of the Southeastern
    United States

fl.   p 133

Asclepias incarnata

Jennings, O. E.
Wild Flowers of Western Pennsylvania
    Vol. II

fl.   Pl. 118

Asclepias incarnata

Massachusetts Hort. Soc.
Horticulture
    vol 53, No. 9   1975

fl.   p. 32

Asclepias incarnata

Moyle, John B.
Northland Wild Flowers

fl. hab.   p 127, Pl. 134

Asclepias incarnata

Wharton, Mary E.
A Guide to the Wildflowers & Ferns
    of Kentucky

fl.   p 130   Pl. 1.70

Asclepias lanceolata

Brown, Clair A.
Wildflowers of Louisiana and
    Adjoining States

fl.   p 143

Asclepias lanceolata

Dean, Blanche E.
Wildflowers of Alabama and
    Adjoining States

fl.   p 139

Asclepias lanceolata

Duncan, Wilbur H.
Wildflowers of the Southeastern
    United States

fl.   p 129

Asclepias lanuginosa

Courtenay, Booth
Wildflowers & Weeds

fl.   p. 27

Asclepias michauxii

Duncan, Wilbur H.
Wildflowers of the Southeastern
    United States

fl.   p 131

Asclepias obovata

Duncan, Wilbur H.
Wildflowers of the Southeastern
    United States

fl.   p 133

Asclepias pedicellata

Fleming, Glenn
Wild Flowers of Florida

fl.   p 41

Asclepias perennis

Duncan, Wilbur H.
Wildflowers of the Southeastern
    United States

fl.   p 129

Asclepias purpurascens

Wharton, Mary E.
A Guide to the Wildflowers & Ferns
    of Kentucky

fl.   p 129   Pl. 1.69

Asclepias quadrifolia

Duncan, Wilbur H.
Wildflowers of the Southeastern
    United States

fl.   p 133

Asclepias quadrifolia

Jennings, O. E.
Wildflowers of Western Pennsylvania
                    Vol. II

fl.   Pl. 4

Asclepias quadrifolia

Wharton, Mary E.
A Guide to the Wildflowers & Ferns
    of Kentucky

fl.   p 146   Pl. 1.103

Asclepias rubra

Brown, Clair A.
Wildflowers of Louisiana and
    Adjoining States

fl.   p 143

Asclepias sp.

American Hort. Soc.
American Horticulturist
    vol 57, No. 5   1978

fr.   cover, backcover

Asclepias sp.

Milne, Lorus
Living Plants of the World

fl.   p 202

Asclepias speciosa

Clark, Lewis J.
Wild Flowers of British Columbia

fl.   p 426

Asclepias speciosa

Curtis's Botanical Magazine 1958-59
                    Vol. 172

fl.   Pl. 308

Asclepias speciosa

Lemmon, Robert S.
Wildflowers of North America in
    Full Color

fl.   pg. 195   pl. 306

Asclepias speciosa

Moyle, John B.
Northland Wild flowers

fl. hab.   p 126, Pl. 133

Asclepias speciosa

Szczawinski, Adam F.
Edible Garden Weeds of Canada

fr.   p. 34

Asclepias speciosa

Walcott, Mary Vaux
North American Wild Flowers

fl.   Pl. 90

Asclepias speciosa

Welsh, Stanley L.
Flowers of the Canyon Country

fl.   p 19

Asclepias subulata

Coyle, Jeanette
A Field Guide to the Common and
Interesting Plants of Baja California

fl. fr.   p 151

Asclepias syriaca

Courtenay, Booth
Wildflowers & Weeds

fl.   p 28

Asclepias syriaca

Ferguson, Mary
Wildflowers

fl.   p 101

Asclepias syriaca

Klimas, John E.
Wildflowers of Eastern America

fl.   Pl. 290

Asclepias syriaca

Klute, Jeannette
Woodland Portraits

fr.   Pl. 47

Asclepias syriaca

Massachusetts Hort. Soc.
Horticulture
    vol 40, No. 12   1962

fr.   p. 600

Asclepias syriaca

Moyle, John B.
Northland Wild Flowers

fl. hab.   p 126, Pl. 132

Asclepias syriaca

Polunin, Oleg
Flowers of Europe

fl.   Pl. 99 #1011

Asclepias syriaca

Wit, H. C. D. de
Plants of the World;
    The Higher Plants, Vol. II

fr.   p. 105   Pl. 60

Asclepias tuberosa

American Hort. Soc.
American Horticulturist
    vol 53, No. 5 1974

fl.   p. 34

Asclepias tuberosa

American Hort. Soc.
American Horticulturist
    vol 55, No. 6   1976

fl.   p. 39

Asclepias tuberosa

American Hort. Soc.
American Horticulturist
    vol 56, No. 3   1977

fl.   p. 15

Asclepias tuberosa

Batson, Wade T.
Wild Flowers in South Carolina

fl.   p 94

Asclepias tuberosa

Brown, Clair A.
Wildflowers of Louisiana and
    Adjoining States

fl. hab.   p 144

Asclepias tuberosa

Campbell, Carlos C.
Great Smoky Mountain Wildflowers

fl.   p 75

Asclepias tuberosa

Courtenay, Booth
Wildflowers & Weeds

fl.   p. 28

Asclepias tuberosa

Dean, Blanche E.
Wildflowers of Alabama and
    Adjoining States

fl.   p 137

Asclepias tuberosa

Duncan, Wilbur H.
Wildflowers of the Southeastern
    United States

fl.   p 131

Asclepias tuberosa

Everett, T.H., ed.
New Illustrated Encyclopedia of
    Gardening vol. 2

fl.   frontispiece

Asclepias tuberosa

Ferguson, Mary
Wildflowers

fl.   p 110

Asclepias tuberosa

Klimas, John E.
Wildflowers of Eastern America

fl.   Pl. 132

Asclepias tuberosa

Lemmon, Robert S.
Wildflowers of North America in
        Full Color

fl.   pg. 41   pl. 69

Asclepias tuberosa

Massachusetts Hort. Soc.
Horticulture
    vol 36, No. 5   1958

fl.   p. 278

Asclepias tuberosa

Massachusetts Hort. Soc.
Horticulture
    vol 41, No. 8   1963

fl.   inside cover

Asclepias tuberosa

Massachusetts Hort. Soc.
Horticulture
    vol 42, No. 5   1964

fl.   p. 16

Asclepias tuberosa

Massachusetts Hort. Soc.
Horticulture
    vol 53, No. 8   1975

fl.   p. 34

Asclepias tuberosa

Moyle, John B.
Northland Wild Flowers

fl. hab.   p 127, Pl. 135

Asclepias tuberosa

Orr, Robert T.
Wildflowers of Western America

fl.   Pl. 123

Asclepias tuberosa

Pond, Barbara
A Sampler of Wayside Herbs

fl.   Pl. XV

Asclepias tuberosa

Tsukamoto, Yotaro
Coloured Illustrations of
        Garden Flowers

fl.   opp. p. 31

Asclepias tuberosa

Walcott, Mary Vaux
North AmericanWild Flowers

fl.   Pl. 36

Asclepias tuberosa

Weeds of the Southern United States

fl.   p. 5

Asclepias tuberosa

Wharton, Mary E.
A Guide to the Wildflowers & Ferns
        of Kentucky

fl. p105 Pl. 1.22

Asclepias tuberosa var.

Fleming, Glenn
Wild Flowers of Florida

fl.   p 63

Asclepias variegata

Batson, Wade T.
Wild Flowers in South Carolina

fl.   p 95

Asclepias variegata

Brown, Clair A.
Wildflowers of Louisiana and
        Adjoining States

fl. hab.   p 144

Asclepias variegata

Dean, Blanche E.
Wildflowers of Alabama and
        Adjoining States

fl.   p 137

Asclepias variegata

Duncan, Wilbur H.
Wildflowers of the Southeastern
        United States

fl.   p 129

Asclepias variegata

Wharton, Mary E.
A Guide to the Wildflowers & Ferns
        of Kentucky

fl.   p. 117   Pl. 1.47

Asclepias verticillata

Courtenay, Booth
Wildflowers & Weeds

fl.   p. 27

Asclepias verticillata

Dean, Blanche E.
Wildflowers of Alabama and
        Adjoining States

fl.   p 137

Asclepias verticillata

Duncan, Wilbur H.
Wildflowers of the Southeastern
        United States

fl.   p 131

Asclepias verticillata

Moyle, John B.
Northland Wild Flowers

fl. hab.   p 128, Pl. 136

Asclepias viridiflora

Brown, Clair A.
Wildflowers of Louisiana and
        Adjoining States

fl. hab.   p 145

Asclepias viridiflora

Wharton, Mary E.
A Guide to the Wildflowers & Ferns
        of Kentucky

fl.   p. 146   Pl. 1.104

Asclepias viridis

Brown, Clair A.
Wildflowers of Louisiana and
        Adjoining States

fl. hab.   p 145

Asclepiodora viridis

Dean, Blanche E.
Wildflowers of Alabama and
        Adjoining States

fl.   p 139

Ascocentrum ampullaceum

Kamemoto, Haruyuki
Beautiful Thai Orchid Species

fl.   p 123

Ascocentrum ampullaceum

Kramer, Jack
Orchids; Flowers of Romance
        and Mystery

fl.   p 291

Ascocentrum curvifolium

Kamemoto, Haruyuki
Beautiful Thai Orchid Species

fl.   p 122

Ascocentrum curvifolium

Kramer, Jack
Orchids; Flowers of Romance
        and Mystery

fl.   p 290

Ascocentrum miniatum

Kamemoto, Haruyuki
Beautiful Thai Orchid Species

fl.   p 123

Ascyrum hypericoides

Klimas, John E.
Wildflowers of Eastern America

fl.   Pl. 130

Ascyrum hypericoides

Wharton, Mary E.
A Guide to the Wildflowers & Ferns
    of Kentucky

fl. p 99 Pl. 1.10

Ascyrum stans

Batson, Wade T.
Wild Flowers in South Carolina

fl.   p 75

Ascyrum stans

Brown, Clair A.
Wildflowers of Louisiana and
    Adjoining States

fl. hab.   p 112

Ash

see

Fraxinus

Ash, Bennett's

see

Flindersia bennettiana

Ash, Blue Berry

see

Elaeocarpus reticulatus

Ash, Chaparral

see

Fraxinus trifoliata

Ash, claret

see

Fraxinus oxycarpa

Ash, Common

see

Fraxinus excelsior

Ash, European

see

Fraxinus excelsior

Ash, Evergreen

see

Fraxinus uhdei

Ash, Flowering

see

Fraxinus ornus

Ash, golden

see

Fraxinus excelsior var.

Ash, Manna

see

Fraxinus ornus

Ash, Modesto

see

Fraxinus velutina var.

Ash, Mountain

see

Eucalyptus regnans

Ash, Mountain

see

Pyrus americana

Ash, Mountain

see

Sorbus

Ash, Mountain American

see

Sorbus americana

Ash, Mountain European

See

Sorbus aucuparia

Ash, Mountain Kashair

see

Sorbus cashmeriana

Ash, Mountain Western

see

Sorbus sambucifolia

Ash, Red

see

Alphitonia excelsa

Ash, Rhodesian

see

Burkea africana

Ash, Wafer

see

Ptelea trifoliata

Ash, Weeping

see

Fraxinus excelsior

Ash, White

see

Fraxinus americana

Ashanti Blood

see

Mussaenda erythrophylla

Asimina reticulata

Fleming, Glenn
Wild Flowers of Florida

fl.   p 20

Asimina triloba

Batson, Wade T.
Wild Flowers in South Carolina

fl.   p. 50

Asimina triloba

Brown, Clair A.
Wildflowers of Louisiana and
    Adjoining States

fl.   p 57

Asimina triloba

Campbell, Carlos C.
Great Smoky Mountain Wildflowers

fl.   p 77

Asimina triloba

Dean, Blanche E.
Wildflowers of Alabama and
Adjoining States

fl.   p 67

Asimina triloba

Walcott, Mary Vaux
North American Wild Flowers
vol. 5

fl.   Pl. 328

Asimina triloba

Wharton, Mary E.
Trees & Shrubs of Kentucky

fl.   p 43, Pl. 1.4
fr.   p 120, Pl. 2.3

Aspalathus lanceifolia

Flowering Plants of Africa
vol 40   1969-70

fl.   Pl. 1570

Aspalathus macrantha

Flowering Plants of Africa
vol 38   1967

fl.   Pl. 1481

Aspalathus macrocarpa

Flowering Plants of Africa
vol 38   1967

fl.   Pl. 1482

Aspalathus rycroftii

Flowering Plants of Africa
vol 40   1969-70

fl., fr.   Pl. 1571

Asparagus

see

Asparagus officinalis

Asparagus acutifolius

Bianchini, F.
The Complete Book of Fruits & Vegetables

hab.   p 89

Asparagus acutifolius

Polunin, Oleg
Flowers of Europe

hab.   Pl. 168 #1650

Asparagus asparagoides var.

Kiaer, Eigil
Indoor Plants in Colour

hab.   p. 18

Asparagus densiflorus var.

Hay, Roy
The Dictionary of House Plants

hab.   Pl. 45

Asparagus falcatus

Kiaer, Eigil
Indoor Plants in Colour

hab.   P. 18

Asparagus, fern

see

Asparagus plumosus var.

Asparagus Fern

see

Asparagus sprengeri

Asparagus juniperiodes

Eliovson, Sima
Namaqualand in Flower

hab.   Pl. 26, 4

Asparagus officinalis

Bianchini, F.
The Complete Book of Fruits & Vegetables

hab.   p 89

Asparagus officinalis

Bianchini, Francesco
Health Plants of the World

fr.   p 131

Asparagus officinalis

Courtenay, Booth
Wildflowers & Weeds

fl.   p. 6

Asparagus officinalis

Hvass, Elsie
Plants That Feed and Serve Us

fl. fr.   p 35, Pl. 65

Asparagus officinalis

Masefield, G. B.
The Oxford Book of Food Plants

hab.   Pl. 163, 3

Asparagus officinalis

Perrot, Emile
Les Plantes Medicinales

fl. fr. hab.   p 121

Asparagus ovatus

Flowering Plants of Africa
vol 29   1952-53

fl.   Pl. 1146

Asparagus pastorianus

Bramwell, David
Wild Flowers of the Canary Islands

hab.   Pl. 323

Asparagus plumosus

Kromdijk, G.
200 House Plants in Colour

hab.   Pl. 22

Asparagus plumosus var.

Kiaer, Eigil
Indoor Plants in Colour

hab.   p. 18

Asparagus prostratus

Martin, W. Keble
The Concise British Flora in Colour

fl. fr. hab.   Pl. 84

Asparagus, sicklethorn

see

Asparagus falcatus

Asparagus sprengeri

Crockett, James Underwood
Lawns & Ground Covers

hab.   p. 123

Asparagus sprengeri

Hersey, Jean
Woman's Day Book of House Plants

hab.   p 31

Asparagus sprengeri

Kiaer, Eigil
Indoor Plants in Colour

fr.   p. 18

Asparagus sprengeri

Kromdijk, G.
200 House Plants in Colour

hab.   Pl. 23

Asparagus sprengeri

Macoby, Stirling
What Flower is That

fr.   P 48

Asparagus Sprengeri

Tsukamoto, Yotaro
Coloured Illustrations of
    Garden Flowers   Vol. 9

hab.   p. 31

Aspasia epidendroides

Ospina, Mariano
Orquideas de las americas

fl.   Pl. 167

Aspasia lunata

Ebel, Friedrich
The Strange and Beautiful
    World of Orchids

fl.   p 111

Aspasia lunata

Pabst, G. F. J.
Orchidaceae Brasilienses,
    Vol 2

fl.   p 259

Aspasia variegata

Pabst, G. F. J.
Orchidaceae Brasilienses,
    Vol 2

fl.   p 259

Aspazoma amplectens

Herre, H.
The Genera of the Mesembryanthemaceae

fl. fr.   p 87

Aspen

    see

Populus

Aspen, Quaking

    see

Populus tremuloides

Asperugo procumbens

Lindman, C. A. M.
Nordens Flora, Vol 8

fl. fr. hab.   Pl. 492

Asperula cynanchica

Ary, S.
The Oxford Book of Wildflowers

fl.   p. 92   Pl. 5

Asperula cynanchica

Barneby, T. P.
European Alpine Flowers in Colour

fl.   Pl. 79, 1

Asperula cynanchica

Martin, W. Keble
The Concise British Flora in Colour

fl. hab.   Pl. 42

Asperula odorata

Ary, S.
The Oxford Book of Wildflowers

fl.   p 92, Pl. 3

Asperula odorata

Bianchini, Francesco
Health Plants of the World

fl. hab.   p 129

Asperula odorata

Crockett, James Underwood
Lawns & Ground Covers

fl.   p. 123

Asperula odorata

Felsko, Elsa
A Book of Wildflowers

fl.   P 144

Asperula odorata

Furrer, D.
Blumen am Wege

fl. hab.   p 101

Asperula odorata

Huxley, Anthony
Garden Perennials and Water Plants

fl.   Pl. 40

Asperula odorata

Loewenfeld, Claire
The Complete Book of Herbs and Spices

fl. hab.   p 193

Asperula odorata

Perrot, Emile
Les Plantes Medicinales

fl. fr. hab.   p 21

Asperula orientalis

Crockett, James Underwood
Annuals

fl.   P 96

Asperula scoparia

Cochrane, G. R.
Flowers and Plants of Victoria

fl. hab.   Pl. 419

Asperula suberosa

Alpine Garden Society
Bulletin
    vol. 39, No. 2   1971

fl., hab.   p. 105

Asperula suberosa

Hay, Roy
The Color Dictionary of Flowers & Plants

fl.   p 2, Pl. 16

Asphodel, Bog

    see

Narthecium ossifragum

Asphodel, False

    see

Tofieldia glutinosa

Asphodel, Scottish

    see

Tofieldia pusilla

Asphodel, Tofield's

    see

Tofieldia calyculata

Asphodel, White

    see

Asphodelus albus

Asphodel, Yellow

    see

Asphodeline lutea

Asphodeline lutea

Goulimis, Constantine N.
Wild Flowers of Greece

fl.   p 137

Asphodeline lutea

Morley, Brian D.
Wild Flowers of the World

fl.,fr.   Pl. 41

Asphodeline lutea

Polunin, Oleg
Flowers of Europe

fl.   Pl. 162 #1593

Asphodeline lutea

Polunin, Oleg
Flowers of the Mediterranean

fl.   Pl. 235

Asphodelus aestivus

Polunin, Oleg
Flowers of Europe

fl.   Pl. 163 #1591

Asphodelus albus

Barneby, T. P.
European Alpine Flowers in Colour

fl.   Pl. 1, 3

Asphodelus albus

Kleijn, H.
Beauty of the Wild Plant

fl.   p. 112   Pl. 159, 159a

Asphodelus albus

Kohlhaupt, Paula
Fleurs des Alpages   Vol. 1

fl.   P. 23

Asphodelus albus

Taylor, A. W.
Wild Flowers of Spain and Portugal

fl.   p. 95

Asphodelus cerasiferus

Wit, H. C. D. de
Plants of the World;
    The Higher Plants, Vol II

fl.fr.   Pl. 123

Asphodelus fistulosus

Polunin, Oleg
Flowers of Europe

fl.   Pl. 163 #1592

Asphodelus fistulosus

Polunin, Oleg
Flowers of the Mediterranean

fl.   Pl. 238

Asphodelus fistulosus

Taylor, A. W.
Wildflowers of Spain and Portugal

fl.   p. 14

Asphodelus microcarpus

Feinbrun-Dothan, Naomi
Wild Plants in the Land
    of Israel

fl.   p 88

Asphodelus microcarpus

Megaw, Elektra
Wild Flowers of Cyprus

fl., hab.   Pl. 30

Asphodelus microcarpus

Polunin, Oleg
Flowers of the Mediterranean

fl.   Pl. 233

Aspidistra elatior

Hay, Roy
The Dictionary of House Plants

hab.   Pl. 46

Aspidistra elatior

Kiaer, Eigil
Indoor Plants in Colour

fl.   p. 19

Aspidistra elatior

Tsukamoto, Yotaro
Coloured Illustrations of Garden
    Flowers   Vol. 9

hab.   p. 31

Aspidistra elatior

Walden, Beryl M.
Wild Flowers of Hong Kong

fl.   hab.   Pl. 82 (245)

Aspidistra lurida

Kromdijk, G.
200 House Plants in Colour

hab.   Pl. 24

Assegai Wood

see

Terminalia sericea

Astartea fascicularis

Blombery, Alec M.
What Wildflower is That

fl.   p. 53   Pl. 108

Astartea heteranthera

Harrison, R. E.
Trees and Shrubs

fl. P 22

Astelia alpina

Cochrane, G. R.
Flowers and Plants of Victoria

fl. hab.   Pl. 534

Astelia chathamica

Royal Hort. Soc.
Journal of the Royal Hort. Soc.
    vol 95, No. 8   1970

fl.   p. 352   Pl. 189

Astelia pumila

Morley, Brian D.
Wild Flowers of the World

fl.   Pl. 7J

Astelia solandri

Harvey, Norman B.
New Zealand Botanical Paintings

fl.   p. 66   Pl. 30

Astelia solandri

Morley, Brian D.
Wild Flowers of the World

fl. fr.   Pl. 141

Aster acris

Hay, Roy
The Color Dictionary of
    Flowers & Plants

hab., fl.   p. 123   Pl. 980

Aster acuminatus

Jennings, O. E.
Wild Flowers of Western Pennsylvania
    Vol. II

fl. fr. hab   Pl. 178

Aster adnatus

Brown, Clair A.
Wildflowers of Louisiana and
    Adjoining States

fl.   p 187

Aster alpigenus

Fries, Mary A.
Wildflowers of Mount Ranier and
the Cascades

fl. P 180

Aster alpigenus

Lemmon, Robert S.
Wildflowers of North America in
    Full Color

fl.   pg. 161   Pl. 258

Aster, Alpine

see

Aster alpigenus

Aster alpinus

Barneby, T. P.
European Alpine Flowers in Colour

fl.   Pl. 86, 1

Aster alpinus

Goulimis, Constantine
Wild Flowers of Greece

fl.   hab. 113

Aster alpinus

Hay, Roy
The Color Dictionary of Flowers & Plants

fl.   p 3, Pl. 17

Aster alpinus

Kleijn, H.
Beauty of the Wild Plant

fl.   p. 93   Pl. 130

Aster alpinus

Kohlhaupt, Paula
Fleurs des Alpages Vol. 1

fl.   P. 106

Aster alpinus

Polunin, Oleg
Flowers of Europe

fl.   Pl. 142 #1363

Aster alpinus

Tosco, Uberto
The World of Mountain Flowers

fl. hab.   p 62

Aster alpinus var.

Huxley, Anthony.
Garden Perennials and Water Plants.

fl.   Pl. 49a,b

Aster amellus

Barneby, T. P.
European Alpine Flowers in Colour

fl.   Pl. 86, 2

Aster amellus

Felsko, Elsa
A Book of Wildflowers

fl.   P 75

Aster amellus

Huxley, Anthony
Garden Perennials and Water Plants

hab.   Pl. 4

Aster amellus

Polunin, Oleg
Flowers of Europe

fl.   Pl. 142 #1364

Aster amellus var.

Hay, Roy
The Color Dictionary of Flowers &
Plants

fl.   p. 123   Pl. 981-82

Aster amellus var.

Huxley, Anthony
Garden Perennials and Water Plants.

fl.   Pl. 50a,b,c.

Aster amellus var.

Macoby, Stirling
What Flower is That

fl.   p 48

Aster amellus var.

Perry, Frances
Flowers of the World

fl.   p. 80

Aster, Annual

see

Callistephus chinensis

Aster, Aromatic

see

Aster oblongifolius

Aster, Arrow-Leaved

see

Aster sagittifolius

Aster azureus

Moyle, John B.
Northland Wild Flowers

fl. hab.   p 175, Pl. 216

Aster baccharoides

Walden, Beryl M.
Wild Flowers of Hong Kong

fl.   Pl. 60 (166)

Aster benthamii

Walden, Beryl M.
Wild Flowers of Hong Kong

fl.   Pl. 61 (178)

Aster, Bog

see

Aster junciformis

Aster Bush

see

Felicia fruticosa

Aster, Calico

see

Aster lateriflorus

Aster campestris

Walcott, Mary Vaux
North American Wild Flowers

fl. hab.   Pl. 118

Aster, Cascade

see

Aster ledophyllus

Aster, China

see

Callistephus chinensis

Aster ciliolatus

Moyle, John B.
Northland Wild Flowers

fl. hab.   p. 174   Pl. 215

Aster conspicuus

Clark, Lewis J.
Wild Flowers of British Columbia

fl.   p 530

Aster cordifolius

Bruggeman, L.
Tropical Plants

fl.   Pl. 75

Aster cordifolius

Jennings, O. E.
Wild Flowers of Western Pennsylvania
Vol. II

fl.   Pl. 169

Aster cordifolius

Lemmon, Robert S.
Wildflowers of North America in
Full Color

fl.   pg. 204   Pl. 323

Aster cordifolius

Moyle, John B.
Northland Wild Flowers

fl. hab.   p 176, Pl. 217

Aster cordifolius

Wharton, Mary E.
A Guide to the Wildflowers & Ferns
of Kentucky

fl.   p 265   Pl. 3.43

Aster, Cornel-leaf

see

Aster infirmus

Aster, Crooked-stem

see

Aster prenanthoides

Aster curtissii

Campbell, Carlos C.
Great Smoky Mountain Wildflowers

fl.   p 101

Aster, Desert

see

Machaeranthera tortifolia

Aster divaricatus

Klimas, John E.
Wildflowers of Eastern America

fl.   Pl. 105

Aster divaricatus

Wharton, Mary E.
A Guide to the Wildflowers & Ferns
of Kentucky

fl.   p 271   Pl. 3.53

Aster dumosus

Duncan, Wilbur H.
Wildflowers of the Southeastern
United States

fl.   p 205

Aster ericoides

Brown, Clair A.
Wildflowers of Louisiana and
Adjoining States

fl.   p 186

Aster ericoides

Courtenay, Booth
Wildflowers & Weeds

fl.   p 121

Aster ericoides

Macoby, Stirling
What Flower is That

fl.   P 48

Aster ericoides

Moyle, John B.
Northland Wild Flowers

fl. hab.   p 181, Pl. 226

Aster ericoides var.

Huxley, Anthony
Garden Perennials and Water Plants

fl.   Pl. 48a, b.

Aster, Fall

see

Aster oblongifolius

Aster, File-Leaved

see

Aster radula

Aster foliaceus

Welsh, Stanley L.
Flowers of the Mountain Country

fl.   p. 73

Aster x frikartii

Hay, Roy
The Color Dictionary of Flowers
& Plants

fl.   p. 123   Pl. 981

Aster x frikartii

Royal Hort. Soc.
Journal of the Royal Hort. Soc.
vol. 97, No. 8   1972

fl.   p. 352   Pl. 178

Aster x frikartii var.

Huxley, Anthony
Garden Perennials and Water Plants

fl.   Pl. 51

Aster, Frost-weed

see

Aster pilosus var.

Aster, Golden

see

Chrysopsis villosa

Aster, Golden

see

Heterotheca scabrella

Aster, Golden

see

Heterotheca subaxillaris

Aster, Grass-Leaved Golden

see

Chrysopsis graminifolia

Aster, Grass-leaved Golden

see

Chrysopsis nervosa

Aster, Guitar

see

Aster panduratus

Aster harveyanus

Moriarty, Audrey
Wild Flowers of Malawi

fl.   Pl. 79; 1

Aster, Heart-leaved

see

Aster cordifolius

Aster, heath

see

Aster incisus

Bruggeman, L.
Tropical Plants

fl.   Pl. 76

Aster infirmus

Wharton, Mary E.
A Guide to the Wildflowers & Ferns
of Kentucky

fl.   p 270   Pl. 3.52

Aster integrifolius

Welsh, Stanley L.
Flowers of the Mountain Country

fl.   p. 26

Aster, italian

see

Aster amellus var.

Aster junciformis

Courtenay, Booth
Wildflowers & Weeds

fl.   p 121

Aster laevis

Courtenay, Booth
Wildflowers & Weeds

fl.   p 121

Aster laevis

Jennings, O. E.
Wild Flowers of Western Pennsylvania
Vol. II

fl. hab.   Pl. 173

Aster laevis

Moyle, John B.
Northland Wild Flowers

fl.   p 177, Pl. 220

Aster laevis var.

Porsild, A. E.
Rocky Mountain Wild Flowers

fl. hab.   p 403

Aster lateriflorus

Brown, Clair A.
Wildflowers of Louisiana and
Adjoining States

fl.   p 186

Aster lateriflorus

Courtenay, Booth
Wildflowers & Weeds

fl.   p 123

Aster lateriflorus

Jennings, O. E.
Wild Flowers of Western Pennsylvania
Vol. II

fl. hab.   Pl. 174

Aster lateriflorus

Moyle, John B.
Northland Wild Flowers

fl.   p 181, Pl. 227

Aster lateriflorus

Wharton, Mary E.
A Guide to the Wildflowers & Ferns
of Kentucky

fl.   p 270   Pl. 3.51

Aster, Leafy

see

Aster foliaceus

Aster ledophyllus

Orr, Robert T.
Wildflowers of Western America

fl.   Pl. 227

Aster ledophyllus

Taylor, Ronald J.
Mountain Wild Flowers

fl.   p 16

Aster linarifolius

Courtenay, Booth
Wildflowers & Weeds

fl.   p 123

Aster linarifolius

Dean, Blanche E.
Wildflowers of Alabama and
Adjoining States

fl.   p. 191

Aster linarifolius

Jennings, O. E.
Wild Flowers of Western Pennsylvania
Vol. II

fl. hab.   Pl. 177

Aster, Lindley's

see

Aster ciliolatus

Aster linosyris

Curtis's Botanical Magazine 1964-65
Vol. 175

fl.   Pl. 471

Aster linosyris

Hay, Roy
The Color Dictionary of Flowers & Plants

fl.   p 123, Pl. 984

Aster linosyris

Lindman, C. A. M.
Nordens Flora, Vol 9

fl. fr. hab.   Pl. 577

Aster lowrieanus

Jennings, O. E.
Wild Flowers of Western Pennsylvania
Vol. II

fl.   Pl. 170, 75

Aster, Lowrie's

see

Aster lowrieanus

Aster macrophyllus

Courtenay, Booth
Wildflowers & Weeds

fl.   p 121

Aster macrophyllus

Morley, Brian D.
Wild Flowers of the World

fl.   Pl. 165

Aster macrophyllus

Moyle, John B.
Northland Wild Flowers

fl. hab.   p 174, Pl. 214

Aster macrophyllus var.

Jennings, O. E.
Wild Flowers of Western Pennsylvania
Vol. II

fl. hab.   Pl. 165

Aster, Mohave

see

Machaeranthera tortifolia

Aster, Mountain

see

Aster acuminatus

Aster multiflorus

Bruggeman, L.
Tropical Plants

fl.   Pl. 77

Aster, New England

see

Aster novaeangliae

Aster, New York

see

Aster novi-belgii

Aster novae-angliae

Bruggeman, L.
Tropical Plants

fl.   Pl. 78

Aster novaeangliae

Courtenay, Booth
Wildflowers & Weeds

fl.   p 123

Aster novae-angliae

Jennings, O. E.
Wild Flowers of Western Pennsylvania
Vol. II

fl. hab.   Pl. 167

Aster novae angliae

Lemmon, Robert S.
Wildflowers of North America in
    Full Color

fl.   pg. 57   Pl. 96

Aster novaeangliae

Moyle, John B.
Northland Wild Flowers

fl. hab.   p 176, Pl. 218

Aster novae-angliae

Wharton, Mary E.
A Guide to the Wildflowers & Ferns
    of Kentucky

fl.   p 267   Pl. 3.47

Aster novae-angliae var.

Hay, Roy
The Color Dictionary of Flowers &
    Plants

fl.   p. 124   Pl. 985-91

Aster novae-angliae var.

Huxley, Anthony
Garden Perennials and Water Plants

fl.   Pl. 53

Aster novae-angliae var.

Perry, Frances
Flowers of the World

fl.   p. 80

Aster novi-belgii

Lemmon, Robert S.
Wildflowers of North America in
    Full Color

fl.   p. 56   Pl. 94

Aster novi-belgii

Milne, Lorus
Living Plants of the World

fl. hab.   p 237

Aster novi-belgii var

Hay, Roy
The Color Dictionary of Flowers
    & Plants

fl.   p. 124   Pl. 966 - 991

Aster novi-belgii var.

Huxley, Anthony
Garden Perennials and Water Plants

fl.   Pl. 52a,b,c,d,

Aster oblongifolius

Courtenay, Booth
Wildflowers & Weeds

fl.   p 123

Aster oblongifolius

Moyle, John B.
Northland Wild Flowers

fl. hab.   p 178, Pl. 221

Aster oblongifolius

Wharton, Mary E.
A Guide to the Wildflowers & Ferns
    of Kentucky

fl.   p. 268   Pl. 3.48

Aster paludosus

Batson, Wade T.
Wild Flowers in South Carolina

fl.   p 116

Aster paludosus

Dean, Blanche E.
Wildflowers of Alabama and
    Adjoining States

fl.   p 189

Aster panduratus

Walden, Beryl M.
Wild Flowers of Hong Kong

fl. hab.   Pl. 45 (130)

Aster, Panicled

see

Aster simplex

Aster pansus

Clark, Lewis J.
Wild Flowers of British Columbia

fl.   p 519

Aster patens

Wharton, Mary E.
A Guide to the Wildflowers & Ferns
    of Kentucky

fl.   p 267   Pl. 3.46

Aster paternus

Duncan, Wilbur H.
Wildflowers of the Southeastern
    United States

fl.   p 203

Aster pilosus

Batson, Wade T.
Wild Flowers in South Carolina

fl.   p 116

Aster pilosus

Courtenay, Booth
Wildflowers & Weeds

fl.   p 121

Aster pilosus

Weeds of the Southern United States

fl.   p. 9

Aster pilosus var.

Wharton, Mary E.
A Guide to the Wildflowers & Ferns
    of Kentucky

fl.   p. 269   Pl. 3.49

Aster, Pineland

see

Aster squarrosus

Aster, Prairie

see

Aster campestris

Aster prealtus

Brown, Clair A.
Wildflowers of Louisiana and
    Adjoining States

fl.   p 188

Aster prenanthoides

Jennings, O. E.
Wild Flowers of Western Pennsylvania
    Vol. II

fl. hab.   Pl. 175

Aster prenanthoides

Wharton, Mary E.
A Guide to the Wildflowers & Ferns
    of Kentucky

fl.   p 266   Pl. 3.45

Aster ptarmicoides

Courtenay, Booth
Wildflowers & Weeds

fl.   p 121

Aster ptarmicoides

Moyle, John B.
Northland Wild Flowers

fl. hab.   p 180, Pl. 224

Aster puniceus

Campbell, Carlos C.
Great Smoky Mountain Wildflowers

fl.   p 101

Aster puniceus

Courtenay, Booth
Wildflowers & Weeds

fl.   p 123

Aster puniceus

Duncan, Wilbur H.
Wildflowers of the Southeastern
     United States

fl.   p 205

Aster puniceus

Jennings, O. E.
Wild Flowers of Western Pennsylvania
          Vol. II

fl. hab.   Pl. a76

Aster puniceus

Moyle, John B.
Northland Wild Flowers

fl.   p 177, Pl. 219

Aster, Purple-Stemmed

     see

Aster puniceus

Aster, Rabbit Ear

     see

Ainsliaea fragrans

Aster Radula

Jennings, O. E.
Wild Flowers of Western Pennsylvania
          Vol. II

fl. hab.   Pl. 166

Aster, Red-Stemmed

     see

Aster puniceus

Aster reticulatus

Batson, Wade T.
Wild Flowers in South Carolina

fl.   p 117

Aster reticulatus

Duncan, Wilbur H.
Wildflowers of the Southeastern
     United States

fl. hab.   p 205

Aster reticulatus

Fleming, Glenn
Wild Flowers of Florida

fl. hab.   p 32

Aster sagittifolius

Courtenay, Booth
Wildflowers & Weeds

fl.   p. 121

Aster sagittifolius

Jennings, O. E.
Wild Flowers of Western Pennsylvania
          Vol. II

fl. hab.   Pl. 172

Aster Savatieri

Tsukamoto, Yotaro
Coloured Illustrations of
     Garden Flowers Vol. 9

fl.   p. 32

Aster, Sea

     see

Aster tripolium

Aster sericeus

Courtenay, Booth
Wildflowers & Weeds

fl.   p 123

Aster sericeus

Moyle, John B.
Northland Wild Flowers

fl.   p 179, Pl. 222

Aster shortii

Jennings, O. E.
Wild Flowers of Western Pennsylvania
          Vol. II

fl. hab.   Pl. 168

Aster shortii

Wharton, Mary E.
A Guide to the Wildflowers & Ferns
     of Kentucky

fl.   p 264   Pl. 3.42

Aster, Short's

     see

Aster shortii

Aster sibiricus

Alaska-Yukon Wild Flower Guide

fl. hab.   p 162

Aster sibiricus

Heller, Christine
Wild Flowers of Alaska

fl.   Pl. 267-268

Aster sibiricus

Lindman, C. A. M.
Nordens Flora, Vol 9

fl. fr. hab.   Pl. 578

Aster sibiricus

Porsild, A. E.
Rocky Mountain Wild Flowers

fl.   p 405

Aster sikkimensis

Hara, Hiroshi, comp.
Photo-Album of Plants of
     Eastern Himalaya

fl.   Pl. 177

Aster, Silky

     see

Aster sericeus

Aster simplex

Courtenay, Booth
Wildflowers & Weeds

fl.   p 121

Aster simplex

Moyle, John B.
Northland Wild Flowers

fl.   p 179, Pl. 223

Aster simplex

Wharton, Mary E.
A Guide to the Wildflowers & Ferns
     of Kentucky

fl.   p 269   Pl. 3.50

Aster sp.

Brown, Clair A.
Wildflowers of Louisiana and
     Adjoining States

fl.   p 188

Aster species

Fries, Mary A.
Wildflowers of Mount Ranier and
     the Cascades

fl.   P 181

Aster spectabilis

Addisonia, Vol. 19, 1935-36

fl.   p. 61   Pl. 639

Aster spectabilis

Hay, Roy
The Color Dictionary of Flowers & Plants

fl.   p 124, Pl. 992

Aster, Spreading

see

Aster, patens

Aster squarrosus

Batson, Wade T.
Wild Flowers in South Carolina

fl.   p 117

Aster squarrosus

Walcott, Mary Vaux
North American Wild Flowers

fl. hab.   Pl. 160

Aster, Stokes'

see

Stokesia caerulea var.

Aster subspicatus

Clark, Lewis J.
Wild Flowers of British Columbia

fl.   p 531

Aster tanacetifolius

American Hort. Soc.
American Horticulturist
vol 57, No. 1   1978

fl.   p. 35

Aster tanacetifolius

Crockett, James Underwood
Annuals

fl.   P 97

Aster tataricus

Kimura, Koiti
Japanese Medicinal Plants, Vol II

fl.   p 235

Aster thomsonii var.

Hay, Roy
The Color Dictionary of Flowers & Plants

fl.   p. 124   Pl. 993

Aster tripolium

Ary, S.
The Oxford Book of Wildflowers

fl.   p. 136   Pl. 3

Aster tripolium

Felsko, Elsa
A Book of Wild Flowers 2nd Ser.

fl.   p. 51

Aster tripolium

Kleijn, H.
Beauty of the Wild Plant

fl.   p. 77   Pl. 98

Aster tripolium

Lindman, C. A. M.
Nordens Flora, Vol 9

fl. hab.   Pl. 579

Aster tripolium

Martin, W. Keble
The Concise British Flora in Colour

fl. hab.   Pl. 44

Aster tripolium

Polunin, Oleg
Flowers of Europe

fl.   Pl. 142 #1365

Aster umbellatus

Courtenay, Booth
Wildflowers & Weeds

fl.   p. 121

Aster umbellatus

Moyle, John B.
Northland Wild Flowers

fl. hab.   p 180, Pl. 225

Aster undulatus

Jennings, O. E.
Wild Flowers of Western Pennsylvania
Vol. II

fl.   Pl. 171

Aster undulatus

Wharton, Mary E.
A Guide to the Wildflowers & Ferns
of Kentucky

fl.   p 266   Pl. 3.44

Aster, Violet Wood

see

Aster macrophyllus var.

Aster, Wavy-Leaved

see

Aster undulatus

Aster, Whiteheath

see

Aster pilosus

Aster, White-top

see

Aster reticulatus

Aster, White-top

see

Sericocarpus asteroides

Aster, White-Topped

see

Sericocarpus linifolius

Aster, White Wood

see

Aster divaricatus

Asteranthera ovata

Curtis's Botanical Magazine
Vol 165   1948

fl.   15

Asteranthera ovata

Harrison, Richmond E.
Climbers and Trailers

fl.   p 20   Pl. 12

Asteranthera ovata

Munoz Pizarro, Carlos
Flores Silvestres de Chile

fl.   Pl. 47

Asteranthera ovata

Royal Hort. Soc.
Journal of the Royal Hort. Soc.
vol 92, No. 8   1967

fl.   p. 344   Pl. 180

Asteranthera ovata

Royal Hort. Soc.
Journal of the Royal Hort. Soc.
vol 100, No. 5   1975

hab.   p. 200   Pl. 81

Asteriscus aquaticus

Polunin, Oleg
Flowers of Europe

fl.   Pl. 146 #1399

Asteriscus maritimus

Polunin, Oleg
Flowers of Europe

fl.   Pl. 146 #1398

Asteriscus sericeus

Bramwell, David
Wild Flowers of the Canary Islands

fl.   Pl. 274

Asteriscus stenophyllus

Bramwell, David
Wild Flowers of the Canary Islands

fl., hab.   Pl. 275

Asterolasia phebalioides

Cochrane, G. R.
Flowers and Plants of Victoria

fl.   Pl. 89

Asterolasia trymilioides

Blombery, Alec M.
What Wildflower is That

fl.   p. 53   Pl. 109

Asteromoea indica

Bruggeman, L.
Tropical Plants

fl.   Pl. 79

Asterotrichion discolor

Curtis, Winifred
The Endemic Flora of Tasmania, Part IV

fl. fr. hab.   Pl. 125

Astianthus viminalis

O'Gorman, Helen
Mexican Flowering Trees and Plants

fl. hab.   p 13

Astilbe x arendsii

Hay, Roy
The Color Dictionary of Flowers
& Plants

fl.   p. 125   Pl. 994

Astilbe x arendsii var.

Huxley, Anthony
Garden Perennials and Water Plants

fl.   Pl. 54a thru 54f

Astilbe atrorosea

Huxley, Anthony
Garden Perennials and Water Plants

fl.   Pl. 55

Astilbe chinensis

Hay, Roy
The Color Dictionary of Flowers & Plants

fl.   p 3, Pl. 19

Astilbe x crispa var.

Huxley, Anthony
Garden Perennials and Water Plants

fl.   Pl. 56

Astilbe hybrid

Hay, Roy
The Color Dictionary of Flowers & Plants

fl.   p 125, Pl. 995 thru 998

Astilbe hybrid

Perry, Frances
Flowers of the World

fl.   p. 272

Astilbe taquetii var.

Royal Hort. Soc.
Journal of the Royal Hort. Soc.
vol. 98, No. 7   1973

hab.,fl.   p. 306   Pl. 160

Astilbe taquetii var.

Royal Hort. Soc.
The Garden
vol 102, No. 6   1977

fl.   p. 265

Astragalus adsurgens

Kimura, Koiti
Japanese Medicinal Plants, Vol II

fl.   p 185

Astragalus adsurgens var.

Ferguson, Mary
Wildflowers

fl.   p 135

Astragalus alopecuroides

Color Treasury of Herbs & Other
Medicinal Plants

fl. p 34   Pl. 39

Astragalus alpinus

Alaska-Yukon Wild Flower Guide

fl.   p 89

Astragalus alpinus

Barneby, T. P.
European Alpine Flowers in Colour

fl.   Pl. 41, 6

Astragalus alpinus

Clark, Lewis J.
Wild Flowers of British Columbia

fl.   p 283

Astragalus alpinus

Heller, Christine
Wild Flowers of Alaska

fl.   Pl. 242

Astragalus alpinus

Martin, W. Keble
The Concise British Flora in Colour

fl. hab.   Pl. 24

Astragalus alpinus

Polunin, Oleg
Flowers of Europe

fl.   Pl. 54 #530

Astragalus alpinus

Porsild, A. E.
Rocky Mountain Wild Flowers

fl.   p 253

Astragalus alpinus

Walcott, Mary Vaux
North American Wild Flowers
vol. 4

fl.   Pl. 295

Astragalus americanus

Alaska-Yukon Wild Flower Guide

fl.   p 87

Astragalus amphioxys

Orr, Robert T.
Wildflowers of Western America

fl.   Pl. 266

Astragalus arenarius

Lindman, C. A. M.
Nordens Flora, Vol 5

fl. fr. hab.   Pl. 347

Astragalus asclepiadoides

Welsh, Stanley L.
Flowers of the Canyon Country

fl.   p 17

Astragalus australis

Barneby, T. P.
European Alpine Flowers in Colour

fl. hab.   Pl. 42, 1

Astragalus balearicus

Royal Hort. Soc.
Journal of the Royal Hort. Soc.
vol 98, No. 7 1973

hab., fl.   P. 306   Pl. 171

Astragalus bourgovii

Walcott, Mary Vaux
North American Wild Flowers

fl. fr. hab.   Pl. 21

Astragalus callichrous

Feinbrun-Dothan, Naomi
Wild Plants in the Land
of Israel

fl. fr. hab.   p 156

Astragalus canadensis

Courtenay, Booth
Wildflowers & Weeds

fl.   p 49

Astragalus canadensis

Moyle, John B.
Northland Wild Flowers

fl. hab.   p 78, Pl. 53

Astragalus centroalpinus var.

Tosco, Uberto
The World of Mountain Flowers

fl.   p 4

Astragalus cicer

Polunin, Oleg
Flowers of Europe

fl.   Pl. 54 #528

Astragalus clycyphyllus

Lindman, C. A. M.
Nordens Flora, Vol 5

fl. fr.   Pl. 346

Astragalus coccineus

Morley, Brian D.
Wild Flowers of the World

fl.   Pl. 153

Astragalus coccineus

Munz, Philip A.
California Desert Wildflowers

fl.   p 36, Pl. 30

Astragalus coccineus

Orr, Robert T.
Wildflowers of Western America

fl.   Pl. 183

Astragalus crassicarpus

Moyle, John B.
Northland Wild Flowers

fr., hab.   p 78, Pl. 54

Astragalus danicus

Lindman, C. A. M.
Nordens Flora, Vol 5

fl. fr. hab.   Pl. 348

Astragalus danicus

Martin, W. Keble
The Concise British Flora in Colour

fl. hab.   Pl. 24

Astragalus durhamii

Curtis's Botanical Magazine
Vol 162   1939

fl.   No. 9579

Astragalus exscapus

Barneby, T. P.
European Alpine Flowers in Colour

fl. hab.   Pl. 42, 2

Astragalus flavus

Welsh, Stanley L.
Flowers of the Canyon Country

fl. hab.   p 39

Astragalus frigidus

Porsild, A. E.
Rocky Mountain Wild Flowers

fr.   p 257

Astragalus glycyphyllos

Ary, S.
The Oxford Book of Wildflowers

fl.   p 22, Pl. 3

Astragalus glycyphyllos

Barneby, T. P.
European Alpine Flowers in Colour

fl. hab.   Pl. 41, 4

Astragalus glycyphyllos

Martin, W. Keble
The Concise British Flora in Colour

fl. fr. hab.   Pl. 24

Astragalus glycyphyllos

Polunin, Oleg
Flowers of Europe

fl.   Pl. 54 #527

Astragalus graecus

Goulimis, Constantine N.
Wild Flowers of Greece

fl.   p 39

Astragalus gummifer

Hvass, Elsie
Plants That Feed and Serve Us

fl.   p 108, Pl. 232

Astragalus larkyaensis

Hara, Hiroshi, comp.
Photo-Album of Plants of
Eastern Himalaya

fl.   Pl. 214

Astragalus lentiginosus

Munz, Philip A.
California Desert Wildflowers

fl.   p. 37   Pl. 32

Astragalus leontinus

Barneby, T. P.
European Alpine Flowers in Colour

fl.   Pl. 41, 5

Astragalus lusitanicus var.

Goulimis, Constantine N.
Wild Flowers of Greece

fl. fr.   p. 41

Astragalus miser

Porsild, A. E.
Rocky Mountain Wild Flowers

fl.   p 255

Astragalus mollissimus

Welsh, Stanley L.
Flowers of the Canyon Country

fl.   p 17

Astragalus monspessulanus

Barneby, T. P.
European Alpine Flowers in Colour

fl. hab.   Pl. 41, 2

Astragalus norvegicus

Lindman, C. A. M.
Nordens Flora, Vol 6

fl. fr. hab.   Pl. 349

Astragalus nubigenus

Hara, Hiroshi, comp.
Photo-Album of Plants of
Eastern Himalaya

fl.   Pl. 215

Astragalus onobrychis

Barneby, T. P.
European Alpine Flowers in Colour

fl. hab.   Pl. 41, 3

Astragalus parnassi

Goulimis, Constantine N.
Wild Flowers of Greece

fl.   p 43

Astragalus penduliflorus

Lindman, C. A. M.
Nordens Flora, Vol 6

fl. fr.   Pl. 350

Astragalus pinetorum

American Hort. Soc.
American Horticulturist
      vol 53, No. 5 1974

fl.   p. 18

Astragalus praelongus

Welsh, Stanley L.
Flowers of the Canyon Country

fl.   p 42

Astragalus purpureus

Vilmorin, Roger de
Plantes Alpines dans les Jardins

fl. hab.   Pl. XXIII

Astragalus sempervirens

Barneby, T. P.
European Alpine Flowers in Colour

fl.   Pl. 41, 1

Astragalus soxmaniorum

Brown, Clair A.
Wildflowers of Louisiana and
      Adjoining States

fl. fr. hab.   p. 73

Astragalus sp.

Coyle, Jeanette
A Field Guide to the Common and
   Interesting Plants of Baja California

fl. fr.   p 97

Astragalus sp/

Royal Hort. Soc.
Journal of the Royal Hort. Soc.
      vol 90, No. 2   1965

fl.   p. 84   Pl. 30

Astragalus striatus

Porsild, A. E.
Rocky Mountain Wild Flowers

fl.   p 259

Astragalus umbellatus

Alaska-Yukon Wild Flower Guide

fl.   p 88

Astragalus umbellatus

Heller, Christine
Wild Flowers of Alaska

fl.   Pl. 32

Astranthium integrifolium

Wharton, Mary E.
A Guide to the Wildflowers & Ferns
      of Kentucky

fl. p 262   Pl. 3.38

Astrantia amethystinum

Polunin, Oleg
Flowers of Europe

fl.   Pl. 83 #856

Astrantia major

Barneby, T. P.
European Alpine Flowers in Colour

fl.   Pl. 53, 1

Astrantia major

Felsko, Elsa
A Book of Wild Flowers 2nd Ser.

fl.   p. 92

Astrantia major

Huxley, Anthony
Garden Perennials and Water Plants

fl.   Pl. 57

Astrantia major

Kleijn, H.
Beauty of the Wild Plant

fl.   pg. 101   Pl. 146

Astrantia major

Martin, W. Keble
The Concise British Flora in Colour

fl. hab.   Pl. 36

Astrantia major

Morley, Brian D.
Wild Flowers of the World

fl.   Pl. 17E

Astrantia major

Perry, Frances
Flowers of the World

fl.   p 301

Astrantia major

Polunin, Oleg
Flowers of Europe

fl.   Pl. 83 #852

Astrantia major

Robert, Paul A
Alpine Flowers

fl.   Pl. 9

Astrantia maxima

Hay, Roy
The Color Dictionary of Flowers & Plants

fl.   p 125, Pl. 999

Astrantia minor

Barneby, T. P.
European Alpine Flowers in Colour

fl.   Pl. 53, 2

Astrantia, Star Umbel

      see

Astrantia major

Astridia longifolia

Herre, H.
The Genera of the Mesembryanthemaceae

fl. fr.   p 89

Astripomoea malvacea

Moriarty, Audrey
Wild Flowers of Malawi

fl. fr.   Pl. 59; 1

Astroloma ciliata

Harrison, R. E.
Trees and Shrubs

fl.   P 23

Astroloma conostephioides

Blombery, Alec M.
What Wildflower is That

fl.   p. 53   Pl. 110

Astroloma pallidum

Blombery, Alec M.
What Wildflower is That

fl.   p 53, Pl. 111

Astrophytum asterias

Cactus & Succulent Society of America
Journal
      vol 19, No. 2   1947

fl., hab.   cover (p. 17)

Astrophytum asterias

Hersey, Jean
Woman's Day Book of House Plants

fl., hab.   p 44

Astrophytum asterias

Kupper, Walter
Cacti

fl., hab.   p. 99   Pl. 46

Astrophytum asterias

Lamb, Edgar
Colorful Cacti of the American Deserts

fl. hab.   Pl. 8, 10

Astrophytum asterias

Lamb, Edgar
The Pocket Encyclopedia of Cacti
   and Succulents in Color

fl.   Pl. 4

Astrophytum asterias

Subik, Rudolf
Decorative Cacti

fl. hab.   p 21

Astrophytum asterias

Van Laren, A. J.
Cactus

fl.   P 56   Fig. 65

Astrophytum capricorne

Kiaer, Eigil
Indoor Plants in Colour

fl.   p. 53

Astrophytum capricorne

Kupper, Walter
Cacti

fl., hab.   p. 95   Pl. 44

Astrophytum capricorne

Lamb, Edgar
The Pocket Encyclopedia of Cacti
   and Succulents in Color

fl. hab.   Pl. 5

Astrophytum capricorne

Lamb, Edgar
Popular Exotic Cacti in Color

fl. hab.   p 35

Astrophytum capricorne

Subik, Rudolf
Decorative Cacti

fl. hab.   p 23

Astrophytum capricorne

Van Laren, A. J.
Cactus

fl.   P 56   Fig. 66

Astrophytum capricorne var.

Lamb, Edgar
The Pocket Encyclopedia of Cacti
   and Succulents in Color

fl. hab.   Pl. 8

Astrophytum myriostigma

Hersey, Jean
Woman's Day Book of House Plants

fl., hab.   p 41

Astrophytum myriostigma

Kromdijk, G.
200 House Plants in Colour

hab.   Pl. 26

Astrophytum myriostigma

Kupper, Walter
Cacti

hab., fl.   p. 97   Pl. 45

Astrophytum myriostigma

O'Gorman, Helen
Mexican Flowering Trees and Plants

fl. hab.   p 203

Astrophytum myriostigma

Subik, Rudolf
Decorative Cacti

fl. hab.   p 25

Astrophytum myriostigma

Van Laren, A. J.
Cactus

fl.   P 56   Fig. 64

Astrophytum myriostigma var.

Lamb, Edgar
The Pocket Encyclopedia of Cacti
   and Succulents in Color

fl. hab.   Pl. 6

Astrophytum ornatum

Hay, Roy
The Dictionary of House Plants

hab.   Pl. 49

Astrophytum ornatum

Kupper, Walter
Cacti

fl., hab.   p. 93   Pl. 43

Astrophytum ornatum

Lamb, Edgar
The Pocket Encyclopedia of Cacti
   and Succulents in Color

fl. hab.   Pl. 7

Astrophytum ornatum

Royal Hort. Soc.
Journal of the Royal Hort. Soc.
   vol 86, No. 8   1961

hab.   p. 354   Pl. 103

Astrophytum ornatum

Van Laren, A. J.
Cactus

hab. P 61   Fig. 67

Astrotricha ledifolia

Cochrane, G. R.
Flowers and Plants of Victoria

fl.   Pl. 397

Astydamia latifolia

Bramwell, David
Wild Flowers of the Canary Islands

fl., hab.   Pl. 202

Asystasia bella

Hay, Roy
The Dictionary of House Plants

fl.   Pl. 50

Asystasia gangetica

Bruggeman, L.
Tropical Plants

fl.   Pl. 80

Asystasia gangetica

Menninger, Edwin A.
Flowering Vines of the World

fl.   Pl. 2

Asystasia gangetica

Moriarty, Audrey
Wild Flowers of Malawi

fl.   Pl. 42; 1

Asystasia gangetica

Perry, Frances
Flowers of the World

fl.   p 14

Atamosco atamasco

Walcott, Mary Vaux
North American Wild Flowers
   vol. 4

fl.   Pl. 255

Athanasia punctata

Flowering Plants of Africa
   vol 43   1974-76

fl.   Pl. 1699

Atherosperma moschatum

Cochrane, G. R.
Flowers and Plants of Victoria

fl.   Pl. 451

Atherosperma moschatum

Curtis's Botanical Magazine
Vol 165   1948

fl.   43

Atherosperma moschatum

Royal Hort. Soc.
Journal of the Royal Hort. Soc.
    vol 100, No. 5   1975

fl.   p. 200   Pl. 85

Athrixia rosmarinifolia

Moriarty, Audrey
Wild Flowers of Malawi

fl.   Pl. 79; 5

Athrotaxis cupressoides

Curtis, Winifred
The Endemic Flora of Tasmania
        Vol. 2

fl. fr.   Pl. 70

Athrotaxis laxifolia

Curtis, Winifred
The Endemic Flora of Tasmania, Part IV

fl. fr. hab.   Pl. 138

Athrotaxis selaginoides

Curtis, Winifred
The Endemic Flora of Tasmania, Part IV

fl. fr. hab.   Pl. 137

Athrotaxis selaginoides

Curtis's Botanical Magazine
Vol 163   1940 - 42

fl. fr.   No. 9639

Atractylodes japonica

Kimura, Koiti
Japanese Medicinal Plants, Vol I

fl.   p 98

Atractylodes japonica

Takatori, Jisuke
Color Atlas of Medicinal Plants of Japan

fl.   Fig. 1A

Atractylodes lancea

Kariyone, Tatsuo
Atlas of Medicinal Plants

fl.   Pl. 27

Atractylodes lancea

Kimura, Koiti
Japanese Medicinal Plants, Vol II

fl.   p 237

Atriplex calotheca

Lindman, C. A. M.
Nordens Flora, Vol 3

hab.   Pl. 187A

Atriplex cinerea

Cochrane, G. R.
Flowers and Plants of Victoria

fl. hab.   Pl. 294

Atriplex glabriuscula var.

Martin, W. Keble
The Concise British Flora in Colour

fl. hab.   Pl. 72

Atriplex halimus

Feinbrun-Dothan, Naomi
Wild Plants in the Land
    of Israel

fr.   p 146

Atriplex halimus

Gault, S. Millar
The Color Dictionary of Shrubs

hab.   Pl. 13

Atriplex halimus

Hellyer, A. G. L.
Shrubs in Colour

hab.   p. 15

Atriplex hastata

Kleijn, H.
Beauty of the Wild Plant

fl.   p. 80   Pl. 105

Atriplex hastata

Martin, W. Keble
The Concise British Flora in Colour

fl. hab.   Pl. 72

Atriplex holocarpa

Blombery, Alec M.
What Wildflower is That

fr.   p. 54   Pl. 112

Atriplex hortensis

Crockett, James Underwood
Annuals

hab. P 97

Atriplex hortensis

Hay, Roy
The Color Dictionary of Flowers & Plants

fl.   p 31, Pl. 241

Atriplex hortensis

Masefield, G. B.
The Oxford Book of Food Plants

fl.   Pl. 161, 4

Atriplex laciniata

Martin, W. Keble
The Concise British Flora in Colour

fl. hab.   Pl. 72

Atriplex latifolia

Lindman, C. A. M.
Nordens Flora, Vol 3

fr. hab.   Pl. 186

Atriplex littoralis

Ary, S.
The Oxford Book of Wildflowers

fl.   p 54, Pl. 2

Atriplex littoralis

Martin, W. Keble
The Concise British Flora in Colour

fl. hab.   Pl. 72

Atriplex patula

Ary, S.
The Oxford Book of Wildflowers

fl.   p 54, Pl. 1

Atriplex patula

Courtenay, Booth
Wildflowers & Weeds

fl.   p 69

Atriplex patula

Lindman, C. A. M.
Nordens Flora, Vol 3

fr.   Pl. 187B

Atriplex patula

Martin, W. Keble
The Concise British Flora in Colour

fl. hab.   Pl. 72

Atriplex patula var.

Clark, Lewis J.
Wild Flowers of British Columbia

fl.   p 122

Atriplex semibaccata

Coyle, Jeanette
A Field Guide to the Common and
  Interesting Plants of Baja California

fl.   p 73

Atriplex vesicaria

Cochrane, G. R.
Flowers and Plants of Victoria

fl.   Pl. 180

Atropa belladonna

Ary, S.
The Oxford Book of Wildflowers

fl. fr.   p. 130   Pl. 4

Atropa belladonna

Bianchini, Francesco
Health Plants of the World

fl.   p. 121

Atropa bella-donna

Color Treasury of Herbs & Other
  Medicinal Plants

fl. fr. p 48   Pl. 69

Atropa belladonna

Felsko, Elsa
A Book of Wildflowers

fl.   P 73

Atropa belladonna

Hvass, Elsie
Plants That Feed and Serve Us

fl. fr.   p 87, Pl. 184

Atropa belladonna

Kimura, Koiti
Japanese Medicinal Plants, Vol II

fl.   p 221

Atropa belladonna

Loewenfeld, Claire
The Complete Book of Herbs and Spices

fl. hab.   p 112

Atropa bella-donna

Martin, W. Keble
The Concise British Flora in Colour

fl. fr. hab.   Pl. 61

Atropa belladonna

Perrot, Emile
Les Plantes Medicinales

fl. fr. hab.   p 29

Atropa bella-donna

Polunin, Oleg
Flowers of Europe

fl.   Pl. 117 #1174

Atropa bella-donna

Webster, Mary
Flora of Moray, Nairn & East
  Inverness

fr.   p. 292   Pl. 16

Attalea funifera

Hvass, Elsie
Plants That Feed and Serve Us

hab.   p 104, Pl. 222

Aubergine

see

Solanum melongena

Aubrieta deltoidea

Huxley, Anthony
Garden Perennials and Water Plants

fl.   Pl. 58

Aubrieta deltoidea

Macoby, Stirling
What Flower is That

fl.   P 48

Aubrieta deltoidea

Perry, Frances
Flowers of the World

fl.   p 98

Aubrieta deltoidea var.

Huxley, Anthony
Garden Perennials and Water Plants

fl.   Pl. 59

Aubrieta deltoidea var.

Vilmorin, Roger de
Plantes Alpines dans les Jardins

fl. hab.   Pl. XXI

Aubrieta hybrid

Hay, Roy
The Color Dictionary of Flowers & Plants

fl.   p. 3   Pl. 20

Aucuba chinensis

Walden, Beryl M.
Wild Flowers of Hong Kong

fl.   Pl. 11(35)

Aucuba japonica

Encke, Fritz
Zimmerpflanzen

fr.   p. 53

Aucuba japonica

Hay, Roy
The Color Dictionary of Flowers & Plants

fr.   p 182, Pl. 1453

Aucuba japonica

Kitamura, Siro
Coloured Illustrations of Trees &
  Shrubs of Japan

fl.   372

Aucuba japonica

Massachusetts Hort. Soc.
Horticulture
vol 40, No. 9   1962

fr.   backcover

Aucuba japonica

Polunin, Oleg
Flowers of Europe

fl.   Pl. 83 #847

Aucuba japonica

Polunin, Oleg
Trees and Bushes of Europe

fl. fr.   p 154

Aucuba japonica var.

Crockett, James Underwood
Evergreens

hab. fr.   p. 113

Aucuba japonica var.

Harrison, E. E.
Trees and Shrubs

fr.   P. 24

Aucuba japonica var.

Hellyer, A. G. L.
Shrubs in Colour

fr.   p. 15

Aucuba japonica var.

Huxley, Anthony
Evergreen Garden Trees and Shrubs

fr. hab.   Pl. 91, 92

Aucuba japonica var.

Kiaer, Eigil
Indoor Plants in Colour

hab.   p. 20

Aucuba japonica var.

Kromdijk, G.
200 House Plants in Colour

hab.   Pl. 27

Aucuba japonica var.

Macoby, Stirling
What Flower is That

hab.   p. 49

Aucuba japonica var.

Perry, Frances
Flowers of the World

hab.   p. 93

Aucuba sp.

Massachusetts Hort. Soc.
Horticulture
    vol 46, No. 4   1968

fr.   p. 29

Aureolaria dispersa

Brown, Clair A.
Wildflowers of Louisiana and
    Adjoining States

fl.   p 166

Aureolaria flava

Klimas, John E.
Wildflowers of Eastern America

fl.   Pl. 171

Aureolaria grandiflora

Courtenay, Booth
Wildflowers & Weeds

fl.   p 100

Aureolaria laevigata

Dean, Blanche E.
Wildflowers of Alabama and
    Adjoining States

Aureolaria laevigata

Jennings, O. E.
Wild Flowers of Western Pennsylvania
    Vol. II

fl.   Pl. 143

Aureolaria pectinata

Duncan, Wilbur H.
Wildflowers of the Southeastern
    United States

fl.   p 177

Aureolaria pedicularia

Wharton, Mary E.
A Guide to the Wildflowers & Ferns
    of Kentucky

fl.   p 213   Pl. 3.6

Aureolaria virginica

Duncan, Wilbur H.
Wildflowers of the Southeastern
    United States

fl.   p 177

Aureolaria virginica

Wharton, Mary E.
A Guide to the Wildflowers & Ferns
    of Kentucky

fl.   p 213   Pl. 3.5

Auricula

    see

Primula auricula

Aurora borealis plant

    see

Kalanchoe fedtschenkoi var.

Australian Nut

    see

Macadamia ternifolia

Austrian Briar

    see

Rosa foetida

Austrocephalocereus dybowskii

Backeberg, Curt
Cactus Lexicon

hab.   p 799

Austrocephalocereus estevesii

Cactus & Succulent Society of America
Journal
    vol 47, No. 6   1975

hab., fl.   p. 267

Austrocephalocereus purpureus

Backeberg, Curt
Cactus Lexicon

fl.   p 799

Austrocylindropuntia inarmata

Backeberg, Curt
Cactus Lexicon

fl.   p 558

Avena glauca var.

Massachusetts Hort. Soc.
Horticulture
    vol 43, No. 8   1965

hab.   p. 15

Avena sativa

Bianchini, F.
The Complete Book of Fruits & Vegetables

fl.   p 19

Avena sativa

Hvass, Elsie
Plants That Feed and Serve Us

fr.   p. 6   Pl. 4

Avena sativa

Masefield, G. B.
The Oxford Book of Food Plants

fr.   Pl. 5, 2

Avena sativa

Perrot, Emile
Les Plantes Medicinales

fr. hab.   p 56

Avena sterilis

Crockett, James Underwood
Annuals

fr.   P 97

Avena sterilis

Polunin, Oleg
Flowers of Europe

fr. Pl. 180 #1265

Avens, Aleutian

    see

Geum pentapetalum

Avens, Alpine

    see

Sieversia montana

Avens, Appalachian

    see

Geum radiatum

Avens, Drummond

    see

Dryas drummondi

Avens, Glacier

    see

Geum glaciale

Avens, Large-Leaved

see

Geum macrophyllum

Avens, Low

see

Geum pentapetalum

Avens, Mountain

see

Dryas octopetala

Avens, Mountain

see

Sieversia montana

Avens, purple

see

Geum rivale

Avens, Ross

see

Geum rosii

Avens, Swamp

see

Geum rivale

Avens, Water

see

Geum rivale

Avens, White

see

Geum canadense

Avens, Yellow

see

Geum aleppicum

Averrhoa carambola

Edlin, Herbert
The Illustrated Encyclopedia
    of Trees

fr.   p 25

Averrhoa carambola

Macoby, Stirling
What Flower is That

fr.   P 49

Averrhoa carambola

Masefield, C. B.
The Oxford Book of Food Plants

fl., fr.   Pl. 103, 1

Avicennia germinans

Brown, Clair A.
Wildflowers of Louisiana and
    Adjoining States

fl. hab.   p 154

Avicennia germinans

Coyle, Jeanette
A Field Guide to the Common and
    Interesting Plants of Baja California

fl. fr.   p 155

Avicennia germinans

Fleming, Glenn
Wild Flowers of Florida

fr. hab.   p 41

Avicennia marina

Cochrane, G. R.
Flowers and Plants of Victoria

fl. hab.   Pl. 190, 191

Avicennia marina

Edlin, Herbert
The Illustrated Encyclopedia
    of Trees

hab.   226

Avicennia resinifolia

Harvey, Norman B.
New Zealand Botanical Paintings

fr.   p. 64   Pl. 29

Avocado

see

Persea americana

Axseed

see

Coronilla varia

Axonopus affinis

Crockett, James Underwood
Lawns & Ground Covers

fr.   p. 115

Ayensua uaipanensis

Bromeliad Society
Journal
    vol 25, No. 6   1975

fl., hab.   p. 215

Aylostera buiningiana

Backeberg, Curt
Cactus Lexicon

fl.   p 800

Aylostera heliosa

Backeberg, Curt
Cactus Lexicon

fl.   p 800

Aylostera jujuyana

Backeberg, Curt
Cactus Lexicon

fl.   p 800

Aylostera kupperiana

Kupper, Walter
Cacti

fl., hab.   p. 69   Pl. 31

Aylostera kupperiana

Subik, Rudolf
Decorative Cacti

fl. hab.   p 27

Azalea

see

Rhododendron

Azalea, Alpine

see

Loiseleuria procumbens

Azalea, Alpine

see

Rhododendron lapponicum

Azalea, Creeping

see

Loiseleuria procumbens

Azalea, Flame

see

Rhododendron calendulaceum

Azalea, Flame

see

Rhododendron lutea

Azalea, Florida

see

Rhododendron austrinum

Azalea, Hino crimson

see

Rhododendron obtusum var.

Azalea, Hoary Pink

see

Rhododendron canescens

Azalea, Indian

see

Rhododendron simsii

Azalea, Korean

see

Rhododendron mucronulatum var.

Azalea, Oregon

see

Rhododendron occidentale

Azalea, Piedmont

see

Rhododendron canescens

Azalea, Pink

see

Rhododendron nudiflorum

Azalea, Pinkshell

see

Rhododendron vaseyi

Azalea, Pinxterbloom

see

Rhododendron nudiflorum

Azalea, Plumleaf

see

Rhododendron prunifolium

Azalea, Sweet

see

Rhododendron arborescens

Azalea, Trailing

see

Loiseleuria procumbens

Azalea, Tree

see

Rhododendron arborescens

Azalea, Western

see

Rhododendron occidentale

Azanza garckeana

Moriarty, Audrey
Wild Flowers of Malawi

fl.   Pl. 35; 1

Azanza garckeana

Palgrave, K. C.
Trees of Central Africa

fl., fr.   p. 209

Azara integrifolia

Curtis's Botanical Magazine
Vol 163   1940-42

fl. fr.   No. 9620

Azara integrifolia var.

Harrison, R. E.
Trees and Shrubs

hab.   P. 24

Azara lanceolata

Harrison, R. E.
Trees and Shrubs

fl.   P. 24

Azara lanceolata

Perry, Frances
Flowers of the World

fl.   p 118

Azara microphylla

Hay, Roy
The Color Dictionary of Flowers & Plants

hab.   p 182, Pl. 1454

Azara microphylla

Hellyer, A. G. L.
Shrubs in Colour

fl.   pg. 15

Azarole

see

Crataegus azarolus

Azorella trifurcata

Huxley, Anthony
Garden Perennials and Water Plants

hab.   Pl. 60

Aztekium ritteri

Lamb, Edgar
Popular Exotic Cacti in Color

fl. hab.   p 17

Babe-in-the-Cradle

see

Epiblema grandiflorum

Babiana cuneifolia

Flowering Plants of South Africa
vol XVIII   1938

fl., hab.   Pl. 685

Babiana dregei

Eliovson, Sima
Namaqualand in Flower

fl.   Pl. 31, 1/2

Babiana geniculata

Eliovson, Sima
Namaqualand in Flower

fl.   Pl. 31, 3

Babiana hypogea

Flowering Plants of Africa
vol XXV   1945-46

fl., hab.   Pl. 962

Babiana pygmaea

Flowering Plants of Africa
vol 44   1977

fl., hab.   Pl. 1731

Babiana, Red

see

Babiana villosa

Babiana sambucina

Hay, Roy
The Dictionary of House Plants

fl.   Pl. 51

Babiana spiralis

Flowering Plants of South Africa
        vol XVIII   1938

fl., hab.   Pl. 686

Babiana stricta

Macoby, Stirling
What Flower is That

fl.   P 51

Babiana villosa

American Hort. Soc.
American Horticulturist
        vol 53, No. 3   1974

fl.   p. 35

Baboon flower

see

Babiana stricta

Baboon root

see

Babiana stricta

Baby blue-eyes

see

Nemophila menziesii

Baby's Breath

see

Gypsophila

Baby's breath, annual

see

Gypsophila elegans

Baby's Breath, Creeping

see

Gypsophila repens

Baby's-Breath, Wild

see

Arenaria patula

Baby's Tears

see

Helxine soleirolii

Baccharis halimifolia

Fleming, Glenn
Wild Flowers of Florida

fl. hab.   p 22

Baccharis pilularis

Crockett, James Underwood
Lawns & Ground Covers

fl.   p. 124

Baccharis sarothroides

Coyle, Jeanette
A Field Guide to the Common and
    Interesting Plants of Baja California

hab.   p 171

Baccharis sp.

Tarver, David P.
Aquatic and Wetland Plants
    of Florida

hab.   p 72

Bachelor's button

see

Centaurea cyanus

Bachelor's Button

see

Craspedia uniflora

Bachelor's Button, Bog

see

Polygala lutea

Bachelor's Button, Low

see

Polygala nana

Bachelor's Button , White

see

Ranunculus aconitifolius

Bachelor's Button, Yellow

see

Polygala lutea

Bachelor's Button, Yellow

see

Polygala rugelii

Bachelor's button , Yellow

see

Ranunculus acris var.

Backebergia chrysomallus

Cactus & Succulent Soc. of America
Journal
        vol 45, No. 1   1973

fl.   cover (p. 1)

Backhousia citriodora

Blombery, Alec M.
What Wildflower is That

fl.   p 54, Pl. 113

Backhousia citriodora

Macoby, Stirling
What Flower is That

fl.   P 52

Backhousia citriodora

Oakman, Harry
Colorful Trees

Hab. Fl.   P. 61

Backhousia myrtifolia

Blombery, Alec M.
What Wilflower is That

fl.   p 55, Pl. 114

Bacopa amplexicaulis

Stodola, Jiri
Encyclopedia of Water Plants

Fl.   P. 238

Bacopa caroliniana

Duncan, Wilbur H.
Wildflowers of the Southeastern
        United States

fl.   p 173

Bacopa caroliniana

Fleming, Glenn
Wild Flowers of Florida

fl. hab.   p 87

Bacopa caroliniana

Tarver, David P.
Aquatic and Wetland Plants
of Florida

fl. hab.   p 36

Bacopa, Fragrant

see

Bacopa caroliniana

Bacopa monnieri

Duncan, Wilbur H.
Wildflowers of the Southeastern
United States

fl.   p 173

Bacopa monnieri

Fleming, Glenn
Wild Flowers of Florida

fl. hab.   p 92

Bacopa monnieri

Stodola, Jiri
Encyclopedia of Water Plants

hab.   P. 238

Bacopa monnieri

Tarver, David P.
Aquatic and Wetland Plants
of Florida

fl. hab.   p 37

Baeckea crassifolia

Cochrane, G. R.
Flowers and Plants of Victoria

fl.   Pl. 92, 165

Baeckea frutescens

Walden, Beryl M.
Wild Flowers of Hong Kong

fl.   Pl. 35(94)

Baeckea gunniana

Cochrane, G. R.
Flowers and Plants of Victoria

fl.   Pl. 520

Baeckea leptocaulis

Curtis, Winifred
The Endemic Flora of Tasmania, Vol. 5

fl. fr.   Pl. 163

Baeckea linifolia

Blombery, Alec M.
What Flower is That

fl.   p 56, Pl. 115

Baeckea ramosissima

Blombery, Alec M.
What Wildflower is That

fl.   p 56, Pl. 116

Baeckea ramosissima

Cochrane, G. R.
Flowers and Plants of Victoria

fl.   Pl. 331

Baeckea, shrubby Chinese

see

Baeckia frutescens

Baeckea virgata

Blombery, Alec M.
What Wildflower is That

fl.   p 56, Pl. 117

Baeria chrysostoma

Munz, Philip A.
California Spring Wildflowers

fl.   p 80, Pl. 66

Baeriopsis guadalupensis

Cactus & Succulent Society of America
Journal
     vol 40, No. 2   1968

fl.   p. 67

Bag flower

see

Clerodendrom thomsoniae

Bahiagrass

see

Paspalum notatum

Baikiaea insignis

Morley, Brian D.
Flowers of the World

fl.   Pl. 54

Baikiaea plurijuga

Palgrave, K. C.
Trees of Central Africa

fl., fr.   p. 67

Baileya multiradiata

American Hort. Soc.
American Horticulturist
     vol 57, No. 1   1978

fl.   p. 34

Baileya multiradiata

Lemmon, Robert S.
Wildflowers of North America
in Full Color

fl.   pg. 108   pl. 174

Baileya multiradiata

Munz, Philip A.
California Desert Wildflowers

fl.   p 56, Pl. 88

Baileya multiradiata

Welsh, Stanley L.
Flowers of the Canyon Country

fl. hab.   p 30

Baissea multiflora

Wit, H. C. D. de
Plants of the World;
The Higher Plants, Vol II

fl.   p 102, Pl. 53

Baissea sp.

Menninger, Edwin A.
Flowering Vines of the World

fl.   Pl. 12

Baked-apple Berry

see

Rubus chamaemorus

Balanophora celebica

Morley, Brian D.
Wild Flowers of the World

fl.   Pl. 117

Balanophora dioica

Hara, Hiroshi, comp.
Photo-Album of Plants of
Eastern Himalaya

fl. hab.   Pl. 20

Balanophora polyandra

Hara, Hiroshi, comp.
Photo-Album of Plants of
Eastern Himalaya

fl. hab.   Pl. 23

Balaustion pulcherrimum

Blombery, Alec M.
What Wildflower is That

fl.   p 56, Pl. 118

Balbisia peduncularis

Munoz Pizarro, Carlos
Flores Silvestres de Chile

fl.   Pl. 5

Baldellia ranunculoides

Felsko, Elsa
A Book of Wild Flowers
                    2nd Ser.
fl. fr.   p 96

Baldellia ranunculoides

Martin, W. Keble
The Concise British Flora in Colour

fl. fr. hab.   Pl. 79

Balduina angustifolia

Duncan, Wilbur H.
Wildflowers of the Southeastern
          United States

fl.   p 221

Balduina uniflora

Brown, Clair A.
Wildflowers of Louisiana and
      Adjoining States

fl.   p. 189

Balisier

see

Heliconia bihai

Ballflower, Common

see

Globularia willkommii

Ball, Golden

see

Echinocactus grusonii

Ballart, Cherry

see

Exocarpos cupressiformis

Ballart, Coast

see

Exocarpos syrticola

Ballart, Leafless

see

Exocarpos aphyllus

Balloon flower

see

Platycodon

Balloon vine

see

Cardiospermum halicacabum

Ballota foetida

Perrot, Emile
Les Plantes Medicinales

fl.,fr.,hab.   p. 26

Ballota nigra

Ary, S.
The Oxford Book of Wildflowers

fl.   p 148, Pl. 5

Ballota nigra

Lindman, C. A. M.
Nordens Flora, Vol 8

fl. fr.   Pl. 507

Ballota nigra

Martin, W. Keble
The Concise British Flora in Colour

fl. hab.   Pl. 70

Ballota nigra

Polunin, Oleg
Flowers of Europe

fl.   Pl. 112 #1134

Ballota pseudo-dictammus

Hay, Roy
The Color Dictionary of Flowers & Plants

fl.   p 125, Pl. 1000

Balm

see

Melissa

Balm, Bastard

see

Melittis melissophyllum

Balm, Horse

see

Collinsonia canadensis

Balm, Lemon

see

Melissa officinalis

Balm, Molucca

see

Moluccella laevis

Balm, Mountain

see

Eriodictyon trichocalyx

Balm of Gilead

see

Populus gileadensis

Balsa Wood

see

Ochroma lagopus

Balsam

see

Impatiens

Balsam, garden

see

Impatiens balsamina

Balsam of Peru

see

Myroxylon pereirae

Balsam, Orange

see

Impatiens capensis

Balsam, pear

see

Momordica charantia

Balsam Root

see

Balsamorhiza

Balsamroot, Arrowleaf

see

Balsamorhiza sagittata

Balsam, small

see

Impatiens parviflora

Balsam, Sultan's

see

Impatiens sultanii

Balsam Tree

see

Colophospermum mopane

Balsam, Yellow

see

Impatiens noli-tangere

Balsam, Zanzibar

see

Impatiens sultanii

Balsamorhiza platylepis

Lemmon, Robert S.
Wildflowers of North America

fl.   p. 272   Pl. 435

Balsamorhiza sagittata

Clark, Lewis J.
Wild Flowers of British Columbia

fl. hab.   p 534, 535

Balsamorhiza sagittata

Ferguson, Mary
Wildflowers

fl.   p 175

Balsamorhiza sagittata

Orr, Robert T.
Wildflowers of Western America

fl.   Pl. 80

Balsamorhiza sagittata

Shaw, Richard J.
Field Guide to the Vascular Plants of
Grand Teton National Park

fl.   Pl. 2

Balsamorhiza sagittata

Taylor, Ronald, J.
Mountain Wild Flowers

fl. hab.   p 17

Balsamorhiza sagittata

Turner, Nancy J.
Wild Coffee and Tea Substitutes of
Canada

fl.   p.34

Balsamorhiza sagittata

Walcott, Mary Vaux
North American Wild Flowers

fl.   Pl. 69

Balsamorhiza sagittata

Welsh, Stanley L,
Flowers of the Mountain Country

fl.   p. 48

Bamboo

see

Bambusa

Bamboo, Chinese sacred

see

Nandina domestica

Bamboo, dwarf

see

Sasa

Bamboo, golden

see

Phyllostachys aurea

Bamboo, sacred

see

Nandina domestica

Bamboo shoots

see

Bambusa vulgaris

Bamburanta

see

Ctenanthe lubbersiana

Bambusa species

Hvass, Elsie
Plants That Feed and Serve Us

fl.   p 131, Pl. 290

Bambusa vulgaris

Masefield, G. B.
The Oxford Book of Food Plants

hab.   Pl. 163, 4

Banana

see

Musa

Banana, Japanese

see

Musa Basjoo

Banana, ladyfinger

see

Musa cavendishii

Banana Passion

see

Passiflora mollissima

Banana, Pink Fruited

see

Musa velutina

Banana, Scarlet

see

Musa coccinea

Banana shrub

see

Michelia fuscata

Banana, Wild

see

Strelitzia nicolai

Baneberry

see

Actaea

Baneberry, Ivory

see

Actaea arguta

Baneberry, Red

see

Actaea arguta

Baneberry, Red

see

Actaea rubra

Baneberry, White

see

Actaea alba

Baneberry, White

see

Actaea pachypoda

Banksia, acorn

see

Banksia prionotes

Banksia ashbyi

Blombery, Alec M.
What Wildflower is That

fl.   p 56, Pl. 119

Banksia ashbyi

Morcombe, M. K.
Australia's Western Wildflowers

fl.   p 34

Banksia attenuata

Morcombe, M. K.
Australia's Western Wildflowers

fl.   p. 38, 49, 53

Banksia baueri

Blombery, Alec M.
What Wildflower is That

fl.   p 57, Pl. 120

Banksia baueri

Harrison, R. E.
Trees and Shrubs

fl.   P 30

Banksia baxteri

Perry, Frances
Flowers of the World

fl.   p 247

Banksia, Bull

see

Banksia grandis

Banksia burdettii

Perry, Frances
Flowers of the World

fl.   p 247

Banksia, Coast

see

Banksia integrifolia

Banksia coccinea

Blombery, Alec M.
What Wildflower is That

fl.   P 58, Pl. 121

Banksia coccinea

Curtis's Botanical Magazine
Vol 179   1972

fl.   630

Banksia coccinea

Harrison, R. E.
Trees and Shrubs

fl.   P 30

Banksia coccinea

Macoby, Stirling
What Flower is That

fl.   P 52

Banksia coccinea

Morcombe, M. K.
Australia's Western Wildflowers

fl.   p 6, 42

Banksia collina

Harrison, R. E.
Trees and Shrubs

fl.   P 30

Banksia collina

Macoby, Stirling
What Flower is That

fl.   P 52

Banksia dryandroides

Morcombe, M. K.
Australia's Western Wildflowers

fl.   p 101

Banksia ericifolia

Blombery, Alec M.
What Wildflower is That

fl.   p 58, Pl. 122

Banksia ericifolia

Harrison, R.E.
Trees and Shrubs

fl.   P 31

Banksia ericifolia

Macoby, Stirling
What Flower is That

fl.   P 52

Banksia ericifolia

Mullins, Barbara
Australian Wildflowers in Colour

fl.   P. 35   Pl. 26

Banksia grandis

Blombery, Alec M.
What Wildflower is That

fl.   p 59, Pl. 123

Banksia, Heath

see

Banksia ericifolia

Banksia hookeriana

Morcombe, M. K.
Australia's Western Wildflowers

fl.   p 73

Banksia ilicifolia

Morcombe, M. K.
Australia's Western Wildflowers

fl.   p 40

Banksia integrifolia

Blombery, Alec M.
What Wildflower is That

fl.   p 59, Pl. 124

Banksia integrifolia

Cochrane, G. R.
Flowers and Plants of Victoria

hab.   Pl. 275

Banksia integrifolia

Harrison, R. E.
Trees and Shrubs

fl.   P 31

Banksia integrifolia

Macoby, Stirling
What Flower is That

fl.   P 52

Banksia integrifolia

Royal Hort. Soc.
Journal of the Royal Hort. Soc.
    vol 90, No. 1   1965

fl.   p. 22   Pl. 20

Banksia laricina

Morcombe, M. K.
Australia's Western Wildflowers

fl.   p 68

Banksia marginata

Blombery, Alec M.
What Wildflower is That

fl.   p 59, Pl. 125

Banksia marginata

Cochrane, C. R.
Flowers and Plants of Victoria

fl.   Pl. 229

Banksia marginata

Mullins, Barbara
Australian Wildflowers in Colour

Hab. & fl.   P 33   Pl. 25

Banksia media

Harrison, R.E.
Trees and Shrubs

fl.   P 32

Banksia media

Massachusetts Hort. Soc.
Horticulture
    vol 44, No. 12 1966

fl.   cover, backcover

Banksia occidentalis

Mathias, Mildred E.
Color for the Landscape

fk.   p 78

Banksia occidentalis

Morcombe, M. K.
Australia's Western Wildflowers

fl.   p 105

Banksia, Orange

see

Banksia prionotes

Banksia ornata

Harrison, R. E.
Trees and Shrubs

fl.   P 30

Banksia prionotes

Blombery, Alec M.
What Wildflower is That

fl.   p 60, Pl. 126

Banksia prionotes

Harrison, R.E.
Trees and Shrubs

fl.   P 32

Banksia prionotes

Mullins, Barbara
Australian Wildflowers in Colour

fl.   P. 37   Pl. 29

Banksia prostrata

Harrison, R. E.
Trees andShrubs

fl.   P 31

Banksia repens

Harrison, R.E.
Trees and Shrubs

fl.   P 32

Banksia robur

Mullins, Barbara
Australian Wildflowers in Colour

fl.   P. 37   Pl. 28

Banksia, Round-fruit

see

Banksia sphaerocarpa

Banksia, Saw

see

Banksia serrata

Banksia, Scarlet

see

Banksia coccinea

Banksia serrata

Blombery, Alec M.
What Wildflower is That

fl.   p 60, Pl. 127

Banksia serrata

Cochrane, C. R.
Flowers and Plants of Victoria

fl.   Pl. 21

Banksia serrata

Curtis's Botanical Magazine
    Vol 163   1940-42

fl.   No. 9642

Banksia serrata

Edlin, Herbert
The Illustrated Encyclopedia
    of Trees

fl.   p 243

Banksia serrata

Morley, Brian D.
Wild Flowers of the World

fl.   pl. 134

Banksia serrata

Mullins, Barbara
Austrailian Wildflowers in Colour

fr.   p. 35   Pl. 27

Banksia serratifolia

Blombery, Alec M.
What Wildflower is That

fl.   p 60, Pl. 128

Banksia, showy

see

Banksia speciosa

Banksia, silver

see

Banksia marginata

Banksia sp.

Milne, Lorus
Living Plants of the World

fl.   P 73

Banksia speciosa

Blombery, Alec M.
What Wildflower is That

fl.   p 60, Pl. 129

Banksia speciosa

Harrison, R.E.
Trees and Shrubs

fl.   P 31

Banksia sphaerocarpa

Blombery, Alec M.
What Wildflower is That

fl.   p 61, Pl. 130

Banksia sphaerocarpa

Morcombe, M. K.
Australia's Western Wildflowers

fl.   p 2, frontispiece

Banksia spinulosa

Blombery, Alec M.
What Wildflower is That

fl.   p 61, Pl. 131

Banksia spinulosa

Cochrane, G. R.
Flowers and Plants of Victoria

fl.   Pl. 387

Banksia spinulosa

Curtis's Botanical Magazine
vol 176   1966-68

fl.   Pl. 498

Banksia spinulosa

Harrison, R. E.
Trees and Shrubs

Banksia, swamp

see

Banksia robur

Banksia, Swamp

see

Banksia verticillata

Banksia verticillata

Blombery, Alec M.
What Wildflower is That

fl.   p 61, Pl. 132

Banyan

see

Ficus benghalensis

Banyan tree, weeping chinese

see

Ficus benjamina

Baobab

see

Adansonia digitata

Baptisia alba

Batson, Wade T.
Wild Flowers in South Carolina

fl. hab.   p 62

Baptisia alba

Duncan, Wilbur H.
Wildflowers of the Southeastern
    United States

fl.   p 71

Baptisia australis

Duncan, Wilbur H.
Wildflowers of the Southeastern
    United States

fl.   p 71

Baptisia australis

Hay, Roy
The Color Dictionary of Flowers & Plants

fl.   p 126, Pl. 1001

Baptisia australis

Klimas, John E.
Wildflowers of Eastern America

fl.   Pl. 252

Baptisia australis

Tsukamoto, Yotaro
Coloured Illustrations of
    Garden Flowers Vol. 9

fl.   opp. p. 33

Baptisia australis

Wharton, Mary E.
A Guide to the Wildflowers & Ferns
    of Kentucky

fl. p 180   Pl. 1.4

Baptisia bracteata

Batson, Wade T.
Wild Flowers in South Carolina

fl. hab.   p 63

Baptisia bracteata

Duncan, Wilbur H.
Wildflowers of the Southeastern
    United States

fl.   p 69

Baptisia lanceolata

Duncan, Wilbur H.
Wildflowers of the Southeastern
    United States

fl. hab.   p 71

Baptisia leucantha

Brown, Clair A.
Wildflowers of Louisiana and
    Adjoining States

fl. hab.   p 74

Baptisia leucantha

Courtenay, Booth
Wildflowers and Weeds

fl.   p. 46

Baptisia leucantha

Dean, Blanche E.
Wildflowers of Alabama and
    Adjoining States

fl.   p 85

Baptisia leucophaea

Brown, Clair A.
Wildflowers of Louisiana and
    Adjoining States

fl. hab.   p 74

Baptisia leucophaea

Courtenay, Booth
Wildflowers and Weeds

fl.   p. 46

Baptisia leucophaea

Wharton, Mary E.
A Guide to the Wildflowers & Ferns
    of Kentucky

fl. p 180   Pl. 1.3

Baptisia nuttaliana

Brown, Clair A.
Wildflowers of Louisiana and
    Adjoining States

fl. hab.   p 75

Baptisia sphaerocarpa

Brown, Clair A.
Wildflowers of Louisiana and
    Adjoining States

fl. hab.   p 75

Baptisia sphaerocarpa

Lemmon, Robert S.
Wildflowers of North America
    in Full Color

fl.   pg. 186   pl. 292

Baptisia tinctoria

Courtenay, Booth
Wildflowers and Weeds

fl.   p. 46

Baptisia tinctoria

Dean, Blanche E.
Wildflowers of Alabama and
    Adjoining States

fl.   p 85

Baptisia tinctoria

Duncan, Wilbur H.
Wildflowers of the Southeastern
United States

fl. p 69

Baptisia tinctoria

Klimas, John E.
Wildflowers of Eastern America

fl. Pl. 139

Baptisia tinctoria

Pond, Barbara
A Sampler of Wayside Herbs

fl. Pl. XI

Baptistonia echinata

Pabst, G. F. J.
Orchidaceae Brasilienses,
Vol 2

fl. p 254

Barberry

see

Berberis

Barberry, Black

see

Berberis gagnepainii

Barberry, common

see

Berberis vulgaris

Barberry, Darwin

see

Berberis Darwinii

Barberry, Dwarf Magellan

see

Berberis buxifolia

Barberry, Japanese

see

Berberis thunbergii

Barberry, Prickly-leaved

see

Berberis darwinii

Barberry, Rosemary

see

Berberis stenophylla

Barberry, Warty

see

Berberis verruculosa

Barberry, Wintergreen

see

Berberis julianae

Barbados flower fence

see

Caesalpina pulcherrima

Barbados-pride

see

Caesalpinia pulcherrima

Barbara's Buttons

see

Marshallia trinervia

Barbarea orthoceras

Clark, Lewis J.
Wild Flowers of British Columbia

fl. p 182

Barbarea stricta

Martin, W. Keble
The Concise British Flora in Colour

fl. fr. hab. Pl. 7

Barbarea verna

Martin, W. Keble
The Concise British Flora in Colour

fl. fr. hab. Pl. 7

Barbarea vulgaris

Ary, S.
The Oxford Book of Wildflowers

fl. p 10, Pl. 2

Barbarea vulgaris

Courtenay, Booth
Wildflowers & Weeds

fl. p 17

Barbarea vulgaris

Duncan, Wilbur H.
Wildflowers of the Southeastern
United States

fl. p 51

Barbarea vulgaris

Klimas, John E.
Wildflowers of Eastern America

fl. Pl. 123

Barbarea vulgaris

Lindman, C. A. M.
Nordens Flora, Vol 4

fl. fr. Pl. 270

Barbarea vulgaris

Martin, W. Keble
The Concise British Flora in Colour

fl. fr. hab. Pl. 7

Barbarea vulgaris

Moyle, John B.
Northland Wild Flowers

fl. hab. p 60, Pl. 25

Barbarea vulgaris

Wharton, Mary E.
A Guide to the Wildflowers & Ferns
of Kentucky

fl. p 148 Pl. 2.2

Barbary Nut

see

Iris sisyrinchium

Barbers' Poles

see

Oxalis versicolor

Barbosella cuculala

Ospina, Mariano
Orquideas de las americas

fl. hab. Pl. 74

Barbosella handroi

Pabst, G. F. J.
Orchidaceae Brasilienses,
Vol 1

fl. p 228

Barbosella miersii

Pabst, G.F.J.
Orchidaceae Brasilienses
vol 1

hab. p. 52

Barclaya longifolia

Stodola, Jiri

Encyclopedia of Water Plants,

fl.   P 215

Barkeria elegans

Ospina, Mariano
Orquideas de las americas

fl.   Pl. 49

Barklya syringifolia

Blombery, Alec M.
What Wildflower is That

fl.   p 62, Pl. 133

Barklya syringifolia

Oakman, Harry
Colorful Trees

fl.   P 63

Barleria albostellata

Flowering Plants of Africa
vol 29   1952-53

fl., fr.   Pl. 1138

Barleria bremekampi

Flowering Plants of South Africa
vol XXIII   1943

fl.   Pl. 893

Barleria cristata

Bruggeman, L.
Tropical Plants

fl.   Pl. 217

Barleria cristata

Macoby, Stirling
What Flower is That

fl.   P 53

Barleria involucrata

Bruggeman, L.
Tropical Plants

fl.   Pl. 218

Barleria involucrata

Perry, Frances
Flowers of the World

fl.   p 14

Barleria kirkii

Curtis's Botanical Magazine
vol 181   1976-77

fl.   Pl. 707

Barleria lupulina

Bruggeman, L.
Tropical Plants

fl.   Pl. 219

Barleria obtusa

Flowering Plants of Africa
vol XXV   1945-46

fl.   Pl. 998

Barleria pretoriensis

Flowering Plants of Africa
vol 41   1970-71

fl.   Pl. 1623

Barleria rotundifolia

Flowering Plants of Africa
vol 36   1963-64

fl.   Pl. 1426

Barleria senensis

Moriarty, Audrey
Wild Flowers of Malawi

fl.   Pl. 43; 1

Barleria spinulosa

Moriarty, Audrey
Wild Flowers of Malawi

fl.   Pl. 43; 3

Barley

see

Hordeum vulgare

Barley, Fox-Tail

see

Hordeum jubatum

Barley, Little

see

Hordeum pusillum

Barley, Six-rowed

see

Hordeum vulgare

Barley, two-rowed

see

Hordeum distichon

Barnyardgrass

see

Echinochloa crusgalli

Barrel, Blue

see

Echinocactus horizonthalonius

Barren Wort

see

Epimedium alpinum

Barrenwort

see

Epimedium rubrum

Barrenwort

see

Epimedium sulphureum

Barringtonia racemosa

Flowering Plants of Africa
vol 43   1974-76

fl., fr.   Pl. 1706

Barringtonia samoensis

Morely, Brian D.
Wild Flowers of the World

fl.   pl. 148

Barthea barthei

Walden, Beryl M.
Wild Flowers of Hong Kong

fl.   Pl. 15(42)

Bartholina burmanniana

Morley, Brian D.
Wild Flowers of the World

fl.,hab.   Pl. 89

Bartholina ethelae

Flowering Plants of Africa
vol 30   1954-55

fl., hab.   Pl. 1178

Bartsia alpina

Barneby, T. P.
European Alpine Flowers in Colour

fl.   Pl. 74, 1

Bartsia alpina

Kohlhaupt, Paula
Fleurs des Alpages vol 1

fl.   P 93

Bartsia alpina

Lindman, C. A. M.
Nordens Flora, Vol 8

fl. fr.   Pl. 535

Bartsia alpina

Martin, W. Keble
The Concise British Flora in Colour

fl. hab.   Pl. 65

Bartsia alpina

Polunin, Oleg
Flowers of Europe

fl.   Pl. 126 #1236

Basella rubra

Menninger, Edwin A.
Flowering Vines of the World

fl.   Pl. 31

Basella rubra

Tsukamoto, Yotaro   Vol 10
Coloured Illustrations of Garden Flowers

fl.   Illus. 13, opp. p. 4

Basil

see

Ocimum

Basil, Red

see

Satureja coccineum

Basil, sweet

see

Ocimum basilicum

Basiphyllaea corallicola

Luer, Carlyle A.
The Native Orchids of Florida

fl. fr. hab.   Pl. 70, 3-7

Baskervillea paranaensis

Pabst, G. F. J.
Orchidaceae Brasilienses,
Vol 1

fl.   p 180

Basket Flower

see

Centaurea americana

Basket Flower

see

Hymenocallis americana

Basket, Indian

see

Echinomastus erectocentrus

Basket of Gold

see

Alyssum saxatile

Basket, Pedlar's

see

Saxifraga stolonifera

Basswood

see

Tilia

Basswood, American

see

Tilia americana

Bassia divaricata

Cochrane, G. R.
Flowers and Plants of Victoria

fl.   Pl. 144

Bassia hirsuta

Lindman, C. A. M.
Nordens Flora, Vol 3

fr. hab.   Pl. 189

Bassia obliquicuspis

Cochrane, G. R.
Flowers and Plants of Victoria

hab.   Pl. 146

Bassia quinquecuspis

Cochrane, G. R.
Flowers and Plants of Victoria

hab.   Pl. 238

Bassia tricuspis

Cochrane, G. R.
Flowers and Plants of Victoria

hab.   Pl. 145

Bast, Cuban

see

Hibiscus elatus

Batemannia colleyi

Ospina, Mariano
Orquideas de las americas

fl.   hab.   Pl. 99

Batemannia colleyi

Pabst, G. F. J.
Orchidaceae Brasilienses,
Vol 2

fl.   p 229

Bats-in-the-Belfry

see

Campanula trachelium

Battleweed

see

Baptisia tinctoria

Bauera rubioides

Blombery, Alec M.
What Wildflower is That

fl.   p 63, Pl. 134

Bauera rubioides

Cochrane, G. R.
Flowers and Plants of Victoria

fl.   Pl. 56

Bauera rubioides

Macoby, Stirling
What Flower is That

fl. P 53

Bauera sessiliflora

Cochrane, G. R.
Flowers and Plants of Victoria

fl.   Pl. 100

Bauera sessiliflora

Harrison, R.E.
Trees and Shrubs

fl. P 33

Bauera, showy

see

Bauera rubioides

Bauhinia blakeana

American Hort. Soc.
American Horticulturist
vol 52, No. 4   1973

fl.   p. 11

Bauhinia blakeana

Crockett, James Underwood
Evergreens

hab. fl.   p. 113

Bauhinia blakeana

Mathias, Mildred E.
Color for the Landscape

fl.   p 8

Bauhinia blakeana

Oakman, Harry
Colorful Trees

Hab. Fl.   P 127

Bauhinia, butterfly

see

Bauhinia monandra

Bauhinia carronii

Blombery, Alec M.
What Wildflower is That

fl.   p 63, Pl. 135

Bauhinia, Climbing

see

Bauhinia corymbosa

Bauhinia corymbosa

Harrison, Richmond E.
Climbers and Trailers

fl.   p 21   Pl. 13

Bauhinia corymbosa

Menninger, Edwin A.
Flowering Vines of the World

fl.   Pl. 86

Bauhinia esculenta

Flowering Plants of Africa
vol 33   1959
fl., fr.   Pl. 1311

Bauhinia flammifera

Menninger, Edwin A.
Flowering Vines of the World

fl.   Pl. 85

Bauhinia galpinii

Mathias, Mildred E.
Color for the Landscape

fl. hab.   p 105

Bauhinia galpinii

Menninger, Edwin A.
Flowering Vines of the World

fl.   Pl. 87

Bauhinia galpinii

Palgrave, K. C.
Trees of Central Africa

fl., fr.   p. 75

Bauhinia hookeri

Oakman, Harry
Colorful Trees

Hab.Fl.   P 56

Bauhinia kockiana

Harrison, Richmond E.
Climbers and Trailers

fl.   p 21   Pl. 14

Bauhinia kockiana

Menninger, Edwin A.
Flowering Vines of the World

fl.   Pl. 104

Bauhinia macrantha

Flowering Plants of Africa
vol 39   1968-69
fl.   Pl. 1531

Bauhinia monandra

Hargreaves, Dorothy
Tropical Blossoms of the Caribbean

fl.   p. 52

Bauhinia monandra

Oakman, Harry
Colorful Trees

Fl.   P 75

Bauhinia, Orchid

see

Bauhinia petersiana

Bauhinia petersiana

Flowering Plants of Africa
vol 39   1968-69

fl.   Pl. 1532

Bauhinia petersiana

Morley, Brian D.
Wild Flowers of the World

fl.,fr.   Pl. 54

Bauhinia petersiana

Palgrave, K. C.
Trees of Central Africa

fl., fr.   p. 71

Bauhinia punctata

Macoby, Stirling
What Flower is That

fl.   P 54

Bauhinia purpurea

Bruggeman, L.
Tropical Plants

fl.   Pl. 220

Bauhinia purpurea

Hall, Clarence E.
Flowers of the Islands in the Sun

fl.   p. 139   Pl. 31

Bauhinia purpurea

Perry, Frances
Flowers of the World

fl.   p 159

Bauhinia tomentosa

Macoby, Stirling
What Flower is That

fl.   P 54

Bauhinia variegata

Flemer, William III
Shade and Ornamental Trees in Color

fl., hab.   p. 81, 100

Bauhinia variegata

Macoby, Stirling
What Flower is That

fl.   p 54

Bauhinia variegata

Massachusetts Hort. Soc.
Horticulture
vol 34, No. 11 1956

fl.   p. 563

Bauhinia variegata

Massachusetts Hort. Soc.
Horticulture
vol 42, No. 11   1964

fl.   p. 20

Bauhinia variegata

Massachusetts Hort. Soc.
Horticulture
vol 51, No. 11   1973

fl.   p. 36

Bauhinia variegata

Mathias, Mildred E.
Color for the Landscape

fl. hab.   p 9

Bauhinia variegata

Oakman, Harry
Colorful Trees

Hab. Fl.   P 131

Bauhinia variegata

Pertchik, Bernard
Flowering Trees of the
   Caribbean

fl. fr.   p 91

Bauhinia variegata

Walden, Beryl M.
Wild Flowers of Hong Kong

fl.   Pl. 17(43)

Bauhinia variegata var.

Flemer, William III
Shade and Ornamental Trees in Color

hab.   p 82

Bauhinia variegata var.

Macoby, Stirling
What Flower is That

fl.   p 54

Bauhinia variegata var.

Oakman, Harry
Colorful Trees

hab., fl.   p. 30, 31.

Bauhinia variegata var.

Perry, Frances
Flowers of the World

fl.   p. 158

Bauhinia, yellow

see

Bauhinia tomentosa

Bay

see

Laurus nobilis

Bay, Bull

see

Magnolia grandiflora

Bay, Loblolly

see

Gordonia lasianthus

Bay, Rose

see

Nerium oleander

Bay, Swamp

see

Magnolia virginiana

Bay, Sweet

see

Laurus nobilis

Bay, Sweet

see

Magnolia virginiana

Be-Still Tree

see

Thevetia peruviana

Beach Grass

See

Uniola paniculata

Bead Plant

see

Crassula nealeana

Bead plant

see

Nertera granadensis

Bead tree

see

Beaks, Brown

see

Lyperanthus suaveolens

Beaks, Red

see

Lyperanthus nigricans

Bean

see

Phaseolus

Bean, Asparagus

see

Vigna unguiculata var.

Bean, Black

see

Castanospermum australe

Bean, Bog

see

Menyanthes trifoliata

Bean, broad

see

Vicia faba

Bean, Buck

see

Menyanthes trifoliata

Bean, butter

see

Phaseolus lunatus

Bean, Castor

see

Ricinus communis

Bean, Cat Tail

see

Uraria macrostachya

Bean, Cherokee

see

Erythrina herbacea

Bean, Chinese Kidney

see

Wistaria sinensis

Bean, Coral

see

Erythrina crista-galli

Bean, Coral

see

Erythrina herbacea

Bean, Dwarf

see

Phaseolus vulgaris

Bean, French

see

Phaseolus vulgaris

Bean, haricot

see

Phaseolus vulgaris

Bean, Hog

see

Hyoscyamus niger

Bean, Horse

see

Parkinsonia aculeata

Bean, Hyacinth

see

Dolichos lablab

Bean, hyacinth

see

Dolichos lignosus

Bean, Indian

see

Catalpa bignonioides

Bean, Jack

see

Canavalia ensiformis

Bean, Jequerity

see

Abrus precatorius

Bean, kidney

see

Phaseolus vulgaris

Bean, Lima

see

Phaseolus lunatus

Bean, Locust

see

Ceratonia siliqua

Bean, Mahogany

see

Afzelia quanzensis

Bean, Mexican black

see

Phaseolus vulgaris

Bean, pea

see

Phaseolus vulgaris

Bean, Pink Wild

see

Strophostyles umbellata

Bean, Precatory

see

Abrus precatorius

Bean, Queensland

see

Bauhinia carronii

Bean, Rabbit

see

Cracca virginiana

Bean, Red

see

Kennedia rubicunda

Bean, Runner

see

Phaseolus multiflorus

Bean, Scarlet Runner

see

Phaseolus coccineus

Bean, Scarlet Runner

see

Phaseolus multiflorus

Bean, Snout

see

Rhynchosia minima

Bean, Trailing Wild

see

Strophostyles umbellata

Bean-tree

see

Aria latifolia

Bean tree, Kaffir

see

Schotia brachypetala

Bean, Wild

see

Phaseolus polystachios

Bean, Yam
see
Pachyrhizus erosus

Bearberry
see
Arctostaphylos uva-ursi

Bearberry, alpine
See
Arctous alpina

Bear Flower
see
Boykinia richardsonii

Bear Grass
see
Camassia esculenta

Bear grass
see
Nolina parryi

Bear Grass
see
Xerophyllum tenax

Bear Grass
see
Yucca louisianensis

Bear Root
see
Hedysarum alpinum

Bear Wood
see
Rhamnus purshiana

Bear
See Also
Bears

Beard , Brown
see
Calochilus robertsonii

Beard , Copper
see
Calochilus campestris

Beard, Crown Virginia
see
Verbesina virginica

Beard, Crown Wing Stem
see
Verbesina walteri

Beard, Crown Yellow
see
Verbesina helianthoides

Beard Flower
see
Pogonia ophioglossoides

Beard, Coats
see
Aruncus sylvester

Beard Grass
see
Erianthus contortus

Beard, Gray Grancy
see
Chionanthus virginica

Beard, Old Man's
see
Clematis aristata

Beard Tongue
see
Penstemon

Beardtongue
see
Penstemon

Beardtongue, Eastern
see
Penstemon hirsutus

Beard tongue, Foxglove
see
Penstemon digitalis

Beardtongue, Hairy
see
Penstemon cobaea

Beardtongue, Small-Sepaled
see
Penstemon sepalulus

Beardtongue, Tall
see
Penstemon procerus

Beardtongue, Whipple
see
Penstemon whippleanus

Beardtongue, Yukon
see
Pentstemon gormanii

Bear's breeches
see
Acanthus mollis

Bear's Ear
see
Primula auricula

Bears Ear, Austral
see
Cymbonotus preissianus

Bearsfoot

see

Acanthus mollis

Bear's Foot

see

Alchemilla vulgaris

Bear's-foot

see

Helleborus foetidus

Bear's-foot

see

Helleborus viridis

Bear's Foot

see

Polymnia uvedalia

Bears, Woolly

see

Trichinium rotundifolium

Beaufortia orbifolia

Blombery, Alec M.
What Wildflower is That

Fl.   p 63, Pl. 136

Beaufortia purpurea

Blombery, Alec M.
What Wildflower is That

fl.   p 64, Pl. 137

Beaufortia purpurea

Harrison, R.E.
Trees andShrubs

fl.   P 33

Beaufortia schaueri

Harrison, R.E.
Trees and Shrubs

fl.   P 33

Beaufortia sparsa

Harrison, R.E.
Trees and Shrubs

fl.   P 33

Beaufortia squarrosa

Blombery, Alec M.
What Wildflower is That

fl.   p 65, Pl. 138

Beaufortia squarrosa

Morcombe, M. K.
Australia's Western Wildflowers

fl.   p 53

Beaumontia grandiflora

Hargreaves, Dorothy
Tropical Blossoms of the Caribbean

fl.   p. 7

Beaumontia grandiflora

Harrison, Richmond E.
Climbers and Trailers

fl.   p 21   Pl. 15

Beaumontia grandiflora

Mathias, Mildred E.
Color for the Landscape

fl. hab.   p 105

Beaumontia grandiflora

Menninger, Edwin A.
Flowering Vines of the World

fl.   Pl. 29

Beaumontia grandiflora

Perry, Frances
Flowers of the World

fl.   p 30

Beaumontia terdoniana

Harrison, Richmond E.
Climbers and Trailers

fl.   p 22   Pl. 16

Beaumontia vine

see

Beaumontia grandiflora

Beautyberry

see

Callicarpa americana

Beauty berry

see

Callicarpa japonica

Beauty Berry, American

see

Callicarpa americana

Beauty Berry, Purple,

see

Callicarpa dichotoma

Beauty bush

see

Kolkwitzia amabilis

Beauty bush, Chinese

see

Kolkwitzia amabilis

Beaver-Tail

see

Opuntia basilaris

Becium obovatum

Moriarty, Audrey
Wild Flowers of Malawi

fl.   Pl. 55; 2

Bedstraw

see

Galium

Bedstraw, Dwarf

see

Galium pumilum

Bedstraw, Hedge

see

Galium mollugo

Bedstraw, Northern

see

Galium boreale

Bedstraw, Shining

see

Galium concinnum

Bedstraw, Swiss

see

Galium helveticum

Bedstraw, Yellow

see

Galium verum

Bedfordia linearis

Curtis, Winifred
The Endemic Flora of Tasmania Vol. 1

fl.   Pl. 30

Bedfordia salicina

Cochrane, G. R.
Flowers and Plants of Victoria

fl.   Pl. 447

Bedfordia salicina

Curtis, Winifred
The Endemic Flora of Tasmania
   Vol VI

fl.   Pl. 214

Bee balm

see

Monarda

Bee Balm, Crimson

see

Monarda didyma

Beebalm, Lemon

see

Monarda citriodora

Beebalm, Spotted

see

Monarda punctata

Bee Plant, Rocky Mountain

see

Cleome serrulata

Bee Plant, Yellow

see

Cleome lutea

Beech

See

Fagus

Beech, African

see

Faurea saligna

Beech, American

see

Fagus grandifolia

Beech, Common

see

Fagus sylvatica

Beech, copper

see

Fagus sylvatica var.

Beech, Dawyck

see

Fagus sylvatica

Beech-drops

see

Epifagus virginiana

Beech, European

see

Fagus sylvatica

Beech, Fastigiate

see

Fagus sylvatica

Beech, Fern-leaved

see

Fagus sylvatica var.

Beech, Mountain

see

Nothofagus solandri

Beech, New Zealand Black

see

Nothofagus solandri

Beech, Purple

see

Fagus sylvatica var.

Beech, Weeping

see

Fagus sylvatica

Beefsteak plant

see

Acalypha wilkesiana

Beefsteak plant

see

Perilla frutescens var.

Beefwood

see

Casuarina stricta

Beefwood, horsetail

see

Casuarina equisetifolia

Beeja

see

Phlogacanthus guttatus

Beet

see

Beta vulgaris var.

Beet, Fodder

see

Beta vulgaris

Beetroot

see

Beta vulgaris

Beet, Sea Kale

see

Beta vulgaris

Beet, spinach

see

Beta vulgaris

Beet, sugar

see

Beta vulgaris var.

Beet, Wild

see

Beta vulgaris

Befaria racemosa

Fleming, Glenn
Wild Flowers of Florida

fl.   p 36

Befaria racemosa

Walcott, Mary Vaux
North American Wild Flowers

fl.   Pl. 17

Beggar Lice

see

Desmodium canadense

Beggarticks

see

Bidens cernua

Beggar-ticks

see

Bidens pilosa

Beggarticks, Tall

see

Bidens vulgata

Beggarweed, Florida

see

Desmodium tortuosum

Begonia acaulis

American Begonia Soc.
Begonian
vol 40, No. 8   1973

fl., hab.   cover (p. 169)

Begonia acaulis

Misono, Isamu
Begonias

fl., hab.   p. 17   Pl. 1

Begonia acetosa

Misono, Isamu
Begonias

fl., hab.   p. 17   Pl. 2

Begonia acida

Massachusetts Hort. Soc.
Horticulture
vol 32, No. 11   1954

hab.   p. 519

Begonia acida

Misano, Isamu
Begonias

fl., hab.   p. 17   Pl. 3

Begonia aconitifolia

Misono, Isamu
Begonias

fl., hab.   p. 17   Pl. 4

Begonia acuminata

Misono, Isamu
Begonias

fl., hab.   p. 20   Pl. 5

Begonia albo-picta

Clark, Alice M
Begonia Portraits

hab., fl.   p. 73

Begonia albo-picta

Misono, Isamu
Begonias

hab., fl.   p. 20 Pl. 7

Begonia alice-clarkae

American Begonia Soc.
Begonian
vol 43, No. 3   1976

fl., hab.   cover (p. 61)

Begonia, angel wing

see

Begonia coccinea

Begonia angularis

Misono, Isamu
Begonias

hab., fl.   p. 20   Pl. 8

Begonia x argenteo-guttata

Kiaer, Eigil
Indoor Plants in Colour

hab.   p. 21

Begonia x argenteo-guttata

Misono, Isamu
Begonias

fl., hab.   p. 21   Pl. 12

Begonia aridicaulis

Misono, Isamu
Begonias

fl., hab.   p. 24 Pl. 13

Begonia auriculata

Misono, Isamu
Begonias

fl., hab.   p. 24   Pl. 14

Begonia barkeri

Clark, Alice M
Begonia Portraits

fl., hab.   p. 47

Begonia, bedding

see

Begonia semperflorens

Begonia, beefsteak

see

Begonia feastii

Begonia boweri

Hersey, Jean
Woman's Day Book of House Plants

fl.   p. 34

Begonia boweri

Misano, Isamu
Begonias

fl., hab.   p. 25   Pl. 18

Begonia boweri var.

American Begonia soc.
Begonian
vol. 40, No. 11   1973

hab.   cover (p. 241)

Begonia boweri var.

Misona, Isamu
Begonias

fl., hab.   p. 25,28   Pl.20,21

Begonia bradei

Misono, Isamu
Begonias

fl., hab.   p. 28   Pl. 23

Begonia caffra

Misono, Isamu
Begonias

fl., hab.   p. 29   Pl. 25

Begonia cathayana

Massachusetts Hort. Soc.
Horticulture
    vol 32, No. 1   1954

hab.   p. 29

Begonia cathayana

Misono, Isamu
Begonias

hab.   p. 29   Pl. 27

Begonia x cheimantha

Everett, T.H., ed.
New Illustrated Encyclopedia of
    Gardening vol. 2

fl.   p. 182

Begonia x cheimantha var.

Misono, Isamu
Begonias

fl.   p. 32-33   Pl. 28-33

Begonia x cheimantha var.

Nicolaisen, Age
Pocket Encyclopedia of Indoor Plants

fl.   Pl. 70

Begonia, Christmas
    see
Begonia hybrida var.

Begonia, Christmas
    see
Begonia socotrana

Begonia cinnabarina

Misono, Isamu
Begonias

fl., hab.   p. 33   Pl. 37

Begonia circumlobata

Misono, Isamu
Begonias

fl., hab.   p. 36   Pl. 38

Begonia, Climbing
    see
Cissus discolor

Begonia coccinea

Hay, Roy
The Dictionary of House Plants

fl.   Pl. 52

Begonia coccinea

Macoby, Stirling
What Flower is That

fl.   P 55

Begonia coccinea

Misono, Isamu
Begonias

fl., hab.   p. 36   Pl. 39

Begonia coccinea

Perry, Frances
Flowers of the World

fl.   p 41

Begonia coccinea var.

Everett, T.H., ed.
New Illustrated Encyclopedia of
    Gardening vol. 2

fl.   p. 182

Begonia coccinea var.

Kiaer, Eigil
Indoor Plants in Colour

fl.   p. 21

Begonia conchifolia var.

Misono, Isamu
Begonias

fl., hab.   p. 176   Pl. 302

Begonia convolvulacea

Kiaer, Eigil
Indoor Plants in Colour

fl.   p. 22

Begonia cooperi

Misono, Isamu
Begonias

hab.   p. 36   Pl. 40

Begonia corallina

Kromdijk, G.
200 House Plants in Colour

hab.   Pl. 28

Begonia x crestabruchii

Clark, Alice M.
Begonia Portraits

fl.   p. 71

Begonia x crestabruchii

Misono, Isamu
Begonias

hab.   p. 37   Pl. 42

Begonia decora

American Begonia Soc.
Begonian
    vol 42, No. 3   1975

hab.   cover (p. 53)

Begonia decora

Misono, Isamu
Begonias

hab.   p. 40   Pl. 47

Begonia deliciosa

Misono, Isamu
Begonias

fl., hab.   p. 41   Pl. 49

Begonia diadema

Clark, Alice M.
Begonia Portraits

fl., hab.   p. 53

Begonia diadema

Misono, Isamu
Begonias

fl., hab.   p. 41   Pl. 50

Begonia dichroa

Misono, Isamu
Begonias

fl., hab.   p. 44   Pl. 52

Begonia x digswelliana

Clark, Alice M.
Begonia Portraits

fl.   p. 95

Begonia dregei

American Begonia Soc.
Begonian
    vol 41, No. 11   1974

fl., hab.   cover (p. 269)

Begonia dregei
Clark, Alice M.
Begonia Portraits
fl., hab.   p. 59

Begonia dregei
Misono, Isamu
Begonias
fl., hab.   p. 45   Pl. 56

Begonia dregei var.
Misono, Isamu
Begonias
fl., hab.   p. 45   Pl. 57

Begonia echinosepala
Misono, Isamu
Begonias
fl., hab.   p. 45   Pl. 58

Begonia egregia
American Begonia Soc.
Begonian
vol 44, No. 9   1977
fl., hab.   cover (p. 233)

Begonia egregia
Misono, Isamu
Begonias
hab.   p. 45   Pl. 59

Begonia eminii
Wit, H. C. D. de
Plants of the World;
The Higher Plants, Vol I
fl.   p 169, Pl. 91

Begonia x erythrophylla
Clark, Alice M.
Begonia Portraits
fl.   p. 75

Begonia erythrophylla
Misono, Isamu
Begonias
fl.,hab.   p. 48   Pl. 61

Begonia x erythrophylla var.
Misono, Isamu
Begonias
fl.,hab.   p. 48   Pl. 62

Begonia evansiana
Miles, Bebe
Bulbs for the Home Gardener
hab.   p 55

Begonia evansiana
Misono, Isamu
Begonias
fl., hab.   p. 49   Pl. 63

Begonia evansiana
Pennsylvania Hort. Soc.
The Green Scene
vol 5, No. 1   1976
hab., fl   p. 11   Pl. 2

Begonia evansiana var.
Misono, Isamu
Begonias
fl., hab.   p. 49   Pl. 64-66

Begonia exotica
Misono, Isamu
Begonias
hab., fl.   p. 52   Pl. 67

Begonia fagifolia
Misono, Isamu
Begonias
fl., hab.   p. 52   Pl. 68

Begonia x feastii
Kiaer, Eigil
Indoor Plants in Colour
fl.   p. 21

Begonia x feastii var.
Massachusetts Hort. Soc.
Horticulture
vol. 32, No. 11   1954
hab.   p. 519

Begonia fernando-costae
Misono, Isamu
Begonias
hab.   p. 52   Pl. 69

Begonia ferruginea
American Begonia Soc.
Begonian
vol 45, No. 3   1978
fl., hab.   p. 75

Begonia, fibrous-rooted
see
Begonia semperflorens

Begonia ficicola
Misono, Isamu
Begonias
fl., hab.   p. 53   Pl. 70

Begonia flava
Flowering Plants of Africa
vol 31   1956
fl.   Pl. 1233

Begonia foliosa
Misono, Isamu
Begonias
hab., fl.   p. 53   Pl. 72

Begonia franconis
Misono, Isamu
Begonias
hab., fl.   p. 56   Pl. 74

Begonia, free-flowering
see
Begonia semperflorens

Begonia fuchsioides
Hay, Roy
The Dictionary of House Plants
fl.   Pl. 53

Begonia fuchsioides
Misono, Isamu
Begonias
fl., hab.   p. 56   Pl. 75

Begonia x fuscomaculata
Misono, Isamu
Begonias
fl.,hab.   p. 56   Pl. 76

Begonia geranioides
Flowering Plants of Africa
vol 43   1974-76
fl.   Pl. 1698

Begonia x gilsonii
Clark, Alice M.
Begonia Portraits
fl.   p. 89

Begonia glabra
Bruggeman, L.
Tropical Plants
fl.   pl. 81

Begonia glabra
Misono, Isamu
Begonias
fl., hab.   p. 56   Pl. 77

Begonia glabra var.

Misono, Isamu
Begonias

fl., hab.   p. 57   Pl. 78

Begonia glaucophylla

Kiaer, Eigil
Indoor Plants in Colour

fl.   p. 22

Begonia goegoensis

Misono, Isamu
Begonias

hab., fl.   p. 57   Pl. 79

Begonia gracilipeteolata

American Begonia Soc.
Begonian
vol 42, No. 5   1975

fl., hab.   cover (p. 101)

Begonia gracilis

O'Gorman, Helen
Mexican Flowering Trees and Plants

fl. hab.   p 131

Begonia grandis

American Begonia Soc.
Begonian
vol 45, No. 3 1978

fl., hab.   cover (p. 61)

Begonia haageana

Hay, Roy
The Dictionary of House Plants

fl.   Pl. 54

Begonia haageana

Hersey, Jean
Woman's Day Book of House Plants

hab.   p. 33

Begonia haageana

Misono, Isamu
Begonias

fl., hab.   p. 57   Pl. 81

Begonia haageana

Morley, Brian D.
Wild Flowers of the World

fl.   pl. 174

Begonia hemsleyana

Misono, Isamu
Begonias

hab.   p. 60   Pl. 84

Begonia hepatica var.

Misono, Isamu
Begonias

hab.   p. 60   Pl. 85

Begonia herbacea

American Begonia Soc.
Begonian
vol 41, No. 1   1974

hab.   cover (p. 1)

Begonia herbacea

American Begonia Soc.
Begonian
vol 44, No. 5   1977

hab.   cover (p. 113)

Begonia herbacea

American Begonia Soc.
Begonian
vol 44, No. 11   1977

hab.   cover (p. 289)

Begonia herbacea

Misono, Isamu
Begonias

fl., hab.   p. 61   Pl. 86

Begonia hidalgensis

Misono, Isamu
Begonias

hab., fl.   p. 40   Pl. 46

Begonia x hiemalis var.

Misono, Isamu
Begonias

fl.   p. 61   Pl. 87-88

Begonia hispida var.

Misono, Isamu
Begonias

hab., fl.   p. 64   Pl. 89

Begonia hybrid

Clark, Alice M.
Begonia Portraits

fl.,hab.   foreward, p.59-61, 64-69,
            72-73, 76-85, 93-94, 96-111

Begonia hybrid

Encke, Fritz
Zimmerpflanzen

fl., hab.   p. 53

Begonia hybrid

Hersey, Jean
Woman's Day Book of House Plants

fl.   p. 34, 35

Begonia hybrid

Kromdijk, G.
200 House Plants in Colour

fl., hab.   Pl. 29,31,33

Begonia hybrid

Macoby, Stirling
What Flower is That

hab.   p. 54

Begonia hybrid

Miles, Bebe
Bulbs for the Home Gardener

fl.   p 153

Begonia hybrid

Nicolaisen, Age
Pocket Encyclopedia of Indoor Plants

fl.   Pl. 72

Begonia hybrid

Perry, Frances
Flowers of the World

fl.   p. 40

Begonia hybrida

Tsukamoto, Yotaro
Coloured Illustrations of Garden
        Flowers   Vol. 9

fl.   opp. p. 33

Begonia hybrida var.

Hay, Roy
The Color Dictionary of Flowers
        & Plants

fl.   p. 52   Pl. 413

Begonia hybrida var.

Kiaer, Eigil
Indoor Plants in Colour

fl.   p. 24

Begonia hydrocotylifolia

Misono, Isamu
Begonias

hab.   p. 64   Pl. 90

Begonia x illustrata

Misono, Isamu
Begonias

hab.,fl.   p. 64   Pl. 91

Begonia imperialis

Misono, Isamu
Begonias

hab., fl.   p. 65   Pl. 93

Begonia imperialis var.

American Begonia Soc.
Begonian
vol 40, No. 9   1973

fl., hab.   cover (p. 193)

Begonia imperialis var.

Misono, Isamu
Begonias

hab., fl.   p. 65   Pl. 94-95

Begonia incana

Misono, Isamu
Begonias

fl., hab.   p. 65   Pl. 96

Begonia incarnata

Clark, Alice M.
Begonia Portraits

fl., hab.   p. 55

Begonia incarnata

Kiaer, Eigil
Indoor Plants in Colour

fl.   p. 22

Begonia incarnata var.

Misono, Isamu
Begonias

fl., hab.   p. 69 Pl. 97

Begonia involucrata

Clark, Alice M
Begonia Portraits

fl., hab.   p. 39

Begonia involucrata

Misono, Isamu
Begonias

fl., hab.   p. 69   Pl. 98

Begonia itaguassuensis

Misono, Isamu
Begonias

fl., hab.   p. 69   Pl. 100

Begonia jussiaeicarpa

Misono, Isamu
Begonias

fl., hab.   p. 72   Pl. 104

Begonia kellermannii

Misono, Isamu
Begonias

fl., hab.   p. 72   Pl. 105

Begonia kenworthyae

Misono, Isamu
Begonias

fl., hab.   p. 72   Pl. 106

Begonia x kewensis

Misono, Isamu
Begonias

fl.,hab.   p. 73   Pl. 107

Begonia laciniata

Walden, Beryl M.
Wild Flowers of Hong Kong

fl.   Pl. 63 (185)

Begonia laciniata var.

Misono, Isamu
Begonias

hab., fl.   p. 73   Pl. 108

Begonia leptotricha

Misono, Isamu
Begonias

fl., hab.   p. 76   Pl. 112

Begonia liebmannii

Misono, Isamu
Begonias

fl., hab.   p. 76   Pl. 114

Begonia limmingheiana

American Begonia Soc.
Begonian
vol 41, No. 12   1974

fl., hab.   p. 315

Begonia limmingheiana

Clark, Alice M.
Begonia Portraits

fl., hab.   p. 37

Begonia limmingheiana

Misono, Isamu
Begonias

fl., hab.   p. 76   Pl. 115

Begonia listida

Misono, Isamu
Begonias

hab., fl.   p. 77   Pl. 116

Begonia x lloydii

Misono, Isamu
Begonias

fl.,hab.   p. 77   Pl. 117

Begonia lobulata

Clark, Alice M.
Begonia Portraits

fl., hab.   p. 45

Begonia lubbersii

American Begonia Soc.
Begonian
vol 44, No. 3   1977

fl.   cover (p. 57)

Begonia lubbersii

Misono, Isamu
Begonias

hab., fl.   p. 80   Pl. 119

Begonia luxurians

Clark, Alice M.
Begonia Portraits

fl., hab.   p. 91

Begonia macdougalii var.

American Begonia Soc.
Begonian
vol 40, No. 5   1973

hab.   cover (p. 97)

Begonia macrocarpa

Misono, Isamu
Begonias

hab., fl.   p. 81   Pl. 123

Begonia maculata

Misono, Isamu
Begonias

fl., hab.   p. 81   Pl. 124

Begonia maculata var.

Misono, Isamu
Begonias

fl., hab.   p. 84   Pl. 125

Begonia malabarica

Misono, Isamu
Begonias

fl., hab.   p. 84   Pl. 126

Begonia manicata

Hay, Roy
The Dictionary of House Plants

fl.   Pl. 55

Begonia manicata var.

American Begonia Soc.
Begonian
vol 42, No. 1   1975

hab.   cover (p. 1)

Begonia manicata var.

Clark, Alice M.
Begonia Portraits

fl., hab.   p. 35

Begonia manicata var.

Massachusetts Hort. Soc.
Horticulture
    vol 32, No. 1   1954

hab.   p. 29

Begonia manicata var.

Misono, Isamu
Begonias

fl., hab.   p. 84-85 Pl. 128-29

Begonia mannii

Misono, Isamu
Begonias

fl., hab.   p. 85   Pl. 130

Begonia x margaritacea

Misono, Isamu
Begonias

fl.,hab.   p. 88   Pl. 133

Begonia x margaritae

Misono, Isamu
Begonias

fl.,hab.   p. 88-89   Pl. 134

Begonia mariae

American Begonia Soc.
Begonian
    vol 41, No. 3   1974

fl., hab.   cover (p. 53)

Begonia masoniana

Hay, Roy
The Color Dictionary of Flowers & Plants

hab.   p 52, Pl. 414

Begonia masoniana

Hay, Roy
The Dictionary of House Plants

fl. hab.   Pl. 56

Begonia masoniana

Misono, Isamu
Begonias

hab., fl.   p. 88   Pl. 135

Begonia masoniana

Perry, Frances
Flowers of the World

hab.   p 39

Begonia masoniana

Royal Hort. Soc.
Journal of the Royal Hort. Soc.
    vol 90, No. 5   1965

hab.   p. 206   Pl. 88

Begonia mazae

Misono, Isamu
Begonias

hab., fl.   p. 89   Pl. 137

Begonia mazae var.

Misono, Isamu
Begonias

fl., hab.   p. 89   Pl. 138

Begonia metallica

Kromdijk, G.
200 House Plants in Colour

fl., hab.   Pl. 30

Begonia metallica

Misono, Isamu
Begonias

hab., fl.   p. 92   Pl. 140

Begonia micranthera var.

Misono, Isamu
Begonias

fl., hab.   p. 92   Pl. 141-42

Begonia morelii

American Begonia Soc.
Begonian
    vol 42, No. 12 1975

fl., hab.   cover (p. 289)

Begonia multiflora

Miles, Bebe
Bulbs for the Home Gardener

fl.   p 152

Begonia multiflora var.

Huxley, Anthony
Garden Annuals and Bulbs

fl.   Pl. 199a,b,c.

Begonia nelumbiifolia

Misono, Isamu
Begonias

fl., hab.   p. 93,   Pl. 146

Begonia nitida var.

Clark, Alice M.
Begonia Portraits

fl., hab.   p. 57

Begonia nurii

American Begonia Soc.
Begonian
    vol 43, No. 10   1976

fl.   cover (p. 25?)

Begonia odeteiantha

American Begonia Soc.
Begonian
    vol 45, No. 4   1978

hab.   cover (p. 89)

Begonia odorata

Misono, Isamu
Begonias

fl.   p. 96   Pl. 149

Begonia olsoniae

American Begonia Soc.
Begonian
    vol 41, No. 7   1974

hab.   cover (p. 165)

Begonia olsoniae

Misono, Isamu
Begonias

fl., hab.   p. 169   Pl. 290

Begonia oxyphylla

Misono, Isamu
Begonias

fl., hab.   p. 97   Pl. 152

Begonia palmifolia

Misono, Isamu
Begonias

fl., hab.   p. 97   Pl. 153

Begonia parilis

Misono, Isamu
Begonias

fl., hab.   p. 97   Pl. 154

Begonia parviflora

Misono, Isamu
Begonias

fl., hab.   p. 97   Pl. 155

Begonia pauciflora

American Begonia Soc.
Begonian
    vol 44, No. 7   1977

fl., hab.   p. 191

Begonia paulensis

Misono, Isamu
Begonias

hab.   p. 100   Pl. 157

Begonia x pearlii

Clark, Alice M.
Begonia Portraits

fl.   p. 87

Begonia peltata

Addisonia
      vol 23   1954-59

fl.   p. 53   Pl. 763

Begonia pendula var.

Perry, Frances
Flowers of the World

fl.   p. 42

Begonia perfectiflora

Misono, Isamu
Begonias

fl., hab.   p. 100   Pl. 158

Begonia phyllomaniaca

Misono, Isamu
Begonias

hab.,fl.   p. 101   Pl. 159

Begonia picta

American Begonia Soc.
Begonian
      vol 43, No. 1   1976

fl., hab.   cover (p. 1)

Begonia picta

American Begonia Soc.
Begonian
      vol 44, No. 1   1977

hab.   cover (p. 1)

Begonia picta

Royal Hort. Soc.
Journal of the Royal Hort. Soc.
      vol 94, No. 5   1969

fl.   p. 214   Pl. 105

Begonia preussen

Massachusetts Hort. Soc.
Horticulture
      vol 32, No. 11   1954

fl., hab.   p. 519

Begonia preussen

Massachusetts Hort. Soc.
Horticulture
      vol 35, No. 11   1957

fl., hab.   p. 556

Begonia preussen

Massachusetts Hort. Soc.
Horticulture
      vol 39, No. 12   1961

fl.   p. 599

Begonia princeae

Misono, Isamu
Begonias

fl., hab.   p. 101   Pl. 162

Begonia prismatocarpa

American Begonia Soc.
Begonian
      vol 40, No. 10   1973

fl., hab.   cover (p. 217)

Begonia prismatocarpa

American Begonia Soc.
Begonian
      vol 43, No. 9   1976

fl.   cover (p. 233)

Begonia prismatocarpa var.

American Begonia Soc.
Begonian
      vol 45, No. 6   1978

fl., hab.   cover (p. 141)

Begonia x pseudophyllomaniaca

Kiaer, Eigil
Indoor Plants in Colour

fl.   p. 22

Begonia x pseudophyllomaniaca

Misono, Isamu
Begonias

fl.   p. 104   Pl. 163

Begonia quadrialata

American Begonia Soc.
Begonian
      vol 42, No. 2   1975

fl., hab.   cover (p. 29)

Begonia quadrialata

Wit, H. C. D. de
Plants of the World;
      The Higher Plants, Vol I

fl.   p 170, Pl. 94

Begonia repens

American Begonia Soc.
Begonia
      vol 44, No. 9   1977

fl., hab.   cover (p. 205)

Begonia rex

Misono, Isamu
Begonias

hab., fl.   p. 105   Pl. 166

Begonia rex

Tsukamoto, Yotaro
Coloured Illustrations of Garden
      Flowers   Vol. 9

fl.   p. 32

Begonia rex var.

Encke, Fritz
Zimmerpflanzen

hab.   p. 56

Begonia rex var.

Everett, T.H., ed.
New Illustrated Encyclopedia of
      Gardening vol. 2

hab.   p. 182

Begonia rex var.

Kiaer, Eigil
Indoor Plants in Colour

hab.   p. 23

Begonia rex var.

Kromdijk, G.
200 House Plants in Colour

hab.   p. 31

Begonia rex var.

Misono, Isamu
Begonias

hab.   p. 105-41   Pl. 167-237

Begonia rex var.

Nicolaisen, Age
Pocket Encyclopedia of Indoor Plants

hab.   Pl. 71.

Begonia rex var.

Perry, Frances
Flowers of the World

hab.   p. 39

Begonia rhizocarpa

Misono, Isamu
Begonias

fl., hab.   p. 140   Pl. 238

Begonia richardsiana

Clark, Alice M.
Begonia Portraits

fl., hab.   p. 59

Begonia richardsiana

Flowering Plants of South Africa
      vol XVII   1937

fl., fr.   Pl. 673

Begonia richii

Misono, Isamu
Begonias

hab.   p. 141   Pl. 240

Begonia x richmondensis

Misono, Isamu
Begonias

fl.,hab.   p. 144   Pl. 241

Begonia x richmondensis var.

Misono, Isamu
Begonias

fl.   p. 144   Pl. 242

Begonia x ricinifolia

Kiaer, Eigil
Indoor Plants in Colour

fl.   p. 23

Begonia rigida

Clark, Alice M.
Begonia Portraits

fl., hab.   p. 51

Begonia x rogeri

Misono, Isamu
Begonias

fl.,hab.   p. 144   Pl. 244

Begonia rotundifolia

Misono, Isamu
Begonias

fl., hab.   p. 145   Pl. 246

Begonia rubro-venia

Clark, Alice M.
Begonia Portraits

fl., hab.   p. 49

Begonia rubro-venia

Misono, Isamu
Begonias

hab., fl.   p. 145   Pl. 247

Begonia rubro-venia var.

Misono, Isamu
Begonias

hab., fl.   p. 145, 148   Pl. 248-49

Begonia rufosericea

Mee, Margaret
Flowers of the Brazilian Forests

fl., hab.   Pl. 9

Begonia rufosericea

Misono, Isamu
Begonias

hab., fl.   p. 148   Pl. 250

Begonia sanguinea

American Begonia Soc.
Begonian
vol 41, No. 5   1974

fl.   cover (p. 105)

Begonia sanguinea

Misono, Isamu
Begonias

fl., hab.   p. 148   Pl. 251

Begonia saxicola

Curtis's Botanical Magazine
Vol. 181   1976-77

fl.   Pl. 739

Begonia scandens

Massachusetts Hort. Soc.
Horticulture
vol 32, No. 1   1954

hab.   p. 29

Begonia scharffiana

Clark, Alice M.
Begonia Portraits

fl., hab.   p. 43

Begonia schmidtiana

Misono, Isamu
Begonias

fl., hab.   p. 148   Pl. 252

Begonia semperflorens

Bruggeman, L.
Tropical Plants

fl.   Pl. 31

Begonia semperflorens

Crockett, James Underwood
Annuals

fl.   p. 98

Begonia semperflorens

Everett, T.H., ed.
New Illustrated Encyclopedia of
    Gardening vol. 2

fl.   p. 182

Begonia semperflorens

Hersey, Jean
Woman's Day Book of House Plants

fl.   p. 35

Begonia semperflorens

Kromdijk, G.
200 House Plants in Colour

fl.   Pl. 32

Begonia semperflorens

Macoby, Stirling
What Flower is That

fl.   p. 55

Begonia semperflorens

Tsukamoto, Yotaro
Coloured Illustrations of Garden
    Flowers   Vol. 9

fl.   p. 32

Begonia semperflorens var.

Hay, Roy
The Color Dictionary of Flowers & Plants

fl.   p. 31

Begonia semperflorens var.

Huxley, Anthony
Garden Annuals and Bulbs

fl.   Pl. 8

Begonia semperflorens var.

Kiaer, Eigil
Indoor Plants in Colour

fl.   p. 25

Begonia semperflorens var.

Massachusetts Hort. Soc.
Horticulture
vol 32, No. 1   1954

fl.   p. 29

Begonia semperflorens var.

Misono, Isamu
Begonias

fl.   p. 148-55   Pl.253-65

Begonia semperflorens var.

Perry, Frances
Flowers of the World

fl.   p. 41

Begonia serratipetala

American Begonia Soc.
Begonian
vol 41, No. 10   1974

hab.   cover (p. 245)

Begonia serratipetala

Misono, Isamu
Begonias

hab.   p. 156   Pl. 266

Begonia, Small-flowered tuberous

see

Begonia multiflora

Begonia socotrana

Addisonia
vol 19   1935-36
fl.   p. 43   Pl 630

Begonia socotrana

Misono, Isamu
Begonias
fl., hab.   p. 157   Pl. 271

Begonia socotrana var.

Hay, Roy
The Color Dictionary of Flowers & Plants
fl.   p 52, 53, Pl. 415, 416, 417,
                     418

Begonia socotrana var.

Massachusetts Hort. Soc.
Horticulture
vol 35, No. 12   1957
fl.   p. 621

Begonia species

Hall, Clarence E.
Flowers of the Islands in the Sun
fl.   p. 23   Pl. 2

Begonia sp.

Milne, Lorus
Living Plants of the World
fl. hab.   p 157

Begonia species

Misono, Isamu
Begonias
fl., hab.   p. 20, 86   Pl. 6, 127

Begonia, strawberry

    see

Saxifraga sarmentosa

Begonia sudjanae

Misono, Isamu
Begonias
hab., fl.   p. 160   Pl. 274

Begonia sutherlandii

Clark, Alice M.
Begonia Portraits
fl., hab.   p. 59

Begonia sutherlandii

Hilliard, O. M.
Streptocarpus
hab.   Pl. 5(b)

Begonia sutherlandii

Misono, Isamu
Begonias
fl., hab.   p. 161   Pl. 276

Begonia tenuifolia

Clark, Alice M
Begonia Portraits
fl., hab.   p. 41

Begonia x thurstonii

Misono, Isamu
Begonias
fl.,hab.   p. 164   Pl. 278

Begonia tomentosa

Misono, Isamu
Begonias
fl., hab.   p. 164   Pl. 280

Begonia x tuberhybrida

Huxley, Anthony
Garden Annuals and Bulbs
hab.   Pl. 148

Begonia x tuberhybrida var.

Hay, Roy
The Color Dictionary of Flowers & Plants
fl.   p. 53-54

Begonia x tuberhybrida var.

Huxley, Anthony
Garden Annuals and Bulbs
fl.   Pl. 200

Begonia x tuberhybrida var.

Kiaer, Eigil
Indoor Plants in Colour
fl.   p. 26-27

Begonia x tuberhybrida var.

Macoby, Stirling
What Flower is That
fl.   p. 55

Begonia x tuberhybrida var.

Misono, Isamu
Begonias
fl.   p.165-70   Pl. 281-87

Begonia tuberosa

Tsukamoto, Yotaro
Coloured Illustrations of Garden
   Flowers   Vol. 9
fl.   p. 4

Begonia ulmifolia

Misono, Isamu
Begonias
fl., hab.   p. 169   Pl. 288

Begonia undulata

Clark, Alice M.
Begonia Portraits
fl., hab.   p. 103

Begonia undulata

Misono, Isamu
Begonias
fl., hab.   p. 169   Pl. 289

Begonia venosa

Misono, Isamu
Begonias
hab.   p. 169   Pl. 291

Begonia x verschaffeltii

Clark, Alice M.
Begonia Portraits
fl.   p. 63

Begonia x verschaffeltii

Misono, Isamu
Begonias
fl.,hab.   p. 172   Pl. 293

Begonia versicolor

American Begonia Soc.
Begonian
vol 39, No. 8   1972
hab.   cover (p. 165)

Begonia versicolor

Misono, Isamu
Begonias
hab., fl.   p. 172   Pl. 294

Begonia x viaudii

Misono, Isamu
Begonias
hab.,fl.   p. 173   Pl. 295

Begonia viscida

Misono, Isamu
Begonias
fl., hab.   p. 173   Pl. 298

Begonia vitifolia

Misono, Isamu
Begonias
fl., hab.   p. 176   Pl. 299

Begonia wallichiana

Misono, Isamu
Begonias

fl., hab.   p. 176   Pl. 300

Begonia, Watermelon

see

Peperomia sandersii var.

Begonia, waved-leaved

see

Begonia undulata

Begonia, wax

see

Begonia semperflorens

Begonia x weltoniensis

Clark, Alice M.
Begonia Portraits

fl.   p. 59

Begonia x weltoniensis

Misono, Isamu
Begonias

hab.,fl.   p. 176   Pl. 301

Belamcanda chinensis

Bruggeman, L.
Tropical Plants

fl.   Pl. 82

Belamcanda chinensis

Dean, Blanche E.
Wildflowers of Alabama and
   Adjoining States

fl.   p 31

Belamcanda chinensis

Duncan, Wilbur H.
Wildflowers of the Southeastern
   United States

fl.   p 267

Belamcanda chinensis

Everett, T.H., ed.
New Illustrated Encyclopedia of
   Gardening vol. 2

fl.   p. 231

Belamcanda chinensis

Kimura, Koiti
Japanese Medicinal Plants, Vol II

fl. fr.   p 145

Belamcanda chinensis

Klimas, John E.
Wildflowers of Eastern America

fl.   Pl. 126

Belamcanda chinensis

Miles, Bebe
Bulbs for the Home Gardener

fl. fr.   p 55

Belamcanda chinensis

Takatori, Jisuke
Color Atlas of Medicinal Plants of Japan

fl.   Fig. 68

Belamcanda chinensis

Wharton, Mary E.
A Guide to the Wildflowers & Ferns
   of Kentucky

fl. fr.   p 49 Pl. 1.5

Bell, Canterbury Wild

see

Phacelia campanularia

Bell, Canterbury Wild

see

Phacelia minor

Bell, Christmas

see

Blandfordia

Bell, Christmas New Zealand

see

Alstroemeria pulchella

Bell Climber, Red

see

Marianthus ringens

Bell, Cranbrook

see

Darwinia meeboldii

Bellflower

see

Campanula

Bellflower, Alpine

see

Campanula allionii

Bellflower, American

see

Campanula americana

Bellflower, Bearded

see

Campanula barbata

Bellflower, Bolognese

see

Campanula bononiensis

Bellflower, Canary Islands

see

Canarina canariensis

Bell Flower, Chilean

see

Lapageria rosea

Bellflower, Chinese

see

Abutilon hybrid

Bell flower, Chinese

see

Enkianthus campanulatus

Bellflower, Chinese

see

Platycodon grandiflorum

Bellflower, Clustered

see

Campanula glomerata

Bellflower, Creeping

see

Campanula rapunculoides

Bellflower, Dwarf
see
Campanula pusilla

Bellflower, Heath
see
Campanula rotundifolia

Bellflower, Irish
see
Moluccella laevis

Bellflower, Italian
see
Campanula isophylla

Bellflower, Marsh
see
Campanula aparinoides

Bellflower, Mt. Cenis
see
Campanula cenisia

Bellflower, Pale Blue
see
Wahlenbergia annularis

Bellflower, Poscharsky
see
Campanula poscharskyana

Bellflower, Rock
see
Campanula portenschlagiana

Bellflower, Rock
see
Campanula poscharskyana

Bellflower, Serbian
see
Campanula poscharskyana

Bellflower, Small
see
Campanula cochleariifolia

Bellflower, Spiked
see
Campanula spicata

Bellflower, Spreading
see
Campanula patula

Bellflower, Stencilled
see
Campanula excisa

Bellflower, Tall
see
Campanula americana

Bellflower, tussock
see
Campanula carpatica

Bellflower, Willow
see
Campanula persicifolia

Bellflower, Yellow
see
Campanula thyrsoides

Bell-fruit Tree
see
Codonocarpus cotinifolius

Bell, Mondurup
see
Darwinia macrostegia

Bell, Mountain
see
Darwinia collina

Bell, Qualap
see
Pimelea physodes

Bell, Red Climber
see
Marianthus ringens

Bell, Silver
see
Halesia carolina

Bell Tree, Scarlet
see
Spathodea campanulata

Bellwort
see
Uvularia

Bellwort, Large-Flowered
see
Uvularia grandiflora

Bellwort, Sessile
see
Uvularia sessilifolia

Bellwort, Smooth
see
Uvularia perfoliata

Bell, Yellow
see
Allamanda cathartica var.

Bell, Yellow
see
Fritillaria pudica

Bell
See also
Bells

Belladonna

see

Atropa belladonna

Bellardia trixago

Polunin, Oleg
Flowers of Europe

fl.   Pl. 126 #1237

Bellardia trixago

Polunin, Oleg
Flowers of the Mediterranean

fl.   Pl. 175

Belle of the night

See Hylocereus undatus

Bellendena montana

Curtis, Winifred
The Endemic Flora of Tasmania
                            vol 3
fl. fr.   p 189

Bellevalia ciliata

Polunin, Oleg
Flowers of the Mediterranean

fl.   Pl. 242

Bellevalia dichroa

Wendelbo, Per
Tulips and Irises of Iran and
    Their Relatives

fl. hab.   p 47, Pl. 49

Bellevalia hermonis

American Hort. Soc.
American Horticulturist
    vol 53, No. 5 1974

fl.   p. 18

Bellevalia longistyla

Wendelbo, Per
Tulips and Irises of Iran and
    Their Relatives

fl. hab.   p 47, Pl. 48

Bellevalia pycnantha

Wendelbo, Per
Tulips and Irises of Iran and
    Their Relatives

fl.   p 47, Pl. 47

Bellidastrum michelii

Barneby, T. P.
European Alpine Flowers in Colour

fl.   Pl. 87, 5

Bellidastrum michelii

Kleijn, H.
Beauty of the Wild Plant

fl.   p. 96   Pl. 136

Bellidastrum michelii

Polunin, Oleg
Flowers of Europe

fl.   Pl. 142 #1362

Bellis perennis

Ary, S.
The Oxford Book of Wildflowers

fl.   p 98, Pl. 3

Bellis perennis

Clark, Lewis J.
Wild Flowers of British Columbia

fl.   p 531

Bellis perennis

Crockett, James Underwood
Annuals

fl.   P 98

Bellis perennis

Everett, T.H., ed.
New Illustrated Encyclopedia of
Gardening vol. 2

fl.   p. 199

Bellis perennis

Felsko, Elsa
A Book of Wildflowers

fl.   P 123

Bellis perennis

Furrer, D.
Blumen am Wege

fl. hab.   p 111

Bellis perennis

Lindman, C. A. M.
Nordens Flora, Vol 9

fl. hab.   Pl. 575

Bellis perennis

Macoby, Stirling
What Flower is That

fl.   P 55

Bellis perennis

Martin, W. Keble
The Concise British Flora in Colour

fl. hab.   Pl. 44

Bellis perennis

Morley, Brian D.
Wild Flowers of the World

fl.,hab.   Pl. 22

Bellis perennis

Tsukamoto, Yotaro
Coloured Illustrations of Garden
    Flowers   Vol. 9

fl.   p. 34

Bellis perennis var.

Hay, Roy
The Color Dictionary of Flowers &
    Plants

fl.   p. 31   Pl. 243

Bellis perennis var.

Huxley, Anthony
Garden Annuals and Bulbs

fl.   Pl. 11

Bellis perennis var.

Huxley, Anthony
Garden Perennials and Water Plants

fl.   Pl. 61

Bellis sylvestris

Feinbrun-Dothan, Naomi
Wild Plants in the Land
    of Israel

fl. hab.   p. 84

Bellium bellidioides

Vilmorin, Roger de
Plantes Alpines dans les Jardins

fl. hab.   Pl. XXXVIII

Bellonia aspera

Elbert, Virginie F.
The Miracle Houseplants

fl.   p 116

Bells, Alpine

see

Cortusa matthioli

Bells, Autumn

see

Gerardia purpurea

Bells, Bronze

see

Stenanthium occidentale

Bells, Canterbury
see
Campanula medium

Bells, Cathedral
see
Bryophyllum pinnatum

Bells, Cathedral
see
Cobaea scandens

Bells, Chiming
see
Mertensia paniculata

Bells, Christmas
see
Blandfordia nobilis

Bells, Christmas
see
Blandfordia punicea

Bells, Cinnamon
see
Gastrodia sesamoides

Bells, Coral
see
Heuchera sp.

Bells, Easter
see
Erythronium grandiflorum

Bells, Golden
see
Forsythia

Bells, Honey
see
Mahernia verticillata

Bells, Oconee
see
Shortia galacifolia

Bells, Oconee
see
Shortia galacifolia

Bells of Ireland
see
Moluccella laevis

Bells, Purple
see
Rhodochiton volubile

Bells, Rabbit
see
Crotolaria spectabilis

Bells, Rock
see
Aquilegia canadensis

Bells, Side
see
Pyrola secunda

Bells, Sunny
See
Schoenolirion croceum

Bells, Temple
see
Smithiantha zebrina

Bells, Wedding
see
Deutzia scabra var.

Bells, Yellow
see
Stenolobium stans

Bells, Yellow
see
Tecoma stans

Bellucia grossularioides
Mee, Margaret
Flowers of the Brazilian Forests
fl.   Pl. 7

Beloperone californica
Coyle, Jeanette
A Field Guide to the Common and
Interesting Plants of Baja California
fl.   p 165

Beloperone californica
Munz, Philip A.
California Desert Wildflowers
fl.   p 51, Pl. 75

Beloperone californica
Orr, Robert T.
Wildflowers of Western America
fl.   Pl. 203

Beloperone comosa
O'Gorman, Helen
Mexican Flowering Trees and Plants
fl. hab.   p 133

Beloperone guttata
Curtis's Botanical Magazine
Vol 163   1940 - 42
fl.   No. 9633

Beloperone guttata
Encke, Fritz
Zimmerpflanzen
fl.   p. 56

Beloperone guttata
Hargreaves, Dorothy
Tropical Blossoms of the Caribbean
fl.   p. 34

Beloperone guttata
Harrison, R. E.
Trees and Shrubs
fl.   p 35

Beloperone guttata
Hay, Roy
The Color Dictionary of Flowers & Plants
fl.   p 54, Pl. 431

Beloperone guttata

Hersey, Jean
Woman's Day Book of House Plants

fl.   p. 95

Beloperone guttata

Kiaer, Eigil
Indoor Plants in Colour

fl.   p. 28

Beloperone guttata

Kromdijk, G.
200 House Plants in Colour

fl.   Pl. 34

Beloperone guttata

Macoby, Stirling
What Flower is That

fl.   P 56

Beloperone guttata

Nicolaisen, Age
Pocket Encyclopedia of Indoor Plants

fl.   pl. 126

Beloperone guttata

Pacific Hort. Foundation
Pacific Horticulture
    vol 39, No. 2   1978

fl.   p. 25

Beloperone guttata

Perry, Frances
Flowers of the World

fl.   p 14

Beloperone guttata

Tsukamoto, Yotaro
Coloured Illustrations of Garden
    Flowers   Vol. 9

fl.   p. 33

Beloperone plumbaginifolia

Bruggeman, L.
Tropical Plants

fl.   Pl. 221

Bent Grass

see

Agrostis humilis

Bent Grass, Colonial

see

Agrostis tenuis

Bent Grass, Reed

see

Calamagrostis purpurascens

Berberidopsis corallina

Harrison, Richmond E.
Climbers and Trailers

fl.   p 22   Pl. 17

Berberidopsis corallina

Perry, Frances
Flowers of the World

fl.   p 118

Berberis aemulans

Curtis's Botanical Magazine
Vol 169   1952-53

fl. fr.   179

Berberis aggregata

Bartels, Andreas
Das Grosse Buch der Gartengehölze

fr.   p 117

Berberis aggregata

Huxley, Anthony
Deciduous Garden Trees and Shrubs

fl. fr.   Pl. 16, 16a

Berberis aquifolium

Clark, Lewis J.
Wild Flowers of British Columbia

fl.   p 170

Berberis aquifolium

Mathias, Mildred E.
Color for the Landscape

fl. hab.   p 174

Berberis aquifolium

Taylor, Ronald J.
Mountain Wild Flowers

fl. hab.   p 41

Berberis aristata

Bartels, Andreas
Das Grosse Buch der Gartengehölze

fl.   p 116

Berberis asiatica

Roxburgh, William
Icones Roxburghianae

fl., fr.   Pl. 6

Berberis buxifolia var.

Huxley, Anthony
Evergreen Garden Trees and Shrubs

hab.   Pl. 93

Berberis calliantha

Curtis's Botanical Magazine
Vol 178   1970-72

fl.   584

Berberis candidula

Huxley, Anthony
Evergreen Garden Trees and Shrubs

fr. hab.   Pl. 94

Berberis chillanensis var.

Curtis's Botanical Magazine
    Vol 160   1937

fl. fr.   No. 9503

Berberis darwinii

Gault, S. Millar
The Color Dictionary of Shrubs

fl.   Pl. 15

Berberis darwinii

Harrison, R. E.
Trees and Shrubs

fl.   P 35

Berberis darwinii

Hay, Roy
The Color Dictionary of Flowers & Plants

fl. fr.   p 182, Pl. 1455, 1456

Berberis darwinii

Hellyer, A.G.L.
Shrubs in Colour

fl., fr.   p. 18

Berberis darwinii

Macoby, Stirling
What Flower is That

fl. P 56

Berberis darwinii

Milne, Lorus
Living Plants of the World

fr.   p 54

Berberis darwinii

Perry, Frances
Flowers of the World

fl.   p 42

Berberis darwinii

Royal Hort. Soc.
Journal of the Royal Hort. Soc.
vol 93, No. 4   1968

fl.   p. 164   Pl. 89

Berberis gagnepainii

Hellyer, A.G.L.
Shrubs in Colour

fl., fr.   p. 18

Berberis gagnepainii

Huxley, Anthony
Evergreen Garden Trees and Shrubs

hab.   Pl. 95

Berberis gagnepainii var.

Curtis's Botanical Magazine
vol 176   1966-68

fl., fr.   Pl. 504

Berberis gyalaica

Curtis's Botanical Magazine
Vol 165   1948

fl. fr.   22

Berberis haematocarpa

Munz, Philip A.
California Desert Wildflowers

fl.   p 32, Pl. 16

Berberis higginsae

Coyle, Jeanette
A Field Guide to the Common and Interesting
Plants of Baja California

fl.   p 75

Berberis hybrid

Gault, S. Millar
The Color Dictionary of Shrubs

fr.   Pl. 14

Berberis johannis

Curtis's Botanical Magazine
Vol 166   1949

fl. fr.   57

Berberis julianae

Crockett, James Underwood
Evergreens

hab. fl.   p. 113

Berberis julianae

Huxley, Anthony
Evergreen Garden Trees and Shrubs

fl. hab.   Pl. 96

Berberis kawakamii

Curtis's Botanical Magazine
Vol 163   1940-42

fl. fr.   No. 9622

Berberis lempergiana

Curtis's Botanical Magazine
Vol 167   1950

fl. fr.   90

Berberis linearifolia

Curtis's Botanical Magazine
Vol 161   1938

fl. fr.   No. 9526

Berberis x lologensis

Gault, S. Millar
The Color Dictionary of Shrubs

fl.   Pl. 16

Berberis x lologensis

Hay, Roy
The Color Dictionary of Flowers &
    Plants

fl.   p.183   Pl. 1457

Berberis mucrifolia

Curtis's Botanical Magazine
Vol 179   1972

fl. fr.   643

Berberis nervosa

Clark, Lewis J.
Wild Flowers of British Columbia

fl. fr.   p 175

Berberis nervosa

Taylor, Ronald J.
Mountain Wild Flowers

fl. hab.   p 41

Berberis nevinii

Mathias, Mildred E.
Color for the Landscape

fl. hab.   p 174

Berberis x ottawensis var.

Gault, S. Millar
The Color Dictionary of Shrubs

fl.   Pl. 17

Berberis parisepala

Curtis's Botanical Magazine
vol 167   1950

fl. fr.   119

Berberis parvifolia

Huxley, Anthony
Deciduous Garden Trees and Shrubs

fr.   Pl. 17

Berberis polyantha

Curtis's Botanical Magazine
Vol 170   1954-55

fl. fr.   236

Berberis prattii

Curtis's Botanical Magazine
Vol 171   1956-57

fl. fr.   286

Berberis repens

Orr, Robert T.
Wildflowers of Western America

fl.   Pl. 87

Berberis repens

Porsild, A.E.
Rocky Mountain Wild Flowers

fl.   p 193

Berberis repens

Walcott, Mary Vaux
North American Wild Flowers

fr. hab.   Pl. 30

Berberis x rubrostilla

Hellyer, A.G.L.
Shrubs in Colour

fr.   p. 18

Berberis sargentiana

Gault, S. Millar
The Color Dictionary of Shrubs

fl.   Pl. 18

Berberis sieboldii

Kitamura, Siro
Coloured Illustrations of Trees &
    Shrubs of Japan

fr.   159

Berberis sikkimensis

Curtis's Botanical Magazine
Vol 168   1951

fl. fr.   173

Berberis x stenophylla

Gault, S. Millar
The Color Dictionary of Shrubs

fl., hab.   Pl. 20

Berberis x stenophylla

Huxley, Anthony
Evergreen Garden Trees and Shrubs

hab.   Pl. 97

Berberis x stenophylla

Royal Hort. Soc.
Journal of the Royal Hort. Soc.
    vol 97, No. 10   1972

fl., hab.   p. 442   Pl. 223

Berberis x stenophylla var.

Hay, Roy
The Color Dictionary of Flowers &
    Plants

fl.   p. 183   Pl. 1458

Berberis thunbergii

Bartels, Andreas
Das Grosse Buch der Gartengeholze

fr.   p 116

Berberis thunbergii

Huxley, Anthony
Deciduous Garden Trees and Shrubs

hab.   Pl. 18

Berberis Thunbergii

Kimura, Koiti
Japanese Medicinal Plants, Vol I

fl.   p 118

Berberis thunbergii

Wit, H. C. D. de
Plants of the World;
    The Higher Plants, Vol I

fl.   p 99, Pl. 42, 43

Berberis thunbergii var.

Everett, T.H., ed.
New Illustrated Encyclopedia of
    Gardening vol. 2

hab.   p. 199

Berberis thunbergii var.

Gault, S. Millar
The Color Dictionary of Shrubs

fl.   Pl. 21

Berberis thunbergii var.

Harrison, R.E.
Trees and Shrubs

hab.   p. 35

Berberis thunbergii var.

Hay, Roy
The Color Dictionary of Flowers & Plants

hab.   p 183, Pl. 1459

Berberis thunbergii var.

Huxley, Anthony
Deciduous Garden Trees and Shrubs

hab.   Pl. 19

Berberis umbellata

Curtis's Botancial Magazine
Vol 168   1951

fl. fr.   145

Berberis valdiviana

Curtis's Botanical Magazine
Vol 168   1951

fl. fr.   139

Berberis verruculosa

Huxley, Anthony
Evergrenn Garden Trees and Shrubs

fl.   Pl. 98

Berberis vulgaris

Ary, S.
The Oxford Book of Wildflowers

fl. fr.   p 190, Pl. 2

Berberis vulgaris

Barneby, T. P.
European Alpine Flowers in Colour

fl.   Pl. 28, 1

Berberis vulgaris

Bianchini, Francesco
Health Plants of the World

fl.   hab.   p 145

Berberis vulgaris

Lindman, C. A. M.
Nordens Flora, Vol 4

fl. fr.   Pl. 244

Berberis vulgaris

Macoby, Stirling
What Flower is That

fr.   P 56

Berberis vulgaris

Martin, W. Keble
The Concise British Flora in Colour

fl. fr. hab.   Pl. 4

Berberis vulgaris

Masefield, G. B.
The Oxford Book of Food Plants

fl. fr.   Pl. 191, 2

Berberis vulgaris

Perrot, Emile
Les Plantes Medicinales

fl. fr. hab.   p 91

Berberis vulgaris

Polunin, Oleg
Flowers of Europe

fl.   Pl. 27 #261

Berberis vulgaris

Polunin, Oleg
Trees and Bushes of Europe

fl. fr.   p 68

Berberis vulgaris

Tosco, Uberto
The World of Mountain Flowers

fl.   p 36

Berberis vulgaris

Vedel, H.
Arbres et Arbustes

fl. fr. hab.   p 64, Pl. 56

Berberis wilsoniae

Harrison, R.E.
Trees and Shrubs

fr.   P 35

Berberis wilsoniae var.

Gault, S. Millar
The Color Dictionary of Shrubs

fr.   Pl. 19

Berchemia racemosa

Kitamura, Siro
Coloured Illustrations of Trees &
    Shrubs of Japan

fr.   325

Bergamot

see

Monarda

Bergamot, Purple

see

Monarda media

Bergamot, Red

see

Monarda didyma

Bergamot, White

see

Monarda russeliana

Bergamot, Wild

see

Monarda fistulosa

Bergenia cordifolia

Huxley, Anthony
Garden Perennials and Water Plants

fl.   pl. 62

Bergenia cordifolia

Macoby, Stirling
What Flower is That

fl.   P 57

Bergenia cordifolia var.

Royal Hort. Soc.
Journal of the Royal Hort. Soc.
vol 92, No. 5   1967

fl.   p. 206   Pl. 94

Bergenia crassifolia

Crockett, James Underwood
Lawns & Ground Covers

fl.   p. 124

Bergenia crassifolia

Everett, T.H., ed.
New Illustrated Encyclopedia of
Gardening vol. 2

fl.   p. 198

Bergenia delavayi

Hay, Roy
The Color Dictionary of Flowers & Plants

fl.   p 126, Pl. 1002

Bergenia, heartleaf

see

Bergenia cordifolia

Bergenia hybrid

Perry, Frances
Flowers of the World

fl.   p. 272

Bergenia Leather

see

Bergenia crassifolia

Bergenia purpurascens

Hay, Roy
The Color Dictionary of Flowers & Plants

fl.   p 126, Pl. 1003

Bergenia purpurascens

Morley, Brian D.
Wild Flowers of the World

fl.fr.   Plate 99

Bergenia purpurascens var.

Curtis's Botanical Magazine
Vol 167   1950

fl.   117

Bergenia stracheyi

Tsukamoto, Yotaro
Coloured Illustrations of Garden
Flowers   Vol. 9

fl.   p. 34

Bergeranthus scapiger

Herre, H.
The Genera of the Mesembryanthemaceae

fl. fr.   p. 91

Bergerocactus emoryi

Coyle, Jeanette
A Field Guide to the Common and
Interesting Plants of Baja California

hab.   p 133

Bergerocactus emoryi

Lamb, Edgar
Colorful Cacti of the American Deserts

fl. hab.   Pl. 114 - 117

Bergia ammanioides

Roxburgh, William
Icones Roxburghianae

fl., fr., hab.   Pl. 11

Berkheya spekeana

Lind, E.M.
Some Common Flowering Plants of
Uganda

fl.   Pl. 12b

Berkheya spinosissima

Eliovson, Sima
Namaqualand in Flower

fl.   Pl. 46, 2

Berkheya zeyheri

Moriarty, Audrey
Wild Flowers of Malawi

fl.   Pl. 78; 1

Berlandiera pumila

Duncan, Wilbur H.
Wildflowers of the Southeastern
United States

fl. hab.   p 209

Berlandiera subacaulis

Fleming, Glenn
Wild Flowers of Florida

fl.   p 58

Berlandiera subacaulis

Lemmon, Robert S.
Wildflowers of North America in Full Color

fl.   p. 50   Pl. 83

Bermuda Grass

see

Cynodon dactylon

Berresfordia khamiesbergensis

Herre, H.
The Genera of the Mesembryanthemaceae

fl.   p 93

Berrigan

see

Eremophila longifolia

Berteroa incana

Courtenay, Booth
Wildflowers & Weeds

fl.   p 17

Berteroa incana

Lindman, C. A. M.
Nordens Flora, Vol 4

fl. fr. hab.   Pl. 264

Bertholletia excelsa

Hvass, Elsie
Plants That Feed and Serve Us

fl. fr.   p 21, Pl. 32

Bertholletia excelsa

Masefield, G. B.
The Oxford Book of Food Plants

fr.   pl. 31, 1

Bertholletia sp.

Bianchini, F.
The Complete Book of Fruits & Vegetables

fr.   p 195

Berula erecta

Ary, S.
The Oxford Book of Wildflowers

fl.   p 88, Pl. 2

Berula erecta

Courtenay, Booth
Wildflowers & Weeds

fl.   p 58

Berula erecta

Martin, W. Keble
The Concise British Flora in Colour

fl. fr. hab.   Pl. 38

Beschorneria yuccoides

Hay, Roy
The Color Dictionary of Flowers & Plants

fl.   p 183, Pl. 1462

Beschorneria yuccoides

Massachusetts Hort. Soc.
Horticulture
   vol 47, No. 7   1969

hab., fl   p. 28

Beschorneria yuccoides

Royal Hort. Soc.
The Garden
   vol 101, No. 5   1976

fl., hab.   p. 279

Bessera elegans

Curtis's Botanical Magazine
Vol 171   1956-57

fl.   270

Bessera elegans

O'Gorman, Helen
Mexican Flowering Trees and Plants

fl. fr.   p 135

Besseya bullii

Courtenay, Booth
Wildflowers & Weeds

fl.   p 100

Beta maritima

Lindman, C. A. M.
Nordens Flora, Vol 3

fl. fr.   Pl. 184

Beta vulgaris

Ary, S.
The Oxford Book of Wildflowers

fl.   p 56, Pl. 4

Beta vulgaris

Bianchini, F.
The Complete Book of Fruits & Vegetables

hab.   p 223

Beta vulgaris

Hvass, Elsie
Plants That Feed and Serve Us

hab.   p 65, Pl. 131

Beta vulgaris

Masefield, G. B.
The Oxford Book of Food Plants

fl.   Pl. 161, 2

Beta vulgaris var.

Bianchini, F.
The Complete Book of Fruits &
Vegetables

hab.   p 77, 81

Beta vulgaris var.

Hvass, Elsie
Plants That Feed and Serve Us

fl. hab.   p 13, Pl. 17
              29, Pl. 51

Beta vulagris var.

Martin, W. Keble
The Concise British Flora in Colour

fl. fr. hab.   Pl. 72

Beta vulgaris var.

Masefield, G. B.
The Oxford Book of Food Plants

fl.   Pl. 15, 2

Beta vulgaris var.

Polunin, Oleg
Flowers of Europe

fl.   Pl. 10, #101

Betonica officinalis

Lindman, C. A. M.
Nordens Flora, Vol 8

fl. fr.   Pl. 508

Betonica officinalis

Martin, W. Keble
The Concise British Flora in Colour

fl. hab.   Pl. 69

Betonica officinalis

Perrot, Emile
Les Plantes Medicinales

fl. hab.   p 31

Betony
   see
Stachys

Betony, Florida
   see
Stachys floridana

Betony Wood
   see
Pedicularis canadensis

Betsy, Little Sweet
   see
Trillium cuneatum

Betula alba

Bianchini, Francesco
Health Plants of the World

fl.   p 161

Betula alba

Oakman, Harry
Colorful Trees

Hab.   P 119

Betula alba

Perrot, Emile
Les Plantes Medicinales

fl. fr. hab.   p 37

Betula albo-sinensis var.

Hay, Roy
The Color Dictionary of Flowers & Plants

hab.   p 183, Pl. 1461

Betula alleghaniensis

Turner, Nancy J.
Wild Coffee and Tea Substitutes of
   Canada

fr.   p. 48

Betula costata

Boom, B. K.
The Glory of the Tree

hab.   opp. p. 41   pl. 60

Betula costata

Edlin, Herbert
The Illustrated Encyclopedia
   of Trees

hab.   p 140

Betula davurica

Kitamura, Siro
Coloured Illustrations of Trees &
  Shrubs of Japan

hab.   103

Betula ermani

Kitamura, Siro
Coloured Illustrations of Trees &
  Shrubs of Japan

fr.   102

Betula glandulosa

Heller, Christine
Wild Flowers of Alaska

hab.   Pl. 281

Betula jacquemontii

Royal Hort. Soc.
Journal of the Royal Hort. Soc.
  vol 91, No. 11   1966

hab.   p. 470   Pl. 243

Betula jacquemontii

Royal Hort. Soc.
The Garden
  vol 102, No. 11 1977

hab.   p. 449

Betula lutea

Edlin, Herbert
The Illustrated Encyclopedia
  of Trees

fr.   p 141

Betula lutea

Wharton, Mary E.
Trees & Shrubs of Kentucky

fl.   p 101, Pl. 3.4a

Betula maximowicziana

Kitamura, Siro
Coloured Illustrations of Trees &
  Shrubs of Japan

fl.   100, 101

Betula maximowicziana

Royal Hort. Soc.
Journal of the Royal Hort. Soc.
  vol 89, No. 2   1964

hab.   p. 84   Pl. 29

Betula medwediewii

Curtis's Botanical Magazine
  Vol 162   1939

fl. fr.   No. 9569

Betula nana

Lindman, C. A. M.
Nordens Flora, Vol 3

fr.   Pl. 167

Betula nana

Martin, W. Keble
The Concise British Flora in Colour

fl. hab.   Pl. 76

Betula nana

Polunin, Oleg
Flowers of Europe

fl.   Pl. 4 #38

Betula nana

Vedel, H.
Arbres et Arbustes

hab.   p 51, Pl. 44

Betula nigra

Boom, B. K.
The glory of the tree

hab.   p. 44   Pl. 63

Betula nigra

Wharton, Mary E.
Trees & Shrubs of Kentucky

fl.   p 101, Pl. 3.4b

Betula occidentalis

Edlin, Herbert
The Illustrated Encyclopedia
  of Trees

fr.   p 141

Betula papyrifera

Edlin, Herbert
The Illustrated Encyclopedia
  of Trees

fr. hab.   p 140-41

Betula pendula

Boom, B. K.
The glory of the tree

hab., fl.   p. 41, 44   Pl. 61,
                                  65

Betula pendula

Edlin, Herbert
The Illustrated Encyclopedia
  of Trees

fr. hab.   p 140

Betula pendula

Everett, Thomas H.
Living Trees of the World

hab.   p. 126

Betula pendula

Everett, T.H., ed.
New Illustrated Encyclopedia of
  Gardening vol. 2

hab.   p. 230

Betula pendula

Hay, Roy
The Color Dictionary of Flowers & Plants

hab.   p 183, Pl. 1462

Betula pendula

Hvass, Elsie
Plants That Feed and Serve Us

fr.   p 123, Pl. 269

Betula pendula

Martin, W. Keble
The Concise British Flora in Colour

fl. hab.   Pl. 76

Betula pendula

Polunin, Oleg
Flowers of Europe

fr. Pl. 4 #36

Betula pendula

Polunin, Oleg
Trees and Bushes of Europe

fl. fr. hab.   p 43, 44

Betula pendula

Wit, H. C. D. de
Plants of the World;
  The Higher Plants, Vol I

fl. hab. p 108, Pl. 61,62

Betula pendula var.

Edlin, Herbert
The Illustrated Encyclopedia
  of Trees

hab.   p 141

Betula pendula var.

Flemer, William III
Shade & Ornamental Trees in Color

hab.   p. 64

Betula pendula var.

Hay, Roy
The Color Dictionary of Flowers & Plants

hab.   p 183, Pl. 1463

Betula pendula var.

Huxley, Anthony
Deciduous Garden Trees & Shrubs

fr.   Pl. 20-22.

Betula platyphylla var.

Kitamura, Siro
Coloured Illustrations of Trees &
  Shrubs of Japan

hab.   97, 98, 99

Betula pubescens

Edlin, Herbert
The Illustrated Encyclopedia
of Trees

fr.   p 141

Betula pubescens

Everett, Thomas H.
Living Trees of the World

hab.   p. 127

Betula pubescens

Huxley, Anthony
Deciduous Garden Trees and Shrubs

fr.   Pl. 23

Betula pubescens

Martin, W. Keble
The Concise British Flora in Colour

fl. hab.   Pl. 76

Betula pubescens

Morley, Brian D.
Wild Flowers of the World

Fl. Fr.   Pl. 12D

Betula pubescens

Polunin, Oleg
Trees and Bushes of Europe

fl. fr. hab.   p 44

Betula pubescens

Vebel, H.
Arbres et Arbustes

fl. fr. hab.   p 51, Pl. 43

Betula utilis

Royal Hort. Soc.
Journal of the Royal Hort. Soc.
vol 99, No. 12   1974

hab.   p. 538   Pl. 254

Betula verrucosa

Lindman, C. A. M.
Nordens Flora, Vol 3

fr.   Pl. 166

Betula verrucosa

Vebel, H.
Arbres et Arbustes

fl. fr. hab.   p 50, Pl. 42

Biarum tenuifolium

Polunin, Oleg
Flowers of Europe

fl.   Pl. 183 #1820

Biarum tenuifolium

Polunin, Oleg
Flowers of the Mediterranean

fl.   Pl. 222

Biarum tenuifolium var.

Coulimis, Constantine N.
Wild Flowers of Greece

fl. hab. p 187

Bidens aristosa

Brown, Clair A.
Wildflowers of Louisiana and
Adjoining States

fl.   p 189

Bidens aristosa

Duncan, Wilbur H.
Wildflowers of the Southeastern
United States

fl.   p 219

Bidens aristosa

Wharton, Mary E.
A Guide to the Wildflowers & Ferns
of Kentucky

fl.   p 252   Pl. 3.16

Bidens bipinnata

Weeds of the Southern United States

fl.   p. 10

Bidens biternata

Walden, Beryl M.
Wild Flowers of Hong Kong

fl.   Pl. 33(99)

Bidens cernua

Ary, S.
The Oxford Book of Wildflowers

fl.   p 32, Pl. 1

Bidens cernua

Brown, Clair A.
Wildflowers of Louisiana and
Adjoining States

fl. hab.   p 190

Bidens cernua

Clark, Lewis J.
Wild Flowers of British Columbia

fl.   p 531

Bidens cernua

Courtenay, Booth
Wildflowers & Weeds

fl.   p 113

Bidens cernua

Martin, W. Keble
The Concise British Flora in Colour

fl. hab.   Pl. 45

Bidens cernua

Wharton, Mary E.
A Guide to the Wildflowers & Ferns
of Kentucky

fl.   p 251   Pl. 3.15

Bidens cernua var.

Polunin, Oleg
Flowers of Europe

fl.   Pl. 146 #1406

Bidens coronata

Courtenay, Booth
Wildflowers & Weeds

fl.   p 113

Bidens grandiflora

O'Gorman, Helen
Mexican Flowering Trees and Plants

fl. hab.   p 137

Bidens laevis

Duncan, Wilbur H.
Wildflowers of the Southeastern
United States

fl.   p 221

Bidens mitis

Fleming, Glenn
Wild Flowers of Florida

fl.   p 55

Bidens pilosa

Brown, Clair A.
Wildflowers of Louisiana and
Adjoining States

fl. fr.   p 190

Bidens pilosa

Chickering, Carol Rogers
Flowers of Guatemala

fl.   p 129

Bidens pilosa

Duncan, Wilbur H.
Wildflowers of the Southeastern
United States

fl.   p 219

Bidens pilosa

Fleming, Glenn
Wild Flowers of Florida

fl. hab.   p 26

Bidens pilosa

Walden, Beryl M.
Wild Flowers of Hong Kong

fl.   Pl. 78 (244)

Bidens steppia

Moriarty, Audrey
Wild Flowers of Malawi

fl.   Pl. 78; 3

Bidens tripartita

Lindman, C. A. M.
Nordens Flora, Vol 9

fl. fr.   Pl. 598

Bidens tripartita

Martin, W. Keble
The Concise British Flora in Colour

fl. hab.   Pl. 45

Bidens vulgata

Courtenay, Booth
Wildflowers & Weeds

fl.   p 113

Bifrenaria atropurpurea

Pabst, G. F. J.
Orchidaceae Brasilienses,
   Vol 2

fl.   p 223

Bifrenaria aurea

Pabst, G. F. J.
Orchidaceae Brasilienses,
   Vol 2

fl.   p 223

Bifrenaria aureo-fulva

Pabst, G. F. J.
Orchidaceae Brasilienses,
   Vol 2

fl.   p 225

Bifrenaria bicornaria

Ospina, Mariano
Orquideas de las americas

fl.   Pl. 100

Bifrenaria harrisoniae

Ebel, Friedrich
The Strange and Beautiful
   World of Orchids

fl.   p 83

Bifrenaria harrisoniae

Kramer, Jack
Orchids; Flowers of Romance
   and Mystery

fl.   p 187

Bifrenaria harrisoniae

Pabst, G.F.J.
Orchidaceae Brasilienses
   vol 1

hab.   p. 59

Bifrenaria harrisoniae

Pabst, G. F. J.
Orchidaceae Brasilienses,
   Vol 2

fl.   p 224

Bifrenaria maguirei

Pabst, G. F. J.
Orchidaceae Brasilienses,
   Vol 2

fl.   p 224

Bifrenaria racemosa

Pabst, G. F. J.
Orchidaceae Brasilienses,
   Vol 2

fl.   p 225

Bifrenaria tetragona

Pabst, G. F. J.
Orchidaceae Brasilienses,
   Vol 2

fl.   p 224

Bifrenaria thyrianthina

Pabst, G. F. J.
Orchidaceae Brasilienses,
   Vol 2

fl.   p 224

Bifrenaria vitellina

Pabst, G. F. J.
Orchidaceae Brasilienses,
   Vol 2

fl.   p 225

Bifrenaria wendlandiana

Pabst, G. F. J.
Orchidaceae Brasilienses,
   Vol 2

fl. hab.   p 225

Big Tree

see

Sequoiadendron giganteum

Bigelowia nudata

Brown, Clair A.
Wildflowers of Louisiana and
   Adjoining States

fl.   p 191

Bigelowia nudata

Duncan, Wilbur H.
Wildflowers of the Southeastern
   United States

fl.   p 203

Bignonia buccinatoria

Royal Hort. Soc.
Journal of the Royal Hort. Soc.
   vol 91, No. 10   1966

fl.   p. 426   Pl. 226

Bignonia capreolata

Batson, Wade T.
Wild Flowers in South Carolina

fl.   p 105

Bignonia capreolata

Brown, Clair A.
Wildflowers of Louisiana and
   Adjoining States

fl.   p 171

Bignonia capreolata

Campbell, Carlos C.
Great Smoky Mountain Wildflowers

fl.   p 85

Bignonia capreolata

Fleming, Glenn
Wild Flowers of Florida

fl.   p 78

Bignonia capreolata

Harrison, Richmond E.
Climbers and Trailers

fl.   p 22   Pl. 18

Bignonia capreolata

Wharton, Mary E.
Trees & Shrubs of Kentucky

fl.   p 67, Pl. 1.32

Bignonia radicans

Walcott, Mary Vaux
North American Wild Flowers

fl. hab.   Pl. 227

Bijlia cana

Herre, H.
The Genera of the Mesembryanthemaceae

fl. fr.   p 95

Bikukulla canadensis

Walcott, Mary Vaux
North American Wild Flowers

fl. hab.   Pl. 136

Bikukulla cucullaria

Walcott, Mary Vaux
North American Wild Flowers
   vol. 4

fl.   Pl. 247

Bilberry

see

Vaccinium myrtillus

Billardiera cymosa

Cochrane, G. R.
Flowers and Plants of Victoria

fl.   Pl. 156

Billardiera longiflora

Cochrane, G. R.
Flowers and Plants of Victoria

fl. fr.   Pl. 428

Billardiera longiflora

Harrison, Richmond E.
Climbers and Trailers

fr.   P 23   Pl. 19

Billardiera scandens

Blombery, Alec M.
What Wildflower is That

Fl.   p 65, Pl. 139

Billardiera scandens

Cochrane, G. R.
Flowers and Plants of Victoria

fl.   Pl. 368

Billbergia alfonsi-joannis

Bromeliad Society
Journal
   vol 24, No. 4   1974

fl.   cover (p. 118)

Billbergia alfonsi-joannis

Bromeliad Society
Journal
   vol 27, No. 3   1977

fl.   p. 120

Billbergia amoena var.

Smith, Lyman B.
The Bromeliads

fl. hab.   Pl. 31

Billbergia amoena var.

Wilson, Robert Gardner
Bromeliads in Cultivation

fl. hab.   Pl. 70, 72, 77

Billbergia brasiliensis

Wilson, Robert Gardner
Bromeliads in Cultivation

fl. hab.   Pl. 74

Billbergia calophylla

Bromeliad Society
Journal
   vol 27, No. 3   1977

fl.   p. 121

Billbergia distachia

Bromeliad Society
Bulletin
   vol 18, No. 1   1968

fl., hab.   p. 13

Billbergia euphemiae

Wilson, Robert Gardner
Bromeliads in Cultivation

fl.   Pl. 73

Billbergia euphemiae var.

Wilson, Robert Gardner
Bromeliads in Cultivation

hab.   Pl. 75

Billbergia fantasia

Wilson, Robert Gardner
Bromeliads in Cultivation

fl.   Pl. 117

Billbergia forgetii

Kiaer, Eigil
Indoor Plants in Colour

hab.   p. 29

Billbergia horrida var.

Padilla, Victoria
Bromeliads

fl.   p. 65   Pl. 3

Billbergia hybrid

Hersey, Jean
Woman's Day Book of House Plants

fl.   p. 36

Billbergia hybrid.

Padilla, Victoria, ed.
Bromeliads in Color and
   Their Culture

fl.   p 30

Billbergia hybrid.

Wilson, Robert Gardner
Bromeliads in Cultivation

fl. hab.   Pl. 92-95

Billbergia iridifolia

Wilson, Robert Gardner
Bromeliads in Cultivation

fl.   Pl. 78

Billbergia iridifolia var.

Wilson, Robert Gardner
Bromeliads in Cultivation

fl.   Pl. 97

Billbergia leptopoda

Bromeliad Society
Journal
   vol 26, No. 4   1976

hab.   p. 176

Billbergia leptopoda

Rauh, Werner
Bromeliads for Home, Garden and
   Greenhouse

fl.   Pl. 86

Billbergia leptopoda var.

Wilson, Robert Gardner
Bromeliads in Cultivation

fl. hab.   Pl. 90

Billbergia lietzei

Wilson, Robert Gardner
Bromeliads in Cultivation

fl.   Pl. 76

Billbergia macrocalyx

Addisonia
   vol 19   1935-36

fl.   p. 19   Pl. 618

Billbergia macrocalyx

Rauh, Werner
Bromeliads for Home, Garden and
   Greenhouse

fl.   Pl. 87

Billbergia meyeri

Wilson, Robert Gardner
Bromeliads in Cultivation

hab.   Pl. 71

Billbergia minarum

Wilson, Robert Gardner
Bromeliads in Cultivation

fl. hab.   Pl. 79

Billbergia morelii

Bromeliad Society
Journal
   vol 27, No. 3   1977

fl.   p. 120

Billbergia morelii

Padilla, Victoria
Bromeliads

fl.   p. 65   Pl. 3

Billbergia nutans

Hay, Roy
The Color Dictionary of Flowers & Plants

fl.   p 54, Pl. 432

Billbergia nutans

Kiaer, Eigil
Indoor Plants in Colour

fl.   p. 28

Billbergia nutans

Macoby, Stirling
What Flower is That

fl.   P 57

Billbergia nutans

Morley, Brian D.
Wild Flowers of the World

fl.   Pl. 188

Billbergia nutans

Wilson, Robert Gardner
Bromeliads in Cultivation

fl.   Pl. 81

Billbergia nutans var.

Rauh, Werner
Bromeliads for Home, Garden and
   Greenhouse

fl.   Pl. 88

Billbergia pallidiflora

Wilson, Robert Gardner
Bromeliads in Cultivation

fl.   Pl. 80

Billbergia porteana

Bromeliad Society
Bulletin
   vol 12, No. 3 1962

fl.   p. 45

Billbergia porteana

Padilla, Victoria, ed.
Bromeliads in Color and
   Their Culture

fl.   p 30

Billbergia pyramidalis

Addisonia
   vol 19   1936

fl.   p. 33 Pl. 625

Billbergia pyramidalis

Hay, Roy
The Dictionary of House Plants

fl.   Pl. 69

Billbergia pyramidalis

Macoby, Stirling
What Flower is That

fl. P 57

Billbergia pyramidalis

Padilla, Victoria
Bromeliads

fl.   p. 65   Pl. 3

Billbergia pyramidalis

Padilla, Victoria, ed.
Bromeliads in Color and
   Their Culture

fl.   p 30

Billbergia pyramidalis var.

Bromeliad Society
Journal
   vol 25, No. 4 1975

fl.   cover (p. 121), p. 144, 160

Billbergia pyramidalis var.

Rauh, Werner
Bromeliads for Home, Garden and
   Greenhouse

fl.   Pl. 89

Billbergia pyramidalis var.

Wilson, Robert Gardner
Bromeliads in Cultivation

fl. hab.   Pl. 82, 83

Billbergia rosea

Bruggeman, L.
Tropical Plants

hab.   Pl. 83

Billbergia sanderana

Bromeliad Society
Bulletin
   Vol. 18, No. 1 1968

fl., hab.   P. 12

Billbergia sanderana

Wilson, Robert Gardner
Bromeliads in Cultivation

fl.,hab.   Pl. 84

Billbergia saundersii

Royal Hort. Soc.
Journal of the Royal Hort. Soc.
   vol 91, No. 8 1966

fl.   p. 338   Pl. 177

Billbergia saundersii

Wilson, Robert Gardner
Bromeliads in Cultivation

fl.   Pl. 85

Billbergia venezuelana

Bromeliad Society
Journal
   vol 27, No. 3 1977

fl.   cover (p. 97)

Billbergia venezuelana

Padilla, Victoria, ed.
Bromeliads in Color and
   Their Culture

fl.   p 111

Billbergia venezuelana

Wilson, Robert Gardner
Bromeliads in Cultivation

fl.   Pl. 86

Billbergia violacea

Mee, Margaret
Flowers of the Brazilian Forests

fl., hab.   Pl. 28

Billbergia vittata

Bromeliad Society
Journal
   vol 23, No. 2 1973

fl., hab.   p. 80

Billbergia vittata

Bromeliad Society
Journal
   vol 24, No. 5 1974

fl.   p. 196

Billbergia vittata

Bromeliad Society
Journal
   vol 27, No. 3 1977

fl., hab.   p. 120

Billbergia vittata

Rauh, Werner
Bromeliads for Home, Garden and
   Greenhouse

fl.   Pl. 90

Billbergia vittata var.

Wilson, Robert Gardner
Bromeliads in Cultivation

fl. hab.   Pl. 87-8

Billbergia x windii

Nicolaisen, Age
Pocket Encyclopedia of Indoor Plants

fl.   Pl. 29

Billbergia x windii

Perry, Frances
Flowers of the World

fl.   p. 54

Billbergia zebrina

Kiaer, Eigil
Indoor Plants in Colour

fl.   p. 29

Billbergia zebrina

Rauh, Werner
Bromeliads for Home, Garden and
          Greenhouse

fl.   Pl. 91

Billbergia zebrina

Wilson, Robert Gardner
Bromeliads in Cultivation

fl.   Pl. 89

Billy Buttons
     see
Craspedia uniflora

Bindweed
     see
Convolvulus

Bindweed, Field
     see
Convolvulus arvensis

Bindweed, Great
     see
Convolvulus sepium

Bindweed, Hedge
     see
Convolvulus sepium

Bindweed, Large
     see
Calystegia sepium

Bindweed, Mallow-leaved
     see
Convolvulus althaeoides

Bindweed, Sea
     see
Calystegia soldanella

Bindweed, Upright
     see
Convolvulus spithamaeus

Biota orientalis

Kimura, Koiti
Japanese Medicinal Plants, Vol II

fl.   p 125

Bipinnula canisii

Pabst, G. F. J.
Orchidaceae Brasilienses,
     Vol 1

fl.   p 177

Birch
See Betula

Birch, Canoe
     see
Betula papyrifera

Birch, Common
     see
Betula pendula

Birch, Common White
     see
Betula pubescens

Birch, Cutleaf European
     see
Betula pendula var.

Birch, Dalecarlian
     see
Betula pendula var.

Birch, Downy
     see
Betula pubescens

Birch, European
     see
Betula pubescens

Birch, European White
     see
Betula pubescens

Birch, Paper-bark
     see
Betula papyrifera

Birch, Scrub
     see
Betula glandulosa

Birch, Silver
     see
Betula pendula

Birch, Swedish
     see
Betula pendula var.

Birch, Sweet
     see
Betula alleghaniensis

Birch, Water
     see
Betula occidentalis

Birch, White
     see
Betula occidentalis

Birch, White
     see
Betula pendula

Birch, White
     see
Betula pubescens

Birch, Yellow
     see
Betula lutea

Bird flower

see

Crotalaria agatiflora

Bird Flower

See

Crotalaria cunninghamii

Bird flower

see

Crotalaria semperflorens

Bird of Paradise

see

Strelitzia reginae

Bird of Paradise, White

see

Strelitzia nicolai

Bird on the Wing

see

Polygala paucifolia

Bird's-eye

see

Veronica persica

Bird's eye bush

see

Ochna serrulata

Bird's Eyes

see

Gilia tricolor

Bird's Foot

see

Trigonella foenum-graecum

Bird's-Foot, Common

see

Ornithopus perpusillus

Birds Nest, Yellow

see

Monotropa hypopitys

Bird's Tongue Flower

see

Strelitzia reginae

Birthroot

see

Trillium recurvatum

Birthwort

see

Aristolochia clematitis

Birthwort

see

Trillium catesbaei

Biscuit Root, Gray's

see

Lomatium grayi

Biscuitroot, Nineleaf

see

Lomatium triternatum

Biscutella laevigata

Barneby, T. P.
European Alpine Flowers in Colour

fl.   Pl. 29, 3

Biscutella laevigata

Kohlhaupt, Paula
Fleurs des Alpages
                            vol 2
fr.   P 73

Biscutella laevigata

Morley, Brian D.
Wild Flowers of the World

fl.,fr.   Pl. 30

Biscutella laevigata

Tosco, Uberto
The World of Mountain Flowers

fl. hab.   p 62

Bishop's Cap

see

Astrophytum myriostigma

Bishop's-cap

see

Mitella diphylla

Bishop's Cap

see

Mitella nuda

Bishop's Hat

see

Astrophytum capricorne

Bishop's Mitre

see

Astrophytum myriostigma

Bishop's Weed, Mock

see

Ptilimnium costatum

Bisquit Root

see

Lomatium triternatum

Bistort

see

Polygonum bistorta

Bistort

see

Polygonum bistortoides

Bistort, Alpine

see

Polygonum viviparum

Bistort, Amphibious

see

Polygonum amphibium

Bistort, Greater

see

Polygonum bistorta

Bistort, Mountain Meadow

see

Polygonum bistorta

Bistorta major

Kimura, Koiti
Japanese Medicinal Plants, Vol II

fl.   p 151

Bitterbloom

see

Sabatia angularis

Bitternut

see

Carya condiformis

Bitterroot

see

Lewisia rediviva

Bitterroot, Pygmy

see

Lewisia pygmaea

Bittersweet

see

Solanum dulcamara

Bittersweet, Climbing

see

Celastrus scandens

Bittersweet, False

see

Celastrus scandens

Bitter-sweet, Shrubby

see

Celastrus scandens

Bitterweed

see

Helenium amarum

Bitterweed, Kaibab

see

Hymenoxys subintegra

Bixa orellana

Bruggeman, L.
Tropical Plants

fl.   pl. 222

Bixa orellana

Chickering, Carol Rogers
Flowers of Guatemala

fl. fr.   p 37

Bixa orellana

Macoby, Stirling
What Flower is That

fl.   P 57

Blackberry

see

Rubus fruticosus

Blackberry

see

Rubus ulmifolius

Blackberry, Highbush

see

Rubus argutus

Blackberry, Sand

see

Rubus cuneifolius

Blackboys

see

Xanthorrhoea hastilis

Black-eyed Susan

see

Gazania splendens

Black-eyed Susan

see

Rudbeckia hirta

Black-eyed Susan

see

Rudbeckia serotina

Black-eyed Susan

see

Rudbeckia speciosa

Black-eyed Susan

See

Tetratheca ericifolia

Black-eyed Susan

see

Thunbergia alata

Blackthorn

see

Bursaria spinosa

Blackthorn

see

Prunus spinosa

Blackwood

see

Acacia melanoxylon

Black Wood

see

Dalbergia latifolia

Blackwood, African

see

Peltophorum africanum

Blackstonia perfoliata

Ary, S.
The Oxford Book of Wildflowers

fl.   p 30, Pl. 3

Blackstonia perfoliata

Martin, W. Keble
The Concise British Flora in Colour

fl. hab.　Pl. 58

Blackstonia perfoliata

Morley, Brian D.
Wild Flowers of the World

Fl.　Pl 15E

Blackstonia perfoliata

Polunin, Oleg
Flowers of Europe

fl.　Pl. 94　#988

Bladder Nut

see

Staphylea holocarpa var.

Bladdernut

see

Staphylea pinnata

Bladderpod

see

Isomeris arborea

Bladder Pod

see

Lesquerella ludoviciana

Bladderpod, Double

see

Physaria didymocarpa

Bladderwort

see

Utricularia

Bladderwort, Arctic

see

Lesquerella arctica

Bladderwort, Greater

see

Utricularia vulgaris

Blaeberry

see

Vaccinium myrtillus

Blaeria kiwuensis

Moriarty, Audrey
Wild Flowers of Malawi

fl.　Pl. 69; 1

Blakea trinervia

Menninger, Edwin A.
Flowering Vines of the World

fl.　Pl. 125

Blancoa canescens

Blombery, Alec M.
What Wildflower is That

fl.　p 65, Pl. 140

Blancoa canescens

Morcombe, M. K.
Australia's Western Wildflowers

fl.　p 7

Blandfordia grandiflora

Blombery, Alec M.
What Wildflower is That

fl.　p 66, Pl. 141, 142

Blandfordia grandiflora

Perry, Frances
Flowers of the World

fl.　p 174

Blandfordia nobilis

Blombery, Alec M.
What Wildflower is That

fl.　p 66, Pl. 143

Blandfordia nobilis

Macoby, Stirling
What Flower is That

fl.　P 57

Blandfordia nobilis

Mullins, Barbara
Australian Wildflowers in Colour

fl.　P. 101　Pl. 99

Blandfordia punicea

Curtis, Winifred
The Endemic Flora of Tasmania　Vol. 1

fl.　Pl. 7

Blandfordia punicea

Morley, Brian D.
Wild Flowers of the World

fl. fr.　pl. 143

Blanket flower

see

Gaillardia

Blanket, Indian

see

Gaillardia pulchella

Blazing Star

see

Chamaelirium luteum

Blazing Star

see

Mentzelia

Blazing Star

see

Tritonia crocata

Blazing Star, Grass Leaved

see

Liatris graminifolia

Blazing Star, Sand

see

Mentzelia involucrata

Bleeding heart

see

Dicentra

Bleeding Heart, Fringed

see

Dicentra eximia

Bleeding Heart, Wild

see

Dicentra eximia

Bleedingheart, Wild

see

Dicentra formosa

Bleeding Hearts

see

Clerodendron thomsonae

Blepharis grandis

Moriarty, Audrey
Wild Flowers of Malawi

fl.   Pl. 42; 2

Blepharis subvolubilis

Flowering Plants of Africa
vol 31   1956

fl.   Pl. 1207

Blephilia ciliata

Courtenay, Booth
Wildflowers & Weeds

fl.   p 92

Blephilia ciliata

Dean, Blanche E.
Wildflowers of Alabama and
     Adjoining States

fl.   p 157

Blephilia ciliata

Duncan, Wilbur H.
Wildflowers of the Southeastern
     United States

fl.   p 161

Blephilia ciliata

Wharton, Mary E.
A Guide to the Wildflowers & Ferns
     of Kentucky

fl.   p 204   Pl. 2.26

Blephilia hirsuta

Courtenay, Booth
Wildflowers & Weeds

fl.   p. 92

Bletia catenulata

Pabst, G. F. J.
Orchidaceae Brasilienses,
     Vol 1

fl.   p 229

Bletia patula

Luer, Carlyle A.
The Native Orchids of Florida

fl.   Pl. 70, 1,2

Bletia patula

Ospina, Mariano
Orquideas de las americas

fl.   Pl. 39

Bletia purpurea

Lemmon, Robert S.
Wildflowers of North America
     in Full Color

fl.   pg. 15   pl. 21

Bletia purpurea

Luer, Carlyle A.
The Native Orchids of Florida

fl. fr. hab.   Pl. 69

Bletilla hyacinthina

Luer, Carlyle A.
The Native Orchids of Florida

fl. hab.   Pl. 1; 6

Bletilla striata

Hay, Roy
The Color Dictionary of Flowers & Plants

fl.   p 126, Pl. 1004

Bletilla striata

Kariyone, Tatsuo.
Atlas of Medicinal Plants

fl., hab.   Pl. 1

Bletilla striata

Kimura, Koiti
Japanese Medicinal Plants, Vol I

fl.   p 18

Bletilla striata

Macoby, Stirling
What Flower is That

fl.   P 58

Bletilla striata

Miles, Bebe
Bulbs for the Home Gardener

fl.   p 56

Bletilla striata

Perry, Frances
Flowers of the World

fl.   p 206

Bletilla striata

Takatori, Jisuke
Color Atlas of Medicinal Plants of Japan

fl.   Fig. 65A

Blighia sapida

Everett, Thomas H.
Living Trees of the World

fr.   p. 225

Blighia sapida

Hargreaves, Dorothy
Tropical Blossoms of the Caribbean

fr.   p. 62

Blighia sapida

Masefield, G. B.
The Oxford Book of Food Plants

fl.   fr.   Pl. 103, 3

Blind Grass, False

see

Agrostocrinum scabrum

Blinks

see

Montia verna

Blinks, Water

see

Montia fontana

Blite, Strawberry

see

Chenopodium capitatum

Bloodberry

see

Rivina humilis

Blood flower

see

Asclepias curassavica

Blood Flower

see

Haemanthus coccineus

Blood Flower, Mexican

see

Phaedranthus buccinatorius

Bloodleaf

see

Iresine herbstii

Bloodroot

see

Sanguinaria canadensis

Bloodwood

see

Eucalyptus corymbosa

Bloodwood

see

Eucalyptus terminalis

Bloodwood

see

Pterocarpus angolensis

Bloodwood, Red

see

Eucalyptus gummifera

Bloodwood, red

see

Eucalyptus ptychocarpa

Bloodwort

see

Sanguisorba menziesii

Bloomeria crocea

Munz, Philip A.
California Spring Wildflowers

fl.   p 88, Pl. 90

Blossfeldia fechseri

Lamb, Edgar
Popular Exotic Cacti in Color

fl. hab.   p 40

Bluebead

see

Clintonia borealis

Blue beard

see

Caryopteris

Bluebell

see

Campanula rotundifolia

Bluebell

see

Endymion non-scriptus

Bluebell

see

Mertensia paniculata

Bluebell

see

Sollya heterophylla

Bluebell

see

Wahlenbergia stricta

Bluebell, Alpine

see

Wahlenbergia sp.

Bluebell, Arizona

see

Mertensia arizonica

Blue bell , California

see

Phacelia campanularia

Blue Bell Creeper

see

Sollya fusiformis

Bluebell, English

see

Scilla nonscripta

Bluebell, Royal

see

Wahlenbergia gloriosa

Bluebell, Spanish

see

Scilla campanulata

Bluebell, Spanish

see

Scilla hispanica

Bluebell, Tall

see

Wahlenbergia stricta

Bluebell, Virginia

see

Mertensia virginica

Bluebell, Wasatch

see

Mertensia brevistyla

Bluebell, Western

see

Mertensia paniculata

Blueberry

see

Vaccinium

Blueberry climber

see

Ampelopsis brevipedunculata

Blueberry, Highbush

see

Vaccinium corymbosum

Blueberry, Low

see

Vaccinium vacillans

Blueberry, Pineland

see

Vaccinium tenellum

Blueberry, Shiny

see

Vaccinium myrsinites

Blue blossom

see

Ceanothus hybrid

Bluebonnet

see

Lupinus havardii

Bluebonnet

see

Lupinus texensis

Bluebonnet, Texas

see

Lupinus subcarnosus

Bluebottle

see

Centaurea cyanus

Blue Bush

see

Eucalyptus macrocarpa

Bluebush

see

Kochia

Bluebush, Rosy

see

Kochia erioclada

Bluebush, Satiny

see

Kochia georgei

Bluebush, Shrubby

see

Kochia pyramidata

Blue curls

see

Trichostema

Blue Curls, Woolly

see

Trichostema lanatum

Blue cushion

see

Houstonia caerulea

Blue dawn flower

see

Ipomoea leari i

Blue Dicks

see

Brodiaea capitata

Blue Dicks

see

Brodiaea pulchella

Blue-eyed Grass

see

Sisyrinchium

Blue-Eyed Grass, Prairie

see

Sisyrinchium campestre

Blue-Eyed Mary

see

Collinsia heterophylla

Blue-Eyed Mary

see

Collinsia verna

Bluegrass

see

Poa

Bluegrass, Alpine

see

Poa alpina

Bluegrass, Annual

see

Poa annua

Bluegrass, Kentucky

see

Poa pratensis

Bluegrass, Roughstalk

see

Poa trivialis

Blue lace flower

see

Trachymene coerulea

Blue Sailors

see

Cichorium intybus

Blueweed

see

Echium vulgare

Bluet

see

Houstonia caerulea

Bluet , Longleaf

see

Houstonia longifolia

Bluet, Mountain

see

Centaurea montana

Bluet , Prostrate

see

Houstonia serpyllifolia

Bluet , Purple

see

Houstonia purpurea

Bluet, Pygmy

see

Houstonia pygmaea

Bluet, Small

see

Houstonia patens

Bluet, Small

see

Houstonia pusilla

Blumea clarkei

Walden, Beryl M.
Wild Flowers of Hong Kong

fl.   Pl. 76 (234)

Blumea megacephala

Walden, Beryl M.
Wild Flowers of Hong Kong

fl.   Pl. 67 (201)

Blushing Bride

see

Serruria florida

Blysmus compressus

Martin, W. Keble
The Concise British Flora in Colour

fl. hab.   Pl. 91

Blyxa echinosperma

Stodola, Jiri
Encyclopedia of Water Plants

hab.   P. 95

Bobartia robusta

Flowering Plants of Africa
    vol 44   1977

fl., fr.   Pl. 1742

Bocconia cordata

Pennsylvania Hort. Soc.
The Green Scene
    vol 2, No. 5   1974

fl.   cover

Boehmeria cylindrica

Courtenay, Booth
Wildflowers & Weeds

fl.   p 64

Boehmeria cylindrica

Klimas, John E.
Wildflowers of Eastern America

fl.   Pl. 55

Boehmeria nivea

Hvass, Elsie
Plants That Feed and Serve Us

fl.   p 102, Pl. 219

Bog Candles

see

Limnorchis dilatata

Bog Candles

see

Platanthera dilatata var.

Bokhara vine

see

Polygonum baldschuanicum

Bolax glebaria

Vilmorin, Roger de
Plantes Alpines dans les Jardins

fl. hab.   Pl. XX

Boldo

see

Peumus boldus

Bollea coelestis

Ospina, Mariano
Orquideas de las americas

fl.   Pl. 101

Bollea lawrenceana

Kramer, Jack
Orchids; Flowers of Romance
    and Mystery

fl.   p 220

Bollea violacea

Pabst, G. F. J.
Orchidaceae Brasilienses,
    Vol 2

fl.   p 231

Boltonia asteroides

Brown, Clair A.
Wildflowers of Louisiana and
    Adjoining States

fl.   p 191

Boltonia asteroides

Bruggeman, L.
Tropical Plants

fl.   pl. 84

Boltonia asteroides

Moyle, John B.
Northland Wild Flowers

fl. hab.   p 182, Pl. 228

Boltonia latisquama

Addisonia
    vol 19   1935-36

fl.   p. 23   Pl. 620

Bolusanthus speciosus

Codd, L.E.W.
Trees & Shrubs of the Kruger
    National Park

fl., fr.   p. 2 Pl. I

Bolusanthus speciosus

Palgrave, K. C.
Trees of Central Africa

fl., fr.   p. 315

Bolusiella iridifolia

Stewart, Joyce
Orchids of Tropical Africa

fl. hab.   Pl. 15

Bomarea acutifolia

O'Gorman, Helen
Mexican Flowering Trees and Plants

fl. fr. hab.   p 89

Bomarea andimarcana

Perry, Frances
Flowers of the World

fl.   p 21

Bomarea calbreyeri

Menninger, Edwin A.
Flowering Vines of the World

fl.   Pl. 15

Bomarea caldasii

Curtis's Botanical Magazine
vol 175   1964-65

fl.   Pl. 465

Bomarea caldasii

Harrison, Richmond E.
Climbers and Trailers

fl.   p 23   Pl. 21

Bomarea caldasii

Morley, Brian D.
Wild Flowers of the World

fl.   pl. 190

Bomarea x cantabrigiensis

Royal Hort. Soc.
The Garden
  vol.103, No. 5   1978

fl.   p. 215

Bomarea carderi

Curtis's Botanical Magazine
Vol 163   1940-42

fl.   No. 9601

Bomarea carderi

Perry, Frances
Flowers of the World

fl.   p 21

Bomarea hybrid

Harrison, Richmond E.
Climbers and Trailers

fl.   p. 24   Pl. 22

Bomarea kalbreyeri

Hay, Roy
The Color Dictionary of Flowers & Plants

fl.   p 55, Pl. 433

Bomarea kalbreyeri

Royal Hort. Soc.
Journal of the Royal Hort. Soc.
  vol 89, No. 12   1964

fl.   p. 498   Pl. 195

Bomarea kalbreyeri

Royal Hort. Soc.
Journal of the Royal Hort. Soc.
  vol 93, No. 10   1968

fl.   p. 426   Pl. 233

Bomarea multiflora

Harrison, Richmond E.
Climbers and Trailers

fl.   p 24   Pl. 23

Bomarea porschiana

Herklots, Geoffrey
Flowering Tropical Climbers

fl.   p. 34   Pl. 2

Bomarea salsilla

Munoz Pizarro, Carlos
Flores Silvestres de Chile

fl., hab.   Pl. 23

Bomarea shuttleworthii

Macoby, Stirling
What Flower is That

fl.   P 58

Bomarea sp.

Royal Hort. Soc.
The Garden
  vol 101, No. 1   1976

fl.   p. 19

Bomarea werklei

Harrison, Richmond E.
Climbers and Trailers

fl.   p 24   Pl. 24

Bombax buonopozense

Wit, H. C. D. de
Plants of the World;
  The Higher Plants, Vol I

fl.   p 218, Pl. 129

Bombax ceiba

Morley, Brian D.
Wild Flowers of the World

fl.   Plate 115

Bombax ceiba

Perry, Frances
Flowers of the World

fl.   p 47

Bombax costatum

Morley, Brian D.
Wild Flowers of the World

fl.   Pl. 57

Bombax ellipticum

Cactus & Succulent Soc. of America
Journal
  vol 44, No. 3   1972

hab.   cover (p. 89)

Bombax ellipticum

Hargreaves, Dorothy
Tropical Blossoms of the Caribbean

fl.   p. 41

Bombax ellipticum

Oakman, Harry
Colorful Trees

Fl.   P 65

Bombax ellipticum

O'Gorman, Helen
Mexican Flowering Trees and Plants

fl.   p 15

Bombax ellipticum

Royal Hort. Soc.
Journal of the Royal Hort. Soc.
  vol 87, No. 10   1962

fl.   p. 456   Pl. 138

Bombax insigne

Morley, Brian D.
Wild Flowers of the World

fl.   Plate 115

Bombax malabaricum

Everett, Thomas H.
Living Trees of the World

hab.,fl.   p. 245

Bombax malabaricum

Oakman, Harry
Colorful Trees

Fl. Hab.   P 22

Bonamia semidygnia

Menninger, Edwin A.
Flowering Vines of the World

fl.   Pl. 64

Bonatea antennifera

Flowering Plants of Africa
  vol 36   1963-64

fl.   Pl. 1405

Bonatea steudneri

Stewart, Joyce
Orchids of Tropical Africa

fl.   Pl. 16

Bone Plant

see

Vanda hookerana

Boneset

see

Eupatorium

Boneset

see

Symphytum officinale

Boneset, False

see

Kuhnia eupatorioides

Boneset, Lindley's

see

Eupatorium lindleyanum

Boneset, Purple

see

Eupatorium perfoliatum var.

Boneset, Upland

see

Eupatorium sessilifolium

Bonewood

see

Emmenosperma alphitonioides

Bongardia chrysogonum

Polunin, Oleg
Flowers of the Mediterranean

fl.   Pl. 33

Bonnet, Mrs. Robb's

see

Euphorbia robbiae

Bonnets, Granny's

see

Aquilegia vulgaris.

Boojum

see

Fouquieria columnaris

Boom, Kaffir

see

Erythrina abyssinica

Boomerang, Trigger-Plant

see

Stylidium breviscapum

Boophone disticha

Flowering Plants of Africa
    vol 29   1952-53

fl.   Pl. 1141

Boophone disticha

Morley, Brian D.
Wild Flowers of the World

fl.   Pl. 86

Boophane disticha

Royal Hort. Soc.
Journal of the Royal Hort. Soc.
    vol 94, No. 1   1969

fl., fr.   p. 22   Pl. 8

Boophone disticha

Royal Hort. Soc.
The Garden
    vol 103, No. 11   1978

fl.   p. 439

Boophone sp.

Moriarty, Audrey
Wild Flowers of Malawi

fl.   Pl. 4; 3

Bootlace Bush

see

Pimelea axiflora

Bootlace, Bushman's

see

Pimelea nivea

Borage

see

Borago officinalis

Borago laxiflora

Perry, Frances
Flowers of the World

fl.   p 51

Borago officinalis

Ary, S.
The Oxford Book of Wildflowers

fl.   p 170, Pl. 5

Borago officinalis

Barneby, T. P.
European Alpine Flowers in Colour

fl.   Pl. 66, 3

Borago officinalis

Bianchini, F.
The Complete Book of Fruits
    & Vegetables

fl   p 59

Borago officinalis

Bianchini, Francesco
Health Plants of the World

fl. hab.   p 133

Borago officinalis

Felsko, Elsa
A Book of Wild Flowers
                2nd Ser.
fl.   p 52

Borago officinalis

Loewenfeld, Claire
The Complete Book of Herbs and Spices

fl. hab.   p 113

Borago officinalis

Macoby, Stirling
What Flower is That

fl.   P 59

Borago officinales

Perrot, Emile
Les Plantes Medicinales

fl. fr. hab.   p 39

Borago officinalis

Polunin, Oleg
Flowers of Europe

fl.   Pl. 103 #1054

Borago officinalis

Polunin, Oleg
Flowers of the Mediterranean

fl.   Pl. 139

Borago officinalis

Royal Hort. Soc.
Journal of the Royal Hort. Soc.
    vol 93, No. 9   1968

fl.   p. 382   Pl. 192

Borassus aethiopium

Wit, H. C. D. de
Plants of the World;
    The Higher Plants, Vol II

hab.   Pl. 167

Borassus flabellifer
Masefield, G. B.
The Oxford Book of Food Plants
fl.   Pl. 17, 3 107, 2

Boronia alata
Curtis's Botanical Magazine
    vol 174   1962-63
fl.   Pl. 426

Boronia algida
Cochrane, G. R.
Flowers and Plants of Victoria
fl.   Pl. 508

Boronia anemonifolia
Cochrane, G. R.
Flowers and Plants of Victoria
fl.   Pl. 309

Boronia, Brown
see
Boronia megastigma

Boronia caerulescens
Cochrane, G. R.
Flowers and Plants of Victoria
fl.   Pl. 162

Boronia citriodora
Curtis, Winifred
The Endemic Flora of Tasmania Vol. 1
fl.   Pl. 9

Boronia crenulata
Harrison, R.E.
Trees and Shrubs
fl. P 34

Boronia cymosa
Blombery, Alec M.
What Wildflower is That
fl.   p. 67, Pl. 145

Boronia denticulata
Blombery, Alec M.
What Wildflower is That
fl.   p. 67, Pl. 146

Boronia denticulata
Harrison, R. E.
Trees and Shrubs
hab. and fl.   P 34

Boronia elatior
Morcombe, M. K.
Australia's Western Wildflowers
fl.   p 106

Boronia floribunda
Blombery, Alec M.
What Wildflower is That
fl.   p 68, Pl. 147

Boronia, Granite
see
Boronia cymosa

Boronia granitica
Morley, Brian D.
Wild Flowers of the World
fl.   pl. 138

Boronia heterophylla
Blombery, Alec M.
What Wildflower is That
fl.   p 68, Pl. 148

Boronia heterophylla
Harrison, R.E.
Trees and Shrubs
fl.   P 34

Boronia ledifolia
Blombery, Alec M.
What Wildflower is That
fl.   p 68, Pl. 149

Boronia ledifolia
Macoby, Stirling
What Flower is That
fl.   P 59

Boronia megastigma
Blombery, Alec M.
What Wildflower is That
fl.   p 68, Pl. 150

Boronia megastigma
Harrison, R. E.
Trees and Shrubs
fl.   P 34

Boronia megastigma
Massachusetts Hort. Soc.
Horticulture
    vol. 41, No. 12 1963
fl.   p. 619

Boronia megastigma
Perry, Frances
Flowers of the World
fl.   p 269

Boronia megastigma var.
Harrison, R.E.
Trees and Shrubs
fl.   P 34

Boronia megastigma var.
Macoby, Stirling
What Flower is That
fl.   P 59

Boronia mollis
Blombery, Alec M.
What Wildflower is That
fl.   p 68, Pl. 151

Boronia mollis
Mullins, Barbara
Australian Wildflowers in Colour
fl.   p. 75   Pl. 72

Boronia muelleri
Cochrane, G. R.
Flowers and Plants of Victoria
fl.   Pl. 423

Boronia nana
Cochrane, G. R.
Flowers and Plants of Victoria
fl.   Pl. 38

Boronia, Pale Pink
see
Boronia floribunda

Boronia pilosa
Cochrane, G. R.
Flowers and Plants of Victoria
fl.   Pl. 106

Boronia pinnata
Blombery, Alec M.
What Wildflower is That
fl.   p 68, Pl. 152

Boronia pinnata
Mullins, Barbara
Australian Wildflowers in Colour
fl.   P. 75   Pl. 73

Boronia polygalifolia

Blombery, Alec M.
What Wildflower is That

fl.   p 69, Pl. 153

Boronia purdeiana

Macoby, Stirling
What Flower is That

fl.   P 59

Boronia, Red

see

Boronia heterophylla

Boronia rhomboidea

Blombery, Alec M.
What Wildflower is That

fl.   p 69, Pl. 155

Boronia serrulata

Blombery, Alec M.
What Wildflower is That

fl.   p 69, Pl. 155

Boronia serrulata

Jones, Paul
Flora Magnifica

fl.   Pl. 2

Boronia, Soft

see

Boronia mollis

Boronia, Sydney

see

Boronia ledifolia

Borreria dibrachiata

Moriarty, Audrey
Wild Flowers of Malawi

fl.   Pl. 49; 3

Borrichia frutescens

Brown, Clair A.
Wildflowers of Louisiana and
   Adjoining States

fl. hab.   p 192

Borrichia frutescens

Fleming, Glenn
Wild Flowers of Florida

fl. hab.   p 56

Borzicactus celsianus

Cactus & Succulent Society of America
Journal
   vol 42, No. 4   1970

fl.   p. 170

Borzicactus humboldtii

Lamb, Edgar
Popular Exotic Cacti in Color

fl. hab.   p 37

Borzicactus sepium

Cactus & Succulent Society of America
Journal
   vol 32, No. 5   1960

fl., hab.   p. 149

Borzicactus sepium

Lamb, Edgar
The Pocket Encyclopedia of Cacti
   and Succulents in Color

fl.   Pl. 2

Borzicactus sp.

Backeberg, Curt
Cactus Lexicon

fl.   p 565

Borzicactus sp.

Cactus & Succulent Society of America
Journal
   vol 40, No. 1   1968

fl.   p. 25

Borzicactus straussii

Van Laren, A. J.
Cactus

hab.   P. 10   Fig. 12

Boschniakia hookeri

Clark, Lewis J.
Wild Flowers of British Columbia

fl. hab.   p 494

Boschniakia rossica

Alaska-Yukon Wild Flower Guide

fl.   p 154

Boschniakia rossica

Heller, Christine
Wild Flowers of Alaska

hab.   Pl. 275

Boschniakia tuberosa

Orr, Robert T.
Wildflowers of Western America

fl.   Pl. 288

Boscia albitrunca

Eliovson, Sima
Namaqualand in Flower

hab.   Pl. 44,1

Boscia albitrunca

Palmer, Eve
Trees of South Africa

hab.   p. 288   Pl. XXVII

Bossiaea aquifolium

Blombery, Alec M.
What Wildflower is That

fl.   p 69, Pl. 156

Bossiaea cinerea

Cochrane, G. R.
Flowers and Plants of Victoria

fl.   Pl. 49

Bossiaea cordigera

Cochrane, G. R.
Flowers and Plants of Victoria

fl.   Pl. 404

Bossiaea eriocarpa

Blombery, Alec M.
What Wildflower is That

fl.   p 70, Pl. 157

Bossiaea foliosa

Cochrane, G. R.
Flowers and Plants of Victoria

fl.   Pl. 507

Bossiaea heterophylla

Blombery, Alec M.
What Wildflower is That

fl.   p 70, Pl. 158

Bossiaea lenticularis

Blombery, Alec M.
What Wildflower is That

fl.   p 70, Pl. 159

Bossiaea rhombifolia

Blombery, Alec M.
What Wildflower is That

fl.   p 70, Pl. 160

Boswellia dalzielii

Wit, H. C. D. de
Plants of the World;
   The Higher Plants, Vol II

hab.   p 54, Pl. 17

Bothriochilus macrostachyus

Ospina, Mariano
Orquideas de las americas

fl. hab.   Pl. 40

Bothriochloa ischaemum

Polunin, Oleg
Flowers of Europe

fr. Pl. 182 #1808

Bottlebrush

see

Callistemon

Bottle Brush

see

Fothergilla monticola

Bottlebrush, Albany

see

Callistemon speciosus

Bottlebrush, Crimson

see

Callistemon citrinus

Bottle-brush grass

see

Hystrix patula

Bottlebrush, green

see

Callistemon pinifolius

Bottle brush, lemon

see

Callistemon citrinus

Bottle-brush, Lemon

see

Callistemon pallidus

Bottle Brush, mountain

see

Greyia radlkoferi

Bottlebrush, Natal

see

Greyia sutherlandii

Bottlebrush, onesided

see

Calothamnus quadrifidus

Bottlebrush, Sand Heath

see

Beaufortia squarrosa

Bottle-brush, Scarlet

see

Callistemon macropunctatus

Bottle-brush tree

See Callistemon lanceolata

Bottle-brush, weeping

see

Callistemon viminalis

Bottlebrush, Willow

see

Callistemon salignus

Bottle Cleaner

see

Oenothera decorticans var.

Bottletree

see

Brachychiton rupestre

Bougainvillea x buttiana

Encke, Fritz
Zimmerpflanzen

fl.   p. 56

Bougainvillea x buttiana

Menninger, Edwin A.
Flowering Vines of the World

fl.   Pl. 139

Bougainvillea x buttiana var.

Hay, Roy
The Dictionary of House Plants

fl.   Pl. 71

Bougainvillea x buttiana var.

Mathias, Mildred E.
Color for the Landscape

fl., hab.   p. 102

Bougainvillea x buttiana var.

Menninger, Edwin A.
Flowering Viles of the World

fl.   Pl. 140

Bougainvillea glabra

Bruggeman, L.
Tropical Plants

fl.   pl. 5

Bougainvillea glabra

Everett, T.H., ed.
New Illustrated Encyclopedia of
    Gardening vol. 2

fl.   p. 231

Bougainvillea glabra

Morley, Brian D.
Wild Flowers of the World

fl.   pl. 178

Bougainvillea glabra

Nicolaisen, Age
Pocket Encyclopedia of Indoor Plants

fl.   Pl. 55

Bougainvillea glabra

O'Gorman, Helen
Mexican Flowering Trees and Plants

fl. hab.   p 91

Bougainvillea glabra

Perry, Frances
Flowers of the World

fl.   p 195

Bougainvillea glabra

Tsukamoto, Yotaro
Coloured Illustrations of Garden
    Flowers   vol 10

fl.   p. 41   Pl. 132

Bougainvillea glabra

Wit, H. C. D. de
Plants of the World;
    The Higher Plants, Vol I

fl.   p 109, Pl. 64

Bougainvillea glabra var.

Hargreaves, Dorothy
Tropical Blossoms of the Caribbean

fl.   p. 9

Bougainvillea glabra var.

Harrison, Richmond E.
Climbers and Trailers

fl.   p. 25   Pl. 25, 27

Bougainvillea glabra var.

Hay, Roy
The Dictionary of House Plants

fl.   Pl. 72

Bougainvillea glabra var

Kiaer, Eigil
Indoor Plants in Colour

fl.   p. 30

Bougainvillea glabra var.

Kromdijk, G.
200 House Plants in Colour

fl.   Pl. 35

Bougainvillea glabra var.

Macoby, Stirling
What Flower is That

fl.   P 60

Bougainvillea hybrid

Crockett, James Underwood
Lawns & Ground Covers

fl.   p. 124

Bougainvillea hybrid

Harrison, Richmond E.
Climbers and Trailers

fl.   p.25-27   Pl.26,28-32

Bougainvillea hybrid

Hersey, Jean
Woman's Day Book of House Plants

fl.   p. 37

Bougainvillea hybrid

Menninger, Edwin A.
Flowering Vines of the World

fl.   Pl. 130,134-35, 137,
                    141-47

Bougainvillea hybrida var.

Kiaer, Eigil
Indoor Plants in Colour

fl.   p. 30

Bougainvillea praetoria

Massachusetts Hort. Soc.
Horticulture
    vol 49, No. 1   1971

fl.   p. 29

Bougainvillea, Sander's

see

Bougainvillea glabra var.

Bougainvillea species

Hall, Clarence E.
Flowers of the Islands in the Sun

fl.   p. 31   Pl. 4

Bougainvillea sp.

Milne, Lorus
Living Plants of the World

fl.   p 84

Bougainvillea spectabilis

Hay, Roy
The Dictionary of House Plants

fl.   Pl. 73

Bougainvillea spectabilis

Macoby, Stirling
What Flower is That

fl.   P 60

Bougainvillea spectabilis

Menninger, Edwin A.
Flowering Vines of the World

fl.   Pl. 131

Bougainvillea spectabilis

Polunin, Oleg
Flowers of the Mediterranean

fl.   Pl. 11

Bougainvillea spectabilis var.

Kiaer, Eigil
Indoor Plants in Colour

fl.   p. 30

Bougainvillea spectabilis var.

Mathias, Mildred E.
Color for the Landscape

fl. hab.   p 102

Bougainvillea spectabilis var.

Menninger, Edwin A.
Flowering Vines of the World

fl.   p. 136

Bougainvillea, variegated

see

Bougainvillea glabra Var.

Bouncing Bet

see

Saponaria officinalis

Boundary mark

see

Cordyline terminalis

Bouteloua oligostachya

Huxley, Anthony
Garden Perennials and Water Plants

fr.   Pl. 316

Bouvardia x domestica var.

Hay, Roy
The Dictionary of House Plants

fl.   Pl. 74, 75

Bouvardia glaberrima

Orr, Robert T.
Wildflowers of Western America

fl.   Pl. 202

Bouvardia hybrid

Harrison, R.E.
Trees and Shrubs.

fl.   p. 36

Bouvardia hybrid

Perry, Frances
Flowers of the World

fl.   p. 267

Bouvardia longiflora

Harrison, R.E.
Trees and Shrubs

fl.   P 36

Bouvardia longiflora

Kiaer, Eigil
Indoor Plants in Colour

fl.   p. 31

Bouvardia longiflora

Macoby, Stirling
What Flower is That

fl.   P 61

Bouvardia longifolia var.

Nicolaisen, Age
Pocket Encyclopedia of
Indoor Plants

fl.   pl. 140

Bouvardia, scarlet

see

Bouvardia ternifolia

Bouvardia, scented

see

Bouvardia longiflora

Bouvardia ternifolia

Chickering, Carol Rogers
Flowers of Guatemala

fl.   p 129

Bouvardia ternifolia

Kiaer, Eigil
Indoor Plants in Colour

fl.   p. 31

Bouvardia ternifolia

Macoby, Stirling
What Flower is That

fl.   P 61

Bouvardia ternifolia

O'Gorman, Helen
Mexican Flowering Trees and Plants

fl. hab.   p 139

Bouvardia ternifolia var.

Hersey, Jean
Woman's Day Book of House Plants

fl.   p. 37

Bouvardia triphylla

Tsukamoto, Yotaro
Coloured Illustrations of Garden
     Flowers   vol 10

fl.   p. 42   Pl. 134

Bouvardia, white

see

Bouvardia longiflora

Bouvardia, Wild

see

Collomia grandiflora

Bow Flower

see

Toxanthes muelleri

Bow Fruit Creeper

see

Toxocarpus wightianus

Bowman's-Root

see

Gillenia trifoliata

Bowmansroot

see

Porteranthus trifoliatus

Bowman's Root

see

Veronicastrum virginicum

Bower of beauty

see

Pandorea jasminoides var.

Bower Plant

see

Pandorea jasminoides

Bowiea volubilis

Flowering Plants of South Africa
     vol XXI   1941

fl.   Pl. 815

Bowkeria citrina

Flowering Plants of Africa
     vol 40   1969-70

fl., fr.   Pl. 1578

Box

see

Buxus sempervirens

Box, Bastard

see

Polygala chamaebuxus

Box, Brush

see

Tristania conferta

Box, Christmas

see

Sarcococca humilis

Box, Common

see

Buxus sempervirens

Box, Red

see

Eucalyptus polyanthemos

Box, Sea

see

Alyxia buxifolia

Box Thorn

see

Lycium barbarum

Boxwood

see

Buxus

Box, Yellow

see

Eucalyptus melliodora

Boykinia elata

Clark, Lewis J.
Wild Flowers of British Columbia

fl.   p 194

Boykinia jamesii

Hay, Roy
The Color Dictionary of Flowers & Plants

fl.   p 3;.Pl. 21

Boykinia richardsonii

Alaska-Yukon Wild Flowers Guide

fl.   p 59

Boys, naked

see

Colchicum autumnale

Brachiaria plantaginea

Burgis, D.S.
Florida Weeds

hab., fr.   p. 5

Brachiaria platyphylla

Weeds of the Southern United States

fr., hab.   p. 23

Brachionidium floribundum

Ospina, Mariano
Orquideas de las americas

fl. hab.   Pl. 75

Brachtia andina

Ospina, Mariano
Orquideas de las americas

fl. hab.   Pl. 168

Brachycarpaea juncea

Eliovson, Sima
Namaqualand in Flower

fl.   Pl. 44, 2

Brachychilus horsfieldii

Wit, H. C. D. de
Plants of the World;
   The Higher Plants, Vol II

fr.   Pl. 151

Brachychiton acerifolius

Blombery, Alec M.
What Wildflower is That

fl.   p 70, Pl. 161

Brachychiton acerifolius

Edlin, Herbert
The Illustrated Encyclopedia
   of Trees

hab.   p 244

Brachychiton acerifolius

Everett, Thomas H.
Living Trees of the World

fl.   p. 246

Brachychiton acerifolius

Macoby, Stirling
What Flower is That

hab.   P 61

Brachychiton acerifolius

Mathias, Mildred E.
Color for the Landscape

fl.   p 10

Brachychiton acerifolius

Mullins, Barbara
Australian Wildflowers in Colour

fl.   P. 45   Pl 40

Brachychiton acerifolius

Oakman, Harry
Colorful Trees

Hab. Fl.   P 59 & 60

Brachychiton acerifolius

Royal Hort. Soc.
Journal of the Royal Hort. Soc.
   vol 98, No. 6   1973

hab., fl.   p. 260   Pl. 147

Brachychiton discolor

Macoby, Stirling
What Flower is That

fl.   P 61

Brachychiton discolor

Mathias, Mildred E.
Color for the Landscape

fl. hab.   p 11

Brachychiton discolor

Oakman, Harry
Colorful Trees

Fl.   P 60

Brachychiton discolor

Perry, Frances
Flowers of the World

fl.   p 288

Brachychiton paradoxum

Blombery, Alec M.
What Wildflower is That

fl.   p 71, Pl. 162

Brachychiton populneus

Blombery, Alec M.
What Wildflowers is That

fl.   p 72   Pl. 163.

Brachychiton populneus

Cochrane, G. R.
Flowers and Plants of Victoria

fl.   Pl. 217

Brachychiton populneus

Everett, T.H.
Living Trees of the World

fr.   p. 247

Brachychiton populneus

Macoby, Stirling
What Flower is That

fl.   p. 61

Brachychiton rupestre

Blombery, Alec M.
What Wildflower is That

hab.   p 73, Pl. 164

Brachychiton species

Mullins, Barbara
Australian Wildflowers in Colour

fl.   P. 47   Pl 41

Brachycome diversifolia

Cochrane, G. R.
Flowers and Plants of Victoria

fl.   Pl. 221

Brachycome iberidifolia

Blombery, Alec M.
What Wildflower is That

fl.   p 73, Pl. 165

Brachycome iberidifolia

Crockett, James Underwood
Annuals

fl.   P 98

Brachycome iberidifolia

Hay, Roy
The Color Dictionary of Flowers & Plants

fl.   p 31, Pl. 244

Brachycome iberidifolia

Huxley, Anthony
Garden Annuals and Bulbs

fl.   Pl. 12

Brachycome iberidifolia

Morley, Brian D.
Wild Flowers of the World

fl.   pl. 140

Brachycome multifida

Blombery, Alec M.
What Wildflower is That

fl.   p 73, Pl. 166

Brachycome nivalis var.

Alpine Flowers of the Kosciusko
State Park

fl.   Pl. 7

Brachycome pusilla

Blombery, Alec M.
What Wildflower is That

fl.   p 73,   Pl. 167

Brachycome rigidula

Cochrane, G. R.
Flowers and Plants of Victoria

fl.   Pl. 542

Brachycome scapigera

Mullins, Barbara
Australian Wildflowers in Colour

fl.   P. 97,   Pl. 97

Brachycome species

Mullins, Barbara
Australian Wildflowers in Colour

fl.   P. 97   Pl. 94

Brachycorythis angolensis var.

Williamson, Graham
The Orchids of South Central Africa

fl.   Pl. 30-31

Brachycorythis buchananii

Williamson, Graham
The Orchids of South Central Africa

fl.   Pl. 32

Brachycorythis congoensis

Williamson, Graham
The Orchids of South Central Africa

fl.   Pl. 29

Brachycorythis conica var.

Williamson, Graham
The Orchids of South Central Africa

fl.   Pl. 28

Brachycorythis friesii

Williamson, Graham
The Orchids of South Central Africa

fl.   Pl. 26

Brachycorythis galeandra

Hu, Shiu-ying
The Genera of Orchidaceae in
Hong Kong

fl.   p 120

Brachycorythis kalbreyeri

Grubb, Roy
Selected Orchidaceous Plants
Vol II

fl.   p 67

Brachycorythis kalbreyeri

Stewart, Joyce
Orchids of Tropical Africa

fl.   Pl. 17

Brachycorythis mixta

Williamson, Graham
The Orchids of South Central Africa

fl.   Pl. 35

Brachycorythis pleistophylla

Moriarty, Audrey
Wild Flowers of Malawi

fl.   Pl. 27; 4

Brachycorythis pleistophylla var.

Williamson, Graham
The Orchids of South Central Africa

fl.   Pl. 33

Brachycorythis pubescens

Williamson, Graham
The Orchids of South Central Africa

fl.   Pl. 34

Brachycorythis tenuior

Williamson, Graham
The Orchids of South Central Africa

fl.   Pl. 27

Brachyglottis repanda

Harvey, Norman B.
New Zealand Botanical Paintings

hab.   p. 54   Pl. 24

Brachyionidium dungaii

Pabst, G. F. J.
Orchidaceae Brasillienses,
Vol. 1

fl., hab.   p. 228

Brachylaena rotundata

Flowering Plants of Africa
vol XXVI   1947

fl.   Pl. 1005

Brachyloma daphnoides

Cochrane, G. R.
Flowers and Plants of Victoria

fl.   Pl. 333

Brachyloma ericoides

Cochrane, G. R.
Flowers and Plants of Victoria

fl.   Pl. 82

Brachypodium pinnatum

Lindman, C. A. M.
Nordens Flora, Vol 2

fr. hab.   Pl. 109

Brachysema aphyllum

Morcombe, M. K.
Australia's Western Wildflowers

fl.   p 43

Brachysema lanceolatum

Blombery, Alec M.
What Wildflower is That

fl.   P. 73   Pl. 168

Brachysema lanceolatum

Harrison, R. E.
Trees and Shrubs

fl.   P. 36

Brachystegia boehmii

Palgrave, K. C.
Trees of Central Africa

fl., fr.   p. 79

Brachystegia spiciformis

Palgrave, K. C.
Trees of Central Africa

fl., fr.   p. 83

Brachystegia tamarindoides

Palgrave, K. C.
Trees of Central Africa

fl., fr.   p. 87

Brachystelma barberiae

Cactus & Succulent Society of America
Journal
vol 43, No. 2   1971

fl.   cover (p. 49)

Brachystelma barberiae

Lamb, Edgar
Popular Exotic Cacti in Color

fl. hab.   p 39

Brachystelma barberiae

Morley, Brian D.
Wild Flowers of the World

fl.,hab.   Pl. 78

Brachystelma campanulatum

Flowering Plants of Africa
vol 38   1967

fl., hab.   Pl. 1483A

Brachystelma cathcartense

Flowering Plants of Africa
vol 42   1972-73

fl., hab.   Pl. 1667

Brachystelma chlorozonum

Flowering Plants of Africa
vol 42   1972-73

fl.   Pl. 1659

Brachystelma coddii

Flowering Plants of Africa
vol 30   1954-55

fl., hab.   Pl. 1181

Brachystelma discoideum

Flowering Plants of Africa
vol 42   1972-73

fl., hab.   Pl. 1668

Brachystelma flavidum

Flowering Plants of Africa
vol 27   1948-49

fl.   Pl. 1067

Brachystelma floribundum

Flowering Plants of Africa
vol 31   1956

fl., hab.   Pl. 1224

Brachystelma foetidum

Flowering Plants of South Africa
vol XXIV   1944

fl.   Pl. 940

Brachystelma gerrardii

Flowering Plants of Africa
vol 43   1974-76

fl.   Pl. 1686

Brachystelma gracilis

Flowering Plants of Africa
vol 27   1948-49

fl., fr.   Pl. 1077

Brachystelma minor

Flowering Plants of Africa
vol 28   1950-51

fl., hab.   Pl. 1096a

Brachystelma modestum

Flowering Plants of Africa
vol 30   1954-55

fl., hab.   Pl. 1165a

Brachystelma pachypodium

Flowering Plants of Africa
vol 32   1957-58

fl.   Pl. 1269

Brachystelma pulchellum

Flowering Plants of Africa
vol 29   1952-53

fl., hab.   Pl. 1121

Brachystelma pygmaeum var.

Flowering Plants of Africa
vol 28   1950-51

fl.   Pl. 1088

Brachystelma ringens

Flowering Plants of Africa
vol 28   1950-51

fl., hab.   Pl. 1096b

Brachystelma stellatum

Flowering Plants of Africa
vol 30   1954-55

fl., hab.   Pl. 1165b

Brachystelma tenellum

Flowering Plants of Africa
vol 42   1972-73

fl., hab.   Pl. 1664

Brachystelma togoense

Moriarty, Audrey
Wild Flowers of Malawi

fl.   Pl. 46; 4

Bramble, Holy

see

Rubus sanguineus

Bramble, Strawberry

see

Rubus pedatus

Brandy bottle

see

Nuphar luteum

Brasenia schreberi

Brown, Clair A.
Wildflowers of Louisiana and
Adjoining States

fl. hab.   p 50

Brasenia schreberi

Clark, Lewis J.
Wild Flowers of British Columbia

fl. hab.   p. 139

Brasenia schreberi

Courtenay, Booth
Wildflowers & Weeds

fl.   p 35

Brasenia schreberi

Stodola, Jiri
Encyclopedia of Water Plants

fl.   P 218

Brasenia schreberi

Tarver, David P.
Aquatic and Wetland Plants
of Florida

fl. hab.   p 23

Brass Buttons, New Zealand

see

Cotula squalida

Brassaia actinophylla

Hargreaves, Dorothy
Tropical Blossoms of the Caribbean

fl.   p. 52

Brassaia actinophylla

Macoby, Stirling
What Flower is That

fl.   p. 61

Brassaia actinophylla

Oakman, Harry
Colorful Trees

Hab. Fl.   P 96

Brassavola cucullata

Grubb, Roy
Selected Orchidaceous Plants
Vol I

fl.   p 91

Brassavola digbyana

Kramer, Jack
Orchids; Flowers of Romance
and Mystery

fl.   p 106

Brassavola digbyana

Kupper, Walter
Orchids

fl.   p 39

Brassavola martiana

Pabst, G. F. J.
Orchidaceae Brasilienses,
    Vol 1

fl.   p 219

Brassavola nodosa

Grubb, Roy
Selected Orchidaceous Plants
    Vol I

fl. hab.   p 29

Brassavola nodosa

Hersey, Jean
Woman's Day Book of House Plants

fl.   p. 83

Brassavola nodosa

Kramer, Jack
Orchids: Flowers of Romance
    and Mystery

fl.   p 107

Brassavola nodosa

Ospina, Mariano
Orquideas de las americas

fl. hab.   Pl. 50

Brassavola nodosa

Wright, Norman Pelham
Orquideas de Mexico

fl.   Pl. 1

Brassavola perrinii

Ebel, Friedrich
The Strange and Beautiful
    World of Orchids

fl.   p 59

Brassavola tuberculata

Pabst, G. F. J.
Orchidaceae Brasilienses,
    Vol 1

fl. hab.   p 219

Brassia caudata

Luer, Carlyle A.
The Native Orchids of Florida

fl. hab.   Pl. 76; 5-6

Brassia chloroleuca

Pabst, G. F. J.
Orchidaceae Brasilienses,
    Vol 2

fl.   p 258

Brassia lawrenceana

American Hort. Soc.
American Horticulturist
    vol 51, No. 4 1972

fl.   p. 25

Brassia lawrenceana

Pabst, G. F. J.
Orchidaceae Brasilienses,
    Vol 2

fl.   p 259

Brassia, Long Tailed

see

Brassia caudata

Brassia longissima

Morley, Brian D.
Wild Flowers of the World

fl.   pl. 192

Brassia verrucosa

Ebel, Friedrich
The Strange and Beautiful
    World of Orchids

fl.   p 81

Brassia verrucosa

Kupper, Walter
Orchids

fl.   p 103

Brassia verrucosa

Ospina, Mariano
Orquideas de las americas

fl.   Pl. 169

Brassica balearica

Curtis's Botanical Magazine
    vol 179   1972-73

fl.   Pl. 641

Brassica campestris

Ary, S.
The Oxford Book of Wildflowers

fl.   p 10, Pl. 5

Brassica campestris

Bianchini, F.
The Complete Book of Fruits & Vegetables

hab.   p 69

Brassica campestris

Lindman, C. A. M.
Nordens Flora, Vol 4

fl. fr. hab.   Pl. 251B

Brassica campestris var.

Bianchini, F.
The Complete Book of Fruits & Vegetables

hab.   p 75

Brassica chinensis

Masefield, G. B.
The Oxford Book of Food Plants

fl.   Pl. 155, 1

Brassica juncea

Kariyone, Tatsuo
Atlas of Medicinal Plants

fl., fr.   Pl. 112

Brassica juncea

Kimura, Koiti
Japanese Medicinal Plants, Vol II

fl.   p 175

Brassica kaber

Pond, Barbara
A Sampler of Wayside Herbs

fl.   Pl. II

Brassica kaber

Weeds of the Southern United States

fl.   p. 18

Brassica napus

Hvass, Elsie
Plants That Feed and Serve Us

fl. fr.   p 15, Pl. 22

Brassica napus

Masefield, G. B.
The Oxford Book of Food Plants

fl. fr. Pl. 25, 3

Brassica napus var.

Bianchini, F.
The Complete Book of Fruits & Vegetables

fr.   p 237

Brassica napus var.

Masefield, G. B.
The Oxford Book of Food Plants

fl. fr.   Pl. 173, 2

Brassica napus var.

Perrot, Emile
Les Plantes Medicinales

fl. fr. hab.   p 75

Brassica nigra

Ary, S.
The Oxford Book of Wildflowers

fl.   p 8, Pl. 1

Brassica nigra

Bianchini, F.
The Complete Book of Fruits & Vegetables

fr.   p 209

Brassica nigra

Courtenay, Booth
Wildflowers & Weeds

fl.   p 17

Brassica nigra

Kimura, Koiti
Japanese Medicinal Plants, Vol I

fl. hab.   p 50

Brassica nigra

Martin, W. Keble
The Concise British Flora in Colour

fl. fr. hab.   Pl. 9

Brassica nigra

Masefield, G. B.
The Oxford Book of Food Plants

fr.   Pl. 133, 4

Brassica nigra

Perrot, Emile
Les Plantes Medicinales

fl. fr. hab.   p 154

Brassica oleracea

Macoby, Stirling
What Flower is That

fl.   P 62

Brassica oleracea

Martin, W. Keble
The Concise British Flora in Colour

fl. fr. hab.   Pl. 9

Brassica oleracea

Masefield, G. B.
The Oxford Book of Food Plants

fl., fr.   Pl. 157, 1-2

Brassica oleracea var.

Bianchini, F.
The Complete Book of Fruits & Vegetables

fl. hab.   p 65, 67, 69

Brassica oleracea var.

Crockett, James Underwood
Annuals

hab. P 99

Brassica oleracea var.

Hvass, Elsie
Plants That Feed and Serve Us

fl.   p 26, Pl. 42

Brassica oleracea var.

Masefield, G. B.
The Oxford Book of Food Plants

hab., fr.   Pl. 159, 1, 2, 3, 4

Brassica oleracea var.

Tsukamoto, Yotaro
Coloured Illustrations of Garden
Flowers   vol 10

fl.   p. 5 Pl. 14.

Brassica pekinensi

Masefield, G. B.
The Oxford Book of Food Plants

fl.   Pl. 155, 2

Brassica rapa

Martin, W. Keble
The Concise British Flora in Colour

fl. fr. hab.   Pl. 9

Brassica rapa

Masefield, G. B.
The Oxford Book of Food Plants

fl. fr.   Pl. 173, 1

Brassica rapa

Perrot, Emile
Les Plantes Medicinales

fl. fr. hab.   p 75

Brassica rapa var.

Bianchini, F.
The Complete Book of Fruits & Vegetables

fr.   p 237

Brassica rapa var.

Hvass, Elsie
Plants That Feed and Serve Us

fl.   p 65, Pl. 132

Brassica Rapa var.

Kimura, Koiti
Japanese Medicinal Plants, Vol I

fl.   p 50

Brassica tenuifolia

Bianchini, Francesco
Health Plants of the World

fl. fr.   p 99

x Brassocattleya cliftonii var.

Hay, Roy
The Color Dictionary of Flowers
& Plants

fl.   p. 55   Pl. 434

x Brassocattleya hybrid

Macoby, Stirling
What Flower is That

fl.   p. 62

x Brassocattleya lindleyana

Pabst, G. F. J.
Orchidaceae Brasilienses,
Vol 1

fl.   p 220

x Brassolaeliocattleya hybrid

Kimber, Sheila
A Handbook of Orchids

fl.   p. 12

x Brassolaeliocattleya hybrid

Kramer, Jack
Orchids; Flowers of Romance
and Mystery

fl.   p 104

x Brassolaeliocattleya hybrid

Massachusetts Hort. Soc.
Horticulture
vol 34, No. 3   1956

fl.   p. 142

x Brassolaeliocattleya hybrid

Perry, Frances
Flowers of the World

fl.   p. 4-5, 214

x Brassosophrolaeliocattleya hybrid

Kramer, Jack
Orchids; Flowers of Romance
and Mystery

fl.   p 207

x Brassovola martiana

American Hort. Soc.
American Horticulturist
vol 51, No. 4   1972

fl.   p. 24

Braunsia apiculata

Herre, H.
The Genera of the Mesembryanthemaceae

fl. fr.   p 97

Brazil Nut

see

Bertholletia excelsa

Brazilian, beautiful blue

see

Jacaranda mimosifolia

Bread-and-jam flowers

see

Darwinia fascicularis

Breadfruit

see

Artocarpus

Breadfruit Tree, Polynesian

see

Artocarpus communis

Bread, Sow

see

Cyclamen persicum

Breath of heaven

see

Coleonema pulchrum

Bredia tuberculata

Curtis's Botanical Magazine
Vol 166   1949

fl.   49

Breynia nivosa var.

Macoby, Stirling
What Flower is That

hab.   P. 62

Briar, Sweet

see

Rosa rubiginosa

Bridal flower

See Stephanotis floribunda

Bridal Wreath

see

Spiraea arguta

Bridal wreath

see

Spiraea prunifolia

Bridewort

see

Spiraea salicifolia

Bride's Bonnet

see

Clintonia uniflora

Bride's Tears

see

Antigonon leptopis

Brier, Sensitive

see

Schrankia microphylla

Brighamia rockii

Morley, Brian D.
Wild Flowers of the World

fl.   pl. 146

Brillantaisia subulugurica

Flowering Plants of Africa
vol 44   1977

fl.   Pl. 1751

Brittlebushes

see

Encelia farinosa

Briza maxima

Crockett, James Underwood
Annuals

fr.   P 99

Briza maxima

Hay, Roy
The Color Dictionary of Flowers & Plants

fl.   p 31, Pl. 245

Briza maxima

Huxley, Anthony
Garden Annuals and Bulbs

fl.   Pl. 10

Briza maxima

Perry, Frances
Flowers of the World

fr.   p 131

Briza maxima

Polunin, Oleg
Flowers of Europe

fr. Pl. 179 #1746

Briza maxima

Royal Hort. Soc.
Journal of the Royal Hort. Soc.
vol 93, No. 11   1968

fr.   cover

Briza maxima

Tsukamoto, Yotaro
Coloured Illustrations of Garden
          Flowers   vol 10

fr.   p. 6   Pl. 17

Briza media

Lindman, C. A. M.
Nordens Flora, Vol 2

fr. hab.   Pl. 85

Briza minor

Huxley, Anthony
Garden Annuals and Bulbs

fl.   Pl. 9

Broccoli

see

Brassica oleracea var.

Brodiaea, Bridges'

see

Triteleia bridgesii

Brodiaea capitata

Addisonia
          vol 21   1939-42

fl.   P. 37 Pl. 691

Brodiaea coronaria

Clark, Lewis J.
Wild Flowers of British Columbia

fl.   p 10

Brodiaea, Golden

see

Brodiaea lutea

Brodiaea ida-maia

Hay, Roy
The Color Dictionary of Flowers & Plants

fl.   p 85, Pl. 677

Brodiaea ida-maia

Lemmon, Robert S.
Wildflowers of North America in
        Full Color   .

fl.   pg. 117   pl. 186

Brodiaea ida-maia

Munz, Philip A.
California Spring Wildflowers

fl.   p 78, Pl. 59

Brodiaea ida-maia

Orr, Robert T.
Wildflowers of Western America

fl.   Pl. 215

Brodiaea lactea

Royal Hort. Soc.
Journal of the Royal Hort. Soc.
        vol 89, No. 4   1964

fl.   p. 158   Pl. 66

Brodiaea laxa

Hay, Roy
The Color Dictionary of Flowers & Plants

fl.   p 85, Pl. 678

Brodiaea laxa

Miles, Bebe
Bulbs for the Home Gardener

fl.   p 56

Brodiaea laxa

Munz, Philip A.
California Spring Wildflowers

fl.   p 83, Pl. 73

Brodiaea laxa

Orr, Robert T.
Wildflowers of Western America

fl.   Pl. 282

Brodiaea lutea

Massachusetts Hort. Soc.
Horticulture
        vol 48, No. 11   1970

fl.   p. 16

Brodiaea lutea

Munz, Philip A.
California Spring Wildflowers

fl.   p 83, Pl. 74

Brodiaea pulchella

Munz, Philip A.
California Spring Wildflowers

fl.   p 83, Pl. 75

Brodiaea pulchella var.

Munz, Philip A.
California Desert Wildflowers

fl.   p 27, Pl. 3

Brodiaea uniflora

Everett, T. H.
New Illustrated Encyclopedia of
        Gardening   Vol. 10

fl.   p. 1847

Brodiaea uniflora

Tsukamoto, Yotaro
Coloured Illustrations of
        Garden Flowers   Vol. 9

fl.   p. 4

Broken Hearts

see

Clerodendron thomsonae

Brome, Drooping

see

Bromus tectorum

Bromelia antiacantha

Bromeliad Society
Journal
        vol. 22, No. 6   1972

hab., fl.   p. 162

Bromelia antiacantha

Bromeliad Society
Journal
        vol 24, No. 1   1974

fl.   p. 40

Bromelia balansae

Bromeliad Society
Bulletin
        vol 13, No. 1   1963

fl., hab.   p. 5

Bromelia balansae

Bromeliad Society
Journal
        vol 22, No. 4   1972

fl.   p. 82

Bromelia balansae

Padilla, Victoria
Bromeliads

fl.   p. 65   Pl. 3

Bromelia balansae

Padilla, Victoria, ed.
Bromeliads in Color and
        Their Culture

fl. hab.   p 78

Bromelia balansae

Rauh, Werner
Bromeliads for Home, Garden and
        Greenhouse

fl.   Pl. 95

Bromelia balansae

Wilson, Robert Gardner
Bromeliads in Cultivation

fl. hab.   Pl. 65

Bromelia, Blue

see

Tillandsia lindeniana

Bromelia humilis

Hargreaves, Dorothy
Tropical Blossoms of the Caribbean

fl.   p. 2

Bromelia humilis

Padilla, Victoria
Bromeliads

hab.   p. 65   Pl. 3

Bromelia humilis

Wilson, Robert Gardner
Bromeliads in Cultivation

hab.   Pl. 64

Bromelia pinguin

Bromeliad Society
Journal
        vol 22, No. 5   1972

fl., hab.   cover (p. 105)

Bromelia pinguin

Rauh, Werner
Bromeliads for Home, Garden and
        Greenhouse

fl.   Pl. 93

Bromelia pinguin

Royal Hort. Soc.
Journal of the Royal Hort. Soc.
        vol 91, No. 8   1966

hab.   p. 338   Pl. 175

Bromelia scarlatina

Bromeliad Society
Journal
vol 22, No. 6 1972

hab., fl.   p. 162

Bromeliad

see

forms of Bromeliaceae

Bromheadia finlaysoniana

Bruggeman, L.
Tropical Plants

fl.   Pl. 85

Bromus benekeni

Lindman, C. A. M.
Nordens Flora, Vol 2

fr. hab.   Pl. 106

Bromus erectus

Polunin, Oleg
Flowers of Europe

fl.   Pl. 178 #1720

Bromus hordeaceus

Lindman, C. A. M.
Nordens Flora, Vol 2

fr. hab.   Pl. 108A

Bromus pumpellianus

Porsild, A. E.
Rocky Mountain Wild Flowers

fl.   p 49

Bromus ramosus

Polunin, Oleg
Flowers of Europe

fr.   Pl. 178 #1722

Bromus secalinus

Lindman, C. A. M.
Nordens Flora, Vol 2

fr.   Pl. 108B

Bromus sterilis

Lindman, C. A. M.
Nordens Flora, Vol 2

fr. hab.   Pl. 107A

Bromus tectorum

Lindman, C. A. M.
Nordens Flora, Vol 2

fr.   Pl. 107B

Bromus tectorum

Morley, Brian D.
Wild Flowers of the World

fl.,hab.   Pl. 25

Bronze-Leaf, Rodgers

see

Rodgersia podophylla

Brook Grass

see

Glyceria aquatica

Brooklime

see

Veronica becabunga

Brooklime, American

see

Veronica americana

Brooklime, Austral

see

Gratiola peruviana

Brookweed

see

Samolus

Broom

see

Cytisus

Broom

see

Sarothamnus scoparius

Broom

see

Spartium junceum

Broom, Black

see

Cytisus nigricans

Broom, Butcher's

see

Ruscus aculeatus

Broom, Canary Island

see

Genista canariensis

Broom, Common

see

Spartium scoparium

Broom, Easter

see

Cytisus racemosus

Broom, Feathered

see

Genista sagittalis

Broom, Golden

see

Adenocarpus bacquei

Broom, Hedgehog

see

Erinacea anthyllis

Broom, Kew

see

Cytisus kewensis

Broom, Montpellier

see

Teline monspessulana

Broom, Mount Etna

see

Genista aetnensis

Broom, Native

see

Viminaria juncea

Broom, New Zealand

see

Carmichaelia aligera

Broom, Purple

see

Cytisus purpureus

Broom Rape

see

Orobanche

Broomrape, Ivy

see

Orobanche hederae

Broom, Scots

see

Cytisus scoparius

Broom, Spanish

see

Spartium junceum

Broom, Warminster

see

Cytisus praecox

Broom, White

see

Retama roetam

Broom, White Spanish

see

Cytisus multiflorus

Broom, white weeping

see

Genista monosperma

Broom, Winged

see

Genista sagittalis

Broom, Woolly-Podded

see

Cytisus grandiflorus

Broughtonia negrilensis

Ospina, Mariano
Orquideas de las americas

fl.   Pl. 51

Broussa

see

Verbascum bombyciferum

Broussonetia kazinoki

Kitamura, Siro
Coloured Illustrations of Trees &
   Shrubs of Japan

fl.   138

Broussonetia papyrifera

Kitamura, Siro
Coloured Illustrations of Trees &
   Shrubs of Japan

hab.   139

Broussonetia papyrifera

Polunin, Oleg
Trees and Bushes of Europe

fl. fr. hab.   p 65

Browallia demissa

Tsukamoto, Yotaro
Coloured Illustrations of
      Garden Flowers   Vol. 10

fl.   Illus. 15,   opp. p. 5

Browallia hybrid

Kromdijk, G.
200 House Plants in Colour

fl.   Pl. 36

Browallia, Orange

see

Streptosolen jamesonii

Browallia speciosa

Everett, T.H., ed.
New Illustrated Encyclopedia of
      Gardening vol. 2

fl.   p. 231

Browallia speciosa

Hay, Roy
The Dictionary of House Plants

fl.   Pl. 76

Browallia speciosa var.

Crockett, James Underwood
Annuals

fl.   P 100

Browallia speciosa var.

Hay, Roy
The Color Dictionary of Flowers & Plants

fl.   p 55, Pl. 435

Browallia speciosa var.

Hersey, Jean
Woman's Day Book of House Plants

fl.   p. 39

Browallia speciosa var.

Massachusetts Hort. Soc.
Horticulture
   vol 34, No. 12   1956

fl.   p. 604

Browallia speciosa var.

Massachusetts Hort. Soc.
Horticulture
   vol 37, No. 12   1959

fl.   p. 632

Browallia speciosa var.

Massachusetts Hort. Soc.
Horticulture
   vol 40, No. 3   1962

fl.   p. 165

Browallia speciosa var.

Tsukamoto, Yotaro
Coloured Illustrations of Garden
      Flowers   vol 10

fl.   p. 5   Pl. 16

Brown-eyed Susan

see

Rudbeckia hirta

Brown-eyed susan

see

Rudbeckia triloba

Brownanthus marlothii

Flowering Plants of Africa
      vol 36   1963-64

fl.   Pl. 1418

Brownea x crawfordii

Perry, Frances
Flowers of the World

fl.   p. 161

Brownea grandiceps

Macoby, Stirling
What Flower is That

fl.   P 63

Brownea grandiceps

Morley, Brian D.
Wild Flowers of the World

fl.   pl. 170

Brownea grandiceps

Oakman, Harry
Colorful Trees

Fl.   P 48

Brownea grandiceps

Pertchik, Bernard
Flowering Trees of the
Caribbean

fl.   p 11

Brownea hybrid

Bruggeman, L.
Tropical Plants

fl.   pl. 202

Brownea macrophylla

Hargreaves, Dorothy
Tropical Blossoms of the Caribbean

fl. hab.   p. 43

Brownea macrophylla

Pacific Hort. Foundation
Pacific Horticulture
     vol 38, No. 2   1977

fl.   p. 28-29

Browningia candelaris

Backeberg, Curt
Cactus Lexicon

hab.   p 569

Browningia candelaris

Lamb, Edgar
The Pocket Encyclopedia of Cacti
     and Succulents in Color

hab.   Pl. 301

Browningia pilleifera

Cactus & Succulent Society of America
Journal
     vol 40, No. 1   1968

fl., hab.   p. 23

Brownleea coerulea

Flowering Plants of South Africa
     vol XVIII   1938

fl.   Pl. 702

Brownleea monophylla

Flowering Plants of South Africa
     vol XIX   1939

fl,   Pl. 740

Brugmansia suaveolens

Bruggeman, L.
Tropical Plants

fl.   pl. 223

Brunella grandiflorum

Felsko, Elsa
A Book of Wildflowers

fl.   P 70

Brunfelsia americana

Bruggeman, L.
Tropical Plants

fl.   Pl. 224

Brunfelsia calycina

Nicolaisen, Age
Pocket Encyclopedia of Indoor Plants

fl.   pl. 114

Brunfelsia calycina

Wit, H. C. D. de
Plants of the World;
     The Higher Plants, Vol II

fl.   Pl. 80

Brunfelsia calycina var.

Crockett, James Underwood
Evergreens

Brunfelsia calycina var.

Hay, Roy
The Color Dictionary of Flowers & Plants

fl.   p 55, Pl. 436

Brunfelsia calycina var.

Hay, Roy
The Dictionary of House Plants

fl.   Pl. 77

Brunfelsia calycina var.

Macoby, Stirling
What Flower is That

fl.   P 63

Brunfelsia calycina var.

Mathias, Mildred E.
Color for the Landscape

fl. hab.   p 61

Brunfelsia calycina var.

Morley, Brian D.
Wild Flowers of the World

fl,   pl. 183

Brunfelsia calycina var.

Perry, Frances
Flowers of the World

fl.   p 286

Brunfelsia pauciflora var.

Encke, Fritz
Zimmerpflanzen

fl.   p. 56

Brunfelsia sp.

Kromdijk, G.
200 House Plants in Colour

fl.   Pl. 37

Brunfelsia uniflora

Bruggeman, L.
Tropical Plants

fl.   pl. 225

Brunia albiflora

Flowering Plants of South Africa
     vol XXIV   1944

fl.   Pl. 928

Brunia albiflora

Harrison, R.E.
Trees and Shrubs

fl.   P 36

Brunia stokoei

Morley, Brian D.
Wild Flowers of the World

fl.   Pl. 72

Brunnera macrophylla

Hay, Roy
The Color Dictionary of Flowers & Plants

fl.   p 126, Pl. 1005

Brunnera macrophylla

Perry, Frances
Flowers of the World

fl.   p 51

Brunnichia cirrhosa

Weeds of the Southern United States

fl., hab.   p. 36

Brunnichia cirrhosa

Wharton, Mary E.
Trees & Shrubs of Kentucky

fl.   p 107, Pl. 4.3

Brunonia australis

Blombery, Alec M.
What Wildflower is That

fl. hab.   p. 74, Pl. 169, 170

Brunonia australis

Cochrane, G. R
Flowers and Plants of Victoria

fl.   Pl. 364

Brunsvigia josephinae

Flowering Plants of Africa
    vol 31   1956

fl.   Pl. 1223

Brunsvigia marginata

Flowering Plants of Africa
    vol 44   1977

fl.   Pl. 1740

Brunsvigia orientalis

Flowering Plants of Africa
    vol 36   1963-64

fl., hab.   Pl. 1440

Brush Box

    see

Tristania conferta

Brussels Sprouts

    see

Brassica oleracea var.

Bryocarpum himalaicum

Hara, Hiroshi, comp.
Photo-Album of Plants of
    Eastern Himalaya

fl. hab.   Pl. 171

Bryonia cretica

Polunin, Oleg
Flowers of Europe

fl.   Pl. 79 #815

Bryonia dioica

Ary, S.
The Oxford Book of Wildflowers

fl. fr.   p 50, Pl. 4

Bryonia dioica

Bianchini, Francesco
Health Plants of the World

fr.   p 73

Bryonia dioica

Kleijn, H.
Beauty of the Wild Plant

fl. fr.   opp. p. 53, pl. 65, 65a

Bryonia dioica

Loewenfeld, Claire
The Complete Book of Herbs and Spices

hab.   p 113

Bryonia dioica

Martin, W. Keble
The Concise British Flora in Colour

fl. fr. hab.   Pl. 36

Bryonia dioica

Perrot, Emile
Les Plantes Medicinales

fl. fr. hab.   p 41

Bryony

    see

Bryonia dioica

Bryony, Black

    see

Tamus communis

Bryony, White

    see

Bryonia dioica

Bryophyllum daigremontianum

Curtis's Botanical Magazine
Vol 168   1951

fl.   16?

Bryophyllum daigremontianum

Encke, Fritz
Zimmerpflanzen

hab.   p. 56

Bryophyllum daigremontianum

Nicolaisen, Age
Pocket Encyclopedia of Indoor Plants

hab.   pl. 87

Bryophyllum pinnatum

Hargreaves, Dorothy
Tropical Blossoms of the Caribbean

fl.   p. 2

Bryophyllum pinnatum

Walden, Beryl M.
Wild Flowers of Hong Kong

fl.   Pl. 3(7)

Bryophyllum pinnatum

Wit, H. C. D. de
Plants of the World;
    The Higher Plants, Vol I

fl.   p 280, Pl. 166

Bryophyllum tubiflorum

Macoby, Stirling
What Flower is That

fl.   P 63

Bryophyllum tubiflorum

Nicolaisen, Age
Pocket Encyclopedia of Indoor Plants

hab pl. 88

Bryophyllum tubiflorum

Wit, H. C. D. de
Plants of the World;
    The Higher Plants, Vol I

fl. hab.   p 281, Pl. 168,
                        169

Bryophyllum uniflorum

Hay, Roy
The Dictionary of House Plants

fl.   Pl. 78

Buchenroedera multiflora

Flowering Plants of South Africa
    vol XXIII   1943

fl.   Pl. 883

Buchloe dactyloides

Crockett, James Underwood
Lawns & Ground Covers

fr.   p. 115

Buchnera cruciata

Walden, Beryl M.
Wild Flowers of Hong Kong

fl.   Pl. 67 (199)

Buchnera floridana

Duncan, Wilbur H.
Wildflowers of the Southeastern
    United States

fl.   p 179

Buchnera hispida

Moriarty, Audrey
Wild Flowers of Malawi

fl.   Pl. 40; 4

Buchnera pulchra

Moriarty, Audrey
Wild Flowers of Malawi

fl.   Pl. 40; 3

Buchnera similis

Moriarty, Audrey
Wild Flowers of Malawi

fl.   Pl. 40; 5

Buckbean

see

Menyanthes

Buckbrush

see

Ceanothus greggii var.

Buckeye

see

Aesculus

Buckeye, Bottle-brush

see

Aesculus parviflora

Buckeye, California

see

Aesculus californica

Buckeye, Dwarf

see

Aesculus pavia

Buckeye, Parry

see

Aesculus parryi

Buckeye, Red

see

Aesculus pavia

Buckeye, Sweet

see

Aesculus flava

Buckeye, Yellow

see

Aesculus flava

Buckeye, Yellow

see

Aesculus octandra

Buckhorn

see

Plantago aristata

Buckthorn

see
Rhamnus

Buckthorn, Alder

see

Frangula alnus

Buckthorn, Alder

see

Rhamnus frangula

Buckthorn, Alpine

see

Rhamnus alpinus

Buckthorn, Dwarf

see

Rhamnus pumila

Buckthorn, Mediterranean

see

Rhamnus alaternus

Buckthorn, Palestine

see

Rhamnus palaestina

Buckthorn, Sea

see

Hippophae rhamnoides

Buckthorn, Southern

see

Bumelia lycioides

Buckwheat

see

Fagopyrum

Buckwheat, Flattop

see

Eriogonum fasciculatum

Buckwheat Tree

see

Cliftonia monophylla

Buckwheat, Umbrella

see

Eriogonum umbellatum

Buckwheat Vine

see

Brunnichia cirrhosa

Buckwheat, Wild

see

Eriogonum fasciculatum var.

Buckwheat, Wild

see

Eriogonum heracleoides

Buckwheat, Wild

see

Eriogonum ovalifolium var.

Buckwheat, wild

see

Erigonum umbellatum

Buckwheat, yellow

see

Eriogonum alleni

Buckinghamia celsissima

Blombery, Alec M.
What Wildflower is That

fl.   p. 75, Pl. 171

Buckinghamia celsissima

Oakman, Harry
Colorful Trees

Hab. Fl.   P 91

Buckleberry

see

Vaccinium myrtillus

Buckleya lanceolata

Kitamura, Siro
Coloured Illustrations of Trees &
    Shrubs of Japan

fr.   150

Buddleia alternifolia

Gault, S. Millar
The Color Dictionary of Shrubs

fl.   Pl. 22

Buddleia alternifolia

Hansen, Richard
Baume und Straucher im Garten

fl.   p. 153

Buddleia alternifolia

Hay, Roy
The Color Dictionary of Flowers & Plants

fl.   p 183, Pl. 1464

Buddleia alternifolia

Hellyer, A.G.L.
Shrubs in Colour

fl.   p. 19

Buddleia alternifolia

Royal Hort. Soc.
Journal of the Royal Hort. Soc.
    vol 100, No. 1 1975

hab.   p. 25 Pl. 9

Buddleia colvilei

Harrison, R.E.
Trees and Shrubs

fl.   p. 37

Buddleia colvilei

Hay, Roy
The Color Dictionary of Flowers & Plants

fl.   p 184, Pl. 1465

Buddleia colvilei

Morley, Brian D.
Wild Flowers of the World

fl., fr.   Pl. 95

Buddleia crispa

Gault, S. Millar
The Color Dictionary of Shrubs

fl.   Pl. 23

Buddleia crispa

Morley, Brian D.
Wild Flowers of the World

fl.   Pl. 95

Buddleia davidii

Huxley, Anthony
Deciduous Garden Trees and Shrubs

fl. Pl. 24a, 24b, 24c

Buddleia davidii

Macoby, Stirling
What Flower is That

fl.   P 64

Buddleia davidii

Polunin, Oleg
Flowers of Europe

fl.   Pl. 121 #1189

Buddleia davidii

Polunin, Oleg
Trees and Bushes of Europe

fl. fr.   p. 172

Buddleia davidii

Wit, H. C. D. de
Plants of the World;
    The Higher Plants, Vol II

fl.   p 101, Pl. 52

Buddleia davidii var.

Gault, S. Millar
The Color Dictionary of Shrubs

fl.   Pl. 24, 25

Buddleia davidii var.

Harrison, R.E.
Trees and Shrubs

fl.   P 37

Buddleia davidii var.

Hay, Roy
The Color Dictionary of Flowers & Plants

fl.   p 184, Pl. 1466

Buddleia davidii var.

Hellyer, A.G.L.
Shrubs in Colour

fl.   pl. 19

Buddleia davidii var.

Perry, Frances
Flowers of the World

fl.   p. 56

Buddleia davidii var.

Royal Hort. Soc.
Journal of the Royal Hort. Soc.
    vol 99, No. 10 1974

fl.   p. 442   Pl. 212

Buddleia fallowiana

Curtis's Botanical Magazine
    Vol. 162   1939

fl.   No. 9564

Buddleia forrestii

Curtis's Botanical Magazine
Vol 167   1950

fl.   93

Buddleia, Fountain

see

Buddleia alternifolia

Buddleia globosa

Gault, S. Millar
The Color Dictionary of Shrubs

fl.   Pl. 26

Buddleia globosa

Hay, Roy
The Color Dictionary of Flowers & Plants

fl.   p 184, Pl. 1467

Buddleia globosa

Huxley, Anthony
Deciduous Garden Trees and Shrubs

fl.   Pl. 25

Buddleia globosa

Perry, Frances
Flowers of the World

fl.   p. 56

Buddleia heliophila

Curtis's Botanical Magazine
Vol 169   1952-53

fl.   193

Buddleia hybrid

Gault, S. Millar
The Color Dictionary of Shrubs

fl.   Pl. 27

Buddleia, sage leaf

see

Buddleia salvifolia

Buddleia salvifolia

Harrison, R.E.
Trees and Shrubs

fl.   P 37

Buddleia salvifolia

Macoby, Stirling
What Flower is That

fl.   P 64

Buddleia tibetica

Hara, Hiroshi, comp.
Photo-Album of Plants of
     Eastern Himalaya

fl.   Pl. 18, 19

Buddleia variabilis

Bruggeman, L.
Tropical Plants

fl.   pl. 226

Buddleia variabilis var.

Tsukamoto, Yotaro
Coloured Illustrations of garden
     Flowers vol 10

fl.   p. 42 Pl. 133

Buddleia x weyeriana

Hay, Roy
The Color Dictionary of Flowers &
     Plants

fl.   p. 184   Pl. 1468

Buddleia x weyeriana

Hellyer, A.G.L.
Shrubs in Colour

fl.   p. 19

Buddleia weyeriana var.

Harrison, R. E.
Trees and Shrubs

fl.   p. 37

Buffaloberry, Canada

see

Lepargyrea canadensis

Buffalo Grass

see

Buchloe dactyloides

Buffalonut

see

Pyrularia pubera

Bugbane

see

Cimicifuga

Bugbane, American

see

Cimicifuga americana

Bugbane, False

see

Trautvetteria caroliniensis

Bugloss

see

Anchusa

Bugloss

see

Lycopsis arvensis

Bugloss, Italian

see

Anchusa azurea

Bugloss, Purple

see

Echium plantagineum

Bugloss, Purple Viper's

see

Echium lycopsis

Bugloss, Siberian

see

Anchusa myosotidiflora

Bugloss, Small

see

Lycopsis arvensis

Bugloss, viper's

see

Echium plantagineum

Bugloss, viper's

see

Echium vulgare

Bugle

see

Ajuga reptans

Bugle, Australian

see

Ajuga australis

Bugle, Blue

see

Ajuga genevensis

Bugle, blue

see

Ajuga reptans

Bugle, carpet

see

Ajuga reptans

Bugle, Common

see

Ajuga reptans

Bugle, Pyramidal

see

Ajuga pyramidalis

Bugle, Red

see

Blancoa sanescens

Bugleweed

see

Ajuga reptans

Bugleweed

see

Lycopus virginicus

Bugler, Scarlet

see

Penstemon centranthifolius

Buiningia aurea

Backeberg, Curt
Cactus Lexicon

hab.   p 801

Buiningia brevicylindrica

Backeberg, Curt
Cactus Lexicon

fl.   p 801

Buiningia purpurea

Backeberg, Curt
Cactus Lexicon

fl.   p 802

Bulbine abyssinica

Flowering Plants of Africa
      vol 34   1960-61

fl.   Pl. 1350

Bulbine abyssinica

Moriarty, Audrey
Wild Flowers of Malawi

fl.   Pl. 11; 5

Bulbine bulbosa

Blombery, Alec M.
What Wildflower is That

fl.   p 76, Pl. 172

Bulbine bulbosa

Cochrane, G. R.
Flowers and Plants of Victoria

fl.   Pl. 211

Bulbine sp.

Eliovson, Sima
Namaqualand in Flower

fl.   Pl. 35, 2

Bulbine stenophylla

Flowering Plants of Africa
      vol 27   1948-49

fl., hab.   Pl. 1044

Bulbine tortifolia

Flowering Plants of Africa
      vol XXVI   1947

fl.   Pl. 1019

Bulbinella floribunda

Eliovson, Sima
Namaqualand in Flower

fl.   Pl. 34, 2

Bulbinella setosa

Flowering Plants of South Africa
      vol XXIV   1944

fl., hab.   Pl. 930

Bulbocodium vernum

Polunin, Oleg
Flowers of Europe

fl.   Pl. 163 #1589

Bulbocodium vernum

Tosco, Uberto
The World of Mountain Flowers

fl.   p 52

Bulbophyllum ambrosia

Walden, Beryl M.
Wild Flowers of Hong Kong

fl. hab.   Pl. 18(53)

Bulbophyllum aristatum

Ospina, Mariano
Orquideas de las americas

fl.   Pl. 91

Bulbophyllum barbigerum

Ebel, Friedrich
The Strange and Beautiful
   World of Orchids

fl.   p 135

Bulbophyllum carssulifolium

Cady, Leo
Australian Native Orchids in Colour

fl. hab.   Pl. 88

Bulbophyllum congolanum

Williamson, Graham
The Orchids of South Central Africa

fl.   Pl. 115

Bulbophyllum elisae

Blombery, Alec M.
What Wildflower is That

fl.   p 76, Pl. 173

Bulbophyllum elisae

Cady, Leo
Australian Native Orchids in Colour

fl. hab.   Pl. 89

Bulbophyllum exigium

Blombery, Alec M.
What Wildflower is That

fl.   p 76, Pl. 174

Bulbophyllum falcatum

Stewart, Joyce
Orchids of Tropical Africa

fl. hab.   Pl. 18

Bulbophyllum frostii

Curtis's Botanical Magazine
Vol 167   1950

fl.   121

Bulbophyllum glutinosum

Pabst, G. F. J.
Orchidaceae Brasilienses,
   Vol 1

fl. hab.   p 184

Bulbophyllum glutinosum

Pabst, G.F.J.
Orchidaceae Brasilienses
   vol 2

hab.   p. 172

Bulbophyllum laciniatum

Pabst, G.F.J.
Orchidaceae Brasilienses
   vol 2

fl., hab p. 172

Bulbophyllum leopardinum

Curtis's Botanical Magazine
Vol 163   1940 - 42

fl. hab.   No. 9631

Bulbophyllum lobbii

Ebel, Friedrich
The Strange and Beautiful
   World of Orchids

fl.   p 151

Bulbophyllum lobbii

Kupper, Walter
Orchids

fl. hab.   p 52

Bulbophyllum macphersonii

Cady, Leo
Australian Native Orchids in Colour

fl. hab.   Pl. 87

Bulbophyllum malawiense

Williamson, Graham
The Orchids of South Central Africa

fl.   Pl. 118

Bulbophyllum medusae

Morley, Brian D.
Wild Flowers of the World

fl.   Plate 127

Bulbophyllum micranthum

Pabst, G. F. J.
Orchidaceae Brasilienses,
    Vol 1

fl.   p 185

Bulbophyllum napellii

Pabst, G. F. J.
Orchidaceae Brasilienses,
    Vol 1

fl.   p 185

Bulbophyllum orthoglossum

Curtis's Botanical Magazine
    Vol 159   1936

fl.   No. 9459

Bulbophyllum oxypterum

Moriarty, Audrey
Wild Flowers of Malawi

fl.   Pl. 28; 1

Bulbophyllum oxypterum

Stewart, Joyce
Orchids of Tropical Africa

fl.   Pl. 19

Bulbophyllum oxypterum var.

Williamson, Graham
The Orchids of South Central Africa

fl.   Pl. 114

Bulbophyllum pachyrhachis

Luer, Carlyle A.
The Native Orchids of Florida

fl. fr. hab.   Pl. 53

Bulbophyllum regnellii

Pabst, G. F. J.
Orchidaceae Brasilienses,
    Vol 1

fl.   p 185

Bulbophyllum rugosibulbum

Williamson, Graham
The Orchids of South Central Africa

fl.   Pl. 119

Bulbophyllum sandersonii

Williamson, Graham
The Orchids of South Central Africa

fl.   Pl. 116

Bulbophyllum sp.

Piers, Frank
Orchids of East Africa

fl.   Pl. IIi

Bulbophyllum warmingianum

Pabst, G. F. J.
Orchidaceae Brasilienses,
    Vol 1

fl.   p 185

Bulbophyllum weddellii

Pabst, G. F. J.
Orchidaceae Brasilienses,
    Vol 1

fl.   p 185

Bulbophyllum youngsayeanum

Hu, Shiu-ying
The Genera of Orchidaceae in
    Hong Kong

fl.   p 122

Bulbophyllum zenkeranum

Williamson, Graham
The Orchids of South Central Africa

fl.   Pl. 117

Bull Hoof

see

Bauhinia monandra

Bullace

see

Prunus institia

Bullock bush

see

Templetonia retusa

Bulrush

see

Scirpus

Bulrush, Great

see

Scirpus validus var.

Bulrush, Small

see

Scirpus atrovirens

Bulrush, weeping

see

Scirpus cernuus var.

Bumble Bee Flower

see

Pedicularis species

Bumelia lycioides

Wharton, Mary E.
Trees & Shrubs of Kentucky

fl.   p 70, Pl. 2.3

Bunchberry

see

Cornus canadensis

Bunch Flower

see

Melanthium virginicum

Bunias erucago

Bianchini, F.
The Complete Book of Fruits & Vegetables

hab.   p 73

Bunias erucago

Polunin, Oleg
Flowers of Europe

fl.   Pl. 32   #293

Bunias orientalis

Lindman, C. A. M.
Nordens Flora, Vol. 4

fl. fr.   Pl. 262

Bunium bulbocastanum

Martin, W. Keble
The Concise British Flora in Colour

fl. fr. hab.   Pl. 37

Bunny ears

see

Opuntia santa-rita

Bunya

see

Araucaria bidwillii

Buphthalmum salicifolium

Felsko, Elsa
A Book of Wild Flowers
            2nd Ser.
fl.   p 19

Buphthalmum salicifolium

Hay, Roy
Color Dictionary of Flowers &
        Plants

fl.   p. 126   Pl. 1006

Buphthalmum speciosissimum

Barneby, T. P.
European Alpine Flowers in Colour

fl.   Pl. 95, 6

Bupleurum baldense

Martin, W. Keble
The Concise British Flora in Colour

fr. hab.   Pl. 36

Bupleurum falcatum

Kariyone, Tatsuo
Atlas of Medicinal Plants

fl., hab.   Pl. 69

Bupleurum falcatum

Kimura, Koiti
Japanese Medicinal Plants, Vol I

fl.   p 66

Bupleurum falcatum

Martin, W. Keble
The Concise British Flora in Colour

fl. fr. hab.   Pl. 36

Bupleurum falcatum

Takatori, Jisuke
Color Atlas of Medicinal Plants of Japan

fl.   Fig. 22

Bupleurum fruticosum

Curtis's Botanical Magazien
        vol 174   1962-63

fl.   Pl. 408

Bupleurum fruticosum

Hay, Roy
The Color Dictionary of Flowers & Plants

fl.   p 184, Pl. 1469

Bupleurum fruticosum

Hellyer, A.G.L.
Shrubs in Colour

fl.   p. 19

Bupleurum fruticosum

Polunin, Oleg
Trees and Bushes of Europe

fl. fr. hab.   p 155

Bupleurum longifolium

Barneby, T. P.
European Alpine Flowers in Colour

fl.   Pl. 53, 6

Bupleurum longifolium var.

Royal Hort. Soc.
The Garden
        vol 103, No. 5   1978

fl.   p. 208

Bupleurum petraeum

Barneby, T. P.
European Alpine Flowers in Colour

fl. hab.   Pl. 54, 1

Bupleurum rotundifolium

Martin, W. Keble
The Concise British Flora in Colour

fl. hab.   Pl. 36

Bupleurum rotundifolium

Wharton, Mary E.
A Guide to the Wildflowers & Ferns
        of Kentucky

fl.   p 158   Pl. 2.21

Bupleurum stellatum

Barneby, T. P.
European Alpine Flowers in Colour

fl. hab.   Pl. 53, 5

Bupleurum stellatum

Kohlhaupt, Paula
Fleurs des Alpages
        vol 1
fl. P 70

Bupleurum stellatum

Polunin, Oleg
Flowers of Europe

fl.   Pl. 84 #883

Bupleurum tenuissimum

Lindman, C. A. M.
Nordens Flora, Vol 7

fl. fr. hab.   Pl. 421

Bupleurum tenuissimum

Martin, W. Keble
The Concise British Flora in Colour

fl. fr. hab.   Pl. 36

Burbush

see

Franseria chenopodifolia

Burdock

see

Arctium

Burdock, Common

see

Arctium minus

Burgrass, Desert

see

Cenchrus palmeri

Burnet

see

Sanguisorba

Burnet

see

Poterium sanguisorba

Burnet, Great

see

Sanguisorba menziesii

Burnet, great

see

Sanguisorba officinalis

Burnet, Japanese

see

Sanguisorba obtusa

Burnet, Salad

see

Sanguisorba minor

Burnet, Sitka Great

see

Sanguisorba sitchensis

Burnet, Thorny

see

Poterium spinosum

Bursage

see

Franseria dumosa

Burchardia umbellata

Blombery, Alec M.
What Wildflower is That

fl.   p 76, Pl. 175

Burchardia umbellata

Cochrane, G. R.
Flowers and Plants of Victoria

fl.   Pl. 26

Burchardia umbellata

Mullins, Barbara
Australian Wildflowers in Colour

fl.   p. 103   Pl. 102

Burchellia bubalina

Morley, Brian D.
Wild Flowers of the World

fl.   Pl. 79

Burchellia bubalina

Perry, Frances
Flowers of the World

fl.   p 267

Burkea africana

Flowering Plants of Africa
vol 38   1967

fl.   Pl. 1505

Burkea africana

Palgrave, K. C.
Trees of Central Africa

fl., fr.   p. 91

Burkea africana

Palmer, Eve
Trees of South Africa

hab.   p. 208   Pl. XIV

Burmannia capitata

Duncan, Wilbur H.
Wildflowers of the Southeastern
United States

fl.   p 269

Burmannia coelestris

Walden, Beryl M.
Wild Flowers of Hong Kong

fl. hab.   Pl. 64 (186)

Burmannia madagascariensis

Flowering Plants of Africa
vol 36   1963-64

fl., hab.   Pl. 1427

Burmannia nepalensis

Hara, Hiroshi, comp.
Photo-Album of Plants of
Eastern Himalaya

fl.   Pl. 22

Burn Plant

see

Aloe vera

Burnweed

see

Erechitites hieracifolia

Burnweed, American

see

Erechtites hieracifolia

Burnettia cuneata

Cady, Leo
Australian Native Orchids in Colour

fl.   Pl. 55

Burnettia cuneata

Cochrane, G. R.
Flowers and Plants of Victoria

fl. hab.   Pl. 71

Burning Bush

see

Dictamnus albus

Burning Bush

see

Kochia scoparia var.

Burro Weed

see

Franseria dumosa

Burro's Tail

see

Sedum morganianum

Bursaria spinosa

Blombery, Alec M.
What Wildflower is That

fl.   p 77, Pl. 176

Bursaria spinosa

Cochrane, G. R.
Flowers and Plants of Victoria

fl.   Pl. 282

Bursera hindsiana

Coyle, Jeanette
A Field Guide to the Common and
Interesting Plants of Baja California

fr.   p 107

Bursera microphylla

Coyle, Jeanette
A Field Guide to the Common and
Interesting Plants of Baja California

fr. hab.   p 107

Burtonia scabra

Blombery, Alec M.
What Wildflower is That

fl.   p 77, Pl. 177

Bush Chinquapin

see

Castanopsis sempervirens

Bush Clover

see

Lespedeza

Bush-Clover, Creeping

see

Lespedeza repens

Bush-clever, Siebold's

see

Lespedeza sieboldii

Bush-Hops

see

Hymenoclea salsola

Bush lawyer

see

Rubus parvus

Bushman's bootlace

see

Pimelea nivea

Bushman's poison

see

Acokanthera spectabilis

Bushman's poison

see

Acokanthera venenata

Bush Nut

see

Macadamia integrifolia

Busy Lizzie

see

Impatiens

But But

see

Eucalyptus bridgesiana

Butea frondosa

Everett, Thomas H.
Living Trees of the World

fl., hab.   p. 178-179

Butea frondosa

Macoby, Stirling
What Flower is That

fl.   P 64

Butea monosperma

American Hort. Soc.
American Horticulturist
vol 57, No. 6   1978

fl.   p. 6

Butea superba

Menninger, Edwin A.
Flowering Vines of the World

fl.   Pl. 110

Butomus umbellatus

Ary, S.
The Oxford Book of Wildflowers

fl.   p 108, Pl. 4

Butomus umbellatus

Courtenay, Booth
Wildflowers and weeds

fl. p. 3

Butomus umbellatus

Felsko, Elsa
A Book of Wild Flowers
                    2nd ser.
fl.   p 130

Butomus umbellatus

Hay, Roy
The Color Dictionary of Flowers & Plants

fl.   p 126, Pl. 1007

Butomus umbellatus

Kleijn, H.
Beauty of the Wild Plant

fl.   opp. p. 17   pl. 10

Butomus umbellatus

Lindman, C. A. M.
Nordens Flora, Vol 1

fl. fr.   Pl. 42

Butomus umbellatus

Martin, W. Keble
The Concise British Flora in Colour

fl. hab.   Pl. 79

Butomus umbellatus

Morley, Brian D.
Wild Flowers of the World

fl.,hab.   Pl. 23

Butomus umbellatus

Perry, Frances
Flowers of the World

fl.   p 56

Butomus umbellatus

Polunin, Oleg
Flowers of Europe

fl.   Pl. 161 #1564

Butter-and-eggs

see

Linaria vulgaris

Butterbur

see

Petasites

Butterbur, Glacial

see

Petasites paradoxus

Butterbur, Northern

see

Petasites hyperboreus

Butterbur, white

see

Petasites albus

Buttercup

see

Ranunculus

Buttercup, Alpine

see

Ranunculus eschscholtzii

Buttercup, Avalanche

see

Ranunculus suksdorfii

Buttercup, Chamisso

see

Ranunculus glacialis

Buttercup, Common
see
Ranunculus acris

Buttercup, Common
see
Ranunculus lappaceus

Buttercup, Cooley
see
Ranunculus cooleyae

Buttercup, creeping
see
Ranunculus repens

Buttercup, Dwarf
see
Ranunculus pygmaeus

Buttercup, Early
see
Ranunculus fascicularis

Buttercup, Eschscholtz
see
Ranunculus eschscholtzii

Buttercup Flower
see
Allamanda cathartica var.

Buttercup, grassy
see
Ranunculus gramineus

Buttercup, Hairy
see
Ranunculus hispidus

Buttercup, Kosciusko
see
Ranunculus dissectifolius

Buttercup, Meadow
see
Ranunculus acer

Buttercup, Meadow
see
Ranunculus acris

Buttercup, Mountain
see
Ranunculus aconitifolius

Buttercup, Mountain
see
Ranunculus eschscholtzii

Buttercup, mountain
see
Ranunculus graniticola

Buttercup, Northern
see
Ranunculus affinis

Buttercup, Persian
see
Ranunculus asiaticus

Buttercup, Plantain-leaf
see
Ranunculus alismaefolius

Buttercup, Poisonous
see
Ranunculus thora

Buttercup, Sagebrush
see
Ranunculus glaberrimus

Buttercup, Small-flowered
see
Ranunculus parviflorus

Buttercup, Subalpine
see
Ranunculus eschscholtzii

Buttercup, Swamp
see
Ranunculus septentrionalis

Buttercup, Tall
see
Ranunculus acris

Buttercup, tongue
see
Ranunculus lingua

Buttercup tree
see
Cassia corymbosa

Buttercup tree
See Cochlospermum vitifolium

Buttercup, Water
see
Ranunculus aquatilis

Buttercup, Western
see
Ranunculus occidentalis

Butterfly bush
see
Buddleia

Butterfly bush
see
Clerodendrum ugandense

Butterfly Bush
see
Psoralea pinnata

Butterfly Flower
see
Hedychium coronarium

Butterfly Flower
see
Schizanthus

Butterfly-Flower Tree
see
Bauhinia monandra

Butterfly weed
see
Asclepias

Butterfly-weed
see
Gaura coccinea

Butternut
see
Juglans cinerea

Butterweed
see
Senecio glabellus

Butterwort
see
Pinguicula

Butterwort, Alpine
see
Pinguicula alpina

Butterwort, Common
see
Pinguicula vulgaris

Butterwort, little blue
see
Pinguicula primuliflora

Butterwort, Northern
see
Pinguicula vulgaris

Butterwort, Pale
see
Pinguicula lusitanica

Butterwort, Purple
see
Pinguicula elatior

Butterwort, Violet
see
Pinguicula caerulea

Butterwort, Yellow
see
Pinguicula lutea

Buttonbush
see
Cephalanthus occidentalis

Button Flower
see
Cotula barbata

Buttonweed
see
Diodia teres

Button Weed
see
Diodia virginiana

Buttonwood
see
Conocarpus erecta

Buttonwood
see
Platanus occidentalis

Buttons, Mescal
see
Lophophora

Buttons, Pink
see
Kunzea capitata

Buttons, Scaly
see
Leptorhynchus squamatus

Buttons, Water
see
Cotula coronopifolia

Butyrospermum parkii

Wit, H. C. D. de
Plants of the World;
  The Higher Plants, Vol I

fr. hab.   p 174, Pl. 103, 104

Buxus balearica

Polunin, Oleg
Trees and Bushes of Europe

fr. hab.   p 132, 133

Buxus microphylla var.

Hersey, Jean
Woman's Day Book of House Plants

hab.   p 38

Buxus microphylla var.

Kitamura, Siro
Coloured Illustrations of Trees &
  Shrubs of Japan

fr.   276

Buxus sempervirens

Edlin, Herbert
The Illustrated Encyclopedia
  of Trees

fl. fr. hab.   p 186-7

Buxus sempervirens

Hay, Roy
The Color Dictionary of Flowers & Plants

hab.   p 184, Pl. 1470

Buxus sempervirens

Huxley, Anthony
Evergreen Garden Trees and Shrubs

hab.   Pl. 99

Buxus sempervirens

Hvass, Elsie
Plants That Feed and Serve Us

fl.   fr.   p 126, Pl. 280

Buxus sempervirens

Martin, W. Keble
The Concise British Flora in Colour

fl. hab.   Pl. 41

Buxus sempervirens

Morley, Brian D.
Wild Flowers of the World

fl.   Pl. 31

Buxus sempervirens

Pennsylvania Hort. Soc.
The Green Scene
      vol 6, No. 4   1978

hab.   p. 29

Buxus sempervirens

Perrot, Emile
Les Plantes Medicinales

fl. fr. hab.   p 42

Buxus sempervirens

Polunin, Oleg
Trees and Bushes of Europe

fl. fr. hab.   p 132

Buxus sempervirens var.

Crockett, James Underwood
Evergreens

hab.   p. 144

Buxus sempervirens var.

Gault, S. Millar
The Color Dictionary of Shrubs

hab.   Pl. 28

Buxus sempervirens var.

Hellyer, A.G.L.
Shrubs in Colour

hab.   p. 22

Buxus sempervirens var.

Huxley, Anthony
Evergreen Garden Trees and Shrubs

hab.   Pl. 101, 102

Byblis gigantea

Blombery, Alec M.
What Wildflower is That

fl.   p. 77, Pl. 178

Byblis gigantea

Morcombe, M. K.
Australia's Western Wildflowers

fl.   p 80

Byblis gigantea

Morley, Brian D.
Wild Flowers of the World

fl.   pl. 131

Byronia alba

Lindman, C. A. M.
Nordens Flora, Vol 6

fl. fr.   Pl. 399

Bystropogon plumosus

Bramwell, David
Wild Flowers of the Canary Islands

fl.   Pl. 242

Cabbage

see

Brassica oleracea var.

Cabbage, Bear Green

see

Veratrum viride

Cabbage, Black

see

Brassica oleracea var.

Cabbage, Field

see

Brassica campestris

Cabbage, Garden

see

Brassica oleracea var.

Cabbage, Hottentot's

see

Trachyandra falcata

Cabbage, Kerguelen

see

Pringlea antiscorbutica

Cabbage, ornamental

see

Brassica oleracea var.

Cabbage, Purple

see

Moricandia arvensis

Cabbage, Savoy

see

Brassica oleracea var.

Cabbage, Skunk

see

Lysichitum americanum

Cabbage, Skunk

see
Spathyema foetida

Cabbage, Skunk

see

Symplocarpus foetidus

Cabbage, Skunk Western

see

Lysichitum camtschatcense

Cabbage, Skunk Yellow

see

Lysichitum americanum

Cabbage tree

see

Cordyline australis

Cabbage tree

see

Cussonia kirkii

Cabbage tree, spiked

see

Cussonia paniculata

Cabbage, wild

see

Brassica oleracea

Cabomba aquatica

Stodola, Jiri
Encyclopedia of Water Plants

fl.   P 239

Cabomba carolinana

Stodola, Jiri
Encyclopedia of Water Plants

fl.   P 242

Cabomba caroliniana

Tarver, David P.
Aquatic and Wetland Plants
    of Florida

fl. hab.   p 61

Cabomba caroliniana var.

Tarver, David P.
Aquatic and Wetland Plants
    of Florida

fl., hab.   p. 62

Cabomba piauhiensis

Wit, H. C. D. de
Plants of the World;
    The Higher Plants, Vol I

fl.   p 63, Pl. 35

Cabomba pulcherrima

Tarver, David P.
Aquatic and Wetland Plants
    of Florida

fl. hab.   p 63

Cacalia atriplicifolia

Courtenay, Booth
Wildflowers & Weeds

fl.   p 126

Cacalia atriplicifolia

Wharton, Mary E.
A Guide to the Wildflowers & Ferns
    of Kentucky

fl.   p 234   Pl. 1.3

Cacalia lanceolata

Duncan, Wilbur H.
Wildflowers of the Southeastern
    United States

fl.   p 225

Cacalia plantaginea

Brown, Clair A.
Wildflowers of Louisiana and
    Adjoining States

fl.   p 192

Cacalia suaveolens

Courtenay, Booth
Wildflowers & Weeds

fl.   p 126

Cacalia tuberosa

Courtenay, Booth
Wildflowers & Weeds

fl.   p 126

Cacalia tuberosa

Dean, Blanche E.
Wildflowers of Alabama and
    Adjoining States

fl.   p 179

Cacao, American

see

Theobroma cacao

Cactus, Bailey's Hedgehog

see

Echinocereus baileyi

Cactus, Barrel

see

Ferocactus acanthodes

Cactus, Barrel

see

Ferocactus peninsulae

Cactus, Barrel

see

Ferocactus wislizenii

Cactus, Beaver Tail

see

Opuntia basilaris

Cactus, Beehive

see

Coryphantha vivipara

Cactus, Bishop's Cap

see

Astrophytum myriostigma

Cactus, Blade

see

Phyllocactus sp.

Cactus, Bonker's Hedgehog

see

Echinocereus fendleri

Cactus, Button

see

Epithelantha bokei

Cactus, Calico

see

Echinocereus viridiflorus

Cactus, California Barrel

see

Ferocactus acanthodes

Cactus, Candelabra

see

Myrtillocactus cochal

Cactus, Candy

see

Homalocephala texensis

Cactus, Cat-Claw

see

Hamatocactus uncinatus

Cactus, Cholla

see

Opuntia sp.

Cactus, Christmas

see

Schlumbergera bridgesii

Cactus, Christmas

see

Zygocactus truncatus

Cactus, Compass

see

Ferocactus acanthodes

Cactus, Coville's Barrel

see

Ferocactus covillei

Cactus, Crab

see

Zygocactus truncatus

Cactus, cream

see

Mammillaria applanata

Cactus, Cream

see

Mammillaria heyderi

Cactus, Dahlia

see

Wilcoxia poselgeri

Cactus, Deer-horn

see

Peniocereus greggii

Cactus, Devil

see

Opuntia schottii

Cactus, Easter

see

Schlumbergera gaertneri

Cactus, Easter lily

see

Echinopsis multiplex

Cactus, empress

see

Nopalxochia phyllanthoides

Cactus, Fish-hook

see

Ancistrocactus scheerii

Cactus, Fish-hook

see

Hamatocactus setispinus

Cactus, Fish-hook

see

Mammillaria microcarpa

Cactus, Fishhook

see

Mammillaria tetrancistra

Cactus, Galloping

see

Machaerocereus gummosus

Cactus, Gherkin

see

Chamaecereus silvestrii

Cactus, Graham Dog

see

Opuntia schottii

Cactus, Gramma Grass

see

Toumeya papyracantha

Cactus, Green-flowered Torch

see

Echinocereus viridiflorus

Cactus, grizzly bear

see

Echinocereus ursinus

Cactus, Hairbrush

see

Pachycereus pecten-aboriginum

Cactus, Hedgehog

see

Coryphantha echinus

Cactus, Hedgehog

see

Echinocereus brandegeei

Cactus, Hedgehog

see

Echinocereus engelmannii

Cactus, Hedgehog

see

Echinocereus fendleri

Cactus, hedgehog

see

Echinopsis multiplex

Cactus, hedgehog

see

Opuntia erinacea

Cactus intortus

Van Laren, A. J.
Cactus

fr.   P 62   Fig. 75

Cactus, Johnson's Pineapple

see

Echinomastus johnsonii

Cactus, Junior Tom Thumb

see

Escobaria runyonii

Cactus, Living Rock

see

Ariocarpus kotschoubeyanus

Cactus, Mojave Mound

see

Echinocereus triglochidiatus var.

Cactus, Nichol's Hedgehog
see
Echinocereus engelmannii

Cactus, Nipple
see
Neobesseya simillis

Cactus, Old-man
see
Cephalocereus senilis

Cactus, Orchid
see
Chiapasia nelsonii

Cactus, orchid
see
Epiphyllum hybrid

Cactus, Organ-pipe
see
Lemaireocereus thurberi

Cactus, Peanut
see
Chamaecereus silvestrii

Cactus, Pencil
see
Wilcoxia poselgeri

Cactus, Pincushion
see
Mammillaria dioica

Cactus, pincushion
see
Mammillaria zeilmanniana

Cactus, Plains
see
Pediocactus simpsonii

Cactus, Rainbow
see
Echinocereus rigidissimus

Cactus, Rat's Tail
see
Aporocactus flagelliformis

Cactus, Rat-tail
see
Mammillaria pottsii

Cactus, Rose
see
Strombocactus disciformis

Cactus, Saguaro
see
Carnegiea gigantea

Cactus, San Diego barrel
see
Echinocactus viridescens

Cactus, Sand-Dollar
see
Astrophytum asterias

Cactus, Sea-Urchin
see
Astrophytum asterias

Cactus, sea urchin
see
Echinopsis multiplex

Cactus, Small Barrel
see
Ferocactus viridescens

Cactus, star
see
Astrophytum capricorne

Cactus, Strawberry
see
Echinocereus engelmannii

Cactus, Strawberry
see
Echinocereus enneacanthus

Cactus, Strawberry
see
Echinocereus mojavensis

Cactus, Strawberry
see
Echinocereus stramineus

Cactus, Strawberry Green
see
Echinocereus viridiflorus

Cactus, Strawberry Hedgehog
see
Echinocereus engelmannii

Cactus, Strawberry Lloyds
see
Echinocereus lloydii

Cactus, Texas Rainbow
see
Echinocereus dasyacanthus

Cactus, Torch
see
Echinocereus engelmannii

Cactus, Velvet
see
Bergerocactus emoryi

Cactus, Velvet
see
Cereus emoryi

Cadaba aphylla

Eliovson, Sima
Namaqualand in Flower

fl.   Pl. 26, 2

Cadaba kirkii

Moriarty, Audrey
Wild Flowers of Malawi

fl.   Pl. 44, 2

Cadaba termitaria

Flowering Plants of South Africa
        vol XVII   1937

fl.   Pl. 656

Cade-Oil

see

Juniperus oxycedrus

Cadia purpurea

Flowering Plants of Africa
        vol 27   1948-49

fl.   Pl. 1070

Caesalpinia ferrea

Oakman, Harry
Colorful Trees

Hab. Fl.   P 97

Caesalpinia gilliesii

American Hort. Soc.
American Horticulturist
        vol 56, No. 4 1977

fl.   p. 23

Caesalpinia gilliesii

Harrison, R. E.
Trees and Shrubs

fl.   p 38

Caesalpinia japonica

Hay, Roy
The Color Dictionary of Flowers & Plants

fl.   p 184, Pl. 1471

Caesalpinia japonica

Kitamura, Siro
Coloured Illustrations of Trees &
        Shrubs of Japan

fl.   246

Caesalpinia japonica

Royal Hort. Soc.
Journal of the Royal Hort. Soc.
        vol 91, No. 10   1966

fl.   p. 426   Pl. 216

Caesalpinia pulcherrima

Bruggeman, L.
Tropical Plants

fl. fr.   pl. 227

Caesalpinia pulcherrima

Chickering, Carol Rogers
Flowers of Guatemala

fl. fr.   p 39

Caesalpinia pulcherrima

Hall, Clarence E.
Flowers of the Islands in the Sun

fl.   p. 115   Pl. 25

Caesalpinia pulcherrima

Macoby, Stirling
What Flower is That

fl.   P 65

Caesalpinia pulcherrima

O'Gorman, Helen
Mexican Flowering Trees and Plants

fl. fr. hab.   p 47

Caesalpinia pulcherrima

Perry, Frances
Flowers of the World

fl.   p 159

Caesalpinia pulcherrima

Wit, H. C. D. de
Plants of the World;
        The Higher Plants, Vol I

fl.   p 282, Pl. 172

Caesalpinia sepiaria

Hara, Hiroshi, comp.
Photo-Album of Plants of
        Eastern Himalaya

fl.   Pl. 66

Caesarweed

see

Urena lobata

Caesia vittata

Blombery, Alec M.
What Wildflower is That

fl.   p 78, Pl. 179

Caiophora cernua

Curtis's Botanical Magazine
Vol 171   1956-57

fl.   273

Cajanus cajan

Masefield, G. B.
The Oxford Book of Food Plants

fl. fr.   Pl. 35, 1

Cajeput

see

Melaleuca quinquenervia

Cajeput-tree

see

Melaleuca leucadendra

Cakile edentula

Clark, Lewis J.
Wild Flowers of British Columbia

fl.   p 183

Cakile edentula

Courtenay, Booth
Wildflowers & Weeds

fl.   p 20

Cakile lanceolata

Fleming, Glenn
Wild Flowers of Florida

Cakile maritima

Ary, S.
The Oxford Book of Wildflowers

fl. fr.   p 68, Pl. 6

Cakile maritima

Bianchini, F.
The Complete Book of Fruits & Vegetables

fl.   p 59

Cakile maritima

Kleijn, H.
Beauty of the Wild Plant

fl.   opp. p. 77   pl. 101

Cakile maritima

Lindman, C. A. M.
Nordens Flora, Vol 4

fl. fr.   Pl. 254

Cakile maritima

Martin, W. Keble
The Concise British Flora in Colour

fl. fr. hab.   Pl. 11

Cakile maritima

Polunin, Oleg
Flowers of Europe

fl.   Pl. 37 #366

Calabazilla

see

Cucurbita foetidissima

Caladenia alba

Blombery, Alec M.
What Wildflower is That

fl.   p 78, Pl. 180

Caladenia angustata

Cochrane, G. R.
Flowers and Plants of Victoria

fl.   Pl. 378

Caladenia caerulea

Cochrane, G. R.
Flowers and Plants of Victoria

fl.   Pl. 343

Caladenia carnea

Blombery, Alec M.
What Wildflower is That

fl.   p 79, Pl. 181

Caladenia carnea

Cady, Leo
Australian Native Orchids in Colour

fl.   Pl. 51

Caladenia carnea

Cochrane, G. R.
Flowers and Plants of Victoria

fl.   Pl. 375

Caladenia caudata

Curtis, Winifred
The Endemic Flora of Tasmania, Part IV

fl. hab.   Pl. 147

Caladenia deformis

Cady, Leo
Australian Native Orchids in Colour

fl.   Pl. 52

Caladenia deformis

Cochrane, G. R.
Flowers and Plants of Victoria

fl.   Pl. 344

Caladenia deformis

Morcombe, M. K.
Australia's Western Wildflowers

fl.   p 92

Caladenia dilatata

Blombery, Alec M.
What Wildflower is That

fl.   p. 79   pl. 182

Caladenia dilatata

Cady, Leo
Australian Native Orchids in Colour

fl.   Pl. 50

Caladenia dilatata

Cochrane, G. R.
Flowers and Plants of Victoria

fl. hab.   Pl. 339

Caladenia dilatata

Morcombe, M. K.
Australia's Western Wildflowers

fl.   p 9

Caladenia dilatata var.

Morcombe, M. K.
Australia's Western Wildflowers

fl.   p 9

Caladenia filamentosa

Morcombe, M. K.
Australia's Western Wildflowers

fl.   p 29

Caladenia filamentosa

Morley, Brian D.
Wild Flowers of the World

fl.   Pl. 145

Caladenia flava

Blombery, Alec M.
What Wildflower is That

fl.   p. 79   Pl. 183

Caladenia flava

Cady, Leo
Australian Native Orchids in Colour

fl. hab.   Pl. 48

Caladenia latifolia

Cady, Leo
Australian Native Orchids in Colour

fl.   Pl. 53

Caladenia latifolia

Cochrane, C. R.
Flowers and Plants of Victoria

fl.   Pl. 277

Caladenia latifolia

Morcombe, M. K.
Australia's Western Wildflowers

fl.   p 8

Caladenia menziesii

Cady, Leo
Australian Native Orchids in Colour

fl.   Pl. 54

Caladenia multiclava

Cady, Leo
Australian Native Orchids in Colour

fl.   Pl. 49

Caladenia patersonii

Cady, Leo
Australian Native Orchids in Colour

fl.   Pl. 47

Caladenia patersonii

Cochrane, G. R.
Flowers and Plants of Victoria

fl.   Pl. 73

Caladenia patersonii var.

Cady, Leo
Australian Native Orchids in Colour

fl.   Pl. 2

Caladenia pectinata

Blombery, Alec M.
What Wildflower is That

fl.   p. 79   Pl. 184

Caladium bicolor

Bruggeman, L.
Tropical Plants

fl.   Pl. 86

Caladium bicolor

Hay, Roy
The Dictionary of House Plants

hab.   Pl. 79

Caladium bicolor

Macoby, Stirling
What Flower is That

hab P 65

Caladium bicolor

Tsukamoto, Yotaro
Coloured Illustrations of
    Garden Flowers

hab.   opp. p. 4

Caladium bicolor var.

Everett, T.H., ed.
New Illustrated Encyclopedia of
    Gardening vol. 2

fl., hab.   p. 231, 278

Caladium bicolor var.

Everett, T. H., ed.
New Illustrated Encyclopedia of
    Gardening   vol. 13

hab.   p. 2406

Caladium candidum

Hay, Roy
The Color Dictionary of Flowers & Plants

hab.   p 55, Pl. 437

Caladium humboldti

Pennsylvania Hort. Soc.
The Green Scene
    vol 5, No. 1   1976

hab.   p. 33

Caladium hybrid

Encke, Fritz
Zimmerpflanzen

hab.   p. 61

Caladium hybrid

Hersey, Jean
Woman's Day Book of House Plants

hab. p. 44

Caladium hybrid

Kromdijk, G.
200 House Plants in Colour

hab.   Pl. 38

Caladium hybrid

Miles, Bebe
Bulbs for the Home Gardener

hab.   p 154

Calamagrostis canescens

Lindman, C. A. M.
Nordens Flora, Vol 2

fr. hab.   Pl. 97

Calamagrostis neglecta

Lindman, C. A. M.
Nordens Flora, Vol 2

fr. hab.   Pl. 98

Calamagrostis purpurascens

Porsild, A. E.
Rocky Mountain Wild Flowers

fl.   p 49

Calamint

see

Satureja calamintha

Calamint, Scarlet

see

Cardoquia coccinea

Calamintha alpina

Barneby, T. P.
European Alpine Flowers in Colour

fl.   Pl. 67, 4

Calamintha alpina

Kohlhaupt, Paula
Fleurs des Alpages
    vol 2
Fl.   P 96

Calamintha alpina var.

Curtis's Botanical Magazine
    Vol 160   1937

fl.   No. 9508

Calamintha ascendens

Ary, S.
The Oxford Book of Wildflowers

fl.   p 144, Pl. 2

Calamintha ascendens

Martin, W. Keble
The Concise British Flora in Colour

fl. hab.   Pl. 68

Calamintha grandiflora

Polunin, Oleg
Flowers of Europe

fl.   Pl. 114 #1154

Calamintha nepeta

Martin, W. Keble
The Concise British Flora in Colour

fl. hab.   Pl. 68

Calamintha nepeta

Polunin, Oleg
Flowers of Europe

fl.   Pl. 114 #1156

Calamintha nepetoides

Hay, Roy
The Color Dictionary of Flowers & Plants

fl.   p 126, Pl. 1008

Calamintha officinales

Perrot, Emile
Les Plantes Medicinales

fl. hab.   p 28

Calamintha sylvatica

Martin, W. Keble
The Concise British Flora in Colour

fl. hab.   Pl. 68

Calamophyllum teretifolium

Herre, H.
The Genera of the Mesembryanthemaceae

fl.   fr.   p 99

Calamus, American

see

Acorus americanus

Calamus rotang

Hvass, Elsie
Plants That Feed and Serve Us

fr.   p 131, Pl. 291

Calandrinia balonensis

Blombery, Alec M.
What Wildflower is That

fl.   p 80, Pl. 185

Calandrinia balonensis

Morley, Brian D.
Wild Flowers of the World

fl.   pl. 132

Calandrinia ciliata

Orr, Robert T
Wildflowers of Western America

fl.   Pl. 206

Calandrinia ciliata var.

Clark, Lewis J.
Wild Flowers of British Columbia

fl.   p. 126

Calandrinia ciliata var.

Lemmon, Robert S.
Wildflowers of North America in
    Full Color

fl.   pg. 18   pl. 28

Calandrinia crassifolia var.

Munoz Pizarro, Carlos
Flores Silvestres de Chile

fl.   Pl. 14a

Calandrinia discolor

Munoz Pizarro, Carlos
Flores Silvestres de Chile

fl.   Pl. 14

Calandrinia feltonii

Royal Hort. Soc.
The Garden
   vol 100, No. 8   1975

fl.   p. 380

Calandrinia polyandra

Perry, Frances
Flowers of the World

fl.   p. 238

Calandrinia species

Mullins, Barbara
Australian Wildflowers in Colour

fl.   P. 81   Pl. 79

Calandrinia umbellata

Crockett, James Underwood
Annuals

fl.   P. 100

Calandrinia umbellata var.

Hay, Roy
The Color Dictionary of Flowers &
        Plants

fl.   p. 31   Pl. 246

Calandrinia umbellata var.

Hay, Roy
The Dictionary of House Plants

fl.   Pl. 80

Calanthe brevicorni

Hara, Hiroshi, comp.
Photo-Album of Plants of
   Eastern Himalaya

fl.   Pl. 124

Calanthe cardioglossa

Kamemoto, Haruyuki
Beautiful Thai Orchid Species

fl.   p 124

Calanthe chloroleuca

Hara, Hiroshi, comp.
Photo-Album of Plants of
   Eastern Himalaya

fl.   Pl. 125

Calanthe corymbosa

Williamson, Graham
The Orchids of South Central Africa

fl.   Pl. 111

Calanthe discolor var.

Tsukamoto, Yotaro
Coloured Illustrations of Garden
        Flowers: vol 9

fl.   p. 34

Calanthe masuca

Kupper, Walter
Orchids

fl.   p 47

Calanthe masuca

Walden, Beryl M.
Wild Flowers of Hong Kong

fl.   Pl. 66

Calanthe mexicana

Ospina, Mariano
Orquideas de las americas

fl. hab.   Pl. 41

Calanthe natalensis

Williamson, Graham
The Orchids of South Central Africa

fl.   Pl. 112

Calanthe patsinensis

Walden, Beryl M.
Wild Flowers of Hong Kong

fl.   Pl. 81 (239)

Calanthe plantaginea

Hara, Hiroshi, comp.
Photo-Album of Plants of
   Eastern Himalaya

fl. hab.   Pl. 122, 123

Calanthe reflexa

Curtis's Botanical Magazine
   Vol 164   1943-48

fl.   No. 9648

Calanthe rubens

Kamemoto, Haruyuki
Beautiful Thai Orchid Species

fl.   p 124

Calanthe striata

Walden, Beryl M.
Wild Flowers of Hong Kong

fl. hab.   Pl. 7(25)

Calanthe triplicata

Blombery, Alec M.
What Wildflower is That

fl.   p. 80   Pl. 186

Calanthe triplicata

Cady, Leo
Australian Native Orchids in Colour

fl. hab.   Pl. 84

Calanthe triplicata

Tsukamoto, Yotaro
Coloured Illustrations of
   Garden Flowers vol 9

fl.   opp. p. 34

Calanthe triplicata

Walden, Beryl M.
Wild Flowers of Hong Kong

fl.   Pl. 45(126)

Calanthe veratrifolia

Bruggeman, L.
Tropical Plants

fl.   Pl. 88

Calanthe vestita

Kamemoto, Haruyuki
Beautiful Thai Orchid Species

fl.   p 124

Calanthe vestita var.

Perry, Frances
Flowers of the World

fl.   p. 214

Calanthe volkensii

Stewart, Joyce
Orchids of Tropical Africa

fl.   Pl. 20

Calathea angustifolia

Morley, Brian D.
Wild Flowers of the World

fl.   Pl. 189

Calathea cylindrica

Wit, H. C. D. de
Plants of the World:
   The Higher Plants, Vol II

fl.   Pl. 152

Calathea insignis

Hay, Roy
The Dictionary of House Plants

hab.   Pl. 81

Calathea insignis

Kiaer, Eigil
Indoor Plants in Colour

hab.　　p. 32

Calathea lietzei

Bruggeman, L.
Tropical Plants

fl.　　Pl. 89

Calathea macleanna

Massachusetts Hort. Soc.
Horticulture
　　vol 35, No. 11　1957

hab.　　p. 556

Calathea makoyana

Hay, Roy
The Dictionary of House Plants

hab.　　Pl. 82

Calathea Makoyana

Kromdijk, C.
200 House Plants in Colour

hab.　　Pl. 39

Calathea makoyana

Macoby, Stirling
What Flower is That

hab P 65

Calathea makoyana

Massachusetts Hort. Soc.
Horticulture
　　vol 46, No. 11　1968

hab.　　cover, backcover

Calathea makoyana

Perry, Frances
Flowers of the World

hab p 186

Calathea ornata var.

Massachusetts Hort. Soc.
Horticulture
　　vol 35, No. 11　1957

hab.　　p. 556

Calathea ornata var.

Perry, Frances
Flowers of the World

hab.　　p. 186

Calathea zebrina

Hay, Roy
The Dictionary of House Plants

hab.　　Pl. 83

Calceolaria crenatiflora

Macoby, Stirling
What Flower is That

fl.　　P 66

Calceolaria darwinii

Hay, Roy
The Color Dictionary of Flowers & Plants

fl.　　　p 3, Pl. 22

Calceolaria darwinii

Morley, Brian D.
Wild Flowers of the World

fl.　　　Pl. 5G

Calceolaria darwinii

Royal Hort. Soc.
Journal of the Royal Hort. Soc.
　　vol 75, No. 3　1950

fl.　　p. 110　　Pl. 39

Calceolaria darwinii

Royal Hort. Soc.
Journal of the Royal Hort. Soc.
　　vol 92, No. 6　1967

fl.　　p. 250　　Pl. 128

Calceolaria darwinii

Royal Hort. Soc.
Journal of the Royal Hort. Soc.
　　vol 93, No. 4　1968

fl.　　p. 164　　Pl. 84

Calceolaria herbeo-hybrida

Kiaer, Eigil
Indoor Plants in Colour

fl.　　p. 32

Calceolaria herbeohybrida

Nicolaisen, Age
Pocket Encyclopedia of Indoor Plants

fl.　　　pl. 118

Calceolaria herbeohybrida

Tsukamoto, Yotaro
Coloured Illustrations of Garden
　　Flowers　　Vol. 10

fl.　　p. 6　　Pl. 19

Calceolaria herbeohybrida var.

Hay, Roy
The Dictionary of House Plants

fl.　　　Pl. 84, 85

Calceolaria hybrid

Kromdijk, C.
200 House Plants in Colour

fl.　　　Pl. 40

Calceolaria hybrida

Perry, Frances
Flowers of the World

fl.　　p. 278

Calceolaria integrifolia

Hay, Roy
The Dictionary of House Plants

fl.　　　Pl. 86

Calceolaria integrifolia

Macoby, Stirling
What Flower is That

fl.　　P 66

Calceolaria integrifolia var.

Hay, Roy
The Color Dictionary of Flowers & Plants

fl.　　　p 184, Pl. 1472

Calceolaria mexicana

Chickering, Carol Rogers
Flowers of Guatemala

fl.　　　p 129

Calceolaria x multiflora var.

Hay, Roy
The Color Dictionary of Flowers &
Plants

fl.　　　P. 55　　Pl.438-39

Calceolaria pavonii

Hay, Roy
The Dictionary of House Plants

fl.　　　Pl. 87

Calceolaria purpurea

Morley, Brian D.
Wild Flowers of the World

fl.　　　pl. 184

Calceolaria purpurea

Munoz Pizarro, Carlos
Flores Silvestres de Chile

fl.　　Pl. 20

Calceolaria scabiosaefolia

Huxley, Anthony
Garden Annuals and Bulbs

fl.　　　Pl. 15

Caleana major

Blombery, Alec M.
What Wildflower is That

fl.　　p. 80　　Pl. 187

Caleana major

Cady, Leo
Australian Native Orchids in Colour

fl.   Pl. 37

Caleana major

Cochrane, G. R.
Flowers and Plants of Victoria

fl.   Pl. 68

Caleana minor

Cady, Leo
Australian Native Orchids in Colour

fl.   Pl. 38

Caleana nigrita

Morley, Brian D.
Wild Flowers of the World

fl.   pl. 145

Calectasia cyanea

Blombery, Alec M.
What Wildflower is that
fl.   p. 81   Pl. 188

Calectasia cyanea

Cochrane, G. R.
Flowers and Plants of Victoria

fl.   Pl. 78

Calectasia cyanea

Morley, Brian D.
Wild Flowers of the World

fl.   pl. 141

Calendula arvensis

Feinbrun-Dothan, Naomi
Wild Plants in the Land
    of Israel

fl. hab.   p 98

Calendula arvensis

Kleijn, H.
Beauty of the Wild Plant

fl.   opp. p. 116   pl. 167

Calendula arvensis

Polunin, Oleg
Flowers of Europe

fl.   Pl. 151 #1460

Calendula arvensis

Polunin, Oleg
Flowers of the Mediterranean

fl.   Pl. 204

Calendula hybrid

Everett, T.H., ed.
New Illustrated Encyclopedia of
    Gardening vol. 2

fl.   p. 294

Calendula officinalis

Bianchini, Francesco
Health Plants of the World

fl.   p 159

Calendula officinalis

Color Treasury of Herbs & Other
    Medicinal Plants

fl.   p 64   Pl. 103

Calendula officinalis

Crockett, James Underwood
Annuals

fl.   P 101

Calendula officinalis

Loewenfeld, Claire
The Complete Book of Herbs and Spices

fl. hab.   p 129

Calendula officinalis

Macoby, Stirling
What Flower is That

fl.   P 66

Calendula officinales

Perrot, Emile
Les Plantes Medicinales

fl. fr. hab.   p 224

Calendula officinalis

Tsukamoto, Yotaro
Coloured Illustrations of Garden
    Flowers vol. 10

fl. Ill. 20   opp. p. 7

Calendula officinalis var.

Hay, Roy
The Color Dictionary of Flowers & Plants

fl.   p 31, 32, Pl. 248, 249, 250

Calendula officinalis var.

Huxley, Anthony
Garden Annuals and Bulbs

fl.   Pl. 13

Calico Bush

see

Kalmia latifolia

Calico, Desert

see

Langloisia matthewsii

Calico Flower

see

Aristolochia elegans

Calicotome spinosa

Curtis's Botanical Magazine
    Vol. 180   1974

fl.   Pl. 683

Calicotome spinosa

Polunin, Oleg
Trees and Bushes of Europe

fl.   p 107

Calicotome villosa

Polunin, Oleg
Flowers of Europe

fl. Pl. 51   #498

Calicotome villosa

Polunin, Oleg
Trees and Bushes of Europe

fl. hab.   p 107

Calla

see

Zantedeschia

Calla, black

see

Arum palaestinum

Calla lily, golden

see

Zantedeschia macreparpa

Calla palustris

Alaska-Yukon Wild Flowers Guide

fl. hab.   p 3

Calla palustris

Courtenay, Booth
Wildflowers and weeds

fl. p. 3

Calla palustris

Felsko, Elsa
A Book of Wildflowers

fl.   P 143

Calla palustris

Ferguson, Mary
Wildflowers

fl. hab.   p 34

Calla palustris

Heller, Christine
Wild Flowers of Alaska

fl.   Pl. 155

Calla palustris

Huxley, Anthony
Garden Perennials and Water Plants

fl. fr.   Pl. 285

Calla palustris

Jennings, O. E.
Wild Flowers of Western Pennsylvania
                                        vol II
fl. hab.   Pl. 145

Calla palustris

Kleijn, H.
Beauty of the Wild Plant

fl.   opp. p. 26   pl. 26

Calla palustris

Klimas, John E.
Wildflowers of Eastern America

fl.   Pl. 28

Calla palustris

Klute, Jeannette
Woodland Portraits

fl.   Pl. 30

Calla palustris

Lindman, C. A. M.
Nordens Flora, Vol 1

fl. fr. hab.   Pl. 32

Calla palustris

Milne, Lorus
Living Plants of the World

fl. hab.   p 291

Calla palustris

Moyle, John B.
Northland Wild Flowers

fl. hab.   p 202, Pl. 263

Calla palustris

Perry, Frances
Flowers of the World

fl., fr.   p. 33

Calla palustris

Polunin, Oleg
Flowers of Europe

fl.   Pl. 183 #1817

Calla palustris

Stodola, Jiri
Encyclopedia of Water Plants

fl.   P. 342

Calla palustris

Walcott, Mary Vaux
North American Wild Flowers

fl. hab.   Pl. 129

Calla palustris

Wit, H. C. D. de
Plants of the World;
    The Higher Plants, Vol II

fl.   Pl. 169

Calla, Wild

    see

Calla palustris

Calla, yellow

    see

Zantedeschia elliettiana

Calliandra anomala

O'Gorman, Helen
Mexican Flowering Trees and Plants

fl. hab.   p 49

Calliandra californica

Coyle, Jeanette
A Field Guide to the Common and
    Interesting Plants of Baja California

fl. fr.   p 89

Calliandra eriophylla

Lemmon, Robert S.
Wildflowers of North America in
    Full Color

Calliandra eriophylla

Munz, Philip A.
California Desert Wildflowers

fl.   p 35, Pl.26

Calliandra eriophylla

Orr, Robert T.
Wildflowers of Western America

fl.   Pl. 181

Calliandra guildingi

Massachusetts Hort. Soc.
Horticulture
    vol 48, No. 12   1970

fl.   cover, backcover

Calliandra haematocephala

Mathias, Mildred E.
Color for the Landscape

fl. hab.   p 62

Calliandra haematocephala

Morley, Brian D.
Wild Flowers of the World

fl. fr.   Pl. 170

Calliandra haematocephala

Perry, Frances
Flowers of the World

fl.   p 159

Calliandra haematona

Bruggeman, L.
Tropical Plants

fl.   Pl. 228-29

Calliandra houstoniana

Chickering, Carol Rogers
Flowers of Guatemala

fl. fr.   p 41

Calliandra inaequilatera

Crockett, James Underwood
Evergreens

fl.   Pl. 115

Calliandra inaequilatera

Hargreaves, Dorothy
Tropical Blossoms of the Caribbean

fl.   P. 57

Calliandra inaequilatera

Macoby, Stirling
What Flower is That

fl.   P 67

Calliandra portoricensis

Bruggeman, L.
Tropical Plants

fl.   Pl. 230

Calliandra surinamensis

Macoby, Stirling
What Flower is That

fl.   P 67

Calliandra surinamensis

Milne, Lorus
Living Plants of the World

fl.   p 107

Calliandra Tweedii

Harrison, R. E.
Trees and Shrubs

fl.   P. 38

Calliandra tweedii

Mathias, Mildred E.
Color for the Landscape

—   — 63

Calliandra tweedii

Royal Hort. Soc.
Journal of the Royal Hort. Soc.
     Vol. 91, No. 10   1966

fl.   p. 426   Pl. 225

Callianthemum coriandrifolium

Barneby, T. P.
European Alpine Flowers in Colour

fl.   Pl. 26, 3

Callianthemum coriandrifolium

Felsko, Elsa
A Book of Wild Flowers
                2nd Ser.
fl. hab.   p 131

Callianthemum kerneranum

Barneby, T. P.
European Alpine Flowers in Colour

fl. hab.   Pl. 26, 4

Callianthemum pimpinelloides

Hara, Hiroshi, comp.
Photo-Album of Plants of
     Eastern Himalaya

fl. hab.   Pl. 220

Callicarpa americana

Batson, Wade T.
Wild Flowers in South Carolina

fr.   p 100

Callicarpa americana

Brown, Clair A.
Wildflowers of Louisiana and
     Adjoining States

fr. hab.   p 155

Callicarpa americana

Walcott, Mary Vaux
North American Wild Flowers

fr.   Pl. 210

Callicarpa bodinieri var.

Bartels, Andreas
Das Grosse Buch der Gartengehölze

fr.   p 120

Callicarpa bodinieri var.

Gault, S. Millar
The Color Dictionary of Shrubs

fr.   Pl. 29

Callicarpa bodinieri var.

Hellyer, A.G.L.
Shrubs in Colour

fl., fr.   p. 22

Callicarpa bodinieri var.

Perry, Frances
Flowers of the World

fr.   p. 304

Callicarpa dichotoma

Kitamura, Siro
Coloured Illustrations of Trees &
     Shrubs of Japan

fr.   426

Callicarpa dichotoma

Pennsylvania Hort. Soc.
The Green Scene
     vol 6, No. 3   1978

fr.   p. 22

Callicarpa japonica

Huxley, Anthony
Deciduous Garden Trees and Shrubs

fr.   Pl. 26

Callicarpa japonica

Kitamura, Siro
Coloured Illustrations of Trees &
     Shrubs of Japan

fl.   424

Callicarpa loureiroi

Walden, Beryl M.
Wild Flowers of Hong Kong

fl.   Pl. 50

Callicarpa rubella

Harrison, R.E.
Trees and Shrubs

fr.   P 39

Callicarpa rubella

Hay, Roy
The Dictionary of House Plants

fl.   Pl. 88

Callicarpa rubella

Morley, Brian D.
Wild Flowers of the World

fl.fr.   Plate 104

Callichlamys latifolia

Herklots, Geoffrey
Flowering Tropical Climbers

fl.   p. 66   Pl. 5

Callicoma serratifolia

Blombery, Alec M.
What Wildflower is That

fl.   p 81, Pl. 189

Callilepis leptophylla

Flowering Plants of Africa
     Vol. 39   1968-69

fl., hab.   Pl. 1554

Callirhoe involucrata

Lemmon, Robert S.
Wildflowers of North America in
     Full Color

fl.   pg. 189   pl. 297

Callirhoe involucrata

Meikle, R. D.
Garden Flowers

fl.   opp. p. 128   Pl. 2

Callirhoe involucrata

Orr, Robert T.
Wildflowers of Western America

fl.   Pl. 207

Callirhoe papaver

Brown, Clair A.
Wildflowers of Louisiana and
     Adjoining States

fl.   p 108

Callirhoe triangulata

Courtenay, Booth
Wildflowers & Weeds

fl.   p 44

Callirhoe triangulata

Dean, Blanche E.
Wildflowers of Alabama and
     Adjoining States

fl.   p 103

Callisia elegans

Kromdijk, C.
200 House Plants in Colour

hab.   Pl. 41

Callistemon brachyandrus

Curtis's Botanical Magazine
       vol 172   1958-59

fl., fr.   Pl. 316

Callistemon brachyandrus

Morley, Brian D.
Wild Flowers of the World

fl. fr.   Pl. 137

Callistemon citrinus

Blombery, Alec M.
What Wildflower is That

fl.   p. 82   Pl. 190

Callistemon citrinus

Crockett, James Underwood
Evergreens

hab.,fl.   p. 115

Callistemon citrinus

Encke, Fritz
Zimmerpflanzen

fl.   p. 61

Callistemon citrinus

Hay, Roy
The Dictionary of House Plants

fl.   Pl. 89

Callistemon citrinus

Macoby, Stirling
What Flower is That

fl. P 67

Callistemon citrinus

Mathias, Mildred E.
Color for the Landscape

fl. hab.   p 13

Callistemon citrinus

Mullins, Barbara
Australian Wildflowers in Colour

fl.   P. 63   Pl. 57

Callistemon citrinus

Perry, Frances
Flowers of the World

fl.   p 192

Callistemon citrinus

Wit, H. C. D. de
Plants of the World;
The Higher Plants, Vol. II

fl.   p. 50   Pl. 3

Callistemon citrinus var.

Harrison, R.E.
Trees and Shrubs

fl. P 39

Callistemon comboynensis

Curtis's Botanical Magazine
Vol 178   1970-72

fl.   602

Callistemon hybrid

Royal Hort. Soc.
Journal of the Royal Hort. Soc.
     vol 98, No. 10   1973

fl.   p. 442   Pl. 228

Callistemon lanceolatus

Hargreaves, Dorothy
Tropical Blossoms of the Caribbean

fl.,hab.   p. 42

Callistemon lilacinus

Blombery, Alec M.
What Wildflower is That

fl.   p 83, Pl. 191

Callistemon macropunctatus

Cochrane, G. R.
Flowers and Plants of Victoria

fl.   Pl. 114

Callistemon macropunctatus

Royal Hort. Soc.
Journal of the Royal Hort. Soc.
     vol 98, No. 10   1973

fl.   p. 442   Pl. 226

Callistemon pallidus

Cochrane, G. R.
Flowers and Plants of Victoria

fl. fr.   Pl. 205

Callistemon paludosus

Cochrane, G. R.
Flowers and Plants of Victoria

fl.   Pl. 273

Callistemon pinifolius

Macoby, Stirling
What Flower is That

fl. P 67

Callistemon rigidus

Blombery, Alec A.
What Wildflower is That

fl.   p 83, Pl. 192

Callistemon rigidus

Curtis's Botanical Magazine
Vol 179   1972

fl. fr.   619

Callistemon rigidus

Tsukamoto, Yotaro
Coloured Illustrations of Garden
     Flowers vol. 10

fl. Illus. 135   opp. p. 42

Callistemon rigidus

Walden, Beryl M.
Wild Flowers of Hong Kong

fl.   Pl. 31(83)

Callistemon rigidus var.

Harrison, R. E.
Trees and Shrubs

fl.   P. 39

Callistemon salignus

Blombery, Alec M.
What Wildflower is That

fl.   p. 83   Pl. 193

Callistemon salignus

Harrison, R. E.
Trees and Shrubs

fl.   P. 39

Callistemon salignus

Hay, Roy
The Color Dictionary of Flowers & Plants

fl.   p 185, Pl. 1473

Callistemon salignus

Macoby, Stirling
What Flower is That

fl. P 67

Callistemon salignus

Royal Hort. Soc.
Journal of the Royal Hort. Soc.
     vol 92, No. 11   1967

fl.   p. 478   Pl. 267

Callistemon salignus

Royal Hort. Soc.
Journal of the Royal Hort. Soc.
     vol 98, No. 10   1973

fl.   p. 442   Pl. 227

Callistemon sieberi

Blombery, Alec M.
What Wildflower is That

fl.   p. 83   Pl. 194

Callistemon speciosus

Addisonia
        vol 19   1935-36

fl.   p. 25   Pl. 621

Callistemon speciosus

Blombery, Alec M.
What Wildflower is That

fl.   p. 84   Pl. 195

Callistemon speciosus

Hay, Roy
The Dictionary of House Plants

fl.   Pl. 90

Callistemon viminalis

Blombery, Alec M.
What Wildflower is That

fl.   p 85, Pl. 196

Callistemon viminalis

Macoby, Stirling
What Flower is That

fl. P 67

Callistemon viminalis

Mathias, Mildred E.
Color for the Landscape

fl. hab.   p 13

Callistemon viminalis

Oakman, Harry
Colorful Trees

Fl. Hab.   P 40

Callistemon viridiflorus

Curtis, Winifred
The Endemic Flora of Tasmania
Vol VI

fl. fr.   Pl. 213

Callistephus chinensis

Bruggeman, L.
Tropical Plants

fl.   Pl. 32 & 33

Callistephus chinensis

Crockett, James Underwood
Annuals

fl. P 101

Callistephus chinensis

Everett, T.H., ed.
New Illustrated Encyclopedia of
        Gardening vol. 2

fl.   p. 247

Callistephus chinensis

Macoby, Stirling
What Flower is That

fl. P 68

Callistephus chinensis

Morley, Brian D.
Wild Flowers of the World

fl. Plate 106

Callistephus chinensis

Tsukamoto, Yotaro
Coloured Illustrations of Garden
        Flowers   vol 10

fl.   p. 7   Pl. 21,22.

Callistephus chinensis var.

Hay, Roy
The Color Dictionary of Flowers & Plants

fl.   p 32, Pl. 251, 252, 253
                                      254

Callistephus chinensis var.

Huxley, Anthony
Garden Annuals and Bulbs

fl.   pl. 16

Callitriche autumnalis

Heller, Christine
Wild Flowers of Alaska

hab.   Pl. 301

Callitriche autumnalis

Lindman, C. A. M.
Nordens Flora, Vol 6

fl. hab.   Pl. 377B

Callitriche autumnalis

Stodola, Jiri
Encyclopedia of Water Plants

hab.   P. 246

Callitriche hermaphroditica

Martin, W. Keble
The Concise British Flora in Colour

fr. hab.   Pl. 34

Callitriche intermedia var.

Martin, W. Keble
The Concise British Flora in Colour

fr. hab.   Pl. 34

Callitriche autumnalis

Martin, W. Keble
The Concise British Flora in Colour

fl. fr. hab.   Pl. 34

Callitriche palustris

Martin, W. Keble
The Concise British Flora in Colour

fr. hab.   Pl. 34

Callitriche palustris

Stodola, Jiri
Encyclopedia of Water Plants

hab.   P. 243

Callitriche platycarpa

Martin, W. Keble
The Concise British Flora in Colour

fr. hab.   Pl. 34

Callitriche stagnalis

Ary, S.
The Oxford Book of Wildflowers

fl.   p 52, Pl. 5

Callitriche stagnalis

Lindman, C. A. M.
Nordens Flora, Vol 6

fl. hab.   Pl. 377A

Callitriche stagnalis

Martin, W. Keble
The Concise British Flora in Colour

fr. hab.   Pl. 34

Callitriche stagnalis

Polunin, Oleg
Flowers of Europe

hab.   Pl. 105 #1088

Callitriche truncata

Martin, W. Keble
The Concise British Flora in Colour

fr. hab.   Pl. 34

Callitris oblonga

Curtis's Botanical Magazine
Vol 161   1938

fr.   No. 9550

Callitris rhomboidea

Blombery, Alec M.
What Wildflower is That

hab.   p 85, Pl. 197

Callitris rhomboidea

Cochrane, G. R.
Flowers and Plants of Victoria

fr.　　　　　Pl. 97

Callitris verrucosa

Cochrane, G. R.
Flowers and Plants of Victoria

fr.　　　　　Pl. 133

Calluna vulgaris

Ary, S.
The Oxford Book of Wildflowers

fl.　　　　p 118, Pl. 5

Calluna vulgaris

Barneby, T. P.
European Alpine Flowers in Colour

fl.　　　　Pl. 56, 6

Calluna vulgaris

Crockett, James Underwood
Lawns & Ground Covers

fl.　p. 125

Calluna vulgaris

Felsko, Elsa
A Book of Wildflowers

fl.　P 119

Calluna vulgaris

Hay, Roy
The Color Dictionary of Flowers & Plants

fl.　　　　p 185, Pl. 1474

Calluna vulgaris

Hellyer, A.G.L.
Shrubs in Colour

fl.　　　p. 22

Calluna vulgaris

Huxley, Anthony
Evergreen Garden Trees and Shrubs

fl. hab.　　　　Pl. 103

Calluna vulgaris

Lindman, C. A. M.
Nordens Flora, Vol 7

fl. hab.　　　Pl. 454

Calluna vulgaris

Martin, W. Keble
The Concise British Flora in Colour

fl. hab.　　　Pl. 55

Calluna vulgaris

Meikle, R. D.
British Trees and Shrubs

fl.　　　　Pl. 10

Calluna vulgaris

Perry, Frances
Flowers of the World

fl.　p 106

Calluna vulgaris

Tosco, Uberto
The World of Mountain Flowers

fl.　　　　p 28

Calluna vulgaris var.

Bartels, Andreas
Das Grosse Buch der Gartengeholze

fl. hab.　　　　p 121

Calluna vulgaris var.

Crockett, James Underwood
Evergreens

hab. fl.　　　p. 116

Calluna vulgaris var.

Gault, S. Millar
The Color Dictionary of Shrubs

fl. hab.　　　Pl. 31, 32, 33, 34, 35,
　　　　　　　36, 37, 38

Calluna vulgaris var.

Harrison, R.E.
Trees and Shrubs

fl.　P 38

Calluna vulgaris var.

Hay, Roy
The Color Dictionary of Flowers & Plants

fl.　　　p 185, Pl. 1475 thru 1478

Calluna vulgaris var.

Hellyer, A.G.L.
Shrubs in Colour

fl.　p. 22

Calluna vulgaris var.

Huxley, Anthony
Evergreen Garden Trees and Shrubs

fl. hab.　　　Pl. 104 thru 107

Calluna vulgaris var.

Perry, Frances
Flowers of the World

fl.　p 107

Calocephalus brownii

Blombery, Alec M.
What Wildflower is That

hab.　p. 85　　　Pl. 198

Calocephalus brownii

Cochrane, G. R.
Flowers and Plants of Victoria

fl. hab.　　Pl. 286

Calocephalus multiflorus

Blombery, Alec M.
What Wildflower is That

fl.　　　p. 85　　　Pl. 199

Calochilus campestris

Blombery, Alec M.
What Wildflower is That

fl.　　　p 86, Pl. 200

Calochilus campestris

Cady, Leo
Australian Native Orchids in Colour

fl.　　　　Pl. 60

Calochilus richae

Cady, Leo
Australian Native Orchids in Colour

fl.　　　front

Calochilus robertsonii

Cady, Leo
Australian Native Orchids in Colour

fl.　　　　Pl. 59

Calochilus robertsonii

Cochrane, G. R.
Flowers and Plants of Victoria

fl.　　　　Pl. 69

Calochilus robertsonii

Morley, Brian D.
Wild Flowers of the World

fl.　　　　Pl. 145

Calochortus albus

Addisonia
　vol 19　1935-36

fl.　　p. 27　　Pl. 622

Calochortus albus

Orr, Robert T.
Wildflowers of Western America

fl.　Pl. 32

Calochortus albus var.

Massachusetts Hort. Soc.
Horticulture
vol 48, No. 11   1970

fl.   p. 14

Calochortus amabilis

Orr, Robert T.
Wildflowers of Western America

fl.   Pl. 101

Calochortus amoenus

Addisonia
     vol 23   1954-59

fl.   p. 13   pl. 743

Calochortus amoenus

Lemmon, Robert S.
Wildflowers of North America in
     Full Color

fl.   p. 232   pl. 367

Calochortus apiculatus

Clark, Lewis J.
Wild Flowers of British Columbia

fl.   p 10

Calochortus barbatus

Mathew, Brian
Dwarf Bulbs

fl.   p. 56   Pl. 16

Calochortus caeruleus var.

Lemmon, Robert S.
Wildflowers of North America in
     Full Color

fl.   pg. 231   pl. 366

Calochortus catalinae

Walcott, Mary Vaux
North American Wild Flowers

fl. fr.   Pl. 205

Calochortus clavatus

Walcott, Mary Vaux
North American Wild Flowers
     vol. 4

fl.   Pl. 314

Calochortus elegans

Orr, Robert T.
Wildflowers of Western America

fl.   Pl. 16

Calochortus elegans

Walcott, Mary Vaux
North American Wild Flowers

fl. fr.   Pl. 2

Calochortus flavus

O'Gorman, Helen
Mexican Flowering Trees and Plants

fl. fr. hab.   p 141

Calochortus flexuosus

Welsh, Stanley L.
Flowers of the Canyon Country

fl.   p 14

Calochortus gunnisonii

Orr, Robert T.
Wildflowers of Western America

fl.   Pl. 141

Calochortus gunnisonii

Welsh, Stanley L.
Flowers of the Mountain Country

fl.   p. 7

Calochortus kennedyi

Alpine Garden Society
Bulletin
     vol. 8, No. 3   1940

fl.   p. 194

Calochortus Kennedyi

Lemmon, Robert S.
Wildflowers of North America in
     Full Color

fl.   pg. 63   pl. 102

Calochortus kennedyi

Mathias, Mildred E.
Color for the Landscape

fl. hab.   194

Calochortus kennedyi

Miles, Bebe
Bulbs for the Home Gardener

fl.   p 57

Calochortus kennedyi

Munz, Philip A.
California Desert Wildflowers

fl.   p 30, Pl. 10

Calochortus kennedyi

Orr, Robert T.
Wildflowers of Western America

fl.   Pl. 111

Calochortus kennedyi

Royal Hort. Soc.
Journal of the Royal Hort. Soc.
     vol 87, No. 4   1962

fl.   p. 174   Pl. 48

Calochortus kennedyi

Walcott, Mary Vaux
North American Wild Flowers
     vol. 5

fl.   Pl. 391

Calochortus leichtlinii

Orr, Robert T.
Wildflowers of Western America

fl.   Pl. 14

Calochortus luteus

Mathew, Brian
Dwarf Bulbs

fl.   p. 56   Pl. 17

Calochortus luteus

Orr, Robert T.
Wildflowers of Western America

fl.   Pl. 92

Calochortus macrocarpus

Clark, Lewis J.
Wild Flowers of British Columbia

fl.   p 11

Calochortus macrocarpus

Pacific Hort. Foundation
Pacific Horticulture
     vol 37, No. 4   1976

fl.   p. 19

Calochortus macrocarpus

Walcott, Mary Vaux
North American Wild Flowers

fl.   Pl. 175

Calochortus nuttallii

Orr, Robert T.
Wildflowers of Western America

fl.   Pl. 40

Calochortus nuttallii

Welsh, Stanley L.
Flowers of the Canyon Country

fl.   p 8

Calochortus nuttallii var.

Lemmon, Robert S.
Wildflowers of North America in
     Full Color

fl.   pg. 64   pl. 103

Calochortus pulchellus

Lemmon, Robert S.
Wildflowers of North America in
     Full Color

fl.   pg. 232   pl. 368

Calochortus pulchellus

Morley, Brian D.
Wild Flowers of the World

fl.   pl. 167

Calochortus, Showy

        see

Calochortus venustus

Calochortus sp.

Massachusetts Hort. Soc.
Horticulture
    vol 47, No.5   1969

fl.   cover, backcover

Calochortus sp.

Royal Hort. Soc.
Journal of the Royal Hort. Soc.
    vol 86, No. 8   1961

fl.   p. 354   Pl. 102

Calochortus sp.

Royal Hort. Soc.
Journal of the Royal Hort. Soc.
    vol 89, No. 4   1964

fl.   p. 158   Pl. 67

Calochortus splendens

Walcott, Mary Vaux
North American Wild Flowers
    vol. 4

fl.   Pl. 310

Calochortus tolmiei

Orr, Robert T.
Wildflowers of Western America

fl.   Pl. 30

Calochortus uniflorus

Hay, Roy
The Color Dictionary of Flowers & Plants

fl.   p 85, Pl. 679

Calochortus uniflorus

Mathew, Brian
Dwarf Bulbs

fl.   p. 56   Pl. 15

Calochortus venustus

Lemmon, Robert S.
Wildflowers of North America in
    Full Color

fl.   pg. 233   pl. 369

Calochortus venustus

Massachusetts Hort. Soc.
Horticulture
    vol 48, No. 11   1970

fl.   p. 14

Calochortus venustus

Mathew, Brian
Dwarf Bulbs

fl.   p. 56   Pl. 18

Calochortus venustus

Munz, Philip A.
California Spring Wildflowers

fl.   P. 81   Pl. 68

Calochortus venustus var.

Royal Hort. Soc.
Journal of the Royal Hort. Soc.
    vol 87, No. 4   1962

fl.   p. 174   Pl. 49

Calochortus weedii

Pacific Hort. Foundation
Pacific Horticulture
    vol 37, No. 4   1976

fl.   p. 18

Calochortus weedii

Walcott, Mary Vaux
North American Wild Flowers

fl.   Pl. 199

Calodendrum capense

Curtis's Botanical Magazine
Vol 165   1948

fl.   34

Calodendrum capense

Flowering Plants of Africa
    vol 27   1948-49

fl., fr.   Pl. 1041

Calodendron capense

Macoby, Stirling
What Flower is That

fl.   P 68

Calodendrum capense

Mathias, Mildred E.
Color for the Landscape

fl. hab.   p 14

Calodendrum capense

Morley, Brian D.
Wild Flowers of the World

fl.   Pl. 77

Calodendrum capense

Oakman, Harry
Colorful Trees

Hab. Fl.   P 53

Calodendrum capense

Pacific Hort. Foundation
California Hort. Journal
    vol 33, No. 4   1972

hab., fl.   p. 147

Calodendrum capense

Palgrave, K. C.
Trees of Central Africa

fl., fr.   p. 397

Calodendrum capense

Palmer, Eve
Trees of South Africa

fl., hab.   p. 304   Pl. XXVIII

Calomeria amaranthoides

Cochrane, G. R.
Flowers and Plants of Victoria

fl.   Pl. 480

Caloncoba gilgiana

Wit, H. C. D. de
Plants of the World;
    The Higher Plants, Vol I

fl.   p 165, Pl. 82

Calonyction aculeatum

Crockett, James Underwood
Annuals

fl.   P 102

Calonyction aculeatum

Royal Hort. Soc.
Journal of the Royal Hort. Soc.
    vol 90, No. 6   1965

fl.   p. 250   Pl. 117

Calonyction aculeatum

Tsukamoto, Yotaro
Coloured Illustrations of Garden
    Flowers vol. 10

fl. 111.42 opp. p. 14

Calonyction album

Bruggeman, L.
Tropical Plants

fl.   Pl. 6

Calonyction album

Blunt, Wilfrid
Flora Superba

fl.,hab.   Pl. VII

Calophyllum membranaceum

Walden, Beryl M.
Wild Flowers of Hong Kong

fl.   Pl. 33(80)

Calopogon barbatus

Duncan, Wilbur H.
Wildflowers of the Southeastern
    United States

fl.   p 277

Calopogon barbatus

Fleming, Glenn
Wild Flowers of Florida

fl.,hab.   p. 68

Calopogon barbatus

Luer, Carlyle A.
The Native Orchids of Florida

fl.   Pl. 16; 7-9

Calopogon multiflorus

Luer, Carlyle A.
The Native Orchids of Florida

fl. hab.   Pl. 15; 1-6

Calopogon pallidus

Lemmon, Robert S.
Wildflowers of North America in
    Full Color

fl.   pg. 173   pl. 171

Calopogon pallidus

Luer, Carlyle A.
The Native Orchids of Florida

fl. fr. hab.   Pl. 16; 1-6

Calopogon pulchellus

Brown, Clair A.
Wildflowers of Louisiana and
    Adjoining States

fl.   p 35

Calopogon pulchellus

Campbell, Carlos C.
Great Smoky Mountain Wildflowers

fl.   p 87

Calopogon pulchellus

Courtenay, Booth
Wildflowers & Weeds

fl.   p 14

Calopogon pulchellus

Dean, Blanche E.
Wildflowers of Alabama and
    Adjoining States

fl.   p 35

Calopogon pulchellus

Ferguson, Mary
Wildflowers

fl.   p 26

Calopogon pulchellus

Jennings, O. E.
Wild Flowers of Western Pennsylvania
                          vol II.
fl.   Pl. 44

Calopogon pulchellus

Klute, Jeannette
Woodland Portraits

fl.   Pl. 33

Calopogon pulchellus

Kramer, Jack
Orchids; Flowers of Romance
    and Mystery

fl.   p 50

Calopogon pulchellus

Lemmon, Robert S.
Wildflowers of North America in
    Full Color

fl.   pg. 173   pl. 172

Calopogon pulchellus

Wharton, Mary E.
A Guide to the Wildflowers & Ferns
    of Kentucky

fl.   p 70   Pl. 2.11

Calopogon tuberosus

American Hort. Soc.
American Horticulturist
    vol 57, No. 4   1978

fl.   p. 2

Calopogon tuberosus

Luer, Carlyle A.
The Native Orchids of Florida

fl.   Pl. 14; 1-8

Calopogon tuberosus

Luer, Carlyle A.
The Native Orchids of the
    United States and Canada

fl. hab.   Pl. 70
                  Pl. 71

Calopogon tuberosus

Ospina, Mariano
Orquideas de las americas

fl.   Pl. 38

Calopogon tuberosus var.

Luer, Carlyle A.
The Native Orchids of Florida

fl. hab.   Pl. 5; 3, 4
                  Pl. 14; 2

Calorhabdos brunoniana

Curtis's Botanical Magazine
    Vol 162   1939

fl.   No. 9580

Calorophus elongatus

Curtis, Winifred
The Endemic Flora of Tasmania, Vol. 5

fl. fr. hab.   Pl. 180

Calostemma purpureum

Blombery, Alec M.
What Wildflower is That

fl.   p. 86   Pl. 201

Calostemma purpureum

Macoby, Stirling
What Flower is That

fl.   P. 69

Calothamnus asper

Blombery, Alec M.
What Wildflower is That

fl.   p 86, Pl. 202

Calothamnus quadrifidus

Macoby, Stirling
What Flower is That

fl.   P 69

Calothamnus quadrifidus

Morcombe, M. K.
Australia's Western Wildflowers

fl.   p 44,

Calothamnus validus

Curtis's Botanical Magazine
Vol 178   1970-72

fl. fr.   614

Calothamnus villosus

Blombery, Alec M.
What Wildflower is That

fl.   p. 86   Pl. 203

Calotis erinacea

Cochrane, G. R.
Flowers and Plants of Victoria

fl.   Pl. 143

Calotis latiuscula

Blombery, Alec M.
What Wildflower is That

fl.   p 86, Pl. 204

Calotropis gigantea

Flowering Plants of Africa
    vol 42   1972-73

fl.   Pl. 1647

Calotropis procera

Morley, Brian D.
Wild Flowers of the World

fl.   Pl. 59

Calpurnia aurea

Morley, Brian D.
Wild Flowers of the World

fl.,fr.   Pl. 54

Calpurnia aurea var.

Flowering Plants of Africa
     vol 44   1977

fl.   Pl. 1759

Caltha asarifolia

Clark, Lewis J.
Wild Flowers of British Columbia

fl.   p 159

Caltha biflora

Clark, Lewis J.
Wild Flowers of British Columbia

fl.   p 159

Caltha biflora

Fries, Mary A.
Wildflowers of Mount Ranier and
the Cascades

fl.   P 85

Caltha biflora

Heller, Christine
Wild Flowers of Alaska

fl.   Pl. 139

Caltha biflora

Taylor, Ronald J.
Mountain Wild Flowers

fl. hab.   p 118

Caltha dionaeifolia

Morley, Brian D.
Wild Flowers of the World

fl.   Pl. 1E

Caltha leptosepala

Alaska-Yukon Wild Flowers Guide

fl. hab.   p 36

Caltha leptosepala

Heller, Christine
Wild Flowers of Alaska

fl.   Pl. 140

Caltha leptosepala

Lemmon, Robert S.
Wildflowers of North America in
   Full Color

fl.   pg. 125   pl. 199

Caltha leptosepala

Orr, Robert T.
Wildflowers of Western America

fl.   Pl. 3

Caltha leptosepala

Porsild, A. E.
Rocky Mountain Wild Flowers

fl. fr. hab.   p 175

Caltha leptosepala

Shaw, Richard J
Field Guide to the Vascular Plants of
   Grand Teton National Park

fl.   Pl. 9

Caltha leptosepala

Walcott, Mary Vaux
North American Wild Flowers
   vol. 4

fl.   Pl. 287

Caltha leptosepala

Welsh, Stanley L.
Flowers of the Mountain Country

fl.   p. 7

Caltha palustris

Alaska-Yukon Wild Flowers Guide

fl.   p 37

Caltha palustris

American Hort. Soc.
American Horticulturist
   vol 53, No. 5 1974

fl.   P. 31

Caltha palustris

Ary, S.
The Oxford Book of Wildflowers

fl.   p 4, Pl. 5

Caltha palustris

Barneby, T. P.
European Alpine Flowers in Colour

fl.   Pl. 21, 2

Caltha palustris

Courtenay, Booth
Wildflowers & Weeds

fl.   p 32

Caltha palustris

Everett, T.H., ed.
New Illustrated Encyclopedia of
   Gardening vol.

fl.   p. 279

Caltha palustris

Felsko, Elsa
A Book of Wildflowers

fl.   P 6

Caltha palustris

Ferguson, Mary
Wildflowers

fl. hab.   p 42

Caltha palustris

Heller, Christine
Wild Flowers of Alaska

fl., hab.   Pl. 18-19

Caltha palustris

Huxley, Anthony
Garden Perennials and Water Plants

fl.   Pl. 287

Caltha palustris

Jennings, O. E.
Wild Flowers of Western Pennsylvania
                         vol II
fl.   Pl. 57

Caltha palustris

Klimas, John E.
Wildflowers of Eastern America

fl.   Pl. 120

Caltha palustris

Lemmon, Robert S.
Wildflowers of North America in
   Full Color

fl.   pg. 176   pl. 276

Caltha palustris

Lindman, C. A. M.
Nordens Flora, Vol 4

fl. fr.   Pl. 221

Caltha palustris

Massachusetts Hort. Soc.,
Horticulture
   vol 38, No. 9 1960

fl.   p. 473

Caltha palustris

Massachusetts Hort. Soc.
Horticulture
   Vol. 55, No. 12 1977

fl., hab.   p. 34

Caltha palustris

Milne, Lorus
Living Plants of the World

fl.   p 53

Caltha palustris

Moyle, John B.
Northland Wild Flowers

fl. hab.   p 50, Pl. 8

Caltha palustris

Neufeld, J.B.
Wild Flowers of the Prairies

fl.   p. 33

Caltha palustris

Royal Hort. Soc.
The Garden
    vol 101, No. 4   1976

fl.   p. 204

Caltha palustris

Tosco, Uberto
The World of Mountain Flowers

fl.   p 120

Caltha palustris

Walcott, Mary Vaux
North American Wild Flowers

fl. hab.   Pl. 208

Caltha palustris var.

Hay, Roy
The Color Dictionary of Flowers & Plants

fl.   p 127, Pl. 1009

Caltha palustris var.

Huxley, Anthony
Garden Perennials and Water Plants

fl.   Pl. 286

Caltha palustris var.

Martin, W. Keble
The Concise British Flora in Colour

fl. hab.   Pl. 4

Caltha palustris var.

Royal Hort. Soc.
The Garden
    vol 101, No. 4   1976

fl., hab.   p. 204

Caltha phylloptera

Curtis, Winifred
The Endemic Flora of Tasmania, Part IV

fl. fr. hab.   Pl. 134

Caltrops

see

Trapa natans

Calycanthus fertilis

Everett, T.H., ed.
New Illustrated Encyclopedia of
    Gardening vol. 2

fl.   p. 295

Calycanthus fertilis

Wharton, Mary E.
Trees & Shrubs of Kentucky

fl.   p 43, Pl. 1.3

Calycanthus floridus

Batson, Wade T.
Wild Flowers in South Carolina

fl.   p 48

Calycanthus floridus

Campbell, Carlos C.
Great Smoky Mountain Wildflowers

fl.   p 71

Calycanthus floridus

Dean, Blanche E.
Wildflowers of Alabama and
    Adjoining States

fl.   p 69

Calycanthus floridus

Hellyer, A.G.L.
Shrubs in Colour

fl.   p. 22

Calycanthus floridus

Huxley, Anthony
Deciduous Garden Trees and Shrubs

fl.   Pl. 27

Calycanthus floridus

Wit, H. C. D. de
Plants of the World;
    The Higher Plants, Vol I

fl.   p 57, Pl. 21

Calycanthus occidentalis

Mathias, Mildred E.
Color for the Landscape

fl. hab.   p 175

Calycanthus occidentalis

Morley, Brian D.
Wild Flowers of the World

fl.   pl. 150

Calycanthus occidentalis

Munz, Philip A.
California Spring Wildflowers

fl.   p 86, Pl. 83

Calycanthus occidentalis

Perry, Frances
Flowers of the World

fl.   p 60

Calycoseris wrightii

Munz, Philip A.
California Desert Wildflowers

fl.   p 58, Pl. 96

Calycotome infesta

Polunin, Oleg
Flowers of the Mediterranean

fl.   Pl. 55

Calycotome villosa

Feinbrun-Dothan, Naomi
Wild Plants in the Land
    of Israel

fl. fr.   p 48

Calypso

see

Cytherea bulbosa

Calypso bulbosa

Clark, Lewis J.
Wild Flowers of British Columbia

fl.   p 79

Calypso bulbosa

Courtenay, Booth
Wildflowers & Weeds

fl.   p 14

Calypso bulbosa

Ferguson, Mary
Wildflowers

fl. hab.   p 150

Calypso bulbosa

Fries, Mary A.
Wildflowers of Mount Ranier and
    the Cascades

fl.   P 32

Calypso bulbosa

Heller, Christine
Wild Flowers of Alaska

fl.   Pl. 61

Calypso bulbosa

Lemmon, Robert S.
Wildflowers of North America in
Full Color

fl.   pg. 244   pl. 384

Calypso bulbosa

Lindman, C. A. M.
Nordens Flora, Vol 3

fl. hab.   Pl. 154

Calypso bulbosa

Moyle, John B.
Northland Wild Flowers

fl.   p 224, Pl. 298

Calypso bulbosa

Orr, Robert T.
Wildflowers of Western America

fl.   Pl. 235

Calypso bulbosa

Ospina, Mariano
Orquideas de las americas

fl.   Pl. 94

Calypso bulbosa

Porsild, A. E.
Rocky Mountain Wild Flowers

fl. hab.   p 111

Calypso bulbosa

Taylor, Ronald J.
Mountain Wild Flowers of the Pacific
Northwest

fl.   p. 28

Calypso bulbosa

Welsh, Stanley L.
Flowers of the Mountain Country

fl.   p. 26

Calypso bulbosa var.

Clark, Lewis J.
Wild Flowers of British Columbia

fl.   p 82, 83

Calypso bulbosa var.

Luer, Carlyle A.
The Native Orchids of the
United States and Canada

fl. hab.   Pl. 95
                Pl. 96

Calyptripidium umbellatum

Orr, Robert T.
Wildflowers of Western America

fl.   Pl. 165

Calystegia marginata

Cochrane, G. R.
Flowers and Plants of Victoria

fl.   Pl. 265

Calystegia sepium

Ary, S.
The Oxford Book of Wildflowers

fl.   p 94, Pl. 5

Calystegia sepium

Brown, Clair A.
Wildflowers of Louisiana and
Adjoining States

fl. hab.   p 146

Calystegia sepium

Dean, Blanche E.
Wildflowers of Alabama and
Adjoining States

fl.   p 143

Calystegia sepium

Felsko, Elsa
A Book of Wild Flowers
                2nd ser.
fl.   p 132

Calystegia sepium

Lindman, C. A. M.
Nordens Flora, Vol 7

fl. fr.   Pl. 480

Calystegia sepium

Martin, W. Keble
The Concise British Flora in Colour

fl. hab.   Pl. 61

Calystegia sepium

Morcombe, M. K.
Australia's Western Wildflowers

fl.   60, 61

Calystegia silvatica

Morley, Brian D.
Wild Flowers of the World

fl.   Pl. 18

Calystegia soldanella

Harvey, Norman B.
New Zealand Botanical Paintings

fl.   p. 60   Pl. 27

Calystegia soldanella

Kleijn, H.
Beauty of the Wild Plant

fl.   opp. p. 80   pl. 104

Calystegia soldanella

Martin, W. Keble
The Concise British Flora in Colour

fl. hab.   Pl. 61

Calystegia soldanella

Polunin, Oleg
Flowers of Europe

fl.   Pl. 100 #1041

Calystegia spithamaea

Duncan, Wilbur H.
Wildflowers of the Southeastern
United States

fl.   p 137

Calytrix alpestris

Cochrane, G. R.
Flowers and Plants of Victoria

fl.   Pl. 99

Calytrix alpestris

Harrison, R.E.
Trees and Shrubs

fl.   P 40

Calytrix brachyphylla

Blombery, Alec M.
What Wildflower is That

fl.   p. 87   Pl. 205

Calytrix brachyphylla

Morcombe, M. K.
Australia's Western Wildflowers

fl.   p 88

Calytrix brevifolia

Morcombe, M. K.
Australia's Western Wildflowers

fl.   p 108

Calytrix fraseri

Blombery, Alec M.
What Wildflower is That

fl.   p 87, Pl. 206

Calytrix fraseri

Morcombe, M. K.
Australia's Western Wildflowers

fl.   p 56

Calytrix glutinosa

Harrison, R.E.
Trees and Shrubs

fl.   P 40

234

Calytrix microphylla

Blombery, Alec M.
What Wildflower is That

fl.   p. 87   Pl. 207

Calytrix microphylla

Mullins, Barbara
Australian Wildflowers in Colour

fl.   P. 67   Pl. 63

Calytrix sullivanii

Harrison, R.E.
Trees and Shrubs

fl.   P 40

Calytrix tetragona

Blombery, Alec M.
What Wildflower is That

fl.   p. 88   Pl. 208

Calytrix tetragona

Cochrane, G. R.
Flowers and Plants of Victoria

fl.   Pl. 125

Calytrix tetragona

Harrison, R.E.
Trees andShrubs

fl.   P 40

Camarotis rostrata

Curtis's Botanical Magazine
Vol 177   1969-70

fl.   537

Camass
see
Camassia

Camass
see
Quamasia quamash

Camass, Blue
see
Camassia quamash

Camass, Death
see
Zigadenus elegans

Camass, Elegant Death
see
Zigadenus elegans

Camass, Leichtlin's
see
Camassia leichtlinii

Camass, white
see
Zigadenus elegans

Camassia cusickii

Curtis's Botanical Magazine
vol 172   1958-59

fl.   Pl. 319

Camassia cusickii

Hay, Roy
The Color Dictionary of Flowers & Plants

fl.   p 85, Pl. 680

Camassia cusickii

Miles, Bebe
Bulbs for the Home Gardener

fl.   p. 12, 58

Camassia esculenta

Royal Hort. Soc.
The Garden
vol 101, No. 7   1976

hab., fl.   p. 351

Camassia leichtlinii

Clark, Lewis J.
Wild Flowers of British Columbia

fl.   p 23

Camassia leichtlinii

Lemmon, Robert S.
Wildflowers of North America in
Full Color

fl.   pg. 114   pl. 182

Camassia leichtlinii

Orr, Robert T.
Wildflowers of Western America

fl.   Pl. 241

Camassia leichtlinii var.

Perry, Frances
Flowers of the World

fl.   p. 175

Camassia leichtlinii var.

Royal Hort. Soc.
Journal of the Royal Hort. Soc.
vol 89, No. 1   1964

fl.   p. 22   Pl. 6

Camassia quamash

Clark, Lewis J.
Wild Flowers of British Columbia

fl. hab.   p 22, 23

Camassia quamash

Ferguson, Mary
Wildflowers

fl.   p 47

Camassia quamash

Hay, Roy
The Color Dictionary of Flowers & Plants

fl.   p 86, Pl. 681

Camassia quamash

Lemmon, Robert S.
Wildflowers of North America in
Full Color

fl.   pg. 114   pl. 181

Camassia quamash

Mathew, Brian
The Larger Bulbs

fl.   p. 64

Camassia quamash

Miles, Bebe
Bulbs for the Home Gardener

fl.   p 59

Camassia quamash

Munz, Philip A.
California Spring Wildflowers

fl.   p 40, Pl. 41

Camassia Quamash

Porsild, A. E.
Rocky Mountain Wild Flowers

fl.   p 95

Camassia scilloides

Brown, Clair A.
Wildflowers of Louisiana and
Adjoining States

fl. hab.   p 17

Camassia scilloides

Courtenay, Booth
Wildflowers & Weeds

fl.   p 6

Camassia scilloides

Dean, Blanche E.
Wildflowers of Alabama and
Adjoining States

fl.   p 21

Camassia scilloides

Jennings, O. E.
Wild Flowers of Western Pennsylvania
Vol. II

fl.   Pl. 20

Camassia scilloides

Wharton, Mary E
A Guide to the Wildflowers & Ferns
of Kentucky

fl.   p. 62   Pl. 1,30

Camelina sativa

Lindman, C. A. M.
Nordens Flora, Vol 5

fl. fr. hab.   Pl. 277

Camellia, common

see

Camellia japonica

Camellia granthamiana

Curtis's Botanical Magazine
Vol 178   1970-72

fl.   597

Camellia granthamiana

Morley, Brian D.
Wild Flowers of the World

fl.   Pl. 101

Camellia granthamiana

Walden, Beryl M.
Wild Flowers of Hong Kong

fl. fr.   Pl. 71 (219), 72 (219)

Camellia hongkongensis

Royal Hort. Soc.
Journal of the Royal Hort. Soc.
vol 74, No. 6   1949

fl.   p. 252   Pl. 80

Camellia hongkongensis

Walden, Beryl M.
Wild Flowers of Hong Kong

fl.   Pl. 1(10), 3(10)

Camellia hybrid

Everett, T.H., ed.
New Illustrated Encyclopedia of
Gardening vol. 2

fl.   p. 294

Camellia hybrid.

Gault, S. Millar
The Color Dictionary of Shrubs

fl.   Pl. 39, 40

Camellia hybrid

Harrison, R. E.
Trees and Shrubs

fl.   P. 41, 42, 45

Camellia hybrid

Hay, Roy
The Color Dictionary of Flowers & Plants

fl.   p 187, Pl. 1489, 1490

Camellia hybrid

Jones, Paul
Flora Magnifica

fl.   frontispiece

Camellia hybrid

Perry, Frances
Flowers of the World

fl.   p. 292-93

Camellia hybrida

Kiaer, Eigil
Indoor Plants in Colour

fl.   p. 33

Camellia japonica

Crockett, James Underwood
Evergreens

hab., fl.   p. 116

Camellia japonica

Encke, Fritz
Zimmerpflanzen

fl.   p. 61

Camellia japonica

Harrison, R. E.
Trees and Shrubs

fl.   P. 42-45

Camellia japonica

Hersey, Jean
Woman's Day Book of House Plants

fl.   p. 46

Camellia japonica

Kitamura, Siro
Coloured Illustrations of Trees &
Shrubs of Japan

fr.   342, 343

Camellia japonica

Massachusetts Hort. Soc.
Horticulture
vol 40, No. 5   1962

hab.   p. 259

Camellia japonica

Nicolaisen, Age
Pocket Encyclopedia of
Indoor Plants

fl.   pl. 73

Camellia japonica

Takatori, Jisuke
Color Atlas of Medicinal Plants of Japan

fl.
fr.   Fig. 27B

Camellia japonica var.

Everett, T. H., ed.
New Illustrated Encyclopedia of
Gardening   Vol. 3

fl.   Frontispiece

Camellia japonica var.

Gault, S. Millar
The Color Dictionary of Shrubs

fl.   Pl. 41 thru 48

Camellia japonica var.

Harrison, R. E.
Trees and Shrubs

fl.   P. 42-45

Camellia japonica var.

Hay, Roy
The Color Dictionary of Flowers & Plants

fl.   p 185, 186, Pl. 1479 thru 1486

Camellia japonica var.

Hay, Roy
The Dictionary of House Plants

fl.   Pl. 91 - 94

Camellia japonica var.

Hellyer, A.G.L.
Shrubs in Colour

fl.   P. 23

Camellia japonica var.

Huxley, Anthony
Evergreen Garden Trees and Shrubs

fl. hab.   Pl. 178

Camellia japonica var.

Kariyone, Tatsuo
Atlas of Medicinal Plants

fl., fr.   Pl. 114

Camellia japonica var.

Kimura, Koiti
Japanese Medicinal Plants, Vol. I

fl.   p. 50

Camellia japonica var.

Macoby, Stirling
What Flower is That

fl.   P. 70

Camellia japonica var.

Perry, Frances
Flowers of the World

fl.   p. 292-94

Camellia japonica var.

Royal Hort. Soc.
Journal of Royal Hort. Soc.
vol 83, No. 9   1958

fl.   p. 382   Pl. 113-14, 119

Camellia japonica var.

Tsukamoto, Yotaro
Coloured Illustrations of Garden
Flowers   Vol. 10

fl.   Illus. 136-39 opp. p. 43,44

Camellia kissi

Walden, Beryl M.
Wild Flowers of Hong Kong

fl.   Pl. 79 (242)

Camellia magnoliaeflora
Macoby, Stirling
What Flower is That

fl.   P 70

Camellia, Mountain

see

Stewartia ovata

Camellia, Mountain Purple Stemmed

see

Malachodendron pentagynum var.

Camellia oleifera

Curtis's Botanical Magazine
Vol. 170   1954-55

fl.   221

Camellia oleifera

Walden, Beryl M.
Wild Flowers of Hong Kong

fl. fr.   Pl. 74 (223)

Camellia pitardii var.

Curtis's Botanical Magazine
Vol 179   1972

fl.   633

Camellia reticulata

Morley, Brian D.
Wild Flowers of the World

fl.   Plate 101

Camellia reticulata var.

Harrison, R. E.
Trees and Shrubs

fl.   P. 41, 43

Camellia reticulata var.

Hay, Roy
The Color Dictionary of Flowers & Plants

fl.   p 186, Pl. 1487, 1488

Camellia reticulata var.

Hay, Roy
The Dictionary of House Plants

fl.   Pl. 95

Camellia reticulata var.

Macoby, Stirling
What Flower is That

fl.   P. 70

Camellia reticulata var.

Massachusetts Hort. Soc.
Horticulture
vol 34, No. 4   1956

fl.   p. 238

Camellia reticulata var.

Royal Hort. Soc.
Journal of Royal Hort. Soc.
vol 83, No. 7   1958

fl.   p. 294   Pl. 80

Camellia reticulata var.

Royal Hort. Soc.
Journal of the Royal Hort. Soc.
vol 93, No. 2   1968

fl.   cover

Camellia salicifolia

Walden, Beryl M.
Wild Flowers of Hong Kong

fl.   Pl. 80 (241)

Camellia saluenensis var.

Curtis's Botanical Magazine
Vol 160   1937

fl.   No. 9505

Camellia sasanqua

Curtis's Botanical Magazine
Vol 162   1939

fl.   No. 9591

Camellia sasanqua

Macoby, Stirling
What Flower is That

fl.   P 70

Camellia sasanqua

Massachusetts Hort. Soc.
Horticulture
vol 53, No. 10   1975

fl.   p. 38

Camellia sasanqua

Takatori, Jisuke
Color Atlas of Medicinal Plants of Japan

fl.
fr.   Fig. 270

Camellia sasanqua var.

Harrison, R. E.
Trees and Shrubs

fl.   p. 41

Camellia sasanqua var.

Massachusetts Hort. Soc.
Horticulture
vol 53, No. 10   1975

fl.   p. 41

Camellia sasanqua var.

Mathias, Mildred E.
Color for the Landscape

fl. hab.   p 64

Camellia sasanqua var.

Pacific Hort. Foundation
Pacific Horticulture
vol 37, No. 1   1976

fl.   p. 40

Camellia sasanqua var.

Tsukamoto, Yotaro
Coloured Illustrations of Garden
Flowers   Vol. 10

fl.   Ill.141,142 opp. p. 44

Camellia, Silky

see

Stewartia malachodendron

Camellia sinensis

Bianchini, F.
The Complete Book of Fruits & Vegetables

fl. fr.   p 217

Camellia sinensis

Hvass, Elsie
Plants That Feed and Serve Us

fl.   fr.   p 78, Pl. 161

Camellia sinensis

Kitamura, Siro
Coloured Illustrations of Trees &
    Shrubs of Japan

fl.   341

Camellia sinensis

Masefield, C. B.
The Oxford Book of Food Plants

fl.   Pl. 113, 2

Camellia sinensis

Milne, Lorus
Living Plants of the World

hab.   p 151

Camellia taliensis

Curtis's Botanical Magazine
    Vol 164   1943-48

fl.   No. 9684

Camellia waldenae

Walden, Beryl M.
Wild Flowers of Hong Kong

fl.   Pl. 79 (240)

Camellia, Wild

    see

Stewartia malacodendron

Camellia x williamsii var.

Gault, S. Millar
The Color Dictionary of Shrubs

fl.   Pl. 49,50,51

Camellia x williamsii var.

Harrison, R. E.
Trees and Shrubs

fl.   P. 43

Camellia x williamsii var.

Hay, Roy
The Color Dictionary of Flowers &
    Plants

fl.   p. 187   Pl.1491 thru
                      1494

Camellia x williamsii var.

Hellyer, A.G.L.
Shrubs in Colour

fl.   p. 23

Camellia x williamsii var.

Macoby, Stirling
What Flower is That

fl.   p. 71

Camellia x williamsii var.

Royal Hort. Soc.
Journal of the Royal Hort. Soc.
    vol 91, No. 7   1966

fl.   p. 294   Pl. 148-49

Camellia x williamsii var.

Royal Hort. Soc.
The Garden
    vol 103, No. 10   1978

fl., hab.   p. 417

Camel's Foot

    see

Bauhinia variegata

Camoensia maxima

Menninger, Edwin A.
Flowering Vines of the World

fl.   Pl. 101

Camomile

    see

Anthemis cotula

Camomile

    see

Matricaria chamomilla

Camomile, Yellow

    see

Anthemis tinctoria

Campanula affinis

Curtis's Botanical Magazine
    Vol 162   1939

fl.   No. 9568

Campanula alliariaefolia

Tsukamoto, Yotaro
Coloured Illustrations of Garden
    Flowers   vol 9

fl.   opp. p. 35

Campanula allionii

Barneby, T. P.
European Alpine Flowers in Colour

fl.   hab.   Pl. 82, 6

Campanula allionii

Kohlhaupt, Paula
Fleurs des Alpages
        vol 2
Fl.   P 106

Campanula americana

Courtenay, Booth
Wildflowers & Weeds

fl.   p 70

Campanula americana

Dean, Blanche E.
Wildflowers of Alabama and
    Adjoining States

fl.   p 177

Campanula americana

Jennings, O. E.
Wild Flowers of Western Pennsylvania
        vol II
fl.   Pl. 151

Campanula americana

Klimas, John E.
Wildflowers of Eastern America

fl.   Pl. 288

Campanula americana

Moyle, John B.
Northland Wild Flowers

fl.   hab.   p 154, Pl. 179

Campanula americana

Wharton, Mary E.
A Guide to the Wildflowers & Ferns
    of Kentucky

fl.   p 139   Pl. 1.90

Campanula aparinoides

Courtenay, Booth
Wildflowers & Weeds

fl.   p 70

Campanula barbata

Barneby, T. P.
European Alpine Flowers in Colour

fl.   Pl. 81, 4

Campanula barbata

Felsko, Elsa
A Book of Wild Flowers
        2nd ser.
fl.   p 54

Campanula barbata

Kohlhaupt, Paula
Fleurs des Alpages
        vol 1
fl.   P 99

Campanula barbata

Polunin, Oleg
Flowers of Europe

fl.   Pl. 139 #1330

Campanula barbata

Robert, Paul A.
Alpine Flowers

fl., hab.   Pl. 14

Campanula barbata

Tosco, Uberto
The World of Mountain Flowers

fl.   p 49

Campanula bononiensis

Barneby, T. P.
European Alpine Flowers in Colour

fl.   Pl. 83, 3

Campanula bononiensis

Polunin, Oleg
Flowers of Europe

fl.   Pl. 140 #1343

Campanula burghaltii

Hay, Roy
The Color Dictionary of Flowers & Plants

fl.   p 127, Pl. 1010

Campanula carpatica

Hay, Roy
The Color Dictionary of Flowers & Plants

fl.   p 3, Pl. 23

Campanula carpatica

Huxley, Anthony
Garden Perennials and Water Plants

fl.   Pl. 64

Campanula carpatica

Macoby, Stirling
What Flower is That

fl. P 71

Campanula carpatica

Perry, Frances
Flowers of the World

fl.   p 62

Campanula cenisia

Alpine Garden Society
Bulletin
    vol. 23, No. 2   1955

fl., hab.   p. 94

Campanula cenisia

Barneby, T. P.
European Alpine Flowers in Colour

fl. hab.   Pl. 82, 3

Campanula cenisia

Kohlhaupt, Paula
Fleurs des Alpages
                vol 1
fl.   P 101

Campanula cervicaria

Lindman, C. A. M.
Nordens Flora, Vol 9

fl.   Pl. 563

Campanula cochleariifolia

Addisonia
    vol 24   1960-64

fl.   p. 3   Pl. 770

Campanula cochleariifolia

Barneby, T.P.
European Alpine Flowers in Colour

fl. hab.   Pl. 82,2

Campanula cochleariifolia

Felsko, Elsa
A Book of Wild Flowers
                2nd ser.
fl. hab.   p 55

Campanula cochleariifolia

Perry, Frances
Flowers of the World

fl.   p 62

Campanula cochleariifolia

Polunin, Oleg
Flowers of Europe

fl.   Pl. 140 #1338

Campanula cochleariifolia

Vilmorin, Roger ce
Plantes Alpines dans les Jardins

fl. hab.   Pl. VIII

Campanula cochleariifolia var.

Hay, Roy
Color Dictionary of Flowers &
    Plants

fl.   p. 3   Pl. 24

Campanula davisii

Curtis's Botanical Magazine
Vol 171   1956-57

fl.   283

Campanula divaricata

Addisonia
vol 21   1939-42

fl., fr,   p. 13   Pl. 679

Campanula divaricata

Dean, Blanche E.
Wildflowers of Alabama and
    Adjoining States

fl.   p 177

Campanula divaricata

Duncan, Wilbur H.
Wildflowers of the Southeastern
    United States

fl.   p 189

Campanula divaricata

Wharton, Mary E.
A Guide to the Wildflowers & Ferns
    of Kentucky

fl.   p 170   Pl. 2.46

Campanula drabifolia

Polunin, Oleg
Flowers of the Mediterranean

fl.   Pl. 193

Campanula ephesia

Curtis's Botanical Magazine
    Vol. 181   1976-77

fl.   Pl. 712

Campanula excisa

Barneby, T. P.
European Alpine Flowers in Colour

fl. hab.   Pl. 82, 4

Campanula foliosa

Goulimis, Constantine N.
Wild Flowers of Greece

fl.   p 101

Campanula formanekiana

Curtis's Botanical Magazine
    Vol 159   1936

fl.   No. 9436

Campanula formanekiana

Goulimis, Constantine N.
Wild Flowers of Greece

fl.   p 103

Campanula garganica

Alpine Garden Society
Bulletin
    vol. 35, No. 1   1967

fl., hab.   p. 45

Campanula garganica

Hay, Roy
The Color Dictionary of Flowers & Plants

fl.   p 4, Pl. 25

Campanula glomerata

Ary, S.
The Oxford Book of Wildflowers

fl.   p 166, Pl. 3

Campanula glomerata

Barneby, T. P.
European Alpine Flowers in Colour

fl.   Pl. 81, 3

Campanula glomerata

Kohlhaupt, Paula
Fleurs des Alpages
                    vol 2
Fl.   P 105

Campanula glomerata

Lindman, C. A. M.
Nordens Flora, Vol 9

fl. hab.   Pl. 564

Campanula glomerata

Martin, W. Keble
The Concise British Flora in Colour

fl. hab.   Pl. 54

Campanula glomerata

Massachusetts Hort. Soc.
Horticulture
    vol 51, No. 6   1973

fl.   p. 24

Campanula glomerata

Perry, Frances
Flowers of the World

fl.   p 61   Pl. 5

Campanula glomerata

Polunin, Oleg
Flowers of Europe

fl.   Pl. 138 #1333

Campanula glomerata

Tsukamoto, Yotaro
Coloured Illustrations of
    Garden Flowers vol 9

fl.   opp. p. 35

Campanula glomerata var.

Hay, Roy
The Color Dictionary of Flowers & Plants

fl.   p 127, Pl. 1011

Campanula glomerata var.

Huxley, Anthony
Garden Perennials and Water Plants

fl.   Pl. 66

Campanula hawkinsiana

Curtis's Botanical Magazine
    vol 176   1966-68

fl.   Pl. 505

Campanula hybrid

Perry, Frances
Flowers of the World

fl.   p. 62

Campanula incurva

Curtis's Botanical Magazine
Vol 161   1938

fl.   No. 9556

Campanula incurva

Goulimis, Constantine N.
Wild Flowers of Greece

fl.   p 105

Campanula isophylla

Hay, Roy
The Color Dictionary of Flowers & Plants

fl.   p 55, Pl. 440

Campanula isophylla

Hay, Roy
The Dictionary of House Plants

fl.   Pl. 96

Campanula isophylla

Hersey, Jean
Woman's Day Book of House Plants

fl.   p 99

Campanula isophylla

Massachusetts Hort. Soc.
Horticulture
vol 34, No. 12   1956

fl.   p. 604

Campanula isophylla

Massachusetts Hort. Soc.
Horticulture
vol 37, No. 12   1959

fl.   p. 632

Campanula isophylla

Massachusetts Hort. Soc.
Horticulture
vol 40, No. 3   1962

fl.   p. 165

Campanula isophylla

Nicolaisen, Age
Pocket Encyclopedia of Indoor Plants

fl.   Pl. 147

Campanula isophylla var.

Encke, Fritz
Zimmerpflanzen

fl.   p. 61

Campanula isophylla var.

Hay, Roy
The Dictionary of House Plants

fl.   Pl. 97

Campanula isophylla var.

Kiaer, Eigil
Indoor Plants in Colour

fl.   p. 33

Campanula isophylla var.

Kromdijk, G.
200 House Plants in Colour

fl.   Pl. 42

Campanula isophylla var.

Massachusetts Hort. Soc.
Horticulture
    vol 34, No. 12   1956

fl.   p. 604

Campanula isophylla var.

Massachusetts Hort. Soc.
Horticulture
    vol 37, No. 12   1959

fl.   p. 632

Campanula isophylla var.

Massachusetts Hort. Soc.
Horticulture
    vol 40, No. 3   1962

fl.   p. 165

Campanula isophylla var.

Nicolaisen, Age
Pocket Encyclopedia of Indoor Plants

fl.   pl. 146

Campanula lactiflora

Meikle, R. D.
Garden Flowers

fl.   opp. p. 257   Pl. 7

Campanula lactiflora var.

Hay, Roy
The Color Dictionary of Flowers & Plants

fl.   p 127, Pl. 1012

Campanula lactiflora var.

Perry, Frances
Flowers of the World

fl.   p. 61   Pl. 1

Campanula lasiocarpa

Alaska-Yukon Wild Flower Guide

fl. hab.   p 159

Campanula lasiocarpa

Heller, Christine
Wild Flowers of Alaska

fl.   Pl. 235

Campanula lasiocarpa

Porsild, A. E.
Rocky Mountain Wild Flowers

fl. hab.   p 375

Campanula lasiocarpa

Walcott, Mary Vaux
North American Wild Flowers
          vol. 5

fl.   Pl. 368

Campanula latifolia

Ary, S.
The Oxford Book of Wildflowers

fl.   p 166, Pl. 5

Campanula latifolia

Lindman, C. A. M.
Nordens Flora, Vol 9

fl. fr.   Pl. 566

Campanula latifolia

Martin, W. Keble
The Concise British Flora in Colour

fl. fr. hab.   Pl. 54

Campanula latifolia

Perry, Frances
Flowers of the World

fl.   p 61   Pl. 2

Campanula latifolia var.

Addisonia
          vol 24   1960-64

fl., fr.   p. 9 Pl. 773

Campanula latifolia var.

Hay, Roy
The Color Dictionary of Flowers & Plants

fl.   p 127, Pl. 1013

Campanula latifolia var.

Huxley, Anthony
Garden Perennials and Water Plants

fl.   pl. 63

Campanula latifolia var.

Tsukamoto, Yotaro
Coloured Illustrations of Garden Flowers
          vol 9

fl.   opp. p. 36

Campanula leutwenii

Vilmoring, Roger de
Plantes Alpines dans les Jardins

fl. hab.   Pl. VII

Campanula linifolia

Barneby, T. P.
European Alpine Flowers in Colour

fl.   Pl. 82, 1

Campanula medium

Crockett, James Underwood
Annuals

fl.   P 102

Campanula medium

Huxley, Anthony
Garden annuals and bulbs

fl.   Pl. 18

Campanula medium

Macoby, Stirling
What Flower is That

fl.   P 71

Campanula medium

Massachusetts Hort. Soc.
Horticulture
          vol 51, No. 1   1973

fl.   p. 40

Campanula medium

Perry, Frances
Flowers of the World

fl.   p 61   Pl. 3

Campanula medium

Tsukamoto, Yotaro
Coloured Illustrations of Garden
          Flowers Vol. 10

fl. Illus. 23   opp. p. 8

Campanula medium var.

Crockett, James Underwood
Annuals

fl.   P 102

Campanula medium var.

Hay, Roy
The Color Dictionary of Flowers &
          Plants

fl.   p. 32   Pl. 255

Campanula medium var.

Massachusetts Hort. Soc.
Horticulture
          vol 44, No. 6   1966

fl.   cover, backcover

Campanula medium var.

Tsukamoto, Yotaro
Coloured Illustrations of garden
          Flowers vol 10

fl.   p. 8   Pl. 24

Campanula morettiana

Alpine Garden Society
Bulletin
          vol. 36, No. 1   1968

fl., hab.   p. 19

Campanula morettiana

Kohlhaupt, Paula
Fleurs des Alpages
          vol 1
fl.   P 102

Campanula myrtifolia

Curtis's Botanical Magazine
Vol 179   1972

fl.   637

Campanula patula

Ary, S.
The Oxford Book of Wildflowers

fl.   p 166, Pl. 1

Campanula patula

Barneby, T. P.
European Alpine Flowers in Colour

fl.   Pl. 83, 5

Campanula patula

Felsko, Elsa
A Book of Wildflowers

fl.   P 63

Campanula patula

Furrer, D.
Blumen am Wege

fl. hab.   p 105

Campanula patula

Kleijn, H.
Beauty of the Wild Plant

fl.   opp. p. 85   Pl. 116

Campanula patula
Lindman, C. A. M.
Nordens Flora, Vol 9

fl. hab.   Pl. 569

Campanula patula
Martin, W. Keble
The Concise British Flora in Colour

fl. fr. hab.   Pl. 54

Campanula persicifolia
Barneby, T. P.
European Alpine Flowers in Colour

fl.   Pl. 83, 4

Campanula persicifolia
Huxley, Anthony
Garden Perennials and Water Plants

fl.   Pl. 68

Campanula persicifolia
Lindman, C. A. M.
Nordens Flora, Vol 9

fl. hab.   Pl. 568

Campanula persicifolia
Pennsylvania Hort. Soc.
The Green Scene
     vol 5, No. 1   1976

fl.   p. 11   Pl. 1

Campanula persicifolia
Perry, Frances
Flowers of the World

fl.   p 61   Pl. 4

Campanula persicifolia
Polunin, Oleg
Flowers of Europe

fl.   Pl. 139 #1336

Campanula persicifolia
Tsukamoto, Yotaro
Coloured Illustrations of Garden Flowers

fl.   opp. p. 36

Campanula persicifolia var
Hay, Roy
The Color Dictionary of Flowers & Plants

fl.   p 127, Pl. 1014-15

Campanula persicifolia var.
Huxley, Anthony
Garden Perennials and Water Plants

fl.   Pl. 69

Campanula portenschlagiana
Curtis's Botanical Magazine
Vol 170   1954-55

fl.   256

Campanula portenschlagiana
Hay, Roy
The Color Dictionary of Flowers & Plants

fl.   p 4, Pl. 26

Campanula portenschlagiana
Huxley, Anthony
Garden Perennials and Water Plants

fl.   Pl. 70

Campanula portenschlagiana
Perry, Frances
Flowers of the World

fl.   p 62

Campanula portenschlagiana
Tsukamoto, Yotaro
Coloured Illustrations of Garden Flowers
                              vol 9
fl.   opp. p. 36

Campanula portenschlagiana
Vilmorin, Roger de
Plantes Alpines dans les Jardins

fl. hab.   Pl. XXXIV

Campanula poscharskyana
Crockett, James Underwood
Lawns & Ground Covers

fl.   p. 125

Campanula poscharskyana
Curtis's Botanical Magazine
vol 172   1958-59

fl.   Pl. 334

Campanula poscharskyana
Macoby, Stirling
What Flower is That

fl.   P 71

Campanula poscharskyana
Tsukamoto, Yotaro
Coloured Illustrations of Garden Flowers
                              vol 9
fl.   opp. p. 36

Campanula poscharskyana var.
Huxley, Anthony
Garden Perennials and Water Plants

fl.   Pl. 67

Campanula propinqua var.
Morley, Brian D.
Wild Flowers of the World

fl.   Pl. 48

Campanula pusilla
Robert, Paul A.
Alpine Flowers

fl.   Pl. 16

Campanula raineri
Barneby, T. P.
European Alpine Flowers in Colour

fl. hab.   Pl. 82, 5

Campanula ramosissima
Polunin, Oleg
Flowers of the Mediterranean

fl.   Pl. 191

Campanula rapunculoides
Ary, S.
The Oxford Book of Wildflowers

fl.   p 166, Pl. 4

Campanula rapunculoides
Barneby, T. P.
European Alpine Flowers in Colour

fl.   Pl. 83, 2

Campanula rapunculoides
Felsko, Elsa
A Book of Wildflowers

Campanula rapunculoides
Lindman, C. A. M.
Nordens Flora, Vol 9

fl. fr. hab.   Pl. 565

Campanula rapunculoides
Martin, W. Keble
The Concise British Flora in Colour

fl. hab.   Pl. 54

Campanula rapunculoides
Moyle, John B.
Northland Wild Flowers

fl.   p 154, Pl. 178

Campanula rapunculoides
Polunin, Oleg
Flowers of Europe

fl.   Pl. 140 #1342

Campanula rapunculus

Barneby, T. P.
European Alpine Flowers in Colour

fl.   Pl. 83, 6

Campanula rapunculus

Color Treasury of Herbs & Other
   Medicinal Plants

fl. p 54 Pl. 82

Campanula rapunculus

Martin, W. Keble
The Concise British Flora in Colour

fl. hab.   Pl. 54

Campanula reiseri

Curtis's Botanical Magazine
Vol 169   1952-53

fl.   178

Campanula rhomboidalis

Barneby, T. P.
European Alpine Flowers in Colour

fl.   Pl. 81, 5

Campanula rhomboidalis

Polunin, Oleg
Flowers of Europe

fl.   Pl. 140 #1344

Campanula rotundifolia

Alaska-Yukon Wild Flower Guide

fl.   p 160

Campanula rotundifolia

American Hort. Soc.
American Horticulturist
   vol 57, No. 5 1978

fl., hab. p. 4

Campanula rotundifolia

Ary, S.
The Oxford Book of Wildflowers

fl.   p 166, Pl. 6

Campanula rotundifolia

Clark, Lewis J.
Wild Flowers of British Columbia

fl.   p 510

Campanula rotundifolia

Courtenay, Booth
Wildflowers & Weeds

fl.   p 70

Campanula rotundifolia

Felsko, Elsa
A Book of Wild Flowers
          2nd ser
fl. hab.   p 56

Campanula rotundifolia

Ferguson, Mary
Wildflowers

fl.   p 77

Campanula rotundifolia

Fries, Mary A.
Wildflowers of Mount Ranier and
the Cascades

fl.   P 148

Campanula rotundifolia

Heller, Christine
Wild Flowers of Alaska

fl.   Pl. 236

Campanula rotundifolia

Klimas, John E.
Wildflowers of Eastern America

fl.   Pl. 286

Campanula rotundifolia

Lemmon, Robert S.
Wildflowers of North America in
    Full Color

fl.   pg. 158   pl. 254

Campanula rotundifolia

Lindman, C. A. M.
Nordens Flora, Vol 9

fl. hab.   Pl. 567

Campanula rotundifolia

Martin, W. Keble
The Concise British Flora in Colour

fl. hab.   Pl. 54

Campanula rotundifolia

Massachusetts Hort. Soc.
Horticulture
   vol 51, No. 1   1973

fl.   p. 40

Campanula rotundifolia

Orr, Robert T.
Wildflowers of Western America

fl.   Pl. 226/228

Campanula rotundifolia

Perry, Frances
Flowers of the World

fl. p 62

Campanula rotundifolia

Porsild, A. E.
Rocky Mountain Wild Flowers

fl.   p 375

Campanula rotundifolia

Taylor, Ronald J.
Mountain Wild Flowers

fl. hab.   p 81

Campanula rotundifolia

Walcott, Mary Vaux
North American Wild Flowers
   vol. 5

fl.   Pl. 369

Campanula rotundifolia

Welsh, Stanley L.
Flowers of the Mountain Country

fl.   p. 74

Campanula rupestris

Morley, Brian D.
Wild Flowers of the World

fl.,hab.   Pl. 33

Campanula rupestris

Polunin, Oleg
Flowers of the Mediterranean

fl.   Pl. 190

Campanula rupestris var.

Alpine Garden Society
Bulletin
   vol. 39, No. 4 1971

fl.   p. 265

Campanula rupestris var.

Polunin, Oleg
Flowers of the Mediterranean

fl.   Pl. 189

Campanula rupestris var.

Royal Hort. Soc.
Journal of the Royal Hort. Soc.
   vol 88, No. 11 1963

fl.   p. 478   Pl. 181

Campanula saxatilis

Goulimis, Constantine N.
Wild Flowers of Greece

fl.   p 107

Campanula saxifraga

Morley, Brian D.
Wild Flowers of the World

fl.   Pl. 15D

Campanula scheuchzeri

Barneby, T. P.
European Alpine Flowers in Colour

fl.   Pl. 81, 6

Campanula scheuchzeri

Kohlhaupt, Paula
Fleurs des Alpages
                    vol 2
Fl.   P 104

Campanula scheuchzeri

Polunin, Oleg
Flowers of Europe

fl.   Pl. 140 #1339

Campanula scheuchzeri

Tosco, Uberto
The World of Mountain Flowers

fl.   p 49

Campanula scouleri

Clark, Lewis J.
Wild Flowers of British Columbia

fl.   p 511

Campanula spicata

Barneby, T. P.
European Alpine Flowers in Colour

fl.   Pl. 82, 2

Campanula spicata

Polunin, Oleg
Flowers of Europe

fl.   Pl. 138 #1332

Campanula thyrsoides

Barneby, T. P.
European Alpine Flowers in Colour

fl.   Pl. 81, 1

Campanula thyrsoides

Kleijn, H.
Beauty of the Wild Plant

fl.   opp. p. 100   Pl. 142

Campanula thyrsoides

Kohlhaupt, Paula
Fleurs des Alpages
                    Vol. 1
fl.   Pl. 100

Campanula thyrsoides

Kohlhaupt, Paula
Fleurs des Alpages
                    Vol. 1
fl.   P. 100

Campanula thyrsoides

Morley, Brian D.
Wild Flowers of the World

fl.   Pl. 15F

Campanula thyrsoides

Polunin, Oleg
Flowers of Europe

fl.   Pl. 139 #1332

Campanula thyrsoides

Tosco, Uberto
The World of Mountain Flowers

fl.   p 7

Campanula trachelium

Ary, S.
The Oxford Book of Wildflowers

fl.   p 166, Pl. 2

Campanula trachelium

Barneby, T. P.
European Alpine Flowers in Colour

fl.   Pl. 83, 1

Campanula trachelium

Martin, W. Keble
The Concise British Flora in Colour

fl. hab.   Pl. 54

Campanula trachelium

Polunin, Oleg
Flowers of Europe

fl.   Pl. 139 #1340

Campanula tridendata var.

Royal Hort. Soc.
Journal of the Royal Hort. Soc.
        vol 86, No. 6   1961
fl.   p. 266   Pl. 75

Campanula uniflora

Heller, Christine
Wild Flowers of Alaska

fl.   Pl. 237

Campanula uniflora

Lindman, C. A. M.
Nordens Flora, Vol 9

fl. hab.   Pl. 570

Campanula uniflora

Porsild, A. E.
Rocky Mountain Wild Flowers

fl. hab.   p 377

Campanula vidalii

Curtis's Botanical Magazine
        vol 176   1966-68

fl.   Pl. 527

Campanula vidalii

Morley, Brian D.
Wild Flowers of the World

fl., hab.   Pl. 43

Campanula zoysii

Alpine Garden Society
Bulletin
        vol. 35, No. 4   1967
fl., hab.   p. 291

Campanula zoysii

Hay, Roy
The Color Dictionary of Flowers & Plants

fl.   p 4, Pl. 27

Campernelle

see

Narcissus odorusvar.

Camphor

see

Cinnamomum camphora

Camphorweed

see

Heterotheca subaxillaris

Camphorweed

see

Pluchea camphorata

Camphorosma monspeliaca

Perrot, Emile
Les Plantes Medicinales

fl. fr. hab.   p 225

Campion

see

Lychnis

Campion

see

Silene

Campion, arctic

See

Lychnis alpina

Campion, Bladder

see

Melandrium attenuatum

Campion, Bladder

See

Silene cucubalus

Campion, Elisabeth's

see

Silene elisabethae

Campion, Moss

see

Silene acaulis

Campion, Nodding

see

Lychnis apetala

Campion, Red

see

Melandrium rubrum

Campion, Red

see

Silene dioica

Campion, Rose

see

Lychnis coronaria

Campion, sea

see

Silene maritima

Campion, Starry

see

Silene stellata

Campion, white

see

Lychnis alba

Campion, white

see

Melandrium album

Campsidium valdivianum

Munoz Pizarro, Carlos
Flores Silvestres de Chile

fl.   Pl. 48

Campsis chinensis

Macoby, Stirling
What Flower is That

fl.   P 72

Campsis grandiflora

Harrison, Richmond E.
Climbers and Trailers

fl.   p 27   Pl. 33

Campsis grandiflora

Kimura, Koiti
Japanese Medicinal Plants, Vol I

fl.   p 90

Campsis grandiflora

Morley, Brian D.
Wild Flowers of the World

fl.   Plate 103

Campsis radicans

Brown, Clair A.
Wildflowers of Louisiana and
Adjoining States

fl. hab.   p 172

Campsis radicans

Campbell, Carlos C.
Great Smoky Mountain Wildflowers

fl.   p 61

Campsis radicans

Dean, Blanche E.
Wildflowers of Alabama and
Adjoining States

fl.   p 167

Campsis radicans

Everett, T.H., ed.
New Illustrated Encyclopedia of
Gardening vol. 2

fl.   p. 294

Campsis radicans

Everett, T.H., ed.
New Illustrated Encyclopedia
of Gardening   vol 13

fl.   p. 2407

Campsis radicans

Fleming, Glenn
Wild Flowers of Florida

fl. hab.   p 80

Campsis radicans

Harrison, Richmond E.
Climbers and Trailers

fl.   p 28   Pl. 34

Campsis radicans

Hellyer, A.G.L.
Shrubs in Colour

fl.   p. 23

Campsis radicans

Huxley, Anthony
Deciduous Garden Trees and Shrubs

fl.   Pl. 282

Campsis radicans

Klimas, John E.
Wildflowers of Eastern America

fl.   Pl. 138

Campsis radicans

Massachusetts Hort. Soc.
Horticulture
vol 51, No. 8   1973

fl.   p. 24

Campsis radicans

Morley, Brian D.
Wild Flowers of the World

fl. fr.   pl. 162

Campsis radicans

Perry, Frances
Flowers of the World

fl.   p 44

Campsis radicans

Weeds of the Southern United States

fl.   p. 5

Campsis radicans

Wharton, Mary E.
Trees & Shrubs of Kentucky

fl.   p 67, Pl. 1.31

Campsis radicans var.

Harrison, Richmond E.
Climbers and Trailers

fl.   p. 28   Pl. 35

Campsis x tagliabuana

Curtis's Botanical Magazine
Vol. 169   1952-53

fl.   198

Campsis x tagliabuana

Hay, Roy
The Color Dictionary of Flowers &
    Plants

fl.   p. 245   Pl. 1958

Campsis x tagliabuana var.

Harrison, Richmond E.
Climbers and Trailers

fl.   p. 28   Pl. 36

Campylanthus salsoloides

Bramwell, David
Wild Flowers of the Canary Islands

fl.   Pl. 265

Campylocentrum aromaticum

Pabst, G. F. J.
Orchidaceae Brasilienses,
    Vol 2

fl.   p 265

Campylocentrum micranthum

Ospina, Mariano
Orquideas de las americas

fl. hab.   Pl. 135

Campylocentrum pachyrrhizum

Luer, Carlyle A.
The Native Orchids of Florida

fl. fr. hab.   Pl. 83

Campylocentrum sellowii

Pabst, G.F.J.
Orchidaceae Brasilienses
    vol. 2

fl.,hab.   p. 174, 265

Campynema lineare

Curtis, Winifred
The Endemic Flora of Tasmania Vol. 2

fl.   Pl. 40

Cananga odorata

Hvass, Elsie
Plants That Feed and Serve Us

fl.   p 116, Pl. 254

Cananga odorata

Morley, Brian D.
Wild Flowers of the World

fl.fr.   Plate 110

Cananga odorata

Wit, H. C. D. de
Plants of the World;
    The Higher Plants, Vol I

fl. fr.   p 58, Pl. 24

Canarina abyssinica

Morley, Brian D.
Wild Flowers of the World

fl.   Pl. 55

Canarina campanula

Harrison, Richmond E.
Climbers and Trailers

fl.   p 29   Pl. 37

Canarina campanula

Menninger, Edwin A.
Flowering Vines of the World

fl.   Pl. 51

Canarina canariensis

Bramwell, David
Wild Flowers of the Canary Islands

fl.   Pl. 273

Canarina canariensis

Morley, Brian D.
Wild Flowers of the World

fl.   Pl. 43

Canarina canariensis

Pacific Hort. Foundation
Pacific Horticulture
    vol 39, No. 3   1978

fl.   p. 32

Canarina canariensis

Perry, Frances
Flowers of the World

fl.   p 64

Canarina eminii

Curtis's Botanical Magazine
Vol 161   1938

fl.   No. 9531

Canary bird bush

see

Crotalaria semperflorens

Canary-bird flower

see

Tropaeolum peregrinum

Canary bird vine

see

Tropaeolum peregrinum

Canary Creeper

see

Senecio tamoides

Canary creeper

see

Tropaeolum peregrinum

Canary Creeper

see

Tropaeolum polyphyllum

Canavalia ensiformis

Masefield, G. B.
The Oxford Book of Food Plants

fl. fr.   Pl. 41, 2

Canavalia kauaiensis

Pacific Tropical Botanical Garden
Bulletin
    vol 3, No. 2   1973

Canavalia rosea

Blombery, Alec M.
What Wildflower is That

fl.   p 88, Pl. 209

Cancer-root

see

Conopholis americana

Cancer root

see

Orobanche fasciculata

Cancer Root, One-Flowered

see

Orobanche uniflora

Candelabra Tree

see

Euphorbia ingens

Candlebarks

see

Eucalyptus rubida

Candle, Christmas

see

Cassia alata

Candle of the Lord

see

Yucca whipplei

Candle, Swamp

see

Lysimachia terrestris

Candy Root

see

Polygala

Candy Root, Orange

see

Polygala lutea

Candystick

see

Allotropa virgata

Candytuft

see

Iberis

Candytuft, annual

see

Iberis amara

Candytuft, Evergreen

see
Iberis sempervirens

Candytuft, globe

see

Iberis umbellata

Candytuft, Perennial

see

Iberis sempervirens

Candytuft, Rock

see

Iberis saxatilis

Candytuft, Wild

see

Thlaspi montanum

Cane, Dumb

see

Dieffenbachia

Cane, Dumb

see

Pedilanthus tithymaloides

Canistrum aurantiacum

Bromeliad Society
Journal
vol 24, No. 3   1974

fl.   cover (p. 81)

Canistrum cyathiforme

Bromeliad Society
Bulletin
vol 17, No. 4 1967

fl., hab.   p. 84

Canistrum fosterianum

Bromeliad Society
Bulletin
vol 17, No. 5 1967

fl.   p. 120

Canistrum fosterianum

Bromeliad Society
Journal
vol 21, No. 2 1971

fl., hab.   cover (p. 25)

Canistrum fosterianum

Padilla, Victoria, ed.
Bromeliads in Color and
Their Culture

fl.   p 110

Canistrum fosterianum

Wilson, Robert Gardner
Bromeliads in Cultivation

fl.   Pl. 98

Canistrum lindenii var.

Rauh, Werner
Bromeliads for Home, Garden and
Greenhouse

fl.   Pl. 94

Canistrum lindenii var.

Wilson, Robert Gardner
Bromeliads in Cultivation

fl. hab.   Pl. 96

Canna, dwarf

see

Canna indica var.

Canna edulis

Chickering, Carol Rogers
Flowers of Guatemala

fl.   p 43

Canna flaccida

Brown, Clair A.
Wildflowers of Louisiana and
Adjoining States

fl. hab.   p 23

Canna flaccida

Duncan, Wilbur H.
Wildflowers of the Southeastern
United States

fl.   p 267

Canna flaccida

Fleming, Glenn
Wild Flowers of Florida

fl.   p 46

Canna x generalis

Tsukamoto, Yotaro
Coloured Illustrations of Garden Flowers
Vol. 9

fl.   opp. p. 6

Canna, Golden

see

Canna flaccida

Canna hortensis

Hay, Roy
The Color Dictionary of Flowers & Plants

fl.   p 56, Pl. 441

247

Canna hybrid

Miles, Bebe
Bulbs for the Home Gardener

fl.   p 154,
                         155

Canna hybrid

Perry, Frances
Flowers of the World

fl.   p. 65

Canna x hybrida

Kiaer, Eigil
Indoor Plants in Colour

fl.   p. 34

Canna x hybrida var.

Hay, Roy
The Dictionary of House Plants

fl.   Pl. 98

Canna indica

Bruggeman, L.
Tropical Plants

fl.   Pl. 87

Canna indica

Macoby, Stirling
What Flower is That

fl.   P 72

Canna indica var.

Huxley, Anthony
Garden Annuals and Bulbs

fl.   Pl. 201

Canna indica var.

Macoby, Stirling
What Flower is That

fl.   P 72

Canna lucifer

Kromdijk, G.
200 House Plants in Colour

fl.   Pl. 43

Cannabis sativa

Bianchini, Francesco
Health Plants of the World

fl. hab.   p 123

Cannabis sativa

Courtenay, Booth
Wildflowers & Weeds

fl.   p 69

Cannabis sativa

Huxley, Anthony
Garden Annuals and Bulbs.

fl.   Pl. 17

Cannabis sativa

Hvass, Elsie
Plants That Feed and Serve Us

fl.   fr.   p 100, Pl. 214

Cannabis sativa

Perrot, Emile
Les Plantes Medicinales

fl. fr. hab.   p 58

Cannabis sativa

Polunin, Oleg
Flowers of Europe

fl.   Pl. 6 #64

Cannabis sativa

Takatori, Jisuke
Color Atlas of Medicinal Plants of
    Japan

fl.   Fig. 61

Cannon Ball Tree

see

Couroupita guianensis

Cantaloupe

see

Cucumis melo

Cantua buxifolia

Harrison, R.E.
Trees and Shrubs

fl.   P 46

Cantua buxifolia

Hay, Roy
The Dictionary of House Plants

fl.   Pl. 99

Cantua buxifolia

Perry, Frances
Flowers of the World

fl.   p 234

Cantua tomentosa

Harrison, R. E.
Trees and Shrubs

fl.   P. 46

Capanemia micromera

Pabst, G. F. J.
Orchidaceae Brasilienses,
    Vol 2

fl.   p 262

Capanemia superflua

Ospina, Mariano
Orquideas de las americas

fl.   Pl. 170

Capanemia superflua

Pabst, G. F. J.
Orchidaceae Brasilienses,
    Vol 2

fl. hab.   p 262

Capanemia thereziae

Pabst, G. F. J.
Orchidaceae Brasilienses,
    Vol 2

fl. hab.   p 262

Cape Gold

see

Arctotheca calendula

Cape Pondweed

see

Aponogeton distachyus

Cape Weed

see

Anchusa capensis

Capeweed

see

Lippia nodiflora

Caper

see

Capparis spinosa

Caper, Hong Kong

see

Capparis membranacea

Capnoides sempervirens

Walcott, Mary Vaux
North American Wild Flowers

fl. hab.   Pl. 5

Capparis membranacea

Walden, Beryl M.
Wild Flowers of Hong Kong

fl.   Pl. 36(92)

Capparis micrantha

Menninger, Edwin A.
Flowering Vines of the World

fl.   Pl. 48

Capparis mitchellii

Blombery, Alec M.
What Wildflower is That

fl.   p 88, Pl. 210

Capparis ovata var.

Coulimis, Constantine N.
Wild Flowers of Greece

fl.   p. 29

Capparis spinosa

Bianchini, F.
The Complete Book of Fruits & Vegetables

fl. fr.   p 209

Capparis spinosa

Color Treasury of Herbs & Other
    Modicinal Plants

fl.   p 30   Pl. 30

Capparis spinosa

Hvass, Elsie
Plants That Feed and Serve Us

fl.   p 71, Pl. 146

Capparis spinosa

Masefield, G. B.
The Oxford Book of Food Plants

fl. fr.   Pl. 133, 3

Capparis spinosa

Megaw, Elektra
Wild Flowers of Cyprus

fl.   p. 8   Pl. 6

Capparis spinosa

Pennsylvania Hort. Soc.
The Green Scene
    vol 6, No. 3   1978

fl., hab.   p. 14-15

Capparis spinosa

Perrot, Emile
Les Plantes Medicinales

fl. fr. hab.   p 15

Capparis spinosa

Perry, Frances
Flowers of the World

fl.   p 66

Capparis spinosa

Polunin, Oleg
Flowers of Europe

fl.   Pl. 31   #283

Capparis spinosa var.

Morley, Brian D.
Wild Flowers of the World

fl. fr.   pl. 148

Capsella bursa-pastoris

Ary, S.
The Oxford Book of Wildflowers

fl. fr.   p 70, Pl. 3

Capsella bursa-pastoris

Bianchini, F.
The Complete Book of Fruits & Vegetables

hab.   p 73

Capsella bursa-pastoris

Bianchini, Francesco
Health Plants of the World

fl. hab.   p 145

Capsella bursa-pastoris

Courtenay, Booth
Wildflowers & Weeds

hab., fl.   p. 17

Capsella bursa-pastoris

Kimura, Koiti
Japanese Medicinal Plants, Vol II

fl.   p 177

Capsella bursa-pastoris

Lindman, C. A. M.
Nordens Flora, Vol 4

fl. fr. hab.   Pl. 259

Capsella bursa-pastoris

Martin, W. Keble
The Concise British Flora in Colour

fl. fr. hab.   Pl. 10

Capsella bursa-pastoris

Perrot, Emile
Les Plantes Medicinales

fl. fr. hab.   p 40

Capsella bursa-pastoris

Weeds of the Southern United States

fl., hab.   p. 18

Capsicum annuum

Bianchini, F.
The Complete Book of Fruits &
    Vegetables

fr.   p. 95

Capsicum annuum

Hay, Roy
The Color Dictionary of Flowers &
    Plants

fr.   p. 32   Pl. 256

Capsicum annuum

Hvass, Elsie
Plants That Feed and Serve Us

fl. fr.   p. 39   Pl. 75

Capsicum annuum

Kariyone, Tatsuo
Atlas of Medicinal Plants

fr.   Pl. 41

Capsicum annuum

Loewenfeld, Claire
The Complete Book of Herbs and Spices

fl. fr. hab.   p 208

Capsicum annuum

Masefield, G. B.
The Oxford Book of Food Plants

fr.   Pl. 129, 1

Capsicum annuum

Perrot, Emile
Les Plantes Medicinales

fl. fr. hab.   p 180

Capiscum annuum

Rosengarten, Frederic, Jr.
The Book of Spices

fl. fr. p 128

Capsicum annuum var.

Hay, Roy
The Dictionary of House Plants

fr.   Pl. 100

Capsicum annuum var.

Macoby, Stirling
What Flower is That

fr.   P 73

Capsicum annuum var.

Perry, Frances
Flowers of the World

fr.   p. 284

Capsicum frutescens

Bianchini, F.
The Complete Book of Fruits &
  Vegetables

fr.   p. 95, 97

Capsicum frutescens

Masefield, G. B.
The Oxford Book of Food Plants

fl. fr.   Pl. 129, 2-3

Capsicum frutescens var.

Hay, Roy
The Dictionary of House Plants

fr.   Pl. 101, 102

Capsicum minimum

Bianchini, Francesco
Health Plants of the World

fr.   p 29

Caragana arborescens

Hellyer, A.G.L.
Shrubs in Colour

fl.   p. 23

Caragana arborescens

Huxley, Anthony
Deciduous Garden Trees and Shrubs

fl.   Pl. 28

Caralluma arenicola

Flowering Plants of South Africa
  vol XIX   1939

fl., hab.   Pl. 760

Caralluma baldratii

Cactus & Succulent Soc. of America
Journal
  vol 45, No. 1   1973

fl.   p. 24

Caralluma baldratii

Flowering Plants of Africa
  vol XXVI   1947

fl.   Pl. 1033

Caralluma burchardii

Bramwell, David
Wild Flowers of the Canary Islands

hab.   Pl. 213, 14

Caralluma burchardii var.

Lamb, Edgar
Stapeliads in Cultivation

fl.   p 36

Caralluma bredae var.

Flowering Plants of Africa
  vol 36   1963-64

fl., hab.   Pl. 1438

Caralluma carnosa

Flowering Plants of Africa
  vol 28   1950-51

fl., hab.   Pl. 1085

Caralluma caudata

Lamb, Edgar
Stapeliads in Cultivation

fl. hab.   p 17

Caralluma chrysostephana

Flowering Plants of Africa
  vol 43   1974-76

fl., hab.   Pl. 1712

Caralluma congestiflora

Flowering Plants of Africa
  vol 43   1974-76

fl., hab.   Pl. 1715

Caralluma dicapuae

Flowering Plants of Africa
  vol XXVI   1947

fl.   Pl. 1032

Caralluma dioscoridis

Flowering Plants of Africa
  vol 42   1972-73

fl., hab.   Pl. 1662

Caralluma distincta

Flowering Plants of Africa
  vol 27   1948-49

fl., fr.   Pl. 1048

Caralluma edithae

Flowering Plants of Africa
  vol 36   1963-64

fl.   Pl. 1430

Caralluma europaea

Hay, Roy
The Dictionary of House Plants

hab.   Pl. 103

Caralluma europaea

Lamb, Edgar
The Pocket Encyclopedia of Cacti
  and Succulents in Color

fl.   Pl. 231

Caralluma europaea

Subik, Rudolf
Decorative Cacti

fl. hab.   p 101

Caralluma foetida

Flowering Plants of Africa
  vol 43   1974-76

fl., hab.   Pl. 1714

Caralluma furta

Flowering Plants of Africa
  vol 37   1965-66

fl., hab.   Pl. 1457

Caralluma gerstneri

Flowering Plants of South Africa
  vol XVI   1936

fl.   Pl. 631

Caralluma gerstneri var.

Flowering Plants of Africa
  vol 40   1969-70

fl., hab.   Pl. 1567

Caralluma incarnata

Flowering Plants of Africa
  vol 39   1968-69

fl., hab.   Pl. 1540

Caralluma lugardi

Lamb, Edgar
Popular Exotic Cacti in Color

fl. hab.   p 41

Caralluma lutea

Flowering Plants of South Africa
  vol XVI   1936

fl.   Pl. 621

Caralluma lutea

Lamb, Edgar
Stapeliads in Cultivation

fl. hab.   p 54

Caralluma marlothii var.

Lamb, Edgar
Stapeliads in Cultivation

fl.   p 36

Caralluma montana

Flowering Plants of Africa
vol XXVI   1947

fl., hab.   Pl. 1034

Caralluma peckii

Flowering Plants of Africa
vol 35   1962

fl.   Pl. 1394

Caralluma pillansii

Flowering Plants of South Africa
vol XXIV   1944

fl.   Pl. 931

Caralluma pruinosa

Flowering Plants of South Africa
vol XIX   1939

fl.   Pl. 722

Caralluma rangeana

Flowering Plants of Africa
vol 38   1967

fl., hab.   Pl. 1484a

Caralluma retrospiciens

Flowering Plants of Africa
vol 27   1948-49

fl., hab.   Pl. 1062

Caralluma retrospiciens var.

Flowering Plants of Africa
vol 27   1948-49

fl., hab.   Pl. 1063

Caralluma retrospiciens var.

Morley, Brian D.
Wild Flowers of the World

fl.,fr.   Pl. 78

Caralluma shadhbana

Flowering Plants of Africa
vol 44   1977

fl., hab.   Pl. 1743

Caralluma socotrana

Cactus & Succulent Society of America
Journal
vol 40, No. 2   1968

fl., hab.   cover (p. 45)

Caralluma socotrana

Flowering Plants of Africa
vol 41   1970-71

fl., hab.   Pl. 1607

Caralluma somalica

Flowering Plants of Africa
vol XXVI   1947

fl., hab.   Pl. 1008

Caralluma speciosa

Flowering Plants of Africa
vol 38   1967

fl.   Pl. 1485

Caralluma sprengeri

Flowering Plants of Africa
vol XXVI   1947

fl., hab.   Pl. 1031

Caralluma tubiformis

Flowering Plants of Africa
vol 27   1948-49

fl., hab.   Pl. 1047

Caralluma umdausensis

Flowering Plants of Africa
vol 43   1974-76

fl., hab.   Pl. 1713

Caralluma umdausensis

Lamb, Edgar
Stapeliads in Cultivation

fl.   p 36

Caralluma vibratilis

Flowering Plants of Africa
vol 27   1948-49

fl., hab.   Pl. 1074

Carambola tree

see

Averrhoa carambola

Carapa procera

Addisonia
vol 19   1935-36

fl.   p. 51   Pl. 634

Caraway

see

Carum carvi

Cardamine amara

Bianchini, F.
The Complete Book of Fruits & Vegetables

hab.   p 53

Cardamine amara

Huxley, Anthony
Garden Perennials and Water Plants

fl.   Pl. 288

Cardamine amara

Martin, W. Keble
The Concise British Flora in Colour

fl. hab.   Pl. 7

Cardamine amara

Polunin, Oleg
Flowers of Europe

fl.   Pl. 34 #310

Cardamine bellidifolia

Lindman, C. A. M.
Nordens Flora, Vol 4

fl. fr. hab.   Pl. 269

Cardamine bellidifolia

Porsild, A.E.
Rocky Mountain Wild Flowers

fl. fr. hab.   p 201

Cardamine bulbifera

Lindman, C. A. M.
Nordens Flora, Vol 4

fl. hab.   Pl. 267

Cardamine bulbifera

Martin, W. Keble
The Concise British Flora in Colour

fl. hab.   Pl. 8

Cardamine bulbifera

Polunin, Oleg
Flowers of Europe

fl.   Pl. 34 #308

Cardamine bulbosa

Courtenay, Booth
Wildflowers & Weeds

fl.   p 19

Cardamine bulbosa

Duncan, Wilbur H.
Wildflowers of the Southeastern
United States

fl. hab.   p 53

Cardamine bulbosa

Jennings, O. E.
Wild Flowers of Western Pennsylvania

fl. hab.   Pl. 15   vol II.

Cardamine bulbosa

Moyle, John B.
Northland Wild Flowers

fl.   p 61, Pl. 27

Cardamine bulbosa

Wharton, Mary E.
A Guide to the Wildflowers & Ferns
      of Kentucky

fl.   p 150   Pl. 2.5

Cardamine concatenata

Dean, Blanche E.
Wildflowers of Alabama and
      Adjoining States

fl.   p 73

Cardamine cordifolia

Welsh, Stanley L.
Flowers of the Mountain Country

fl.   p. 8

Cardamine depressa var.

Morley, Brian D.
Wild Flowers of the World

fl.   Pl. 1F

Cardamine diphylla

Dean, Blanche E.
Wildflowers of Alabama and
      Adjoining States

fl.   p 73

Cardamine douglassii

Jennings, O. E.
Wild Flowers of Western Pennsylvania
                              vol II
fl. hab.   Pl. 15

Cardamine douglassii

Klimas, John E.
Wildflowers of Eastern America

fl.   Pl. 238

Cardamine douglassii

Roberts, June Carver
Born in the Spring

fl. hab.   p 31

Cardamine douglassii

Wharton, Mary E.
A Guide to the Wildflowers & Ferns
      of Kentucky

fl.   p 142   Pl. 1.95

Cardamine enneaphyllos

Polunin, Oleg
Flowers of Europe

fl.   Pl. 33   #309

Cardamine ferrarii

Tosco, Uberto
The World of Mountain Flowers

fl. hab.   p 52

Cardamine flexuosa

Martin, W. Keble
The Concise British Flora in Colour

fl. fr. hab.   Pl. 7

Cardamine heptaphylla

Barneby, T. P.
European Alpine Flowers in Colour

fl.   Pl. 29, 2

Cardamine hirsuta

Ary, S.
The Oxford Book of Wildflowers

fl.   p 68, Pl. 4

Cardamine hirsuta

Bianchini, F.
The Complete Book of Fruits & Vegetables

fl. hab.   p 73

Cardamine hirsuta

Martin, W. Keble
The Concise British Flora in Colour

fl. fr. hab.   Pl. 7

Cardamine impatiens

Martin, W. Keble
The Concise British Flora in Colour

fl. fr. hab.   Pl. 8

Cardamine lyrata

Stodola, Jiri
Encyclopedia of Water Plants

hab.   P. 250

Cardamine palustris

Lindman, C. A. M.
Nordens Flora, Vol 4

fl. fr.   Pl. 268B

Cardamine parviflora

Duncan, Wilbur H.
Wildflowers of the Southeastern
      United States

fl. hab.   p 53

Cardamine pensylvanica

Courtenay, Booth
Wildflowers & Weeds

fl.   p 19

Cardamine pentaphyllos

Barneby, T. P.
European Alpine Flowers in Colour

fl.   Pl. 29 1

Cardamine pentaphyllos

Polunin, Oleg
Flowers of Europe

fl. Pl. 34 #309

Cardamine pentaphyllos

Wit, H. C. D. de
Plants of the World;
      The Higher Plants, Vol I

fl.   p 170, Pl. 96

Cardamine pratensis

Ary, S.
The Oxford Book of Wildflowers

fl.   p 112, Pl. 1

Cardamine pratensis

Courtenay, Booth
Wildflowers & Weeds

fl.   p 19

Cardamine pratensis

Felsko, Elsa
A Book of Wildflowers

fl.   P 84

Cardamine pratensis

Furrer, D.
Blumen am Wege

fl. hab.   p 33

Cardamine pratensis

Kleijn, H.
Beauty of the Wild Plant

fl.   opp. p. 32   pl. 33

Cardamine pratensis

Lindman, C. A. M.
Nordens Flora, Vol 4

fl. hab.   Pl. 268A

Cardamine pratensis

Martin, W. Keble
The Concise British Flora in Colour

fl. fr. hab.   Pl. 7

Cardamine pratensis

Polunin, Oleg
Flowers of Europe

fl. Pl. 34 #310

Cardamine pulcherrima

Clark, Lewis J.
Wild Flowers of British Columbia

fl.   p 183

Cardamine purpurea

Alaska-Yukon Wild Flowers Guide

fl. hab.   p 54

Cardamine purpurea

Heller, Christine
Wild Flowers of Alaska

fl.   Pl. 233

Cardamine resedifolia

Barneby, T. P.
European Alpine Flowers in Colour

fl. hab.   Pl. 30, 5

Cardaminopsis arenosa

Polunin, Oleg
Flowers of Europe

fl. Pl. 35 #314

Cardaminopsis petraea

Martin, W. Keble
The Concise British Flora in Colour

fl. fr. hab.   Pl. 7

Cardamom

see

Elettaria cardamomum

Cardaria draba

Martin, W. Keble
The Concise British Flora in Colour

fl. fr. hab.   Pl. 10

Cardaria draba

Polunin, Oleg
Flowers of Europe

fl. Pl. 36 #353

Cardinal climber

see

Quamoclit sloteri

Cardinal-Flower

see

Lobelia cardinalis

Cardinal-Flower

see

Lobelia fulgens

Cardinal's Guard

see

Pachystachys coccinea

Cardiocrinum giganteum

Hay, Roy
The Color Dictionary of Flowers & Plants

fl.   p 86, Pl. 682

Cardiospermum corindum

Coyle, Jeanette
A Field Guide to the Common and
Interesting Plants of Baja California

fr.   p 119

Cardiospermum halicacabum

Crockett, James Underwood
Annuals

fl.fr.   P 103

Cardiospermum halicacabum

Dean, Blanche E.
Wildflowers of Alabama and
Adjoining States

fr.   p 101

Cardiospermum halicacabum

Tsukamoto, Yotaro
Coloured Illustrations of Garden
Flowers Vol. 10

fr. Ill.25   p. 8

Cardiospermum halicacabum

Wit, H. C. D. de
Plants of the World;
The Higher Plants, Vol II

fr.   p 59, Pl. 27

Cardoon

see

Cynara cardunculus

Cardopatium corymbosum

Morley, Brian D.
Wild Flowers of the World

fl.,hab.   Pl. 40

Carduncellus rhaponticoides

Morley, Brian D.
Wild Flowers of the World

fl.   Pl. 39

Carduus acanthoides

Courtenay, Booth
Wildflowers & Weeds

fl.   p 109

Carduus acanthoides

Lindman, C. A. M.
Nordens Flora, Vol 9

fl. fr.   Pl. 625

Carduus acanthoides

Martin, W. Keble
The Concise British Flora in Colour

fl. hab.   Pl. 48

Carduus carlinoides

Kohlhaupt, Paula
Fleurs des Alpages
vol 2
Pl.   P 119

Carduus crispus

Ary, S.
The Oxford Book of Wildflowers

fl.   p 150, Pl. 1

Carduus crispus

Lindman, C. A. M.
Nordens Flora, Vol 9

fl. fr.   Pl. 626

Carduus defloratus

Barneby, T. P.
European Alpine Flowers in Colour

fl.   Pl. 90, 2

Carduus defloratus

Felsko, Elsa
A Book of Wild Flowers
2nd ser.
fl.   p 93

Carduus discolor

Dean, Blanche E.
Wildflowers of Alabama and
Adjoining States

fl.   p 181

Carduus marianus

Huxley, Anthony
Garden Annuals and Bulbs

fl.   Pl. 19

Carduus nutans

Ary, S.
The Oxford Book of Wildflowers

fl.   p 150, Pl. 3

Carduus nutans

Barneby, T. P.
European Alpine Flowers in Colour

fl.   Pl. 90, 1

Carduus nutans

Courtenay, Booth
Wildflowers & Weeds

fl.   p 109

Carduus nutans

Kleijn, H.
Beauty of the Wild Plant

fl.   opp. p. 72   Pl. 92

Carduus nutans

Klimas, John E.
Wildflowers of Eastern America

fl.   Pl. 223

Carduus nutans

Lemmon, Robert S.
Wildflowers of North America in
     Full Color

fl.   pg. 215   pl. 341

Carduus nutans

Martin, W. Keble
The Concise British Flora in Colour

fl. hab.   Pl. 48

Carduus nutans

Moyle, John B.
Northland Wild Flowers

fl. hab.   p 189, Pl. 241

Carduus nutans

Polunin, Oleg
Flowers of Europe

fl.   Pl. 152 #1478

Carduus nutans

Welsh Stanley L.
Flowers of the Mountain Country

fl.   p. 27

Carduus nutans

Wharton, Mary E.
A Guide to the Wildflowers & Ferns
     of Kentucky

fl.   p 238   Pl. 1.13

Carduus personatus

Barneby, T. P.
European Alpine Flowers in Colour

fl.   Pl. 90, 3

Carduus personatus

Polunin, Oleg
Flowers of Europe

fl.   Pl. 153 #1479

Carduus tenuiflorus

Ary, S.
The Oxford Book of Wildflowers

fl.   p 150, Pl. 5

Carduus tenuiflorus

Martin, W. Keble
The Concise British Flora in Colour

fl. fr. hab.   Pl. 48

Carex acuta

Lindman, C. A. M.
Nordens Flora, Vol 2

fr. hab.   Pl. 125

Carex acutiformis

Martin, W. Keble
The Concise British Flora in Colour

fl. fr. hab.   Pl. 94

Carex appropinquata

Martin, W. Keble
The Concise British Flora in Colour

fr. hab.   Pl. 92

Carex aquatilis

Martin, W. Keble
The Concise British Flora in Colour

fr. hab.   Pl. 93

Carex aquatilis

Welsh, Stanley L.
Flowers of the Mountain Country

fr.   p. 9

Carex arenaria

Lindman, C. A. M.
Nordens Flora, Vol 2

fr. hab.   Pl. 123

Carex arenaria

Martin, W. Keble
The Concise British Flora in Colour

fr. hab.   Pl. 92

Carex atrata

Lindman, C. A. M.
Nordens Flora, Vol 2

fr. hab.   Pl. 126B

Carex atrata

Martin, W. Keble
The Concise British Flora in Colour

fr. hab.   Pl. 93

Carex atrata

Morley, Brian D.
Wild Flowers of the World

fl. fr.   Pl. 7M

Carex atrata

Polunin, Oleg
Flowers of Europe

fr. Pl. 186 #1869

Carex atrofusca

Lindman, C. A. M.
Nordens Flora, Vol 2

fr. hab.   Pl. 126A

Carex atrofusca

Martin, W. Keble
The Concise British Flora in Colour

fr. hab.   Pl. 93

Carex atrosquama

Porsild, A. E.
Rocky Mountain Wild Flowers

fl.   p 61

Carex aurea

Walcott, Mary Vaux
North American Wild Flowers
     vol. 4

fr., hab.   Pl. 281

Carex baldensis

Kohlhaupt, Paula
Fleurs des Alpages
                    vol 2
fl.   P 20

Carex bigelowii

Martin, W. Keble
The Concise British Flora in Colour

fr. hab.   Pl. 93

Carex binervis

Martin, W. Keble
The Concise British Flora in Colour

fl. fr. hab.   Pl. 94

Carex buxbaumii

Martin, W. Keble
The Concise British Flora in Colour

fr. hab.   Pl. 93

Carex buxbaumii

Webster, Mary
Flora of Moray, Nairn & East
Inverness

fl.   p. 292   Pl. 19

Carex caespitosa

Lindman, C. A. M.
Nordens Flora, Vol 2

fr. hab.   Pl. 124A

Carex canescens

Lindman, C. A. M.
Nordens Flora, Vol 2

fr. hab.   Pl. 121A

Carex capillaris

Martin, W. Keble
The Concise British Flora in Colour

fr. hab.   Pl. 93

Carex caryophyllea

Martin, W. Keble
The Concise British Flora in Colour

fr. hab.   Pl. 93

Carex concinna

Porsild, A. E.
Rocky Mountain Wild Flowers

fl. hab.   p 63

Carex contigua

Lindman, C. A. M.
Nordens Flora, Vol 2

fr. hab.   Pl. 122A

Carex crinita

Jennings, O.E.
Wild Flowers of Western Pennsylvania
vol II
fl. fr.   Pl. 2

Carex crinita

Wharton, Mary E.
A Guide to the Wildflowers & Ferns
of Kentucky

fl. p 86 Pl. 2.7

Carex curta

Martin, W. Keble
The Concise British Flora in Colour

fr. hab.   Pl. 92

Carex demissa

Martin, W. Keble
The Concise British Flora in Colour

fl. fr. hab.   Pl. 94

Carex diandra

Martin, W. Keble
The Concise British Flora in Colour

fr. hab.   Pl. 92

Carex digitata

Martin, W. Keble
The Concise British Flora in Colour

fr. hab.   Pl. 93

Carex dioica

Lindman, C. A. M.
Nordens Flora, Vol 2

fr. hab.   Pl. 120

Carex dioica

Martin, W. Keble
The Concise British Flora in Colour

fr. hab.   Pl. 92

Carex distans

Martin, W. Keble
The Concise British Flora in Colour

fr. hab.   Pl. 94

Carex disticha

Lindman, C. A. M.
Nordens Flora, Vol 2

fr. hab.   Pl. 122C

Carex disticha

Martin, W. Keble
The Concise British Flora in Colour

fr. hab.   Pl. 92

Carex divisa

Martin, W. Keble
The Concise British Flora in Colour

fr. hab.   Pl. 92

Carex divulsa

Martin, W. Keble
The Concise British Flora in Colour

fr. hab.   Pl. 92

Carex echinata

Martin, W. Keble
The Concise British Flora in Colour

fr. hab.   Pl. 92

Carex elata

Martin, W. Keble
The Concise British Flora in Colour

fl. fr. hab.   Pl. 93

Carex elongata

Martin, W. Keble
The Concise British Flora in Colour

fr. hab.   Pl. 92

Carex ericetorum

Martin, W. Keble
The Concise British Flora in Colour

fr. hab.   Pl. 93

Carex extensa

Martin, W. Keble
The Concise British Flora in Colour

fr. hab.   Pl. 94

Carex extensa

Polunin, Oleg
Flowers of Europe

fr.   Pl. 185 #1851

Carex fascicularis

Cochrane, G. R.
Flowers and Plants of Victoria

fl.   Pl. 263

Carex festivella

Porsild, A. E.
Rocky Mountain Wild Flowers

fl.   p 65

Carex flacca

Martin, W. Keble
The Concise British Flora in Colour

fr. hab.   Pl. 93

Carex flacca

Polunin, Oleg
Flowers of Europe

fr.   Pl. 186 #1859

Carex flava

Lindman, C. A. M.
Nordens Flora, Vol 2

fr. hab.   Pl. 128A

Carex flava

Martin, W. Keble
The Concise British Flora in Colour

fr.   Pl. 94

Carex gaudichaudiana

Cochrane, G. R.
Flowers and Plants of Victoria

fl.   Pl. 260

Carex glareosa

Lindman, C. A. M.
Nordens Flora, Vol 2

fr.   hab.   Pl. 121B

Carex grayi

Huxley, Anthony
Garden Perennials and Water Plants

fr.   Pl. 317

Carex grayi

Jennings, O. E.
Wild Flowers of Western Pennsylvania
                                vol II
fl. fr.   Pl. 2

Carex hirta

Martin, W. Keble
The Concise British Flora in Colour

fr. hab.   Pl. 94

Carex hirta

Polunin, Oleg
Flowers of Europe

fr. Pl. 186 #1860

Carex humilis

Martin, W. Keble
The Concise British Flora in Colour

fr. hab.   Pl. 93

Carex laevigata

Martin, W. Keble
The Concise British Flora in Colour

fl. fr. hab.   Pl. 94

Carex lasiocarpa

Martin, W. Keble
The Concise British Flora in Colour

fr. hab.   Pl. 94

Carex lepidocarpa

Martin, W. Keble
The Concise British Flora in Colour

fl. fr. hab.   Pl. 94

Carex leporina

Lindman, C. A. M.
Nordens Flora, Vol 2

fr.   Pl. 122B

Carex limosa

Lindman, C. A. M.
Nordens Flora, Vol 2

fr. hab.   Pl. 127A

Carex limosa

Martin, W. Keble
The Concise British Flora in Colour

fr. hab.   Pl. 93

Carex lupuliformis

Duncan, Wilbur H.
Wildflowers of the Southeastern
    United States

fl.   p 237

Carex lupulina

Wharton, Mary E.
A Guide to the Wildflowers & Ferns
    of Kentucky

fl.   p 86 Pl. 2.8

Carex maritima

Martin, W. Keble
The Concise British Flora in Colour

fr. hab.   Pl. 92

Carex media

Porsild, A. E.
Rocky Mountain Wild Flowers

fl.   p 65

Carex microglochin

Martin, W. Keble
The Concise British Flora in Colour

fr. hab.   Pl. 92

Carex microglochin

Porsild, A. E.
Rocky Mountain Wild Flowers

fr. hab.   p 67

Carex montana

Martin, W. Keble
The Concise British Flora in Colour

fr. hab.   Pl. 93

Carex muricata

Martin, W. Keble
The Concise British Flora in Colour

fr. hab.   Pl. 92

Carex nardina

Morley, Brian D.
Wild Flowers of the World

fl. fr.   Pl. 7F

Carex nardina var.

Porsild, A. E.
Rocky Mountain Wild Flowers

fr. hab.   p. 69

Carex nigra

Lindman, C. A. M.
Nordens Flora, Vol 2

fr. hab.   Pl. 124B

Carex nigra

Martin, W. Keble
The Concise British Flora in Colour

fl. fr. hab.   Pl. 93

Carex nigra

Polunin, Oleg
Flowers of Europe

fr. Pl. 186 #1861

Carex norvegica

Martin, W. Keble
The Concise British Flora in Colour

fr. hab.   Pl. 93

Carex obtusata

Lindman, C. A. M.
Nordens Flora, Vol 2

fr. hab.   Pl. 131A

Carex obtusata

Porsild, A. E.
Rocky Mountain Wild Flowers

fr. hab.   p 71

Carex otrubae

Martin, W. Keble
The Concise British Flora in Colour

fr. hab.   Pl. 92

Carex ovalis

Martin, W. Keble
The Concise British Flora in Colour

fr. hab.   Pl. 92

Carex ovalis

Polunin, Oleg
Flowers of Europe

fr. Pl. 186 #1867

Carex pallescens

Lindman, C. A. M.
Nordens Flora, Vol 2

fr. hab.   Pl. 128B

Carex pallescens

Martin, W. Keble
The Concise British Flora in Colour

fl. fr. hab.   Pl. 93

Carex panicea

Lindman, C. A. M.
Nordens Flora, Vol 2

fr.   Pl. 127B

Carex panicea

Martin, W. Keble
The Concise British Flora in Colour

fr. hab.   Pl. 93

Carex paniculata

Martin, W. Keble
The Concise British Flora in Colour

fr. hab.   Pl. 92

Carex pauciflora

Martin, W. Keble
The Concise British Flora in Colour

fr. hab.   Pl. 92

Carex paupercula

Martin, W. Keble
The Concise British Flora in Colour

fl. fr. hab.   Pl. 93

Carex pendula

Martin, W. Keble
The Concise British Flora in Colour

fl. fr. hab.   Pl. 94

Carex pendula

Polunin, Oleg
Flowers of Europe

fr. Pl. 186 #1854

Carex petricosa

Porsild, A. E.
Rocky Mountain Wild Flowers

fl. hab.   p 73

Carex pilulifera

Martin, W. Keble
The Concise British Flora in Colour

fr. hab.   Pl. 93

Carex praticola

Porsild, A. E.
Rocky Mountain Wild Flowers

fl.   p 73

Carex pseudocyperus

Martin, W. Keble
The Concise British Flora in Colour

fr. hab.   Pl.. 94

Carex pseudocyperus

Polunin, Oleg
Flowers of Europe

fr.   P. 185 #1853

Carex pulicaris

Lindman, C. A. M.
Nordens Flora, Vol 2

fr. hab.   Pl. 131B

Carex pulicaris

Martin, W. Keble
The Concise British Flora in Colour

fr. hab.   Pl. 92

Carex pyrenaica

Porsild, A. E.
Rocky Mountain Wildflowers

fl.   p. 75

Carex remota

Martin, W. Keble
The Concise British Flora in Colour

fr. hab.   Pl. 92

Carex riparia

Martin, W. Keble
The Concise British Flora in Colour

fl. fr. hab.   Pl. 94

Carex riparia

Morley, Brian D.
Wild Flowers of the World

fl.,fr.   Pl. 25

Carex riparia

Polunin, Oleg
Flowers of Europe

fl.   Pl. 186 #1858

Carex rostrata

Lindman, C. A. M.
Nordens Flora, Vol 2

fr. hab.   Pl. 130

Carex rostrata

Martin, W. Keble
The Concise British Flora in Colour

fl. fr. hab.   Pl. 94

Carex rostrata

Polunin, Oleg
Flowers of Europe

fr. Pl. 186 #1857

Carex rupestris

Martin, W. Keble
The Concise British Flora in Colour

fr. hab.   Pl. 92

Carex rupestris

Morley, Brian D.
Wild Flowers of the World

fl. fr.   Pl. 7H

Carex saxatilis

Martin, W. Keble
The Concise British Flora in Colour

fl. fr. hab.   Pl. 94

Carex scirpoidea

Porsild, A. E.
Rocky Mountain Wild Flowers

fl. hab.   p 77

Carex sp.

Tarver, David P.
Aquatic and Wetland Plants
   of Florida

fr. hab.   p 80

Carex spicata

Martin, W. Keble
The Concise British Flora in Colour

fr.   Pl. 92

Carex stenolepis

Martin, W. Keble
The Concise British Flora in Colour

fr. hab.   Pl. 94

Carex stricta

Royal Hort. Soc.
The Garden
   vol 101, No. 9   1976

hab.   p. 450

Carex strigosa

Martin, W. Keble
The Concise British Flora in Colour

fl. fr. hab.   Pl. 93

Carex sylvatica

Martin, W. Keble
The Concise British Flora in Colour

fl. fr. hab.   Pl. 94

Carex vaginata

Martin, W. Keble
The Concise British Flora in Colour

fr. hab.   Pl. 93

Carex vesicaria

Lindman, C. A. M.
Nordens Flora, Vol 2

fr.   Pl. 129

Carex vulpina

Martin, W. Keble
The Concise British Flora in Colour

fr.   Pl. 92

Careya species

Mullins, Barbara
Australian Wildflowers in Colour

fl.   P. 67   Pl. 65

Carica papaya

Bianchini, F.
The Complete Book of Fruits & Vegetables

fr.   p 173

Carica papaya

Hvass, Elsie
Plants That Feed and Serve Us

fl. fr.   p 52, Pl. 97

Carica papaya

Kariyone, Tatsuo
Atlas of Medicinal Plants

fr.   Pl. 81

Carica papaya

Kimura, Koiti
Japanese Medicinal Plants, Vol II

fr.   p 201

Carica papaya

Macoby, Stirling
What Flower is That

fl.   P 73

Carica papaya

Masefield, G. B.
The Oxford Book of Food Plants

fr. fr.   Pl. 115, 1

Carica papaya

Massachusetts Hort. Soc.
Horticulture
   vol 32, No. 11   1954

fr.   cover

Carica papaya

Massachusetts Hort. Soc.
Horticulture
   vol 50, No. 5   1972

fr., hab.   p. 28

Carica papaya

Milne, Lorus
Living Plants of the World

fr.   p 152

Carissa edulis

Lind, E.M.
Some Common Flowering Plants of
   Uganda

fl., fr.   Pl. 7

Carissa grandiflora var.

Crockett, James Underwood
Evergreens

hab. fl. fr.   p. 117

Carissa macrocarpa

Mathias, Mildred E.
Color for the Landscape

fl. fr. hab.   p 63

Carissa Macrocarpa var.

Crockett, James Underwood
Lawns & Ground Covers

fl.   p. 125

Carlina acanthifolia

Goulimis, Constantine N.
Wild Flowers of Greece

fl., hab.   p. 115

Carlina acaulis

Color Treasury of Herbs & Other
   Medicinal Plants

fl.   p 61   Pl. 99

Carlina acaulis

Huxley, Anthony
Garden Perennials and Water Plants

fl.   pl. 65

Carlina acaulis

Kohlhaupt, Paula
Fleurs des Alpages
             vol 2
Fl.   P 120

Carlina acaulis

Polunin, Oleg
Flowers of Europe

fl.   Pl. 151 #1467

Carlina acaulis

Robert, Paul A.
Alpine Flowers

fl., hab.   Pl. 10

Carlina acaulis

Royal Hort. Soc.
Journal of the Royal Hort. Soc.
   vol 90, No. 11   1965

fl.   p. 470   Pl. 218

Carlina acaulis

Royal Hort. Soc.
The Garden
   vol 101, No. 3   1976

fl.   p. 125

Carlina acaulis

Tosco, Uberto
The World of Mountain Flowers

fl.   p 65

Carlina acaulis var.

Barneby, T. P.
European Alpine Flowers in Colour

fl. hab.   Pl. 91;3

Carlina corymbosa

Polunin, Oleg
Flowers of Europe

fl. Pl. 152 #1469

Carlina utzka

Goulimis, Constantine N.
Wild Flowers of Greece

fl.   p 115

Carlina vulgaris

Ary, S.
The Oxford Book of Wildflowers

fl.   p 38, Pl. 5

Carlina vulgaris

Kleijn, H.
Beauty of the Wild Plant

fl.   opp. p. 72   Pl. 90

Carlina vulgaris

Lindman, C. A. M.
Nordens Flora, Vol 9

fl. hab.   Pl. 622

Carlina vulgaris

Martin, W. Keble
The Concise British Flora in Colour

fl. hab.   Pl. 47

Carlina vulgaris

Morley, Brian D.
Wild Flowers of the World

fl.,hab.   Pl. 22

Carlina vulgaris

Webster, Mary
Flora of Moray, Nairn & East
Inverness

fl.   p. 292   Pl. 18

Carludovica palmata

Hvass, Elsie
Plants That Feed and Serve Us

hab.   p 112, Pl. 243

Carmel Creeper

see

Ceanothus griseus var.

Carmichaelia aligera

Harvey, Norman B.
New Zealand Botanical Paintings

fl.   p. 44   Pl. 19

Carmichaelia odorata

Curtis's Botanical Magazine
Vol 160   1937

fl.   No. 9479

Carmichaelia williamsii

Curtis's Botanical Magazine
Vol 166   1949

fl. fr.   70

Carnation

see

Dianthus

Carnation, annual

see

Dianthus caryophyllus

Carnation, Chabaud

see

Dianthus caryophyllus

Carnation, wheat-ear

see

Dianthus caryophyllus var.

Carnauba

see

Copernicia cerifera

Carnegiea gigantea

Backeberg, Curt
Cactus Lexicon

hab.   p 571

Carnegiea gigantea

Lamb, Edgar
Colorful Cacti of the American Deserts

fl. hab.   Pl. 69 - 72

Carnegiea gigantea

Lamb, Edgar
The Pocket Encyclopedia of Cacti
and Succulents in Color

hab.   Pl. 304

Carnegiea gigantea

Lemmon, Robert S.
Wildflowers of North America in
Full Color

fl.   p. 78   Pl. 126

Carnegiea gigantea

Morley, Brian D.
Wild Flowers of the World

fl.   pl. 175

Carnegiea gigantea

Van Laren, A.J.
Cactus

hab.   frontispiece

Carnival bush

see

Ochna serrulata

Carob

see

Ceratonia siliqua

Carpanthea pomeridiana

Flowering Plants of Africa
vol 32   1957-58

fl., hab.   Pl. 1264

Carpanthea pomeridiana

Herre, H.
The Genera of the Mesembryanthemaceae

fl.   fr.   p 101

Carpanthea pomeridiana

Morley, Brian D.
Wild Flowers of the World

fl.,fr.,hab.   Pl. 73

Carpenteria californica

Gault, S. Millar
The Color Dictionary of Shrubs

fl.   Pl. 52

Carpenteria californica

Hay, Roy
The Color Dictionary of Flowers & Plants

fl.   p 187, Pl. 1495

Carpenteria californica

Hellyer, A.G.L.
Shrubs in Color

fl.   p. 23

Carpenteria californica

Mathias, Mildred E.
Color for the Landscape

fl.   p 186

Carpenteria californica

Munz, Philip A.
California Spring Wildflowers

fl.   p 31, Pl. 13

Carpenteria californica

Perry, Frances
Flowers of the World

fl.   p 229

Carpenteria californica

Royal Hort. Soc.
The Garden
vol 100, No. 7   1975

fl.   p. 317

Carpenteria californica

Royal Hort. Soc.
The Garden
vol 102, No. 6   1977

fl.   p. 271

Carpet Grass

see

Axonopus affinis

Carpet weed

see

Mollugo verticillata

Carphephorus bellidifolius

Batson, Wade T.
Wild Flowers in South Carolina

fl.   p 112

Carphephorus corymbosus

Duncan, Wilbur H.
Wildflowers of the Southeastern
   United States

fl.   p 199

Carphephorus odoratissimus

Duncan, Wilbur H.
Wildflowers of the Southeastern
   United States

fl. hab.   p 199

Carphephorus pseudo-liatris

Brown, Clair A.
Wildflowers of Louisiana and
   Adjoining States

fl.   p 194

Carphephorus pseudo-liatris

Dean, Blanche E.
Wildflowers of Alabama and
   Adjoining States

fl.   p 185

Carpinus betulus

Ary, S.
The Oxford Book of Wildflowers

fl.   p 188, Pl. 3

Carpinus betulus

Boom, B. K.
The glory of the tree

fl.   opp. p. 45   pl. 68

Carpinus betulus

Edlin, Herbert
The Illustrated Encyclopedia
   of Trees

fl. fr. hab.   p 143

Carpinus betulus

Huxley, Anthony
Deciduous Garden Trees and Shrubs

fr.   Pl. 29

Carpinus betulus

Lindman, C. A. M.
Nordens Flora, Vol 3

fl. fr.   Pl. 164

Carpinus betulus

Martin, W. Keble
The Concise British Flora in Colour

fl. hab.   Pl. 77

Carpinus betulus

Perrot, Emile
Les Plantes Medicinales

fl. fr. hab.   p 167

Carpinus betulus

Polunin, Oleg
Flowers of Europe

fr. Pl. 5 #42

Carpinus betulus

Polunin, Oleg
Trees and Bushes of Europe

fl. fr. hab.   p 47

Carpinus betulus

Vedel, H.
Arbres et Arbustes

fl. fr. hab.   p 54, No. 47

Carpinus caroliniana

Wharton, Mary E.
Trees & Shrubs of Kentucky

fl.   p 100, Pl. 3.3

Carpinus japonica

Addisonia
   vol 24   1960-64

fl., fr.   p. 43 Pl. 790

Carpinus japonica

Kitamura, Siro
Coloured Illustrations of Trees &
   Shrubs of Japan

fl.   91

Carpinus laxiflora

Kitamura, Siro
Coloured Illustrations of Trees &
   Shrubs of Japan

fl.   93

Carpinus orientalis

Boom, B. K.
The glory of the tree

hab.   opp. p. 45   pl. 67

Carpinus orientalis

Polunin, Oleg
Trees and Bushes of Europe

hab.   p 48

Carpinus tschonoskii

Kitamura, Siro
Coloured Illustrations of Trees &
   Shrubs of Japan

fl.   92

Carpinus turczaninovii

Royal Hort. Soc.
Journal of the Royal Hort. Soc.
   vol 92, No. 3   1967

hab.   p. 116   Pl. 51

Carpobrotus acinaciformis

Kleijn, H.
Beauty of the Wild Plant

fl.   opp. p. 113   Pl. 165

Carpobrotus acinaciformis

Polunin, Oleg
Flowers of Europe

fl.   Pl. 11 # 120

Carpobrotus acinaciformis

Polunin, Oleg
Flowers of the Mediterranean

fl.   No. 12

Carpobrotus acinaciformis

Taylor, A. W.
Wild Flowers of Spain and Portugal

fl.   p. 39

Carpobrotus aequilateralis

Morcombe, M. K.
Australia's Western Wildflowers

fl.   p 89

Carpobrotus chilensis

Macoby, Stirling
What Flower is That

fl.   P 73

Carpobrotus chilensis

Massachusetts Hort. Soc.
Horticulture
   vol 52, No. 8   1974

fl.   p. 37

Carpobrotus chilensis

Menninger, Edwin A.
Flowering Vines of the World

fl.   Pl. 13

Carpobrotus chilensis

Munoz Pizarro, Carlos
Flores Silvestres de Chile

fl.   Pl. 18

Carpobrotus cinaciformis

Polunin, Oleg
Flowers of Europe

fl.   Pl. 11 #120

Carpobrotus deliciosus

Herre, H.
The Genera of the Mesembryanthemaceae

fl.   fr.   p 103

Carpobrotus edulis

Crockett, James Underwood
Lawns & Ground Covers

fl.   p. 126

Carpobrotus edulis

Kleijn, H.
Beauty of the Wild Plant

fl.   opp. p. 113   Pl. 166

Carpobrotus edulis

Morcombe, M. K.
Australia's Western Wildflowers

fl.   p 108

Carpobrotus edulis

Polunin, Oleg
Flowers of the Mediterranean

fl.   No. 13

Carpobrotus glaucescens

Blombery, Alec M.
What Wildflower is That

fl.   p 89, Pl. 211

Carpobrotus quadrifidus

Eliovson, Sima
Namaqualand in Flower

fl.   Pl. 53

Carrierea calycina

Curtis's Botanical Magazine
Vol 166   1949

fl.   53

Carrion flower

see

Smilax herbacea

Carrion flower

see

Stapelia

Carrion tree

See Couroupita guianensis

Carrot

see

Daucus carota

Carrot, Deadly

see

Thapsia garganica

Carrot, Deadly

see

Thapsia villosa

Carrot, Spanish

see

Ammi visnaga

Carrot, Wild

see

Daucus carota

Carrotwood

see

Cupaniopsis anacardioides

Carruanthus sp.

Herre, H.
The Genera of the Mesembryanthemaceae

fl. fr.   p 105

Carruthersia scandens

Morley, Brian D.
Wild Flowers of the World

fl.   Pl. 148

Carthamus arborescens

Polunin, Oleg
Flowers of the Mediterranean

fl.   No. 210

Carthamus arborescens

Taylor, A. W.
Wild Flowers of Spain and Portugal

fl.   p. 79

Carthamus lanatus

Polunin, Oleg
Flowers of Europe

fl.   Pl. 157 #1508

Carthamus lanatus

Polunin, Oleg
Flowers of the Mediterranean

fl.   No. 211

Carthamus tenuis

Feinbrun-Dothan, Naomi
Wild Plants in the Land
of Israel

fl.   p 116

Carthamus tinctorius

Kariyone, Tatsuo
Atlas of Medicinal Plants

fl.   Pl. 28

Carthamus tinctorius

Kimura, Koiti
Japanese Medicinal Plants, Vol I

fl.   p 100

Carthamus tinctorius

Takatori, Jisuke
Color Atlas of Medicinal Plants of Japan

fl.   Fig. 1B

Carthamus tinctorius

Wit, H. C. D. de
Plants of the World;
The Higher Plants, Vol II

fl.   Pl. 116

Carum carvi

Bianchini, F.
The Complete Book of Fruits & Vegetables

fr.   p 101

Carum carvi

Hvass, Elsie
Plants That Feed and Serve Us

fl. fr.   p 67, Pl. 135

Carum carvi

Lindman, C. A. M.
Nordens Flora, Vol 7

fl. fr.   Pl. 424

Carum carvi

Loewenfeld, Claire
The Complete Book of Herbs and Spices

fl. fr. hab.   p 208

Carum carvi

Martin, W. Keble
The Concise British Flora in Colour

fl. fr. hab.   Pl. 37

Carum carvi

Masefield, G. B.
The Oxford Book of Food Plants

fl. fr.   Pl. 139, 1

Carum carvi

Perrot, Emile
Les Plantes Medicinales

fl. fr. hab.   p 78

Carum carvi

Rosengarten, Frederic, Jr.
The Book of Spices

fl. fr.   p 148

Carum verticillatum

Martin, W. Keble
The Concise British Flora in Colour

fl. fr. hab.   Pl. 37

Carya cordiformis

Boom, B. K.
The glory of the tree

hab.   opp. p. 57   pl. 85

Carya cordiformis

Edlin, Herbert
The Illustrated Encyclopedia
of Trees

fr. hab.   p 138

Carya cordiformis

Polunin, Oleg
Trees and Bushes of Europe

hab.   p 42

Carya illinoinensis

Bianchini, F.
The Complete Book of Fruits &
Vegetables

fr.   P. 195

Carya illinoinensis

Edlin, Herbert
The Illustrated Encyclopedia
of Trees

fr.   p. 139

Carya illinoinensis

Masefield, G. B.
The Oxford Book of Food Plants

fl. fr.   Pl. 29, 5

Carya illinoinensis var.

Massachusetts Hort. Soc.
Horticulture
vol 51, No. 9   1973

fr.   p. 43

Carya laciniosa

Polunin, Oleg
Trees and Bushes of Europe

fr.   p 42

Carya ovata

Edlin, Herbert
The Illustrated Encyclopedia
of Trees

fr.   p 139

Carya ovata

Hvass, Elsie
Plants That Feed and Serve Us

fl. fr.   p 21, Pl. 31
125, Pl. 278

Carya ovata

Massachusetts Hort. Soc.
Horticulture
vol 50, No. 11   1972

fr.   p. 32-33

Carya ovata

Wharton, Mary E.
Trees & Shrubs of Kentucky

fl.   p 100, Pl. 3.2

Carya tomentosa

Boom, B. K.
The glory of the tree

fr.   opp. p. 60   pl. 91

Carya tomentosa

Edlin, Herbert
The Illustrated Encyclopedia
of Trees

fr.   p 139

Caryopteris x clandonensis

Curtis's Botanical Magazine
Vol. 166   1949

fl.   75

Caryopteris x clandonensis

Gault, S. Millar
The Color Dictionary of Shrubs

fl.   Pl. 53

Caryopteris x clandonensis

Hay, Roy
The Color Dictionary of Flowers &
Plants

fl.   p. 187   Pl. 1496

Caryopteris x clandonensis

Hellyer, A.G.L.
Shrubs in Colour

fl.   p. 27

Caryopteris x clandonensis

Royal Hort. Soc.
The Garden
vol 100, No. 10   1975

fl.   p. 473

Caryopteris x clandonensis var

Hay, Roy
The Color Dictionary of Flowers &
Plants

fl.   p. 188   Pl. 1497

Caryopteris incana

Huxley, Anthony
Deciduous Garden Trees and Shrubs

fl. Pl. 30

Caryopteris incana

Walden, Beryl M.
Wild Flowers of Hong Kong

fl.   Pl. 63 (191)

Caryopteris incana var.

Harrison, R.E.
Trees and Shrubs

fl. P 47

Caryotophora skiatophytoides

Herre, H.
The Genera of the Mesembryanthemaceae

fl. fr.   p 107

Cashew
see
Anacardium occidentale

Cassava
see
Manihot utilissima

Cassia
see
Cinnamomum cassia

Cassia abbreviata

Palgrave, K. C.
Trees of Central Africa

fl., fr.   p. 95

Cassia abbreviata var.

Codd, L.E.W.
Trees & Shrubs of the Kruger
National Park

fl.   p. 34   Pl. II

Cassia acutifolia

Hvass, Elsie
Plants That Feed and Serve Us

fl. fr.   p 95, Pl. 202

Cassia acutifolia

Perrot, Emile
Les Plantes Medicinales

fl. fr. hab.   p 222

Cassia alata

American Hort. Soc.
Americah Horticulturist
    vol 56, No. 5   1977

fl.   p. 11

Cassia alata

Bruggeman, L.
Tropical Plants

fl.   pl. 231

Cassia alata

Hall, Clarence E.
Flowers of the Islands in the Sun

fl.   p. 123   Pl. 27

Cassia alata

Wit, H. C. D. de
Plants of the World;
    The Higher Plants, Vol I

fl.   p 284, Pl. 176

Cassia angustifolia

Perrot, Emile
Les Plantes Medicinales

fl. fr. hab.   p 222

Cassia, apple blossom

See Cassia javanica

Cassia armata

Lemmon, Robert S.
Wildflowers of North America in
    Full Color

fl.   pg. 71   pl. 114

Cassia armenta

Munz, Philip A.
California Desert Wildflowers

fl.   p. 36   Pl. 28

Cassia artemisioides

Blombery, Alec M.
What Wildflower is That

fl.   p. 90   Pl. 212

Cassia artemisioides

Curtis's Botanical Magazine
Vol 178   1970-72

fl.   599

Cassia artemisioides

Macoby, Stirling
What Flower is That

fl. P 74

Cassia artemisioides

Mathias, Mildred E.
Color for the Landscape

fl. hab.   p 41

Cassia artemisioides

Morley, Brian D.
Wild Flowers of the World

fl.   Plate 130

Cassia brewsteri

Blombery, Alec M.
What Wildflower is That

fl.   p. 90   Pl. 213

Cassia carnaval

Mathias, Mildred E.
Color for the Landscape

fl. hab.   p 39

Cassia chatelainiana

Morcombe, M. K.
Australia's Western Wildflowers

fl.   p 20

Cassia coluteoides

Fleming, Glenn
Wild Flowers of Florida

fl. hab.   p 48

Cassia corymbosa

Hay, Roy
The Dictionary of House Plants

fl.   Pl. 104

Cassia corymbosa

Macoby, Stirling
What Flower is That

fl. P 74

Cassia corymbosa var.

Harrison, R. E.
Trees and Shrubs

fl. hab. P. 46

Cassia covesii

Orr, Robert T.
Wildflowers of Western America

fl.   Pl. 116

Cassia didymobotrya

Lind, E.M.
Some Common Flowering Plants of
    Uganda

fl., fr.   Pl. 1

Cassia emarginata

Coyle, Jeanette
A Field Guide to the Common and
    Interesting Plants of Baja California

fl.   p 95

Cassia fasciculata

Batson, Wade T.
Wild Flowers in South Carolina

fl.   p 61

Cassia fasciculata

Brown, Clair A.
Wildflowers of Louisiana and
    Adjoining States

fl. fr. hab.   p 76

Cassia fasciculata

Courtenay, Booth
Wildflowers & Weeds

fl.   p 52

Cassia fasciculata

Dean, Blanche E.
Wildflowers of Alabama and
    Adjoining States

fl.   p 83

Cassia fasciculata

Duncan, Wilbur H.
Wildflowers of the Southeastern
    United States

fl. fr.   p 67

Cassia fasciculata

Fleming, Glenn
Wild Flowers of Florida

fl.   p 61

Cassia fasciculata

Klimas, John E.
Wildflowers of Eastern America

fl.   Pl. 157

Cassia fasciculata

Moyle, John B.
Northland Wild Flowers

fl. hab.   p 70, Pl. 43

Cassia fasciculata

Weeds of the Southern United States

fl., hab.   p. 29

Cassia fasciculata

Wharton, Mary E.
A Guide to the Wildflowers & Ferns
of Kentucky

fl.   p 211   Pl. 3.2

Cassia, feathery

see

Cassia artemisioides

Cassia fistula

Everett, Thomas H.
Living Trees of the World

fl.   p. 178

Cassia fistula

Hargreaves, Dorothy
Tropical Blossoms of the Caribbean

fl.   p. 45

Cassia fistula

Macoby, Stirling
What Flower is That

fl.   P 74

Cassia fistula

Morley, Brian D.
Wild Flowers of the World

fl.fr.   Plate 112

Cassia fistula

Oakman, Harry
Colorful Trees

Hab. Fl.   P 78

Cassia fistula

Perry, Frances
Flowers of the World

fl.   p 161

Cassia fistula

Pertchik, Bernard
Flowering Trees of the
Caribbean

fl. fr.   p 67

Cassia glutinosa

Blombery, Alec M.
What Wildflower is That

fl.   p. 90   Pl. 214

Cassia hebecarpa

Courtenay, Booth
Wildflowers & Weeds

fl.   p 52

Cassia hebecarpa

Klimas, John E.
Wildflowers of Eastern America

fl.   Pl. 159

Cassia hybrid

Massachusetts Hort. Soc.
Horticulture
vol 46, No. 10   1968

fl., hab.   p. 27

Cassia indecora

Chickering, Carol Rogers
Flowers of Guatemala

fl. fr.   p 45

Cassia javanica

Bruggeman, L.
Tropical Plants

fl.   Pl. 203

Cassia javanica

Hargreaves, Dorothy
Tropical Blossoms of the Caribbean

fl.   p. 46

Cassia javanica

Oakman, Harry
Colorful Trees

Fl. P 50

Cassia javanica

Pertchik, Bernard
Flowering Trees of the
Caribbean

fl.   p 115

Cassia laevigata

O'Gorman, Helen
Mexican Flowering Trees and Plants

fl. fr. hab.   p 51

Cassia lechenaultiana

Wit, H. C. D. de
Plants of the World;
The Higher Plants, Vol I

fl.   p 285, Pl. 178

Cassia leptophylla

Mathias, Mildred E.
Color for the Landscape

fl. hab.   p 40

Cassia, Long-Tail

see

Cassia abbreviata var.

Cassia marginata

Bruggeman, L.
Tropical Plants

fl.   Pl. 204

Cassia marilandica

Duncan, Wilbur H.
Wildflowers of the Southeastern
United States

fl.   p 67

Cassia marilandica

Wharton, Mary E.
A Guide to the Wildflowers & Ferns
of Kentucky

fl. p 211 Pl. 3.1

Cassia mimosoides

Walden, Beryl M.
Wild Flowers of Hong Kong

fl. fr.   Pl. 16(44)

Cassia mimosoides var.

Kimura, Koiti
Japanese Medicinal Plants, Vol II

fl. fr.   p 187

Cassia multijuga

Bruggeman, L.
Tropical Plants

fl.   Pl. 205

Cassia multijuga

Macoby, Stirling
What Flower is That

fl. P 74

Cassia multijuga

Oakman, Harry
Colorful Trees

Hab. Fl.   P 79

Cassia nemophila

Cochrane, G. R.
Flowers and Plants of Victoria

fl.   Pl. 159

Cassia obtusifolia

Duncan, Wilbur H.
Wildflowers of the Southeastern
United States

fl. fr.   p 67

Cassia obtusifolia

Kariyone, Tatsuo
Atlas of Medicinal Plants

fl.   Pl. 98

Cassia obtusifolia

Kimura, Koiti
Japanese Medicinal Plants, Vol II

fl. fr.   p 187

Cassia obtusifolia

Takatori, Jisuke
Color Atlas of Medicinal Plants of Japan

fl.
fr.   Fig. 38A

Cassia obtusifolia

Weeds of the Southern United States

fl., fr., hab.   p. 29

Cassia occidentalis

Duncan, Wilbur H.
Wildflowers of the Southeastern
      United States

fl. fr.   p 67

Cassia odorata

Blombery, Alec M.
What Wildflower is That

fl.   p 90, Pl. 215

Cassia, pink

See Cassia javanica

Cassia polytricha

Moriarty, Audrey
Wild Flowers of Malawi

fl.   Pl. 58; 4

Cassia, purging

See Cassia fistula

Cassia siamea

Oakman, Harry
Colorful Trees

Hab. Fl.   P 100

Cassia, silvery

see

Cassia artemisioides

Cassia singueana

Palgrave, K. C.
Trees of Central Africa

fl., fr.   p. 99.

Cassia stipulacea

Curtis's Botanical Magazine
      vol 172   1958-59

fl., fr.   Pl. 345

Cassia tomentosa

Harrison, R.E.
Trees and Shrubs

fl.   P 46

Cassia torosa

Kimura, Koiti
Japanese Medicinal Plants, Vol I

fl.   p 92

Cassia torosa

Takatori, Jisuke
Color Atlas of Medicinal Plants of Japan

fl.
fr.   Fig. 38B

Cassinia aculeata

Blombery, Alec M.
What Wildflower is That

fl.   p. 90   Pl. 216

Cassinia arcuata

Cochrane, G. R.
Flowers and Plants of Victoria

fl.   Pl. 317

Cassinia denticulata

Blombery, Alec M.
What Wildflower is That

fl.   p 90, Pl. 217

Cassinia, Showy

see

Apalochlamys spectabilis

Cassinia vauvilliersii

Curtis's Botanical Magazine
Vol 177   1969-70

fl.   549

Cassiope hybrid

Gault, S. Millar
The Color Dictionary of Shrubs

fl.   Pl. 54

Cassiope hybrid

Hay, Roy
The Color Dictionary of Flowers &
      Plants

fl.   p. 5   Pl. 28

Cassiope hybrid

Perry, Frances
Flowers of the World

fl.   p. 108

Cassiope hypnoides

Lindman, C. A. M.
Nordens Flora, Vol 7

fl. fr. hab.   Pl. 445B

Cassiope, Lapland

see

Cassiope tetragona

Cassiope lycopodioides

Curtis's Botanical Magazine
Vol 171   1956-57

fl.   298

Cassiope lycopodioides

Hay, Roy
The Color Dictionary of Flowers & Plants

fl.   p 4, Pl. 29

Cassiope lycopodioides

Morley, Brian D.
Wild Flowers of the World

fl.   Pl. 161

Cassiope lycopodioides

Royal Hort. Soc.
Journal of the Royal Hort Soc.
      Vol. 86, No. 7   1961

fl.   p. 310   Pl. 88

Cassiope mertensiana

Clark, Lewis J.
Wild Flowers of British Columbia

fl.   p. 386

Cassiope mertensiana

Fries, Mary A.
Wildflowers of Mount Ranier and
the Cascades

fl.   P 116

Cassiope mertensiana

Munz, Philip A.
California Mountain Wildflowers

fl.   p 45, Pl. 56

Cassiope mertensiana

Orr, Robert T.
Wildflowers of Western America

fl.   Pl. 4

Cassiope Mertensiana

Porsild, A. E.
Rocky Mountain Wild Flowers

fr.   p 307

Cassiope mertensiana

Taylor, Ronald J.
Mountain Wild Flowers

fl. hab.   p 107

Cassiope mertensiana

Walcott, Mary Vaux
North American Wild Flowers

fl. hab.   Pl. 75

Cassiope, Rocky Mountain
         see
Cassiope mertensiana

Cassiope stelleriana

Alaska-Yukon Wild Flower Guide

fl. hab.   p 116

Cassiope stellariana

Heller, Christine
Wild Flowers of Alaska

fl.   Pl. 200

Cassiope tetragona

Alaska-Yukon Wild Flower Guide

fl. hab.   p 115

Cassiope tetragona

Heller, Christine
Wild Flowers of Alaska

fl.   Pl. 199

Cassiope tetragona

Lindman, C. A. M.
Nordens Flora, Vol 7

fl. fr. hab.   Pl. 445A

Cassiope tetragona var.

Porsild, A. E.
Rocky Mountain Wild Flowers

fl.   p 309

Cassiope wardii

Curtis's Botanical Magazine
Vol 168   1951

fl.   151

Cassiope wardii

Royal Hort. Soc.
Journal of the Royal Hort. Soc.
vol 92, No. 2   1967

fl.   p. 86   Pl. 36

Cassytha glabella

Cochrane, G. R.
Flowers and Plants of Victoria

fl. fr.   Pl. 44

Cast-Iron Plant
         see
Aspidistra elatier

Castalia odorata

Walcott, Mary Vaux
North American Wild Flowers

fl. hab.   Pl. 223

Castanea crenata

Kimura, Koiti
Japanese Medicinal Plants, Vol II

fl.   p 149

Castanea crenata

Kitamura, Siro
Coloured Illustrations of Trees &
   Shrubs of Japan

fl.   127

Castanea dentata

Royal Hort. Soc.
Journal of the Royal Hort. Soc.
vol 99, No. 4   1974

hab., fl.   p. 156 Pl. 73

Castanea dentata

Wharton, Mary E.
Trees & Shrubs of Kentucky

fl.   p 104, Pl. 3.10a

Castanea mollissima

Massachusetts Hort. Soc.
Horticulture
vol 39, No. 10   1961

fr.   p. 508

Castanea mollissima

Massachusetts Hort. Soc.
Horticulture
vol 42, No. 9   1964

fr.   p. 16

Castanea pumila

Wharton, Mary E.
Trees & Shrubs of Kentucky

fl.   p 104, Pl. 3.10b
fr.   p 114, Pl. 1.1

Castanea sativa

Bianchini, F.
The Complete Book of Fruits & Vegetables

fl. fr.   p 191

Castanea sativa

Bianchini, Francesco
Health Plants of the World

fr.   p]23

Castanea sativa

Boom, B. K.
The glory of the tree

fl., fr.   p.56 & 57   Pl. 82-83

Castanea sativa

Edlin, Herbert
The Illustrated Encyclopedia
   of Trees

fl. fr. hab.   p 148-9

Castanea sativa

Huxley, Anthony
Deciduous Garden Trees and Shrubs

fl. fr.   Pl. 31, 31a, 31b

Castanea sativa

Hvass, Elsie
Plants That Feed and Serve Us

fl. fr.   p 12, Pl. 14

Castanea sativa

Masefield, G. B.
The Oxford Book of Food Plants

fl. fr.   Pl. 27, 3

Castanea sativa

Polunin, Oleg
Flowers of Europe

fl. Pl. 5 #46

Castanea sativa

Polunin, Oleg
Trees and Bushes of Europe

fl. fr. hab.   p 51

Castanea sativa

Riefel, Carlos von
A Folio of Fruit

fr.   Pl. 10

Castanea sativa

Wit, H. C. D. de
Plants of the World;
   The Higher Plants, Vol I

fr.   p 107, Pl. 60

Castanea sp.

Everett, T.H., ed.
New Illustrated Encyclopedia of
    Gardening vol. 2

fr.   p. 295

Castanea vulgaris

Perrot, Emile
Les Plantes Medicinales

fl. fr. hab.   p 60

Castanopsis cuspidata

Kitamura, Siro
Coloured Illustrations of Trees &
    Shrubs of Japan

fl.   128

Castanopsis indica

Hara, Hiroshi, comp.
Photo-Album of Plants of
    Eastern Himalaya

fl.   Pl. 11

Castanopsis sempervirens

Munz, Philip A.
California Mountain Wildflowers

fl.,fr.   p. 39   Pl. 39

Castanospermum australe

Blombery, Alec M.
What Flower is That

fl.   p. 91   Pl. 218

Castanospermum australe

Macoboy, Stirling
What Flower is That

fl.   P 74

Castanospermum australe

Masefield, G. B.
The Oxford Book of Food Plants

fl. fr.   Pl. 31, 5

Castanospermum australe

Mathias, Mildred E.
Color for the Landscape

fl. hab.   p 15

Castanospermum australe

Mullins, Barbara
Australian Wildflowers in Colour

fl.   p. 21   Pl. 12

Castanospermum australe

Oakman, Harry
Colorful Trees

fl.   P 46

Castilleja californica

Massachusetts Hort. Soc.
Horticulture
vol 49, No. 8   1971

fl.   p. 28

Castilleja caudata

Alaska-Yukon Wild Flower Guide

fl.   p 146

Castilleja chromosa

Munz, Philip A.
California Desert Wildflowers

fl.   p 51, Pl. 74

Castilleja chromosa

Royal Hort. Soc.
Journal of the Royal Hort. Soc.
vol 89, No. 4   1964

fl.   p. 158   Pl. 62

Castilleja chromosa

Welsh, Stanley L.
Flowers of the Canyon Country

fl.   p 26

Castilleja coccinea

Addisonia
vol 23   1954-59

fl., fr.   p. 37   Pl. 755

Castilleja coccinea

Campbell, Carlos C.
Great Smoky Mountain Wildflowers

fl.   p 83

Castilleja coccinea

Courtenay, Booth
Wildflowers & Weeds

fl.   p 100

Castilleja coccinea

Duncan, Wilbur H.
Wildflowers of the Southeastern
    United States

fl.   p 179

Castilleja coccinea

Klimas, John E.
Wildflowers of Eastern America

fl.   Pl. 204

Castilleja coccinea

Lemmon, Robert S.
Wildflowers of North America in
    Full Color

fl.   pg. 200   pl. 314

Castilleja coccinea

Wit, H. C. D. de
Plants of the World;
    The Higher Plants, Vol II

fl.   Pl. 90

Castilleja elegans

Alaska-Yukon Wild Flower Guide

fl. hab.   p 147

Castilleja gracilis

O'Gorman, Helen
Mexican Flowering Trees and Plants

fl. fr. hab.   p 143

Castilleja hispida

Taylor, Ronald J.
Mountain Wild Flowers

fl.   p 110

Castilleja hyetophilia

Heller, Christine
Wild Flowers of Alaska

fl.   Pl. 114

Castilleja indivisa

Brown, Clair A.
Wildflowers of Louisiana and
    Adjoining States

fl.   p 167

Castilleja integra

Welsh, Stanley L.
Flowers of the Mountain Country

fl.   p. 47

Castilleja lancifolia

Walcott, Mary Vaux
North American Wild Flowers

fl. hab.   Pl. 102

Castilleja lemmonii

Munz, Philip A.
California Mountain Wildflowers

fl.   p 38, Pl. 34

Castilleja levisecta

Clark, Lewis J.
Wild Flowers of British Columbia

fl.   p 462

Castilleja miniata

Clark, Lewis J.
Wild Flowers of British Columbia

fl. hab.   p 459, 463

Castilleja miniata

Ferguson, Mary
Wildflowers

fl.   p 138

Castilleja miniata

Munz, Philip A.
California Mountain Wildflowers

fl.   p 37, Pl. 32

Castilleja miniata

Orr, Robert T.
Wildflowers of Western America

fl.   Pl. 173/174

Castilleja miniata

Porsild, A. E.
Rocky Mountain Wild Flowers

fl.   p 345

Castilleja miniata

Walcott, Mary Vaux
North American Wild Flowers
          vol. 5
fl.   Pl. 372

Castilleja occidentalis

Porsild, A. E.
Rocky Mountain Wild Flowers

fl.   p 347

Castilleja occidentalis

Welsh, Stanley L.
Flowers of the Mountain Country

fl.   p. 49

Castilleja pallida

Heller, Christine
Wild Flowers of Alaska

fl.   Pl. 115a-b

Castilleja pallida

Walcott, Mary Vaux
North American Wild Flowers

fl.   Pl. 48

Castilleja parviflora

Alaska-Yukon Wild Flower Guide

fl.   p 148

Castilleja parviflora

Clark, Lewis J.
Wild Flowers of British Columbia

fl.   p 470

Castilleja parviflora

Taylor, Ronald J.
Mountain Wild Flowers

fl.   p 110

Castilleja parviflora var.

Fries, Mary A.
Wildflowers of Mount Ranier and
the Cascades

fl.   P 136-37

Castilleja pulchella

Lemmon, Robert S.
Wildflowers of North America in
     Full Color

fl.   p. 153   pl. 244

Castilleja rhexifolia

Clark, Lewis J.
Wild Flowers of British Columbia

fl.   p 470

Castilleja rhexifolia

Shaw, Richard J.
Field Guide to the Vascular Plants of
     Grand Teton National Park

fl.   Pl. 5

Castilleja rhexifolia

Welsh, Stanley L.
Flowers of the Mountain Country

fl.   p. 27

Castilleja rupicola

Taylor, Ronald J.
Mountain Wild Flowers

fl.   p 110

Castilleja scabrida

Welsh, Stanley L.
Flowers of the Canyon Country

fl. hab.   p 23

Castilleja sessiliflora

Courtenay, Booth
Wildflowers & Weeds

fl.   p 100

Castilleja sp.

Coyle, Jeanette
A Field Guide to the Common and
     Interesting Plants of Baja California

fl.   p 161

Castilleja species

Fries, Mary A.
Wildflowers of Mount Ranier and
     the Cascades

hab.   P. 132-133

Castilleja sp.

Heller, Christine
Wild Flowers of Alaska

fl.   Pl. 33-34

Castilleja sp.

Royal Hort. Soc.
Journal of the Royal Hort. Soc.
     vol 89, No. 4   1964

fl.   p. 158   Pl. 63

Castilleja sulphurea

Clark, Lewis J.
Wild Flowers of British Columbia

fl.   p 463

Castilleja tenuifolia

Chickering, Carol Rogers
Flowers of Guatemala

fl. fr.   p 47

Castilleja unalaschcensis

Alaska-Yukon Wild Flower Guide

fl.   p 144, 145

Castor Bean

see

Ricinus communis

Castor Oil

see

Ricinus communis

Casuarina decaisneana

Blombery, Alec M.
What Wildflower is That

hab.   p. 91   Pl. 219

Casuarina distyla

Blombery, Alec M.
What Wildflower is That

hab.   p 91, Pl. 220

Casuarina distyla

Perry, Frances
Flowers of the World

fl.   p 73

Casuarina equisetifolia

Crockett, James Underwood
Evergreens

hab. fr.   p. 117

Casuarina equisetifolia

Tarver, David P.
Aquatic and Wetland Plants
of Florida

fl. fr. hab.   p 71

Casuarina paludosa

Cochrane, G. R.
Flowers and Plants of Victoria

fl. fr.   Pl. 41

Casuarina sp.

Edlin, Herbert
The Illustrated Encyclopedia
of Trees

hab.   p 243

Casuarina stricta

Cochrane, G. R.
Flowers and Plants of Victoria

hab.   Pl. 199

Catbrier

see

Smilax glauca

Catclaw

see

Acacia greggii

Catfoot

see

Gnaphalium obtusifolium

Catmint

see

Nepeta faassenii

Catmint, Mauve

see

Nepeta mussinii

Catnip

see

Nepeta cataria

Cat-tail

see

Typha

Cattail, Narrow-leaved

see

Typha angustifolia

Cattail, Red Hot

see

Acalypha hispida

Cat

see Also

Cats

Catabrosa aquatica

Lindman, C. A. M.
Nordens Flora, Vol 2

fr. hab.   Pl. 84

Catalpa bignonioides

Batson, Wade T.
Wild Flowers in South Carolina

fl.   p 104

Catalpa bignonioides

Bianchini, Francesco
Health Plants of the World

fl. fr.   p 101

Catalpa bignonioides

Boom, B. K.
The glory of the tree

fr. fl. opp. p. 105   pl. 182-83

Catalpa bignonioides

Edlin, Herbert
The Illustrated Encyclopedia
of Trees

fl. fr. hab.   p 197

Catalpa bignonioides

Hay, Roy
The Color Dictionary of Flowers & Plants

fl.   p 188, Pl. 1498

Catalpa bignonioides

Huxley, Anthony
Deciduous Garden Trees and Shrubs

fl.   Pl. 32

Catalpa bignonioides

Macoby, Stirling
What Flower is That

fl.   P 75

Catalpa bignonioides

Morley, Brian D.
Wild Flowers of the World

fl. fr.   Pl. 162

Catalpa bignonioides

Oakman, Harry
Colorful Trees

Hab. Fl.   P 71

Catalpa bignonioides

Perry, Frances
Flowers of the World

fl.   p 45

Catalpa bignonioides

Polunin, Oleg
Flowers of Europe

fl.   Pl. 129 #1265

Catalpa bignonioides

Polunin, Oleg
Trees and Bushes of Europe

fl. fr. hab.   p 173

Catalpa fargesii var.

Curtis's Botanical Magazine
Vol 159   1936

fl.   No. 9458

Catalpa ovata

American Hort. Soc.
American Horticulturist
vol 54, No. 2 1975

fr., hab.   p. 18

Catalpa ovata

Kariyone, Tatsuo
Atlas of Medicinal Plants

fl., fr.   Pl. 38

Catalpa ovata

Kimura, Koiti
Japanese Medicinal Plants, Vol I

fl.   p 90

Catalpa ovata

Takatori, Jisuke
Color Atlas of Medicinal Plants of Japan

fl.
fr.   Fig. 10

Catalpa speciosa

Brown, Clair A.
Wildflowers of Louisiana and
Adjoining States

fl. hab.   p 172

Catalpa speciosa

Wharton, Mary E.
Trees & Shrubs of Kentucky

fl.   p 66, Pl. 1.30

Catananche caerulea

Hay, Roy
The Color Dictionary of Flowers &
    Plants

fl.   p. 127   Pl. 1016

Catananche caerulea

Macoby, Stirling
What Flower is That

fl. P 75

Catananche caerulea

Perry, Frances
Flowers of the World

fl.   p. 85

Catananche caerulea var.

Crockett, James Underwood
Annuals

fl. P 103

Catapodium rigidum

Polunin, Oleg
Flowers of Europe

fr Pl. 180 #1759

Catasetum appendiculatum

Pabst, G. F. J.
Orchidaceae Brasilienses,
   Vol 1

fl.   p 237

Catasetum atratum

Kupper, Walter
Orchids

fl.   p 59

Catasetum atratum

Pabst, G. F. J.
Orchidaceae Brasilienses,
   Vol 1

fl.   p 236

Catasetum barbatum

Mee, Margaret
Flowers of the Brazilian Forests

fl., hab. Pl. 15

Catasetum barbatum

Pabst, G. F. J.
Orchidaceae Brasilienses,
   Vol 1

fl.   p 237

Catasetum brichtae

Pabst, G. F. J.
Orchidaceae Brasilienses,
   Vol 1

fl.   p 235

Catasetum callosum

Mee, Margaret
Flowers of the Brazilian Forests

fl., hab. Pl. 14

Catasetum callosum

Pabst, G. F. J.
Orchidaceae Brasilienses,
   Vol 1

fl.   p 236,
      237

Catasetum cernum

Pabst, G. F. J.
Orchidaceae Brasilienses,
   Vol 1

fl.   p 233

Catasetum ciliatum

Pabst, G. F. J.
Orchidaceae Brasilienses,
   Vol 1

fl.   p 230

Catasetum cirrhaeoides

Pabst, G. F. J.
Orchidaceae Brasilienses,
   Vol 1

fl.   p 235

Catasetum cristatum

Pabst, G. F. J.
Orchidaceae Brasilienses,
   Vol 1

fl.   p 237,
      238

Catasetum discolor var.

Pabst, G. F. J.
Orchidaceae Brasilienses,
   Vol 1

fl.   p 230

Catasetum ferox

Pabst, G. F. J.
Orchidaceae Brasilienses,
   Vol 1

fl.   p 235

Catasetum fimbriatum

Ebel, Friedrich
The Strange and Beautiful
   World of Orchids

fl.   p 67

Catasetum fimbriatum

Kupper, Walter
Orchids

fl.   p 61

Catasetum fimbriatum

Pabst, G. F. J.
Orchidaceae Brasilienses,
   Vol 1

fl.   p 236

Catasetum globiflorum

Pabst, G. F. J.
Orchidaceae Brasilienses,
   Vol 1

fl.   p 234

Catasetum gnomus

Kupper, Walter
Orchids

fl.   p 63

Catasetum gnomus

Pabst, G. F. J.
Orchidaceae Brasilienses,
   Vol 1

fl.   p 231

Catasetum hookerii

Pabst, G. F. J.
Orchidaceae Brasilienses,
   Vol 2

fl.   p 213

Catasetum x issanensis

Pabst, G. F. J.
Orchidaceae Brasilienses,
   Vol. 1

fl.   p. 238

Catasetum juruense

Pabst, G. F. J.
Orchidaceae Brasilienses,
   Vol 1

fl.   p 237

Catasetum luridum

Pabst, G. F. J.
Orchidaceae Brasilienses,
   Vol 1

fl.   p 234

Catasetum macrocarpum

Pabst, G. F. J.
Orchidaceae Brasilienses,
   Vol 1

fl.   p 231

Catasetum mattosianum

Pabst, G. F. J.
Orchidaceae Brasilienses,
   Vol 1

fl.   p 235

Catasetum meeae

Mee, Margaret
Flowers of the Brazilian Forests

fl., hab. Pl. 16

Catasetum meeae

Pabst, G. F. J.
Orchidaceae Brasilienses,
Vol 1

fl.   p 237

Catasetum pileatum

Dunsterville, G. C. K.
Introduction to the World
of Orchids

fl.   Pl. 2

Catasetum pileatum

Kramer, Jack
Orchids; Flowers of Romance
and Mystery

fl.   p 171

Catasetum pileatum

Ospina, Mariano
Orquideas de las americas

fl.   Pl. 144

Catasetum pileatum

Pabst, G. F. J.
Orchidaceae Brasilienses,
Vol 1

fl.   p 231

Catasetum pileatum var.

Pabst, G. F. J.
Orchidaceae Brasilienses,
Vol 1

fl.   p 232

Catasetum punctatum

Pabst, G. F. J.
Orchidaceae Brasilienses,
Vol 1

fl.   p 234

Catasetum richteri

Pabst, G. F. J.
Orchidaceae Brasilienses,
Vol 1

fl.   p 235

Catasetum rooseveltianum

Pabst, G. F. J.
Orchidaceae Brasilienses,
Vol 1

fl.   p 236

Catasetum roseum

Pabst, G. F. J.
Orchidaceae Brasilienses,
Vol 1

fl.   p 230

Catasetum saccatum

Pabst, G. F. J.
Orchidaceae Brasilienses,
Vol 1

fl.   p 232

Catasetum saccatum var.

Mee, Margaret
Flowers of the Brazilian Forests

fl.   frontispiece

Catasetum saccatum var.

Pabst, G. F. J.
Orchidaceae Brasilienses,
Vol 1

fl.   p 233

Catasetum species

Mee, Margaret
Flowers of the Brazilian Forests

fl., hab.   Pl. 16

Catasetum spitzii

Pabst, G. F. J.
Orchidaceae Brasilienses,
Vol 1

fl.   p 233

Catasetum splendens

Dunsterville, G. C. K.
Introduction to the World
of Orchids

fl.   Pl. 3

Catasetum tabulare

Ospina, Mariano
Orquideas de las americas

fl.   Pl. 145

Catasetum triodon

Pabst, G. F. J.
Orchidaceae Brasilienses,
Vol 1

fl.   p 233

Catasetum trulla

Pabst, G. F. J.
Orchidaceae Brasilienses,
Vol 1

fl.   p 234

Catasetum vinaceum

Pabst, G. F. J.
Orchidaceae Brasilienses,
Vol 1

fl.   p 234

Catasetum warczewitzii

Curtis's Botanical Magazine
Vol 163   1940-42

fl.   No. 9619

Catasetum warczewitzii

Pabst, G. F. J.
Orchidaceae Brasilienses
Vol. 1

fl.   p. 230

Catawba

see

Catalpa bignonioides

Catch Fly

see

Drosera

Catchfly

see

Silene

Catchfly, Alpine Red

see

Lychnis alpina

Catchfly, drooping

see

Silene pendula

Catchfly, forked

see

Silene dichotoma

Catchfly, French

see

Silene gallica var.

Catchfly, garden

see

Silene armeria

Catchfly, German

see

Viscaria vulgaris

Catchfly, Nottingham

see

Silene nutans

Catchfly, Red

see

Silene virginica

Catchfly, Red

see

Viscaria vulgaris

Catchfly, Red Alpine

See

Lychnis alpina

Catchfly, Red Alpine

see

Viscaria alpina

Catchfly, Red German

see

Lychnis viscaria

Catchfly, red german

see

Viscaria vulgaris

Catchfly, Round-leaved

see

Silene rotundifolia

Catchfly, sleepy

see

Silene antirrhina

Catchfly, Sweet William

see

Silene maritima

Catchfly, White

see

Silene parryi

Catchweed

see

Galium aparine

Catha edulis

Flowering Plants of Africa
vol 43   1974-76

fl., fr.   Pl. 1685

Catharanthus roseus

Bianchini, Francesco
Health Plants of the World

fl.   p 177

Catharanthus roseus

Encke, Fritz
Zimmerpflanzen

fl.   p. 64

Catharanthus roseus

Macoby, Stirling
What Flower is That

fl.   P 76

Catharanthus roseus

Mathias, Mildred E.
Color for the Landscape

fl. hab.   p 132

Catharanthus roseus

Morley, Brian D.
Wild Flowers of the World

fl.   Pl. 71

Catharanthus roseus

Perry, Frances
Flowers of the World

fl.   p 30

Cathcartia villosa

Hara, Hiroshi, comp.
Photo-Album of Plants of
     Eastern Himalaya

fl.   Pl. 174, 175

Catimbium speciosum

Hall, Clarence E.
Flowers of the Islands in the Sun

fl.   p. 19   Pl. 1

Catophractes alexandri

Flowering Plants of Africa
     vol 27   1948-49

fl., fr.   Pl. 1060

Cat's claw

see

Doxantha unguis-cati

Cat's-claw

see

Fouquiera splendens

Cat's Ear, One Flowered

see

Hypochoeris uniflora

Cat's-foot

see

Antennaria

Cat's Paw

see

Anigozanthos humilis

Cat's paw creeper

see

Doxantha unguis-cati

Cat's Tail

see

Acalypha hispida

Cat's Tail

see

Phleum pratense

Cat's Tail

see

Struthiola leptantha

Cat's Tail, Red-hot

see

Acalypha hispida

Cats

see Also

Cat

Cattleya aclandiae

Fowlie, J. A.
The Brasilian Bifoliate Cattleyas

fl.   p 32, 37

Cattleya aclandiae

Pabst, G. F. J.
Orchidaceae Brasilienses,
     Vol. 1

fl.   p. 192

Cattleya amethystoglossa

Fowlie, J. A.
The Brasilian Bifoliate Cattleyas

fl.   p 85, 90, 94

Cattleya amethystoglossa

Pabst, G. F. J.
Orchidaceae Brasilienses,
   Vol 1

fl.   p 195

Cattleya amethystoglossa var.

Addisonia
   vol 24   1960-64

fl.   p. 25   Pl. 781

Cattleya amethystoglossa var.

Fowlie, J. A.
The Brasilian Bifoliate Cattleyas

fl.   p 90, 94

Cattleya araguaiensis

Pabst, G. F. J.
Orchidaceae Brasilienses,
   Vol 1

fl.   p 199

Cattleya aurantiaca

Chickering, Carol Rogers
Flowers of Guatemala

fl.   p 49

Cattleya aurantiaca

Ebel, Friedrich
The Strange and Beautiful
   World of Orchids

fl.   p 137

Cattleya aurantiaca

Ospina, Mariano
Oruqideas de las americas

fl.   Pl. 53

Cattleya aurantiaca

Wright, Norman Pelham
Orquideas de Mexico

fl.   Pl. 2

Cattleya bicolor

Fowlie, J. A.
The Brasilian Bifoliate Cattleyas

fl.   p 23

Cattleya bicolor

Pabst, G. F. J.
Orchidaceae Brasilienses,
   Vol 1

fl.   p 192

Cattleya bicolor var.

Fowlie, J. A.
The Brasilian Bifoliate Cattleyas

fl.   p 23, 26, 27, 30

Cattleya bicolor var.

Pabst, G. F. J.
Orchidaceae Brasilienses,
   Vol 1

fl.   p 193

Cattleya bowringiana

Curtis's Botanical Magazine
   1964-65   vol 175

fl.   Pl. 451

Cattleya bowringiana

Kramer, Jack
Orchids; Flowers of Romance
   and Mystery

fl.   p 18

Cattleya bowringiana

Perry, Frances
Flowers of the World

fl.   p 209

Cattleya x brasiliensis

Fowlie, J. A.
The Brasilian Bifoliate Cattleyas

fl.   p 64

Cattleya x brasiliensis

Pabst, G. F. J.
Orchidaceae Brasilienses,
   Vol. 1

fl.   p. 203

Cattleya x brymeriana

Pabst, G. F. J.
Orchidaceae Brasilienses,
   Vol. 1

fl.   p. 203

Cattleya citrina

Ebel, Friedrich
The Strange and Beautiful
   World of Orchids

fl. hab.   p 27

Cattleya citrina

Kupper, Walter
Orchids

fl. hab.   p 31

Cattleya citrina

Wright, Norman Pelham
Orquideas de Mexico

fl.   Pl. 3

Cattleya x dolosa

Fowlie, J. A.
The Brasilian Bifoliate Cattleyas

fl.   p. 53, 57, 60

Cattleya x dolosa

Pabst, G. F. J.
Orchidaceae Brasilienses,
   Vol. 1

fl.   p. 205

Cattleya dormaniana

Fowlie, J. A.
The Brasilian Bifoliate Cattleyas

fl.   p 16, 17

Cattleya dormaniana

Pabst, G. F. J.
Orchidaceae Brasilienses,
   Vol 1

fl.   p 195

Cattleya dowiana var.

Kupper, Walter
Orchids

fl.   p 33

Cattleya x duveenii

Fowlie, J. A.
The Brasilian Bifoliate Cattleyas

fl.   p 94

Cattleya x duveenii

Pabst, G. F. J.
Orchidaceae Brasilienses,
   Vol. 2

fl.   p. 212

Cattleya eldorado

Pabst, G. F. J.
Orchidaceae Brasilienses,
   Vol 1

fl.   p 200

Cattleya elongata

Fowlie, J. A.
The Brasilian Bifoliate Cattleyas

fl.   p 85, 87

Cattleya elongata

Pabst, G. F. J.
Orchidaceae Brasilienses,
   Vol 1

fl.   p 193

Cattleya elongata var.

Fowlie, J. A.
The Brasilian Bifoliate Cattleyas

fl.   p 87

Cattleya elongata var.

Pabst, G. F. J.
Orchidaceae Brasilienses,
Vol 1

fl.   p 193

Cattleya forbesii

Fowlie, J. A.
The Brasilian Bifoliate Cattleyas

fl.   p 107, 119

Cattleya forbesii

Pabst, G. F. J.
Orchidaceae Brasilienses,
Vol 1

fl.   p 198

Cattleya granulosa

Fowlie, J. A.
The Brasilian Bifoliate Cattleyas

fl.   p 71, 74

Cattleya granulosa

Pabst, G. F. J.
Orchidaceae Brasilienses,
Vol 1

fl.   p 194

Cattleya granulosa var.

Fowlie, J. A.
The Brasilian Bifoliate Cattleyas

fl.   p 71, 74

Cattleya granulosa var.

Pabst, G. F. J.
Orchidaceae Brasilienses,
Vol 1

fl.   p 194

Cattleya x guatemalensis

Wright, Norman Pelham
Orquideas de Mexico

fl.   Pl. 4

Cattleya guttata

Fowlie, J. A.
The Brasilian Bifoliate Cattleyas

fl.   p 102

Cattleya guttata var.

Fowlie, J. A.
The Brasilian Bifoliate Cattleyas

fl.   p 85, 99,102

Cattleya guttata var.

Pabst, G. F. J.
Orchidaceae Brasilienses,
Vol 1

fl.   p 195, 196

Cattleya harrisoniana

Fowlie, J. A.
The Brasilian Bifoliate Cattleyas

fl. hab.   p 107, 126

Cattleya harrisoniana

Pabst, G. F. J.
Orchidaceae Brasilienses,
Vol 1

fl.   p 198

Cattleya harrisoniana var.

Fowlie, J. A.
The Brasilian Bifoliate Cattleyas

fl.   p 126

Cattleya hybrid

Everett, T. H., ed.
New Illustrated Encyclopedia
of Gardening vol. 3

fl.   p. 358

Cattleya hybrid

Fowlie, J. A.
The Brasilian Bifoliate Cattelyas

fl.   p 37

Cattleya hybrid

Hay, Roy
The Color Dictionary of Flowers &
Plants

fl.   p. 56   Pl. 442

Cattleya hybrid

Hay, Roy
The Dictionary of House Plants

fl.   Pl. 105

Cattleya hybrid

Hersey, Jean
Woman's Day Book of House Plants

fl.   p 81

Cattleya hybrid

Kramer, Jack
Orchids: Flowers of Romance
and Mystery

fl.   p 81,87

Cattleya hybrid

Macoby, Stirling
What Flower is That

fl.   p. 76

Cattleya hybrid

Perry, Frances
Flowers of the World

fl.   p. 205

Cattleya x hybrida

Kiaer, Eigil
Indoor Plants in Colour

fl.   p. 35

Cattleya x hybrida

Pabst, G. F. J.
Orchidaceae Brasilienses,
Vol. 1

fl.   p. 204

Cattleya x inter-guttata

Fowlie, J. A.
The Brasilian Bifoliate Cattleyas

fl.   p 106

Cattleya x inter-leopoldii

Fowlie, J. A.
The Brasilian Bifoliate Cattleyas

fl.   p 106

Cattleya intermedia

Fowlie, J. A.
The Brasilian Bifoliate Cattleyas

fl.   p 107, 110

Cattleya intermedia

Pabst, G. F. J.
Orchidaceae Brasilienses,
Vol 1

fl.   p 198

Cattleya intermedia var.

Fowlie, J. A.
The Brasilian Bifoliate Cattleyas

fl.   p 106, 110

Cattleya intermedia var.

Kramer, Jack
Orchids: Flowers of Romance
and Mystery

fl.   p 82

Cattleya intermedia var.

Pabst, G. F. J.
Orchidaceae Brasilienses,
Vol 1

fl.   p 198

Cattleya x isabellae

Pabst, G. F. J.
Orchidaceae Brasilienses,
Vol. 1

fl.   p. 203

Cattleya x kautskyi

Fowlie, J. A.
The Brasilian Bifoliate Cattleyas

fl.   p 122

Cattleya x kautskyi

Pabst, G. F. J.
Orchidaceae Brasilienses,
Vol. 1

fl.   p. 205

Cattleya x Krameriana

Fowlie, J. A.
The Brasilian Bifoliate Cattleyas

fl.   p 106

Cattleya labiata

Luer, Carlyle A.
The Native Orchids of Florida

fl. hab.   Pl. 1; 9

Cattleya labiata

Nicolaisen, Age
Pocket Encyclopedia of
Indoor Plants

fl., hab.   Pl. 41

Cattleya labiata

Pabst, G. F. J.
Orchidaceae Brasilienses,
Vol 1

fl.   p 200, 201

Cattleya labiata var.

Ospina, Mariano
Orquideas de las americas

fl.   Pl. 54

Cattleya lawrenciana

Pabst, G. F. J.
Orchidaceae Brasilienses,
Vol 1

fl.   p 202

Cattleya x leczar

Pabst, G. F. J.
Orchidaceae Brasilienses,
Vol. 1

fl.   p. 204

Cattleya leopoldii

Fowlie, J. A.
The Brasilian Bifoliate Cattleyas

fl.   p 98, 99

Cattleya leopoldii var.

Fowlie, J. A.
The Brasilian Bifoliate Cattleyas

fl.   p 85, 94, 98-9

Cattleya leopoldii var.

Pabst, G. F. J.
Orchidaceae Brasilienses,
Vol. 1

fl.   p. 196, 197

Cattleya leopoldii var.

Pabst, G. F. J.
Orchidaceae Brasilienses,
Vol. 2

fl.   p. 211

Cattleya loddigesii

Fowlie, J. A.
The Brasilian Bifoliate Cattleyas

fl.   p 107, 123

Cattleya loddigesii

Pabst, G. F. J.
Orchidaceae Brasilienses,
Vol 1

fl.   p 199

Cattleya loddigesii var.

Fowlie, J. A.
The Brasilian Bifoliate Cattleyas

fl.   p 122

Cattleya lueddemanniana

Dunsterville, G. C. K.
Introduction to the World
of Orchids

fl.   Pl. 4

Cattleya luteola

Pabst, G. F. J.
Orchidaceae Brasilienses,
Vol 1

fl.   p 202

Cattleya nobilior

Pabst, G. F. J.
Orchidaceae Brasilienses,
Vol 1

fl.   p 192

Cattleya nobilior var.

Fowlie, J. A.
The Brasilian Bifoliate Cattleyas

fl.   p 57, 60, 68

Cattleya x o'brieniana

Pabst, G. F. J.
Orchidaceae Brasilienses,
Vol. 1

fl.   p. 205

Cattleya x o'brieniana var.

Pabst, G. F. J.
Orchidaceae Brasilienses,
Vol. 1

fl.   p. 205

Cattleya x patrocinii

Pabst, G. F. J.
Orchidaceae Brasilienses,
Vol. 2

fl.   p. 212

Cattleya x picturata

Pabst, G. F. J.
Orchidaceae Brasilienses,
Vol. 1

fl.   p. 204

Cattleya x picturata

Pabst, G. F. J.
Orchidaceae Brasilienses
Vol. 2

fl.   p. 49

Cattleya porphyroglossa

Fowlie, J. A.
The Brasilian Bifoliate Cattleyas

fl.   p 74, 80

Cattleya porphyroglossa

Pabst, G. F. J.
Orchidaceae Brasilienses,
Vol 1

fl.   p 194

Cattleya x sancheziana

Fowlie, J. A.
The Brasilian Bifoliate Cattleyas

fl.   p 122

Cattleya schilleriana

Fowlie, J. A.
The Brasilian Bifoliate Cattleyas

fl.   p 37

Cattleya schilleriana

Pabst, G. F. J.
Orchidaceae Brasilienses,
Vol 1

fl. fr.   p 197

Cattleya schilleriana var.

Fowlie, J. A.
The Brasilian Bifoliate Cattleyas

fl.   p. 36, 37

Cattleya schilleriana var.

Pabst, G. F. J.
Orchidaceae Brasilienses,
Vol 1

fl.   p 197

Cattleya schofieldiana

Fowlie, J. A.
The Brasilian Bifoliate Cattleyas

fl.   p 74, 75

Cattleya schofieldiana var.

Fowlie, J. A.
The Brasilian Bifoliate Cattleyas

fl.   p 74, 75

Cattleya silvana var.

Pabst, G. F. J.
Orchidaceae Brasilienses,
Vol 2

fl.   p 211

Cattleya skinneri

Kramer, Jack
Orchids; Flowers of Romance
and Mystery

fl.   p 83

Cattleya skinneri

Pacific Hort. Foundation
Pacific Horticulture
vol 38, No. 2   1977

fl., hab.   p. 28

Cattleya skinneri var.

Wright, Norman Pelham
Orquideas de Mexico

fl.   Pl. 5

Cattleya x sororia

Pabst, G. F. J.
Orchidaceae Brasilienses,
Vol. 1

fl.   p. 203

Cattleya trianaei

Ebel, Friedrich
The Strange and Beuatiful
World of Orchids

fl.   p 15

Cattleya velutina

Fowlie, J. A.
The Brasilian Bifoliate Cattleyas

fl.   p 21

Cattleya velutina

Kramer, Jack
Orchids; Flowers of Romance
and Mystery

fl.   p 88

Cattleya velutina

Pabst, G. F. J.
Orchidaceae Brasilienses,
Vol 1

fl.   p 193

Cattleya velutina var.

Fowlie, J. A.
The Brasilian Bifoliate Cattleyas

fl.   p 20, 21

Cattleya x venosa

Fowlie, J. A.
The Brasilian Bifoliate Cattleyas

fl.   p 106

Cattleya x venosa

Pabst, G. F. J.
Orchidaceae Brasilienses,
Vol. 1

fl.   p. 204

Cattleya violacea

Fowlie, J. A.
The Brasilian Bifoliate Cattleyas

fl.   p 48, 60

Cattleya violacea

Ospina, Mariano'
Orquideas de las americas

fl.   Pl. 52

Cattleya violacea

Pabst, G. F. J.
Orchidaceae Brasilienses,
Vol 1

fl.   p 199

Cattleya violacea var.

Fowlie, J. A.
The Brasilian Bifoliate Cattleyas

fl.   p 60

Cattleya walkeriana

Kramer, Jack
Orchids; Flowers of Romance
and Mystery

fl.   p 90

Cattleya walkeriana

Pabst, G. F. J.
Orchidaceae Brasilienses,
Vol 1

fl.   p 192

Cattleya walkeriana var.

Fowlie, J. A.
The Brasilian Bifoliate Cattleyas

fl. hab.   p 56, 57, 60-1

Cattleya x whitei

Fowlie, J. A.
The Brasilian Bifoliate Cattleyas

fl.   p 37

Cattleya x whitei

Pabst, G. F. J.
Orchidaceae Brasilienses,
Vol. 1

fl.   p. 206

x Cattleytonia hybrid

Grubb, Roy
Selected Orchidaceous Plants

fl.   p. 47

Caucalis platycarpos

Martin, W. Keble
The Concise British Flora in Colour

fl. fr. hab.   Pl. 40

Caularthron bicornutum

Ospina, Mariano
Orquideas de las americas

fl.   Pl. 55

Caularthron bicornutum

Pabst, G. F. J,
Orchidaceae Brasilienses,
Vol 1

fl.   p 186

Cauliflower

see

Brassica oleracea var,

Caulophyllum thalictroides

Campbell, Carlos C.
Great Smoky Mountain Wildflowers

fl.   p 19

Caulophyllum thalictroides

Courtenay, Booth
Wildflowers & Weeds

fl.   p 20

Caulophyllum thalictroides

Duncan, Wilbur H.
Wildflowers of the Southeastern
United States

fr.   p 49

Caulophyllum thalictroides

Ferguson, Mary
Wildflowers

fl.   p 157

Caulophyllum thalictroides

Klimas, John E.
Wildflowers of Eastern America

fl.   Pl. 299

Caulophyllum thalictroides

Moyle, John B.
Northland Wild Flowers

fl. fr. hab.   p 56, Pl. 18, 18a

Caulophyllum thalictroides

Wharton, Mary E.
A Guide to the Wildflowers & Ferns
of Kentucky

fl. fr.   p 276 Pl. 2.4

Cautleya robusta

Hay, Roy
The Color Dictionary of Flowers & Plants

fl.   p 128, Pl. 1017

Cavendishia acuminata

Harrison, Richmond E.
Climbers and Trailers

fl.   p 29   Pl. 38

Ceanothus americanus

Brown, Clair A.
Wildflowers of Louisiana and
    Adjoining States

fl. hab.   p 107

Ceanothus americanus

Campbell, Carlos C.
Great Smoky Mountain Wildflowers

fl.   p 53

Ceanothus americanus

Courtenay, Booth
Wildflowers & Weeds

fl.   p 60

Ceanothus americanus

Dean, Blanche E.
Wildflowers of Alabama and
    Adjoining States

fl.   p 103

Ceanothus americanus

Turner, Nancy J.
Wild Coffee and Tea Substitutes of
    Canada

fl.   p. 80

Ceanothus americanus

Wharton, Mary E.
Trees & Shrubs of Kentucky

fl.   p 75, Pl. 2.12

Ceanothus arboreus

Hay, Roy
The Color Dictionary of Flowers & Plants

fl.   p 188, Pl. 1499

Ceanothus burkwoodii

Hellyer, A.G.L.
Shrubs in Colour

fl.   p. 26

Ceanothus x delilianus

Huxley, Anthony
Deciduous Garden Trees and Shrubs

fl.   Pl. 33

Ceanothus x delilianus var.

Harrison, R.E.
Trees and Shrubs

fl.   P 47

Ceanothus dentatus

Gault, S. Millar
The Color Dictionary of Shrubs

fl.   Pl. 56

Ceanothus foliosus

Curtis's Botanical Magazine
     Vol 161   1938

fl.   No. 9540

Ceanothus greggii var.

Coyle, Jeanette
A Field Guide to the Common and
    Interesting Plants of Baja California

fl.   p 121

Ceanothus griseus var.

Crockett, James Underwood
Lawns & Ground Covers

fl.   p. 126

Ceanothus griseus var.

Menninger, Edwin A.
Flowering Vines of the World

fl.   Pl. 176

Ceanothus hybrid

Gault, S. Millar
The Color Dictionary of Shrubs

fl.   Pl. 55, 57, 61

Ceanothus hybrid

Hay, Roy
The Color Dictionary of Flowers & Plants

fl.   p 188, Pl. 1500 thru 1504

Ceanothus hybrid

Hellyer, A.G.L.
Shrubs in Colour

fl.   p. 26

Ceanothus hybrid

Macoby, Stirling
What Flower is That

fl.   P 76

Ceanothus hybrid

Royal Hort. Soc.
Journal of the Royal Hort. Soc.
     vol 89, No. 8   1964

fl.   p. 334   Pl. 120

Ceanothus impressus

Gault, S. Millar
The Color Dictionary of Shrubs

fl.   Pl. 58

Ceanothus impressus

Mathias, Mildred E.
Color for the Landscape

hab.   p 183

Ceanothus impressus

Royal Hort. Soc.
Journal of the Royal Hort. Soc.
     vol 95, No. 2   1970

fl.   p. 22   Pl. 45

Ceanothus impressus

Royal Hort. Soc.
The Garden
     vol 100, No. 11   1975

hab., fl.   p. 526-27.

Ceanothus integerrimus

Munz, Philip A.
California Mountain Wildflowers

fl.   p 42, Pl. 47

Ceanothus integerrimus

Orr, Robert T.
Wildflowers of Western America

fl.   Pl. 31

Ceanothus leucodermis

Mathias, Mildred E.
Color for the Landscape

hab.   p 183

Ceanothus x lobbianus

Royal Hort. Soc.
Journal of the Royal Hort. Soc.
     vol 89, No. 8   1964

fl.   p. 334   Pl. 120

Ceanothus mendocinensis

Gault, S. Millar
The Color Dictionary of Shrubs

fl.   Pl. 59

Ceanothus papillosus var.

Harrison, R. E.
Trees and Shrubs

fl.   p. 47

Ceanothus prostratus

Munz, Philip A.
California Mountain Wildflowers

fl.   p 49, Pl. 68

Ceanothus purpureus

Curtis's Botanical Magazine
Vol 165   1948

fl. fr.   37

Ceanothus sangiuneus

Clark, Lewis J.
Wild Flowers of British Columbia

fl.   p 315

Ceanothus thrysiflorus

Crockett, James Underwood
Evergreens

hab. fl.   p. 118

Ceanothus thrysiflorus

Hay, Roy
The Color Dictionary of Flowers
& Plants

fl.   p. 188

Ceanothus thyrsiflorus

Perry, Frances
Flowers of the World

fl.   p 257

Ceanothus thyrsiflorus var.

Gault, S. Millar
The Color Dictionary of Shrubs

fl.   Pl. 60

Ceanothus thyrsiflorus var.

Royal Hort. Soc.
Journal of the Royal Hort. Soc.
vol 97, No. 12   1972

fl., hab.   p. 528   Pl. 283

Ceanothus velutinus

Clark, Lewis J.
Wild Flowers of British Columbia

fl.   p 315

Ceanothus velutinus

Fries, Mary A.
Wildflowers of Mount Ranier and
the Cascades

fl.   P 81

Ceanothus velutinus

Shaw, Richard J.
Field Guide to the Vascular Plants of
Grand Teton National Park

fl.   Pl. 8

Ceasar Weed

see

Urena lobata

Cecropia sp.

Royal Hort. Soc.
The Garden
Vol. 101, No. 1   1976

hab.   p. 18

Cedar

see

Cedrus

Cedar, Atlantic

see

Cedrus atlantica

Cedar, Atlas

see

Cedrus atlantica

Cedar, California incense

see

Libocedrus decurrens

Cedar, Clanwilliam

see

Widdringtonia juniperoides

Cedar, Cyprus

see

Cedrus brevifolia

Cedar, deodar

see

Cedrus deodara

Cedar, Incense

see

Libocedrus decurrens

Cedar, Indian

see

Cedrus deodara

Cedar, Japanese

see

Cryptomeria japonica

Cedar, Mount Atlas

see

Cedrus atlantica var.

Cedar of Lebanon

see

Cedrus libani

Cedar, Pencil

see

Juniperus virginiana

Cedar, Plume

see

Cryptomeria japonica

Cedar, Port Oxford

see

Chamaecyparis lawsoniana

Cedar, Red cannaert

see

Juniperus virginiana var.

Cedar, Salt

see

Tamarix

Cedar, Seaside

see

Tamarix gallica

Cedar, West Indian

see

Cedrela odorata

Cedar, Western Red

see

Juniperus scopulorum

Cedar, Western Red

see

Thuja plicata

Cedar, white

see

Melia azederach

Cedrela odorata

Hvass, Elsie
Plants That Feed and Serve Us

fl.   p. 121   Pl. 265

Cedrela sinensis

Boom, B. K.
The glory of the tree

hab.   opp. p. 92   pl. 154

Cedrela sinensis

Royal Hort. Soc.
The Garden
        vol 101, No. 4   1976

hab.   p. 217

Cedronella canariensis

Bramwell, David
Wild Flowers of the Canary Islands

fl.   Pl. 241

Cedrus atlantica

Boom, B. K.
The glory of the tree

hab., fr. opp. p. 29   pl. 39-40

Cedrus atlantica

Edlin, Herbert
The Illustrated Encyclopedia
of Trees

fr.   hab.   p 104-5

Cedrus atlantica

Huxley, Anthony
Evergreen Garden Trees and Shrubs

fr.   hab.   Pl. 9

Cedrus atlantica

Perrot, Emile
Les Plantes Medicinales

fr.   hab.   p 54

Cedrus atlantica

Polunin, Oleg
Trees and Bushes of Europe

fr.   hab.   p 10, 11

Cedrus atlantica

Wit, H. C. D. de
Plants of the World;
    The Higher Plants, Vol I

hab.   p 54, Pl. 12

Cedrus atlantica var.

Bartels, Andreas
Das Grosse Buch der Gartengeholze

fr.   p 248

Cedrus atlantica var.

Crockett, James Underwood
Evergreens

hab.   fr.   p. 91

Cedrus atlantica var.

Flemer, William III
Shade and Ornamental Trees in Color

hab.   p. 95

Cedrus atlantica var.

Harrison, R. E.
Trees and Shrubs

hab.   Pl. 566

Cedrus atlantica var.

Hay, Roy
The Color Dictionary of Flowers & Plants

hab.   p 251, Pl. 2003

Cedrus atlantica var.

Hvass, Elsie
Plants That Feed and Serve Us

fr.   p 119, Pl. 259

Cedrus atlantica var.

Macoby, Stirling
What Flower is That

fr.   P 77

Cedrus atlantica var.

Royal Hort. Soc.
Journal of the Royal Hort. Soc.
    vol 98, No. 3   1973

fr.   cover

Cedrus brevifolia

Edlin, Herbert
The Illustrated Encyclopedia
of Trees

hab.   p 105

Cedrus brevifolia

Megaw, Elektra
Wild Flowers of Cyprus

fl., hab.   Pl. 40

Cedrus deodara

Bartels, Andreas
Das Grosse Buch der Gartengeholze

hab.   p 248

Cedrus deodara

Crockett, James Underwood
Evergreens

hab. fr.   p. 91

Cedrus deodara

Edlin, Herbert
The Illustrated Encyclopedia
of Trees

fr. hab.   p 105

Cedrus deodara

Flemer, William III
Shade and Ornamental Trees in Color

hab.   p. 96

Cedrus deodara

Huxley, Anthony
Evergreen Garden Trees and Shrubs

hab.   Pl. 10

Cedrus deodara

Kitamura, Siro
Coloured Illustrations of Trees &
    Shrubs of Japan

fr.   30, 31

Cedrus deodara

Macoby, Stirling
What Flower is That

hab.   P. 77

Cedrus deodara

Pennsylvania Hort. Soc.
The Green Scene
    vol 6, No. 1   1977

hab., fr.   p. 26

Cedrus deodara

Polunin, Oleg
Trees and Bushes of Europe

hab.   p 11

Cedrus libani

Boom, B. K.
The glory of the tree

fr.   opp. p. 29   pl. 39, 41

Cedrus libani

Crockett, James Underwood
Evergreens

hab. fr.   p. 92

Cedrus libani

Edlin, Herbert
The Illustrated Encyclopedia
of Trees

fr. hab.   p 104-5

Cedrus libani

Hvass, Elsie
Plants That Feed and Serve Us

fr.   p 119, Pl. 260

Cedrus libani

Perrot, Emile
Les Plantes Medicinales

fl. fr. hab.   p 54

Cedrus libani

Polunin, Oleg
Trees and Bushes of Europe

hab.   p 11

Cedrus libani

Royal Hort. Soc.
Journal of the Royal Hort. Soc.
   vol 89, No. 3   1964

hab.   p. 114   Pl. 41

Ceiba casearia

Hvass, Elsie
Plants That Feed and Serve Us

fl.   fr.   p 105, Pl. 225

Celandine, greater

see

Chelidonium majus

Celandine, Lesser

see

Ranunculus ficaria

Celastrus orbiculatus

Bartels, Andreas
Das Grosse Buch der Gartengeholze

fr. hab.   p 128

Celastrus orbiculatus

Gault, S. Millar
The Color Dictionary of Shrubs

fr.   Pl. 62

Celastrus orbiculatus

Harrison, Richmond E.
Climbers and Trailers

fr.   p 29   Pl. 39

Celastrus orbiculatus

Hay, Roy
The Color Dictionary of Flowers & Plants

fr.   p 245, Pl. 1959

Celastrus orbiculatus

Hellyer, A.G.L.
Shrubs in Colour

fr.   p. 26

Celastrus orbiculatus

Kitamura, Siro
Coloured Illustrations of Trees &
Shrubs of Japan

fr.   295

Celastrus orbiculatus

Perry, Frances
Flowers of the World

fr.   p 73

Celastrus orbiculatus

Royal Hort. Soc.
Journal of the Royal Hort. Soc.
   vol 85, No. 12   1960

fr.   p. 522   Pl. 159

Celastrus orbiculatus

Royal Hort. Soc.
Journal of the Royal Hort. Soc.
   vol 96, No. 1   1971

fr.   p. 22   Pl. 9

Celastrus scandens

Harrison, Richmond E.
Climbers and Trailers

fr.   p 30   Pl. 40

Celastrus scandens

Huxley, Anthony
Deciduous Garden Trees and Shrubs

fr.   Pl. 283

Celastrus scandens

Massachusetts Hort. Soc.
Horticulture
   vol 50, No. 11   1972

fr.   p. 33

Celastrus scandens

Milne, Lorus
Living Plants of the World

fr.   p 128

Celastrus scandens

Wharton, Mary E.
Trees & Shrubs of Kentucky

fl.   p 73, Pl. 2.9
fr.   p 125, Pl. 2.11

Celeriac

see

Apium graveolens var.

Celery

see

Apium graveolens

Celery, Alpine

see

Aciphylla glacialis

Celestials

see

Herbertia drummondii

Celmisia bellidifolia

Royal Hort. Soc.
The Garden
   vol 100, No. 9   1975

fl.   p. 414

Celmisia coriacea

Hay, Roy
The Color Dictionary of Flowers & Plants

fl.   p 4, Pl. 30

Celmisia coriacea

Royal Hort. Soc.
Journal of the Royal Hort. Soc.
   vol 99, No. 4   1974

fl.   p. 156   Pl. 77

Celmisia hookeri

Morley, Brian D.
Wild Flowers of the World

fl.   pl. 140

Celmisia hookeri

Royal Hort. Soc.
Journal of the Royal Hort. Soc.
   vol 99, No. 4   1974

fl.   p. 156   Pl. 79

Celmisia longifolia

Alpine Flowers of the Kosciusko
   State Park

fl.   Pl. 3

Celmisia longifolia

Blombery, Alec M.
What Wildflower is That

fl. hab.   p. 92   pl. 221

Celmisia longifolia

Cochrane, G. R.
Flowers and Plants of Victoria

fl. hab.   Pl. 498

Celmisia longifolia

Macoby, Stirling
What Flower is That

fl.   P 77

Celmisia nova-zealandiae

Royal Hort. Soc.
Journal of the Royal Hort. Soc.
vol 99, No. 4   1974

fl.   p. 156   pl. 80

Celmisia petiolata

Hay, Roy
The Color Dictionary of Flowers & Plants

fl.   p 4, Pl. 31

Celmisia saxifraga

Curtis, Winifred
The Endemic Flora of Tasmania Vol. 2

fl.   Pl. 55

Celmisia sessiliflora

Royal Hort. Soc.
Journal of the Royal Hort. Soc.
vol 99, No. 4   1974

fl.   p. 156   Pl. 78

Celmisia spectabilis

Hay, Roy
The Color Dictionary of Flowers & Plants

fl.   p 4, Pl. 32

Celmisia spectabilis var.

Royal Hort. Soc.
Journal of the Royal Hort. Soc.
vol 86, No. 7   1961

fl.   p. 310   Pl. 90

Celosia argentea

Everett, T.H., ed.
New Illustrated Encyclopedia of
Gardening   vol 3

fl.   p. 359

Celosia argentea

Takatori, Jisuke
Color Atlas of Medicinal Plants of Japan

fl.   Fig. 58

Celosia argentea var.

Crockett, James Underwood
Annuals

fl.   P 103

Celosia argentea var.

Everett, T.H., ed.
New Illustrated Encyclopedia of
Gardening vol. 3

fl.   p. 359

Celosia argentea var.

Hay, Roy
The Color Dictionary of Flowers & Plants

fl.   p 33, Pl. 257-58

Celosia argentea var.

Hay, Roy
The Dictionary of House Plants

fl.   Pl. 106, 107

Celosia argentea var.

Macoby, Stirling
What Flower is That

fl.   P 77

Celosia cristata

Macoby, Stirling
What Flower is That

fl.   P 77

Celosia cristata

Tsukamoto, Yotaro
Coloured Illustrations of Garden
Flowers Vol. 10

fl. Illus. 26,27   opp. p. 9

Celosia cristata var.

Tsukamoto, Yotaro
Coloured Illustrations of Garden
Flowers Vol. 10

fl. Illus. 28   opp. p. 9

Celosia plumosa var.

Huxley, Anthony
Garden annuals and bulbs

fl.   Pl. 20

Celosia scabra

Flowering Plants of Africa
vol 31   1956

fl., fr.   Pl. 1236

Celsia acaulis

Hay, Roy
The Color Dictionary of Flowers & Plants

fl.   p 5, Pl. 33

Celsia acaulis var.

Royal Hort. Soc.
Journal of the Royal Hort. Soc.
vol 86, No. 7   1961

fl.   p. 310   Pl. 80

Celsia arcturus

Hay, Roy
The Dictionary of House Plants

fl.   Pl. 108

Celsia brevipedicellata

Flowering Plants of Africa
vol 28   1950-51

fl.   Pl. 1105

Celsia sinuata

Curtis's Botanical Magazine
Vol 168   1951

fl.   175

x Celsioverbascum hybrid

Alpine Garden Society
Bulletin
vol. 37, No. 2   1969

fl., hab.   p. 139

x Celsioverbascum hybrid

Royal Hort. Soc.
Journal of the Royal Hort. Soc.
vol 93, No. 9   1968

fl.   p. 382   Pl. 196

Celtis africana

Flowering Plants of Africa
vol 31   1956

fl., fr.   Pl. 1210

Celtis africana

Palmer, Eve
Trees of South Africa

hab.   Pl. III p. 144

Celtis australis

Polunin, Oleg
Trees and Bushes of Europe

fl. fr. hab.   p 63

Celtis sinensis var.

Kitamura, Siro
Coloured Illustrations of Trees &
Shrubs of Japan

fr.   136

Celtis tournefortii

Polunin, Oleg
Trees and Bushes of Europe

fr. hab.   p 64

Cenarrhenes nitida

Curtis, Winifred
The Endemic Flora of Tasmania, Part IV

fl. fr. hab.   Pl. 153

Cenchrus palmeri

Coyle, Jeanette
A Field Guide to the Common and
Interesting Plants of Baja California

fr. hab.   p 47

Cenchrus tribuloides

Duncan, Wilbur H.
Wildflowers of the Southeastern
United States

fl.   p 235

Cenia barbata

Crockett, James Underwood
Annuals

fl.   P 104

Cenia turbinata

Morley, Brian D.
Wild Flowers of the World

fl.   Pl. 83

Centaurea americana

Brown, Clair A.
Wildflowers of Louisiana and
Adjoining States

fl.   p 195

Centaurea americana

Bruggeman, L.
Tropical Plants

fl.   Pl. 34

Centaurea americana

Tsukamoto, Yotaro
Coloured Illustrations of Garden
Flowers Vol. 10

fl. Illus. 29   opp. p. 10

Centaurea arbutifolia

Bramwell, David
Wild Flowers of the Canary Islands

fl.   Pl. 298

Centaurea aspera

Martin, W. Keble
The Concise British Flora in Colour

fl. hab.   Pl. 49

Centaurea bella

Vilmorin, Roger de
Plantes Alpines dans les Jardins

fl. hab.   Pl. V

Centaurea calcitrapa

Martin, W. Keble
The Concise British Flora in Colour

fl. hab.   Pl. 49

Centaurea calcitrapa

Polunin, Oleg
Flowers of Europe

fl.   Pl. 156 #1500

Centaurea chilensis

Munoz Pizarro, Carlos
Flores Silvestres de Chile

fl.   Pl. 10

Centaurea conifera

Polunin, Oleg
Flowers of Europe

fl.   Pl. 157 #1507

Centaurea cyanoides

Feinbrun-Dothan, Naomi
Wild Plants in the Land
of Israel

fl. hab.   p 102

Centaurea cyanus

American Hort. Soc.
American Horticulturist
vol 55, No. 4   1976

fl.   p. 6

Centaurea cyanus

Ary, S.
The Oxford Book of Wildflowers

fl.   p 178, P. 1

Centaurea cyanus

Barneby, T. P.
European Alpine Flowers in Colour

fl.   Pl. 91, 4

Centaurea cyanus

Batson, Wade T.
Wild Flowers in South Carolina

fl.   p 124

Centaurea cyanus

Color Treasury of Herbs & Other
Medicinal Plants

fl.   p. 62   Pl. 100

Centaurea cyanus

Crockett, James Underwood
Annuals

fl.   P 104

Centaurea cyanus

Felsko, Elsa
A Book of Wildflowers

fl.   P 55

Centaurea cyanus

Furrer, D.
Blumen am Wege

fl. hab.   p 113

Centaurea cyanus

Kleijn, H.
Beauty of the Wild Plant

fl.   opp. p. 38   Pl. 39

Centaurea cyanus

Lindman, C. A. M.
Nordens Flora, Vol 10

fl.   Pl. 633

Centaurea cyanus

Macoby, Stirling
What Flower is That

fl.   P 78

Centaurea cyanus

Martin, W. Keble
The Concise British Flora in Colour

fl. hab.   Pl. 49

Centaurea cyanus

Perrot, Emile
Les Plantes Medicinales

fl. hab.   p 34

Centaurea cyanus

Polunin, Oleg
Flowers of Europe

fl.   Pl. 156 #1501

Centaurea cyanus

Tsukamoto, Yotaro
Coloured Illustrations of Garden Flowers

fl.   Pl. 30, 31

Centaurea cyanus

Weeds of the Southern United States

fl.   p. 10

Centaurea cyanus var.

Hay, Roy
The Color Dictionary of Flowers &
Plants

fl.   p. 33   Pl. 259

Centaurea cyanus var.

Huxley, Anthony
Garden Annuals and Bulbs

fl.   Pl. 21-22

Centaurea dealbata

Huxley, Anthony
Garden Perennials and Water Plants

fl.   Pl. 72

Centaurea dealbata

Tsukamoto, Yotaro
Coloured Illustrations of Garden
    Flowers   Vol. 9

fl.   opp. p. 37

Centaurea dealbata var.

Hay, Roy
The Color Dictionary of Flowers & Plants

fl.   p 128, Pl. 1018, 1019

Centaurea ebenoides

Goulimis, Constantine N.
Wild Flowers of Greece

fl. hab.   p 117

Centaurea graeca

Goulimis, Constantine N.
Wild Flowers of Greece

fl. hab.   p 119

Centaurea hypoleuca

Curtis's Botanical Magazine
Vol 167   1950

fl.   95

Centaurea hypoleuca

Morley, Brian D.
Wild Flowers of the World

fl.   Pl. 39

Centaurea hypoleuca

Royal Hort. Soc.
The Garden
    vol 103, No. 3   1978

fl.   p. 113

Centaurea jacea

Felsko, Elsa
A Book of Wildflowers

fl.   P 112

Centaurea jacea

Furrer, D.
Blumen am Wege

fl. hab.   p 115

Centaurea jacea

Lemmon, Robert S.
Wildflowers of North America in
    Full Color

fl.   p. 216   pl. 342

Centaurea jacea

Martin, W. Keble
The Concise British Flora in Colour

fl. hab.   Pl. 49

Centaurea junoniana

Bramwell, David
Wild Flowers of the Canary Islands

fl.   Pl. 297

Centaurea lactucifolia

Goulimis, Constantine N.
Wild Flowers of Greece

fl. hab.   121

Centaurea macrocephala

Addisonia
    vol 20   1937-38

Fl.   p. 23   Pl. 652

Centaurea macrocephala

Everett, T.H., ed.
New Illustrated Encyclopedia of
    Gardening vol 8

fl.   p. 1399

Centaurea macrocephala

Huxley, Anthony
Garden Perennials and Water Plants

fl.   Pl. 74

Centaurea macrocephala

Tsukamoto, Yotaro
Coloured Illustrations of Garden
    Flowers   Vol. 9

fl.   opp. p. 37

Centaurea maculosa

Clark, Lewis J.
Wild Flowers of British Columbia

fl.   p 531

Centaurea maculosa

Courtenay, Booth
Wildflowers & Weeds

fl.   p 107

Centaurea maculosa

Wharton, Mary E.
A Guide to the Wildflowers & Ferns
    of Kentucky

fl.   p 240   Pl. 1.17

Centaurea montana

Barneby, T. P.
European Alpine Flowers in Colour

fl.   Pl. 91, 5

Centaurea montana

Felsko, Elsa
A Book of Wild Flowers
                    2nd ser.
fl.   p 41

Centaurea montana

Kohlhaupt, Paula
Fleurs des Alpages   vol 1

fl.   P 119

Centaurea montana

Massachusetts Hort. Soc.
Horticulture
    vol 43, No. 6   1965

fl.   cover

Centaurea montana var.

Hay, Roy
The Color Dictionary of Flowers & Plants

fl.   p 128, Pl. 1020

Centaurea montana var.

Huxley, Anthony
Garden Perennials and Water Plants

fl.   Pl. 73A

Centaurea montana var.

Royal Hort. Soc.
The Garden
    vol 103, No. 3   1978

fl.   p. 113

Centaurea moschata

Hay, Roy
The Color Dictionary of Flowers & Plants

fl.   p 33, Pl. 260

Centaurea moschata

Huxley, Anthony
Garden Annuals and Bulbs

fl.   Pl. 23

Centaurea moschata

Tsukamoto, Yotaro
Coloured Illustrations of Garden Flowers

fl.   Pl. 32

Centaurea nemoralis

Martin, W. Keble
The Concise British Flora in Colour

fl. hab.   Pl. 49

Centaurea nervosa

Kohlhaupt, Paula
Fleurs des Alpages   vol 1
fl.   P 120

Centaurea nervosa

Polunin, Oleg
Flowers of Europe

fl.   Pl. 156 #1504

Centaurea nervosa

Tosco, Uberto
The World of Mountain Flowers

fl. hab.   p 24

Centaurea nigra

Ary, S.
The Oxford Book of Wildflowers

fl.   p 154, Pl. 1

Centaurea nigra

Lindman, C. A. M.
Nordens Flora, Vol 9

fl.   Pl. 632

Centaurea nigra

Martin, W. Keble
The Concise British Flora in Colour

fl. hab.   Pl. 49

Centaurea phrygia

Polunin, Oleg
Flowers of Europe

fl.   Pl. 156 #1504

Centaurea phrygia .var.

Goulimis, Constantine N.
Wild Flowers of Greece

fl. hab. p 123

Centaurea pulcherrima

Huxley, Anthony
Garden Perennials and Water Plants

fl.   Pl. 75

Centaurea pullata

Polunin, Oleg
Flowers of the Mediterranean

fl.   Pl. 212

Centaurea rhapontica

Barneby, T. P.
European Alpine Flowers in Colour

fl.   Pl. 92, 2

Centaurea rhapontica

Polunin, Oleg
Flowers of Europe

fl.   Pl. 156 #1506

Centaurea rhapontica

Vilmorin, Roger de
Plantes Alpines dans les Jardins

fl. hab.   Pl. XXX

Centaurea salonitana

Goulimis, Constantine N.
Wild Flowers of Greece

fl.   p 125

Centaurea salonitana

Morley, Brian D.
Wild Flowers of the World

fl.   Pl. 39

Centaurea salonitana

Polunin, Oleg
Flowers of Europe

fl.   Pl. 156 #1503

Centaurea scabiosa

American Hort. Soc.
American Horticulturist
    vol 55, No. 4   1976

fl.   p. 6

Centaurea scabiosa

Ary, S.
The Oxford Book of Wildflowers

fl.   p 154, Pl. 3

Centaurea scabiosa

Barneby, T. P.
European Alpine Flowers in Colour

fl.   Pl. 91, 6

Centaurea scabiosa

Courtenay, Booth
Wildflowers & Weeds

fl.   p 107

Centaurea scabiosa

Lindman, C. A. M.
Nordens Flora, Vol 10

fl.   Pl. 634

Centaurea scabiosa

Martin, W. Keble
The Concise British Flora in Colour

fl. hab.   Pl. 49

Centaurea solstitialis

Polunin, Oleg
Flowers of Europe

fl.   Pl. 155 #1499

Centaurea sp.

Klimas, John E.
Wildflowers of Eastern America

fl.   Pl. 220

Centaurea thracica

Goulimis, Constantine N.
Wild Flowers of Greece

fl.   hab. p 127

Centaurea triumfetti var.

Goulimis, Constantine N.
Wild Flowers of Greece

fl., hab.   p. 129

Centaurea uniflora

Barneby, T. P.
European Alpine Flowers in Colour

fl.   Pl. 92, 1

Centaurea variegata

American Hort. Soc.
American Horticulturist
    vol 53, No. 5 1974

fl.   p. 16

Centaurium capitatum

Ary, S.
The Oxford Book of Wildflowers

fl.   p 124, Pl. 2

Centaurium capitatum

Martin, W. Keble
The Concise British Flora in Colour

fl. hab.   Pl. 58

Centaurium erythraea

Bianchini, Francesco
Health Plants of the World

fl. hab.   p 19

Centaurium erythraea

Color Treasury of Herbs & Other
    Medicinal Plants

fl.   p 42   Pl. 56

Centaurium erythraea

Polunin, Oleg
Flowers of Europe

fl.   Pl. 95 #986

Centaurium erythraea var.

Martin, W. Keble
The Concise British Flora in Colour

fl. hab.   Pl. 58

Centaurium littorale

Martin, W. Keble
The Concise British Flora in Colour

fl. hab.   Pl. 58

Centaurium maritimum

Kleijn, H.
Beauty of the Wild Plant

fl.   opp. p. 117   Pl. 172

Centaurium minus

Ary, S.
The Oxford Book of Wildflowers

fl.   p 124, Pl. 1

Centaurium portense

Martin, W. Keble
The Concise British Flora in Colour

fl. hab.   Pl. 58

Centaurium pulchellum

Lindman, C. A. M.
Nordens Flora, Vol 7

fl. hab.   Pl. 471B

Centaurium pulchellum

Martin, W. Keble
The Concise British Flora in Colour

fl. hab.   Pl. 58

Centaurium umbellatum

Felsko, Elsa
A Book of Wildflowers

fl.   P 118

Centaurium umbellatum

Furrer, D.
Blumen am Wege

fl. hab.   p 75

Centaurium venustum

Orr, Robert T.
Wildflowers of Western America

fl.   Pl. 255

Centaurium venustum

Walcott, Mary Vaux
North American Wild Flowers
    vol. 4

fl.   Pl. 303

Centaurium vulgare

Lindman, C. A. M.
Nordens Flora, Vol 7

fl. hab.   Pl. 471A

Centaury

    see

Centaurium erythraea

Centaury, Caucasian

    See

Centaurea ocholeuca

Centaury, common

    see

Centaurium umbellatum

Centaury, Dumpy

    see

Centaurium capitatum

Centaury, Sea

    see

Centaurium maritimum

Centipede Grass

    see

Eremochloa ophiuroides

Centradenia floribunda

Perry, Frances
Flowers of the World

fl.   p 188

Centradenia floribunda

Wit, H. C. D. de
Plants of the World;
    The Higher Plants, Vol II

fl.   p 51, Pl. 6

Centradenia grandifolia

Addisonia
    vol 20   1937-38

fl., fr.   p. 1   Pl. 641

Centranthus angustifolius

Barneby, T. P.
European Alpine Flowers in Colour

fl.   Pl. 80, 3

Centranthus angustifolius

Polunin, Oleg
Flowers of Europe

fl.   Pl. 137 #1316

Centranthus ruber

Barneby, T. P.
European Alpine Flowers in Colour

fl.   Pl. 80, 2

Centranthus ruber

Bianchini, Francesco
Health Plants of the World

fl.   p 111

Centranthus ruber

Huxley, Anthony
Garden Perennials and Water Plants

hab.   Pl. 5

Centranthus ruber

Kleijn, H.
Beauty of the Wild Plant

fl.   opp. p. 117   Pl. 170

Centranthus ruber

Mathias, Mildred E.
Color for the Landscape

fl. hab.   p 141

Centranthus ruber

Morley, Brian D.
Wild Flowers of the World

fl.   Pl. 17A

Centranthus ruber

Perry, Frances
Flowers of the World

fl.   p 303

Centranthus ruber

Polunin, Oleg
Flowers of Europe

fl.   Pl. 136 #1316

Centranthus ruber

Polunin, Oleg
Flowers of the Mediterranean

fl.   Pl. 184

Centranthus ruber var.

Huxley, Anthony
Garden Perennials and Water Plants

fl.   Pl. 71

Centrogenium roseo-album

Ospina, Mariano
Orquideas de las americas

fl.   Pl. 19

Centrogenium setaceum

Luer, Carlyle A.
The Native Orchids of Florida

fl. hab.   Pl. 34

Centrolepis monogyna

Curtis, Winifred
The Endemic Flora of Tasmania, Part IV

fl. hab.   Pl. 144

Centrolepis muscoides

Curtis, Winifred
The Endemic Flora of Tasmania
     Vol VI

fl. hab.   Pl. 229

Centrolepis pulvinata

Curtis, Winifred
The Endemic Flora of Tasmania, Part IV

fl. fr. hab.   Pl. 145

Centropetalum sanguineum

Ospina, Mariano
Orquideas de las americas

fl. hab.   Pl. 148

Centropogon cornutus

Morley, Brian D.
Wild Flowers of the World

fl.   Pl. 172

Centropogon hybridus var.

Addisonia
     vol. 23   1954-59

fl.   p. 19   Pl. 746

Centrosema pubescens

Menninger, Edwin A.
Flowering Vines of the World

fl.   Pl. 96

Centrosema virginianum

Brown, Clair A.
Wildflowers of Louisiana and
     Adjoining States

fl. fr. hab.   p 77

Centrosema virginianum

Dean, Blanche E.
Wildflowers of Alabama and
     Adjoining States

fl.   p 91

Centrosema virginianum

Duncan, Wilbur H.
Wildflowers of the Southeastern
     United States

fl.   p 85

Centunculus minimus

Ary, S.
The Oxford Book of Wildflowers

fl.   p 124, Pl. 4

Century Plant
     see
Agave americana

Century Plant
     see
Agave deserti

Century Plant
     see
Agave shawii

Century Plant, Blue
     see
Agave palmeri

Cephaelis ipecacuanha

Hvass, Elsie
Plants That Feed and Serve Us

fl. fr.   p 92, Pl. 196

Cephaelis ipecacuanha

Perrot, Emile
Les Plantes Medicinales

fl. fr. hab.   p 124

Cephalanthera alba

Barneby, T. P.
European Alpine Flowers in Colour

fl.   Pl. 11, 5

Cephalanthera austinae

Luer, Carlyle A.
The Native Orchids of the
     United States and Canada

fl. fr. hab.   Pl. 14

Cephalanthera damasonium

Ary, S.
The Oxford Book of Wildflowers

fl.   p 100, Pl. 2

Cephalanthera damasonium

Felsko, Elsa
A Book of Wild Flowers
                    2nd ser.
fl.   p 10

Cephalanthera damasonium

Lindman, C. A. M.
Nordens Flora, Vol 3

fl. hab.   Pl. 146

Cephalanthera damasonium

Martin, W. Keble
The Concise British Flora in Colour

fl. hab.   Pl. 80

Cephalanthera damasonium

Morley, Brian D.
Wild Flowers of the World

fl., hab.   Pl. 26

Cephalanthera ensifolia

Brooke, Jocelyn
The Wild Orchids of Britain

fl.   Pl. 3

Cephalanthera grandiflora

Brooke, Jocelyn
The Wild Orchids of Britain

fl.   Pl. 2

Cephalanthera longifolia

Barneby, T. P.
European Alpine Flowers in Colour

fl.   Pl. 11, 6

Cephalanthera longifolia

Kleijn, H.
Beauty of the Wild Plant

fl.   opp. p. 120   Pl. 176

Cephalanthera longifolia

Lindman, C. A. M.
Nordens Flora, Vol 3

fl. hab.   Pl. 147

Cephalanthera longifolia

Martin, W. Keble
The Concise British Flora in Colour

fl. hab.   Pl. 80

Cephalanthera longifolia

Polunin, Oleg
Flowers of Europe

fl.   Pl. 191 #1919

Cephalanthera longifolia

Webster, Mary
Flora of Moray, Nairn & East
     Inverness

fl.   p. 292   Pl. 19

Cephalanthera rubra

Barneby, T. P.
European Alpine Flowers in Colour

fl.   Pl. 11, 4

Cephalanthera rubra

Brooke, Jocelyn
The Wild Orchids of Britain

fl.   Pl. 4

Cephalanthera rubra

Kohlhaupt, Paula
Fleurs des Alpages
                    vol 1
fl. P 28

Cephalanthera rubra

Martin, W. Keble
The Concise British Flora in Colour

fl. hab.   Pl. 80

Cephalanthera rubra

Polunin, Oleg
Flowers of Europe

fl.   Pl. 191 #1920

Cephalantheropsis gracilis

Walden, Beryl M.
Wild Flowers of Hong Kong

fl.   Pl. 81 (254)

Cephalanthus occidentalis

Batson, Wade T.
Wild Flowers in South Carolina

fl.   p 107

Cephalanthus occidentalis

Brown, Clair A.
Wildflowers of Louisiana and
    Adjoining States

fl. hab.   p 175

Cephalanthus occidentalis

Dean, Blanche E.
Wildflowers of Alabama and
    Adjoining States

fl.   p 173

Cephalanthus occidentalis

Ferguson, Mary
Wildflowers

fl.   p 16

Cephalanthus occidentalis

Fleming, Glenn
Wild Flowers of Florida

fl. fr.   p 35

Cephalanthus occidentalis

Tarver, David P.
Aquatic and Wetland Plants
    of Florida

fl. hab.   p 78

Cephalanthus occidentalis

Wharton, Mary E.
Trees & Shrubs of Kentucky

fl.   p 70, Pl. 2.4
fr.   p 117, Pl. 1.6

Cephalaria alpina

Barneby, T. P.
European Alpine Flowers in Colour

fl.   Pl. 80, 6

Cephalaria pungens

Moriarty, Audrey
Wild Flowers of Malawi

fl.   Pl. 56; 4

Cephalipterum drummondii

Morcombe, M. K.
Australia's Western Wildflowers

fl.   p 25

Cephalocereus chrysacanthus

Lamb, Edgar
The Pocket Encyclopedia of Cacti
    and Succulents in Color

hab.   Pl. 9

Cephalocereus hoppenstedtii

Van Laren, A. J.
Cactus

hab P 10   Fig. 10

Cephalocereus leucocephalus

Van Laren, A. J.
Cactus

fl.   P 15   Fig. B

Cephalocereus palmeri

Cactus & Succulent Society of America
Journal
    vol 40, No. 6   1968

fl.   p. 256

Cephalocereus palmeri

Lamb, Edgar
The Pocket Encyclopedia of Cacti
    and Succulents in Color

hab.   Pl. 10

Cephalocereus palmeri

Lamb, Edgar
Popular Exotic Cacti in Color

fl. hab.   p 43

Cephalocereus senilis

Cactus & Succulent Society of America
Journal
    vol 46, No. 3   1974

hab.   cover

Cephalocereus senilis

Hay, Roy
The Color Dictionary of Flowers & Plants

hab.   p 56, Pl. 443

Cephalocereus senilis

Hay, Roy
The Dictionary of House Plants

hab.   Pl. 109

Cephalocereus senilis

Hersey, Jean
Woman's Day Book of House Plants

hab.   p 43

Cephalocereus senilis

Kromdijk, G.
200 House Plants in Colour

hab.   Pl. 44

Cephalocereus senilis

Kupper, Walter
Cacti

hab.   P. 91   Pl. 42

Cephalocereus senilis

Lamb, Edgar
The Pocket Encyclopedia of Cacti
    and Succulents in Color

hab.   Pl. 11

Cephalocereus senilis

Subik, Rudolf
Decorative Cacti

hab.   p 29

Cephalocereus senilis

Van Laren, A. J.
Cactus

hab P 10   Fig. 9

Cephalocleistocactus schattatianus

Backeberg, Curt
Cactus Lexicon

hab.   p 574

Cephalophyllum albertiniense

Flowering Plants of Africa
    vol 35   1962

fl., hab.   Pl. 1398

Cephalophyllum loreum

Flowering Plants of Africa
    vol 35   1962

fl., hab.   Pl. 1396

Cephalophyllum loreum

Herre, H.
The Genera of the Mesembryanthemaceae

fl. fr.   p 109

Cephalophyllum maritimum

Flowering Plants of Africa
     vol 35   1962

fl., hab.   Pl. 1397

Cephalotaxus harringtonia

Boom, B. K.
The glory of the tree

fr.   p. 12   Pl. 7

Cephalotaxus harringtonia

Kitamura, Siro
Coloured Illustrations of Trees &
     Shrubs of Japan

fr.   10

Cephalotaxus harringtonia var.

Huxley, Anthony
Evergreen Garden Trees and Shrubs

hab.   Pl. 11

Cephalotus follicularis

Blombery, Alec M.
What Wildflower is That

hab.   p. 92   Pl. 222

Cephalotus follicularis

Milne, Lorus
Living Plants of the World

hab.   p 93

Cephalotus follicularis

Morley, Brian D.
Wild Flowers of the World

fl.   Pl. 133

Cerastium alpinum

Martin, W. Keble
The Concise British Flora in Colour

fl. hab.   Pl. 14

Cerastium alpinum

Polunin, Oleg
Flowers of Europe

fl.   Pl. 12 #142

Cerastium arcticum

Martin, W. Keble
The Concise British Flora in Colour

fl. hab.   Pl. 14

Cerastium arvense

Clark, Lewis J.
Wild Flowers of British Columbia

fl. hab.   p 138

Cerastium arvense

Courtenay, Booth
Wildflowers & Weeds

fl.   p 22

Cerastium arvense

Ferguson, Mary
Wildflowers

fl.   p 95

Cerastium arvense

Kleijn, H.
Beauty of the Wild Plant

fl.   opp. p. 73   Pl. 96

Cerastium arvense

Martin, W. Keble
The Concise British Flora in Colour

fl. hab.   Pl. 14

Cerastium arvense

Neufeld, J.B.
Wild Flowers of the Prairies

fl.   p. 23

Cerastium arvense

Orr, Robert T.
Wildflowers of Western America

fl.   Pl. 66

Cerastium arvense

Taylor, Ronald J.
Mountain Wild Flowers

fl.   p 87

Cerastium arvense var.

Barneby, T. P.
European Alpine Flowers in Colour

fl. hab.   Pl. 19,1

Cerastium berringianum

Heller, Christine
Wild Flowers of Alaska

fl.   Pl. 157

Cerastium biebersteinii

Huxley, Anthony
Garden Perennials and Water Plants

fl.   Pl. 76

Cerastium cerastoides

Martin, W. Keble
The Concise British Flora in Colour

fl. hab.   Pl. 14

Cerastium diffusum

Martin, W. Keble
The Concise British Flora in Colour

fl. hab.   Pl. 14

Cerastium Earlei

Porsild, A. E.
Rocky Mountain Wild Flowers

fl. hab.   p 159

Cerastium fontanum

Polunin, Oleg
Flowers of Europe

fl.   Pl. 12 #144

Cerastium fontanum var.

Martin, W. Keble
The Concise British Flora in Colour

fl. hab.   Pl. 14

Cerastium glomeratum

Ary, S.
The Oxford Book of Wildflowers

fl.   p 74, Pl. 4

Cerastium glomeratum

Martin, W. Keble
The Concise British Flora in Colour

fl. hab.   Pl. 14

Cerastium holosteoides

Lindman, C. A. M.
Nordens Flora, Vol 3

fl. hab.   Pl. 200B

Cerastium latifolium

Barneby, T. P.
European Alpine Flowers in Colour

fl.   Pl. 19, 2

Cerastium latifolium

Kohlhaupt, Paula
Fleurs des Alpages
          vol 2
fl.   P 32

Cerastium pumilum

Martin, W. Keble
The Concise British Flora in Colour

fl. hab.   Pl. 14

Cerastium regelii

Morley, Brian D.
Wild Flowers of the World

fl.   Pl. 3F

Cerastium semidecandrum

Martin, W. Keble
The Concise British Flora in Colour

fl. hab.   Pl. 14

Cerastium tetrandrum

Ary, S.
The Oxford Book of Wildflowers

fl.   p 74, Pl. 5

Cerastium tomentosum

Crockett, James Underwood
Lawns & Ground Covers

fl.   p. 126

Cerastium tomentosum

Macoby, Stirling
What Flower is That

fl.   P 78

Cerastium tomentosum

Mathias, Mildred E.
Color for the Landscape

fl. hab.   p 141

Cerastium uniflorum

Barneby, T. P.
European Alpine Flowers in Colour

fl. hab.   Pl. 19, 3

Cerastium uniflorum

Kohlhaupt, Paula
Fleurs des Alpages
                    vol 1
fl. P 42

Cerastium vulgatum

Ary, S.
The Oxford Book of Wildflowers

fl.   p 74, Pl. 3

Cerastium vulgatum

Courtenay, Booth
Wildflowers & Weeds

fl.   p 23

Cerastium vulgatum

Crockett, James Underwood
Lawns & Ground Covers

fl.   p. 70

Cerastium vulgatum

Weeds of the Southern United States

fl., hab.   p. 6

Ceratiola ericoides

Batson, Wade T.
Wild Flowers in South Carolina

fl.   p 70

Ceratonia siliqua

Bianchini, F.
The Complete Book of Fruits & Vegetables

fr.   p 199

Ceratonia siliqua

Bianchini, Francesco
Health Plants of the World

fr.   p 59

Ceratonia siliqua

Edlin, Herbert
The Illustrated Encyclopedia
    of Trees

fr. hab.   p 183

Ceratonia siliqua

Flemer, William III
Shade and Ornamental Trees in Color

hab.   p. 82

Ceratonia siliqua

Hvass, Elsie
Plants That Feed and Serve Us

fr.   p 49, Pl. 92

Ceratonia siliqua

Perrot, Emile
Les Plantes Medicinales

fl., fr. hab.   p 50

Ceratonia siliqua

Polunin, Oleg
Flowers of Europe

fr.   Pl. 49 #486

Ceratonia siliqua

Polunin, Oleg
Trees and Bushes of Europe

fl. fr. hab.   p 100

Ceratopetalum apetalum

Blombery, Alec M.
What Wildflower is That

fl.   p 93, Pl. 223

Ceratopetalum gummiferum

Blombery, Alec M.
What Wildflower is That

fl.   p 93, Pl. 224

Ceratopetalum gummiferum

Curtis's Botanical Magazine
    vol 172   1958-59

fl.   Pl. 312

Ceratopetalum gummiferum

Harrison, R.E.
Trees and Shrubs

fl. P 48

Ceratopetalum gummiferum

Macoby, Stirling
What Flower is That

fl.   P 78

Ceratopetalum gummiferum

Morley, Brian D.
Wild Flowers of the World

fl.   Pl. 133

Ceratopetalum gummiferum

Oakman, Harry
Colorful Trees

Hab. Fl.   P 82

Ceratophyllum demersum

Ary, S.
The Oxford Book of Wildflowers

fl.   p 52, Pl. 6

Ceratophyllum demersum

Lindman, C. A. M.
Nordens Flora, Vol 4

fl. fr. hab.   Pl. 217

Ceratophyllum demersum

Martin, W. Keble
The Concise British Flora in Colour

hab.   Pl. 34

Ceratophyllum demersum

Stodola, Jiri
Encyclopedia of Water Plants

hab.   P. 70

Ceratophyllum demersum

Tarver, David P.
Aquatic and Wetland Plants
    of Florida

hab.   p 45

Ceratophyllum submersum

Martin, W. Keble
The Concise British Flora in Colour

fl. hab.   Pl. 34

Ceratophyllum submersum

Stodola, Jiri
Encyclopedia of Water Plants

hab.   P. 71

Ceratostigma plumbaginoides

Crockett, James Underwood
Lawns & Ground Covers

fl.   p. 127

Ceratostigma plumbaginoides

Curtis's Botanical Magazine
Vol 169   1952-53

fl.   210

Ceratostigma plumbaginoides

Hay, Roy
The Color Dictionary of Flowers & Plants

fl.   p 189, Pl. 1505

Ceratostigma plumbaginoides

Meikle, R. D.
Garden Flowers

fl.   opp. p. 288   Pl. 8

Ceratostigma plumbaginoides

Morley, Brian D.
Wild Flowers of the World

fl.   Plate 105

Ceratostigma willmottianum

Gault, S. Millar
The Color Dictionary of Shrubs

fl.   Pl. 63

Ceratostigma willmottianum

Harrison, R.E.
Trees and Shrubs

fl.   P 48

Ceratostigma willmottianum

Hay, Roy
The Color Dictionary of Flowers & Plants

fl.   p 189, Pl. 1506

Ceratostigma willmottianum

Hellyer, A.G.L.
Shrubs in Colour

fl.   p. 26

Ceratostigma willmottianum

Perry, Frances
Flowers of the World

fl.   p 232

Ceratostylis retisquama

Curtis's Botanical Magazine
Vol 179   1972

fl.   640

Ceratotheca sesamoides

Moriarty, Audrey
Wild Flowers of Malawi

fl.   Pl. 63; 2

Ceratozamia mexicana

Wit, H. C. D. de
Plants of the World;
   The Higher Plants, Vol I

fr.   p 50, Pl. 2

Cerbera manghas

Walden, Beryl M.
Wild Flowers of Hong Kong

fr.   Pl. 46 (135)

Cercidiphyllum japonicum

American Hort. Soc.
American Horticulturist
   vol 56, No. 6   1977

hab.   p. 22-23

Cercidiphyllum japonicum

Huxley, Anthony
Deciduous Garden Trees and Shrubs

hab.   Pl. 34

Cercidiphyllum japonicum

Kitamura, Siro
Coloured Illustrations of Trees &
   Shrubs of Japan

hab.   p. 62,152

Cercidium floridum

Mathias, Mildred E.
Color for the Landscape

hab.   p 199

Cercidium torreyanum

Everett, Thomas H.
Living Trees of the World

hab.   p. 190

Cercidium undulatum

Lemmon, Robert S.
Wildflowers of North America in
   Full Color

fl. hab.   p. 73   pl. 117

Cercis canadensis

Batson, Wade T.
Wild Flowers in South Carolina

fl. hab.   p 62

Cercis canadensis

Brown, Clair A.
Wildflowers of Louisiana and
   Adjoining States

fl.   p 76

Cercis canadensis

Campbell, Carlos C.
Great Smoky Mountain Wildflowers

fl.   p 39

Cercis canadensis

Dean, Blanche E.
Wildflowers of Alabama and
   Adjoining States

fl.   p 91

Cercis canadensis

Edlin, Herbert
The Illustrated Encyclopedia
   of Trees

hab.   p 181

Cercis canadensis

Flemer, William III
Shade and Ornamental Trees in Color

hab.   p. 77

Cercis canadensis

Jennings, O. E.
Wild Flowers of Western Pennsylvania
                              vol II
fl.   Pl. 89

Cercis canadensis

Massachusetts Hort. Soc.
Horticulture
   vol 32, No. 4   1954

fl.   backcover

Cercis canadensis

Massachusetts Hort. Soc.
Horticulture
   vol 35, No. 2   1957

fl.   p. 76

Cercis canadensis

Walcott, Mary Vaux
North American Wild Flowers

fl.   Pl. 26

Cercis canadensis

Wharton, Mary E.
Trees & Shrubs of Kentucky

fl.   p 58, Pl. 1.24

Cercis chinensis
American Hort. Soc.
American Horticulturist
vol 56, No. 2 1977
fl. p. 22

Cercis occidentalis
Mathias, Mildred E.
Color for the Landscape
fl. hab. p 175

Cercis occidentalis
Munz, Philip A.
California Spring Wildflowers
fl. p 78, Pl. 58

Cercis occidentalis
Orr, Robert T.
Wildflowers of Western America
fl. Pl. 190

Cercis occidentalis
Pacific Hort. Foundation
California Hort. Journal
vol 33, No. 4 1972
hab., fl. p. 146-47

Cercis siliquastrum
Bianchini, F.
The Complete Book of Fruits & Vegetables
fl. p 55

Cercis siliquastrum
Boom, B. K.
The glory of the tree
fl. p. 84 Pl. 134

Cercis siliquastrum
Edlin, Herbert
The Illustrated Encyclopedia of Trees
fl. fr. hab. p 180-1

Cercis siliquastrum
Feinbrun-Dothan, Naomi
Wild Plants in the Land of Israel
fl. fr. p 38

Cercis siliquastrum
Gault, S. Millar
The Color Dictionary of Shrubs
fl. Pl. 64

Cercis siliquastrum
Harrison, R. E.
Trees and Shrubs
fl. fr. P. 48

Cercis siliquastrum
Hay, Roy
The Color Dictionary of Flowers & Plants
fl. p 189, Pl. 1507

Cercis siliquastrum
Macoby, Stirling
What Flower is That
fl. P 79

Cercis siliquastrum
Morley, Brian D.
Wild Flowers of the World
fl., fr. Pl. 28

Cercis siliquastrum
Perry, Frances
Flowers of the World
fl. p 160

Cercis siliquastrum
Polunin, Oleg
Flowers of Europe
fl. Pl. 50 #485

Cercis siliquastrum
Polunin, Oleg
Flowers of the Mediterranean
fl. Pl. 54

Cercis siliquastrum
Polunin, Oleg
Trees and Bushes of Europe
fl. fr. hab. p 99, 100

Cercis siliquastrum
Royal Hort. Soc.
Journal of the Royal Hort. Soc.
vol 92, No. 6 1967
fl. p. 250 Pl. 132

Cercis siliquastrum
Taylor, A. W.
Wild Flowers of Spain and Portugal
fl. p. 59

Cerefolium sativum
Perrot, Emile
Les Plantes Medicinales
fl. fr. hab. p 57

Cereus chalybaeus
Lamb, Edgar
The Pocket Encyclopedia of Cacti and Succulents in Color
hab. Pl. 12

Cereus emoryi
Munz, Philip A.
California Spring Wildflowers
fl. p. 30 Pl. 10

Cereus giganteus
Massachusetts Hort. Soc.
Horticulture
vol 50, No. 3 1972
fl. p. 43

Cereus giganteus
Milne, Lorus
Living Plants of the World
fl. hab. p 163

Cereus jamacaru
Kupper, Walter
Cacti
fl., fr. P. 29 Pl. 11

Cereus jamacaru
Van Laren, A. J.
Cactus
hab P 10 Fig. 8

Cereus jamacaru var.
Lamb, Edgar
Popular Exotic Cacti in Color
hab. p 45

Cereus monstrosus
Hersey, Jean
Woman's Day Book of House Plants
fl. p. 40

Cereus, Nightblooming
see
Hylocereus undatus

Cereus, Night-Blooming
see
Selenicereus grandiflorus

Cereus, Nightblooming
see
Selenicereus hamatus

Cereus peruvianus
Hersey, Jean
Woman's Day Book of House Plants
hab. p. 41

Cereus peruvianus

Kromdijk, G.
200 House Plants in Colour

hab.   Pl. 45

Cereus peruvianus

Lamb, Edgar
The Pocket Encyclopedia of Cacti
    and Succulents in Color

hab.   Pl. 13

Cereus peruvianus var.

Van Laren, A. J.
Cactus

hab P 94   Fig. 112

Cereus, Senita

see

Lophocereus schottii

Cereus validus

Van Laren, A. J.
Cactus

hab. P 19   Fig. D

Cereus variabilis

Cactus & Succulent Society of America
Journal
    vol 32, No. 5   1960

fr.   p. 149

Cereus variabilis

Everett, T.H., ed.
New Illustrated Encyclopedia of
    Gardening   vol 3

fl.   p. 391

Cereus weberi

Royal Hort. Soc.
Journal of the Royal Hort. Soc.
    vol 86, No. 8   1961

hab., fl.   p. 354   Pl. 101

Cereus, Wool

see

Harrisia bonplandi

Ceriman

see

Monstera deliciosa

Cerinthe aspera

Crockett, James Underwood
Annuals

fl.   P 105

Cerinthe glabra

Barneby, T. P.
European Alpine Flowers in Colour

fl.   Pl. 67, 2

Cerinthe major

Morley, Brian D.
Wild Flowers of the World

fl.   Pl. 37

Cerinthe major

Polunin, Oleg
Flowers of Europe

fl.   Pl. 105 #1079

Cerinthe major

Polunin, Oleg
Flowers of the Mediterranean

fl.   Pl. 148

Cerinthe major var.

Polunin, Oleg
Flowers of the Mediterranean

fl.   Pl. 149

Cerinthe minor

Polunin, Oleg
Flowers of Europe

fl.   Pl. 105 #1078

Cerinthe retorta

Polunin, Oleg
Flowers of the Mediterranean

fl.   Pl. 151

Cerocarpus betuloides

Coyle, Jeanette
A Field Guide to the Common and
    Interesting Plants of Baja California

fr.   p 83

Cerochlamys pachyphylla

Herre, H.
The Genera of the Mesembryanthemaceae

fl. fr.   p 111

Cerochlamys pachyphylla

Lamb, Edgar
Popular Exotic Cacti in Color

fl. hab.   p 47

Ceropegia africana

Flowering Plants of Africa
    vol 39   1968-69

fl.   Pl. 1544

Ceropegia ampliata

Harrison, Richmond
Climbers and Trailers

fl.   p. 30   Pl. 41

Ceropegia ballyana

Flowering Plants of Africa
    vol 41   1970-71

fl.   Pl. 1614

Ceropegia ballyana

Morley, Brian D.
Wild Flowers of the World

fl.   Pl. 60

Ceropegia barkleyi

Flowering Plants of Africa
    Vol. 29   1952-53

fl., hab.   Pl. 1156

Ceropegia barkleyi

Lamb, Edgar
The Pocket Encyclopedia of Cacti
    and Succulents in Color

fl. hab.   Pl. 217

Ceropegia conrathii

Flowering Plants of Africa
    vol 32   1957-58

fl., hab.   Pl. 1246

Ceropegia crassifolia

Flowering Plants of South Africa
    vol XXIV   1944

fl.   Pl. 924

Ceropegia cufodontis

Flowering Plants of Africa
    vol 35   1962

fl.   Pl. 1370

Ceropegia de-vechii

Flowering Plants of Africa
    vol 35   1962

fl.   Pl. 1368

Ceropegia dichotoma

Bramwell, David
Wild Flowers of the Canary Islands

fl.   Pl. 211

Ceropegia dimorpha

Flowering Plants of Africa
    vol 36   1963-64

fl., hab.   Pl. 1437

Ceropegia distincta var.

Flowering Plants of Africa
    Vol. 44   1977

fl.   Pl. 1735

Ceropegia filiformis

Flowering Plants of Africa
    vol 40   1969-70

fl.   Pl. 1580

Ceropegia filipendula

Moriarty, Audrey
Wild Flowers of Malawi

fl.   Pl. 45; 5

Ceropegia floribunda

Cacti & Succulent Society of America
Journal
    vol 13, No. 6   1941

fl.   p. 90-91

Ceropegia fortuita

Flowering Plants of South Africa
    vol XXIV   1944

fl.   Pl. 925

Ceropegia fusca

Bramwell, David
Wild Flowers of the Canary Islands

hab.   Pl. 210

Ceropegia galeata

Flowering Plants of Africa
    vol 37   1965-66

fl.   Pl. 1443

Ceropegia gemmifera

Wit, H. C. D. de
Plants of the World;
    The Higher Plants, Vol II

fl.   p 106, Pl. 61

Ceropegia grandis

Flowering Plants of Africa
    vol 28   1950-51

fl.   Pl. 1113

Ceropegia haygarthii

Harrison , Richmond E.
Climbers and Trailers

fl.   p 30   Pl. 42

Ceropegia hians

Bramwell, David
Wild Flowers of the Canary Islands

hab.   Pl. 212

Ceropegia insignis

Flowering Plants of South Africa
    vol XXIII   1943

fl.   Pl. 902

Ceropegia linearis

Kromdijk, G.
200 House Plants in Colour

hab.   Pl. 46

Ceropegia meyeri-johannis

Flowering Plants of Africa
    vol 35   1962

fl.   Pl. 1371

Ceropegia meyeri-johannis var.

Flowering Plants of Africa
    vol 36   1963-64

fl., hab.   Pl. 1410

Ceropegia multiflora

Flowering Plants of South Africa
    vol XXIII   1943

fl.   Pl. 909

Ceropegia nilotica

Lamb, Edgar
Popular Exotic Cacti in Color

fl. hab.   p 49

Ceropegia papillata

Moriarty, Audrey
Wild Flowers of Malawi

fl.   Pl. 45; 3

Ceropegia papillata var.

Flowering Plants of Africa
    vol 43   1974-76

fl.   Pl. 1716

Ceropegia patersoniae

Flowering Plants of Africa
    vol 33   1959

fl.   Pl. 1296

Ceropegia plicata

Flowering Plants of South Africa
    vol XVII   1937

fl.   Pl. 675

Ceropegia radicans

Flowering Plants of Africa
    vol XXV   1945-46

fl., hab.   Pl. 970

Ceropegia sandersonii

Harrison, Richmond E.
Climbers and Trailers

fl.   p 31   Pl. 43

Ceropegia sandersonii

Hersey, Jean
Woman's Day Book of House Plants

fl.   p. 46

Ceropegia sandersonii

Subik, Rudolf
Decorative Cacti

fl. hab.   p 103

Ceropegia serpentina

Flowering Plants of Africa
    vol 27   1948-49

fl., hab.   Pl. 1072

Ceropegia seticorona var.

Flowering Plants of Africa
    Vol. 41   1970-71

fl.   Pl. 1616

Ceropegia smithii

Flowering Plants of Africa
    vol 31   1956

fl.   Pl. 1240

Ceropegia stapeliiformis

Cactus & Succulent Society of America
Journal
    vol 37, No. 3   1965

fl.   cover (p. 61)

Ceropegia stapeliiformis

Cactus & Succulent Society of America
Journal
    Vol. 49, No. 4   1977

fl.   cover

Ceropegia stapeliiformis

Flowering Plants of South Africa
    vol XXI   1941

fl.   Pl. 809

Ceropegia stapeliiformis

Lamb, Edgar
The Pocket Encyclopedia of Cacti
    and Succulents in Color

fl.   Pl. 218

Ceropegia succulenta

Flowering Plants of Africa
    vol 36   1963-64

fl.   Pl. 1431

Ceropegia turricula

Flowering Plants of Africa
vol 27   1948-49

fl., hab.   Pl. 1045

Ceropegia woodii

Encke, Fritz
Zimmerpflanzen

fl., hab.   p. 64

Ceropegia woodii

Hay, Roy
The Dictionary of House Plants

hab.   Pl. 110

Ceropegia woodii

Hersey, Jean
Woman's Day Book of House Plants

hab.   p. 100

Ceropegia woodii

Kiaer, Eigil
Indoor Plants in Colour

hab.   p. 36

Ceropegia woodii

Perry, Frances
Flowers of the World

fl.   p 38

Cestrum aurantiacum

Chickering, Carol Rogers
Flowers of Guatemala

fl.   p 51

Cestrum aurantiacum

Hay, Roy
The Color Dictionary of Flowers & Plants

fl.   p 56, Pl. 444

Cestrum aurantiacum

Hay, Roy
The Dictionary of House Plants

fl.   Pl. 111

Cestrum aurantiacum

Macoby, Stirling
What Flower is That

fl.   P 79

Cestrum elegans

Kiaer, Eigil
Indoor Plants in Colour

fl.   p. 36

Cestrum elegans var.

Curtis's Botanical Magazine
Vol 170   1954-55

fl. fr.   249

Cestrum fasciculatum

Macoby, Stirling
What Flower is That

fl. P 79

Cestrum fasciculatum

O'Gorman, Helen
Mexican Flowering Trees and Plants

fl. hab.   p 53

Cestrum fasciculatum

Royal Hort. Soc.
Journal of the Royal Hort. Soc.
vol 93, No. 4   1968

fl.   p. 164   Pl. 81

Cestrum hybrid.

Curtis's Botanical Magazine
Vol. 180   1974

fl.   Pl. 661

Cestrum hybrid

Perry, Frances
Flowers of the World

fl.   p. 283

Cestrum newellii

Harrison, R.E.
Trees and Shrubs

fl.   P 50

Cestrum newellii

Hay, Roy
The Color Dictionary of Flowers & Plants

fl.   p 56, Pl. 445

Cestrum nocturnum

Macoby, Stirling
What Flower is That

fl. P 79

Cestrum nocturnum

Tsukamoto, Yotaro
Coloured Illustrations of Garden
Flowers vol. 10

fl. Illus. 143   opp. p. 45

Cestrum, orange

see

Cestrum aurantiacum

Cestrum parqui

Hersey, Jean
Woman's Day Book of House Plants

fl.   p 72

Cestrum parqui

Morley, Brian D.
Wild Flowers of the World

fl.   Pl. 183

Cestrum parqui

Polunin, Oleg
Trees and Bushes of Europe

fl. fr.   p 172

Cestrum, purple

see

Cestrum elegans

Cestrum purpureum

Nicolaisen, Age
Pocket Encyclopedia of Indoor
Plants

fl.   Pl. 115

Cestrum purpureum

Wit, H. C. D. de
Plants of the World;
The Higher Plants, Vol II

fl.   Pl. 82

Chaenactis douglasii

Clark, Lewis J.
Wild Flowers of British Columbia

Chaenactis douglasii

Welsh, Stanley L.
Flowers of the Canyon Country

fl. hab.   p 3

Chaenomeles hybrid.

Bartels, Andreas
Das Grosse Buch der Gartengeholze

fl.   p 128

Chaenomeles hybrid

Cault, S. Millar
The Color Dictionary of Shrubs

fl.   Pl. 65-67

Chaenomeles hybrid

Harrison, R.E.
Trees and Shrubs

fl., hab.   p. 49

Chaenomeles hybrid

Massachusetts Hort. Soc.
Horticulture
vol 44, No. 3   1966

fl.   p. 32-33

Chaenomeles japonica

Huxley, Anthony
Deciduous Garden Trees and Shrubs

fl.   Pl. 35

Chaenomeles japonica

Kimura, Koiti
Japanese Medicinal Plants, Vol II

fl.   p 179

Chaenomeles lagenaria

Kimura, Koiti
Japanese Medicinal Plants, Vol II

fl.   p 181

Chaenomeles lagenaria

Tsukamoto, Yotaro   vol 10
Coloured Illustrations of Garden Flowers

fl.   Pl. 145

Chaenomeles lagenaria var.

Everett, T.H., ed.
New Illustrated Encyclopedia of
    Gardening   vol 3

fl.   p. 502

Chaenomeles lagenaria var.

Macoby, Stirling
What Flower is That

fl.   P 80

Chaenomeles sinensis

Kariyone, Tatsuo
Atlas of Medicinal Plants

fl., fr.   Pl. 105

Chaenomeles sinensis

Kitamura, Siro
Coloured Illustrations of Trees &
    Shrubs of Japan

fl.   233

Chaenomeles sp.

Gault, S. Millar
The Color Dictionary of Shrubs

fl.   Pl. 65, 66, 67

Chaenomeles speciosa

American Hort. Soc.
American Horticulturist
vol 56, No. 6   1977

fl.   p. 15

Chaenomeles speciosa

Hellyer, A.G.L.
Shrubs in Colour

fl.   p. 27

Chaenomeles speciosa

Huxley, Anthony
Deciduous Garden Trees and Shrubs

fl. fr. Pl. 36, 36a

Chaenomeles speciosa

Morley, Brian D.
Wild Flowers of the World

fl.   Plate 92

Chaenomeles speciosa

Perry, Frances
Flowers of the World

fl.   p 262

Chaenomeles speciosa var.

Hay, Roy
The Color Dictionary of Flowers & Plants

fl.   p 189, Pl. 1508

Chaenomeles speciosa var.

Huxley, Anthony
Deciduous Garden Trees and Shrubs

fl.   Pl. 37

Chaenomeles speciosa var.

Royal Hort. Soc.
Journal of Royal Hort. Soc.
vol 83, No. 11   1958

fl.   p. 470   Pl. 136

Chaenomeles x superba var.

Hay, Roy
The Color Dictionary of Flowers &
    Plants

fl.   p. 189   Pl. 1509

Chaenomeles x superba var.

Royal Hort. Soc.
Journal of the Royal Hort. Soc.
vol 94, No. 11   1969

fr.   p. 480   Pl. 261

Chaenorhinum minus

Ary, S.
The Oxford Book of Wildflowers

fl. fr.   p. 140   Pl. 1

Chaenorhinum minus

Martin, W. Keble
The Concise British Flora in Colour

fl. hab.   Pl. 62

Chaerophyllum procumbens

Courtenay, Booth
Wildflowers & Weeds

fl.   p 58

Chaerophyllum temulum

Ary, S.
The Oxford Book of Wildflowers

fl.   p 90, Pl. 3

Chaerophyllum tenulentum

Martin, W. Keble
The Concise British Flora in Colour

fl. fr. hab.   Pl. 38

Chaffweed

see

Centunculus mininus

Chain, Love's

See Antigonon leptopus

Chalice Flower

see

Anemone occidentalis

Chalice, golden

See Solandra nitida

Chalice Vine

see

Solandra nitida

Chalk Plant

see

Gypsophila cerastioides

Chalk plant

see

Gypsophila elegans

Chalk Plant

see

Gypsophila paniculata var.

Chalk Plant

see

Gyphophila repens var.

Chamaeangis orientalis

Stewart, Joyce
Orchids of Tropical Africa

fl.   Pl. 21

Chamaebatia australis

Coyle, Jeanette
A Field Guide to the Common and
  Interesting Plants of Baja California

fl.   p 83

Chamaebatia foliososa

Orr, Robert T.
Wildflowers of Western America

fl.   Pl. 13

Chamaebatiaria millefolium

Royal Hort. Soc.
Journal of the Royal Hort. Soc.
  vol 95, No. 5   1970

fl.   p. 214   Pl. 125

Chamaecereus sylvestri

Hay, Roy
The Color Dictionary of Flowers
  & Plants

fl.   p. 56   Pl. 446

Chamaecereus sylvestri

Kromdijk, G.
200 House Plants in Colour

fl., hab.   Pl. 47

Chamaecereus sylvestri

Kupper, Walter
Cacti

fl., hab.   p. 35   Pl. 14

Chamaecereus sylvestri

Lamb, Edgar
The Pocket Encyclopedia of Cacti
  and Succulents in Color

fl. hab.   Pl. 15

Chamaecereus sylvestri

Lamb, Edgar
Popular Exotic Cacti in Color

fl. hab.   p. 50

Chamaecereus sylvestri

Subik, Rudolf
Decorative Cacti

fl. hab.   p. 30

Chamaecereus sylvestri

Susik, Rudol
Decorative Cacti

fl.   p. 31

Chamaecereus sylvestri

Van Laren, A. J.
Cactus

fl.   p. 35   Fig. 29

Chamaecereus sylvestri var.

Lamb, Edgar
The Pocket Encyclopedia of Cacti
  and Succulents in Color

fl. hab.   Pl. 16

Chamaecyparis formosensis

Edlin, Herbert
The Illustrated Encyclopedia
  of Trees

hab.   p 89

Chamaecyparis lawsoniana

Polunin, Oleg
Trees and Bushes of Europe

fr. hab.   p 25

Chamaecyparis lawsoniana

Vedel, H.
Arbres et Arbustes

fr. hab.   p 34, Pl. 22

Chamaecyparis lawsoniana var.

Bartels, Andreas
Das Grosse Buch der Gartengeholze

hab.   p 248, 249

Chamaecyparis lawsoniana var.

Edlin, Herbert
The Illustrated Encyclopedia
  of Trees

hab.   p 88

Chamaecyparis lawsoniana var.

Gault, S. Millar
The Color Dictionary of Shrubs

hab.   Pl. 68, 69, 70

Chamaecyparis lawsoniana var.

Harrison, R. E.
Trees and Shrubs

hab.   Pl. 565, 571

Chamaecyparis lawsoniana var.

Hay, Roy
The Color Dictionary of Flowers &
  Plants

hab.   p. 251   Pl. 2004 thru
                          2008

Chamaecyparis lawsoniana var.

Huxley, Anthony
Evergreen Garden Trees and Shrubs

hab.   Pl. 13, 14, 15, 16, 17

Chamaecyparis nootkatensis

Boom, B. K.
The glory of the tree

hab., fr.   p. 40   pl. 53, 55

Chamaecyparis nootkatensis

Edlin, Herbert
The Illustrated Encyclopedia
  of Trees

fr. hab.   p 88-9

Chamaecyparis nootkatensis var.

Hansen, Richard
Baume und Straucher im Carten

hab.   p. 207

Chamaecyparis nootkatensis var.

Huxley, Anthony
Evergreen Garden Trees and Shrubs

hab.   Pl. 18

Chamaecyparis obtusa

Edlin, Herbert
The Illustrated Encyclopedia
  of Trees

fr.   p 89

Chamaecyparis obtusa

Everett, T.H., ed.
New Illustrated Encyclopedia of
  Gardening   vol 3

hab.   p. 390

Chamaecyparis obtusa

Kitamura, Siro
Coloured Illustrations of Trees &
  Shrubs of Japan

fr.   60

Chamaecyparis obtusa var.

Crockett, James Underwood
Evergreens

hab.   p. 93

Chamaecyparis obtusa var.

Gault, S. Millar
The Color Dictionary of Shrubs

hab.   Pl. 71

Chamaecyparis obtusa var.

Harrison, R. E.
Trees and Shrubs

hab.   Pl. 562, 564, 567

Chamaecyparis obtusa var.

Hay, Roy
The Color Dictionary of Flowers & Plants

hab. p 252, Pl. 2009, 2010

Chamaecyparis obtusa var.

Huxley, Anthony
Evergreen Garden Trees and Shrubs

hab. Pl. 19

Chamaecyparis obtusa var.

Royal Hort. Soc.
Journal of the Royal Hort. Soc.
vol 92, No. 9 1967

hab. p. 388 Pl. 211

Chamaecyparis pisifera

Edlin, Herbert
The Illustrated Encyclopedia
of Trees

hab. p 88

Chamaecyparis pisifera

Kitamura, Siro
Coloured Illustrations of Trees &
Shrubs of Japan

fr. 61

Chamaecyparis pisifera var.

Crockett, James Underwood
Evergreens.

hab. p. 93-94

Chamaecyparis pisifera var.

Edlin, Herbert
The Illustrated Encyclopedia
of Trees

hab. p 88

Chamaecyparis pisifera var.

Gault, S. Millar
The Color Dictionary of Shrubs

hab. Pl. 72, 73

Chamaecyparis pisifera var.

Harrison, R. E.
Trees and Shrubs

hab. Pl. 568

Chamaecyparis pisifera var.

Huxley, Anthony
Evergreen Garden Trees and Shrubs

hab. Pl. 12, 20, 21

Chamaecyparis thyoides var.

Gault, S. Millar
The Color Dictionary of Shrubs

hab. Pl. 74

Chamaecytisus hirsutus

Polunin, Oleg
Flowers of Europe

fl. Pl. 51 #506

Chamaecytisus proliferus

Bramwell, David
Wild Flowers of the Canary Islands

fl. Pl. 170

Chamaecytisus purpureus

Polunin, Oleg
Flowers of Europe

fl. Pl. 51 #506

Chamaedaphne calyculata

Courtenay, Booth
Wildflowers & Weeds

fl. p 77

Chamaedaphne calyculata

Ferguson, Mary
Wildflowers

fl. p 21

Chamaedaphne calyculata

Heller, Christine
Wild Flowers of Alaska

fl. Pl. 202

Chamaedaphne calyculata

Lindman, C. A. M.
Nordens Flora, Vol 7

fl. fr. hab. Pl. 447

Chamaedaphne calyculata

Moyle, John B.
Northland Wild Flowers

fl. p 117, Pl. 117

Chamaedorea oblongata

Morley, Brian D.
Wild Flowers of the World

fr. Pl. 191

Chamaedorea oreophila

Morley, Brian D.
Wild Flowers of the World

fl. pl. 191

Chamaedorea sp.

Kromdijk, G.
200 House Plants in Colour

hab. Pl. 48

Chamaelaucium uncinatum

Blombery, Alec M.
What Wildflower is That

fl. p 92, Pl. 225

Chamaelaucium uncinatum

Macoby, Stirling
What Flower is That

fl. P 81

Chamaelaucium uncinatum

Mathias, Mildred E.
Color for the Landscape

fl. hab. p 66

Chamaelaucium uncinatum

Morcombe, M. K.
Australia's Western Wildflowers

fl. p 102

Chamaelirium luteum

American Hort. Soc.
American Horticulturist
vol 53, No. 5 1974

fl. p. 24

Chamaelirium luteum

Campbell, Carlos C.
Great Smoky Mountain Wildflowers

fl. p 51

Chamaelirium luteum

Courtenay, Booth
Wildflowers & Weeds

fl. p 9

Chamaelirium luteum

Dean, Blanche E.
Wildflowers of Alabama and
Adjoining States

fl. hab. p 9

Chamaelirium luteum

Duncan, Wilbur H.
Wildflowers of the Southeastern
United States

fl. hab. p 247

Chamaelirium luteum

Jennings, O. E.
Wild Flowers of Western Pennsylvania
vol II

fl. Pl. 11

Chamaelirium luteum

Klimas, John E.
Wildflowers of Eastern America

fl. Pl. 37

Chamaelirium luteum

Wharton, Mary E.
A Guide to the Wildflowers & Ferns
of Kentucky

fl.   p 55   Pl. 1.16

Chamaemelum nobile

Martin, W. Keble
The Concise British Flora in Colour

fl. hab.   Pl. 45

Chamaemelum nobile

Polunin, Oleg
Flowers of Europe

fl.   Pl. 147 #1413

Chamaenerion angustifolium

Ary, S.
The Oxford Book of Wildflowers

fl.   p 110, Pl. 3

Chamaenerion angustifolium

Lindman, C. A. M.
Nordens Flora, Vol 6

fl. fr.   Pl. 407

Chamaenerion angustifolium

Martin, W. Keble
The Concise British Flora in Colour

fl. hab.   Pl. 35

Chamaepericlymenum canadense

Perry, Frances
Flowers of the World

fl.   p 93

Chamaepericlymenum suecicum

Martin, W. Keble
The Concise British Flora in Colour

fl. fr. hab.   Pl. 41

Chamaepericlymenum suecicum

Raven, John
Mountain Flowers

fl.   p. 178   Pl. 13

Chamaerops humilis

Hay, Roy
The Color Dictionary of Flowers & Plants

hab.   p 189, Pl. 1510

Chamaerops humilis

Huxley, Anthony
Evergreen Garden Trees and Shrubs

hab.   Pl. 192

Chamaerops humilis

Hvass, Elsie
Plants That Feed and Serve Us

hab.   p 105, Pl. 226

Chamaerops humilis

Polunin, Oleg
Flowers of Europe

hab Pl. 176 #1719

Chamaerops humilis

Polunin, Oleg
Flowers of the Mediterranean

hab.   Pl. 219

Chamaerops humilis

Polunin, Oleg
Trees and Bushes of Europe

fr. hab.   p 184

Chamaespartium sagittale

Polunin, Oleg
Flowers of Europe

fl.   Pl. 52 #513

Chamise

see

Adenostoma fasciculatum

Chamomile

see

Anthemis cretica

Chamomile

see

Anthemis nobilis

Chamomile

see

Anthemis tinctoria

Chamomile

see

Matricaria chamomilla

Chamomile, Common

see

Anthemis nobilis

Chamomile, Corn

see

Anthemis arvensis

Chamomile, German

see

Matricaria chamomilla

Chamomile, Roman

See

Anthemis nobilis

Chamomile, wild

see

Matricaria chamomilla

Chamomile, Wild

see

Matricaria recutita

Chamomile, Yellow

see

Anthemis tinctoria

Chamorchis alpina

Lindman, C. A. M.
Nordens Flora, Vol 3

fl. hab.   Pl. 143

Chaplet flower, Madagascar

see

Stephanotis floribunda

Chaptalia tomentosa

Brown, Clair A.
Wildflowers of Louisiana and
Adjoining States

fl. hab.   p 195

Chaptalia tomentosa

Dean, Blanche E.
Wildflowers of Alabama and
Adjoining States

fl. hab.   p 183

Charlock

see

Brassica kaber

Charlock

see

Sinapis arvensis

Chard

see

Beta vulgaris

Charieis heterophylla

Crockett, James Underwood
Annuals

fl.   P 105

Charieis sp.

Eliovson, Sima
Namaqualand in Flower

fl.   Pl. 6

Chasmanthe caffra

Curtis's Botanical Magazine
    Vol 160   1937

fl.   No. 9470

Chasmatophyllum musculinum

Herre, H.
The Genera of the Mesembryanthemaceae

fl.   fr.   p 113

Chaste Tree

see

Vitex agnus-castus

Chatterbox

see

Epipactis gigantea

Chaubardia chasmatochila

Ospina, Mariano
Orquideas de las americas

fl.   Pl. 102

Chayote

see

Sechium edule

Checkerberry

see

Gaultheria procumbens

Checker Bloom

see

Sidalcea malvaeflora

Cheeseberry

see

Cyathodes glauca

Cheese Bush

see

Hymenoclea salsola

Cheesemania radicata

Curtis, Winifred
The Endemic Flora of Tasmania, Vol. 5

fl. hab.   Pl. 192

Cheeses

see

Malva neglecta

Cheeses

see

Modiola caroliniana

Cheiranthera cyanea

Morley, Brian D.
Wild Flowers of the World

fl.   pl. 131

Cheiranthera linearis

Blombery, Alec M.
What Wildflower is That

fl.   p 94, Pl. 227

Cheiranthera linearis

Cochrane, G. R.
Flowers and Plants of Victoria

fl.   Pl. 318

Cheiranthus allionii

Crockett, James Underwood
Annuals

fl.   p. 105

Cheiranthus allionii var.

Hay, Roy
The Color Dictionary of Flowers &
    Plants

fl.   p. 33   Pl. 261

Cheiranthus cheiri

Barneby, T. P.
European Alpine Flowers in Colour

fl.   Pl. 29, 4

Cheiranthus cheiri

Bianchini, Francesco
Health Plants of the World

fl.   p 151

Cheiranthus cheiri

Crockett, James Underwood
Annuals

fl.   P 105

Cheiranthus cheiri

Hay, Roy
The Color Dictionary of Flowers & Plants

fl.   p 33, Pl. 262

Cheiranthus cheiri

Macoby, Stirling
What Flower is That

fl.   P 82

Cheiranthus cheiri

Perry, Frances
Flowers of the World

fl.   p 97

Cheiranthus cheiri

Polunin, Oleg
Flowers of Europe

fl.   Pl. 32   #299

Cheiranthus cheiri

Tsukamoto, Yotaro
Coloured Illustrations of Garden
    Flowers Vol. 10

fl. Illus. 123   opp. p. 39

Cheiranthus cheiri var.

Hay, Roy
The Color Dictionary of Flowers & Plants

fl.   p 128, Pl. 1021 thru 1023

Cheiranthus cheiri var.

Huxley, Anthony
Garden annuals and bulbs

fl.   pl. 24

Cheiranthus hybrid

Hay, Roy
The Color Dictionary of Flowers &
    Plants

fl.   p. 5   Pl. 34

Cheiranthus hybrid

Royal Hort. Soc.
Journal of the Royal Hort. Soc.
vol 99, No. 2   1974

fl.   p. 72   Pl. 33

Cheiranthus scoparius

Bramwell, David
Wild Flowers of the Canary Islands

fl.   Pl. 137

Cheiranthus scoparius var.

Bramwell, David
Wild Flowers of the Canary Islands

fl., hab.   Pl. 138

Cheiranthus virescens

Bramwell, David
Wild Flowers of the Canary Islands

fl.   Pl. 139

Cheiridopsis candidissima

Flowering Plants of Africa
vol 38   1967

fl.   Pl. 1507

Cheiridopsis sp.

Cactus & Succulent Society of America
Journal
vol 28, No. 5   1956

fl., hab.   p. 151

Cheiridopsis tuberculata

Flowering Plants of Africa
vol 39   1968-69

fl., hab.   Pl. 1529

Cheiridopsis tuberculata

Herre, H.
The Genera of the Mesembryanthemaceae

fl. fr.   p 115

Cheirostylis chinensis

Hu, Shiu-ying
The Genera of Orchidaceae in
Hong Kong

fl.   p 121

Chelidonium majus

Ary, S.
The Oxford Book of Wildflowers

fl.   p 6, Pl. 3

Chelidonium majus

Bianchini, Francesco
Health Plants of the World

fl. hab.   p 177

Chelidonium majus

Color Treasury of Herbs & Other Medicinal
Plants

fl.   p 28   Pl. 25

Chelidonium majus

Courtenay, Booth
Wildflowers & Weeds

fl.   p 20

Chelidonium majus

Felsko, Elsa
A Book of Wildflowers

fl.   P 24

Chelidonium majus

Furrer, D.
Blumen am Wege

fl. fr. hab.   p 27

Chelidonium majus

Kimura, Koiti
Japanese Medicinal Plants, Vol II

fl.   p 175

Chelidonium majus

Klimas, John E.
Wildflowers of Eastern America

fl.   Pl. 117

Chelidonium majus

Lemmon, Robert S.
Wildflowers of North America in
Full Color

fl.   pg. 255   pl. 402

Chelidonium majus

Lindman, C. A. M.
Nordens Flora, Vol 4

fl. fr.   Pl. 246

Chelidonium majus

Martin, W. Keble
The Concise British Flora in Colour

fl. hab.   Pl. 5

Chelidonium majus

Perrot, Emile
Les Plantes Medicinales

fl. fr. hab.   p 61

Chelidonium majus

Pond, Barbara
A Sampler of Wayside Herbs

fl. fr.   Pl. VII

Chelidonium majus

Takatori, Jinuke
Color Atlas of Medicinal Plants of Japan

fl.
fr.   Fig. 45A

Chelone glabra

American Hort. Soc.
American Horticulturist
vol 57, No. 5   1978

fl.   p. 11

Chelone glabra

Batson, Wade T.
Wild Flowers in South Carolina

fl.   p 102

Chelone glabra

Courtenay, Booth
Wildflowers & Weeds

fl.   p 99

Chelone glabra

Dean, Blanche E.
Wildflowers of Alabama and
Adjoining States

fl.   p 161

Chelone glabra

Duncan, Wilbur H.
Wildflowers of the Southeastern
United States

fl.   p 169

Chelone glabra

Jennings, O. E.
Wild Flowers of Western Pennsylvania
vol II
fl.   Pl. 137

Chelone glabra

Klimas, John E.
Wildflowers of Eastern America

fl.   Pl. 59

Chelone glabra

Moyle, John B.
Northland Wild Flowers

fl.   p 144, Pl. 163

Chelone glabra

Walcott, Mary Vaux
North American Wild Flowers
vol. 4

fl.   Pl. 259

Chelone glabra

Wharton, Mary E.
A Guide to the Wildflowers & Ferns
of Kentucky

fl.   p 196   Pl. 2.10

Chelone glabra var.

Jennings, O. E.
Wild Flowers of Western Pennsylvania
vol II
fl.   Pl. 137

Chelone lyoni

Campbell, Carlos C.
Great Smoky Mountain Wildflowers
fl.   p 97

Chelone lyoni

Massachusetts Hort. Soc.
Horticulture
vol 53, No. 9   1975
fl.   p. 31

Chelone obliqua

Duncan, Wilbur H.
Wildflowers of the Southeastern
United States
fl.   p 169

Chelone obliqua

Hay, Roy
The Color Dictionary of Flowers & Plants
fl.   p 128, Pl. 1024

Chelone obliqua

Lemmon, Robert S.
Wildflowers of North America in
Full Color
fl.   pg. 46   pl. 76

Chenille Plant

    see

Acalypha hispida

Chenopodium album

Ary, S.
The Oxford Book of Wildflowers
fl.   p 54, Pl. 6

Chenopodium album

Courtenay, Booth
Wildflowers & Weeds
fl.   p. 67

Chenopodium album

Crockett, James Underwood
Lawns & Ground Covers
fl., hab.   p. 72

Chenopodium album

Lindman, C. A. M.
Nordens Flora, Vol 3
fl. fr.   Pl. 185

Chenopodium album

Martin, W. Keble
The Concise British Flora in Colour
fl. fr. hab.   Pl. 71

Chenopodium album

Polunin, Oleg
Flowers of Europe
fl. Pl. 10 #105

Chenopodium album

Szczawinski, Adam F.
Edible Garden Weeds of Canada
fr.   p. 103

Chenopodium album

Weeds of the Southern United States
hab.   p. 7

Chenopodium amaranticolor

Hay, Roy
The Color Dictionary of Flowers &
Plants
hab.   p. 33   Pl. 263

Chenopodium ambrosioides

Bianchini, Francesco
Health Plants of the World
fl. fr.   P 55

Chenopodium ambrosioides

Burgis, D. S.
Florida Weeds
fr.   p. 3

Chenopodium ambrosioides var.

Kimura, Koiti
Japanese Medicinal Plants, Vol. I
fl.   p. 22

Chenopodium anthelminthicum

Perrot, Emile
Les Plantes Medicinales
fl. fr. hab.   p 64

Chenopodium bonus-henricus

Ary, S.
The Oxford Book of Wildflowers
fl.   p 54, Pl. 5

Chenopodium bonus-henricus

Bianchini, F.
The Complete Book of Fruits & Vegetables
hab.   p 79

Chenopodium bonus-henricus

Martin, W. Keble
The Concise British Flora in Colour
fl. hab.   Pl. 72

Chenopodium bonus-henricus

Masefield, G. B.
The Oxford Book of Food Plants
fl.   Pl. 191, 3

Chenopodium bonus-henricus

Perrot, Emile
Les Plantes Medicinales
'fl. fr. hab.   p 64

Chenopodium bonus-henricus

Polunin, Oleg
Flowers of Europe
fl. Pl. 10 #103

Chenopodium botryodes

Martin, W. Keble
The Concise British Flora in Colour
fl. hab.   Pl. 72

Chenopodium capitatum

Alaska-Yukon Wild Flowers Guide
fl. hab.   p 30

Chenopodium capitatum

Clark, Lewis J.
Wild Flowers of British Columbia
fl.   p 123

Chenopodium capitatum

Courtenay, Booth
Wildflowers & Weeds
fl.   p 69

Chenopodium capitatum

Heller, Christine
Wild Flowers of Alaska
fl.   Pl. 83

Chenopodium capitatum

Porsild, A. E.
Rocky Mountain Wild Flowers
fr.   p 151

Chenopodium capitatum

Walcott, Mary Vaux
North American Wild Flowers
vol. 5
fl.   Pl. 349

Chenopodium ficifolium

Martin, W. Keble
The Concise British Flora in Colour

fl. hab.   Pl. 72

Chenopodium foliosum

Kleijn, H.
Beauty of the Wild Plant

fr.   opp. p. 39   Pl. 45

Chenopodium foliosum

Polunin, Oleg
Flowers of Europe

fr. Pl. 10 #104

Chenopodium glaucum

Martin, W. keble
The Concise British Flora in Colour

fl. hab.   Pl. 72

Chenopodium hybridum

Courtenay, Booth
Wildflowers & Weeds

fl.   p 67

Chenopodium murale

Martin, W. Keble
The Concise British Flora in Colour

fl. hab.   Pl. 72

Chenopodium polyspermum

Martin, W. Keble
The Concise British Flora in Colour

fl. hab.   Pl. 71

Chenopodium polyspermum

Morley, Brian D.
Wild Flowers of the World

fl. fr.   Pl. 14E

Chenopodium quinoa

Wit, H. C. D. de
Plants of the World;
    The Higher Plants, Vol I

fl.   p 161, Pl. 73

Chenopodium rubrum

Ary, S.
The Oxford Book of Wildflowers

fl.   p 54, Pl. 4

Chenopodium rubrum

Martin, W. keble
The Concise British Flora in Colour

fl. fr. hab.   Pl. 72

Chenopodium vulvaria

Ary, S.
The Oxford Book of Wildflowers

fl.   p 54, Pl. 3

Chenopodium vulvaria

Martin, W. Keble
The Concise British Flora in Colour

fl. fr. hab.   Pl. 71

Cherimoya

see

Annona cherimolia

Cherleria sedoides

Martin, W. Keble
The Concise British Flora in Colour

fl. hab.   Pl. 16

Cherry

see

Prunus

Cherry, Amanogawa

see

Prunus serrulata var.

Cherry, Barbados

see

Malpighia coccigera

Cherry, Barbados

see

Malpighia glabra

Cherry, Barbadoes wedge-leaved

see

Malpighia lucida

Cherry, Bell-Flowered

see

Prunus campanulata

Cherry, Bird

see

Prunus padus

Cherry, Bitter

see

Prunus emarginata

Cherry, Black

see

Prunus serotina

Cherry, Bladder

see

Physalis alkekengi

Cherry, brush

see

Eugenia australis

Cherry, Carolina Laurel

see

Prunus caroliniana

Cherry, Christmas

see

Solanum capsicastrum

Cherry, Christmas

see

Solanum pseudo-capsicum

Cherry, Conradine's

see

Prunus conradinae

Cherry, European Bird

see

Prunus padus

Cherry, Flowering

see

Prunus serrulata

Cherry, Flowering Almond

see

Prunus glandulosa var.

Cherry, Flowering Oriental
see
Prunus serrulata

Cherry, Ground
see
Physalis angulata

Cherry, Ground
sue
Physalis crassifolia

Cherry, ground
see
Physalis pruinosa

Cherry, Ground
see
Quincula lobata

Cherry, Ground Smooth
see
Physalis subglabrata

Cherry, Holly-Leaved
see
Prunus ilicifolia

Cherry in a lantern
see
Physalis franchetii

Cherry, Indian
see
Rhamnus carolinianas

Cherry, Japanese
see
Prunus amanogawa

Cherry, Japanese
see
Prunus kiku-shidare var.

Cherry, Japanese
see
Prunus shirotae

Cherry, Japanese flowering
see
Prunus serrulata var.

Cherry, Jerusalem
see
Solanum capsicastrum

Cherry, Jerusalem
see
Solanum pseudo-capsicum

Cherry, Jerusalem yellow-fruited
see
Solanum pseudo-capsicum

Cherry, Kwanzan
see
Prunus serrulata var.

Cherry, Laurel
see
Prunus caroliniana

Cherry, Mount Fuji
see
Prunus serrulata var.

Cherry, Native
see
Exocarpus cupressiformis

Cherry Pie
see
Heliotropium arborescens

Cherry Pie
see
Heliotropium peruvianum

Cherry, pink rush
see
Prunus glandulosa var.

Cherry, Rosebud
see
Prunus subhirtella

Cherry, Rum
see
Prunus serotina

Cherry, St. Lucie
see
Prunus mahaleb

Cherry, Sargent's
see
Prunus sargentii

Cherry, scrub
see
Eugenia australis

Cherry, Shiro-Fugen
see
Prunus serrulata var.

Cherry, Sour
see
Prunus cerasus

Cherry, Spring
see
Prunus subhirtella

Cherry, Sweet
see
Prunus avium

Cherry, weeping
see
Prunus subhirtella var.

Cherry, Wild

see

Prunus avium

Cherry, winter

see

Physalis franchetii

Cherry, Winter

see

Solanum capsicastrum

Cherry, winter

see

Solanum pseudo-capsicum

Cherry, Yoshino

see

Prunus yedoensis

Chervil

see

Anthriscus cerefolium

Chervil, American

see

Chaerophyllum procumbens

Chervil, Rough

see

Chaerophyllum temulum

Chestnut

see

Castanea

Chestnut, Cape

see

Calodendron capense

Chestnut, Chinese water

see

Eleocharis tuberosa

Chestnut, Horse

see

Aesculus hippocastanum

Chestnut, Indian Horse

see

Aesculus indica

Chestnut, Lowveld

see

Sterculia murex

Chestnut, Moreton Bay

see

Castanospermum australe

Chestnut, Rhodesian

see

Baikiaea plurijuga

Chestnut, Ruby Horse

see

Aesculus carnea var.

Chestnut, Spanish

see

Castanea sativa

Chestnut, Spanish

see

Castanospermum australe

Chestnut, Sweet

see

Castanea sativa

Chestnut, water

see

Trapa natans

Chestnut, Wild

see

Calodendrum capense

Chestnut, Wild

see

Pachira insignis

Chevelure

see

Adiantum capillis-veneris

Chewing Gum Tree

See

Achras sapota

Chiapasia nelsonii

Lamb, Edgar
Popular Exotic Cacti in Color

fl.   hab.   p 51

Chiastophyllum oppositifolium

Huxley, Anthony
Garden Perennials and Water Plants

fl.   pl. 77

Chicalote

see

Argemone platyceras

Chicele

see

Achras sapota

Chickweed

see

Cerastium

Chickweed

see

Stellaria media

Chickweed, Bering

see

Cerastium berringianum

Chickweed, common

see

Stellaria media

Chickweed, Field

see

Cerastium arvense

Chickweed, giant

see

Stellaria aquatica

Chickweed, Great

see

Stellaria pubera

Chickweed, Mouse-ear

see

Cerastium arvense

Chickweed, mouse-ear

see

Cerastium vulgatum

Chickweed, Mouse-Ear Sticky

see

Cerastium glomeratum

Chickweed, Polar

see

Cerastium regelii

Chickweed, Sea

see

Honckenya peploides

Chickweed, Star

see

Stellaria pubera

Chickweed, starry

see

Cerastium arvense

Chickweed, Upright

see

Moenchia erecta

Chickweed, Water

see

Myosoton aquaticum

Chickweed, Water

see

Stellaria aquatica

Chicory

see

Cichorium intybus

Chicory, Desert

see

Rafinesquia neomixicana

Chicory, Red Treviso

see

Cichorium intybus var.

Chicary, Red Verona

see

Cichorium intybus var.

Chigger Weed

see

Asclepias tuberosa

Chili

see

Capsicum anuum

Chili

see

Capsicum frutescens

Chiloglottis gunnii

Cady, Leo
Australian Native Orchids in Colour

fl.   Pl. 36

Chiloglottis gunnii

Cochrane, G. R.
Flowers and Plants of Victoria

fl. hab.   Pl. 434

Chiloglottis reflexa

Cady, Leo
Australian Native Orchids in Color

fl.   Pl. 35

Chilopsis linearis

Coyle, Jeanette
A Field Guide to the Common and
Interesting Plants of Baja California

fl.   p 163

Chilopsis linearis

Mathias, Mildred E.
Color for the Landscape

fl. hab.   p 176

Chilopsis linearis

Orr, Robert T.
Wildflowers of Western America

fl.   Pl. 201

Chimaphila maculata

Courtenay, Booth
Wildflowers & Weeds

fl.   p 79

Chimaphila maculata

Dean, Blanche E.
Wildflowers of Alabama and
Adjoining States

fl.   p 125

Chimaphila maculata

Duncan, Wilbur H.
Wildflowers of the Southeastern
United States

fl. hab.   p 117

Chimaphila maculata

Klimas, John E.
Wildflowers of Eastern America

fl.   Pl. 95

Chimaphila maculata

Lemmon, Robert S.
Wildflowers of North America in
Full Color

fl.   pg. 267   pl. 423

Chimaphila maculata

Pennsylvania Hort. Soc.
The Green Scene
vol 3, No. 3   1975

fl.   backcover

Chimaphila maculata

Wharton, Mary E.
A Guide to the Wildflowers & Ferns
of Kentucky

fl.   p 116   Pl. 1.44

Chimaphila menziesii

Clark, Lewis J.
Wild Flowers of British Columbia

fl.   p 387

Chimaphila umbellata

Courtenay, Booth
Wildflowers & Weeds

fl.   p 79

Chimaphila umbellata

Ferguson, Mary
Wildflowers

fl.   p 181

Chimaphila umbellata

Fries, Mary A.
Wildflowers of Mount Ranier and
the Cascades

fl.   P 65

Chimaphila umbellata

Klimas, John E.
Wildflowers of Eastern America

fl.   Pl. 228

Chimaphila umbellata

Lindman, C. A. M.
Nordens Flora, Vol 7

fl. fr. hab.   Pl. 435

Chimaphila umbellata

Taylor, Ronald J.
Mountain Wild Flowers

fl. hab.   p 47

Chimaphila umbellata var.

Clark, Lewis J.
Wild Flowers of British Columbia

fl.   p 390

Chimaphila umbellata var.

Munz, Philip A.
California Mountain Wildflowers

fl.   p 33, Pl. 19

Chimaphila umbellata var.

Porsild, A. E.
Rocky Mountain Wild Flowers

fr.   p 297

Chimaphila umbellata var

Walcott, Mary Vaux
North American Wild Flowers
vol. 5

fl.   Pl. 360

Chimonanthus praecox

Curtis's Botanical Magazine
Vol 169   1952-53

fl.   184

Chimonanthus praecox

Hellyer, A.G.L.
Shrubs in Colour

fl.   p. 26

Chimonanthus praecox

Morley, Brian D.
Wild Flowers of the World

fl.   Plate 92

Chimonanthus praecox

Perry, Frances
Flowers of the World

fl.   p 60

Chimonanthus praecox var.

Gault, S. Millar
The Color Dictionary of Shrubs

fl.   Pl. 75

Chimonanthus praecox var.

Harrison, R.E.
Trees and Shrubs

fl.   P 50

Chimonanthus praecox var.

Hay, Roy
The Color Dictionary of Flowers & Plants

fl.   p 189, Pl. 1511

Chimonanthus praecox var.

Perry, Frances
Flowers of the World

fl.   p 60

Chimonanthus praecox var.

Royal Hort. Soc.
Journal of the Royal Hort. Soc.
vol 77, No. 3   1952

fl.   p. 86   Pl. 49

China berry

see

Melia azederach

China Root

see

Smilax glabra

China Tree

see

Koelreuteria paniculata

China Tree

see

Melia azedarach

Chinaman's hat

see

Holmskioldia sanguinea

Chinese Houses

see

Collinsia bicolor

Chinese Houses

see

Collinisia heterophylla

Chinese New Year Flower

see

Enkianthus quinqueflorus

Chinese Paper Plant

see

Tetrapanax papyriferus

Chinese Scholar Tree

see

Sophora japonica

Chinquapin, Water

see

Nelumbo pentapetala

Chionanthus retusus

Kitamura, Siro
Coloured Illustrations of Trees &
Shrubs of Japan

fl.   413

Chionanthus retusus

Macoby, Stirling
What Flower is That

fl.   p. 82

Chionanthus retusus

Mathias, Mildred E.
Color for the Landscape

fl.,hab.   p. 16

Chionanthus sardensis

Hay, Roy
The Color Dictionary of Flowers
& Plants

fl.   p. 86   Pl. 685

Chionanthus virginicus

Batson, Wade T.
Wild Flowers in South Carolina

fl.   p 78

Chionanthus virginicus

Brown, Clair A.
Wildflowers of Louisiana and
Adjoining States

fl. hab.   p. 135

Chionanthus virginicus

Dean, Blanche E.
Wildflowers of Alabama and
Adjoining States

fl.   p 131

Chionanthus virginicus

Everett, T.H., ed.
New Illustrated Encyclopedia of
Gardening vol. 3

hab., fl.   p. 390

Chionanthus virginicus

Royal Hort. Soc.
Journal of the Royal Hort. Soc.
vol 87, No. 8   1962

hab., fl.   p. 358   Pl. 106

Chionanthus virginicus

Walcott, Mary Vaux
North American Wild Flowers

fl. hab.   Pl. 140

Chionanthus virginicus

Wharton, Mary E.
Trees & Shrubs of Kentucky

fl.   p 57, Pl. 1.20
fr.   p 140, Pl. 2.34

Chionodoxa hybrid

Miles, Bebe
Bulbs for the Home Gardener

fl.   p 59

Chionodoxa lochiae

Curtis's Botanical Magazine
Nol. 171   1956-57

fl.   281

Chionodoxa lochiae

Megaw, Elektra
Wild Flowers of Cyprus

fl., hab.   Pl. 35

Chionodoxa luciliae

American Hort. Soc.
American Horticulturist
vol 55, No. 5   1976

fl.   p. 30

Chionodoxa luciliae

Curtis's Botanical Magazine
Vol. 181   1976-77

fl.   Pl. 730

Chinodoxa luciliae

Hay, Roy
The Color Dictionary of Flowers & Plants

fl.   p 86, Pl. 683, 684

Chionodoxa luciliae

Huxley, Anthony
Garden Annuals and Bulbs

hab., fl.   Pl. 149, 202

Chionodoxa luciliae

Macoby, Stirling
What Flower is That

fl.   P 83

Chionodoxa luciliae

Massachusetts Hort. Soc.
Horticulture
vol 42, No. 3   1964

fl.   p. 14

Chionodoxa luciliae

Massachusetts Hort. Soc.
Horticulture
vol 45, No. 2   1967

fl.   cover, backcover

Chionodoxa luciliae

Mathew, Brian
Dwarf Bulbs

fl.   p. 40   Pl. 14

Chionodoxa luciliae

Perry, Frances
Flowers of the World

fl.   p 172

Chionodoxa luciliae

Royal Hort. Soc.
Journal of the Royal Hort. Soc.
vol 80, No. 10   1955

fl., hab.   p. 466   Pl. 97

Chionodoxa luciliae

Royal Hort. Soc.
Journal of the Royal Hort. Soc.
vol 95, No. 1   1970

fl.   p. 22   Pl. 19

Chionodoxa luciliae var.

Hay, Roy
The Color Dictionary of Flowers &
Plants

fl.   p. 86

Chionodoxa luciliae var.

Royal Hort. Soc.
Journal of the Royal Hort. Soc.
vol 95, No. 1   1970

fl.   p. 22   Pl. 20

Chionodoxa sardensis

American Hort. Soc.
American Horticulturist
vol 55, No. 5   1976

fl.   p. 30

Chionodoxa sardensis

Curtis's Botanical Magazine
Vol 166   1949

fl.   50

x Chionoscilla allenii

Curtis's Botanical Magazine
Vol. 169   1952-53

fl.   207

Chiranthodendron pentadactylon

Chickering, Carol Rogers
Flowers of Guatemala

fl. fr.   p 53

Chiranthodendron pentadactylon

Morley, Brian D.
Wild Flowers of the World

fl.   pl. 176

Chiranthodendron pentadactylon

O'Gorman, Helen
Mexican Flowering Trees and Plants

fl.   p 17

Chirita lavandulacea

Morley, Brian D.
Wild Flowers of the World

fl.   Plate 122

Chirita micromusa

American Gloxinia & Gesneriad Society
The Gloxinian
vol 23, No. 2   1973

fl., hab.   cover

Chirita sinensis

American Gloxinia & Gesneriad Society
The Gloxinian
vol. 28, No. 3   1978

fl.,hab.   cover, backcover

Chirita sinensis

Walden, Beryl M.
Wild Flowers of Hong Kong

fl.   Pl. 71 (218)

Chirita sinensis var.

American Gloxinia & Gesneriad Society
The Gloxinian
vol 28, No. 3   1978

fl., hab.   backcover

Chirita urticifolia

Hara, Hiroshi, comp.
Photo-Album of Plants of
Eastern Himalaya

fl.   Pl. 95

Chironia humilis

Flowering Plants of South Africa
vol XXIV   1944

fl.   Pl. 951

Chironia krebsii

Moriarty, Audrey
Wild Flowers of Malawi

fl.   Pl. 50; 1

Chironia laxiflora

Moriarty, Audrey
Wild Flowers of Malawi

fl.   Pl. 50; 2

Chironia transvaalensis

Flowering Plants of South Africa
vol XXI   1941

fl.   Pl. 814

Chittick

see

Lambertia inermis

Chives

see

Allium schoenoprasum

Chives, Garlic

see

Allium tuberosum

Chlamydocarya macrocarpa

Wit, H. C. D. de
Plants of the World;
The Higher Plants, Vol. II

fr.   p. 63   Pl. 37

Chlidanthus fragrans

Miles, Bebe
Bulbs for the Home Gardener

fl.   p 156

Chloanthes coccinea

Blombery, Alec M.
What Wildflower is That

fl.   p. 94   Pl. 228

Chloanthes parviflora

Morley, Brian D.
Wild Flowers of the World

fl.   pl. 138

Chloanthes stoechadis

Blombery, Alec M.
What Wildflower is That

fl.   p 94, Pl. 229

Chloraea ulanthoides

Munoz Pizarro, Carlos
Flores Silvestres de Chile

fl., hab Pl. 16

Chloranthus glaber

Kitamura, Siro
Coloured Illustrations of Trees &
Shrubs of Japan

fr.   74

Chlorophytum aridum

Flowering Plants of Africa
vol 35   1962

fl., fr.   Pl. 1399

Chlorophytum bichetii

Bruggeman, L.
Tropical Plants

hab.   Pl. 90

Chlorophytum comosum

Kromdijk, G.
200 House Plants in Colour

hab.   Pl. 49

Chlorophytum comosum var.

Hay, Roy
The Color Dictionary of Flowers &
Plants

hab.   p. 56   Pl. 447

Chlorophytum comosum var.

Kiaer, Eigil
Indoor Plants in Colour

hab.   p. 36

Chlorophytum comosum var.

Nicolaisen, Age
Pocket Encyclopedia of Indoor Plants

hab.   Pl. 20

Chlorophytum elatum var.

Hersey, Jean
Woman's Day Book of House Plants

hab.   p 105

Chlorophytum longipedunculatum

Flowering Plants of South Africa
vol XXII   1942

fl., fr.   Pl. 861

Chlorophytum sp.

Moriarty, Audrey
Flowers of Malawi

fl. fr.   Pl. 12; 3

Choananthus cyrtanthiflorus

Flowering Plants of Africa
vol 34   1960-61

fl., hab.   Pl. 1340

Chocolate Berry

see

Vitex payos

Choisya ternata

Curtis's Botanical Magazine
vol 172   1958-59

fl.   Pl. 318

Choisya ternata

Gault, S. Millar
The Color Dictionary of Shrubs

fl.   Pl. 76

Choisya ternata

Harrison, R.E.
Trees and Shrubs

fl.   P 48

Choisya ternata

Hay, Roy
The Color Dictionary of Flowers & Plants

fl.   p 189, Pl. 1512

Choisya ternata

Hellyer, A.G.L.
Shrubs in Colour

fl.   p. 27

Choisya ternata

Macoby, Stirling
What Flower is That

fl.   P 83

Choisya ternata

Perry, Frances
Flowers of the World

fl.   p 269

Chokeberry

see

Aronia arbutifolia

Chokeberry, Black

see

Aronia melanocarpa

Chokeberry, Red

see

Pyrus arbutifolia

Cholla

see

Opuntia

Cholla, Buckhorn

see

Opuntia acanthocarpa

Cholla, cane

see

Opuntia spinosior

Cholla, Golden

see

Opuntia echinocarpa

Cholla, Jumping

see

Opuntia bigelovii

Cholla, purple

see

Opuntia versicolor

Cholla, Silver

see

Opuntia echinocarpa

Cholla, Teddy Bear

see

Opuntia bigelovii

Cholla, Yellow

see

Opuntia spinosior

Chondrorhyncha chestertoni

Kupper, Walter
Orchids

fl.   p. 87

Chondrorhyncha fimbriata

Ospina, Mariano
Orquideas de las americas

fl. hab.   Pl. 103

Chonemorpha macrophylla

Menninger, Edwin A.
Flowering Vines of the World

fl.   Pl. 16

Chordospartium stevensonii

Curtis's Botanical Magazine
Vol 164   1943-48

fl.   No. 9654

Chorisia insignis

Everett, Thomas H.
Living Trees of the World

fl.   p. 247

Chorisia speciosa

American Hort. Soc.
American Horticulturist
vol 56, No. 4 1977

fl., hab.   p. 42

Chorisia speciosa

Massachusetts Hort. Soc.
Horticulture
vol 52, No. 2 1974

fl.   cover, backcover

Chorisia speciosa

Mathias, Mildred E.
Color for the Landscape

fl. hab.   p 17

Chorisia speciosa

Milne, Lorus
Living Plants of the World

hab.   p 141

Chorisia speciosa

Pacific Hort. Foundation
California Hort. Journal
vol 33, No. 4 1972

hab., fl.   p. 146

Chorizema aciculare

Menninger, Edwin A.
Flowering Vines of the World

fl.   Pl. 100

Chorizema cordatum

Curtis's Botanical Magazine
Vol 170   1954-55

fl.   237

Chorizema cordatum

Hersey, Jean
Woman's Day Book of House Plants

fl.   p 32

Chorizema cordatum

Macoby, Stirling
What Flower is That

fl.   P 83

Chorizema cordatum

Morcombe, M. K.
Australia's Western Wildflowers

fl.   p 101

Chorizema cordatum

Perry, Frances
Flowers of the World

fl.   p 165

Chorizema dicksonii

Harrison, R.E.
Trees and Shrubs

fl.   P 51

Chorizema diversifolia

Harrison, Richmond E.
Climbers and Trailers

fl.   p 42   Pl. 81

Chorizema diversifolia var.

Harrison, R.E.
Trees and Shrubs

fl.   P 51

Chorizema ilicifolia

Blombery, Alec M.
What Wildflower is That

fl.   p 95, Pl. 230

Chorizema ilicifolia

Morley, Brian D.
Wild Flowers of the World

fl.   Pl. 130

Chorizema varium

Harrison, R.E.
Trees andShrubs

fl.   P 51

Chortolirion stenophyllum

Flowering Plants of South Africa
    vol XXIV   1944
fl., hab. Pl. 932

Christmas Berry

See

Heteromeles arbutifolia

Christmas Berry

See

Lycium carolinianum

Christmas bush

see

Ceratopetalum gummiferum

Christmas bush, Victorian

see

Prostanthera lasianthos

Christmas-cheer

see

Sedum rubrotinctum

Christmas jewels

see

Aechmea racinae

Christmas Pride

see

Ruellia devosiana

Christmas Tree

see

Metrosideros excelsa

Christmas-Tree

see

Nuytsia floribunda

Christmas Tree, New Zealand

see

Metrosideros excelsa

Christopher, Herb

see

Actaea spicata

Christ's Thorn

see

Euphorbia milii

Chrosperma muscaetoxicum

Walcott, Mary Vaux
North American Wild Flowers

fl. hab.   Pl. 147

Chrozophora tinctoria

Polunin, Oleg
Flowers of Europe

fl. Pl. 65 #665

Chrysamphora californica

Walcott, Mary Vaux
North American Wild Flowers
        vol. 5

fl., hab.   Pl. 390

Chrysanthemum alpinum

Barneby, T. P.
European Alpine Flowers in Colour

fl.   Pl. 87, 6

Chrysanthemum alpinum

Kohlhaupt, Paula
Fleurs des Alpages
            vol 1
fl. P 105

Chrysanthemum, annual

see

Chrysanthemum carinatum

Chrysanthemum, annual

see

Chrysanthemum coronarium

Chrysanthemum arcticum

Addisonia
    vol 24   1960-64

fl.   p. 37   Pl. 787

Chrysanthemum arcticum

Heller, Christine
Wild Flowers of Alaska

fl.   Pl. 215-216

Chrysanthemum arcticum var.

Alaska-Yukon Wild Flower Guide

fl.   p 171

Chrysanthemum balsamita

Masefield, G. B.
The Oxford Book of Food Plants

hab.   Pl. 145, 4

Chrysanthemum burbankii

Tsukamoto, Yotaro
Coloured Illustrations of Garden Flowers
            Vol. 9

fl.   p. 37-38

Chrysanthemum carinatum

Crockett, James Underwood
Annuals

fl.   P 106

Chrysanthemum carinatum

Huxley, Anthony
Garden annuals and bulbs

fl.   pl. 26

Chrysanthemum carinatum

Tsukamoto, Yotaro
Coloured Illustrations of Garden Flowers
            Vol. 10

fl. Illus. 33   opp. p. 11

Chrysanthemum carinatum var.

Hay, Roy
The Color Dictionary of Flowers &
    Plants

fl.   p. 33   Pl. 264

Chrysanthemum catananche

Morley, Brian D.
Wild Flowers of the World

fl.   Pl. 39

Chrysanthemum catananche

Royal Hort. Soc.
Journal of the Royal Hort. Soc.
    vol 90, No. 9   1965

    fl.   p. 382   Pl. 175

Chrysanthemum cinerariaefolium

Bianchini, Francesco
Health Plants of the World

fl. hab.   p 179

Chrysanthemum cinerariaefolium

Kariyone, Tatsuo
Atlas of Medicinal Plants

fl.   Pl. 29

Chrysanthemum cinerariaefolium

Kimura, Koiti
Japanese Medicinal Plants, Vol II

fl.   p 237

Chrysanthemum cinerariaefolium

Perrot, Emile
Les Plantes Medicinales

fl. fr. hab.   p 67

Chrysanthemum cinerariaefolium

Takatori, Jisuke
Color Atlas of Medicinal Plants of Japan

fl.   Fig. 3A

Chrysanthemum coccineum

Kimura, Koiti
Japanese Medicinal Plants, Vol I

fl.   p 100

Chrysanthemum coccineum var.

Huxley, Anthony
Garden Perennials and Water Plants

fl.   Pl. 81, 82

Chrysanthemum coronarium

Crockett, James Underwood
Annuals

fl.   P 106

Chrysanthemum coronarium

Feinbrun-Dothan, Naomi
Wild Plants in the Land
   of Israel

fl. hab.   p 104

Chrysanthemum coronarium

Kleijn, H.
Beauty of the Wild Plant

fl. opp. p. 116   Pl. 168

Chrysanthemum coronarium

Masefield, G. B.
The Oxford Book of Food Plants

fl.   Pl. 155, 5

Chrysanthemum coronarium

Polunin, Oleg
Flowers of Europe

fl. Pl. 148 #1425

Chrysanthemum coronarium

Polunin, Oleg
Flowers of the Mediterranean

fl.   Pl.199

Chrysanthemum coronarium

Royal Hort. Soc.
Journal of the Royal Hort. Soc.
    vol 88, No. 9   1963

    fl.   p. 388   Pl. 140

Chrysanthemum coronarium

Taylor, A. W.
Wild Flowers of Spain and Portugal

fl.   p. 78

Chrysanthemum coronarium var.

Hay, Roy
The Color Dictionary of Flowers &
   Plants

fl.   p. 34, Pl. 265

Chrysanthemum coronarium var.

Huxley, Anthony
Garden Annuals and Bulbs.

fl.   Pl. 25, 27

Chrysanthemum coronarium var.

Polunin, Oleg
Flowers of the Mediterranean

fl.   Pl. 200

Chrysanthemum ferulaceum

Curtis's Botanical Magazine
Vol 178   1970-72

fl.   595

Chrysanthemum frutescens

Huxley, Anthony
Garden annuals and bulbs

fl.   pl. 28

Chrysanthemum frutescens

Macoby, Stirling
What Flower is That

fl.   P 84

Chrysanthemum frutescens

Tsukamoto, Yotaro
Coloured Illustrations of Garden Flowers
                    Vol. 9

fl.   opp. p. 38

Chrysanthemum frutescens var.

Harrison, R. E.
Trees and Shrubs

fl.   p. 50

Chrysanthemum gayanua

Curtis's Botanical Magazine
Vol 169   1952-53

fl.   183

Chrysanthemum haradjanii

Hay, Roy
The Color Dictionary of Flowers & Plants

hab.   p 5, Pl. 35

Chrysanthemum hybrid

Hay, Roy
The Color Dictionary of Flowers &
   Plants

fl.   p.56-59, 129-131 Pl.448-467,
                    1025-1048

Chrysanthemum hybrid

Huxley, Anthony
Garden Perennials and Water Plants.

fl.   Pl 79, 80

Chrysanthemum hybrid

Macoby, Stirling
What Flower is That

fl.   p. 84

Chrysanthemum hybrid

Perry, Frances
Flowers of the World

fl.   p. 78-79

Chrysanthemum hybridum var.

Kiaer, Eigil
Indoor Plants in Colour

fl.   p. 37, 38

Chrysanthemum indicum

Bruggeman, L.
Tropical Plants

fl.   Pl. 91

Chrysanthemum indicum

Kariyone, Tatsuo
Atlas of Medicinal Plants

fl.   Pl. 30

Chrysanthemum indicum

Walden, Beryl M.
Wild Flowers of Hong Kong

fl.   Pl. 74 (229)

Chrysanthemum integrifolium

Alaska-Yukon Wild Flower Guide

fl.   p 172

Chrysanthemum integrifolium

Heller, Christine
Wild Flowers of Alaska

fl.   Pl. 214

Chrysanthemum leucanthemum

Ary, S.
The Oxford Book of Wildflowers

fl.   p. 98   Pl. 2

Chrysanthemum leucanthemum

Batson, Wade T.
Wild Flowers in South Carolina

fl.   p 123

Chrysanthemum leucanthemum

Clark, Lewis J.
Wild Flowers of British Columbia

fl.   p 534

Chrysanthemum leucanthemum

Courtenay, Booth
Wildflowers & Weeds

fl.   p 110

Chrysanthemum leucanthemum

Crockett, James Underwood
Lawns & Ground Covers

fl.   p. 70

Chrysanthemum leucanthemum

Dean, Blanche E.
Wildflowers of Alabama and
Adjoining States

fl.   p 201

Chrysanthemum leucanthemum

Duncan, Wilbur H.
Wildflowers of the Southeastern
United States

fl.   p 225

Chrysanthemum leucanthemum

Felsko, Elsa
A Book of Wildflowers

fl.   P 140

Chrysanthemum leucanthemum

Ferguson, Mary
Wildflowers

fl. hab.   p 97

Chrysanthemum leucanthemum

Furrer, D.
Blumen am Wege

fl. hab.   p 119

Chrysanthemum leucanthemum

Heller, Christine
Wild Flowers of Alaska

fl.   Pl. 217

Chrysanthemum leucanthemum

Klimas, John E.
Wildflowers of Eastern America

fl.   Pl. 70

Chrysanthemum leucanthemum

Lemmon, Robert S.
Wildflowers of North America in
Full Color

fl.   pg. 213   pl. 338

Chrysanthemum leucanthemum

Lindman, C. A. M.
Nordens Flora, Vol 9

fl. hab.   Pl. 603

Chrysanthemum leucanthemum

Martin, W. Keble
The Concise British Flora in Colour

fl. hab.   Pl. 46

Chrysanthemum leucanthemum

Moyle, John B.
Northland Wild Flowers

fl.   p 165, Pl. 198

Chrysanthemum leucanthemum

Polunin, Oleg
Flowers of Europe

fl.   Pl. 148 #1427

Chrysanthemum leucanthemum

Roberts, June Carver
Born in the Spring

fl. hab.   p 137

Chrysanthemum leucanthemum

Szczawinski, Adam F.
Edible Garden Weeds of Canada

fl.   p. 46

Chrysanthemum leucanthemum

Taylor, Ronald J.
Mountain Wild Flowers

fl.   p 42

Chrysanthemum leucanthemum

Tsukamoto, Yotaro
Coloured Illustrations of Garden
Flowers

fl.   opp. p. 38

Chrysanthemum leucanthemum var.

Wharton, Mary E.
A Guide to the Wildflowers &
Ferns of Kentucky

fl.   p. 261   Pl. 337

Chrysanthemum maresii var.

Curtis's Botanical Magazine
1964-65

fl.   Pl. 454

Chrysanthemum maximum

Bruggeman, L.
Tropical Plants

fl.   Pl. 92

Chrysanthemum maximum

Everett, T. H.
New Illustrated Encyclopedia of
Gardening vol 3

fl.   Pl. (3-6a)

Chrysanthemum maximum

Macoby, Stirling
What Flower is That

fl.   P 84

Chrysanthemum maximum

Massachusetts Hort. Soc.
Horticulture
vol 42, No. 6   1964

fl.   p. 28-29

Chrysanthemum maximum var.

Hay, Roy
The Color Dictionary of Flowers &
Plants

fl.   p. 132   Pl. 1049

Chrysanthemum maximum var.

Huxley, Anthony
Garden Perennials and Water Plants

fl.   Pl. 83a,b,c.

Chrysanthemum, Mexican

see

Tridax procumbens

Chrysanthemum x morifolium var.

Tsukamoto, Yotaro
Coloured Illustrations of Garden
Flowers   Vol. 9

fl.   opp. p. 39, 40

Chrysanthemum myconis

Polunin, Oleg
Flowers of Europe

fl.   Pl. 148 #1424

Chrysanthemum nipponicum

Everett, T.H., ed.
New Illustrated Encyclopedia of
Gardening vol. 3

fl.   frontispiece

Chrysanthemum ochroleucum

Curtis's Botanical Magazine
Vol 166   1949

fl.   67

Chrysanthemum parthenium

Ary, S.
The Oxford Book of Wildflowers

fl.   p 98, Pl. 5

Chrysanthemum parthenium

Bianchini, Francesco
Health Plants of the World

fl.   p 57

Chrysanthemum parthenium

Crockett, James Underwood
Annuals

fl.   P 106

Chrysanthemum parthenium

Huxley, Anthony
Garden annuals and bulbs

fl.   pl. 29

Chrysanthemum parthenium

Macoby, Stirling
What Flower is That

fl.   P 84

Chrysanthemum parthenium

Martin, W. Keble
The Concise British Flora in Colour

fl. hab.   Pl. 46

Chrysanthemum parthenium

Polunin, Oleg
Flowers of Europe

fl.   Pl. 148 #1429

Chrysanthemum parthenium

Tsukamoto, Yotaro   vol 10
Coloured Illustrations of Garden Flowers

fl.   No. 34

Chrysanthemum ptarmiciflorum

Curtis's Botanical Magazine
Vol 178   1970-72

fl.   583

Chrysanthemum ptariciflorum

Macoby, Stirling
What Flower is That

hab.   p. 85

Chrysanthemum, red-flowered

See

Chrysanthemum coccineum

Chrysanthemum roseum

Hvass, Elsie
Plants That Feed and Serve Us

fl.   p 99, Pl. 211

Chrysanthemum x rubellum

Curtis's Botanical Magazine
Vol. 162   1939

fl.   No. 9566

Chrysanthemum segetum

Ary, S.
The Oxford Book of Wildflowers

fl.   p 38, Pl. 1

Chrysanthemum segetum

Crockett, James Underwood
Annuals

fl.   P 106

Chrysanthemum segetum

Lindman, C. A. M.
Nordens Flora, Vol 9

fl. hab.   Pl. 604

Chrysanthemum segetum

Martin, W. Keble
The Concise British Flora in Colour

fl. hab.   Pl. 46

Chrysanthemum segetum

Morley, Brian D.
Wild Flowers of the World

fl.   Pl. 39

Chrysanthemum segetum

Polunin, Oleg
Flowers of the Mediterranean

fl.   Pl. 198

Chrysanthemum segetum var.

Hay, Roy
The Color Dictionary of Flowers
& Plants.

fl   Pl. 266 p. 34

Chrysanthemum segetum var.

Huxley, Anthony
Garden Annuals and Bulbs.

fl.   Pl. 30

Chrysanthemum x superbum var.

Royal Hort. Soc.
The Garden
vol 103, No. 12   1978

fl.   p. 474

Chrysanthemum virginianum

Hay, Roy
The Color Dictionary of Flowers
& Plants

fl.   p. 132

Chrysanthemum vulgare

Color Treasury of Herbs & Other
Medicinal Plants

fl.   p 56   Pl. 88

Chrysanthemum vulgare

Felsko, Elsa
A Book of Wildflowers

fl.   P 40

Chrysanthemum vulgare

Lindman, C. A. M.
Nordens Flora, Vol 9

fl.   Pl. 602

Chrysanthemum vulgare

Martin, W. Keble
The Concise British Flora in Colour

fl. hab.   Pl. 46

Chrysanthemum vulgare

Polunin, Oleg
Flowers of Europe

fl.   Pl. 147 #1426

Chrysobactron rossii

Morley, Brian D.
Wild Flowers of the World

fl.   Pl. 7L

Chrysobalanus icaco

Massachusetts Hort. Soc.
Horticulture
   vol 53, No. 7   1975

fr., hab.   p. 28

Chrysocoma coma-aurea

Harrison, R.E.
Trees and Shrubs

fl.   P 51

Chrysocychnis schlimii

Ospina, Mariana
Orquideas de las Americas

fl. hab.   Pl. 120

Chrysogonum virginianum

Batson, Wade T.
Wild Flowers in South Carolina

fl.   p 119

Chrysogonum virginianum

Duncan, Wilbur H.
Wildflowers of the Southeastern
   United States

fl. hab.   p 211

Chrysogonum virginianum

Everett, T.H., ed.
New Illustrated Encyclopedia of
   Gardening   vol 3

fl.   p. 407

Chrysogonum virginianum

Hay, Roy
The Color Dictionary of Flowers & Plants

fl.   p 132, Pl. 1050

Chrysogonum virginianum

Massachusetts Hort. Soc.
Horticulture
   vol 39, No. 7   1961

fl.   inside cover

Chrysogonum virginianum

Vilmorin, Roger de
Plantes Alpines dans les Jardins

fl. hab.   Pl. XVI

Chrysogonum virginianum

Walcott, Mary Vaux
North American Wild Flowers

fl. hab.   Pl. 145

Chrysoma pauciflosculosa

Batson, Wade T.
Wild Flowers in South Carolina

fl.   p 115

Chrysophyllum magalismontanum

Palgrave, K. C.
Trees of Central Africa

fl., fr.   p. 409

Chrysopsis gossypina

Batson, Wade T.
Wild Flowers in South Carolina

fl. hab.   p 113

Chrysopsis graminifolia

Batson, Wade T.
Wild Flowers in South Carolina

fl.   p 113

Chrysopsis graminifolia

Brown, Clair A.
Wildflowers of Louisiana and
   Adjoining States

fl.   p 196

Chrysopsis hyssopifolia

Addisonia
   vol 21   1939-42

.   fl.   p. 11   Pl. 678

Chrysopsis mariana

Wharton, Mary E.
A Guide to the Wildflowers & Ferns
   of Kentucky

Chrysopsis nervosa

Wharton, Mary E.
A Guide to the Wildflowers & Ferns
   of Kentucky

fl.   p 278   Pl. 2.7

Chrysopsis villosa

Clark, Lewis J.
Wild Flowers of British Columbia

fl.   p 535

Chrysopsis villosa

Moyle, John B.
Northland Wild Flowers

fl. hab.   p 169, Pl. 205

Chrysopsis villosa

Porsild, A. E.
Rocky Mountain Wild Flowers

fl. hab.   p 407

Chrysosplenium alternifolium

Barneby, T. P.
European Alpine Flowers in Colour

fl.   Pl. 36; 2

Chrysosplenium alternifolium

Lindman, C. A. M.
Nordens Flora, Vol 5

fl. fr. hab.   Pl. 294

Chrysosplenium alternifolium

Martin, W. Keble
The Concise British Flora in Colour

fl. hab.   Pl. 32

Chrysosplenium americanum

Courtenay, Booth
Wildflowers & Weeds

fl.   p 42

Chrysosplenium Forrestii

Hara, Hiroshi, comp.
Photo-Album of Plants of
   Eastern Himalaya

fl.   Pl. 225

Chrysosplenium oppositifolium

Ary, S.
The Oxford Book of Wildflowers

fl.   p 14, Pl. 1

Chrysosplenium oppositifolium

Martin, W. Keble
The Concise British Flora in Colour

fl. hab.   Pl. 32

Chrysosplenium oppositifolium

Polunin, Oleg
Flowers of Europe

fl. Pl. 42   #412

Chrysosplenium tetrandrum

Heller, Christine
Wild Flowers of Alaska

fl.   Pl. 306

Chrysothamnus nauseosus

Clark, Lewis J.
Wild Flowers of British Columbia

fl.   p 511

Chrysothamnus nauseosus

Orr, Robert T.
Wildflowers of Western America

fl.   Pl. 93

Chrysothamnus nauseosus

Welsh, Stanley L.
Flowers of the Canyon Country

fl.   p 34

Chrysothamnus sp.

Mathias, Mildred E.
Color for the Landscape

fl. hab.   p 177

Chrysothamnus sp.

Pacific Hort. Foundation
California Hort. Journal
    vol 33, No. 4   1972

fl., hab.   backcover

Chrysothemis villosa

Elbert, Virginie F.
The Miracle Houseplants

fl.   p 116

Chuparosa

see

Beloperone californica

Chysis aurea

Wright, Norman Pelham
Orquideas de Mexico

fl.   Pl. 6

Chysis bractescens

Ebel, Friedrich
The Strange and Beautiful
    World of Orchids

fl.   p 117

Chysis bractescens

Kupper, Walter
Orchids

fl. hab.   p 51

Chysis bractescens

Wright, Norman Pelham
Orquideas de Mexico

fl.   Pl. 7

Chysis laevis

Ospina, Mariano
Orquideas de las americas

fl.   Pl. 42

Chytroglossa aurata

Pabst, G. F. J.
Orchidaceae Brasilienses,
    Vol 2

fl.   p 264

Cicely, sweet

see

Myrrhis odorata

Cicely, Sweet

see

Osmorhiza claytonia

Cicely, Sweet

see

Osmorhiza longistylis

Cicendia filiformis

Martin, W. Keble
The Concise British Flora in Colour

fl. hab.   Pl. 58

Cicer arietinum

Bianchini, F.
The Complete Book of Fruits & Vegetables

fr.   p 41

Cicer arietinum

Masefield, G. B.
The Oxford Book of Food Plants

fl. fr.   Pl. 39, 2

Cicerbita alpina

Barneby, T. P.
European Alpine Flowers in Colour

fl.   Pl. 92, 5

Cicerbita alpina

Kleijn, H.
Beauty of the Wild Plant

fl.   opp. p. 104   Pl. 151

Cicerbita alpina

Martin, W. Keble
The Concise British Flora in Colour

fl. hab.   Pl. 53

Cicerbita alpina

Morley, Brian D.
Wild Flowers of the World

fl.,hab.   Pl. 22

Cicerbita alpina

Polunin, Oleg
Flowers of Europe

fl.   Pl. 159 #1537

Cichorium endivia

Hvass, Elsie
Plants That Feed and Serve Us

hab.   p. 30   Pl. 55

Cichorium endivia

Masefield, G. B.
The Oxford Book of Food Plants

hab.   Pl. 151, 4

Cichorium endivia var.

Bianchini, F.
The Complete Book of Fruits & Vegetables

hab.   p 47

Cichorium intybus

Ary, S.
The Oxford Book of Wildflowers

fl.   p 178, Pl. 6

Cichorium intybus

Bianchini, Francesco
Health Plants of the World

fl. hab.   p 37

Cichorium intybus

Clark, Lewis J.
Wild Flowers of British Columbia

fl.   p. 542

Cichorium intybus

Color Treasury of Herbs & Other
    Medicinal Plants

fl.   p 64 Pl. 104

Cichorium intybus

Courtenay, Booth
Wildflowers & Weeds

fl.   p 107

Cichorium intybus

Dean, Blanche E.
Wildflowers of Alabama and
    Adjoining States

fl.   p 179

Cichorium intybus

Duncan, Wilbur H.
Wildflowers of the Southeastern
    United States

fl.   p 227

Cichorium intybus

Felsko, Elsa
A Book of Wildflowers

fl.   p. 58

Cichorium intybus

Furrer, D.
Blumen am Wege

fl. hab.   p 117

Cichorium intybus

Hvass, Elsie
Plants That Feed and Serve Us

fl.   p 30, Pl. 53

Cichorium intybus

Kleijn, H.
Beauty of the Wild Plant

fl.   opp. p. 69   Pl. 88

Cichorium intybus

Klimas, John E.
Wildflowers of Eastern America

fl.   Pl. 283

Cichorium intybus

Lemmon, Robert S.
Wildflowers of North America in
        Full Color

fl.   pg. 216   pl. 343

Cichorium intybus

Lindman, C. A. M.
Nordens Flora, Vol 10

fl.   Pl. 635

Cichorium intybus

Masefield, G. B.
The Oxford Book of Food Plants

fl.   Pl. 111, 3

Cichorium intybus

Morley, Brian D.
Wild Flowers of the World

fl.,hab.   Pl. 22

Cichorium intybus

Moyle, John B.
Northland Wild Flowers

fl.   p 195, Pl. 252

Cichorium intybus

Perrot, Emile
Les Plantes Medicinales

fl. fr. hab.   p 65

Cichorium intybus

Polunin, Oleg
Flowers of Europe

fl. Pl. 158 #1512

Cichorium intybus

Pond, Barbara
A Sampler of Wayside Herbs

fl. fr.   Pl. XX

Cichorium intybus

Roberts, June Carver
Born in the Spring

fl. hab.   p 147

Cichorium intybus

Welsh, Stanley L.
Flowers of the Canyon Country

fl.   p 47

Cichorium intybus

Wharton, Mary E.
A Guide to the Wildflowers & Ferns
    of Kentucky

fl.   p 241   Pl. 2.1

Cichorium intybus var.

Bianchini, F.
The Complete Book of Fruits & Vegetables

hab.   p 43, 45

Cichorium pumilum

Feinbrun-Dothan, Naomi
Wild Plants in the Land
    of Israel

fl. hab.   p 122

Cicuta bulbifera

Courtenay, Booth
Wildflowers & Weeds

fl.   p 58

Cicuta douglasii

Alaska-Yukon Wild Flower Guide

fl. hab.   p 102, 103

Cicuta douglasii

Clark, Lewis J.
Wild Flowers of British Columbia

fl.   p 351

Cicuta mackenzieana

Alaska-Yukon Wild Flower Guide

fl.   p 104

Cicuta mackenzieana

Heller, Christine
Wild Flowers of Alaska

fl.   Pl. 194-195

Cicuta maculata

Courtenay, Booth
Wildflowers & Weeds

fl.   p 58

Cicuta maculata

Klimas, John E.
Wildflowers of Eastern America

fl.   Pl. 84

Cicuta maculata

Moyle, John B.
Northland Wild Flowers

fl. hab.   p 101, Pl. 93

Cicuta maculata

Wharton, Mary E.
A Guide to the Wildflowers & Ferns
    of Kentucky

fl.   p 162   Pl. 2.31

Cicuta virosa

Lindman, C. A. M.
Nordens Flora, Vol 7

fl. fr.   Pl. 423

Cicuta virosa

Martin, W. Keble
The Concise British Flora in Colour

fl. fr. hab.   Pl. 37

Cicuta virosa

Perrot, Emile
Les Plantes Medicinales

fl. fr. hab.   p 69

Cigar bush

see

Cuphea platycentra

Cigar, Indian

see

Catalpa bignonioides

Cigar plant

see

Cuphea ignea

Cigar plant

see

Cuphea miniata

Cigar tree

see

Catalpa bignonioides

Cimicifuga americana

Campbell, Carlos C.
Great Smoky Mountain Wildflowers

fl.   p 83

Cimicifuga americana

Jennings, O. E.
Wild Flowers of Western Pennsylvania
vol II
fl. fr.   Pl. 60

Cimicifuga dahurica

Huxley, Anthony
Garden Perennials and Water
    Plants
fl.   Pl. 85

Cimicifuga racemosa

Courtenay, Booth
Wildflowers & Weeds

fl.   p 31

Cimicifuga racemosa

Dean, Blanche E.
Wildflowers of Alabama and
    Adjoining States
fl.   p 61

Cimicifuga racemosa

Hay, Roy
The Color Dictionary of Flowers & Plants

fl.   p 132, Pl. 1051

Cimicifuga racemosa

Jennings, O. E.
Wild Flowers of Western Pennsylvania
vol II
fl.   Pl. 59

Cimicifuga racemosa

Klimas, John E.
Wildflowers of Eastern America

fl.   Pl. 33

Cimicifuga racemosa

Perrot, Emile
Les Plantes Medicinales

fl. fr. hab.   p 70

Cimicifuga racemosa

Royal Hort. Soc.
The Garden
vol 100, No. 6 1975
fl.   p. 233

Cimicifuga racemosa

Wharton, Mary E.
A Guide to the Wildflowers & Ferns
    of Kentucky
fl.   p 152   Pl. 2.10

Cimicifuga simplex var.

Royal Hort. Soc.
The Garden
vol 100, No. 6 1975
fl.   p. 233

Cinchona calisaya

Hvass, Elsie
Plants That Feed and Serve Us

fl.   p 90, Pl. 191

Cinchona succirubra

Kariyone, Tatsuo
Atlas of Medicinal Plants

fl.   Pl. 54

Cinchona succirubra

Perrot, Emile
Les Plantes Medicinales

fl. fr. hab.   p 195

Cinderella Plant

see

Hylocereus undatus

Cineraria

see

Senecio hybrid

Cineraria maritima var.

Hay, Roy
The Color Dictionary of Flowers &
    Plants
hab.   p. 34   Pl. 267

Cineraria multiflora

Hay, Roy
The Color Dictionary of Flowers &
    Plants
fl.   p. 34   Pl. 268

Cinnamomum camphora

Crockett, James Underwood
Evergreens

hab.   p. 118

Cinnamomum camphora

Flemer, William III
Shade and Ornamental Trees in Color

hab.   p 92

Cinnamomum camphora

Hvass, Elsie
Plants That Feed and Serve Us

fl. fr. p 91, Pl. 193

Cinnamomum camphora

Kariyone, Tatsuo
Atlas of Medicinal Plants

fr.   Pl. 128

Cinnamomum Camphora

Kimura, Koiti
Japanese Medicinal Plants, Vol II

fl. fr.   p 171

Cinnamomum camphora

Kitamura, Siro
Coloured Illustrations of Trees &
    Shrubs of Japan
fl.   179

Cinnamomum camphora

Perrot, Emile
Les Plantes Medicinales

fl. fr. hab.   p 47

Cinnamomum camphora

Takatori, Jisuke
Color Atlas of Medicinal Plants of Japan

fl.
fr.   Fig. 47

Cinnamomum cassia

Perrot, Emile
Les Plantes Medicinales

fl. hab.   p 47

Cinnamomum cassia

Rosengarten, Frederic, Jr.
The Book of Spices

fl. fr.   p 194

Cinnamomum cassia

Takatori, Jisuke
Color Atlas Medicinal Plants of Japan

hab.   Fig 46B

Cinnamomum japonicum

Kitamura, Siro
Coloured Illustrations of Trees &
    Shrubs of Japan
fr.   177

Cinnamomum loureirii

Kitamura, Siro
Coloured Illustrations of Trees &
    Shrubs of Japan
hab.   178

Cinnamomum loureirii

Takatori, Jisuke
Color Atlas of Medicinal Plants of Japan

fl.   Fig. 46A

Cinnamomum sieboldii

Kariyone, Tatsuo
Atlas of Medicinal Plants

fr.   Pl. 129

Cinnamomum zeylanicum

Bianchini, F.
The Complete Book of Fruits & Vegetables

hab.   p 211

Cinnamomum zeylanicum

Bianchini, Francesco
Health Plants of the World

hab.   p 21

Cinnamomum zeylanicum

Hvass, Elsie
Plants That Feed and Serve Us

fl.   p 74, Pl. 153

Cinnamomum zeylanicum

Kimura, Koiti
Japanese Medicinal Plants, Vol II

fl.   p 173

Cinnamomum zeylanicum

Loewenfeld, Claire
The Complete Book of Herbs and Spices

fl. hab.   p 208

Cinnamomum zeylanicum

Masefield, G. B.
The Oxford Book of Food Plants

hab.   Pl. 131, 3

Cinnamomum zeylanicum

Perrot, Emile
Les Plantes Medicinales

fl. fr. hab.   p 47

Cinnamomum zeylanicum

Rosengarten, Frederic, Jr.
The Book of Spices

fl. fr.   p 186

Cinnamomum zeylanicum

Takatori, Jisuke
Color Atlas Medicinal Plants of Japan

hab.   Fig 46C

Cinnamon

see

Cinnamomum zeylanicum

Cinnamon Vine

see

Apios americana

Cinquefoil

see

Potentilla

Cinquefoil, Alpine

see

Potentilla crantsii

Cinquefoil, Arctic

see

Potentilla hyparctica

Cinquefoil, Beach

see

Potentilla pacifica

Cinquefoil, Bush

see

Potentilla fruticosa

Cinquefoil, Common

see

Potentilla simplex

Cinquefoil, Dwarf

see

Potentilla canadensis

Cinquefoil, Early

see

Potentilla canadensis

Cinquefoil, Fanleaf

see

Potentilla flabellifolia

Cinquefoil, Hairy

see

Potentilla villosa

Cinquefoil, hoary

see

Potentilla argentea

Cinquefoil, Marsh

see

Potentilla palustris

Cinquefoil, Norwegian

see

Potentilla norvegica var.

Cinquefoil, Purple

see

Potentilla palustris

Cinquefoil, Rosette

see

Potentilla uniflora

Cinquefoil, Rough-Fruited

see

Potentilla Recta

Cinquefoil, Shrubby

See

Potentilla Fruiticosa

Cinquefoil, Sticky

see

Potentilla glandulosa

Cinquefoil, Sulphur

see

Potentilla recta

Circaea alpina

Clark, Lewis J.
Wild Flowers of British Columbia

fl.   p 339

Circaea alpina

Lindman, C. A. M.
Nordens Flora, Vol 6

fl. hab.   Pl. 408A

Circaea alpina

Martin, W. Keble
The Concise British Flora in Colour

fl. hab.   Pl. 35

Circaea lutetiana

Ary, S.
The Oxford Book of Wildflowers

fl.   p 94, Pl. 4

Circaea lutetiana

Furrer, D.
Blumen am Wege

fl. hab.   p 67

Circaea lutetiana

Lindman, C. A. M.
Nordens Flora, Vol 6

fl. fr.   Pl. 408B

Circaea lutetiana

Martin, W. Keble
The Concise British Flora in Colour

fl. fr. hab.   Pl. 35

Circaea lutetiana

Polunin, Oleg
Flowers of Europe

fl.   Pl. 80 #829

Circaea quadrisulcata

Courtenay, Booth
Wildflowers & Weeds

fl.   p 75

Circaea quadrisulcata

Moyle, John B.
Northland Wild Flowers

fl.   p 98, Pl. 88

Circaea quadrisulcata

Wharton, Mary E.
A Guide to the Wildflowers & Ferns
  of Kentucky

fl.   p 150   Pl. 2.9

Cirio

see

Idria columnaris

Cirrhaea dependens

Pabst, G. F. J.
Orchidaceae Brasilienses,
  Vol 2

fl., fr.   p 222

Cirrhaea loddigesii

Pabst, G. F. J.
Orchidaceae Brasilienses,
  Vol 2

fl.   p 222

Cirrhaea saccata

Pabst, G. F. J.
Orchidaceae Brasilienses,
  Vol 2

fl.   p 223

Cirrhopetalum fascinator

Kupper, Walter
Orchids

fl. hab.   p 55

Cirrhopetalum makoyanum

Ebel, Friedrich
The Strange and Beautiful
  World of Orchids

fl. hab.   p 85

Cirrhopetalum mastersianum

Perry, Frances
Flowers of the World

fl.   p. 209

Cirrhopetalum medusae

Ebel, Friedrich
The Strange and Beautiful
  World of Orchids

fl.   p 65

Cirrhopetalum medusae

Kupper, Walter
Orchids

fl. hab.   p 57

Cirrhopetalum ornatissimum

Ebel, Friedrich
The Strange and Beautiful
  World of Orchids

fl.   p 25

Cirrhopetalum umbellatum

Perry, Frances
Flowers of the World

fl.   p. 209

Cirrhopetalum umbellatum

Stewart, Joyce
Orchids of Tropical Africa

fl.   Pl. 22

Cirsium acarna

Polunin, Oleg
Flowers of Europe

fl.   Pl. 153 #1488

Cirsium acaule

Ary, S.
The Oxford Book of Wildflowers

fl.   p 152, Pl. 4

Cirsium acaule

Barneby, T. P.
European Alpine Flowers in Colour

fl. hab.   Pl. 91, 2

Cirsium acaule

Felsko, Elsa
A Book of Wildflowers

fl.   P 159

Cirsium acaule

Felsko, Elsa
A Book of Wild Flowers
              2nd ser
fl.   p 106

Cirsium acaule

Kohlhaupt, Paula
Fleurs des Alpages
              vol 2
Fl.   P 117

Cirsium acaule

Martin, W. Keble
The Concise British Flora in Colour

fl. hab.   Pl. 48

Cirsium altissimum

Jennings, O. E.
Wild Flowers of Western Pennsylvania
              Vol. II

fl. hab.   Pl. 194

Cirsium arvense

Ary, S.
The Oxford Book of Wildflowers

fl.   p 152, Pl. 2

Cirsium arvense

Courtenay, Booth
Wildflowers & Weeds

fl.   p 109

Cirsium arvense

Crockett, James Underwood
Lawns & Ground Covers

fl.   p. 73

Cirsium arvense

Felsko, Elsa
A Book of Wildflowers

fl.   P 76

Cirsium arvense

Klimas, John E.
Wildflowers of Eastern America

fl.   Pl. 272

Cirsium arvense

Martin, W. Keble
The Concise British Flora in Colour

fl. hab.   Pl. 48

Cirsium arvense

Welsh, Stanley L.
Flowers of the Mountain Country

fl.   p. 28

Cirsium candelabrum

Polunin, Oleg
Flowers of Europe

fl.   Pl. 153 #1484

Cirsium carolinianum

Duncan, Wilbur H.
Wildflowers of the Southeastern
    United States

fl.   p 227

Cirsium coulteri

Munz, Philip A.
California Spring Wildflowers

fl.   p 81, Pl. 67

Cirsium discolor

Courtenay, Booth
Wildflowers & Weeds

fl.   p 109

Cirsium discolor

Lemmon, Robert S.
Wildflowers of North America in
    Full Color

fl.   pg. 214   pl. 340

Cirsium discolor

Moyle, John B.
Northland Wild Flowers

fl.   p 190, Pl. 243

Cirsium discolor

Wharton, Mary E.
A Guide to the Wildflowers & Ferns
    of Kentucky

fl.   p 239   Pl. 1.15

Cirsium dissectum

Ary, S.
The Oxford Book of Wildflowers

fl.   p 152, Pl. 5

Cirsium dissectum

Martin, W. Keble
The Concise British Flora in Colour

fl. hab.   Pl. 48

Cirsium edule

Clark, Lewis J.
Wild Flowers of British Columbia

fl.   p 542

Cirsium edule

Ferguson, Mary
Wildflowers

fl.   p 93

Cirsium edule

Fries, Mary A.
Wildflowers of Mount Ranier and
the Cascades

fl.   P 152

Cirsium edule

Taylor, Ronald J.
Mountain Wild Flowers

fl. hab.   p 144

Cirsium eriophorum

Ary, S.
The Oxford Book of Wildflowers

fl.   p 150, Pl. 2

Cirsium eriophorum

Barneby, T. P.
European Alpine Flowers in Colour

fl.   Pl. 90, 5

Cirsium eriophorum

Kleijn, H.
Beauty of the Wild Plant

fl.   opp. p. 69   pl. 89

Cirsium eriophorum

Martin, W. Keble
The Concise British Flora in Colour

fl. hab.   Pl. 48

Cirsium eriophorum

Polunin, Oleg
Flowers of Europe

fl.   Pl. 154 #1485

Cirsium eriophorum

Royal Hort. Soc.
The Garden
    vol.103, No. 2   1978

fr.   p. 53

Cirsium eriophorum

Tosco, Uberto
The World of Mountain Flowers

fl.   p 67

Cirsium eristhales

Barneby, T. P.
European Alpine Flowers in Colour

fl.   Pl. 91, 1

Cirsium eristhales

Royal Hort. Soc.
The Garden
    vol 103, No. 2   1978

fl.   p. 55

Cirsium flodmani

Moyle, John B.
Northland Wild Flowers

fl. hab.   p 191, Pl. 245

Cirsium foliosum

Welsh, Stanley L.
Flowers of the Mountain Country

fl., hab.   p. 28

Cirsium helenioides

Webster, Mary
Flora of Moray, Nairn & East
    Inverness

fl.   p. 292   Pl. 18

Cirsium heterophyllum

Barneby, T. P.
European Alpine Flowers in Colour

fl.   Pl. 92, 4

Cirsium heterophyllum

Kohlhaupt, Paula
Fleurs des Alpages
                vol 1
fl. P 118

Cirsium heterophyllum

Martin, W. Keble
The Concise British Flora in Colour

fl. hab.   Pl. 48

Cirsium hookerianum

Clark, Lewis J.
Wild Flowers of British Columbia

fl.   p 542

Cirsium hookerianum

Walcott, Mary Vaux
North American Wild Flowers

fl. hab.   Pl. 103

Cirsium horridulum

Brown, Clair A.
Wildflowers of Louisiana and
     Adjoining States

fl. hab.   p 194

Cirsium horridulum

Duncan, Wilbur H.
Wildflowers of the Southeastern
     United States

fl.   p 227

Cirsium horridulum

Lemmon, Robert S.
Wildflowers of North America in
     Full Color

fl.   pg. 52   pl. 87

Cirsium japonicum

Crockett, James Underwood
Annuals

fl.   P 106

Cirsium japonicum

Walden, Beryl M.
Wild Flowers of Hong Kong

fl.   Pl. 33(96)

Cirsium kamtschaticum

Alaska-Yukon Wild Flower Guide

fl.   p 186

Cirsium mexicanum

O'Gorman, Helen
Mexican Flowering Trees and Plants

fl. fr. hab.   p 145

Cirsium muticum

Brown, Clair A.
Wildflowers of Louisiana and
     Adjoining States

fl.   p 193

Cirsium muticum

Campbell, Carlos C.
Great Smoky Mountain Wildflowers

fl.   p 93

Cirsium muticum

Courtenay, Booth
Wildflowers & Weeds

fl.   p 109

Cirsium muticum

Jennings, O. E.
Wild Flowers of Western Pennsylvania
                              vol II
fl. hab.   Pl. 193

Cirsium muticum

Moyle, John B.
Northland Wild Flowers

fl.   p 190, Pl. 244

Cirsium muticum

Wharton, Mary E.
A Guide to the Wildflowers & Ferns
     of Kentucky

fl.   p 240   Pl. 1.16

Cirsium neomexicanum

Munz, Philip A.
California Desert Wildflowers

fl.   p 57, Pl. 92

Cirsium nutans

Brown, Clair A.
Wildflowers of Louisiana and
     Adjoining States

fl.   p 193

Cirsium oleraceum

Barneby, T. P.
European Alpine Flowers in Colour

fl.   Pl. 90, 6

Cirsium oleraceum

Lindman, C. A. M.
Nordens Flora, Vol 9

fl. fr.   Pl. 629

Cirsium oleraceum

Polunin, Oleg
Flowers of Europe

fl.   Pl. 154 #1482

Cirsium palustre

Ary, S.
The Oxford Book of Wildflowers

fl.   p 152, Pl. 3

Cirsium palustre

Lindman, C. A. M.
Nordens Flora, Vol 9

fl. fr.   Pl. 628

Cirsium palustre

Martin, W. Keble
The Concise British Flora in Colour

fl. hab.   Pl. 48

Cirsium pitcheri

Courtenay, Booth
Wildflowers & Weeds

fl.   p 109

Cirsium pumilum

Jennings, O. E.
Wild Flowers of Western Pennsylvania
                              vol II
fl. hab.   Pl. 192

Cirsium repandum

Batson, Wade T.
Wild Flowers in South Carolina

fl. hab.   p. 125

Cirsium rivulare var.

Curtis's Botanical Magazine
Vol 169   1952-53

fl.   217

Cirsium rivulare var.

Hay, Roy
The Color Dictionary of Flowers & Plants

fl.   p 132, Pl. 1052

Cirsium rivulare var.

Huxley, Anthony
Garden Perennials and Water Plants

fl.   pl. 84

Cirsium spinosissimum

Barneby, T. P.
European Alpine Flowers in Colour

fl.   Pl. 90. 4

Cirsium spinosissimum

Kleijn, H.
Beauty of the Wild Plant

fl.   opp. p. 104   pl. 154

Cirsium spinosissimum

Kohlhaupt, Paula
Fleurs des Alpages
                              vol 1
fl.   P 117

Cirsium spinosissimum

Polunin, Oleg
Flowers of Europe

fl.   Pl. 154 #1483

Cirsium tuberosum

Polunin, Oleg
Flowers of Europe

fl.   Pl. 153 #1489

Cirsium undulatum

Clark, Lewis J.
Wild Flowers of British Columbia

fl.   p 542

Cirsium undulatum

Neufeld, J.B.
Wild Flowers of the Prairies

fl.   p. 9

Cirsium undulatum

Walcott, Mary Vaux
North American Wild Flowers
            vol. 4

fl.   Pl. 309

Cirsium undulatum

Welsh, Stanley L.
Flowers of the Mountain Country

fl.   p. 29

Cirsium utahense

Welsh, Stanley L.
Flowers of the Canyon Country

fl.   p 10

Cirsium vulgare

Ary, S.
The Oxford Book of Wildflowers

fl.   p 150, Pl. 4

Cirsium vulgare

Clark, Lewis J.
Wild Flowers of British Columbia

fl.   p 543

Cirsium vulgare

Courtenay, Booth
Wildflowers & Weeds

fl.   p 109

Cirsium vulgare

Ferguson, Mary
Wildflowers

fl.   p 130

Cirsium vulgare

Jennings, O. E.
Wild Flowers of Western Pennsylvania
                    vol II
fl. hab.   Pl. 191

Cirsium vulgare

Klimas, John E.
Wildflowers of Eastern America

fl.   Pl. 273

Cirsium vulgare

Lindman, C. A. M.
Nordens Flora, Vol 9

fl. fr.   Pl. 627

Cirsium vulgare

Martin, W. Keble
The Concise British Flora in Colour

fl. hab.   Pl. 48

Cirsium vulgare

Milne, Lorus
Living Plants of the World

fl.   238

Cirsium vulgare

Moyle, John B.
Northland Wild Flowers

fl.   p 189, Pl. 242

Cirsium vulgare

Szczawinski, Adam F.
Edible Garden Weeds of Canada

fl., fr.   p. 49

Cirsium vulgare

Webster, Mary
Flora of Moray, Nairn & East
        Inverness

fl.   p. 292   Pl. 18

Cirsium vulgare

Welsh, Stanley L.
Flowers of the Canyon Country

fl. fr.   p 12

Cirsium vulgare

Wharton, Mary E.
A Guide to the Wildflowers & Ferns
    of Kentucky

fl.   p 239   Pl. 1.14

Cirsium wheeleri

Orr, Robert T.
Wildflowers of Western America

fl.   Pl. 172

Cissus adenopodus

Morley, Brian D.
Wild Flowers of the World

fl.   Pl. 58

Cissus antarctica

Hay, Roy
The Color Dictionary of Flowers & Plants

hab.   p 59, Pl. 469

Cissus antarctica

Hay, Roy
The Dictionary of House Plants

hab.   Pl. 129

Cissus antarctica

Hersey, Jean
Woman's Day Book of House Plants

hab.   p. 75

Cissus antarctica

Kromdijk, G.
200 House Plants in Colour

hab.   Pl. 50

Cissus cucurbitina

O'Gorman, Helen
Mexican Flowering Trees and Plants

fl. hab.   p 93

Cissus discolor

Harrison, Richmond E.
Climbers and Trailers

hab p 31   Pl. 44

Cissus discolor

Kiaer, Eigil
Indoor Plants in Colour

hab.   p. 39

Cissus discolor

Perry, Frances
Flowers of the World

hab. P 307

Cissus hypoglauca

Blombery, Alec M.
What Wildflower is That

fr.   p. 95   Pl. 231

Cissus rhombifolia

Hersey, Jean
Womans Day Book of House Plants

hab p. 69

Cissus rhombifolia

Macoby, Stirling
What Flower is That

hab P 85

Cissus striata

Kiaer, Eigil
Indoor Plants in Colour

hab.   p. 39

Cissus unguiformifolius

Flowering Plants of Africa
vol XXV   1945-46

fl., fr., hab.   Pl. 972

Cistanche tubulosa

Royal Hort. Soc.
Journal of the Royal Hort. Soc.
vol 96, No. 9   1971

fl.   p. 404   Pl. 168

Cistus albidus

Barneby, T. P.
European Alpine Flowers in Colour

fl.   Pl. 50, 4

Cistus albidus

Morley, Brian D.
Wild Flowers of the World

fl.   Pl. 34

Cistus albidus

Polunin, Oleg
Flowers of Europe

fl.   Pl. 77 #788

Cistus albidus

Polunin, Oleg
Flowers of the Mediterranean

fl.   Pl. 99

Cistus albiflorus

Harrison, R.E.
Trees and Shrubs

fl.   P 52

Cistus creticus

Morley, Brian D.
Wild Flowers of the World

fl.,fr.   Pl. 34

Cistus creticus

Pacific Hort. Foundation
Pacific Horticulture
vol 39, No. 3   1978

fl.   cover

Cistus creticus

Royal Hort. Soc.
Journal of the Royal Hort. Soc.
vol 97, No. 6   1972

fl.   p. 258   Pl. 115

Cistus crispus

Polunin, Oleg
Flowers of the Mediterranean

fl.   Pl. 106

Cistus x cyprius

Hellyer, A.G.L.
Shrubs in Colour

fl.   p. 27

Cistus x florentinus

Gault, S. Millar
The Color Dictionary of Shrubs

fl.   Pl. 77

Cistus, Gum

see

Cistus ladaniferus

Cistus hybrid

Gault, S. Millar
The Color Dictionary of Shrubs

fl.   Pl. 79, 83

Cistus hybrid

Harrison, R. E.
Trees and Shrubs

fl.   P. 52

Cistus hybrid

Hellyer, A.G.L.
Shrubs in Color

fl.   p. 27

Cistus hybrid

Mathias, Mildred E.
Color for the Landscape

fl. hab.   p 67

Cistus hybrid

Pacific Hort. Foundation
Pacific Horticulture
vol 39, No. 3   1978

fl.   p. 36-37

Cistus hybrid

Perry, Frances
Flowers of the World

fl.   p. 75

Cistus incanus

Goulimis, Constantine N.
Wild Flowers of Greece

fl.   p 60

Cistus incanus

Perry, Frances
Flowers of the World

fl.   p 75

Cistus incanus

Polunin, Oleg
Flowers of Europe

fl.   Pl. 77 #787

Cistus ladanifer

Massachusetts Hort. Soc.
Horticulture
vol 50, No. 8   1972

fl.   p. 35

Cistus ladanifer

Morley, Brian D.
Wild Flowers of the World

fl.   Pl. 34

Cistus ladanifer

Perrot, Emile
Les Plantes Medicinales

fl. fr. hab.   p 158

Cistus ladanifer

Polunin, Oleg
Flowers of Europe

fl.   Pl. 77 #793

Cistus ladanifer

Polunin, Oleg
Flowers of the Mediterranean

fl.   Pl. 104, 105

Cistus ladanifer

Polunin, Oleg
Trees and Bushes of Europe

fl. fr.   p 144

Cistus ladanifer

Royal Hort. Soc.
Journal of the Royal Hort. Soc.
vol 97, No. 6   1972

fl.   p. 258   Pl. 114

Cistus ladanifer

Royal Hort. Soc.
The Garden
vol 103, No. 5   1978

fl.   p. 216

Cistus ladanifer

Taylor, A. W.
Wild Flowers of Spain and Portugal

fl.   p. 98

Cistus laurifolius

Goulimis, Constantine N.
Wild Flowers of Greece

fl.   p 63

Cistus laurifolius

Hay, Roy
The Color Dictionary of Flowers & Plants

fl.   p 190, Pl. 1514

Cistus laurifolius

Polunin, Oleg
Flowers of Europe

fl. Pl. 77 #794

Cistus x loretii

Gault, S. Millar
The Color Dictionary of Shrubs

fl.   Pl. 78

Cistus x lusitanicus var.

Perry, Frances
Flowers of the World

fl.   p. 75

Cistus monspeliensis

Bramwell, David
Wild Flowers of the Canary Islands

fl.   Pl. 200

Cistus monspeliensis

Polunin, Oleg
Flowers of Europe

fl. Pl. 77 #791

Cistus monspeliensis

Polunin, Oleg
Flowers of the Mediterranean

fl.   Pl. 103

Cistus palhinhaii

Curtis's Botanical Magazine
Vol 168   1951

fl.   157

Cistus populifolius

Polunin, Oleg
Flowers of the Mediterranean

fl. fr.   Pl. 107

Cistus x pulverulentus

Gault, S. Millar
The Color Dictionary of Shrubs

fl.   Pl. 80

Cistus x pulverulentus

Hay, Roy
The Color Dictionary of Flowers & Plants

fl.   p. 190   Pl. 1515

Cistus x pulverulentus var.

Harrison, R.E.
Trees and Shrubs

fl.   p. 52

Cistus x purpureus

Gault, S. Millar
The Color Dictionary of Shrubs

fl.   Pl. 82

Cistus x purpureus

Hay, Roy
The Color Dictionary of Flowers &
  Plants

fl.   p. 190   Pl. 1516

Cistus x purpureus

Macoby, Stirling
What Flower is That

fl.   p. 85

Cistus x purpureus

Perry, Frances
Flowers of the World

fl.   p. 75

Cistus x purpureus

Royal Hort. Soc.
Journal of the Royal Hort. Soc,
  Vol. 95, No. 2   1970

fl,   p. 54   Pl. 46

Cistus x purpureus var.

Harrison, R. E.
Trees and Shrubs

fl.   p. 52

Cistus rosmarinifolius

Macoby, Stirling
What Flower is That

fl. P 85

Cistus salviifolius

Feinbrun-Dothan, Naomi
Wild Plants in the Land of
  Israel

fl. fr.   P. 58

Cistus salviifolius

Megaw, Elektra
Wild Flowers of Cyprus

fl.   p. 8   Pl. 7

Cistus salviifolius

Polunin, Oleg
Flowers of Europe

fl.   Pl. 77 #790

Cistus salviifolius

Polunin, Oleg
Flowers of the Mediterranean

fl.   Pl. 102

Cistus skanbergii

Curtis's Botanical Magazine
Vol 161   1938

fl.   No. 9514

Cistus symphytifolius

Bramwell, David
Wild Flowers of the Canary Islands

fl.   Pl. 200a

Cistus villosus

Feinbrun-Dothan, Naomi
Wild Plants in the Land
  of Israel

fl. fr.   p 56

Cistus villosus

Polunin, Oleg
Flowers of the Mediterranean

fl.   Pl. 100

Cistus villosus

Wit, H. C. D. de
Plants of the World;
  The Higher Plants, Vol I

fl.   p 166, Pl. 83

Cistus villosus var.

Polunin, Oleg
Flowers of the Mediterranean

fl.   Pl. 101

Citron

see

Citrus medica

Citrullus colocynthis

Hvass, Elsie
Plants That Feed and Serve Us

fl. fr.   p 98, Pl. 209

Citrullus colocynthis

Kimura, Koiti
Japanese Medicinal Plants, Vol II

fr.   p 233

Citrullus colocynthis

Polunin, Oleg
Flowers of Europe

fl. Pl. 78 #812

Citrullus sp.

Milne, Lorus
Living Plants of the World

fr. hab.   p 232

Citrullus vulgaris

Bianchini, F.
The Complete Book of Fruits & Vegetables

fr.   p 139

Citrullus vulgaris

Hvass, Elsie
Plants That Feed and Serve Us

fl. fr.   p 55, Pl. 102

Citrullus vulgaris

Masefield, G. B.
The Oxford Book of Food Plants

fl. fr.   Pl. 121, 1

Citrullus vulgaris

Weeds of the Southern United States

fr., fl., hab.   p. 19

Citrus aurantifolia

Hersey, Jean
Woman's Day Book of House Plants

fr.   p. 78

Citrus aurantifolia

Masefield, G. B.
The Oxford Book of Food Plants

fl. fr.   Pl. 87, 2

Citrus aurantifolia

Morley, Brian D.
Wild Flowers of the World

fl.fr.   Plate 119

Citrus aurantium

Bianchini, F.
The Complete Book of Fruits & Vegetables

fr.   p 183

Citrus aurantium

Boom, B. K.
The glory of the tree

fr.   opp. p. 89   pl. 150

Citrus aurantium

Linnell, T.
Plantes Utiles du Monde entier

fr.   p. 76

Citrus aurantium

Macoby, Stirling
What Flower is That

fr.   P 86

Citrus aurantium

Masefield, G. B.
The Oxford Book of Food Plants

fr.   Pl. 85, 3

Citrus aurantium

Morley, Brian D.
Wild Flowers of the World

fl.fr.   Plate 119

Citrus aurantium

Polunin, Oleg
Trees and Bushes of Europe

fl. fr. hab.   p 114

Citrus aurantium var.

Bianchini, F.
The Complete Book of Fruits &
    Vegetables

fr.   p. 183, 187

Citrus aurantium var.

Bianchini, Francesco
Health Plants of the World

fr.   p 41

Citrus aurantium var.

Macoby, Stirling
What Flower is That

fr.   P 86

Citrus aurantium var.

Takatori, Jisuke
Color Atlas of Medicinal Plants of Japan

fl.
fr.   Fig. 55A

Citrus bergamia

Polunin, Oleg
Trees and Bushes of Europe

fr.   p 116

Citrus deliciosa

Polunin, Oleg
Trees and Bushes of Europe

fr.   p 115

Citrus grandis

Polunin, Oleg
Trees and Bushes of Europe

fr.   p 115

Citrus japonica

Bianchini, F.
The Complete Book of Fruits & Vegetables

fr.   p 187

Citrus japonica

Kromdijk, G.
200 House Plants in Colour

fr.   Pl. 51

Citrus limetta

Polunin, Oleg
Trees and Bushes of Europe

fr.   p 116

Citrus limon

Edlin, Herbert
The Illustrated Encyclopedia
    of Trees

fr. hab.   p 222

Citrus limon

Masefield, G. B.
The Oxford Book of Food Plants

fl. fr.   Pl. 85, 4

Citrus limon

Polunin, Oleg
Flowers of Europe

fr.   Pl. 67 #693

Citrus limon

Polunin, Oleg
Trees and Bushes of Europe

fl. fr.   p 116

Citrus x limonia

Huxley, Anthony
Evergreen Garden Trees and Shrubs

fl.,fr.,hab.   Pl. 180

Citrus x limonia

Hvass, Elsie
Plants That Feed and Serve Us

fl., fr.   p. 48   Pl. 91

Citrus x limonia

Macoby, Stirling
What Flower is That

fr.   p. 86

Citrus x limonia

Perrot, Emile
Les Plantes Medicinales

fl.,fr.,hab.   p. 71

Citrus x limonia var.

Hersey, Jean
Woman's Day Book of House Plants

fr.   p. 76

Citrus medica

Masefield, G. B.
The Oxford Book of Food Plants

fl. fr.   Pl. 89, 1

Citrus medica

Polunin, Oleg
Trees and Bushes of Europe

fr.   p 115

Citrus medica var.

Bianchini, F.
The Complete Book of Fruits & Vegetables

fl. fr.   p 185

Citrus medica var.

Bianchini, Francesco
Health Plants of the World

fr.   p 29

Citrus microcarpa

Encke, Fritz
Zimmerpflanzen

fl.   p. 64

Citrus mitis

American Hort. Soc.
American Horticulturist
vol 57, No. 6 1978

fr.   p. 30

Citrus mitis

Hay, Roy
The Color Dictionary of Flowers & Plants

fr.   p 59, Pl. 470

Citrus mitis

Hay, Roy
The Dictionary of House Plants

fr.   Pl. 130

Citrus x nobilis

Macoby, Stirling
What Flower is That

fl.   p. 86

Citrus x nobilis var.

Everett, T.H., ed.
New Illustrated Encyclopedia of
    Gardening   Vol. 3

fl., fr.   p. 438

Citrus x nobilis var.

Huxley, Anthony
Evergreen Garden Trees and Shrubs

fr. hab.   Pl. 181

Citrus x nobilis var.

Hvass, Elsie
Plants That Feed and Serve Us

Fr.   p. 47   Pl. 89

Citrus notsudaidai

Kimura, Koiti
Japanese Medicinal Plants, Vol I

fl.   p 54

Citrus x paradisi

Hvass, Elsie
Plants That Feed and Serve Us

fr.   p. 48   Pl. 90

Citrus x paradisi

Masefield, G. B.
The Oxford Book of Food Plants

fr.   Pl. 87, 1

Citrus x paradisi

Polunin, Oleg
Trees and Bushes of Europe

fr.   p. 115

Citrus reticulata

Masefield, G. B.
The Oxford Book of Food Plants

fr.   Pl. 87, 3

Citrus reticulata var.

Bianchini, F.
The Complete Book of Fruits & Vegetables

fr.   p 187

Citrus sinensis

Crockett, James Underwood
Evergreens

hab., fr., fl.   p. 119

Citrus sinensis

Flemer, William III
Shade and Ornamental Trees in Color

hab.   p. 57

Citrus sinensis

Huxley, Anthony
Evergreen Garden Trees and Shrubs

fl. fr. hab.   Pl. 179

Citrus sinensis

Hvass, Elsie
Plants That Feed and Serve Us

fl. fr.   p 47, Pl. 88

Citrus sinensis

Masefield, G. B.
The Oxford Book of Food Plants

fl., fr.   Pl. 85, 1-2

Citrus sinensis

Perry, Frances
Flowers of the World

fl.   p 268

Citrus sinensis

Polunin, Oleg
Flowers of Europe

fl. fr. Pl. 67 #690

Citrus sinensis

Polunin, Oleg
Trees and Bushes of Europe

fr. hab.   p 114

Citrus sinensis

Wit, H. C. D. de
Plants of the World;
The Higher Plants, Vol II

fl. fr.   p 57, Pl. 21, 23

Citrus taitensis

Hersey, Jean
Woman's Day Book of House Plants

fr.   frontispiece, p. 81

Citrus unshu

Kariyone, Tatsuo
Atlas of Medicinal Plants

fl., fr.   Pl. 91

Citrus vulgaris

Perrot, Emile
Les Plantes Medicinales

fl. fr. hab.   p 32

Cladanthus arabicus

Crockett, James Underwood
Annuals

fl.   P 107

Cladanthus arabicus

Curtis's Botanical Magazine
vol 176   1966-68

fl.   Pl. 490

Cladanthus arabicus

Polunin, Oleg
Flowers of the Mediterranean

fl.   Pl. 201

Cladium mariscus

Lindman, C. A. M.
Nordens Flora, Vol 2

fr. hab.   Pl. 117

Cladium mariscus

Martin, W. Keble
The Concise British Flora in Colour

fl. hab.   Pl. 91

Cladium sinclairi

Harvey, Norman B.
New Zealand Botanical Paintings

fr.   p. 84   Pl. 39

Cladothamnus pyroliflorus

Clark, Lewis J.
Wild Flowers of British Columbia

fl.   p. 378

Cladrastis lutea

American Hort. Soc.
American Horticulturist
    vol 56, No. 3 1977

fl., hab.   p. 42

Cladrastis lutea

Huxley, Anthony
Deciduous Garden Trees and Shrubs

fl.   Pl. 38

Cladrastis lutea

Wharton, Mary E.
Trees & Shrubs of Kentucky

fl.   p 57, Pl. 1.22

Cladrastis sikokiana

Kitamura, Siro
Coloured Illustrations of Trees
    & Shrubs of Japan

hab.   251

Clammy Weed

    see

Polanisia dodecandra

Clappertonia ficifolia

Morley, Brian D.
Wild Flowers of the World

fl.,fr.   Pl. 55

Claret Cup

    see

Echinocereus triglochidiatus

Clarkia amoena

Clark, Lewis J.
Wild Flowers of British Columbia

fl.   p 339

Clarkia amoena

Morley, Brian D.
Wild Flowers of the World

fl.   Pl. 156

Clarkia amoena

Munz, Philip A.
California Spring Wildflowers

fl.   p 30, Pl. 12

Clarkia amoena

Orr, Robert T.
Wildflowers of Western America

fl.   Pl. 211

Clarkia bottae

Munz, Philip A.
California Spring Wildflowers

fl.   p 42, Pl. 48

Clarkia elegans

Crockett, James Underwood
Annuals

fl.   P 107

Clarkia elegans var.

Hay, Roy
The Color Dictionary of Flowers &
    Plants

fl.   p. 34   Pl. 269

Clarkia elegans

Huxley, Anthony
Garden Annuals and bulbs

fl.   Pl. 31

Clarkia elegans

Macoby, Stirling
What Flower is That

fl.   P 86

Clarkia grandiflora

Lemmon, Robert S.
Wildflowers of North America in
    Full Color

fl.   pg. 36   pl. 61

Clarkia hybrid

Perry, Frances
Flowers of the World

fl.   p. 204

Clarkia pulchella

Clark, Lewis J.
Wild Flowers of British Columbia

fl.   p 339

Clarkia pulchella var.

Crockett, James Underwood
Annuals

fl   P 107

Clarkia rubicunda

Orr, Robert T.
Wildflowers of Western America

fl.   Pl. 213

Clarkia unguiculata

Perry, Frances
Flowers of the World

fl.   p 204

Clary

    see

Salvia horminum

Clary

    see

Salvia sclarea

Clary, meadow

    see

Salvia pratensis

Clary, Yellow

    see

Salvia glutinosa

Claviceps purpurea

Bianchini, Francesco
Health Plants of the World

fr.   p 151

Claviceps purpurea

Kariyone, Tatsuo
Atlas of Medicinal Plants

fr.   Pl. 148

Claytonia lanceolata
Porsild, A. E.
Rocky Mountain Wild Flowers
fl. hab.   p 153

Cleisostoma fordii

Walden, Beryl M.
Wild Flowers of Hong Kong

fl.   Pl. 54 (156)

Cleisostoma teres

Hu, Shiu-ying
The Genera of Orchidaceae in
    Hong Kong

fl.   p. 123

Cleisostoma teres

Walden, Beryl M
Wild Flowers of Hong Kong

fl. hab.   Pl. 71 (220)

Cleistes divaricata

Brown, Clair A.
Wildflowers of Louisiana and
    Adjoining States

fl.   p 35

Cleistes divaricata

Campbell, Carlos C.
Great Smoky Mountain Wildflowers

fl.   p 59

Cleistes divaricata

Dean, Blanche E.
Wildflowers of Alabama and
    Adjoining States

fl.   p 39

Cleistes divaricata

Duncan, Wilbur H.
Wildflowers of the Southeastern
    United States

fl.   p 275

Cleistes divaricata

Klimas, John E.
Wildflowers of Eastern America

fl.   Pl. 230

Cleistes divaricata

Luer, Carlyle A.
The Native Orchids of Florida

fl.   Pl. 7, 8

Cleistes divaricata

Luer, Carlyle A.
The Native Orchids of the
    United States and Canada

fl. hab.   Pl. 66

Cleistes divaricata

Wharton, Mary E.
A Guide to the Wildflowers & Ferns
    of Kentucky

fl. p 69 Pl. 2.10

Cleistes revoluta

Pabst, G. F. J.
Orchidaceae Brasilienses,
    Vol 1

fl.   p 178

Cleistes rosea

Ospina, Mariano
Orquideas de las americas

fl.   Pl. 35

Cleistocactus azerensis

Backeberg, Curt
Cactus Lexicon

fl.   p 576

Cleistocactus baumannii

Van Laren, A. J.
Cactus

fl.   P 20   Fig. 17

Cleistocactus brookei

Backeberg, Curt
Cactus Lexicon

fl.   p 576

Cleistocactus candelilla

Lamb, Edgar
The Pocket Encyclopedia of Cacti
    and Succulents in Color

fl. hab.   Pl. 18

Cleistocactus dependens

Backeberg, Curt
Cactus Lexicon

fl.   p 576

Cleistocactus jujuyensis

Lamb, Edgar
The Pocket Encyclopedia of Cacti
    and Succulents in Color

fl.   Pl. 14

Cleistocactus strausii

Hay, Roy
The Dictionary of House Plants

hab.   Pl. 131

Cleistocactus strausii

Kupper, Walter
Cacti

fl., hab.   p. 33   Pl. 13

Cleistocactus strausii

Lamb, Edgar
The Pocket Encyclopedia of Cacti
    and Succulents in Color

hab.   Pl. 300

Cleistocactus vallegrandensis

Backeberg, Curt
Cactus Lexicon

fl.   p 576

Cleistocactus viridiflorus

Backeberg, Curt
Cactus Lexicon

fl.   p 577

Cleistocactus vulpis-cauda

Backeberg, Curt
Cactus Lexicon

fl.   p 577

Cleistocactus wendlandiorum

Lamb, Edgar
Popular Exotic Cacti in Color

fl. hab.   p 53

Clematis alpina

Barneby, T. P.
European Alpine Flowers in Colour

fl.   Pl. 23, 3

Clematis alpina

Bartels, Andreas
Das Grosse Buch der Gartengeholze

fl.   p 130

Clematis alpina

Felsko, Elsa
A Book of Wild Flowers
            2nd ser.
fl.   p 42

Clematis alpina

Harrison, Richmond E.
Climbers and Trailers

fl. p 31   Pl. 45

Clematis alpina

Kohlhaupt, Paula
Fleurs des Alpages
            vol 1
fl. P 11

Clematis alpina

Polunin, Oleg
Flowers of Europe

fl. Pl. 23 #227

Clematis alpina

Robert, Paul A.
Alpine Flowers

fl., hab.   Pl. 3

Clematis alpina var.

Gault, S. Millar
The Color Dictionary of Shrubs

fl.   Pl. 81

Clematis aristata

Blombery, Alec M.
What Wildflower is That

fl.   p. 97   Pl. 233

Clematis aristata

Cochrane, G. R.
Flowers and Plants of Victoria

fl.   Pl. 424

Clematis aristata

Harrison, Richmond E.
Climbers and Trailers

fl.   fr.   p 32   Pl. 47

Clematis aristata

Mullins, Barbara
Australian Wildflowers in Colour

fl.   p. 79   Pl. 75

Clematis armandii

Harrison, Richmond E.
Climbers and Trailers

fl.   p 32   Pl. 46

Clematis armandii

Hay, Roy
The Color Dictionary of Flowers & Plants

fl.   p 245, Pl. 1960

Clematis armandii

Mathias, Mildred E.
Color for the Landscape

fl.   hab.   p 106

Clematis armandii

Menninger, Edwin A.
Flowering Vines of the World

fl.   Pl. 175

Clematis armandii

Royal Hort. Soc.
Journal of the Royal Hort. Soc.
vol 91, No. 10 1966

fl.   p. 426   Pl. 212

Clematis armandii

Royal Hort. Soc.
The Garden
vol 101, No. 10   1976

fl.   p. 500

Clematis armandii var.

Royal Hort. Soc.
Journal of the Royal Hort. Soc.
vol 88, No. 11   1963

fl.   p. 478   Pl. 189

Clematis armandii var.

Royal Hort. Soc.
Journal of the Royal Hort. Soc.
vol 91, No. 1   1966

fl.   p. 22   Pl. 14

Clematis baldwinii

Fleming, Glenn
Wild Flowers of Florida

fl. fr.   p 85

Clematis brachiata

Flowering Plants of Africa
vol 30   1954-55

fl., fr.   Pl. 1197

Clematis chrysantha

Royal Hort. Soc.
Journal of the Royal Hort. Soc.
vol 98, No. 8   1973

fl.   p. 354   Pl. 177

Clematis chrysocoma var.

Harrison, Richmond E.
Climbers and Trailers

fl.   p 35   Pl. 57

Clematis cirrhosa

Harrison, Richmond E.
Climbers and Trailers

fl.   p 32   Pl. 48

Clematis cirrhosa

Megaw, Elektra
Wild Flowers of Cyprus

fl.   p. 7   Pl. 1

Clematis cirrhosa

Polunin, Oleg
Trees and Bushes of Europe

fl.   p 68

Clematis cirrhosa var.

Royal Hort. Soc.
The Garden
vol 102, No. 6   1977

fl.   p. 271

Clematis colensoi var.

Curtis's Botanical Magazine
Vol. 170   1954-55

fl.   250

Clematis columbiana

Clark, Lewis J.
Wild Flowers of British Columbia

fl.   p 166

Clematis columbiana

Porsild, A.E.
Rocky Mountain Wild Flowers

fl.   p 177

Clematis columbiana

Walcott, Mary Vaux
North American Wild Flowers

fl. fr. hab.   Pl. 99, 100

Clematis columbiana

Welsh, Stanley L.
Flowers of the Canyon Country

fl.   p 50

Clematis crispa

Addisonia
vol 24   1960-64

fl., fr.   p. 59   Pl. 798

Clematis crispa

Brown, Clair A.
Wildflowers of Louisiana and
Adjoining States

fl. hab.   p 51

Clematis crispa

Walcott, Mary Vaux
North American Wild Flowers

fl. hab.   Pl. 150

Clematis dioscoreifolia

Brown, Clair A.
Wildflowers of Louisiana and
Adjoining States

fl. hab.   p 51

Clematis durandii

Harrison, Richmond E.
Climbers and Trailers

fl.   p. 36   Pl. 61

Clematis, dwarf

see

Clematis gentianoides

Clematis x eriostemon var.

Gault, S. Millar
The Color Dictionary of Shrubs

fl.   Pl. 84

Clematis, Evergreen
see
Clematis armandii

Clematis, Father David's
see
Clematis heracleifolia var.

Clematis flammula

Harrison, Richmond E.
Climbers and Trailers

fl.   p 33   Pl. 49

Clematis flammula

Polunin, Oleg
Flowers of Europe

fl.   Pl. 24 #225

Clematis flammula

Polunin, Oleg
Trees and Bushes of Europe

fl. fr.   p 67

Clematis florida

Kimura, Koiti
Japanese Medicinal Plants, Vol I

fl.   p 114

Clematis florida var.

Harrison, Richmond E.
Climbers and Trailers

fl.   p 33   Pl. 50

Clematis florida var.

Menninger, Edwin A.
Flowering Vines of the World

fl.   Pl. 172

Clematis gentianoides

Curtis, Winifred
The Endemic Flora of Tasmania
Vol. 1

fl. fr.   Pl. 8

Clematis glycinoides

Blombery, Alec M.
What Wildflower is That

fl.   p 97, Pl. 234

Clematis, Golden
see
Clematis tangutica

Clematis gouriana

Roxburgh, William
Icones Roxburghianae

fl., fr.   Pl. 1

Clematis grewiaeflora

Hara, Hiroshi, comp.
Photo-Album of Plants of
Eastern Himalaya

fl.   Pl. 70

Clematis heracleifolia var.

Addisonia
vol 22   1943-46

fl., fr.   p. 11   Pl. 710

Clematis heracleifolia var.

Hay, Roy
The Color Dictionary of Flowers &
Plants

fl.   p. 132   Pl. 1053

Clematis hirsutissima

Lemmon, Robert S.
Wildflowers of North America in
Full Color

fl.   pg. 177   pl. 279

Clematis hirsutissima

Orr, Robert T.
Wildflowers of Western America

fl.   Pl. 217

Clematis hirsutissima

Welsh, Stanley L.
Flowers of the Mountain Country

fl.   p. 75

Clematis hybrid.

Bartels, Andreas
Das Grosse Buch der Gartengeholze

fl.   p   131

Clematis hybrid

Blunt, Wilfrid
Flora Superba

fl.   Pl. XVI

Clematis hybrid

Gault, S. Millar
The Color Dictionary of Shrubs

fl.   Pl. 93 thru 99

Clematis hybrid

Hansen, Richard
Baume und Straucher im Garten

fl.   p. 208

Clematis hybrid

Harrison, Richmond E.
Climbers and Trailers

fl.   p. 37-42   Pl. 65-80

Clematis hybrid.

Hay, Roy
The Color Dictionary of Flowers & Plants

fl.   p 246, 247; Pl. 1967 thru 1973

Clematis hybrid

Huxley, Anthony
Deciduous Garden Trees and Shrubs

fl.   Pl. 286

Clematis hybrid

Menninger, Edwin A.
Flowering Vines of the World

fl.   Pl. 169-70, 173-74

Clematis hybrid

Perry, Frances
Flowers of the World

fl.   p. 254-55

Clematis hybrid

Tsukamoto, Yotaro
Coloured Illustrations of Garden
Flowers   Vol. 9

fl.   p. 40

Clematis indivisa

Royal Hort. Soc.
Journal of the Royal Hort. Soc.
vol 91, No. 1   1966

fl.   p. 22   Pl. 20

Clematis integrifolia

Polunin, Oleg
Flowers of Europe

fl.   Pl. 24 #229

Clematis x jackmanii

Harrison, Richmond E.
Climbers and Trailers

fl.   p. 40   Pl. 73

Clematis x jackmanii

Hay, Roy
The Color Dictionary of Flowers &
Plants

fl.   p. 247   Pl. 1969

Clematis x jackmanii

Huxley, Anthony
Deciduous Garden Trees and Shrubs

fl.   Pl. 286b

Clematis x jackmanii
Menninger, Edwin A.
Flowering Vines of the World

fl.   Pl. 171

Clematis x jackmanii
Perry, Frances
Flowers of the World

fl.   p. 254

Clematis x jackmanii var.
Bartels, Andreas
Das Gross Buch der Gartengeholze

fl.   p. 130

Clematis x jackmanii var.
Hellyer, A.G.L.
Shrubs in Colour

fl.   p. 30

Clematis x jouiniana
Harrison, Richmond E.
Climbers and Trailers

fl.   p. 36   Pl. 62

Clematis x jouiniana
Hay, Roy
The Color Dictionary of Flowers &
Plants

fl.   p. 246   Pl. 1961

Clematis x jouiniana var.
Gault, S. Millar
The Color Dictionary of Shrubs

fl.   Pl. 85

Clematis lanuginosa var.
Macoby, Stirling
What Flower is That

fl.   P. 87

Clematis lanuginosa var.
Massachusetts Hort. Soc.
Horticulture
     vol 42, No. 7   1964

fl.   cover, backcover

Clematis lasiantha
Munz, Philip A.
California Spring Wildflowers

fl.   p 87, Pl. 87

Clematis x lawsoniana
Massachusetts Hort. Soc.
Horticulture
     Vol. 38, No. 6   1960

fl.   p. 319

Clematis x lawsoniana var.
Massachusetts Hort. Soc.
Horticulture
     Vol. 39, No. 5   1962

fl.   backcover

Clematis ligusticifolia
Clark, Lewis J.
Wild Flowers of British Columbia

fr.   p 163

Clematis macropetala
Harrison, Richmond E.
Climbers and Trailers

fl.   p 33   Pl. 51

Clematis macropetala
Hay, Roy
The Color Dictionary of Flowers & Plants

fl.   p 246, Pl. 1962

Clematis macropetala
Hellyer, A.G.L.
Shrubs in Colour

fl.   p. 30

Clematis meyeniana
Walden, Beryl M.
Wild Flowers of Hong Kong

fl.   Pl. 48 (143), 51 (143)

Clematis microphylla
Cochrane, G. R.
Flowers and Plants of Victoria

fl.   Pl. 210

Clematis microphylla
Morley, Brian D.
Wild Flowers of the World

fl.fr.   Plate 128

Clematis montana
Bartels, Andreas
Das Grosse Buch der Gartengeholze

fl.   p 130

Clematis montana
Macoby, Stirling
What Flower is That

fl.   P 87

Clematis montana var.
Gault, S. Millar
The Color Dictionary of Shrubs

fl.   Pl. 86

Clematis montana var.
Hansen, Richard
Baume und Straucher im Garten

fl.   p. 208

Clematis montana var.
Harrison, Richmond E.
Climbers and Trailers

fl.   p. 35, 36   Pl. 58,59,60

Clematis montana var.
Hay, Roy
The Color Dictionary of Flowers & Plants

fl.   p 246, Pl. 1963

Clematis montana var.
Huxley, Anthony
Deciduous Garden Trees and Shrubs

fl.   Pl. 284

Clematis montana var.
Massachusetts Hort. Soc.
Horticulture
     vol 35, No. 8   1957

fl.   p. 421

Clematis montana var.
Massachusetts Hort. Soc.
Horticulture
     vol 44, No. 1   1966

fl.   p. 37

Clematis montana var.
Royal Hort. Soc.
The Garden
     vol 101, No. 10   1976

fl.   p. 500

Clematis nannophylla
Curtis's Botanical Magazine
     Vol 163   1940 - 42

fl.   No. 9641

Clematis orientalis
Gault, S. Millar
The Color Dictionary of Shrubs

fl. fr.   Pl. 87

Clematis orientalis
Harrison, Richmond E.
Climbers and Trailers

fl.   p 33   Pl. 52

Clematis orientalis
Hay, Roy
The Color Dictionary of Flowers & Plants

fl.   p 246, Pl. 1964

Clematis orientalis

Perry, Frances
Flowers of the World

fl.   p 255

Clematis orientalis

Royal Hort. Soc.
Journal of the Royal Hort. Soc.
vol 91, No. 1   1966

fl.   p. 22   Pl. 21

Clematis paniculata

Bruggeman, L.
Tropical Plants

fl.   Pl. 7

Clematis paniculata

Harrison, Richmond E.
Climbers and Trailers

fl.   p 34   Pl. 53

Clematis paniculata

Harvey, Norman B.
New Zealand Botanical Paintings

fl.   p. 22   Pl. 8

Clematis paniculata

Royal Hort. Soc.
The Garden
vol 101, No. 10   1976

fl.   p. 515

Clematis paniculata

Takatori, Jisuke
Color Atlas of Medicinal Plants of Japan

fl.   Fig. 54B

Clematis patens

Kimura, Koiti
Japanese Medicinal Plants, Vol II

fl.   p 159

Clematis patens var.

Hellyer, A.G.L.
Shrubs in Colour

fl.   p. 30-31

Clematis patens var.

Tsukamoto, Yotaro
Coloured Illustrations of Garden
        Flowers Vol. 9

fl.   p. 40

Clematis phlebantha

Curtis's Botanical Magazine
Vol 178   1970-72

fl.   574

Clematis pseudoalpina

Welsh, Stanley L.
Flowers of the Mountain Country

fl.   p. 76

Clematis pubescens

Blomberg, Alec M.
What Wildflower is That

fl. hab.   p 97, Pl. 235

Clematis recta

Hay, Roy
The Color Dictionary of Flowers & Plants

fl.   p 132, Pl. 1054

Clematis rehderiana

Curtis's Botanical Magazine
    vol 176   1966-68

fl.   Pl. 523

Clematis rehderiana

Gault, S. Millar
The Color Dictionary of Shrubs.

fl.   Pl. 88

Clematis rehderiana

Harrison, Richmond E.
Climbers and Trailers

fl.   p 34   Pl. 54

Clematis rehderiana

Hay, Roy
The Color Dictionary of Flowers & Plants

fl.   p 246, Pl. 1965

Clematis reticulata

Duncan, Wilbur H.
Wildflowers of the Southeastern
    United States

fl.   p 45

Clematis, Scarlet

see

Clematis texensis

Clematis sp.

Everett, T.H., ed.
New Illustrated Encyclopedia of
    Gardening vol 13

fl.   p. 2407

Clematis tangutica

Gault, S. Millar
The Color Dictionary of Shrubs

fl. fr.   Pl. 89

Clematis tangutica

Harrison, Richmond E.
Climbers and Trailers

fl. fr.   p. 34   Pl. 55

Clematis tangutica

Hay, Roy
The Color Dictionary of Flowers & Plants

fl. fr.   p 246, Pl. 1966

Clematis tangutica

Hellyer, A.G.L.
Shrubs in Colour

fl.   p. 30

Clematis tangutica

Huxley, Anthony
Deciduous Garden Trees and Shrubs

fl.   Pl. 287

Clematis terniflora

Kimura, Koiti
Japanese Medicinal Plants, Vol I

fl.   p 28

Clematis texensis

Addisonia
    vol 21   1939-42

fl., fr.   p. 5   Pl. 675

Clematis texensis

Harrison , Richmond E.
Climbers and Trailers

fl. p 34   Pl. 56

Clematis texensis

Royal Hort. Soc.
Journal of the Royal Hort. Soc.
vol 95, No. 4   1970

fl.   p. 166   Pl. 112

Clematis texensis var.

Gault, S. Millar
The Color Dictionary of Shrubs

fl.   Pl. 90

Clematis texensis var.

Hellyer, A.G.L.
Shrubs in Colour

fl.   p. 30

Clematis verticillaris

Courtenay, Booth
Wildflowers & Weeds

fl.   p 30

Clematis verticillaris

Ferguson, Mary
Wildflowers

fl.   p 122

Clematis viorna

Batson, Wade T.
Wild Flowers in South Carolina

fl.   p 46

Clematis viorna

Dean, Blanche E.
Wildflowers of Alabama and
   Adjoining States

fl.   p 63

Clematis viorna

Jennings, O. E.
Wild Flowers of Western Pennsylvania
                    vol II
fl. fr.   Pl. 64

Clematis viorna

Klimas, John E.
Wildflowers of Eastern America

fl.   Pl. 265

Clematis viorna

Walcott, Mary Vaux
North American Wild Flowers

fl. hab.   Pl. 41

Clematis viorna

Wharton, Mary E.
A Guide to the Wildflowers & Ferns
   of Kentucky

fl. fr.   p. 125   Pl. 1.61

Clematis viorna var.

Royal Hort. Soc.
Journal of the Royal Hort. Soc.
   vol 95, No. 4   1970

fl.   p. 166   Pl. 113

Clematis virginiana

Campbell, Carlos G.
Great Smoky Mountain Wildflowers

fl.   p 89

Clematis virginiana

Courtenay, Booth
Wildflowers & Weeds

fl.   p 30

Clematis virginiana

Dean, Blanche E.
Wildflowers of Alabama and
   Adjoining States

fl.   p 61

Clematis virginiana

Klimas, John E.
Wildflowers of Eastern America

fl.   Pl. 75

Clematis virginiana

Moyle, John B.
Northland Wild Flowers

fl. hab.   p 52, Pl. 12

Clematis virginiana

Wharton, Mary E.
A Guide to the Wildflowers & Ferns
   of Kentucky

fl.   p 111   Pl. 1.34

Clematis vitalba

Ary, S.
The Oxford Book of Wildflowers

fl.   p 50, Pl. 1

Clematis vitalba

Bartels, Andreas
Das Grosse Buch der Gartengeholze

fr.   p 131

Clematis vitalba

Felsko, Elsa
A Book of Wildflowers

fl.   P 155

Clematis vitalba

Huxley, Anthony
Deciduous Garden Trees and Shrubs

fl. fr.   Pl. 285, 285a

Clematis vitalba

Martin, W. Keble
The Concise British Flora in Colour

fl. fr. hab.   Pl. 1

Clematis vitalba

Perrot, Emile
Les Plantes Medicinales

fl. fr. hab.   p 72

Clematis vitalba

Perry, Frances
Flowers of the World

fr.   p. 255

Clematis vitalba

Polunin, Oleg
Flowers of Europe

fl.   Pl. 24   #224

Clematis vitalba

Polunin, Oleg
Trees and Bushes of Europe

fl. fr. hab.   p 67

Clematis viticella

Huxley, Anthony
Deciduous Garden Trees and Shrubs

fl.   Pl. 288

Clematis viticella

Massachusetts Hort. Soc.
Horticulture
   vol 46, No. 4   1968

fl.   p. 35

Clematis viticella

Polunin, Oleg
Trees and Bushes of Europe

fl.   p 68

Clematis viticella var.

Gault, S. Millar
The Color Dictionary of Shrubs

fl.   Pl. 91, 92

Clematis viticella var.

Harrison, Richmond E.
Climbers and Trailers

fl.   p 37   Pl. 64

Clematis viticella var.

Hellyer, A.G.L.
Shrubs in Colour

fl.   p. 30

Clematis viticella var.

Huxley, Anthony
Deciduous Garden Trees and Shrubs

fl.   Pl. 289

Clematis welwitschii

Moriarty, Audrey
Wild Flowers of Malawi

fl.   Pl. 34; 4

Clematis, White
   see
Clematis paniculata

Clematoclethra actinidioides

Curtis's Botanical Magazine
Vol 159   1936

fl. fr.   No. 9439

Clematopsis kirkii

Flowering Plants of Africa
vol XXVI   1947

fl.   Pl. 1026

Clematopsis scabiosifolia

Moriarty, Audrey
Wild Flowers of Malawi

fl.   Pl. 34; 2

Clematopsis uhehensis

Moriarty, Audrey
Wild Flowers of Malawi

fl.   Pl. 34; 3

Clematopsis uhehensis

Morley, Brian D.
Wild Flowers of the World

fl.,fr.   Pl. 52

Cleome angustifolia

Morley, Brian D.
Wild Flowers of the World

fl.,fr.   Pl. 52

Cleome foliosa

Flowering Plants of Africa
vol 34   · 1960-61

fl.   Pl. 1330

Cleome hasslerana

Morley, Brian D.
Wild Flowers of the World

fl. fr.   Pl. 174

Cleome hasslerana

Perry, Frances
Flowers of the World

fl.   p 75

Cleome hirta

Moriarty, Audrey
Wild Flowers of Malawi

fl.   Pl. 44; 1

Cleome houtteana

Brown, Clair A.
Wildflowers of Louisiana and
    Adjoining States

fl. fr.   p 59

Cleome lutea

Welsh, Stanley L.
Flowers of the Canyon Country

fl. hab.   p 41

Cleome serrulata

Orr, Robert T.
Wildflowers of Western America

fl.   Pl. 182

Cleome spinosa

Crockett, James Underwood
Annuals

fl.   P 107

Cleome spinosa

Macoby, Stirling
What Flower is That

fl.   P 87

Cleome spinosa

Tsukamoto, Yotaro
Coloured Illustrations of Garden
    Flowers   vol 10

fl.   p. 13   Pl. 38

Cleome spinosa

Wit, H. C. D. de
Plants of the World;
    The Higher Plants, Vol I

fl.   p 171, Pl. 97

Cleome spinosa var.

Hay, Roy
The Color Dictionary of Flowers &
    Plants

Cleome spinosa var.

Huxley, Anthony
Garden Annuals and Bulbs

fl.   Pl. 32

Cleomella palmerana

Welsh, Stanley L.
Flowers of the Canyon Country

fl. hab.   p 36

Clermontia arborescens

Morley, Brian D.
Wild Flowers of the World

fl.   Pl. 146

Clermontia parviflora

Morley, Brian D.
Wild Flowers of the World

fl.   Pl. 146

Clerodendrum bungei

Gault, S. Millar
The Color Dictionary of Shrubs.

fl.   Pl. 100

Clerodendrum Bungei

Hay, Roy
The Color Dictionary of Flowers & Plants

fl   p. 190, Pl. 1517

Clerodendrum bungei

Hellyer, A.G.L.
Shrubs in Colour

fl.   p. 31

Clerodendrum bungei

Morley, Brian D.
Wild Flowers of the World

fl.   Plate 104

Clerodendrum capitatum

Morley, Brian D.
Wild Flowers of the World

fl.,fr.   Pl. 60

Clerodendrum capitatum

Perry, Frances
Flowers of the World

fl.   p 304

Clerodendrum fargesii

Harrison, R. E.
Trees and Shrubs

fr.   p. 53

Clerodendrum fortunatum

Walden, Beryl M.
Wild Flowers of Hong Kong

fl.   Pl. 40(110)

Clerodendrum fragrans

Walden, Beryl M.
Wild Flowers of Hong Kong

fl.   Pl. 10(32)

Clerodendrum glabrum var.

Palgrave, K. C.
Trees of Central Africa

fl., fr.   p. 429

Clerodendrum hybrid

Moriarty, Audrey
Wild Flowers of Malawi

fl.   Pl. 70; 2

Clerodendrum incisum

Bruggeman, L.
Tropical Plants

fl.   Pl. 232

Clerodendrum inerme

Walden, Beryl M.
Wild Flowers of Hong Kong

fl.   Pl. 43(111)

Clerodendrum makanjanum

Flowering Plants of Africa
vol 32   1957-58

fl., fr.   Pl. 1274

Clerodendrum myricoides

Moriarty, Audrey
Wild Flowers of Malawi

fl.   Pl. 70; 1

Clerodendrum nutans

Bruggeman, L.
Tropical Plants

fl.   Pl. 233

Clerodendrum nutans

Macoby, Stirling
What Flower is That

fl.   P. 88

Clerodendrum rotundifolium

Moriarty, Audrey
Wild Flowers of Malawi

fl.   Pl. 70; 3

Clerodendrum scandens

Wit, H. C. D. de
Plants of the World;
   The Higher Plants, Vol II

fl.   Pl. 107

Clerodendrum speciosissimum

Hargreaves, Dorothy
Tropical Blossoms of the Caribbean

fl.   p. 29

Clerodendrum speciosissimum

Hay, Roy
The Color Dictionary of Flowers & Plants

fl.   p 59, Pl. 471

Clerodendrum speciosissimum

Hay, Roy
The Dictionary of House Plants

fl.   Pl. 132

Clerodendrum speciosissimum

Macoby, Stirling
What Flower is That

fl.   P 88

Clerodendrum speciosissimum

Nicolaisen, Age
Pocket Encyclopedia of Indoor Plants

fl.   Pl. 131

Clerodendrum splendens

Curtis's Botanical Magazine
   vol 174   1962-63

fl.   Pl. 414

Clerodendrum splendens

Harrison, Richmond E.
Climbers and Trailers

fl.   p 43   Pl. 82

Clerodendrum splendens

Kiaer, Eigil
Indoor Plants in Colour

fl.   p. 40

Clerodendrum splendens

Morley, Brian D.
Wild Flowers of the World

fl.   Pl. 60

Clerodendrum splendens

Perry, Frances
Flowers of the World

fl.   p 305

Clerodendrum thomsoniae

Bruggeman, L.
Tropical Plants

fl.   pl. 8

Clerodendrum thomsoniae

Encke, Fritz
Zimmerpflanzen

fl.   p. 64

Clerodendrum thomsoniae

Harrison, Richmond E.
Climbers and Trailers

fl.   p 43   Pl. 83

Clerodendrum Thomsoniae

Harrison, R.E.
Trees and Shrubs

fl.   P 53

Clerodendrum thomsoniae

Hay, Roy
The Color Dictionary of Flowers &
   Plants

fl.   p. 59   Pl. 472

Clerodendrum thomsoniae

Hay, Roy
The Dictionary of House Plants

fl.   Pl. 133

Clerodendrum thomsoniae

Kromdijk, G.
200 House Plants in Colour

fl.   Pl. 52

Clerodendrum thomsoniae

Macoby, Stirling
What Flower is That

fl.   Pl. 88

Clerodendrum thomsoniae

Menninger, Edwin A.
Flowering Vines of the World

fl.   Pl. 195

Clerodendrum thomsoniae

Perry, Frances
Flowers of the World

fl.   titlepage, p.305

Clerodendrum thomsoniae

Tsukamoto, Yotaro
Coloured Illustrations of Garden Flowers
   vol 9 Bulbs and Perennials.

fl.   p. 41

Clerodendrum thomsoniae

Wit, H. C. D. de
Plants of the World;
   The Higher Plants, Vol. II

fl.   Pl. 104

Clerodendrum tomentosum

Blombery, Alec M.
What Wildflower is That

fl. fr.   p 97, Pl. 236, 237

Clerodendrum trichotomum

Boom, B.K.
The Glory of the Tree.

fl.   p. 105   Pl. 184

Clerodendrum trichotomum

Hay, Roy
The Color Dictionary of Flowers & Plants

fl.   p 190, Pl. 1519

Clerodendrum trichotomum

Hellyer, A.G.L.
Shrubs in Colour

fl.   p. 31

Clerodendrum trichotomum

Kitamura, Siro
Coloured Illustrations of Trees &
    Shrubs of Japan

fl.   427

Clerodendrum trichotomum

Massachusetts Hort. Soc.
Horticulture
    vol. 32, No. 7   1954

fl.   p. 338

Clerodendrum trichotomum

Massachusetts Hort. Soc.
Horticulture
    vol 43, No. 12   1965

fl.   inside backcover

Clerodendrum trichotomum

Royal Hort. Soc.
Journal of the Royal Hort. Soc.
    Vol. 86, No. 3   1961

fl.   p. 118   Pl. 30

Clerodendrum trichotomum

Royal Hort. Soc.
The Garden
    vol 102, No. 11   1977

fl.   p. 442

Clerodendrum trichotomum var.

Bartels, Andreas
Das Grosse Buch der Gartengeholze

fr.   p 134

Clerodendrum trichotomum var.

Perry, Frances
Flowers of the World

fl.   fr.   p 305

Clerodendrum trichotomum var.

Royal Hort. Soc.
The Garden
    vol 101, No. 9   1976

fl.   p. 479

Clerodendrum ugandense

Harrison, R.E.
Trees and Shrubs

fl.   P 53

Clerodendrum ugandense

Hay, Roy
The Dictionary of House Plants

fl.   Pl. 134

Clerodendrum viscosum

Walden, Beryl M.
Wild Flowers of Hong Kong

fl.   fr.   Pl. 42(112)

Clethra acuminata

Campbell, Carlos C.
Great Smoky Mountain Wildflowers

fl.   p 79

Clethra acuminata

Wharton, Mary E.
Trees & Shrubs of Kentucky

fl.   p 75, Pl. 2.13

Clethra alnifolia

Batson, Wade T.
Wild Flowers in South Carolina

fl.   p 81

Clethra alnifolia

Brown, Clair A.
Wildflowers of Louisiana and
    Adjoining States

fl.   hab.   p 127

Clethra alnifolia

Everett, T.H., ed.
New Illustrated Encyclopedia of
    Gardening   vol 3

fl., hab.   p. 454

Clethra alnifolia

Hellyer, A.G.L.
Shrubs in Colour

fl.   p. 31

Clethra alnifolia

Huxley, Anthony
Deciduous Garden Trees and Shrubs

fl.   Pl. 39

Clethra arborea

Harrison, R.E.
Trees and Shrubs

fl.   P 53

Clethra arborea

Royal Hort. Soc.
Journal of the Royal Hort. Soc.
    vol 91, No. 1   1966

hab., fl.   p. 22   Pl. 3

Clethra arborea

Royal Hort. Soc.
Journal of the Royal Hort. Soc.
    vol 93, No. 8   1968

fl.   p. 336   Pl. 170

Clethra barbinervis

Curtis's Botanical Magazine
    Vol 174   1962-63

fl.   Pl. 398

Clethra barbinervis

Hay, Roy
The Color Dictionary of Flowers & Plants

fl.   p 190, Pl. 1520

Clethra barbinervis

Kitamura, Siro
Coloured Illustrations of Trees &
    Shrubs of Japan

fl.   376

Clethra delavayi

Royal Hort. Soc.
The Garden
    vol 101, No. 2   1976

fl., hab.   p. 86-87

Clethra tomentosa

Morley, Brian D.
Wild Flowers of the World

fl. fr.   pl. 161

Cleyera fortunei

Royal Hort. Soc.
Journal of the Royal Hort. Soc.
    vol 97, No. 11   1972

hab.   p. 484   Pl. 245

Cleyera japonica

Kitamura, Siro
Coloured Illustrations of Trees &
    Shrubs of Japan

fr.   348

Cleyera japonica

Kromdijk, G.
200 House Plants in Colour

hab.   Pl. 53

Cleyera japonica var.

Curtis's Botanical Magazine
    Vol 163   1940 - 42

fl.   No. 9606

Clianthus dampieri

Kiaer, Eigil
Indoor Plants in Colour

fl.   p. 41

Clianthus formosus

Blombery, Alec M.
What Wildflower is That

fl.   p 96, Pl. 232

Clianthus formosus

Hay, Roy
The Color Dictionary of Flowers & Plants

fl.   p 60, Pl. 473

Clianthus formosus

Hay, Roy
The Dictionary of House Plants

fl.   Pl. 135

Clianthus formosus

Milne, Lorus
Living Plants of the World

fl.   p 107

Clianthus formosus

Morley, Brian D.
Wild Flowers of the World

fl.fr.   Plate 129

Clianthus formosus

Mullins, Barbara
Australian Wildflowers in Colour

fl.   p. 17   Pl. 6

Clianthus formosus

Perry, Frances
Flowers of the World

fl.   p 164

Clianthus formosus

Royal Hort. Soc.
Journal of the Royal Hort. Soc.
    vol 95, No. 7   1970

fl.   p. 304   Pl. 171

Clianthus puniceus

Harrison, R.E.
Trees and Shrubs

fl.   P 54

Clianthus puniceus

Harvey, Norman B.
New Zealand Botanical Paintings

fl.   p. 42   Pl. 18

Clianthus puniceus

Hay, Roy
The Color Dictionary of Flowers & Plants

fl.   p 247, Pl. 1974

Clianthus puniceus

Hay, Roy
The Dictionary of House Plants

fl.   Pl. 136

Clianthus puniceus

Hellyer, A.G.L.
Shrubs in Colour

fl.   p. 31

Clianthus puniceus

Royal Hort. Soc.
Journal of Royal Hort. Soc.
    vol 84, No. 3   1959

fl.   p. 122   Pl. 38

Clianthus puniceus

Royal Hort. Soc.
Journal of the Royal Hort. Soc.
    vol 95, No. 2   1970

fl.   p. 54   Pl. 47

Clianthus puniceus

Royal Hort. Soc.
The Garden
    vol 102, No. 11   1977

fl.   p. 443

Cliftonia monophylla

Brown, Clair A.
Wildflowers of Louisiana and
    Adjoining States

fl. hab.   p 104

Cliftonia monophylla

Dean, Blanche E.
Wildflowers of Alabama and
    Adjoining States

fl.   p 101

Clinopodium vulgare

Ary, S.
The Oxford Book of Wildflowers

fl.   p 146, Pl. 1

Clinopodium vulgare

Martin, W. Keble
The Concise British Flora in Colour

fl. hab.   Pl. 68

Clinopodium vulgare

Polunin, Oleg
Flowers of Europe

fl.   Pl. 115 #1158

Clintonia andrewsiana

Lemmon, Robert S.
Wildflowers of North America in
    Full Color

fl.   pg. 226   pl. 360

Clintonia andrewsiana

Royal Hort. Soc.
The Garden
    vol 102, No. 3   1977

fl.   p. 125

Clintonia borealis

American Hort. Soc.
American Horticulturist
    vol 55, No. 6   1976

fr.   p. 14

Clintonia borealis

Campbell, Carlos C.
Great Smoky Mountain Wildflowers

fl. hab.   p 45

Clintonia borealis

Courtenay, Booth
Wildflowers and weeds

fl.   p. 4

Clintonia borealis

Jennings, O. E.
Wild Flowers of Western Pennsylvania
                            vol II
fl. hab.   Pl. 21

Clintonia borealis

Klimas, John E.
Wildflowers of Eastern America

fl.   Pl. 131

Clintonia borealis

Lemmon, Robert S.
Wildflowers of North America in
    Full Color

fl. fr.   p. 226   pl. 358

Clintonia borealis

Lemmon, Robert S.
Wildflowers of North America in
    Full Color

fr.   p. 226   pl. 360

Clintonia borealis

Moyle, John B.
Northland Wild Flowers

fl. hab.   p 210, Pl. 276

Clintonia borealis

Walcott, Mary Vaux
North American Wild Flowers
    vol. 5

fl.   Pl. 338

Clintonia, red

see

Clintonia andrewsiana

Clintonia umbellulata

Campbell, Carlos C.
Great Smoky Mountain Wildflowers

fl. hab.   p 55

Clintonia umbellulata

Duncan, Wilbur H.
Wildflowers of the Southeastern
    United States

fl. hab.   p 255

Clintonia umbellulata

Jennings, O. E.
Wild Flowers of Western Pennsylvania
vol II
fl.   Pl. 22

Clintonia umbellulata

Klimas, John E.
Wildflowers of Eastern America

fl.   Pl. 52

Clintonia umbellulata

Wharton, Mary E.
A Guide to the Wildflowers & Ferns
of Kentucky

fl. p 52 Pl. 1.11

Clintonia uniflora

Clark, Lewis J.
Wild Flowers of British Columbia

fl. fr.   p 22, 26

Clintonia uniflora

Ferguson, Mary
Wildflowers

fl.   p 149

Clintonia uniflora

Fries, Mary A.
Wildflowers of Mount Ranier and
the Cascades

fl.   P 41

Clintonia uniflora

Orr, Robert T.
Wildflowers of Western America

fl.   Pl. 23

Clintonia uniflora

Porsild, A. E.
Rocky Mountain Wild Flowers

fr. hab.   p 97

Clintonia uniflora

Taylor, Ronald J.
Mountain Wild Flowers

fl. hab.   p 48

Clintonia uniflora

Walcott, Mary Vaux
North American Wild Flowers

fl. fr. hab.   Pl. 203, 204

Clintonia, White

see

Clintonia umbellulata

Clintonia, Yellow

see

Clintonia borealis

Clitoria mariana

Dean, Blanche E.
Wildflowers of Alabama and
Adjoining States

fl.   p 91

Clitoria mariana

Duncan, Wilbur H.
Wildflowers of the Southeastern
United States

fl.   p 85

Clitoria mariana

Klimas, John E.
Wildflowers of Eastern America

fl.   Pl. 258

Clitoria mariana

Wharton, Mary E.
A Guide to the Wildflowers & Ferns
of Kentucky

fl. p 185 Pl. 1.13

Clitoria ternatea

Bruggeman, L.
Tropical Plants

fl.,fr.   Pl. 10

Clitoria ternatea

Harrison, Richmond E.
Climbers and Trailers

fl. p 43 Pl. 84

Clitoria ternatea

Morley, Brian D.
Wild Flowers of the World

fl.   Pl. 53

Clitoria ternatea

Tsukamoto, Yotaro
Coloured Illustrations of Garden
Flowers Vol. 10

fl. Illus. 41 opp. p. 14

Clivia, Cape

see

Clivia nobilis

Clivia caulescens

Flowering Plants of South Africa
vol XXIII   1943

fl.   Pl. 891

Clivia x cyrtanthiflora

Hay, Roy
The Dictionary of House Plants

fl.   Pl. 137

Clivia gardenii

Flowering Plants of Africa
vol 42   1972-73

fl.   Pl. 1641

Clivia miniata

Encke, Fritz
Zimmerpflanzen

fl.   p. 70

Clivia miniata

Hay, Roy
The Color Dictionary of Flowers & Plants

fl.   p 60, Pl. 474

Clivia miniata

Hay, Roy
The Dictionary of House Plants

fl.   Pl. 138

Clivia miniata

Hersey, Jean
Woman's Day Book of House Plants

fl.   p. 77

Clivia miniata

Kiaer, Eigil
Indoor Plants in Colour

fl.   p. 41

Clivia miniata

Macoby, Stirling
What Flower is That

fl.   P 88

Clivia miniata

Massachusetts Hort. Soc.
Horticulture
vol 38, No. 4   1960

fl.   p. 215

Clivia miniata

Massachusetts Hort. Soc.
Horticulture
vol 49, No. 10   1971

fl.   p. 32

Clivia miniata

Massachusetts Hort. Soc.
Horticulture
vol 56, No. 12   1978

fl.   p. 27

Clivia miniata

Miles, Bebe
Bulbs for the Home Gardener

fl.   p 157

Clivia miniata

Nicolaisen, Age
Pocket Encyclopedia of Indoor Plants

fl.   Pl. 34

Clivia miniata

Perry, Frances
Flowers of the World

fl.   p 24

Clivia miniata

Tsukamoto, Yotaro
Coloured Illustrations of Garden
        Flowers vol 9

fl.   p. 41

Clivia miniata var.

Pacific Hort. Foundation
Pacific Horticulture
    vol 37, No. 2   1976

fl., fr.   p. 9

Clivia nobilis

Everett, T.H., ed.
New Illustrated Encyclopedia of
        Gardening   vol 3

fl.   p. 407

Clivia nobilis

Hay, Roy
The Dictionary of House Plants

fl.   Pl. 139

Clivia nobilis

Macoby, Stirling
What Flower is That

fl. P 88

Clivia sp.

Kromdijk, G.
200 House Plants in Colour

fl.   Pl. 54

Clockvine, Bengal

see

Thunbergia grandiflora

Clock Vine, Orange

see

Thunbergia alata

Clock Vine, Scarlet

see

Thunbergia coccinea

Clog Plant

see

Hypocyrta radicans

Cloudberry

see

Rubus chamaemorus

Cloud Grass

see

Agrostis nebulosa

Clover

see

Trifolium

Clover, Alpine

see

Trifolium alpinum

Clover, Alsike

see

Trifolium hybridum

Clover, Brown

see

Trifolium badium

Clover, Buffalo

see

Trifolium reflexum

Clover, Bur Spotted

see

Medicago arabica

Clover, Bush

see

Lespedeza virginica

Clover, Common

see

Trifolium pratense

Clover, Crimson

see

Trifolium incarnatum

Clover, Dutch

see

Trifolium repens

Clover, Hare's Foot

see

Trifolium arvense

Clover, Holy

see

Onobrychis viciifolia

Clover, Hop

see

Trifolium agrarium

Clover, Korean

see

Lespedeza stipulacea

Clover, Ladino

see

Trifolium repens

Clover, Mustang

see

Linanthus montanus

Clover, Owl's

see

Orthocarpus purpurascens

Clover, Owl's

see

Orthocarpus tenuifolius

Clover, Parry

see

Trifolium parryi

Clover, Persian

see

Trifolium resupinatum

Clover, Prairie Purple

see

Petalostemum purpureum

Clover, Prairie White

see

Petalostemum candidum

Clover, Rabbit-foot

see

Trifolium arvense

Clover, Red

see

Trifolium pratense

Clover, Sea

see

Trifolium squamosum

Clover, Snow

see

Trifolium pratense var.

Clover, Sour Yellow

see

Melilotus indica

Clover, Strawberry

see

Trifolium fragiferum

Clover, White

see

Trifolium repens

Clover, White Sweet

see

Melilotus alba

Clover, Yellow Sweet

see

Melilotus officinalis

Clover, Zigzag

see

Trifolium medium

Cloves

see

Eugenia aromatica

Cloves

see

Eugenia caryophyllata

Cloves

see

Syzygium aromaticum

Clusia grandiflora

Morley, Brian D.
Wild Flowers of the World

fl.   Pl. 177

Clusia rosea

Everett, Thomas H.
Living Trees of the World

fl.   p. 247 bottom

Clusia schomburgkiana

Mee, Margaret
Flowers of the Brazilian Forests

fl.   Pl. 2

Clusia viscida

Mee, Margaret
Flowers of the Brazilian Forests

fl.,   Pl 3, 4

Clusterberry

see

Cotoneaster frigidus

Clytostoma callistegioides

Harrison, Richmond E.
Climbers and Trailers

fl. p 44 Pl. 85

Clytostoma callistegioides

Macoby, Stirling
What Flower is That

fl.   P 89

Clytostoma callistegioides

Mathias, Mildred E.
Color for the Landscape

fl. hab.   p 107

Clytostoma callistegioides

Menninger, Edwin A.
Flowering Vines of the World

fl.   Pl. 34

Cneoridium dumosum

Munz, Philip A.
California Spring Wildflowers

fl., fr.   p. 76   Pl. 54

Cneorum pulverulentum

Bramwell, David
Wild Flowers of the Canary Islands

hab.   Pl. 191

Cnestis ferruginea

Wit, H. C. D. de
Plants of the World;
The Higher Plants, Vol II

fr.   p 60, Pl. 30

Cnicus benedictus

Bianchini, Francesco
Health Plants of the World

fl. hab.   p 17

Cnicus benedictus

Perrot, Emile
Les Plantes Medicinales

fl. fr. hab.   p 59

Cnicus benedictus

Polunin, Oleg
Flowers of Europe

fl.   Pl. 160 #1509

Cnicus benedictus

Weeds of the Southern United States

hab., fl.   p. 10

Cnicus calcitrapa

Bianchini, Francesco
Health Plants of the World

fl. hab.   p 17

Cnidium officinale

Kariyone, Tatsuo
Atlas of Medicinal Plants

fl., hab.   Pl. 70

Cnidoscolus angustidens

Coyle, Jeanette
A Field Guide to the Common and
  Interesting Plants of Baja California

fl. hab.   p 109

Cnidoscolus stimulosus

Brown, Clair A.
Wildflowers of Louisiana and
  Adjoining States

fl. hab.   p 100

Cnidoscolus stimulosus

Duncan, Wilbur H.
Wildflowers of the Southeastern
  United States

fl. hab.   p 95

Cnidoscolus stimulosus

Lemmon, Robert S.
Wildflowers of North America in
  Full Color

fl.   pg. 26   pl. 43

Cnidoscolus texanus

Brown, Clair A.
Wildflowers of Louisiana and
  Adjoining States

fl. hab.   p 101

Coachwhip

see

Fouquiera splendens

Coachwood

see

Ceratopetalum apetalum

Cobnut

see

Corylus avellana

Cobaea, Purple bell

see

Cobaea scandens

Cobaea scandens

Crockett, James Underwood
Annuals

fl.   P 108

Cobaea scandens

Harrison, Richmond E.
Climbers and Trailers

fl.   p 44   Pl. 86

Cobaea scandens

Hay, Roy
The Color Dictionary of Flowers & Plants

fl.   p 247, Pl. 1975

Cobaea scandens

Hay, Roy
The Dictionary of House Plants

fl.   Pl. 140

Cobaea scandens

Huxley, Anthony
Garden annuals and bulbs

fl.   pl. 33

Cobaea Scandens

Kiaer, Eigil
Indoor Plants in Colour

fl.   p. 40

Cobaea scandens

Macoby, Stirling
What Flower is That

fl.   p. 89

Cobaea scandens

Menninger, Edwin A.
Flowering Vines of the World

fl.   Pl. 35

Cobaea scandens

Morley, Brian D.
Wild Flowers of the World

fl. fr.   pl. 181

Cobaea scandens

Nicolaisen, Age
Pocket Encyclopedia of Indoor Plants

fl.   Pl. 112

Cobaea scandens

O'Gorman, Helen
Mexican Flowering Trees and Plants

fl. fr. hab.   p 95

Cobaea scandens

Perry, Frances
Flowers of the World

fl.   p 76

Cobaea villosa

Chickering, Carol Rogers
Flowers of Guatemala

fl.   p 55

Cobra plant

see

Darlingtonia californica

Coca-Cola

see

Cola acuminata

Cocaine

see

Erythroxylon coca

Coccinia adoensis

Moriarty, Audrey
Wild Flowers of Malawi

fl.   Pl. 74; 3

Coccinia quinqueloba

Morley, Brian D.
Wild Flowers of the World

fl.,fr.   Pl. 79

Coccoloba uvifera

Fleming, Glenn
Wild Flowers of Florida

fl. hab.   p 40

Coccoloba uvifera

Hargreaves, Dorothy
Tropical Blossoms of the Caribbean

fr.   p. 61

Coccoloba uvifera

Hay, Roy
The Color Dictionary of Flowers & Plants

hab.   p 60, Pl. 475

Coccoloba uvifera

Hay, Roy
The Dictionary of House Plants

hab.   Pl. 141

Coccoloba uvifera

Hersey, Jean
Woman's Day Book of House Plants

hab.   p. 93

Cocculus carolinus

Brown, Clair A.
Wildflowers of Louisiana and
Adjoining States

fr. hab.   p 54

Cocculus carolinus

Dean, Blanche E.
Wildflowers of Alabama and
Adjoining States

fl.   p 69

Cocculus carolinus

Wharton, Mary E.
Trees & Shrubs of Kentucky

fr.   p 123, Pl. 2.7

Cocculus trilobus

Kimura, Koiti
Japanese Medicinal Plants, Vol II

fl.   p 167

Cocculus trilobus

Kitamura, Siro
Coloured Illustrations of Trees &
Shrubs of Japan

fr.   160

Cocculus trilobus

Takatori, Jisuke
Color Atlas of Medicinal Plants of Japan

fl.
fr.   Fig. 50

Cochemiea poselgeri

Cactus & Succulent Society of America
Journal
vol 39, No. 6   1967

fl.   cover (p. 201)

Cochiseia robbinsorum

Cactus & Succulent Soc. of America
Journal
vol 50, No. 6   1978

fl., hab.   p. 293

Cochleanthes candida

Pabst, G. F. J.
Orchidaceae Brasilienses,
Vol 2

fl.   p 232

Cochleanthes discolor

Ebel, Friedrich
The Strange and Beautiful
World of Orchids

fl.   p 129

Cochleanthes discolor

Ospina, Mariano
Orquideas de las americas

fl. hab.   Pl. 104

Cochleanthes flabelliformis

Pabst, G. F. J.
Orchidaceae Brasilienses,
Vol 2

fl.   p 232

Cochleanthes marginata

Dunsterville, G. C. K
Introduction to the World
of Orchids

fl.   Pl. 5

Cochleanthes marginata

Grubb, Roy
Selected Orchidaceous Plants
Vol I

fl.   p 135

Cochleanthes wailesiana

Pabst, G. F. J.
Orchidaceae Brasilienses,
Vol 2

fl.   p 232

Cochlearia anglica

Martin, W. Keble
The Concise British Flora in Colour

fl. fr. hab.   Pl. 8

Cochlearia armoracia

Hvass, Elsie
Plants That Feed and Serve Us

fl.   hab.   p 70, Pl. 144

Cochlearia armoracia

Perrot, Emile
Les Plantes Medicinales

fl. fr. hab.   p 196

Cochlearia armoracia

Rosengarten, Frederic, Jr.
The Book of Spices

fl.   p 266

Cochlearia danica

Martin, W. Keble
The Concise British Flora in Colour

fl. hab.   Pl. 8

Cochlearia danica

Polunin, Oleg
Flowers of Europe

fl.   Pl. 36 #337

Cochlearia alcacea

Martin, W. Keble
The Concise British Flora in Colour

fl. fr. hab.   Pl. 8

Cochlearia officinalis

Ary, S.
The Oxford Book of Wildflowers

fl. fr.   p 68, Pl. 5

Cochlearia officinalis

Lindman, C. A. M.
Nordens Flora, Vol 4

fl. fr. hab.   Pl. 260

Cochlearia officinalis

Perrot, Emile
Les Plantes Medicinales

fl. fr. hab.   p 196

Cochlearia officinalis var.

Martin, W. Keble
The Concise British Flora in Colour

fl. fr. hab.   Pl. 8

Cochlioda sanguinea

Ospina, Mariano
Orquideas de las americas

fl.   Pl. 171

Cochliostema jacobianum

Morley, Brian D.
Wild Flowers of the World

fl.   pl. 187

Cochliostema jacobianum

Perry, Frances
Flowers of the World

fll   p 77

Cochlospermum fraseri

Mullins, Barbara
Australian Wildflowers in Colour

fl.   P. 23   Pl. 15

Cochlospermum gillivraei

Blombery, Alec M.
What Wildflower is That

fl. hab.   p 98, Pl. 238

Cochlospermum planchoni

Wit, H. C. D. de
Plants of the World;
The Higher Plants, Vol I

fl.   p 165, Pl. 81

Cochlospermum religiosum

Wit, H. C. D. de
Plants of the World;
   The Higher Plants, Vol I

fl.   p 166, Pl. 85

Cochlospermum vitifolium

Chickering, Carol Rogers
Flowers of Guatemala

fl. fr.   p. 57

Cochlospermum vitifolium

Hall, Clarence E.
Flowers of the Islands in the Sun

fl.   p. 127   Pl. 28

Cochlospermum vitifolium

Hargreaves, Dorothy
Tropical Blossoms of the Caribbean

fl.,hab.   p. 60

Cochlospermum vitifolium

O'Gorman, Helen
Mexican Flowering Trees and Plants

fl. fr. hab.   p 19

Cochlospermum vitifolium

Pertchik, Bernard
Flowering Trees of the
   Caribbean

fl. fr.   p 39

Cockroach plant

see

Drejerella guttata

Cockspur flower

see

Plectranthus eckloni

Cockade, Rose

see

Leucadendron grandiflorum

Cockle

see

Agrostema githago

Cocklebur, Common

see

Xanthium pensylvanicum

Cockle, Corn

see

Agrostemma githago

Cockscomb

see

Celosia argentea var.

Cockscomb

see

Celosia cristata

Cockscomb, Feather

see

Celosia argentea var.

Cockscomb, feathered

see

Celosia plumosa

Cock's foot

see

Dactylis glomerata

Coco Palm

see

Cocos nucifera

Coco, Wild

see

Eulophia alta

Cocoa

see

Theobroma cacao

Cocos nucifera

Bianchini, F.
The Complete Book of Fruits & Vegetables

fr.   p 231

Cocos nucifera

Coyle, Jeanette
A Field Guide to the Common and
   Interesting Plants of Baja California

fr.   p 51

Cocos nucifera

Edlin, Herbert
The Illustrated Encyclopedia
   of Trees

fr. hab.   p 232-3

Cocos nucifera

Everett, Thomas H.
Living Trees of the World

fr. hab.   p. 70

Cocos nucifera

Hvass, Elsie
Plants That Feed and Serve Us

fr.   p 17, Pl. 25

Cocos nucifera

Masefield, G. B.
The Oxford Book of Food Plants

hab., fr.   Pl. 19

Cocos nucifera

Massachusetts Hort. Soc.
Horticulture
   Vol. 47, No. 1   1969

hab.   cover, backcover

Cocos nucifera

Massachusetts Hort. Soc.
Horticulture
   Vol. 52, No. 1   1974

hab.   P. 52

Cocos weddeliana

Kromdijk, G.
200 House Plants in Colour

hab.   Pl. 55

Cocoxochitl

see

Dahlia coccinea

Codiaeum hybrid

Kromdijk, G.
200 House Plants in Colour

hab.   Pl. 56

Codiaeum sp.

Hersey, Jean
Woman's Day Book of House Plants

hab.   p. 51

Codiaeum variegatum

Bruggeman, L.
Tropical Plants

hab.   Pl. 280, 282, 283, 284,
                285

Codiaeum variegatum

Crockett, James Underwood
Evergreens

hab.   p. 119

Codiaeum variegatum

Nicolaisen, Age
Pocket Encyclopedia of Indoor Plants

hab.   Pl. 83

Codiaeum variegatum var.

Encke, Fritz
Zimmerpflanzen

hab.   p. 70

Codiaeum variegatum var.

Hay, Roy
The Color Dictionary of Flowers & Plants

hab.   p 60, Pl. 476

Codiaeum variegatum var.

Hay, Roy
The Dictionary of House Plants

hab.   Pl. 142,143

Codiaeum variegatum var.

Kiaer, Eigil
Indoor Plants in Colour

hab.   p. 42

Codiaeum variegatum var.

Macoby, Stirling
What Flower is That

hab. P 89

Codiaeum variegataum var.

Tsukamoto, Yotaro
Coloured Illustrations of Garden
    Flowers   Vol. 10

hab.   p. 46   Pl. 146

Codlins-and-cream

    see

Epilobium hirsutum

Codonacanthus pauciflorus

Walden, Beryl M.
Wild Flowers of Hong Kong

fl.   Pl. 74 (224)

Codonanthe gracilis

American Gloxinia & Gesneriad Society
The Gloxinian
    vol 27, No. 5   1977

fl., hab.   p. 20

Codonanthe hybrid

American Gloxinia & Gesneriad Society
The Gloxinian
    vol 27, No. 5   1977

fl., fr., hab.   p. 21

Codonanthe macradenia

American Gloxinia & Gesneriad Society
The Gloxinian
    vol 27, No. 5   1977

fl.   p. 21

Codonanthe uleana

American Gloxinia & Gesneriad Society
The Gloxinian
    vol 27, No. 5   1977

fl.   p. 20

Codonocarpus cotinifolius

Cochrane, G. R.
Flowers and Plants of Victoria

fr.   Pl. 161

Codonopsis clematidea

Royal Hort. Soc.
Journal of the Royal Hort. Soc.
    vol 93, No. 3   1968

fl.   p. 122   Pl. 55

Codonopsis convolvulacea

Royal Hort. Soc.
Journal of the Royal Hort. Soc.
    vol 97, No. 2   1972

fl.   p. 88   Pl. 37

Codonopsis convolvulacea var.

Curtis's Botanical Magazine
Vol 162   1939

fl.   No. 9581

Codonopsis dicentrifolia

Royal Hort. Soc.
Journal of the Royal Hort. Soc.
    vol 97, No. 2   1972

fl.   p. 56   Pl. 34

Codonopsis macrocalyx

Curtis's Botanical Magazine
Vol 167   1950

fl.   94

Codonopsis meleagris var.

Royal Hort. Soc.
Journal of the Royal Hort. Soc.
    vol 97, No. 2   1972

fl.   p. 56   Pl. 33

Codonopsis mollis

Curtis's Botanical Magazine
Vol 164   1943-48

fl.   No. 9677

Codonopsis ovata

Hay, Roy
The Color Dictionary of Flowers & Plants

fl.   p 5, Pl. 36

Codonopsis ovata

Massachusetts Hort. Soc.
Horticulture
    vol 51, No. 1   1973

fl.   p. 37

Codonopsis ovata

Morley, Brian D.
Wild Flowers of the World

fl.fr.   Plate 97

Codonopsis ovata

Perry, Frances
Flowers of the World

fl.   p 64

Codonopsis rotundifolia var.

Curtis's Botanical Magazine
Vol 167   1950

fl.   131

Codonopsis vinciflora

Curtis's Botanical Magazine
Vol 166   1949

fl.   59

Codonopsis vinciflora

Royal Hort. Soc.
Journal of the Royal Hort. Soc.
    vol 97, No. 2   1972

fl.   p. 88   Pl. 35

Codonopsis viridis

Hara, Hiroshi, comp.
Photo-Album of Plants of
    Eastern Himalaya

fl.   Pl. 132

Codonorchis canisioi

Pabst, G. F. J.
Orchidaceae Brasilienses,
    Vol 1

fl.   p 177

Coeloglossum viride

Ary, S.
The Oxford Book of Wildflowers

fl.   p 44, Pl. 1

Coeloglossum viride

Barneby, T. P.
European Alpine Flowers in Colour

fl.   Pl. 10, 3

Coeloglossum viride

Brooke, Jocelyn
The Wild Orchids of Britain

fl.   Pl. 16

Coeloglossum viride

Lindman, C. A. M.
Nordens Flora, Vol 2

fl. hab.   Pl. 134A

Coeloglossum viride

Martin, W. Keble
The Concise British Flora in Colour

fl. hab.   Pl. 82

Coeloglossum viride var.

Luer, Carlyle A.
The Native Orchids of the
United States and Canada

fl. hab.   Pl. 42

Coelogyne asperata

Bruggeman, L.
Tropical Plants

fl.   Pl. 93

Coelogyne cristata

Ebel, Friedrich
The Strange and Beautiful
World of Orchids

fl.   p 99

Coelogyne cristata

Encke, Fritz
Zimmerpflanzen

fl.   p. 104

Coelogyne cristata

Hay, Roy
The Dictionary of House Plants

fl.   Pl. 144

Coelogyne cristata

Kimber, Sheila
A Handbook of Orchids

fl., hab.   p. 14

Coelogyne cristata

Macoby, Stirling
What Flower is That

fl.   P 90

Coelogyne cristata

Royal Hort. Soc.
Journal of the Royal Hort. Soc.
vol 94, No. 7   1969

fl.   p. 302   Pl. 161

Coelogyne cumingii

Kamemoto, Haruyuki
Beautiful Thai Orchid Species

fl.   p 125

Coelogyne dayana

Curtis's Botanical Magazine
vol 172   1958-59

fl.   Pl. 309

Coelogyne elata

Kimber, Sheila
A Handbook of Orchids

fl., hab.   p. 16

Coelogyne fimbriata

Walden, Beryl M.
Wild Flowers of Hong Kong

fl.   Pl. 56 (175)

Coelogyne flaccida

Ebel, Friedrich
The Strange and Beautiful
World of Orchids

fl.   p 105

Coelogyne lawrenceana

Kramer, Jack
Orchids; Flowers of Romance
and Mystery

fl.   p 208

Coelogyne massangeana

Ebel, Friedrich
The Strange and Beautiful
World of Orchids

fl.   p 61

Coelogyne nitida

Perry, Frances
Flowers of the World

fl.   p 210

Coelogyne ochracea

Hay, Roy
The Color Dictionary of Flowers & Plants

fl.   p 60, Pl. 477

Coelogyne ochracea

Hay, Roy
The Dictionary of House Plants

fl.   Pl. 145

Coelogyne pandurata

Everett, T.H., ed.
New Illustrated Encyclopedia of
Gardening vol 3

fl.   p. 358

Coelogyne pandurata

Kramer, Jack
Orchids; Flowers of Romance
and Mystery

fl.   p 62

Coelogyne pulchella

Curtis's Botanical Magazine
Vol 165   1948

fl.   28

Coelogyne speciosa var.

Curtis's Botanical Magazine
Vol 161   1938

fl.   No. 9539

Coelogyne virescens

Kamemoto, Haruyuki
Beautiful Thai Orchid Species

fl.   p 125

Coffea arabica

Bianchini, F.
The Complete Book of Fruits & Vegetables

fl. fr.   p 217

Coffea arabica

Bianchini, Francesco
Health Plants of the World

fr.   p 127

Coffea arabica

Encke, Fritz
Zimmerpflanzen

fl.   p. 70

Coffea arabica

Hvass, Elsie
Plants That Feed and Serve Us

fl. fr.   p 77, Pl. 159

Coffea arabica

Kariyone, Tatsuo
Atlas of Medicinal Plants

fr.   Pl. 55

Coffea arabica

Masefield, G. B.
The Oxford Book of Food Plants

fl. fr.   Pl. 111, 1

Coffea arabica

Massachusetts Hort. Soc.
Horticulture
vol 40, No. 10   1962

fl., fr.   cover

Coffea arabica

Morley, Brian D.
Wild Flowers of the World

fl.,fr.   Pl. 60

Coffea arabica

Nicolaisen, Age
Pocket Encyclopedia of Indoor Plants

fl.   Pl. 141

Coffea arabica

Perrot, Emile
Les Plantes Medicinales

fl. fr. hab.   p 45

Coffea arabica

Wit, H. C. D. de
Plants of the World;
   The Higher Plants, Vol II

fr.   p 107, Pl. 64

Coffea bengalensis

Hara, Hiroshi, comp.
Photo-Album of Plants of
   Eastern Himalaya

fl.   Pl. 4

Coffea canephora

Masefield, G. B.
The Oxford Book of Food Plants

fr.   Pl. 111, 2

Coffea liberica

Perrot, Emile
Les Plantes Medicinales

fl. fr.   p 45

Coffee

see

Coffea

Coffee, Arabian

see

Coffea arabica

Coffee Bean Tree

see

Vitex payos

Coffeeberry

see

Rhamnus californica

Coffee Weed

see

Cassia obtusifolia

Coffee, Wild

see

Psychotria rubra

Cogniauxiocharis glazioviana

Pabst, G. F. J.
Orchidaceae Brasilienses,
   Vol 1

fl.   p 182

Cohosh, Black

see

Cimicifuga americana

Cohosh, Black

see

Cimicifuga racemosa

Cohosh, blue

see

Caulophyllum thalictroides

Coix lacryma-jobi

Crockett, James Underwood
Annuals

fl.   P 108

Coix lacryma-jobi var.

Kariyone, Tatsuo
Atlas of Medicinal Plants

fl., hab.   Pl. 8

Coix lacryma-jobi var.

Kimura, Koiti
Japanese Medicinal Plants, Vol. I

fl.   p. 2

Coix lacryma-jobi var.

Takatori, Jisuke
Color Atlas of Medicinal Plants of Japan

fl.,fr.   Fig. 78

Cola acuminata

Hvass, Elsie
Plants That Feed and Serve Us

fl. fr. p 82, Pl. 170

Cola attiensis

Wit, H. C. D. de
Plants of the World;
   The Higher Plants, Vol I

fl.   p 214, Pl. 119

Cola caricifolia

Wit, H. C. D. de
Plants of the World;
   The Higher Plants, Vol I

fl. fr.   p 214, 215,
             Pl. 120 - 122

Cola nitida

Bianchini, Francesco
Health Plants of the World

fr.   p 127

Cola nitida

Perrot, Emile
Les Plantes Medicinales

fl. fr. hab.   p 130

Colchicum autumnale

American Hort. Soc.
American Horticulturist
vol 54, No. 4   1975

fl.   p. 5

Colchicum autumnale

Ary, S.
The Oxford Book of Wildflowers

fl. fr.   p 162, Pl. 1

Colchicum autumnale

Barneby, T. P.
European Alpine Flowers in Colour

fl.   Pl. 3, 2

Colchicum autumnale

Bianchini, Francesco
Health Plants of the World

fl. hab.   p. 177

Colchicum autumnale

Color Treasury of Herbs & Other
   Medicinal Plants

fl. p 22 Pl. 11

Colchicum autumnale

Felsko, Elsa
A Book of Wildflowers

fl.   P 79

Colchicum autumnale

Hay, Roy
The Color Dictionary of Flowers & Plants

fl.   p 86, Pl. 686

Colchicum autumnale

Hvass, Elsie
Plants That Feed and Serve Us

fl.   p 89, Pl. 190

Colchicum autumnale

Kariyone, Tatsuo
Atlas of Medicinal Plants

fl., fr., hab.   Pl. 18

Colchicum autumnale

Kiaer, Eigil
Indoor Plants in Colour

fl.   p. 42

Colchicum autumnale

Kimura, Koiti
Japanese Medicinal Plants, Vol I

fl. hab.   p 104

Colchicum autumnale

Kleijn, H.
Beauty of the Wild Plant

fl.   opp. p. 32   Pl. 32

Colchicum autumnale

Macoby, Stirling
What Flower is That

fl.   P 90

Colchicum autumnale

Martin, W. Keble
The Concise British Flora in Colour

fl. hab.   Pl. 86

Colchicum autumnale

Miles, Bebe
Bulbs for the Home Gardener

fl.   p, 61, 62

Colchicum autumnale

Perrot, Emile
Les Plantes Medicinales

fl. fr. hab.   p 74

Colchicum autumnale

Polunin, Oleg
Flowers of Europe

fl.   Pl. 163 #1588

Colchicum autumnale

Polunin, Oleg
Flowers of the Mediterranean

fl.   Pl. 230

Colchicum autumnale

Tsukamoto, Yotaro
Coloured Illustrations of Garden
Flowers   Vol. 9

fl.   opp. p. 6

Colchicum autumnale var.

Massachusetts Hort. Soc.
Horticulture
vol 52, No. 8   1974

fl.   p. 30

Colchicum boissieri

Goulimis, Constantine N.
Wild Flowers of Greece

fl. hab. p 139

Colchicum bornmuelleri

Massachusetts Hort. Soc.
Horticulture
vol 35, No. 8   1957

fl.   p. 403

Colchicum bornmuelleri

Massachusetts Hort. Soc.
Horticulture
vol 45, No. 4   1967

fl.   p. 20

Colchicum bowlesianum

Goulimis, Constantine N.
Wild Flowers of Greece

fl. hab. p 141

Colchicum burttii

Curtis's Botanical Magazine
Vol. 181   1976-77

fl. hab.   Pl. 735

Colchicum catacuzenium

Curtis's Botanical Magazine
Vol 164   1943-48

fl. hab.   No. 9652

Colchicum cilicicum

Hay, Roy
The Color Dictionary of Flowers & Plants

fl.   p 86, Pl. 687

Colchicum cilicicum

Royal Hort. Soc.
Journal of the Royal Hort. Soc.
vol 76, No. 3   1951

fl.   p. 90   Pl. 40

Colchicum cilicicum var.

Royal Hort. Soc.
Journal of the Royal Hort. Soc.
vol 76, No. 3   1951

fl.   p. 90   Pl. 41

Colchicum hierosolymitanum

Feinbrun-Dothan, Naomi
Wild Plants in the Land
of Israel

fl. hab.   p 78

Colchicum hungaricum

Curtis's Botanical Magazine
vol 173   1960

fl.   Pl. 373

Colchicum hybrid

Miles, Bebe
Bulbs for the Home Gardener

fl.   p 60

Colchicum, Jerusalem

see

Colchicum hierosolymitanum

Colchicum jezdianum

Wendelbo, Per
Tulips and Irises of Iran and
Their Relatives

fl.   p 13, Pl. 3

Colchicum Kotschyi

Curtis's Botanical Magazine
Vol. 176   1966-68

fl.   Pl. 520

Colchicum kotschyi

Wendelbo, Per
Tulips and Irises of Iran and
Their Relatives

fl.   p 13, Pl. 2

Colchicum lusitanum

Curtis's Botanical Magazine
Vol 165   1948

fl.   21

Colchicum macrophyllum

Goulimis, Constantine N.
Wild Flowers of Greece

fl. hab. p 143

Colchicum parnassicum

Curtis's Botanical Magazine
vol 174   1962-63

fl.   Pl. 403

Colchicum persicum

Wendelbo, Per
Tulips and Irises of Iran and
Their Relatives

fl.   p 13, Pl. 4

Colchicum speciosum

Hay, Roy
The Color Dictionary of Flowers & Plants

fl.   p 86, 87, Pl. 688 thru 690

Colchicum speciosum

Huxley, Anthony
Garden Annuals and Bulbs

fl.   Pl. 204

Colchicum speciosum

Massachusetts Hort. Soc.
Horticulture
   vol 32, No. 8   1954

fl.   p. 381

Colchicum speciosum

Wendelbo, Per
Tulips and Irises of Iran and
   Their Relatives

fl. hab.   p 11, Pl. 1

Colchicum speciosum var.

Hay, Roy
The Color Dictionary of Flowers
   & Plants

fl.   p. 87

Colchicum speciosum var.

Royal Hort. Soc.
Journal of the Royal Hort. Soc.
   vol 99, No. 7   1974

fl.   p. 296   Pl. 132

Colchicum steveni

Feinbrun-Dothan, Naomi
Wild Flowers in the Land of
   Israel

fl. hab.   p. 78

Colchicum troodi

Megaw, Elektra
Wild Flowers of Cyprus

fl., hab.   Pl. 37

Coleonema pulchrum

Harrison, R.E.
Trees and Shrubs

fl.   P 55

Coleonema pulchrum

Macoby, Stirling
What Flower is That

fl. P 90

Coleotrype natalensis

Flowering Plants of Africa
   vol 37   1965-66

fl.   Pl. 1465

Coleus amboinicus

Addisonia
   vol 20   1937-38

fl.   p. 11   Pl. 646

Coleus autrani

Curtis's Botanical Magazine
Vol 163   1940 - 42

fl.   No. 9605

Coleus barbatus

Lind, E.M.
Some Common Flowering Plants of
   Uganda

fl.   Pl. 9

Coleus blumei

Crockett, James Underwood
Annuals

hab.   P 108

Coleus blumei

Hay, Roy
The Color Dictionary of Flowers & Plants

hab.   p 60, Pl. 478

Coleus blumei

Hay, Roy
The Dictionary of House Plants

hab.   Pl. 146, 148

Coleus blumei

Macoby, Stirling
What Flower is That

hab P 91

Coleus blumei

Nicolaisen, Age
Pocket Encyclopedia of
   Indoor Plants

hab.   Pl. 129

Coleus blumei

Perry, Frances
Flowers of the World

hab.   p 153

Coleus blumei

Tsukamoto, Yotaro
Coloured Illustrations of Garden
   Flowers   Vol. 10

hab. Illus. 39   opp. p. 13

Coleus blumei var.

Macoby, Stirling
What Flower is That

hab. P 91

Coleus, creeping

see

Coleus rehneltianus var.

Coleus frederici

Morley, Brian D.
Wild Flowers of the World

fl.   Pl. 63

Coleus hybrid

Encke, Fritz
Zimmerpflanzen

hab.   p. 70

Coleus hybrid

Hersey, Jean
Woman's Day Book of House Plants

fl.,hab.   p. 48

Coleus hybrid

Kromdijk, G.
200 House Plants in Colour

hab.   Pl. 57

Coleus hybrid

Perry, Frances
Flowers of the World

hab.   p. 153

Coleus x hybridus

Kiaer, Eigil
Indoor Plants in Colour

hab., fl.   p. 43

Coleus mirabilis

Flowering Plants of Africa
   vol 36   1963-64

fl.   Pl. 1417

Coleus rehneltianus var.

Kiaer, Eigil
Indoor Plants in Colour

hab.   p. 43

Coleus rehneltianus var.

Macoby, Stirling
What Flower is That

hab.   p. 91

Coleus thyrsoideus

Hay, Roy
The Color Dictionary of Flowers & Plants

fl.   p 60, Pl. 479

Coleus thyrsoideus

Hay, Roy
The Dictionary of House Plants

fl.   Pl. 147

Coleus thyrsoideus

Macoby, Stirling
What Flower is That

hab P 91

Coleus thyrsoideus

Perry, Frances
Flowers of the World

fl.   p 152

Colic root

see

Aletris

Colletia armata

Curtis's Botanical Magazine
Vol 178   1970-72

fl.   586

Colletia armata

Hellyer, A.G.L.
Shrubs in Colour

fl.   p. 31

Colletia armata var.

Gault, S. Millar
The Color Dictionary of Shrubs

fl.   Pl. 101

Collinsia bicolor

Hay, Roy
The Color Dictionary of Flowers & Plants

fl.   p 34, Pl. 271

Collinsia bicolor

Orr, Robert T.
Wildflowers of Western America

fl.   Pl. 258

Collinsia grandiflora

Clark, Lewis J.
Wild Flowers of British Columbia

fl.   p 463

Collinsia heterophylla

Crockett, James Underwood
Annuals

fl.   P. 109

Collinsia heterophylla

Klimas, John E.
Wildflowers of Eastern America

fl.   Pl. 244

Collinsia heterophylla

Munz, Philip A.
California Spring Wildflowers

fl.   p 30, Pl. 11

Collinsia heterophylla

Wit, H. C. D. de
Plants of the World;
   The Higher Plants, Vol II

fl.   Pl. 88

Collinsia verna

Jennings, O. E.
Wild Flowers of Western Pennsylvania   vol II

fl.   Pl. 138

Collinsia verna

Roberts, June Carver
Born in the Spring

fl. hab.   p 103

Collinsia verna

Walcott, Mary Vaux
North American Wild Flowers
   vol. 4

fl.   Pl. 246

Collinsia verna

Wharton, Mary E.
A Guide to the Wildflowers & Ferns
   of Kentucky

fl.   p 200   Pl. 2.17

Collinsonia canadensis

Courtenay, Booth
Wildflowers & Weeds

fl.   p 97

Collinsonia canadensis

Dean, Blanche E.
Wildflowers of Alabama and
   Adjoining States

fl.   p 157

Collinsonia canadensis

Duncan, Wilbur H.
Wildflowers of the Southeastern
   United States

fl.   p 165

Collinsonia canadensis

Wharton, Mary E.
A Guide to the Wildflowers & Ferns
   of Kentucky

fl.   p 191   Pl. 2.2

Collomia grandiflora

Clark, Lewis J.
Wild Flowers of British Columbia

fl.   p 435

Collomia grandiflora

Munz, Philip A.
California Spring Wildflowers

fl.   p 89, Pl. 93

Colobanthus affinis

Cochrane, G. R.
Flowers and Plants of Victoria

fl. hab.   Pl. 522

Colocasia antiquorum

Brown, Clair A.
Wildflowers of Louisiana and
   Adjoining States

fl. hab.   p 8

Colocasia antiquorum

Hvass, Elsie
Plants That Feed and Serve Us

fl. fr.   p 11, Pl. 12

Colocasia antiquorum

Kiaer, Eigil
Indoor Plants in Colour

hab.   p. 44

Colocasia antiquorum

Masefield, G. B.
The Oxford Book of Food Plants

hab.   Pl. 181, 3

Colocasia esculenta

Bianchini, F.
The Complete Book of Fruits & Vegetables

hab.   p 221

Colocasia esculenta

Miles, Bebe
Bulbs for the Home Gardener

hab.   p 154

Colocasia esculenta

Tarver, David P.
Aquatic and Wetland Plants
   of Florida

hab.   p. 69

Colocasia esculenta var.

Pacific Tropical Botanical Garden
Bulletin
   vol 6, No. 4   1976

hab.   cover

Colombo

see

Sabatia angularis

Colophospermum mopane

Palgrave, K. C.
Trees of Central Africa

fl., fr.   p. 105

Colophospermum mopane

Palmer, Eve
Trees of South Africa

hab.   p. 208   Pl. XV

Coloradoa mesae var.

Cacti & Succulent Society of America
Journal
     vol 13, No. 3   1941

fl., hab.   op. p. 41

Colquhounia coccinea

Curtis's Botanical Magazine
Vol 167   1950

fl.   115

Colquhounia coccinea

Morley, Brian D.
Wild Flowers of the World

fl.   Plate 105

Coltsfoot

see

Petasites frigidus

Coltsfoot

see

Tussilago farfara

Coltsfoot, Alpine

see

Homogyne alpina

Coltsfoot, Sweet-Scented

see

Tussilago fragrans

Coltsfoot, Western

see

Petasites palmatus

Columbine

see

Aquilegia

Columbine, Blue

see

Aquilegia caerulea

Columbine, Eastern

see

Aquilegia canadensis

Columbine, Golden

see

Aquilegia chrysantha

Columbine, Lemon

see

Aquilegia flavescens

Columbine, Red

see

Aquilegia formosa

Columbine, Shortspur

see

Aquilegia brevistyla

Columbine, Western

see

Aquilegia formosa

Columbine, Wild

see

Aquilegia canadensis

Columbine, Wild Blue

see

Aquilegia caerulea var.

Columbine, Yellow

see

Aquilegia flavescens

Columbo, American

see

Swertia caroliniensis

Column flower, showy

see

Columnea gloriosa

Columnea argentea

Morley, Brian D.
Wild Flowers of the World

fl.   Pl. 185

Columnea x banksii

Curtis's Botanical Magazine
Vol. 171   1956-57

fl.   300

Columnea x banksii

Hay, Roy
The Dictionary of House Plants

fl.   Pl. 149

Columnea x banksii

Menninger, Edwin A.
Flowering Vines of the World

fl.   Pl. 83

Columnea, Costa Rica

see

Columnea gloriosa

Columnea erythrophaea

Royal Hort. Soc.
The Garden
     vol 100, No. 11   1975

fl.   p. 535

Columnea fendleri

Morley, Brian D.
Wild Flowers of the World

fl.   Pl. 185

Columnea flaccida

American Gloxinia & Gesneriad Society
The Gloxinian
     vol 27, No. 2   1977

fl.   p. 20

Columnea gloriosa

Kiaer, Eigil
Indoor Plants in Colour

fl.   p. 44

Columnea gloriosa

Macoby, Stirling
What Flower is That

fl.   P 91

Columnea gloriosa var.

Hay, Roy
The Color Dictionary of Flowers &
    Plants

fl.   p. 60   Pl. 480

Columnea gloriosa var.

Hay, Roy
The Dictionary of House Plants

fl.   Pl. 150

Columnea hirsutissima

American Gloxinia & Gesneriad Society
The Gloxinian
    vol 27, No. 2   1977

fl.   p. 21

Columnea hirta

American Gloxinia & Gesneriad Society
The Gloxinian
    vol 23, No. 1   1973

fl.   cover

Columnea hirta

Curtis's Botanical Magazine
Vol 161   1938

fl.   No. 9542

Columnea hybrid

Elbert, Virginie F.
The Miracle Houseplants

fl.   p 116

Columnea hybrid

Encke, Fritz
Zimmerpflanzen

fl.   p. 71

Columnea hybrida

Nicolaisen, Age
Pocket Encyclopedia of Indoor Plants

fl.   Pl. 121

Columnea kewensis

Kiaer, Eigil
Indoor Plants in Colour

fl.   p. 44

Columnea lepidocaula

American Gloxinia & Gesneriad Society
The Gloxinian
    vol 25, No. 1   1975

fl.   cover

Columnea lepidocaula

Curtis's Botanical Magazine
    Vol 176   1966-68

fl.   Pl. 519

Columnea microphylla

American Gloxinia & Gesneriad Society
The Gloxinian
    vol 27, No. 2   1977

fl.   p. 21

Columnea microphylla

Hersey, Jean
Woman's Day Book of House Plants

fl.   p. 49

Columnea mortonii

American Gloxinia & Gesneriad Society
The Gloxinian
    vol 27, No. 2   1977

fl., hab.   p. 20

Columnea percrassa

Curtis's Botanical Magazine
Vol 178   1970-72

fl.   607

Columnea picta

American Gloxinia & Gesneriad Society
The Gloxinian
    vol 27, No. 2   1977

hab.   cover

Columnea purpureovittata

Royal Hort. Soc.
The Garden
    vol 100, No. 9   1975

fl., hab.   p. 438

Columnea schiedeana

American Gloxinian & Gesneriad Society
The Gloxinian
    vol 25, No. 4   1975

fl.   cover

Columnea schiedeana

O'Gorman, Helen
Mexican Flowering Trees and Plants

fl. hab.   p 115

Columnea sp.

Kromdijk, G.
200 House Plants in Colour

fl.   Pl. 58

Columnea teuscheri

Elbert, Virginie F.
The Miracle Houseplants

fl.   p 116

Columnea urbanii

Curtis's Botanical Magazine
    Vol. 181   1976-77

fl.   Pl. 734

Colutea arborescens

Barneby, T. P.
European Alpine Flowers in Colour

fl.   Pl. 40, 3

Colutea arborescens

Hellyer, A.G.L.
Shrubs in Colour

fl., fr.   p. 34

Colutea arborescens

Huxley, Anthony
Deciduous Garden Trees and Shrubs

fl. fr.   Pl. 40, 40a

Colutea arborsecens

Polunin, Oleg
Flowers of Europe

fr.   Pl. 53 #525

Colutea arborescens

Polunin, Oleg
Trees and Bushes of Europe

fl. fr.   p 112

Colvillea racomosa

Oakman, Harry
Colorful Trees

fl.   P 128

Comandra livida

Clark, Lewis J.
Wild Flowers of British Columbia

fr.   p 110

Comandra livida

Walcott, Mary Vaux
North American Wild Flowers
    vol. 5

fr.   Pl. 361

Comandra, Red

see

Comandra livida

Comandra umbellata

Clark, Lewis J.
Wild Flowers of British Columbia

fl.   p 110

Comandra umbellata

Courtenay, Booth
Wildflowers & Weeds

fl.   p 60

Combretun aubletti

American Hort. Soc.
American Horticulturist
     vol 57, No. 6   1978

fl.   p. 6

Combretum coccineum

Menninger, Edwin A.
Flowering Vines of the World

fl.   Pl. 54

Combretum farinosum

Menninger, Edwin A.
Flowering Vines of the World

fl.   Pl. 52

Combretum farinosum

O'Gorman, Helen
Mexican Flowering Trees and Plants

fl. hab.   p 97

Combretum fruticosum

Mathias, Mildred E.
Color for the Landscape

fl.   p. 107

Combretum grandiflorum

Bruggeman, L.
Tropical Plants

fl.,fr.   Pl. 9

Combretum grandiflorum

Menninger, Edwin A.
Flowering Vines of the World

fl.   Pl. 53

Combretum grandiflorum

Morley, Brian D.
Wild Flowers of the World

fl.,fr.   Pl. 61

Combretum loeflingi

Macoby, Stirling
What Flower is That

fl. P 92

Combretum micranthum

Perrot, Emile
Les Plantes Medicinales

fl. fr. hab.   p 129

Combretum microphyllum

Flowering Plants of Africa
     vol XXV   1945-46

fl.   Pl. 978

Combretum microphyllum

Menninger, Edwin A.
Flowering Vines of the World

fl.   Pl. 50

Combretum molle

Palgrave, K. C.
Trees of Central Africa

fl., fr.   p. 133

Combretum platypterum

Menninger, Edwin A.
Flowering Vines of the World

fl.   Pl. 55

Combretum transvaalense

Cobb, L.E.W.
Trees & Shrubs of the Kruger
     National Park

fr.   p. 98   Pl. IVb

Combretum transvaalense

Palmer, Eve.
Trees of South Africa

fr.   Pl. Ib   p. 144

Combretum zeyheri

Flowering Plants of Africa
     vol 31   1956

fl., fr.   Pl. 1230

Combretum zeyheri

Palgrave, K. C.
Trees of Central Africa

fl., fr.   p. 137

Comesperma calymega

Cochrane, G. R.
Flowers and Plants of Victoria

fl.   Pl. 42

Comesperma defoliatum

Blombery, Alec M.
What Wildflower is That

fl.   p 98, Pl. 239

Comesperma ericinum

Blombery, Alec M.
What Wildflower is That

fl.   p 99, Pl. 240

Comesperma ericinum

Mullins, Barbara
Australian Wildflowers in Colour

fl.   P. 43   Pl. 39

Comesperma retusum

Cochrane, G. R.
Flowers and Plants of Victoria

fl.   Pl. 518

Comesperma volubile

Cochrane, G. R.
Flowers and Plants of Victoria

fl.   Pl. 370

Comfrey

see

Symphytum officinale

Comfrey, Tuberous

see

Symphytum tuberosum

Comfrey, Wild

see

Cynoglossum virginianum

Commandra, Northern

see

Geocaulon lividum

Commelina africana

Moriarty, Audrey
Wild Flowers of Malawi

fl.   Pl. 19; 3

Commelina aspera

Moriarty, Audrey
Wild Flowers of Malawi

fl.   Pl. 19; 5

Commelina benghalensis

Lind, E.M.
Some Common Flowering Plants of
     Uganda

fl.   Pl. 14

Commelina benghalensis

Wit, H. C. D. de
Plants of the World;
     The Higher Plants, Vol II

fl.   Pl. 158

Commelina coelestis

O'Gorman, Helen
Mexican Flowering Trees and Plants

fl. hab.   p 147

Commelina communis

Courtenay, Booth
Wildflowers and weeds

fl. p. 2

Commelina communis

Klimas, John E.
Wildflowers of Eastern America

fl.   Pl. 254

Commelina communis

Weeds of the Southern United States

hab., fl.   p. 7

Commelina communis

Wharton, Mary E
A Guide to the Wildflowers & Ferns
of Kentucky

fl.   p 63 Pl. 2.1

Commelina debilis

Jennings, O. E.   vol II
Wild Flowers of Western Pennsylvania

fl.   Pl. 10

Commelina diffusa

Moriarty, Audrey
Wild Flowers of Malawi

fl.   Pl. 19; 4

Commelina erecta

Brown, Clair A.
Wildflowers of Louisiana and
Adjoining States

fl.   p 12

Commelina erecta

Dean, Blanche E.
Wildflowers of Alabama and
Adjoining States

fl.   p 7

Commelina erecta

Duncan, Wilbur H.
Wildflowers of the Southeastern
United States

fl.   p 243

Commelina erecta

Fleming, Glenn
Wild Flowers of Florida

fl.   p 93

Commelina erecta

Lemmon, Robert S.
Wildflowers of North America in
Full Color

fl.   pg. 167   pl. 263

Commelina neurophylla

Moriarty, Audrey
Wild Flowers of Malawi

fl.   Pl. 19; 2

Commelina nudiflora

Walden, Beryl M.
Wild Flowers of Hong Kong

fl.   Pl. 18(47)

Commelina pohliana

Bruggeman, L.
Tropical Plants

fl.   Pl. 94

Commelina virginiana

Duncan, Wilbur H.
Wildflowers of the Southeastern
United States

fl.   p 243

Commelina zambesiaca

Moriarty, Audrey
Wild Flowers of Malawi

fl.   Pl. 19; 1

Commelinantia anomala

Addisonia
vol 21   1939-42

fl.   Pl. 700 p. 55

Commersonia fraseri

Blombery, Alec M.
What Wildflower is That

fl.   p 99, Pl. 241

Commiphora marlothii

Palgrave, K. C.
Trees of Central Africa

fl., fr.   p. 56

Commiphora pilosa

Palgrave, K. C.
Trees of Central Africa

fl., fr.   p. 59

Comparettia coccinea

Pabst, G. F. J.
Orchidaceae Brasilienses,
Vol 2

fl.   p 260

Comparettia macroplectron

Ospina, Mariano
Orquideas de las americas

fl.   Pl. 172

Compass plant

see

Silphium laciniatum

Conebush

see

Ispogon anemonifolius

Cone Bush, Narrow-leaf

see

Petrophila linearis

Cone Bush, Rose

see

Isopogon dubius

Coneflower

see

Rudbeckia

Coneflower

see

Ratibida columnifera

Coneflower, Branched

see

Rudbeckia triloba

Coneflower, Cutleaf

see

Rudbeckia laciniata

Coneflower, Gray-headed

see

Ratibida pinnata

Coneflower, Green

see

Rudbeckia laciniata

Coneflower, Orange

see

Rudbeckia fulgida

Coneflower, Prairie

see

Ratibida pinnata

Coneflower, purple

see

Echinacea pallida
var.

Coneflower, Purple

see

Echinacea purpurea

Coneflower, purple

see

Rudbeckia purpurea

Coneflower, Tall

see

Rudbeckia laciniata

Coneflower, Western

see

Rudbeckia occidentalis

Cone, Horny

see

Isopogon ceratophyllus

Confederate vine

see

Antigonon leptopus

Confetti tree

see

Maytenus senegalensis

Congea tomentosa

Hargreaves, Dorothy
Tropical Blossoms of the Caribbean

fl.   p. 12

Congea tomentosa

Harrison, Richmond E.
Climbers and Trailers

fl.   p 44   Pl. 87

Congea tomentosa

Menninger, Edwin A.
Flowering Vines of the World

fl.   Pl. 189

Congea velutina

Bruggeman, L.
Tropical Plants

fl.   Pl. 11

Congo Root

see

Psoralea psoralioides

Conicosia pugioniformis

Eliovson, Sima
Namaqualand in Flower

fl.   Pl. 52

Conicosia pugioniformis

Herre, H.
The Genera of the Mesembryanthemaceae

fl.   fr.   p 117

Conium maculatum

Ary, S.
The Oxford Book of Wildflowers

fl.   p 86, Pl. 1

Conium maculatum

Bianchini, Francesco
Health Plants of the World

fl.   p 163

Conium maculatum

Brown, Clair A.
Wildflowers of Louisiana and
Adjoining States

fl. fr. hab.   p 123

Conium maculatum

Clark, Lewis J.
Wild Flowers of British Columbia

fl.   p 351

Conium maculatum

Color Treasury of Herbs & Other
Medicinal Plants

hab. p 40   Pl. 53

Conium maculatum

Jennings, O.E.   vol II
Wild Flowers of Western Pennsylvania

fl.   Pl. 105

Conium maculatum

Kimura, Koiti
Japanese Medicinal Plants, Vol I

fl.   p 68

Conium maculatum

Lindman, C. A. M.
Nordens Flora, Vol 7

fl. fr.   Pl. 420

Conium maculatum

Loewenfeld, Claire
The Complete Book of Herbs and Spices

fl. hab.   p 128

Conium maculatum

Martin, W. Keble
The Concise British Flora in Colour

fl. fr. hab.   Pl. 36

Conium maculatum

Perrot, Emile
Les Plantes Medicinales

fl. fr. hab.   p 68

Conium maculatum

Polunin, Oleg
Flowers of Europe

fl. Pl. 85 #879

Conium maculatum

Wharton, Mary E.
A Guide to the Wildflowers & Ferns
of Kentucky

fl.   p 162   Pl. 2.30

Conocarpus erecta

Fleming, Glenn
Wild Flowers of Florida

fl.   p 39

Conopharyngia elegans

Palgrave, K. C.
Trees of Central Africa

fl., fr.   p. 19

Conopholis americana

Campbell, Carlos C.
Great Smoky Mountain Wildflowers

fl. hab.   p 13

355

Conopholis americana

Courtenay, Booth
Wildflowers & Weeds

fl.   p 103

Conopholis americana

Duncan, Wilbur H.
Wildflowers of the Southeastern
   United States

fl. hab.   p 181

Conopholis americana

Jennings, O. E.   vol II
Wild Flowers of Western Pennsylvania

fl. hab.   Pl. 145

Conopholis americana

Klimas, John E.
Wildflowers of Eastern America

fl.   Pl. 298

Conopholis americana

Walcott, Mary Vaux
North American Wild Flowers

fl. hab.   Pl. 214

Conopholis americana

Wharton, Mary E.
A Guide to the Wildflowers & Ferns
   of Kentucky

fl.   p 280   Pl. 3.5

Conophyllum dissitum

Herre, H.
The Genera of the Mesembryanthemaceae

fl. fr.   p. 119

Conophytum bicarinatum

Lamb, Edgar
The Pocket Encyclopedia of Cacti
   and Succulents in Color

fl. hab.   Pl. 220

Conophytum bilobum

Herre, H.
The Genera of the Mesembryanthemaceae

fl.   fr.   p 121

Conophytum cupreiflorum

Cactus & Succulent Society of America
Journal
   vol 35, No. 5   1963

fl., hab.   cover (p. 129)

Conophytum ernianum

Lamb, Edgar
The Pocket Encyclopedia of Cacti
   and Succulents in Color

fl. hab.   Pl. 221

Conophytum fenestratum

Lamb, Edgar
The Pocket Encyclopedia of Cacti
and Succulents in Color

fl. hab.   Pl. 223

Conophytum flavum

Lamb, Edgar
Popular Exotic Cacti in Color

fl. hab.   p 58

Conophytum fraternum

Lamb, Edgar
The Pocket Encyclopedia of Cacti
and Succulents in Color

fl. hab.   Pl. 224

Conophytum meyeri

Lamb, Edgar
The Pocket Encyclopedia of Cacti
   and Succulents in Color

fl. hab.   Pl. 227

Conophytum minusculum

Morley, Brian D.
Wild Flowers of the World

fl.,hab.   Pl. 73

Conophytum minutum

Herre, H.
The Genera of the Mesembryanthemaceae

fl. fr.   p 121

Conophytum minutum

Lamb, Edgar
The Pocket Encyclopedia of Cacti
and Succulents in Color

fl. hab.   Pl. 225

Conophytum minutum

Morley, Brian D.
Wild Flowers of the World

fl.,hab.   Pl. 73

Conophytum praecox

Perry, Frances
Flowers of the World

fl.   p 18

Conophytum proximum

Lamb, Edgar
The Pocket Encyclopedia of Cacti
   and Succulents in Color

fl. hab.   Pl. 222

Conophytum regale

Lamb, Edgar
The Pocket Encyclopedia of Cacti
   and Succulents in Color

fl. hab.   Pl. 226

Conophytum springbokensis

Nicolaisen, Age
Pocket Encyclopedia of Indoor Plants

fl.   pl. 57

Conophytum wettsteinii

Cactus & Succulent Society of America
Journal
   vol 22, No. 1   1950

fl., hab.   cover (p. 1)

Conopodium majus

Ary, S.
The Oxford Book of Wildflowers

fl.   p 86, Pl. 3

Conopodium majus

Martin, W. Keble
The Concise British Flora in Colour

fl. fr. hab.   Pl. 38

Conospermum amoenum

Blombery, Alec M.
What Wildflower is That

fl.   p 99, Pl. 242

Conospermum caeruleum

Blombery, Alec M.
What Wildflower is That

fl.   p 99, Pl. 243

Conospermum mitchellii

Cochrane, G. R.
Flowers and Plants of Victoria

fl.   Pl. 101

Conospermum species

Mullins, Barbara
Australian Wildflowers in Colour

fl.   P. 41   Pl. 35

Conospermum stoechadis

Blombery, Alec M.
What Wildflower is That

fl. hab.   p 99, Pl. 244

Conospermum taxifolium

Blombery, Alec M.
What Wildflower is That

fl.   p 99, Pl. 245

Conospermum tenuifolium

Blombery, Alec M.
What Wildflower is That

fl.   p 100, Pl. 246

Conospermum triplinervium

Royal Hort. Soc.
Journal of the Royal Hort. Soc.
vol 95, No. 7   1970

hab.   p. 304   Pl. 164

Conostylis setigera

Blombery, Alec M.
What Wildflower is That

fl.   p 100, Pl. 247

Consolida ambigua

Crockett, James Underwood
Annuals

fl.   P 109

Consolida ambigua

Polunin, Oleg
Flowers of Europe

fl.   Pl. 20 #213

Consolida regalis

Polunin, Oleg
Flowers of Europe

fl.   Pl. 20 #213

Consound

    see

Symphytum officinale

Constantia rupestris

Pabst, C. F. J.
Orchidaceae Brasilienses,
    Vol 1

fl. hab.   p 222

Convallaria keiskei

Takatori, Jisuke
Color Atlas of Medicinal Plants of Japan

fl.   Fig. 74A

Convallaria majalis

Barneby, T. P.
European Alpine Flowers in Colour

fl.   Pl. 6, 2

Convallaria majalis

Bianchini, Francesco
Health Plants of the World

fl.   p 77

Convallaria majalis

Color Treasury of Herbs & Other
    Medicinal Plants

fl. p 24   Pl. 16

Convallaria majalis

Crockett, James Underwood
Lawns & Ground Covers

fl.   P. 127

Convallaria majalis

Everett, T.H., ed.
New Illustrated Encyclopedia of
    Gardening   Vol. 3

fl.   p. 502

Convallaria majalis

Felsko, Elsa
A Book of Wildflowers

fl.   P 141

Convallaria majalis

Huxley, Anthony
Garden Perennials and Water Plants

fl.   Pl. 87

Convallaria majalis

Hvass, Elsie
Plants That Feed and Serve Us

fl.   p 115, Pl. 252

Convallaria majalis

Kiaer, Eigil
Indoor Plants in Colour

fl.   p. 44

Convallaria majalis

Kimura, Koiti
Japanese Medicinal Plants, Vol I

fl.   p 106

Convallaria majalis

Kleijn, H.
Beauty of the Wild Plant

fl.   opp. p. 52   pl. 63

Convallaria majalis

Kromdijk, G.
200 House Plants in Colour

fl.   Pl. 59

Convallaria majalis

Lemmon, Robert S.
Wildflowers of North America in
    Full Color

fl.   pg. 113   pl. 179

Convallaria majalis

Lindman, C. A. M.
Nordens Flora, Vol 1

fl. fr. hab.   Pl. 60

Convallaria majalis

Macoby, Stirling
What Flower is That

fl.   P 93

Convallaria majalis

Martin, W. Keble
The Concise British Flora in Colour

fl. fr. hab.   Pl. 84

Convallaria majalis

Miles, Bebe
Bulbs for the Home Gardener

fl.   p 63

Convallaria majalis

Perrot, Emile
Les Plantes Medicinales

fl. fr. hab.   p 155

Convallaria majalis

Perry, Frances
Flowers of the World

fl.   p 170

Convallaria majalis

Polunin, Oleg
Flowers of Europe

fl.   Pl. 171 #1657

Convallaria majalis

Tosco, Uberto
The World of Mountain Flowers

fl. hab.   p 99

Convallaria montana

Duncan, Wilbur H.
Wildflowers of the Southeastern
    United States

fl.   p 257

Convolvulus althaeoides

Hay, Roy
The Color Dictionary of Flowers & Plants

fl.   p 5, Pl. 37

Convolvulus althaeoides

Morley, Brian D.
Wild Flowers of the World

fl.   Pl. 36

Convolvulus althaeoides

Polunin, Oleg
Flowers of Europe

fl.   Pl. 100 #1039

Convolvulus althaeoides

Polunin, Oleg
Flowers of the Mediterranean

fl.   Pl. 136

Convolvulus althaeoides

Royal Hort. Soc.
Journal of the Royal Hort. Soc.
    vol 88, No. 9   1963

fl.   p. 388   Pl. 145

Convolvulus althaeoides

Royal Hort. Soc.
The Garden
    vol 101, No. 2   1976

fl.   p. 104

Convolvulus althaeoides

Taylor, A. W.
Wild Flowers of Spain and Portugal

hab., fl.   p. 86

Convolvulus arvensis

Ary, S.
The Oxford Book of Wildflowers

fl.   p 124, Pl. 6

Convolvulus arvensis

Clark, Lewis J.
Wild Flowers of British Columbia

fl.   p 426

Convolvulus arvensis

Color Treasury of Herbs & Other
    Medicinal Plants

fl. p 46   Pl. 65

Convolvulus arvensis

Courtenay, Booth
Wildflowers & Weeds

fl.   p 91

Convolvulus arvensis

Felsko, Elsa
A Book of Wildflowers

fl.   P 106

Convolvulus arvensis

Furrer, D.
Blumen am Wege

fl. hab.   p 79

Convolvulus arvensis

Lindman, C. A. M.
Nordens Flora, Vol 7

fl. fr.   Pl. 479

Convolvulus arvensis

Martin, W. Keble
The Concise British Flora in Colour

fl. hab.   Pl. 61

Convolvulus arvensis

Perry, Frances
Flowers of the World

fl.   90

Convolvulus arvensis

Weeds of the Southern United States

hab.   p. 16

Convolvulus arvensis

Wit, H. C. D. de
Plants of the World;
    The Higher Plants, Vol. II

fl.   p. 108   Pl. 71

Convolvulus cantabricus

Polunin, Oleg
Flowers of the Mediterranean

fl.   Pl. 137

Convolvulus caput-medusae

Bramwell, David
Wild Flowers of the Canary Islands

hab.   Pl. 220

Convolvulus cneorum

Addisonia
    vol 20   1937-38

fl.   p. 45   Pl. 663

Convolvulus cneorum

Hay, Roy
The Color Dictionary of Flowers & Plants

fl.   p 191, Pl. 1521

Convolvulus cneorum

Hellyer, A.G.L.
Shrubs in Colour

fl.   p. 34

Convolvulus cneorum

Perry, Frances
Flowers of the World

fl.   p 89

Convolvulus cneorum

Royal Hort. Soc.
Journal of the Royal Hort. Soc.
    vol 93, No. 7   1968

fl.   cover

Convolvulus cyclostegius

Orr, Robert T.
Wildflowers of Western America

fl.   Pl. 192

Convolvulus elegantissimus

Polunin, Oleg
Flowers of Europe

fl.   Pl. 100 #1039

Convolvulus elegantissimus

Polunin, Oleg
Flowers of the Mediterranean

fl.   Pl. 135

Convolvulus erubescens

Blombery, Alec M.
What Wildflower is That

fl.   p. 100   Pl. 248

Convolvulus erubescens

Cochrane, G. R.
Flowers and Plants of Victoria

fl.   Pl. 227

Convolvulus floridus

Bramwell, David
Wild Flowers of the Canary Islands

fl.   Pl. 217

Convolvulus fruticulosus

Bramwell, David
Wild Flowers of the Canary Islands

fl.   Pl. 218

Convolvulus glandulosus

Bramwell, David
Wild Flowers of the Canary Islands

fl.   Pl. 219

Convolvulus lanuginosus

Polunin, Oleg
Flowers of the Mediterranean

fl.   Pl 138

Convolvulus, Large White

see
Convolvulus sepium

Convolvulus leiocalycinus

Royal Hort. Soc.
Journal of the Royal Hort. Soc.
    vol 98, No. 7   1973

fl.   p. 306   Pl. 153

Convolvulus mauritanicus

Crockett, James Underwood
Lawns & Ground Covers

fl.   p. 128

Convolvulus mauritanicus

Harrison, Richmond E.
Climbers and Trailers

fl.   p 92   Pl. 228

Convolvulus mauritanicus

Hay, Roy
The Color Dictionary of Flowers & Plants

fl.   p 5, Pl. 38

Convolvulus mauritanicus

Macoby, Stirling
What Flower is That

fl.   p. 92

Convolvulus mauritanicus

Vilmorin, Roger de
Plantes Alpines dans les Jardins

fl. hab.   Pl. XXXVI

Convolvulus minor

Huxley, Anthony
Garden annuals and bulbs

fl.   pl. 34

Convolvulus occidentalis

Orr, Robert T.
Wildflowers of Western America

fl.   Pl. 44

Convolvulus oleifolius

Coulimis, Constantine N.
Wild Flowers of Greece

fl.   p 87

Convolvulus oleifolius

Megaw, Elektra
Wild Flowers of Cyprus

fl.   p. 10   Pl. 15

Convolvulus oleifolius var.

Curtis's Botanical Magazine
vol 172   1958-59

fl.   Pl. 324

Convolvulus pitardii var.

Curtis's Botanical Magazine
Vol. 170   1954-55

fl.   240

Convolvulus sabatius var.

Mathias, Mildred E.
Color for the Landscape

fl. hab.   p 137

Convolvulus, Sand
see
Calystegia soldanella

Convolvulus scoparius

Bramwell, David
Wild Flowers of the Canary Islands

fl.   Pl. 221

Convolvulus sepium

Bianchini, Francesco
Health Plants of the World

fl. hab.   p 71

Convolvulus sepium

Courtenay, Booth
Wildflowers & Weeds

fl.   p 91

Convolvulus sepium

Ferguson, Mary
Wildflowers

fl.   p 109

Convolvulus sepium

Klimas, John E.
Wildflowers of Eastern America

fl.   Pl. 60

Convolvulus sepium

Lemmon, Robert S.
Wildflowers of North America in Full Color

fl.   pg. 196   pl. 307

Convolvulus sepium

Moyle, John B.
Northland Wild Flowers

fl. hab.   p 129, Pl. 138

Convolvulus sepium

Pond, Barbara
A Sampler of Wayside Herbs

fl.   Pl. XXVII

Convolvulus sepium var.

Clark, Lewis J.
Wild Flowers of British Columbia

fl.   p 422

Convolvulus soldanella

Clark, Lewis J.
Wild Flowers of British Columbia

fl.   p 198

Convolvulus spithamaeus

Courtenay, Booth
Wildflowers & Weeds

fl.   p 91

Convolvulus spithamaeus

Moyle, John B.
Northland Wild Flowers

fl.   p 129, Pl. 139

Convolvulus spithamaeus

Wharton, Mary E.
A Guide to the Wildflowers & Ferns of Kentucky

fl.   p 118   Pl. 1.48

Convolvulus tricolor

Crockett, James Underwood
Annuals

fl.   P 110

Convolvulus tricolor

Everett, T.H., ed.
New Illustrated Enccylopedia of Gardening   vol 3

fl.   p. 438

Convolvulus tricolor

Morley, Brian D.
Wild Flowers of the World

fl.   Pl. 36

Convolvulus tricolor

Polunin, Oleg
Flowers of Europe

fl.   Pl. 100 #1035

Convolvulus tricolor

Polunin, Oleg
Flowers of the Mediterranean

fl.   Pl. 133

Convolvulus tricolor

Tsukamoto, Yotaro
Coloured Illustrations of Garden Flowers Vol. 10

fl. Illus. 43   opp. p. 14

Convolvulus tricolor var.

Hay, Roy
The Color Dictionary of Flowers & Plants

fl.   p. 34-35   Pl.272-73

Convolvulus tricolor var.

Perry, Frances
Flowers of the World

fl.   p. 90

Conyza canadensis

Clark, Lewis J.
Wild Flowers of British Columbia

fl.   p 543

Conyza canadensis

Courtenay, Booth
Wildflowers & Weeds

fl.   p 118

Conyza canadensis

Martin, W. Keble
The Concise British Flora in Colour

fl. hab.   Pl. 44

Coolabah Tree

see

Eucalyptus microtheca

Cooperia drummondii

Brown, Clair A.
Wildflowers of Louisiana and
          Adjoining States

fl. hab.   p 25

Cooperia drummondii

Bruggeman, L.
Tropical Plants

fl.   Pl. 95

Cooperia smallii

Addisonia
     vol 21   1939-42

fl., fr.   p. 7   Pl. 676

Cooperia, yellow

see

Cooperia smallii

Copal

see

Bursera hindsiana

Copal, Manila

see

Agathis dammara

Copernicia cerifera

Hvass, Elsie
Plants That Feed and Serve Us

hab.   p 109, Pl. 235

Copernicia torreana

Milne, Lorus
Living Plants of the World

hab.   p 274

Copey

see

Clusia rosea

Copiapoa echinoides

Lamb, Edgar
The Pocket Encyclopedia of Cacti
     and Succulents in Color

fl. hab.   Pl. 19

Copiapoa echinoides

Lamb, Edgar
Popular Exotic Cacti in Color

fl. hab.   p 55

Copiapoa humilis var.

Lamb, Edgar
The Pocket Encyclopedia of Cacti
     and Succulents in Color

fl. hab.   Pl. 20

Copiapoa longispina

Backeberg, Curt
Cactus Lexicon

fl.   p 582

Copiapoa wagenknechtii

Backeberg, Curt
Cactus Lexicon

hab.   p 582

Copper Cups

see

Pileanthus peduncularis

Copperleaf

see

Acalypha hispida

Copper leaf

see

Acalypha wilkesiana

Copperleaf, banana

see

Acalypha wilkesiana var.

Copperleaf, painted

see

Acalypha wilkesiana var.

Copperleaf, Virginia

see

Acalypha virginica

Coprosma baueri var.

Macoby, Stirling
What Flower is That

hab.   P. 93

Coprosma hirtella

Cochrane, G. R.
Flowers and Plants of Victoria

fr.   Pl. 395

Coprosma hybrid

Harrison, Richmond E.
Climbers and Trailers

fl.   p. 23   Pl. 20

Coprosma x kirkii

Harrison, Richmond E.
Climbers and Trailers

hab.   p. 91   Pl. 225

Coprosma nitida

Blombery, Alec M.
What Wildflower is That

fr.   p. 100   Pl. 249

Coprosma nitida

Cochrane, G. R.
Flowers and Plants of Victoria

fl.   Pl. 430

Coprosma nitida

Curtis's Botanical Magazine
Vol 166   1949

fl. fr.   88

Coprosma pumila

Morley, Brian D.
Wild Flowers of the World

fl. fr.   Pl. 2J

Coprosma quadrifida

Cochrane, G. R.
Flowers and Plants of Victoria

hab.   Pl. 458

Coprosma repens

Milne, Lorus
Living Plants of the World

fr.   p 231

Coprosma repens var.

Harrison, R.E.
Trees and Shrubs

hab.   p 55

Coprosma repens var.

Royal Hort. Soc.
The Garden
      vol 100, No. 12 1975

hab.   p. 576

Coprosma robusta var.

Harrison, R.E.
Trees and Shrubs

hab. P 54

Coptis asplenifolia

Clark, Lewis J.
Wild Flowers of British Columbia

fl.   p 159

Coptis groenlandica

Klimas, John E.
Wildflowers of Eastern America

fl.   Pl. 69

Coptis japonica

Kariyone, Tatsuo
Atlas of Medicinal Plants

fl., hab.   Pl. 126

Coptis japonica

Kimura, Koiti
Japanese Medicinal Plants, Vol I

fl.   p 26

Coptis trifolia

Courtenay, Booth
Wildflowers & Weeds

fl.   p 31

Coptis trifolia

Heller, Christine
Wild Flowers of Alaska

fl.   Pl. 141

Coptis trifolia

Lemmon, Robert S.
Wildflowers of North America in
   Full Color

fl.   pg. 250   pl. 394

Coral
   see
Erythrina americana

Coral, Batswing
   see
Erythrina vespertilio

Coral Beads
   see
Cocculus carolinus

Coral berry
   see
Ardisia crenata

Coral blow
   see
Russellia juncea

Coralbush
   see
Russelia juncea

Coral, cockscomb
   see
Erythrina crista-galli

Coral Drops, Mexican
   see
Bessera elegans

Coral gem
   see
Lotus bertholetii

Coral, Indian
   see
Erythrina indica

Coral Root
   see
Corallorhiza

Coralroot, Autumn
   see
Corallorhiza odontorhiza

Coralroot, Crested
   see
Hexalectris spicata

Coralroot, Spotted
   see
Corallorhiza maculata

Coral Root, Spring
   see
Corallorhiza wisteriana

Coral Root, Western
   see
Corallorhiza mertensiana

Coral, South African
   see
Erythrina caffra

Coral Tree
   see
Erythrina

Coral-tree, Abyssinian
   see
Erythrina absyssinica

Coral Tree, Bat's Wing
   see
Erythrina verspertilio

Coral tree, coral bean
   see
Erythrina indica

Coral Tree, Crabclaw

see

Erythrina indica

Coral Tree, Indian

see

Erythrina indica

Coral tree, purple

see

Erythrina fusca

Coral tree, variegated

see

Erythrina parcellii

Coral Vine

see

Antigonon leptopus

Coral vine

see

Berberidopsis corallina

Coral Vine

see

Cissus cucurbitina

Coral, weeping

see

Erythrina ovalifolia

Corallita

see

Antigonon leptopis

Corallodiscus kingianus

Curtis's Botanical Magazine
Vol 165   1948

fl.   6

Corallodiscus kingianus

Royal Hort. Soc.
Journal of the Royal Hort. Soc.
     Vol. 73, No. 9   1948

fl., hab.   p. 292   Pl. 98

Corallorhiza innata

Brooke, Jocelyn
The Wild Orchids of Britain

fl., hab.   Pl. 14

Corallorhiza maculata

Alaska-Yukon Wild Flowers Guide

fl.   p. 23

Corallorhiza maculata

Courtenay, Booth
Wildflowers & Weeds

fl.   p 14

Corallorhiza maculata

Jennings, O. E.   vol II
Wild Flowers of Western Pennsylvania

fl.   Pl. 45

Corallorhiza maculata

Luer, Carlyle A.
The Native Orchids of the
     United States and Canada

fl. fr. hab.   Pl. 89

Corallorhiza maculata

Moyle, John B.
Northland Wild Flowers

fl. hab.   p 225, Pl. 300

Corallorhiza maculata

Orr, Robert T.
Wildflowers of Western America

fl.   Pl. 7

Corallorhiza maculata

Ospina, Mariano
Orquideas de las americas

fl.   Pl. 95

Corallorhiza maculata

Welsh, Stanley L.
Flowers of the Mountain Country

fl.   p. 31

Corallorhiza maculata var.

Clark, Lewis J.
Wild Flowers of British Columbia

fl. hab.   p 63, 83

Corallorhiza mertensiana

Fries, Mary A.
Wildflowers of Mount Ranier and
the Cascades

fl. P 36 & 37

Corallorhiza mertensiana

Luer, Carlyle A.
The Native Orchids of the
     United States and Canada

fl. hab.   Pl. 90

Corallorhiza odontorhiza

Klimas, John E.
Wildflowers of Eastern America

fl.   Pl. 302

Corallorhiza odontorhiza

Luer, Carlyle A.
The Native Orchids of Florida

fl. hab.   Pl. 52; 6-9

Corallorhiza odontorhiza

Luer, Carlyle A.
The Native Orchids of the
     United States and Canada

fl. hab.   Pl. 93; 6-9

Corallorhiza sp.

Taylor, Ronald J.
Mountain Wild Flowers

fl. hab.   p 22, 23

Corallorhiza striata

Clark, Lewis J.
Wild Flowers of British Columbia

fl.   p 91

Corallorhiza striata

Courtenay, Booth
Wildflowers & Weeds

fl.   p 14

Corallorhiza striata

Luer, Carlyle A.
The Native Orchids of the United States
and Canada

fl.   frontispiece

Corallorhiza striata var.

Luer, Carlyle A.
The Native Orchids of the
     United States and Canada

fl. hab.   Pl. 92

Corallorhiza trifida

Ary, S.
The Oxford Book of Wildflowers

fl.   p 44, Pl. 7

Corallorhiza trifida

Barneby, T. P.
European Alpine Flowers in Colour

fl.   Pl. 13, 6

Corallorhiza trifida

Lindman, C. A. M.
Nordens Flora, Vol 3

fl. hab.   Pl. 155

Corallorhiza trifida

Luer, Carlyle A.
The Native Orchids of the
   United States and Canada

fl. hab.   Pl. 91

Corallorhiza trifida

Martin, W. Keble
The Concise British Flora in Colour

fl. hab.   Pl. 80

Corallorhiza trifida

Morley, Brian D.
Wild Flowers of the World

fl.   Pl. 7A

Corallorhiza trifida

Polunin, Oleg
Flowers of Europe

fl. Pl. 192 #1926

Corallorhiza trifida

Porsild, A.E.
Rocky Mountain Wild Flowers

fl. hab.   p. 113

Corallorhiza trifida

Webster, Mary
Flora of Moray, Nairn & East
   Inverness

fl.   p. 292   Pl. 19

Corallorhiza wisteriana

Duncan, Wilbur H.
Wildflowers of the Southeastern
   United States

fl. hab.   p 279

Corallorhiza wisteriana

Luer, Carlyle A.
The Native Orchids of Florida

fl. hab.   Pl. 52; 1-5

Corallorhiza wisteriana

Luer, Carlyle A.
The Native Orchids of the
   United States and Canada

fl. hab.   Pl. 93; 1-5

Corallorhiza wisteriana

Wharton, Mary E.
A Guide to the Wildflowers & Ferns
   of Kentucky

fl. p 92 Pl. 3.5

Corchorus capsularis

Hvass, Elsie
Plants That Feed and Serve Us

fl. fr.   p 102, Pl. 218

Cordia alba

Chickering, Carol Rogers
Flowers of Guatemala

fl. fr.   p 59

Cordia dodecandra

O'Gorman, Helen
Mexican Flowering Trees and Plants

fl. fr. hab.   p 21

Cordia sebestena

Hall, Clarence E.
Flowers of the Islands in the Sun

fl.   p. 51   Pl. 9

Cordia sebestena

Hargreaves, Dorothy
Tropical Blossoms of the Caribbean

fl.,hab.   p. 47

Cordia sebestena

Perry, Frances
Flowers of the World

fl.   p 48

Cordia sebestena

Pertchik, Bernard
Flowering Trees of the
   Caribbean

fl.   p 103

Cordyline australis

Boom, B. K.
The glory of the tree

hab.   opp. p. 111   pl. 192

Cordyline australis

Harvey, Norman B.
New Zealand Botanical Paintings

fl.   p. 72   Pl. 33

Cordyline australis

Hay, Roy
The Color Dictionary of Flowers & Plants

hab.   p 191, Pl. 1522

Cordyline australis

Hay, Roy
The Dictionary of House Plants

hab.   Pl. 151

Cordyline australis

Macoby, Stirling
What Flower is That

fl. P 93

Cordyline australis

Polunin, Oleg
Trees and Bushes of Europe

hab.   p 181

Cordyline australis

Royal Hort. Soc.
Journal of the Royal Hort. Soc.
   vol 93, No. 8   1968

fl.   p. 336   Pl. 169

Cordyline australis

Royal Hort. Soc.
The Garden
   vol 103, No. 7   1978

hab.   p. 269

Cordyline cannifolia

Blombery, Alec M.
What Wildflower is That

fl.   p 101, Pl. 250

Cordyline congesta

Royal Hort. Soc.
The Garden
   vol.103, No. 1 1978

hab.   p. 24

Cordyline hybrid

Kromdijk, G.
200 House Plants in Colour

hab.   Pl. 60

Cordyline indivisa

Hay, Roy
The Color Dictionary of Flowers & Plants

fl.   p 191, Pl. 1523

Cordyline stricta

Blombery, Alec M.
What Wildflower is That

fl.   p. 101   Pl. 251

Cordyline stricta

Morley, Brian D.
Wild Flowers of the World

fl.   pl. 142

Cordyline terminalis

Everett, T.H., ed.
New Illustrated Encyclopedia of
   Gardening   vol 3

hab.   p. 391

Cordyline terminalis

Hargreaves, Dorothy
Tropical Blossoms of the Caribbean

hab.   p. 63-64

Cordyline terminalis

Hay, Roy
The Color Dictionary of Flowers & Plants

hab.   p 61, Pl. 481

Cordyline terminalis

Kiaer, Eigil
Indoor Plants in Colour

hab.   p. 45

Cordyline terminalis

Macoby, Stirling
What Flower is That

hab.   P. 93

Cordyline terminalis

Massachusetts Hort. Soc.
Horticulture
    vol 41, No. 2   1963

fl., fr.   backcover

Cordyline terminalis

Tsukamoto, Yotaro
Coloured Illustrations of Garden
    Flowers   Vol. 9

hab.   p. 45-46

Cordyline terminalis var.

American Hort. Soc.
American Horticulturist
    vol 55, No. 5   1976

hab., fr.   p. 6-7

Cordyline terminalis var.

Encke, Fritz
Zimmerpflanzen

hab.   p. 71

Cordyline terminalis var.

Hay, Roy
The Dictionary of House Plants

hab.   Pl. 152

Coreopsis angustifolia

Duncan, Wilbur H.
Wildflowers of the Southeastern
    United States

fl.   p 217

Coreopsis auriculata

Dean, Blanche E.
Wildflowers of Alabama and
    Adjoining States

fl.   p 199

Coreopsis basalis

Duncan, Wilbur H.
Wildflowers of the Southeastern
    United States

fl.   p 219

Coreopsis calliopsidea

Munz, Philip A.
California Desert Wildflowers

fl.   p 54, Pl. 82

Coreopsis drummondii

Bruggeman, L.
Tropical Plants

fl.   Pl. 35

Coreopsis drummondii

Crockett, James Underwood
Annuals

fl.   P 110

Coreopsis drummondii

Tsukamoto, Yotaro
Coloured Illustrations of Garden
    Flowers   vol. 10

fl. Illus. 84   opp. p. 27

Coreopsis gigantea

Mathias, Mildred E.
Color for the Landscape

fl. hab.   p 181

Coreopsis gigantea

Munz, Philip A.
California Spring Wildflowers

fl.   p 37, Pl. 33

Coreopsis grandiflora var.

Hay, Roy
The Color Dictionary of Flowers & Plants

fl.   p 132, Pl. 1055

Coreopsis lanceolata

Batson, Wade T.
Wild Flowers in South Carolina

fl.   p 121

Coreopsis lanceolata

Brown, Clair A.
Wildflowers of Louisiana and
    Adjoining States

fl. hab.   p 196

Coreopsis lanceolata

Klimas, John E.
Wildflowers of Eastern America

fl.   Pl. 152

Coreopsis lanceolata

Lemmon, Robert S.
Wildflowers of North America in
    Full Color

fl.   pg. 211   pl. 446

Coreopsis lanceolata

Macoby, Stirling
What Flower is That

fl.   P 93

Coreopsis lanceolata

Tsukamoto, Yotaro
Coloured Illustrations of Garden
    Flowers   Vol. 10

fl. Illus. 40   opp. p. 13

Coreopsis, Large

    see

Coreopsis major

Coreopsis leavenworthii

Fleming, Glenn
Wild Flowers of Florida

fl. hab.   p 45

Coreopsis linifolia

Addisonia
    vol 23   1954-59

fl.   p. 55   Pl. 764

Coreopsis major

Brown, Clair A.
Wildflowers of Louisiana and
    Adjoining States

fl.   p 197

Coreopsis major

Campbell, Carlos C.
Great Smoky Mountain Wildflowers

fl.   p 93

Coreopsis major

Dean, Blanche E.
Wildflowers of Alabama and
    Adjoining States

fl.   p 199

Coreopsis major

Duncan, Wilbur H.
Wildflowers of the Southeastern
    United States

fl.   p 217

Coreopsis major

Wharton, Mary E.
A Guide to the Wildflowers & Ferns
    of Kentucky

fl.   p 252   Pl. 3.17

Coreopsis maritima

Lemmon, Robert S.
Wildflowers of North America in
Full Color

fl.   pg. 53   pl. 89

Coreopsis maritima

Massachusetts Hort. Soc.
Horticulture
vol 47, No. 9   1969

fl.   p. 39

Coreopsis nudata

Dean, Blanche E.
Wildflowers of Alabama and
Adjoining States

fl.   p 197

Coreopsis nudata

Duncan, Wilbur H.
Wildflowers of the Southeastern
United States

fl.   p 219

Coreopsis nudata

Lemmon, Robert S.
Wildflowers of North America in
Full Color

fl.   pg. 55   pl. 92

Coreopsis palmata

Courtenay, Booth
Wildflowers & Weeds

fl.   p 115

Coreopsis palmata

Moyle, John B.
Northland Wild Flowers

fl.   p 161, Pl. 192

Coreopsis saxicola

Addisonia
vol 20   1937-38

fl., fr.   p. 17   Pl. 649

Coreopsis, Smooth

see

Coreopsis lanceolata

Coreopsis, Tall

see

Coreopsis tripteris

Coreopsis tinctoria

Brown, Clair A.
Wildflowers of Louisiana and
Adjoining States

fl.   p 197

Coreopsis tinctoria

Bruggeman, L.
Tropical Plants

fl.   Pl. 36

Coreopsis tinctoria

Crockett, James Underwood
Annuals

fl.   P 110

Coreopsis tinctoria

Dean, Blanche E.
Wildflowers of Alabama and
Adjoining States

fl.   p 199

Coreopsis tinctoria

Huxley, Anthony
Garden annuals and bulbs

fl.   pl. 39

Coreopsis tinctoria var.

Hay, Roy
The Color Dictionary of Flowers &
Plants

fl.   p. 35   Pl. 274

Coreopsis tripteris

Duncan, Wilbur H.
Wildflowers of the Southeastern
United States

fl.   p 217

Coreopsis tripteris

Wharton, Mary E.
A Guide to the Wildflowers & Ferns
of Kentucky

fl.   p 253   Pl. 3.18

Coreopsis verticillata

Everett, T.H., ed.
New Illustrated Encyclopedia of
Gardening   vol 8

fl.   p. 1399

Coreopsis verticillata

Meikle, R. D.
Garden Flowers

fl.   opp. p. 256   Pl. 6

Coreopsis verticillata var.

Hay, Roy
The Color Dictionary of Flowers &
Plants

fl.   p. 132   Pl. 1056

Coreopsis verticillata var.

Huxley, Anthony
Garden Perennials and Water Plants

fl.   Pl. 88

Coriander

see

Coriandrum sativum

Coriandrum sativum

Bianchini, F.
The Complete Book of Fruits &
Vegetables

fr.   p 101

Coriandrum sativum

Kimura, Koiti
Japanese Medicinal Plants, Vol I

fl.   p 68

Coriandrum sativum

Loewenfeld, Claire
The Complete Book of Herbs and Spices

fl. fr. hab.   p 208

Coriandrum sativum

Masefield, G. B.
The Oxford Book of Food Plants

fl. fr.   Pl. 139, 2

Coriandrum sativum

Perrot, Emile
Les Plantes Medicinales

fl. fr. hab.   p 78

Coriandrum sativum

Rosengarten, Frederic, Jr.
The Book of Spices

fl. fr.   p. 216

Coriaria arborea

Harvey, Norman B.
New Zealand Botanical Paintings

fl.   p. 28   Pl. 11

Coriaria japonica

Kitamura, Siro
Coloured Illustrations of Trees &
Shrubs of Japan

fr.   277

Coriaria myrtifolia

Polunin, Oleg
Flowers of Europe

fr.   Pl. 68 #702

Coriaria myrtifolia

Polunin, Oleg
Trees and Bushes of Europe

fr. hab.   p 118

Coriaria nepalensis

Hara, Hiroshi, comp.
Photo-Album of Plants of
Eastern Himalaya

fl.   Pl. 72

Coridothymus capitatus

Feinbrun-Dothan, Naomi
Wild Plants in the Land
of Israel

fl.   p 66

Coris monspeliensis

Polunin, Oleg
Flowers of the Mediterranean

fl. fr.   Pl. 122

Corkbush

see

Euonymus

Corkscrew Flower

see

Phaseolus caracalla

Cork Tree

see

Parinari mobola

Corkwood

see

Hakea sp.

Corn

see

Zea mays

Cornflower

see

Centaurea cyanus

Cornflower, Mountain

see

Centaurea montana

Corn, Guinea

see

Sorghum vulgare

Corn, Indian

see

Zea mays

Corn, Kaffir

see

Sorghum vulgare

Corn Plant

see

Dracaena fragrans var.

Corn, Rainbow

see

Zea mays var.

Corn-salad

see

Valerianella intermedia

Corn Salad

see

Valerianella olitoria

Corn, Sand

see

Zigadenus paniculatus

Corn, Squirrel

see

Bikukulla canadensis

Corn, Squirrel

see

Dicentra canadensis

Cornel

see

Cornus

Cornel, Dwarf

see

Chamaepericlymenum suecicum

Cornel, Dwarf

see

Cornus canadensis

Cornel, Dwarf

see

Cornus suecica

Cornus alba

American Hort. Soc.
American Horticulturist
vol 56, No. 2   1977

hab.   p. 18

Cornus alba

Polunin, Oleg
Trees and Bushes of Europe

hab.   p 153

Cornus alba var.

American Hort. Soc.
American Horticulturist
vol 56, No. 2   1977

hab.   p. 18

Cornus alba var.

Gault, S. Millar
The Color Dictionary of Shrubs

hab.   Pl. 102

Cornus alba var.

Hay, Roy
The Color Dictionary of Flowers & Plants

fr. hab.   p 191, Pl. 1524, 1525

Cornus alba var.

Hellyer, A.G.L.
Shrubs in Colour

hab.   p. 34

Cornus alba var.

Huxley, Anthony
Deciduous Garden Trees and Shrubs

fr., hab.   Pl. 41-42

Cornus alba var.

Massachusetts Hort. Soc.
Horticulture
vol 43, No. 2   1965

hab.   p. 32-33

Cornus alba var.

Royal Hort. Soc.
Journal of the Royal Hort. Soc.
vol 94, No. 12   1969

hab.   cover

Cornus alternifolia

Campbell, Carlos C.
Great Smoky Mountain Wildflowers

fl.   p 27

Cornus alternifolia

Wharton, Mary E.
Trees & Shrubs of Kentucky

fl. fr.   p. 93, 144   Pl. 2.29a &
                                      239a

Cornus baileyi

Gault, S. Millar
The Color Dictionary of Shrubs

fr.   Pl. 103

Cornus canadensis

Alaska-Yukon Wild Flower Guide

fl. fr. hab.   p 29, 106,
                                107

Cornus canadensis

American Hort. Soc.
American Horticulturist
vol 55, No. 6   1976

fr., hab.   p. 15

Cornus canadensis

Clark, Lewis J.
Wild Flowers of British Columbia

fl. fr.   p 363, 366

Cornus canadensis

Courtenay, Booth
Wildflowers & Weeds

fl.   p 79

Cornus canadensis

Ferguson, Mary
Wildflowers

fl.   p 124

Cornus canadensis

Fries, Mary A.
Wildflowers of Mount Ranier and
    the Cascades

fl., hab.   P. 48-49

Cornus canadensis

Hay, Roy
The Color Dictionary of Flowers & Plants

fl.   p 5, Pl. 39

Cornus canadensis

Heller, Christine
Wild Flowers of Alaska

fl., fr.   Pl. 190-191

Cornus canadensis

Jennings, O. E.   vol II
Wild Flowers of Western Pennsylvania

fl.   Pl. 65

Cornus canadensis

Klimas, John E.
Wildflowers of Eastern America

fl.   Pl. 94

Cornus canadensis

Lemmon, Robert S.
Wildflowers of North America in
    Full Color

fl. fr.   p. 138,139   pl. 220,221

Cornus canadensis

Neufeld, J.B.
Wild Flowers of the Prairies

fl.   p. 19

Cornus canadensis

Orr, Robert T.
Wildflowers of Western America

fl.   Pl. 65

Cornus canadensis

Porsild, A. E.
Rocky Mountain Wild Flowers

fl.   p 295

Cornus canadensis

Royal Hort. Soc.
Journal of the Royal Hort. Soc.
vol 92, No. 4   1967

fl., hab.   p. 160   Pl. 73

Cornus canadensis

Taylor, Ronald J.
Mountain Wild Flowers

fl. fr. hab.   p 20

Cornus canadensis

Walcott, Mary Vaux
North American Wild Flowers
vol. 4

fl., fr.   Pl. 271-72

Cornus canadensis

Welsh, Stanley L.
Flowers of the Mountain Country

fl.   p. 10

Cornus capitata

Macoby, Stirling
What Flower is That

fr.   P 94

Cornus chinensis

Royal Hort. Soc.
Journal of the Royal Hort. Soc.
vol 87, No. 6   1962

fl.   p. 266   Pl. 72

Cornus controversa

Kitamura, Siro
Coloured Illustrations of Trees &
    Shrubs of Japan

fl.   373

Cornus drummondi

Wharton, Mary E.
Trees & Shrubs of Kentucky

fl.   p 92, Pl. 2.29b
fr.   p 144, Pl. 2.39c

Cornus florida

Batson, Wade T.
Wild Flowers in South Carolina

fl.   p 80

Cornus florida

Boom, B. K.
The glory of the tree

fl.   opp. p. 104   pl. 172

Cornus florida

Brown, Clair A.
Wildflowers of Louisiana and
    Adjoining States

fl.   p 126

Cornus florida

Campbell, Carlos C.
Great Smoky Mountain Wildflowers

fl.   p 39

Cornus florida

Flemer, William III
Shade and Ornamental Trees in Color

hab., fl.   p. 33

Cornus florida

Gault, S. Millar
The Color Dictionary of Shrubs

hab.   Pl. 105

Cornus florida

Massachusetts Hort. Soc.
Horticulture
vol 37, No. 9   1959

fr.   cover

Cornus florida

Royal Hort. Soc.
Journal of the Royal Hort. Soc.
vol 87, No. 7   1962

hab., fl.   p. 310   Pl. 89

Cornus florida

Walcott, Mary Vaux
North American Wild Flowers
vol. 5

fl., fr.   Pl. 321-22

Cornus florida

Wharton, Mary E.
Trees & Shrubs of Kentucky

fl.   p 94, Pl. 2.29f
fr.   p 123, Pl. 2.9

Cornus florida var.

Bartels, Andreas
Das Grosse Buch der Gartengeholze

fl. hab.   p 135

Cornus florida var.

Everett, T.H., ed.
New Illustrated Encyclopedia of
    Gardening   vol 3

hab.   p. 487

Cornus florida var.

Flemer, William III
Shade and Ornamental Trees in Color

hab., fl.   p. 34

Cornus florida var.

Harrison, R.E.
Trees and Shrubs

fl.   p. 54

Cornus florida var.

Hay, Roy
The Color Dictionary of Flowers &
    Plants

fl.   p. 191   Pl. 1526

Cornus florida var.

Macoby, Stirling
What Flower is That

fl.   P 94

Cornus florida var.

Massachusetts Hort. Soc.
Horticulture
    vol 33, No. 1   1955

fl.   p. 12

Cornus florida var.

Massachusetts Hort. Soc.
Horticulture
    vol 35, No. 2   1957

fl.   p. 76

Cornus florida var.

Massachusetts Hort. Soc.
Horticulture
    vol 40, No. 5   1962

fl.   inside backcover

Cornus hybrid.

Gault, S. Millar
The Color Dictionary of Shrubs

fl.   Pl. 104

Cornus kousa

American Hort. Soc.
American Horticulturist
    vol 55, No. 2   1976

fl., fr., hab.   p. 30

Cornus kousa

Bartels, Andreas
Das Grosse Buch der Gartengeholze

fr.   p 135

Cornus kousa

Boom, B. K.
The glory of the tree

fr.   opp. p. 104   pl. 173

Cornus kousa

Edlin, Herbert
The Illustrated Encyclopedia
    of Trees

fr.   p 199

Cornus kousa

Everett, T.H., ed.
New Illustrated Encyclopedia of
    Gardening   vol 3

fl.   frontis.

Cornus kousa

Flemer, William III
Shade and Ornamental Trees in Color

fl.   p. 35

Cornus kousa

Gault, S. Millar
The Color Dictionary of Shrubs

fl.   Pl. 106

Cornus kousa

Hansen, Richard
Baume und Straucher im Garten

hab.   P. 99

Cornus kousa

Kitamura, Siro
Coloured Illustrations of Trees &
    Shrubs of Japan

fl.   375

Cornus kousa

Massachusetts Hort. Soc.
Horticulture
    vol 39, No. 10   1961

fr.   p. 526

Cornus kousa

Pennsylvania Hort. Soc.
The Green Scene
    vol 6, No. 6   1978

fl.   p. 25

Cornus kousa var.

Gault, S. Millar
The Color Dictionary of Shrubs

fl.   Pl. 107, 108

Cornus kousa var.

Hay, Roy
The Color Dictionary of Flowers & Plants

fl.   p 191, Pl. 1527

Cornus kousa var.

Massachusetts Hort. Soc.
Horticulture
    vol 46, No. 4   1968

fr.   p. 48

Cornus kousa var.

Perry, Frances
Flowers of the World

fl.   p 92

Cornus kousa var.

Royal Hort. Soc.
The Garden
    vol 101, No. 10   1976

fl.   p. 500

Cornus macrophylla

Kitamura, Siro
Coloured Illustrations of Trees &
    Shrubs of Japan

fl.   374

Cornus mas

American Hort. Soc.
American Horticulturist
    vol 56, No. 2   1977

fl.   p. 22

Cornus mas

Bianchini, F.
The Complete Book of Fruits & Vegetables

fr.   p 165

Cornus mas

Bianchini, Francesco
Health Plants of the World

fr.   p 63

Cornus mas

Boom, B. K.
The glory of the tree

fr.   opp. p. 104   pl. 174

Cornus mas

Color Treasury of Herbs & Other
   Medicinal Plants

fr.   p 39   Pl. 51

Cornus mas

Everett, T. H.
New Illustrated Encyclopedia
   of Gardening   vol 3

fl.   Pl. (3-7c)

Cornus mas

Gault, S. Millar
The Color Dictionary of Shrubs

fl.   Pl. 109

Cornus mas

Hansen, Richard
Baume und Straucher im Garten

fl.   p. 99

Cornus mas

Hay, Roy
The Color Dictionary of Flowers & Plants

fl.   p 191, Pl. 1528

Cornus mas

Huxley, Anthony
Deciduous Garden Trees and Shrubs

fl.   fr.   Pl. 43, 43a, 43b

Cornus mas

Perry, Frances
Flowers of the World

fl.   p 93

Cornus mas

Polunin, Oleg
Flowers of Europe

fl.   Pl. 82 #845

Cornus mas

Polunin, Oleg
Trees and Bushes of Europe

fl. fr.   p 153

Cornus nuttallii

Clark, Lewis J.
Wild Flowers of British Columbia

fl.   p 367

Cornus nuttallii

Edlin, Herbert
The Illustrated Encyclopedia
   of Trees

fl. hab.   p 198

Cornus nuttallii

Ferguson, Mary
Wildflowers

fl.   p 156

Cornus nuttallii

Flemer, William III
Shade and Ornamental Trees in Color

hab.   p. 35

Cornus nuttallii

Morley, Brian D.
Wild Flowers of the World

fl. fr.   pl. 155

Cornus nuttallii

Munz, Philip A.
California Mountain Wildflowers

fl.   p 43, Pl. 51

Cornus nuttallii

Orr, Robert T.
Wildflowers of Western America

fl.   Pl. 29

Cornus nuttallii

Royal Hort. Soc.
Journal of the Royal Hort. Soc.
   vol 97 No. 1 1972

fl.   p. 22   Pl. 12

Cornus nuttallii

Royal Hort. Soc.
The Garden
   vol 103, No. 5   1978

fl.   p. 186

Cornus officinalis

Kariyone, Tatsuo
Atlas of Medicinal Plants

fl., fr.   Pl. 75

Cornus racemosa

Wharton, Mary E.
Trees & Shrubs of Kentucky

fl.   p 94, Pl. 2.29c
fr.   p 145, Pl. 2.39d

Cornus sangiunea

Edlin, Herbert
The Illustrated Encyclopedia
   of Trees

fr. hab.   p 198-9

Cornus sanguinea

Lindman, C. A. M.
Nordens Flora, Vol 6

fl. fr.   Pl. 411

Cornus sanguinea

Meikle, R. D.
British Trees and Shrubs

fl. fr.   Pl. 8

Cornus sanguinea

Polunin, Oleg
Flowers of Europe

fl.   Pl. 82 #846

Cornus sanguinea

Polunin, Oleg
Trees and Bushes of Europe

fl. fr. hab.   p 152

Cornus sanguinea

Vedel, H.
Arbres et Arbustes

fl. fr. p 103, Pl. 109

Cornus sericea

Polunin, Oleg
Trees and Bushes of Europe

fl.   p 153

Cornus stolonifera

American Hort. Soc.
American Horticulturist
   vol 56, No. 2 1977

hab.   p. 18

Cornus stolonifera

Gault, S. Millar
The Color Dictionary of Shrubs

hab.   Pl. 110

Cornus stolonifera

Heller, Christine
Wild Flowers of Alaska

fr.   Pl. 193

Cornus stolonifera

Walcott, Mary Vaux
North American Wild Flowers

fr. hab.   Pl. 38

Cornus stolonifera var.

Clark, Lewis J.
Wild Flowers of British Columbia

fr.   p 370

Cornus suecica

Heller, Christine
Wild Flowers of Alaska

fl.   Pl. 192

Cornus suecica

Kleijn, H.
Beauty of the Wild Plant

fl. fr.   opp. p. 105   pl. 155,155a

Cornus suecica

Lindman, C. A. M.
Nordens Flora, Vol 6

fl. fr. hab.   Pl. 412

Cornus suecica

Morley, Brian D.
Wild Flowers of the World

fl. fr.   Pl. 3E

Cornus suecica

Polunin, Oleg
Flowers of Europe

fl. Pl. 82 #847

Cornus suecica

Webster, Mary
Flora of Moray, Nairn & East
   Inverness

fl., fr.   p. 164   Pl. 10

Corokia cotoneaster

Hellyer, A.G.L.
Shrubs in Colour

fl.   p. 35

Corokia cotoneaster

Menninger, Edwin A.
Flowering Vines of the World

fr.   Pl. 82

Corokia cotoneaster var.

Harrison, R.E.
Trees and Shrubs

hab.   P 55

Coronilla coronata

Barneby, T. P.
European Alpine Flowers in Colour

fl.   Pl. 45, 4

Coronilla coronata

Vilmorin, Roger de
Plantes Alpines dans les Jardins

fl. hab.   Pl. XXXVII

Coronilla emerus

Lindman, C. A. M.
Nordens Flora, Vol 6

fl. fr.   Pl. 351

Coronilla emerus

Polunin, Oleg
Flowers of Europe

fl. Pl. 62 #624

Coronilla emerus var.

Goulimis, Constantine N.
Wild Flowers of Greece

fl.   fr.   p 45

Coronilla emerus var.

Polunin, Oleg
Flowers of the Mediterranean

fl.   Pl. 68

Coronilla glauca

Gault, S. Millar
The Color Dictionary of Shrubs

fl.   Pl. 111

Coronilla glauca

Hay, Roy
The Color Dictionary of Flowers & Plants

fl.   p 192, Pl. 1529

Coronilla glauca

Hellyer, A.G.L.
Shrubs in Colour

fl.   p. 35

Coronilla glauca var.

Hay, Roy
The Dictionary of House Plants

fl.   Pl. 153

Coronilla juncea

Polunin, Oleg
Flowers of the Mediterranean

fl.   Pl. 71

Coronilla minima

Vilmorin, Roger de
Plantes Alpines dans les Jardins

fl. hab.   Pl. XXXVII

Coronilla varia

Barneby, T. P.
European Alpine Flowers in Colour

fl.   Pl. 45, 5

Coronilla varia

Courtenay, Booth
Wildflowers & Weeds

fl.   p 51

Coronilla varia

Crockett, James Underwood
Lawns & Ground Covers

fl. p. 128

Coronilla varia

Felsko, Elsa
A Book of Wildflowers

fl.   P 92

Coronilla varia

Harrison, Richmond E.
Climbers and Trailers

fl.   p 91   Pl. 226, 227

Coronilla varia

Kleijn, H.
Beauty of the Wild Plant

fl.   opp. p. 84   pl. 110

Coronilla varia

Klimas, John E.
Wildflowers of Eastern America

fl.   Pl. 227

Coronilla varia

Massachusetts Hort. Soc.
Horticulture
   vol 40, No. 6   1962

fl.   inside backcover

Coronilla varia

Moyle, John B.
Northland Wild Flowers

fl. hab.   p 80, Pl. 56

Coronilla varia

Perrot, Emile
Les Plantes Medicinales

fl. hab.   p 245

Coronilla varia

Polunin, Oleg
Flowers of Europe

fl. Pl. 62 #627

Coronilla varia

Wharton, Mary E.
A Guide to the Wildflowers & Ferns
   of Kentucky

fl.   p 183   Pl. 1.10

Coronopus didymus

Martin, W. Keble
The Concise British Flora in Colour

fr. hab.   Pl. 10

Coronopus didymus

Polunin, Oleg
Flowers of Europe

fr.   Pl. 36 #354

Coronopus didymus

Weeds of the Southern United States

hab.   p. 19

Coronopus squamatus

Ary, S.
The Oxford Book of Wildflowers

fl. fr.   p 70, Pl. 6

Coronopus squamatus

Martin, W. Keble
The Concise British Flora in Colour

fl. fr. hab.   Pl. 10

Correa aemula

Cochrane, G. R.
Flowers and Plants of Victoria

fl.   Pl. 113

Correa alba

Cochrane, G. R.
Flowers and Plants of Victoria

fl.   Pl. 284

Correa backhousiana

Curtis, Winifred
The Endemic Flora of Tasmania
Vol VI

fl. fr.   Pl. 252

Correa backhousiana

Curtis's Botanical Magazine
Vol 171   1956-57

fl.   289

Correa backhousiana

Morley, Brian D.
Wild Flowers of the World

fl.   pl. 138

Correa decumbens

Curtis's Botanical Magazine
Vol 177   1969-70

fl.   538

Correa lawrenciana

Cochrane, G. R.
Flowers and Plants of Victoria

fl.   Pl. 431

Correa pulchella

Blombery, Alec M.
What Wildflower is That

fl.   p. 101   Pl. 252

Correa pulchella

Harrison, R.E.
Trees and Shrubs

fl. P 55

Correa reflexa

Blombery, Alec M.
What Wildflower is That

fl.   p 101, Pl. 253

Correa reflexa

Cochrane, G. R.
Flowers and Plants of Victoria

fl.   Pl. 24, 25

Correa reflexa

Harrison, R.E.
Trees and Shrubs

fl. P 55

Correa reflexa

Mullins, Barbara
Australian Wildflowers in Colour

fl.   P. 73   Pl. 70

Correa reflexa var.

Morley, Brian D.
Wild Flowers of the World

fl.   Pl. 138

Correa reflexa var.

Perry, Frances
Flowers of the World

fl.   p 269

Corrigiola litoralis

Martin, W. Keble
The Concise British Flora in Colour

fl. hab.   Pl. 71

Corrigiola telephiifolia

Perrot, Emile
Les Plantes Medicinales

fl. fr. hab.   p 192

Corryocactus ayopayanus

Lamb, Edgar
Popular Exotic Cacti in Color

fl. hab.   p 57

Cortaderia argentea

Macoby, Stirling
What Flower is That

fr P 94

Cortaderia argentea

Tsukamoto, Yotaro
Coloured Illustrations of Garden
    Flowers   Vol. 9

fr. hab.   opp. p. 35

Cortaderia argentea var.

Royal Hort. Soc.
Journal of the Royal Hort. Soc.
    vol 93, No. 10   1968

fl., hab.   p. 426   Pl. 229

Cortaderia selloana

Hay, Roy
The Color Dictionary of Flowers & Plants

fr.   p 133, Pl. 1057

Cortaderia sp.

Harvey, Norman B.
New Zealand Botanical Paintings

fr., hab.   Pl. 40

Cortusa matthioli

Felsko, Elsa
A Book of Wild Flowers
          2nd ser
fl.   p 80

Cortusa matthioli

Kohlhaupt, Paula
Fleurs des Alpages
          vol 1
fl.   P 32

Cortusa matthioli

Polunin, Oleg
Flowers of Europe

fl. Pl. 91 #955

Coryanthes albertinae

Kupper, Walter
Orchids

fl.   p 79

Coryanthes albertinae

Pabst, G. F. J.
Orchidaceae Brasilienses,
    Vol 2

fl.   p 221

Coryanthes alborosea

Ospina, Mariano
Orquideas de las americas

fl.   Pl. 152

Coryanthes elegantium

Pabst, G. F. J.
Orchidaceae Brasilienses,
Vol 2

fl.   p 221

Coryanthes macrantha

American Hort. Soc.
American Horticulturist
   vol 51, No. 4   1972

fl., hab.   p. 28-29

Coryanthes speciosa

Pabst, G. F. J.
Orchidaceae Brasiliensis
   vol 2

fl.   p. 91, 221

Corybas acontiflorus

Cady, Leo
Australian Native Orchids in Colour

fl.   Pl. 68

Corybas dilatatus

Cady, Leo
Australian Native Orchids in Colour

fl.   Pl. 67

Corybas dilatatus

Cochrane, G. R.
Flowers and Plants of Victoria

fl.   Pl. 440

Corybas rivularis

Morley, Brian D.
Wild Flowers of the World

fl.   Pl. 145

Corybas undulatum

Blombery, Alec M.
What Wildflower is That

fl.   p. 102   Pl. 254

Corybas unguiculatus

Cochrane, G. R.
Flowers and Plants of Victoria

fl. hab.   Pl. 70

Corydalis aurea

Alaska-Yukon Wild Flowers Guide

fl.   p 53

Corydalis aurea

Clark, Lewis J.
Wild Flowers of British Columbia

fl.   p 175

Corydalis aurea

Courtenay, Booth
Wildflowers & Weeds

fl.   p 21

Corydalis aurea

Heller, Christine
Wild Flowers of Alaska

fl.   Pl. 28

Corydalis aurea

Moyle, John B.
Northland Wild Flowers

fl. hab.   p 59, Pl. 23

Corydalis aurea

Porsild, A.E.
Rocky Mountain Wild Flowers

fl.   p 197

Corydalis bulbosa

Perrot, Emile
Les Plantes Medicinales

fl. fr. hab.   p 72

Corydalis cashmeriana

Curtis's Botanical Magazine
   vol 176   1966-68

fl.   Pl. 522

Corydalis cashmeriana

Hay, Roy
The Color Dictionary of Flowers & Plants

fl.   p 5, Pl. 40

Corydalis cashmeriana

Perry, Frances
Flowers of the World

fl.   p 119

Corydalis cashmeriana

Royal Hort. Soc.
Journal of the Royal Hort. Soc.
   vol 74, No. 10   1949

fl.   p. 446   Pl. 170

Corydalis cashmeriana

Royal Hort. Soc.
Journal of the Royal Hort. Soc.
   vol 91, No. 7   1966

fl.   p. 294   Pl. 155

Corydalis cashmeriana

Royal Hort. Soc.
Journal of the Royal Hort. Soc.
   vol 93, No. 1   1968

hab., fl.   p. 24   Pl. 21

Corydalis cava

Bianchini, Francesco
Health Plants of the World

fl. hab.   p 125

Corydalis cava

Felsko, Elsa
A Book of Wild Flowers
                    2nd ser.
fl.   p 68

Corydalis claviculata

Ary, S.
The Oxford Book of Wildflowers

fl.   p 82, Pl. 3

Corydalis claviculata

Martin, W. Keble
The Concise British Flora in Colour

fl. fr. hab.   Pl. 6

Corydalis claviculata

Polunin, Oleg
Flowers of Europe

fl.   Pl. 30 #277

Corydalis decumbens

Takatori, Jisuke
Color Atlas of Medicinal Plants of Japan

fl.
fr.   Fig. 45B

Corydalis fabacea

Lindman, C. A. M.
Nordens Flora, Vol 4

fl. hab.   Pl. 249

Corydalis flavula

Jennings, O. E.
Wild Flowers of Western Pennsylvania
                              vol II
fl. hab.   Pl. 67

Corydalis flavula

Wharton, Mary E.
A Guide to the Wildflowers & Ferns
   of Kentucky

fl.   p 216   Pl. 3.12

Corydalis lutea

Ary, S.
The Oxford Book of Wildflowers

fl.   p 30, Pl. 4

Corydalis lutea

Barneby, T. P.
European Alpine Flowers in Colour

fl. hab.   Pl. 28, 2

Corydalis lutea
Felsko, Elsa
A Book of Wildflowers
fl.   P 26

Corydalis lutea
Huxley, Anthony
Garden Perennials and Water Plants
fl.   Pl. 89

Corydalis lutea
Martin, W. Keble
The Concise British Flora in Colour
fl. fr. hab.   Pl. 6

Corydalis lutea
Polunin, Oleg
Flowers of Europe
fl. Pl. 30 #278

Corydalis micrantha
Brown, Clair A.
Wildflowers of Louisiana and
   Adjoining States
fl. hab.   p 58

Corydalis micrantha
Dean, Blanche E.
Wildflowers of Alabama and
   Adjoining States
fl.   p 71

Corydalis micrantha
Fleming, Glenn
Wild Flowers of Florida
fl.   p 55

Corydalis ochroleuca
Polunin, Oleg
Flowers of Europe
fl. Pl. 30 #278

Corydalis, Pale
   see
Corydalis sempervirens

Corydalis, Pale Yellow
   see
Corydalis flavula

Corydalis pauciflora
Heller, Christine
Wild Flowers of Alaska
fl.   Pl. 257

Corydalis, Pink
   see
Corydalis sempervirens

Corydalis rutifolia
Morley, Brian D.
Wild Flowers of the World
fl.   Plate 91

Corydalis scouleri
Clark, Lewis J.
Wild Flowers of British Columbia
fl.   p 182

Corydalis scouleri
Fries, Mary A.
Wildflowers of Mount Ranier and
the Cascades
fl.   P 44 & 45

Corydalis scouleri
Orr, Robert T.
Wildflowers of Western America
fl.   Pl. 146

Corydalis sempervirens
Courtenay, Booth
Wildflowers & Weeds
fl.   p 21

Corydalis sempervirens
Ferguson, Mary
Wildflowers
fl.   p 102

Corydalis sempervirens
Heller, Christine
Wild Flowers of Alaska
fl.   Pl. 63

Corydalis sempervirens
Klimas, John E.
Wildflowers of Eastern America
fl.   Pl. 194

Corydalis sempervirens
Lemmon, Robert S.
Wildflowers of North America in
   Full Color
fl.   pg. 129   pl. 205

Corydalis sempervirens
Wharton, Mary E.
A Guide to the Wildflowers & Ferns
   of Kentucky
fl.   p 216   Pl. 3.11

Corydalis solida
Hay, Roy
The Color Dictionary of Flowers & Plants
fl.   p 87, Pl. 691

Corydalis solida
Morley, Brian D.
Wild Flowers of the World
fl.   Pl. 11C

Corydalis solida
Polunin, Oleg
Flowers of Europe
fl. Pl. 30 #279

Corydalis solida
Tosco, Uberto
The World of Mountain Flowers
fl. hab.   p 98

Corydalis verticillaris
Curtis's Botanical Magazine
   Vol 160   1937
fl. hab.   No. 9486

Corydalis, Western
   see
Corydalis scouleri

Corydalis, Yellow
   see
Corydalis flavula

Corydalis, Yellow
   see
Corydalis lutea

Corylopsis himalayana
Hara, Hiroshi, comp.
Photo-Album of Plants of
   Eastern Himalaya
fl.   Pl. 75, 76

Corylopsis pauciflora
Gault, S. Millar
The Color Dictionary of Shrubs
fl.   Pl. 112

Corylopsis pauciflora
Hay, Roy
The Color Dictionary of Flowers & Plants
fl.   p 192, Pl. 1530

Corylopsis pauciflora

Kitamura, Siro
Coloured Illustrations of Trees &
    Shrubs of Japan

fl.   212

Corylopsis pauciflora

Royal Hort. Soc.
Journal of the Royal Hort. Soc.
    vol 98, No. 11   1973

hab., fl   p. 486   Pl. 244

Corylopsis sinensis

Hay, Roy
The Color Dictionary of Flowers & Plants

fl.   p 192, Pl. 1531

Corylopsis sinensis

Morley, Brian D.
Wild Flowers of the World

fl.   Plate 94

Corylopsis sinensis var.

Royal Hort. Soc.
Journal of the Royal Hort. Soc.
    vol 92, No. 6   1967

fl.   p. 250   Pl. 142

Corylopsis sinensis var.

Royal Hort. Soc.
The Garden
    vol 102, No. 3   1977

fl.   p. 106

Corylopsis spicata

Harrison, R.E.
Trees and Shrubs

fl.   P 55

Corylopsis spicata

Hay, Roy
The Color Dictionary of Flowers & Plants

fl.   p 192, Pl. 1532

Corylopsis spicata

Hellyer, A.G.L.
Shrubs in Colour

fl.   p. 35

Corylopsis spicata

Kitamura, Siro
Coloured Illustrations of Trees &
    Shrubs of Japan

fl. fr.   211

Corylopsis spicata

Royal Hort. Soc.
The Garden
    vol 102, No. 3   1977

fl.   p. 106

Corylopsis veitchiana

Gault, S. Millar
The Color Dictionary of Shrubs

fl.   Pl. 113

Corylopsis willmottiae

Curtis's Botanical Magazine
    vol 174   1962-63

fl.   Pl. 438

Corylopsis willmottiae

Everett, T.H., ed.
New Illustrated Encyclopedia of
    Gardening   vol 3

fl.   p. 391

Corylopsis willmottiae

Huxley, Anthony
Deciduous Garden Trees and Shrubs

fl Pl. 44

Corylopsis willmottiae var.

Royal Hort. Soc.
The Garden
    vol 102, No. 9   1977

hab.   p. 390

Corylus americana

Wharton, Mary E.
Trees & Shrubs of Kentucky

fl.   p 102, Pl. 3.6

Corylus avellana

Ary, S.
The Oxford Book of Wildflowers

fl. fr.   p 188, Pl. 4

Corylus avellana

Bianchini, F.
The Complete Book of Fruits &
    Vegetables

fr.   p. 193

Corylus avellana

Edlin, Herbert
The Illustrated Encyclopedia
    of Trees

fl. fr. hab.   p 144-5

Corylus avellana

Hvass, Elsie
Plants That Feed and Serve Us

fl. fr.   p 20, Pl. 29

Corylus avellana

Lindman, C. A. M.
Nordens Flora, Vol 3

fr.   Pl. 165

Corylus avellana

Martin, W. Keble
The Concise British Flora in Colour

fl. fr. hab.   Pl. 77

Corylus avellana

Masefield, G. B.
The Oxford Book of Food Plants

fl. fr.   Pl. 27, 1

Corylus avellana

Meikle, R. D.
British Trees and Shrubs

fl. fr.   Pl. 13

Corylus avellana

Perrot, Emile
Les Plantes Medicinales

fl. fr. hab.   p 161

Corylus avellana

Polunin, Oleg
Trees and Bushes of Europe

fl. fr. hab.   p 49

Corylus avellana

Vedel, H.
Arbres et Arbustes

fl. fr. hab.   p 55, Pl. 48

Corylus avellana var.

Gault, S. Millar
The Color Dictionary of Shrubs

fr.   Pl. 114

Corylus avellana var.

Hay, Roy
The Color Dictionary of Flowers & Plants

fl.   p 192, Pl. 1533

Corylus avellana var.

Huxley, Anthony
Deciduous Garden Trees and Shrubs

fr.   Pl. 45

Corylus colurna

Curtis's Botanical Magazine
    Vol 160   1937

fl. fr.   No. 9469

Corylus colurna

Polunin, Oleg
Trees and Bushes of Europe

hab.   p 49

Corylus cornuta

Edlin, Herbert
The Illustrated Encyclopedia
of Trees

fr. hab.   p 144

Corylus cornuta var.

Clark, Lewis J.
Wild Flowers of British Columbia

fl.   p 103

Corylus ferox

Hara, Hiroshi, comp.
Photo-Album of Plants of
Eastern Himalaya

fl.   Pl. 28

Corylus heterophylla var.

Kitamura, Siro
Coloured Illustrations of Trees &
Shrubs of Japan

hab.   95

Corylus jacquemontii

Curtis's Botanical Magazine
vol 173   1960

fl., fr.   Pl. 391

Corylus maxima

Edlin, Herbert
The Illustrated Encyclopedia
of Trees

fr. hab.   p 144

Corylus maxima

Masefield, G. B.
The Oxford Book of Food Plants

fr.   Pl. 27, 2

Corylus maxima

Morley, Brian D.
Wild Flowers of the World

fl. fr.   Pl. 12C

Corylus maxima

Polunin, Oleg
Trees and Bushes of Europe

fr.   p 49

Corylus maxima var.

Curtis's Botanical Magazine
Vol 171   1956-57

fl.   268

Corylus maxima var.

Edlin, Herbert
The Illustrated Encyclopedia
of Trees

hab.   p 144

Corylus maxima var.

Gault, S. Millar
The Color Dictionary of Shrubs

hab.   Pl. 115

Corylus maxima var.

Harrison, R.E.
Trees and Shrubs

hab P 56

Corylus maxima var.

Huxley, Anthony
Deciduous Garden Trees and Shrubs

hab Pl. 46

Corylus sieboldiana

Kitamura, Siro
Coloured Illustrations of Trees &
Shrubs of Japan

fr.   96

Corymborchis forcipigera

Ospina, Mariano
Orquideas de las Americas

fl., hab.   Pl. 18

Corynabutilon vitifolium

Morley, Grian D.
Wild Flowers of the World

fl.   Pl. 177

Corynabutilon vitifolium

Munoz Pizarro, Carlos
Flores Silvestres de Chile

fl.   Pl. 39

Corynephorus canescens

Lindman, C. A. M.
Nordens Flora, Vol 2

fr. hab.   Pl. 94

Corynocarpus laevigatus

Harvey, Norman B.
New Zealand Botanical Paintings

fr.   p. 46   Pl. 20

Corynopuntia planibulbispina

Backeberg, Curt
Cactus Lexicon

fl.   P. 584

Corypha umbraculifera

Royal Hort. Soc.
Journal of the Royal Hort. Soc.
vol 94, No. 6   1969

hab., fl.   p. 258   Pl. 129

Coryphantha arizonica

Hay, Roy
The Color Dictionary of Flowers & Plants

fl.   p 61, Pl. 482

Coryphantha bumamma

Van Laren, A. J.
Cactus

fl.   P 67   Fig. 81

Coryphantha clava

Lamb, Edgar
The Pocket Encyclopedia of Cacti
and Succulents in Color

fl. hab.   Pl. 21

Coryphantha difficilis

Cactus & Succulent Society of America
Journal
vol 47, No. 1   1975

fl., hab.   cover

Coryphantha echinus

Hay, Roy
The Dictionary of House Plants

hab.   Pl. 154

Coryphantha echinus

Lamb, Edgar
Colorful Cacti of the American Deserts

fl. hab.   Pl. 15, 16

Coryphantha elephantidens

Van Laren, A. J.
Cactus

fl.   P 67   Fig. 80

Coryphantha erecta

Van Laren, A. J.
Cactus

hab P 67   Fig. 79

Coryphantha gracilis

Cactus & Succulent Society of America
Journal
vol 49, No. 2   1977

hab., fl.   p. 71

Coryphantha grandis

Cactus & Succulent Society of America
Journal
vol 50, No. 3   1978

hab., fl.   p. 134

Coryphantha hesteri

Lamb, Edgar
The Pocket Encyclopedia of Cacti
and Succulents in Color

fl. hab.   Pl. 22

Coryphantha nelliae

Lamb, Edgar
The Pocket Encyclopedia of Cacti
    and Succulents in Color

fl. hab.   Pl. 26

Coryphantha organensis

Cactus & Succulent Soc. of America
Journal
    vol 44, No. 3   1972

fl.   p. 114

Coryphantha pallida

Lamb, Edgar
The Pocket Encyclopedia of Cacti
    and Succulents in Color

fl. hab.   Pl. 23

Coryphantha palmeri

Van Laren, A. J.
Cactus

hab P 67   Fig. 84

Coryphantha poselgeriana

Van Laren, A. J.
Cactus

hab P 67   Fig. 82

Coryphantha radians

Lamb, Edgar
The Pocket Encyclopedia of Cacti
    and Succulents in Color

fl.   Pl. 27

Coryphantha radians

Van Laren, A. J.
Cactus

hab P 67   Fig. 83

Coryphantha radians var.

Lamb, Edgar
The Pocket Encyclopedia of Cacti
    and Succulents in Color

fl. hab.   Pl. 24

Coryphantha vivipara

Cactus & Succulent Society of America
Journal
    vol 40, No. 6   1968

fl., hab.   p. 217

Coryphantha vivipara var.

Lamb, Edgar
Colorful Cacti of the American Deserts

fl.   Pl. 107

Coryphantha vivipara var.

Lamb, Edgar
The Pocket Encyclopedia of Cacti
    and Succulents in Color

fl. hab.   Pl. 25

Coryphantha vivipara var.

Lamb, Edgar
Popular Exotic Cacti in Color

fl. hab.   p 59

Corytholoma sp.

Kromdijk, G.
200 House Plants in Colour

fl.   Pl. 61

Cosmetic plant

see

Morus nigra

Cosmos bipinnatus

Bruggeman, L.
Tropical Plants

fl.   Pl. 37

Cosmos bipinnatus

Crockett, James Underwood
Annuals

fl   P 111

Cosmos bipinnatus

Macoby, Stirling
What Flower is That

fl.   p. 94

Cosmos bipinnatus

O'Gorman, Helen
Mexican Flowering Trees and Plants

fl. hab.   p 149

Cosmos bipinnatus

Tsukamoto, Yotaro
Coloured Illustrations of Garden
    Flowers   vol. 10

fl. Illus. 44,45   opp. p. 15

Cosmos bipinnatus var.

Hay, Roy
The Color Dictionary of Flowers & Plants

fl.   p 35, Pl. 275, 276

Cosmos bipinnatus var.

Huxley, Anthony
Garden Annuals and Bulbs

fl.   Pl. 35-36

Cosmos ocellatus

O'Gorman, Helen
Mexican Flowering Trees and Plants

fl.   p 149

Cosmos scabiosioides

O'Gorman, Helen
Mexican Flowering Trees and Plants

fl.   p 149

Cosmos sulphureus

Bruggeman, L.
Tropical Plants

fl.   Pl. 38

Cosmos sulphureus

Crockett, James Underwood
Annuals

fl.   P 111

Cosmos sulphureus

Macoby, Stirling
What Flower is That

fl.   P 94

Cosmos sulphureus

O'Gorman, Helen
Mexican Flowering Trees and Plants

fl.   p 149

Cosmos sulphureus

Tsukamoto, Yotaro
Coloured Illustrations of Garden
    Flowers

fl. Ill. 46   opp. p. 15

Cosmos sulphureus var.

Huxley, Anthony
Garden Annuals and Bulbs

fl.   Pl. 37a,b

Cosmos, Yellow

see

Cosmos sulphureus

Costus igneus

American Hort. Soc.
American Horticulturist
    vol 53, No. 4   1974

fl.   p. 24

Costus igneus

Bruggeman, L.
Tropical Plants

fl.   Pl. 96

Costus igneus

Hersey, Jean
Woman's Day Book of House Plants

fl.   p. 98

Costus igneus

Perry, Frances
Flowers of the World

fl.   p 94

Costus longibracteata

Royal Hort. Soc.
The Garden
vol 101, No. 1   1976

fl.   p. 25

Costus malortieanus

Addisonia
vol 19   1935-36

fl.   p. 39   Pl. 628

Costus malortieanus

Wit, H. C. D. de
Plants of the World;
The Higher Plants, Vol II

fl.   Pl. 150

Costus speciosus

American Hort. Soc.
American Horticulturist
vol 53, No. 4   1974

fl.   p. 23

Costus speciosus

Milne, Lorus
Living Plants of the World

fl.   p 320

Costus spectabilis

Moriarty, Audrey
Wild Flowers of Malawi

fl.   Pl. 18; 5

Costus spectabilis

Morley, Brian D.
Wild Flowers of the World

fl.,hab.   Pl. 68

Costus spicatus

American Hort. Soc.
American Horticulturist
vol 53, No. 4   1974

fl.   p. 24

Costus tappenbeckianus

Addisonia
vol 19   1935-36

fl.   p. 15   Pl. 616

Cotinus americanus

Harrison, R.E.
Trees and Shrubs

hab.   P 56

Cotinus americanus

Hay, Roy
The Color Dictionary of Flowers & Plants

hab.   p 192, Pl. 1534

Cotinus coggygria

Edlin, Herbert
The Illustrated Encyclopedia
of Trees

fl. hab.   p 185

Cotinus coggygria

Gault, S. Millar
The Color Dictionary of Shrubs

fl.   Pl. 116, 117

Cotinus coggygria

Hay, Roy
The Color Dictionary of Flowers & Plants

hab.   p 192, Pl. 1535

Cotinus coggygria

Hellyer, A.G.L.
Shrubs in Colour

fr.   p. 35

Cotinus coggygria

Huxley, Anthony
Deciduous Garden Trees and Shrubs

fl.   Pl. 47

Cotinus coggygria

Macoby, Stirling
What Flower is That

fl.   P. 95

Cotinus coggygria

Polunin, Oleg
Flowers of Europe

fl.   Pl. 69 #706

Cotinus coggygria

Polunin, Oleg
Trees and Bushes of Europe

fl. fr. hab.   p. 119

Cotinus coggygria var.

Edlin, Herbert
The Illustrated Encyclopedia
of Trees

hab.   p 185

Cotinus coggygria var.

Gault, S. Millar
The Color Dictionary of Shrubs

fl.   Pl. 118

Cotinus coggygria var.

Harrison, R. E.
Trees and Shrubs

fl. hab.   p. 56

Cotinus coggygria var.

Hellyer, A.G.L.
Shrubs in Colour

fl.   p. 35

Cotinus coggygria var.

Perry, Frances
Flowers of the World

fl.   p. 27

Cotinus obovatus

Milne, Lorus
Living Plants of the World

hab.   p 128

Cotinus sp.

Pennsylvania Hort. Soc.
The Green Scene
vol 6, No. 3   1978

fl., hab.   p. 32

Cotoneaster acutifolius

Huxley, Anthony
Deciduous Garden Trees and Shrubs

fr.   Pl. 48

Cotoneaster, bearberry

see

Cotoneaster dammeri

Cotoneaster bullatus

Bartels, Andreas
Das Grosse Buch der Gartengeholze

fr.   p 139

Cotoneaster conspicuus

Curtis's Botanical Magazine
Vol. 161   1938

fl. fr.   No. 9554

Cotoneaster conspicuus

Gault, S. Millar
The Color Dictionary of Shrubs

fl. fr.   Pl. 119, 120

Cotoneaster conspicuus var.

Hay, Roy
The Color Dictionary of Flowers & Plants

fr.   p 193, Pl. 1537

Cotoneaster conspicuus var.

Hellyer, A.G.L.
Shrubs in Colour

fr.   p. 38

Cotoneaster cooperi var.

Curtis's Botanical Magazine
Vol 160   1937

fl. fr.   No. 9478

Cotoneaster, creeping

see

Cotoneaster, horizontalis

Cotoneaster dammeri

Crockett, James Underwood
Evergreens

hab.,fr.   p. 119

Cotoneaster dammeri

Crockett, James Underwood
Lawns & Ground Covers

fr.   p. 128

Cotoneaster dammeri

Huxley, Anthony
Evergreen Garden Trees and Shrubs

fr. hab.   Pl. 108

Cotoneaster dielsianus

Hellyer, A.G.L.
Shrubs in Colour

fr.   p. 38

Cotoneaster divaricatus

Hansen, Richard

Baume und Straucher im Garten

hab.   p. 136

Cotoneaster divaricatus

Huxley, Anthony
Deciduous Garden Trees and Shrubs

fr.   Pl. 49

Cotoneaster x exburyensis

Hay, Roy
The Color Dictionary of Flowers &
    Plants

fr.   p. 193   Pl. 1539

Cotoneaster, fishbone

see

Cotoneaster horizontalis

Cotoneaster franchetii

Harrison, R. E.
Trees and Shrubs

fr.   P. 57

Cotoneaster franchetii var.

Curtis's Botanical Magazine
Vol 167   1950

fl. fr.   130

Cotoneaster franchetii var.

Hay, Roy
The Color Dictionary of Flowers & Plants

fr.   p 193, Pl. 1540

Cotoneaster frigidus

Macoby, Stirling
What Flower is That

fr.   P. 95

Cotoneaster frigidus var.

Harrison, R.E.
Trees and Shrubs

fr.   P 57

Cotoneaster horizontalis

American Hort. Soc.
American Horticulturist
    vol 57, No. 1   1978

fr.   p. 27

Cotoneaster horizontalis

Gault, S. Millar
The Color Dictionary of Shrubs

fr.   Pl. 123

Cotoneaster horizontalis

Harrison, R.E.
Trees and Shrubs

fr.   P 57

Cotoneaster horizontalis

Hay, Roy
The Color Dictionary of Flowers & Plants

fr.   p 193, Pl. 1541

Cotoneaster horizontalis

Hellyer, A.G.L.
Shrubs in Colour

fr.   p. 38

Cotoneaster horizontalis

Huxley, Anthony
Deciduous Garden Trees and Shrubs

fr.   Pl. 50

Cotoneaster horizontalis

Macoby, Stirling
What Flower is That

fr.   P 95

Cotoneaster horizontalis

Tsukamoto, Yotaro
Coloured Illustrations of Garden
    Flowers   Vol. 10

fr. Illus. 147   opp. p. 46

Cotoneaster horizontalis var.

Gault, S. Millar
The Color Dictionary of Shrubs

hab. fr.   Pl. 124

Cotoneaster hupehensis

Curtis's Botanical Magazine
Vol 170   1954-55

fl. fr.   245

Cotoneaster hupehensis

Hansen, Richard
Baume und Straucher im Garten

fl., fr.   p. 135

Cotoneaster hupehensis

Royal Hort. Soc.
Journal of the Royal Hort. Soc.
    vol 80, No. 6   1955

fl., fr.   p. 270   Pl. 50

Cotoneaster hybrid.

Gault, S. Millar
The Color Dictionary of Shrubs

fr.   Pl. 121, 122, 125, 126

Cotoneaster hybrid

Harrison, R. E.
Trees and Shrubs

fr.   p. 57

Cotoneaster hybrid.

Hay, Roy
The Color Dictionary of Flowers & Plants

fr.   p 193, Pl. 1538

Cotoneaster hybrid

Hellyer, A.G.L.
Shrubs in Colour

fr.   p. 38

Cotoneaster hybrid

Huxley, Anthony
Evergreen Garden Trees and Shrubs

fr. hab.   Pl. 112

Cotoneaster hybrid

Perry, Frances
Flowers of the World

fr.   p. 260

Cotoneaster hybrid

Royal Hort. Soc.
Journal of the Royal Hort. Soc.
   vol 95, No. 11   1970

fr.   cover

Cotoneaster integerrimus

Lindman, C. A. M.
Nordens Flora, Vol 5

fl. fr.   Pl. 331

Cotoneaster integerrimus

Martin, W. Keble
The Concise British Flora in Colour

fr. hab.   Pl. 31

Cotoneaster integerrimus

Polunin, Oleg
Flowers of Europe

fl.   Pl. 48 #473

Cotoneaster integerrimus

Vedel, H.
Arbres et Arbustes

fl. fr.   p 67, Pl. 63

Cotoneaster lacteus

Curtis's Botanical Magazine
   Vol. 159   1936

fl. fr.   No. 9454

Cotoneaster lacteus

Gault, S. Millar
The Color Dictionary of Shrubs

fr.   Pl. 127

Cotoneaster lacteus

Harrison, R.E.
Trees and Shrubs

fr.   P 58

Cotoneaster microphyllus

Huxley, Anthony
Evergreen Garden Trees and Shrubs

fr. hab.   Pl. 109

Cotoneaster multiflorus

Huxley, Anthony
Deciduous Garden Trees and Shrubs

fl.   Pl. 51

Cotoneaster multiflorus var.

Hansen, Richard
Baume und Straucher im Garten

fr.   p. 135

Cotoneaster nebrodensis

Polunin, Oleg
Trees and Bushes of Europe

fl. fr. hab.   p 85

Cotoneaster praecox

Bartels, Andreas
Das Grosse Buch der Gartengeholze

fr.   p 138

Cotoneaster purpusilla

Macoby, Stirling
What Flower is That

fr.   P 95

Cotoneaster, rock

see

Cotoneaster purpusilla

Cotoneaster salicifolius var.

Huxley, Anthony
Evergreen Garden Trees and Shrubs

fr. hab.   Pl. 110

Cotoneaster simonsii

Polunin, Oleg
Trees and Bushes of Europe

fl. fr. hab.   p 86

Cotoneaster x watereri

Curtis's Botanical Magazine
Vol. 171   1956-57

fl., fr.   282

Cotoneaster x watereri

Huxley, Anthony
Evergreen Garden Trees and Shrubs

fr., hab.   Pl. 111

Cotton

see

Gossypium

Cotton bush

see

Pimelea nivea

Cotton Grass

see

Eriophorum

Cottongrass, Tassel

see

Eriophorum angustifolium

Cotton, Lavender

see

Santolina chamaecyparissus

Cotton, Red Silk

see

Bombax ceiba

Cotton, Sea Island

see

Gossypium barbadense

Cotton, Short-Staple

see

Gossypium hirsutum

Cotton Thorn

see

Tetradymia axillaris

Cotton, Tree

see

Gossypium arboreum

Cotton tree, red

see

Bombat malabaricum

Cotton Tree, Silk

see

Chorisia speciosa

Cotton, Upland

see

Gossypium hirsutum

Cotton, wild

see

Cochlospermum fraseri

Cotton, Wild

see

Cochlospermum vitifolium

Cotton, Wild

see

Hibiscus palustris

Cottonweed

see

Froelichia floridana

Cottonwood

see

Hibiscus tiliaceus

Cottonwood

see

Populus deltoides

Cottonwood

see

Populus fremontii

Cottonwood, Black

see

Populus trichocarpa

Cotula barbata

Eliovson, Sima
Namaqualand in Flower

fl.   Pl. 7

Cotula barbata

Hay, Roy
The Color Dictionary of Flowers & Plants

fl.   p 35, Pl. 277

Cotula coronopifolia

Clark, Lewis J.
Wild Flowers of British Columbia

fl.   p 543

Cotula coronopifolia

Cochrane, G. R.
Flowers and Plants of Victoria

fl. hab.   Pl. 252

Cotula coronopifolia

Polunin, Oleg
Flowers of Europe

fl.   Pl. 149 #1433

Cotula filicula

Cochrane, G. R.
Flowers and Plants of Victoria

fl. hab.   Pl. 540

Cotula squalida

Crockett, James Underwood
Lawns & Ground Covers

fl.   p. 129

Cotyledon ascendens

Flowering Plants of Africa
vol 27   1948-49

fl.   Pl. 1080

Cotyledon companulata

Flowering Plants of South Africa
vol XXI   1941

fl.   Pl. 808

Cotyledon fragilis

Flowering Plants of Africa
vol 41   1970-71

fl., hab.   Pl. 1631

Cotyledon grandiflora

Flowering Plants of Africa
vol 27   1948-49

fl., hab.   Pl. 1046

Cotyledon hirtifolium

Flowering Plants of South Africa
vol XVIII   1938

fl., hab.   Pl. 690

Cotyledon leucophylla

Flowering Plants of South Africa
vol XVII   1937

fl.   Pl. 652

Cotyledon mucronata

Flowering Plants of South Africa
vol XVIII   1938

fl.   Pl. 711

Cotyledon oppositifolia

Hay, Roy
The Color Dictionary of Flowers & Plants

fl.   p 6, Pl. 41

Cotyledon orbiculata

Kiaer, Eigil
Indoor Plants in Colour

hab.   p. 45

Cotyledon orbiculata

Lamb, Edgar
The Pocket Encyclopedia of Cacti
and Succulents in Color

hab.   Pl. 228

Cotyledon orbiculata

Lamb, Edgar
Popular Exotic Cacti in Color

fl. hab.   p 61

Cotyledon paniculata

Flowering Plants of Africa
vol 29   1952-53

fl.   Pl. 1142

Cotyledon racemosa

Flowering Plants of South Africa
vol XXII   1942

fl.   Pl. 848

Cotyledon reticulata

Flowering Plants of Africa
vol 36   1963-64

fl.   Pl. 1411

Cotyledon singularis

Flowering Plants of Africa
vol 41   1970-71

fl.   Pl. 1606

Cotyledon sp.

Lamb, Edgar
The Pocket Encyclopedia of Cacti
and Succulents in Color

fl. hab.   Pl. 229

Cotyledon undulata

Lamb, Edgar
The Pocket Encyclopedia of Cacti
and Succulents in Color

hab.   Pl. 230

Cotyledon undulata

Subik, Rudolf
Decorative Cacti

hab.   p 105

Coughwart

see

Tussilago farfara

Coulterella capitata

Cactus & Succulent Society of America
Journal
vol 42, No. 4   1970

fl.   p. 167

Couroupita guianensis

Hargreaves, Dorothy
Tropical Blossoms of the Caribbean

fl.   p. 44

Couroupita guianensis

Milne, Lorus
Living Plants of the World

fl.   p 175

Couroupita guianensis

Morley, Brian D.
Wild Flowers of the World

fl. fr.   pl. 179

Couroupita guianensis

Pertchik, Bernard
Flowering Trees of the
   Caribbean

fl. fr.   p 31

Couroupita guianensis

Royal Hort. Soc.
Journal of the Royal Hort. Soc.
   vol 95, No. 7   1970

fl.   p. 304   Pl. 170

Cowbane

see

Oxypolis rigidior

Cowberry

see

Vaccinium vitis-idaea

Cow Itch

see

Campsis radicans

Cow-itch tree

see

Lagunaria patersoni

Cowslip

see

Caltha palustris

Cowslip

see

Mertensia virginica

Cowslip

see

Primula veris

Cowslip, American

see

Dodecatheon hendersonii

Cowslip, Virginia

see

Mertensia virginica

Cow Tongue

see

Opuntia linguiformis

Cowania mexicana

Welsh, Stanley L.
Flowers of the Canyon Country

fl.   p 4

Cowania mexicana var.

Munz, Philip A.
California Desert Wildflowers

fl.   p 34, Pl. 24

Cowania stansburiana

Lemmon, Robert S.
Wildflowers of North America in
   Full Color

fl.   pg. 71   pl. 115

Cow's udder plant

see

Solanum mammosum

Coyote Bush

see

Baccharis pilularis

Crabgrass, Large

see

Digitaria sanguinalis

Crab grass, smooth

see

Digitaria ischaemum

Crab's Eye

see

Abrus precatorius

Cracca virginiana

Walcott, Mary Vaux
North American Wild Flowers

fl. hab.   Pl. 44

Crambe cordifolia

Hay, Roy
The Color Dictionary of Flowers & Plants

fl.   p 133, Pl. 1058

Crambe laevigata

Bramwell, David
Wild Flowers of the Canary Islands

fl., hab.   Pl. 133

Crambe maritima

Bianchini, F.
The Complete Book of Fruits & Vegetables

hab.   p 69

Crambe maritima

Lindman, C. A. M.
Nordens Flora, Vol 4

fl. fr.   Pl. 253

Crambe maritima

Martin, W. Keble
The Concise British Flora in Colour

fl. fr. hab.   Pl. 11

Crambe maritima

Masefield, G. B.
The Oxford Book of Food Plants

hab.   Pl. 163, 2

Crambe maritima

Perrot, Emile
Les Plantes Medicinales

fl. fr. hab.   p 81

Crambe maritima

Polunin, Oleg
Flowers of Europe

fl.   Pl. 37 #368

Crambe maritima

Royal Hort. Soc.
The Journal of the Royal Hort. Soc.
   vol 99, No. 11   1974

hab.   p. 490   Pl. 231

Crambe sventenii

Braswell, David
Wild Flowers of the Canary Islands

fl., hab.   Pl. 134

Cranberry

see

Vaccinium

Cranberry, American

see

Vaccinium macrocarpon

Cranberry, American

see

Viburnum trilobum

Cranberry, Bog

see

Oxycoccus microcarpus

Cranberry, Dwarf

see

Oxycoccus microcarpus

Cranberry, European

see

Viburnum opulus

Cranberry, High-Bush

see

Viburnum edule

Cranberry, Large

see

Vaccinium macrocarpon

Cranberry, Lowbrush

see

Vaccinium vitis

Cranberry, Mountain

see

Vaccinium vitis-idaea

Cranberry, Rock

see

Vaccinium vitis-idaea

Cranberry, Scots

see

Vaccinium vitis-idaea

Cranberry, Small

see

Oxycoccus palustris

Cranberry, Southern Mountain

see

Hugeria erythrocarpa

Crane Flower

see

Strelitzia reginae

Cranesbill

see

Geranium

Cranesbill, Ashby

see

Geranium argenteum var.

Cranesbill, Bloody

see

Geranium sanguineum

Cranesbill, Cut-leaved

see

Geranium dissectum

Cranesbill, Dove's Foot

see

Geranium molle

Cranesbill, Dusky

see

Geranium phaeum

Cranesbill, Knotted

see

Geranium nodosum

Cranesbill, Livid

see

Geranium phaeum var.

Cranesbill, Marsh

see

Geranium palustre

Cranesbill, Meadow

see

Geranium pratense

Cranesbill, Mountain

see

Geranium pyrenaicum

Cranesbill, Northern

see

Geranium erianthum

Cranesbill, Rock

see

Geranium macrorrhizum

Cranesbill, Shining

see

Geranium lucidum

Cranesbill, Silver

see

Geranium argenteum

Cranesbill, Western

see

Geranium viscosissimum

Cranesbill, White

see

Geranium rivulare

Cranesbill, Wood

see

Geranium sylvaticum

Cranichis candida

Pabst, G. F. J.
Orchidaceae Brasilienses,
Vol 1

fl.   p 180

Cranichis muscosa

Luer, Carlyle A.
The Native Orchids of Florida

fl. hab.   Pl. 22; 6-8

Cranichis schaffneri

Ospina, Mariano
Orquideas de las americas

fl.   Pl. 26

Craspedia uniflora

Alpine Flowers of the Kosciusko
State Park

fl.   Pl. 14

Craspedia uniflora

Blombery, Alec M.
What Wildflower is That

fl.   p 102, Pl. 255

Craspedia uniflora

Morley, Brian D.
Wild Flowers of the World

fl.   pl. 140

Craspedia uniflora

Mullins, Barbara
Australian Wildflowers in Colour

fl.   pg.95   Pl. 93

Crassocephalum rubens

Moriarty, Audrey
Wild Flowers of Malawi

fl.   Pl. 79; 3

Crassula alba

Moriarty, Audrey
Wild Flowers of Malawi

fl.   Pl. 36; 4

Crassula aquatica

Lindman, C. A. M.
Nordens Flora, Vol 5

fl. fr. hab.   Pl. 285A

Crassula aquatica

Martin, W. Keble
The Concise British Flora in Colour

hab.   Pl. 33

Crassula aquatica

Stodola, Jiri
Encyclopedia of Water Plants

fl.   P. 246

Crassula arborescens

Hay, Roy
The Color Dictionary of Flowers & Plants

hab.   p 61, Pl. 483

Crassula arborescens

Hersey, Jean
Woman's Day Book of House Plants

hab. p. 72

Crassula arborescens

Lamb, Edgar
The Pocket Encyclopedia of Cacti
and Succulents in Color

hab.   Pl. 234

Crassula arborescens

Subik, Rudolf
Decorative Cacti

hab.   p 107

Crassula argentea

Hay, Roy
The Dictionary of House Plants

hab.   Pl. 155

Crassula argentea

Lamb, Edgar
The Pocket Encyclopedia of Cacti
and Succulents in Color

fl.   Pl. 233

Crassula argentea var.

Lamb, Edgar
The Pocket Encyclopedia of Cacti
and Succulents in Color

hab.   Pl. 232

Crassula argyrophylla

Flowering Plants of South Africa
vol XIX   1939

fl.   Pl. 754

Crassula argyrophylla

Moriarty, Audrey
Wild Flowers of Malawi

fl.   Pl. 36; 2

Crassula barbata

Flowering Plants of South Africa
vol XXIII   1943

fl., hab.   Pl. 881

Crassula barbata

Lamb, Edgar
The Pocket Encyclopedia of Cacti
and Succulents in Color

fl. hab.   Pl. 235

Crassula compacta

Flowering Plants of Africa
vo 27   1948-49

fl., hab.   Pl. 1059a

Crassula deceptrix

Lamb, Edgar
The Pocket Encyclopedia of Cacti
and Succulents in Color

fl. hab.   Pl. 238

Crassula ericoides

Lamb, Edgar
The Pocket Encyclopedia of Cacti
and Succulents in Color

hab.   Pl. 236

Crassula falcata

Hay, Roy
The Dictionary of House Plants

fl.   Pl. 156

Crassula falcata

Kromdijk, G.
200 House Plants in Colour

hab.   Pl. 62

Crassula falcata

Mathias, Mildred E.
Color for the Landscape

fl., hab.   p. 158

Crassula falcata

Nicolaisen, Age
Pocket Encyclopedia of Indoor Plants

hab.   Pl. 90

Crassula globularioides

Moriarty, Auirey
Wild Flowers of Malawi

fl.   Pl. 36; 3

Crassula hemisphaerica

Flowering Plants of South Africa
vol XXIII   1943

fl., hab.   Pl. 892

Crassula lactea

Addisonia
vol 23   1954-59

Crassula lactea

Flowering Plants of South Africa
vol XXIII   1943

fl.   Pl. 888

Crassula lactea

Kiaer, Eigil
Indoor Plants in Colour

fl.   p. 45

Crassula lycopodioides

Kromdijk, G.
200 House Plants in Colour

hab.   Pl. 63

Crassula milfordae

Wit, H. C. D. de
Plants of the World;
The Higher Plants, Vol I

fl. hab.   p 280, Pl. 165

Crassula, milky

see

Crassula lactea

Crassula multicava

Flowering Plants of South Africa
vol XXII   1942

fl.   Pl. 871

Crassula namaquensis

Flowering Plants of South Africa
vol XXIII   1943

fl.   Pl. 908

Crassula natans

Eliovson, Sima
Namaqualand in Flower

hab.   Pl. 56, 1

Crassula nealeana

Lamb, Edgar
Popular Exotic Cacti in Color

fl. hab.   p 63

Crassula perfoliata

Addisonia
vol 19   1935-36

fl.   p. 57   Pl. 637

Crassula perfoliata

Encke, Fritz
Zimmerpflanzen

fl.   p. 71

Crassula perforata

Subik, Rudolf
Decorative Cacti

hab.   p 109

Crassula portulacea

Kiaer, Eigil
Indoor Plants in Colour

fl.   p. 45

Crassula portulacea

Nicolaisen, Age
Pocket Encyclopedia of
Indoor Plants

hab.   Pl. 89

Crassula pyramidalis

Flowering Plants of Africa
vol 34   1960-61

fl.   Pl. 1334b

Crassula pyramidalis

Lamb, Edgar
The Pocket Encyclopedia of Cacti
and Succulents in Color

hab.   Pl. 237

Crassula quadrangula

Flowering Plants of Africa
vol 34   1960-61

fl., hab.   Pl. 1334a

Crassula recurva

Stodola, Jiri
Encyclopedia of Water Plants

hab.   P. 247

Crassula rubicunda

Flowering Plants of Africa
vol 38   1967

fl., hab.   Pl. 1520

Crassula rupestris

Flowering Plants of South Africa
vol XXI   1941

fl.   Pl. 839

Crassula sarcocaulis

Hay, Roy
The Color Dictionary of Flowers & Plants

fl.   p 61, Pl. 484

Crassula socialis

Flowering Plants of Africa
vol 27   1948-49

fl., hab.   Pl. 1059b

Crassula sp.

Milne, Lorus
Living Plants of the World

fl.   p 113

Crassula streyi

Flowering Plants of Africa
vol 42   1972-73

fl.   Pl. 1672

Crassula tecta

Flowering Plants of Africa
vol 32   1957-58

fl., hab.   Pl. 1243

Crassula tillaea

Martin, W. Keble
The Concise British Flora in Colour

fl. hab.   Pl. 33

Crassula volkensii

Lamb, Edgar
The Pocket Encyclopedia of Cacti
and Succulents in Color

hab.   Pl. 239

Crataegus agglestoni

Addisonia
vol 22   1943-46

fl., fr.   p. 47   Pl. 728

Crataegus azarolus

Bianchini, F.
The Complete Book of Fruits & Vegetables

fr.   p 137

Crataegus azarolus

Masefield, G. B.
The Oxford Book of Food Plants

fl. fr.   Pl. 63, 5

Crataegus azarolus

Polunin, Oleg
Trees and Bushes of Europe

fl. fr.   p 89

Crataegus calycina

Lindman, C. A. M.
Nordens Flora, Vol 5

fr.   Pl. 330

Crataegus carrierei

Harrison, R. E.
Trees and Shrubs

fr.   P. 58

Crataegus chlorosarca

Kitamura, Siro
Coloured Illustrations of Trees &
    Shrubs of Japan

fr.   231

Crataegus chrysocarpa var.

Addisonia
    Vol. 24   1960-64

fl., fr.   p. 7   Pl. 772

Crataegus columbiana

Clark, Lewis J.
Wild Flowers of British Columbia

fr.   p 235

Crataegus columbiana

Neufeld, J.B.
Wild Flowers of the Prairies

fl.   p. 15

Crataegus crus-galli

Boom, B. K.
The glory of the tree

hab.   p. 80   Pl. 124

Crataegus crus-galli

Edlin, Herbert
The Illustrated Encyclopedia
    of Trees

fr. hab.   p 167

Crataegus crus-galli

Morley, Brian D.
Wild Flowers of the World

fl. fr.   pl. 152

Crataegus crus-galli

Oakman, Harry
Colorful Trees

fr.   P 108

Crataegus crus-galli

Polunin, Oleg
Trees and Bushes of Europe

fr. hab.   p 90

Crataegus crus-galli

Wharton, Mary E.
Trees & Shrubs of Kentucky

fl.   p 49, Pl. 1.10a
fr.   p 131, Pl. 2.19b

Crataegus douglasii

Clark, Lewis J.
Wild Flowers of British Columbia

fr.   p 235

Crataegus douglasii

Welsh, Stanley L.
Flowers of the Mountain Country

fl.   p. 11

Crataegus egglestoni

Addisonia
    Vol. 22   1943-46

fl., fr.   p. 47   Pl. 728

Crataegus ellwangeriana

Curtis's Botanical Magazine
Vol 167   1950

fl. fr.   105

Crataegus ellwangeriana

Macoby, Stirling
What Flower is That

fr.   P 96

Crataegus harbisoni

Addisonia
    vol 21   1939-42

fl., fr.   p. 41   Pl. 693

Crataegus heldreichii

Polunin, Oleg
Trees and Bushes of Europe

fl.   p 90

Crataegus hybrid

Macoby, Stirling
What Flower is That

fr.   P. 96

Crataegus laciniata

Polunin, Oleg
Trees and Bushes of Europe

fr. hab.   p 89

Crataegus laevigata

Polunin, Oleg
Trees and Bushes of Europe

fr. hab.   p 88

Crataegus x lavallei

Hay, Roy
The Color Dictionary of Flowers &
    Plants

fr.   p. 193   Pl. 1542

Crataegus mexicana

Oakman, Harry
Colorful Trees

fr.   P 108

Crataegus mexicana var.

Macoby, Stirling
What Flower is That

fr.   P 96

Crataegus mollis

Edlin, Herbert
The Illustrated Encyclopedia
    of Trees

fr. hab.   p 167

Crataegus mollis

Flemer, William III
Shade and Ornamental Trees in Color

hab.   p. 37

Crataegus mollis

Polunin, Oleg
Trees and Bushes of Europe

fr.   p 90

Crataegus mollis

Wharton, Mary E.
Trees & Shrubs of Kentucky

fl.   p 50, Pl. 1.10c
fr.   p 132, Pl. 2.19c

Crataegus monogyna

Ary, S.
The Oxford Book of Wildflowers

fl. fr.   p 180, Pl. 3

Crataegus monogyna

Boom, B. K.
The glory of the tree

hab.   opp. p. 73   pl. 119

Crataegus monogyna

Edlin, Herbert
The Illustrated Encyclopedia
    of Trees

fl. fr. hab.   p 166-7

Crataegus monogyna

Huxley, Anthony
Deciduous Garden Trees and Shrubs

fr.   Pl. 52

Crataegus monogyna

Martin, W. Keble
The Concise British Flora in Colour

fl. fr. hab.   Pl. 31

Crataegus monogyna

Polunin, Oleg
Trees and Bushes of Europe

fl. fr. hab.   p 88

Crataegus monogyna

Vedel, H.
Arbres et Arbustes

fl. fr.   p 69, Pl. 66

Crataegus nigra

Polunin, Oleg
Trees and Bushes of Europe

fr.   p 89

Crataegus opaca

Brown, Clair A.
Wildflowers of Louisiana and
Adjoining States

fl.   p 66

Crataegus orientalis

Edlin, Herbert
The Illustrated Encyclopedia
of Trees

fr.   p 167

Crataegus orientalis

Hay, Roy
The Color Dictionary of Flowers & Plants

fr.   p 193, Pl. 1543

Crataegus orientalis

Huxley, Anthony
Deciduous Garden Trees and Shrubs

fr.   Pl. 53

Crataegus oxyacantha

Bianchini, Francesco
Health Plants of the World

fl.   p. 77

Crataegus oxyacantha

Edlin, Herbert
The Illustrated Encyclopedia
of Trees

hab.   p 167

Crataegus oxyacantha

Everett, T. H.
New Illustrated Encyclopedia of
Gardening Vol. 3

hab.   (3-13a)

Crataegus oxyacantha

Huxley, Anthony
Deciduous Garden Trees and Shrubs

fl. fr.   Pl. 54

Crataegus oxyacantha

Lindman, C. A. M.
Nordens Flora, Vol 5

fl. fr.   Pl. 329

Crataegus oxyacantha

Oakman, Harry
Colorful Trees

Hab. Fl.   P. 38

Crataegus oxyacantha

Perrot, Emile
Les Plantes Medicinales

fl. fr. hab.   p 22

Crataegus oxyacantha

Vedel, H.
Arbres et Arbustes

fl. fr.   p. 69   Pl. 65

Crataegus oxycantha var.

Everett, T.H., ed.
New Illustrated Encyclopedia of
Gardening vol. 3

hab., fl.   p. 487

Crataegus oxyacantha var.

Flemer, William III
Shade and Ornamental Trees in Color

hab.   p. 36

Crataegus oxyacantha var.

Harrison, R. E.
Trees and Shrubs

fl.   p. 58

Crataegus oxyacantha var.

Hay, Roy
The Color Dictionary of Flowers & Plants

fr.   p 193, Pl. 1544

Crataegus oxyacantha var.

Huxley, Anthony
Deciduous Garden Trees and Shrubs

fl.   Pl. 55-56

Crataegus oxyacantha var.

Macoby, Stirling
What Flower is that

fr.   P 96

Crataegus oxyacanthoides

Martin, W. Keble
The Concise British Flora in Colour

fl. fr. hab.   Pl. 31

Crataegus pedicellata

Hansen, Richard
Baume und Straucher im Garten

fr.   p. 100

Crataegus pentagyna

Polunin, Oleg
Trees and Bushes of Europe

fl.   p 89

Crataegus phaenopyrum

Flemer, William III
Shade and Ornamental Trees in Color

hab.   p. 38

Crataegus phaenopyrum

Oakman, Harry
Colorful Trees

fr.   P 108

Crataegus x prunifolia

Boom, B. K.
The glory of the tree

fr.   opp. p. 80   Pl. 122

Crataegus schraderana

Polunin, Oleg
Trees and Bushes of Europe

fr.   p 90

Crataegus submollis

Polunin, Oleg
Trees and Bushes of Europe

fl.   p 90

Craterostigma lanceolatum

Moriarty, Audrey
Wild Flowers of Malawi

fl.   Pl. 38; 7

Craterostigma wilmsii

Flowering Plants of South Africa
vol XIX   1939

fl., hab.   Pl. 730

Cratoxylum ligustrinum

Walden, Beryl M.
Wild Flowers of Hong Kong

fl.   Pl. 39(120)

Crawfurdia fasciculata

Walden, Beryl M.
Wild Flowers of Hong Kong

fl.   Pl. 73 (21?)

Crazyweed, Slender

see

Oxytropis campestris

Cream Cups

see

Platystemon californicus

Cream of Tartar Tree

see

Adansonia digitata

Creeping Charlie

see

Glechoma hederacea

Creeping Charlie

see

Lippia nodiflora

Creeping Charlie

see

Pilea nummularifolia

Creeping Jenny

see

Lysimachia nummularia

Cremanthodium reniforme

Hara, Hiroshi, comp.
Photo-Album of Plants of
   Eastern Himalaya

fl.   Pl. 176

Creosote

see

Larrea tridentata

Creosote Bush

See

Larrea divaricata

Crepis atrabarba var.

Clark, Lewis J.
Wild Flowers of British Columbia

fl.   p 535

Crepis aurea

Barneby, T. P.
European Alpine Flowers in Colour

fl. hab.   Pl. 94, 3

Crepis aurea

Kleijn, H.
Beauty of the Wild Plant

fl.   opp. p. 100   pl. 144

Crepis aurea

Kohlhaupt, Paula
Fleurs des Alpages
               vol 2
Fl.   P 115

Crepis aurea

Polunin, Oleg
Flowers of Europe

fl.   Pl. 160 #1549

Crepis biennis

Lindman, C. A. M.
Nordens Flora, Vol 10

fl. fr. hab.   Pl. 656

Crepis biennis

Martin, W. Keble
The Concise British Flora in Colour

fl. hab.   Pl. 50

Crepis capillaris

Ary, S.
The Oxford Book of Wildflowers

fl.   p 36, Pl. 2

Crepis capillaris

Clark, Lewis J.
Wild Flowers of British Columbia

fl.   p 543

Crepis capillaris

Felsko, Elsa
A Book of Wild Flowers
               2nd ser.
fl.   p 20

Crepis capillaris

Martin, W. Keble
The Concise British Flora in Colour

fl. hab.   Pl. 50

Crepis foetida

Martin, W. Keble
The Concise British Flora in Colour

fl. hab.   Pl. 50

Crepis incana

Morley, Brian D.
Wild Flowers of the World

fl.,hab.   Pl. 40

Crepis mollis

Martin, W. Keble
The Concise British Flora in Colour

fl. hab.   Pl. 50

Crepis moschata

Tsukamoto, Yotaro
Coloured Illustrations of Garden
   Flowers   vol 10

fl.   Pl. 32

Crepis multicaulis

Lindman, C. A. M.
Nordens Flora, Vol 10

fl.fr. hab.   Pl. 657

Crepis nana

Alaska-Yukon Wild Flower Guide

fl. hab.   p 187

Crepis nana

Heller, Christine
Wild Flowers of Alaska

fl.   Pl. 46

Crepis nana

Porsild, A. E.
Rocky Mountain Wild Flowers

fl.   p 409

Crepis paludosa

Lindman, C. A. M.
Nordens Flora, Vol 10

fl. fr.   Pl. 659

Crepis paludosa

Martin, W. Keble
The Concise British Flora in Colour

fl. hab.   Pl. 50

Crepis praemorsa

Lindman, C. A. M.
Nordens Flora, Vol. 10

fl. fr. hab.   Pl. 658

Crepis rubra

Crockett, James Underwood
Annuals

fl.   P 111

Crepis rubra

Hay, Roy
The Color Dictionary of Flowers & Plants

fl.   p 35, Pl. 279

Crepis rubra

Polunin, Oleg
Flowers of the Mediterranean

fl.   Pl. 216

Crepis rubra

Tsukamoto, Yotaro
Coloured Illustrations of Garden
    Flowers   Vol. 10

fl. Illus. 47   opp. p. 16

Crepis taraxacifolia

Ary, S.
The Oxford Book of Wildflowers

fl.   p 36, Pl. 1

Crepis tectorum

Courtenay, Booth
Wildflowers & Weeds

fl.   p 105

Crepis tectorum

Moyle, John B.
Northland Wild Flowers

fl.   p 198, Pl. 258

Crepis vesicaria

Polunin, Oleg
Flowers of Europe

fl.   Pl. 160 #1546

Crepis vesicaria var.

Martin, W. Keble
The Concise British Flora in Colour

fl. hab.   Pl. 50

Cress

see

Lepidium sativum

Cress, Alpine

see

Hutchinsia alpina

Cress, bitter

see

Cardamine pensylvanica

Cress, Bitter

see

Cardamine purpurea

Cress, Bitter Bulbous

see

Cardamine bulbosa

Cress, Bitter Hairy

see

Cardamine hirsuta

Cress, Bitter Heart-leaved

see

Cardamine cordifolia

Cress, Bitter Large

see

Cardamine amara

Cress, Bitter Lavender

see

Cardamine douglassii

Cress, blister

see

Erysimum kotschyanum

Cress, Blister Swiss

see

Erysimum sylvestris var.

Cress, Brazil

see

Spilanthes oleracea

Cress, Cushion

see

Draba lemmonii

Cress, Garden

see

Lepidum sativum

Cress, Indian

see

Tropaeolum majus

Cress, Penny

see

Thlaspi arvense

Cress, Purple

see

Cardamine douglasii

Cress, Purple

see

Cardamine purpurea

Cress, Rock

see

Arabis albida var.

Cress, Rock

see

Arabis lyrata

Cress, Rock

see

Aubrieta deltoides

Cress, Rock Hairy

see

Arabis hirsuta

Cress, Rock Lyall's

see

Arabis lyallii

Cress, Rock Pyrenean

see

Petrocallis pyrenaica

Cress, Rock Smooth

see

Arabis laevigata

Cress, Rock Wall

see

Arabis albida

Cress, Spring

see

Cardamine bulbosa

Cress, Swine

see

Coronopus didymus

Cress, Thale

see

Arabidopsis thaliana

Cress, Wart

see

Coronopus squamatus

Cress, Winter

see

Barbarea vulgaris

Cress, Yellow

see

Rorippa islandica

Cress, Yellow Creeping

see

Rorippa sylvestris

Cress, Yellow Marsh

see

Rorippa islandica

Crest, Gold

see

Lophiola americana

Crinitaria linosyris

Martin, W. Keble
The Concise British Flora in Colour

fl. hab.   Pl. 44

Crinkleroot

see

Dentaria diphylla

Crinkleroot

see

Dentaria laciniata

Crinkleroot, Two-Leaved

see

Dentaria diphylla

Crinodendron hookeranum

Gault, S. Millar
The Color Dictionary of Shrubs

fr.   Pl. 128

Crinodendron hookeranum

Hay, Roy
The Color Dictionary of Flowers & Plants

fl.   p 194, Pl. 1545

Crinodendron hookeranum

Munoz Pizarro, Carlos
Flores Silvestres de Chile

fl., fr.   Pl. 46

Crinodendron hookeranum

Perry, Frances
Flowers of the World

fl.   p 104

x Crinodonna corsii

Hay, Roy
The Color Dictionary of Flowers & Plants

fl.   p 87, Pl. 692

Crinum americanum

Brown, Clair A.
Wildflowers of Louisiana and Adjoining States

fl.   p 25

Crinum americanum

Bruggeman, L.
Tropical Plants

fl.   Pl. 97

Crinum americanum

Dean, Blanche E.
Wildflowers of Alabama and Adjoining States

fl.   p 29

Crinum americanum

Fleming, Glenn
Wild Flowers of Florida

fl. hab.   p 19

Crinum amoenum

Curtis's Botanical Magazine
Vol 177   1969-70

fl.   528

Crinum asiaticum

Bruggeman, L.
Tropical Plants

fl.   Pl. 98

Crinum augustum

Everett, T.H., ed.
New Illustrated Encyclopedia of Gardening   vol 3

fl.   p. 391

Crinum baumii

Flowering Plants of Africa
vol 36   1963-64

fl., hab.   Pl. 1432

Crinum bulbispermum

Flowering Plants of Africa
vol 29   1952-53

fl., hab.   Pl. 1150

Crinum bulbispermum

Hersey, Jean
Woman's Day Book of House Plants

fl.   p. 51

Crinum buphanoides

Flowering Plants of South Africa
vol XXIII   1943

fl.   Pl. 878

Crinum campanulatum

Flowering Plants of Africa
vol 37   1965-66

fl.   Pl. 1455

Crinum carolo-schmidtii
Flowering Plants of Africa
vol 41   1970-71
fl.   Pl. 1629

Crinum crassicaule
Flowering Plants of Africa
vol 42   1972-73
fl.   Pl. 1676

Crinum delagoense
Flowering Plants of Africa
vol 35   1962
fl., hab.   Pl. 1389

Crinum distichum
Royal Hort. Soc.
Journal of the Royal Hort. Soc.
vol 92, No. 5   1967
fl.   p. 206   Pl. 109

Crinum eucrophyllum
Flowering Plants of Africa
vol 42   1972-73
fl.   Pl. 1642

Crinum flaccidum
Blombery, Alec M.
What Wildflower is That
fl.   p 102, Pl. 256

Crinum flaccidum
Morley, Brian D.
Wild Flowers of the World
fl.   Pl. 143

Crinum giganteum
Bruggeman, L.
Tropical Plants
fl.   Pl. 99

Crinum glaucum
Royal Hort. Soc.
Journal of the Royal Hort. Soc.
vol 92, No. 5   1967
fl.   p. 206   Pl. 110

Crinum graminicolum
Flowering Plants of Africa
vol 29   1952-53
fl., hab.   Pl. 1155

Crinum jagus
Royal Hort. Soc.
Journal of the Royal Hort. Soc.
vol 92, No. 5   1967
fl.   p. 206   Pl. 108

Crinum kirkii
Bruggeman, L.
Tropical Plants
fl.   Pl. 100

Crinum latifolium
Tsukamoto, Yotaro
Coloured Illustrations of Garden Flowers
fl.   opp. p. 14

Crinum lineare
Flowering Plants of Africa
vol 37   1965-66
fl.   Pl. 1471

Crinum minimum
Flowering Plants of Africa
vol 40   1969-70
fl., hab.   Pl. 1577

Crinum moorei
Flowering Plants of Africa
vol 34   1960-61
fl.   Pl. 1351

Crinum moorei
Macoby, Stirling
What Flower is That
fl.   P 96

Crinum natans
Royal Hort. Soc.
Journal of the Royal Hort. Soc.
vol 91, No. 3   1966
fl.   p. 116   Pl. 77-78

Crinum ornatum
Morley, Brian D.
Wild Flowers of the World
fl.   Pl. 66

Crinum paludosum
Flowering Plants of Africa
vol 39   1968-69
fl.   Pl. 1523

Crinum pedicellatum
Moriarty, Audrey
Wild Flowers of Malawi
fl.   Pl. 2

Crinum pedunculatum
Blombery, Alec M.
What Wildflower is That
fl.   p. 103   Pl. 257

Crinum x powellii
Hay, Roy
The Color Dictionary of Flowers & Plants
fl.   p. 87   Pl. 693

Crinum x powellii
Hay, Roy
The Dictionary of House Plants
fl.   Pl. 157

Crinum x powellii
Huxley, Anthony
Garden annuals and bulbs
fl.   Pl. 206

Crinum x powellii
Perry, Frances
Flowers of the World
fl.   p. 24

Crinum rautanenianum
Flowering Plants of Africa
vol 42   1972-73
fl.   Pl. 1643

Crinum variabile
Flowering Plants of Africa
vol 36   1963-64
fl.   Pl. 1433

Crinum zeylanicum
Hargreaves, Dorothy
Tropical Blossoms of the Caribbean
fl.   p. 5

Crithmum maritimum
Ary, S.
The Oxford Book of Wildflowers
fl.   p 46, Pl. 2

Crithmum maritimum
Bianchini, F.
The Complete Book of Fruits & Vegetables
hab.   p 105

Crithmum maritimum
Martin, W. Keble
The Concise British Flora in Colour
fl. hab.   Pl. 39

Crithmum maritimum
Masefield, G. B.
The Oxford Book of Food Plants
fl.   Pl. 147, 4

Crocidium multicaule

Clark, Lewis J.
Wild Flowers of British Columbia

fl.   p 535

Crocodile Bark Tree

see

Diospyros quiloensis

Crocosmia aurea

Moriarty, Audrey
Wild Flowers of Malawi

fl. fr.   Pl. 6; 1

Crocosmia x crocosmiflora

Wit, H. C. D. de
Plants of the World;
    The Higher Plants, Vol. II

fl.   Pl. 139

Crocosmia hybrid

Hay, Roy
The Color Dictionary of Flowers & Plants

fl.   p 87, Pl. 694

Crocosmia hybrid

Royal Hort. Soc.
The Garden
    vol 101, No. 9   1976

fl.   p. 463

Crocosmia masoniorum

Hay, Roy
The Color Dictionary of Flowers &
    Plants

fl.   P. 87   Pl. 695

Crocosmia masoniorum

Massachusetts Hort. Soc.
Horticulture
    vol. 41, No. 1   1963

fl.   p. 20

Crocosmia masoniorum

Miles, Bebe
Bulbs for the Home Gardener

fl.   p. 157

Crocosmia masoniorum

Royal Hort. Soc.
Journal of the Royal Hort. Soc.
    Vol. 89, No. 1   1964

fl.   p. 22   Pl. 7

Crocosmia masoniorum

Royal Hort. Soc.
The Garden
    Vol. 101, No. 9   1976

fl.   P. 463

Crocosmia pottsii

Royal Hort. Soc.
The Garden
    vol 101, No. 9   1976

fl.   p. 463

Crocus aerius

Royal Hort. Soc.
Journal of the Royal Hort. Soc.
    vol 91, No. 8   1966

fl.   p. 338   Pl. 184

Crocus aerius var.

Alpine Garden Society
Bulletin
    vol. 45, No. 2   1977

fl.   p. 114

Crocus alatavicus

Curtis's Botanical Magazine
    Vol. 180   1974

fl. hab.   Pl. 692

Crocus albiflorus

Barneby, T. P.
European Alpine Flowers in Colour

fl.   Pl. 7, 1

Crocus albiflorus

Kohlhaupt, Paula
Fleurs des Alpages
                    vol 1
fl. P 18

Crocus albiflorus

Polunin, Oleg
Flowers of Europe

fl.   Pl. 174 #1678

Crocus almehensis

Wendelbo, Per
Tulips and Irises of Iran and
    Their Relatives

fl. hab.   p 61, Pl. 63

Crocus ancyrensis

Curtis's Botanical Magazine
Vol 167   1950

fl.   99

Crocus ancyrensis

Miles, Bebe
Bulbs for the Home Gardener

fl.   p 65

Crocus ancyrensis

Pennsylvania Hort. Soc.
The Green Scene
    vol 6, No. 1   1977

fl.   p. 15

Crocus antalyensis

Curtis's Botanical Magazine
Vol 179   1972

fl. hab.   629

Crocus aureus

Hay, Roy
The Color Dictionary of Flowers & Plants

fl.   p 87, Pl. 696

Crocus, autumn

see

Colchicum autumnale

Crocus, autumn

see

Sternbergia lutea

Crocus, Autumn

see

Zephranthes candida

Crocus balansae

Goulimis, Constantine N.
Wild Flowers of Greece

fl. hab. p 171

Crocus banaticus

Mathew, Brian
Dwarf Bulbs

fl.   p. 88   Pl. 29

Crocus banaticus

Royal Hort. Soc.
Journal of the Royal Hort. Soc.
    vol 90, No. 4   1965

fl.   p. 160   Pl. 86

Crocus banaticus

Royal Hort. Soc.
Journal of the Royal Hort. Soc.
    vol 94, No. 10   1969

fl.   p. 434   Pl. 230

Crocus baytopiorum

Curtis's Botanical Magazine
    Vol. 180   1974

fl. hab.   Pl. 664

Crocus biflorus

Mathew, Brian
Dwarf Bulbs

fl.   p. 72   Pl. 23

Crocus biflorus

Miles, Bebe
Bulbs for the Home Gardener

fl.   p 65

Crocus biflorus var.

Alpine Garden Society
Bulletin
       vol. 45, No. 2   1977

fl.   p. 114

Crocus biflorus var.

Miles, Bebe
Bulbs for the Home Gardener

fl.   p. 191

Crocus cancellatus

Wendelbo, Per
Tulips and Irises of Iran and
   Their Relatives

fl. hab.   p 63, Pl. 67

Crocus candidus

Royal Hort. Soc.
Journal of the Royal Hort. Soc.
    vol 98, No. 5   1973

fl.   p. 216   Pl. 114

Crocus carpetanus

Curtis's Botanical Magazine
   Vol. 181   1976-77

fl. hab.   Pl. 711

Crocus caspius

Mathew, Brian
Dwarf Bulbs

fl.   p. 72   Pl. 27

Crocus caspius

Royal Hort. Soc.
Journal of the Royal Hort. Soc.
   vol 91, No. 3   1966

fl.   p. 116   Pl. 75

Crocus caspius

Wendelbo, Per
Tulips and Irises of Iran and
   Their Relatives

fl. hab.   p 63, Pl. 65

Crocus chrysanthus

Royal Hort. Soc.
Journal of the Royal Hort. Soc.
   vol 78, No. 3   1953

fl., hab.   p. 100   Pl. 39

Crocus chrysanthus var.

Alpine Garden Society
Bulletin
     vol. 45, No. 2   1977

fl.   p. 114

Crocus chrysanthus var.

Curtis's Botanical Magazine
Vol 169   1952-53

fl.   182

Crocus chrysanthus var.

Hay, Roy
The Color Dictionary of Flowers & Plants

fl.   p 88, Pl. 697 thru 702

Crocus chrysanthus var.

Massachusetts Hort. Soc.
Horticulture
   vol 53, No. 9   1975

fl.   p. 46-47

Crocus chrysanthus var.

Miles, Bebe
Bulbs for the Home Gardener

fl.   p 64, 65

Crocus chrysanthus var.

Pennsylvania Hort. Soc.
The Green Scene
   vol 6, No. 1   1977

fl.   p. 14

Crocus chrysanthus var.

Perry, Frances
Flowers of the World

fl.   p. 148

Crocus chrysanthus var.

Royal Hort. Soc.
Journal of the Royal Hort. Soc.
   vol 91, No. 6   1966

fl.   p. 250   Pl. 129

Crocus cvijicii

Curtis's Botanical Magazine
   Vol. 181   1976-77

fl. hab.   Pl. 721

Crocus cyprius

Curtis's Botanical Magazine
   Vol. 180   1974

fl. hab.   Pl. 675

Crocus cyprius

Megaw, Elektra
Wild Flowers of Cyprus

fl., hab.   Pl. 27

Crocus cyprius

Royal Hort. Soc.
Journal of the Royal Hort. Soc.
   vol 94, No. 10   1969

fl.   p. 434   Pl. 231

Crocus dalmaticus

Curtis's Botanical Magazine
Vol 179   1972

fl. hab.   617

Crocus etruscus

Hay, Roy
The Color Dictionary of Flowers & Plants

fl.   p 88, Pl. 703

Crocus flavus

Polunin, Oleg
Flowers of the Mediterranean

fl.   Pl. 271

Crocus fleischeri

Goulimis, Constantine N.
Wild Flowers of Greece

fl. hab. p 171

Crocus gargaricus

Curtis's Botanical Magazine
   Vol. 180   1974

fl. hab.   Pl. 703

Crocus, golden

see

Crocus aureus

Crocus goulimyi

Curtis's Botanical Magazine
   vol 173   1960

fl.   Pl. 354

Crocus goulimyi

Goulimis, Constantine N.
Wild Flowers of Greece

fl. hab. p 173

Crocus hartmannianus

Megaw, Elektra
Wild Flowers of Cyprus

fl., hab.   Pl. 27

Crocus hybrid

Hay, Roy
The Color Dictionary of Flowers &
Plants

fl.   p.89-90   Pl.711-14

Crocus hybrid

Huxley, Anthony
Garden Annuals and Bulbs

hab., fl.   Pl. 150, 205

Crocus hybrid

Kromdijk, G.
200 House Plants in Colour

fl. Pl. 64

Crocus hybrid

Miles, Bebe
Bulbs for the Home Gardener

fl. p. 4, 5, 64

Crocus imperati

Hay, Roy
The Color Dictionary of Flowers & Plants

fl. p 88, Pl. 704

Crocus imperati

Perry, Frances
Flowers of the World

fl. p 149

Crocus karduchorum

Miles, Bebe
Bulbs for the Home Gardener

fl. p. 193

Crocus korolkowii

Mathew, Brian
Dwarf Bulbs

fl. p. 72 Pl. 24

Crocus korolkowii

Morley, Brian D.
Wild Flowers of the World

fl.,hab. Pl. 51

Crocus korolkowii

Royal Hort. Soc.
Journal of the Royal Hort. Soc.
vol 93, No. 2 1968

fl. p. 54 Pl. 34

Crocus kotschyanus var.

Miles, Bebe
Bulbs for the Home Gardener

fl. p 67

Crocus laevigatus

Curtis's Botanical Magazine
Vol 161 1938

fl. No. 9515

Crocus laevigatus

Coulimis, Constantine N.
Wild Flowers of Greece

fl. hab. p 175

Crocus laevigatus

Hay, Roy
The Color Dictionary of Flowers & Plants

fl. p 89, Pl. 705

Crocus laevigatus

Royal Hort. Soc.
Journal of the Royal Hort. Soc.
vol 78, No. 3 1953

fl., hab. p. 100 Pl. 27

Crocus lazicus

Royal Hort. Soc.
Journal of the Royal Hort. Soc.
vol 92, No. 12 1967

fl. p. 514 Pl. 287

Crocus longiflorus

Massachusetts Hort. Soc.
Horticulture
vol 53, No. 4 1975

fl. p. 54

Crocus maesiacus

Tsukamoto, Yotaro
Coloured Illustrations of Garden
Flowers Vol. 9

fl. opp. p. 7

Crocus medius

Miles, Bebe
Bulbs for the Home Gardener

fl. p 67

Crocus michelsonii

Curtis's Botanical Magazine
Vol 178 1970-72

fl. 606

Crocus michelsonii

Wendelbo, Per
Tulips and Irises of Iran and
Their Relatives

fl. hab. p 61, Pl. 64

Crocus minimus

Hay, Roy
The Color Dictionary of Flowers & Plants

fl. p 89, Pl. 706

Crocus nevadensis

Curtis's Botanical Magazine
vol 174 1962-63

fl. Pl. 439

Crocus nevadensis

Taylor, A. W.
Wild Flowers of Spain and Portugal

fl. p. 26

Crocus niveus

Curtis's Botanical Magazine
Vol. 168 1951

fl. 146

Crocus niveus

Coulimis, Constantine N.
Wild Flowers of Greece

fl. hab. p 177

Crocus niveus

Morley, Brian D.
Wild Flowers of the World

fl.,hab. Pl. 42

Crocus nubigena

Curtis's Botanical Magazine
Vol 170 1954-55

fl. 235

Crocus nudiflorus

Ary, S.
The Oxford Book of Wildflowers

fl. p 162, Pl. 3

Crocus nudiflorus

Curtis's Botanical Magazine
Vol 168 1951

fl. 169

Crocus nudiflorus

Hay, Roy
The Color Dictionary of Flowers & Plants

fl. p 90, Pl. 715

Crocus nudiflorus

Polunin, Oleg
Flowers of Europe

fl. Pl. 174 #1676

Crocus ochroleucus

Miles, Bebe
Bulbs for the Home Gardener

fl. p 67

Crocus, old cloth of gold

see

Crocus sulphureus

Crocus olivieri

Curtis's Botanical Magazine
Vol 179 1972

fl. hab. 639

Crocus pallasii var.

Polunin, Oleg
Flowers of the Mediterranean

fl.   Pl. 270

Crocus, Prairie

see

Anemone patens

Crocus, Prairie

see

Pulsatilla ludoviciana

Crocus purpureus

Martin, W. Keble
The Concise British Flora in Colour

fl. hab.   Pl. 83

Crocus reticulatus

Curtis's Botanical Magazine
      Vol. 181   1976-77

fl. hab.   Pl. 736

Crocus, saffron

see

Crocus sativus

Crocus sativus

Bianchini, F.
The Complete Book of Fruits & Vegetables

fl. hab.   p 215

Crocus sativus

Bianchini, Francesco
Health Plants of the World

fr.   p 21

Crocus sativus

Felsko, Elsa
A Book of Wildflowers

fl.   P. 78

Crocus sativus

Kariyone, Tatsuo
Atlas of Medicinal Plants

fl., hab.   Pl. 9

Crocus sativus

Kimura, Koiti
Japanese Medicinal Plants, Vol I

fl.   p 16

Crocus sativus

Loewenfeld, Claire
The Complete Book of Herbs and Spices

fl. fr.   p 209

Crocus sativus

Masefield, G. B.
The Oxford Book of Food Plants

fl.   Pl. 133, 2

Crocus sativus

Perrot, Emile
Les Plantes Medicinales

fl. fr. hab.   p 210

Crocus sativus

Perry, Frances
Flowers of the World

fl.   p 148

Crocus sativus

Rosengarten, Federic, Jr.
The Book of Spices

fl.   p. 388

Crocus sativus

Royal Hort. Soc.
Journal of the Royal Hort. Soc.
      vol 94, No. 10   1969

fl.   p. 434   Pl. 232

Crocus sativus

Takatori, Jisuke
Color Atlas of Medicinal Plants of Japan

fl.   Fig. 67B

Crocus scepusiensis

Mathew, Brian
Dwarf Bulbs

fl.   p. 88   Pl. 28

Crocus scepusiensis

Royal Hort. Soc.
Journal of the Royal Hort. Soc.
      vol 92, No. 9   1967

fl.   p. 388   Pl. 203

Crocus sieberi

Curtis's Botanical Magazine
      vol 172   1958-59

fl.   Pl. 340

Crocus sieberi

Hay, Roy
The Color Dictionary of Flowers & Plants

fl.   p 89, Pl. 707

Crocus sieberi

Perry, Frances
Flowers of the World

fl.   p. 149

Crocus sieberi var.

Goulimis, Constantine N.
Wild Flowers of Greece

fl. hab. p 179

Crocus sieberi var.

Royal Hort. Soc.
Journal of the Royal Hort. Soc.
      vol 74, No. 1   1949

fl., hab.   p. 24   Pl. 1, 6-7

Crocus sieberi var.

Royal Hort. Soc.
Journal of the Royal Hort. Soc.
      vol 92, No. 6   1967

fl.   p. 250   Pl. 123

Crocus sieheanus

Curtis's Botanical Magazine
      Vol 162   1939

fl. hab.   No. 9583

Crocus speciosus

Hay, Roy
The Color Dictionary of Flowers & Plants

fl.   p 90, Pl. 716

Crocus speciosus

Huxley, Anthony
Garden Annuals and Bulbs

hab., fl.   Pl. 151, 204

Crocus speciosus

Miles, Bebe
Bulbs for the Home Gardener

fl.   p 66

Crocus speciosus

Perry, Frances
Flowers of the World

fl.   p 148

Crocus speciosus

Royal Hort. Soc.
Journal of the Royal Hort. Soc.
      vol 76, No. 3   1951

fl.   p. 90   Pl. 59

Crocus speciosus

Wendelbo, Per
Tulips and Irises of Iran and
      Their Relatives

fl.   p 63, Pl. 66

Crocus speciosus var.

Hay, Roy
The Color Dictionary of Flowers &
Plants

fl.   p. 90   Pl. 717

Crocus susianus

Miles, Bebe
Bulbs for the Home Gardener

fl.   p 64

Crocus susianus

Vilmorin, Roger de
Plantes Alpines dans les Jardins

fl. hab.   Pl. XXX

Crocus tomasinianus

Hay, Roy
The Color Dictionary of Flowers & Plants

fl.   p 89, Pl. 708

Crocus tomasinianus

Macoby, Stirling
What Flower is That

fl.   P 97

Cocus tomasinianus

Mahew, Brian
Dwarf Bulbs

fl.   p. 72   Pl. 25

Crocus omasinianus

Miles, Bebe
Bulbs for the Home Gardener

fl.   p 65

Crocus toasinianus

Royal Hort. Soc.
Journal of he Royal Hort. Soc.
vol 97, No. 2   1972

fl.   p. 8   Pl. 46

Crocus tomasinianus var.

Hay, Roy
Tho Color Dictionary f Flowers &
Plants

fl.   p. 89   Pl. 709

Crocus tomasinianus var.

Massachusetts Hort. Soc.
Horticulture
vol 48, No. 10   1970

fl.   p. 19

Crocus tomasinianas var.

Massachusetts Hort. Soc.
Horticulture
vol 53, No. 9   1975

fl.   p. 47

Crocus vallicola

Curtis's Botanical Magazine
vol 174   1962-63

fl., fr.   Pl. 424

Crocus vallicola

Mathew, Brian
Dwarf Bulbs

fl.   p. 72   Pl. 26

Crocus vallicola

Royal Hort. Soc.
Journal of the Royal Hort. Soc.
vol 92, No. 12   1967

fl.   p. 514   Pl. 286

Crocus veluchensis

Royal Hort. Soc.
The Garden
vol 103, No. 5   1978

fl.   p. 208

Crocus vernus

Perry, Frances
Flowers of the World

fl.   p 149

Crocus vernus

Tosco, Uberto
The World of Mountain Flowers

fl. hab.   p 66

Crocus vernus

Wit, H. C. D. de
Plants of the World;
The Higher Plants, Vol II

fl.   Pl. 142

Crocus vernus var.

Hay, Roy
The Color Dictionary of Flowers &
Plants

fl.   p. 89   Pl. 710

Crocus vernus var.

Hay, Roy
The Dictionary of House Plants

fl.   Pl. 158

Crocus vernus var.

Massachusetts Hort. Soc.
Horticulture
vol 48, No. 10   1970

fl.   p. 20

Crocus zonatus

Tsukamoto, Yotaro
Coloured Illustrations of
Garden Flowers   Vol. 9

fl.   opp. p. 6

Cross Flower

see

Polygala serpyllifolia

Cross, Jerusalem

see

Lychnis chalcedonica

Cross, Southern

see

Xanthosia rotundifolia

Crossvine

see

Anisostichus caprelatus

Cross-Vine

see

Bignonia capreolata

Crosswort

see

Galium caproolata

Crossandra flava

Wit, H. C. D. de
Plants of the World;
The Higher Plants, Vol II

fl.   Pl. 97

Crossandra friesiorum

Curtis's Botanical Magazine
Vol 179   1972

fl.   647

Crossandra greenstockii

Moriarty, Audrey
Wild Flowers of Malawi

fl.   Pl. 43; 6

Crossandra infundibuliformis

Hersey, Jean
Woman's Day Book of House Plants

fl.   p. 50

Crossandra infundibuliformis

Kromdijk, G.
200 House Plants in Colour

fl.   Pl. 65

Crossandra infundibuliformis

Massachusetts Hort. Soc.
Horticulture
vol 37, No. 12   1959

fl.   p. 632

Crossandra infundibuliformis

Massachusetts Hort. Soc.
Horticulture
vol 40, No. 3   1962

fl.   p. 165

Crossandra infundibuliformis

Perry, Frances
Flowers of the World

fl.   p 14

Crossandra infundibuliformis var.

Encke, Fritz
Zimmerpflanzen

fl.   p. 71

Crossandra infundibuliformis var.

Macoby, Stirling
What Flower is That

fl.   p. 97

Crossandra massaica

Curtis's Botanical Magazine
vol 174   1962-63

fl.   Pl. 404

Crossandra nilotica var.

Morley, Brian D.
Wild Flowers of the World

fl.   Pl. 64

Crossandra puberula
Flowering Plants of Africa
vol 28   1950-51

fl.   Pl. 1098

Crossandra undulifolia

Bruggeman, L.
Tropical Plants

fl.   Pl. 234

Crossandra undulifolia

Massachusetts Hort. Soc.
Horticulture
vol. 34, No. 12   1956

fl.   p. 604

Crossandra undulifolia var.

Hay, Roy
The Dictionary of House Plants

fl.   Pl. 159

Crotalaria agatiflora

Harrison, R.E.
Trees and Shrubs

fl.   P 58

Crotalaria agatiflora var.

Morley, Brian D.
Wild Flowers of the World

fl.,fr.   Pl. 53

Crotalaria calycina

Walden, Beryl M.
Wild Flowers of Hong Kong

fl.   Pl. 62 (181)

Crotalaria cunninghamii

Blombery, Alec M.
What Wildflower is That

fl.   p. 104   Pl.148

Crotalaria laburnifolia

Blombery, Alec M.
What Wildflower is That

fl.   p 104, Pl. 259

Crotalaria laburnifolia

Moriarty, Audrey
Wild Flowers of Malawi

fl.   Pl. 64;5

Crotalaria laburnifolia var.

Flowering Plants of Africa
vol 43   1974-76

fl., fr.   Pl. 1689

Crotalaria lachnorphora

Moriarty, Audrey
Wild Flowers of Malawi

fl.   Pl. 64; 4

Crotalaria lanceolata

Moriarty, Audrey
Wild Flowers of Malawi

fl.   Pl. 64; 1

Crotalaria pallida

Moriarty, Audrey
Wild Flowers of Malawi

fl.   Pl. 64; 3

Crotalaria raffillii

Curtis's Botanical Magazine
Vol 167   1950

fl.   111

Crotalaria recta
Flowering Plants of Africa
vol 28   1950-51

fl.   Pl. 1104

Crotalaria retusa

Crockett, James Underwood
Annuals

fl.   P 111

Crotalaria rotundifolia

Duncan, Wilbur H.
Wildflowers of the Southeastern
United States

fl. hab.   p 73

Crotalaria rura
Flowering Plants of South Africa
vol XXII   1942

fl., fr.   Pl. 878

Crotalaria sagittalis

Courtenay, Booth
Wildflowers and Weeds

fl.   p. 46

Crotalaria semperflorens

Macoby, Stirling
What Flower is That

fl.   P 97

Crotalaria sessiliflora

Walden, Beryl M.
Wild Flowers of Hong Kong

fl.   Pl. 78 (248)

Crotalaria shirensis

Moriarty, Audrey
Wild Flowers of Malawi

fl.   Pl. 64;2

Crotalaria spectabilis

Batson, Wade T.
Wild Flowers in South Carolina

fl.   p 63

Crotalaria spectabilis

Dean, Blanche E.
Wildflowers of Alabama and
Adjoining States

fl.   p 85

Crotalaria spectabilis

Duncan, Wilbur H.
Wildflowers of the Southeastern
United States

fl.   p 73

Crotalaria spectabilis

Fleming, Glenn
Wild Flowers of Florida

fl. hab.   p 44

Crotalaria spectabilis

Weeds of the Southern United States

fl.   p. 30

Crotalaria verrucosa

Walden, Beryl M.
Wild Flowers of Hong Kong

fl.   Pl. 48 (148)

Crotalaria zanzibarica

Wit, H. C.D. de
Plants of the World
   The Higher Plants, Vol. I

fl.   p. 288   Pl. 185

Croton

see

Codiaeum variegatum

Croton alabamensis

Dean, Blanche E.
Wildflowers of Alabama and
     Adjoining States

fl. fr.   p 97

Croton capitatus

Weeds of the Southern United States

fl., hab.   p. 21

Croton glandulosus

Weeds of the Southern United States

fl.   p. 22

Croton Oil Plant

see

Croton tiglium

Croton tiglium

Walden, Beryl M.
Wild Flowers of Hong Kong

fl. fr.   Pl. 46 (136)

Croton, Tropic

see

Croton glandulosus

Croton, Woolly

see

Croton capitatus

Crowberry

see

Empetrum

Crow Flower

see

Plumeria rubra

Crowfoot

see

Ranunculus

Crowfoot, Alpine

see

Ranunculus glacialis

Crowfoot, Celery-leaved

see

Ranunculus sceleratus

Crowfoot, Cursed

see

Ranunculus sceleratus

Crowfootgrass

see

Dactyloctenium aegyptium

Crowfoot, Hooked

see

Ranunculus recurvatus

Crowfoot, round-leaved

see

Ranunculus cymbalaria

Crowfoot, Stiff

see

Ranunculus circinatus

Crowfoot, Water

see

Ranunculus trichophyllus

Crowpoison

see

Chrosperma muscaetoxicum

Crow Poison

see

Zigadenus densus

Crowea angustifolia

Blombery, Alec M.
What Wildflower is That

fl.   p 104, Pl. 260

Crowea exalata

Blombery, Alec M.
What Wildflower is That

fl.   p 104, Pl. 261

Crowea exalata

Cochrane, G. R.
Flowers and Plants of Victoria

fl.   Pl. 213

Crowea exalata

Harrison, R.E.
Trees and Shrubs

fl.   P 59

Crowea saligna

Blombery, Alec M.
What Wildflower is That

fl.   p 105, Pl. 262

Crowea saligna

Morley, Brian D.
Wild Flowers of the World

fl.   Pl. 138

Crownbeard

see

Verbesina

Crownbeard, golden

see

Verbesina encelioides

Crownbeard, Sunflower

see

Verbesina helianthoides

Crown imperial

see

Fritillaria imperialis

Crown of Gold

see

Cassia carnaval

Crown of Thorns

see

Euphorbia milii

Crown, Rose

see

Sedum rhodanthum

Cruciata laevipes

Martin, W. Keble
The Concise British Flora in Colour

fl. hab.   Pl. 42

Cruciata laevipes

Polunin, Oleg
Flowers of Europe

fl.   Pl. 99 #1029

Cruel Plant

see

Araujia sericifera

Crusea calocephala

Chickering, Carol Rogers
Flowers of Guatemala

fl.   p 129

Cryptandra alpina

Curtis, Winifred
The Endemic Flora of Tasmania Vol. 2

fl.   Pl. 71

Cryptandra amara

Blombery, Alec M.
What Wildflower is That

fl.   p. 105   Pl. 263

Cryptandra propinqua

Cochrane, G. R.
Flowers and Plants of Victoria

fl.   Pl. 186

Cryptandra, Spring

see

Cryptandra amara

Cryptandra tomentosa

Cochrane, G. R.
Flowers and Plants of Victoria

fl.   Pl. 123

Cryptandra tomentosa

Cochrane, G. R.
Flowers and Plants of Victoria

fl.   Pl. 313

Cryptantha confertiflora

Welsh, Stanley L.
Flowers of the Canyon Country

fl. hab.   p 33

Cryptantha flava

Welsh, Stanley L.
Flowers of the Canyon Country

fl. hab.   p 43

Cryptantha humilis

Welsh, Stanley L.
Flowers of the Canyon Country

fl.   p 11

Cryptantha nubigena

Lemmon, Robert S.
Wildflowers of North America in
      Full Color

fl.   p. 151   pl. 241

Cryptanthemis slateri

Cady, Leo
Australian Native Orchids in Colour

fl.   Pl. 75

Cryptanthus acaulis

Bruggeman, L.
Tropical Plants

fl.   Pl. 101

Cryptanthus bahianus

Wilson, Robert Gardner
Bromeliads in Cultivation

fl. hab.   Pl. 100

Cryptanthus bahianus var.

Bromeliad Society
Journal
      vol 27, No. 5   1977

hab.   p. 216

Cryptanthus bivittatus

Hay, Roy
The Dictionary of House Plants

hab.   Pl. 160

Cryptanthus bivittatus

Wilson, Robert Gardner
Bromeliads in Cultivation

hab.   Pl. 106

Cryptanthus bivittatus var.

Wilson, Robert Gardner
Bromeliads in Cultivation

hab.   Pl. 106

Cryptanthus bromelioides

Padilla, Victoria
Bromeliads

hab.   p. 65   Pl. 3

Cryptanthus bromelioides

Wilson, Robert Gardner
Bromeliads in Cultivation

hab.   Pl. 101

Cryptanthus bromelioides var.

Pacific Hort. Foundation
Pacific Horticulture
      vol 39, No. 4   1978-79

hab.   p. 32

Cryptanthus bromelioides var

Rauh, Werner
Bromeliads for Home, Garden and
      Greenhouse

hab., fl.   Pl. 98

Cryptanthus fosteranus

Bromeliad Society
Journal
      Vol. 27, No. 5   1977

hab.   p. 216

Cryptanthus fosteranus

Massachusetts Hort. Soc.
Horticulture
      Vol. 43, No. 6   1965

hab.   P. 25

Cryptanthus fosteranus

Padilla, Victoria
Bromeliads

hab.   p. 66   Pl. 4

Cryptanthus fosteranus

Perry, Frances
Flowers of the World

hab., fl.   p. 54

Cryptanthus fosteranus

Rauh, Werner
Bromeliads for Home, Garden and
    Greenhouse

hab.   Pl. 99

Cryptanthus fosteranus

Wilson, Robert Gardner
Bromeliads in Cultivation

hab.   Pl. 103

Cryptanthus fosteranus var.

Hay, Roy
The Color Dictionary of Flowers &
    Plants

hab.   p. 61   Pl. 485

Cryptanthus glaziovii

Addisonia
    vol 20   1937-38

fl.   p. 9   Pl. 645

Cryptanthus hybrid.

Padilla, Victoria, ed.
Bromeliads in Color and
    Their Culture

hab.   p 78

Cryptanthus hybrid.

Wilson, Robert Gardner
Bromeliads in Cultivation

hab.   Pl. 91, 99, 102

Cryptanthus sp.

Kromdijk, G.
200 House Plants in Color

hab.   Pl. 66

Cryptanthus sp.

Wilson, Robert Gardner
Bromeliads in Cultivation

hab.   Pl. 104, 105,
                107

Cryptanthus zonatus

Bruggeman, L.
Tropical Plants

fl.   Pl. 102

Cryptanthus zonatus

Hay, Roy
The Dictionary of House Plants

hab.   Pl. 161

Cryptanthus zonatus

Wilson, Robert Gardner
Bromeliads in Cultivation

hab.   Pl. 103

Cryptanthus zonatus var.

Massachusetts Hort. Soc.
Horticulture
    vol 35, No. 1 1957

hab.   p. 37 (see errata p. 108)

Cryptanthus zonatus var.

Massachusetts Hort. Soc.
Horticulture
    vol 35, No. 11 1957

hab.   p. 556

Cryptanthus zonatus var.

Massachusetts Hort. Soc.
Horticulture
    vol 36, No. 11   1958

hab., fl p. 577

Cryptocentrum latifolium

Ospina, Mariano
Orquideas de las americas

fl., hab.   Pl. 126

Cryptocereus anthonyanus

Cactus & Succulent Society of America
Journal
    vol 28, No. 1   1956

fl.   cover (p. 1)

Cryptocoryne affinis

Stodola, Jiri
Encyclopedia of Water Plants

fl.   P. 64

Cryptocoryne balansae

Stodola, Jiri
Encyclopedia of Water Plants

fl.   p. 126

Cryptocoryne beckettii

Stodola, Jiri
Encyclopedia of Water Plants

fl.   p. 127

Cryptocoryne blassii

Stodola, Jiri
Encyclopedia of Water Plants

hab.   P. 130

Cryptocoryne ciliata

Stodola, Jiri
Encyclopedia of Water Plants

fl.   P. 131

Cryptocoryne cordata

Stodola, Jiri
Encyclopedia of Water Plants

fl.   P. 134

Cryptocoryne grandis

Stodola, Jiri
Encyclopedia of Water Plants

fl.   P. 135

Cryptocoryne griffithii

Stodola, Jiri
Encyclopedia of Water Plants

fl.   P. 138

Cryptocoryne griffithii

Wit, H. C. D. de
Plants of the World:
    The Higher Plants, Vol II

fl.   Pl. 170, 171

Cryptocoryne johorensis

Stodola, Jiri
Encyclopedia of Water Plants

fl.   P. 139

Cryptocoryne longicauda

Stodola, Jiri
Encyclopedia of Water Plants

fl.   P. 142

Cryptocoryne lutea

Stodola, Jiri
Encyclopedia of Water Plants

fl.   P. 143

Cryptocoryne nevillii

Stodola, Jiri
Encyclopedia of Water Plants

fl.   P. 146

Cryptocoryne purpurea

Wit, H. C. D. de
Plants of the World:
    The Higher Plants, Vol II

fl.   Pl. 172

Cryptocoryne retrosiralis

Stodola, Jiri
Encyclopedia of Water Plants

fl.   P. 147

Cryptocoryne spiralis

Stodola, Jiri
Encyclopedia of Water Plants

fl.   P. 147

Cryptocoryne undulata

Stodola, Jiri
Encyclopedia of Water Plants

fl.   P. 150 & 151

Cryptocoryne versteegii

Stodola, Jiri
Encyclopedia of Water Plants

fl.   P. 154

Cryptocoryne wendtii

Stodola, Jiri
Encyclopedia of Water Plants

fl.   P. 155

Cryptocoryne willisii

Stodola, Jiri
Encyclopedia of Water Plants

fl.(159)   P. 158 & 159

Cryptomeria japonica

Boom, B. K.
The glory of the tree

fr.   pl. 16
           opp. p. 16

Cryptomeria japonica

Crockett, James Underwood
Evergreens

hab.   p. 94

Cryptomeria japonica

Edlin, Herbert
The Illustrated Encyclopedia
    of Trees

fr. hab.   p 96-7

Cryptomeria japonica

Hay, Roy
The Color Dictionary of Flowers & Plants

hab.   p 252, Pl. 2011

Cryptomeria japonica

Kitamura, Siro
Coloured Illustrations of Trees &
    Shrubs of Japan

fl. fr. hab.   53, 54, 55

Cryptomeria japonica

Polunin, Oleg
Trees and Bushes of Europe

fr. hab.   p 23

Cryptomeria japonica

Wit, H. C. D. de
Plants of the World;
    The Higher Plants, Vol I

fl. fr.   p 55, Pl. 16

Cryptomeria japonica var.

Edlin, Herbert
The Illustrated Encyclopedia
    of Trees

hab.   p 97

Cryptomeria japonica var.

Gault, S. Millar
The Color Dictionary of Shrubs

hab.   Pl. 129-130

Cryptomeria japonica var.

Harrison, R. E.
Trees and Shrubs

hab.   Pl. 572

Cryptomeria japonica var.

Huxley, Anthony
Evergreen Garden Trees and Shrubs

hab.   Pl. 22, 23, 24

Cryptophoranthus dayanus

Ospina, Mariano
Orquideas de las americas

fl. hab.   Pl. 76

Cryptophoranthus fenestratus

Pabst, G. F. J.
Orchidaceae Brasilienses,
    Vol 1

fl.   p 225

Cryptostegia grandiflora

Menninger, Edwin A.
Flowering Vines of the World

fl.   Pl. 27

Cryptostegia grandiflora var.

Menninger, Edwin A.
Flowering Vines of the World

fl.   Pl. 28

Cryptostephanus vansonii

Flowering Plants of South Africa
    vol XXIII   1943

fl.   Pl. 885

Cryptostylis arachnites

Walden, Beryl M.
Wild Flowers of Hong Kong

fl. hab.   Pl. 27(72)

Cryptostylis erecta

Blombery, Alec M.
What Wildflower is That

fl.   p. 105   Pl. 264

Cryptostylis hunteriana

Cady, Leo
Australian Native Orchids in Colour

fl.   Pl. 70

Cryptostylis ovata

Morcombe, M. K.
Australia's Western Wildflowers

fl.   p 65

Cryptostylis subulata

Cady, Leo
Australian Native Orchids in Colour

fl.   Pl. 69

Cryptostylis subulata

Curtis's Botanical Magazine
Vol 178   1970-72

fl.   573

Cryptotaenia canadensis

Courtenay, Booth
Wildflowers & Weeds

fl.   p 57

Ctenanthe lubbersiana

Macoby, Stirling
What Flower is That

hab P 98

Ctenanthe oppenheimiana var.

Hay, Roy
The Dictionary of House Plants

hab.   Pl. 162

Cubelium concolor

Courtenay, Booth
Wildflowers & Weeds

fl.   p 56

Cuckoo flower

see

Cardamine pratensis

Cuckoo pint

see

Arum maculatum

Cucubalus baccifer

Polunin, Oleg
Flowers of Europe

fl. Pl. 15 #177

Cucumber

see

Cucumis sativus

Cucumber, Bitter

see

Citrullus colocynthis

Cucumber, Bur

see

Sicyos angulatus

Cucumber, Cow

see

Magnolia macrophylla

Cucumber, Creeping

see

Melothria pendula

Cucumber, Mock

see

Echinocystis lobata

Cucumber Root, Indian

see

Medeola virginiana

Cucumber, Squirting

see

Echallium elaterium

Cucumber Tree

see

Kigelia pinnata

Cucumbertree

see

Magnolia acuminata

Cucumbertree, Yellow

see

Magnolia cordata

Cucumber, wild

see

Echinocystis lobata

Cucumis africanus

American Hort. Soc.
American Horticulturist
vol 52, No. 3 1973

fr.   p. 24

Cucumis anguria

Masefield, G. B.
The Oxford Book of Food Plants

fr.   Pl. 117, 7

Cucumis dipsaceus

American Hort. Soc.
American Horticulturist
vol 52, No. 3 1973

fr.   p. 25

Cucumis melo

Bianchini, F.
The Complete Book of Fruits & Vegetables

fr.   p 141

Cucumis melo

Crockett, James Underwood
Annuals

fr.   p. 122

Cucumis melo

Masefield, G. B.
The Oxford Book of Food Plants

fl. fr.   Pl. 119

Cucumis melo var.

Brown, Clair A.
Wildflowers of Louisiana and
Adjoining States

fl. fr. hab.   p 181

Cucumis melo var.

Hvass, Elsie
Plants That Feed and Serve Us

fl. fr.   p 55, Pl. 101

Cucumis metuliferus

American Hort. Soc.
American Horticulturist
vol 52, No. 3 1973

fr.   p. 25

Cucumis myriocarpus

American Hort. Soc.
American Horticulturist
vol 52, No. 3 1973

fr.   p. 24

Cucumis sativus

Bianchini, F.
The Complete Book of Fruits & Vegetables

fl. fr.   p 117

Cucumis sativus

Hvass, Elsie
Plants That Feed and Serve Us

fl. fr.   p 38, Pl. 71

Cucumis sativus

Masefield, G. B.
The Oxford Book of Food Plants

fl., fr.   Pl.117,1-6

Cucurbita foetidissima

Morley, Brian D.
Wild Flowers of the World

fl. fr.   Pl. 159

Cucurbita foetidissima

Orr, Robert T.
Wildflowers of Western America

fl.   Pl. 119

Cucurbita maxima

Bianchini, F.
The Complete Book of Fruits &
Vegetables

fr.   p. 113

Cucurbita maxima

Crockett, James Underwood
Annuals

fr.   p. 122

Cucurbita maxima

Masefield, G. B.
The Oxford Book of Food Plants

fl. fr.   Pl. 123, 7

Cucurbita moschata

Bianchini, F.
The Complete Book of Fruits &
Vegetables

fr.   p. 107, 109

Cucurbita palmata

Coyle, Jeanette
A Field Guide to the Common and
Interesting Plants of Baja California

fl. fr.   p 167

Cucurbita pepo

Crockett, James Underwood
Annuals

fr.   P 122

Cucurbita pepo

Huxley, Anthony
Garden annuals and bulbs

fr.   Pl. 40

Cucurbita pepo

Hvass, Elsie
Plants That Feed and Serve Us

fr.   p 38, Pl. 72

Cucurbita pepo

Masefield, G. B.
The Oxford Book of Food Plants

fr.   Pl. 123, 6

Cucurbita pepo var.

Bianchini, F.
The Complete Book of Fruits &
Vegetables

fr.   p 109, 111, 115

Curcurbita pepo var.

Crockett, James Underwood
Annuals

fr.   p. 122

Cucurbita pepo var.

Macoby, Stirling
What Flower is That

fr.   P 98

Cudweed

see

Gnaphalium

Cudweed, Common

see

Filago germanica

Cudweed, Dwarf

see

Gnaphalium supinum

Cudweed, Flannel

see

Actinobole uliginosum

Cudweed, Highland

see

Gnaphalium norvegicum

Cudweed, Jersey

see

Gnaphalium luteo-album

Cudweed, Least

see

Gnaphalium supinum

Cudweed, Marsh

see

Gnaphalium uliginosum

Cudweed, Slender

see

Filago minima

Cudweed, Wayside

see

Gnaphalium uliginosum

Cudweed, Wood

see

Gnaphalium sylvaticum

Cudrania cochinchinensis var.

Kitamura, Siro
Coloured Illustrations of Trees &
Shrubs of Japan

hab.   140

Cudrania tricuspidata

Kitamura, Siro
Coloured Illustrations of Trees &
Shrubs of Japan

fr.   141

Cullowhee

see

Zephyranthes atamasco

Culver's-Root

see

Veronicastrum virginicum

Cumin

see

Cuminum cyminum

Cuminum cyminum

Bianchini, F.
The Complete Book of Fruits &
Vegetables

fr.   p. 101

Cuminum cyminum

Loewenfeld, Claire
The Complete Book of Herbs and Spices

fr. hab.   p 209

Cuminum cyminum

Masefield, G. B.
The Oxford Book of Food Plants

fl. fr.   Pl. 139, 3

Cuminum cyminum

Perrot, Emile
Les Plantes Medicinales

fl. fr. hab.   p 49

Cuminum cyminum

Rosengarten, Frederic, Jr.
The Book of Spices

fr.   p 224

Cunila origanoides

Duncan, Wilbur H.
Wildflowers of the Southeastern
United States

fl.   p 163

Cunila origanoides

Wharton, Mary E.
A Guide to the Wildflowers & Ferns
of Kentucky

fl. p 205 Pl. 2.28

Cunninghamia konishii

Kitamura, Siro
Coloured Illustrations of Trees &
Shrubs of Japan

fr.   56

Cunninghamia lanceolata

Boom, B. K.
The glory of the tree

hab.   opp. p. 16   Pl. 14

Cunninghamia lanceolata

Crockett, James Underwood
Evergreens

hab.,fr.   p. 94

Cunninghamia lanceolata

Huxley, Anthony
Evergreen Garden Trees and Shrubs

hab.   Pl. 25

402

Cunninghamia lanceolata

Kitamura, Siro
Coloured Illustrations of Trees &
  Shrubs of Japan

hab.   57

Cunonia capensis

Hay, Roy
The Dictionary of House Plants

fl.   Pl. 163

Cup-and-Saucer vine

see

Cobaea scandens

Cup Flower

see

Angianthus preissianus

Cup flower

see

Nierembergia caerulea

Cup, Frnge

see

Tellima gandiflora

Cup, Gold

see

Solandra nitida

Cup, Golden

see

Adonis chrusocyanthus

Cup, Golden

see

Solandra guttata

Cup-of-gold

see

Allamanda cathartica

Cup-of-gold

see

Solandra guttata

Cup of Gold

see

Solandra nitida

Cup of Gold Vine

see

Solandra hartwegii

Cup Plant

see

Silphium perfoliatum

Cupania

see

   Cupaniopsis anacardioides

Cupaniopsis anacardioides

Crockett, James Underwood
Evergreens

hab.   p. 120

Cupaniopsis anacardioides

Oakman, Harry
Colorful Trees

Hab. Fr.   P 52

Cuphea aequipetala

O'Gorman, Helen
Mexican Flowering Trees and Plants

fl. hab.   p 151

Cuphea cyanea

Royal Hort. Soc.
The Garden
    vol 102, No. 7   1977

fl.   p. 315

Cuphea ignea

Macoby, Stirling
What Flower is That

fl.   P 98

Cuphea lanceolata

Morley, Brian D.
Wild Flowers of the World

fl. fr.   Pl. 156

Cuphea miniata

Crockett, James Underwood
Annuals

fl.   P 112

Cuphea miniata var.

Hay, Roy
The Color Dictionary of Flowers
  & Plants

fl.   p. 35

Cuphea miniata var.

Hay, Roy
The Dictionary of House Plants

fl.   Pl. 164

Cuphea pinetorum

Chickering, Carol Rogers
Flowers of Guatemala

fl.   p 129

Cuphea platycentra

Harrison, R. E.
Trees and Shrubs

fl.   P. 59

Cuphea platycentra

Hay, Roy
The Dictionary of House Plants

fl.   Pl. 165

Cuphea procumbens

Bruggeman, L.
Tropical Plants

fl.   Pl. 39

Cupidone, Blue

see

Catananche caerulea

Cupid's dart

see

Catananche caerulea

x Cupressocyparis leylandii

Boom, B. K.
The glory of the tree

hab.   opp. p. 40   pl. 52

x Cupressocyparis leylandii

Crockett, James Underwood
Evergreens

hab.   p. 95

xCupressocyparis leylandii

Edlin, Herbert
The Illustrated Encyclopedia
  of Trees

fr. hab.   p 91

xCupressocyparis leylandii

Hay, Roy
The Color Dictionary of Flowers & Plants

hab.   p 252, Pl. 2012

xCupressocyparis leylandii var.

Royal Hort. Soc.
The Garden
    vol 102, No. 12   1977

hab.   p. 505

Cupressus arizonica

Edlin, Herbert
The Illustrated Encyclopedia
of Trees

fr. hab.   p 91

Cupressus arizonica var.

Crockett, James Underwood
Evergreens

hab.   p. 95

Cupressus cashmeriana

Hay, Roy
The Dictionary of House Plants

hab.   Pl. 166

Cupressus duprezzyana

Tosco, Uberto
The World of Mountain Flowers

hab.   p 94

Cupressus glabra var.

Edlin, Herbert
The Illustrated Encyclopedia
of Trees

hab.   p 91

Cupressus glabra var.

Hay, Roy
The Color Dictionary of Flowers & Plants

hab.   p 252, Pl. 2013

Cupressus lusitanica

Curtis's Botanical Magazine
    Vol 159   1936

fl. fr.   No. 9434

Cupressus macrocarpa

Edlin, Herbert
The Illustrated Encyclopedia
of Trees

fr. hab.   p 90

Cupressus macrocarpa

Massachusetts Hort. Soc.
Horticulture
    vol 41, No. 4   1963

hab.   backcover

Cupressus macrocarpa

Polunin, Oleg
Trees and Bushes of Europe

fr. hab.   p 24, 25

Cupressus macrocarpa var.

Harrison, R. E.
Trees and Shrubs

hab.   Pl. 570

Cupressus macrocarpa var.

Oakman, Harry
Colorful Trees

Hab.   P 88

Cupressus sempervirens

Bianchini, Francesco
Health Plants of the World

fr.   p 175

Cupressus sempervirens

Boom, B. K.
The glory of the tree

hab.,fr.   p. 37, 40   Pl. 51, 54

Cupressus sempervirens

Edlin, Herbert
The Illustrated Encyclopedia
of Trees

fr. hab.   p 90-1

Cupressus sempervirens

Huxley, Anthony
Evergreen Garden Trees and Shrubs

fr. hab.   Pl. 182

Cupressus sempervirens

Perrot, Emile
Les Plantes Medicinales

fl. fr. hab.   p 107

Cupressus sempervirens

Polunin, Oleg
Flowers of Europe

fr. Pl. 2 #11

Cupressus sempervirens

Polunin, Oleg
Trees and Bushes of Europe

fr. hab.   p 24

Cupressus sempervirens

Royal Hort. Soc.
The Garden
    vol 102, No. 9   1977

hab.   p. 368

Cupressus sempervirens var.

Crockett, James Underwood
Evergreens

hab. fr.   p. 96

Cupressus sempervirens var.

Harrison, R. E.
Trees and Shrubs

hab.   Pl. 569

Cupressus sempervirens var.

Macoby, Stirling
What Flower is That

hab.   P 99

Curcuma domestica

American Hort. Soc.
American Horticulturist
    vol 53, No. 4 1974

fl.   p. 25

Curcuma domestica

Kimura, Koiti
Japanese Medicinal Plants, Vol I

fl.   p 14

Curcuma longa

Kariyone, Tatsuo
Atlas of Medicinal Plants

fl., hab.   Pl. 4

Curcuma longa

Loewenfeld, Claire
The Complete Book of Herbs and Spices

hab.   p 209

Curcuma longa

Masefield, G. B.
The Oxford Book of Food Plants

hab.   Pl. 135, 2

Curcuma longa

Rosengarten, Frederic, Jr.
The Book of Spices

fl. fr.   p 444

Curcuma longa

Takatori, Jisuke
Color Atlas of Medicinal Plants of Japan

fl.   Fig. 66

Curcuma roscoeana

Massachusetts Hort. Soc.
Horticulture
    vol 52, No. 2   1974

fl.   p. 46

Curcuma roscoeana
Royal Hort. Soc.
The Garden
Vol. 100, No. 12   1975

fl.   P. 582

Curcuma xanthorrhiza
Perrot, Emile
Les Plantes Medicinales

fl. fr. hab.   p 230

Curlewberry
see
Empetrum nigrum

Currant, Alpine Prickly
see
Ribes montigenum

Currant, black
see
Ribes nigrum

Currant, Buffalo
see
Ribes odoratum

Currant, Bush Leafless
see
Leptomeria aphylla

Currant, Flowering
see
Ribes sanguineum

Currant, Golden
see
Ribes aureum

Currant, Mapleleaf
see
Ribes howellii

Currant, Mountain
see
Ribes alpinum

Currant, Prickly
see
Ribes lacustre

Currant, Red
see
Ribes rubrum

Currant, red
see
Ribes sativum

Currant, Red-flowering
see
Ribes sanguineum

Currant, Sierra
see
Ribes nevadense

Currant, Swamp
see
Ribes lacustre

Currant, Wild Red
see
Ribes triste

Curtain bush
see
Garrya elliptica

Cuscuta campestris
Flowering Plants of Africa
vol 33   1959
hab.   Pl. 1308

Cuscuta epithymum
Ary, S.
The Oxford Book of Wildflowers

fl.   p 124, Pl. 5

Cuscuta epithymum
Martin, W. Keble
The Concise British Flora in Colour

fl. hab.   Pl. 61

Cuscuta epithymum
Perrot, Emile
Les Plantes Medicinales

fl. fr. hab.   p 82

Cuscuta epithymum
Polunin, Oleg
Flowers of Europe

fl.   Pl. 101 #1014

Cuscuta europaea
Felsko, Elsa
A Book of Wild Flowers
2nd Ser.

fl.   p. 135

Cuscuta europaea
Lindman, C. A. M.
Nordens Flora, Vol 7

fl. fr.   Pl. 481

Cuscuta europaea
Martin, W. Keble
The Concise British Flora in Colour

fl. hab.   Pl. 61

Cuscuta gronovii
Brown, Clair A.
Wildflowers of Louisiana and
Adjoining States

fl. hab.   p 146

Cuscuta gronovii
Courtenay, Booth
Wildflowers & Weeds

fl.   p 91

Cuscuta japonica
Kimura, Koiti
Japanese Medicinal Plants, Vol II

fr.   p 217

Cuscuta japonica
Walden, Beryl M.
Wild Flowers of Hong Kong

fl.   Pl. 71 (213)

Cuscuta salina
Clark, Lewis J.
Wild Flowers of British Columbia

fl.   p 414

Cuscuta sp.
Klimas, John E.
Wildflowers of Eastern America

fl.   Pl. 67

Cuscuta sp.

Moyle, John B.
Northland Wild Flowers

fl. hab.   p 130, Pl. 140

Cuscuta sp.

Weeds of the Southern United States

fl.   p. 16

Cuscuta veatchii

Coyle, Jeanette
A Field Guide to the Common and
   Interesting Plants of Baja California

hab.   p 153

Cushion, Blue

      see

Houstonia caerulea

Cushion, Bush

      see

Calocephalus brownii

Cushion Bush

   see

Scleranthus biflorus

Cushionflower

   see

Hakea sp.

Cussonia barteri

Wit, H. C. D. de
Plants of the World;
   The Higher Plants, Vol II

hab.   p 63, Pl. 38

Cussonia kirkii

Palgrave, K. C.
Trees of Central Africa

fl., fr.   p. 35

Cussonia paniculata

Macoby, Stirling
What Flower is That

hab P 99

Cutweed, Everlasting

   see

Gnaphalium obtusifolium

Cuthbertia ornata

Fleming, Glenn
Wild Flowers of Florida

fl.   p 70

Cyanaeorchis arundinae

Pabst, G. F. J.
Orchidaceae Brasilienses,
   Vol 1

fl.   p 241

Cyananthus lobatus

Hay, Roy
The Color Dictionary of Flowers & Plants

fl.   p 6, Pl. 42

Cyananthus macrocalyx

Curtis's Botanical Magazine
   Vol 162   1939

fl. fr. hab.   No. 9562

Cyananthus microphyllus

Curtis's Botanical Magazine
   Vol 162   1939

fl.   No. 9598

Cyananthus microphyllus

Hay, Roy
The Color Dictionary of Flowers & Plants

fl.   p 6, Pl. 43

Cyananthus sherriffii

Curtis's Botanical Magazine
   Vol 164   1943-48

fl.   No. 9655

Cyanella orchidiformis

Eliovson, Sima
Namaqualand in Flower

fl.   Pl. 28, 2

Cyanostegia lanceolata

Blombery, Alec M.
What Wildflower is That

fl.   p 106, Pl. 265

Cyanotis foecunda

Moriarty, Audrey
Wild Flowers of Malawi

fl.   Pl. 20; 6

Cyanotis longifolia

Moriarty, Audrey
Wild Flowers of Malawi

fl.   Pl. 20; 7

Cyanotis speciosa

Moriarty, Audrey
Wild Flowers of Malawi

fl.   Pl. 20; 2

Cyanotis veldthoutiana

Hersey, Jean
Woman's Day Book of House Plants

hab   p. 105

Cyathodes abietina

Curtis, Winifred
The Endemic Flora of Tasmania
                    vol 3

fl. fr. p 185

Cyathodes dealbata

Curtis, Winifred
The Endemic Flora of Tasmania
                    Vol. 2

fl. fr.   Pl. 57

Cyathodes divaricata

Curtis, Winifred
The Endemic Flora of Tasmania
                    Vol. 2

fl. fr.   Pl. 58

Cyathodes glauca

Curtis, Winifred
The Endemic Flora of Tasmania
   vol 1

fl., fr.   Pl. 36
            (see errata, vol 6)

Cyathodes nitida

Curtis, Winifred
The Endemic Flora of Tasmania
   Vol VI

fl. fr.   Pl. 211

Cyathodes parvifolia

Curtis, Winifred
The Endemic Flora of Tasmania Vol. 2

fl. fr.   Pl. 59

Cyathodes petiolaris

Curtis, Winifred
The Endemic Flora of Tasmania
   Vol VI

fl. fr.   Pl. 210

Cyathodes straminea

Curtis, Winifred
The Endemic Flora of Tasmania
   vol 1

fl., fr.   Pl. 37
            (see errata, vol 6)

Cybistetes longifolia

Flowering Plants of Africa
   vol XXV   1945-46

fl., hab.   Pl. 1000

Cycad

see

Cycas revoluta

Cycas circinalis

Wit, H. C. D. de
Plants of the World;
    The Higher Plants, Vol I

fl.   p 49, Pl. 1

Cycas media

Blombery, Alec M.
What Wildflower is That

fr. hab.   p 106, Pl. 266

Cycas revoluta

Hay, Roy
The Dictionary of House Plants

hab.   Pl. 167

Cycas revoluta

Kimura, Koiti
Japanese Medicinal Plants, Vol II

fl. fr.   p 123, 125

Cycas revoluta

Kitamura, Siro
Coloured Illustrations of Trees &
    Shrubs of Japan

fr. hab.   1, 2

Cycas revoluta

Macoby, Stirling
What Flower is That

hab. P 100

Cycas revoluta

Milne, Lorus
Living Plants of the World

fr.   p 18

Cyclamen alpinum

Curtis's Botanical Magazine
    vol 174   1962-63

fl.   Pl. 437

Cyclamen alpinum

Morley, Brian D.
Wild Flowers of the World

fl.   Pl. 15H

Cyclamen alpinum var.

Royal Hort. Soc.
Journal of the Royal Hort. Soc.
vol 90, No. 7   1965

fl.   p. 294   Pl. 128

Cyclamen x atkinsi

Massachusetts Hort. Soc.
Horticulture
    vol. 34, No. 7   1956

fl.   p. 372

Cyclamen x atkinsi

Massachusetts Hort. Soc.
Horticulture
    vol. 37, No. 7   1959

fl.   p. 390

Cyclamen x atkinsi

Massachusetts Hort. Soc.
Horticulture
    vol. 44, No. 2   1966

fl.   p. 36

Cyclamen balearicum

Royal Hort. Soc.
Journal of the Royal Hort. Soc.
    vol 96, No. 4   1971

fl.   p. 170   Pl. 75

Cyclamen cilicium

Hay, Roy
The Color Dictionary of Flowers & Plants

fl.   p 90, Pl. 718

Cyclamen cilicium

Mathew, Brian
Dwarf Bulbs

fl.   p. 96   Pl.34

Cyclamen cilicium

Royal Hort. Soc.
Journal of the Royal Hort. Soc.
    vol 85, No. 5   1960

fl.   p. 222   Pl. 62

Cyclamen cilicium

Royal Hort. Soc.
Journal of the Royal Hort. Soc.
    vol 91, No. 6   1966

fl.   p. 250   Pl. 124

Cyclamen cilicium

Royal Hort. Soc.
Journal of the Royal Hort. Soc.
    vol 99, No. 11   1974

fl.   p. 490   Pl. 222

Cyclamen cilicium var.

Curtis's Botanical Magazine
Vol. 171   1956-57

fl.   307

Cyclamen colchicum

Royal Hort. Soc.
Journal of the Royal Hort. Soc.
    vol 99, No. 1   1974

fl., hab.   p. 22   Pl. 21

Cyclamen coum

Hay, Roy
The Color Dictionary of Flowers & Plants

fl.   p 90, Pl. 719

Cyclamen coum

Mathew, Brian
Dwarf Bulbs

fl.   p. 96   Pl. 33

Cyclamen coum

Perry, Frances
Flowers of the World

fl.   p 240

Cyclamen coum

Royal Hort. Soc.
Journal of the Royal Hort. Soc.
    vol 97, No. 2   1972

fl.   p. 88   Pl. 43

Cyclamen coum var.

Mathew, Brian
Dwarf Bulbs

fl.   p. 96   Pl. 35

Cyclamen coum var.

Royal Hort. Soc.
Journal of the Royal Hort. Soc.
    vol 90, No. 7   1965

fl.   p. 294   Pl. 127

Cyclamen creticum

Curtis's Botanical Magazine
    vol 175   1964-65

fl.   Pl. 450

Cyclamen creticum

Goulimis, Constantine N.
Wild Flowers of Greece

fl.   p69

Cyclamen creticum

Royal Hort. Soc.
Journal of the Royal Hort. Soc.
    vol 96, No. 4   1971

fl.   p. 170   Pl. 71

Cyclamen cyprium

Megaw, Elektra
Wild Flowers of Cyprus

fl., hab.   p. 10   Pl. 13

Cyclamen elegans

Royal Hort. Soc.
Journal of the Royal Hort. Soc.
    vol 99, No. 3   1974

fl., hab.   p. 110   Pl. 64

Cyclamen europaeum

Barneby, T. P.
European Alpine Flowers in Colour

fl. hab.   Pl. 61, 4

Cyclamen europaeum

Felsko, Elsa
A Book of Wildflowers

fl.   P 104

Cyclamen europaeum

Kleijn, H.
Beauty of the Wild Plant

fl.   opp. p. 84   Pl. 113

Cyclamen europaeum

Miles, Bebe
Bulbs for the Home Gardener

fl.   p. 68

Cyclamen europaeum

Perrot, Emile
Les Plantes Medicinales

fl. hab.   p 243

Cyclamen europaeum

Perry, Frances
Flowers of the World

fl.   p 240

Cyclamen europaeum

Tosco, Uberto
The World of Mountain Flowers

fl.   p. 31

Cyclamen, Florist's

see

Cyclamen persicum

Cyclamen graecum

Curtis's Botanical Magazine
Vol 178   1970-72

fl.   603

Cyclamen graecum

Goulimis, Constantine N.
Wild Flowers of Greece

fl.   p 71

Cyclamen graecum

Polunin, Oleg
Flowers of Europe

fl. Pl. 92 #959

Cyclamen graecum

Royal Hort. Soc.
The Garden
          vol 102, No. 10   1977

hab.   p. 432

Cyclamen graecum var.

Royal Hort. Soc.
Journal of the Royal Hort. Soc.
          vol 85, No. 5   1960

fl.   p. 222   Pl. 64-65

Cyclamen hederifolium

Martin, W. Keble
The Concise British Flora in Colour

fl. fr. hab.   Pl. 57

Cyclamen hederifolium

Perry, Frances
Flowers of the World

fl.   p 240

Cyclamen hederifolium

Polunin, Oleg
Flowers of Europe

fl. Pl. 92 #958

Cyclamen hederifolium

Royal Hort. Soc.
The Garden
          vol 102, No. 6   1977

fl., hab.   p. 260

Cyclamen hybrid

Miles, Bebe
Bulbs for the Home Garden

fl. hab.   p 158

Cyclamen libanoticum

Curtis's Botanical Magazine
Vol 171   1956-57

fl.   296

Cyclamen libanoticum

Hay, Roy
The Color Dictionary of Flowers & Plants

fl.   p 90, Pl. 720

Cyclamen libanoticum

Mathew, Brian
Dwarf Bulbs

fl.   p. 96   Pl. 32

Cyclamen libanoticum

Royal Hort. Soc.
Journal of Royal Hort. Soc.
          vol 83, No. 7   1958

fl.   p. 294   Pl. 92

Cyclamen libanoticum

Royal Hort. Soc.
Journal of the Royal Hort. Soc.
          vol 85, No. 5   1960

fl.   p. 222   Pl. 61

Cyclamen libanoticum

Royal Hort. Soc.
Journal of the Royal Hort. Soc.
          vol 92, No. 1   1967

fl.   p. 22   Pl. 8

Cyclamen mirabile

Curtis's Botanical Magazine
Vol 178   1970-72

fl.   579

Cyclamen mirabile

Royal Hort. Soc.
Journal of the Royal Hort. Soc.
          vol 92, No. 1   1967

fl.   p. 54   Pl. 31

Cyclamen mirabile

Royal Hort. Soc.
Journal of the Royal Hort. Soc.
          vol 96, No. 4   1971

fl.   p. 170   Pl. 70

Cyclamen neapolitanum

Addisonia
          vol 22   1943-46

fl.   p. 51   Pl. 730

Cyclamen neapolitanum

Barneby, T. P.
European Alpine Flowers in Colour

fl. hab.   Pl. 61, 5

Cyclamen neapolitanum

Hay, Roy
The Color Dictionary of Flowers & Plants

fl.   p 91, Pl. 721

Cyclamen neapolitanum

Huxley, Anthony
Garden Annuals and Bulbs

hab., fl.   Pl. 152, 207

Cyclamen neapolitanum

Polunin, Oleg
Flowers of the Mediterranean

fl.   Pl. 124

Cyclamen neapolitanum

Royal Hort. Soc.
Journal of the Royal Hort. Soc.
          vol 92, No. 6   1967

fl.   p. 250   Pl. 124

Cyclamen neapolitanum var.

Macoby, Stirling
What Flower is That

fl.   P 100

Cyclamen orbiculatum

Polunin, Oleg
Flowers of the Mediterranean

fl.   Pl. 125

Cyclamen orbiculatum var.

Royal Hort. Soc.
Journal of the Royal Hort. Soc.
     vol 85, No. 5   1960

fl.   p. 222   Pl. 63

Cyclamen orbiculatum var.

Royal Hort. Soc.
Journal of the Royal Hort. Soc.
     vol 89, No. 5   1964

fl.   p. 202   Pl. 75

Cyclamen persicum

Feinbrun-Dothan, Naomi
Wild Plants in the Land
   of Israel

fl. hab.   p 80

Cyclamen persicum

Goulimis, Constantine N.
Wild Flowers of Greece

fl.   p 73

Cyclamen persicum

Hay, Roy
The Dictionary of House Plants

fl.   Pl. 168, 169

Cyclamen persicum

Hersey, Jean
Woman's Day Book of House Plants

fl.   p. 51

Cyclamen persicum

Kiaer, Eigil
Indoor Plants in Colour

fl.   p. 46

Cyclamen persicum

Kromdijk, G.
200 House Plants in Colour

fl.   Pl. 67

Cyclamen persicum

Megaw, Elektra
Wild Flowers of Cyprus

fl., hab.   p. 10   Pl. 12

Cyclamen persicum

Morley, Brian D.
Wild Flowers of the World

fl.,hab.   Pl. 48

Cyclamen persicum

Pennsylvania Hort. Soc.
The Green Scene
    vol 6, No. 1   1977

fl.   p. 33

Cyclamen persicum

Polunin, Oleg
Flowers of the Mediterranean

fl.   Pl. 126

Cyclamen persicum

Royal Hort. Soc.
Journal of the Royal Hort. Soc.
     vol 92, No. 1   1967

fl.   p. 22   Pl. 9

Cyclamen persicum

Royal Hort. Soc.
Journal of the Royal Hort. Soc.
     vol 96, No. 3   1971

fl.   cover

Cyclamen persicum

Royal Hort. Soc.
The Garden
     vol 102, No. 10   1977

fl., hab.   p. 432

Cyclamen persicum

Tsukamoto, Yotaro
Coloured Illustrations of Garden
    Flowers   Vol. 9

fl.   opp. p. 7

Cyclamen persicum var.

Hay, Roy
The Color Dictionary of Flowers &
   Plants

fl.   p. 61,91   Pl.486,722

Cyclamen persicum var.

Hay, Roy
The Dictionary of House Plants

fl.   Pl. 170

Cyclamen persicum var.

Macoby, Stirling
What Flower is That

fl.   P 100

Cyclamen persicum var.

Nicolaisen, Age
Pocket Encyclopedia of Indoor Plants

fl.   pl. 108

Cyclamen persicum var.

Perry, Frances
Flowers of the World

fl.   p. 239

Cyclamen persicum var.

Royal Hort. Soc.
Journal of the Royal Hort. Soc.
     vol 85, No. 5   1960

fl.   p. 222   Pl. 56-57

Cyclamen pseudibericum

Curtis's Botanical Magazine
     vol 174   1962-63

fl.   Pl. 417

Cyclamen pseudibericum

Hay, Roy
The Color Dictionary of Flowers & Plants

fl.   p 91, Pl. 723

Cyclamen pseudibericum

Mathew, Brian
Dwarf Bulbs

fl.   p. 96   Pl. 36

Cyclamen pseudibericum

Royal Hort. Soc.
Journal of Royal Hort. Soc.
     vol 82, No. 4   1957

fl.   p. 162   Pl. 58

Cyclamen pseudibericum

Royal Hort. Soc.
Journal of the Royal Hort. Soc.
     vol 85, No. 5   1960

fl.   p. 222   Pl. 60

Cyclamen pseudibericum

Royal Hort. Soc.
Journal of the Royal Hort. Soc.
     vol 91, No. 6   1966

fl.   p. 250   Pl. 127

Cyclamen purpurascens

Kohlhaupt, Paula
Fleurs des Alpages
                vol 1
fl. P 31

Cyclamen purpurascens

Polunin, Oleg
Flowers of Europe

fl.   Pl. 92 #959

Cyclamen repandum

Goulimis, Constantine N.
Wild Flowers of Greece

fl. hab. p 75

Cyclamen repandum

Hay, Roy
The Color Dictionary of Flowers & Plants

fl.   p 91, Pl. 724

Cyclamen repandum

Massachusetts Hort. Soc.
Horticulture
vol 34, No. 7   1956

fl.   p. 372

Cyclamen repandum

Massachusetts Hort. Soc.
Horticulture
vol 48, No. 8   1970

fl.   p. 25

Cyclamen repandum

Morley, Brian D.
Wild Flowers of the World

fl., hab.   Pl. 33

Cyclamen repandum

Polunin, Oleg
Flowers of Europe

fl. Pl. 92 #960

Cyclamen repandum

Polunin, Oleg
Flowers of the Mediterranean

fl.   Pl. 123

Cyclamen repandum

Royal Hort. Soc.
Journal of the Royal Hort. Soc.
vol 96, No. 4   1971

fl.   p. 170   Pl. 72

Cyclamen repandum var.

Royal Hort. Soc.
Journal of the Royal Hort. Soc.
vol 88, No. 10   1963

fl.   p. 434   Pl. 163

Cyclamen repandum var.

Royal Hort. Soc.
Journal of the Royal Hort. Soc.
vol 90, No. 7   1965

fl., hab.   p. 294   Pl. 121

Cyclamen repandum var.

Royal Hort. Soc.
Journal of the Royal Hort. Soc.
vol 96, No. 4   1971

fl.   p. 170   Pl. 73-74

Cyclamen rohlfsianum

Curtis's Botanical Magazine
Vol 169   1952-53

fl.   192

Cyclamen vernum

Curtis's Botanical Magazine
Vol 165   1948

fl.   29

Cyclanthus bipartitus

Bruggeman, L.
Tropical Plants

hab.   Pl. 103

Cycloloma atriplicifolia

Courtenay, Booth
Wildflowers & Weeds

hab.   p. 69

Cyclopogon argyrifolius

Pabst, G. F. J.
Orchidaceae Brasilienses,
Vol 1

fl., hab.   p 180, 181

Cycnium adonense

Moriarty, Audrey
Wild Flowers of Malawi

fl.   Pl. 39; 5

Cycnoches chlorochilon

Everett, T. H., ed.
New Illustrated Encyclopedia of
    Gardening   vol. 3

fl.   Pl. (3-2b)

Cycnoches guttulatum

Ospina, Mariano
Orquideas de las americas

fl.   Pl. 146

Cycnoches haagii

Mee, Margaret
Flowers of the Brazilian Forests

fl., hab.   Pl. 17

Cycnoches haagii

Pabst, G. F. J.
Orchidaceae Brasilienses
Vol 1

fl.   p. 238

Cycnoches loddigesii

Pabst, G. F. J.
Orchidaceae Brasilienses,
Vol 1

fl.   p 238

Cycnoches pentadactylon

Kupper, Walter
Orchids

fl.   p 65

Cycnoches ventricosum var.

Kramer, Jack
Orchids: Flowers of Romance
    and Mystery

fl.   p 174

Cydista aequinoctialis

Menninger, Edwin A.
Flowering Vines of the World

fl.   Pl. 14

Cydonia oblonga

Bianchini, F.
The Complete Book of Fruits & Vegetables

fr.   p 135

Cydonia oblonga

Massachusetts Hort. Soc.
Horticulture
vol 51, No. 4   1973

fr.   p. 52

Cydonia oblonga

Polunin, Oleg
Flowers of Europe

fl. Pl. 47 #463

Cydonia oblonga

Polunin, Oleg
Trees and Bushes of Europe

fl. fr. hab.   p 74

Cydonia oblonga var.

Hvass, Elsie
Plants That Feed and Serve Us

fl. fr.   p 42, Pl. 78

Cydonia vulgaris

Masefield, G. B.
The Oxford Book of Food Plants

fl. fr.   Pl. 63, 1

Cydonia vulgaris

Perrot, Emile
Les Plantes Medicinales

fl. fr. hab.   p 149

Cylindrophyllum tugwelliae

Herre, H.
The Genera of the Mesembryanthemaceae

fl. fr.   p 123

Cylindropuntia rosea var.

Backeberg, Curt
Cactus Lexicon

fl.   p 585

Cylindropuntia vestita

Kupper, Walter
Cacti

hab.   p. 13   Pl. 3

Cymbalaria muralis

Ary, S.
The Oxford Book of Wildflowers

fl.   p. 140   Pl. 6

Cymbalaria muralis

Martin, W. Keble
The Concise British Flora in Colour

fl. hab.   Pl. 62

Cymbalaria muralis

Pennsylvania Hort. Soc.
The Green Scene
vol 6, No. 5   1978

fl., hab.   p. 18

Cymbalaria muralis

Polunin, Oleg
Flowers of Europe

fl.   Pl. 122 #1210

Cymbalaria pallida

Vilmorin, Roger de
Plantes Alpines dans les Jardins

fl. hab.   Pl. IV

Cymbidiella rhodochila

Grubb, Roy
Selected Orchidaceous Plants
Vol I

fl.   p 115

Cymbidiella rhodochila

Stewart, Joyce
Orchids of Tropical Africa

fl.   Pl. 23

Cymbidium aloifolium

Ebel, Friedrich
The Strange and Beautiful
World of Orchids

fl.   p 51

Cymbidium atropurpureum

Morley, Brian D.
Wild Flowers of the World

fl.   Plate 126

Cymbidium canaliculatum

Blombery, Alec M.
What Wildflower is That

fl.   p 106, Pl. 267

Cymbidium canaliculatum

Cady, Leo
Australian Native Orchids in Colour

fl.   Pl. 91

Cymbidium canaliculatum

Macoby, Stirling
What Flower is That

fl.   P 101

Cymbidium canaliculatum

Morley, Brian D.
Wild Flowers of the World

fl.   Pl. 145

Cymbidium devonianum

Kramer, Jack
Orchids; Flowers of Romance
and Mystery

fl.   p 159

Cymbidium ensifolium

Kamemoto, Haruyuki
Beautiful Thai Orchid Species

fl.   p 127

Cymbidium ensifolium

Morley, Brian D.
Wild Flowers of the World

fl.   Plate 109

Cymbidium ensifolium

Royal Hort. Soc.
Journal of the Royal Hort. Soc.
vol 98, No. 3   1973

fl.   p. 122   Pl. 79

Cymbidium erythrostylum

Massachusetts Hort. Soc.
Horticulture
vol 34, No. 3   1956

fl.   p. 142

Cymbidium finlaysonianum

Kamemoto, Haruyuki
Beautiful Thai Orchid Species

fl.   p 128

Cymbidium giganteum

Royal Hort. Soc.
The Garden
vol 101, No. 4   1976

fl.   p. 178

Cymbidium giganteum var.

Curtis's Botanical Magazine
Vol 181   1976-77

fl. hab.   Pl. 704

Cymbidium goeringii

Curtis's Botanical Magazine
vol 174   1962-63

fl.   Pl. 413

Cymbidium grandiflorum

Kimber, Sheila
A Handbook of Orchids

fl., hab.   p. 18

Cymbidium hybrid

Grubb, Roy
Selected Orchidaceous Plants
Vol. I

fl.   p. 59, 83, 87

Cymbidium hybrid

Hay, Roy
The Color Dictionary of Flowers & Plants

fl.   p 61, 62; Pl. 487 thru 491

Cymbidium hybrid.

Hay, Roy
The Dictionary of House Plants

fl.   Pl. 171 - 174

Cymbidium hybrid

Kimber, Sheila
A Handbook of Orchids

fl., hab.   p. 20

Cymbidium hybrid

Kramer, Jack
Orchids; Flowers of Romance
and Mystery

fl.   p 158, 162

Cymbidium hybrid

Macoby, Stirling
What Flower is That

fl.   P 101

Cymbidium hybrid

Perry, Frances
Flowers of the World

fl.   p. 210

Cymbidium lancifolium

Grubb, Roy
Selected Orchidaceous Plants
Vol I

fl.   p 51

Cymbidium lowianum

Hay, Roy
The Dictionary of House Plants

fl.   Pl. 175

Cymbidium lowianum

Kamemoto, Haruyuki
Beautiful Thai Orchid Species

fl.   p 126

Cymbidium lowianum

Kimber, Sheila
A Handbook of Orchids

fl., hab.   p. 22

Cymbidium maclehoseae

Walden, Beryl M.
Wild Flowers of Hong Kong

fl. hab.   Pl. 83

Cymbidium madidum

Cady, Leo
Australian Native Orchids in Colour

fl.   Pl. 92

Cymbidium pendulum

Walden, Beryl M.
Wild Flowers of Hong Kong

fl.   Pl. 27(71)

Cymbidium pumilum

Nicolaisen, Age
Pocket Encyclopedia of Indoor
Plants

fl.   Pl. 42

Cymbidium pumilum var.

Royal Hort. Soc.
Journal of the Royal Hort. Soc.
vol 98, No. 3   1973

fl.   p. 122   Pl. 80

Cymbidium schroederi

Curtis's Botanical Magazine
Vol 163   1940 - 42

fl.   No. 9637

Cymbidium siamense

Kamemoto, Haruyuki
Beautiful Thai Orchid Species

fl.   p 127

Cymbidium simulans

Kamemoto, Haruyuki
Beautiful Thai Orchid Species

fl.   p 128

Cymbidium simulans

Kimber, Sheila
A Handbook of Orchids

fl., hab.   p. 24

Cymbidium suave

Bedford, Roger B.
A Guide to Native Australian Orchids

fl., hab.   p. 62   Pl. I

Cymbidium suave

Blombery, Alec M.
What Wildflower 's That

fl.   p. 107   Pl. 268

Cymbidium suave

Cady, Leo
Australian Native Orchids in Colour

fl. hab.   Pl. 93

Cymbidium suavissimum

Curtis's Botanical Magazine
Vol. 180   1974

fl. hab.   Pl. 671

Cymbidium tracyanum

Curtis's Botanical Magazine
Vol 166   1949

fl.   56

Cymbidium tracyanum

Kamemoto, Haruyuki
Beautiful Thai Orchid Species

fl.   p 126

Cymbonotus preissianus

Cochrane, G. R.
Flowers and Plants of Victoria

fl. hab.   Pl. 349

Cynanchum acutum

Polunin, Oleg
Trees and Bushes of Europe

fl. fr.   p 169

Cynanchum marnierianum

Cactus & Succulent Society of America
Journal
vol 42, No. 3   1970

fl.   cover (p. 97)

Cynanchum vincetoxicum

Lindman, C. A. M.
Nordens Flora, Vol 7

fl. fr.   Pl. 476

Cynara cardunculus

Bianchini, F.
The Complete Book of Fruits & Vegetables

hab.   p 61

Cynara cardunculus

Masefield, G. B.
The Oxford Book of Food Plants

fl.   Pl. 165, 2

Cynara cardunculus

Perrot, Emile
Les Plantes Medicinales

fl. hab.   p 20

Cynara cardunculus var.

Color Treasury of Herbs & Other
Medicinal Plants

fl.   p 63   Pl. 101-02

Cynara hystrix

Curtis's Botanical Magazine
Vol 168   1951

fl.   172

Cynara hystrix

Royal Hort. Soc.
The Garden
Vol. 102, No. 6   1977

fl.   p. 242

Cynara hystrix

Royal Hort. Soc.
The Garden
vol 103, No. 2   1978

fl.   p. 56

Cynara scolymus

Bianchini, F.
The Complete Book of Fruits & Vegetables

fr.   p 61, 63

Cynara scolymus

Bianchini, Francesco
Health Plants of the World

fl.   p 31

Cynara scolymus

Hvass, Elsie
Plants That Feed and Serve Us

fl.   p 34, Pl. 64

Cynara scolymus

Macoby, Stirling
What Flower is That

fl.   P 101

Cynara scolymus

Masefield, G. B.
The Oxford Book of Food Plants

fl.   Pl. 165, 1

Cynara scolymus

Morley, Brian D.
Wild Flowers of the World

fl.   Pl. 40

Cynara scolymus

Perrot, Emile
Les Plantes Medicinales

fl. fr. hab.   p 20

Cynara scolymus

Perry, Frances
Flowers of the World

fl.   p 85

Cynodon dactylon

Crockett, James Underwood
Lawns & Ground Covers

hab., fr.   p. 76, 116

Cynodon dactylon

Perrot, Emile
Les Plantes Medicinales

fl. fr. hab.   p 66

Cynodon dactylon

Weeds of the Southern United States

hab.   p. 24

Cynoglossum amabile

Crockett, James Underwood
Annuals

fl.   P 112

Cynoglossum amabile

Tsukamoto, Yotaro
Coloured Illustrations of Garden
    Flowers   vol. 10

fl. Illus. 124 opp. p. 39

Cynoglossum amabile var.

Curtis's Botanical Magazine
Vol 166   1949

fl.   82

Cynoglossum creticum

Polunin, Oleg
Flowers of Europe

fl.   Pl. 101 #1049

Cynoglossum geometricum

Moriarty, Audrey
Wild Flowers of Malawi

fl.   Pl. 52; 1

Cynoglossum germanicum

Martin, W. Keble
The Concise British Flora in Colour

fl. fr. hab.   Pl. 59

Cynoglossum grande

Clark, Lewis J.
Wild Flowers of British Columbia

fl.   p 446

Cynoglossum grande

Lemmon, Robert S.
Wildflowers of North America in
    Full Color

fl.   pg. 150   pl. 240

Cynoglossum grande

Orr, Robert T.
Wildflowers of Western America

fl.   Pl. 247

Cynoglossum grande

Royal Hort. Soc.
Journal of the Royal Hort. Soc.
    vol 95, No. 5   1970

fl.   p. 214   Pl. 134

Cynoglossum nervosum

Hay, Roy
The Color Dictionary of Flowers & Plants

fl.   p 133, Pl. 1059

Cynoglossum officinale

Ary, S.
The Oxford Book of Wildflowers

fl. fr.   p 104, Pl. 2

Cynoglossum officinale

Clark, Lewis J.
Wild Flowers of British Columbia

fl.   p 434

Cynoglossum officinale

Courtenay, Booth
Wildflowers & Weeds

fl.   p 86

Cynoglossum officinale

Lindman, C. A. M.
Nordens Flora, Vol 8

fl. fr.   Pl. 494

Cynoglossum officinale

Martin, W. Keble
The Concise British Flora in Colour

fl. fr. hab.   Pl. 59

Cynoglossum officinale

Perrot, Emile
Les Plantes Medicinales

fl. fr. hab.   p 76

Cynoglossum officinale

Polunin, Oleg
Flowers of Europe

fl.   Pl. 102 #1049

Cynoglossum suaveolens

Cochrane, G. R.
Flowers and Plants of Victoria

fl. fr.   Pl. 230

Cynoglossum virginianum

Jennings, O. E.
Wild Flowers of Western Pennsylvania
                              vol II
fl.   Pl. 128

Cynoglossum virginianum

Klimas, John E.
Wildflowers of Eastern America

fl.   Pl. 237

Cynoglossum virginianum

Wharton, Mary E.
A Guide to the Wildflowers & Ferns
    of Kentucky

fl.   p 147 Pl. 1.106

Cynorkis fastigiata

Stewart, Joyce
Orchids of Tropical Africa

fl.   Pl. 24a

Cynorkis hanningtonii

Williamson, Graham
The Orchids of South Central Africa

fl.   Pl. 37

Cynorkis kassnerana

Moriarty, Audrey
Wild Flowers of Malawi

fl.   Pl. 27; 2

Cynorkis kirkii

Moriarty, Audrey
Wild Flowers of Malawi

fl.   Pl. 27; 1

Cynorkis purpurascens

Stewart, Joyce
Orchids of Tropical Africa

fl.   Pl. 24B

Cynosurus cristatus

Lindman, C. A. M.
Nordens Flora, Vol 2

fr. hab.   Pl. 86

Cynosurus echinatus

Polunin, Oleg
Flowers of Europe

fl.   Pl. 179 #1748

Cynoxylon floridum

Tsukamoto, Yotaro
Coloured Illustrations of Garden
    Flowers'   vol 10

fl.   Pl. 144

Cynthia

    see

Krigia biflora

Cyperus alternifolius

Bruggeman, L.
Tropical Plants

fl.   Pl. 171

Cyperus alternifolius

Hersey, Jean
Woman's Day Book of House Plants

    hab.   p. 102

Cyperus alternifolius

Kiaer, Eigil
Indoor Plants in Colour

fl.   p. 47

Cyperus alternifolius

Kromdijk, G.
200 House Plants in Colour

hab.   Pl. 68

Cyperus alternifolius

Nicolaisen, Age
Pocket Encyclopedia of Indoor
    Plants

fl.   Pl. 4

Cyperus alternifolius

Stodola, Jiri
Encyclopedia of Water Plants

hab.   P. 339

Cyperus compressus

Burgis, D.S.
Florida Weeds

hab.   p. 4

Cyperus esculentus

Crockett, James Underwood
Lawns & Ground Covers

fl.   p. 72

Cyperus esculentus

Weeds of the Southern United States

hab., fr.   p. 20

Cyperus fuscus

Lindman, C. A. M.
Nordens Flora, Vol 2

fr. hab.   Pl. 112

Cyperus fuscus

Martin, W. Keble
The Concise British Flora in Colour

fl. hab.   Pl. 91

Cyperus gymnocaulos

Cochrane, G. R.
Flowers and Plants of Victoria

fl.   Pl. 267

Cyperus longus

Martin, W. Keble
The Concise British Flora in Colour

fl. hab.   Pl. 91

Cyperus obtusiflorus var.

Flowering Plants of Africa
    vol 42   1972-73

fr., hab.   Pl. 1666

Cyperus odoratus

Tarver, David P.
Aquatic and Wetland Plants
    of Florida

fr. hab.   p 82

Cyperus papyrus

Hay, Roy
The Color Dictionary of Flowers & Plants

hab.   p 62, Pl. 492

Cyperus papyrus

Hay, Roy
The Dictionary of House Plants

hab.   Pl. 176

Cyperus papyrus

Hvass, Elsie
Plants That Feed and Serve Us

    hab.   p 112, Pl. 241

Cyperus papyrus

Macoby, Stirling
What Flower is That

hab P 102

Cyperus rotundus

Weeds of the Southern United States

hab., fr.   p. 21

Cyperus strigosus

Wharton, Mary E.
A Guide to the Wildflowers & Ferns
    of Kentucky

fl.   p 84 Pl. 2.3

Cyphel, Mossy

    see

Minuartia sedoides

Cyphia sp.

Moriarty, Audrey
Wild Flowers of Malawi

fl.   Pl. 37; 1

Cyphomandra betacea

Macoby, Stirling
What Flower is That

fr.   P 102

Cyphomandra betacea

Masefield, G. B.
The Oxford Book of Food Plants

fl. fr.   Pl. 125, 5

Cyphostemma uter

Cactus & Succulent Society of America
Journal
    vol 46, No. 6   1974

hab.   p. 263

Cypress

    see

Cupressus

Cypress, Allum's

    see

Chamaecyparis lawsoniana var.

Cypress, Arizona

    see

Cupressus arizonica

Cypress, Bald

see

Taxodium distichum

Cypress, column

see

Cupressus sempervirens

Cypress, Dwarf False

see

Chamaecyparis obtusa var.

Cypress, False

see

Chamaecyparis pisifera

Cypress, Flowering

see

Tamarix

Cypress, Formosan

see

Chamaecyparis

Cypress, Garee Arizona

see

Cupressus arizonica var.

Cypress, golden Lambert

see

Cupressus macrocarpa var.

Cypress, Hinoki

see

Chamaecyparis obtusa

Cypress, Italian

see

Cupressus sempervirens

Cypress, Lawson

see

Chamaecyparis lawsonia

Cypress, Leyland

see

x Cupressocyparis leylandii

Cypress, Mexican

see

Cupressus lusitanica

Cypress, Monterey

see

Cupressus macrocarpa

Cypress, Nootka

see

Chamaecyparis

Cypress, Pond

see

Taxodium ascendens

Cypress, Port Jackson

see

Callitris rhomboidea

Cypress, San Pedro Martir

see

Cupressus montana

Cypress, Sawara moss false

see

Chamaecyparis pisifera var.

Cypress, slender Hinoki false

see

Chamaecyparis obtusa var.

Cypress, Standing

see

Ipomopsis rubra

Cypress, summer

see

Kochia scoparia

Cypress, Swamp

see

Taxodium distichum

Cypress, thread Sawara false

see

Chamaecyparis pisifera var.

Cypress Vine

see

Ipomoea quamoclit

Cypress vine

see

Quamoclit pennata

Cypress Weed

see

Eupatorium capillifolium

Cypripedium acaule

American Hort. Soc.
American Horticulturist
vol 56, No. 2   1977

fl.   p. 15

Cypripedium acaule

Batson, Wade T.
Wild Flowers in South Carolina

fl. hab.   p 38

Cypripedium acaule

Campbell, Carlos C.
Great Smoky Mountain Wildflowers

fl. hab.   p 43

Cypripedium acaule

Courtenay, Booth
Wildflowers & Weeds

fl.   p 11

Cypripedium acaule

Dean, Blanche E.
Wildflowers of Alabama and
    Adjoining States

fl.   p 35

Cypripedium acaule

Duncan, Wilbur H.
Wildflowers of the Southeastern
    United States

fl. hab.   p 269

Cypripedium acaule

Everett, T.H., ed.
New Illustrated Encyclopedia of
Gardening   vol 3

fl.   p. 503

Cypripedium acaule

Ferguson, Mary
Wildflowers

fl. hab.   p 155

Cypripedium acaule

Jennings, O. E.   vol II
Wild Flowers of Western Pennsylvania

fl.   Pl. 35

Cypripedium acaule

Klimas, John E.
Wildflowers of Eastern America

fl.   Pl. 206

Cypripedium acaule

Klute, Jeannette
Woodland Portraits

fl.   Pl. 24

Cypripedium acaule

Lemmon, Robert S.
Wildflowers of North America in
Full Color

fl.   Pg. 238   pl. 375

Cypripedium acaule

Luer, Carlyle A.
The Native Orchids of Florida

fl.   Pl. 4; 2

Cypripedium acaule

Luer, Carlyle A.
The Native Orchids of the
United States and Canada

fl. fr. hab.   Pl. 1

Cypripedium acaule

Massachusetts Hort. Soc.
Horticulture
vol 47, No. 3 1969

fl.   cover, backcover

Cypripedium acaule

Massachusetts Hort. Soc.
Horticulture
vol 49, No. 3 1971

fl.   p. 35

Cypripedium acaule

Massachusetts Hort. Soc.
Horticulture
vol 53, No. 2 1975

fl.   p. 47

Cypripedium acaule

Massachusetts Hort. Soc.
Horticulture
vol 55, No. 12 1977

fl., hab.   p. 33

Cypripedium acaule

Moyle, John B.
Northland Wild Flowers

fl. hab.   p 218, Pl. 289

Cypripedium acaule

Oepina, Mariano
Orquideas de las americas

fl. hab.   Pl. 13

Cypripedium acaule

Walcott, Mary Vaux
North American Wild Flowers

fl. hab.   Pl. 58

Cypripedium acaule

Walcott, Mary Vaux
North American Wild Flowers
vol. 5

fl.   Pl. 327

Cypripedium acaule

Wharton, Mary E.
A Guide to the Wildflowers & Ferns
of Kentucky

fl. p 67 Pl. 2,7

Cypripedium acaule var.

Massachusetts Hort. Soc.
Horticulture
vol 49, No. 3 1971

fl.   p. 35

Cypripedium x andrewsii

Luer, Carlyle A.
The Native Orchids of the
United States and Canada

fl.   Pl. 6; 6

Cypripedium arietinum

Courtenay, Booth
Wildflowers & Weeds

fl.   p 11

Cypripedium arietinum

Lemmon, Robert S.
Wildflowers of North America in
Full Color

fl.   pg. 236   pl. 373

Cypripedium arietinum

Luer, Carlyle A.
The Native Orchids of the
United States and Canada

fl. hab.   Pl. 2

Cypripedium arietinum

Walcott, Mary Vaux
North American Wild Flowers

fl. hab.   Pl. 216

Cypripedium calceolus

Alaska-Yukon Wild Flowers Guide

fl.   p 14

Cypripedium calceolus

American Hort. Soc.
American Horticulturist
vol 56, No. 2 1977

fl. cover

Cypripedium calceolus

Brooke, Jocelyn
The Wild Orchids of Britain

fl.   Pl. 1

Cypripedium calceolus

Brown, Clair A.
Wildflowers of Louisiana and
Adjoining States

fl. hab.   p 36

Cypripedium calceolus

Campbell, Carlos C.
Great Smoky Mountain Wildflowers

fl.   p 47

Cypripedium calceolus

Dean, Blanche E.
Wildflowers of Alabama and
Adjoining States

fl.   p 33

Cypripedium calceolus

Duncan, Wilbur H.
Wildflowers of the Southeastern
United States

fl.   p 269

Cypripedium calceolus

Ebel, Friedrich
The Strange and Beautiful
World of Orchids

fl.   p 101

Cypripedium calceolus

Felsko, Elsa
A Book of Wildflowers

fl.   P 15

Cypripedium calceolus

Ferguson, Mary
Wildflowers

fl.   p 158

Cypripedium calceolus

Hay, Roy
The Color Dictionary of Flowers & Plants

fl.   p 6, Pl. 44

Cypripedium calceolus

Huxley, Anthony
Garden Perennials and Water Plants

fl.   Pl. 90

Cypripedium calceolus

Klimas, John E.
Wildflowers of Eastern America

fl.   Pl. 156

Cypripedium calceolus

Kohlhaupt, Paula
Fleurs des Alpages
                    vol 1
fl. P 27

Cypripedium calceolus

Lindman, C. A. M.
Nordens Flora, Vol 2

fl. hab.   Pl. 132

Cypripedium calceolus

Martin, W. Keble
The Concise British Flora in Colour

fl. hab.   Pl. 82

Cypripedium calceolus

Milne, Lorus
Living Plants of the World

fl.   p 323

Cypripedium calceolus

Moyle, John B.
Northland Wild Flowers

fl. hab.   p 220, Pl. 291

Cypripedium calceolus

Orr, Robert T.
Wildflowers of Western America

fl.   Pl. 75

Cypripedium calceolus

Perry, Frances
Flowers of the World

fl.   p 206

Cypripedium calceolus

Royal Hort. Soc.
Journal of Royal Hort. Soc.
    vol 83, No. 7   1958

fl.   p. 294   Pl. 85

Cypripedium calceolus var.

Batson, Wade T.
Wild Flowers in South Carolina

fl. hab.   p 38

Cypripedium calceolus var.

Clark, Lewis J.
Wild Flowers of British Columbia

fl.   p 90

Cypripedium calceolus var.

Everett, T.H., ed.
New Illustrated Encyclopedia of
    Gardening   vol 3

fl.   p. 503

Cypripedium calceolus var.

Jennings, O. E.   vol II
Wild Flowers of Western Pennsylvania

fl.   Pl. 36

Cypripedium Calceolus var.

Klute, Jeannette
Woodland Portraits

fl.   Pl. 29

Cypripedium calceolus var.

Kramer, Jack
Orchids; Flowers of Romance
    and Mystery

fl.   Pl. 3, p 27

Cypripedium calceolus var.

Lemmon, Robert S.
Wildflowers of North America in
    Full Color

fl.   pg. 236   pl. 374

Cypripedium calceolus var.

Luer, Carlyle A.
The Native Orchids of Florida

fl.   Pl. 1; 4

Cypripedium calceolus var.

Luer, Carlyle A.
The Native Orchids of the
    United States and Canada

fl.   Pl. 3,
                    4, 5

Cypripedium calceolus var.

Massachusetts Hort. Soc.
Horticulture
    vol 41, No. 5   1963

fl.   backcover

Cypripedium calceolus var.

Massachusetts Hort. Soc.
Horticulture
    vol 42, No. 5   1964

fl.   p. 16

Cypripedium calceolus var.

Massachusetts Hort. Soc.
Horticulture
    vol 52, No. 5   1974

fl.   p. 24

Cypripedium calceolus var.

Wharton, Mary E.
A Guide to the Wildflowers & Ferns
    of Kentucky

fl. p 66 Pl. 2.6

Cypripedium calceolus var.

Wit, H. C. D. de
Plants of the World;
    The Higher Plants, Vol II

fl.   Pl. 157

Cypripedium californicum

Luer, Carlyle A.
The Native Orchids of the
    United States and Canada

fl. hab.   Pl. 11

Cypripedium californicum

Munz, Philip A.
California Spring Wildflowers

fl.   p 32, Pl. 16

Cypripedium californicum

Royal Hort. Soc.
Journal of the Royal Hort. Soc.
    vol 89, No. 4   1964

fl.   p. 158   Pl. 68

Cypripedium candidum

Klute, Jeannette
Woodland Portraits

fl.   Pl. 16

Cypripedium candidum

Lemmon, Robert S.
Wildflowers of North America in
    Full Color

fl.   pg. 238   pl. 376

Cypripedium candidum

Luer, Carlyle A.
The Native Orchids of the
    United States and Canada

fl. fr. hab.   Pl. 6; 1-4

Cypripedium candidum

Moyle, John B.
Northland Wild Flowers

fl. hab.   p 220, Pl. 292

Cypripedium cordigerum

Royal Hort. Soc.
Journal of the Royal Hort. Soc.
    vol 90, No. 3   1965

fl.   p. 114   Pl. 53

Cypripedium delenatii

Grubb, Roy
Selected Orchidaceous Plants
Vol I

fl.   p. 75

Cypripedium fasciculatum

Luer, Carlyle A.
The Native Orchids of the
United States and Canada

fl. hab.   Pl. 12

Cypripedium x favillianum

Luer, Carlyle A.
The Native Orchids of the
United States and Canada

fl.   Pl. 6, 5

Cypripedium guttatum

Alaska-Yukon Wild Flowers Guide

fl.   p 13, 28

Cypripedium guttatum

Heller, Christine
Wild Flowers of Alaska

fl.   Pl. 62

Cypripedium guttatum

Massachusetts Hort. Soc.
Horticulture
vol 46, No. 6   1968

fl., hab.   p. 20

Cypripedium guttatum

Massachusetts Hort. Soc.
Horticulture
vol 52, No. 5   1974

fl.   p. 25

Cypripedium guttatum var.

Luer, Carlyle A.
The Native Orchids of the
United States and Canada

fl. hab.   Pl. 13

Cypripedium himalaicum

Morley, Brian D.
Wild Flowers of the World

fl.   Plate 109

Cypripedium hybrid

Grubb, Roy
Selected Orchidaceous Plants
Vol I

fl.   p 39

Cypripedium hybrid.

Hay, Roy
The Color Dictionary of Flowers & Plants

fl.   p 62, Pl. 493, 494

Cypripedium irapeanum

Luer, Carlyle A.
The Native Orchids of the
United States and Canada

fl.   Pl. 9

Cypripedium irapeanum

O'Gorman, Helen
Mexican Flowering Trees and Plants

fl. hab.   p 153

Cypripedium japonicum

Curtis's Botanical Magazine
Vol 161   1938

fl.   No. 9520

Cypripedium montanum

Clark, Lewis J.
Wild Flowers of British Columbia

fl.   p 87

Cypripedium montanum

Luer, Carlyle A.
The Native Orchids of the
United States and Canada

fl. hab.   Pl. 7

Cypripedium montanum

Massachusetts Hort. Soc.
Horticulture
vol 52, No. 5   1974

fl.   p. 24

Cypripedium montanum

Orr, Robert T.
Wildflowers of Western America

fl.   Pl. 21

Cypripedium montanum

Royal Hort. Soc.
Journal of the Royal Hort. Soc.
vol 89, No. 4   1964

fl.   p. 158   Pl. 69

Cypripedium montanum

Walcott, Mary Vaux
North American Wild Flowers

fl.   Pl. 1

Cypripedium parviflorum

Neufeld, J.B.
Wild Flowers of the Prairies

fl.   p. 21

Cypripedium parviflorum

Walcott, Mary Vaux
North American Wild Flowers

fl. hab.   Pl. 92

Cypripedium passerinum

Alaska-Yukon Wild Flowers Guide

fl.   p 15

Cypripedium passerinum

Clark, Lewis J.
Wild Flowers of British Columbia

fl.   p 78

Cypripedium passerinum

Heller, Christine
Wild Flowers of Alaska

fl.   Pl. 124

Cypripedium passerinum

Luer, Carlyle A.
The Native Orchids of the
United States and Canada

fl. hab.   Pl. 10

Cypripedium passerinum

Porsild, A. E.
Rocky Mountain Wild Flowers

fl. hab.   p 115

Cypripedium passerinum

Walcott, Mary Vuax
North American Wild Flowers

fl. hab.   Pl. 91

Cypripedium pubescens

Courtenay, Booth
Wildflowers & Weeds

fl.   p 11

Cypripedium pubescens

Massachusetts Hort. Soc.
Horticulture
vol 33, No. 9   1955

fl.   p. 441

Cypripedium pubescens

Massachusetts Hort. Soc.
Horticulture
vol 35, No. 7   1957

fl.   inside backcover

Cypripedium pubescens

Massachusetts Hort. Soc.
Horticulture
vol 36, No. 5   1958

fl.   p. 278

Cypripedium reginae

Courtenay, Booth
Wildflowers & Weeds

fl.   p 11

Cypripedium reginae

Dean, Blanche E.
Wildflowers of Alabama and
   Adjoining States

fl.   p 35

Cypripedium reginae

Duncan, Wilbur H.
Wildflowers of the Southeastern
   United States

fl.   p 269

Cypripedium reginae

Ferguson, Mary
Wildflowers

fl.   p 41

Cypripedium reginae

Jennings, O. E.   vol II
Wild Flowers of Western Pennsylvania

fl.   Pl. 37

Cypripedium reginae

Klute, Jeannette
Woodland Portraits

fl.   Pl. 20

Cypripedium reginae

Lemmon, Robert S.
Wildflowers of North America in
   Full Color

fl.   p. 239   Pl. 377

Cypripedium reginae

Luer, Carlyle A.
The Native Orchids of the
   United States and Canada

fl. hab.   Pl. 8

Cypripedium reginae

Massachusetts Hort. Soc.
Horticulture
   vol 43, No. 4   1965

fl.   p. 20

Cypripedium reginae

Massachusetts Hort. Soc.
Horticulture
   vol 52, No. 5   1974

fl.   p. 25

Cypripedium reginae

Moyle, John B.
Northland Wild Flowers

fl. hab.   p 219, Pl. 290

Cypripedium reginae

Perry, Frances
Flowers of the World

fl.   p 206

Cypripedium reginae

Royal Hort. Soc.
Journal of the Royal Hort. Soc.
   vol 95, No. 1   1970

fl.   cover

Cypripedium reginae

Walcott, Mary Vaux
North American Wild Flowers

fl. hab.   Pl. 217

Cypripedium species

Everett, T.H., ed.
New Illustrated Encyclopedia of
   Gardening   vol 3

fl.   p. 358

Cypripedium venustum

Kimber, Sheila
A Handbook of Orchids

fl., hab.   p. 26

Cyrilla arida

Addisonia
   vol 19   1935-36

fl., fr.   p. 45   Pl. 631

Cyrilla racemiflora

Batson, Wade T.
Wild Flowers in South Carolina

fl.   p 71

Cyrilla racemiflora

Brown, Clair A.
Wildflowers of Louisiana and
   Adjoining States

fl. hab.   p 104

Cyrilla racemiflora

Dean, Blanche E.
Wildflowers of Alabama and
   Adjoining States

fl.   p 101

Cyrtanthus affinis

Flowering Plants of South Africa
   vol XXII   1942

fl., hab.   Pl. 867

Cyrtanthus affinis

Royal Hort. Soc.
Journal of the Royal Hort. Soc.
   vol 92, No. 4   1967

fl.   p. 160   Pl. 85

Cyrtanthus angustifolius

Royal Hort. Soc.
Journal of the Royal Hort. Soc.
   Vol. 90, No. 8   1965

fl.   p. 338   Pl. 150

Cyrtanthus brachyscyphus

Flowering Plants of Africa
   vol 37   1965-66

fl., hab.   Pl. 1444

Cyrtanthus breviflorus

Moriarty, Audrey
Wild Flowers of Malawi

fl.   Pl. 4; 1

Cyrtanthus carneus

Royal Hort. Soc.
Journal of the Royal Hort. Soc.
   vol 92, No. 4   1967

fl.   p. 160   Pl. 79

Cyrtanthus contractus

Royal Hort. Soc.
Journal of the Royal Hort. Soc.
   vol 92, No. 4   1967

fl.   p. 160   Pl. 78

Cyrtanthus erubescens

Flowering Plants of Africa
   vol 37   1965-66

fl.   Pl. 1442

Cyrtanthus eucallus

Royal Hort. Soc.
Journal of the Royal Hort. Soc.
   vol 92, No. 4   1967

fl.   p. 160   Pl. 87

Cyrtanthus falcatus

Curtis's Botanical Magazine
Vol 169   1952-53

fl.   202

Cyrtanthus falcatus

Flowering Plants of Africa
   vol 41   1970-71

fl., hab.   Pl. 1632

Cyrtanthus fergusoniae

Flowering Plants of Africa
   vol 39   1968-69

fl., hab.   Pl. 1545

Cyrtanthus flanaganii

Curtis's Botanical Magazine
Vol. 180   1974

fl, hab.   Pl. 666

Cyrtanthus flanaganii

Flowering Plants of South Africa
   vol XVIII   1938

fl., hab.   Pl. 693

Cyrtanthus galpinii var.

Royal Hort. Soc.
Journal of the Royal Hort. Soc.
vol 92, No. 4   1967

fl.   p. 160   Pl. 88

Cyrtanthus guthrieae

Flowering Plants of South Africa
vol XXI   1941

fl., hab.   Pl. 807

Cyrtanthus herrei

Flowering Plants of Africa
vol 33   1959

fl.   Pl. 1281

Cyrtanthus huttonii

Flowering Plants of Africa
vol 42   1972-73

fl., hab.   Pl. 1660

Cyrtanthus mackenii

Curtis's Botanical Magazine
vol 173   1960

fl.   Pl. 368

Cyrtanthus mackenii

Macoby, Stirling
What Flower is That

fl.   P 102

Cyrtanthus mackenii var.

Addisonia
vol 21   1939-42

fl.   p. 39   Pl. 692

Cyrtanthus montanus

Flowering Plants of Africa
vol 44   1977

fl., hab.   Pl. 1756

Cyrtanthus nutans

Flowering Plants of Africa
vol 30   1954-55

fl., hab.   Pl. 1182

Cyrtanthus nutans

Royal Hort. Soc.
Journal of the Royal Hort. Soc.
vol 92, No. 4   1967

fl.   p. 160   Pl. 86

Cyrtanthus salmonoides

Flowering Plants of Africa
vol 35   1962

fl., hab.   Pl. 1393

Cyrtanthus sanguineus var.

Curtis's Botanical Magazine
vol 175   1964-65

fl.   Pl. 481

Cyrtanthus smithiae

Royal Hort. Soc.
Journal of the Royal Hort. Soc.
vol 92, No. 4   1967

fl.   p. 160   Pl. 77

Cyrtanthus speciosus

Flowering Plants of South Africa
vol XXII   1942

fl., hab.   Pl. 868

Cyrtanthus spiralis

Flowering Plants of Africa
vol 37   1965-66

fl., hab.   Pl. 1460

Cyrtanthus thorncroftii

Flowering Plants of Africa
vol 33   1959

fl., hab.   Pl. 1318

Cyrtanthus thorncroftii

Royal Hort. Soc.
Journal of the Royal Hort. Soc.
vol 92, No. 4   1967

fl.   p. 160   Pl. 84

Cyrtanthus tuckii var.

Flowering Plants of South Africa
vol XVII   1937

fl., hab.   Pl. 680

Cyrtanthus welwitschii

Moriarty, Audrey
Wild Flowers of Malawi

fl.   Pl. 4; 2

Cyrtidium frontinoense

Ospina, Mariano
Orquideas de las americas

fl.   Pl. 121

Cyrtocarpa edulis

Coyle, Jeanette
A Field Guide to the Common and
Interesting Plants of Baja California

fr.   p 117

Cyrtopodium andersonii

Luer, Carlyle A.
The Native Orchids of Florida

fl.   Pl. 71, 1,2

Cyrtopodium andersonii

Pabst, G. F. J.
Orchidaceae Brasilienses,
Vol 1

fl.   p 240

Cyrtopodium eugenii

Pabst, G. F. J.
Orchidaceae Brasilienses,
Vol 1

fl.   p 240

Cyrtopodium falcilobum

Pabst, G. F. J.
Orchidaceae Brasilienses,
Vol 1

fl.   p 240

Cyrtopodium gigas

Pabst, G. F. J.
Orchidaceae Brasilienses,
Vol 1

fl.   p 240

Cyrtopodium poecilum

Pabst, G. F. J.
Orchidaceae Brasilienses,
Vol 1

fl.   p 240

Cyrtopodium punctatum

Lemmon, Robert S.
Wildflowers of North America in
Full Color

fl.   p. 13   Pl. 18

Cyrtopodium punctatum

Luer, Carlyle A.
The Native Orchids of Florida

fl.,fr.,hab.   Pl. 3;3
7;3-7

Cyrtopodium punctatum

Ospina, Mariano
Orquideas de las americas

fl.   Pl. 139

Cyrtopodium punctatum

Walcott, Mary Vaux
North American Wild Flowers

fl.   Pl. 212

Cyrtopodium, Spotted

see

Cyrtopodium punctatum

Cyrtorchis arcuata var.

Stewart, Joyce
Orchids of Tropical Africa

fl.   Pl. 25

Cyrtorchis arcuata var.

Williamson, Graham
The Orchids of South Central Africa

fl.   Pl. 179-81

Cyrtorchis crassifolia

Williamson, Graham
The Orchids of South Central Africa

fl.   Pl. 183

Cyrtorchis praetermissa

Flowering Plants of Africa
      vol 33   1959

fl., hab.   Pl. 1313

Cyrtorchis praetermissa

Williamson, Graham
The Orchids of South Central Africa

fl.   Pl. 182

Cyrtostachys lakko

Everett, Thomas H
Living Trees of the World

hab.   p. 71

Cytherea bulbosa

Walcott, Mary Vaux
North American Wild Flowers

fl.   Pl. 105

Cytinus hypocistis

Kleijn, H.
Beauty of the Wild Plant

fl.   opp. p. 113   pl. 163

Cytinus hypocistis

Megaw, Elektra
Wild Flowers of Cyprus

fl.   p. 8   Pl. 7

Cytinus hypocistis

Morley, Brian D.
Wild Flowers of the World

fl.   Pl. 31

Cytinus hypocistis

Polunin, Oleg
Flowers of Europe

fl. Pl. 8 #78

Cytinus hypocistis

Polunin, Oleg
Flowers of the Mediterranean

fl.   Pl. 10

Cytinus hypocistis

Royal Hort. Soc.
Journal of the Royal Hort. Soc.
      vol 88, No. 111   1963

hab., fl.   p. 478   Pl. 183

Cytinus hypocistus

Wit, H. C. D. de
Plants of the World;
    The Higher Plants, Vol I

fl.   p 161, Pl. 71

Cytinus sanguineus

Flowering Plants of Africa
      vol 38   1967

fl.   Pl. 1486

Cytisus absinthioides

Curtis's Botanical Magazine
Vol 166   1949

fl. fr.   87

Cytisus albus

Hellyer, A.G.L.
Shrubs in Colour

fl.   p. 39

Cytisus ardoinii

Hay, Roy
The Color Dictionary of Flowers & Plants

fl.   p 6, Pl. 45

Cytisus battandieri

Curtis's Botanical Magazine
Vol 161   1938

fl.   No. 9528

Cytisus battandieri

Gault, S. Millar
The Color Dictionary of Shrubs

fl.   Pl. 131

Cytisus battandieri

Hay, Roy
The Color Dictionary of Flowers & Plants

fl.   p 194, Pl. 1546

Cytisus battandieri

Hellyer, A.G.L.
Shrubs in Colour

fl.   p. 39

Cytisus x beanii

Curtis's Botanical Magazine
      vol 173   1960

fl.   Pl. 366

Cytisus x beanii

Gault, S. Millar
The Color Dictionary of Shrubs

fl.   Pl. 132

Cytisus x beanii

Hay, Roy
The Color Dictionary of Flowers &
    Plants

fl.   p. 6   Pl. 46

Cytisus canariensis

Addisonia
      vol 22   1943-46

fl., fr.   p. 53   Pl. 731

Cytisus canariensis

Nicolaisen, Age
Pocket Encyclopedia of Indoor Plants

fl.   Pl. 97

Cytisus canariensis

Royal Hort. Soc.
The Garden
      vol 100, No. 11   1975

hab., fl.   p. 526-27

Cytisus decumbens

Bartels, Andreas
Das Grosse Buch der Gartengeholze

fl.   p 142

Cytisus fragrans

Everett, T.H., ed.
New Illustrated Encyclopedia of
    Gardening   vol 3

fl.   p. 486

Cytisus grandiflorus

Polunin, Oleg
Trees and Bushes of Europe

fl.   p 108

Cytisus hirsutus

Barneby, T. P.
European Alpine Flowers in Colour

fl.   Pl. 40, 2

Cytisus hybrid

Gault, S. Millar
The Color Dictionary of Shrubs

fl.   Pl. 134, 137, 138, 139,
                140, 141

Cytisus hybrid

Harrison, R.E.
Trees and Shrubs

fl.   p. 59

Cytisus ingrami

Curtis's Botanical Magazine
Vol 169   1952-53

fl.   211

Cytisus x kewensis

Crockett, James Underwood
Lawns & Ground Covers

fl.   p. 129

Cytisus x kewensis

Curtis's Botanical Magazine
Vol. 171   1956-57

fl.   299

Cytisus x kewensis

Gault, S. Millar
The Color Dictionary of Shrubs

fl., hab.   Pl. 133

Cytisus x kewensis

Hay, Roy
The Color Dictionary of Flowers &
Plants

fl.   p. 194   Pl. 1547

Cytisus laburnum

Perrot, Emile
Les Plantes Medicinales

fl. fr. hab.   p 83

Cytisus maderensis

Harrison, R.E.
Trees and Shrubs

fl.   P 59

Cytisus multiflorus

Pacific Hort. Foundation
Pacific Horticulture
vol 39, No. 2 1978

fl., hab.   p. 40

Cytisus multiflorus

Polunin, Oleg
Trees and Bushes of Europe

fl. hab.   p 108

Cytisus multiflorus

Tsukamoto, Yotaro
Coloured Illustrations of Garden
Flowers   Vol. 10

fl. Illus. 148   opp. p. 46

Cytisus nigricans

Barneby, T. P.
European Alpine Flowers in Colour

fl.   Pl. 39, 5

Cytisus x praecox

Hansen, Richard
Baume und Straucher im Garten

fl.   p. 153

Cytisus x praecox

Hellyer, A.G.L.
Shrubs in Colour

fl.   p. 39

Cytisus x praecox

Huxley, Anthony
Deciduous Garden Trees and Shrubs

fl.   Pl. 58

Cytisus x praecox

Massachusetts Hort. Soc.
Horticulture
vol. 33, No. 4   1955

fl., hab.   backcover

Cytisus x praecox var.

Gault, S. Millar
The Color Dictionary of Shrubs

fl., hab.   Pl. 135

Cytisus x praecox var.

Huxley, Anthony
Deciduous Garden Trees and Shrubs

fl.   Pl. 59, 60

Cytisus purpureus

Gault, S. Millar
The Color Dictionary of Shrubs

fl. hab.   Pl. 136

Cytisus purpureus

Huxley, Anthony
Deciduous Garden Trees and Shrubs

fl.   Pl. 61

Cytisus purpureus var.

Huxley, Anthony
Deciduous Garden Trees and Shrubs

fl.   Pl. 62

Cytisus racemosus

Kiaer, Eigil
Indoor Plants in Colour

fl.   p. 48

Cytisus racemosus

Kromdijk, G.
200 House Plants in Colour

fl., hab.   Pl. 69

Cytisus radiatus

Barneby, T. P.
European Alpine Flowers in Colour

fl. hab.   Pl. 40, 1

Cytisus scoparius

Bianchini, Francesco
Health Plants of the World

fl. hab.   p 81

Cytisus scoparius

Clark, Lewis J.
Wild Flowers of British Columbia

fl.   p 283

Cytisus scoparius

Hay, Roy
The Color Dictionary of Flowers & Plants

fl.   p 194, Pl. 1548

Cytisus scoparius

Kimura, Koiti
Japanese Medicinal Plants, Vol I

fl.   p 38

Cytisus scoparius

Lindman, C. A. M.
Nordens Flora, Vol 5

fl. fr.   Pl. 336

Cytisus scoparius

Macoby, Stirling
What Flower is That

fl.   p. 103

Cytisus scoparius

Massachusetts Hort. Soc.
Horticulture
vol 38, No. 5   1960

fl.   cover

Cytisus scoparius

Massachusetts Hort. Soc.
Horticulture
vol 50, No. 6 1972

fl.   p. 38

Cytisus scoparius

Morley, Brian D.
Wild Flowers of the World

fl. fr.   Pl. 10F

Cytisus scoparius

Polunin, Oleg
Flowers of Europe

fl. Pl. 51 #505

Cytisus scoparius

Polunin, Oleg
Trees and Bushes of Europe

fl. fr. bab.   p 108

Cytisus scoparius var.

Bartels, Andreas
Das Grosse Buch der Gartengeholze

fl.   p 142

Cytisus scoparius var.

Gault, S. Millar
The Color Dictionary of Shrubs

fl. hab.   Pl. 142

Cytisus scoparius var.

Hellyer, A.G.L.
Shrubs in Colour

fl.   p. 39

Cytisus sessilifolius

Barneby, T. P.
European Alpine Flowers in Colour

fl.   Pl. 39, 6

Cytisus sessilifolius

Polunin, Oleg
Flowers of Europe

fl.   Pl. 51 #504

Cytisus sessilifolius

Royal Hort. Soc.
Journal of the Royal Hort. Soc.
    vol 96, No. 11   1971

hab., fl.   p. 494   Pl. 216

Cytisus stenopetalus

Curtis's Botanical Magazine
    vol 172   1958-59

fl.   Pl. 327

Cytisus triflorus

Polunin, Oleg
Flowers of the Mediterranean

fl.   Pl. 51

Cyttarium gunnii

Cockrane, G. R.
Flowers and Plants of Victoria

fr.   Pl. 435

Daboecia azorica

Curtis's Botanical Magazine
Vol 166   1949

fl.   46

Daboecia azorica

Royal Hort. Soc.
Journal of the Royal Hort. Soc.
    vol 94, No. 10   1969

fl.   p. 434   Pl. 238

Daboecia azorica var.

Royal Hort. Soc.
Journal of the Royal Hort. Soc.
    vol 94, No. 10   1969

fl.   p. 434   Pl. 239

Daboecia cantabrica

Crockett, James Underwood
Evergreens

hab. fl.   p. 120

Daboecia cantabrica

Hay, Roy
The Color Dictionary of Flowers & Plants

fl.   p 194, Pl. 1549

Daboecia cantabrica

Hellyer, A.G.L.
Shrubs in Colour

fl.   p. 42

Daboecia cantabrica

Macoby, Stirling
What Flower is That

fl.   P 103

Daboecia cantabrica

Martin, W. Keble
The Concise British Flora in Colour

fl. hab.   Pl. 55

Daboecia cantabrica

Polunin, Oleg
Flowers of Europe

fl.   Pl. 88 #922

Daboecia cantabrica

Wit, H. C. D. de
Plants of the World;
    The Higher Plants, Vol I

fl.   p 173, Pl. 101

Daboecia cantabrica var.

Gault, S. Millar
The Color Dictionary of Shrubs

fl.   Pl. 143, 144

Daboecia cantabrica var.

Hellyer, A.G.L.
Shrubs in Colour

fl.   p. 42

Daboecia cantabrica var.

Royal Hort. Soc.
Journal of the Royal Hort. Soc.
    vol. 96, No. 4   1971

fl.   p. 404   Pl. 183

Daboecia hybrid.

Gault, S. Millar
The Color Dictionary of Shrubs

fl. hab.   Pl. 145

Daboecia x scotica var.

Royal Hort. Soc.
The Garden
    vol 103, No. 3   1978

fl.   p. 114

Dacrydium cupressinum

Edlin, Herbert
The Illustrated Encyclopedia
    of Trees

hab.   p 247

Dacrydium cupressinum

Royal Hort. Soc.
The Garden
    vol 101, No. 3   1976

hab., fr.   p. 163

Dacrydium franklinii

Curtis, Winifred
The Endemic Flora of Tasmania
                vol 3

fl.   fr.   p 201

Dactylis glomerata

Hvass, Else
Plants That Feed and Serve Us

fr.   p. 62   Pl. 123

Dactylis glomerata

Lindman, C. A. M.
Nordens Flora, Vol 2

fr. hab.   Pl. 87

Dactyloctenium aegyptium

Weeds of the Southern United States

hab., fr.   p. 24

Dactylopsis digitata

Herre, H.
The Genera of the Mesembryanthemaceae

fl.   fr.   p 125

Dactylorchis maculata

Kleijn, H.
The Beauty of the Wild Plant

fl.   p. 57   Pl. 73

Dactylorchis maculata

Lindman, C. A. M.
Nordens Flora, Vol 2

fl. fr. hab.   Pl. 139

Dactylorchis majalis

Lindman, C. A. M.
Nordens Flora, Vol 2

fl.   Pl. 138

Dactylorchis romana

Polunin, Oleg
Flowers of the Mediterranean

fl.   No. 303, 304

Dactylorchis sambucina

Lindman, C. A. M.
Nordens Flora, Vol 2

fl.   Pl. 137

Dactylorhiza aristata

Alaska-Yukon Wild Flowers Guide

fl.   p 16

Dactylorhiza aristata var.

Luer, Carlyle A.
The Native Orchids of the
   United States and Canada

fl. hab.   Pl. 38
             Pl. 39; 1-7

Dactylorhiza elata

Royal Hort. Soc.
The Garden
   vol 101, No. 7   1976

fl., hab.   p. 359

Dactylorhiza fuchsii

Martin, W. Keble
The Concise British Flora in Colour

fl. hab.   Pl. 81

Dactylorhiza fuchsii

Polunin, Oleg
Flowers of Europe

fl.   Pl. 189 #1900

Dactylorhiza hybrid

Perry, Frances
Flowers of the World

fl.   p. 212

Dactylorhiza incarnata

Martin, W. Keble
The Concise British Flora in Colour

fl. hab.   Pl. 81

Dactylorhiza maculata

Luer, Carlyle A.
The Native Orchids of the
   United States and Canada

fl.   Pl. 39; 8-9

Dactylorhiza maculata var.

Martin, W. Keble
The Concise British Flora in Colour

fl.   Pl. 81

Dactylorhiza maculata   var.

Webster, Mary
Flora of Moray, Nairn & East
   Inverness

fl.   p. 453   Pl. 20

Dactylorhiza majalis

Ebel, Friedrich
The Strange and Beautiful
   World of Orchids

fl.   p 45

Dactylorhiza majalis

Polunin, Oleg
Flowers of Europe

fl.   Pl. 187 #1898

Dactylorhiza praetermissa

Martin, W. Keble
The Concise British Flora in Colour

fl. hab.   Pl. 81

Dactylorhiza purpurella

Martin, W. Keble
The Concise British Flora in Colour

fl. hab.   Pl. 81

Dactylorhiza purpurella

Webster, Mary
Flora of Moray, Nairn & East
   Inverness

fl.   p. 453   Pl. 20

Dactylorhiza romana var.

Royal Hort. Soc.
Journal of the Royal Hort. Soc.
   vol 92, No. 7   1967

fl.   p. 296   Pl. 165-66.

Dactylorhiza sambucina

Polunin, Oleg
Flowers of Europe

fl.   Pl. 189 #1899

Daffodil

see

Narcissus

Daffodil, autumn

see

Sternbergia lutea

Daffodil, Hoop Petticoat

see

Narcissus bulbocodium

Daffodil, Peruvian

see

Hymenocallis calathina

Daffodil, Sand

see

Pancratium maritimum

Daffodil, wild

see

Narcissus pseudonarcissus

Dagga, Wild

see

Leonotis leonurus

Dagger Bush

wee

Hakea teretifolia

Dagger, Giant

see

Yucca carnerosana

Dahlia coccinea

Morley, Brian D.
Wild Flowers of the World

fl.   pl. 186

Dahlia coccinea

O'Gorman, Helen
Mexican Flowering Trees and Plants

fl. hab.   p 155

Dahlia excelsa

O'Gorman, Helen
Mexican Flowering Trees and Plants

fl. hab.   p 157

Dahlia excelsa

Royal Hort. Soc.
The Garden
vol 101, No. 5   1976

fl.   p. 278

Dahlia, Flat Tree

see

Dahlia excelsa

Dahlia hybrid

Bruggeman, L.
Tropical Plants

fl.   Pl. 104

Dahlia hybrid

Hay, Roy
The Color Dictionary of Flowers &
        Plants

fl.   p.133-136   Pl.1060-1083

Dahlia hybrid

Huxley, Anthony
Garden Annuals and Bulbs

fl.   Pl. 207-211

Dahlia hybrid

Macoby, Stirling
What Flower is That

fl.   p. 105

Dahlia hybrid

Miles, Bebe
Bulbs for the Home Garden

fl.   p 160-161

Dahlia hybrid

Perry, Frances
Flowers of the World

fl.   p. 81

Dahlia hybrida var.

Hay, Roy
The Color Dictionary of Flowers &
        Plants

fl.   p. 35-36   Pl.280-281

Dahlia imperialis

Royal Hort. Soc.
Journal of the Royal Hort. Soc.
    vol 91, No. 6   1966

fl.   p. 250   Pl. 136

Dahlia maxonii

Chickering, Carol Rogers
Flowers of Guatemala

fl.   p 61

Dahlia popenovii

Chickering, Carol Rogers
Flowers of Guatemala

fl.   p 63

Dahlia, Sea

see

Coreopsis maritima

Dahlia sp.

Huxley, Anthony
Garden Annuals and Bulbs

hab.   Pl. 153

Dahlia variabilis

Tsukamoto, Yotaro
Coloured Illustrations of Garden
        Flowers   Vol. 10

fl. Illus. 48   opp. p. 16

Dahoon

see

Ilex cassine

Dais cotinifolia

Flowering Plants of South Africa
    vol XXII   1942

fl.   Pl. 869

Dais cotinifolia

Harrison, R.E.
Trees and Shrubs

fl.   P 60

Dais cotinifolia

Oakman, Harry
Colorful Trees

Hab. Fl.   P 90

Dais cotinifolia

Palmer, Eve
Trees of South Africa

fl.   p. 256   Pl. XXIII

Daisy, African

see

Arctotis

Daisy, African

see

Dimorphotheca aurantiaca

Daisy, African

see

Gerbera

Daisy, African

see

Helianthus simulans

Daisy, Albany

see

Actinidium cunninghamii

Daisy, Alpine

see

Bellidiastrum michelii

Daisy, Alpine

see

Brachycome nivalis var.

Daisy, alpine

see

Brachycome scapigera

Daisy, Alpine

see
Erigeron grandiflorus

Daisy, Alpine Moon

see

Chrysanthemum alpinum

Daisy, Arctic

see

Chrysanthemum arcticum

Daisy, aurora

see

Arctotis   hybrid

Daisy, Barberton

see

Gerbera jamesonii

Daisy, Beetle

see

Corteria diffusa var.

Daisy, blue

see

Felicia amelloides

Daisy, Burr

see

Calotis latiuscula

Daisy Bush

see

Olearia tomentosa

Daisy-Bush, Heath

see

Olearia floribunda

Daisy-Bush, Snowy

see

Olearia lirata

Daisy Bush, Starry

see

Olearia subrepanda

Daisy bush, Tasmanian

see

Olearia phlogopappa

Daisy-bush, Twiggy

see

Olearia ramulosa

Daisy, crown

see

Chrysanthemum coronarium

Daisy, Dahlberg

see

Thymophylla tenuiloba

Daisy, Desert

see

Erigeron pumilus var.

Daisy, desert

see

Melampodium leucanthum

Daisy, Double Namaqualand

See

Arctotis fastuosa

Daisy, Early Morning

See

Osteospermum hyoseroides

Daisy, English

see

Bellis perennis

Daisy, everlasting

see

Acroclinium roseum

Daisy, False

see

Bellidastrum michelii

Daisy, Field

see

Chrysanthemum leucanthemum . var.

Daisy Fleabane

see

Erigeron annuus

Daisy, Fleabane

see

Erigeron pumilus

Daisy, Globe

see

Globularia cordifolia

Daisy, gloriosa

see

Rudbeckia hirta var.

Daisy, gloriosa

see

Rudbeckia hybrid

Daisy, golden paper

see

Helichrysum bracteatum

Daisy, kingfisher

see

Felicia angustifolia

Daisy, lawn

see

Bellis perennis

Daisy, lazy

see

Aphanostephus skirrhobasis

Daisy, Livingston

see

Mesembryanthemum criniflorum

Daisy, Livingstone

see

Dorotheanthus bellidiformis

Daisy, Marguerite

see

Chrysanthemum frutescens

Daisy, michaelmus

see

Aster amellus

Daisy, michaelmus

see

Aster ericoides

Daisy, michaelmus
see
Aster frikartii

Daisy, Minnie
see
Minuria leptophylla

Daisy, moon
see
Chrysanthemum leucanthemum

Daisy, Mountain
see
Erigeron compositus

Daisy, mountain
see
Erigeron peregrinus

Daisy, Namaqualand
see
Dimorphotheca aurantiaca

Daisy, Namaqualand
see
Dimorphotheca sinuata

Daisy, New Holland
see
Vittadinia triloba

Daisy, Ox-eye
see
Chrysanthemum leucanthemum

Daisy, Painted
see
Chrysanthemum coccineum

Daisy, Painted
see
Chrysanthemum coronarium

Daisy, Panamint
see
Enceliopsis argophylla

Daisy, paper
see
Acroclinium roseum

Daisy, Paper
see
Helipterum albicans

Daisy, Paris
see
Chrysanthemum frutescens

Daisy Paris
see
Euryops abrotanifolius

Daisy, Paris
see
Gamolepis chrysanthemoides

Daisy, Pink
see
Erigeron peregrinus

Daisy, Pink Paper
see
Helipterum roseum

Daisy, Poached Egg
see
Myriocephalus stuartii

Daisy, purple
see
Brachycome species

Daisy, Rock
see
Perityle emoryi

Daisy, Rottnest Island
see
Trachymene coerulea

Daisy, sailor boy
see
Dimorphotheca ecklonis

Daisy, Sea
see
Borrichia frutescens

Daisy, Seaside
see
Erigeron glaucus

Daisy, seaside
see
Erigeron karvinskianus

Daisy, Shasta
see
Chrysanthemum maximum

Daisy, Showy
see
Erigeron speciosus

Daisy, Silver Snow
see
Celmisia longifolia

Daisy, Smooth
see
Erigeron leiomeris

Daisy, snow
see
Celmisia longifolia

Daisy, Subalpine
see
Erigeron peregrinus

Daisy, sunshine

see

Gamolepis tagetes

Daisy, swamp

see

Actinodium cunninghamii

Daisy, Swan River

see

Brachycome ibiridifolia

Daisy, tahoka

see

Aster tanacetifolius

Daisy, Tall

see

Brachycome diversifolia

Daisy, Trailing African

see

Osteospermum fruticosum

Daisy, Transvaal

see

Gerbera jamesonii

Daisy Tree

see

Montanoa arborescens

Daisy, Tree

see

Olearia

Daisy, Western

see

Astranthium integrifolium

Daisy, White

see

Chrysanthemum leucanthemum

Daisy, White Paper

see

Helichrysum elatum

Daisy, White Rain

see

Dimorphotheca pluvialis

Daisy, Wild

see

Erigeron formosissimus

Daisy, Wood

see

Bellis sylvestris

Daisy, Woolly

see

Eriophyllum wallacei

Daisy, Woolly yellow

see

Eriophyllum lanatum

Daisy, Yam

see

Microseris scapigera

Daisy, Yellow Everlasting

see

Helichrysum bracteatum

Daisy, Yellow Paper

see

Helichrysum bracteatum

Dalbergia latifolia

Hvass, Eslie
Plants That Feed and Serve Us
fl.   p 128, Pl. 283

Dalea emoryi

Munz, Philip A.
California Desert Wildflowers

fl.   p 37, Pl. 31

Dalea fremontii

American Hort. Soc.
American Horticulturist
vol 53. No. 5 1974

fl.   p. 20

Dalea schottii

Coyle, Jeanette
A Field Guide to the Common and
Interesting Plants of Baja California

fl.   p 97

Dalea spinosa

Coyle, Jeanette
A Field Guide to the Common and
Interesting Plants of Baja California

hab.   p 99

Dalea spinosa

Mathias, Mildred E.
Color for the Landscape

fl. hab.   p 193

Dalea spinosa

Wit, H. C. D. de
Plants of the World;
The Higher Plants, Vol I

hab.   p 286, Pl. 181

Dalechampia roezliana

Morley, Brian D.
Wild Flowers of the World

fl. fr.   Pl. 178

Dalechampia spathulata

Wit, H. C. D. de
Plants of the World;
The Higher Plants, Vol I

fl. fr.   p 275, Pl. 151,
                        152

Dalibarda repens

Courtenay, Booth
Wildflowers & Weeds

fl.   p 40

Dallis Grass

see

Paspalum dilatatum

Damasonium alisma

Martin, W. Keble
The Concise British Flora in Colour

fl. fr. hab.   Pl. 79

Damnacanthus indicus

Kitamura, Siro
Coloured Illustrations of Trees &
Shrubs of Japan

fr.   435

Dampiera, blue

see

Dampiera stricta

Dampiera diversifolia

Blombery, Alec M.
What Wildflower is That

fl.   p 108, Pl. 269

Dampiera diversifolia

Harrison, Richmond E.
Climbers and Trailers

fl.   p. 92   Pl. 229

Dampiera diversifolia

Royal Hort. Soc.
The Garden
     vol 102, No. 3   1977

fl.   p. 109

Dampiera lanceolata

Cochrane, G. R.
Flowers and Plants of Victoria

fl.   Pl. 184

Dampiera linearis

Blombery, Alec M.
What Wildflower is That

fl.   p 108, Pl. 270

Dampiera marifolia

Cochrane, G. R.
Flowers and Plants of Victoria

fl.   Pl. 153

Dampiera purpurea

Blombery, Alec M.
What Wildflower is That

fl.   p. 108   Pl. 271

Dampiera stricta

Blombery, Alec M.
What Wildflower is That

fl.   p. 109   Pl. 272

Dampiera stricta

Cochrane, G. R.
Flowers and Plants of Victoria

fl.   Pl. 393

Dampiera stricta

Mullins, Barbara
Australian Wildflowers in Colour

fl.   P. 91   Pl. 88

Dampiera wellsiana

Blombery, Alec M.
What Wildflower is That

fl. hab.   p 109, Pl. 273

Damson

see

Prunus damascena

Danae racemosa

Hellyer, A.G.L.
Shrubs in Colour

hab.   p. 42

Dandelion

see

Taraxacum

Dandelion, Desert

see

Malacothrix californica

Dandelion, Desert

see

Malacothrix glabrata

Dandelion, Dwarf

see

Krigia biflora

Dandelion, Dwarf

see

Krigia dandelion

Dandelion, False

see

Agoseris aurantiaca

Dandelion, False

see

Pyrrhopappus carolinianus

Dandelion, False Carolina

see

Pyrrhopappus carolinianus

Dandelion, goat

see

Krigia biflora

Dandelion, Mountain

see

Agoseris aurantiaca

Dandelion, Mountain

see

Agroseris glauca

Dandelion, Potato

see

Krigia dandelion

Dandelion, Prairie

see

Microseris cuspidata

Dandelion, Rubber

see

Taraxacum bicorne

Danewort

see

Sambucus ebulus

Dangleberry

see

Gaylussacia frondosa

Danthonia dimidiata

Curtis, Winifred
The Endemic Flora of Tasmania
     Vol VI

fl. fr. hab.   Pl. 233

Danthonia fortunae-hibernae

Curtis, Winifred
The Endemic Flora of Tasmania
     Vol VI

fl. fr. hab.   Pl. 234

Danthonia intermedia

Porsild, A. E.
Rocky Mountain Wild Flowers

fl. hab.   p 51

Daphne alpina

Felsko, Elsa
A Book of Wild Flowers
2nd ser
fl.   p 122

Daphne arbuscula

Hay, Roy
The Color Dictionary of Flowers & Plants

fl.   p 194, Pl. 1550

Daphne bholua

American Hort. Soc.
American Horticulturist
Vol. 55, No. 2   1976

fl.   p. 15

Daphne bholua

Curtis's Botanical Magazine
Vol. 180   1974

fl.   Pl. 681

Daphne Bholua

Hara, Hiroshi, comp.
Photo-Album of Plants of
Eastern Himalaya

fl.   Pl. 80

Daphne bholua

Royal Hort. Soc.
The Garden
vol 101, No. 9 1976

fl.   p. 455

Daphne blagayana

Hay, Roy
The Color Dictionary of Flowers & Plants

fl.   p 194, Pl. 1551

Daphne blagayana

Perry, Frances
Flowers of the World

fl.   p 296

Daphne x burkwoodii

Curtis's Botanical Magazine
Vol. 166   1949

fl.   55

Daphne x burkwoodii

Hellyer, A.G.L.
Shrubs in Colour

fl.   p. 42

Daphne x burkwoodii

Perry, Frances
Flowers of the World

fl.   p. 296

Daphne x burkwoodii var.

Gault, S. Millar
The Color Dictionary of Shrubs

fl.   Pl. 146

Daphne x burkwoodii var.

Harrison, R. E.
Trees and Shrubs

fl.   p. 61

Daphne cneorum

Barneby, T. P.
European Alpine Flowers in Colour

fl.   Pl. 52, 5

Daphne cneorum

Bartels, Andreas
Das Grosse Buch der Gartengehölze

fl.   p 142

Daphne cneorum

Crockett, James Underwood
Evergreens

hab. fl.   p. 121

Daphne cneorum

Everett, T.H., ed.
New Illustrated Encyclopedia of
Gardening   vol 4

fl., hab.   p. 550

Daphne cneorum

Felsko, Elsa
A Book of Wild Flowers
2nd ser
fl.   p 81

Daphne cneorum

Hansen, Richard
Bäume und Sträucher im Garten

fl.   p. 172

Daphne cneorum

Harrison, R.E.
Trees and Shrubs

fl. P 61

Daphne cneorum

Hay, Roy
The Color Dictionary of Flowers & Plants

fl.   p 6, Pl. 47

Daphne cneorum

Huxley, Anthony
Evergreen Garden Trees and Shrubs

fl. hab.   Pl. 113

Daphne cneorum

Massachusetts Hort. Soc.
Horticulture
vol 33, No. 3   1955

fl.   backcover

Daphne cneorum

Polunin, Oleg
Flowers of Europe

fl.   Pl. 74 #756

Daphne cneorum

Vilmorin, Roger de
Plantes Alpines dans les Jardins

fl. hab.   Pl. XXVII

Daphne cneorum var.

Hellyer, A.G.L.
Shrubs in Colour

fl.   p. 42

Daphne cneorum var.

Pacific Hort. Foundation
Pacific Horticulture
vol 38, No. 2   1977

fl.   p. 37

Daphne cneorum var.

Perry, Frances
Flowers of the World

fl.   p. 296

Daphne Cneorum var.

Royal Hort. Soc.
Journal of Royal Hort. Soc.
vol 82, No. 8   1957

hab., fl.   p. 340   Pl. 97

Daphne cneorum var.

Royal Hort. Soc.
Journal of the Royal Hort. Soc.
vol 86, No. 11   1961

hab., fl.   p. 486   Pl. 149

Daphne collina

Hay, Roy
The Color Dictionary of Flowers & Plants

fl.   p 194, Pl. 1552

Daphne collina

Pacific Hort. Foundation
Pacific Horticulture
vol 38, No. 2   1977

fl.   p. 36

Daphne collina

Royal Hort. Soc.
Journal of Royal Hort. Soc.
vol 82, No. 8   1957

hab., fl.   p. 340   Pl. 96

Daphne collina

Royal Hort. Soc.
Journal of the Royal Hort. Soc.
vol 94, No. 11   1969

hab., fl.   p. 480   Pl. 254

Daphne genkwa

Addisonia
vol 19   1935-36

fl.   p. 37   Pl. 627

Daphne Genkwa

Curtis's Botanical Magazine
vol 173   1960

fl.   Pl. 360

Daphne Genkwa

Everett, T.H., ed.
New Illustrated Encyclopedia of
Gardening   vol 4

fl.   p. 551

Daphne genkwa

Harrison, R.E.
Trees and Shrubs

fl. P 61

Daphne genkwa

Hay, Roy
The Color Dictionary of Flowers & Plants

fl.   p 195, Pl. 1553

Daphne genkwa

Kariyone, Tatsuo
Atlas of Medicinal Plants

fl.   Pl. 83

Daphne genkwa

Kimura, Koiti
Japanese Medicinal Plants, Vol II

fl.   p 203

Daphne genkwa

Royal Hort. Soc.
Journal of the Royal Hort. Soc.
vol 87, No. 12   1962

fl.   p. 538   Pl. 153

Daphne giraldii

Royal Hort. Soc.
Journal of the Royal Hort. Soc.
vol 86, No. 11   1961

fl.   p. 486   Pl. 148

Daphne glomerata

Royal Hort. Soc.
The Garden
vol 102, No. 12   1977

fl.   p. 498

Daphne gnidium

Perrot, Emile
Les Plantes Medicizales

fl. fr. hab.   p 84

Daphne gnidium

Polunin, Oleg
Flowers of the Mediterranean

fl. fr.   No. 113

Daphne gnidium

Wit, H. C. D. de
Plants of the World;
The Higher Plants, Vol I

fl. fr.   p 276, Pl. 154

Daphne hybrid

Pacific Hort. Foundation
Pacific Horticulture
vol 38, No. 2   1977

fl.   p. 36

Daphne x hybrida

Curtis's Botanical Magazine
Vol. 172   1958-59

fl.   Pl. 320

Daphne jezoensis

Curtis's Botanical Magazine
Vol. 178   1970-72

fl.   613

Daphne julia

Royal Hort. Soc.
Journal of the Royal Hort. Soc.
vol 99, No. 8   1974

hab., fl.   p. 346 Pl. 164

Daphne kiusiana

Kitamura, Siro
Coloured Illustrations of Trees &
Shrubs of Japan

fl.   354

Daphne laureola

Ary, S.
The Oxford Book of Wildflowers

fl.   p 190, Pl. 4

Daphne laureola

Martin, W. Keble
The Concise British Flora in Colour

fl. Hab.   Pl. 76

Daphne laureola

Perrot, Emile
Les Plantes Medicinales

fl. fr. hab.   p 84

Daphne laureola

Perry, Frances
Flowers of the World

fl.   p. 296

Daphne laureola

Polunin, Oleg
Flowers of Europe

fl.   Pl. 74 #760

Daphne longilobata

Curtis's Botanical Magazine
vol 172   1958-59

fl., fr.   Pl. 344

Daphne x mantensiana

Pacific Hort. Foundation
Pacific Horticulture
vol 38, No. 2   1977

fl.   p. 36

Daphne mezereum

American Hort. Soc.
American Horticulturist
vol 57, No. 1   1978

fr.   p. 26

Daphne mezereum

Barneby, T. P.
European Alpine Flowers in Colour

fl.   Pl. 52, 4

Daphne mezereum

Bartels, Andreas
Das Grosse Buch der Gartengeholze

fr.   p 143

Daphne mezereum

Bianchini, Francesco
Health Plants of the World

fr.   p 163

Daphne mezereum

Color Treasury of Herbs & Other
Medicinal Plants

fr. p 37 Pl. 46

Daphne mezereum

Curtis's Botanical Magazine
Vol 171   1956-57

fl. fr.   272

Daphne mezereum

Felsko, Elsa
A Book of Wildflowers

fl.   P 81

Daphne mezereum

Gault, S. Millar
The Color Dictionary of Shrubs

fl.   Pl. 147

Daphne mezereum

Hansen, Richard
Baume und Straucher im Garten

fl.   p. 172

Daphne mezereum

Hellyer, A.G.L.
Shrubs in Colour

fl.   p. 42

Daphne mezereum

Huxley, Anthony
Deciduous Garden Trees and Shrubs

fl. fr.   Pl. 63, 63a

Daphne mezereum

Kohlhaupt, Paula
Fleurs des Alpages   Vol. 2

fl. fr.   P. 65-66

Daphne mezereum

Lindman, C. A. M.
Nordens Flora, Vol 6

fl. fr.   Pl. 400

Daphne mezereum

Martin, W. Keble
The Concise British Flora in Colour

fl. fr. hab.   Pl. 76

Daphne mezereum

Meikle, R. D.
British Trees and Shrubs

fl. fr.   Pl. 12

Daphne mezereum

Morley, Brian D.
Wild Flowers of the World

fl. fr.   Pl. 18G

Daphne mezereum

Perrot, Emile
Les Plantes Medicinales

fl. fr. hab.   p 84

Daphne mezereum

Polunin, Oleg
Flowers of Europe

fl. Pl. 74 #759

Daphne mezereum

Vedel, H.
Arbres et Arbustes

fl. fr.   p 100, No. 105

Daphne mezereum var.

Everett, T.H., ed.
New Illustrated Encyclopedia of
     Gardening   vol 4

fl.   p. 551

Daphne mezereum var.

Hay, Roy
The Color Dictionary of Flowers & Plants

fl.   p 195, Pl. 1554

Daphne, Native

see

Pittosporum undulatum

Daphne odora

Macoby, Stirling
What Flower is That

fl. P 106

Daphne odora var.

Gault, S. Millar
The Color Dictionary of Shrubs

fl.   Pl. 148

Daphne odora var.

Harrison, R.E.
Trees and Shrubs

fl. P 61

Daphne odora var.

Hay, Roy
The Color Dictionary of Flowers & Plants

fl.   p 195, Pl. 1555

Daphne odora var.

Macoby, Stirling
What Flower is That

fl. P 106

Daphne odorata var.

Perry, Frances
Flowers of the World

fl.   p. 296

Daphne papyracea

Walden, Beryl M.
Wild Flowers of Hong Kong

fl.   Pl. 1(1), Pl. 2(1)

Daphne petraea

Royal Hort. Soc.
Journal of the Royal Hort. Soc.
     vol 78, No. 1   1953

fl.   p. 24   Pl. 1

Daphne petraea

Royal Hort. Soc.
Journal of the Royal Hort. Soc.
     vol 92, No. 7   1967

fl.   p. 296   Pl. 157

Daphne petraea var.

Hay, Roy
The Color Dictionary of Flowers &
     Plants

fl.   p. 6   Pl. 48

Daphne petraea var.

Royal Hort. Soc.
Journal of the Royal Hort. Soc.
     vol 91, No. 6   1966

fl.   p. 250   Pl. 117

Daphne retusa

Gault, S. Millar
The Color Dictionary of Shrubs

fl.   Pl. 149

Daphne retusa

Hay, Roy
The Color Dictionary of Flowers & Plants

fl.   p 195, Pl. 1556

Daphne retusa

Royal Hort. Soc.
Journal of the Royal Hort. Soc.
     vol 97, No. 4   1972

fl.   p. 170   Pl. 76

Daphne retusa

Royal Hort. Soc.
The Garden
     vol 100, No. 10 1975

fl.   p. 503

Daphne, Rock

see

Daphne petraea

Daphne, rose

see

Daphne cneorum

Daphne, South African

see

Dais cotinifolia

Daphne striata

Barneby, T. P.
European Alpine Flowers in Colour

fl.   Pl. 52, 6

Daphne striata

Hay, Roy
The Color Dictionary of Flowers & Plants

fl.   p 195, Pl. 1557

Daphne striata

Kohlhaupt, Paula
Fleurs des Alpages
          vol 1
fl.fr.   P 61-62

Daphne, sweet

see

Daphne odora

Daphne tangutica

Cault, S. Millar
The Color Dictionary of Shrubs

fl.   Pl. 150

Daphne, variegated

see

Daphne odora var.

Daphne, White

see

Daphne papyracea

Daphne, yellow

see

Edgeworthia papyrifera

Daphniphyllum macropodum

Kitamura, Siro
Coloured Illustrations of Trees &
     Shrubs of Japan

fr.   269

Darlingtonia californica

Hersey, Jean
Woman's Day Book of House Plants

hab.   p. 52

Darlingtonia californica

Lemmon, Robert S.
Wildflowers of North America in
      Full Color

hab., fl.   p. 23   Pl. 37

Darlingtonia californica

Massachusetts Hort. Soc.
Horticulture
    vol 48, No. 1   1964

hab.   p. 38

Darlingtonia californica

Munz, Philip A.
California Mountain Wildflowers

hab.   p. 54   Pl. 84

Darlingtonia californica

Orr, Robert T.
Wildflowers of Western America

hab., fl.   Pl. 79

Darlingtonia californica

Pacific Hort. Foundation
Pacific Horticulture
    vol 39, No. 3 1978

hab.   p. 29

Darlingtonia californica

Perry, Frances
Flowers of the World

hab.   p. 271

Darlingtonia californica

Royal Hort. Soc.
Journal of the Royal Hort. Soc.
    vol 87, No. 4   1962

hab.   p. 174 Pl. 43

Darlingtonia californica

Schnell, Donald E.
Carnivorous Plants of the U. S. and Canada

fl. hab.   p 53, 55

Darwinia citriodora

Blombery, Alec M.
What Wildflower is That

fl.   p. 109   Pl. 274

Darwinia collina

Morcombe, M. K.
Australia's Western Wildflowers

fl.   p 100

Darwinia fascicularis

Blombery, Alec M.
What Wildflower is That

fl.   p 110, Pl. 275

Darwinia fascicularis

Macoby, Stirling
What Flower is That

fl.   P 106

Darwinia fascicularis

Mullins, Barbara
Australian Wildflowers in Colour

fl.   P. 65   Pl. 59

Darwinia leiostyla

Blombery, Alec M.
What Wildflower is That

fl.   p 111, Pl. 276

Darwinia leiostyla

Morcombe, M. K.
Australia's Western Wildflowers

fl.   p 98

Darwinia macrostegia

Morcombe, M. K.
Australia's Western Wildflowers

fl.   p 17

Darwinia meeboldii

Morcombe, M. K.
Australia's Western Wildflowers

fl.   p 96

Darwinia meeboldii

Morley, Brian D.
Wild Flowers of the World

fl.   Pl. 137

Darwinia nieldiana

Morcombe, M. K.
Australia's Western Wildflowers

fl.   p 81

Darwinia oldfieldii

Blombery, Alec M.
What Wildflower is That

fl.   p 111, Pl. 277

Daaistoma macrophylla

Wharton, Mary E.
A Guide to the Wildflowers & Ferns
    of Kentucky

fl.   p 214   Pl. 3.7

Dasylirion leiophyllum

Lamb, Edgar
Colorful Cacti of the American Deserts

hab.   Pl. 17

Dasypogon bromaliaefolius

Blombery, Alec M.
What Wildflower is That

fl. hab.   p 111, Pl. 278

Dasypogon, Pineapple-leaved

see

Dasypogon bromaliaefolius

Dasystachys campanulata

Moriarty, Audrey
Wild Flowers of Malawi

fl.   Pl. 12; 1

Date

see

Phoenix dactylifera

Datillo

see

Yucca valida

Datura alba

Kimura, Koiti
Japanese Medicinal Plants, Vol I

fl.   p 88

Datura alba

Tsukamoto, Yotaro
Coloured Illustrations of Garden
        Flowers   Vol. 10

fl. Illus. 37   opp. p. 12

Datura arborea

Blunt, Wilfrid
Flora Superba

fl.   frontispiece, Pl. 1

Datura arborea

Kiaer, Eigil
Indoor Plants in Colour

fl.   p. 49

Datura arborea

Massachusetts Hort. Soc.
Horticulture
  vol 46, No. 8   1968

fl.   cover, backcover

Datura aurea

Curtis's Botanical Magazine
  vol 176   1966-68

fl.   Pl. 484

Datura x candida

Hargreaves, Dorothy
Tropical Blossoms of the Caribbean

fl., hab.   p. 40

Datura x candida

Harrison, R. E.
Trees and Shrubs

fl.   p. 60

Datura x candida

O'Gorman, Helen
Mexican Flowering Trees and Plants

fl., hab.   p. 55

Datura cornigera

Macoby, Stirling
What Flower is That

fl.   P 106

Datura cornigera var.

Hay, Roy
The Dictionary of House Plants

fl.   Pl. 177

Datura cornucopia

American Hort. Soc.
American Horticulturist
    vol 54, No. 4   1975

fl.   p. 5

Datura discolor

Coyle, Jeanette
A Field Guide to the Common and
  Interesting Plants of Baja California

fl.   p 157

Datura innoxia

Moriarty, Audrey
Wild Flowers of Malawi

fl. fr.   Pl. 41; 2, 2a

Datura metel

Crockett, James Underwood
Annuals

fl.   P 113

Datura metel

Polunin, Oleg
Flowers of Europe

fl.   Pl. 118 #1186

Datura metel

Polunin, Oleg
Flowers of the Mediterranean

fl.   No. 168

Datura meteloides

Orr, Robert T.
Wildflowers of Western America

fl.   Pl. 37

Datura meteloides

Welsh, Stanley L.
Flowers of the Canyon Country

fl.   p 5

Datura rosei

Harrison, R.E.
Trees and Shrubs

fl.   P 60

Datura rosei

Morley, Brian D.
Wild Flowers of the World

fl.   Pl. 183

Datura, Sacred

see

Datura meteloides

Datura sanguinea

Hay, Roy
The Dictionary of House Plants

fl.   Pl. 178

Datura sanguinea

Kiaer, Eigil
Indoor Plants in Colour

fl.   p. 49

Datura sanguinea

Royal Hort. Soc.
The Garden
    vol 100, No. 11   1975

fl.   p. 525

Datura stramonium

Bianchini, Francesco
Health Plants of the World

fl. fr. hab.   p 119

Datura stramonium

Brown, Clair A.
Wildflowers of Louisiana and
    Adjoining States

fl. hab.   p 163

Datura stramonium

Color Treasury of Herbs & Other
Medicinal Plants

fl. fr.   p 49 Pl. 71

Datura stramonium

Courtenay, Booth
Wildflowers & Weeds

fl., fr.   p. 91

Datura stramonium

Duncan, Wilbur H.
Wildflowers of the Southeastern
United States

fl.   p 167

Datura stramonium

Felsko, Elsa
A Book of Wildflowers

fl.fr.   P 157

Datura stramonium

Hvass, Elsie
Plants That Feed and Serve Us

fl.   fr.   p 87, Pl. 185

Datura stramonium

Kariyone, Tatsuo
Atlas of Medicinal Plants

fl., fr.   Pl. 42

Datura stramonium

Kleijn, H.
Beauty of the Wild Plant

fl. fr.   opp. p. 39   pl. 44

Datura stramonium

Klimas, John E.
Wildfowers of Eastern America

fl.   Pl. 263

Datura stramonium

Milne, Lorus
Living Plants of the World

fl. hab.   p 214

Datura stramonium

Perrot, Emile
Les Plantes Medicinales

fl. fr. hab.   p 85

Datura stramonium

Polunin, Oleg
Flowers of Europe

fl.   Pl. 118 #1186

Datura stramonium

Weeds of the Southern United States

fl., fr., hab.   p. 40

Datura stramonium **var.**

Wharton, Mary E.
A Guide to the Wildflowers & Ferns
of Kentucky

fl. p 138   Pl. 1.87

Datura suaveolens

Hay, Roy
The Color Dictionary of Flowers & Plants

fl.   p 62, Pl. 495

Datura suaveolens

Hay, Roy
The Dictionary of House Plants

fl.   Pl. 179

Datura suaveolens

Macoby, Stirling
What Flower is That

fl. P 106

Datura suaveolens

Nicolaisen, Age
Pocket Encyclopedia of Indoor Plants

fl.   Pl. 116

Datura suaveolens

Perry, Frances
Flowers of the World

fl. p 283

Datura tatula

Kariyone, Tatsuo
Atlas of Medicinal Plants

fl., fr.   Pl. 42

Datura tatula

Takatori, Jisuke
Color Atlas of Medicinal Plants of Japan
fl.
fr.   Fig. 13A

Daubentonia punicea

Brown, Clair A.
Wildflowers of Louisiana and
Adjoining States

fl. hab.   p 77

Daubentonia texana

Brown, Clair A.
Wildflowers of Louisiana and
Adjoining States

fl. hab.   p 78

Daubenya aurea

Curtis's Botanical Magazine
Vol. 180   1974

fl.   Pl. 700

Daucus carota

Ary, S.
The Oxford Book of Wildflowers

fl.   p 90, Pl. 1

Daucus carota

Batson, Wade T.
Wild Flowers in South Carolina

fl.   p 80

Daucus carota

Bianchini, F.
The Complete Book of Fruits & Vegetables

hab.   p 105

Daucus carota

Campbell, Carlos C.
Great Smoky Mountain Wildflowers

fl.   p 99

Daucus carota

Clark, Lewis J.
Wild Flowers of British Columbia

fl.   p 342

Daucus carota

Courtenay, Booth
Wildflowers & Weeds

fl.   p 58

Daucus carota

Dean, Blanche E.
Wildflowers of Alabama and
Adjoining States

fl.   p 123

Daucus carota

Duncan, Wilbur H.
Wildflowers of the Southeastern
United States

fl.   p 117

Daucus carota

Hvass, Elsie
Plants That Feed and Serve Us

fl.   hab.   p 22, Pl. 33

Daucus carota

Klimas, John E.
Wildflowers of Eastern America

fl.   Pl. 83

Daucus carota

Lemmon, Robert S.
Wildflowers of North America in
Full Color

fl.   pg. 193   pl. 304

Daucus carota

Lindman, C. A. M.
Nordens Flora, Vol 7

fl. fr. hab.   Pl. 434

Daucus carota

Masefield, G. B.
The Oxford Book of Food Plants

fl.   Pl. 175, 1

Daucus carota

Massachusetts Hort. Soc.
Horticulture
  vol 54, No. 6 1976

fl., fr., hab.   p. 41

Daucus carota

Moyle, John B.
Northland Wild Flowers

fl.   p 101, Pl. 92

Daucus carota

Perrot, Emile
Les Plantes Medicinales

fl. fr. hab.   p 49

Daucus carota

Polunin, Oleg
Flowers of Europe

fl. Pl. 87 #910

Daucus carota

Pond, Barbara
A Sampler of Wayside Herbs

fl.   Pl. XX

Daucus carota

Weeds of the Southern United States

fl., hab.   p. 41

Daucus carota

Wharton, Mary E.
A Guide to the Wildflowers & Ferns
  of Kentucky

fl. fr. p 161 Pl. 2.28

Daucus carota var.

Martin, W. Keble
The Concise British Flora in Colour

fl. fr. hab.   Pl. 40

Davidia involucrata

American Hort. Soc.
American Horticulturist
  vol 57, No. 1 1978

fl., hab.   p. 6-7

Davidia involucrata

Boom, B. K.
The glory of the tree

fl.   opp. p. 88   pl. 148

Davidia involucrata

Edlin, Herbert
The Illustrated Encyclopedia
  of Trees

fl. hab.   p 196

Davidia involucrata

Everett, T.H., ed.
New Illustrated Encyclopedia of
  Gardening   vol 4

fl.   Frontispiece

Davidia involucrata

Hay, Roy
The Color Dictionary of Flowers & Plants

fl.   p 195, Pl. 1558

Davidia involucrata

Huxley, Anthony
Deciduous Garden Trees and Shrubs

fl. Pl. 64

Davidia involucrata

Perry, Frances
Flowers of the World

fr. p 100

David's Harp

see

Polygonatum multiflorum

Daviesia acicularis

Blombery, Alec M.
What Wildflower is That

fl.   p 112, Pl. 279

Daviesia alata

Blombery, Alec M.
What Wildflower is That

fl.   p 112, Pl. 280

Daviesia brevifolia

Cochrane, G. R.
Flowers and Plants of Victoria

fl.   Pl. 119

Daviesia corymbosa

Blombery, Alec M.
What Wildflower is That

fl.   p 112, Pl. 281

Daviesia genistifolia

Cochrane, G. R.
Flowers and Plants of Victoria

fl.   Pl. 104

Daviesia latifolia

Blombery, Alec M.
What Wildflower is That

fl.   p 112, Pl. 282

Daviesia latifolia

Cochrane, G. R.
Flowers and Plants of Victoria

fl.   Pl. 400

Dayflower

see

Commelina

Dayflower, Asiatic

see

Commelina communis

Day Flower, Dark Blue

see

Commelina debilis

Daylily

see

Hemerocallis

Day-Lily, Common

see

Hemerocallis fulva

Daylily, Orange

see

Hemerocallis fulva

Deamia testudo

Backeberg, Curt
Cactus Lexicon

fl.   p 586

Debregeasia edulis

Kitamura, Siro
Coloured Illustrations of Trees &
Shrubs

hab.   148

Debregeasia longifolia

Wit, H. C. D. de
Plants of the World:
  The Higher Plants, Vol I

hab.   p 105, Pl. 56

Decabelone grandiflora

Morley, Brian D.
Wild Flowers of the World

fl.,hab.   Pl. 78

Decabelone meintjesii

Flowering Plants of Africa
vol 36   1963-64

fl., hab.   Pl. 1420

Decaisnea fargesii

Bartels, Andreas
Das Grosse Buch der Gartengeholze

fr.   p 143

Decaisnea fargesii

Gault, S. Millar
The Color Dictionary of Shrubs

fr.   Pl. 151

Decaisnea fargesii

Hay, Roy
The Color Dictionary of Flowers & Plants

fr.   p 195, Pl. 1559

Decaisnea fargesii

Hellyer, A.G.L.
Shrubs in Colour

fl., fr.   p. 43

Decaisnea fargesii

Huxley, Anthony
Deciduous Garden Trees and Shrubs

fr.   Pl. 65

Decaisnea fargesii

Perry, Frances
Flowers of the World

fr.   p 156

Decaisnea fargesii

Royal Hort. Soc.
Journal of the Royal Hort. Soc.
vol 92, No. 1   1967

fr.   p. 22   Pl. 24

Decaspermum fruticosum

Parham, J.W.
Plants of the Fiji Islands

fl.   p. 196

Decaspermum rubrum

Bruggeman, L.
Tropical Plants

fl.   Pl. 235

Decodon verticillatus

Courtenay, Booth
Wildflowers & Weeds

fl.   p 26

Decodon verticillatus

Wharton, Mary E.
Trees & Shrubs of Kentucky

fl.   p 53, Pl. 1.13

Decumaria barbara

Batson, Wade T.
Wild Flowers in South Carolina

fl.   p 53

Decumaria barbara

Menninger, Edwin A.
Flowering Vines of the World

fl.   Pl. 187

Decumaria sinensis

Curtis's Botanical Magazine
Vol 159   1936

fl.   No. 9429

Deerberry

see

Maianthemum dilitatum

Deerberry

see

Polycodium stamineum

Deerberry

see

Vaccinium stamineum

Deer Brush

see

Ceanothus integerrimus

Deergrass

see

Rhexia sp.

Deer Tongue

see

Frasera speciosa

Deer Tongue

see

Trilisa odoratissima

Deerweed

see

Lotus scoparius

Deherainia smaragdina

Morley, Brian D.
Wild Flowers of the World

fl.   pl. 178

Deinanthe caerulea

Morley, Brian D.
Wild Flowers of the World

fl.   Plate 99

Delaetia woutersiana

Backeberg, Curt
Cactus Lexicon

fl.   p 586

Delight, Single

see

Moneses uniflora

Delonix regia

Bruggeman, L.
Tropical Plants

fl.   Pl. 206

Delonix regia

Edlin, Herbert
The Illustrated Encyclopedia
of Trees

fl. hab.   224-5

Delonix regia

Everett, Thomas H.
Living Trees of the World

fl. hab.   p. 189 top

Delonix regia

Everett, T.H., ed.
New Illustrated Encyclopedia of
Gardening vol. 4

fl.   frontispiece

Delonix regia

Hall, Clarence E.
Flowers of the Islands in the Sun

fl.   P. 59   Pl. 11

Delonix regia

Hargreaves, Dorothy
Tropical Blossoms of the Caribbean

fl.   p. 55

Delonix regia

Macoby, Stirling
What Flower is That

fl.   P 107

Delonix regia

Morley, Brian D.
Wild Flowers of the World

fl.fr.   Plate 113

Delonix regia

Oakman, Harry
Colorful Trees

Hab. Fl.   P 72

Delonix regia

O'Gorman, Helen
Mexican Flowering Trees and Plants

fl.   p 23

Delonix regia

Perry, Frances
Flowers of the World

fl.   p 158

Delonix regia

Pertchik, Bernard
Flowering Trees of the
    Caribbean

fl.   p 95

Delonix regia

Royal Hort. Soc.
Journal of the Royal Hort. Soc.
    vol 95, No. 11   1970

fl.   p. 488   Pl. 226

Delosperma ashtonii

Flowering Plants of Africa
    vol XXVI   1947

fl.   Pl. 1023

Delosperma jansei

Flowering Plants of Africa
    vol XXVI   1947

fl.   Pl. 1024

Delosperma tradescantioides

Herre, H.
The Genera of the Mesembryanthemaceae

fl.   fr.   p 127

Delphinium ajacis

Crockett, James Underwood
Annuals

fl.   P 109

Delphinium ajacis

Macoby, Stirling
What Flower is That

fl.   P 107

Delphinium ajacis

Tsukamoto, Yotaro
Coloured Illustrations of Garden
    Flowers   Vol. 10

fl. Illus 49   opp. p. 16

Delphinium ajacis

Wharton, Mary E.
A Guide to the Wildflowers & Ferns
    of Kentucky

fl.   p 225   Pl. 3.30

Delphinium ajacis var.

Hay, Roy
The Color Dictionary of Flowers &
    Plants

fl.   p. 36   Pl. 282

Delphinium ambiguum

Martin, W. Keble
The Concise British Flora in Colour

fl. fr.   Pl. 4

Delphinium x belladona var.

Huxley, Anthony
Garden Perennials and Water Plants

fl.   Pl. 92

Delphinium bicolor

Porsild, A.E.
Rocky Mountain Wild Flowers

fl. hab.   p 179

Delphinium brachycentrum

Alaska-Yukon Wild Flowers Guide

fl.   p 39

Delphinium brachycentrum

Heller, Christine
Wild Flowers of Alaska

fl.   Pl. 230

Delphinium brunonianum

Hay, Roy
The Color Dictionary of Flowers & Plants

fl.   p 136, Pl. 1084

Delphinium, butterfly

see

Delphinium grandiflorum

Delphinium cardinale

Munz, Philip A.
California Spring Wildflowers

fl.   p 39, Pl. 39

Delphinium carolinianum

Brown, Clair A.
Wildflowers of Louisiana and
    Adjoining States

fl.   P 53

Delphinium, Chinese

see

Delphinium grandiflorum

Delphinium consolida

Barneby, T. P.
European Alpine Flowers in Colour

fl.   Pl. 22, 4

Delphinium consolida

Felsko, Elsa
A Book of Wildflowers

fl.   P 50

Delphinium consolida

Lindman, C. A. M.
Nordens Flora, Vol 4

fl. fr.   Pl. 223

Delphinium consolida var.

Curtis's Botanical Magazine
    Vol 159   1936

fl. fr.   No. 9435

Delphinium consolida var.

Huxley, Anthony
Garden Annuals and Bulbs

fl.   Pl. 38

Delphinium x cultorum

Tsukamoto, Yotaro
Coloured Illustrations of Garden
    Flowers   Vol. 9

fl.   opp. p. 41

Delphinium dasyanthum

Royal Hort. Soc.
Journal of the Royal Hort. Soc.
    vol 93, No. 3   1968

fl.   p. 122   Pl. 59

Delphinium dasycaulon

Moriarty, Audrey
Wild Flowers of Malawi

fl.   Pl. 33; 2

Delphinium delavayi

Curtis's Botanical Magazine
Vol 166   1949

fl.   68

Delphinium depauperatum

Walcott, Mary Vaux
North American Wild Flowers
        vol. 5
fl.   Pl. 384

Delphinium elatum

Kohlhaupt, Paula
Fleurs des Alpages
                vol 2
fl.   P 5

Delphinium elatum

Polunin, Oleg
Flowers of Europe

fl.   Pl. 20 #211

Delphinium elatum var.

Huxley, Anthony
Garden Perennials and Water Plants

fl.   Pl. 93

Delphinium elatum var.

Macoby, Stirling
What Flower is That

fl.   p. 107

Delphinium elongatum

Walcott, Mary Vuax
North American Wild Flowers

fl.   Pl. 27

Delphinium exaltum

Dean, Blanche E.
Wildflowers of Alabama and
        Adjoining States

fl.   p 59

Delphinium glaucum

Clark, Lewis J.
Wild Flowers of British Columbia

fl.   p 163

Delphinium glaucum

Heller, Christine
Wild Flowers of Alaska

fl.   Pl. 231

Delphinium glaucum

Runz, Philip A.
California Mountain Wildflowers

fl.   p 47, Pl. 63

Delphinium glaucum

Porsild, A.E.
Rocky Mountain Wild Flowers

fl.   p 181

Delphinium grandiflorum

Macoby, Stirling
What Flower is That

fl.   P 107

Delphinium grandiflorum

Tsukamoto, Yotaro
Coloured Illustrations of Garden
        Flowers   Vol. 10

fl. Illus. 50   opp. p. 17

Delphinium halteratum

Polunin, Oleg
Flowers of the Mediterranean

fl.   No. 20

Delphinium hesperium

Orr, Robert T.
Wildflowers of Western America

fl.   Pl. 265

Delphinium huetianum

Curtis's Botanical Magazine
Vol. 177   1969-70

fl.   566

Delphinium hybrid

Hay, Roy
The Color Dictionary of Flowers & Plants

fl.   p 136, 137; Pl. 1085 thru 1091

Delphinium hybrid

Huxley, Anthony
Garden Perennials and Water Plants

hab.   Pl. 6

Delphinium hybrid

Perry, Frances
Flowers of the World

fl.   p. 249

Delphinium lacostei

Morley, Brian D.
Wild Flowers of the World

Fl.   Plate 91

Delphinium leroyi

Moriarty, Audrey
Wild Flowers of Malawi

fl.   Pl. 33; 1

Delphinium macrocentron

Flowering Plants of Africa
        vol 35   1962

fl.   Pl. 1377

Delphinium macrocentron

Morley, Brian D.
Wild Flowers of the World

fl.,fr.   Pl. 52

Delphinium x magnificum

Curtis's Botanical Magazine
        Vol. 181   1976-77

fl.   Pl. 705

Delphinium menziesii

Clark, Lewis J.
Wild Flowers of British Columbia

fl.   p 166

Delphinium menziesii

Lemmon, Robert S.
Wildflowers of North America in
        Full Color

fl.   pg. 125   pl. 198

Delphinium menziesii

Taylor, Ronald J.
Mountain Wild Flowers

fl.   p 113

Delphinium menziesii

Welsh, Stanley L.
Flowers of the Mountain Country

fl.   p. 76

Delphinium muscosum

Royal Hort. Soc.
Journal of the Royal Hort. Soc.
        vol 87, No. 1   1962

fl.   p. 22   Pl. 6

Delphinium nudicaule

Huxley, Anthony
Garden Perennials and Water Plants

fl.   Pl. 91

Delphinium nudicaule

Lemmon, Robert S.
Wildflowers of North America in
        Full Color

fl.   pg. 124   pl. 197

Delphinium nudicaule
Orr, Robert T.
Wildflowers of Western America

fl.   Pl. 176

Delphinium nuttallianum
Taylor, Ronald J.
Mountain Wild Flowers

fl.   p 113

Delphinium orientale
Curtis's Botanical Magazine
Vol 169   1952-53

fl.   186

Delphinium parishii
Lemmon, Robert S.
Wildflowers of North America in
Full Color

fl.   pg. 66   pl. 106

Delphinium parryi var.
Munz, Philip A.
California Spring Wildflowers

fl.   p 84, Pl. 77

Delphinium peregrinum
Morley, Brian D.
Wild Flowers of the World

fl.   Pl. 27

Delphinium peregrinum
Polunin, Oleg
Flowers of Europe

fl. Pl. 20 #211

Delphinium polycladon
Orr, Robert T.
Wildflowers of Western America

fl.   Pl. 237

Delphinium sp.
Moriarty, Audrey
Wild Flowers of Malawi

fl.   Pl. 33; 3

Delphinium stapeliosmum
Curtis's Botanical Magazine
vol 174   1962-63

fl.   Pl. 402

Delphinium staphisagria
Goulimis, Constantine N.
Wild Flowers of Greece

fl.   p 21

Delphinium staphisagria
Perrot, Emile
Les Plantes Medicinales

fl. fr. hab.   p 86

Delphinium tenuisectum
Orr, Robert T.
Wildflowers of Western America

fl.   Pl. 249

Delphinium tricorne
Courtenay, Booth
Wildflowers & Weeds

fl.   p 31

Delphinium tricorne
Dean, Blanche E.
Wildflowers of Alabama and
Adjoining States

fl.   p 59

Delphinium tricorne
Duncan, Wilbur H.
Wildflowers of the Southeastern
United States

fl. hab.   p 43

Delphinium tricorne
Jennings, O. E.
Wild Flowers of Western Pennsylvania
vol II

fl.   Pl. 62

Delphinium tricorne
Klimas, John E.
Wildflowers of Eastern America

fl.   Pl. 243

Delphinium tricorne
Roberts, June Carver
Born in the Spring

fl. fr. hab.   p 73

Delphinium tricorne
Wharton, Mary E.
A Guide to the Wildflowers & Ferns
of Kentucky

fl. p 225 Pl. 3.29

Delphinium virescens
Moyle, John B.
Northland Wild Flowers

fl. hab.   p 53, Pl. 13

Delphinium wellbyi
Royal Hort. Soc.
The Garden
vol 101, No. 7 1976

fl.   p. 359

Dendriopoterium menendezii
Bramwell, David
Wild Flowers of the Canary Islands

fl.   Pl. 162

Dendrobium aemulum
Blombery, Alec M.
What Wildflower is That

fl.   p 113, Pl. 283

Dendrobium aggregatum
Kamemoto, Haruyuki
Beautiful Thai Orchid Species

fl.   p 129

Dendrobium aggregatum var.
Kramer, Jack
Orchids; Flowers of Romance
and Mystery

fl.   p 126

Dendrobium agrostophyllum
Bedford, Roger B.
A Guide to Native Australian Orchids

fl., hab.   p. 62   Pl. II

Dendrobium amoenum
Kimber, Sheila
A Handbook of Orchids

fl.   p. 28

Dendrobium anosmum
Morley, Brian D.
Wild Flowers of the World

fl. Plate 127

Dendrobium aphrodite
Kamemoto, Haruyuki
Beautiful Thai Orchid Species

fl.   p 133

Dendrobium aphyllumb
Morley, Brian D.
Wild Flowers of the World

fl. Plate 109

Dendrobium aureum
Hay, Roy
The Dictionary of House Plants

fl.   Pl. 180

Dendrobium beckleri
Blombery, Alec M.
What Wildflower is That

fl.   p 113, Pl. 284

Dendrobium bullatulum

Kamemoto, Haruyuki
Beautiful Thai Orchid Species

fl.   p 145

Dendrobium bigibbum
.
Blombery, Alec M.
What Wildflower is That

fl.   p 113, Pl. 285

Dendrobium bigibbum

Blunt, Wilfrid
Flora Superba

fl., hab.   Pl. IX

Dendrobium bigibbum

Macoby, Stirling
What Flower is That

fl.   P 108

Dendrobium bigibbum

Mullins, Barbara
Australian Wildflowers in Colour

fl.   P. 109   Pl. 107

Dendrobium bigibbum

Royal Hort. Soc.
Journal of the Royal Hort. Soc.
   vol 95, No. 11   1970

fl.   p. 488   Pl. 229

Dendrobium bracteosum

Curtis's Botanical Magazine
      vol 173   1960

fl.   Pl. 389

Dendrobium caespitificum

Royal Hort. Soc.
Journal of the Royal Hort. Soc.
   vol 92, No. 3   1967

fl.   p. 116   Pl. 61

Dendrobium canaliculatum

Bedford, Roger B.
A Guide to Native Australian Orchids

fl., hab.   p. 62   Pl. III

Dendrobium canaliculatum

Blombery, Alec M.
What Wildflower is That

fl. hab.   p 113, 114, Pl. 286, 287

Dendrobium canaliculatum

Cady, Leo
Australian Native Orchids in Colour

fl.   Pl. 78

Dendrobium capillipes

Kamemoto, Haruyuki
Beautiful Thai Orchid Species

fl.   p 134

Dendrobium cariniferum

Kamemoto, Haruyuki
Beautiful Thai Orchid Species

fl.   p 144

Dendrobium chrysanthum

Kamemoto, Haruyuki
Beautiful Thai Orchid Species

fl.   p 133

Dendrobium chrysotoxum

Addisonia
      vol 21   1939-42

fl.   p. 43   Pl. 694

Dendrobium chrysotoxum

Kamemoto, Haruyuki
Beautiful Thai Orchid Species

fl.   p 129

Dendrobium crassinode

Kamemoto, Haruyuki
Beautiful Thai Orchid Species

fl.   p 134

Dendrobium crepidatum

Kamemoto, Haruyuki
Beautiful Thai Orchid Species

fl.   p 133

Dendrobium crispatum

Parham, J.W.
Plants of the Fiji Islands

fl., hab.   p. 381

Dendrobium cruentum

Kamemoto, Haruyuki
Beautiful Thai Orchid Species

fl.   p 146

Dendrobium crystallinum

Kamemoto, Haruyuki
Beautiful Thai Orchid Species

fl.   p 134

Dendrobium cucumerinum

Bedford, Roger B.
A Guide to Native Australian Orchids

fl., hab.   p. 62   Pl. IV

Dendrobium cucumerinum

Blombery, Alec M.
What Wildflower is That

fl.   p 115, Pl. 288

Dendrobium delacourii

Kamemoto, Haruyuki
Beautiful Thai Orchid Species

fl.   p 147

Dendrobium delicatum

Blombery, Alec M.
What Wildflower is That

fl.   p 115, Pl. 289

Dendrobium delicatum var.

Grubb, Roy
Selected Orchidaceous Plants
   Vol I

fl.   p 139

Dendrobium densiflorum

Ebel, Friedrich
The Strange and Beautiful
   World of Orchids

fl.   p 41

Dendrobium densiflorum

Hara, Hiroshi, comp.
Photo-Album of Plants of
   Eastern Himalaya

fl.   Pl. 118, 119

Dendrobium densiflorum

Kamemoto, Haruyuki
Beautiful Thai Orchid Species

fl.   p 131

Dendrobium densiflorum

Kramer, Jack
Orchids; Flowers of Romance
   and Mystery

fl.   p 127

Dendrobium densiflorum

Morley, Brian D.
Wild Flowers of the World

fl.   Plate 109

Dendrobium devonianum

Kamemoto, Haruyuki
Beautiful Thai Orchid Species

fl.   p 135

Dendrobium dicuphum

Cady, Leo
Australian Native Orchids in Colour

fl. hab.   Pl. 81

Dendrobium discolor

Blombery, Alec M.
What Wildflower is That

fl.   p 115, Pl. 290

Dendrobium discolor

Morley, Brian D.
Wild Flowers of the World

fl.   Pl. 145

Dendrobium dixanthum

Kamemoto, Haruyuki
Beautiful Thai Orchid Species

fl.   p 135

Dendrobium draconis

Kamemoto, Haruyuki
Beautiful Thai Orchid Species

fl.   p 144

Dendrobium falconeri

Kamemoto, Haruyuki
Beautiful Thai Orchid Species

fl.   p 135

Dendrobium falcorostrum

Blombery, Alec M.
What Wildflower is That

fl.   p 115, Pl. 291

Dendrobium fantasticum

Curtis's Botanical Magazine
       Vol. 181   1976-77

fl.   Pl. 727

Dendrobium farmeri

Kamemoto, Haruyuki
Beautiful Thai Orchid Species

fl.   p 130

Dendrobium farmeri var.

Kamemoto, Haruyuki
Beautiful Thai Orchid Species

fl.   p 132

Dendrobium fimbriatum

Ebel, Friedrich
The Strange and Beautiful
    World of Orchids

fl.   p 133

Dendrobium fimbriatum

Kamemoto, Haruyuki
Beautiful Thai Orchid Species

fl.   p 137

Dendrobium fimbriatum

Milne, Lorus
Living Plants of the World

fl.   p 327

Dendrobium fimbriatum var.

Kamemoto, Haruyuki
Beautiful Thai Orchid Species

fl.   p 137

Dendrobium fimbriatum var.

Kimber, Sheila
A Handbook of Orchids

fl.   p. 30

Dendrobium fimbriatum var.

Royal Hort. Soc.
The Garden
       vol 102, No. 3   1977

fl.   p. 104

Dendrobium findlayanum

Kamemoto, Haruyuki
Beautiful Thai Orchid Species

fl.   p 136

Dendrobium formosum

Kimber, Sheila
A Handbook of Orchids

fl.   p. 32

Dendrobium formosum var.

Kamemoto, Haruyuki
Beautiful Thai Orchid Species

fl.   p 142

Dendrobium friedericksianum

Kamemoto, Haruyuki
Beautiful Tahi Orchid Species

fl.   p 136

Dendrobium, golden

see

Dendrobium thyrsiflorum

Dendrobium, golden-arched

see

Dendrobium chrysotoxum

Dendrobium hercoglossum

Curtis's Botancial Magazine
     Vol 159   1936

fl. hab.   No. 9428

Dendrobium hercoglossum

Hu, Shiu-ying
The Genera of Orchidaceae in
      Hong Kong

fl.   p. 122

Dendrobium hercoglossum

Kamemoto, Haruyuki
Beautiful Thai Orchid Species

fl.   p 137

Dendrobium heterocarpum

Kamemoto, Haruyuki
Beautiful Thai Orchid Species

fl.   p 138

Dendrobium heterocarpum

Kimber, Sheila
A Handbook of Orchids

fl.   p. 34

Dendrobium hybrid

Encke, Fritz
Zimmerpflanzen

fl., hab.   p. 104

Dendrobium hybrid

Hay, Roy
The Color Dictionary of Flowers & Plants

fl.   p 62, 63, Pl. 496,
                       497, 498

Dendrobium hybrid

Kramer, Jack
Orchids; Flowers of Romance
     and Mystery

fl.   p 131

Dendrobium infundibulum

Kamemoto, Haruyuki
Beautiful Thai Orchid Species

fl.   p 142

Dendrobium johnsoniae

Curtis's Botanical Magazine
Vol 177   1969-70

fl.   560

Dendrobium johnsoniae

Royal Hort. Soc.
Journal of the Royal Hort. Soc.
     vol 87, No. 2   1962

fl., hab.   p. 86   Pl. 23

Dendrobium kingianum

Bedford, Roger B.
A Guide to Native Australian Orchids

fl., hab.   p. 62   Pl. V

Dendrobium kingianum

Blombery, Alec M.
What Wildflower is That

fl.   p 115, Pl. 292

Dendrobium lancifolium

Bruggeman, L.
Tropical Plants

fl.   Pl. 105

Dendrobium lawesii

Curtis's Botanical Magazine
    vol 176   1966-68

fl.   Pl. 524

Dendrobium linguiforme

Blombery, Alec M.
What Wildflower is That

fl.   p 115, Pl. 293

Dendrobium linguiforme

Macoby, Stirling
What Flower is That

fl.   P 108

Dendrobium lituiflorum

Kamemoto, Haruyuki
Beautiful Thai Orchid Species

fl.   p 138

Dendrobium loddigesii

Hu, Shiu-ying
The Genera of Orchidaceae in
    Hong Kong

fl.   p 122

Dendrobium loddigesii

Milne, Lorus
Living Plants of the World

fl.   p. 327

Dendrobium loddigesii

Walden, Beryl M.
Wild Flowers of Hong Kong

fl.   Pl. 24(70)

Dendrobium margaritaceum

Kamemoto, Haruyuki
Beautiful Thai Orchid Species

fl.   p 145

Dendrobium moniliforme

Kimura, Koiti
Japanese Medicinal Plants, Vol I

fl.   p 110

Dendrobium monophyllum

Blombery, Alec M.
What Wildflower is That

fl.   p 116, Pl. 294

Dendrobium monophyllum

Cady, Leo
Australian Native Orchids in Colour

fl.   Pl. 77

Dendrobium moschatum

Kamemoto, Haruyuki
Beautiful Thai Orchid Species

fl.   p 138

Dendrobium moschatum

Kramer, Jack
Orchids: Flowers of Romance
    and Mystery

fl.   p 134

Dendrobium nobile

Ebel, Friedrich
The Strange and Beuatiful
    World of Orchids

fl.   p 75

Dendrobium nobile

Kamemoto, Haruyuki
Beautiful Thai Orchid Species

fl.   p 141

Dendrobium nobile

Kariyone, Tatsuo
Atlas of Medicinal Plants

fl.   Pl. 2

Dendrobium nobile

Kimber, Sheila
A Handbook of Orchids

fl.   p. 36

Dendrobium nobile

Kramer, Jack
Orchids: Flowers of Romance
    and Mystery

fl.   p 135

Dendrobium nobile

Macoby, Stirling
What Flower is That

fl.   P 108

Dendrobium nobile

Perry, Frances
Flowers of the World

fl.   titlepage

Dendrobium nobile var.

Hay, Roy
The Dictionary of House Plants

fl.   Pl. 181

Dendrobium parishii

Kamemoto, Haruyuki
Beautiful Thai Orchid Species

fl.   p 140

Dendrobium phalaenopsis

Kramer, Jack
Orchids: Flowers of Romance
    and Mystery

fl.   p 130

Dendrobium phalaenopsis

Macoby, Stirling
What Flower is That

fl.   pl 108

Dendrobium phalaenopsis var.

Everett, T.H., ed.
New Illustrated Encyclopedia of
    Gardening   vol 4

fl.   p. 551

Dendrobium pierardii

Kamemoto, Haruyuki
Beautiful Thai Orchid Species

fl.   p 139

Dendrobium pierardii

Kimber, Sheila
A Handbook of Orchids

fl.   p. 38

Dendrobium pierardii

Macoby, Stirling
What Flower is That

fl.   P 108

Dendrobium primulinum

Kamemoto, Haruyuki
Beautiful Thai Orchid Species

fl.   p 139

Dendrobium pulchellum

Kamemoto, Haruyuki
Beautiful Thai Orchid Species

fl.   p 140

Dendrobium pulchellum

Kupper, Walter
Orchids

fl.   p 43

Dendrobium scabrilingue

Kamemoto, Haruyuki
Beautiful Thai Orchid Species

fl.   p 143

Dendrobium secundum

Kamemoto, Haruyuki
Beautiful Thai Orchid Species

fl.   p 147

Dendrobium senile

Kamemoto, Haruyuki
Beautiful Thai Orchid Species

fl.   p 141

Dendrobium speciosum

Blombery, Alec M.
What Wildflower is That

fl.   p 117, Pl. 295

Dendrobium speciosum

Cady, Leo
Australian Native Orchids in Colour

fl.   Pl. 80

Dendrobium speciosum

Macoby, Stirling
What Flower is That

fl.   P 108

Dendrobium speciosum

Mullins, Barbara
Australian Wildflowers in Colour

fl.   P. 109   Pl. 108

Dendrobium speciosum var.

Blombery, Alec M.
What Wildflower is That

fl.   p 117, Pl. 296

Dendrobium stratiotes

Curtis's Botanical Magazine
        vol 174   1962-63

fl.   Pl. 436

Dendrobium striolatum

Cady, Leo
Australian Native Orchids in Colour

fl. hab.   Pl. 79

Dendrobium striolatum

Cochrane, G. R.
Flowers and Plants of Victoria

fl.   Pl. 208

Dendrobium x superbiens var.

Kimber, Sheila
A Handbook of Orchids

fl.   p. 40-42

Dendrobium superbum

Kramer, Jack
Orchids; Flowers of Romance
    and Mystery

fl.   p 138

Dendrobium superbum var.

Kramer, Jack
Orchids; Flowers of Romance
    and Mystery

fl.   p 139

Dendrobium sutepense

Kamemoto, Haruyuki
Beautiful Thai Orchid Species

fl.   p 143

Dendrobium teretifolium

Blombery, Alec M.
What Wildflower is That

fl.   p 117, Pl. 297

Dendrobium tetragonum

Blombery, Alec M.
What Wildflower is That

fl.   p 117, Pl. 298

Dendrobium tetragonum

Grubb, Roy
Selected Orchidaceous Plants
    Vol I

fl.   p 25

Dendrobium tetragonum var.

Bedford, Roger B.
A Guide to Native Australian Orchids

fl., hab.   p. 62   Pl. VI

Dendrobium thyrsiflorum

Everett, T.H., ed.
New Illustrated Encyclopedia of
    Gardening   vol 4

fl.   p. 583

Dendrobium thyrsiflorum

Kamemoto, Haruyuki
Beautiful Thai Orchid Species

fl.   p 131

Dendrobium thyrsiflorum

Kupper, Walter
Orchids

fl. hab.   p 45

Dendrobium thyrsiflorum

Macoby, Stirling
What Flower is That

fl.   P 109

Dendrobium tortile

Kamemoto, Haruyuki
Beautiful Thai Orchid Species

fl.   p 141

Dendrobium tosaense

Kimura, Koiti
Japanese Medicinal Plants, Vol II

fl. hab.   p 149

Dendrobium trigonopus

Kamemoto, Haruyuki
Beautiful Thai Orchid Species

fl.   p 146

Dendrobium unicum

Curtis's Botanical Magazine
Vol 179   1972

fl.   616

Dendrobium victoriae-reginae

Grubb, Roy
Selected Orchidaceous Plants
    Vol I

fl.   p 71

Dendrobium wardianum

Kamemoto, Haruyuki
Beautiful Thai Orchid Species

fl.   p 139

Dendrocereus nudiflorus

Backeberg, Curt
Cactus Lexicon

fl.   p 587

Dendrochilum cobbianum

Addisonia
        vol 22   1943-46

fl.   p. 25   Pl. 717

Dendrochilum cobbianum

Everett, T.H., ed.
New Illustrated Encyclopedia of
    Gardening   vol 4

fl.   p. 551

Dendromecon harfordii

Lemmon, Robert S.
Wildflowers of North America in
    Full Color

fl.   pg. 21   pl. 33

Dendromecon harfordii var.

Royal Hort. Soc.
Journal of the Royal Hort. Soc.
Vol. 90, No. 11   1965

fl.   p. 470   Pl. 231

Dendromecon rigida

Gault, S. Millar
The Color Dictionary of Shrubs

fl.   Pl. 152

Dendromecon rigida

Hay, Roy
The Color Dictionary of Flowers & Plants

fl.   p 195, Pl. 1560

Dendromecon rigida

Macoby, Stirling
What Flower is That

fl.   P 109

Dendromecon rigida

Mathias, Mildred E.
Color for the Landscape

fl. hab.   p 180

Dendromecon rigida

Royal Hort. Soc.
The Garden
vol 101, No. 6   1976

fl.   p. 330

Dendromecon rigida

Walcott, Mary Vaux
North American Wild Flowers
vol. 5

fl.   Pl. 392

Dendromecon rigida var.

American Hort. Soc.
American Horticulturist
vol 56, No. 4   1977

hab., fl.   p. 27

Dendropanax trifidus

Kitamura, Siro
Coloured Illustrations of Trees &
Shrubs of Japan

fr.   366

Dendrophthoe vitellina

Cochrane, G. R.
Flowers and Plants of Victoria

fl.   Pl. 477

Dendrophylax barrettiae

Ospina, Mariano
Orquideas de las americas

fl. hab.   Pl. 136

Denmoza erythrocephala

Lamb, Edgar
Popular Exotic Cacti in Color

fl. hab.   p 64

Dentaria diphylla

Campbell, Carlos C.
Great Smoky Mountain Wildflowers

fl.   p 17

Dentaria diphylla

Courtenay, Booth
Wildflowers & Weeds

fl.   p 18

Dentaria diphylla

Duncan, Wilbur H.
Wildflowers of the Southeastern
United States

fl. hab.   p 53

Dentaria diphylla

Jennings, O. E.
Wild Flowers of Western Pennsylvania
vol II
fl. hab.   Pl. 71

Dentaria diphylla

Lemmon, Robert S.
Wildflowers of North America in
Full Color

fl.   p. 256   Pl. 404

Dentaria diphylla

Wharton, Mary E.
A Guide to the Wildflowers & Ferns
of Kentucky

fl.   p 113 Pl. 1.39

Dentaria heterophylla

Wharton, Mary E.
A Guide to the Wildflowers & Ferns
of Kentucky

fl.   p 114   Pl. 1.40

Dentaria integrifolia

Orr, Robert T.
Wildflowers of Western America

fl.   Pl. 39

Dentaria laciniata

Courtenay, Booth
Wildflowers & Weeds

fl.   p 18

Dentaria laciniata

Ferguson, Mary
Wildflowers

fl.   p 177

Dentaria laciniata

Jennings, O. E.
Wild Flowers of Western Pennsylvania
vol II
fl. hab.   Pl. 104

Dentaria laciniata

Klimas, John E.
Wildflowers of Eastern America

fl.   Pl. 14

Dentaria laciniata

Moyle, John B.
Northland Wild Flowers

fl. hab.   p 60, Pl. 26

Dentaria laciniata

Roberts, June Carver
Born in the Spring

fl. hab.   p 31

Dentaria laciniata

Walcott, Mary Vaux
North American Wild Flowers
vol. 4

fl.   Pl. 249

Dentaria laciniata

Wharton, Mary E.
A Guide to the Wildflowers & Ferns
of Kentucky

fl.   p 113   Pl. 1.38

Dentaria multifida

Duncan, Wilbur H.
Wildflowers of the Southeastern
United States

fl.   p 53

Derris elliptica

Hvass, Elsie
Plants That Feed and Serve Us

fl.   fr.   p 99, Pl. 212

Derris elliptica

Kariyone, Tatsuo
Atlas of Medicinal Plants

fl., fr.   Pl. 99

Derris microphylla

Bruggeman, L.
Tropical Plants

fl.   Pl. 190

Deschampsia caespitosa

Lindman, C. A. M.
Nordens Flora, Vol 2

fr. hab.   Pl. 92B

445

Deschampsia flexuosa

Lindman, C. A. M.
Nordens Flora, Vol 2

fr. hab.   Pl. 92A

Deschampsia flexuosa

Polunin, Oleg
Flowers of Europe

fr. hab.   Pl. 180 #1772

Descurainia bourgaeana

Bramwell, David
Wild Flowers of the Canary Islands

fl.   Pl. 135

Descurainia sophia

Ary, S.
The Oxford Book of Wildflowers

fl.   p 10, Pl. 4

Descurainia sophia

Lindman, C. A. M.
nordens Flora, Vol 4

fl. fr. hab.   Pl. 276B

Descurainia sophia

Martin, W. Keble
The Concise British Flora in Colour

fl. fr. hab.   Pl. 8

Desert Candle

see

Dasylirion leiophyllum

Desert Candle

see

Eremurus robustus

Desert Candle

see

Streptanthus inflatus

Desert Plume

see

Stanleya pinnata

Desert Savior

see

Dudleya lanceolata

Desert Spray

see

Cadaba aphylla

Desert Velvet

see

Psathyrotes ramosissima

Desfontainea spinosa

Harrison, R.E.
Trees and Shrubs

fl.   P 60

Desfontainea spinosa

Hay, Roy
The Color Dictionary of Flowers & Plants

fl.   p 196, Pl. 1561

Desfontainea spinosa

Hellyer, A.G.L.
Shrubs in Colour

fl.   p. 43

Desfontainea spinosa

Wit, H. C. D. de
Plants of the World;
The Higher Plants, Vol II

fl.   p 100, Pl. 48

Desfontainea spinosa var.

Munoz Pizarro, Carlos
Flores Silvestres de Chile

fl.   Pl. 42

Desmanthus illinoensis

Brown, Clair A.
Wildflowers of Louisiana and
Adjoining States

fl. hab.   p 78

Desmanthus illinoensis

Wharton, Mary E.
A Guide to the Wildflowers & Ferns
of Kentucky

fl. fr. p 155   Pl. 2.15

Desmodium canadense

Courtenay, Booth
Wildflowers & Weeds

fl.   p. 49

Desmodium canadense

Massachusetts Hort. Soc.
Horticulture
vol 51, No. 10   1973

fr.   p. 26

Desmodium canadense

Moyle, John B.
Northland Wild Flowers

fl. hab.   p 81, Pl. 58

Desmodium caudatum

Kitamura, Siro
Coloured Illustrations of Trees &
Shrubs of Japan

fr.   254

Desmodium cuspidatum

Duncan, Wilbur H.
Wildflowers of the Southeastern
United States

fl.   p 79

Desmodium glutinosum

Courtenay, Booth
Wildflowers & Weeds

fl.   p. 48

Desmodium glutinosum

Moyle, John B.
Northland Wild Flowers

fl. hab.   p 80, Pl. 57

Desmodium glutinosum

Wharton, Mary E.
A Guide to the Wildflowers & Ferns
of Kentucky

fl. p 187   Pl. 1.18

Desmodium heterocarpum

Walden, Beryl M.
Wild Flowers of Hong Kong

fl.   Pl. 70 (214)

Desmodium nudiflorum

Jennings, O. E.
Wild Flowers of Western Pennsylvania
                              vol II
fl.   Pl. 91

Desmodium perplexum

Wharton, Mary E.
A Guide to the Wildflowers & Ferns
of Kentucky

fl. p 188   Pl. 1.20

Desmodium praestans

Curtis's Botanical Magazine
vol 174   1962-63

fl.   Pl. 407

Desmodium repandum

Moriarty, Audrey
Wild Flowers of Malawi

fl. fr.   Pl. 68; 4

Desmodium rotundifolium

Wharton, Mary E.
A Guide to the Wildflowers & Ferns
    of Kentucky

fl.   p 188   Pl. 1.19

Desmodium sp.

Klimas, John E.
Wildflowers of Eastern America

fl.   Pl. 229

Desmodium tortuosum

Duncan, Wilbur H.
Wildflowers of the Southeastern
    United States

fl. fr.   p 79

Desmodium tortuosum

Weeds of the Southern United States

hab., fr.   p. 30

Desmos chinensis

Herklots, Geoffrey
Flowering Tropical Climbers

fl., fr.   p. 50   Pl. 3

Desplatsia chrysochlamys

Wit, H. C. D. de
Plants of the World;
    The Higher Plants, Vol. I

fr.   p. 213   Pl. 116

Deuterocohnia chrysantha

Munoz Pizarro, Carlos
Flores Silvestres de Chile

fl., hab.   Pl. 2

Deuterocohnia meziana

Wilson, Robert Gardner
Bromeliads in Cultivation

hab.   Pl. 108

Deuterocohnia schreiteri

Bromeliad Society
Journal
        vol 26, No. 1   1976

fl.   p. 26

Deuterocohnia schreiteri

Padilla, Victoria
Bromeliads

hab.   p. 66   Pl. 4

Deutzia crenata

Kitamura, Siro
Coloured Illustrations of Trees &
    Shrubs of Japan

fl.   201

Deutzia x elegantissima

Gault, S. Millar
The Color Dictionary of Shrubs

fl.   Pl. 153

Deutzia x elegantissima var.

Perry, Frances
Flowers of the World

fl.   p. 229

Deutzia gracilis

Macoby, Stirling
What Flower is That

fl.   P 109

Deutzia hybrid

Harrison, R.E.
Trees and Shrubs

fl.   p. 62

Deutzia hybrid

Hay, Roy
The Color Dictionary of Flowers &
    Plants

fl.   p. 196   Pl. 1562

Deutzia x lemoinei

Everett, T.H., ed.
New Illustrated Encyclopedia of
    Gardening vol. 4

fl., hab.   p. 583

Deutzia longifolia

Morley, Brian D.
Wild Flowers of the World

fl.   Plate 99

Deutzia longifolia var.

Curtis's Botanical Magazine
    Vol 161   1938

fl.   No. 9532

Deutzia x magnifica

Hellyer, A.G.L.
Shrubs in Colour

fl.   p. 43

Deutzia x magnifica

Huxley, Anthony
Deciduous Garden Trees and Shrubs

fl.   Pl. 66

Deutzia maximowiczii

Kitamura, Siro
Coloured Illustrations of Trees & Shrubs
    of Japan

fl.   202

Deutzia monbeigii

Curtis's Botanical Magazine
Vol 167   1950

fl.   123

Deutzia x rosea

Curtis's Botanical Magazine
Vol. 169   1952-53

fl.   189

Deutzia x rosea

Hay, Roy
The Color Dictionary of Flowers &
    Plants

fl.   p. 196   Pl. 1563

Deutzia x rosea var.

Gault, S. Millar
The Color Dictionary of Shrubs

fl.   Pl. 154

Deutzia x rosea var.

Huxley, Anthony
Deciduous Garden Trees and Shrubs

fl.   Pl. 67

Deutzia scabra

Hay, Roy
The Color Dictionary of Flowers & Plants

fl.   p 196, Pl. 1564

Deutzia scabra

Kitamura, Siro
Coloured Illustrations of Trees &
    Shrubs of Japan

fl.   203

Deutzia scabra var.

Harrison, R.E.
Trees and Shrubs

fl. P 62

Deutzia scabra var.

Hellyer, A. G. L.
Shrubs in Colour

fl.   p. 43

Deutzia scabra var.

Huxley, Anthony
Deciduous Garden Trees and Shrubs

fl.   Pl. 68

Deutzia scabra var.

Macoby, Stirling
What Flower is That

fl.   P 109

Deutzia setchuenensis

Hay, Roy
The Color Dictionary of Flowers & Plants

fl.   p 196, Pl. 1565

Deutzia, slender

see

Deutzia gracilis

Devil, Blue

see

Echium vulgare

Devil, Blue

see

Eryingium rostratum

Devil, Creeping

see

Machaerocereus eruca

Devil, Mountain

see

Lambertia formosa

Devil's bit

see

Chamaelirium luteum

Devil's Boots

see

Sarracenia purpurea

Devil's Claw

see

Phyteuma comosum

Devil's Claw

see

Proboscidea altheaefolia

Devil's Club

see

Echinopax horridum

Devil's club

see

Oplopanax horridum

Devil's Ivy

see

Scindapsus aureus

Devil's Lantern

see

Oenothera deltoides

Devil's Pins

see

Hovea pungens

Devil's Root

see

Lophophora lewinii

Devil's Shoe String

see

Tephrosia virginiana

Dewberry

see

Rubus acaulis

Dewberry

see

Rubus caesius

Dewberry, Red

see

Rubus pedatus

Dewberry, Southern

see

Rubus enslenii

Dewberry, Southern

see

Rubus trivialis

Dewberry, Swamp

see

Rubus hispidus

Dewdrop, Golden

see

Duranta repens

Dew Drops

see

Dalibarda repens

Dew Flowers

see

Drosanthemum hispidum

Dew plant

see

Mesembryanthemum crassifolium

Deyeuxia accedens

Curtis, Winifred
The Endemic Flora of Tasmania
Vol VI

fl. fr. hab.   Pl. 228

Dhak tree

see

Butea frondosa

Diadenium barkeri

Pabst, G. F. J.
Orchidaceae Brasilienses,
Vol 2

fl.   p 262

Dialiopsis africana

Palgrave, K. C.
Trees of Central Africa

fl., fr.   p. 405

Diamorpha smallii

Duncan, Wilbur H.
Wildflowers of the Southeastern
United States

fl.   p 59

Dianella laevis

Blombery, Alec M.
What Wildflower is That

fl.   p 118, Pl. 300

Dianella tasmanica

Cochrane, G. R.
Flowers and Plants of Victoria

fr.   Pl. 454

Dianella tasmanica

Hay, Roy
The Dictionary of House Plants

fr.   Pl. 182

Dianella tasmanica

Morley, Brian D.
Wild Flowers of the World

fl. fr.   Pl. 142

Dianthus x allwoodii var.

Hay, Roy
The Color Dictionary of Flowers & Plants

fl.   p. 138,139   Pl.1104 thru
                                    1107

Dianthus x allwoodii var.

Tsukamoto, Yotaro
Coloured Illustrations of Garden
    Flowers   Vol. 9

fl.   p. 41

Dianthus alpinus

Barneby, T. P.
European Alpine Flowers in Colour

fl. hab.   Pl. 18, 4

Dianthus alpinus

Hay, Roy
The Color Dictionary of Flowers & Plants

fl.   p 7, Pl. 49

Dianthus alpinus

Kohlhaupt, Paula
Fleurs des Alpages
                    vol 1
fl. P 46

Dianthus alpinus

Royal Hort. Soc.
Journal of the Royal Hort. Soc.
    vol 87, No. 1   1962

fl.   p. 22   Pl. 10

Dianthus alpinus

Royal Hort. Soc.
Journal of the Royal Hort. Soc.
    vol 92, No. 10   1967

fl.   p. 432   Pl. 224

Dianthus alpinus

Vilmorin, Roger de
Plantes Alpines dans les Jardins

fl. hab.   Pl. XXIX

Dianthus alpinus var.

Royal Hort. Soc.
The Garden
    vol 101, No. 7   1976

fl.   p. 375

Dianthus arenarius

Macoby, Stirling
What Flower is That

fl.   P 110

Dianthus armeria

American Hort. Soc.
American Horticulturist
    vol 57, No. 4   1978

fl.   p. 34

Dianthus armeria

Clark, Lewis J.
Wild Flowers of British Columbia

fl.   p 146

Dianthus armeria

Courtenay, Booth
Wildflowers & Weeds

fl.   p 25

Dianthus armeria

Klimas, John E.
Wildflowers of Eastern America

fl.   Pl. 185

Dianthus armeria

Martin, W. Keble
The Concise British Flora in Colour

fl. hab.   Pl. 13

Dianthus armeria

Polunin, Oleg
Flowers of Europe

fl.   Pl. 17 #186

Dianthus armeria

Roberts, June Carver
Born in the Spring

fl. hab.   p 129

Dianthus armeria

Wharton, Mary E.
A Guide to the Wildflowers & Ferns
    of Kentucky

fl. p 124   Pl. 1.58

Dianthus barbatus

Barneby, T. P.
European Alpine Flowers in Colour

fl.   Pl. 16, 6

Dianthus barbatus

Crockett, James Underwood
Annuals

fl.   P 114

Dianthus barbatus

Hay, Roy
The Color Dictionary of Flowers & Plants

fl.   p 36, Pl. 283

Dianthus barbatus

Macoby, Stirling
What Flower is That

fl.   P 110

Dianthus barbatus

Perry, Frances
Flowers of the World

fl.   p 71

Dianthus barbatus

Tsukamoto, Yotaro
Coloured Illustrations of Garden
    Flowers   Vol. 10

fl. Illus. 57   opp. p. 18

Dianthus barbatus var.

Huxley, Anthony
Garden Perennials & Water Plants

fl.   Pl. 94

Dianthus callizonus

Alpine Garden Society
Bulletin
    vol. 39, No. 1   1971

fl., hab.   p. 36

Dianthus carthusianorum

Barneby, T. P.
European Alpine Flowers in Colour

fl.   Pl. 17, 1

Dianthus carthusianorum

Felsko, Elsa
A Book of Wildflowers

fl.   p. 100

Dianthus carthusianorum

Kleijn, H.
Beauty of the Wild Plant

fl.   opp. p. 62   Pl. 77

Dianthus carthusianorum

Kohlhaupt, Paula
Fleurs des Alpages
                    vol 1
fl. P 45

Dianthus carthusianorum

Polunin, Oleg
Flowers of Europe

fl.   Pl. 17 #187

Dianthus carthusianorum

Tosco, Uberto
The World of Mountain Flowers

fl. hab.   p 112

Dianthus carthusianorum

Wit, H.C.D. de
Plants of the World
    The Higher Plants vol 1

fl.   p. 110   Pl. 66

Dianthus caryophyllus

Crockett, James Underwood
Annuals

fl.   P 114

Dianthus caryophyllus

Huxley, Anthony
Garden annuals and bulbs

fl.   Pl. 41

Dianthus caryophyllus

Kiaer, Eigil
Indoor Plants in Colour

fl.   p. 50

Dianthus caryophyllus

Martin, W. Keble
The Concise British Flora in Colour

fl.   Pl. 13

Dianthus caryophyllus

Nicolaisen, Age
Pocket Encyclopedia of Indoor Plants

fl.   pl. 54

Dianthus caryophyllus

Perrot, Emile
Les Plantes Medicinales

fl. fr. hab.   p 118

Dianthus caryophyllus

Tsukamoto, Yotaro
Coloured Illustrations of Garden
    Flowers   Vol. 10

fl. Illus. 54   opp. p. 18

Dianthus caryophyllus var.

Barneby, T. P.
European Alpine Flowers in Colour

fl.   Pl. 17, 5

Dianthus caryophyllus var.

Hay, Roy
The Color Dictionary of Flowers &
    Plants

fl.   p. 36   Pl. 284

Dianthus carophyllus var.

Huxley, Anthony
Garden Perennials and Water Plants

fl.   Pl. 95

Dianthus caryophyllus var.

Macoby, Stirling
What Flower is That

fl.   p. 110

Dianthus chinensis

Crockett, James Underwood
Annuals

fl.   P 114

Dianthus chinensis

Macoby, Stirling
What Flower is That

fl.   P 110

Dianthus chinensis var.

Huxley, Anthony
Garden Annuals and Bulbs

fl.   Pl. 42

Dianthus chinensis var.

Tsukamoto, Yotaro
Coloured Illustrations of Garden
    Flowers   Vol. 9

fl.   p. 42

Dianthus chinensis var.

Tsukamoto, Yotaro
Coloured Illustrations of Garden
    Flowers   Vol. 10

fl. Illus. 51,52,53 opp. p. 17

Dianthus creticus

Goulimis, Constantine N.
Wild Flowers of Greece

fl.   p 7

Dianthus crinitus

Goulimis, Constantine N.
Wild Flowers of Greece

fl.   p 9

Dianthus deltoides

Batson, Wade T.
Wild Flowers in South Carolina

fl.   p 44

Dianthus deltoides

Crockett, James Underwood
Lawns & Ground Covers

fl. p. 129

Dianthus deltoides

Hay, Roy
The Color Dictionary of Flowers & Plants

fl.   p 7, Pl. 50

Dianthus deltoides

Huxley, Anthony
Garden Perennials and Water Plants

fl.   pl. 97

Dianthus deltoides

Lindman, C. A. M.
Nordens Flora, Vol 4

fl. hab.   Pl. 214B

Dianthus deltoides

Martin, W. Keble
The Concise British Flora in Colour

fl. hab.   Pl. 13

Dianthus deltoides

Tsukamoto, Yotaro
Coloured Illustrations of Garden
    Flowers   Vol. 9

fl.   opp. p. 44

Dianthus deltoides

Vilmorin, Roger de
Plantes Alpines dans les Jardins

fl. hab.   Pl. XII

Dianthus glacialis

Barneby, T. P.
European Alpine Flowers in Colour

fl. hab.   Pl. 18, 3

Dianthus glacialis

Kohlhaupt, Paula
Fleurs des Alpages
    vol 2

Fl.   P 36

Dianthus gratianopolitanus

Barneby, T. P.
European Alpine Flowers in Colour

fl. hab.   Pl. 18, 1

Dianthus gratianopolitanus

Hay, Roy
The Color Dictionary of Flowers & Plants

fl.   p 7, Pl. 51

Dianthus gratianopolitanus

Martin, W. Keble
The Concise British Flora in Colour

fl. hab.   Pl. 13

Dianthus gratianopolitanus

Royal Hort. Soc.
The Garden
    vol 102, No. 5   1977

fl., hab.   p. 197

Dianthus gratianopolitanus

Vilmorin, Roger de
Plantes Alpines dans les Jardins

fl. hab.   Pl. IX

Dianthus haematocalyx var.

Curtis's Botanical Magazine
Vol 165   1948

fl.   9

Dianthus hybrid

Everett, T.H., ed.
New Illustrated Encyclopedia of
    Gardening.   vol 4

fl.   p. 598

Dianthus hybrid

Hay, Roy
The Color Dictionary of Flowers &
    Plants

fl.   p.36, 63, 64,   Pl.286,499-510,
            137-38   1092-1103

Dianthus hybrid.

Hay, Roy
The Dictionary of House Plants

fl.   Pl. 183 - 187

Dianthus hybrid

Perry, Frances
Flowers of the World

fl.   p. 70-71

Dianthus hyssopifolius

Barneby, T. P.
European Alpine Flowers in Colour

fl.   Pl. 17, 4

Dianthus japonicus

Tsukamoto, Yotaro
Coloured Illustrations of Garden
    Flowers   Vol. 10

fl. Illus. 55   opp. p. 18

Dianthus kirkii

Flowering Plants of South Africa
    vol XVI   1936

fl.   Pl. 626

Dianthus knappii

Addisonia
    vol 19   1935-36

fl.   p. 3   Pl. 610

Dianthus knappii var.

Tsukamoto, Yotaro
Coloured Illustrations of
    Garden Flowers   Vol. 9

fl.   p. 44

Dianthus monspessulanus

Polunin, Oleg
Flowers of Europe

fl.   Pl. 17 #190

Dianthus monspessulanus

Royal Hort. Soc.
The Garden
    vol 101, No. 11   1976

fl.   p. 555

Dianthus neglectus

Alpine Garden Society
Bulletin
    vol. 35, No. 1   1967

fl., hab.   p. 45

Dianthus neglectus

Barneby, T. P.
European Alpine Flowers in Colour

fl. hab.   Pl. 18, 2

Dianthus neglectus

Hay, Roy
The Color Dictionary of Flowers & Plants

fl.   p 7, Pl. 52

Dianthus neglectus

Royal Hort. Soc.
The Garden
    vol 103, No. 5   1978

fl.   p. 177

Dianthus pancicii var.

Tsukamoto, Yotaro
Coloured Illustrations of
    Garden Flowers   Vol. 9

fl.   p. 44

Dianthus plumarius

Everett, T.H., ed.
New Illustrated Encyclopedia of
    Gardening vol. 4

fl.   p. 598

Dianthus plumarius

Huxley, Anthony
Garden Perennials and Water Plants

fl.   pl. 96

Dianthus plumarius

Macoby, Stirling
What Flower is That

fl.   P 110

Dianthus plumarius

Martin, W. Keble
The Concise British Flora in Colour

fl.   Pl. 13

Dianthus plumarius

Morley, Brian D.
Wild Flowers of the World

fl. fr.   Pl. 13D

Dianthus seguieri

Barneby, T. P.
European Alpine Flowers in Colour

fl.   Pl. 17, 2

Dianthus sinensis var.

Hay, Roy
The Color Dictionary of Flowers &
    Plants

fl.   p. 36   Pl. 285

Dianthus subacaulis var.

Morley, Brian D.
Wild Flowers of the World

fl.,fr.,hab.   Pl. 32

Dianthus superbus

Barneby, T. P.
European Alpine Flowers in Colour

fl.   Pl. 17, 3

Dianthus superbus

Felsko, Elsa
A Book of Wild Flowers
        2nd ser.
fl.   p 95

Dianthus superbus

Kleijn, H.
Beauty of the Wild Plant

fl.   opp. p. 57   Pl. 139

Dianthus superbus

Lindman, C. A. M.
Nordens Flora, Vol 4

fl.   Pl. 214A

Dianthus superbus

Polunin, Oleg
Flowers of Eurppe

fl.   Pl. 17 #188

Dianthus superbus

Tsukamoto, Yotaro
Coloured Illustrations of Garden
Flowers   Vol. 10

fl. Illus. 56   opp. p. 18

Dianthus superbus var.

Kimura, Koiti
Japanese Medicinal Plants, Vol II

fl.   p 157

Dianthus superbus var.

Robert, Paul A
Alpine Flowers

fl.   Pl. 15

Dianthus sylvester

Kohlhaupt, Paula
Fleurs des Alpages
Vol. 2

fl.   p. 35

Dianthus sylvestris

Polunin, Oleg
Flowers of Europe

fl.   Pl. 17 #194

Dianthus virgineus

Vilmorin, Roger de
Plantes Alpines dans Les Jardins

fl. hab.   Pl. XXV

Dianthus, yellow

see

Dianthus knappii

Diapensia himalaica

Royal Hort. Soc.
Journal of the Royal Hort. Soc.
vol 77, No. 7   1952

fl.   p. 248   Pl. 109

Diapensia lapponica

Alaska-Yukon Wild Flower Guide

fl. hab.   p 120

Diapensia lapponica

Alpine Garden Society
Bulletin
vol. 36, No. 1   1968

fl., hab.   p. 19

Diapensia lapponica

Ferguson, Mary
Wildflowers

fl.   p 79

Diapensia lapponica

Heller, Christine
Wild Flowers of Alaska

fl.   Pl. 205

Diapensia lapponica

Lemmon, Robert S.
Wildflowers of North America in
Full Color

fl.   pg. 142   pl. 227

Diapensia lapponica

Lindman, C. A. M.
Nordens Flora, Vol 7

fl. fr.   Pl. 457

Diapensia lapponica

Martin, W. Keble
The Concise British Flora in Colour

fl. hab.   Pl. 57

Diapensia lapponica

Milne, Lorus
Living Plants of the World

fl.   p 186

Diapensia lapponica

Morley, Brian D.
Wild Flowers of the World

fl., hab.   Pl. 6A

Diapensia lapponica

Polunin, Oleg
Flowers of Europe

fl.   Pl. 87 #911

Diapensia lapponica

Raven, John
Mountain Flowers

fl.   p. 125 Pl. 12

Diapensia lapponica

Royal Hort. Soc.
The Garden
vol 102, No. 5   1977

fl., hab.   p. 201

Diaphananthe fragrantissima

Williamson, Graham
The Orchids of South Central Africa

fl., hab.   Pl. 170-71

Diaphananthe pulchella var.

Williamson, Graham
The Orchids of South Central Africa

fl.   Pl. 172

Diaphananthe rutila

Williamson, Graham
The Orchids of South Central Africa

fl.   Pl. 173

Diaphananthe xanthopollinia

Williamson, Graham
The Orchids of South Central Africa

fl.   Pl. 174

Diascia barberae

Crockett, James Underwood
Annuals

fl   P 114

Diascia barberae

Hay, Roy
The Color Dictionary of Flowers & Plants

fl.   p 36, Pl. 287

Diascia thunbergiana

Eliovson, Sima
Namaqualand in Flower

fl.   Pl. 27

Diastema vexans

Elbert, Virginie F.
The Miracle Houseplants

fl. hab.   p 116

Dicentra canadensis

Campbell, Carlos C.
Great Smoky Mountain Wildflowers

fl.   p 29

Dicentra canadensis

Courtenay, Booth
Wildflowers & Weeds

fl.   p 22

Dicentra canadensis

Duncan, Wilbur H.
Wildflowers of the Southeastern
United States

fl.   p 51

Dicentra canadensis

Jennings, O. E.
Wild Flowers of Western Pennsylvania
vol II

fl. hab.   Pl. 26

Dicentra canadensis

Klute, Jeannette
Woodland Portraits

fl.   Pl. 14

Dicentra canadensis

Wharton, Mary E.
A Guide to the Wildflowers & Ferns
   of Kentucky

fl.   p 217   Pl. 3.14

Dicentra chrysantha

Royal Hort. Soc.
The Garden
   vol 101, No. 8 1976

fl.   p. 433

Dicentra cucullaria

Addisonia
   vol 20   1937-38

fl., fr.   p. 21   Pl. 651

Dicentra cucullaria

Campbell, Carlos C.
Great Smoky Mountain Wildflowers

fl.   p 29

Dicentra cucullaria

Courtenay, Booth
Wildflowers & Weeds

fl.   p 22

Dicentra cucullaria

Dean, Blanche E.
Wildflowers of Alabama and
   Adjoining States

fl.   p 71

Dicentra cucullaria

Duncan, Wilbur H.
Wildflowers of the Southeastern
   United States

fl.   p 51

Dicentra cucullaria

Ferguson, Mary
Wildflowers

fl.   p 162

Dicentra cucullaria

Jennings, O. E.
Wild Flowers of Western Pennsylvania
   vol II
fl.   hab.   Pl. 68

Dicentra cucullaria

Klimas, John E.
Wildflowers of Eastern America

fl.   Pl. 25

Dicentra cucullaria

Lemmon, Robert S.
Wildflowers of North America in
   Full Color

fl.   pg. 255   pl. 403

Dicentra cucullaria

Massachusetts Hort. Soc.
Horticulture
   vol 50, No. 3   1972

fl.   p. 40-41

Dicentra cucullaria

Moyle, John B.
Northland Wild Flowers

fl. hab.   p 59, Pl. 24

Dicentra cucullaria

Perry, Frances
Flowers of the World

fl.   p 119

Dicentra cucullaria

Royal Hort. Soc.
The Garden
   vol 101, No. 8   1976

fl.   p. 433

Dicentra cucullaria

Wharton, Mary E.
A Guide to the Wildflowers & Ferns
   of Kentucky

fl.   p 217   Pl. 3.13

Dicentra eximia

Campbell, Carlos C.
Great Smoky Mountain Wildflowers

fl.   p. 29

Dicentra eximia

Huxley, Anthony
Garden Perennials and Water Plants

fl.   Pl. 99

Dicentra eximia

Klimas, John E.
Wildflowers of Eastern America

fl.   Pl. 193

Dicentra eximia

Lemmon, Robert S.
Wildflowers of North America in
   Full Color

fl.   pg. 129   pl. 206

Dicentra eximia var.

Hay, Roy
The Color Dictionary of Flowers & Plants

fl.   p 139, Pl. 1108

Dicentra formosa

Clark, Lewis J.
Wild Flowers of British Columbia

fl.   p 182

Dicentra formosa

Fries, Mary A.
Wildflowers of Mount Ranier and
the Cascades

Dicentra formosa

Orr, Robert T.
Wildflowers of Western America

fl.   Pl. 153

Dicentra formosa

Taylor, Ronald J.
Mountain Wild Flowers

fl. hab.   p 19

Dicentra formosa var.

Hay, Roy
The Color Dictionary of Flowers & Plants

fl.   p 139, Pl. 1109

Dicentra peregrina

Alpine Garden Society
Bulletin
   vol. 36, No. 4   1968

fl., hab.   p. 313

Dicentra peregrina

Curtis's Botanical Magazine
   vol 175   1964-65

fl.   Pl. 461

Dicentra peregrina var.

Hay, Roy
The Color Dictionary of Flowers & Plants

fl.   p 7, Pl. 53

Dicentra spectabilis

American Hort. Soc.
American Horticulturist
   vol 54, No. 1 1975

fl.   p. 30

Dicentra spectabilis

Everett, T.H., ed.
New Illustrated Encyclopedia of
   Gardening   vol 4

fl.   p. 599

Dicentra spectabilis

Everett, T. H., ed.
New Illustrated Encyclopedia of
   Gardening   Vol. 12

fl.   p. 2118

Dicentra spectabilis

Hay, Roy
The Color Dictionary of Flowers & Plants

fl.   p 139, Pl. 1110

Dicentra spectabilis

Hay, Roy
The Dictionary of House Plants

fl.   Pl. 183

Dicentra spectabilis

Huxley, Anthony
Garden Perennials and Water Plants

hab., fl.   Pl. 7, 98

Dicentra spectabilis

Macoby, Stirling
What Flower is That

fl.   p 111

Dicentra spectabilis

Milne, Lorus
Living Plants of the World

fl.   p 90

Dicentra spectabilis

Perry, Frances
Flowers of the World

fl.   p 119

Dicentra spectabilis

Royal Hort. Soc.
The Garden
          Vol. 101, No. 7   1976

fl.   p. 378

Dicerandra linearifolia

Duncan, Wilbur H.
Wildflowers of the Southeastern
          United States

fl.   p 163

Dicerandra odoratissima

Duncan, Wilbur H.
Wildflowers of the Southeastern
          United States

fl.   p 161

Dichaea anchoraelabia

Pabst, G. F. J.
Orchidaceae Brasilienses
          Vol. 2

fl.   p. 240

Dichaea brachyphylla

Pabst, G. F. J.
Orchidaceae Brasilienses
          Vol. 2

fl.   p. 240

Dichaea pendula

Pabst, G.F.J.
Orchidaceae Brasilienses
          vol 2

hab.   p. 175

Dichaea samenkapsel

Pabst, G. F. J.
Orchidaceae Brasilienses,
          Vol 2

          fr.   p 240

Dichaea trulla

Ospina, Mariano
Orquideas de las Americas

fl. hab.   Pl. 127

Dicheranthus plocamoides

Bramwell, David
Wild Flowers of the Canary Islands

hab.   Pl. 131

Dichondra repens

Cochrane, G. R.
Flowers and Plants of Victoria

fl. hab.   Pl. 31

Dichondra repens

Crockett, James Underwood
Lawns & Ground Covers

hab P. 130

Dichondra repens

Weeds of the Southern United States

hab.   p. 16

Dichopogon strictus

Blombery, Alec M.
What Wildflower is That

fl.   p. 118   Pl. 301

Dichopogon strictus

Cochrane, G. R.
Flowers and Plants of Victoria

fl.   Pl. 336

Dichorisandra mosaica var.

Massachusetts Hort. Soc.
Horticulture
          vol 32, No. 2   1954

hab.   p. 77

Dichorisandra reginae

Morley, Brian D.
Wild Flowers of the World

fl.   Pl. 187

Dichorisandra thyrsiflora

Curtis's Botanical Magazine
Vol 178   1970-72

fl.   590

Dichorisandra thyrsiflora

Macoby, Stirling
What Flower is That

fl.   P 111

Dichorisandra thyrsiflora

Menninger, Edwin A.
Flowering Vines of the World

fl.   Pl. 59

Dichromena colorata

Batson, Wade T.
Wild Flowers in South Carolina

fl. hab.   p 25

Dichromena colorata

Brown, Clair A.
Wildflowers of Louisiana and
          Adjoining States

fl.   p 6

Dichromena colorata

Dean, Blanche E.
Wildflowers of Alabama and
          Adjoining States

fl.   p 5

Dichromena colorata

Fleming, Glenn
Wild Flowers of Florida

fl.   p 26

Dichromena latifolia

Duncan, Wilbur H.
Wildflowers of the Southeastern
          United States

fl.   p 237

Dichrostachys glomerata

Flowering Plants of South Africa
          vol XXIII   1943

fl.   Pl. 894

Dichrostachys glomerata

Palgrave, K. C.
Trees of Central Africa

fl., fr.   p. 273

Dicliptera leonotis

Moriarty, Audrey
Wild Flowers of Malawi

fl.   Pl. 42; 3

Dicliptera suberecta

Macoby, Stirling
What Flower is That

fl.   P 112

Dicoma anomala

Flowering Plants of South Africa
vol XVI   1936

fl.   Pl. 625

Dicoma anomala var.

Moriarty, Audrey
Wild Flowers of Malawi

fl.   Pl. 79; 7

Dicoma sessiliflora

Moriarty, Audrey
Wild Flowers of Malawi

fl.   Pl. 79; 6

Dicrocaulon brevifolium

Herre, H.
The Genera of the Mesembryanthemaceae

fl. fr.   p 129

Dictamnus albus

Barneby, T. P.
European Alpine Flowers in Colour

fr.   Pl. 57, 5

Dictamnus albus

Everett, T.H., ed.
New Illustrated Encyclopedia of
    Gardening   vol 4

fl.   p. 599

Dictamnus albus

Felsko, Elsa
A Book of Wild Flowers
    2nd Ser.

fl.   p. 82

Dictamnus albus

Huxley, Anthony
Garden Perennials and Water Plants

fl.   Pl. 100

Dictamnus albus

Kohlhaupt, Paula
Fleurs des Alpages
    vol 2

Fl.   P 71

Dictamnus albus

Morley, Brian D.
Wild Flowers of the World

fl.,fr.   Pl. 35

Dictamnus albus

Perry, Frances
Flowers of the World

fl.   p 269

Dictamnus albus

Polunin, Oleg
Flowers of Europe

fl.   Pl. 67 #688

Dictamnus albus var.

Hay, Roy
The Color Dictionary of Flowers & Plants

fl.   p 139, Pl. 1111

Didelta spinosa

Eliovson, Sima
Namaqualand in Flower

fl.   Pl. 46, 1

Didierea madagascariensis

de Wit, H. C. D.
Plants of the World, The Higher Plants I
                        vol II
fl.   p 110, Pl. 67

Didierea madagascariensis

Morley, Brian D.
Wild Flowers of the World

fl.   Pl. 70

Didierea trollii

Cactus & Succulent Society of America
Journal
    vol 50, No. 1   1978

hab.   p. 14

Didiscus caeruleus

Hay, Roy
The Color Dictionary of Flowers & Plants

fl.   p 36, Pl. 288

Didissandra atrocyanea

Morley, Brian D.
Wild Flowers of the World

fl.   Plate 122

Didymaotus lapidiformis

Herre, H.
The Genera of the Mesembryanthemaceae

fl.   fr.   p 131

Dieffenbachia amoena

Hay, Roy
The Color Dictionary of Flowers & Plants

hab.   p 64, Pl. 511

Dieffenbachia arvida var.

Hay, Roy
The Color Dictionary of Flowers & Plants

hab.   p 64, Pl. 512

Dieffenbachia hybrid

Encke, Fritz
Zimmerpflanzen

hab.   p. 74

Dieffenbachia jenmannii

Kiaer, Eigil
Indoor Plants in Colour

hab.   p. 51

Dieffenbachia leoniae

Macoby, Stirling
What Flower is That

hab.   P. 113

Dieffenbachia leopoldii

Kiaer, Eigil
Indoor Plants in Colour

hab.   p. 51

Dieffenbachia leopoldii

Nicolaisen, Age
Pocket Encyclopedia of Indoor Plants

hab.   Pl. 11

Dieffenbachia maculata

Hersey, Jean
Woman's Day Book of House Plants

hab.   P. 53

Dieffenbachia picta

Nicolaisen, Age
Pocket Encyclopedia of Indoor Plants

hab.   Pl. 10

Dieffenbachia picta var.

Hay, Roy
The Color Dictionary of Flowers & Plants

hab.   p 65, Pl. 513

Dieffenbachia picta var.

Hay, Roy
The Dictionary of House Plants

hab.   Pl. 190, 191

Dieffenbachia picta var.

Massachusetts Hort. Soc.
Horticulture
    vol 35, No. 1   1957

hab.   p. 37 (see errata p. 108)

Dieffenbachia picta var.

Massachusetts Hort. Soc.
Horticulture
    vol 35, No. 11   1957

hab.   p. 556

Dieffenbachia seguine

Bruggeman, L.
Tropical Plants

hab.   Pl. 106

Dieffenbachia sp.

Everett, T. H., ed.
New Illustrated Encyclopedia of
        Gardening   Vol. 4

hab.   frontispiece

Dieffenbachia sp.

Kromdijk, G.
200 House Plants in Colour

hab.   Pl. 70

Dierama igneum

Flowering Plants of South Africa
    vol XXII   1942

fl.   Pl. 874

Dierama luteo-albidum

Flowering Plants of South Africa
    vol XXII   1942

fl.   Pl. 845

Dierama medium var.

Flowering Plants of South Africa
    vol XXII   1942

fl.   Pl. 855

Dierama pendulum

Mathew, Brian
The Larger Bulbs

fl.   p. 64

Dierama pendulum

Moriarty, Audrey
Wild Flowers of Malawi

fl. fr.   Pl. 9; 2

Dierama pulcherrimum

Hay, Roy
The Color Dictionary of Flowers & Plants

fl.   p 91, Pl. 725

Dierama pulcherrimum

Macoby, Stirling
What Flower is That

fl.   P 113

Dierama pulcherrimum

Perry, Frances
Flowers of the World

fl.   p 150

Dierama reynoldsii

Flowering Plants of South Africa
    vol XXI   1941

fl.   Pl. 836

Diervilla lonicera

Courtenay, Booth
Wildflowers & Weeds

fl.   p 73

Diervilla sessilifolia

Campbell, Carlos, C.
Great Smoky Mountain Wildflowers

fl.   p 47

Diervilla sessilifolia

Morley, Brian D.
Wild Flowers of the World

fl. fr.   Pl. 155

Dietes bicolor

Flowering Plants of Africa
    vol 39   1968-69

fl., fr.   Pl. 1525

Dietes bicolor

Macoby, Stirling
What Flower is That

fl. P 113

Dietes butcheriana

Flowering Plants of Africa
    vol 38   1967

fl., fr.   Pl. 1487

Dietes flavida

Flowering Plants of Africa
    vol 38   1967

fl., fr.   Pl. 1488

Dietes grandiflora

Macoby, Stirling
What Flower is That

fl. P 113

Dietes vegeta

Flowering Plants of Africa
    vol 39   1968-69

fl., fr.   Pl. 1524

Dietes vegeta

Moriarty, Audrey
Wild Flowers of Malawi

fl.   Pl. 6; 3

Digitalis ambigua

Tosco, Uberto
The World of Mountain Flowers

fl.   p 31

Digitalis ambigua

Vilmorin, Roger de
Plantes Alpines dans les Jardins

fl. hab.   Pl. XI

Digitalis ferruginea

Polunin, Oleg
Flowers of Europe

fl. Pl. 125 #1230

Digitalis grandiflora

Barneby, T. P.
European Alpine Flowers in Colour

fl.   Pl. 71, 3

Digitalis grandiflora

Goulimis, Constantine N.
Wild Flowers of Greece

fl.   p 95

Digitalis grandiflora

Kohlhaupt, Paula
Fleurs des Alpages
        vol 1
fl. P 98

Digitalis grandiflora

Morley, Brian D.
Wild Flowers of the World

fl.   Pl. 21B

Digitalis grandiflora

Polunin, Oleg
Flowers of Europe

fl. Pl. 125 #1232

Digitalis lanata

Kimura, Koiti
Japanese Medicinal Plants, Vol II

fl.   p 227

Digitalis lanata

Perrot, Emile
Les Plantes Medicinales

fl. fr. hab.   p 87

Digitalis lanata

Takatori, Jisuke
Color Atlas of Medicinal Plants of Japan

fl.   Fig. 12

Digitalis lutea

Barneby, T. P.
European Alpine Flowers in Colour

fl.   Pl. 71, 4

Digitalis lutea

Perrot, Emile
Les Plantes Medicinales

fl. fr. hab.   p 87

Digitalis lutea

Polunin, Oleg
Flowers of Europe

fl.   Pl. 125 #1233

Digitalis obscura

Polunin, Oleg
Flowers of Europe

fl.   Pl. 125 #1233

Digitalis obscura

Royal Hort. Soc.
Journal of the Royal Hort. Soc.
    vol 90, No. 5 1965

fl.   p. 206   Pl. 103

Digitalis obscura

Taylor, A. W.
Wild Flowers of Spain and Portugal

fl.   p. 71

Digitalis purpurea

Ary, S.
The Oxford Book of Wildflowers

fl.   p 122, Pl. 5

Digitalis purpurea

Bianchini, Francesco
Health Plants of the World

fl. hab.   p 75

Digitalis purpurea

Color Treasury of Herbs & Other
    Medicinal Plants

fl.   p 50   Pl. 74

Digitalis purpurea

Crockett, James Underwood
Annuals

fl.   P 115

Digitalis purpurea

Everett, T.H., ed.
New Illustrated Encyclopedia of
    Gardening   vol 4

fl.   p. 630

Digitalis purpurea

Felsko, Elsa
A Book of Wildflowers

fl.   P 107

Digitalis purpurea

Hvass, Eslie
Plants That Feed and Serve Us

fl.   p 86, Pl. 182

Digitalis purpurea

Kariyone, Tatsuo
Atlas of Medicinal Plants

fl.   Pl. 39

Digitalis purpurea

Kimura, Koiti
Japanese Medicinal Plants, Vol II

fl.   p 225

Digitalis purpurea

Kleijn, H.
Beauty of the Wild Plant

fl.   opp. p. 92   Pl. 125

Digitalis purpurea

Lindman, C. A. M.
Nordens Flora, Vol 8

fl. hab.   Pl. 529

Digitalis purpurea

Macoby, Stirling
What Flower is That

fl.   P 114

Digitalis purpurea

Martin, W. Keble
The Concise British Flora in Colour

fl. hab.   Pl. 63

Digitalis purpurea'

Orr, Robert T.
Wildflowers of Western America

fl.   Pl. 218

Digitalis purpurea

Perrot, Emile
Les Plantes Medicinales

fl. fr. hab.   p 88

Digitalis purpurea

Perry, Frances
Flowers of the World

fl.   p 276

Digitalis purpurea

Polunin, Oleg
Flowers of Europe

fl.   Pl. 125 #1234

Digitalis purpurea

Takatori, Jisuke
Color Atlas of Medicinal Plants of Japan

fl.   Fig. 11

Digitalis purpurea

Taylor, Ronald J.
Mountain Wild Flowers

fl.   p 32

Digitalis purpurea

Tsukamoto, Yotaro
Coloured Illustrations of Garden
    Flowers   vol 9

fl.   opp. p. 45

Digitalis purpurea var.

Hay, Roy
The Color Dictionary of Flowers & Plants

fl.   P 37, Pl. 289, 290

Digitalis purpurea var.

Huxley, Anthony
Garden Perennials and Water Plants

fl.   Pl. 101

Digitalis purpurea var.

Wit, H. C. D. de
Plants of the World;
    The Higher Plants, Vol II

fl.   Pl. 84

Digitalis purpurea var.

Wit, H. C. D. de
Plants of the World;
    The Higher Plants, Vol II

fl.   Pl. 86

Digitalis sp.

Edlin, Herbert
The Illustrated Encyclopedia
    of Trees

fl. hab.   p 38, 39

Digitaria decumbens

Burgis, D.S.
Florida Weeds

hab., fr.   p. 7

Digitaria ischaemum

Crockett, James Underwood
Lawns & Ground Covers

hab., fr.   P. 70

Digitaria sanguinalis

Weeds of the Southern United States

hab., fr.   p. 24

Dilatris viscosa

Flowering Plants of Africa
vol 41   1970-71

fl., hab.   Pl. 1621

Dill

see

Anethum graveolens

Dill

see

Peucedanum graveolens

Dillenia indica

Oakman, Harry
Colorful Trees

Hab. Fl.   P 92

Dillenia indica

Perry, Frances
Flowers of the World

fl.   p 101

Dillenia obovata

Morley, Brian D.
Wild Flowers of the World

fl.   Plate 112

Dillenia pentagyna

Roxburgh, William
Icones Roxburghianae

fl., fr.   Pl. 2

Dillwynia glaberrima

Cochrane, G. R.
Flowers and Plants of Victoria

fl.   Pl. 362

Dillwynia hispida

Cochrane, G. R.
Flowers and Plants of Victoria

fl.   Pl. 242

Dillwynia retorta

Blombery, Alec M.
What Wildflower is That

fl.   p 119, Pl. 303

Dillwynia sericea

Blombery, Alec M.
What Wildflower is That

fl.   p 119, Pl. 304

Dillwynia sericea

Cochrane, G. R.
Flowers and Plants of Victoria

fl.   Pl. 86

Dimerandra emarginata

Pabst, G. F. J.
Orchidaceae Brasilienses,
Vol 1

fl.   p 186

Dimorphanthera sp.

Royal Hort. Soc.
Journal of the Royal Hort. Soc.
vol 91, No. 3   1966

fl.   p. 116   Pl. 74

Dimorphotheca aurantiaca

American Hort. Soc.
American Horticulturist
vol 53, No. 3   1974

fl.   p. 35

Dimorphotheca aurantiaca

Macoby, Stirling
What Flower is That

fl.   p. 114

Dimorphotheca aurantiaca

Massachusetts Hort. Soc.
Horticulture
vol 52, No. 1   1974

fl.   p. 39

Dimorphotheca aurantiaca

Tsukamoto, Yotaro
Coloured Illustrations of Garden
Flowers   Vol. 10

fl.   Illus. 59   opp. p. 19

Dimorphotheca aurantiaca var.

Hay, Roy
The Color Dictionary of Flowers & Plants

fl.   p 37, Pl. 291

Dimorphotheca barberiae

Hay, Roy
The Color Dictionary of Flowers & Plants

fl.   p 7, Pl. 54

Dimorphotheca chrysanthemifolia

Royal Hort. Soc.
Journal of the Royal Hort. Soc.
vol 91, No. 12   1966

fl.   p. 508   Pl. 265

Dimorphotheca ecklonis

Addisonia
vol 23   1954-59

fl.   p. 63   Pl. 768

Dimorphoteca ecklonis

Macoby, Stirling
What Flower is That

fl.   P 115

Dimorphotheca ecklonis

Tsukamoto, Yotaro
Coloured Illustrations of Garden
Flowers   Vol. 9

fl.   opp. p. 45

Dimorphotheca ecklonis var.

Macoby, Stirling
What Flower is That

fl.   P 115

Dimorphotheca hybrid

Crockett, James Underwood
Annuals

fl.   P 115

Dimorphotheca hybrid

Mathias, Mildred E.
Color for the Landscape

fl. hab.   p 134

Dimorphotheca jucunda

Flowering Plants of South Africa
vol XVI   1936

fl.   Pl. 629

Dimorphotheca pluvialis

American Hort. Soc.
American Horticulturist
vol 53, No. 3   1974

fl.   p. 34

Dimorphotheca pluvialis

Curtis's Botanical Magazine
Vol 177   1969-70

fl.   532

Dimorphotheca pluvialis

Flowering Plants of Africa
vol 32   1957-58

fl.   Pl. 1249

Dimorphotheca sinuata

Eliovson, Sima
Namaqualand in Flower

fl. hab.   Pl. 1, Pl. 9

Dinema polybulbon

Ebel, Friedrich
The Strange and Beautiful
World of Orchids

fl.   p 43

Dinteranthus microspermus

Herre, H.
The Genera of the Mesembryanthemaceae

fl.   fr.   p 133

Dioclea reflexa

Menninger, Edwin A.
Flowering Vines of the World

fl.   Pl. 102

Diodia teres

Courtenay, Booth
Wildflowers & Weeds

fl.   p 70

Diodia teres

Weeds of the Southern United States

hab., fl.   p. 38

Diodia virginiana

Brown, Clair A.
Wildflowers of Louisiana and
Adjoining States

fl. hab.   p 176

Diodia virginiana

Duncan, Wilbur H.
Wildflowers of the Southeastern
United States

fl.   p 189

Diodia virginiana

Fleming, Glenn
Wild Flowers of Florida

fl. hab.   p 24

Diodia virginiana

Wharton, Mary E.
A Guide to the Wildflowers & Ferns
of Kentucky

fl. p 172   Pl. 2.52

Dionaea muscipula

Hay, Roy
The Dictionary of House Plants

hab.   Pl. 192

Dionaea muscipula

Hersey, Jean
Woman's Day Book of House Plants

hab.   P. 103

Dionaea muscipula

Morley, Brian D.
Wild Flowers of the World

fl.   Pl. 158

Dionaea muscipula

Pacific Hort. Foundation
Pacific Horticulture
vol 39, No. 3   1978

hab.   p. 29

Dionaea muscipula

Perry, Frances
Flowers of the World

fl.   p 103

Dionaea muscipula

Schnell, Donald E.
Carnivorous Plants of the U. S. and Canada

fl. fr. hab.   p 5, 16-21

Dionaea muscipula

Walcott, Mary Vaux
North American Wild Flowers

fl. hab.   Pl. 219

Dionaea muscipula

Wit, H. C. D. de
Plants of the World;
The Higher Plants, Vol I

hab.   p 284, Pl. 177

Dionysia aretioides

Curtis's Botanical Magazine
Vol 178   1970-72

fl.   581

Dionysia aretioides

Hay, Roy
The Color Dictionary of Flowers & Plants

fl.   p 7, Pl. 55

Dionysia aretioides

Royal Hort. Soc.
Journal of the Royal Hort. Soc.
vol 90, No. 11   1965

fl.   p. 470   Pl. 205

Dionysia aretioides var.

Royal Hort. Soc.
Journal of the Royal Hort. Soc.
Vol. 94, No. 7   1969

fl.   p. 302   Pl. 137

Dionysia bryoides

Royal Hort. Soc.
The Garden
vol 100, No. 8   1975

fl.   p. 381

Dionysia curviflora

Alpine Garden Society
Bulletin
vol. 36, No. 2   1968

fl., hab.   p. 149

Dionysia curviflora

Royal Hort. Soc.
Journal of the Royal Hort. Soc.
vol 90, No. 7   1965

fl.   p. 294   Pl. 132

Dionysia hedgei

Royal Hort. Soc.
Journal of the Royal Hort. Soc.
vol 96, No. 9   1971

fl.   p. 404   Pl. 172

Dionysia hybrid

Royal Hort. Soc.
The Garden
vol 100, No. 6   1975

fl.   p. 256

Dionysia lamingtonii

Royal Hort. Soc.
The Garden
Vol. 100, No. 6   1975

fl.   p. 258

Dionysia odora

Morley, Brian D.
Wild Flowers of the World

fl.   Pl. 48

Dionysia sp.

Alpine Garden Society
Bulletin
vol. 31, No. 4   1963

fl.   p. 302

Dionysia tapetodes

Curtis's Botanical Magazine
Vol 178   1970-72

fl.   589

Dionysia tapetodes

Royal Hort. Soc.
Journal of the Royal Hort. Soc.
vol 90, No. 11   1965

fl.   p. 470   Pl. 206

Dionysia termeana

Royal Hort. Soc.
The Garden
vol 100, No. 6   1975

fl.   p. 256

Dioon edule

American Hort. Soc.
American Horticulturist
vol 56, No. 3   1977

fr., hab.   p. 23

Dioon edule

Milne, Lorus
Living Plants of the World

fr.   p 18

Dioscorea batatas

Hvass, Elsie
Plants That Feed and Serve Us

fl. fr.   p 11, Pl. 13

Dioscorea benthamii

Walden, Beryl M.
Wild Flowers of Hong Kong

fl. fr.   Pl. 47 (137)

Dioscorea bulbifera var.

Kimura, Koiti
Japanese Medicinal Plants, Vol II

fl.   p 143

Dioscorea elephantipes

Eliovson, Sima
Namaqualand in Flower

hab.   Pl. 56, 2

Dioscorea japonica

Kariyone, Tatsuo
Atlas of Medicinal Plants

fl., hab.   Pl. 10

Dioscorea japonica

Takatori, Jisuke
Color Atlas of Medicinal Plants of Japan

fl.
fr.   Fig. 69

Dioscorea quartiniana

Moriarty, Audrey
Wild Flowers of Malawi

fl.   Pl. 58; 2

Dioscorea quaternata

Wharton, Mary E.
A Guide to the Wildflowers & Ferns
of Kentucky

fl. fr. p 81 Pl. 1.12

Dioscorea rotundata

Masefield, G. B.
The Oxford Book of Food Plants

hab.   Pl. 183, 2

Dioscorea tokoro

Kariyone, Tatsuo
Atlas of Medicinal Plants

fr.   Pl. 11

Dioscorea villosa

Courtenay, Booth
Wildflowers & Weeds

fl.   p 16

Diosma, Pink

see

Coleonema pulchrum

Diosphaera asperuloides

Morley, Brian D.
Wild Flowers of the World

fl.,hab.   Pl. 33

Diospyros ebenum

Hvass, Elsie
Plants That Feed and Serve Us

fl. fr.   p. 129   Pl. 285

Diospyros kaki

Bianchini, F.
The Complete Book of Fruits & Vegetables

fr.   p 179

Diospyros kaki

Edlin, Herbert
The Illustrated Encyclopedia
of Trees

fr.   p 203

Diospyros Kaki

Everett, Thomas H.
Living Trees of the World

hab. fr.   p. 293

Diospyros kaki

Hvass, Elsie
Plants That Feed and Serve Us

fl. fr.   p 51, Pl. 96

Diospyros kaki

Kimura, Koiti
Japanese Medicinal Plants, Vol II

fl.   p 211

Diospyros kaki

Kitamura, Siro
Coloured Illustrations of Trees &
Shrubs of Japan

fr.   402

Diospyros kaki

Macoby, Stirling
What Flower is That

fr.   P 115

Diospyros kaki

Masefield, G. B.
The Oxford Book of Food Plants

fl. fr.   Pl. 105, 2

Diospyros kaki

Massachusetts Hort. Soc.
Horticulture
vol 49, No. 9 1971

fr.   p. 32-33

Diospyros kaki

Massachusetts Hort. Soc.
Horticulture
vol 50, No. 9 1972

hab., fr.   p. 41

Diospyros kaki

Massachusetts Hort. Soc.
Horticulture
Vol. 53, No. 7   1975

fr.   p. 36

Diospyros kaki

Polunin, Oleg
Trees and Bushes of Europe

fr. hab.   p 159

Diospyros lotus

Curtis's Botanical Magazine
Vol. 180   1974

fl. fr.   Pl. 696

Diospyros lotus

Polunin, Oleg
Trees and Bushes of Europe

fr. hab.   p 159

Diospyros malabarica

Oakman, Harry
Colorful Trees

Hab.   P 94

Diospyros morrisiana

Kitamura, Siro
Coloured Illustrations of Trees &
Shrubs of Japan

fr.   403

Diospyros nummularia

Palgrave, K. C.
Trees of Central Africa

fl., fr.   p. 167

Diospyros quiloensis

Palgrave, K. C.
Trees of Central Africa

fl., fr.   p. 170

Diospyros soubreana

Wit, H. C. D. de
Plants of the World;
    The Higher Plants, Vol I

fr.   p 174, Pl. 102

Diospyros virginiana

American Hort. Soc.
American Horticulturist
    vol 57, No. 4 1978

fr.   p. 26

Diospyros virginiana

Edlin, Herbert
The Illustrated Encyclopedia
    of Trees

fr. fl. hab.   p 202-3

Diospyros virginiana

Wharton, Mary E.
Trees & Shrubs of Kentucky

fl.   p 99, Pl. 2.36
fr.   p 122, Pl. 2.5

Diothonea cottoniaeflora

Ospina, Mariano
Orquideas de las americas

fl.   Pl. 56

Dipcadi oligotrichum

Flowering Plants of South Africa
    vol XXIV   1944

fl., fr., hab.   Pl. 934

Dipcadi papillatum

Flowering Plants of Africa
    vol 36   1963-64

fl., hab.   Pl. 1429

Dipcadi sp.

Moriarty, Audrey
Wild Flowers of Malawi

fl.   Pl. 15; 2

Dipcadi tortile

Flowering Plants of South Africa
    vol XXIV   1944

fl., hab.   Pl. 956

Dipelta floribunda

Hellyer, A.G.L.
Shrubs in Colour

fl.   p. 43

Dipelta floribunda

Morley, Brian D.
Wild Flowers of the World

fl.   Plate 94

Diphylleia cymona

Campbell, Carlos C.
Great Smoky Mountain Wildflowers

fl.   p 13

Diphylleia cymosa

Duncan, Wilbur H.
Wildflowers of the Southeastern
    United States

fl.   p 49

Dipidax triquetra

Eliovson, Sima
Namaqualand in Flower

fl.   Pl. 34, 3

Dipidax triquetra

Perry, Frances
Flowers of the World

fl.   p 176

Diplacus longiflorus

Walcott, Mary Vaux
North American Wild Flowers

fl. hab.   Pl. 200

Diplacus puniceus

Walcott, Mary Vaux
North American Wild Flowers
    vol. 4

fl.   Pl. 315

Dipladenia hybrid

Kromdijk, G.
200 House Plants in Colour

fl.   Pl. 71

Dipladenia sanderii

Encke, Fritz
Zimmerpflanzen

fl.   p. 74

Dipladenia sanderii

Nicolaisen, Age
Pocket Encyclopedia of Indoor Plants

fl.   pl. 135

Dipladenia sanderii var.

Harrison, Richmond E.
Climbers and Trailers

fl.   p. 45   Pl. 88

Dipladenia splendens

Everett, T.H., ed.
New Illustrated Encyclopedia of
    Gardening   vol 4

fl.   frontispiece

Dipladenia splendens

Hay, Roy
The Dictionary of House Plants

fl.   Pl. 193

Diplarrhena latifolia

Curtis, Winifred
The Endemic Flora of Tasmania
    Vol. VI

fl. hab.   Pl. 200

Diplarrhena moraea

Blombery, Alec M.
What Wildflower is That

fl.   p. 119   Pl. 305

Diplaspis cordifolia

Curtis, Winifred
The Endemic Flora of Tasmania, Vol. 5

fl. fr. hab.   Pl. 179

Diplocyatha ciliata

Flowering Plants of South Africa
    vol IV   1924

fl.   Pl. 137

Diplolaena angustifolia

Blombery, Alec M.
What Wildflower is That

fl.   p 120, Pl. 306

Diplolaena dampieri

Mullins, Barbara
Australian Wildflowers in Colour

fl.   p. 71   Pl. 67

Diplolaena grandiflora

Harrison, R.E.
Trees andShrubs

fl. P 62

Diplolophium buchananii

Moriarty, Audrey
Wild Flowers of Malawi

fl.   Pl. 71; 1

Diploprora championii

Hu, Shiu-ying
The Genera of Orchidaceae in
    Hong Kong

fl.   p 123

Diploprora championii

Walden, Beryl M.
Wild Flowers of Hong Kong

fl.   Pl. 27(68)

Diplorhynchus condylocarpon var.

Palgrave, K. C.
Trees of Central Africa

fl., fr.   p. 23

Diplosoma retroversum

Herre, H.
The Genera of the Mesembryanthemaceae

fl.   fr.   p 135

Diplotaxis muralis

Bianchini, F.
The Complete Book of Fruits & Vegetables

hab.   p 59

Diplotaxis muralis

Felsko, Elsa
A Book of Wild Flowers
            2nd ser
fl. fr. hab. p 21

Diplotaxis muralis

Martin, W. Keble
The Concise British Flora in Colour

fl. fr. hab.   Pl. 10

Diplotaxis tenuifolia

Ary, S.
The Oxford Book of Wildflowers

fl.   p 8, Pl. 3

Diplotaxis tenuifolia

Martin, W. Keble
The Concise British Flora in Colour

fl. fr. hab.   Pl. 9

Dipodium punctatum

Blombery, Alec M.
What Wildflower is That

fl.   p 120, Pl. 307

Dipodium punctatum

Cady, Leo
Australian Native Orchids in Colour

fl.   Pl. 74

Dipodium punctatum

Cochrane, G. R.
Flowers and Plants of Victoria

fl.   Pl. 438

Dipsacus fullonum

Ary, S.
The Oxford Book of Wildflowers

fl.   p 156, Pl. 5

Dipsacus fullonum

Color Treasury of Herbs & Other
    Medicinal Plants

fl.   p 52   Pl. 79

Dipsacus fullonum

Martin, W. Keble
The Concise British Flora in Colour

fl. hab.   Pl. 43

Dipsacus fullonum

Morley, Brian
Wild Flowers of the World

fl.   Pl. 17B

Dipsacus fullonum

Perrot, Emile
Les Plantes Medicinales

fr.   p 235

Dipsacus fullonum

Perry, Frances
Flowers of the World

fl.   p 101

Dipsacus laciniatus

Courtenay, Booth
Wildflowers & Weeds

fl.   p 79

Dipsacus laciniatus

Polunin, Oleg
Flowers of Europe

fl.   Pl. 137 #1318

Dipsacus pilosus

Martin, W. Keble
The Concise British Flora in Colour

fl. hab.   Pl. 43

Dipsacus sylvestris

Clark, Lewis J.
Wild Flowers of British Columbia

fl.   p 507

Dipsacus sylvestris

Ferguson, Mary
Wildflowers

fl.   p 100

Dipsacus sylvestris

Kleijn, H.
Beauty of the Wild Plant

fl.   opp. p. 63   pl. 82

Dipsacus sylvestris

Klimas, John E.
Wildflowers of Eastern America

fl.   Pl. 224

Dipsacus sylvestris

Perrot, Emile
Les Plantes Medicinales

fl. hab.   p 235

Dipsacus sylvestris

Pond, Barbara
A Sampler of Wayside Herbs

fl.   Pl. XXIII

Dipsacus sylvestris

Wit, H. C. D. de
Plants of the World:
    The Higher Plants, Vol II

fl.   p 108, Pl. 73

Dipteranthus grandiflorus

Pabst, G. F. J.
Orchidaceae Brasilienses,
    Vol 2

fl.   p 263

Dipteranthus planifolius

Dunsterville, G. C. K.
Introduction to the World
    of Orchids

fl. hab.   Pl. 6

Dipteranthus pustulatus

Pabst, G. F. J.
Orchidaceae Brasilienses,
    Vol 2

fl.   p 263

Dipterocarpus alatus

Roxburgh, William
Icones Roxburghianae

fr.   Pl. 19

Dipterocarpus trinervis

Morley, Brian D.
Wild Flowers of the World

fl.fr.   Plate 119

Dipterocarpus tuberculatus

Roxburgh, William
Icones Roxburghianae

fr.   Pl. 20

Dirca palustris

Wharton, Mary E.
Trees & Shrubs of Kentucky

fl.   p 108, Pl. 4.6

Disa aequiloba

Williamson, Graham
The Orchids of South Central Africa

fl.   Pl. 56

Disa amblyopetala

Williamson, Graham
The Orchids of South Central Africa

fl.   Pl. 60

Disa aperta

Williamson, Graham
The Orchids of South Central Africa

fl.   Pl. 67

Disa begleyi

Flowering Plants of South Africa
vol XIX   1939

fl.   Pl. 736

Disa celata

Williamson, Graham
The Orchids of South Central Africa

fl.   Pl. 58

Disa compta

Williamson, Graham
The Orchids of South Central Africa

fl.   Pl. 64

Disa concinna

Moriarty, Audrey
Wild Flowers of Malawi

fl.   Pl. 25; 3

Disa cryptantha

Williamson, Graham
The Orchids of South Central Africa

fl.   Pl. 57

Disa dichroa

Williamson, Graham
The Orchids of South Central Africa

fl.   Pl. 68

Disa eminii

Williamson, Graham
The Orchids of South Central Africa

fl.   Pl. 59

Disa englerana var.

Williamson, Graham
The Orchids of South Central Africa

fl.   Pl. 52-53

Disa erubescens

Moriarty, Audrey
Wild Flowers of Malawi

fl.   Pl. 24; 1

Disa hamatopetala

Moriarty, Audrey
Wild Flowers of Malawi

fl.   Pl. 25; 4

Disa hamatopetala

Williamson, Graham
The Orchids of South Central Africa

fl.   Pl. 71

Disa hiricornis

Moriarty, Audrey
Wild Flowers of Malawi

fl.   Pl. 25; 1

Disa hircicornis

Williamson, Graham
The Orchids of South Central Africa

fl.   Pl. 61

Disa hiricornis var.

Williamson, Graham
The Orchids of South Central Africa

fl.   Pl. 62-63

Disa longilabris

Williamson, Graham
The Orchids of South Central Africa

fl.   Pl. 70

Disa maculata

Flowering Plants of Africa
vol 42   1972-73

fl., hab.   Pl. 1678

Disa nervosa

Flowering Plants of Africa
vol 30   1954-55

fl.   Pl. 1173

Disa ornithantha

Moriarty, Audrey
Wild Flowers of Malawi

fl.   Pl. 24; 2

Disa ornithantha

Williamson, Graham
The Orchids of South Central Africa

fl.   Pl. 50

Disa patula var.

Flowering Plants of Africa
vol 30   1954-55

fl.   Pl. 1174

Disa pillansii

Flowering Plants of South Africa
vol XIX   1939

fl., hab.   Pl. 737

Disa robusta

Moriarty, Audrey
Wild Flowers of Malawi

fl.   Pl. 24; 4

Disa roeperocharoides

Williamson, Graham
The Orchids of South Central Africa

fl.   Pl. 55

Disa rungweensis

Williamson, Graham
The Orchids of South Central Africa

fl.   Pl. 69

Disa saxicola

Moriarty, Audrey
Wild Flowers of Malawi

fl.   Pl. 24; 3

Disa similis

Williamson, Graham
The Orchids of South Central Africa

fl.   Pl. 66

Disa stolzii

Williamson, Graham
The Orchids of South Central Africa

fl.   Pl. 49

Disa triloba

Flowering Plants of South Africa
vol XIX   1939

fl., hab.   Pl. 738

Disa uniflora

American Hort. Soc.
American Horticulturist
vol 53, No. 3 1974

fl.   p. 35

Disa uniflora

Flowering Plants of Africa
vol 30   1954-55

fl., hab.   Pl. 1180

Disa uniflora

Morley, Brian D.
Wild Flowers of the World

fl.,hab.   Pl. 89

Disa uniflora

Royal Hort. Soc.
Journal of the Royal Hort. Soc.
    vol 88, No. 8   1963

fl.   p. 342   Pl. 138

Disa uniflora

Royal Hort. Soc.
Journal of the Royal Hort. Soc.
    vol 94, No. 7   1969

fl.   p. 302   Pl. 166

Disa uniflora

Royal Hort. Soc.
Journal of the Royal Hort. Soc.
    vol 96, No. 3   1971

fl.   p. 122   Pl. 50

Disa verdickii

Williamson, Graham
The Orchids of South Central Africa

fl.   Pl. 51

Disa walleri

Williamson, Graham
The Orchids of South Central Africa

fl.   Pl. 65

Disa Welwitschii

Moriarty, Audrey
Wild Flowers of Malawi

fl.   Pl. 25; 5

Disa welwitschii var.

Williamson, Graham
The Orchids of South Central Africa

fl.   Pl. 54

Disanthus cercidifolius

Gault, S. Millar
The Color Dictionary of Shrubs

hab.   Pl. 155

Disanthus cercidifolius

Hay, Roy
The Color Dictionary of Flowers &
    Plants

hab.   p. 196   Pl. 1566

Disanthus cercidifolius

Hellyer, A.G.L.
Shrubs in Colour

hab.   p. 46

Disanthus cercidifolius

Kitamura, Siro
Coloured Illustrations of Trees &
    Shrubs of Japan

fl. fr.   207

Discocactus estevesii

Cactus & Succulent Society of America
Journal
    vol 50, No. ?   1978

fl., hab.   p. 83

Discocactus horstii

Cactus & Succulent Soc. of America
Journal
    vol 45, No. 5   1973

hab.   cover (p. 201)

Discocactus silicicola

Cactus & Succulent Society of America
Journal
    vol 47, No. 5   1975

fl., hab.   p. 215

Diselma archeri

Curtis, Winifred
The Endemic Flora of Tasmania, Vol. 5

fr.   Pl. 191

Disocactus macrantha

Lamb, Edgar
The Pocket Encyclopedia of Cacti
    and Succulents in Color

fl. hab.   Pl. 28

Disocactus pachythele

Cactus & Succulent Society of America
Journal
    vol 47, No. 4   1975

hab.   p. 163

Disocactus semicampaniflorus

Cactus & Succulent Society of America
Journal
    vol 47, No. 3   1975

fl., hab.   p. 123

Disparago ericoides

Flowering Plants of Africa
    vol 28   1950-51

fl.   Pl. 1102

Disperis anthoceros var.

Williamson, Graham
The Orchids of South Central Africa

fl.   Pl. 87

Disperis aphylla

Williamson, Graham
The Orchids of South Central Africa

fl.   Pl. 89

Disperis capensis

Flowering Plants of Africa
    vol 40   1969-70

fl., hab.   Pl. 1573

Disperis capensis

Morley, Brian D.
Wild Flowers of the World

fl.   Pl. 89

Disperis dicerochila

Moriarty, Audrey
Wild Flowers of Malawi

fl.   Pl. 27; 3

Disperis katangensis

Williamson, Graham
The Orchids of South Central Africa

fl.   Pl. 91

Disperis macowani

Flowering Plants of South Africa
    vol XXIV   1944

fl., hab.   Pl. 929

Disperis parvifolia var.

Williamson, Graham
The Orchids of South Central Africa

fl.   Pl. 88

Disperis reichenbachiana

Williamson, Graham
The Orchids of South Central Africa

fl.   Pl. 90

Disperis thorncroftii

Flowering Plants of Africa
    vol XXV   1945-46

fl., hab.   Pl. 963

Disphyma australe

Blombery, Alec M.
What Wildflower is That

fl. hab.   p 121, Pl. 308

Disphyma australe

Cochrane, G. R.
Flowers and Plants of Victoria

fl. hab.   Pl. 174, 175

Disphyma australe

Mullins, Barbara
Australian Wildflowers in Colour

fl.   p. 83   Pl. 80

Disphyma crassifolium

Herre, H.
The Genera of the Mesembryanthemaceae

fl.   fr.   p 137

Disporum hookeri

Taylor, Ronald J.
Mountain Wild Flowers

fl. hab.   p 99

Disporum lanuginosum

Dean, Blanche E.
Wildflowers of Alabama and
   Adjoining States

fl.   p 21

Disporum lanuginosum

Jennings, O. E.
Wild Flowers of Western Pennsylvania
                             vol II
fl.   hab.   Pl. 179

Disporum lanuginosum

Wharton, Mary E.
A Guide to the Wildflowers & Ferns
   of Kentucky

fl.   p 57   Pl. 1.20

Disporum maculatum

Klimas, John E.
Wildflowers of Eastern America

fl.   Pl. 137

Disporum maculatum

Wharton, Mary E.
A Guide to the Wildflowers & Ferns
   of Kentucky

fl.   p 56   Pl. 1.19

Disporum oreganum

Porsild, A. E.
Rocky Mountain Wild Flowers

fr.   p 97

Disporum smithii

Clark, Lewis J.
Wild Flowers of British Columbia

fl.   p 27

Disporum smithii

Lemmon, Robert S.
Wildflowers of North America in
   Full Color

fl.   pg. 227   pl. 361

Disporum smithii

Royal Hort. Soc.
Journal of the Royal Hort. Soc.
      vol 98, No. 2   1973

fr., hab.   p. 56   Pl. 27

Dissotis canescens

Moriarty, Audrey
Wild Flowers of Malawi

fl.   Pl. 72; 3

Dissotis debilis

Moriarty, Audrey
Wild Flowers of Malawi

fl.   Pl. 72; 2

Dissotis macrocarpa

Lind, E. M.
Some Common Flowering Plants of
   Uganda

fl.   Pl. 4

Dissotis melleri

Moriarty, Audrey
Wild Flowers of Malawi

fl.   Pl. 72; 1

Dissotis princeps

Flowering Plants of Africa
      vol 32   1957-58

fl.   Pl. 1250

Dissotis princeps

Morley, Brian D.
Wild Flowers of the World

fl.   Pl. 59

Dissotis senegambiensis

Moriarty, Audrey
Wild Flowers of Malawi

fl.   Pl. 72; 4

Distictella elongata

Menninger, Edwin A.
Flowering Vines of the World

fl.   Pl. 38

Distictella magnoliifolia

Mee, Margaret
Flowers of the Brazilian Forests

fl.   Pl. 12

Distictis buccinatoria

Mathias, Mildred E.
Color for the Landscape

fl. hab.   p 118

Distictis x riversii

Mathias, Mildred E.
Color for the Landscape

fl., hab.   p. 110

Distictis x riversii

Menninger, Edwin A.
Flowering Vines of the World

fl.   Pl. 42

Distylium racemosum

Curtis's Botanical Magazine
      Vol 160   1937

fl.   No. 9501

Distylium racemosum

Kitamura, Siro
Coloured Illustrations of Trees &
   Shrubs of Japan

fl.   210

Dithyrea wislizenii

Welsh, Stanley L.
Flowers of the Canyon Country

fl. hab.   p 3

Dittany

see

Cunila origanoides

Dittany

see

Dictamnus albus

Dittany of Crete

see

Origanum dictamnus

Diuris aurea

Blombery, Alec M.
What Wildflower is That

fl.   p 121, Pl. 309

Diuris aurea

Cady, Leo
Australian Native Orchids in Colour

fl.   Pl. 16

Diuris carinata

Morcombe, M. K.
Australia's Western Wildflowers

fl.   p 14

Diuris longifolia

Blombery, Alec M.
What Wildflower is That

fl.   p 121, Pl. 310

Diuris longifolia

Cady, Leo
Australian Native Orchids in Colour

fl.   Pl. 15

Diuris longifolia

Morcombe, M. K.
Australia's Western Wildflowers

fl.   p 15

Diuris maculata

Cady, Leo
Australian Native Orchids in Colour

fl.   Pl. 17

Diuris pedunculata

Curtis's Botanical Magazine
        vol 176   1966-68

fl.   Pl. 513

Diuris pedunculata

Perry, Frances
Flowers of the World

fl.   p 211

Diuris punctata

Blombery, Alec M.
What Wildflower is That

fl.   p. 121   Pl. 311

Diuris punctata

Cady, Leo
Australian Native Orchids in Colour

fl.   Pl. 18

Diuris punctata

Cochrane, G. R.
Flowers and Plants of Victoria

fl.   Pl. 66

Diuris punctata var.

Cochrane, G. R.
Flowers and Plants of Victoria

fl.   Pl. 228

Dizygotheca elegantissima

Hay, Roy
The Color Dictionary of Flowers & Plants

hab.   p 65, Pl. 514

Dizygotheca elegantissima

Hay, Roy
The Dictionary of House Plants

hab.   Pl. 194

Dizygotheca elegantissima

Hersey, Jean
Woman's Day Book of House Plants

hab.   p 62

Dizygotheca elegantissima

Kiaer, Eigil
Indoor Plants in Colour

hab.   p. 51

Dizygotheca elegantissima

Kromdijk, G.
200 House Plants in Colour

hab.   Pl. 72

Dock
see
Rumex

Dock, Arctic
see
Rumex arctica

Dock, Broad Leaved
see
Rumex obtusifolius

Dock, Curled
see
Rumex crispus

Dock, Great Water
see
Rumex orbiculatus

Dock, mountain
see
Polygonum bistortoides

Dock, Pale
see
Rumex altissimus

Dock, Prairie
see
Silphium terebinthinaceum

Dock, Rubble
see
Rumex scutatus

Dock, Swamp
see
Rumex verticillatus

Dock, Veined
see
Rumex venosus

Dock, Water
see
Rumex verticillatus

Dock, Yellow
see
Rumex crispus

Dockmackie
see
Viburnum acerifolium

Dodder
see
Cuscuta

Dodder, Common
see
Cuscata epithymum

Dodder, Japanese
see
Cuscuta japonica

Dodder, Large
see
Cuscata europaea

Dodecatheon alpinum

Hay, Roy
The Color Dictionary of Flowers & Plants

fl.   p 7, Pl. 56

Dodecatheon alpinum

Royal Hort. Soc.
The Garden
vol 101, No. 5   1976

fl., hab.   p. 263

Dodecatheon alpinum

Welsh, Stanley L.
Flowers of the Mountain Country

fl.   p. 31

Dodecatheon cylindrocarpum

Neufeld, J.B.
Wild Flowers of the Prairies

fl.   p. 31

Dodecatheon dentatum

Clark, Lewis J.
Wild Flowers of British Columbia

fl.   p 415

Dodecatheon frigidum

Alaska-Yukon Wild Flower Guide

fl. hab.   p 126

Dodecatheon frigidum

Heller, Christine
Wild Flowers of Alaska

fl., hab.   Pl. 75

Dodecatheon hendersonii

Clark, Lewis J.
Wild Flowers of British Columbia

fl.   p 418

Dodecatheon hendersonii

Ferguson, Mary
Wildflowers

fl.   P 75

Dodecatheon hendersonii

Milne, Lorus
Living Plants of the World

fl.   p. 192

Dodecatheon jeffreyi

Clark, Lewis J.
Wild Flowers of British Columbia

fl.   p 415

Dodecatheon jeffreyi

Fries, Mary A.
Wildflowers of Mount Ranier and
the Cascades

fl.   P 77

Dodecatheon jeffreyi

Munz, Philip A.
California Mountain Wildflowers

fl.   p 34, Pl. 24

Dodecatheon jeffreyi

Orr, Robert T.
Wildflowers of Western America

fl.   Pl. 151

Dodecatheon littorale

Clark, Lewis J.
Wild Flowers of British Columbia

fl.   p 419

Dodecatheon meadia

Courtenay, Booth
Wildflowers & Weeds

fl.   p 84

Dodecatheon meadia

Dean, Blanche E.
Wildflowers of Alabama and
Adjoining States

fl. hab.   p 127

Dodecatheon meadia

Duncan, Wilbur H.
Wildflowers of the Southeastern
United States

fl. hab.   p 121

Dodecatheon meadia

Klimas, John E.
Wildflowers of Eastern America

fl.   Pl. 192

Dodecatheon meadia

Lemmon, Robert S.
Wildflowers of North America in
Full Color

fl.   pg. 194   pl. 305

Dodecatheon meadia

Perry, Frances
Flowers of the World

fl.   p 243

Dodecatheon meadia

Vilmorin, Roger de
Plantes Alpines dans les Jardins

fl. hab.   Pl. IV

Dodecatheon meadia

Walcott, Mary Vaux
North American Wild Flowers

fl. hab.   Pl. 49

Dodecatheon meadia

Wharton, Mary E.
A Guide to the Wildflowers & Ferns
of Kentucky

fl.   p 116   Pl. 1.45

Dodecatheon meadia var.

Hay, Roy
The Color Dictionary of Flowers &
Plants

fl.   p. 139   Pl. 1112

Dodecatheon meadia var.

Massachusetts Hort. Soc.
Horticulture
vol 45, No. 4   1967

fl.   p. 29

Dodecatheon pauciflorum

Porsild, A. E.
Rocky Mountain Wild Flowers

fl. hab.   p 327

Dodecatheon pauciflorum

Taylor, Ronald J.
Mountain Wild Flowers

fl.   p 139

Dodecatheon pauciflorum

Walcott, Mary Vaux
North American Wild Flowers
vol. 4

fl., hab.   Pl. 276

Dodecatheon pulchellum

Alaska-Yukon Wild Flowers Guide

fl. hab.   p 4, 5

Dodecatheon pulchellum

Clark, Lewis J.
Wild Flowers of British Columbia

fl.   p 422

Dodecatheon pulchellum

Morley, Brian D.
Wild Flowers of the World

fl. fr.   Pl. 156

Dodecatheon radicatum

Lemmon, Robert S.
Wildflowers of North America in
Full Color

fl.   pg. 144   pl. 229

Dodecatheon radicatum

Welsh, Stanley L.
Flowers of the Canyon Country

fl.   P. 19

Dodecatheon alinum

Massachusetts Hort. Soc.
Horticulture
vol 52, No. 8   1974

fl., hab.   p. 35

Dodocatheon viviparum

Heller, Christine
Wild Flowers of Alaska

fl., hab.   Pl. 73-74

Dodonaea cuneata

Cochrane, G. R.
Flowers and Plants of Victoria

fl.   Pl. 200

Dodonaea ericifolia

Curtis, Winifred
The Endemic Flora of Tasmania Vol. 2

fl, fr.   Pl. 77

Dodonaea pinnata

Blombery, Alec M.
What Wildflower is That

fr.   p. 122   Pl. 312

Dodonaea triquetra

Blombery, Alec M.
What Wildflower is That

fr.   p 122, Pl. 313

Dodonaea viscosa var.

Harrison, R.E.
Trees and Shrubs

fl.   P 62

Dogbane

see

Apocynum

Dogbane, Blue

see

Amsonia tabernaemontana

Dogbane, Climbing

see

Trachelospermum difforme

Dogbane, Hemp

see

Apocynum cannabinum

Dogbane, Spreading

see

Apocynum androsaemifolium

Dogbane, star

see

Amsonia rigida

Dogberry

see

Geocaulon lividum

Dogwood

see

Cornus

Dogwood

see

Jacksonia scoparia

Dogwood, Canadian

see

Cornus canadensis

Dogwood, eastern

see

Cornus florida var.

Dogwood, evergreen

see

Cornus capitata

Dogwood, Japanese

see

Cornus kousa

Dogwood, Japanese Flowering

see

Cornus kousa

Dogwood, Mountain

see

Cornus nuttallii

Dogwood, Pacific

see

Cornus nuttallii

Dogwood, pink

see

Cornus florida var.

Dogwood, Red-barked

see

Cornus alba var.

Dogwood, Red-osier

see

Cornus stolonifera

Dogwood, White Flowering

see

Cornus florida

Dogwood, Yellow

see

Pomaderris elliptica

Dolichandrone alba

Flowering Plants of Africa
vol 28   1950-51

fl.   Pl. 1092

Dolichos kilimandscharicus

Moriarty, Audrey
Wild Flowers of Malawi

fl.   Pl. 65; 3

Dolichos lablab

Crockett, James Underwood
Annuals

fl.   P 116

Dolichos lablab

Masefield, G. B.
The Oxford Book of Food Plants

fl. fr.   Pl. 45, 2

Dolichos lignosus

Harrison, Richmond E.
Climbers and Trailers

fl.   p 45   Pl. 89

Dolichos lignosus

Macoby, Stirling
What Flower is That

fl.   P 115

Dolichos peglerae

Flowering Plants of Africa
vol XXVI   1947

fl.   Pl. 1028

Dolichothele longimamma

Kupper, Walter
Cacti

fl., hab.   p. 113 Pl. 53

Dolichothele longimamma

Lamb, Edgar
Popular Exotic Cacti in Color

fl. hab.   p 65

Dolichothele longimamma

Van Laren, A. J.
Cactus

fl.   P 68   Fig. 86

Dolichothele melaleuca

Kupper, Walter
Cacti

fl., hab.   p. 59 Pl. 26

Dolichothele sphaerica

Lamb, Edgar
Colorful Cacti of the American Deserts

fl.   Pl. 6

Dolichothele sphaerica

Lamb, Edgar
The Pocket Encyclopedia of Cacti
and Succulents in Color

fl. hab.   Pl. 29

Dollar Grass

see

Hydrocotyle umbellata

Dollar, Sand

see

Astrophytum asterias

Dollar Weed

see

Rhynchosia simplicifolia

Doll's Eye

see

Actaea alba

Doll's Eye

see

Actaea pachypoda

Dombeya cacuminum

American Hort. Soc.
American Horticulturist
vol 56, No. 5   1977

fl.   p. 14

Dombeya cacuminum

Mathias, Mildred E.
Color for the Landscape

fl. hab.   p 18

Dombeya x cayeuxii

Addisonia
vol. 23   1954-59

fl.   p. 1   Pl. 737

Dombeya x cayeuxii

Curtis's Botanical Magazine
vol 175   1964-65

fl.   Pl. 473

Dombeya mastersii

Morley, Brian D.
Wild Flowers of the World

fl.   Pl. 57

Dombeya natalensis

Macoby, Stirling
What Flower is That

fl.   P 116

Dombeya pulchra

Flowering Plants of Africa
vol 38   1967

fl., fr.   Pl. 1489

Dombeya rotundifolia

Flowering Plants of Africa
vol 29   1952-53

fl.   Pl. 1143

Dombeya rotundifolia

Palgrave, K. C.
Trees of Central Africa

fl.   p. 417

Dombeya rotundifolia

Palmer, Eve
Trees of South Africa

fl.   p. 288   Pl. XXVI

Donax canniformis

Bruggeman, L.
Tropical Plants

fl.,fr.   Pl. 107

Donkey's Tail

see

Sedum morganianum

x Doritaenopsis hybrid

Kramer, Jack
Orchids; Flowers of Romance
and Mystery

fl.   p 274

Doritis pulcherrima

Kamemoto, Haruyuki
Beautiful Thai Orchid Species

fl.   frontispiece, p. 148

Doritis pulcherrima var.

Kamemoto, Haruyuki
Beautiful Thai Orchid Species

fl.   p 148

Doronicum austriacum var.

Goulimis, Constantine N.
Wild Flowers of Greece

fl.   p. 131

Doronicum caucasicum

Everett, T.H., ed.
New Illustrated Encyclopedia of
Gardening   vol 4

fl.   p. 630

Doronicum caucasicum

Massachusetts Hort. Soc.
Horticulture
vol 49, No. 11   1971

fl.   p. 39

Doronicum caucasicum var.

Hay, Roy
The Color Dictionary of Flowers & Plants

fl.   p 140, Pl. 1113

Doronicum caucasicum var.

Huxley, Anthony
Garden Perennials and Water Plants

fl.   Pl. 102

Doronicum clusii

Barneby, T. P.
European Alpine Flowers in Colour

fl. hab.   Pl. 95, 4

Doronicum cordatum

Barneby, T. P.
European Alpine Flowers in Colour

fl. hab.   Pl. 95, 5

Doronicum cordatum

Hay, Roy
The Color Dictionary of Flowers & Plants

fl.   p 140, Pl. 1114

Doronicum grandiflorum

Barneby, T. P.
European Alpine Flowers in Colour

fl. hab.   Pl. 95, 3

Doronicum grandiflorum

Kleijn, H.
Beauty of the Wild Plant

fl.   opp. p. 97   Pl. 137

Doronicum grandiflorum

Kohlhaupt, Paula
Fleurs des Alpages
                vol 1
fl.   P 115-116

Doronicum grandiflorum

Polunin, Oleg
Flowers of Europe

fl.   Pl. 150 #1448

Doronicum grandiflorum

Tosco, Uberto
The World of Mountain Flowers

fl. hab.   p 64

Doronicum hybridum

Macoby, Stirling
What Flower is That

fl.   P 116

Doronicum pardalianches

Felsko, Elsa
A Book of Wildflowers

fl.   P 41

Doronicum pardalianches

Polunin, Oleg
Flowers of Europe

fl.   Pl. 151 #1447

Dorotheanthus bellidiformis

Crockett, James Underwood
Annuals

fl.   p 136

Dorotheanthus bellidiformis

Eliovson, Sima
Namaqualand in Flower

fl.   Pl. 54, 1

Dorotheanthus bellidiformis

Flowering Plants of Africa
            vol 35   1962

fl., hab.   Pl. 1365

Dorotheanthus bellidiformis

Herre, H.
The Genera of the Mesembryanthemaceae

fl.   fr.   p 139

Dorotheanthus bellidiformis

Macoby, Stirling
What Flower is That

fl. P 116

Dorotheanthus bellidiformis

Perry, Frances
Flowers of the World

fl.   p 18

Dorstenia crispa

Flowering Plants of Africa
            vol 35   1962

fl., hab.   Pl. 1372

Dorstenia crispa

Morley, Brian D.
Wild Flowers of the World

fl.,hab.   Pl. 53

Dorstenia gigas

Cactus & Succulent Society of America
Journal
            vol 46, No. 5   1974

fl., hab.   p. 224

Dorstenia gigas

Curtis's Botanical Magazine
            vol 178   1970-72

fl.   Pl. 596

Dorstenia gypsophila

Cactus & Succulent Soc. of America
Journal
            vol 44, No. 6   1972

fl.   p. 259

Dorstenia poinsettifolia

Wit, H. C. D. de
Plants of the World;
    The Higher Plants, Vol. I

fl.   p. 102   Pl. 50

Doryanthes excelsa

Blombery, Alec M.
What Wildflower is That

fl. hab.   p. 122   Pl. 314, 315

Doryanthes excelsa

Macoby, Stirling
What Flower is That

fl.   P 117

Doryanthes excelsa

Mullins, Barbara
Australian Wildflowers in Colour

fl.   P. 107   Pl. 105

Dorycnium hirsutum

Polunin, Oleg
Flowers of Europe

fl.   Pl. 60 #609

Dorycnium hirsutum

Royal Hort. Soc.
Journal of the Royal Hort. Soc.
            vol 99, No. 2   1974

fl.   p. 72   Pl. 34

Dorycnium rectum

Polunin, Oleg
Flowers of Europe

fl.   Pl. 60 #610

Doryphora sassafras

Curtis's Botanical Magazine
Vol 165   1948

fl.   30

Doubah, Sweet-scented

see

Marsdenia suavolens

Douglasia gormanii

Alaska-Yukon Wild Flower Guide

fl. hab.   p 125

Douglasia laevigata

Taylor, Ronald J.
Mountain Wild Flowers

fl. hab.   p 97

Douglasia laevigata var.

Clark, Lewis J.
Wild Flowers of British Columbia

fl.   p 426

Douglasia ochotensis

Alaska-Yukon Wild Flower Guide

fl. hab.   p 124

Douglasia, Ochotsk

see

Douglasia ochotensis

Douglasia, Smooth

see

Douglasia laevigata

Douglasia vitaliana

Barneby, T. P.
European Alpine Flowers in Colour

fl. hab.   Pl. 60, 4

Douglasia vitaliana

Hay, Roy
The Color Dictionary of Flowers & Plants

fl.   p 8, Pl. 57

Dove-Tree

see

Davidia involucrata

Dovyalis caffra

Flowering Plants of Africa
vol 41   1970-71

fl., fr.   Pl. 1608

Dovyalis zeyheri

Flowering Plants of Africa
vol 39   1968-69

fl., fr.   Pl. 1546

Doxantha unguis-cati

Harrison, Richmond E.
Climbers and Trailers

fl. p 45   Pl. 90

Doxantha unguis-cati

Macoby, Stirling
What Flower is That

fl.   P 117

Doxantha unguis-cati

Mathias, Mildred E.
Color for the Landscape

fl. hab.   p 109

Doxantha unguis-cati

Menninger, Edwin A.
Flowering Vines of the World

fl.   Pl. 32

Draba aizoides

Barneby, T. P.
European Alpine Flowers in Colour

fl. hab.   Pl. 30, 3

Draba aizoides

Felsko, Elsa
A Book of Wild Flowers
                2nd ser.
fl. fr. hab. p 5

Draba aizoides

Kohlhaupt, Paula
Fleurs des Alpages
                vol 1
fl.   P 68

Draba aizoides

Martin, W. Keble
The Concise British Flora in Colour

fl. hab.   Pl. 8

Draba aizoides

Morley, Brian D.
Wild Flowers of the World

fl. fr.   Pl. 11E

Draba aizoides

Polunin, Oleg
Flowers of Europe

fl.   Pl. 34 #331

Draba alpina

Lindman, C. A. M.
Nordens Flora, Vol 4

fl. fr. hab.   Pl. 266A

Draba bryoides var.

Hay, Roy
The Color Dictionary of Flowers & Plants

fl.   p 8, Pl. 58

Draba crassifolia

Morley, Brian D.
Wild Flowers of the World

fl. fr.   Pl. 1C

Draba hirta

Alaska-Yukon Wild Flower Guide

fl. hab.   p 184

Draba imbricata

Royal Hort. Soc.
Journal of the Royal Hort. Soc.
vol 89, No. 7   1964

fl.   p. 290   Pl. 107

Draba incana

Martin, W. Keble
The Concise British Flora in Colour

fl. hab.   Pl. 8

Draba incana

Polunin, Oleg
Flowers of Europe

fl. Pl. 35 #333

Draba incerta

Clark, Lewis J.
Wild Flowers of British Columbia

fl.   p 183

Draba incerta

Porsild, A.E.
Rocky Mountain Wild Flowers

fl. hab.   p 201

Draba lemmonii

Munz, Philip A.
California Mountain Wildflowers

fl.   p 54, Pl. 83

Draba mollissima

Hay, Roy
The Color Dictionary of Flowers & Plants

fl.   p 8, Pl. 59

Draba muralis

Martin, W. Keble
The Concise British Flora in Colour

fl. fr. hab.   Pl. 8

Draba nivalis

Lindman, C. A. M.
Nordens Flora, Vol 4

fl. fr. hab.   Pl. 266B

Draba norvegica

Lindman, C. A. M.
Nordens Flora, Vol 4

fl. fr. hab.   Pl. 265B

Draba norvegica

Martin, W. Keble
The Concise British Flora in Colour

fl. fr. hab.   Pl. 8

Draba oligosperma

Clark, Lewis J.
Wild Flowers of British Columbia

fl.   p 186

Draba oligosperma

Porsild, A.E.
Rocky Mountain Wild Flowers

fl. fr. hab.   p 203

Draba paysonii

Taylor, Ronald J.
Mountain Wild Flowers of the Pacific
Northwest

fl.   p. 148

Draba repens

Vilmorin, Roger de
Plantes Alpines dans les Jardins

fl. hab.   Pl. XX

Draba reptans

Courtenay, Booth
Wildflowers & Weeds

fl.   p 18

Draba rigida

Alpine Garden Society
Bulletin
     vol. 35, No. 3   1967

fl., hab.   p. 225

Draba tomentosa

Barneby, T. P.
European Alpine Flowers in Colour

fl. hab.   Pl. 31, 5

Draba ventosa

Porsild, A.E.
Rocky Mountain Wild Flowers

fl. fr.   p 203

Dracaena cinnabari

Lamb, Edgar
The Pocket Encyclopedia of Cacti
     and Succulents in Color

hab.   Pl. 317

Dracaena deremensis

Hay, Roy
The Dictionary of House Plants

hab.   Pl. 195

Dracaena deremensis

Nicolaisen, Age
Pocket Encyclopedia of Indoor Plants

hab.   Pl. 21

Dracaena deremensis

Perry, Frances
Flowers of the World

hab. p 17

Dracaena deremensis var.

Encke, Fritz
Zimmerpflanzen
hab.   p. 74

Dracaena deremensis var.

Hay, Roy
The Color Dictionary of Flowers &
Plants

hab.   p. 65   Pl. 515

Dracaena deremensis var.

Kiaer, Eigil
Indoor Plants in Colour

hab.   p. 52

Dracaena deremensis var.

Massachusetts Hort. Soc.
Horticulture
     vol 35, No. 1   1957

hab.   p. 37 (see errata p. 108)

Dracaena deremensis var.

Massachusetts Hort. Soc.
Horticulture
     vol 35, No. 11   1957

hab.   p. 556

Dracaena draco

Bramwell, David
Wild Flowers of the Canary Islands
hab.   Pl. 322

Dracaena draco

Milne, Lorus
Living Plants of the World

hab.   p. 261

Dracaena draco

Perry, Frances
Flowers of the World

hab.   P. 15

Dracaena fragrans

Hay, Roy
The Dictionary of House Plants

hab.   Pl. 196

Dracaena fragrans

Kiaer, Eigil
Indoor Plants in Colour

hab.   p. 52

Dracaena fragrans

Kromdijk, G.
200 House Plants in Colour

hab.   Pl. 73

Dracaena fragrans var.

Hay, Roy
The Color Dictionary of Flowers &
Plants

hab.   p. 65   Pl. 516

Dracaena fragrans var.

Hersey, Jean
Woman's Day Book of House Plants

hab.   p. 49

Dracaena godseffiana

Hay, Roy
The Dictionary of House Plants

hab.   Pl. 197

Dracaena godseffiana

Kromdijk, G.
200 House Plants in Colour

hab.   Pl. 74

Dracaena godseffiana

Macoby, Stirling
What Flower is That

hab P 117

Dracaena hookeriana var.

Kiaer, Eigil
Indoor Plants in Colour

hab.   p. 52

Dracaena kindtiana

Flowering Plants of Africa
     vol 33   1959

fl., fr.   Pl. 1309

Dracaena marginata

Hay, Roy
The Color Dictionary of Flowers
& Plants

hab.   p 65   Pl. 517

Dracaena marginata

Nicolaisen, Age
Pocket Encyclopedia of Indoor Plants

hab.   Pl. 22

Dracaena marginata var.

Hay, Roy
The Dictionary of House Plants

hab.   Pl. 198

Dracaena massangeana

Royal Hort. Soc.
The Garden
    vol 103, No. 1   1978

hab.   p. 22

Dracaena sanderiana

Hay, Roy
The Color Dictionary of Flowers
    & Plants

hab.   p 65   Pl. 518

Dracaena sanderiana

Kromdijk, G.
200 House Plants in Colour

hab.   Pl. 75

Dracaena sanderiana

Macoby, Stirling
What Flower is That

hab. P 117

Dracaena sp.

Polunin, Oleg
Trees and Bushes of Europe

hab.   p 181

Dracaena surculosa

Perry, Frances
Flowers of the World

hab. p 17

Dracaena surculosa var.

Wit, H. C. D. de
Plants of the World;
    The Higher Plants, Vol. II

fl. fr.   Pl. 124

Dracaena terminalis

Royal Hort. Soc.
Journal of the Royal Hort. Soc.
    vol 90, No. 5   1965

hab.   p. 206   Pl. 93

Dracaena terminalis var.

Massachusetts Hort. Soc.
Horticulture
    vol 32, No. 2   1954

hab.   p. 77

Dracocephalum bullatum

Curtis's Botanical Magazine
    Vol 164   1943-48

fl.   No. 9657

Dracocephalum hemsleyanum

Curtis's Botanical Magazine
    Vol 161   1938

fl.   No. 9547

Dracocephalum ruyschianum

Barneby, T. P.
European Alpine Flowers in Colour

fl.   Pl. 69, 4

Dracocephalum sibiricum

Curtis's Botanical Magazine
    Vol 164   1943-48

fl.   No. 0646

Dracophilus próximus

Herre, H.
The Genera of the Mesembryanthemaceae

fl. fr.   p 141

Dracophyllum latifolium

Harvey, Norman B.
New Zealand Botanical Paintings

hab.   p. 52   Pl. 23

Dracophyllum milliganii

Curtis, Winifred
The Endemic Flora of Tasmania, Vol. 5

fl. fr.   Pl. 169

Dracophyllum minimum

Curtis, Winifred
The Endemic Flora of Tasmania, Vol. 5

fl. fr. hab.   Pl. 186

Dracophyllum secundum

Blombery, Alec M.
What Wildflower is That

fl.   p 123, Pl. 316, 317

Dracopis amplexicaulis

Dean, Blanche E.
Wildflowers of Alabama and
    Adjoining States

fl.   p 197

Dracunculus vulgaris

Hay, Roy
The Color Dictionary of Flowers & Plants

fl.   p 91, Pl. 726

Dracunculus vulgaris

Perry, Frances
Flowers of the World

fl.   p 32

Dracunculus vulgaris

Polunin, Oleg
Flowers of Europe

fl. Pl. 183 #1819

Dracunculus vulgaris

Polunin, Oleg
Flowers of the Mediterranean

fl.   No. 223

Dragon, Green

see

Arisaema dracontium

Dragonhead

see

Physostegia virginiana

Dragonhead, False

see

Physostegia virginiana

Dragonmouth

see

Arethusa bulbosa

Dragon Tree

see

Dracaena draco

Dragon's Blood

see

Cordyline terminalis

Dragon's Head, Ruysch's

see

Dracocephalum ruyschianum

Dragon's Head, Southern

see

Dracocephalum austriacum

Dragon's-Mouth

see

Arethusa bulbosa

Drakaea glyptodon

Cady, Leo
Australian Native Orchids in Colour

fl.   Pl. 39

Drakaea jeanensis

Blombery, Alec M.
What Wildflower is That

fl.   p. 123   Pl. 318

Drapetes tasmanicus

Curtis, Winifred
The Endemic Flora of Tasmania
            vol 3

fl.   p 187

Dregea sinensis

Harrison, Richmond E.
Climbers and Trailers

fl. p 46 Pl. 91

Drejerella guttata

Hall, Clarence E.
Flowers of the Islands in the Sun

fl   p. 131   Pl. 29

Drejerella guttata

Morley, Brian D.
Wild Flowers of the World

fl.   Pl. 184

Drepanocarpus lunatus

Wit, H. C. D. de
Plants of the World;
   The Higher Plants, Vol I

fr.   p 286, Pl. 182

Drimia alta

Flowering Plants of South Africa
    vol XXIII   1943

fl.   Pl. 890

Drimia anomala

Flowering Plants of Africa
    vol 28   1950-51

fl.   Pl. 1117

Drimia zombensis

Moriarty, Audrey
Wild Flowers of Malawi

fl.   Pl. 12; 5

Drimiopsis crenata

Flowering Plants of Africa
    vol XXV   1945-46

fl., hab.   Pl. 975

Drimiopsis maculata

Flowering Plants of South Africa
    vol XXIV   1944

fl., hab.   Pl. 957

Drimiopsis purpurea

Flowering Plants of Africa
    vol XXV   1945-46

fl., hab.   Pl. 976

Drimiopsis woodii

Flowering Plants of Africa
    vol XXV   1945-46

fl., hab.   Pl. 988

Drimys lanceolata

Cochrane, G. R.
Flowers and Plants of Victoria

fr.   Pl. 394

Drimys winteri

Gault, S. Millar
The Color Dictionary of Shrubs

fl.   Pl. 156

Drimys winteri

Hellyer, A.G.L.
Shrubs in Colour

fl.   p. 43

Drimys winteri

Perry, Frances
Flowers of the World

fl.   p 308

Drimys winteri

Royal Hort. Soc.
Journal of the Royal Hort. Soc.
    vol 97, No. 11   1972

fl.   p. 484   Pl. 248

Drimys winteri var.

Curtis's Botanical Magazine
Vol 169   1952-53

fl. fr.   200

Drimys winteri var.

Hay, Roy
The Color Dictionary of Flowers & Plants

fl.   p 196, Pl. 1567

Drimys xerophila

Cochrane, G. R.
Flowers and Plants of Victoria

fl.   Pl. 494

Dropwort
  see
Filipendula vulgaris

Dropwort, Parsley-Water
  see
Oenanthe lachenalii

Dropwort, Water
  see
Oenanthe fistulosa

Drosanthemum bellum

Flowering Plants of Africa
    vol 28   1950-51

fl.   Pl. 1106

Drosanthemum bicolor

Royal Hort. Soc.
Journal of the Royal Hort. Soc.
    vol 90, No. 8   1965

fl.   p. 338   Pl. 143

Drosanthemum floribundum

Harrison, Richmond E.
Climbers and Trailers

fl.   p. 92   Pl. 230

Drosanthemum floribundum

Macoby, Stirling
What Flower is That

fl. P 118

Drosanthemum floribundum

Menninger, Edwin A.
Flowering Vines of the World

fl.   Pl. 9

Drosanthemum hispidum

Crockett, James Underwood
Lawns & Ground Covers

fl. p. 130

Drosanthemum hispidum

Eliovson, Sima
Namaqualand in Flower

fl.   Pl. 48, 1

Drosanthemum lavisii

Flowering Plants of Africa
    vol 28   1950-51

fl., hab.   Pl. 1107

Drosanthemum speciosum

Cactus & Succulent Society of America
    Journal
    vol 26, No. 5   1954

fl.   p. 150

Drosanthemum speciosum

Eliovson, Sima
Namaqualand in Flower

fl.   Pl. 48, 3

Drosanthemum speciosum

Herre, H.
The Genera of the Mesembryanthemaceae

fl. fr.   p. 143

Drosanthemum splendens

Curtis's Botanical Magazine
    Vol. 165   1948

fl.   44

Drosanthemum splendens

Morley, Brian D.
Wild Flowers of the World

fl.   Pl. 73

Drosanthemum strictifolium

Flowering Plants of Africa
    vol 29   1952-53

fl.   Pl. 1152

Drosera adelae

American Hort. Soc.
American Horticulturist
    vol 53, No. 4   1974

fl., hab.   p. 16

Drosera anglica

Alaska-Yukon Wild Flowers Guide

hab.   p 57

Drosera anglica

Clark, Lewis J.
Wild Flowers of British Columbia

hab.   p 186

Drosera anglica

Martin, W. Keble
The Concise British Flora in Colour

fl. hab.   Pl. 34

Drosera anglica

Polunin, Oleg
Flowers of Europe

fl. Pl. 38 #377

Drosera anglica

Schnell, Donald E.
Carnivorous Plants of the U.S. and Canada

hab.   p 62

Drosera arcturi

Cochrane, G. R.
Flowers and Plants of Victoria

fl. hab.   Pl. 509

Drosera auriculata

Blombery, Alec M.
What Wildflower is That

fl.   p 123, Pl. 319

Drosera auriculata

Cochrane, G. R.
Flowers and Plants of Victoria

fl.   Pl. 367

Drosera binata

Blombery, Alec M.
What Wildflower is That

fl.   p 123, Pl. 320

Drosera binata

Cochrane, G. R.
Flowers and Plants of Victoria

fl.   Pl. 4

Drosera binata

Wit, H. C. D. de
Plants of the World;
    The Higher Plants, Vol I

hab.   p 282, Pl. 170

Drosera brevifolia

Brown, Clair A.
Wildflowers of Louisiana and
    Adjoining States

fl. hab.   p 62

Drosera brevifolia

Lemmon, Robert S.
Wildflowers of North America in
    Full Color

fl.   pg. 181   Pl. 284

Drosera brevifolia

Schnell, Donald E.
Carnivorous Plants of the U.S. and Canada

fl. hab.   p 69

Drosera burmannii

Walden, Beryl M.
Wild Flowers of Hong Kong

fl. hab.   Pl. 2(6)

Drosera capensis

Hay, Roy
The Dictionary of House Plants

hab.   Pl. 199

Drosera capensis

Perry, Frances
Flowers of the World

fl.   p 102

Drosera capillaris

Schnell, Donald E.
Carnivorous Plants of the U.S. and Canada

fl. hab.   p 6, 68

Drosera cistiflora

Morley, Brian D.
Wild Flowers of the World

fl.,hab.   Pl. 75

Drosera dichrosepala

American Hort. Soc.
American Horticulturist
    vol 53, No. 4 1974

fl., hab.   p. 16

Drosera filiformis

Addisonia
    vol22   1943-46

fl.   p. 39   Pl. 724

Drosera filiformis

Brown, Clair A.
Wildflowers of Louisiana and
    Adjoining States

fl. hab.   p 63

Drosera filiformis

Dean, Blanche E.
Wildflowers of Alabama and
    Adjoining States

fl. hab.   p 77

Drosera filiformis var.

Schnell, Donald E.
Carnivorous Plants of the U.S. and Canada

fl. hab.   p 65, 66

Drosera intermedia

Brown, Clair A.
Wildflowers of Louisiana and
    Adjoining States

hab.   p 63

Drosera intermedia

Courtenay, Booth
Wildflowers & Weeds

fl.   p 45

Drosera intermedia

Dean, Blanche E.
Wildflowers of Alabama and
  Adjoining States

fl.   p 77

Drosera intermedia

Martin, W. Keble
The Concise British Flora in Colour

fl. hab.   Pl. 34

Drosera intermedia

Perrot, Emile
Les Plantes Medicinales

fl. fr. hab.   p 90

Drosera intermedia

Polunin, Oleg
Flowers of Europe

fl.   Pl. 39 #378

Drosera intermedia

Schnell, Donald E.
Carnivorous Plants of the U.S. and Canada

fl. hab.   p 63, 64

Drosera leucantha

Duncan, Wilbur H.
Wildflowers of the Southeastern
  United States

fl. hab.   p 57

Drosera linearis

Schnell, Donald E.
Carnivorous Plants of the U.S. and Canada

hab.   p 60

Drosera longifolia

Perrot, Emile
Les Plantes Medicinales

fl. fr. hab.   p 90

Drosera madagascariensis

Moriarty, Audrey
Wild Flowers of Malawi

fl.   Pl. 38; 1

Drosera rotundifolia

Ary, S.
The Oxford Book of Wildflowers

fl.   p 82, Pl. 5

Drosera rotundifolia

Felsko, Elsa
A Book of Wildflowers

fl.   P 152

Drosera rotundifolia

Heller, Christine
Wild Flowers of Alaska

hab.   Pl. 164

Drosera rotundifolia

Jennings, O. E.
Wild Flowers of Western Pennsylvania
                                      vol II
fl. hab.   Pl. 73

Drosera rotundifolia

Kleijn, H.
Beauty of the Wild Plant

fl.   opp. p. 56   Pl. 69

Drosera rotundifolia

Lindman, C. A. M.
Nordens Flora, Vol 6

fl. hab.   Pl. 393

Drosera rotundifolia

Martin, W. Keble
The Concise British Flora in Colour

fl. hab.   Pl. 34

Drosera rotundifolia

Morley, Brian D.
Wild Flowers of the World

fl.   Pl. 16E

Drosera rotundifolia

Perrot, Emile
Les Plantes Medicinales

fl. fr. hab.   p 90

Drosera rotundifolia

Perry, Frances
Flowers of the World

hab.   p. 102

Drosera rotundifolia

Schnell, Donald E.
Carnivorous Plants of the U.S. and Canada

fl. hab.   p 58, 59

Drosera sp.

Klimas, John E.
Wildflowers of Eastern America

fl.   Pl. 86

Drosera sp.

Milne, Lorus
Living Plants of the World

fl.   p 94

Drosera sp.

Morcombe, M. K.
Australia's Western Wildflowers

fl.   p 10

Drosera whittakeri

Cochrane, G. R.
Flowers and Plants of Victoria

fl. hab.   Pl. 62

Drosophyllum lusitanicum

Perry, Frances
Flowers of the World

fl.   p 103

Drosophyllum lusitanicum

Polunin, Oleg
Flowers of Europe

fl.   Pl. 38 #378

Drumsticks

see

Ispogon anemonifolius

Drumsticks

see

Petrophila linearis

Drymary, Heartleaf

see

Drymaria cordata

Dryad, Mountain

see

Dryas octopetala

Dryad, White

see

Dryas octopetala

Dryad, Yellow

see

Dryas drummondi

Dryandra, Cut-leaf

see

Dryandra praemorsa

Dryandra drummondii

Morley, Brian D.
Wild Flowers of the World

fl.   Pl. 134

Dryandra formosa

Blombery, Alec M.
What Wildflower is That

fl.   p 124, Pl. 321

Dryandra formosa

Morcombe, M. K.
Australia's Western Wildflowers

fl.   p 3

Dryandra, great

see

Dryandra nobilis

Dryandra hewardiana

Morcombe, M. K.
Australia's Western Wildflowers

fl.   p 41

Dryandra nobilis

Harrison, R.E.
Trees and Shrubs

fl. P 63

Dryandra nobilis

Macoby, Stirling
What Flower is That

fl.   P 118

Dryandra nobilis

Morcombe, M. K.
Australia's Western Wildflowers

fl.   p 18

Dryandra, Oak-leafed

see

Dryandra quercifolia

Dryandra patens

Blombery, Alec M.
What Wildflower is That

fl.   p 124, Pl. 322

Dryandra polycephala

Harrison, R. E.
Trees and Shrubs

fl.   P. 63

Dryandra polycephala

Macoby, Stirling
What Flower is That

fl. P 118

Dryandra polycephala

Mullins, Barbara
Australian Wildflowers in Colour

fl.   p. 27   Pl. 17

Dryandra praemorsa

Blombery, Alec M.
What Wildflower is That

fl.   p 125, Pl. 323

Dryandra praemorsa

Harrison, R.E.
Trees and Shrubs

fl. P 63

Dryandra quercifolia

Blombery, Alec M.
What Wildflower is That

fl.   p 125, Pl. 324

Dryandra sessilis

Blombery, Alec M.
What Wildflower is That

fl.   p 125, Pl. 325

Dryandra, showy

see

Dryandra formosa

Dryandra sp.

Morcombe, M. K.
Australia's Western Wildflowers

fl.   p 28

Dryas drummondii

Alaska-Yukon Wild Flower Guide

fl. hab.   p 82

Dryas drummondii

Clark, Lewis J.
Wild Flowers of British Columbia

fl.   p 242

Dryas drummondii

Heller, Christine
Wild Flowers of Alaska

fl. fr.   Pl. 12

Dryas drummondii

Porsild, A. E.
Rocky Mountain Wild Flowers

fl. fr. hab.   p. 233

Dryas drummondii

Walcott, Mary Vaux
North American Wild Flowers
vol. 5

fl., fr.   Pl. 364-65

Dryas Hookeriana

Porsild, A.E.
Rocky Mountain Wild Flowers

fl. hab.   p 235

Dryas integrifolia

Alaska-Yukon Wild Flowers Guide

fl. hab.   p 31

Dryas octopetala

Alaska-Yukon Wild Flower Guide

fl. hab.   p 83

Dryas octopetala

Ary, S.
The Oxford Book of Wildflowers

fl.   p 80, Pl. 2

Dryas octopetala

Barneby, T. P.
European Alpine Flowers in Colour

fl.   Pl. 38, 6

Dryas octopetala

Clark, Lewis J.
Wild Flowers of British Columbia

fl.   p 243

Dryas octopetala

Felsko, Elsa
A Book of Wildflowers

fl.   P 153

Dryas octopetala

Ferguson, Mary
Wildflowers

fl.   p 82

Dryas octopetala

Hay, Roy
The Color Dictionary of Flowers & Plants

fl.   p 8, Pl. 60

Dryas octopetala

Heller, Christine
Wild Flowers of Alaska

fl.   Pl. 152-153

Dryas octopetala

Huxley, Anthony
Garden Perennials and Water Plants

fl.   Pl. 103

Dryas octopetala

Kleijn, H.
Beauty of the Wild Plant

fl.   opp. p. 105   pl. 156

Dryas octopetala

Kohlhaupt, Paula
Fleurs des Alpages
                  vol 1
fl. P 53-54

Dryas octopetala

Lindman, C. A. M.
Nordens Flora, Vol 5

fl. fr. hab.   Pl. 332

Dryas octopetala

Martin, W. Keble
The Concise British Flora in Colour

fl. fr. hab.   Pl. 29

Dryas octopetala

Morley, Brian D.
Wild Flowers of the World

fl. fr.   Pl. 2C

Dryas octopetala

Orr, Robert T.
Wildflowers of Western America

fl.   Pl. 1

Dryas octopetala

Polunin, Oleg
Flowers of Europe

fl.   Pl. 45   #441

Dryas octopetala

Taylor, Ronald J.
Mountain Wild Flowers

fl. hab.   p 124

Dryas octopetala

Tosco, Uberto
The World of Mountain Flowers

fl. hab.   p 8, 70, 79

Dryas octopetala

Vilmorin, Roger de
Plantes Alpines dans les Jardins

fl. hab.   Pl. XXIV

Dryas octopetala

Walcott, Mary Vaux
North American Wild Flowers

fl. fr. hab.   Bl. 176, 177

Dryas octopetala

Webster, Mary
Flora of Moray, Nairn & East
Inverness

fl.   p. 164   Pl. 7

Dryas x suendermanni

Massachusetts Hort. Soc.
Horticulture
      vol. 52, No. 7   1974

fl.   p. 31

Drymaria cordata

Weeds of the Southern United States

hab.   p. 6

Drymonia macrantha

American Gloxinia & Gesneriad Society
The Gloxinian
      vol 28, No. 6   1978

fl.   p. 20

Drymonia serrulata

American Gloxinia & Gesneriad Society
The Gloxinian
      vol 28, No. 6   1978

fl.   cover

Drymonia strigosa

American Gloxinia & Gesneriad Society
The Gloxinian
      vol 28, No. 6   1978

fl., fr.   p. 21

Drymonia turrialvae

American Gloxinia & Gesneriad Society
The Gloxinian
      vol 28, No. 6   1978

fl.   p. 20

Drymophila cyanocarpa

Blombery, Alec M.
What Wildflower is That

fl. fr.   p 125, Pl. 326, 327

Drypetes chevalieri

Wit, H. C. D. de
Plants of the World;
   The Higher Plants, Vol I

fr.   p 222, Pl. 139

Duchesnea indica

Batson, Wade T.
Wild Flowers in South Carolina

fl. fr.   p 56

Duchesnea indica

Crockett, James Underwood
Lawns & Ground Covers

fl. p. 130

Duchesnea indica

Walden, Beryl M.
Wild Flowers of Hong Kong

fl.   Pl. 46 (132)

Duchesnea indica

Wharton, Mary E.
A Guide to the Wildflowers & Ferns
   of Kentucky

fl. fr. p 96 Pl. 1.5

Duckweed

see

Lemna minor

Duckling, Ugly

see

Massonia latifolia

Ducks, Flying

see

Caleana nigrita

Dudleya arizonica

Lemmon, Robert S.
Wildflowers of North America in
   Full Color

hab., fl pg. 71   pl. 113

Dudleya attenuata

Coyle, Jeanette
A Field Guide to the Common and
   Interesting Plants of Baja California

fl. hab.   p 78

Dudleya brittonii

Cactus & Succulent Society of America
Journal
      vol 50, No. 2   1978

fl., hab.   p. 82

Dudleya brittonii

Lamb, Edgar
Colorful Cacti of the American Deserts

hab.   Pl. 139

Dudleya campanulata

Cactus & Succulent Society of America
Journal
   vol 50, No. 1   1/78

fl., hab.   p. 20-21

Dudleya cymosa

Munz, Philip A.
California Mountain Wildflowers

fl.   p 55, Pl. 87

Dudleya farinosa

Orr, Robert T.
Wildflowers of Western America

fl.   Pl. 127

Dudleya formosa

Lamb, Edgar
Popular Exotic Cacti in Color

fl. hab.   p 67

Dudleya greenei

Morley, Brian D.
Wild Flowers of the World

fl.   pl. 157

Dudleya lanceolata

Lamb, Edgar
Colorful Cacti of the American Deserts

fl. hab.   Pl. 116

Dudleya lanceolata

Lemmon, Robert S.
Wildflowers of North America in
   Full Color

fl.   pg. 128   pl. 204

Dudleya pulverulenta

Coyle, Jeanette
A Field Guide to the Common and
   Interesting Plants of Baja California

fl. hab.   p 79

Dudleya saxosa

Munz, Philip A.
California Desert Wildflowers

fl.   p 34, Pl. 22

Dudleya saxosa var.

Lamb, Edgar
Colorful Cacti of the American Deserts

hab.   Pl. 100, 140

Duiker Tree

see

Pseudolachnostylis maprouneifolia

Duparquetia orchidaea

Menninger, Edwin A.
Flowering Vines of the World

fl.   Pl. 99

Duranta repens

Bruggeman, L.
Tropical Plants

fl.   Pl. 236 & 237

Duranta repens

Macoby, Stirling
WhatFlower is That

fr.   P 118

Duranta repens

O'Gorman, Helen
Mexican Flowering Trees and Plants

fl. fr. hab.   p 57

Durian

see

Durio sibethinus

Durio zibethinus

Masefield, G. B.
The Oxford Book of Food Plants

fr.   Pl. 103, 2

Durio Zibethinus

Morley, Brian D.
Wild Flowers of the World

fl.fr.   Plate 115

Durra

see

Sorghum vulgare var.

Dusty Miller

see

Lychnis coronaria

Dusty miller

see

Senecio cineraria

Dutchman's Breeches

see

Dicentra cucullaria

Dutchman's breeches

see

Dicentra spectabilis

Dutchman's Pipe

see

Aeginetia indica

Dutchman's-Pipe

see

Aristolochia

Duvalia modesta

Cactus & Succulent Society of America
Journal
   vol 42, No. 1   1970

fl., hab.   p. 23

Duvalia modesta

Lamb, Edgar
Popular Exotic Cacti in Color

fl. hab.   p. 68

Duvalia parviflora

Lamb, Edgar
The Pocket Encyclopedia of Cacti
   and Succulents in Color

fl. hab.   Pl. 240

Duvalia procumbens

Flowering Plants of Africa
   vol 31   1956

fl.   Pl. 1218

Duvalia reclinata

Lamb, Edgar
Stapeliads in Cultivation

fl. hab.   p 53

Duvalia sulcata

Cactus & Succulent Soc. of America
Journal
   vol 45, No. 1   1973

fl.   p. 25

Duvalia sulcata

Flowering Plants of Africa
   vol 44   1977

fl., hab.   Pl. 1734

Duvalia tanganyikensis

Flowering Plants of Africa
   vol 28   1950-51

fl., hab.   Pl. 1082

Duvernoya aconitiflora

Flowering Plants of Africa
vol. 31   1956

fl.   Pl. 1216

Duvernoya adhatodoides

Flowering Plants of Africa
Vol. 35   1962

fl., fr.   Pl. 1375

Dwale

see

Atropa belladonna

Dyckia alt ssima

Macoby, Stirling
What Flower is That

fl.   P 119

Dyckia altissima

Padilla, Victoria
Bromeliads

fl.   p. 66   Pl. 4

Dyckia altissima

Wilson, Robert Gardner
Bromeliads in Cultivation

fl. hab.   Pl. 109, 110

Dyckia fosteriana

Wilson, Robert Gardner
Bromeliads in Cultivation

hab.   Pl. 112

Dyckia leptostachya

Wilson, Robert Gardner
Bromeliads in Cultivation

hab.   Pl. 111

Dyeweed

see

Dalea emoryi

Dyer's Weed

see

Solidago nemoralis

Dyschoriste hildebrantii

Moriarty, Audrey
Wild Flowers of Malawi

fl.   Pl. 43; 2

Dyschoriste oblongifolia

Duncan, Wilbur H.
Wildflowers of the Southeastern
United States

fl.   p 185

Dyssodia speciosa

Coyle, Jeanette
A Field Guide to the Common and
Interesting Plants of Baja California

fl.   p 169

Eagle's Claw

see

Echinocactus horizonthalonius

Ebenus cretica

Goulimis, Constantine N.
Wild Flowers of Greece

fl.   p 47

Eberlanzia spinosa

Herre, H.
The Genera of the Mesembryanthemaceae

fl. fr.   p 145

Ebony

see

Diospyros ebenum

Ebony, Malabar

see

Diospyros malabarica

Ebony, mountain

see

Bauhinia variegata

Ebracteola montis-moltkei

Herre, H.
The Genera of the Mesembryanthemaceae

fl. fr.   p 147

Ecballium elaterium

Bianchini, Francesco
Health Plants of the World

fr. hab.   p 69

Ecballium elaterium

Color Treasury of Herbs & Other
Medicinal Plants

fl. fr. p 53   Pl. 80

Ecballium elaterium

Morley, Brian D.
Wild Flowers of the World

fl.,fr.   Pl. 34

Ecballium elaterium

Polunin, Oleg
Flowers of Europe

fl. fr.   Pl. 79 #811

Ecballium elaterium

Polunin, Oleg
Flowers of the Mediterranean

fl. fr.   No. 188

Ecbolium aruiculatum

Curtis's Botanical Magazine
vol 176   1966-68

fl.   Pl. 516

Eccremocarpus scaber

Harrison , Richmond E.
Climbers and Trailers

Eccremocarpus scaber

Hay, Roy
The Color Dictionary of Flowers & Plants

fl.   p 247, Pl. 1976

Eccremocarpus scaber

Hay, Roy
The Dictionary of House Plants

fl.   Pl. 200

Eccremocarpus scaber

Hellyer, A.G.L.
Shrubs in Colour

fl.   p. 46

Eccremocarpus scaber

Morley, Brian D.
Wild Flowers of the World

fl. fr.   pl. 181

Eccremocarpus scaber

Munoz Pizarro, Carlos
Flores Silvestres de Chile

fl., fr.   Pl. 24

Eccremocarpus scaber

Perry, Frances
Flowers of the World

fl.   p 46

Echeveria agavoides

Lamb, Edgar
The Pocket Encyclopedia of Cacti
and Succulents in Color

hab.   Pl. 241

Echeveria amphoralis

Walther, Eric
Echeveria

fl.   p. 241   Pl. 15

Echeveria atropurpurea

Walther, Eric
Echeveria

fl.   p 232

Echeveria australis

Walther, Eric
Echeveria

fl.   p 225

Echeveria bifida

Walther, Eric
Echeveria

fl.   p 217

Echeveria calycosa

Cactus & Succulent Society of America
Journal
     vol 39, No. 1 1967

fl.   cover (p. 1)

Echeveria ciliata

Walther, Eric
Echeveria

fl.   p. 240   Pl. 14

Echeveria coccinea

Cactus & Succulent Society of America
Journal
     vol 48, No. 5   1976

fl.   p. 225

Echeveria dactylifera

Lamb, Edgar
The Pocket Encyclopedia of Cacti
and Succulents in Color

hab.   Pl. 242

Echeveria elegans

Cactus & Succulent Society of America
Journal
     vol 28, No. 5   1956

hab.   p. 151

Echeveria elegans

Kiaer, Eigil
Indoor Plants in Colour

fl.   p. 55

Echeveria elegans

Massachusetts Hort. Soc.
Horticulture
     vol 41, No. 5   1963

hab.   p. 278

Echeveria elegans

Massachusetts Hort. Soc.
Horticulture
     vol 47, No. 11   1969

hab.   p. 24-25

Echeveria elegans

Nicolaisen, Age
Pocket Encyclopedia of Indoor Plants

hab.   pl. 92

Echeveria elegans

Pacific Hort. Foundation
Pacific Horticulture
     vol 39, No. 4   1978-79

hab.   p. 28

Echeveria gibbiflora

O'Gorman, Helen
Mexican Flowering Trees and Plants

fl. hab.   p 195

Echeveria gibbiflora

Perry, Frances
Flowers of the World

fl.   p 95

Echeveria gibbiflora var.

Hay, Roy
The Color Dictionary of Flowers & Plants

hab.   p 65, Pl. 519

Echeveria gibbiflora var.

Nicolaisen, Age
Pocket Encyclopedia of Indoor Plants

fl.   pl. 91

Echeveria glauca

Hersey, Jean
Woman's Day Book of House Plants

fl.   p. 53

Echeveria glauca

Perry, Frances
Flowers of the World

fl.   p 95

Echeveria halbingeri

Walther, Eric
Echeveria

fl.   p. 213   Pl. 1

Echeveria harmsii

Hay, Roy
The Dictionary of House Plants

fl.   Pl. 201

Echeveria harmsii

Pacific Hort. Foundation
Pacific Horticulture
     vol 39, No. 4   1978-79

fl.   p. 29

Echeveria harmsii

Walther, Eric
Echeveria

fl.   p. 244   Pl. 16

Echeveria hybrid

Lamb, Edgar
The Pocket Encyclopedia of Cacti
and Succulents in Color

hab.   Pl. 244 , 246

Echeveria hybrid

Macoby, Stirling
What Flower is That

fl. P 119

Echeveria hybrid

Pacific Hort. Foundation
Pacific Horticulture
     vol 39, No. 4   1978-79

hab.   p. 28

Echeveria longissima

Walther, Eric
Echeveria

fl.   p. 244   Pl. 16

Echeveria montana

Walther, Eric
Echeveria

fl.   p 224

Echeveria multicaulis

Walther, Eric
Echeveria

fl.   p 228

Echeveria nodulosa

Curtis's Botanical Magazine
     Vol. 181   1976-77

fl.   Pl. 719

Echeveria nodulosa

Walther, Eric
Echeveria

fl.   p 229

Echeveria nuda

Walther, Eric
Echeveria

fl.   p 221

Echeveria obtusifolia

Walther, Eric
Echeveria

fl. hab.   p 216

Echeveria perelegans

Kiaer, Eigil
Indoor Plants in Colour

fl.,hab.   p. 55

Echeveria platyphylla

Walther, Eric
Echeveria

fl., hab.   p. 236   Pl. 12

Echeveria pringlei

Walther, Eric
Echeveria

fl.   p. 241   Pl. 15

Echeveria prolifica

Cactus & Succulent Soc. of America
Journal
      vol 50, No. 6   1978

fl.   p. 289

Echeveria pulvinata

Cactus & Succulent Society of America
Journal
      vol 17, No. 6   1945

hab.   cover (p. 81)

Echeveria pulvinata

Lamb, Edgar
The Pocket Encyclopedia of Cacti
    and Succulents in Color

hab.   Pl. 243

Echeveria pulvinata

Lamb, Edgar
Popular Exotic Cacti in Color

fl. hab.   p 71

Echeveria pulvinata

Walther, Eric
Echeveria

fl., hab.   p. 237   Pl. 13

Echeveria racemosa

Walther, Eric
Echeveria

fl.   p 233

Echeveria retusa var.

Kiaer, Eigil
Indoor Plants in Colour

fl.   p. 55

Echeveria rosea

Walther, Eric
Echeveria

fl.   p 229

Echeveria sanches-mejoradae

Walther, Eric
Echeveria

fl.   p. 312   Pl. 1

Echeveria setosa

Cactus & Succulent Society of America
Journal
      vol 8, No. 10   1937

fl., hab.   cover (p. 165)

Echeveria setosa

Lamb, Edgar
The Pocket Encyclopedia of Cacti
    and Succulents in Color

hab.   Pl. 245

Echeveria setosa

Morley, Brian D.
Wild Flowers of the World

fl.   Pl. 157

Echeveria setosa

Walther, Eric
Echeveria

fl., hab.   p. 240   Pl. 14

Echeveria shaviana

Pacific Hort. Foundation
Pacific Horticulture
      vol 39, No. 4   1978-79

hab.   p. 29

Echeveria shaviana

Walther, Eric
Echeveria

hab.   p 221

Echeveria sprucei

Walther, Eric
Echeveria

fl.   p 224

Echeveria strictiflora

Walther, Eric
Echeveria

fl. hab.   p 217

Echeveria subrigida

Morley, Brian D.
Wild Flowers of the World

fl.   pl. 157

Echeveria walpoleana

Walther, Eric
Echeveria

fl. hab.   p 220

Echidnopsis angustiloba

Flowering Plants of Africa
vol XXVI   1947

fl.   Pl. 1003

Echidnopsis columnaris

Flowering Plants of Africa
vol 40   1969-70

fl., hab.   Pl. 1563

Echidnopsis radians

Cactus & Succulent Society of America
Journal
      vol 49, No. 6   1977

fl.   p. 263

Echidnopsis repens

Flowering Plants of Africa
vol XXV   1945-46

fl.   Pl. 993

Echidnopsis sharpei

Flowering Plants of Africa
vol XXVI   1947

fl.   Pl. 1003

Echinacanthus attenuatus

Hara, Hiroshi, comp.
Photo-Album of Plants of
    Eastern Himalaya

fl.   Pl. 92

Echinacea angustifolia

Moyle, John B.
Northland Wild Flowers

fl. hab.   p 160, Pl. 190

Echinacea pallida

Brown, Clair A.
Wildflowers of Louisiana and
    Adjoining States

fl.   p 198

Echinacea pallida

Duncan, Wilbur H.
Wildflowers of the Southeastern
    United States

fl.   p 213

Echinacea pallida var.

Lemmon, Robert S.
Wildflowers of North America in
    Full Color

fl.   pg. 209   pl. 332

Echinacea purpurea

American Hort. Soc.
American Horticulturist
    vol 57, No. 3   1978

fl.   p. 7

Echinacea purpurea

Courtenay, Booth
Wildflowers & Weeds

fl.   p 115

Echinacea purpurea

Dean, Blanche E.
Wildflowers of Alabama and
    Adjoining States

fl.   p 195

Echinacea purpurea

Everett, T.H., ed.
New Illustrated Encyclopedia of
    Gardening   vol 4

fl.   p. 694

Echinacea purpurea

Macoby, Stirling
What Flower is That

fl.   P 120

Echinacea purpurea

Wharton, Mary E.
A Guide to the Wildflowers & Ferns
    of Kentucky

fl.   p 261   Pl. 3.36

Echinacea purpurea var.

American Hort. Soc.
American Horticulturist
    vol 56, No. 3   1977

fl.   p. 15

Echinacea purpurea var.

Perry, Frances
Flowers of the World

fl.   p. 83

Echinocactus asterias

Weniger, Del
Cacti of the Southwest

fl.   P 19

Echinocactus bicolor var.

Weniger, Del
Cacti of the Southwest

fl.   P 23

Echinocactus brevihamatus

Weniger, Del
Cacti of the Southwest

fl.   P 21

Echinocactus conoideus

Weniger, Del
Cacti of the Southwest

fl.   P 25

Echinocactus erectocentrus var.

Weniger, Del
Cacti of the Southwest

fl.   P 24

Echinocactus flavidispinus

Weniger, Del
Cacti of the Southwest

fl.   P 23

Echinocactus grusonii

Backeberg, Curt
Cactus Lexicon

fl.   p 591

Echinocactus grusonii

Hay, Roy
The Color Dictionary of Flowers & Plants

hab.   p 65, Pl. 520

Echinocactus grusonii

Hay, Roy
The Dictionary of House Plants

hab.   Pl. 202

Echinocactus grusonii

Kromdijk, G.
200 House Plants in Colour

hab.   Pl. 78

Echinocactus grusonii

Kupper, Walter
Cacti

hab.   p. 109   Pl. 51

Echinocactus grusonii

Lamb, Edgar
Colorful Cacti of the American Deserts

hab.   Pl. 135

Echinocactus grusonii

Lamb, Edgar
The Pocket Encyclopedia of Cacti
    and Succulents in Color

hab.   Pl. 42

Echinocactus grusonii

Massachusetts Hort. Soc.
Horticulture
    vol 43, No. 10   1965

hab.   p. 18

Echinocactus grusonii

Nicolaisen, Age
Pocket Encyclopedia of Indoor Plants

hab.   pl. 62

Echinocactus grusonii

Perry, Frances
Flowers of the World

hab. p 58

Echinocactus grusonii

Royal Hort. Soc.
Journal of the Royal Hort. Soc.
    vol 91, No. 10   1966

hab.   p. 426   Pl. 219

Echinocactus grusonii

Royal Hort. Soc.
Journal of the Royal Hort. Soc.
    vol 95, No. 12   1970

hab.   p. 524   Pl. 256

Echinocactus grusonii

Van Laren, A. J.
Cactus

hab P 55   Fig. 60

Echinocactus hamatacanthus

Weniger, Del
Cacti of the Southwest

fl.   P 23

Echinocactus horizonthalonius

Lamb, Edgar
Colorful Cacti of the American Deserts

fl. hab.   Pl. 34, 36

Echinocactus horizonthalonius

Lamb, Edgar
Popular Exotic Cacti in Color

fl. hab.   p 60

Echinocactus horizonthalonius

Van Laren, A. J.
Cactus

fl.   P 56   Fig. 62

Echinocactus horizonthalonius

Weniger, Del
Cacti of the Southwest

fl.   P 18

Echinocactus horizonthalonius var.

Weniger, Del
Cacti of the Southwest

fl.   P 19

Echinocactus ingens

Van Laren, A. J.
Cactus

hab. P 56   Fig. 61

Echinocactus intertextus var.

Weniger, Del
Cacti of the Southwest

fl.   P 24

Echinocactus marioposensis

Weniger, Del
Cacti of the Southwest

fl.   P 25

Echinocactus mesae-verdae

Weniger, Del
Cacti of the Southwest

fl.   P 21

Echinocactus pectinatus var.

Lemmon, Robert S.
Wildflowers of North America in
    Full Color

fl.   pg. 88   pl. 141

Echinocactus sanguiniflorus

Kiaer, Eigil
Indoor Plants in Colour

fl.   p. 53

Echinocactus scheeri

Hay, Roy
The Color Dictionary of Flowers & Plants

fl.   p 66, Pl. 521

Echinocactus scheeri

Weniger, Del
Cacti of the Southwest

fl.   P 21

Echinocactus setispinus var.

Weniger, Del
Cacti of the Southwest

fl.   P 22

Echinocactus sinuatus

Weniger, Del
Cacti of the Southwest

fl.   P 23

Echinocactus texensis

Weniger, Del
Cacti of the Southwest

fl.   P 19

Echinocactus tobuschii

Weniger, Del
Cacti of the Southwest

fl.   P 22

Echinocactus uncinatus var.

Weniger, Del
Cacti of the Southwest

fl.   P 20

Echinocactus viridescens

Lemmon, Robert S.
Wildflowers of North America in
    Full Color

fl.   pg. 97   pl. 153

Echinocactus whipplei

Weniger, Del
Cacti of the Southwest

fl.   P 20

Echinocactus wislizeni

Weniger, Del
Cacti of the Southwest

fl.   P 20

Echinocereus albispinus

Weniger, Del
Cacti of the Southwest

fl.   P 6

Echinocereus baileyi

Lamb, Edgar
Colorful Cacti of the American Deserts

hab.   Pl. 139

Echinocereus baileyi

Weniger, Del
Cacti of the Southwest

fl.   P 6

Echinocereus berlandieri

Weniger, Del
Cacti of the Southwest

fl.   P 17

Echinocereus blankii

Van Laren, A. J.
Cactus

fl.   P 31   Fig. 22

Echinocereus blankii

Weniger, Del
Cacti of the Southwest

fl.   P 17

Echinocereus brandegeei

Coyle, Jeanette
A Field Guide to the Common and
    Interesting Plants of Baja California

fl. hab.   p 133

Echinocereus caespitosus var.

Weniger, Del
Cacti of the Southwest

fl. Pl 3-5, 7

Echinocereus chisoensis

Weniger, Del
Cacti of the Southwest

fl.   P 7

Echinocereus chloranthus

Van Laren, A. J.
Cactus

fl.   P 31   Fig. 19

Echinocereus chloranthus var.

Weniger, Del
Cacti of the Southwest

fl.   P 2

Echinocereus coccineus

Van Laren, A. J.
Cactus

fl.   P 20   Fig. 18

Echinocereus coccineus var.

Weniger, Del
Cacti of the Southwest

hab. fl.   P. 12-13

Echinocereus dasyacanthus

Lamb, Edgar
Colorful Cacti of the American Deserts

fl. hab.   Pl. 46, 47

Echinocereus dasyacanthus

Lamb, Edgar
The Pocket Encyclopedia of Cacti
    and Succulents in Color

fl,   Pl. 38

Echinocereus dasyacanthus

Lemmon, Robert S.
Wildflowers of North America in
    Full Color

fl.   p. 89   Pl. 142

Echinocereus dasyacanthus

Van Laren, A. J.
Cactus

fl.   P 31   Fig. 20

Echinocereus dasyacanthus var.

Lamb, Edgar
The Pocket Encyclopedia of Cacti
    and Succulents in Color

fl. hab.   Pl. 35

Echinocereus dasyacanthus var.

Weniger, Del
Cacti of the Southwest

fl.   P 9

Echinocereus davisii

Weniger, Del
Cacti of the Southwest

fl.   P 2

Echinocereus decapetalus

Lemmon, Robert S.
Wildflowers of North America in
    Full Color

fl.   pg. 89   pl. 142

Echinocereus delaetii

Van Laren, A. J.
Cactus

fl.   P 37   Fig. 1

Echinocereus dubius

Lamb, Edgar
Colorful Cacti of the American Deserts

fl. hab.   Pl. 53, 54

Echinocereus dubius

Weniger, Del
Cacti of the Southwest

fl.   P 15

Echinocereus engelmannii

Coyle, Jeanette
A Field Guide to the Common and
    Interesting Plants of Baja California

fl. hab.   p 133

Echinocereus engelmannii

Lamb, Edgar
Colorful Cacti of the American Deserts

fl. fr. hab.   Pl. 92, 93, 111, 113

Echinocereus engelmannii

Lemmon, Robert S.
Wildflowers of North America in
    Full Color

fl.   p. 79   pl. 127

Echinocereus engelmannii

Lemmon, Robert S.
Wildflowers of North America in
    Full Color

fl.   p. 95   pl. 149

Echinocereus engelmannii

Mathias, Mildred E.
Color for the Landscape

fl. hab.   p 200

Echinocereus engelmannii

Munz, Philip A.
California Desert Wildflowers

fl.   p 41, Pl. 43

Echinocereus engelmannii var.

Lamb, Edgar
Colorful Cacti of the American Deserts

fl. hab.   Pl. 75, 76

Echinocereus enneacanthus

Kupper, Walter
Cacti

fl., hab.   p. 85   Pl. 39

Echinocereus enneacanthus

Lamb, Edgar
The Pocket Encyclopedia of Cacti
    and Succulents in Color

fl. hab.   Pl. 34

Echinocereus enneacanthus var.

Weniger, Del
Cacti of the Southwest

fl.   P 14

Echinocereus fendleri

Lamb, Edgar
The Pocket Encyclopedia of Cacti
    and Succulents in Color

fl. hab.   Pl. 30

Echinocereus fendleri

Lemmon, Robert S.
Wildflowers of North America in
    Full Color

fl.   pg. 94   pl. 148

Echinocereus fendleri

Orr, Robert T.
Wildflowers of Western America

fl.   Pl. 194

Echinocereus fendleri

Welsh, Stanley L.
Flowers of the Canyon Country

fl.   p 22

Echinocereus fendleri var.

Lamb, Edgar
Colorful Cacti of the American Deserts

fl. hab.   Pl. 97-99, 123-24

Echinocereus fendleri var.

Lamb, Edgar
Popular Exotic Cacti in Color

fl. hab.   p 73

Echinocereus fendleri var.

Weniger, Del
Cacti of the Southwest

fl.   P 15

Echinocereus ferreirianus

Cactus & Succulent Soc. of America
Journal
    vol 44, No. 4   1972

fl.   p. 162

Echinocereus fitchii

Kupper, Walter
Cacti

fl., hab.   p. 57   Pl. 25

Echinocereus fitchii

Lamb, Edgar
The Pocket Encyclopedia of Cacti
    and Succulents in Color

fl. hab.   Pl. 36, 37

Echinocereus fitchii

Weniger, Del
Cacti of the Southwest

fl.   P 5

Echinocereus knippelianus

Van Laren, A. J.
Cactus

fl. hab.   P. 35   Fig. 26

Echinocereus knippelianus var.

Cactus & Succulent Society of America
Journal
    vol 50, No. 2   1978

fl.   p. 79

Echinocereus llayaii

Weniger, Del
Cacti of the Southwest

fl.   P 11

Echinocereus lloydii

Walcott, Mary Vaux
North American Wild Flowers

fl. hab.   Pl. 155

Echinocereus lloydii

Weniger, Del
Cacti of the Southwest

fl.   frontispiece, p. 11

Echinocereus maritimus

Coyle, Jeanette
A Field Guide to the Common and
  Interesting Plants of Baja California

fl. hab.   p 135

Echinocereus melanocentrus

Lamb, Edgar
Colorful Cacti of the American Deserts

hab.   Pl. 140

Echinocereus melanocentrus

Weniger, Del
Cacti of the Southwest

fl.   P 5

Echinocereus mojavensis

Munz, Philip A.
California Desert Wildflowers

fl.   p 41, Pl. 44

Echinocereus mojavensis

Orr, Robert T.
Wildflowers of Western America

fl.   Pl. 184

Echinocereus nivosus

Cactus & Succulent Society of America
Journal
    vol 50, No. 1   1978

fl., hab.   p. 18

Echinocereus papillosus

Lamb, Edgar
The Pocket Encyclopedia of Cacti
  and Succulents in Color

fl. hab.   Pl. 31

Echinocereus papillosus var.

Weniger, Del
Cacti of the Southwest

fl.   P 16 -17

Echinocereus pectinatus

Lamb, Edgar
The Pocket Encyclopedia of Cacti
  and Succulents in Color

fl. hab.   Pl. 39

Echinocereus pectinatus var.

Backeberg, Curt
Cactus Lexicon

fl.   p 593

Echinocereus pectinatus var.

Lemmon, Robert S.
Wildflowers of North America in
    Full Color

fl.   p. 88   Pl. 141

Echinocereus pectinatus var.

Weniger, Del
Cacti of the Southwest

fl. hab.   P. 7-8

Echinocereus pensilis

Cactus & Succulent Society of America
Journal
    vol 41, No. 4   1969

fl.   p. 192

Echinocereus pentalophus

Van Laren, A. J.
Cactus

fl.   P 31   Fig. 21

Echinocereus pentalophus

Weniger, Del
Cacti of the Southwest

fl.   P 16

Echinocereus polyacanthus

Cactus & Succulent Society of America
Journal
    vol 41, No. 6   1969

fl.   p. 260

Echinocereus polyacanthus var.

Weniger, Del
Cacti of the Southwest

fl.   P 13

Echinocereus pulchellus

Cactus & Succulent Society of America
Journal
    vol 42, No. 3   1970

fl., hab.   p. 124

Echinocereus pulchellus

Subik, Rudolf
Decorative Cacti

fl. hab.   p 33

Echinocereus pulchellus

Van Laren, A. J.
Cactus

fl.   P 35   Fig. 27

Echinocereus reichenbachii

Cactus & Succulent Society of America
Journal
    vol 21, No. 3   1949

fl., hab.   p. 81

Echinocereus reichenbachii

Lamb, Edgar
The Pocket Encyclopedia of Cacti
  and Succulents in Color

fl. hab.   Pl. 32, 33

Echinocereus reichenbachii

Van Laren, A. J.
Cactus

hab P 31   Fig. 24

Echinocereus rigidissimus

Cactus & Succulent Society of America
Journal
    vol 18, No. 4   1946

fl., hab.   cover (p. 49)

Echinocereus rigidissimus

Kupper, Walter
Cacti

fl., hab.   p. 87   Pl. 40

Echinocereus rigidissimus

Lamb, Edgar
Colorful Cacti of the American Deserts

fl. hab.   Pl. 56 - 59

Echinocereus rigidissimus

Van Laren, A. J.
Cactus

fl.   P 31   Fig. 23

Echinocereus roetteri

Weniger, Del
Cacti of the Southwest

fl.   P 10

Echinocereus rosei

Hay, Roy
The Dictionary of House Plants

hab.   Pl. 203

Echinocereus russanthus

Weniger, Del
Cacti of the Southwest

fl.   P 3

Echinocereus salm-dyckianus

Cactus & Succulent Society of America
Journal
    vol 14, No. 10-11   1942

fl., hab.   cover (p. 137)

Echinocereus salm-dyckianus

Kupper, Walter
Cacti

fl., hab.   p. 89   Pl. 41

Echinocereus stramineus

Lamb, Edgar
Colorful Cacti of the American Deserts

fl. fr. hab.   Pl. 27 - 30

Echinocereus stramineus

Van Laren, A. J.
Cactus

hab.   p. 35   Fig. 28

Echinocereus stramineus

Weniger, Del
Cacti of the Southwest

fl.   P 14

Echinocereus subinermis

Cactus & Succulent Soc. of America
Journal
    vol 44, No. 5   1972

fl., hab.   cover (p. 185)

Echinocereus triglochidiatus

Cactus & Succulent Society of America
Journal
    vol 21, No. 6   1949

fl.   p. 169

Echinocereus triglochidiatus

Lemmon, Robert S.
Wildflowers of North America in
Full Color

fl.   pg. 96   pl. 152

Echinocereus triglocidiatus

Massachusetts Hort. Soc.
Horticulture
    vol 54, No. 12   1976

fl., hab.   p. 44

Echinocereus triglochidiatus var.

Cactus & Succulent Society of America
Journal
    vol 39, No. 5   1967

fl.   p. 181

Echinocereus triglochidiatus var.

Lamb, Edgar
Colorful Cacti of the American Deserts

fl. hab.   Pl. 18, 19, 103-04

Echinocereus triglochidiatus var.

Lamb, Edgar
The Pocket Encyclopedia of Cacti
    and Succulents in Color

fl. hab.   Pl. 40, 41

Echinocereus triglochidiatus var.

Lemmon, Robert S.
Wildflowers of North America in
Full Color

fl.   pg. 96   pl. 151

Echinocereus triglochidiatus var.

Weniger, Del
Cacti of the Southwest

fl.   P 11-12

Echinocereus ursinus

Lemmon, Robert S.
Wildflowers of North America in
Full Color

fl.   p. 95   pl. 150

Echinocereus viridiflorus

Lamb, Edgar
Colorful Cacti of the American Desert

fl. hab.   Pl. 125

Echinocereus viridiflorus

Lemmon, Robert S.
Wildflowers of North America in
Full Color

fl.   pg. 91   pl. 144

Echinocereus viridiflorus

Walcott, Mary Vaux
North American Wild Flowers
    vol. 4

hab., fl.   Pl. 308

Echinocereus viridiflorus var.

Weniger, Del
Cacti of the Southwest

fl.   P 1

Echinocereus weinbergii

Van Laren, A. J.
Cactus

hab. P 35   Fig. 25

Echinochloa crus-galli

Polunin, Oleg
Flowers of Europe

fr. Pl. 182 #1804

Echinochloa crusgalli

Tarver, David P.
Aquatic and Wetland Plants
    of Florida

fr. hab.   p 87

Echinochloa crusgalli

Weeds of the Southern United States

hab., fr.   p. 25

Echinochloa frumentacea

Masefield, G. B.
The Oxford Book of Food Plants

fr.   Pl. 13, 4

Echinocystis lobata

Courtenay, Booth
Wildflowers & Weeds

fl.   p 73

Echinocystis lobata

Crockett, James Underwood
Annuals

fl.fr.   P 116

Echinocystis lobata

Jennings, O. E.
Wild Flowers of Western Pennsylvania
    vol II
fl. fr. hab.   Pl. 150

Echinocystis lobata

Weeds of the Southern United States

fr., fl., hab.   p. 20

Echinodorus berteroi

Stodola, Jiri
Encyclopedia of Water Plants

hab.   P. 194

Echinodorus brevipedicellatus

Stodola, Jiri
Encyclopedia of Water Plants

hab.   P. 195

Echinodorus cordifolius

Stodola, Jiri
Encyclopedia of Water Plants

fl.   P. 198

Echinodorus longistylus

Stodola, Jiri
Encyclopedia of Water Plants

hab.   P. 199

Echinodorus magdalenensis

Stodola, Jiri
Encyclopedia of Water Plants

fl.   P. 203

Echinodorus martii

Stodola, Jiri
Encyclopedia of Water Plants

fl.   P. 207

Echinodorus nymphaeifolius

Stodola, Jiri
Encyclopedia of Water Plants

hab.   P. 210

Echinodorus paniculatus

Stodola, Jiri
Encyclopedia of Water Plants

fl.   P. 211

Echinodorus sp.

Stodola, Jiri
Encyclopedia of Water Plants

hab.   P. 202

Echinodorus tenellus

Stodola, Jiri
Encyclopedia of Water Plants

hab.   P. 206

Echinofossulocactus crispatis

Van Laren, A. J.
Cactus

fl.   P 45   Fig. 48

Echinofossulocactus lancifer

Kupper, Walter
Cacti

fl., hab.   p. 111   Pl. 52

Echinofossulocactus multicostatus

Van Laren, A. J.
Cactus

hab P 45   Fig. 47

Echinofossulocactus pentacanthus

Lamb, Edgar
Popular Exotic Cacti in Color

fl. hab.   p 74

Echinofossulocactus pentacanthus

Subik, Rudolf
Decorative Cacti

fl. hab.   p 35

Echinomastus erectocentrus

Lemmon, Robert S.
Wildflowers of North America in
    Full Color

fl.   pg. 90   pl. 143

Echinomastus johnsonii

Lamb, Edgar
Colorful Cacti of the American Desert

fl.   Pl. 122

Echinomastus laui

Cactus & Succulent Society of America
Journal
    vol 50, No. 4   1978

hab., fl.   p. 188-89

Echinomastus macdowellii

Lamb, Edgar
The Pocket Encyclopedia of Cacti
    and Succulents in Color

fl. hab.   Pl. 43

Echinomastus macdowellii

Van Laren, A. J.
Cactus

hab.   P. 46   Fig. 54

Echinomastus warnockii

Cactus & Succulent Society of America
Journal
    vol 47, No. 5   1975

fl., hab.   p. 218

Echinopanax horridum

Heller, Christine
Wild Flowers of Alaska

fr.   Pl. 288

Echinopanax horridum

Walcott, Mary Vaux
North American Wild Flowers

fr. hab.   Pl. 32

Echinops amplexicaulis

Flowering Plants of South Africa
    vol XX   1940

fl.   Pl. 796

Echinops dahurica

Morley, Brian D.
Wild Flowers of the World

fl.   Plate 106

Echinops ritro

Huxley, Anthony
Garden Perennials and Water Plants

fl.   Pl. 104

Echinops ritro

Perry, Frances
Flowers of the World

fl.   p 85

Echinops ritro

Polunin, Oleg
Flowers of Europe

fl.   Pl. 152 #1462

Echinops ritro

Polunin, Oleg
Flowers of the Mediterranean

fl.   No. 205

Echinops ritro

Royal Hort. Soc.
The Garden
    vol 103, No. 2   1978

fl.   p. 56

Echinops ritro var.

Hay, Roy
The Color Dictionary of Flowers & Plants

fl.   p 140, Pl. 1115

Echinops ruthenicus

Curtis's Botanical Magazine
    Vol. 180   1974

fl.   Pl. 677

Echinopsis ancistrophora

Lamb, Edgar
The Pocket Encyclopedia of Cacti
    and Succulents in Color

fl. hab.   Pl. 44

Echinopsis aurea

Van Laren, A. J.
Cactus

fl.   P 50   Fig. J

Echinopsis eyriesii

Subik, Rudolf
Decorative Cacti

fl. hab.   p 37

Echinopsis eyriesii

Van Laren, A. J.
Cactus

fl.   P 41   Fig. 35

Echinopsis kermesina

Lamb, Edgar
Popular Exotic Cacti in Color

fl. hab.   p 77

Echinopsis longispina

Lamb, Edgar
The Pocket Encyclopedia of Cacti
    and Succulents in Color

fl. hab.   Pl. 45

Echinopsis multiplex

Kromdijk, G.
200 House Plants in Colour

hab.   Pl. 79

Echinopsis multiplex

Macoby, Stirling
What Flower is That

fl.   P 120

Echinopsis multiplex

Massachusetts Hort. Soc.
Horticulture
vol 33, No. 2   1955

fl.   p. 82

Echinopsis multiplex

Van Laren, A. J.
Cactus

fl.   P 41   Fig. 34

Echinopsis tubiflora

Van Laren, A. J.
Cactus

fl. P 41   Fig. 36

Echites peltata

Wit, H. C. D. de
Plants of the World;
  The Higher Plants, Vol II

fl.   p 103, Pl. 57

Echium aculeatum

Bramwell, David
Wild Flowers of the Canary Islands

hab., fl.   Pl. 224

Echium auberianum

Bramwell, David
Wild Flowers of the Canary Islands

fl.   Pl. 232

Echium brevirame

Bramwell, David
Wild Flowers of the Canary Islands

fl., hab.   Pl. 223

Echium decaisnei

Bramwell, David
Wild Flowers of the Canary Islands

fl., hab.   Pl. 234

Echium diffusum

Polunin, Oleg
Flowers of the Mediterranean

fl.   No. 142

Echium fastuosum

Harrison, R.E.
Trees andShrubs

fl.   P 64

Echium fastuosum

Hay, Roy
The Dictionary of House Plants

fl.   Pl. 204

Echium fastuosum

Macoby, Stirling
What Flower is That

fl.   P 121

Echium fastuosum

Mathias, Mildred E.
Color for the Landscape

fl. hab.   p 142

Echium fastuosum

Royal Hort. Soc.
Journal of the Royal Hort. Soc.
  vol 91, No. 10   1966

fl.   p. 426   Pl. 213

Echium giganteum

Bramwell, David
Wild Flowers of the Canary Islands

fl.   Pl. 222

Echium hierrense

Bramwell, David
Wild Flowers of the Canary Islands

fl.   Pl. 225

Echium italicum

Polunin, Oleg
Flowers of Europe

fl. Pl. 105 #1081

Echium judaicum

Feinbrun-Dothan, Naomi
Wild Plants in the Land of
  Israel

fl. hab.   P. 104

Echium judaicum

Polunin, Oleg
Flowers of the Mediterranean

fl.   No. 145

Echium leucophaeum

Bramwell, David
Wild Flowers of the Canary Islands

fl.   Pl. 233

Echium lycopsis

Morley, Brian D.
Wild Flowers of the World

fl.   Pl. 37

Echium lycopsis

Polunin, Oleg
Flowers of Europe

fl. Pl. 106 #1083

Echium lycopsis

Polunin, Oleg
Flowers of the Mediterranean

fl.   No. 141

Echium lycopsis

Taylor, A. W.
Wild Flowers of Spain and Portugal

fl.   p. 91

Echium onosmifolium

Bramwell, David
Wild Flowers of the Canary Islands

fl.   Pl. 230, 231

Echium pininana

Bramwell, David
Wild Flowers of the Canary Islands

fl.   Pl. 229

Echium pininana

Curtis's Botanical Magazine
Vol 171   1956-57

fl.   269

Echium plantagineum

Crockett, James Underwood
Annuals

fl.   P 116

Echium plantagineum

Wit, H. C. D. de
Plants of the World;
  The Higher Plants, Vol II

fl.   Pl. 79

Echium plantagineum var.

Hay, Roy
The Color Dictionary of Flowers &
  Plants

fl.   p. 37

Echium simplex

Bramwell, David
Wild Flowers of the Canary Islands

fl., hab.   Pl. 228

Echium strictum

Bramwell, David
Wild Flowers of the Canary Islands

fl.   Pl. 226

Echium vulgare

Ary, S.
The Oxford Book of Wildflowers

fl.   p 170, Pl. 4

Echium vulgare

Barneby, T. P.
European Alpine Flowers in Colour

fl.   Pl. 66, 4

Echium vulgare

Courtenay, Booth
Wildflowers & Weeds

fl.   p 86

Echium vulgare

Dean, Blanche E.
Wildflowers of Alabama and
    Adjoining States

fl.   p 149

Echium vulgare

Felsko, Elsa
A Book of Wildflowers

fl.   P 54

Echium vulgare

Ferguson, Mary
Wildflowers

fl.   p 139

Echium vulgare

Kleijn, H.
Beauty of the Wild Plant

fl. opp. p. 68   Pl. 83

Echium vulgare

Klimas, John E.
Wildflowers of Eastern America

fl.   Pl. 285

Echium vulgare

Lindman, C. A. M.
Nordens Flora, Vol 7

fl. fr.   Pl. 484

Echium vulgare

Macoby, Stirling
What Flower is That

fl.   P 121

Echium vulgare

Martin, W. Keble
The Concise British Flora in Colour

fl. hab.   Pl. 61

Echium vulgare

Morley, Brian D.
Wild Flowers of the World

fl.   Pl. 20F

Echium vulgare

Polunin, Oleg
Flowers of Europe

fl.   Pl. 104 #1082

Echium vulgare

Wharton, Mary E.
A Guide to the Wildflowers & Ferns
    of Kentucky

fl.   p 226   Pl. 3.31

Echium webbii

Braxwell, David
Wild Flowers of the Canary Islands

fl.   Pl. 235

Echium wildpretii

Braxwell, David
Wild Flowers of the Canary Islands

hab., fl.   Pl. 227

Echium wildpretii

Morley, Brian D.
Wild Flowers of the World

fl.,hab.   Pl. 44

Echium wildpretii

Royal Hort. Soc.
Journal of the Royal Hort. Soc.
    Vol. 94, No. 3   1969

hab., fl.   P. 118   Pl. 65

Eclipta alba

Weeds of the Southern United States

hab.   p. 11

Eclipta prostrata

Walden, Beryl M.
Wild Flowers of Hong Kong

fl.   Pl. 6(19)

Edelweiss

see

Leontopodium alpinum

Edelweiss, Brazilian

see

Rechsteineria leucotricha

Edgeworthia Gardneri

Hara, Hiroshi, comp.
Photo-Album of Plants of
    Eastern Himalaya

fl.   Pl. 79

Edgeworthia papyrifera

Gault, S. Millar
The Color Dictionary of Shrubs

Edgeworthia papyrifera

Harrison, R.E.
Trees and Shrubs

fl.   P 64

Edgeworthia papyrifera

Hay, Roy
The Color Dictionary of Flowers & Plants

fl.   p 196, Pl. 1568

Edgeworthia papyrifera

Kimura, Koiti
Japanese Medicinal Plants, Vol II

fl.   p 203

Edgeworthia papyrifera

Kitamura, Siro
Coloured Illustrations of Trees &
    Shrubs of Japan

fl.   355

Edgeworthia papyrifera

Perry, Frances
Flowers of the World

fl.   p 296

Edithcolea grandis

Curtis's Botanical Magazine
Vol 177   1969-70

fl.   562

Edithcolea grandis

Flowering Plants of Africa
    vol 40   1969-70

fl.   Pl. 1600

Edraianthus pumilio

Hay, Roy
The Color Dictionary of Flowers & Plants

fl.   p 8, Pl. 61

Edraianthus serpyllifolius var.

Hay, Roy
The Color Dictionary of Flowers &
    Plants

fl.   p. 8   Pl. 62

Egeria densa

Tarver, David P.
Aquatic and Wetland Plants
    of Florida

fl. hab.   p 49

Eggplant

see

Solanum melongena

Eggs and bacon

see

Aotus ericoides

Eggs and Bacon

see

Dillwynia retorta

Eggs and bacon

see

Linaria maroccana

Eggs, Poached

see

Myriocephalus stuartii

Eggs, Scrambled

see

Cassia corymbosa

Ehretia dicksonii var.

Kitamura, Siro
Coloured Illustrations of Trees &
　Shrubs of Japan

fl.　　　　422

Ehretia longiflora

Walden, Beryl M.
Wild Flowers of Hong Kong

fl.　　　　Pl. 20(60), 21(60)

Ehretia ovalifolia

Kitamura, Siro
Coloured Illustrations of Trees &
　Shrubs of Japan

fl.　　　　423

Ehretia thyrsiflora

Curtis's Botanical Magazine
　vol 175　　1964-65

fl.　　Pl. 440

Eichhornia azurea

Stodola, Jiri
Encyclopedia of Water Plants

fl.　　　p 51

Eichhornia azurea

Tarver, David P.
Aquatic and Wetland Plants
　of Florida

fl. hab.　　　p 32

Eichhornia crassipes

Brown, Clair A.
Wildflowers of Louisiana and
　Adjoining States

fl. hab.　　p 13

Eichhornia crassipes

Bruggeman, L.
Tropical Plants

fl.　　Pl. 172

Eichhornia crassipes

Dean, Blanche E.
Wildflowers of Alabama and
　Adjoining States

fl.　　　　p 9

Eichhornia crassipes

Duncan, Wilbur H.
Wildflowers of the Southeastern
　United States

hab.　　　p 245

Eichhornia crassipes

Everett, T.H.ed.
New Illustrated Encyclopedia of
　Gardening　　vol 4

fl.　　p. 694

Eichhornia crassipes

Fleming, Glenn
Wild Flowers of Florida

fl. hab.　　p 91

Eichhornia crassipes

Lemmon, Robert S.
Wildflowers of North America in
　Full Color

fl.　　pg. 6　　　pl. 8

Eichhornia crassipes

Macoby, Stirling
What Flower is That

fl.　　P 121

Eichhornia crassipes

Perry, Frances
Flowers of the World

fl.　p 237

Eichhornia crassipes

Stodola, Jiri
Encyclopedia of Water Plants

Fl.　　p 54

Eichhornia crassipes

Tarver, David P.
Aquatic and Wetland Plants
　of Florida

fl. hab.　　　p 7

Eichhornia crassipes

Tsukamoto, Yotaro
Coloured Illustrations of Garden
　Flowers　　　Vol. 9

fl.　　opp. p. 46

Eichhornia crassipes

Wit, H. C. D. de
Plants of the World;
　The Higher Plants, Vol II

fl. hab.　　　Pl. 132,
　　　　　　　133

Ekebergia benguelensis

Palgrave, K. C.
Trees of Central Africa

fl., fr.　　　p. 217

Ekebergia sp.

Palgrave, K. C.
Trees of Central Africa

fl., fr.　　p. 213

Elaeagnus angustifolia

American Hort. Soc.
American Horticulturist
　vol 54, No. 2　1975

fr.　p. 18

Elaeagnus angustifolia

Flemer, William III
Shade and Ornamental Trees in Color

hab.　　p. 77

Elaeagnus angustifolia

Pennsylvania Hort. Soc.
The Green Scene
　Vol. 6, No. 1　1977

hab., fl.　　P. 23

Elaeagnus angustifolia

Polunin, Oleg
Trees and Bushes of Europe

fl. fr. hab.　　p 142, 143

Elaeagnus commutata

Clark, Lewis J.
Wild Flowers of British Columbia

fl.　　p. 318

Elaeagnus commutata

Ferguson, Mary
Wildflowers

fl.　　　p 65

Elaeagnus commutata

Heller, Christine
Wild Flowers of Alaska

fl., fr.   Pl. 30

Elaeagnus commutata

Huxley, Anthony
Deciduous Garden Trees and Shrubs

fl.   Pl. 69

Elaeagnus commutata

Neufeld, J.B.
Wild Flowers of the Prairies

fl.   p. 27

Elaeagnus commutata

Walcott, Mary Vaux
North American Wild Flowers

fl. fr. hab.   Pl. 70, 71

Elaeagnus x ebbingei

Hay, Roy
The Color Dictionary of Flowers &
  Plants

fl.   p. 197   Pl. 1569

Elaeagnus glabra

Kitamura, Siro
Coloured Illustrations of Trees &
  Shrubs of Japan

fl.   356

Elaeagnus multiflora

Hansen, Richard
Baume und Straucher im Garten

hab.   p. 135

Elaeagnus multiflora

Pennsylvania Hort. Soc.
The Green Scene
  vol 6, No. 1   1977

fl., fr. p. 23

Elaeagnus multiflora var.

Kitamura, Siro
Coloured Illustrations of Trees &
  Shrubs of Japan

fr.   358

Elaeagnus murakamiana

Kitamura, Siro
Coloured Illustrations of Trees &
  Shrubs of Japan

fl.   359

Elaeagnus pungens

Crockett, James Underwood
Evergreens

hab.,fr.   p. 121

Elaeagnus pungens

Pennsylvania Hort. Soc.
The Green Scene
  vol 6, No. 1   1977

fl., fr.   p. 22

Elaeagnus pungens

Royal Hort. Soc.
Journal of the Royal Hort. Soc.
  vol 99, No. 12   1974

fl.   cover

Elaeagnus pungens var.

Gault, S. Millar
The Color Dictionary of Shrubs

hab.   Pl. 158, 159

Elaeagnus pungens var.

Harrison, R.E.
Trees and Shrubs

fr.   p 65

Elaeagnus pungens var.

Hay, Roy
The Color Dictionary of Flowers & Plants

hab.   p 197, Pl. 1570, 1571

Elaeagnus pungens var.

Hellyer, A.G.L.
Shrubs in Colour

hab p. 46

Elaeagnus pungens var.

Perry, Frances
Flowers of the World

hab.   p. 104

Elaeagnus, thorny

see

Elaeagnus pungens

Elaeagnus tutcheri

Walden, Beryl M.
Wild Flowers of Hong Kong

fl.   Pl. 80 (252)

Elaeagnus umbellata

Kitamura, Siro
Coloured Illustrations of Trees &
  Shrubs of Japan

fr.   357

Elaeis guineensis

Edlin, Herbert
The Illustrated Encyclopedia
  of Trees

fr. hab.   p 236

Elaeis guineensis

Hvass, Elsie
Plants That Feed and Serve Us

fl. fr.   p 18, Pl. 26

Elaeis guineensis

Masefield, G. B.
The Oxford Book of Food Plants

hab., fl., fr.   Pl. 21

Elaeis guineensis

Wit, H. C. D. de
Plants of the World;
  The Higher Plants, Vol. II

fl. fr. hab.   Pl. 165, 166

Elaeocarpus dentatus

Harvey, Norman B.
New Zealand Botanical Paintings

fl.   p. 34   Pl. 14

Elaeocarpus grandiflorus

Bruggeman, L.
Tropical Plants

fl.   Pl. 207

Elaeocarpus grandiflorus

Morley, Brian D.
Wild Flowers of the World

fl.   Plate 116

Elaeocarpus reticulatus

Blombery, Alec M.
What Wildflower is That

fl.   p 126, Pl. 328

Elaeocarpus storckii

Morley, Brian D.
Wild Flowers of the World

fr.   Pl. 146

Elatine hexandra

Martin, W. Keble
The Concise British Flora in Colour

fl. hab.   Pl. 17

Elatine hydropiper

Lindman, C. A. M.
Nordens Flora, Vol 6

fl. hab.   Pl. 390A

Elatine hydropiper

Martin, W. Keble
The Concise British Flora in Colour

hab.   Pl. 17

Elatine hydropiper

Stodola, Jiri
Encyclopedia of Water Plants

hab.   P. 255

Elatine macropoda

Stodola, Jiri
Encyclopedia of Water Plants

hab.   P. 255

Elatine triandra

Lindman, C. A. M.
Nordens Flora, Vol 6

fl. hab.   Pl. 390B

Elderberry

see

Sambucus

Elderberry, American

see

Sambucus canadensis

Elderberry, Mexican

see

Sambucus mexicana

Elderberry, Red

see

Sambucus racemosa

Elderberry, Southern

see

Sambucus simpsonii

Elderberry, Sweet

see

Sambucus canadensis

Elderberry, White

see

Sambucus gaudichaudiana

Elder, Box

see

Acer negundo

Elder, Common

see

Sambucus nigra

Elder, Dwarf

see

Sambucus.ebulus

Elder, European

see

Sambucus nigra

Elder, Golden

see

Sambucus nigra var.

Elder, Marsh

see

Iva xanthifolia

Elder, Native

see

Sambuccus gaudichaudiana

Elder, Pacific red

see

Sambucus callicarpa

Elder, Red-berried

see

Sambucus racemosa

Elder, yellow

see

Stenolobium stans

Elder, Yellow

see

Tecoma stans

Elecampane

see

Inula ensifolic

Elecampane

see

Inula helenium

Elegia juncea

Morley, Brian D.
Wild Flowers of the World

fl.,hab.   Pl. 88

Eleocharis acicularis

Lindman, C. A. M.
Nordens Flora, Vol 2

fr. hab.   Pl. 115B

Eleocharis acicularis

Martin, W. Keble
The Concise British Flora in Colour

hab.   Pl. 91

Eleocharis acicularis

Stodola, Jiri
Encyclopedia of Water Plants

hab.   P. 82

Eleocharis baldwinii

Tarver, David P.
Aquatic and Wetland Plants
of Florida

hab.   p 18

Eleocharis multicaulis

Martin, W. Keble
The Concise British Flora in Colour

fr. hab.   Pl. 91

Eleocharis palustris

Lindman, C. A. M.
Nordens Flora, Vol 2

fr. hab.   Pl. 115A

Eleocharis palustris

Martin, W. Keble
The Concise British Flora in Colour

fr. hab.   Pl. 91

Eleocharis palustris

Polunin, Oleg
Flowers of Europe

fl.   Pl. 185 #1846

Eleocharis palustris

Stodola, Jiri
Encyclopedia of Water Plants

fl.   P. 83

Eleocharis parvula

Stodola, Jiri
Encyclopedia of Water Plants

hab.   P. 83

Eleocharis quinqueflora

Martin, W. Keble
The Concise British Flora in Colour

fr. hab.   Pl. 91

Eleocharis tuberosa

Masefield, G. B.
The Oxford Book of Food Plants

fl. fr.   Pl. 33, 3

Eleocharis uniglumis

Martin, W. Keble
The Concise British Flora in Colour

fl. hab.   Pl. 91

Eleocharis vivipara

Stodola, Jiri
Encyclopedia of Water Plants

hab.   P. 86

Elephant heads
    see
Pedicularis groenlandica

Elephant Tree
see
Pachycormus discolor

Elephantwood
    see
Pachycormus discolor

Elephant
see also
Elephants

Elephantella
    see
Pedicularis groenlandica

Elephantopus tomentosus

Duncan, Wilbur H.
Wildflowers of the Southeastern
    United States

fl. hab.   p 193

Elephant's ear
    see
Alocasia macrorhiza var.

Elephant's Ear
    see
Bergenia cordifolia

Elephants-Ear
    see
Whiteheadia bifolia

Elephant's food
    see
Portulacaria afra

Elephant's Foot
    see
Dioscorea elephantipes

Elephant's Head
    see
Pedicularis groenlandica

Elephant's Head, Little
    see
Pedicularis attolens

Elephants
    see also
Elephant

Elettaria cardamomum

Bianchini, Francesco
Health Plants of the World

fr.   p 43

Elettaria cardamomum

Hvass, Elsie
Plants That Feed and Serve Us

fl.   fr.   p 75, Pl. 156

Elettaria cardamomum

Loewenfeld, Claire
The Complete Book of Herbs and Spices

fl. fr. hab.   p 208

Elettaria cardamomum

Masefield, G. B.
The Oxford Book of Food Plants

fl. fr.   Pl. 131, 4

Elettaria cardamomum

Rosengarten, Frederic, Jr.
The Book of Spices

fl.   fr.   p 160

Eleusine coracana

Masefield, G. B.
The Oxford Book of Food Plants

fr.   Pl. 11, 3-4

Eleusine indica

Crockett, James Underwood
Lawns & Ground Covers

hab. fr.   p. 71

Eleusine indica

Weeds of the Southern United States

hab., fr.   p. 25

Elisma natans

Stodola, Jiri
Encyclopedia of Water Plants

fl.   P. 178

Elkslip
    see
Caltha leptosepala

Elleanthus aurantiacus

Ospina, Mariano
Orquideas de las americas

fl.   Pl. 44

Elm
    see
Ulmus

Elm, American
    see
Ulmus Americana

Elm, Chinese
    see
Ulmus parvifolia

Elm, English

see

Ulmus campestris

Elm, English

see

Ulmus procera

Elm, evergreen

see

Ulmus parvifolia var.

Elm, Jersey

see

Ulmus carpinifolia var.

Elm, Scotch

see

Ulmus glabra

Elm, Smooth-leafed

see

Ulmus procera

Elm, Wheatley

see

Ulmus carpinifolia var.

Elm, Wych

see

Ulmus glabra

Elodea canadensis

Ary, S.
The Oxford Book of Wildflowers

fl.   p 52, Pl. 7

Elodea canadensis

Lindman, C. A. M.
Nordens Flora, Vol 1

fl. hab.   Pl.43

Elodea canadensis

Martin, W. Keble
The Concise British Flora in Colour

fl. hab.   Pl. 79

Elodea canadensis

Stodola, Jiri
Encyclopedia of Water Plants

hab.   P. 258

Elodea densa

Stodola, Jiri
Encyclopedia of Water Plants

hab.   P. 258

Elodea nuttali

Stodola, Jiri
Encyclopedia of Water Plants

fl.   P. 262

Elodea occidentalis

Stodola, Jiri
Encyclopedia of Water Plants

hab.   P. 259

Elsholtzia ciliata

Walden, Beryl M.
Wild Flowers of Hong Kong

fl.   Pl. 73 (222)

Elsholtzia stauntonii

Hellyer, A.G.L.
Shrubs in Colour

fl.   p. 46

Eltroplectris roseo-alba

Pabst, G. F. J.
Orchidaceae Brasilienses,
   Vol 1

fl.   p 183

Eltroplectris triloba

Pabst, G. F. J.
Orchidaceae Brasilienses,
   Vol 1

fl.   p 183

Elymus arenarius

Hvass, Elsie
Plants That Feed and Serve Us

hab.   p 132, Pl. 292

Elymus arenarius

Lindman, C. A. M.
Nordens Flora, Vol 2

fr.   Pl. 111

Elymus arenarius

Polunin, Oleg
Flowers of Europe

fr. hab. Pl. 179 #1736

Elymus arenarius

Royal Hort. Soc.
The Garden
   vol 101, No. 9 1976

hab.   p. 450

Elymus innovatus

Porsild, A. E.
Rocky Mountain Wild Flowers

fl.   p 51

Elythranthera brunonis

Blombery, Alec M.
What Wildflower is That

fl.   p. 162   Pl. 329

Elythranthera brunonis

Cady, Leo
Australian Native Orchids in Colour

fl.   Pl. 41

Elythranthera brunonis

Morley, Brian D.
Wild Flowers of the World

fl.   pl. 145

Elythranthera emarginata

Blombery, Alec M.
What Wildflower is That

fl.   p 126, Pl. 330

Elytrigia juncea

Lindman, C. A. M.
Nordens Flora, Vol 2

fr. hab.   Pl. 110B

Elytrigia repens

Lindman, C. A. M.
Nordens Flora, Vol 2

fr. hab.   Pl. 110A

Embothrium coccineum

American Hort. Soc.
American Horticulturist
   vol 55, No. 2 1976

fl.   p. 23

Embothrium coccineum

Hay, Roy
The Color Dictionary of Flowers & Plants

fl.   p 197, Pl. 1572

Embothrium coccineum

Munoz Pizarro, Carlos
Flores Silvestres de Chile

fl., fr.   Pl. 36

Embothrium coccineum

Royal Hort. Soc.
The Garden
vol 102, No. 11   1977

fl.   p. 472

Embothrium coccineum var.

Cault, S. Millar
The Color Dictionary of Shrubs

fl.   Pl. 160, 161

Embothrium coccineum var.

Harrison, R.E.
Trees and Shrubs

fl.   P 64

Embothrium coccineum var.

Hay, Roy
The Color Dictionary of Flowers &
Plants

fl.   p. 197   Pl. 1573

Embothrium coccineum var.

Hellyer, A. G. L.
Shrubs in Colour

fl.   p. 47

Embothrium coccineum var.

Royal Hort. Soc.
Journal of the Royal Hort. Soc.
vol. 93, No. 4   1968

fl.   p. 164   Pl. 73

Embothrium lanceolatum var.

Royal Hort. Soc.
Journal of the Royal Hort. Soc.
Vol. 73, No. 11   1948

fl.   p. 388   Pl. 139

Embothrium lanceolatum var.

Royal Hort. Soc.
Journal of the Royal Hort. Soc.
vol 87, No. 9   1962

fl.   p. 408   Pl. 126

Embothrium wichamii

Harrison, R.E.
Trees and Shrubs

fl.   P 64

Emilia coccinea

Morley, Brian D.
Wild Flowers of the World

fl.,fr.   Pl. 65

Emilia flammea

Crockett, James Underwood
Annuals

fl.   P. 117

Emilia flammea

Tsukamoto, Yotaro
Coloured Illustrations of Garden
Flowers   Vol. 10

fl.   Pl. 18

Emilia javanica

Bruggeman, L.
Tropical Plants

fl.   Pl. 40

Emilia javanica

Moriarty, Audrey
Wild Flowers of Malawi

fl.   Pl. 78; 7

Emilia sonchifolia

Walden, Beryl M.
Wild Flowers of Hong Kong

fl.   Pl. 76 (232)

Emmenopterys henryi

Royal Hort. Soc.
Journal of the Royal Hort. Soc.
vol 96, No. 11   1971

fl.   p. 494   Pl. 220

Emmenosperma alphitonioides

Blombery, Alec M.
What Wildflower is That

fr.   p 127, Pl. 331

Emmer

see

Triticum dicoccum

Empetrum nigrum

Ary, S.
The Oxford Book of Wildflowers

fl. fr.   p 120, Pl. 4

Empetrum nigrum

Barneby, T. P.
European Alpine Flowers in Colour

fr.   Pl. 57, 4

Empetrum nigrum

Heller, Christine
Wild Flowers of Alaska

fr.   Pl. 304

Empetrum nigrum

Lindman, C. A. M.
Nordens Flora, Vol 7

fl. fr.   Pl. 456

Empetrum nigrum

Martin, W. Keble
The Concise British Flora in Colour

fr. hab.   Pl. 55

Empetrum nigrum

Walcott, Mary Vaux
North American Wild Flowers
vol. 5   Pl. 382

Empetrum nigrum var.

Porsild, A. E.
Rocky Mountain Wild Flowers

fr.   p 275

Empodium namaquensis

Flowering Plants of Africa
vol 44   1977

fl., hab.   Pl. 1727

Empress, German

see

Nopalxochia phyllanthoides

Empress Tree

see

Paulownia tomentosa

Emu-Bush

see

Eremophila

Enarganthe octonaria

Herre, H.
The Genera of the Mesembryanthemaceae

fl.   fr.   p 149

Encelia californica

Coyle, Jeanette
A Field Guide to the Common and
Interesting Plants of Baja California

fl.   p 169

Encelia californica

Munz, Philip A.
California Spring Wildflowers

fl.   p 32, Pl. 17

Encelia farinosa

Coyle, Jeanette
A Field Guide to the Common and
Interesting Plants of Baja California

fl.   p 171

Encelia farinosa

Lemmon, Robert S.
Wildflowers of North America in
    Full Color

fl. hab.   pg. 106   pl. 171

Encelia farinosa

Mathias, Mildred E.
Color for the Landscape

fl. hab.   p 188

Enceliopis argophylla

Orr, Robert T.
Wildflowers of Western America

fl.   Pl. 112

Enceliopsis argophylla var.

Munz, Philip A.
California Desert Wildflowers

fl.   p 53, Pl. 81

Encephalartos ferox

American Hort. Soc.
American Horticulturist
vol 57, No. 6   1978

fr.   p. 7

Encephalartos latifrons

Cactus & Succulent Society of America
Journal
    vol 47, No. 2   1975

hab.   p. 72-73

Encephalartos lebomboensis

Flowering Plants of Africa
    vol 27   1948-49

fr.   Pl. 1078, 1079

Encephalartos ngoyanus

Flowering Plants of Africa
    vol 27   1948-49

fr.   Pl. 1053, 1054

Encephalartos umbeluziensis

Flowering Plants of Africa
    vol 28   1950-51

fr.   Pl. 1100

Encephalartos villosus

Flowering Plants of Africa
    vol XXVI   1947

fl., fr.   Pl. 1001, 1002

Encephalocarpus strobiliformis

Kupper, Walter
Cacti

fl., hab.   p. 105   Pl. 49

Encephalocarpus strobiliformis

Lamb, Edgar
Popular Exotic Cacti in Color

fl. hab.   p 79

Encheiridion macrorrhynchium

Williamson, Graham
The Orchids of South Central Africa

fl.   Pl. 168

Encholirium spectabile

Smith, Lyman B.
The Bromeliads

fl. hab.   Pl. 1

Enchylaena tomentosa

Cochrane, G. R.
Flowers and Plants of Victoria

fl. fr.   Pl. 160

Encyclia allemanii

Pabst, G. F. J.
Orchidaceae Brasilienses,
    Vol 1

fl.   P 187

Encyclia atropurpurea

Ebel, Friedrich
The Strange and Beautiful
    World of Orchids

fl.   p 55

Encyclia atropurpurea var.

Curtis's Botanical Magazine
Vol 171   1956-57

fl.   290

Encyclia boothiana var.

Luer, Carlyle A.
The Native Orchids of Florida

fl. hab.   Pl. 61

Encyclia bracteata

Pabst, G. F. J.
Orchidaceae Brasilienses,
    Vol 1

fl. hab.   p 186

Encyclia bulbosa

Pabst, G. F. J.
Orchidaceae Brasilienses,
    Vol 1

fl.   p 187

Encyclia cochleata

Luer, Carlyle A.
The Native Orchids of Florida

fl.   Pl. 60; 3-4

Encyclia cochleata var.

Luer, Carlyle A.
The Native Orchids of Florida

fl. hab.   Pl. 60; 1, 2,
        5, 6

Encyclia dichroma

Pabst, G. F. J.
Orchidaceae Brasilienses,
    Vol 1

fl.   p 186

Encyclia farosiana

Pabst, G. F. J.
Orchidaceae Brasilienses,
    Vol 2

Encyclia mariae

Ospina, Mariano
Orquideas de las americas

fl.   Pl. 57

Encyclia odoratissima

Pabst, G. F. J.
Orchidaceae Brasilienses,
    Vol 1

fl.   p 187

Encyclia pygmaea

Luer, Carlyle A.
The Native Orchids of Florida

fl. hab.   Pl. 59

Encyclia tampensis

Fleming, Glenn
Wild Flowers of Florida

fl.   p 42

Encyclia tampensis

Luer, Carlyle A.
The Native Orchids of Florida

fl., hab.   Pl. 4, 6, 57, 58

Encyclia vespa

Pabst, G. F. J.
Orchidaceae Brasilienses,
    Vol 1

fl. fr.   p 187

Endive

see

Cichorium endivia

Endive, Belgian

see

Cichorium intybus var.

Endonema retzioides

Morley, Brian D.
Wild Flowers of the World

fl.   Pl. 75

Endymion hispanicus

Curtis's Botanical Magazine
Vol. 169   1952-53

fl.   176

Endymion hispanicus

Polunin, Oleg
Flowers of Europe

fl.   Pl. 169 #1638

Endymion hybrid

Miles, Bebe
Bulbs for the Home Gardener

fl.   p 69

Endymion nonscriptus

Ary, S.
The Oxford Book of Wildflowers

fl.   p 168, Pl. 3

Endymion non-scriptus

Hay, Roy
The Color Dictionary of Flowers & Plants

fl.   p 91, Pl. 727

Endymion non-scriptus

Kleijn, H.
Beauty of the Wild Plant

fl.   opp. p. 52   pl. 61

Endymion non-scriptus

Martin, W. Keble
The Concise British Flora in Colour

fl. hab.   Pl. 85

Endymion non-scriptus

Morley, Brian D.
Wild Flowers of the World

fl.,hab.   Pl. 24

Endymion non-scriptus

Polunin, Oleg
Flowers of Europe

fl.   Pl. 170 #1638

Enkianthus campanulatus

Curtis's Botanical Magazine
vol 176   1966-68

fl.   Pl. 512

Enkianthus campanulatus

Everett, T.H., ed.
New Illustrated Encyclopedia of
Gardening vol 4

fl.   p. 631

Enkianthus campanulatus

Gault, S. Millar
The Color Dictionary of Shrubs

fl.   Pl. 162

Enkianthus campanulatus

Harrison, R.E.
Trees and Shrubs

fl.   P 74

Enkianthus campanulatus

Hay, Roy
The Color Dictionary of Flowers & Plants

fl.   p 197, Pl. 1574

Enkianthus campanulatus

Hellyer, A.G.L.
Shrubs in Colour

fl.   p. 46

Enkianthus campanulatus

Huxley, Anthony
Deciduous Garden Trees and Shrubs

fl.   Pl. 70

Enkianthus campanulatus

Kitamura, Siro
Coloured Illustrations of Trees &
Shrubs of Japan

fl.   393

Enkianthus campanulatus

Macoby, Stirling
What Flower is That

fl.   P 121

Enkianthus campanulatus

Massachusetts Hort. Soc.
Horticulture
vol 38, No. 5   1960

fl.   p. 282

Enkianthus campanulatus

Morley, Brian D.
Wild Flowers of the World

fl.fr.   Plate 101

Enkianthus campanulatus

Perry, Frances
Flowers of the World

fl.   p 108

Enkianthus cernuus

Hay, Roy
The Color Dictionary of Flowers & Plants

hab.   p 197, Pl. 1575

Enkianthus cernuus var.

Hay, Roy
The Color Dictionary of Flowers &
Plants

hab.   p. 197   Pl. 1576

Enkianthus cernuus var.

Kitamura, Siro
Coloured Illustrations of Trees &
Shrubs of Japan

fl.   396

Enkianthus deflexus

Hara, Hiroshi, comp.
Photo-Album of Plants of
Eastern Himalaya

fl.   Pl. 89

Enkianthus perulatus

Gault, S. Millar
The Color Dictionary of Shrubs

fl.   Pl. 163

Enkianthus perulatus

Kitamura, Siro
Coloured Illustrations of Trees &
Shrubs of Japan

fl.   395

Enkianthus perulatus

Pennsylvania Hort. Soc.
The Green Scene
vol 5, No. 4   1977

hab.   p. 14, Pl. 2

Enkianthus quinqueflorus

Royal Hort. Soc.
The Garden
vol 102, No. 2   1977

fl.   p. 67

Enkianthus quinqueflorus

Walden, Beryl M.
Wild Flowers of Hong Kong

fl.   Pl. 6(16)

Entandrophragma caudatum

Palgrave, K. C.
Trees of Central Africa

fl., fr.   p. 222

Entandrophragma utile

Wit, H. C. D. de
Plants of the World;
The Higher Plants, Vol II

fr.   p 57, Pl. 22

Entelea arborescens

Harvey, Norman B.
New Zealand Botanical Paintings

fl.   p. 36   Pl. 15

Enterospermum borbonicum

Royal Hort. Soc.
Journal of the Royal Hort. Soc.
        vol 96, No. 7   1971

hab.   p. 304   Pl. 133

Eomecon chionantha

Royal Hort. Soc.
The Garden
        vol 102, No. 3   1977

fl., hab.   cover

Epacris acuminata

Curtis, Winifred
The Endemic Flora of Tasmania, Part IV

fl. fr. hab.   Pl. 126

Epacris barbata

Curtis, Winifred
The Endemic Flora of Tasmania Vol. 2

fl.   Pl. 69

Epacris corymbiflora

Curtis, Winifred
The Endemic Flora of Tasmania
        vol 3
fl.   p 175

Epacris crassifolia

Blombery, Alec M.
What Wildflower is That

fl.   p 127, Pl. 332

Epacris exserta

Curtis, Winifred
The Endemic Flora of Tasmania
        vol 3
fl.   p 175

Epacris gunnii

Curtis, Winifred
The Endemic Flora of Tasmania Vol. 2

fl.   Pl. 68

Epacris impressa

Blombery, Alec M.
What Wildflower is That

fl.   p 128, Pl. 333

Epacris impressa

Cochrane, G. R.
Flowers and Plants of Victoria

fl.   Pl. 91

Epacris impressa

Morley, Brian D.
Wild Flowers of the World

fl.   pl. 136

Epacris impressa var.

Perry, Frances
Flowers of the World

fl.   p 105

Epacris lanuginosa

Cochrane, G. R.
Flowers and Plants of Victoria

fl.   Pl. 88

Epacris longiflora

Blombery, Alec M.
What Wildflower is That

fl.   p 128, Pl. 334

Epacris longiflora

Harrison, R.E.
Trees and Shrubs

fl.   P 65

Epacris longiflora

Macoby, Stirling
What Flower is That

fl.   P 122

Epacris longiflora

Mullins, Barbara
Australian Wildflowers in Colour

fl.   P. 51   Pl. 45

Epacris longiflora

Perry, Frances
Flowers of the World

fl.   p 105

Epacris marginata

Curtis, Winifred
The Endemic Flora of Tasmania, Part IV

fl. fr. hab.   Pl. 127

Epacris microphylla

Cochrane, G. R.
Flowers and Plants of Victoria

fl.   Pl. 10

Epacris microphylla

Morley, Brian D.
Wild Flowers of the World

fl.   Pl. 136

Epacris microphylla

Mullins, Barbara
Australian Wildflowers in Colour

fl.   P. 53   Pl. 47

Epacris mucronulata

Curtis, Winifred
The Endemic Flora of Tasmania
        Vol VI
fl.   Pl. 235

Epacris myrtifolia

Curtis, Winifred
The Endemic Flora of Tasmania
        vol 3
fl.   p 177

Epacris obtusifolia

Morley, Brian D.
Wild Flowers of the World

fl.   pl. 136

Epacris paludosa

Cochrane, G. R.
Flowers and Plants of Victoria

fl.   Pl. 422

Epacris pulchella

Blombery, Alec M.
What Wildflower is That

fl.   p 128, Pl. 335

Epacris serpyllifolia

Alpine Flowers of the Kosciusko
        State Park

fl.   Pl. 6

Epacris stuartii

Curtis, Winifred
The Endemic Flora of Tasmania
        Vol VI
fl.   Pl. 236

Epacris tasmanica

Curtis, Winifred
The Endemic Flora of Tasmania Vol. 2

fl.   Pl. 67a, b

Epacris virgata

Curtis, Winifred
The Endemic Flora of Tasmania
        vol 3
fl.   p 177

Ephedra andina

Curtis's Botanical Magazine
Vol 168   1951

fl. fr.   142

Ephedra californica
Coyle, Jeanette
A Field Guide to the Common and
   Interesting Plants of Baja California
fl.   p 45

Ephedra distachya
Kariyone, Tatsuo
Atlas of Medicinal Plants
fl., hab.   Pl. 144

Ephedra distachya
Kimura, Koiti
Japanese Medicinal Plants, Vol I
fl.   p 74

Ephedra distachya
Perrot, Emile
Les Plantes Medicinales
fl. fr. hab.   p 72

Ephedra distachya
Takatori, Jisuke
Color Atlas of Medicinal Plants of Japan
fl.   Fig. 80

Ephedra fragilis
Wit, H. C. D. de
Plants of the World;
   The Higher Plants, Vol I
fr.   p 98, Pl. 40

Ephedra fragilis var.
Polunin, Oleg
Flowers of Europe
fl.   Pl. 2 #18

Ephedra sp.
Bianchini, Francesco
Health Plants of the World
fl. hab.   p 97

Epiblema grandiflora
Blombery, Alec M.
What Wildflower is That
fl.   p 129, Pl. 336

Epiblema grandiflora
Cady, Leo
Australian Native Orchids in Colour
fl.   Pl. 28

Epiblema grandiflora
Morcombe, M. K.
Australia's Western Wildflowers
fl.   p 101

Epidendrum acunae
Luer, Carlyle A.
The Native Orchids of Florida
fl. hab.   Pl. 67, 5 - 7

Epidendrum addac
Pabst, G. F. J.
Orchidaceae Brasilienses,
   Vol. 1
fl.   p. 188

Epidendrum anceps
Luer, Carlyle A.
The Native Orchids of Florida
fl. hab.   Pl. 63

Epidendrum atropurpureum
Wright, Norman Pelham
Orquideas de Mexico
fl.   Pl. 8

Epidendrum baculus
Curtis's Botanical Magazine
   Vol. 180   1974
fl. hab.   Pl. 702

Epidendrum brassavolae
Hay, Roy
The Dictionary of House Plants
fl.   Pl. 205

Epidendrum chondylobolbon
Curtis's Botanical Magazine
Vol 168   1951
fl.   160

Epidendrum ciliare
Morley, Brian D.
Wild Flowers of the World
fl.   Pl. 192

Epidendrum ciliare
Pabst, G. F. J.
Orchidaceae Brasilienses,
   Vol 1
fl.   p 187

Epidendrum cochleatum
Ebel, Friedrich
The Strange and Beautiful
   World of Orchids
fl.   p 37

Epidendrum cochleatum
Milne, Lorus
Living Plants of the World
fl.   p 328

Epidendrum conopseum
Brown, Clair A.
Wildflowers of Louisiana and
   Adjoining States
fl. hab.   p 36

Epidendrum conopseum
Fleming, Glenn
Wild Flowers of Florida
fl.   p 42

Epidendrum conopseum
Luer, Carlyle A.
The Native Orchids of Florida
fl. hab.   Pl. 62

Epidendrum conspicum
Addisonia
   vol 23   1954-59
fl.   p. 7   Pl. 740

Epidendrum cooperianum
Pabst, G. F. J.
Orchidaceae Brasilienses,
   Vol 1
fl.   p 188

Epidendrum coriifolium
Curtis's Botanical Magazine
   Vol 160   1937
fl.   No. 9477

Epidendrum coronatum
Pabst, G. F. J.
Orchidaceae Brasilienses,
   Vol 1
fl.   p 188

Epidendrum crassifolium
Pabst, G. F. J.
Orchidaceae Brasilienses,
   Vol 1
fl.   p 189

Epidendrum dendrobioides
Pabst, G. F. J.
Orchidaceae Brasilienses,
   Vol 1
fl.   p 190

Epidendrum difforme
Addisonia
   vol 24   1960-64
fl., fr.   p. 47   Pl. 792

Epidendrum difforme
Dunsterville, G. C. K.
Introduction to the World
   of Orchids
fl.   Pl. 7

Epidendrum difforme

Luer, Carlyle A.
The Native Orchids of Florida

fl. fr. hab.   Pl. 64

Epidendrum, Dwarf

see

Encyclia pygmaea

Epidendrum endresii

Grubb, Roy
Selected Orchidaceous Plants
Vol II

fl.   p 71

Epidendrum erubescens

Wright, Norman Pelham
Orquideas de Mexico

fl.   Pl. 9

Epidendrum eximium

Grubb, Roy
Selected Orchidaceous Plants
Vol I

fl.   p 131

Epidendrum falcatum

Kupper, Walter
Orchids

fl. hab.   p 25

Epidendrum fragrans

American Hort. Soc.
American Horticulturist
vol 51, No. 4 1972

fl.   p. 25

Epidendrum fulgens

Pabst, G. F. J.
Orchidaceae Brasilienses,
Vol 1

fl.   p 189

Epidendrum ibaguense

Hay, Roy
The Color Dictionary of Flowers & Plants

fl.   p 66, Pl. 522

Epidendrum ibaguense

Hay, Roy
The Dictionary of House Plants

fl.   Pl. 206

Epidendrum ibaguense

Macoby, Stirling
What Flower is That

fl. P 122

Epidendrum imatophyllum

Pabst, G. F. J.
Orchidaceae Brasilienses,
Vol 1

fl.   p 189

Epidendrum klueppelianum

Pabst, G. F. J.
Orchidaceae Brasilienses,
Vol. 1

fl.   p. 188

Epidendrum latilabre

Pabst, G. F. J.
Orchidaceae Brasilienses,
Vol 1

fl. fr.   p 190

Epidendrum longihastatum

Pabst, G. F. J.
Orchidaceae Brasilienses,
Vol 1

fl.   p 189

Epidendrum mariae

Grubb, Roy
Selected Orchidaceous Plants
Vol I

fl.   p 67

Epidendrum mariae

Wright, Norman Pelham
Orquideas de Mexico

fl.   Pl. 10

Epidendrum medusae

Kupper, Walter
Orchids

fl. hab.   p 27

Epidendrum nemorale

Kramer, Jack
Orchids; Flowers of Romance
and Mystery

fl.   p 74

Epidendrum nemorale

Wright, Norman Pelham
Orquideas de Mexico

fl.   Pl.11

Epidendrum, Night Smelling

see

Epidendrum nocturnum

Epidendrum nocturnum

Luer, Carlyle A.
The Native Orchids of Florida

fl. fr. hab.   Pl. 65

Epidendrum nocturnum

Ospina, Mariano
Orquideas de las americas

fl. hab.   Pl. 59

Epidendrum nocturnum

Walcott, Mary Vaux
North American Wild Flowers
vol. 5

fl.   Pl. 337

Epidendrum nocturnum var.

Pabst, G. F. J.
Orchidaceae Brasilienses,
Vol 1

fl. fr.   p 189

Epidendrum x obrienianum

Luer, Carlyle A.
The Native Orchids of the United States
and Canada

fl. hab.   Pl. 78; 1, 4

Epidendrum x o'brienianum

Macoby, Stirling
What Flower is That

fl.   p. 122

Epidendrum oerstedii

Curtis's Botanical Magazine
vol 173 1960

fl.   Pl. 375

Epidendrum paniculatum

Pabst, G. F. J.
Orchidaceae Brasilienses,
Vol 1

fl.   p 191

Epidendrum parkinsonianum

Wright, Norman Pelham
Orquideas de Mexico

fl.   Pl. 12

Epidendrum peperomia

Pabst, G. F. J.
Orchidaceae Brasilienses,
Vol 1

fl.   p 190

Epidendrum porpax

Dunsterville, G. C. K.
Introduction to the World
of Orchids

fl. hab.   Pl. 8

Epidendrum prismatocarpum

Kupper, Walter
Orchids

fl.   p 29

Epidendrum proligerum

Pabst, G. F. J.
Orchidaceae Brasilienses,
Vol 1

fl. hab.   p 188

Epidendrum purpurascens

Pabst, G. F. J.
Orchidaceae Brasilienses,
Vol 1

fl.   p 191

Epidendrum radiatum

Curtis's Botanical Magazine
Vol. 180   1974

fl.   Pl. 662

Epidendrum radicans

Hersey, Jean
Woman's Day Book of House Plants

fl.   p. 82

Epidendrum rigidum

Luer, Carlyle A.
The Native Orchids of Florida

fl. hab.   Pl. 66

Epidendrum robustum

Pabst, G. F. J.
Orchidaceae Brasilienses,
Vol 1

fl.   p 188

Epidendrum secundum

Ebel, Friedrich
The Strange and Beautiful
World of Orchids

fl.   p 39

Epidendrum seudospidendrum

Ospina, Mariano
Orquideas de las americas

fl.   Pl. 58

Epidendrum sprucianum

Pabst, G. F. J.
Orchidaceae Brasilienses,
Vol 1

fl.   p 190

Epidendrum stamfordianum

Kramer, Jack
Orchids: Flowers of Romance
and Mystery

fl.   p 70

Epidendrum stamfordianum

Wright, Norman Pelham
Orquideas de Mexico

fl.   Pl. 13

Epidendrum strobiliferum

Luer, Carlyle A.
The Native Orchids of Florida

fl. hab.   Pl. 67, 1 - 4

Epidendrum tampense

Lemmon, Robert S.
Wildflowers of North America in
Full Color

fl.   pg. 15   pl. 20

Epidendrum tampense

Walcott, Mary Vaux
North American Wild Flowers

fl. hab.   Pl. 152

Epidendrum vesicatum

Pabst, G. F. J.
Orchidaceae Brasilienses,
Vol 1

fl. hab.   p 191

Epidendrum vitellinum

Wright, Norman Pelham
Orquideas de Mexico

fl.   Pl. 14

Epidendrum, White

see

Epidendrum nocturnum

Epidendrum xanthinus

Pabst, G. F. J.
Orchidaceae Brasilienses,
Vol 1

fl.   p 189

Epifagus virginiana

Courtenay, Booth
Wildflowers & Weeds

fl.   p 103

Epifagus virginiana

Klimas, John E.
Wildflowers of Eastern America

fl.   Pl. 303

Epifagus virginiana

Wharton, Mary E.
A Guide to the Wildflowers & Ferns
of Kentucky

fl.   p 280   Pl. 3,4

Epigaea gaultherioides

Royal Hort. Soc.
Journal of the Royal Hort. Soc.
vol 97, No. 5   1972

fl.   p. 216   Pl. 94

Epigaea repens

Campbell, Carlos C.
Great Smoky Mountain Wildflowers

fl.   p 11

Epigaea repens

Courtenay, Booth
Wildflowers & Weeds

fl.   p 79

Epigaea repens

Dean, Blanche E.
Wildflowers of Alabama and
Adjoining States

fl.   p 127

Epigaea repens

Duncan, Wilbur H.
Wildflowers of the Southeastern
United States

fl.   p 119

Epigaea repens

Everett, T.H., ed.
New Illustrated Encyclopedia of
Gardening   vol 4

fl.   p. 631

Epigaea repens

Ferguson, Mary
Wildflowers

fl.   p 171

Epigaea repens

Jennings, O. E.
Wild Flowers of Western Pennsylvania
vol II

fl.   Pl. 112

Epigaea repens

Klimas, John E.
Wildflowers of Eastern America

fl.   Pl. 4

Epigaea repens

Lemmon, Robert S.
Wildflowers of North America in
Full Color

fl.   p. 265   Pl. 420

Epigaea repens

Menninger, Edwin A.
Flowering Vines of the World

fl.   Pl. 75

Epigaea repens

Morley, Brian D.
Wild Flowers of the World

fl.   pl. 161

Epigaea repens

Roberts, June Carver
Born in the Spring

fl. hab.   p 67

Epigaea repens

Walcott, Mary Vaux
North American Wild Flowers

fl. hab.   Pl. 126

Epigaea repens

Wharton, Mary E.
A Guide to the Wildflowers & Ferns
    of Kentucky

fl.   p 144   Pl. 1.99

Epigaea repens

Wharton, Mary E.
Trees & Shrubs of Kentucky

fl. hab.   p 55, Pl. 1.16

Epilobium alpinum

Taylor, Ronald J.
Mountain Wild Flowers of the Pacific
    Northwest

fl.   p. 149

Epilobium alsinifolium

Lindman, C. A. M.
Nordens Flora, Vol 6

fl. hab.   Pl. 406B

Epilobium alsinifolium

Martin, W. Keble
The Concise British Flora in Colour

fl. hab.   Pl. 35

Epilobium anagallidifolium

Lindman, C. A. M.
Nordens Flora, Vol 6

fl. hab.   Pl. 406C

Epilobium anagallidifolium

Martin, W. Keble
The Concise British Flora in Colour

fl. hab.   Pl. 35

Epilobium anagallidifolium

Porsild, A. E.
Rocky Mountain Wild Flowers

fl. hab.   p 283

Epilobium angustifolium

Alaska-Yukon Wild Flowers Guide

fl. hab.   p 51, 100

Epilobium angustifolium

Barneby, T. P.
European Alpine Flowers in Colour

fl.   Pl. 52, 1

Epilobium angustifolium

Clark, Lewis J.
Wild Flowers of British Columbia

fl.   p 338

Epilobium angustifolium

Courtenay, Booth
Wildflowers & Weeds

fl.   p 75

Epilobium angustifolium

Ferguson, Mary
Wildflowers

fl.   p 107

Epilobium angustifolium

Fries, Mary A.
Wildflowers of Mount Ranier and
the Cascades

fl.   P 73

Epilobium angustifolium

Goulimis, Constantine N.
Wild Flowers of Greece

fl.   p. 67

Epilobium angustifolium

Heller, Christine
Wild Flowers of Alaska

fl., hab.   Pl. 96-97

Epilobium angustifolium

Kleijn, H.
Beauty of the Wild Plant

fl. opp. p. 89   pl. 121

Epilobium angustifolium

Klimas, John E.
Wildflowers of Eastern America

fl.   Pl. 212

Epilobium angustifolium

Massachusetts Hort. Soc.
Horticulture
vol 43, No. 8   1965

hab., fl.   p. 23

Epilobium angustifolium

Milne, Lorus
Living Plants of the World

fl. hab.   p 169

Epilobium angustifolium

Morley, Brian D.
Wild Flowers of the World

fl. fr.   Pl. 4E

Epilobium angustifolium

Munz, Philip A.
California Mountain Wildflowers

fl.   p 32, Pl. 17

Epilobium angustifolium

Neufeld, J.B.
Wild Flowers of the Prairies

fl.   p. 17

Epilobium angustifolium

Orr, Robert T.
Wildflowers of Western America

fl.   Pl. 159

Epilobium angustifolium

Polunin, Oleg
Flowers of Europe

fl.   Pl. 80 #835

Epilobium angustifolium

Porsild, A. E.
Rocky Mountain Wild Flowers

fl.   p 283

Epilobium angustifolium

Taylor, Ronald J.
Mountain Wild Flowers

fl. hab.   p. 101

Epilobium angustifolium

Tosco, Uberto
The World of Mountain Flowers

fl.   p 52

Epilobium angustifolium

Walcott, Mary Vaux
North American Wild Flowers
    vol. 4

fl.   Pl. 301

Epilobium angustifolium

Welsh, Stanley L.
Flowers of the Mountain Country

fl.   p. 32

Epilobium angustifolium

Wit, H. C. D. de
Plants of the World;
    The Higher Plants, Vol II

fl.   p 50, Pl. 5

Epilobium chloraefolium var.

Curtis's Botanical Magazine
    vol 175   1964-65

fl.   Pl. 456

Epilobium coloratum

Courtenay, Booth
Wildflowers & Weeds

fl.   p 75

Epilobium davuricum

Morley, Brian D.
Wild Flowers of the World

fl. fr.   Pl. 4C

Epilobium dodonaei

Barneby, T. P.
European Alpine Flowers in Colour

fl.   Pl. 52, 2

Epilobium dodonaei

Felsko, Elsa
A Book of Wild Flowers
            2nd ser.
fl. fr.   p 107

Epilobium fleischeri

Barneby, T. P.
European Alpine Flowers in Colour

fl.   Pl. 52, 3

Epilobium fleischeri

Kohlhaupt, Paula
Fleurs des Alpages
            vol 2
Fl.   P 72

Epilobium fleischeri

Tosco, Uberto
The World of Mountain Flowers

fl. hab.   p 70

Epilobium fugitivum

Curtis, Winifred
The Endemic Flora of Tasmania, Vol. 5

fl. fr.   Pl. 173

Epilobium hirsutum

Ary, S.
The Oxford Book of Wildflowers

fl.   p 110, Pl. 5

Epilobium hirsutum

Lindman, C. A. M.
Nordens Flora, Vol 6

fl. fr.   Pl. 405

Epilobium hirsutum

Martin, W. Keble
The Concise British Flora in Colour

fl. hab.   Pl.35

Epilobium hirsutum

Polunin, Oleg
Flowers of Europe

fl.   Pl. 81 #837

Epilobium Hornemannii

Porsild, A. E.
Rocky Mountain Wild Flowers

fl. fr. hab.   p 285

Epilobium lanceolatum

Martin, W. Keble
The Concise British Flora in Colour

fl. fr. hab.   Pl. 35

Epilobium latifolium

Alaska-Yukon Wild Flower Guide

fl.   p 101

Epilobium latifolium

Clark, Lewis J.
Wild Flowers of British Columbia

fl.   p 350

Epilobium latifolium

Ferguson, Mary
Wildflowers

fl.   p 61

Epilobium latifolium

Fries, Mary A.
Wildflowers of Mount Ranier and
the Cascades

fl.   P 187

Epilobium latifolium

Heller, Christine
Wild Flowers of Alaska

fl.   Pl. 98-99

Epilobium latifolium

Massachusetts Hort. Soc.
Horticulture
    vol 43, No. 8   1965
fl.   p. 22

Epilobium latifolium

Massachusetts Hort. Soc.
Horticulture
    vol 46, No. 6   1968
hab., fl.   p. 18

Epilobium latifolium

Porsild, A. E.
Rocky Mountain Wild Flowers

fl.   p 287

Epilobium latifolium

Taylor, Ronald J.
Mountain Wild Flowers

fl. hab.   p 150

Epilobium latifolium

Walcott, Mary Vaux
North American Wild Flowers
    vol. 5
fl.   Pl. 370

Epilobium luteum

Clark, Lewis J.
Wild Flowers of British Columbia

fl.   p 339

Epilobium luteum

Heller, Christine
Wild Flowers of Alaska

fl., hab.   Pl. 31

Epilobium luteum

Taylor, Ronald J.
Mountain Wild Flowers

fl. hab.   p 151

Epilobium luteum

Walcott, Mary Vaux
North American Wild Flowers
    vol. 4
fl.   Pl. 300

Epilobium montanum

Ary, S.
The Oxford Book of Wildflowers

fl.   p 110, Pl. 1

Epilobium montanum

Felsko, Elsa
A Book of Wildflowers

fl.   P 94

Epilobium montanum

Lindman, C. A. M.
Nordens Flora, Vol. 6

fl.   Pl. 406A

Epilobium montanum

Martin, W. Keble
The Concise British Flora in Colour

fl. hab.   Pl. 35

Epilobium montanum

Polunin, Oleg
Flowers of Europe

fl.   Pl. 81 #839

Epilobium obcordatum

Munz, Philip A.
California Mountain Wildflowers

fl.   p 32, Pl. 18

Epilobium obcordatum

Orr, Robert T.
Wildflowers of Western America

fl.   Pl. 138

Epilobium obcordatum

Royal Hort. Soc.
Journal of the Royal Hort. Soc.
     vol 89, No. 4   1964

fl.   p. 158   Pl. 65

Epilobium obscurum

Martin, W. Keble
The Concise British Flora in Colour

fl. hab.   Pl. 35

Epilobium palustre

Ary, S.
The Oxford Book of Wildflowers

fl.   p 110, Pl. 4

Epilobium palustre

Martin, W. Keble
The Concise British Flora in Colour

fl. hab.   Pl. 35

Epilobium parviflorum

Martin, W. Keble
The Concise British Flora in Colour

fl. hab.   Pl. 35

Epilobium pedunculare

Ary, S.
The Oxford Book of Wildflowers

fl.   p 110, Pl. 2

Epilobium, Rose

     see

Epilobium obcordatum

Epilobium roseum

Martin, W. Keble
The Concise British Flora in Colour

fl. hab.   Pl. 35

Epilobium tetragonum var.

Martin, W. Keble
The Concise British Flora in Colour

fl. hab.   Pl. 35

Epilobium treleaseanum

Heller, Christine
Wild Flowers of Alaska

fl.   Pl. 95

Epimedium alpinum

Felsko, Elsa
A Book of Wild Flowers
               2nd ser.
fl.   p 69

Epimedium alpinum

Martin, W. Keble
The Concise British Flora in Colour

fl. hab.   Pl. 4

Epimedium grandiflorum

Crockett, James Underwood
Lawns & Ground Covers

fl.   P. 131

Epimedium x rubrum

Huxley, Anthony
Garden Perennials and Water Plants

fl.   Pl. 105

Epimedium rugosum

Kimura, Koiti
Japanese Medicinal Plants, Vol I

fl.   p 36

Epimedium sagittatum

Kimura, Koiti
Japanese Medicinal Plants, Vol I

fl.   p 34

Epimedium sempervirens

Kariyone, Tatsuo
Atlas of Medicinal Plants

fl., hab.   Pl. 122

Epimedium sempervirens

Kimura, Koiti
Japanese Medicinal Plants, Vol I

fl.   p 36

Epimedium sp.

American Hort. Soc.
American Horticulturist
     vol 52, No. 2   1973

fl.   p. 38

Epimedium sulphureum

Huxley, Anthony
Garden Perennials and Water Plants

fl.   pl. 106

Epimedium x versicolor var.

Addisonia
     vol. 23   1954-59

fl.   p. 35   Pl. 754

Epimedium violaceum

Kimura, Koiti
Japanese Medicinal Plants, Vol I

fl.   p 34

Epipactis atropurpurea

Barneby, T. P.
European Alpine Flowers in Colour

fl.   Pl. 12, 5

Epipactis atrorubens

Ary, S.
The Oxford Book of Wildflowers

fl.   p 160, Pl. 6

Epipactis atrorubens

Felsko, Elsa
A Book of Wild Flowers
               2nd ser.
fl.   p 97

Epipactis atrorubens

Martin, W. Keble
The Concise British Flora in Colour

fl. hab.   Pl. 80

Epipactis cambrensis

Brooke, Jocelyn
The Wild Orchids of Britain

fl.   Pl. 7

Epipactis consimilis

Kimber, Sheila
A Handbook of Orchids

fl., hab.   p. 44

Epipactis dunensis

Brooke, Jocelyn
The Wild Orchids of Britain

fl.   Pl. 7

Epipactis gigantea

Lemmon, Robert S.
Wildflowers of North America in
     Full Color

fl.   pg. 121   pl. 192

Epipactis gigantea

Luer, Carlyle A.
The Native Orchids of the
     United States and Canada

fl. hab.   Pl. 16

Epipactis gigantea

Munz, Philip A.
California Spring Wildflowers

fl.   p. 89   Pl. 91

Epipactis gigantea

Orr, Robert T.
Wildflowers of Western America

fl.   Pl. 219

Epipactis gigantea

Ospina, Mariano
Orquideas de las americas

fl.   Pl. 16

Epipactis gigantea

Welsh, Stanley L.
Flowers of the Canyon Country

fl..   p 14

Epipactis helleborine

Ary, S.
The Oxford Book of Wildflowers

fl.   p 160, Pl. 1

Epipactis helleborine

Clark, Lewis J.
Wild Flowers of British Columbia

fl. hab.   p 82

Epipactis helleborine

Courtenay, Booth
Wildflowers & Weeds

fl.   p 14

Epipactis helleborine

Klimas, John E.
Wildflowers of Eastern America

fl.   Pl. 41

Epipactis helleborine

Lindman, C. A. M.
Nordens Flora, Vol 2

fl. hab.   Pl. 149

Epipactis helleborine

Luer, Carlyle A.
The Native Orchids of the
    United States and Canada

fl. hab.   Pl. 15

Epipactis helleborine

Martin, W. Keble
The Concise British Flora in Colour

fl. hab.   Pl. 80

Epipactis latifolia

Barneby, T. P.
European Alpine Flowers in Colour

fl.   Pl. 12, 6

Epipactis latifolia

Brooke, Jocelyn
The Wild Orchids of Britain

fl.   Pl. 5

Epipactis leptochila

Brooke, Jocelyn
The Wild Orchids of Britain

fl.   Pl. 6

Epipactis leptochila

Martin, W. Keble
The Concise British Flora in Colour

fl.   Pl. 80

Epipactis palustris

Ary, S.
The Oxford Book of Wildflowers

fl.   p 160, Pl. 4

Epipactis palustris

Barneby, T. P.
European Alpine Flowers in Colour

fl.   Pl. 12, 4

Epipactis palustris

Brooke, Jocelyn
The Wild Orchids of Britain

fl.   Pl. 10

Epipactis palustris

Felsko, Elsa
A Book of Wild Flowers
                 2nd ser.
fl.   p 98

Epipactis palustris

Kleijn, H.
Beauty of the Wild Plant

fl.   opp. p. 26   pl. 25

Epipactis palustris

Lindman, C. A. M.
Nordens Flora, Vol 3

fl. hab.   Pl. 148

Epipactis palustris

Martin, W. Keble
The Concise British Flora in Colour

fl. hab.   Pl. 81

Epipactis palustris

Polunin, Oleg
Flowers of Europe

fl.   Pl. 191 #1916

Epipactis pendula

Brooke, Jocelyn
The Wild Orchids of Britain

fl.   Pl. 10

Epipactis rubiginosa

Brooke, Jocelyn
The Wild Orchids of Britain

fl.   Pl. 9

Epipactis sessilifolia

Ary, S.
The Oxford Book of Wildflowers

fl.   p 160, Pl. 2

Epipactis vectensis

Brooke, Jocelyn
The Wild Orchids of Britain

fl.   Pl. 6

Epipactis violacea

Brooke, Jocelyn
The Wild Orchids of Britain

fl.   Pl. 8

Epiphyllum ackermannii

Hay, Roy
The Color Dictionary of Flowers & Plants

fl.   p 66, Pl. 523

Epiphyllum ackermannii

Hay, Roy
The Dictionary of House Plants

fl.   Pl. 207

Epiphyllum ackermannii

Massachusetts Hort. Soc.
Horticulture
    vol 41, No. 12   1963

fl.   p. 602

Epiphyllum ackermannii

Perry, Frances
Flowers of the World

fl.   p 59

Epiphyllum ackermannii

Van Laren, A. J.
Cactus

fl.   P 86   Fig. Q

Epiphyllum crenatum

Van Laren, A. J.
Cactus

fl.   P 85   Fig. 106

Epiphyllum hybrid

Cactus & Succulent Society of America
Journal
vol 50, No. 3   1978

fl.   p. 130-31

Epiphyllum hybrid

Hay, Roy
The Color Dictionary of Flowers & Plants

fl.   p 66, Pl. 524

Epiphyllum hybrid

Hersey, Jean
Woman's Day Book of House Plants

fl.   p 43

Epiphyllum hybrid

Lamb, Edgar
The Pocket Encyclopedia of Cacti
    and Succulents in Color

fl. hab.   Pl. 46, 47

Epiphyllum hybrid

Lamb, Edgar
Popular Exotic Cacti in Color

fl. hab.   p 75

Epiphyllum hybrid

Macoby, Stirling
What Flower is That

fl.   P 122

Epiphyllum hybrid

Nicolaisen, Age
Pocket Encyclopedia of Indoor Plants

fl.   pl. 68

Epiphyllum hybrid

O'Gorman, Helen
Mexican Flowering Trees and Plants

fl. hab.   p 205

Epiphyllum hybridum

Kiaer, Eigil
Indoor Plants in Colour

fl.   p. 54

Epiphyllum hybridum

Subik, Rudolf
Decorative Cacti

fl. hab.   p 39

Epipogium aphyllum

Barneby, T. P.
European Alpine Flowers in Colour

fl.   Pl. 13, 5

Epipogium aphyllum

Brooke, Jocelyn
The Wild Orchids of Britain

fl., hab.   Pl. 26

Epipogium aphyllum

Lindman, C. A. M.
Nordens Flora, Vol 3

fl. hab.   Pl. 150

Episcia acajou

Massachusetts Hort. Soc.
Horticulture
    vol 32, No. 2   1954

hab.   p. 77

Episcia coccinea

Massachusetts Hort. Soc.
Horticulture
    vol 32, No. 2   1954

hab.   p. 77

Episcia cupreata

Massachusetts Hort. Soc.
Horticulture
    vol 32, No. 2   1954

hab.   p. 77

Episcia cupreata

Massachusetts Hort. Soc.
Horticulture
    vol 55, No. 2   1977

fl.   p. 60

Episcia cupreata var.

American Gloxinia & Gesneriad Society
The Gloxinian
    vol 22, No. 3   1972

hab., fl.   cover

Episcia cupreata var.

Massachusetts Hort. Soc.
Horticulture
    vol 32, No. 9   1954

hab., fl.   p. 412

Episcia cupreata var.

Massachusetts Hort. Soc.
Horticulture
    vol 36, No. 12   1958

fl., hab.   p. 621

Episcia cupreata var.

Perry, Frances
Flowers of the World

fl.   p. 130

Episcia dianthiflora

Massachusetts Hort. Soc.
Horticulture
    vol 51, No. 6   1973

fl.   p. 22

Episcia fulgida

Bruggeman, L.
Tropical Plants

fl.   Pl. 108

Episcia fulgida

Kiaer, Eigil
Indoor Plants in Colour

fl.   p. 55

Episcia hybrid

American Gloxinia & Gesneriad Society
The Gloxinian
    vol 27, No. 3   1977

hab., fl.   cover, p. 20-21

Episcia hybrid

Elbert, Virginie F.
The Miracle Houseplants

fl.   p 116

Episcia lilacina

Curtis's Botanical Magazine
Vol 171   1956-57

fl.   265

Episcia lilacina

Elbert, Virginie F.
The Miracle Houseplants

fl.   p 116

Episcia lilacina var.

Massachusetts Hort. Soc.
Horticulture
    vol 32, No. 2   1954

hab.   p. 77

Episcia punctata

American Gloxinia & Gesneriad Society
The Gloxinian
    vol 22, No. 4   1972

fl.   cover

Episcia reptans

Encke, Fritz
Zimmerpflanzen

fl.   p. 74

Episcia reptans

Kromdijk, G.
200 House Plants in Colour

hab.   Pl. 80

Episcia sp.

Milne, Lorus
Living Plants of the World

fl.   p 210

Episcia splendens

Massachusetts Hort. Soc.
Horticulture
vol 32, No. 9   1954

hab., fl.   p. 412

Episcia tessellata

Massachusetts Hort. Soc.
Horticulture
vol 32, No. 2   1954

hab.   p. 77

Epistephium duckei

Pabst, G. F. J.
Orchidaceae Brasilienses,
Vol 1

fl.   p 178

Epistephium elatum

Ospina, Mariano
Orquideas de las americas

fl.   Pl. 33

Epithelantha bokei

Lamb, Edgar
Colorful Cacti of the American Deserts

fl. hab.   Pl. 39, 40

Epithelantha micromeris

Kupper, Walter
Cacti

hab.   p. 87   Pl. 40

Epithelantha micromeris

Lamb, Edgar
Colorful Cacti of the American Deserts

fl. hab.   Pl. 41

Epithelantha micromeris

Van Laren, A. J.
Cactus

fl.   P 42   Fig. 42

Epithelantha micromeris

Weniger, Del
Cacti of the Southwest

hab.   P 27

Epithelantha micromeris var.

Cactus & Succulent Society of America
Journal
vol 50, No. 4   1978

fl., hab.   p. 184

Epithelantha micromeris var.

Lamb, Edgar
The Pocket Encyclopedia of Cacti
and Succulents in Color

fl. hab.   Pl. 48

Eranthemum nervosum

Bruggeman, L.
Tropical Plants

fl.   Pl. 238-239

Eranthemum nervosum

Perry, Frances
Flowers of the World

fl.   p 13

Eranthis cilicia

Hay, Roy
The Color Dictionary of Flowers & Plants

fl.   p 91, Pl. 728

Eranthis hyemalis

Bianchini, Francesco
Health Plants of the World

fr.   p 53

Eranthis hyemalis

Everett, T.H., ed.
New Illustrated Encyclopedia of
Gardening vol. 4

fl.   p. 631

Eranthis hyemalis

Felsko, Elsa
A Book of Wild Flowers
2nd ser.
fl. hab.   p 1

Eranthis hyemalis

Hay, Roy
The Color Dictionary of Flowers & Plants

fl.   p 92, Pl. 729

Eranthis hyemalis

Huxley, Anthony
Garden Annuals and Bulbs

hab., fl.   Pl. 154, 214

Eranthis hyemalis

Martin, W. Keble
The Concise British Flora in Colour

fl. fr. hab.   Pl. 4

Eranthis hyemalis

Massachusetts Hort. Soc.
Horticulture
vol 42, No. 3   1964

fl.   p. 14

Eranthis hyemalis

Mathew, Brian
Dwarf Bulbs

fl.   p. 112   Pl. 39

Eranthis hyemalis

Miles, Bebe
Bulbs for the Home Gardener

fl. hab.   p 70

Eranthis hyemalis

Morley, Brian D.
Wild Flowers of the World

fl., hab.   Pl. 27

Eranthis hyemalis

Perry, Frances
Flowers of the World

fl.   p 250

Eranthis hyemalis

Polunin, Oleg
Flowers of Europe

fl.   Pl. 18 #202

Eranthis hyemalis

Royal Hort. Soc.
Journal of the Royal Hort. Soc.
vol 97, No. 2   1972

fl.   p. 88   Pl. 44

Eranthis hyemalis

Royal Hort. Soc.
The Garden
vol 101, No. 7   1976

fl.   p. 351

Eranthis hyemalis

Vilmorin, Roger de
Plantes Alpines dans les Jardins

fl. hab.   Pl. XXIX

Eranthis hyemalis

Wit, H. C. D. de
Plants of the World;
The Higher Plants, Vol I

fl.   p 59, Pl. 26

Eranthis x tubergenii

Curtis's Botanical Magazine
Vol. 169   1952-53

fl.   196

Eranthis x tubergenii

Hay, Roy
The Color Dictionary of Flowers &
Plants

fl.   p. 92   Pl. 730

Erblichia odorata

Royal Hort. Soc.
Journal of the Royal Hort. Soc.
vol 96, No. 11   1971

fl.   p. 494   Pl. 227

Ercilla volubilis

Royal Hort. Soc.
Journal of the Royal Hort. Soc.
    vol 100, No. 5 1975

    fl., hab.   p. 200   Pl. 83

Erechtites hieracifolia

Burgis, D.S.
Florida Weeds

    hab.   p. 4

Erechtites hieracifolia

Courtenay, Booth
Wildflowers & Weeds

    fl.   p. 126

Eremaea beaufortioides

Blombery, Alec M.
What Wildflower is That

    fl.   p 129, Pl. 337

Eremaea fimbriata

Blombery, Alec M.
What Wildflower is That

    fl.   p 129, Pl. 338

Eremaea fimbriata

Harrison, R.E.
Trees and Shrubs

    fl. P 65

Eremaea fimbriata

Morcombe, M. K.
Australia's Western Wildflowers

    fl.   p 8

Eremochloa ophiuroides

Crockett, James Underwood
Lawns & Ground Covers

    fr., hab.   p. 77, 116

Eremocrinum albomarginatum

Orr, Robert T.
Wildflowers of Western America

    fl.   Pl. 55

Eremophila alternifolia

Blombery, Alec M.
What Wildflower is That

    fl.   p 130, Pl. 339

Eremophila calorhabdos

Blombery, Alec M.
What Wildflower is That

    fl.   p 130, Pl. 340

Eremophila clarkei

Morcombe, M. K.
Australia's Western Wildflowers

    fl.   p 36

Eremophila fraseri

Blombery, Alec M.
What Wildflower is That

    fl.   p 130, Pl. 341

Eremophila latrobei

Morcombe, M. K.
Australia's Western Wildflowers

    fl.   p 36

Eremophila longifolia

Cochrane, G. R.
Flowers and Plants of Victoria

    fl.   Pl. 142

Eremophila maculata

Blombery, Alec M.
What Wildflower is That

    fl.   p 131, Pl. 342

Eremophila maculata

Morcombe, M. K.
Australia's Western Wildflowers

    fl.   p 37

Eremophila maculata

Morley, Brian D.
Wild Flowers of the World

    fl.   pl. 132

Eremophila oldfieldii

Morcombe, M. K.
Australia's Western Wildflowers

    fl.   p 36

Eremophila oppositifolia

Cochrane, G. R.
Flowers and Plants of Victoria

    fl.   Pl. 140

Eremophila scoparia

Cochrane, G. R.
Flowers and Plants of Victoria

    fl.   Pl. 141

Eremophila scoparia

Morcombe, M. K.
Australia's Western Wildflowers

    fl.   p 27

Eremophila subfloccosa

Blombery, Alec M.
What Wildflower is That

    fl.   p 132, Pl. 343

Eremostachys laciniata

Morley, Brian D.
Wild Flowers of the World

    fl.   Pl. 49

Eremurus bungei

Hay, Roy
The Color Dictionary of Flowers & Plants

    fl.   p 140, Pl. 1116

Eremurus himalaicus

Everett, T.H., ed.
New Illustrated Encyclopedia of
    Gardening   vol 4

    fl.   p. 646

Eremurus hybrid

Perry, Frances
Flowers of the World

    fl.   p. 171

Eremurus kaufmannii

Royal Hort. Soc.
Journal of the Royal Hort. Soc.
    vol 88, No. 9 1963

    fl., hab.   p. 388   Pl. 156

Eremurus kaufmannii

Royal Hort. Soc.
Journal of the Royal Hort. Soc.
    vol 93, No. 3 1968

    fl.   p. 122   Pl. 58

Eremurus olgae

Morley, Brian D.
Wild Flowers of the World

    fl.,hab.   Pl. 50

Eremurus olgae

Wendelbo, Per
Tulips and Irises of Iran and
    Their Relatives

    fl. hab.   p 15, Pl. 5

Eremurus persicus

Wendelbo, Per
Tulips and Irises of Iran and
    Their Relatives

    fl. hab.   p 17, Pl. 8

Eremurus robustus

Hay, Roy
The Color Dictionary of Flowers & Plants

    fl.   p 140, Pl. 1117

Eremurus robustus

Huxley, Anthony
Garden Perennials and Water Plants

fl.   pl. 107

Eremurus robustus

Macoby, Stirling
What Flower is That

fl.   P 123

Eremurus sp.

Huxley, Anthony
Garden Perennials and Water Plants

hab.   Pl. 9

Eremurus spectabilis

Wendelbo, Per
Tulips and Irises of Iran and
    Their Relatives

fl.   p 17, Pl. 7

Eremurus stenophyllus

Morley, Brian D.
Wild Flowers of the World

fl.,fr.,hab.   Pl. 50

Eremurus stenophyllus

Wendelbo, Per
Tulips and Irises of Iran and
    Their Relatives

fl. hab.   p 17, Pl. 6

Eremurus stenophyllus var.

Huxley, Anthony
Garden Perennials and Water Plants

fl.   pl. 108

Erepsia bracteata

Herre, H.
The Genera of the Mesembryanthemaceae

fl.   fr.   p 151

Erepsia pentagona

Flowering Plants of Africa
    vol 27   1948-49

fl.   Pl. 1065

Ergot

   see

Claviceps purpurea

Eria amica

Curtis's Botanical Magazine
    Vol 159   1936

fl. hab.   No. 9453

Eria corneri

Walden, Beryl M.
Wild Flowers of Hong Kong

fl. hab.   Pl. 60 (173)

Eria fitzalani

Cady, Leo
Australian Native Orchids in Colour

fl. hab.   Pl. 82

Eria globifera

Curtis's Botanical Magazine
    Vol 166   1949

fl.   83

Eria pannea

Curtis's Botanical Magazine
    Vol 177   1969-70

fl.   570

Eria rosea

Hu, Shiu-ying
The Genera of Orchidaceae in
    Hong Kong

fl. hab.   p 123

Eria rosea

Walden, Beryl M.
Wild Flowers of Hong Kong

fl. hab.   Pl. 8(26)

Erianthus contortus

Batson, Wade T.
Wild Flowers in South Carolina

hab.   p 24

Erianthus giganteus

Duncan, Wilbur H.
Wildflowers of the Southeastern
    United States

fl.   p 233

Erianthus ravennae

Polunin, Oleg
Flowers of Europe

fr. Pl. 182 #1807

Erica alfredii

Baker, H. A.
Ericas in Southern Africa

fl. fr. hab.   Pl. 144

Erica alopecurus

Baker, H. A.
Ericas in Southern Africa

fl. fr. hab.   Pl. 61

Erica amoena

Baker, H. A.
Ericas in Southern Africa

fl. hab.   Pl. 111

Erica ampullacea

Baker, H. A.
Ericas in Southern Africa

fl. fr. hab.   Pl. 47

Erica annectens

Baker, H. A.
Ericas in Southern Africa

fl. fr. hab.   Pl. 16

Erica arborea

Bramwell, David
Wild Flowers of the Canary Islands

fl.   Pl. 203

Erica arborea

Hellyer, A.G.L.
Shrubs in Colour

fl.   p. 47

Erica arborea

Hvass, Elsie
Plants That Feed and Serve Us

fl.   p 121, Pl. 266

Erica arborea

Polunin, Oleg
Flowers of Europe

fl.   Pl. 89 #927

Erica arborea

Polunin, Oleg
Flowers of the Mediterranean

fl.   No. 121

Erica arborea

Polunin, Oleg
Trees and Bushes of Europe

fl. hab.   p 156

Erica arborea var.

Gault, S. Millar
The Color Dictionary of Shrubs

fl.   Pl. 164, 165

Erica arborea var.

Hay, Roy
The Color Dictionary of Flowers & Plants

fl.   p 198, Pl. 1577

Erica aristata

Baker, H. A.
Ericas in Southern Africa

fl. fr. hab.   Pl. 44

Erica atrovinosa

Baker, H. A.
Ericas in Southern Africa

fl. fr. hab.   Pl. 51

Erica atrovinosa

Morley, Brian D.
Wild Flowers of the World

fl.   Pl. 77

Erica australis

Royal Hort. Soc.
Journal of the Royal Hort. Soc.
      vol 87, No. 8   1962

fl.   p. 358   Pl. 108

Erica australis var.

Hay, Roy
The Color Dictionary of Flowers & Plants

fl.   p 198, Pl. 1578

Erica autumnalis

Harrison, R.E.
Trees and Shrubs

fl.   P 66

Erica axilliflora

Flowering Plants of Africa
      vol 39   1968-69

fl.   Pl. 1548

Erica azalaefolia

Baker, H. A.
Ericas in Southern Africa

fl. fr. hab.   Pl. 135

Erica baccans

Harrison, R.E.
Trees and Shrubs

fl.   P 66

Erica bakeri

Baker, H. A.
Ericas in Southern Africa

fl. fr. hab.   Pl. 74

Erica banksia

Baker, H. A.
Ericas in Southern Africa

fl. fr. hab.   Pl. 5

Erica bauera

Baker, H. A.
Ericas in Southern Africa

fl. fr. hab.   Pl. 10

Erica bauera

Curtis's Botanical Magazine
Vol 170   1954-55

fl.   222

Erica bauera

Harrison, R.E.
Trees and Shrubs

fl.   P 66

Erica bergiana

Baker, H. A.
Ericas in Southern Africa

fl. fr. hab.   Pl. 88

Erica bicolor

Baker, H. A.
Ericas in Southern Africa

fl. fr. hab.   Pl. 85

Erica blandfordia

Baker, H. A.
Ericas in Southern Africa

fl. fr. hab.   Pl. 95

Erica blandfordia

Flowering Plants of Africa
      vol 39   1968-69

fl.   Pl. 1535

Erica blenna

Baker, H. A.
Ericas in Southern Africa

fl. fr. hab.   Pl. 99

Erica blenna

Harrison, R.E.
Trees and Shrubs

fl.   P 66

Erica bodkinii

Baker, H. A.
Ericas in Southern Africa

fl. fr. hab.   Pl. 152

Erica borboniaefolia

Baker, H. A.
Ericas in Southern Africa

fl. fr. hab.   Pl. 143

Erica boscowiana

Harrison, R.E.
Trees and Shrubs

fl.   P 67

Erica bowieana

Flowering Plants of Africa
      vol XXV   1945-46

fl.   Pl. 982

Erica brachialis

Baker, H. A.
Ericas in Southern Africa

fl. fr. hab.   Pl. 25

Erica brownleeae

Baker, H. A.
Ericas in Southern Africa

fl. fr. hab.   Pl. 141

Erica bruniades

Baker, H. A.
Ericas in Southern Africa

fl. hab.   Pl. 130

Erica bruniaefolia

Baker, H. A.
Ericas in Southern Africa

fl. fr. hab.   Pl. 129

Erica calcareophila

Baker, H. A.
Ericas in Southern Africa

fl. fr. hab.   Pl. 155

Erica calycina

Baker, H. A.
Ericas in Southern Africa

fl. hab.   Pl. 161

Erica cameronii

Baker, H. A.
Ericas in Southern Africa

fl. fr. hab.   Pl. 32

Erica campanularis

Baker, H. A.
Ericas in Southern Africa

fl. fr. hab.   Pl. 98

Erica campanularis

Royal Hort. Soc.
Journal of the Royal Hort. Soc.
      vol 94, No. 8   1969

fl.   p. 346   Pl. 188

Erica canaliculata

Baker, H. A.
Ericas in Southern Africa

fl. fr. hab.   Pl. 165

Erica canaliculata

Curtis's Botanical Magazine
vol 172   1958-59

fl.   Pl. 339

Erica canaliculata

Harrison, R.E.
Trees and Shrubs

fl.   P 67

Erica canaliculata

Hay, Roy
The Dictionary of House Plants

fl.   Pl. 217

Erica capitata

Baker, H. A.
Ericas in Southern Africa

fl. fr. hab.   Pl. 131

Erica carduifolia

Baker, H. A.
Ericas in Southern Africa

fl. fr. hab.   Pl. 103

Erica carnea

Barneby, T. P.
European Alpine Flowers in Colour

fl.   Pl. 56, 5

Erica carnea

Bartels, Andreas
Das Grosse Buch der Gartengeholze

fl. hab.   p 148

Erica carnea

Crockett, James Underwood
Lawns & Ground Covers

fl.   p. 131

Erica carnea

Felsko, Elsa
A Book of Wildflowers

fl.   P 80

Erica carnea

Perry, Frances
Flowers of the World

fl. p 106

Erica carnea

Tosco, Uberto
The World of Mountain Flowers

fl.   p 36

Erica carnea

Vilmorin, Roger de
Plantes Alpines dans les Jardins

fl.   Pl. XXXIII, XLI

Erica carnea var.

Bartels, Andreas
Das Grosse Buch der Gartengeholze

fl.   p 148

Erica carnea var.

Harrison, R.E.
Trees and Shrubs

fl., hab.   p. 72

Erica carnea var.

Hay, Roy
The Color Dictionary of Flowers & Plants

fl.   p 198, Pl. 1579, 1580

Erica carnea var.

Hellyer, A.G.L.
Shrubs in Colour

fl.   p. 47

Erica carnea var.

Huxley, Anthony
Evergreen Garden Trees and Shrubs

fl. hab.   Pl. 116, 117

Erica cerinthoides

Baker, H. A.
Ericas in Southern Africa

fl. hab.   Pl. 29

Erica cerinthoides

Flowering Plants of Africa
vol 28   1950-51

fl.   Pl. 1099

Erica cerinthoides

Harrison, R.E.
Trees and Shrubs

fl.   P 67

Erica cerinthoides

Royal Hort. Soc.
Journal of the Royal Hort. Soc.
vol 94, No. 8   1969

fl.   p. 346   Pl. 179

Erica chamissonis

Baker, H. A.
Ericas in Southern Africa

fl. fr. hab.   Pl. 163

Erica chionophila

Baker, H. A.
Ericas in Southern Africa

fl. fr. hab.   Pl. 92

Erica chloroloma

Baker, H. A.
Ericas in Southern Africa

fl. fr. hab.   Pl. 22

Erica chrysocodon

Baker, H. A.
Ericas in Southern Africa

fl. fr. hab.   Pl. 78

Erica ciliaris

Ary, S.
The Oxford Book of Wildflowers

fl.   p 118, Pl. 2

Erica ciliaris

Martin, W. Keble
The Concise British Flora in Colour

fl. hab.   Pl. 55

Erica ciliaris

Polunin, Oleg
Flowers of Europe

fl.   Pl. 90 #329

Erica cinerea

Ary, S.
The Oxford Book of Wildflowers

fl.   p 118, Pl. 1

Erica cinerea

Hay, Roy
The Color Dictionary of Flowers &
Plants

fl.   p. 198   Pl. 1581

Erica cinerea

Martin, W. Keble
The Concise British Flora in Colour

fl. hab.   Pl. 55

Erica cinerea

Meikle, R. D.
British Trees and Shrubs

fl.   Pl. 10

Erica cinerea

Perry, Frances
Flowers of the World

fl.   p 106

Erica cinerea var.

Gault, S. Millar
The Color Dictionary of Shrubs

fl.   Pl. 166, 167, 168

Erica cinerea var.

Harrison, R.E.
Trees and Shrubs

fl.   P 72

Erica cinerea var.

Hay, Roy
The Color Dictionary of Flowers & Plants

fl.   p 198, Pl.   1582, 1583

Erica cinera var.

Royal Hort. Soc.
Journal of the Royal Hort. Soc.
    vol 96, No. 9   1971

fl.   p. 404   Pl. 181

Erica clavisepala

Baker, H. A.
Ericas in Southern Africa

fl. fr. hab.   Pl. 123

Erica coarctata

Baker, H. A.
Ericas in Southern Africa

fl. fr. hab.   Pl. 118

Erica coccinea

Baker, H. A.
Ericas in Southern Africa

fl. fr. hab.   Pl. 1

Erica coccinea var.

Royal Hort. Soc.
Journal of the Royal Hort. Soc.
    vol 94, No. 8   1969

fl.   p. 346   Pl. 176

Erica comptonii

Baker, H. A.
Ericas in Southern Africa

fl. fr. hab.   Pl. 4

Erica conica

Harrison, R. E.
Trees and Shrubs

fl.   P. 68

Erica copiosa

Baker, H. A.
Ericas in Southern Africa

fl. fr. hab.   Pl. 120

Erica coricolia

Baker, H. A.
Ericas in Southern Africa

fl. fr. hab.   Pl. 145

Erica corydalis

Baker, H. A.
Ericas in Southern Africa

fl. fr. hab.   Pl. 167

Erica crenata

Baker, H. A.
Ericas in Southern Africa

fl. fr. hab.   Pl. 108

Erica cristata

Baker, H. A.
Ericas in Southern Africa

fl. fr. hab.   Pl. 48

Erica cruenta

Harrison, R.E.
Trees and Shrubs

fl.   P 68

Erica cubica

Baker, H. A.
Ericas in Southern Africa

fl. fr. hab.   Pl. 162

Erica cumuliflora

Baker, H. A.
Ericas in Southern Africa

fl. fr. hab.   Pl. 129

Erica curviflora

Baker, H. A.
Ericas in Southern Africa

fl. fr. hab.   Pl. 28

Erica curviflora var.

Curtis's Botanical Magazine
    vol 172   1958-59

fl.   Pl. 350

Erica curvirostris

Baker, H. A.
Ericas in Southern Africa

fl. fr. hab.   Pl. 91

Erica cylindrica

Baker, H. A.
Ericas in Southern Africa

fl. fr. hab.   Pl. 55

Erica cyrillaeflora

Baker, H. A.
Ericas in Southern Africa

fl. fr. hab.   Pl. 62

Erica daphniflora

Baker, H. A.
Ericas in Southern Africa

fl. fr. hab.   Pl. 54

Erica daphniflora

Royal Hort. Soc.
Journal of the Royal Hort. Soc.
    vol 94, No. 8   1969

fl.   p. 346   Pl. 177

Erica daphniflora var.

Royal Hort. Soc.
Journal of the Royal Hort. Soc.
    vol 94, No. 8   1969

fl., fr.   p. 346   Pl. 177

Erica x darleyensis

Hay, Roy
The Color Dictionary of Flowers &
    Plants

fl.   p. 198   Pl. 1584

Erica x darleyensis

Hellyer, A.G.L.
Shrubs in Colour

fl.   p. 47

Erica x darleyensis

Huxley, Anthony
Evergreen Garden Trees and Shrubs

fl.   Pl. 114

Erica x darleyensis var.

Gault, S. Millar
The Color Dictionary of Shrubs

fl.   Pl.160,170,171

Erica x darleyensis var.

Huxley, Anthony
Evergreen Garden Trees and Shrubs

fl.   Pl. 115

Erica decora

Baker, H. A.
Ericas in Southern Africa

fl. fr. hab.   Pl. 114

Erica deliciosa

Baker, H. A.
Ericas in Southern Africa

fl. fr. hab.   Pl. 81

Erica densifolia

Baker, H. A.
Ericas in Southern Africa

fl. fr. hab.   Pl. 19

Erica denticulata

Baker, H. A.
Ericas in Southern Africa

fl. fr. hab.   Pl. 53

Erica depressa

Baker, H. A.
Ericas in Southern Africa

fl. fr. hab.   Pl. 157

Erica desmantha

Baker, H. A.
Ericas in Southern Africa

fl. fr. hab.   Pl. 134

Erica diaphana

Flowering Plants of Africa
     vol 27   1948-49

fl.   Pl. 1042

Erica diaphana

Harrison, R.E.
Trees and Shrubs

fl.   P 68

Erica discolor

Flowering Plants of Africa
     vol XXV   1945-46

fl.   Pl. 986

Erica doliiformis

Baker, H. A.
Ericas in Southern Africa

fl. fr. hab.   Pl. 33

Erica drakensbergensis

Baker, H. A.
Ericas in Southern Africa

fl. fr. hab.   Pl. 86

Erica drakensbergensis

Flowering Plants of Africa
     vol 30   1954-55

fl.   Pl. 1161

Erica embothriifolia

Baker, H. A.
Ericas in Southern Africa

fl. fr. hab.   Pl. 37

Erica erigena

Gault, S. Millar
The Color Dictionary of Shrubs

fl. hab.   Pl. 172

Erica fairii

Baker, H. A.
Ericas in Southern Africa

fl. fr. hab.   Pl. 102

Erica fascicularis

Baker, H. A.
Ericas in Southern Africa

fl. fr. hab.   Pl. 38

Erica fastigiata

Baker, H. A.
Ericas in Southern Africa

fl. fr. hab.   Pl. 52

Erica filiformis

Baker, H. A.
Ericas in Southern Africa

fl. fr. hab.   Pl. 76

Erica filipendula

Baker, H. A.
Ericas in Southern Africa

fl. fr. hab.   Pl. 13A

Erica filipendula

Royal Hort. Soc.
Journal of the Royal Hort. Soc.
vol 94, No. 8   1969

fl.   p. 346   Pl. 181

Erica filipendula var.

Baker, H. A.
Ericas in Southern Africa

fl. fr. hab.   Pl. 13B

Erica filipendula var.

Royal Hort. Soc.
Journal of the Royal Hort. Soc.
vol 94, No. 8   1969

fl.   p. 346   Pl. 181

Erica foliacea

Baker, H. A.
Ericas in Southern Africa

fl. fr. hab.   Pl. 23

Erica fontana

Baker, H. A.
Ericas in Southern Africa

fl. fr. hab.   Pl. 64

Erica formosa

Baker, H. A.
Ericas in Southern Africa

fl. fr. hab.   Pl. 107

Erica formosa

Flowering Plants of Africa
     vol 39   1968-69

fl.   Pl. 1533

Erica genistaefolia

Baker, H. A.
Ericas in Southern Africa

fl. fr. hab.   Pl. 128

Erica georgica

Baker, H. A.
Ericas in Southern Africa

fl. fr. hab.   Pl. 57

Erica gibbosa

Baker, H. A.
Ericas in Southern Africa

fl. fr. hab.   Pl. 96

Erica glandulosa

Baker, H. A.
Ericas in Southern Africa

fl. fr. hab.   Pl. 20

Erica glaphyra

Baker, H. A.
Ericas in Southern Africa

fl. fr. hab.   Pl. 116

Erica glauca

Baker, H. A.
Ericas in Southern Africa

fl. fr. hab.   Pl. 150(a)

Erica glauca var.

Baker, H. A.
Ericas in Southern Africa

fl. fr. hab.   Pl. 150(b)

Erica glauca var.

Harrison, R. E.
Trees and Shrubs

fl.   p. 68

Erica glumaeflora

Baker, H. A.
Ericas in Southern Africa

fl. fr. hab.   Pl. 140

Erica glutinosa

Baker, H. A.
Ericas in Southern Africa

fl. fr. hab.   Pl. 59

Erica glutinosa

Flowering Plants of Africa
      vol 34   1960-61

fl.   Pl. 1357

Erica gossypioides

Baker, H. A.
Ericas in Southern Africa

fl. fr. hab.   Pl. 132

Erica gracilis

Kroadijk, G.
200 House Plants in Colour

hab.   Pl. 81

Erica gracilis

Nicolaisen, Age
Pocket Encyclopedia of Indoor Plants

fl.   Pl. 106

Erica grandiflora

Baker, H. A.
Ericas in Southern Africa

fl. fr. hab.   Pl. 14

Erica grandiflora var.

Perry, Frances
Flowers of the World

fl.   p 107

Erica grisbrookii

Baker, H. A.
Ericas in Southern Africa

fl. fr. hab.   Pl. 153

Erica haematocodon

Baker, H. A.
Ericas in Southern Africa

fl. fr. hab.   Pl. 66

Erica haematosiphon

Baker, H. A.
Ericas in Southern Africa

fl. fr. hab.   Pl. 26

Erica halicacaba

Baker, H. A.
Ericas in Southern Africa

fl. fr. hab.   Pl. 151

Erica heleogena

Baker, H. A.
Ericas in Southern Africa

fl. hab.   Pl. 63

Erica herbacea

Crockett, James Underwood
Evergreens

hab. fl.   p. 122

Erica herbacea

Polunin, Oleg
Flowers of Europe

fl.   Pl. 89 #933

Erica herbacea var.

Gault, S. Millar
The Color Dictionary of Shrubs

fl.   Pl. 173, 174, 175

Erica hibbertia

Baker, H. A.
Ericas in Southern Africa

fl. fr. hab.   Pl. 18

Erica hirtiflora

Harrison, R. E.
Trees and Shrubs

fl.   P. 69

Erica hispidula

Baker, H. A.
Ericas in Southern Africa

fl. fr. hab.   Pl. 119

Erica holosericea

Baker, H. A.
Ericas in Southern Africa

fl. hab.   Pl. 149

Erica holosericea

Flowering Plants of Africa
      vol 34   1960-61

fl.   Pl. 1358

Erica holosericea

Harrison, R.E.
Trees and Shrubs

fl.   P 71

Erica holosericea

Macoby, Stirling
What Flower is That

fl.   P 123

Erica holtii

Baker, H. A.
Ericas in Southern Africa

fl. fr. hab.   Pl. 122

Erica holtii

Flowering Plants of Africa
      vol 31   1956

fl.   Pl. 1202

Erica hybrid

Harrison, R. E.
Trees and Shrubs

fl.   p. 67

Erica hybrid

Macoby, Stirling
What Flower is That

fl.   P 122

Erica hybrida

Harrison, R. E.
Trees and Shrubs

fl.   p. 69

Erica imbricata

Baker, H. A.
Ericas in Southern Africa

fl. hab.   Pl. 137

Erica inflata

Baker, H. A.
Ericas in Southern Africa

fl. fr. hab.   Pl. 49

Erica infundibuliformis

Baker, H. A.
Ericas in Southern Africa

fl. fr. hab.   Pl. 56

Erica involucrata

Baker, H. A.
Ericas in Southern Africa

fl. hab.   Pl. 127

Erica irbyana

Baker, H. A.
Ericas in Southern Africa

fl. hab.   Pl. 43

Erica irregularis

Baker, H. A.
Ericas in Southern Africa

fl. fr. hab.   Pl. 158

Erica irrorata

Baker, H. A.
Ericas in Southern Africa

fl. fr. hab.   Pl. 60

Erica jasminiflora

Baker, H. A.
Ericas in Southern Africa

fl. hab.   Pl. 40

Erica johnstoniana

Moriarty, Audrey
Wild Flowers of Malawi

fl.   Pl. 69; 2

Erica junonia

Baker, H. A.
Ericas in Southern Africa

fl. fr. hab.   Pl. 41

Erica junonia

Flowering Plants of Africa
        vol 39   1968-69

fl.   Pl. 1542

Erica junonia

Morley, Brian D.
Wild Flowers of the World

fl.   Pl. 77

Erica junonia

Royal Hort. Soc.
Journal of the Royal Hort. Soc.
        vol 94, No. 8   1969

fl.   p. 346   Pl. 180

Erica lananthera

Baker, H. A.
Ericas in Southern Africa

fl. fr. hab.   Pl. 46

Erica lanuginosa

Baker, H. A.
Ericas in Southern Africa

fl. fr. hab.   Pl. 154

Erica lasciva

Baker, H. A.
Ericas in Southern Africa

fl. fr. hab.   Pl. 139

Erica lateralis

Baker, H. A.
Ericas in Southern Africa

fl. fr. hab.   Pl. 94

Erica leptopus

Baker, H. A.
Ericas in Southern Africa

fl. fr. hab.   Pl. 117

Erica leucanthera

Baker, H. A.
Ericas in Southern Africa

fl. fr. hab.   Pl. 159

Erica limosa

Baker, H. A.
Ericas in Southern Africa

fl. hab.   Pl. 65

Erica lineata

Baker, H. A.
Ericas in Southern Africa

fl. fr. hab.   Pl. 3

Erica longifolia

Baker, H. A.
Ericas in Southern Africa

fl. fr. hab.   Pl. 15

Erica longipedunculata

Baker, H. A.
Ericas in Southern Africa

fl. fr. hab.   Pl. 80

Erica lusitanica

Gault, S. Millar
The Color Dictionary of Shrubs

fl.   Pl. 176

Erica lusitanica

Hay, Roy
The Color Dictionary of Flowers & Plants

fl.   p 199, Pl. 1585

Erica lusitanica

Macoby, Stirling
What Flower is That

fl.   P 123

Erica lusitanica

Perry, Frances
Flowers of the World

fl.   p 107

Erica lusitanica

Polunin, Oleg
Flowers of Europe

fl.   Pl. 89 #927

Erica lutea

Baker, H. A.
Ericas in Southern Africa

fl. fr. hab.   Pl. 148

Erica mackaiana

Martin, W. Keble
The Concise British Flora in Colour

fl. hab.   Pl. 55

Erica mackaiana var.

Royal Hort. Soc.
Journal of the Royal Hort. Soc.
        vol 96, No. 9   1971

fl.   p. 404   Pl. 184, 186

Erica mammosa

Baker, H. A.
Ericas in Southern Africa

fl. fr. hab.   Pl. 9

Erica mammosa

Curtis's Botanical Magazine
Vol 167   1950

fl.   100

Erica mammosa

Harrison, R.E.
Trees and Shrubs

fl.   P 69

Erica mammosa var.

Royal Hort. Soc.
Journal of the Royal Hort. Soc.
        vol 94, No. 8   1969

fl.   p. 346   Pl. 182

Erica margaritacea

Baker, H. A.
Ericas in Southern Africa

fl. fr. hab.   Pl. 89

Erica margaritacea

Flowering Plants of Africa
        vol 34   1960-61

fl.   Pl. 1345a

Erica marifolia

Baker, H. A.
Ericas in Southern Africa

fl. fr. hab.   Pl. 70

Erica massonii

Baker, H. A.
Ericas in Southern Africa

fl. fr. hab.   Pl. 39

Erica massonii

Royal Hort. Soc.
Journal of the Royal Hort. Soc.
  vol 94, No. 8   1969

fl., fr.   p. 346   Pl. 178

Erica mediterranea

Hellyer, A.G.L.
Shrubs in Colour

fl.   p. 47

Erica mediterranea

Martin, W. Keble
The Concise British Flora in Colour

fl. hab.   Pl. 55

Erica mediterranea var.

Harrison, R.E.
Trees and Shrubs

fl. hab.   P 72

Erica melanacme

Baker, H. A.
Ericas in Southern Africa

fl. fr. hab.   Pl. 147

Erica melanthera

Baker, H. A.
Ericas in Southern Africa

fl. fr. hab.   Pl. 164

Erica melanthera

Tsukamoto, Yotaro
Coloured Illustrations of Garden
  Flowers   Vol. 10

fl. Illus. 149   opp. p. 46

Erica monadelpha

Baker, H. A.
Ericas in Southern Africa

fl. fr. hab.   Pl. 6

Erica multiflora

Polunin, Oleg
Flowers of Europe

fl.   Pl. 89 #934

Erica multumbellifera

Baker, H. A.
Ericas in Southern Africa

fl. fr. hab.   Pl. 106

Erica nabea

Baker, H. A.
Ericas in Southern Africa

fl. fr. hab.   Pl. 156

Erica nana

Baker, H. A.
Ericas in Southern Africa

fl. fr. hab.   Pl. 24

Erica nudiflora

Baker, H. A.
Ericas in Southern Africa

fl. fr. hab.   Pl. 82

Erica oatesii

Baker, H. A.
Ericas in Southern Africa

fl. fr. hab.   Pl. 30

Erica oatesii

Flowering Plants of Africa
  vol 42   1972-73

fl.   Pl. 1644

Erica oatesii

Harrison, R.E.
Trees and Shrubs

fl.   P 69

Erica obliqua

Baker, H. A.
Ericas in Southern Africa

fl. fr. hab.   Pl. 104

Erica obtusata

Baker, H. A.
Ericas in Southern Africa

fl. fr. hab.   Pl. 78

Erica odorata

Baker, H. A.
Ericas in Southern Africa

fl. fr. hab.   Pl. 105

Erica oliveri

Baker, H. A.
Ericas in Southern Africa

fl. fr. hab.   Pl. 79

Erica oreophila

Baker, H. A.
Ericas in Southern Africa

fl. fr. hab.   Pl. 72

Erica ovina

Baker, H. A.
Ericas in Southern Africa

fl. fr. hab.   Pl. 69

Erica oxycoccifolia

Baker, H. A.
Ericas in Southern Africa

fl. fr. hab.   Pl. 75

Erica parilia

Baker, H. A.
Ericas in Southern Africa

fl. fr. hab.   Pl. 113

Erica parviflora

Flowering Plants of Africa
  vol XXV   1945-46

fl.   Pl. 990

Erica patersonia

Baker, H. A.
Ericas in Southern Africa

fl. fr. hab.   Pl. 21

Erica patersonia

Flowering Plants of Africa
  vol 34   1960-61

fl.   Pl. 1346

Erica pectinifolia

Baker, H. A.
Ericas in Southern Africa

fl. fr. hab.   Pl. 36

Erica peltata

Baker, H. A.
Ericas in Southern Africa

fl. fr. hab.   Pl. 160

Erica perlata

Baker, H. A.
Ericas in Southern Africa

fl. fr. hab.   Pl. 67

Erica peziza

Harrison, R.E.
Trees and Shrubs

fl.   P 70

Erica physodes

Baker, H. A.
Ericas in Southern Africa

fl. hab.   Pl. 101

Erica physophylla
Baker, H. A.
Ericas in Southern Africa

fl. fr. hab.   Pl. 73

Erica pillansii
Baker, H. A.
Ericas in Southern Africa

fl. fr. hab.   Pl. 27

Erica pillansii
Curtis's Botanical Magazine
Vol 164   1943-48

fl.   No. 9676

Erica pillansii
Perry, Frances
Flowers of the World

fl.   p 107

Erica placentaeflora
Baker, H. A.
Ericas in Southern Africa

fl. fr. hab.   Pl. 136

Erica plukeneti
Baker, H. A.
Ericas in Southern Africa

fl. fr. hab.   Pl. 2

Erica porteri
Baker, H. A.
Ericas in Southern Africa

fl. hab.   Pl. 17

Erica porteri
Flowering Plants of Africa
vol 31   1956

fl.   Pl. 1212

Erica x praegeri var.
Royal Hort Soc.
Journal of the Royal Hort. Soc.
vol 96, No. 9   1971

fl.   p. 404   Pl. 185

Erica pubescens
Baker, H. A.
Ericas in Southern Africa

fl. fr. hab.   Pl. 68

Erica pudens
Baker, H. A.
Ericas in Southern Africa

fl. fr. hab.   Pl. 124

Erica pulchella
Baker, H. A.
Ericas in Southern Africa

fl. hab.   Pl. 112

Erica pulchella
Flowering Plants of Africa
vol 34   1960-61

fl.   Pl. 1345b

Erica pyxidiflora
Baker, H. A.
Ericas in Southern Africa

fl. fr. hab.   Pl. 110

Erica quadrangularis
Baker, H. A.
Ericas in Southern Africa

fl. fr. hab.   Pl. 93

Erica quadrangularis
Harrison, R.E.
Trees and Shrubs

fl.   P 70

Erica regia var.
Harrison, R.E.
Trees and Shrubs

fl.   P 70

Erica retorta
Baker, H. A.
Ericas in Southern Africa

fl. fr. hab.   Pl. 45

Erica rhopalantha
Baker, H. A.
Ericas in Southern Africa

fl. fr. hab.   Pl. 146

Erica rubiginosa
Baker, H. A.
Ericas in Southern Africa

fl. fr. hab.   Pl. 83

Erica scabriuscula
Curtis's Botanical Magazine
Vol 169   1952-53

fl.   206

Erica scoparia
Polunin, Oleg
Trees and Bushes of Europe

fl. hab.   p 156

Erica scytophylla
Baker, H. A.
Ericas in Southern Africa

fl. fr. hab.   Pl. 84

Erica sessiliflora
Baker, H. A.
Ericas in Southern Africa

fl. hab.   Pl. 8

Erica sessiliflora
Harrison, R.E.
Trees and Shrubs

fl.   P 70

Erica shannonea
Baker, H. A.
Ericas in Southern Africa

fl. fr. hab.   Pl. 42

Erica sitiens
Baker, H. A.
Ericas in Southern Africa

fl. fr. hab.   Pl. 87

Erica sitiens
Flowering Plants of Africa
vol 27   1948-49

fl.   Pl. 1061

Erica sonderiana
Baker, H. A.
Ericas in Southern Africa

fl. fr. hab.   Pl. 138

Erica sparrmannii
Baker, H. A.
Ericas in Southern Africa

fl. fr. hab.   Pl. 34

Erica spumosa
Baker, H. A.
Ericas in Southern Africa

fl. fr. hab.   Pl. 133

Erica strigilifolia
Baker, H. A.
Ericas in Southern Africa

fl. fr. hab.   Pl. 35

Erica taxifolia
Baker, H. A.
Ericas in Southern Africa

fl. fr. hab.   Pl. 142

Erica tenella
Baker, H. A.
Ericas in Southern Africa
fl. fr. hab.   Pl. 90

Erica tenella
Flowering Plants of Africa
vol 39   1968-69
fl.   Pl. 1552

Erica tenuis
Baker, H. A.
Ericas in Southern Africa
fl. fr. hab.   Pl. 121

Erica terminalis
Polunin, Oleg
Trees and Bushes of Europe
fl.   p 157

Erica tetragona
Baker, H. A.
Ericas in Southern Africa
fl. fr. hab.   Pl. 58

Erica tetralix
Ary, S.
The Oxford Book of Wildflowers
fl.   p 118, Pl. 3

Erica tetralix
Felsko, Elsa
A Book of Wildflowers
fl.   P 116

Erica tetralix
Kleijn, H.
Beauty of the Wild Plant
fl. opp. p. 57   Pl. 72

Erica tetralix
Lindman, C. A. M.
Nordens Flora, Vol 7
fl. fr. hab.   Pl. 455

Erica tetralix
Martin, W. Keble
The Concise British Flora in Colour
fl. hab.   Pl. 55

Erica tetralix
Meikle, R. D.
British Trees and Shrubs
fl.   Pl. 10

Erica tetralix
Milne, Lorus
Living Plants of the World
fl.   p 186

Erica tetralix var.
Gault, S. Millar
The Color Dictionary of Shrubs
fl.   Pl. 177

Erica tetralix var.
Hay, Roy
The Color Dictionary of Flowers & Plants
fl. hab.   p 199, Pl. 1586

Erica tetralix var.
Huxley, Anthony
Evergreen Garden Trees and Shrubs
fl. hab.   Pl. 118

Erica tetralix var.
Massachusetts Hort. Soc.
Horticulture
Vol. 50, No. 8   1972
fl.   P. 30

Erica tetralix var.
Royal Hort. Soc.
Journal of the Royal Hort. Soc.
vol 96, No. 9   1971
fl.   p. 404   Pl. 182

Erica thimifolia
Baker, H. A.
Ericas in Southern Africa
fl. fr. hab.   Pl. 77

Erica thodei
Baker, H. A.
Ericas in Southern Africa
fl. fr. hab.   Pl. 109

Erica thunbergii
Baker, H. A.
Ericas in Southern Africa
fl. fr. hab.   Pl. 166

Erica tumida
Baker, H. A.
Ericas in Southern Africa
fl.,fr.,hab. frontispiece   Pl. 31

Erica tumida
Royal Hort. Soc.
Journal of the Royal Hort. Soc.
vol 94, No. 8   1969
fl., fr. p. 346   Pl. 175, 187

Erica umbellata
Hay, Roy
The Color Dictionary of Flowers & Plants
fl.   p 199, Pl. 1587

Erica urna-viridis
Baker, H. A.
Ericas in Southern Africa
fl. fr. hab.   Pl. 100

Erica urna-viridis
Morley, Brian D.
Wild Flowers of the World
fl.   Pl. 77

Erica ustulescens
Baker, H. A.
Ericas in Southern Africa
fl. fr. hab.   Pl. 126

Erica vagans
Martin, W. Keble
The Concise British Flora in Colour
fl. hab.   Pl. 55

Erica vagans var.
Gault, S. Millar
The Color Dictionary of Shrubs
fl.   Pl. 178, 179

Erica vagans var.
Harrison, R.E.
Trees and Shrubs
fl.   P 72

Erica vagans var.
Hay, Roy
The Color Dictionary of Flowers & Plants
fl.   p 199, Pl. 1588, 1589

Erica vagans var.
Hellyer, A.G.L.
Shrubs in Colour
fl.   p. 47

Erica vagans var.
Huxley, Anthony
Evergreen Garden Trees and Shrubs
fl. hab.   Pl. 119

Erica vagans var.
Massachusetts Hort. Soc.
Horticulture
vol 56, No. 4   1978
fl.   p. 54

Erica vallis-araneanum

Flowering Plants of Africa
vol 42   1972-73

fl.   Pl. 1680

Erica vanhuerckii

Baker, H. A.
Ericas in Southern Africa

fl. fr. hab.   Pl. 97

Erica x veitchii

Gault, S. Millar
The Color Dictionary of Shrubs

fl.,hab.   Pl. 180

Erica x veitchii

Hay, Roy
The Color Dictionary of Flowers &
    Plants

fl.   p. 199   Pl. 1590

Erica ventricosa

Baker, H. A.
Ericas in Southern Africa

fl. fr. hab.   Pl. 50

Erica ventricosa

Flowering Plants of Africa
vol 39   1968-69

fl.   Pl. 1555

Erica ventricosa

Harrison, R. E.
Trees and Shrubs

fl.   P. 71

Erica vestita

Flowering Plants of Africa
vol XXVI   1947

fl.   Pl. 1040

Erica viridiflora

Baker, H. A.
Ericas in Southern Africa

fl. fr. hab.   Pl. 7

Erica walkeria

Harrison, R.E.
Trees and Shrubs

fl. P 71

Erica x wilmorei

Harrison, R. E.
Trees and Shrubs

fl.   p. 71

Erica woodii

Baker, H. A.
Ericas in Southern Africa

fl. hab.   Pl. 115

Erica woodii

Flowering Plants of Africa
vol 27   1948-49

fl.   Pl. 1071

Erigenia bulbosa

Courtenay, Booth
Wildflowers & Weeds

fl.   p 58

Erigenia bulbosa

Dean, Blanche E.
Wildflowers of Alabama and
    Adjoining States

fl.   p 123

Erigenia bulbosa

Jennings, O. E.
Wild Flowers of Western Pennsylvania
    Vol II
fl. hab.   Pl. 104

Erigenia bulbosa

Wharton, Mary E.
A Guide to the Wildflowers & Ferns
    of Kentucky

fl.   p 163   Pl. 2.32

Erigeron acris

Ary, S.
The Oxford Book of Wildflowers

fl.   p 154, Pl. 5

Erigeron acris

Heller, Christine
Wild Flowers of Alaska

fl.   Pl. 120

Erigeron acris

Lindman, C. A. M.
Nordens Flora, Vol. 9

fl. fr. hab.   Pl. 580

Erigeron acris

Martin, W. Keble
The Concise British Flora in Colour

fl. hab.   Pl. 44

Erigeron acris

Polunin, Oleg
Flowers of Europe

fl.   Pl. 143 #1369

Erigeron alpinus

Barneby, T. P.
European Alpine Flowers in Colour

fl. hab.   Pl. 85, 5

Erigeron annuus

Klimas, John E.
Wildflowers of Eastern America

fl.   Pl. 72

Erigeron annuus

Weeds of the Southern United States

fl.   p. 11

Erigeron aurantiacus

Curtis's Botanical Magazine
vol 172   1958-59

fl.   Pl. 351

Erigeron aureus

Hay, Roy
The Color Dictionary of Flowers & Plants

fl.   p 8, Pl. 63

Erigeron aureus

Porsild, A. E.
Rocky Mountain Wild Flowers

fl. hab.   p 411

Erigeron aureus

Taylor, Ronald J.
Mountain Wild Flowers

fl.   p 104

Erigeron aureus

Walcott, Mary Vaux
North American Wild Flowers
    vol. 4

fl., hab.   Pl. 280

Erigeron borealis

Martin, W. Keble
The Concise British Flora in Colour

fl. hab.   Pl. 44

Erigeron borealis

Raven, John
Mountain Flowers

fl.   p. 179   Pl. 14

Erigeron caespitosus

Alaska-Yukon Wild Flower Guide

fl.   p 170

Erigeron caespitosus

Ferguson, Mary
Wildflowers

fl.   p 62

Erigeron caespitosus

Walcott, Mary Vaux
North American Wild Flowers

fl. hab.   Pl. 61

Erigeron canadensis

Weeds of the Southern United States

hab., fr.   p. 11

Erigeron compositus

Alaska-Yukon Wild Flower Guide

fl. hab.   p 163

Erigeron compositus

Heller, Christine
Wild Flowers of Alaska

fl.   Pl. 218

Erigeron compositus

Porsild, A. E.
Rocky Mountain Wild Flowers

fl. hab.   p 411

Erigeron compositus

Taylor, Ronald J.
Mountain Wild Flowers

fl. hab.   p 94

Erigeron compositus var.

Clark, Lewis J.
Wild Flowers of British Columbia

fl.   p 547, 550

Erigeron compositus var.

Curtis's Botanical Magazine
Vol 165   1948

fl.   2

Erigeron divergens

Orr, Robert T.
Wildflowers of Western America

fl.   Pl. 54

Erigeron dureus

Fries, Mary A.
Wildflowers of Mount Ranier and
the Cascades

fl.   P 200

Erigeron eriocephalum

Lindman, C. A. M.
Nordens Flora, Vol 9

fl. hab.   Pl. 582

Erigeron foliosus var.

Curtis's Botanical Magazine
Vol 162   1939

fl.   No. 9572

Erigeron formosissimus

Orr, Robert T.
Wildflowers of Western America

fl.   Pl. 246

Erigeron glabellus var.

Alaska-Yukon Wild Flower Guide

fl.   p 169

Erigeron glaucus

Lemmon, Robert S.
Wildflowers of North America in
Full Color

fl.   pg. 56   pl. 93

Erigeron glaucus

Munz, Philip A.
California Spring Wildflowers

fl.   p 79, Pl. 63

Erigeron glaucus

Orr, Robert T.
Wildflowers of Western America

fl.   Pl. 278

Erigeron grandiflorus

Heller, Christine
Wild Flowers of Alaska

fl.   Pl. 122

Erigeron grandiflorus

Porsild, A. E.
Rocky Mountain Wild Flowers

fl. hab.   p 413

Erigeron humilis

Alaska-Yukon Wild Flower Guide

fl.   p 166

Erigeron humilis

Heller, Christine
Wild Flowers of Alaska

fl.   Pl. 220

Erigeron humilis

Lindman, C. A. M.
Nordens Flora, Vol. 9

fl. fr. hab.   Pl. 583

Erigeron hybrid

Hay, Roy
The Color Dictionary of Flowers & Plants

fl.   p 140, 141; Pl. 1118 thru 1123

Erigeron hybrid

Huxley, Anthony
Garden Perennials and Water Plants

fl., hab.   Pl. 8, 109

Erigeron hyperboreus

Alaska-Yukon Wild Flower Guide

fl.   p 167

Erigeron hyperboreus

Heller, Christine
Wild Flowers of Alaska

fl.   Pl. 266

Erigeron karvinskianus

Macoby, Stirling
What Flower is That

fl.   P 124

Erigeron karvinskianus

Vilmorin, Roger de
Plantes Alpines dans les Jardins

fl. hab.   Pl. IX

Erigeron lanatus

Porsild, A. E.
Rocky Mountain Wild Flowers

fl. hab.   p 415

Erigeron leiomeris

Welsh, Stanley L.
Flowers of the Mountain Country

fl.   p. 33

Erigeron linearis

Clark, Lewis J.
Wild Flowers of British Columbia

fl.   p 546

Erigeron margaritacea

Moyle, John B.
Northland Wild Flowers

fl. hab.   p 184, Pl. 231

Erigeron mucronatus

Hay, Roy
The Color Dictionary of Flowers & Plants

fl.   p 8, Pl. 64

Erigeron mucronatus

Royal Hort. Soc.
The Garden
    vol 103, No. 5   1978

fl.   p. 178

Erigeron pappochroma

Cochrane, C. R.
Flowers and Plants of Victoria

fr. hab.   Pl. 499

Erigeron peregrinus

Alaska-Yukon Wild Flower Guide

fl. hab.   p 168

Erigeron peregrinus

Fries, Mary A.
Wildflowers of Mount Ranier and
the Cascades

fl.   P 144-145

Erigeron peregrinus

Heller, Christine
Wild Flowers of Alaska

fl.   Pl. 121

Erigeron peregrinus

Porsild, A. E.
Rocky Mountain Wild Flowers

fl.   p 415

Erigeron peregrinus

Taylor, Ronald J.
Mountain Wild Flowers

fl. hab.   p 95

Erigeron peregrinus var.

Clark, Lewis J.
Wild Flowers of British Columbia

fl.   p 546

Erigeron philadelphicus

Brown, Clair A.
Wildflowers of Louisiana and
    Adjoining States

fl. hab.   p 198

Erigeron philadelphicus

Campbell, Carlos C.
Great Smoky Mountain Wildflowers

fl.   p 89

Erigeron philadelphicus

Clark, Lewis J.
Wild Flowers of British Columbia

fl.   p 547

Erigeron philadelphicus

Courtenay, Booth
Wildflowers & Weeds

fl.   p 118

Erigeron philadelphicus

Dean, Blanche E.
Wildflowers of Alabama and
    Adjoining States

fl.   p 189

Erigeron philadelphicus

Duncan, Wilbur H.
Wildflowers of the Southeastern
    United States

fl.   p 207

Erigeron philadelphicus

Jennings, O. E.
Wild Flowers of Western Pennsylvania
                    Vol II
fl. hab.   Pl. 180

Erigeron philadelphicus

Moyle, John B.
Northland Wild Flowers

fl.   p 182, Pl. 229

Erigeron philadelphicus

Neufeld, J.B.
Wild Flowers of the Prairies

fl.   p. 33

Erigeron philadelphicus

Wharton, Mary E.
A Guide to the Wildflowers & Ferns
    of Kentucky

fl.   p 263   Pl. 3.41

Erigeron pulchellus

Courtenay, Booth
Wildflowers & Weeds

fl.   p 118

Erigeron pulchellus

Jennings, O. E.
Wild Flowers of Western Pennsylvania
                    Vol II
fl.   hab.   Pl. 179

Erigeron pulchellus

Klimas, John E.
Wildflowers of Eastern America

fl.   Pl. 190

Erigeron pulchellus

Wharton, Mary E.
A Guide to the Wildflowers & Ferns
    of Kentucky

fl.   p 263   Pl. 3.40

Erigeron pumilus

Clark, Lewis J.
Wild Flowers of British Columbia

fl.   p 550

Erigeron pumilus

Welsh, Stanley L.
Flowers of the Canyon Country

fl.   p 50

Erigeron pumilus var.

Munz, Philip A.
California Desert Wildflowers

fl.   p 52, Pl. 79

Erigeron purpuratus

Heller, Christine
Wild Flowers of Alaska

fl.   Pl. 219

Erigeron salsuginosus

Walcott, Mary Vaux
North American Wild Flowers
    vol. 4

fl.   Pl. 290

Erigeron speciosus

Macoby, Stirling
What Flower is That

fl.   p. 124

Erigeron speciosus

Walcott, Mary Vaux
North American Wild Flowers

fl. hab.   Pl. 165

Erigeron speciosus

Welsh, Stanley L.
Flowers of the Mountain Country

fl.   p. 34

Erigeron stellatus

Curtis, Winifred
The Endemic Flora of Tasmania, Vol. 5

fl. fr. hab.   Pl. 183

Erigeron strigosus

Courtenay, Booth
Wildflowers & Weeds

fl.   p 118

Erigeron unalaschensis

Walcott, Mary Vaux
North American Wild Flowers
vol. 5

fl.   Pl. 376

Erigeron uniflorus

Barneby, T. P.
European Alpine Flowers in Colour

fl. hab.   Pl. 85, 6

Erigeron uniflorus

Lindman, C. A. M.
Nordens Flora, Vol. 9

fl. fr. hab.   Pl. 581

Erigeron vernus

Duncan, Wilbur H.
Wildflowers of the Southeastern
     United States

fl. hab.   p 205

Erinacea anthyllis

Hellyer, A.G.L.
Shrubs in Colour

fl.   p. 50

Erinacea anthyllis

Polunin, Oleg
Flowers of Europe

fl. Pl. 52 #515

Erinacea anthyllis

Taylor, A. W.
Wild Flowers of Spain and Portugal

fl.   p. 27

Erinacea pungens

Alpine Garden Society
Bulletin
     vol. 36, No. 3   1968

fl., hab.   p. 245

Erinacea pungens

Hay, Roy
The Color Dictionary of Flowers & Plants

fl.   p 9, Pl. 65

Erinus alpinus

Barneby, T. P.
European Alpine Flowers in Colour

fl.   Pl. 73, 6

Erinus alpinus

Kohlhaupt, Paula
Fleurs des Alpages
               Vol. 2

fl.   p. 93

Erinus alpinus

Polunin, Oleg
Flowers of Europe

fl. Pl. 127 #1235

Erinus alpinus

Royal Hort. Soc.
The Garden
     vol 103, No. 5   1978

fl.   p. 178

Erinus alpinus

Vilmorin, Roger de
Plantes Alpines dans les Jardins

fl. hab.   Pl. XXXIV

Erinus alpinus

Wit, H. C. D. de
Plants of the World;
     The Higher Plants, Vol II

fl.   p 161, Pl. 78

Eriobotrya fragrans

Walden, Beryl M.
Wild Flowers of Hong Kong

fl.   Pl. 9(30), 10(30)

Eriobotrya japonica

Bianchini, F.
The Complete Book of Fruits & Vegetables

fr.   p 135

Eriobotrya japonica

Crockett, James Underwood
Evergreens

hab. fr.   p. 122

Eriobotrya japonica

Flemer, William III
Shade and Ornamental Trees in Color

hab.   p. 82

Eriobotrya japonica

Hay, Roy
The Color Dictionary of Flowers & Plants

fr.   p 199, Pl. 1591

Eriobotrya japonica

Huxley, Anthony
Evergreen Garden Trees and Shrubs

fr. hab.   Pl. 183

Eriobotrya japonica

Kitamura, Siro
Coloured Illustrations of Trees &
     Shrubs of Japan

fl.   230

Eriobotrya japonica

Macoby, Stirling
What Flower is That

fr.   P 124

Eriobotrya japonica

Masefield, G. B.
The Oxford Book of Food Plants

fl. fr.   Pl. 105, 3

Eriobotrya japonica

Polunin, Oleg
Flowers of Europe

fr. Pl. 48 #470

Eriobotrya japonica

Polunin, Oleg
Flowers of the Mediterranean

fr.   No. 47

Eriobotrya japonica

Polunin, Oleg
Trees and Bushes of Europe

fl. fr. hab.   p 84

Eriocactus leninghausii

Kupper, Walter
Cacti

fl., hab.   p. 51   Pl. 22

Eriocactus leninghausii

Subik, Rudolf
Decorative Cacti

fl. hab.   p 41

Eriocaulon decangulare

Batson, Wade T.
Wild Flowers in South Carolina

fl.   p 28

Eriocaulon decangulare

Brown, Clair A.
Wildflowers of Louisiana and
     Adjoining States

fl. hab.   p 11

Eriocaulon decangulare

Dean, Blanche E.
Wildflowers of Alabama and
     Adjoining States

fl.   p 7

Eriocaulon decangulare

Duncan, Wilbur H.
Wildflowers of the Southeastern
     United States

fl.   p 241

Eriocaulon decangulare

Fleming, Glenn
Wild Flowers of Florida

fl. hab.   p 25

Eriocaulon schimperi

Moriarty, Audrey
Wild Flowers of Malawi

fl.   Pl. 17; 2

Eriocaulon septangulare

Courtenay, Booth
Wildflowers and weeds

fl.   p. 2

Eriocaulon septangulare

Martin, W. Keble
The Concise British Flora in Colour

fr. hab.   Pl. 70

Eriocephalus africanus

Macoby, Stirling
What Flower is That

fl.   P 124

Eriocephalus glaber

Harrison. R.E.
Trees and Shrubs

fl.   ? 65

Eriochilus cucullatus

Blombery, Alec M.
What Wildflower is That

fl.   p 132, Pl. 344

Eriochilus cucullatus

Cady, Leo
Australian Native Orchids in Colour

fl.   Pl. 46

Eriochilus cucullatus

Cochrane, C. R.
Flowers and Plants of Victoria

fl.   Pl. 377

Eriochilus dilatatus

Cady, Leo
Australian Native Orchids in Colour

fl.   Pl. 45

Eriochloa sericea

Massachusetts Hort. Soc.
Horticulture
     vol 55, No. 4   1977

fl.   p. 41

Eriodictyon trichocalyx

Coyle, Jeanette
A Field Guide to the Common and
  Interesting Plants of Baja California

fl.   p 153

Eriogonum alleni

Addisonia
     vol 20   1937-38

fl., fr.   p. 43   Pl. 662

Eriogonum alleni

Curtis's Botanical Magazine
Vol 166   1949

fl.   65

Eriogonum brevicaule

Welsh, Stanley L.
Flowers of the Canyon Country

fl. hab.   p 38

Eriogonum fasciculatum

Coyle, Jeanette
A Field Guide to the Common and
  Interesting Plants of Baja California

fl.   p 71

Eriogonum fasciculatum

Munz, Philip A.
California Spring Wildflowers

fl.   p 31, Pl. 14

Eriogonum fasciculatum var.

Munz, Philip A.
California Desert Wildflowers

fl.   p 30, Pl. 11

Eriogonum flavum

Clark, Lewis J.
Wild Flowers of British Columbia

fl.   p 111

Eriogonum flavum

Fries, Mary A.
Wildflowers of Mount Ranier and the
  Cascades

fl.   p. 200

Eriogonum giganteum

Munz, Philip A.
California Spring Wildflowers

fl.   p. 33   Pl. 19

Eriogonum grandis var.

Royal Hort. Soc.
The Garden
     vol 102, No. 6   1977

fl.   p. 242

Eriogonum heracleoides

Clark, Lewis J.
Wild Flowers of British Columbia

fl.   p. 111

Eriogonum heracleoides

Welsh, Stanley L.
Flowers of the Mountain Country

fl.   p. 12

Eriogonum inflatum

Orr, Robert T.
Wildflowers of Western America

fl.   Pl. 108

Eriogonum nudum var.

Curtis's Botanical Magazine
     Vol 162   1939

fl.   No. 9571

Eriogonum ovalifolium

Porsild, A. E.
Rocky Mountain Wild Flowers

fl. hab.   p 143

Eriogonum ovalifolium

Welsh, Stanley L.
Flowers of the Canyon Country

fl.   p 10

Eriogonum ovalifolium var.

Munz, Philip A.
California Mountain Wildflowers

fl.   p 30, Pl. 10

Eriogonum piperi

Massachusetts Hort. Soc.
Horticulture
     vol 52, No. 7   1974

hab., fl.   p. 31

Eriogonum subalpinum

Porsild, A. E.
Rocky Mountain Wild Flowers

fl. hab.   p 145

Eriogonum subalpinum

Royal Hort. Soc.
The Garden
     vol 103, No. 11   1978

fl.   p. 442

Eriogonum tomentosum

Duncan, Wilbur H.
Wildflowers of the Southeastern
  United States

fl. hab.   p 27

Eriogonum umbellatum

Clark, Lewis J.
Wild Flowers of British Columbia

fl.   p 114

Eriogonum umbellatum

Lemmon, Robert S.
Wildflowers of North America
    in Full Color

fl.   p. 122   Pl. 194

Eriogonum umbellatum

Munz, Philip A.
California Mountain Wildflowers

fl.   p 52, Fl. 79

Eriogonum umbellatum

Orr, Robert T.
Wildflowers of Western America

fl.   Pl. 94

Eriogonum umbellatum

Royal Hort. Soc.
Journal of the Royal Hort. Soc.
    vol 93, No. 7   1968

fl.   p. 290   Pl. 153

Eriogonum umbellatum

Taylor, Ronald J.
Mountain Wild Flowers

fl. hab.   p 84

Eriolaena candollei

Morley, Brian D.
Wild Flowers of the World

fl. fr.   Pl. 116

Eriophorum angustifolium

Heller, Christine
Wild Flowers of Alaska

fr.   Pl. 277

Eriophorum angustifolium

Lindman, C. A. M.
Nordens Flora, Vol 2

fr. hab.   Pl. 113A

Eriophorum angustifolium

Martin, W. Keble
The Concise British Flora in Colour

fl. hab.   Pl. 91

Eriophorum angustifolium

Tosco, Uberto
The World of Mountain Flowers

fl. hab.   p 52

Eriophorum angustifolium

Walcott, Mary Vaux
North American Wild Flowers
    vol. 4

fr.   Pl. 312

Eriophorum chamissonis

Alaska-Yukon Wild Flowers Guide

fr.   p 1

Eriophorum chamissonis

Heller, Christine
Wild Flowers of Alaska

fr.   Pl. 278

Eriophorum chamissonis

Walcott, Mary Vaux
North American Wild Flowers

fr.   Pl. 12

Eriophorum latifolium

Martin, W. Keble
The Concise British Flora in Colour

fl. hab.   Pl. 91

Eriophorum latifolium

Polunin, Oleg
Flowers of Europe

fr.   Pl. 184 #1834

Eriophorum scheuchzeri

Alaska-Yukon Wild Flowers Guide

hab.   p viii

Eriophorum scheuchzeri

Kohlhaupt, Paula
Fleurs des Alpages
    vol 2
fl.   P 21-22

Eriophorum scheuchzeri

Tosco, Uberto
The World of Mountain Flowers

hab.   p 75

Eriophorum sp.

Milne, Lorus
Living Plants of the World

fl.   p 315

Eriophorum spissum

Ferguson, Mary
Wildflowers

fr.   p. 73

Eriophorum vaginatum

Lindman, C. A. M.
Nordens Flora, Vol 2

fr. hab.   Pl. 113B

Eriophorum vaginatum

Martin, W. Keble
The Concise British Flora in Colour

fl. hab.   Pl. 91

Eriophorum vaginatum

Polunin, Oleg
Flowers of Europe

fr.   Pl. 184 #1835

Eriophyllum confertiflorum

Munz, Philip A.
California Spring Wildflowers

fl.   p 34, Pl. 22

Eriophyllum lanatum

Clark, Lewis J.
Wild Flowers of British Columbia

fl.   p 551

Eriophyllum lanatum

Lemmon, Robert S.
Wildflowers of North America in
    Full Color

fl.   pg. 159   pl. 255

Eriophyllum lanatum

Shaw, Richard J.
Field Guide to the Vascular Plants of
    Grand Teton National Park

fl.   Pl. 4

Eriophyllum wallacei

Coyle, Jeanette
A Field Guide to the Common and
    Interesting Plants of Baja California

fl.   p 173

Eriophyllum wallacei

Lemmon, Robert S.
Wildflowers of North America in
    Full Color

fl.   pg. 105   pl. 170

Eriophyllum wallacei

Munz, Philip A.
California Desert Wildflowers

fl.   p 55, Pl. 87

Eriophyllum wallacei

Welsh, Stanley L.
Flowers of the Canyon Country

fl. hab.   p 39

Eriopsis biloba

Curtis's Botanical Magazine
Vol 178   1970-72

fl.   611

Eriopsis biloba

Ospina, Mariano
Orquideas de las americas

fl.   Pl. 140

Eriopsis biloba

Pabst, G. F. J.
Orchidaceae Brasilienses,
   Vol 2

fl.   p 214

Eriopsis sceptrum

Dunsterville, G. C. K.
Introduction to the World
   of Orchids

fl.   Pl. 9

Eriopsis sceptrum

Pabst, G. F. J.
Orchidaceae Brasilienses,
   Vol 2

fl.   p 214

Eriospermum abyssinicum

Moriarty, Audrey
Wild Flowers of Malawi

fl. fr.   Pl. 15; 1

Eriostemon australasius

Blombery, Alec M.
What Wildflower is That

fl.   p 133, Pl. 345

Eriostemon hispidulus

Blombery, Alec M.
What Wildflower is That

fl.   p 133, Pl. 347

Eriostemon lanceolatus

Macoby, Stirling
What Flower is That

fl.   P 125

Eriostemon lanceolatus

Mullins, Barbara
Australian Wildflowers in Colour

fl.   p. 73   pl. 69

Eriostemon myoporoides

Blombery, Alec M.
What Wildflower is That

fl.   p. 133   Pl. 348

Eriostemon myoporoides

Cochrane, G. R.
Flowers and Plants of Victoria

fl.   Pl. 429

Eriostemon myoporoides

Harrison, R.E.
Trees and Shrubs

fl.   P 74

Eriostemon myoporoides

Macoby, Stirling
What Flower is That

fl.   P 125

Eriostemon obovalis

Blombery, Alec M.
What Wildflower is That

fl.   p 133, Pl. 349

Eriostemon pungens

Cochrane, G. R.
Flowers and Plants of Victoria

fl.   Pl. 132

Eriostemon scaber

Blombery, Alec M.
What Wildflower is That

fl.   p 133, Pl. 350

Eriostemon spicatus

Blombery, Alec M.
What Wildflower is That

fl.   p 133, Pl. 351

Eriostemon spicatus

Milne, Lorus
Living Plants of the World

fl.   p 114

Eriostemon spicatus

Morcombe, M. K.
Australia's Western Wildflowers

fl.   p 13

Eriostemon verrucosus

Blombery, Alec M.
What Wildflower is That

fl.   p 133, Pl. 352

Eriostemon verrucosus

Cochrane, G. R.
Flowers and Plants fo Victoria

fl.   Pl. 324

Eriostemon virgatus

Curtis, Winifred
The Endemic Flora of Tasmania Vol. 1

fl.   Pl. 10

Eriosyce ceratistes

Backeberg, Curt
Cactus Lexicon

fl.   p 603

Eritrichium aretioides

Alaska-Yukon Wild Flower Guide

fl. hab.   p 135

Eritrichium aretioides

Heller, Christine
Wild Flowers of Alaska

fl.   Pl. 265

Eritrichium elongatum

Walcott, Mary Vaux
North American Wild Flowers
   vol. 5

fl.   Pl. 359

Eritrichium nanum

Alpine Garden Society
Bulletin
   vol. 6, No. 4   1938

fl., hab.   p. 341

Eritrichium nanum

Barneby, T. P.
European Alpine Flowers in Colour

fl.   Pl. 67, 1

Eritrichium nanum

Hay, Roy
The Color Dictionary of Flowers & Plants

fl.   p 9, Pl. 66

Eritrichium nanum

Kohlhaupt, Paula
Fleurs des Alpages
   vol 1

fl.   P 81-82

Eritrichium nanum

Lemmon, Robert S.
Wildflowers of North America in
   Full Color

fl.   pg. 151   pl. 242

Eritrichium nanum

Polunin, Oleg
Flowers of Europe

fl. Pl. 104 #1070

Eritrichium nanum

Royal Hort. Soc.
Journal of the Royal Hort. Soc.
vol 78, No. 8   1953

fl.   p. 284   Pl. 92

Eritrichium nanum

Royal Hort. Soc.
The Garden
vol 101, No. 5   1976

fl., hab.   p. 263

Eritrichium nanum

Welsh, Stanley L.
Flowers of the Mountain Country

fl.   p. 67

Erlangea sp.

Moriarty, Audrey
Wild Flowers of Malawi

fl.   Pl. 80; 5

Ernestimeyera magna

Morley, Brian D.
Wild Flowers of the World

fl.   Pl. 79

Erodium chamaedryoides var.

Royal Hort. Soc.
The Garden
vol 100, No. 9   1975

fl.   p. 416

Erodium cicutarium

Ary, S.
The Oxford Book of Wildflowers

fl.   p 128, Pl. 5

Erodium cicutarium

Clark, Lewis J.
Wild Flowers of British Columbia

fl.   p 310

Erodium cicutarium

Coyle, Jeanette
A Field Guide to the Common and
  Interesting Plants of Baja California

fl. hab.   p 103

Erodium cicutarium

Felsko, Elsa
A Book of Wildflowers

fl.fr.   P 83

Erodium cicutarium

Lindman, C. A. M.
Nordens Flora, Vol 6

fl. fr.   Pl. 372

Erodium cicutarium

Orr, Robert T.
Wildflowers of Western America

fl.   Pl. 285

Erodium cicutarium

Perrot, Emile
Les Plantes Medicinales

fl. hab.   p 109

Erodium cicutarium

Weeds of the Southern United States

hab., fl.   p. 22

Erodium cicutarium

Welsh, Stanley L.
Flowers of the Canyon Country

fl. hab.   p 49

Erodium cicutarium var.

Martin, W. Keble
The Concise British Flora in Colour

fl. fr. hab.   Pl. 20

Erodium corsicum var.

Hay, Roy
The Color Dictionary of Flowers &
  Plants

fl.   p. 9   Pl. 67

Erodium crinitum

Cochrane, G. R.
Flowers and Plants of Victoria

fl.   Pl. 244

Erodium gruinum

Morley, Brian D.
Wild Flowers of the World

fl.,fr.   Pl. 32

Erodium gruinum

Polunin, Oleg
Flowers of Europe

fl. fr.   Pl. 64 #652

Erodium guttatum

Hay, Roy
The Color Dictionary of Flowers & Plants

fl.   p 9, Pl. 68

Erodium macradenum

Vilmorin, Roger de
Plantes Alpines dans les Jardins

fl. hab.   Pl. XVI

Erodium malacoides

Polunin, Oleg
Flowers of Europe

fl.   Pl. 64 #652

Erodium maritimum

Martin, W. Keble
The Concise British Flora in Colour

fl. fr. hab.   Pl. 20

Erodium moschatum

Martin, W. Keble
The Concise British Flora in Colour

fl. fr. hab.   Pl. 20

Erodium moschatum

Perrot, Emile
Les Plantes Medicinales

fl. fr. hab.   p 109

Erodium petraeum var.

Morley, Brian D.
Wild Flowers of the World

fr.,hab.   Pl. 32

Erodium texanum

Orr, Robert T.
Wildflowers of Western America

fl.   Pl. 270

Erophila spathulata

Martin, W. Keble
The Concise British Flora in Colour

fl. fr. hab.   Pl. 8

Erophila verna

Ary, S.
The Oxford Book of Wildflowers

fl.   p 70, Pl. 7

Erophila verna

Lindman, C. A. M.
Nordens Flora, Vol 4

fl. fr. hab.   Pl. 265A

Erophila verna

Martin, W. Keble
The Concise British Flora in Colour

fl. fr. hab.   Pl. 8

Eruca sativa

Masefield, G. B.
The Oxford Book of Food Plants

fl., fr., hab.   Pl. 153. 5

Eruca sativa

Morley, Brian D.
Wild Flowers of the World

fl.,fr.   Pl. 30

Eruca vesicaria

Polunin, Oleg
Flowers of Europe

fl.   Pl. 37 #363

Ervatamia coronaria

Macoby, Stirling
What Flower is That

fl   P 125

Erycina diaphana

Wright, Norman Pelham
Orquideas de Mexico

fl.   Pl. 15

Erycina echinata

Ospina, Mariano
Orquideas de las americas

fl.   Pl. 173

Eryngium alpinum

Barneby, T. P.
European Alpine Flowers in Colour

fl   Pl. 53, 3

Eryngium alpinum

Kohlhaupt, Paula
Fleurs des Alpages
                    vol 2
fl   p 79

Eryngium alpinum

Morley, Brian D.
Wild Flowers of the World

fl   Pl. 170

Eryngium alpinum

Robert, Paul A.
Alpine Flowers

fl.   Pl. 7

Eryngium alpinum

Royal Hort. Soc.
The Garden
    vol 101, No. 2   1976

fl.   p. 78-9

Eryngium alpinum

Tosco, Uberto
The World of Mountain Flowers

fl.   p 62

Eryngium alpinum var.

American Hort. Soc.
American Horticulturist
    vol 51, No. 4   1972

fl.   cover

Eryngium alpinum var.

Hay, Roy
The Color Dictionary of Flowers & Plants

fl.   p 141, Pl. 1124

Eryngium amethystinum

Huxley, Anthony
Garden Perennials and Water Plants

fl.   pl. 110

Eryngium aromaticum

Fleming, Glenn
Wild Flowers of Florida

fl.   p 38

Eryngium bourgatii

Kohlhaupt, Paula
Fleurs des Alpages
                    v-1 2
Fl.   P 80

Eryngium bourgatii

Royal Hort. Soc.
The Garden
    vol 101, No. 2   1976

fl., hab.   p. 80

Eryngium campestre

Kleijn, H.
Beauty of the Wild Plant

fl.   opp. p. 73   pl. 93

Eryngium campestre

Martin, W. Keble
The Concise British Flora in Colour

fl. hab.   Pl. 36

Eryngium campestre

Polunin, Oleg
Flowers of Europe

fl.   Pl. 83 #853

Eryngium creticum

Feinbrun-Dothan, Naomi
Wild Plants in the Land
    of Israel

fl.   p 114

Eryngium giganteum

Hay, Roy
The Color Dictionary of Flowers & Plants

fl.   p 141, Pl. 1125

Eryngium giganteum

Perry, Frances
Flowers of the World

fl.   p 301

Eryngium integrifolium

Brown, Clair A.
Wildflowers of Louisiana and
    Adjoining States

fl.   p 123

Eryngium maritimum

Ary, S.
The Oxford Book of Wildflowers

fl.   p 178, Pl. 5

Eryngium maritimum

Felsko, Elsa
A Book of Wild Flowers
                    2nd ser
fl.   p 57

Eryngium maritimum

Kleijn, H.
Beauty of the Wild Plant

fl.   opp. p. 81   pl. 106

Eryngium maritimum

Lindman, C. A. M.
Nordens Flora. Vol 6

fl. fr. hab.   Pl. 416

Eryngium maritimum

Martin, W. Keble
The Concise British Flora in Colour

fl. hab.   Pl. 36

Eryngium maritimum

Polunin, Oleg
Flowers of Europe

fl.   Pl. 83 #853

Eryngium maritimum

Royal Hort. Soc.
The Garden
    vol 101, No. 2   1976

fl., hab.   p. 80

Eryngium planum

Felsko, Elsa
A Book of Wildflowers

fl.   p 56

Eryngium planum

Huxley, Anthony
Garden Perennials and Water Plants

fl.   Pl. 111

Eryngium prostratum

Duncan, Wilbur H.
Wildflowers of the Southeastern
United States

fl.   p 115

Eryngium prostratum

Wharton, Mary E.
A Guide to the Wildflowers & Ferns
of Kentucky

fl.   p 152   Pl. 2.20

Eryngium rostratum

Cochrane, G. R.
Flowers and Plants of Victoria

fl.   Pl. 235

Eryngium synchaetum

Addisonia
vol 22   1943-46

fl.   p. 13   Pl. 711

Eryngium tripartitum

Hay, Roy
The Color Dictionary of Flowers & Plants

fl.   p 141, Pl. 1126

Eryngium tripartitum

Royal Hort. Soc.
The Garden
vol 101, No. 2   1976

fl.   p. 80

Eryngium variifolium

Hay, Roy
The Color Dictionary of Flowers & Plants

fl.   p 141, Pl. 1127

Eryngium yuccifolium

Batson, Wade T.
Wild Flowers in South Carolina

fl.   p 79

Eryngium yuccifolium

Brown, Clair A.
Wildflowers of Louisiana and
Adjoining States

fl. hab.   p 124

Eryngium yuccifolium

Courtenay, Booth
Wildflowers & Weeds

fl.   p 58

Eryngium yuccifolium

Dean, Blanche E.
Wildflowers of Alabama and
Adjoining States

fl.   p 123

Eryngium yuccifolium

Duncan, Wilbur H.
Wildflowers of the Southeastern
United States

fl. hab.   p 115

Eryngium yuccifolium

Wharton, Mary E.
A Guide to the Wildflowers & Ferns
of Kentucky

fl.   p 277   Pl. 2.6

Erysimum asperum

Clark, Lewis J.
Wild Flowers of British Columbia

fl.   p 187

Erysimum asperum

Crockett, James Underwood
Annuals

fl.   P 105

Erysimum asperum

Orr, Robert T.
Wildflowers of Western America

fl.   Pl. 135

Erysimum asperum

Shaw, Richard J.
Field Guide to the Vascular Plants of
Grand Teton National Park

fl.   Pl. 2

Erysimum asperum

Welsh, Stanley L.
Flowers of the Canyon Country

fl.   p 29

Erysimum capitatum

Lemmon, Robert S.
Wildflowers of North America in
Full Color

fl.   pg. 178   pl. 281

Erysimum capitatum var.

Curtis's Botanical Magazine
Vol 178   1970-72

fl.   593

Erysimum cheiranthoides

Ary, S.
The Oxford Book of Wildflowers

fl.   p 8, Pl. 5

Erysimum cheiranthoides

Courtenay, Booth
Wildflowers & Weeds

fl.   p 17

Erysimum cheiranthoides

Martin, W. Keble
The Concise British Flora in Colour

fl. fr. hab.   Pl. 9

Erysimum helveticum

Curtis's Botanical Magazine
Vol 177   1969-70

fl.   535

Erysimum kotschyanum

Macoby, Stirling
What Flower is That

fl.   P 125

Erysimum x marshallii

Perry, Frances
Flowers of the World

fl.   p. 97

Erysimum nivale

Welsh, Stanley L.
Flowers of the Mountain Country

fl.   50

Erysimum pallasii

Heller, Christine
Wild Flowers of Alaska

fl.   Pl. 67

Erysimum perenne

Munz, Philip A.
California Mountain Wildflowers

fl.   p. 54   Pl. 82

Erysimum perofskianum

Morley, Brian D.
Wild Flowers of the World

fl. fr.   Pl. 11A

Erysimum purpureum

American Hort. Soc.
American Horticulturist
vol 53, No. 5 1974

fl.   p. 18

Erysimum sylvestris var.

Barneby, T. P.
European Alpine Flowers in Colour

fl.   Pl. 30, 4

Erythea armata

Boom, B. K.
The glory of the tree

fl. hab.   p. 110   Pl. 191

Erythea armata

Coyle, Jeanette
A Field Guide to the Common and
Interesting Plants of Baja California

fl. hab.   p 49

Erythea armata

Massachusetts Hort. Soc.
Horticulture
   vol 47, No. 1   1969

fl.   p. 22

Erythea brandegeei

Coyle, Jeanette
A Field Guide to the Common and
Interesting Plants of Baja California

hab.   p 49

Erythraea centaurium

Perrot, Emile
Les Plantes Medicinales

fl. fr. hab.   p 55

Erythrina abyssinica

Flowering Plants of Africa
   vol 44   1977

fl., fr.   Pl. 1738

Erythrina abyssinica

Oakman, Harry
Colorful Trees

Hab. Fl.   P 33

Erythrina abyssinica

Palgrave, K. C.
Trees of Central Africa

fl., fr.   p. 319

Erythrina acanthocarpa

Harrison, R.E.
Trees and Shrubs

fl.   P 73

Erythrina acanthocarpa

Macoby, Stirling
What Flower is That

fl.   P 126

Erythrina americana

O'Gorman, Helen
Mexican Flowering Trees and Plants

fl. fr. hab.   p 25

Erythrina baumii

Flowering Plants of Africa
   vol 36   1963-64

fl.   Pl. 1412

Erythrina x bidwillii

Mathias, Mildred E.
Color for the Landscape

fl.   p. 51

Erythrina caffra

Flowering Plants of Africa
   vol 43   1974-76

fl., fr.   Pl. 1707

Erythrina caffra

Massachusetts Hort. Soc.
Horticulture
   vol 39, No. 12   1961

fl.   p. 622

Erythrina caffra

Mathias, Mildred E.
Color for the Landscape

fl.   cover, p. 5

Erythrina caffra

Oakman, Harry
Colorful Trees

Hab. Fl.   P 21

Erythrina coralloides

Mathias, Mildred E.
Color for the Landscape

fl.   p 50

Erythrina crista-galli

Cault, S. Millar
The Color Dictionary of Shrubs

fl.   Pl. 181

Erythrina crista-galli

Harrison, R.E.
Trees and Shrubs

fl.   P 73

Erythrina crista-galli

Hay, Roy
The Color Dictionary of Flowers & Plants

fl.   p 199, Pl. 1592

Erythrina crista-galli

Hay, Roy
The Dictionary of House Plants

fl.   Pl. 218

Erythrina crista-galli

Macoby, Stirling
What Flower is That

fl.   P 126

Erythrina crista-galli

Oakman, Harry
Colorful Trees

Hab. Fl.   P 43

Erythrina crista-galli

Perry, Frances
Flowers of the World

fl.   p 165

Erythrina crista-galli

Royal Hort. Soc.
Journal of Royal Hort. Soc.
   vol 84, No. 3   1959

fl.   p. 122   Pl. 39

Erythrina crista-galli

Wit, H. C. D. de
Plants of the World;
   The Higher Plants, Vol I

fl.   p 287, Pl. 184

Erythrina crista-galli var.

Tsukamoto, Yotaro
Coloured Illustrations of Garden
   Flowers   vol 10

fl.   p. 47   Pl. 150

Erythrina falcata

Mathias, Mildred E.
Color for the Landscape

fl. hab.   p 52

Erythrina flabelliformis

Coyle, Jeanette
A Field Guide to the Common and Interesting
Plants of Baja California

fl. fr.   p 99

Erythrina flabelliformis

Orr, Robert T.
Wildflowers of Western America

fl.   Pl. 204

Erythrina fusca

Flowering Plants of Africa
   vol 44   1977

fl.   Pl. 1754

Erythrina fusca

Oakman, Harry
Colorful Trees

Hab. Fl.   P 32

Erythrina glauca

Pertchik, Bernard
Flowering Trees of the Caribbean

fl.   p 79

Erythrina guatemalensis

Pacific Tropical Botanical Garden
Bulletin
vol 8, No. 4   1978

fl.   cover

Erythrina herbacea

Batson, Wade T.
Wild Flowers in South Carolina

fl.   p 60

Erythrina herbacea

Brown, Clair A.
Wildflowers of Louisiana and
   Adjoining States

fl. hab.   p 79

Erythrina herbacea

Dean, Blanche E.
Wildflowers of Alabama and
   Adjoining States

fl.   p 87

Erythrina herbacea

Duncan, Wilbur H.
Wildflowers of the Southeastern
   United States

fl.   p 85

Erythrina herbacea

Fleming, Glenn
Wild Flowers of Florida

fl.   p 72

Erythrina humeana

Mathias, Mildred E.
Color for the Landscape

fl. hab.   p 52

Erythrina indica

Hargreaves, Dorothy
Tropical Blossoms of the Caribbean

fl.,hab.   p. 59

Erythrina indica

Macoby, Stirling
What Flower is That

fl.   P 126

Erythrina indica

Oakman, Harry
Colorful Trees

Hab. Fl.   P 20

Erythrina indica

Tsukamoto, Yotaro
Coloured Illustrations of Garden
   Flowers   Vol. 10

fl. Illus. 151 opp. p. 47

Erythrina latissima

Flowering Plants of Africa
vol 43   1974-76

fl., fr.   Pl. 1710

Erythrina livingstoniana

Flowering Plants of Africa
vol 44   1977

fl., fr.   Pl. 1737

Erythrina lysistemon

Massachusetts Hort. Soc.
Horticulture
vol 40, No. 2   1962

fl.   p. 86

Erythrina lysistemon

Massachusetts Hort. Soc.
Horticulture
vol 44, No. 12   1966

fl.   inside cover

Erythrina lysistemon

Morley, Brian D.
Wild Flowers of the World

fl.,fr.   Pl. 72

Erythrina lysistemon

Palgrave, K. C.
Trees of Central Africa

fl., fr.   p. 323

Erythrina lysistemon

Palmer, Eve.
Trees of South Africa

fl.,fr.   p. 144,208   Pl. Ia, XIII

Erythrina ovalifolia

Oakman, Harry
Colorful Trees

Fl. hab.   P 42

Erythrina parcellii

Macoby, Stirling
What Flower is That

fl.   P 126

Erythrina parcellii

Oakman, Harry
Colorful Trees

Hab.   P 112

Erythrina phlebocarpa

American Hort. Soc.
American Horticulturist
vol 56, No. 5   1977

fl.   p. 14

Erythrina poeppigiana

Everett, Thomas H.
Living Trees of the World

Erythrina poeppigiana

Milne, Lorus
Living Plants of the World

hab., fl.   p 292

Erythrina poeppigiana

Morley, Brian D.
Wild Flowers of the World

fl.   pl. 170

Erythrina poeppigiana

Pertchik, Bernard
Flowering Trees of the
   Caribbean

fl.   p 75

Erythrina princeps

Mathias, Mildred E.
Color for the Landscape

fl. hab.   p 54

Erythrina sp.

Everett, T.H.
Living Trees of the World

fl.   p. 180

Erythrina sp.

Pacific Tropical Botanical Garden
Bulletin
vol 7, No. 2   1977

fl., hab.   cover

Erythrina tahitensis

Pacific Tropical Botanical Garden
Bulletin
vol 3, No. 1   1973

fl., fr.   cover

Erythrina vespertilio

Blombery, Alec M.
What Wildflower is That

fl.   p. 134   Pl. 353

Erythrina vespertilio

Oakman, Harry
Colorful Trees

fl. hab   p. 58

Erythrina zeyheri

Flowering Plants of Africa
vol XXVI   1947

fl., fr.   Pl. 1011

Erythrocephalum zambesianum

Moriarty, Audrey
Wild Flowers of Malawi

fl.   Pl. 79; 2

Erythrodes lindleyana

Pabst, G. F. J.
Orchidaceae Brasilienses,
  Vol 1

fl. hab.   p 183

Erythrodes, Low

see

Erythrodes querceticola

Erythrodes nobilis var.

Grubb, Roy
Selected Orchidaceous Plants
  Vol II

hab.   p 49

Erythrodes nobilis var.

Pabst, G.F.J.
Orchidaceae Brasilienses
  vol 2

hab.   p. 171

Erythrodes querceticola

Luer, Carlyle A.
The Native Orchids of Florida

fl. hab.   Pl. 35

Erythrodes querceticola

Ospina, Mariano
Orquideas de las americas

fl.   Pl. 23

Erythronium albidum

American Hort. Soc.
American Horticulturist
  vol 53, No. 5   1974

fl.   p. 35

Erythronium albidum

Dean, Blanche E.
Wildflowers of Alabama and
  Adjoining States

fl.   p 23

Erythronium albidum

Jennings, O. E.
Wild Flowers of Western Pennsylvania
                        Vol. II

fl. hab.   Pl. 19

Erythronium albidum

Moyle, John B.
Northland Wild Flowers

fl. hab.   p 209, Pl. 24

Erythronium albidum

Walcott, Mary Vaux
North American Wild Flowers

fl. hab.   Pl. 15

Erythronium albidum

Wharton, Mary E.
A Guide to the Wildflowers & Ferns
  of Kentucky

fl.   p 52 Pl. 1.10

Erythronium americanum

Campbell, Carlos C.
Great Smoky Mountain Wildflowers

fl.   p 97

Erythronium americanum

Courtenay, Booth
Wildflowers and weeds

fl.   p. 4

Erythronium americanum

Duncan, Wilbur H.
Wildflowers of the Southeastern
  United States

fl. hab.   p 253

Erythronium americanum

Jennings, O. E.
Wild Flowers of Western Pennsylvania
                        vol II
fl. hab.   Pl. 19

Erythronium americanum

Klimas, John E.
Wildflowers of Eastern America

fl.   Pl. 113

Erythronium americanum

Klute, Jeannette
Woodland Portraits

fl. hab.   Pl. 7

Erythronium americanum

Lemmon, Robert S.
Wildflowers of North America in
  Full Color

fl.   p. 221   pl. 348

Erythronium americanum

Miles, Bebe
Bulbs for the Home Gardener

fl.   p 71

Erythronium americanum

Roberts, June Carver
Born in the Spring

fl. fr. hab.   p 51

Erythronium americanum

Royal Hort. Soc.
The Garden
  Vol. 102, No. 1   1977

fl.   p. 18

Erythronium americanum

Walcott, Mary Vaux
North American Wild Flowers
  vol. 5

fl.   Pl. 339

Erythronium americanum

Wharton, Mary E.
A Guide to the Wildflowers & Ferns
  of Kentucky

fl. hab. p 51 Pl. 1.9

Erythronium californicum

Royal Hort. Soc.
Journal of the Royal Hort. Soc.
  vol 99, No. 8   1974
fl., hab.   p. 346 Pl. 165

Erythronium caucasicum

Wendelbo, Per
Tulips and Irises of Iran and
  Their Relatives

fl.   p 41, Pl. 38

Erythronium dens-canis

Felsko, Elsa
A Book of Wild Flowers
                        2nd ser
fl. hab.   p 67

Erythronium dens-canis

Huxley, Anthony
Garden Annuals and Bulbs

hab., fl.   Pl. 155, 216

Erythronium dens-canis

Macoby, Stirling
What Flower is That

fl.   P 126

Erythronium dens-canis

Perry, Frances
Flowers of the World

fl.   p 175

Erythronium dens-canis

Polunin, Oleg
Flowers of Europe

fl.   Pl. 168 #1628

Erythronium dens-canis

Royal Hort. Soc.
The Garden
  vol 101, No. 7   1976

fl.   p. 351

Erythronium dens-canis var.

Hay, Roy
The Color Dictionary of Flowers &
Plants

fl.   p. 92   Pl. 731

Erythronium giganteum

Mathew, Brian
Dwarf Bulbs

fl.   p. 112   Pl. 40

Erythronium giganteum

Royal Hort. Soc.
Journal of the Royal Hort. Soc.
    Vol. 92, No. 11   1967

fl.   p. 478   Pl. 254

Erythronium grandiflorum

Clark, Lewis J.
Wild Flowers of British Columbia

fl. hab.   p 42

Erythronium grandiflorum

Ferguson, Mary
Wildflowers

fl.   p 64

Erythronium grandiflorum

Fries, Mary A.
Wildflowers of Mount Ranier and
the Cascades

fl.   P 88 & 89

Erythronium grandiflorum

Lemmon, Robert S.
Wildflowers of North America in
    Full Color

fl.   pg. 113   pl. 180

Erythronium grandiflorum

Porsild, A. E.
Rocky Mountain Wild Flowers

fl. hab.   p 99

Erythronium grandiflorum

Shaw, Richard J.
Field Guide to the Vascular Plants of
    Grand Teton National Park

fl.   Pl. 8

Erythronium grandiflorum

Taylor, Ronald J.
Mountain Wild Flowers

fl. hab.   p 102

Erythronium grandiflorum

Walcott, Mary Vaux
North American Wild Flowers

fl. hab.   Pl. 68

Erythronium grandiflorum

Welsh, Stanley L.
Flowers of the Mountain Country

fl.   p. 51

Erythronium helenae

Lemmon, Robert S.
Wildflowers of North America in
    Full Color

fl.   p. 221   pl. 349

Erythronium hybrid

Hay, Roy
The Color Dictionary of Flowers
    & Plants

fl.   p. 92

Erythronium hybrid

Miles, Bebe
Bulbs for the Home Gardener

fl.   p 71

Erythronium hybrid

Royal Hort. Soc.
Journal of the Royal Hort. Soc.
    vol 97, No. 3   1972

fl.   cover

Erythronium japonicum

Kimura, Koiti
Japanese Medicinal Plants, Vol II

fl.   p 135

Erythronium montanum

Clark, Lewis J.
Wild Flowers of British Columbia

fl.   p 27

Erythronium montanum

Fries, Mary A.
Wildflowers of Mount Ranier and
the Cascades

fl.   P 93, 97, 98

Erythronium montanum

Orr, Robert T.
Wildflowers of Western America

fl.   Pl. 18

Erythronium montanum

Walcott, Mary Vaux
North American Wild Flowers

fl. hab.   Pl. 202

Erythronium multiscapoideum

Addisonia
    vol 20   1937-38

fl., fr.   p. 37   Pl. 659

Erythronium multiscapoideum

Morley, Brian D.
Wild Flowers of the World

fl.   pl. 167

Erythronium oregonum

Clark, Lewis J.
Wild Flowers of British Columbia

fl. hab.   p 30, 31

Erythronium oregonum

Massachusetts Hort. Soc.
Horticulture
    vol 45, No. 3   1967

fl.   p. 19

Erythronium oregonum

Pacific Hort. Foundation
Pacific Horticulture
    vol 39, No. 2   1978

fl.   p. 33

Erythronium oregonum var.

Royal Hort. Soc.
The Garden
    vol 101, No. 8   1976

fl.   p. 409

Erythronium propullans

Moyle, John B.
Northland Wild Flowers

fl. hab.   p 209. Pl. 275

Erythronium revolutum

Clark, Lewis J.
Wild Flowers of British Columbia

fl. hab.   p 31, 38

Erythronium revolutum

Ferguson, Mary
Wildflowers

fl.   p 180

Erythronium revolutum

Miles, Bebe
Bulbs for the Home Gardener

fl.   p 71

Erythronium revolutum

Perry, Frances
Flowers of the World

fl.   p 175

Erythronium revolutum var.

Huxley, Anthony
Garden annuals and bulbs

fl.   pl. 215

Erythronium rostratum

Dean, Blanche E.
Wildflowers of Alabama and
Adjoining States

fl.   p 21

Erythronium tuolumnense

Huxley, Anthony
Garden annuals and bulbs

fl.   pl. 217

Erythronium tuolumnense

Miles, Bebe
Bulbs for the Home Gardener

fl.   p 71

Erythronium tuolumnense

Royal Hort. Soc.
Journal of the Royal Hort. Soc.
vol 89, No. 7   1964

fl.   p. 290   Pl. 106

Erythrophysa transvaalensis

Flowering Plants of South Africa
vol XXII   1942

fl., fr.   Pl. 857

Erythrorhipsalis pilocarpus

Lamb, Edgar
Popular Exotic Cacti in Color

fl. hab.   p 80

Erythroxylon coca

Bianchini, Francesco
Health Plants of the World

hab.   p 117

Erythroxylon coca

Hvass, Else
Plants That Feed and Serve Us

fl. fr.   p 90, Pl. 192

Erythroxylon Coca

Kimura, Koiti
Japanese Medicinal Plants, Vol I

fl. fr.   p 52

Erythroxylon coca

Perrot, Emile
Les Plantes Medicinales

fl. fr. hab.   p 73

Erythroxylon sp.

Wit, H. C. D. de
Plants of the World;
The Higher Plants, Vol II

fl. fr.   p 60, Pl. 32

Erythroxylon vovogranatense

Kariyone, Tatsuo
Atlas of Medicinal Plants

fl., fr.   Pl. 96

Escallonia alpina

Curtis's Botanical Magazine
Vol 179   1972

fl.   642

Escallonia fradesi

Crockett, James Underwood
Evergreens

hab., fl.   p. 123

Escallonia hybrid.

Gault, S. Millar
The Color Dictionary of Shrubs

fl.   Pl. 182

Escallonia hybrid

Harrison, R. E.
Trees and Shrubs

fl.   p. 73

Escallonia hybrid

Hay, Roy
The Color Dictionary of Flowers & Plants

fl.   p 200, Pl. 1593, 1594, 1595

Escallonia hybrid

Perry, Frances
Flowers of the World

fl.   p. 113

Escallonia x langleyensis

Hellyer, A.G.L.
Shrubs in Colour

fl.   p. 50

Escallonia macrantha

Hay, Roy
The Color Dictionary of Flowers & Plants

fl.   p 200, Pl. 1596

Escallonia macrantha

Macoby, Stirling
What Flower is That

fl.   P 127

Escallonia, Pink Princess

see

Escallonia fradesi

Escallonia rubra

Munoz Pizarro, Carlos
Flores Silvestres de Chile

fl.   Pl. 35

Escallonia tucumanensis

Curtis's Botanical Magazine
Vol 177   1969-70

fl.   565

Escarole

see

Cichorium endivia var.

Eschscholzia californica

Crockett, James Underwood
Annuals

fl.   P 117

Eschscholzia californica

Huxley, Anthony
Garden annuals and bulbs

fl.   pl. 43

Eschscholzia californica

Lemmon, Robert S.
Wildflowers of North America in
Full Color

fl.   Pg. 20   pl. 32

Eschscholzia californica

Macoby, Stirling
What Flower is That

fl.   P 127

Eschscholzia californica

Mathias, Mildred E.
Color for the Landscape

fl. hab.   p 170, 171

Eschscholzia californica

Milne, Lorus
Living Plants of the World

fl. hab.   p 89

Eschscholzia californica

Morley, Brian D.
Wild Flowers of the World

fl. fr.   pl. 154

Eschscholzia californica

Munz, Philip A.
California Spring Wildflowers

fl.   p 82, Pl. 70

Eschscholzia californica

Orr, Robert T.
Wildflowers of Western America

fl.   Pl. 120

Eschscholsia californica

Pacific Hort. Foundation
Pacific Horticulture
    vol 37, No. 1   1976

fl.   cover

Eschscholzia californica

Perry, Frances
Flowers of the World

fl.   p 220

Eschscholzia californica

Polunin, Oleg
Flowers of Europe

fl.   Pl. 29   #275

Eschscholzia californica

Royal Hort. Soc.
The Garden
    vol 102, No. 1   1977

fl.   p. 11

Eschscholzia californica

Tsukamoto, Yotaro
Coloured Illustrations of Garden
    Flowers   Vol. 10

fl. Illus 58 opp. p. 19

Eschscholzia californica var.

Hay, Roy
The Color Dictionary of Flowers &
    Plants

fl.   P. 37   Pl. 294

Eschscholzia californica var.

Perry, Frances
Flowers of the World

fl.   p. 224

Eschscholzia lobbii

Orr, Robert T.
Wildflowers of Western America

fl.   Pl. 121

Eschscholsia mexicana

Lemmon, Robert S.
Wildflowers of North America
    in Full Color

fl.   p. 68   Pl. 108

Eschscholzia mexicana

Walcott, Mary Vaux
North American Wild Flowers
    vol. 5

fl.   Pl. 393

Eschscholzia parishii

Munz, Philip A.
California Desert Wildflowers

fl.   p 32, Pl. 18

Escobaria henricksonii

Cactus & Succulent Society of America
Journal
    vol 49, No. 5   1977

fl., hab.   p. 195

Escobaria runyonii

Lamb, Edgar
Colorful Cacti of the American Deserts

fl. hab.   Pl. 4

Escobaria runyonii

Lamb, Edgar
Popular Exotic Cacti in Color

fl. hab.   p 81

Escobaria tuberculosa

Lamb, Edgar
Colorful Cacti of the American Deserts

hab.   Pl. 27

Escobita

see

Orthocarpus purpurascens

Espeletia grandiflora

Tosco, Uberto
The World of Mountain Flowers

hab.   p 85

Espostoa lanata

Lamb, Edgar
The Pocket Encyclopedia of Cacti
    and Succulents in Color

hab.   Pl. 49

Espostoa lanata

Subik, Rudolf
Decorative Cacti

hab.   p 43

Espostoa lanata

Van Laren, A. J.
Cactus

hab. P 10   Fig. 11

Espostoa lanata var.

Lamb, Edgar
The Pocket Encyclopedia of Cacti
    and Succulents in Color

hab.   Pl. 50

Espostoa lanata var.

Lamb, Edgar
Popular Exotic Cacti in Color

hab.   p 83

Espostoa melanostele

Lamb, Edgar
The Pocket Encyclopedia of Cacti
    and Succulents in Color

hab.   Pl. 51

Espostoa sericata

Kupper, Walter
Cacti

hab.   p. 79   Pl. 36

Esterhuysenia alpina

Herre, H.
The Genera of the Mesembryanthemaceae

fl.   fr.   p 153

Euadenia eminens

Morley, Brian D.
Wild Flowers of the World

fl.   Pl. 52

Eucalyptus alpina

Holliday, Ivan
Eucalypts in Colour

hab.   p 6

Eucalyptus amygdalina

Curtis, Winifred
The Endemic Flora of Tasmania Vol. 2

fl. fr.   Pl. 39

Eucalyptus annulata

Holliday, Ivan
Eucalypts in Colour

fl.   p. 29

Eucalyptus barberi

Curtis, Winifred
The Endemic Flora of Tasmania
    Vol VI

fl. fr.   Pl. 208

Eucalyptus bicolor

Holliday, Ivan
Eucalypts in Colour

hab.   p 17

Eucalyptus bridgesiana

Cochrane, G. R.
Flowers and Plants of Victoria

fl. fr.   Pl. 418

Eucalyptus caesia

Blombery, Alec M.
What Wildflower is That

fl.   p 134, Pl. 354

Eucalyptus caesia

Holliday, Ivan
Eucalypts in Colour

hab.   p 15

Eucalyptus caesia

Morcombe, M. K.
Australia's Western Wildflowers

fl.   p 16

Eucalyptus calophylla

Edlin, Herbert
The Illustrated Encyclopedia
    of Trees

fr.   p 240

Eucalyptus calophylla

Holliday, Ivan
Eucalypts in Colour

fl. fr.   p 20

Eucalyptus camuldulensis

Blombery, Alec M.
What Wildflower is That

hab.   p. 27   Pl. 41

Eucalyptus camaldulensis

Cochrane, G. R.
Flowers and Plants of Victoria

hab.   Pl. 226

Eucalyptus camaldulensis

Edlin, Herbert
The Illustrated Encyclopedia
    of Trees

hab.   p 239

Eucalyptus camaldulensis

Holliday, Ivan
Eucalypts in Colour

hab.   p 10

Eucalyptus camaldulensis

Polunin, Oleg
Trees and Bushes of Europe

fl. fr. hab.   p 150, 148

Eucalyptus chapmaniana

Cochrane, G. R.
Flowers and Plants of Victoria

fl. fr.   Pl. 421

Eucalyptus cinerea

Holliday, Ivan
Eucalypts in Colour

hab.   p 25

Eucalyptus citriodora

Holliday, Ivan
Eucalypts in Colour

hab.   p 19

Eucalyptus coccifera

Curtis, Winifred
The Endemic Flora of Tasmania
                        vol 3

fl. fr. p 159

Eucalyptus coccifera var.

Curtis's Botanical Magazine
    Vol 160   1937

fl. fr.   No. 9511

Eucalyptus cordata

Curtis, Winifred
The Endemic Flora of Tasmania
        Vol. 2

fl. fr.   Pl. 51

Eucalyptus cordata

Royal Hort. Soc.
Journal of the Royal Hort. Soc.
vol 91, No. 5   1966

hab.   p. 206   Pl. 116

Eucalyptus cornuta

Morley, Brian D.
Wild Flowers of the World

fl. fr.   pl. 137

Eucalyptus corymbosa

Macoby, Stirling
What Flower is That

fl.   P 128

Eucalyptus cosmophylla

Mullins, Barbara
Australian Wildflowers in Colour

fl.   p. 59   Pl. 52

Eucalyptus, Crimson

    see

Eucalyptus ficifolia

Eucalyptus dalrympleana

Royal Hort. Soc.
Journal of the Royal Hort. Soc.
    vol 91, No. 6   1966

hab.   p. 250   Pl. 134-35

Eucalyptus diversicolor

Edlin, Herbert
The Illustrated Encyclopedia
    of Trees

hab.   p 241

Eucalyptus diversicolor

Holliday, Ivan
Eucalypts in Colour

hab.   p 2

Eucalyptus dives

Cochrane, G. R.
Flowers and Plants of Victoria

fl.   Pl. 359

Eucalyptus dives

Holliday, Ivan
Eucalypts in Colour

hab.   p 17

Eucalyptus drepanophylla

Oakman, Harry
Colourful Trees

hab.   p. 136

Eucalyptus dumosa

Holliday, Ivan
Eucalypts in Colour

hab.   p 13

Eucalyptus eremophila

Blombery, Alec M.
What Wildflower is That

fl.   p 135, Pl. 355

Eucalyptus erythrocorys

Blombery, Alec M.
What Wildflower is That

fl.   p 135, Pl. 356

Eucalyptus erythrocorys

Holliday, Ivan
Eucalypts in Colour

fl. hab.   p 27

Eucalyptus erythrocorys

Mathias, Mildred E.
Color for the Landscape

fl. hab.   p 47

Eucalyptus erythronema

Holliday, Ivan
Eucalypts in Colour

hab.   p 22

Eucalyptus erythronema

Mullins, Barbara
Australian Wildflowers in Colour

fr.   P. 57   Pl. 50

Eucalyptus fasciculosa

Holliday, Ivan
Eucalypts in Colour

hab.   p 24

Eucalyptus ficifolia

Blombery, Alec M.
What Wildflower is That

fl.   p 135, Pl. 357, 358

Eucalyptus ficifolia

Crockett, James Underwood
Evergreens

hab. fl.   p. 123

Eucalyptus ficifolia

Edlin, Herbert
The Illustrated Encyclopedia
   of Trees

fl.   p 238

Eucalyptus ficifolia

Flemer, William III
Shade and Ornamental Trees in Color

fl., hab.   p. 15, 100

Eucalyptus ficifolia

Harrison, R.E.
Trees and Shrubs

fl.   P 75

Eucalyptus ficifolia

Holliday, Ivan
Eucalypts in Colour

fl. hab.   p 20

Eucalyptus ficifolia

Macoby, Stirling
What Flower is That

fl.   P. 128

Eucalyptus ficifolia

Mathias, Mildred E.
Color for the Landscape

fl. hab.   p 48

Eucalyptus ficifolia

Morley, Brian D.
Wild Flowers of the World

fl.   Pl. 137

Eucalyptus ficifolia

Mullins, Barbara
Australian Wildflowers in Colour

fl.   P. 55   Pl. 48

Eucalyptus ficifolia

Oakman, Harry
Colorful Trees

Fl. P 70

Eucalyptus ficifolia

Royal Hort. Soc.
Journal of the Royal Hort. Soc.
   vol 90, No. 1   1965

fl.   p. 22   Pl. 19

Eucalyptus ficifolia

Royal Hort. Soc.
Journal of the Royal Hort. Soc.
   vol 93, No. 8   1968

fl.   p. 336   Pl. 161.

Eucalyptus ficifolia

Royal Hort. Soc.
The Garden
   vol 103, No. 3   1978

fl.   p. 105

Eucalyptus, Flame

see

Eucalyptus ficifolia

Eucalyptus foecunda

Holliday, Ivan
Eucalypts in Colour

fl. hab.   p 14

Eucalyptus forrestiana

Edlin, Herbert
The Illustrated Encyclopedia
   of Trees

fl.   p 240

Eucalyptus forrestiana

Holliday, Ivan
Eucalypts in Colour

fl.   p 30

Eucalyptus forrestiana

Mullins, Barbara
Australian Wildflowers in Colour

fl.   Front Cover

Eucalyptus gillii

Holliday, Ivan
Eucalypts in Colour

fl.   p 11

Eucalyptus globulus

Cochrane, G. R.
Flowers and Plants of Victoria

hab.   Pl. 350

Eucalyptus globulus

Hay, Roy
The Dictionary of House Plants

hab.   Pl. 219

Eucalyptus globulus

Holliday, Ivan
Eucalypts in Colour

fl.   p 2

Eucalyptus globulus

Huxley, Anthony
Evergreen Garden Trees and Shrubs

fr. hab.   Pl. 164

Eucalyptus globulus

Hvass, Elsie
Plants That Feed and Serve Us

fl.   p 95, Pl. 201

Eucalyptus globulus

Kariyone, Tatsuo
Atlas of Medicinal Plants

fl., fr.   Pl. 78

Eucalyptus globulus

Kiaer, Eigil
Indoor Plants in Colour

hab., fr.   p. 56

Eucalyptus globulus

Kimura, Koiti
Japanese Medicinal Plants, Vol I

fl. fr.   p 62

Eucalyptus globulus

Macoby, Stirling
What Flower is That

hab. P 128

Eucalyptus globulus

Nicolaisen, Age
Pocket Encyclopedia of Indoor Plants

hab.   pl. 102

Eucalyptus globulus

Perrot, Emile
Les Plantes Medicinales

fl. fr. hab.   p 94

Eucalyptus globulus

Polunin, Oleg
Trees and Bushes of Europe

fl. fr. hab.   p 148, 149

Eucalyptus globulus

Wit, H. C. D. de
Plants of the World;
    The Higher Plants, Vol II

hab.   p 50, Pl. 4

Eucalyptus gracilis

Cochrane, G.R.
Flowers and Plants of Victoria

fl., fr.   Pl. 178

Eucalyptus gracilis

Holliday, Ivan
Eucalypts in Colour

hab.   p 13

Eucalyptus grandis

Holliday, Ivan
Eucalypts in Colour

hab.   p 3

Eucalyptus grandis

Oakman, Harry
Colourful Trees

hab.   p. 136

Eucalyptus gummifera

Blombery, Alec M.
What Wildflower is That

fl.   p 135, Pl. 359

Eucalyptus gunnii

Curtis, Winifred
The Endemic Flora of Tasmania Vol. 2

fl. fr.   Pl. 66

Eucalyptus gunnii

Hay, Roy
The Color Dictionary of Flowers & Plants

hab.   p 200, Pl. 1597

Eucalyptus gunnii

Royal Hort. Soc.
Journal of the Royal Hort. Soc.
    vol 91, No. 5   1966

hab.   p. 206   Pl. 114-15

Eucalyptus Gunnii

Tsukamoto, Yotaro
Coloured Illustrations of Garden
    Flowers   Vol. 10

hab. Illus. 152 opp. p. 47

Eucalyptus haemastoma

Holliday, Ivan
Eucalypts in Colour

hab.   p. 16

Eucalyptus haemastoma

Oakman, Harry
Colourful Trees

hab.   p. 135

Eucalyptus hybrid

Blombery, Alec M.
What Wildflower is That

fl. fr.   p 139, Pl. 369

Eucalyptus hybrid

Holliday, Ivan
Eucalypts in Colour

fl.   p 19

Eucalyptus hybrid

Morcombe, M. K.
Australia's Western Wildflowers

fl.   p 80

Eucalyptus incrassata

Cochrane, G. R.
Flowers and Plants of Victoria

fl. fr.   Pl. 157

Eucalyptus johnstonii

Curtis, Winifred
The Endemic Flora of Tasmania, Part IV

fl. fr. hab.   Pl. 156

Eucalyptus kruseana

Milne, Lorus
Living Plants of the World

fl.   p 170

Eucalyptus lehmannii

Harrison, R.E.
Trees and Shrubs

fl.   P 75

Eucalyptus lehmannii

Morcombe, M. K.
Australia's Western Wildflowers

fr.   p 77

Eucalyptus lehmannii var.

Holliday, Ivan
Eucalypts in Colour

fl. hab.   p 26

Eucalyptus leucoxylon

Holliday, Ivan
Eucalypts in Colour

hab.   p 9

Eucalyptus leucoxylon

Milne, Lorus
Living Plants of the World

fl.   p 175

Eucalyptus leucoxylon

Mullins, Barbara
Australian Wildflowers in Colour

fl.   P. 59   Pl. 54

Eucalyptus leucoxylon

Perry, Frances
Flowers of the World

fl.   titlepage, p. 191

Eucalyptus leucoxylon var.

Blombery, Alec M.
What Wildflower is That

fl.   p. 135   Pl. 360

Eucalyptus leucoxylon var.

Harrison, R. E.
Trees and Shrubs

fl.   p. 75

Eucalyptus leucoxylon var.

Holliday, Ivan
Eucalypts in Colour

fl. hab.   p 10

Eucalyptus leucoxylon var.

Mathias, Mildred E.
Color for the Landscape

fl. hab.   p 49

Eucalyptus longifolia

Everett, Thomas H.
Living Trees of the World

fl.   p. 274 bottom left

Eucalyptus macrocarpa

Blombery, Alec M.
What Wildflower is That

fl.   p 136, Pl. 361

Eucalyptus macrocarpa

Edlin, Herbert
The Illustrated Encyclopedia
    of Trees

fl.   p 240

Eucalyptus macrocarpa

Holliday, Ivan
Eucalypts in Colour

fl. hab.   p 23

Eucalyptus macrocarpa

Morley, Brian D.
Wild Flowers of the World

fl.   Pl. 137

Eucalyptus macrorhyncha

Holliday, Ivan
Eucalypts in Colour

hab.   p 18

Eucalyptus maculata

Edlin, Herbert
The Illustrated Encyclopedia
    of Trees

hab.   p 241

Eucalyptus maculata

Holliday, Ivan
Eucalypts in Colour

hab.   p 4

Eucalyptus maculata

Oakman, Harry
Colourful Trees

hab.   p. 136

Eucalyptus maculosa

Oakman, Harry
Colourful Trees

hab.   p. 135

Eucalyptus maidenii

Polunin, Oleg
Trees and Bushes of Europe

fr. hab.   p 148, 149

Eucalyptus mannifera var.

Holliday, Ivan
Eucalypts in Colour

hab.   p 21

Eucalyptus melliodora

Cochrane, C. R.
Flowers and Plants of Victoria

fl.   Pl. 327

Eucalyptus microcorys

Holliday, Ivan
Eucalypts in Colour

hab.   p 3

Eucalyptus microtheca

Everett, Thomas H.
Living Trees of the World

hab.   p. 276

Eucalyptus microtheca

Holliday, Ivan
Eucalypts in Colour

hab.   p 11

Eucalyptus miniata

Blombery, Alec M.
What Wildflower is That

fl.   p 136, Pl. 362

Eucalyptus miniata

Mullins, Barbara
Australian Wildflowers in Colour

fl.   P. 61   Pl. 56

Eucalyptus morrisbyi

Curtis, Winifred
The Endemic Flora of Tasmania
    Vol VI

fl. fr.   Pl. 241

Eucalyptus niphophila

Alpine Flowers of the Kosciusko
    State Park

fl.   Pl. 16

Eucalyptus niphophila

Ray, Roy
The Color Dictionary of Flowers & Plants

hab.   p 200, Pl. 1598

Eucalyptus niphophila

Royal Hort. Soc.
Journal of the Royal Hort. Soc.
    vol 89, No. 9   1964

hab.   p. 376   Pl. 151,

Eucalyptus niphophila

Royal Hort. Soc.
Journal of the Royal Hort. Soc.
    vol 91, No. 5   1966

fl.   p. 206   Pl. 112

Eucalyptus obliqua

Holliday, Ivan
Eucalypts in Colour

hab.   p 10

Eucalyptus obliqua

Oakman, Harry
Colourful Trees

hab.   p. 135

Eucalyptus obtusiflora

Blombery, Alec M.
What Wildflower is That

fl.   p 137, Pl. 363

Eucalyptus odorata

Holliday, Ivan
Eucalypts in Colour

hab.   p 8

Eucalyptus oleosa

Holliday, Ivan
Eucalypts in Colour

fl.   p 14

Eucalyptus x orpetii

Mathias, Mildred E.
Color for the Landscape

fl.,fr.,hab.   p. 45

Eucalyptus pachyphylla

Blombery, Alec M.
What Wildflower is That

fl.   p 138, Pl. 364

Eucalyptus pachyphylla

Holliday, Ivan
Eucalypts in Colour

fl.   p 11

Eucalyptus papuana

Edlin, Herbert
The Illustrated Encyclopedia
    of Trees

hab.   p 206

Eucalyptus papuana

Everett, Thomas H.
Living Trees of the World

hab.   p. 274

Eucalyptus papuana

Holliday, Ivan
Eucalypts in Colour

hab.   p 12

Eucalyptus pauciflora

Edlin, Herbert
The Illustrated Encyclopedia
    of Trees

hab.   p 241

Eucalyptus pauciflora

Holliday, Ivan
Eucalypts in Colour

hab.   p 5

Eucalyptus pauciflora var.

Cochrane, G. R.
Flowers and Plants of Victoria

hab.   Pl. 513

Eucalyptus pauciflora var.

Holliday, Ivan
Eucalypts in Colour

hab.   p 6

Eucalyptus perriniana

Hay, Roy
The Color Dictionary of Flowers & Plants

hab.   p 200, Pl. 1599

Eucalyptus polyanthemos

Cochrane, G. R.
Flowers and Plants of Victoria

fl.   Pl. 330

Eucalyptus preissiana

American Hort. Soc.
American Horticulturist
    vol 57, No. 5   1978

fl., fr.   p. 26-27

Eucalyptus preissiana

Blombery, Alec M.
What Wildflower is That

fl. fr.   p 138, Pl. 365

Eucalyptus preissiana

Harrison, R.E.
Trees and Shrubs

fl.fr.   P 75

Eucalyptus preissiana

Morcombe, M. K.
Australia's Western Wildflowers

fl.   p 5

Eucalyptus ptychocarpa

Oakman, Harry
Colorful Trees

Fl.   P 109

Eucalyptus pulchelia

Curtis, Winifred
The Endemic Flora of Tasmania Vol. 3

fl.,fr.   p. 181

Eucalyptus pulverulenta

Curtis's Botanical Magazine
    vol 176   1966-68

fl., fr.   Pl. 525

Eucalyptus pyriformis

Holliday, Ivan
Eucalypts in Colour

fl. hab.   p 26

Eucalyptus racemosa

Macoby, Stirling
What Flower is That

fl.   P 128

Eucalyptus radiata

Cochrane, G. R.
Flowers and Plants of Victoria

hab., fl.   Pl. 345, 358

Eucalyptus regnans

Cochrane, G. R.
Flowers and Plants of Victoria

hab.   Pl. 382

Eucalyptus regnans

Holliday, Ivan
Eucalypts in Colour

hab.   titlepage

Eucalyptus resinifer

Polunin, Oleg
Trees and Bushes of Europe

fl. fr.   p 148, 150

Eucalyptus x rhodantha

American Hort. Soc.
American Horticulturist
    vol. 56, No. 4   1977

fl.   p. 15

Eucalyptus x rhodantha

Holliday, Ivan
Eucalypts in Colour

fl.   p. 27

Eucalyptus x rhodantha

Mathias, Mildred E.
Color for the Landscape

fl.,hab.   p. 43

Eucalyptus x rhodantha

Morcombe, M. K.
Australia's Western Wildflowers

fl.   p. 52, 76

Eucalyptus x rhodantha

Mullins, Barbara
Australian Wildflowers in Colour

fr.   p. 57   Pl. 51

Eucalyptus risdonii

Curtis, Winifred
The Endemic Flora of Tasmania Vol. 2

fl. fr.   Pl. 78

Eucalyptus robertsonii

Holliday, Ivan
Eucalypts in Colour

hab.   p 18

Eucalyptus robustus

Polunin, Oleg
Trees and Bushes of Europe

fl. fr. hab.   p 148, 150

Eucalyptus rodwayi

Curtis, Winifred
The Endemic Flora of Tasmania
    Vol VI

fl. fr.   Pl. 224

Eucalyptus rossii

Holliday, Ivan
Eucalypts in Colour

hab.   p 22

Eucalyptus rubida

Holliday, Ivan
Eucalypts in Colour

hab.   p 21

Eucalyptus rubida

Royal Hort. Soc.
Journal of the Royal Hort. Soc.
    vol 91, No. 5   1966

hab.   p. 206   Pl. 113

Eucalyptus rubra

Oakman, Harry
Colourful Trees

hab.   p. 136

Eucalyptus saligna

Holliday, Ivan
Eucalypts in Colour

hab.   p 22

Eucalyptus salubris

Holliday, Ivan
Eucalypts in Colour

hab.   p 15

Eucalyptus siderophloia

Holliday, Ivan
Eucalypts in Colour

hab.   p 25

Eucalyptus sideroxylon

Cochrane, G. R.
Flowers and Plants of Victoria

fl.   Pl. 329

Eucalyptus sideroxylon

Holliday, Ivan
Eucalypts in Colour

hab.   p 16

Eucalyptus sideroxylon

Mathias, Mildred E.
Color for the Landscape

fl. fr. hab.   p 46

Eucalyptus sideroxylon

Mullins, Barbara
Australian Wildflowers in Colour

fl.   P. 57   Pl. 49

Eucalyptus simmondsii

Hay, Roy
The Color Dictionary of Flowers & Plants

fl.   p 200, Pl. 1600

Eucalyptus sp.

Bianchini, Francesco
Health Plants of the World

fl.   p 101

Eucalyptus stoatei

Holliday, Ivan
Eucalypts in Colour

fl.   P. 28

Eucalyptus stricklandii

Mullins, Barbara
Australian Wildflowers in Colour

fl.   P. 59   Pl. 55

Eucalyptus subcrenulata

Curtis, Winifred
The Endemic Flora of Tasmania, Vol. 5

fl. fr.   Pl. 185

Eucalyptus tasmanica

Curtis, Winifred
The Endemic Flora of Tasmania Vol. 1

fl. fr.   Pl. 1

Eucalyptus tasmanica

Curtis, Winifred
The Endemic Flora of Tasmania Vol. 2

fl.   Pl. 48

Eucalyptus terminalis

Blombery, Alec M.
What Wildflower is That

fl.   p 138, Pl. 366

Eucalyptus tessalaris

Holliday, Ivan
Eucalypts in Colour

hab.   p. 15

Eucalyptus tessalaris

Oakman, Harry
Colourful Trees

hab.   p. 136

Eucalyptus tetraptera

American Hort. Soc.
American Horticulturist
vol 57, No. 5   1978

fl.   p. 27

Eucalyptus tetraptera

Blombery, Alec M.
What Wildflower is That

fl.   p 138, Pl. 367

Eucalyptus tetraptera

Holliday, Ivan
Eucalyptus in Colour

fl. hab.   p 28

Eucalyptus tetraptera

Mullins, Barbara
Australian Wildflowers in Colour

fl.   p. 59   Pl. 53

Eucalyptus torquata

Blombery, Alec M.
What Wildflower is That

fl.   p 139, Pl. 368

Eucalyptus torquata

Morcombe, M. K.
Australia's Western Wildflowers

fl.   p 80

Eucalyptus transcontinentalis

Holliday, Ivan
Eucalypts in Colour

fl.   p 31

Eucalyptus urnigera

Curtis, Winifred
The Endemic Flora of Tasmania
                vol 3
fl.   fr.   p 205

Eucalyptus urnigera

Curtis's Botanical Magazine
Vol 177   1969-70

fl. fr.   536

Eucalyptus vernicosa

Curtis, Winifred
The Endemic Flora of Tasmania, Vol. 5

fl. fr.   Pl. 195

Eucalyptus viminalis

Edlin, Herbert
The Illustrated Encyclopedia
  of Trees

hab.   p 238-9

Eucalyptus viminalis

Holliday, Ivan
Eucalypts in Colour

hab.   p 7

Eucalyptus viminalis

Polunin, Oleg
Trees and Bushes of Europe

fl. fr. hab.   p 148, 149

Eucalyptus woodwardii

Morcombe, M. K.
Australia's Western Wildflowers

fl.   p 80

Eucalyptus youngiana

Blombery, Alec M.
What Wildflower is That

fl.   p 139, Pl. 370

Eucalyptus youngiana

Holliday, Ivan
Eucalypts in Colour

fl.   p 14

Eucharidium breweri var.

Royal Hort. Soc.
The Garden
  vol 101, No. 10   1976

fl.   p. 518

Eucharidium concinnum var.

Hay, Roy
The Color Dictionary of Flowers &
  Plants

fl.   P. 37   Pl. 295

Eucharis amazonica

Bruggeman, L.
Tropical Plants

fl.   Pl. 109

Eucharis grandiflora

Hay, Roy
The Color Dictionary of Flowers & Plants

fl.   p 66, Pl. 525

Eucharis grandiflora

Hay, Roy
The Dictionary of House Plants

fl.   Pl. 220

Eucharis grandiflora

Hersey, Jean
Woman's Day Book of House Plants

fl.   p 28

Eucharis grandiflora

Kromdijk, G.
200 House Plants in Colour

fl.   Pl. 82

Eucharis grandiflora

Miles, Bebe
Bulbs for the Home Gardener

fl. hab.   p 163

Eucharis grandiflora

Perry, Frances
Flowers of the World

fl.   p 26

Eucharis grandiflora

Tsukamoto, Yotaro
Coloured Illustrations of Garden
    Flowers   Vol. 9

fl.   opp. p. 46

Euchresta japonica

Kitamura, Siro
Coloured Illustrations of Trees &
    Shrubs of Japan

fr.   250

Eucnide urens

Munz, Philip A.
California Desert Wildflowers

fl.   p 38, Pl. 35

Eucomis bicolor

Hay, Roy
The Color Dictionary of Flowers & Plants

fl.   p 92, Pl. 733

Eucomis bicolor

Miles, Bebe
Bulbs for the Home Gardener

fl.   p 165

Eucomis bicolor

Morley, Brian D.
Wild Flowers of the World

fl.   Pl. 85

Eucomis comosa

Hay, Roy
The Dictionary of House Plants

fl.   Pl. 221

Eucomis comosa

Macoby, Stirling
What Flower is That

fl.   P 128

Eucomis humilis

Flowering Plants of South Africa
    vol XXIV   1944

fl.   Pl. 954

Eucomis pallidiflora

Curtis's Botanical Magazine
Vol 177   1969-70

fl.   561

Eucomis pole-evansii

Flowering Plants of South Africa
    vol XXIV   1944

fl.   Pl. 953

Eucomis punctata

Royal Hort. Soc.
Journal of the Royal Hort. Soc.
    vol 97, No. 4   1972

fl.   p. 170   Pl. 77

Eucomis punctata

Royal Hort. Soc.
The Garden
    vol 101, No. 7   1976

fl.   p. 361

Eucomis punctata var.

Royal Hort. Soc.
Journal of the Royal Hort. Soc.
    vol 97, No. 8   1972

fl.   p. 352   Pl. 188

Eucomis undulata

Moriarty, Audrey
Wild Flowers of Malawi

fl.   Pl. 11; 1

Eucomis vandermerwei

Flowering Plants of South Africa
    vol XXIV   1944

fl.   Pl. 955

Eucomis zambesiaca

American Hort. Soc.
American Horticulturist
    vol 53, No. 3   1974

fl.   p. 33

Eucommia ulmoides

Kariyone, Tatsuo
Atlas of Medicinal Plants

fr.   Pl. 142

Eucryphia cordifolia

Munos Pizarro, Carlos
Flores Silvestres de Chile

fl.   Pl. 41

Eucryphia glutinosa

Hay, Roy
The Color Dictionary of Flowers & Plants

fl.   p 201, Pl. 1601

Eucryphia glutinosa

Hellyer, A.G.L.
Shrubs in Colour

fl.   p. 50

Eucryphia glutinosa

Munos Pizarro, Carlos
Flores Silvestres de Chile

fl.   Pl. 32

Eucryphia glutinosa

Perry, Frances
Flowers of the World

fl.   p 113

Eucryphia glutinosa

Royal Hort. Soc.
The Garden
    vol 100, No. 7   1975

fl.   p. 286

Eucryphia glutinosa var.

Gault, S. Millar
The Color Dictionary of Shrubs

fl.   Pl. 184

Eucryphia x intermedia

Curtis's Botanical Magazine
Vol. 177   1969-70

fl.   534

Eucryphia x intermedia var.

Gault, S. Millar
The Color Dictionary of Shrubs

fl.   Pl. 185

Eucryphia lucida

Blombery, Alec M.
What Wildflower is That

fl.　　　　　p 140, Pl. 371

Eucryphia lucida

Curtis, Winifred
The Endemic Flora of Tasmania Vol. 1

fl. fr.　　　　　Pl. 33

Eucryphia milliganii

Curtis, Winifred
The Endemic Flora of Tasmania Vol. 1

fl. fr.　　　　　Pl. 34

Eucryphia x nymansensis

Hellyer, A.G.L.
Shrubs in Colour

fl.　　　　　p. 50

Eucryphia x nymansensis

Royal Hort. Soc.
Journal of the Royal Hort. Soc.
　　vol 91, No. 1　1966

fl., hab.　　p. 54　Pl. 29

Eucryphia x nymansensis

Royal Hort. Soc.
Journal of the Royal Hort. Soc.
　　vol 97, No. 7　1972

fl., hab.　　p. 304　Pl. 158

Eucryphia x nymansensis

Royal Hort. Soc.
Journal of the Royal Hort. Soc.
　　vol. 99, No. 8　1974

fl.　　p. 346　　Pl. 172

Eucryphia x nymansensis

Royal Hort. Soc.
The Garden
　　vol 103, No. 8　1978

fl.　　p. 335

Eucryphia x nymansensis var.

Hay, Roy
The Color Dictionary of Flowers &
　Plants

fl.　　p. 201　　Pl. 1602

Eugenia aromatica

Hvass, Elsie
Plants That Feed and Serve Us

fl.　p 73, Pl. 151

Eugenia aromatica

Loewenfeld, Claire
The Complete Book of Herbs and Spices

fl. fr. hab.　　p. 208

Eugenia australis

Harrison, R.E.
Trees and Shrubs

fr.　P 74

Eugenia australis

Macoby, Stirling
What Flower is That

fr.　P 129

Eugenia caryophyllata

Bianchini, F.
The Complete Book of Fruits & Vegetables

fr.　　　　p 215

Eugenia caryophyllata

Bianchini, Francesco
Health Plants of the World

fr.　　　　p 117

Eugenia caryophyllata

Edlin, Herbert
The Illustrated Encyclopedia
　of Trees

fl. hab.　　p. 224

Eugenia caryophyllata

Masefield, G. B.
The Oxford Book of Food Plants

fl.　　　　Pl. 133, 6

Eugenia caryophyllata

Perrot, Emile
Les Plantes Medicinales

fl. fr. hab.　　p 157

Eugenia cyanocarpa

Curtis's Botanical Magazine
　Vol 163　　1940-42

fl. fr.　　No. 9602

Eugenia gracilipes

Morley, Brian D.
Wild Flowers of the World

fl. fr.　　pl. 147

Eugenia luehmannii

Macoby, Stirling
What Flower is That

fr.　　p. 129

Eugenia luehmannii

Oakman, Harry
Colorful Trees

Hab. Fl. Fr.　P 76

Eugenia malaccensis

Everett, Thomas H.
Living Trees of the World

fl.　p. 274

Eugenia malaccensis

Macoby, Stirling
What Flower is That

fl.　P 129

Eugenia malaccensis

Pertchik, Bernard
Flowering Trees of the Caribbean

fl.　　　　p 15

Eugenia myrtifolia

Royal Hort. Soc.
Journal of the Royal Hort. Soc.
vol 100, No. 1　1975

fr.　　p. 25　Pl. 15

Eugenia ramosissima

Hara, Hiroshi, comp.
Photo-Album of Plants of Eastern
　Himalaya

fl.　　　Pl. 6

Eugenia smithii

Harrison, R.E.
Trees and Shrubs

fr.　P 74

Eulalia

see

Miscanthus sinensis var.

Eulophia acutilabra

Williamson, Graham
The Orchids of South Central Africa

fl.　　　　Pl. 142

Eulophia alta

Luer, Carlyle A.
The Native Orchids of Florida

fl. fr. hab.　　Pl. 72

Eulophia alta

Ospina, Mariano
Orquideas de las americas

fl.　　　　Pl. 141

Eulophia alta

Pabst, G. F. J.
Orchidaceae Brasilienses.
　Vol 1

fl.　　　p 239

Eulophia angolensis var.

Williamson, Graham
The Orchids of South Central Africa

fl.   Pl. 152-153

Eulophia antennata

Flowering Plants of South Africa
vol XIX   1939

fl., hab.   Pl. 728

Eulophia antennisepala

Williamson, Graham
The Orchids of South Central Africa

fl.   Pl. 150

Eulophia arenicola

Williamson, Graham
The Orchids of South Central Africa

fl.   Pl. 146

Eulophia aurantiaca

Williamson, Graham
The Orchids of South Central Africa

fl.   Pl. 131

Eulophia calantha

Williamson, Graham
The Orchids of South Central Africa

fl.   Pl. 147

Eulophia carunculifera

Flowering Plants of Africa
vol XXV   1945-46

fl.   Pl. 966

Eulophia chrysops

Williamson, Graham
The Orchids of South Central Africa

fl.   Pl. 163

Eulophia clitellifera

Flowering Plants of Africa
vol 31   1956

fl., hab.   Pl. 1235

Eulophia coeloglossa

Moriarty, Audrey
Wild Flowers of Malawi

fl.   Pl. 21; 3

Eulophia coeloglossa

Williamson, Graham
The Orchids of South Central Africa

fl.   Pl. 154

Eulophia complanata

Flowering Plants of Africa
vol 27   1948-49

fl.   Pl. 1056

Eulophia complanata

Moriarty, Audrey
Wild Flowers of Malawi

fl.   Pl. 23; 2

Eulophia cristata

Royal Hort. Soc.
Journal of the Royal Hort. Soc.
vol 92, No. 5   1967

fl.   p. 206   Pl. 111

Eulophia cucullata

Flowering Plants of Africa
vol 30   1954-55

fl., hab.   Pl. 1171

Eulophia cucullata

Lind, E.M.
Some Common Flowering Plants of
Uganda

fl.   Pl. 5

Eulophia cucullata

Moriarty, Audrey
Wild Flowers of Malawi

fl.   Pl. 22; 5

Eulophia cucullata

Williamson, Graham
The Orchids of South Central Africa

fl.   Pl. 155

Eulophia dissimilis

Flowering Plants of Africa
vol 27   1948-49

fl.   Pl. 1066

Eulophia ecristata

Luer, Carlyle A.
The Native Orchids of Florida

fl. fr. hab.   Pl. 73

Eulophia euantha

Moriarty, Audrey
Wild Flowers of Malawi

fl.   Pl. 21; 1

Eulophia euantha

Williamson, Graham
The Orchids of South Central Africa

fl.   Pl. 129

Eulophia flava

Walden, Beryl M.
Wild Flowers of Hong Kong

fl.   Pl. 37

Eulophia flavopurpurea

Williamson, Graham
The Orchids of South Central Africa

fl.   Pl. 148

Eulophia gonychila

Williamson, Graham
The Orchids of South Central Africa

fl.   Pl. 161

Eulophia hereroensis

Flowering Plants of Africa
vol 38   1967

fl., hab.   Pl. 1518

Eulophia hians

Flowering Plants of Africa
vol 29   1952-53

fl., fr., hab.   Pl. 1132

Eulophia ischna

Williamson, Graham
The Orchids of South Central Africa

fl.   Pl. 134

Eulophia katangensis

Williamson, Graham
The Orchids of South Central Africa

fl.   Pl. 139

Eulophia kirkii

Moriarty, Audrey
Wild Flowers of Malawi

fl.   Pl. 23; 1

Eulophia latilabris

Williamson, Graham
The Orchids of South Central Africa

fl.   Pl. 157

Eulophia litoralis

Flowering Plants of Africa
vol 30   1954-55

fl., hab.   Pl. 1176

Eulophia livingstoniana

Moriarty, Audrey
Wild Flowers of Malawi

fl.   Pl. 22; 1

Eulophia livingstoniana

Williamson, Graham
The Orchids of South Central Africa

fl.   Pl. 138

Eulophia mackenii

Flowering Plants of Africa
    vol XXVI   1947

fl., hab.   Pl. 1020

Eulophia macrantha

Moriarty, Audrey
Wild Flowers of Malawi

fl.   Pl. 23; 4

Eulophia monotropis

Williamson, Graham
The Orchids of South Central Africa

fl.   Pl. 141

Eulophia nunbwaensis

Williamson, Graham
The Orchids of South Central Africa

fl.   Pl. 140

Eulophia nuttii

Williamson, Graham
The Orchids of South Central Africa

fl.   Pl. 130

Eulophia orthoplectra

Moriarty, Audrey
Wild Flowers of Malawi

fl.   Pl. 21; 5

Eulophia orthoplectra

Williamson, Graham
The Orchids of South Central Africa

fl.   Pl. 162

Eulophia paivaeana

Moriarty, Audrey
Wild Flowers of Malawi

fl.   Pl. 22; 3

Eulophia paivaeana

Williamson, Graham
The Orchids of South Central Africa

fl.   Pl. 159

Eulophia paivaeana var.

Flowering Plants of Africa
    vol 38   1967

fl.   Pl. 1490

Eulophia papillosa

Flowering Plants of South Africa
    vol XXIV   1944

fl.   Pl. 949

Eulophia paradoxa

Williamson, Graham
The Orchids of South Central Africa

fl.   Pl. 137

Eulophia petersii

Stewart, Joyce
Orchids of Tropical Africa

fl.   Pl. 26

Eulophia porphyroglossa

Stewart, Joyce
Orchids of Tropical Africa

fl.   Pl. 27

Eulophia quartiniana

Curtis's Botanical Magazine
Vol 171   1956-57

fl.   267

Eulophia quartiniana

Morley, Brian D.
Wild Flowers of the West

fl., hab.   **Pl. 69**

Eulophia quartiniana

Perry, Frances
Flowers of the World

fl.   p 212

Eulophia quartiniana

Stewart, Joyce
Orchids of Tropical Africa

fl.   Pl. 28

Eulophia quartiniana

Williamson, Graham
The Orchids of South Central Africa

fl.   Pl. 127

Eulophia rhodesiaca

Williamson, Graham
The Orchids of South Central Africa

fl.   Pl. 143

Eulophia schnelliae

Flowering Plants of Africa
    vol XXV   1945-46

fl., hab.   Pl. 965

Eulophia siaensis

Walden, Beryl M.
Wild Flowers of Hong Kong

fl.   Pl. 38(125)

Eulophia sp.

Moriarty, Audrey
Wild Flowers of Malawi

fl.   Pl. 21; 2

Eulophia sp.

Williamson, Graham
The Orchids of South Central Africa

fl.   Pl. 136, 139
                144-45, 151

Eulophia speciosa

Moriarty, Audrey
Wild Flowers of Malawi

fl.   Pl. 22; 4

Eulophia subulata

Lind, E.M.
Some Common Flowering Plants of
    Uganda

fl.   Pl. 13b

Eulophia tabularis

Flowering Plants of Africa
    vol 38   1967

fl.   Pl. 1491

Eulophia tanganyikensis

Williamson, Graham
The Orchids of South Central Africa

fl.   Pl. 149

Eulophia thomsonii

Moriarty, Audrey
Wild Flowers of Malawi

fl.   Pl. 21; 6

Eulophia thomsonii

Williamson, Graham
The Orchids of South Central Africa

fl.   Pl. 128

Eulophia tristriata

Williamson, Graham
The Orchids of South Central Africa

fl.   Pl. 156

Eulophia tuberculata

Williamson, Graham
The Orchids of South Central Africa

fl.   Pl. 160

Eulophia venosa

Cady, Leo
Australian Native Orchids in Colour

fl.   Pl. 71

Eulophia walleri

Moriarty, Audrey
Wild Flowers of Malawi

fl.   Pl. 23; 3

Eulophia walleri

Williamson, Graham
The Orchids of South Central Africa

fl.   Pl. 133

Eulophia warneckeana

Williamson, Graham
The Orchids of South Central Africa

fl.   Pl. 135

Eulophia watkinsonii

Flowering Plants of Africa
     vol 29   1952-53

fl., hab.   Pl. 1127

Eulophia welwitschii

Williamson, Graham
The Orchids of South Central Africa

fl.   Pl. 132

Eulophia, Yellow

    see

Eulophia flava

Eulophia zeyheri

Moriarty, Audrey
Wild Flowers of Malawi

fl.   Pl. 22; 2

Eulophidium maculatum

Ospina, Mariano
Orquideas de las americas

fl.   Pl. 142

Eulophidium maculatum

Pabst, G. F. J.
Orchidaceae Brasilienses,
Vol 1

fl., hab.   p 239

Eulophidium maculatum

Williamson, Graham
The Orchids of South Central Africa

fl.   Pl. 126

Eulophidium saundersianum

Williamson, Graham
The Orchids of South Central Africa

fl.   Pl. 125

Eulophiella elizabethae

Curtis's Botanical Magazine
Vol 179   1972

fl.   656

Eulychnia saint-pieana

Lamb, Edgar
Popular Exotic Cacti in Color

hab.   p 85

Eumorphia sericea

Curtis's Botanical Magazine
Vol 178   1970-72

fl.   608

Euonymus alatus

American Hort. Soc.
American Horticulturist
     vol 56, No. 2 1977

Hab.   p. 30

Euonymus alatus

Everett, T.H., ed.
New Illustrated Encyclopedia of
     Gardening   vol 4

hab.   p. 646

Euonymus alatus

Hansen, Richard
Baume und Straucher in Garten

hab.   p. 136

Euonymus alatus

Hay, Roy
The Color Dictionary of Flowers & Plants

hab.   p 201, Pl. 1603

Euonymus alatus

Hellyer, A.G.L.
Shrubs in Colour

hab.   p. 51

Euonymus alatus

Kimura, Koiti
Japanese Medicinal Plants, Vol I

fl.   p 52

Euonymus alatus var.

Kitamura, Siro
Coloured Illustrations of Trees &
     Shrubs of Japan

fr.   296, 297

Euonymus alatus var.

Massachusetts Hort. Soc.
Horticulture
     vol 35, No. 2 1957

hab.   p. 75

Euonymus americanus

Batson, Wade T.
Wild Flowers in South Carolina

fr.   p 72

Euonymus americanus

Campbell, Carlos C.
Great Smoky Mountain Wildflowers

fr.   p 103

Euonymus americanus

Wharton, Mary E.
Trees & Shrubs of Kentucky

fl.   p 96, Pl. 2.32b
fr.   p 126, Pl. 2.12a

Euonymus atropurpureus

Perrot, Emile
Les Plantes Medicinales

fl. fr. hab.   p 101

Euonymus atropurpureus

Wharton, Mary E.
Trees & Shrubs of Kentucky

fl.   p 96, Pl. 2.32a
fr.   p 126, Pl. 2.12c

Euonymus bungeanus

Addisonia
     vol 24   1960-64

fl., fr.   p. 15   Pl. 776

Euonymus chibai

Kitamura, Siro
Coloured Illustrations of Trees &
     Shrubs of Japan

fr.   300

Euonymus europaeus

Ary, S.
The Oxford Book of Wildflowers

fl.   p 180, Pl. 4

Euonymus europaeus

Bianchini, Francesco
Health Plants of the World

fr.   p 73

Euonymus europaeus

Gault, S. Millar
The Color Dictionary of Shrubs

fr.   Pl. 186

Euonymus europaeus

Huxley, Anthony
Deciduous Garden Trees and Shrubs

fl.   Pl. 71

Euonymus europaeus

Lindman, C. A. M.
Nordens Flora, Vol 6

fl. fr.   Pl. 379

Euonymus europaeus

Martin, W. Keble
The Concise British Flora in Colour

fl. fr. hab.   Pl. 20

Euonymus europaeus

Meikle, R.D.
British Trees and Shrubs

fl. fr.   Pl. 2

Euonymus europaeus

Perrot, Emile
Les Plantes Medicinales

fl. fr. hab.   p 101

Euonymus europaeus

Perry, Frances
Flowers of the World

fr.   p 73

Euonymus europaeus

Polunin, Oleg
Flowers of Europe

fl.   Pl. 71 #718

Euonymus Europaeus

Polunin, Oleg
Trees and Bushes of Europe

fl. fr.   p 130

Euonymus europaeus

Vedel, H.
Arbres et Arbustes

fl. fr. Illus. 94   p. 91

Euonymus europaeus var.

Bartels, Andreas
Das Grosse Buch der Gartengehölze

fl.   p 149

Euonymus europaeus var.

Harrison, R.E.
Trees and Shrubs

fr.   P 76

Euonymus europaeus var.

Hay, Roy
The Color Dictionary of Flowers & Plants

fr.   p 201, Pl. 1604

Euonymus europaeus var.

Hellyer, A.G.L.
Shrubs in Colour

fr.   p. 51

Euonymus fortunei

American Hort. Soc.
American Horticulturist
vol 57, No. 3   1978

hab.   p. 31

Euonymus fortunei var.

Crockett, James Underwood
Evergreens.

hab.   p. 124

Euonymus fortunei var.

Crockett, James Underwood
Lawns & Ground Covers

hab.   p. 131

Euonymus fortunei var.

Gault, S. Millar
The Color Dictionary of Shrubs

hab.   Pl. 187, 188, 189

Euonymus fortunei var.

Hellyer, A.G.L.
Shrubs in Colour

hab.   p. 51

Euonymus fortunei var.

Huxley, Anthony
Evergreen Garden Trees and Shrubs

hab.   Pl. 120, 121, 122

Euonymus fortunei var.

Kitamura, Siro
Coloured Illustrations of Trees &
   Shrubs of Japan

fr.   299

Euonymus frigidus

Curtis's Botanical Magazine
Vol 168   1951

fl.   161

Euonymus hamiltonianus

Curtis's Botanical Magazine
Vol 169   1952-53

fl.   181

Euonymus japonicus

Kiaer, Eigil
Indoor Plants in Colour

hab.   p. 56

Euonymus japonicus

Kitamura, Siro
Coloured Illustrations of Trees &
   Shrubs of Japan

fr.   298

Euonymus japonicus

Kromdijk, G.
200 House Plants in Colour

hab.   Pl. 83

Euonymus japonicus

Polunin, Oleg
Trees and Bushes of Europe

fl. fr. hab.   p 131

Euonymus japonicus var.

Gault, S. Millar
The Color Dictionary of Shrubs

hab.   Pl. 190

Euonymus japonicus var.

Harrison, R.E.
Trees and Shrubs

hab.   P 76

Euonymus japonicus var.

Hersey, Jean
Woman's Day Book of House Plants

hab.   p. 52

Euonymus latifolius

Polunin, Oleg
Trees and Bushes of Europe

fl. fr.   p 130

Euonymus laxiflora

Walden, Beryl M.
Wild Flowers of Hong Kong

fl.   Pl. 39(123)

Euonymus maackii

Wit, H. C. D. de
Plants of the World;
   The Higher Plants, Vol II

fr.   p 64, Pl. 41

Euonymus myrianthus

Curtis's Botanical Magazine
Vol 166   1949

fl. fr.   64

Euonymus obovatus

Wharton, Mary E.
A Guide to the Wildflowers & Ferns
of Kentucky

fl.   fr.   p 173   Pl. 2.54

Euonymus obovatus

Wharton, Mary E.
Trees & Shrubs of Kentucky

fl.   p 98, Pl. 2.32c
fr.   p 126, Pl. 2.12b

Euonymus oxyphyllus

Kitamura, Siro
Coloured Illustrations of Trees &
Shrubs of Japan

fl.   303

Euonymus radicans

Everett, T.H., ed.
New Illustrated Encyclopedia of
Gardening   vol 13

hab.   p. 2407

Euonymus radicans var.

Kroadijk, G.
200 House Plants in Colour

hab.   Pl. 84

Euonymus sachalinensis

Hansen, Richard
Baume und Straucher im Garten

fr.   p. 136

Euonymus sachalinensis

Hay, Roy
The Color Dictionary of Flowers & Plants

fl.   p 201, Pl. 1605

Euonymus sachalinensis

Morley, Brian D.
Wild Flowers of the World

fr.   Plate 101

Euonymus sieboldianus

Curtis's Botanical Magazine
Vol 177   1969-70

fl.   548

Euonymus sieboldianus

Kimura, Koiti
Japanese Medicinal Plants, Vol II

fl.   p.195

Euonymus sieboldianus

Kitamura, Siro
Coloured Illustrations of Trees &
Shrubs of Japan

fr.   302

Euonymus, Star

see

Euonymus laxiflora

Euonymus tanakae

Kitamura, Siro
Coloured Illustrations of Trees &
Shrubs of Japan

fl.   301

Euonymus verrucosus

Polunin, Oleg
Trees and Bushes of Europe

fl. fr.   p 131

Euonymus yedoensis

Hay, Roy
The Color Dictionary of Flowers & Plants

fr.   p 201, Pl. 1606

Eupatorium album

Brown, Clair A.
Wildflowers of Louisiana and
Adjoining States

fr.   p 199

Eupatorium atrorubens

Hay, Roy
The Dictionary of House Plants

fl.   Pl. 222

Eupatorium cannabinum

Ary, S.
The Oxford Book of Wildflowers

fl.   p. 154   Pl. 4

Eupatorium cannabinum

Bianchini, Francesco
Health Plants of the World

fl. hab.   p 35

Eupatorium cannabinum

Lindman, C. A. M.
Nordens Flora, Vol 9

fl. fr. hab.   Pl. 574

Eupatorium cannabinum

Martin, W. Keble
The Concise British Flora in Colour

fl. fr.   Pl. 24

Eupatorium cannabinum

Polunin, Oleg
Flowers of Europe

fl.   Pl. 141 #1357

Eupatorium capillifolium

Brown, Clair A.
Wildflowers of Louisiana and
Adjoining States

hab.   p 202

Eupatorium capillifolium

Weeds of the Southern United States

hab., fr.   p. 12

Eupatorium coelestinum

Brown, Clair A.
Wildflowers of Louisiana and
Adjoining States

fl. hab.   p 199

Eupatorium coelestinum

Dean, Blanche E.
Wildflowers of Alabama and
Adjoining States

fl.   p 187

Eupatorium coelestinum

Klimas, John E.
Wildflowers of Eastern America

fl.   Pl. 268

Eupatorium coelestinum

Massachusetts Hort. Soc.
Horticulture
vol 53, No. 9   1975

fl.   p. 32

Eupatorium coelestinum

Wharton, Mary E.
A Guide to the Wildflowers & Ferns
of Kentucky

fl.   p 236   Pl. 1.8

Eupatorium compositifolium

Batson, Wade T.
Wild Flowers in South Carolina

hab.   p 110

Eupatorium compositifolium

Duncan, Wilbur H.
Wildflowers of the Southeastern
United States

fl. hab.   p 195

Eupatorium dubium

Pond, Barbara
A Sampler of Wayside Herbs

fl.   Pl. XXIV

Eupatorium fistulosum

Brown, Clair A.
Wildflowers of Louisiana and
Adjoining States

fl.   p 200

Eupatorium fistulosum

Duncan, Wilbur H.
Wildflowers of the Southeastern
United States

fl.   p 197

Eupatorium fistulosum

Jennings, O. E.
Wild Flowers of Western Pennsylvania
                              vol II
fl.   Pl. 155

Eupatorium fistulosum

Klimas, John E.
Wildflowers of Eastern America

fl.   Pl. 267

Eupatorium fistulosum

Wharton, Mary E.
A Guide to the Wildflowers & Ferns
    of Kentucky

fl.   p 236   Pl. 1.9

Eupatorium fortunei

Kimura, Koiti
Japanese Medicinal Plants, Vol II

fl.   p 239

Eupatorium incarnatum

Duncan, Wilbur H.
Wildflowers of the Southeastern
United States

fl.   p 195

Eupatorium lindleyanum

Walden, Baryl M.
Wild Flowers of Hong Kong

fl   Pl. 45 (131)

Eupatorium maculatum

Campbell, Carlos C.
Great Smoky Mountain Wildflowers

fl.   p 99

Eupatorium maculatum

Courtenay, Booth
Wildflowers & Weeds

fl.   p 110

Eupatorium maculatum

Moyle, John B.
Northland Wild Flowers

fl. hab.   p 185, Pl. 233

Eupatorium perfoliatum

Brown, Clair A.
Wildflowers of Louisiana and
    Adjoining States

fl. hab.   p 200

Eupatorium perfoliatum

Courtenay, Booth
Wildflowers & Weeds

fl.   p 110

Eupatorium perfoliatum

Dean, Blanche E.
Wildflowers of Alabama and
    Adjoining States

fl.   p 187

Eupatorium perfoliatum

Jennings, O. E.
Wild Flowers of Western Pennsylvania
                              vol II
fl.   Pl. 156

Eupatorium perfoliatum

Klimas, John E.
Wildflowers of Eastern America

fl.   Pl. 107

Eupatorium perfoliatum

Moyle, John B.
Northland Wild Flowers

fl. hab.   p 186, Pl. 235

Eupatorium perfoliatum

Pond, Barbara
A Sampler of Wayside Herbs

fl.   Pl. XXIV

Eupatorium perfoliatum

Wharton, Mary E.
A Guide to the Wildflowers & Ferns
    of Kentucky

fl.   p 235   Pl. 1.6

Eupatorium perfoliatum var.

Jennings, O. E.
Wild Flowers of Western Pennsylvania
                              vol II
fl.   Pl. 156

Eupatorium purpureum

Batson, Wade T.
Wild Flowers in South Carolina

fl.   p 111

Eupatorium purpureum

Dean, Blanche E.
Wildflowers of Alabama and
    Adjoining States

fl.   p 187

Eupatorium purpureum

Hay, Roy
The Color Dictionary of Flowers & Plants

fl.   p 141, Pl. 1128

Eupatorium purpureum

Huxley, Anthony
Garden Perennials and Water Plants

fl.   pl. 117

Eupatorium purpureum

Lemmon, Robert S.
Wildflowers of North America in
    Full Color

fl.   pg. 205   pl. 324

Eupatorium purpureum

Moyle, John B.
Northland Wild Flowers

fl. hab.   p 185, Pl. 234

Eupatorium rotundifolium

Brown, Clair A.
Wildflowers of Louisiana and
    Adjoining States

fl. hab.   p 201

Eupatorium rotundifolium

Duncan, Wilbur H.
Wildflowers of the Southeastern
    United States

fl.   p 195

Eupatorium rugosum

Brown, Clair A.
Wildflowers of Louisiana and
    Adjoining States

fl.   p 201

Eupatorium rugosum

Courtenay, Booth
Wildflowers & Weeds

fl.   p 110

Eupatorium rugosum

Dean, Blanche E.
Wildflowers of Alabama and
    Adjoining States

fl.   p 187

Eupatorium rugosum

Duncan, Wilbur H.
Wildflowers of the Southeastern
    United States

fl.   p 195

Eupatorium rugosum

Jennings, O. E.
Wild Flowers of Western Pennsylvania
                              vol II
fl.   Pl. 157

Eupatorium rugosum

Klimas, John E.
Wildflowers of Eastern America

fl.   Pl. 82

Eupatorium rugosum

Moyle, John B.
Northland Wild Flowers

fl. hab.   p 186, Pl. 236

Eupatorium rugosum

Wharton, Mary E.
A Guide to the Wildflowers & Ferns
    of Kentucky

fl.   p 234   Pl. 1.4

Eupatorium rugosum var.

Campbell, Carlos C.
Great Smoky Mountain Wildflowers

fl.   p 101

Eupatorium serotinum

Wharton, Mary E.
A Guide to the Wildflowers & Ferns
    of Kentucky

fl.   p 235   Pl. 1.5

Eupatorium sessilifolium

Wharton, Mary E.
A Guide to the Wildflowers & Ferns
    of Kentucky

fl.   p 235   Pl. 1.7

Eupatorium sordidum

Macoby, Stirling
What Flower is That

fl.   P 129

Euphorbia acanthothamnos

Polunin, Oleg
Flowers of Europe

fl.   Pl. 66 #670

Euphorbia acanthothamnos

Polunin, Oleg
Flowers of the Mediterranean

fr.   No. 89

Euphorbia acanthothamnos

Royal Hort. Soc.
Journal of the Royal Hort. Soc.
    Vol. 88, No. 9   1963

hab.   p. 388   Pl. 144

Euphorbia ammak

Edlin, Herbert
The Illustrated Encyclopedia
    of Trees

hab.   p 230

Euphorbia amygdaloides

Ary, S.
The Oxford Book of Wildflowers

fl.   p 62, Pl. 2

Euphorbia amygdaloides

Martin, W. Keble
The Concise British Flora in Colour

fl. fr. hab.   Pl. 25

Euphorbia amygdaloides

Polunin, Oleg
Flowers of Europe

fl.   Pl. 66 #677

Euphorbia anoplia

Lamb, Edgar
The Pocket Encyclopedia of Cacti
    and Succulents in Color

fl. hab.   Pl. 253

Euphorbia aphylla

Bramwell, David
Wild Flowers of the Canary Islands

hab., fl.   Pl. 185

Euphorbia ballyi

Flowering Plants of Africa
    vol 36   1963-64

fl., hab.   Pl. 1408

Euphorbia balsamifera

Bramwell, David
Wild Flowers of the Canary Islands

hab.   Pl. 184

Euphorbia balsamifera

Lamb, Edgar
The Pocket Encyclopedia of Cacti
    and Succulents in Color

hab.   Pl. 247

Euphorbia barnardii

Flowering Plants of South Africa
    vol XXII   1942

fl.   Pl. 877

Euphorbia bevilaniensis

Lamb, Edgar
Popular Exotic Cacti in Color

fl. hab.   p 87

Euphorbia bicolor

Brown, Clair A.
Wildflowers of Louisiana and
    Adjoining States

fl. hab.   p 101

Euphorbia biglandulosa

Polunin, Oleg
Flowers of the Mediterranean

fl.   No. 92

Euphorbia bourgaeana

Bramwell, David
Wild Flowers of the Canary Islands

fl.   Pl. 188

Euphorbia bravoana

Bramwell, David
Wild Flowers of the Canary Islands

fl.   Pl. 190

Euphorbia breoni

Lamb, Edgar
The Pocket Encyclopedia of Cacti
    and Succulents in Color

fl. hab.   Pl. 249, 321

Euphorbia brevitorta

Flowering Plants of Africa
    vol 33   1959

fl., hab.   Pl. 1288

Euphorbia bupleurifolia

Flowering Plants of South Africa
    vol XVII   1937

fl., hab.   Pl. 650

Euphorbia bupleurifolia

Morley, Brian D.
Wild Flowers of the World

fl., hab.   Pl. 58

Euphorbia canariensis

Bramwell, David
Wild Flowers of the Canary Islands

hab.   Pl. 180, 181

Euphorbia canariensis

Lamb, Edgar
The Pocket Encyclopedia of Cacti
    and Succulents in Color

fl. hab.   Pl. 318

Euphorbia caput-medusae

Macoby, Stirling
What Flower is That

hab.   P. 130

Euphorbia characias

Gault, S. Millar
The Color Dictionary of Shrubs

fl.   Pl. 191

Euphorbia characias

Hay, Roy
The Color Dictionary of Flowers & Plants

fl.   p 142, Pl. 1129

Euphorbia characias

Kleijn, H.
Beauty of the Wild Plant

fl.   p. 121   Pl. 179

Euphorbia characias

Polunin, Oleg
Flowers of Europe

fl.   Pl. 66 #678

Euphorbia characias

Polunin, Oleg
Flowers of the Mediterranean

fl.   No. 93

Euphorbia characias var.

Goulimis, Constantine N.
Wild Flowers of Greece

fl.   p 57

Euphorbia characias var.

Polunin, Oleg
Flowers of Europe

fl.   Pl. 66 #678

Euphorbia clavarioides

Flowering Plants of Africa
vol 33   1959

fl.   Pl. 1307

Euphorbia coerulescens

Flowering Plants of Africa
vol XXVI   1947

fl.   Pl. 1025

Euphorbia columnaris

Cactus & Succulent Society of America
Journal
vol 48, No. 4   1976

hab.   p. 181

Euphorbia commutata

Wharton, Mary E.
A Guide to the Wildflowers & Ferns
of Kentucky

fl.   p 273   Pl. 1.2

Euphorbia complexa

Flowering Plants of South Africa
vol XVII   1937

fl.   Pl. 643

Euphorbia cooperi

Lamb, Edgar
The Pocket Encyclopedia of Cacti
and Succulents in Color

hab.   Pl. 319

Euphorbia corniculata

Flowering Plants of Africa
vol 27   1948-49

fl.   Pl. 1076

Euphorbia corollata

Courtenay, Booth
Wildflowers & Weeds

fl.   p 67

Euphorbia corollata

Dean, Blanche E.
Wildflowers of Alabama and
Adjoining States

fl.   p 95

Euphorbia corollata

Duncan, Wilbur H.
Wildflowers of the Southeastern
United States

fl.   p 95

Euphorbia corollata

Klimas, John E.
Wildflowers of Eastern America

fl.   Pl. 62

Euphorbia corollata

Moyle, John B.
Northland Wild Flowers

fl. hab.   p 87, Pl. 69

Euphorbia corollata

Tsukamoto, Yotaro
Coloured Illustrations of Garden
Flowers   Vol. 9

fl.   opp. p. 47

Euphorbia corollata

Wharton, Mary E.
A Guide to the Wildflowers & Ferns
of Kentucky

fl.   p 167   Pl. 2.41

Euphorbia corsica

Vilmorin, Roger de
Plantes Alpines dans les Jardins

fl. hab.   Pl. XXXVIII

Euphorbia cyathophora

Duncan, Wilbur H.
Wildflowers of the Southeastern
United States

fr.   p 97

Euphorbia cyparissias

Barneby, T. P.
European Alpine Flowers in Colour

fl. hab.   Pl. 49, 1

Euphorbia cyparissias

Crockett, James Underwood
Lawns & Ground Covers

fl.   p. 132

Euphorbia cyparissias

Felsko, Elsa
A Book of Wildflowers

fl.   P 10

Euphorbia cyparissias

Martin, W. Keble
The Concise British Flora in Colour

fl. hab.   Pl. 75

Euphorbia cyparissias

Pond, Barbara
A Sampler of Wayside Herbs

fl.   Pl. II

Euphorbia dendroides

Polinun, Oleg
Flowers of Europe

fl. hab.   Pl. 66 #669

Euphorbia dendroides

Polunin, Oleg
Flowers of the Mediterranean

fl.   No. 90

Euphorbia depauperata

Moriarty, Audrey
Wild Flowers of Malawi

fl.   Pl. 51, 1

Euphorbia dregeana

Eliovson, Sima
Namaqualand in Flower

hab.   Pl. 54, 2

Euphorbia epithymoides

Hay, Roy
The Color Dictionary of Flowers & Plants

fl.   p 142, Pl. 1130

Euphorbia epithymoides

Huxley, Anthony
Garden Perennials and Water Plants

fl., hab.   Pl. 10, 112

Euphorbia epithymoides

Perry, Frances
Flowers of the World

fl.   p 115

Euphorbia epitymoides

Royal Hort. Soc.
Journal of the Royal Hort. Soc.
vol 91, No. 6   1966

fl.   p. 250   Pl. 137

Euphorbia esula

Courtenay, Booth
Wildflowers & Weeds

fl.   p 67

Euphorbia esula

Moyle, John B.
Northland Wild Flowers

fl. hab.   p 88, Pl. 70

Euphorbia excelsa

Flowering Plants of South Africa
vol XXIII   1943

fl.   Pl. 886

Euphorbia exigua

Martin, W. Keble
The Concise British Flora in Colour

fl. hab.   Pl. 75

Euphorbia fascicaulis

Cactus & Succulent Society of America
Journal
vol 49, No. 4   1977

fl., hab.   p. 177

Euphorbia fimbriata

Flowering Plants of Africa
vol 27   1948-49

fl., hab.   Pl. 1068

Euphorbia flanaganii var.

Lamb, Edgar
The Pocket Encyclopedia of Cacti
and Succulents in Color

hab.   Pl. 248

Euphorbia franksiae

Flowering Plants of South Africa
vol XVII   1937

fl., hab.   Pl. 648

Euphorbia fulgens

Harrison, R.E.
Trees and Shrubs

fl.   P 77

Euphorbia fulgens

Hay, Roy
The Color Dictionary of Flowers & Plants

fl.   p 66, Pl. 526

Euphorbia fulgens

Hay, Roy
The Dictionary of House Plants

fl.   Pl. 223

Euphorbia fulgens

Kiaer, Eigil
Indoor Plants in Colour

fl.   p. 56

Euphorbia fulgens

Kromdijk, G.
200 House Plants in Colour

fl.   Pl. 85

Euphorbia fulgens

Perry, Frances
Flowers of the World

fl.   p 115

Euphorbia gatbergensis

Flowering Plants of South Africa
vol XVII   1937

fl., hab.   Pl. 649

Euphorbia gillettii var.

Cactus & Succulent Society of America
Journal
vol 49, No. 4   1977

hab., fl.   177

Euphorbia globosa

Flowering Plants of South Africa
vol XVII   1937

fl., hab.   Pl. 647

Euphorbia grandialata

Flowering Plants of South Africa
vol XVII   1937

fl., fr.   Pl. 641

Euphorbia grandicornis

Flowering Plants of South Africa
vol XVII   1937

fl.   Pl. 642

Euphorbia grandicornis

Subik, Rudolf
Decorative Cacti

hab.   p 111

Euphorbia grandicornis

Wit, H. C. D. de
Plants of the World:
The Higher Plants, Vol. I

hab.   p. 221   Pl. 138

Euphorbia grandidens

Lamb, Edgar
The Pocket Encyclopedia of Cacti
and Succulents in Color

hab.   Pl. 300

Euphorbia Griffithii

Hara, Hiroshi, comp.
Photo-Album of Plants of
Eastern Himalaya

fl.   Pl. 91

Euphorbia griffithii

Perry, Frances
Flowers of the World

fl.   p 115

Euphorbia griffithii var.

Hay, Roy
The Color Dictionary of Flowers & Plants

fl.   p 142, Pl. 1131

Euphorbia groenewaldii

Flowering Plants of South Africa
vol XVIII   1938

fl., hab.   Pl. 714

Euphorbia hadramautica

Flowering Plants of Africa
vol 34   1960-61

fl., hab.   Pl. 1343

Euphorbia handiensis

Bramwell, David
Wild Flowers of the Canary Islands

hab.   Pl. 182

Euphorbia helioscopia

Ary, S.
The Oxford Book of Wildflowers

fl.   p 62, Pl. 1

Euphorbia helioscopia

Lindman, C. A. M.
Nordens Flora, Vol 6

fl. hab.   Pl. 375

Euphorbia helioscopia

Martin, W. Keble
The Concise British Flora in Colour

fl. hab.   Pl. 75

Euphorbia helioscopia

Polunin, Oleg
Flowers of Europe

fl.   Pl. 67 #671

Euphorbia heterophylla

Brown, Clair A.
Wildflowers of Louisiana and
    Adjoining States

fr.   p 102

Euphorbia **heterophylla**

Crockett, James Underwood
Annuals

fl.   P 118

Euphorbia heterophylla

Dean, Blanche E.
Wildflowers of Alabama and
    Adjoining States

fl.   p 95

Euphorbia heterophylla

Orr, Robert T.
Wildflowers of Western America

fl.   Pl. 208

Euphorbia heterophylla

Tsukamoto, Yotaro
Coloured Illustrations of Garden
    Flowers   Vol. 10

hab. Illus. 61   p. 20

Euphorbia hirta

Burgis, D.S.
Florida Weeds

hab.   p. 5

Euphorbia hirta

Moriarty, Audrey
Wild Flowers of Malawi

fl.   Pl. 51; 2

Euphorbia horrida

Hay, Roy
The Dictionary of House Plants

fr. hab.   Pl. 224

Euphorbia horrida

Subik, Rudolf
Decorative Cacti

hab.   p 113

Euphorbia horwoodii

Cactus & Succulent Society of America
Journal
    vol 50, No. 1   1978

hab., fl.   p. 26-27

Euphorbia hyberna

Martin, W. Keble
The Concise British Flora in Colour

fl. hab.   Pl. 75

Euphorbia ingens

Everett, Thomas H.
Living Trees of the World

hab.   p. 191 bottom

Euphorbia ingens

Morley, Brian D.
Wild Flowers of the World

fl.   Pl. 58

Euphorbia ingens

Palgrave, K. C.
Trees of Central Africa

fl., fr.   p. 174

Euphorbia lactea

Hall, Clarence E.
Flowers of the Islands in the Sun

fl., hab.   p. 87   Pl. 18

Euphorbia lactea

Hersey, Jean
Woman's Day Book of House Plants

hab.   p. 54

Euphorbia lathyris

Kimura, Koiti
Japanese Medicinal Plants, Vol I

fl.   p 54

Euphorbia lathyris

Martin, W. Keble
The Concise British Flora in Colour

fr. hab.   Pl. 75

Euphorbia lathyris

Polunin, Oleg
Flowers of Europe

fl. fr. Pl. 67 #676

Euphorbia leucochlamys

Flowering Plants of Africa
    vol 34   1960-61

fl.   Pl. 1344

Euphorbia leucophylla

Coyle, Jeanette
A Field Guide to the Common and
    Interesting Plants of Baja California

fl. hab.   p 109

Euphorbia lophogona

Flowering Plants of Africa
    vol 43   1974-76

fl.   Pl. 1688

Euphorbia lophogona

Lamb, Edgar
The Pocket Encyclopedia of Cacti
    and Succulents in Color

fl. hab.   Pl. 252

Euphorbia maculata

Courtenay, Booth
Wildflowers & Weeds

fl.   p 67

Euphorbia maculata

Weeds of the Southern United States

fl., hab.   p. 22

Euphorbia marginata

Crockett, James Underwood
Annuals

hab.,fl.   p. 118

Euphorbia marginata

Royal Hort. Soc.
Journal of the Royal Hort. Soc.
    vol 99, No. 10   1974

hab.   p. 442   Pl. 205

Euphorbia marginata

Tsukamoto, Yotaro
Coloured Illustrations of Garden
    Flowers   Vol. 10

hab. Illus. 60   opp. p. 19

Euphorbia mellifera

Bramwell, David
Wild Flowers of the Canary Islands

hab.   Pl. 189

Euphorbia mellifera

Royal Hort. Soc.
The Garden
    vol 100, No. 12   1975

hab.   p. 608

Euphorbia memoralis

Flowering Plants of Africa
    vol 29   1952-53

fl.   Pl. 1129

Euphorbia milii

Encke, Fritz
Zimmerpflanzen

fl.   p. 75

Euphorbia milii

Harrison, R.E.
Trees and Shrubs

fl.   P 77

Euphorbia milii

Hersey, Jean
Woman's Day Book of House Plants

hab., fl. P 50

Euphorbia milii

Kromdijk, G.
200 House Plants in Colour

fl.   Pl. 86

Euphorbia milii

Macoby, Stirling
What Flower is That

fl.   P 130

Euphorbia milii

Morley, Brian D.
Wild Flowers of the World

fl.   Pl. 58

Euphorbia milii

Nicolaisen, Age
Pocket Encyclopedia of Indoor Plants

fl.   Pl. 85

Euphorbia milii var.

Hay, Roy
The Dictionary of House Plants

fl.   Pl. 225

Euphorbia milii var.

Lamb, Edgar
The Pocket Encyclopedia of Cacti
   and Succulents in Color

fl. hab.   Pl. 250

Euphorbia milii var.

Perry, Frances
Flowers of the World

fl.   p 115

Euphorbia misera

Coyle, Jeanette
A Field Guide to the Common and
   Interesting Plants of Baja California

fl.   p 111

Euphorbia misera

Orr, Robert T.
Wildflowers of Western America

fl.   Pl. 279

Euphorbia mitriformis

Cactus & Succulent Society of America
Journal
     vol 48, No. 3   1976

hab.   p. 128

Euphorbia mosaica

Cactus & Succulent Society of America
Journal
     vol 48, No. 3   1976

hab.   p. 125

Euphorbia myrsinites

Hay, Roy
The Color Dictionary of Flowers & Plants

fl.   p 9, Pl. 69

Euphorbia myrsinites

Polunin, Oleg
Flowers of the Mediterranean

fl.   No. 94

Euphorbia myrsinites

Royal Hort. Soc.
Journal of the Royal Hort. Soc.
   vol 91, No. 6   1966

fl.   p. 250   Pl. 138

Euphorbia myrsinites

Royal Hort. Soc.
Journal of the Royal Hort. Soc.
   vol 96, No. 12   1971

fl., hab.   p. 540   Pl. 241

Euphorbia myrsinites

Royal Hort. Soc.
Journal of the Royal Hort. Soc.
   vol 97, No. 8   1972

fl.   p. 352   Pl. 180

Euphorbia myrsinites

Royal Hort. Soc.
The Garden
   vol 103, No. 3   1978

fl.   cover

Euphorbia neriifolia var.

Lamb, Edgar
The Pocket Encyclopedia of Cacti
   and Succulents in Color

fl. hab.   Pl. 254

Euphorbia nigrispina

Flowering Plants of Africa
   vol 38   1967

fl.   Pl. 1510

Euphorbia obesa

Cactus & Succulent Society of America
Journal
     vol 49, No. 5   1977

hab., fl.   cover

Euphorbia obesa

Flowering Plants of South Africa
   vol XX   1940

fl., hab.   Pl. 788

Euphorbia obesa

Lamb, Edgar
The Pocket Encyclopedia of Cacti
   and Succulents in Color

fl. hab.   Pl. 256

Euphorbia obesa

Subik, Rudolf
Decorative Cacti

fl. hab.   p 115

Euphorbia obtusifolia

Bramwell, David
Wild Flowers of the Canary Islands

fl.   Pl. 186

Euphorbia palustris

Hay, Roy
The Color Dictionary of Flowers & Plants

fl.   p 142, Pl. 1132

Euphorbia paralias

Bramwell, David
Wild Flowers of the Canary Islands

hab.   Pl. 183

Euphorbia paralias

Kleijn, H.
Beauty of the Wild Plant

fl.   opp. p. 81   Pl. 107

Euphorbia paralias

Martin, W. Keble
The Concise British Flora in Colour

fl. hab.   Pl. 75

Euphorbia paralias

Polunin, Oleg
Flowers of Europe

fl.   Pl. 66 #684

Euphorbia peplus

Martin, W. Keble
The Concise British Flora in Colour

fl. fr. hab.   Pl. 75

Euphorbia perangusta

Flowering Plants of South Africa
   vol XVIII   1938

fl.   Pl. 716

Euphorbia persistens

Flowering Plants of South Africa
   vol XVIII   1938

fl.   Pl. 713

Euphorbia phillipsiae

Flowering Plants of Africa
vol 35   1962

fl., hab.   Pl. 1395

Euphorbia platyphyllos

Martin, W. Keble
The Concise British Flora in Colour

fl. hab.   Pl. 75

Euphorbia polygona

Flowering Plants of South Africa
vol XVII   1937

fl.   Pl. 645

Euphorbia portlandica

Martin, W. Keble
The Concise British Flora in Colour

fl. hab.   Pl. 75

Euphorbia proballyana

Cactus & Succulent Society of America
Journal
vol 50, No. 3   1978

fl., hab.   cover

Euphorbia pseudocactus

Flowering Plants of South Africa
vol XX   1940

fl.   Pl. 778

Euphorbia pseudocactus

Lamb, Edgar
The Pocket Encyclopedia of Cacti
   and Succulents in Color

hab.   Pl. 255

Euphorbia pugniformis

Nicolaisen, Age
Pocket Encyclopedia of Indoor Plants

fl., hab.   Pl. 86

Euphorbia pulcherrima

Bruggeman, L.
Tropical Plants

fl.   Pl. 286

Euphorbia pulcherrima

Crockett, James Underwood
Evergreens

hab. fl.   p. 124

Euphorbia pulcherrima

Encke, Fritz
Zimmerpflanzen

fl.   p. 75

Euphorbia pulcherrima

Hall, Clarence E.
Flowers of the Islands in the Sun

fl.   p. 107   Pl. 23

Euphorbia pulcherrima

Hargreaves, Dorothy
Tropical Blossoms of the Caribbean

fl.,hab.   p. 33

Euphorbia pulcherrima

Harrison, R.E.
Trees and Shrubs

fl.   P 77

Euphorbia pulcherrima

Hay, Roy
The Dictionary of House Plants

fl.   Pl. 226

Euphorbia pulcherrima

Kromdijk, G.
200 House Plants in Colour

fl.   Pl. 87

Euphorbia pulcherrima

Macoby, Stirling
What Flower is That

fl.   P 130

Euphorbia pulcherrima

Massachusetts Hort. Soc.
Horticulture
   vol. 56, No. 12   1978

fl.   p. 21

Euphorbia pulcherrima

Nicolaisen, Age
Pocket Encyclopedia of Indoor Plants

fl.   Pl. 84

Euphorbia pulcherrima

O'Gorman, Helen
Mexican Flowering Trees and Plants

fl. hab.   p 59

Euphorbia pulcherrima

Tsukamoto, Yotaro
Coloured Illustrations of Garden Flowers
                    vol 10
fl.   No. 153

Euphorbia pulcherrima

Wit, H. C. D. de
Plants of the World;
   The Higher Plants, Vol I

fl. hab.   p 224, Pl. 142,
                    143

Euphorbia pulcherrima var.

American Hort. Soc.
American Horticulturist
vol 54, No. 6   1975

fl.   cover, p. 5-7

Euphorbia pulcherrima var.

Everett, T.H., ed.
New Illustrated Encyclopedia of
   Gardening vol. 4

fl.   p. 631

Euphorbia pulcherrima var.

Hay, Roy
The Dictionary of House Plants

fl.   Pl. 227

Euphorbia pulcherrima var.

Hersey, Jean
Woman's Day Book of House Plants

fl.   p. 90

Euphorbia pulcherrima var.

Macoby, Stirling
What Flower is That

fl P 130

Euphorbia pulcherrima var.

Massachusetts Hort. Soc.
Horticulture
   vol 50, No. 12   1972

fl.   p. 22-23

Euphorbia pulcherrima var.

Perry, Frances
Flowers of the World

fl.   p. 114

Euphorbia razafinjohanii

Lamb, Edgar
The Pocket Encyclopedia of Cacti
   and Succulents in Color

fl. hab.   Pl. 251

Euphorbia regis-jubae

Bramwell, David
Wild Flowers of the Canary Islands

hab.   Pl. 187

Euphorbia resinifera

Hay, Roy
The Dictionary of House Plants

hab.   Pl. 228

Euphorbia resinifera

Perrot, Emile
Les Plantes Medicinales

fl. fr. hab.   p 95

Euphorbia rigida

Mathias, Mildred E.
Color for the Landscape

fl. hab.   p 144

Euphorbia robbiae

Curtis's Botanical Magazine
Vol 169   1952-53

fl.   208

Euphorbia robbiae

Hay, Roy
The Color Dictionary of Flowers & Plants

fl.   p 142, Pl. 1133

Euphorbia robbiae

Morley, Brian D.
Wild Flowers of the World

fl.   Pl. 35

Euphorbia robbiae

Royal Hort. Soc.
Journal of the Royal Hort. Soc.
vol 93, No. 3   1968

fl.   p. 122   Pl. 47

Euphorbia schoenlandii

Flowering Plants of South Africa
vol XX   1940

fl., hab.   Pl. 772

Euphorbia sekukuniensis

Flowering Plants of South Africa
vol XX   1940

fl.   Pl. 775

Euphorbia sepulta

Cactus & Succulent Society of America
Journal
vol 48, No. 3   1976

hab.   p. 125

Euphorbia sikkimensis

Royal Hort. Soc.
The Garden
vol 103, No. 8   1978

fl.   p. 311

Euphorbia sp.

Lemmon, Robert S.
Wildflowers of North America in
Full Color

fl.   p. 74   Pl. 120

Euphorbia spinosa

Polunin, Oleg
Flowers of the Mediterranean

fl.   No. 88

Euphorbia spiralis

Lamb, Edgar
The Pocket Encyclopedia of Cacti
and Succulents in Color

hab.   Pl. 320

Euphorbia splendens

Cactus & Succulent Society of America
Journal
vol 8, No. 7   1937

fl.   cover (p. 97)

Euphorbia splendens

Milne, Lorus
Living Plants of the World

fl.   p 114

Euphorbia squarrosa

Flowering Plants of South Africa
vol XX   1940

fl.   Pl. 789

Euphorbia stolonifera

Flowering Plants of South Africa
vol XXII   1942

fl.   Pl. 863

Euphorbia stolonifera

Morley, Brian D.
Wild Flowers of the World

fl.,hab.   Pl. 58

Euphorbia supina

Duncan, Wilbur H.
Wildflowers of the Southeastern
United States

hab.   p 97

Euphorbia tetragona

Royal Hort. Soc.
Journal of the Royal Hort. Soc.
vol 90, No. 8   1965

hab.   p. 338   Pl. 149

Euphorbia tortirama

Flowering Plants of South Africa
vol XVII   1937

fl.   Pl. 644

Euphorbia triangularis

Flowering Plants of Africa
vol 43   1974-76

fl., fr.   Pl. 1687

Euphorbia vandermerwei

Flowering Plants of South Africa
vol XVII   1937

fl.   Pl. 660

Euphorbia veneta

Curtis's Botanical Magazine
vol 175   1964-65

fl.   Pl. 482

Euphorbia veneta

Harrison, R. E.
Trees and Shrubs

fl. hab.   P. 77

Euphorbia veneta

Polunin, Oleg
Flowers of the Mediterranean

fl.   No. 91

Euphorbia wallichii

Curtis's Botanical Magazine
vol 175   1964-65

fl.   Pl. 442

Euphorbia wallichii

Royal Hort. Soc.
The Garden
vol 102, No. 11   1977

hab., fl.   p. 449

Euphorbia waterbergensis

Flowering Plants of Africa
vol 28   1950-51

fl.   Pl. 1095

Euphorbia woodii

Flowering Plants of South Africa
vol XIX   1939

fl.   Pl. 723

Euphorbia wulfenii

Macoby, Stirling
What Flower is That

fl.   p. 130

Euphorbia wulfenii

Royal Hort. Soc.
Journal of Royal Hort. Soc.
vol 84, No. 9   1959

hab., fl.   p. 414   Pl. 127

Euphorbia zambesiana

Moriarty, Audrey
Wild Flowers of Malawi

fl.   Pl. 51; 3

Euphorbia zoutpansbergensis

Flowering Plants of South Africa
vol XVIII   1938

fl.   Pl. 715

Euphrasia anglica
Martin, W. Keble
The Concise British Flora in Colour

fl. fr. hab.   Pl. 64

Euphrasia brevipila
Martin, W. Keble
The Concise British Flora in Colour

fl. hab.   Pl. 64

Euphrasia collina
Alpine Flowers of the Kosciusko
State Park

fl.   Pl. 11

Euphrasia collina
Cochrane, C. R.
Flowers and Plants of Victoria

fl.   Pl. 3

Euphrasia collina var.
Curtis, Winifred
The Endemic Flora of Tasmania
Vol VI

fl. fr.   Pl. 226

Euphrasia confusa
Martin, W. Keble
The Concise British Flora in Colour

fl. fr. hab.   Pl. 64

Euphrasia curta
Lindman, C. A. M.
Nordens Flora, Vol 8

fl.   Pl. 532B

Euphrasia diemenica
Curtis, Winifred
The Endemic Flora of Tasmania
Vol. 1

fl. fr.   Pl. 3

Euphrasia glacialis
Cochrane, G. R.
Flowers and Plants of Victoria

fl.   Pl. 497

Euphrasia hookeri
Curtis, Winifred
The Endemic Flora of Tasmania, Part IV

fl. hab.   Pl. 132

Euphrasia kingii
Curtis, Winifred
The Endemic Flora of Tasmania, Part IV

fl. hab.   Pl. 131

Euphrasia micrantha
Martin, W. Keble
The Concise British Flora in Colour

fl. hab.   Pl. 64

Euphrasia minima
Barneby, T. P.
European Alpine Flowers in Colour

fl.   Pl. 74, 3

Euphrasia minima
Polunin, Oleg
Flowers of Europe

fl. Pl. 127 #1244

Euphrasia nemorosa
Martin, W. Keble
The Concise British Flora in Colour

fl. hab.   Pl. 64

Euphrasia officinalis
Ary, S.
The Oxford Book of Wildflowers

fl.   p 92, Pl. 7

Euphrasia officinalis
Loewenfeld, Claire
The Complete Book of Herbs and Spices

fl. hab.   p 128

Euphrasia officinalis
Perrot, Emile
Les Plantes Medicinales

fl. fr. hab.   p 82

Euphrasia officinalis
Tosco, Uberto
The World of Mountain Flowers

fl.   p 49

Euphrasia phragmostoma
Curtis, Winifred
The Endemic Flora of Tasmania
Vol VI

fl. fr.   Pl. 227

Euphrasia rostkoviana
Barneby, T. P.
European Alpine Flowers in Colour

fl.   Pl. 74, 2

Euphrasia rostkoviana
Lindman, C. A. M.
Nordens Flora, Vol 8

fl. fr. hab.   Pl. 532A

Euphrasia rostkoviana
Polunin, Oleg
Flowers of Europe

fl. Pl. 127 #1242

Euphrasia salisburgensis
Martin, W. Keble
The Concise British Flora in Colour

fl. hab.   Pl. 64

Euphrasia scottica
Martin, W. Keble
The Concise British Flora in Colour

fl. hab.   Pl. 64

Euphrasia speciosa
Blombery, Alec M.
What Wildflower is That

fl.   p 140, Pl. 372

Euphrasia striata
Curtis, Winifred
The Endemic Flora of Tasmania
Vol. 1

fl.   Pl. 2

Euptelea polyandra
Kitamura, Siro
Coloured Illustrations of Trees &
Shrubs of Japan

fl.   154

Eurotia lanata
Munz, Philip A.
California Desert Wildflowers

fl.   p 31, Pl. 13

Eurya chinensis
Walden, Beryl M.
Wild Flowers of Hong Kong

fl.   Pl. 78 (243)

Eurya emarginata
Kitamura, Siro
Coloured Illustrations of Trees &
Shrubs of Japan

fr.   350

Eurya japonica
Curtis's Botanical Magazine
Vol 178   1970-72

fl. fr.   588

Eurya japonica
Kitamura, Siro
Coloured Illustrations of Trees &
Shrubs of Japan

fr.   349

Eurya macartneyi

Walden, Beryl M.
Wild Flowers of Hong Kong

fl.   Pl. 76 (227)

Euryale ferox

Morley, Brian D.
Wild Flowers of the World

fl.fr.   Plate 91

Eurycentrum obscurum

Grubb, Roy
Selected Orchidaceous Plants
    Vol II

hab.   p 27

Eurychone rothschildiana

Stewart, Joyce
Orchids of Tropical Africa

fl. hab.   Pl. 29

Eurycles amboinensis

Blombery, Alec M.
What Wildflower is That

fl. hab.   p 141, Pl. 373

Eurycles cunninghamii

Blombery, Alec M.
What Wildflower is That

fl.   p 141, Pl. 374

Euryops abrotanifolius

Harrison, R.E.
Trees and Shrubs

fl.   P 78

Euryops acraeus

Hay, Roy
The Color Dictionary of Flowers & Plants

fl.   p 9, Pl. 70

Euryops acraeus

Royal Hort. Soc.
Journal of the Royal Hort. Soc.
    vol 95, No. 2   1970

fl.   cover

Euryops athanasiae

Flowering Plants of Africa
    vol 39   1968-69

fl.   Pl. 1553

Euryops athanasiae

Macoby, Stirling
What Flower is That

fl.   P 131

Euryops evansii

Royal Hort. Soc.
Journal of the Royal Hort. Soc.
    vol 88, No. 1   1963

fl.   p. 22   Pl. 17

Euryops evansii

Royal Hort. Soc.
Journal of the Royal Hort. Soc.
    vol 89, No. 10   1964

fl.   p. 418   Pl. 163

Euryops sulcatus

Flowering Plants of Africa
    vol 33   1959

fl.   Pl. 1305

Eurystigma clavata

Herre, H.
The Genera of the Mesembryanthemaceae

fl.   fr.   p 155

Eurystyles actinosophila

Pabst, G.F.J.
Orchidaceae Brasilienses
    vol 1

hab., fl.   p. 53

Euscaphis japonica

Kitamura, Siro
Coloured Illustrations of Trees &
    Shrubs of Japan

fr.   305

Eustoma exaltatum

Brown, Clair A.
Wildflowers of Louisiana and
    Adjoining States

fl.   p 137

Eustoma exaltatum

Munz, Philip A.
California Desert Wildflowers

fl.   p 43, Pl. 51

Eustoma exaltatum

Pacific Hort. Foundation
Pacific Horticulture
    vol 39, No. 2   1978

fl.   p. 29

Eustoma grandiflorum

Orr, Robert T.
Wildflowers of Western America

fl.   Pl. 272

Eustoma russellianum

Tsukamoto, Yotaro
Coloured Illustrations of Garden
    Flowers Vol. 10

fl. Illus. 62 opp. p. 20

Eustrephus latifolius

Blombery, Alec M.
What Wildflower is That

fl. fr.   p 141, Pl. 375

Eustrephus latifolius

Cochrane, G. R.
Flowers and Plants of Victoria

fr.   Pl. 473

Eustrephus latifolius

Harrison, Richmond E.
Climbers and Trailers

fr. p 46   Pl. 93

Eustrephus latifolius

Mullins, Barbara
Australian Wildflowers in Colour

fr.   P. 105   Pl. 103

Eustylis purpurea

Addisonia
    vol 20   1937-38

fl., fr.   p. 13   Pl. 647

Eustylis purpurea

Brown, Clair A.
Wildflowers of Louisiana and
    Adjoining States

fl.   p 29

Eutaxia microphylla

Cochrane, G. R.
Flowers and Plants of Victoria

fl. hab.   Pl. 337

Eutaxia obovata

Harrison, R.E.
Trees and Shrubs

fl. P 78

Evax pygmaea

Polunin, Oleg
Flowers of Europe

fl.   Pl. 143 #1374

Evergreen, Chinese

see

Aglaonema

Everlasting

see

Antennaria

Everlasting

see

Helichrysum

Everlasting, Clustered

see

Helichrysum semipapposum

Everlasting flower

see

Helichrysum bracteatum

Everlasting flower

see

Helipterum roseum

Everlasting Flower, Australian

see

Helipterum albicans var.

Everlasting, Pearly

see

Anaphalis margaritacea

Everlasting, Swan River

see

Helipterum manglesii

Everlasting, Sweet

see

Gnaphalium obtusifolium

Everlasting, winged

see

Ammobium alatum

Evodia rutaecarpa

Kariyone, Tatsuo
Atlas of Medicinal Plants

fr.   Pl. 92

Evodia rutaecarpa

Kimura, Koiti
Japanese Medicinal Plants, Vol II

fl.   p 191

Evodia rutaecarpa

Inkatori, Jisuke
Color Atlas of Medicinal Plants of Japan

fl.   Fig. 34

Evodiopanax innovans

Kitamura, Siro
Coloured Illustrations of Trees &
Shrubs of Japan

fl.   370

Evolvulus arizonicus

Lemmon, Robert S.
Wildflowers of North America in
Full Color

fl.   pg. 100   pl. 163

Ewartia catipes

Curtis, Winifred
The Endemic Flora of Tasmania Vol. 1

fl.   Pl. 32

Ewartia meredithae

Curtis, Winifred
The Endemic Flora of Tasmania
Vol. 1

fl.   Pl. 31

Ewartia nubigena

Cochrane, G. R.
Flowers and Plants of Victoria

fl. hab.   Pl. 516, 517

Ewartia planchonii

Curtis, Winifred
The Endemic Flora of Tasmania, Vol. 5

fl. hab.   Pl. 165

Exaculum pusillum

Martin, W. Keble
The Concise British Flora in Colour

fl. hab.   Pl. 58

Exacum affine

Flowering Plants of Africa
vol 40   1969-70

fl., fr., hab.   Pl. 1576

Exacum affine

Kromdijk, G.
200 House Plants in Colour

fl.   Pl. 88

Exacum affine

Perry, Frances
Flowers of the World

fl.   p 122

Exacum affine var.

Hay, Roy
The Dictionary of House Plants

fl.   Pl. 229

Exocarpos aphyllus

Cochrane, G. R.
Flowers and Plants of Victoria

fr.   Pl. 135

Exocarpos cupressiformis

Blombery, Alec M.
What Wildflower is That

fr.   p 141, Pl. 376

Exocarpos cupressiformis

Cochrane, G. R.
Flowers and Plants of Victoria

fr.   Pl. 206

Exocarpos humifusus

Curtis, Winifred
The Endemic Flora of Tasmania, Vol. 5

fl. fr.   Pl. 177

Exocarpos nanus

Cochrane, G. R.
Flowers and Plants of Victoria

fr.   Pl. 526

Exocarpos syrticola

Cochrane, G. R.
Flowers and Plants of Victoria

fr.   Pl. 293

Exochorda giraldii

Hay, Roy
The Color Dictionary of Flowers & Plants

fl.   p 201, Pl. 160?

Exochorda korolkowii

Royal Hort. Soc.
Journal of the Royal Hort. Soc.
vol 100, No. 3 1975

fl.   p. 116 Pl. 34

Exochorda x macrantha

Royal Hort. Soc.
Journal of the Royal Hort. Soc.
vol 94, No. 2 1969

fl.   p. 86 Pl. 40

Exorchorda x macrantha var.

Gault, S. Millar
The Color Dictionary of Shrubs

fl.   Pl. 192

Exochorda racemosa

Harrison, R.E.
Trees and Shrubs

fl.   P 76

Exochorda racemosa

Hellyer, A.G.L.
Shrubs in Colour

fl.   p. 50

Exochorda racemosa

Huxley, Anthony
Deciduous Garden Trees and Shrubs

fl.   Pl. 72

Exochorda racemosa

Perry, Frances
Flowers of the World

fl.   p. 265

Exogonium bracteatum

O'Gorman, Helen
Mexican Flowering Trees and Plants

fl. hab.   p 99

Eyebright

see

Euphrasia officinalis

Eye-bright

see

Euphrasia speciosa

Eyebright, Large Flowered Sticky

see

Euphrasia rostkoviana

Eyebright, Little Kneeling

see

Euphrasia minima

Eyebright, Purple

see

Euphrasia collina

Eye, Crab's

see

Abrus precatorius

Eye, Pink

see

Tetratheca ciliata

Eye, Sore

see

Boophone disticha

Eyes, Doll's

see

Actaea pachypoda

Eyes, Green

see

Berlandiera subacaulis

Eyes, Pink

see

Tetratheca ericifolia

Eyes, Pixie

see

Primula cuneifolia

Fabiana imbricata

Gault, S. Millar
The Color Dictionary of Shrubs

fl.   Pl. 193

Fabiana imbricata

Harrison, R.E.
Trees and Shrubs

fl.   P 79

Fabiana imbricata

Hay, Roy
The Color Dictionary of Flowers & Plants

fl.   p 201, Pl. 1608

Fabiana imbricata

Hay, Roy
The Dictionary of House Plants

fl.   Pl. 230

Fabiana imbricata var.

Hellyer, A.G.L.
Shrubs in Color

fl.   p. 51

Fagara capensis

Flowering Plants of Africa
vol 42   1972-73

fl., fr.   Pl. 1671

Fagopyrum cymosum

Kimura, Koiti
Japanese Medicinal Plants, Vol I

fl.   p 112

Fagopyrum esculentum

Polunin, Oleg
Flowers of Europe

fl.   Pl. 9 #89

Fagopyrum sagittatum

Bianchini, F.
The Complete Book of Fruits & Vegetables

fl. fr.   p 29

Fagraea auriculata

Morley, Brian D.
Wild Flowers of the World

fl.   Pl. 121

Fagus crenata

Kitamura, Siro
Coloured Illustrations of Trees &
     Shrubs of Japan

hab.   112, 113

Fagus grandifolia

Edlin, Herbert
The Illustrated Encyclopedia
     of Trees

hab.   p 147

Fagus grandifolia

Turner, Nancy J.
Wild Coffee and Tea Substitutes of
     Canada

fr.   p. 65

Fagus grandifolia

Wharton, Mary E.
Trees & Shrubs of Kentucky

fl.   p 104, Pl. 3.9

Fagus orientalis

Polunin, Oleg
Trees and Bushes of Europe

hab.   p 51

Fagus sylvatica

Boom, B. K.
The glory of the tree

fl.,hab.   p. 56   Pl. 81, 84

Fagus sylvatica

Edlin, Herbert
The Illustrated Encyclopedia
of Trees

fl. fr. hab.   p 146-7

Fagus sylvatica

Huxley, Anthony
Deciduous Garden Trees and Shrubs

fr.   Pl. 74, 74a

Fagus sylvatica

Hvass, Elsie
Plants That Feed and Serve Us

fr.   p. 122   Pl. 268

Fagus sylvatica

Lindman, C. A. M.
Nordens Flora, Vol. 3

fr.   Pl. 170

Fagus sylvatica

Martin, W. Keble
The Concise British Flora in Colour

fl. fr. hab.   Pl. 77

Fagus sylvatica

Milne, Lorus
Living Plants of the World

hab.   p 59

Fagus sylvatica

Morley, Brian D.
Wild Flowers of the World

fl. fr.   Pl. 12A

Fagus sylvatica

Perrot, Emile
Les Plantes Medicinales

fl. fr. hab.   p 119

Fagus sylvatica

Polunin, Oleg
Trees and Bushes of Europe

fl. fr. hab.   p 50

Fagus sylvatica

Vedel, H.
Arbres et Arbustes

fl. fr. hab.   p 56, No. 49

Fagus sylvatica var.

Edlin, Herbert
The Illustrated Encyclopedia
of Trees

hab.   p 147

Fagus sylvatica var.

Everett, T.H., ed.
New Illustrated Encyclopedia of
Gardening vol. 4

hab.   p. 679

Fagus sylvatica var.

Flemer, William III
Shade and Ornamental Trees in Color

hab.   p 88

Fagus sylvatica var.

Harrison, R. E.
Trees and shrubs

hab.   P. 78

Fagus sylvatica var.

Hay, Roy
The Color Dictionary of Flowers
& Plants

hab.   p. 202   Pl. 1609

Fagus sylvatica var.

Huxley, Anthony
Deciduous Garden Trees & Shrubs

fr.   Pl. 75-77

Fagus sylvatica var.

Macoby, Stirling
What Flower is That

hab.   P 131

Fagus sylvatica var.

Royal Hort. Soc.
Journal of the Royal Hort. Soc.
Vol. 94, No. 2   1969

fl.   p. 22   Pl. 54

Fair Maids of Kent

see

Ranunculus aconitifolius

Fairies Aprons

see

Utricularia dichotoma

Fairies, Blue

see

Caladenia deformis

Fairies, Pink

see

Caladenia latifolia

Fairybell

see

Disporum

Fairy Bells

see

Sarcochilus ceciliae

Fairy Duster

see

Calliandra californica

Fairy Duster

see

Calliandra eriophylla

Fairy duster

see

Calliandra inaequilatera

Fairy Fishing Flower

see

Dierama pulcherrima

Fairy lantern

see

Calochortus albus

Fairy lantern, pink

see

Calochortus amoenus

Fairy lantern, yellow

see

Calochortus pulchellus

Fairy Slipper

see

Calypso bulbosa

Fairy Wand

see

Chamaelirium luteum

Fairy Wand

see

Dierama pulcherrima

Fallugia paradoxa

American Hort. Soc.
American Horticulturist
vol 54, No. 2 1975

fl.   p. 18

Fallugia paradoxa

Massachusetts Hort. Soc.
Horticulture
vol 46, No. 12   1968

fr.   p. 32-33

Fallugia paradoxa

Mathias, Mildred E.
Color for the Landscape

fl. hab.   p 176

Fallugia paradoxa

Munz, Philip A.
California Desert Wildflowers

fl.   p 34, Pl. 23

Fallugia paradoxa

Orr, Robert T.
Wildflowers of Western America

fl.   Pl. 38

Fame Flower

see

Talinum rugospermum

Fan Flower, Coast

see

Scaevola pallida

Fan-flower, purple

see

Scaevola ramosissima

Fanunculus sceleratus

Heller, Christine
Wild Flowers of Alaska

fl.   Pl. 6

Paradaya splendida

Menninger, Edwin A.
Flowering Vines of the World

fl.   Pl. 196

Farewell-to-Spring

see

Clarkia

Farewell-to-Spring

see

Godetia amoena

Farkleberry

see

Vaccinium arboreum

Fascicularia bicolor

Bromeliad Society
Bulletin
vol 19, No. 5   1969

fl.   cover (p. 97)

Fascicularia bicolor

Padilla, Victoria
Bromeliads

fl.   p. 66   Pl. 4

Fascicularia bicolor

Rauh, Werner
Bromeliads for Home, Garden and
Greenhouse

fl., hab.   Pl. 92

Fascicularia kirchhoffiana

Munoz Pizarro, Carlos
Flores Silvestres de Chile

fl., hab.   Pl. 28

Fascicularia pitcairniifolia

Bromeliad Society
Bulletin
vol 16, No. 3   1966

fl.   cover (p. 49)

Fascicularia pitcairniifolia

Padilla, Victoria
Bromeliads

fl.   p. 66   Pl. 4

x Fatshedera lizei

Everett, T.H., ed.
New Illustrated Encyclopedia
of Gardening   vol 4

hab   p. 694

x Fatshedera lizei

Kiaer, Eigil
Indoor Plants in Colour

hab.   p. 57

x Fatshedera lizei

Macoby, Stirling
What Flower is That

hab.   p. 132

x Fatshedera lizei

Nicolaisen, Age
Pocket Encyclopedia of Indoor
        Plants

hab.   Pl. 103

x Fatshedera lizei var.

Harrison, Richmond E.
Climbers and Trailers

hab. p 94   Pl. 234

x Fatshedera lizei var.

Hay, Roy
The Color Dictionary of Flowers & Plants

hab.   p 66, Pl. 527

x Fatshedera lizei var.

Hay, Roy
The Dictionary of House Plants

hab.   Pl. 231

x Fatshedera sp.

Kromdijk, G.
200 House Plants in Colour

hab.   Pl. 89

Fatsia japonica

Crockett, James Underwood
Evergreens

hab. fl.   p. 125

Fatsia japonica

Gault, S. Millar
The Color Dictionary of Shrubs

fl.   Pl. 194

Fatsia japonica

Hay, Roy
The Color Dictionary of Flowers & Plants

fl.   p 202, Pl. 1610

Fatsia japonica

Hay, Roy
The Dictionary of House Plants

fl.   Pl. 232

Fatsia japonica

Hellyer, A.G.K.
Shrubs in Colour

fl. fr.   p. 54

Fatsia japonica

Kiaer, Eigil
Indoor Plants in Colour

hab.   p. 57

Fatsia japonica

Kimura, Koiti
Japanese Medicinal Plants, Vol I

fl.   p 64

Fatsia japonica

Kitamura, Siro
Coloured Illustrations of Trees &
    Shrubs of Japan

fr.   365

Fatsia japonica

Kromdijk, G.
200 House Plants in Colour

hab.   Pl. 90

Fatsia japonica

Takatori, Jisuke
Color Atlas of Medicinal Plants of Japan

fl.   Fig. 25

Fatsia japonica var.

Gault, S. Millar
The Color Dictionary of Shrubs

fl.   Pl. 195

Fatsia japonica Var.

Macoby, Stirling
What Flower is That

hab P 132

Faucaria tigrina

Hay, Roy
The Dictionary of House Plants

hab.   Pl. 233

Faucaria tigrina

Herre, H.
The Genera of the Mesembryanthemaceae

fl.   fr.   p 157

Faucaria tigrina

Morley, Brian D.
Wild Flowers of the World

fl.,hab.   Pl. 73

Faucaria tigrina

Nicolaisen, Age
Pocket Encyclopedia of Indoor
    Plants

fl.   Pl. 58

Faucaria tuberculosa

Lamb, Edgar
Popular Exotic Cacti in Color

fl. hab.   p 88

Faurea saligna

Palgrave, K. C.
Trees of Central Africa

fl., fr.   p. 361

Faurea speciosa

Palgrave, K. C.
Trees of Central Africa

fl., fr.   p. 365

Fauria crista-galli

Clark, Lewis J.
Wild Flowers of British Columbia

fl.   p 418

Featherbells

see

Stenanthium gramineum

Feather, Emerald

see

Asparagus sprengeri

Feather-Fleece

see

Stenanthium gramineum

Featherflower, Golden

see

Verticordia nitens

Feather flower, pink

see

Verticordia picta

Feather Flower, Pink Woolly

see

Verticordia monodelpha

Feather-flower, Scarlet

see

Verticordia grandis

Feather, Gay

see

Liatris elegans

Featherhead

see

Phylica pubescens

Featherhead

see

Ptilotus exaltatus

Feather, prince's

see

Amaranthus hypochondriacus

Featherling, White

see

Tofieldia glabra

Feathers, Prince of Wales

see

Celosia argentea

Feathers Tree, Prince of Wales

see

Brachystegia boehmii

Fedia cornucopiae

Polunin, Oleg
Flowers of Europe

fl.   Pl.137 #1312

Fedia cornucopiae

Polunin, Oleg
Flowers of the Mediterranean

fl.   No. 185

Feijoa sellowiana

Crockett, James Underwood
Evergreens

hab. fl.   p. 126

Feijoa sellowiana

Harrison, R.E.
Trees and Shrubs

fl.   p. 79

Feijoa sellowiana

Macoby, Stirling
What Flower is That

fl.   P 132

Feijoa sellowiana

Morley, Brian D.
Wild Flowers of the World

fl.   Pl. 179

Feijoa sellowiana

Tsukamoto, Yotaro
Coloured Illustrations of Garden
    Flowers   Vol. 10

fl. Illus. 154 opp. p. 48

Felicia amelloides

Crockett, James Underwood
Annuals

fl.   P 118

Felicia amelloides

Flowering Plants of Africa
    vol 42   1972-73

fl.   Pl. 1645

Felicia amelloides

Hersey, Jean
Woman's Day Book of House Plants

fl.   p. 79

Felicia amelloides

Huxley, Anthony
Garden annuals and bulbs

fl.   Pl. 44

Felicia amelloides

Macoby, Stirling
What Flower is That

fl.   P 133

Felicia amelloides

Massachusetts Hort. Soc.
Horticulture
    vol 52, No. 1   1974

fl.   p. 39

Felicia amelloides var.

Harrison, R.E.
Trees and Shrubs

fl. P 80

Felicia amelloides var.

Mathias, Mildred E.
Color for the Landscape

fl. hab.   p 143

Felicia angustifolia

Harrison, R.E.
Trees and Shrubs

fl.hab.   P 80

Felicia angustifolia

Macoby, Stirling
What Flower is That

fl.   P 133

Felicia chariels

Eliovson, Sima
Namaqualand in Flower

fl.   Pl. 6, 2

Felicia fruticosa

Mathias, Mildred
Color for the Landscape

fl. hab.   p 143

Felicia namaquana

Eliovson, Sima
Namaqualand in Flower

fl.   Pl. 6, 3/4

Felicia pappei var.

Curtis's Botanical Magazine
Vol 170   1954-55

fl.   239

Felicia tenella

Eliovson, Sima
Namaqualand in Flower

fl.   Pl. 6, 1

Fellerbush

see

Desmothamnus lucidus

Felwort

see

Gentianella amarella

Felwort, Field

see

Gentianella campestris

Felwort, German

see

Gentiana germanica

Felwort, Marsh

see

Lomatogonium rotatum

Felwort, Marsh

see

Swertia perennis

Fence, Flower

see

Caesalpinia pulcherrima

Fenestraria aurantiaca

Herre, H.
The Genera of the Mesembryanthemaceae

fl.   fr.   p 159

Fenestraria rhopalophylla

Lamb, Edgar
Popular Exotic Cacti in Color

fl. hab.   p 89

Fenestraria rhopalophylla

Nicolaisen, Age
Pocket Encyclopedia of Indoor
    Plants

hab.   Pl. 59

Fenestraria rhopalophylla

Subik, Rudolf
Decorative Cacti

fl. hab.   p 117

Fennel

see

Foeniculum vulgare

Fennel, Dog

see

Anthemis cotula

Fennel, Dog

see

Eupatorium capillifolium

Fennel, Dog

see

Eupatorium compositifolium

Fennel flower

see

Nigella hispanica

Fennel, Sea

see

Crithmum maritimum

Fennel, Sweet

see

Foeniculum vulgare

Fennel, Wild

see

Foeniculum officinale

Fenugreek

see

Trigonella foenum-graecum

Ferdinandusa speciosa

Morley, Brian D.
Wild Flowers of the World

fl. fr.   pl. 182

Fern, Asparagus

see

Asparagus sprengeri

Fernbush

see

Chamaebatiaria millefolium

Fernleaf, Alpine

see

Pedicularis contorta

Fern tree

see

Jacaranda acutifolia

Fern Tree

see

Jacaranda filicifolia

Fern tree

see

Jacaranda mimosaefolia

Fernwood

see

Pedicularis sudetica

Fernseea itatiaia

Bromeliad Society
Bulletin
vol 16, No. 4   1966

hab.   p. 96

x Ferobergia hybrid

Cactus & Succulent Society of America
Journal
    vol 38, No. 5   1966

hab.   cover (p. 149)

Ferocactus acanthodes

Backeberg, Curt
Cactus Lexicon

hab.   p 608

Ferocactus acanthodes

Cactus & Succulent Society of America
Journal
vol 47, No. 5   1975

hab.   cover

Ferocactus acanthodes

Coyle, Jeanette
A Field Guide to the Common and
  Interesting Plants of Baja California

fl. hab.   p 135

Ferocactus acanthodes

Lamb, Edgar
Colorful Cacti of the American Deserts

fl. hab.   73, 74, 95

Ferocactus acanthodes

Lamb, Edgar
The Pocket Encyclopedia of Cacti
  and Succulents in Color

hab.   Pl. 52

Ferocactus acanthodes

Mathias, Mildred E.
Color for the Landscape

fl. hab.   p 202

Ferocactus covillei

Cactus & Succulent Society of America
Journal
vol 48, No. 5   1976

hab.   cover

Ferocactus covillei

Lamb, Edgar
Colorful Cacti of the American Deserts

fl. hab.   Pl. 65, 66

Ferocactus diguetti

Lamb, Edgar
The Pocket Encyclopedia of Cacti
  and Succulents in Color

hab.   Pl. 302

Ferocactus echidne var.

Cactus & Succulent Society of America
Journal
vol 40, No. 4   1968

fl., hab.   p. 172

Ferocactus fordii

Lamb, Edgar
Popular Exotic Cacti in Color

fl. hab.   p 91

Ferocactus glaucescens

Van Laren, A. J.
Cactus

hab P 46   Fig. 52

Ferocactus hamatacanthus

Van Laren, A. J.
Cactus

fl.   p 46   Fig. 53

Ferocactus latispinus

Hay, Roy
The Dictionary of House Plants

hab.   Pl. 234

Ferocactus latispinus

Kupper, Walter
Cacti

hab.   p. 109   Pl. 51

Ferocactus latispinus

Lamb, Edgar
The Pocket Encyclopedia of Cacti
  and Succulents in Color

fl. hab.   Pl. 53

Ferocactus latispinus

Van Laren, A. J.
Cactus

fl.   P 79   Fig. M

Ferocactus nobilis

Van Laren, A. J.
Cactus

hab P 46   Fig. 50

Ferocactus peninsulae

Coyle, Jeanette
A Field Guide to the Common and
Interesting Plants of Baja California

fl.   p 135

Ferocactus pilosus

Cactus & Succulent Society of America
Journal
    vol 38, No. 4   1966

hab.   p. 127

Ferocactus pringlei

Van Laren, A. J.
Cactus

hab P 46   Fig. 49

Ferocactus stainesii

Van Laren, A. J.
Cactus

hab. P 79   Fig. L

Ferocactus uncinatus

Van Laren, A. J.
Cactus

hab P 46   Fig. 51

Ferocactus viridescens

Lamb, Edgar
Colorful Cacti of the American
    Deserts

fl. hab.   Pl. 118-120

Ferocactus wislizenii

Lamb, Edgar
Colorful Cacti of the American Deserts

fl. fr. hab.   Pl. 80 - 82

Ferocactus wislizenii

Massachusetts Hort. Soc.
Horticulture
    vol. 34, No. 4   1956

fl.   p. 227

Ferocactus wislizenii

Orr, Robert T.
Wildflowers of Western America

fl.   Pl. 197

Ferraria longa

Eliovson, Sima
Namaqualand in Flower

fl.   Pl. 35, 3

Ferraria tigridia

Huxley, Anthony
Garden annuals and bulbs

fl.   Pl. 212

Ferraria undulata

Addisonia
    vol 19   1935-36

fl.   p. 55   Pl. 636

Ferraria undulata

Flowering Plants of Africa
    vol 33   1959

fl.   Pl. 1316

Ferula chiliantha

Polunin, Oleg
Flowers of the Mediterranean

fl.   No. 116

Ferula chiliantha

Royal Hort. Soc.
Journal of the Royal Hort. Soc.
    Vol. 88, No. 9   1963

fl.   p. 388   Pl. 146

Ferula communis

Polunin, Oleg
Flowers of Europe

fl.   Pl. 85 #897

Ferula communis

Polunin, Oleg
Flowers of the Mediterranean

fl.   No. 115

Ferula communis

Wit, H. C. D. de
Plants of the World;
    The Higher Plants, Vol II

fl.   p 62, Pl. 35

Ferula linkii

Bramwell, David
Wild Flowers of the Canary Islands

fl.   Pl. 201

Fescue, Blue

see

Festuca ovina var.

Fescue, Red

see

Festuca rubra

Fescue, Rocky Mountain

see

Festuca saximontana

Fescue, short-leaved

see

Festuca brachyphylla

Fescue, Tall

see

Festuca arundinacea

Festuca arundinacea

Crockett, James Underwood
Lawns & Ground Covers

fr.   p. 117

Festuca arundinacea

Polunin, Oleg
Flowers of Europe

fr.   Pl. 180 #1757

Festuca brachyphylla

Morley, Brian D.
Wild Flowers of the World

fr.   P. 73

Festuca eskia

Huxley, Anthony
Garden Perennials and Water Plants

hab.   Pl. 319

Festuca ovina

Lindman, C. A. M.
Nordens Flora, Vol 2

fr. hab.   Pl. 75A

Festuca ovina var.

Crockett, James Underwood
Lawns & Ground Covers

fr. hab.   p. 30-31, 132

Festuca ovina var.

Hay, Roy
The Color Dictionary of Flowers & Plants

fr.   p 142, Pl. 1134

Festuca ovina var.

Huxley, Anthony
Garden Perennials and Water Plants

hab.   Pl. 318

Festuca pratensis

Lindman, C. A. M.
Nordens Flora, Vol 2

fr. hab.   Pl. 76

Festuca pumila

Kohlhaupt, Paula
Fleurs des Alpages
Vol. 2

fr.   Pl. 17-18

Festuca rubra

Crockett, James Underwood
Lawns & Ground Covers

fr., hab.   p. 79, 117

Festuca rubra

Kohlhaupt, Paula
Fleurs des Alpages Vol. 2

fr.   p. 17-18

Festuca rubra

Lindman, C. A. M.
Nordens Flora, Vol 2

fr. hab.   Pl. 75B

Festuca saximontana

Porsild, A. E.
Rocky Mountain Wild Flowers

fr. hab.   p. 53

Festuca varia

Kohlhaupt, Paula
Fleurs des Alpages
Vol. 2

fr.   Pl. 17-18

Festuca vivipara

Morley, Brian D.
Wild Flowers of the World

fl.,hab.   Pl. 25

Fetter bush
    see
Leucothoe axillaris

Fetter-Bush
    see
Lyonia lucida

Feverfew
    see
Chrysanthemum parthenium

Feverfew
    see
    Matricaria eximia

Fever Tree
    see
Pinckneya pubens

Fibigia clypeata

Polunin, Oleg
Flowers of Europe

fl. Pl. 35 #327

Ficinia radiata

Morley, Brian D.
Wild Flowers of the World

fl.,hab.   Pl. 88

Ficus altissima

Kiaer, Eigil
Indoor Plants in Colour

hab.   p. 57

Ficus aurea

Edlin, Herbert
The Illustrated Encyclopedia
of Trees

hab.   p 34-35

Ficus benghalensis

Massachusetts Hort. Soc.
Horticulture
    vol 51, No. 2   1973

hab.   p. 40

Ficus benghalensis

Milne, Lorus
Living Plants of the World

hab.   p 67

Ficus benjamina

Bruggeman, L.
Tropical Plants

fr.   Pl. 191

Ficus benjamina

Crockett, James Underwood
Evergreens

hab.,fr.   p. 126

Ficus benjamina

Hay, Roy
The Color Dictionary of Flowers & Plants

hab.   p 66, Pl. 528

Ficus benjamina

Hay, Roy
The Dictionary of House Plants

hab.   Pl. 235

Ficus benjamina

Hersey, Jean
Woman's Day Book of House Plants

hab.   p. 107

Ficus benjamina

Kiaer, Eigil
Indoor Plants in Colour

hab.   p. 58

Ficus benjamina

Macoby, Stirling
What Flower is That

hab P 133

Ficus benjamina

Nicolaisen, Age
Pocket Encyclopedia of Indoor Plants

hab.   Pl. 47

Fig, Cape
    see
    Ficus capensis

Ficus capensis

Palgrave, K. C.
Trees of Central Africa

fr.   p. 277

Ficus capensis

Palmer, Eve
Trees of South Africa

hab.   p. 256   Pl. XXI

Ficus carica

Boom, B. K.
The glory of the tree

fr.   opp. p. 64   Pl. 97

Ficus carica

Edlin, Herbert
The Illustrated Encyclopedia
of Trees

fr.   p 221

Ficus carica

Huxley, Anthony
Deciduous Garden Trees and Shrubs

fr.   Pl. 73

Ficus carica

Hvass, Elsie
Plants That Feed and Serve Us

fr.   p 46, Pl. 87

Ficus carica

Kiaer, Eigil
Indoor Plants in Colour

fr.   p. 59

Ficus carica

Kimura, Koiti
Japanese Medicinal Plants, Vol II

fr.   p 151

Ficus carica

Masefield, C. B.
The Oxford Book of Food Plants

fr.   Pl. 95, 1

Ficus carica

Massachusetts Hort. Soc.
Horticulture
      vol 51, No. 2   1973

fr.   p. 39

Ficus carica

Nicolaisen, Age
Pocket Encyclopedia of Indoor Plants

hab.   Pl. 48

Ficus carica

Perrot, Emile
Les Plantes Medicinales

fl. fr. hab.   p 97

Ficus carica

Polunin, Oleg
Flowers of the Mediterranean

fr.   No. 7

Ficus carica

Polunin, Oleg
Trees and Bushes of Europe

fl. fr. hab.   p 66

Ficus carica var.

Bianchini, F.
The Complete Book of Fruits & Vegetables

fr.   p 167

Ficus deltoides

Wit, H. C. D. de
Plants of the World;
    The Higher Plants, Vol I

fl. fr.   p 102, Pl. 51

Ficus diversifolia

Hay, Roy
The Dictionary of House Plants

hab. fr.   Pl. 236

Ficus diversifolia

Hersey, Jean
Woman's Day Book of House Plants

fr.   p. 61

Ficus diversifolia

Kromdijk, G.
200 House Plants in Colour

hab.   Pl. 91

Ficus diversifolia

Tsukamoto, Yotaro
Coloured Illustrations of Garden
      Flowers Vol. 10

fr. Illus. 156   opp. p. 49

Ficus elastica

Bruggeman, L.
Tropical Plants

hab.   Pl. 192

Ficus elastica

Hersey, Jean
Woman's Day Book of House Plants

hab.   p. 92

Ficus elastica

Kiaer, Eigil
Indoor Plants in Colour

hab.   p. 58

Ficus elastica

Nicolaisen, Age
Pocket Encyclopedia of Indoor Plants

hab.   Pl. 49

Ficus elastica

Tsukamoto, Yotaro
Coloured Illustrations of Garden
      Flowers Vol. 10

hab. Illus. 155   opp. p. 48

Ficus elastica var.

Everett, T.H., ed.
New Illustrated Encyclopedia of
      Gardening   vol 4

hab.   p. 679

Ficus elastica var.

Hay, Roy
The Color Dictionary of Flowers & Plants

hab.   p 67, Pl. 529, 530

Ficus elastica var.

Hay, Roy
The Dictionary of House Plants

hab.   Pl. 237, 238

Ficus elastica var.

Kromdijk, G.
200 House Plants in Colour

hab.   Pl. 92

Ficus elastica var.

Oakman, Harry
Colorful Trees

Hab.   P 110

Ficus elastica var.

Royal Hort. Soc.
Journal of the Royal Hort. Soc.
      vol 90, No. 5   1965

hab.   p. 206   Pl. 94

Ficus erecta

Kitamura, Siro
Coloured Illustrations of Trees &
      Shrubs of Japan

fr.   142

Ficus lyrata

Hersey, Jean
Woman's Day Book of House Plants

hab.   frontispiece, p.60

Ficus lyrata

Kiaer, Eigil
Indoor Plants in Colour

hab.   p. 58

Ficus lyrata

Kromdijk, G.
200 House Plants in Colour

hab.   Pl. 93

Ficus lyrata

Lowenmo, Runo
Plantes d'Appartement

hab.   p. 58

Ficus lyrata

Macoby, Stirling
What Flower is That

hab.   P. 133

Ficus nipponica

Kitamura, Siro
Coloured Illustrations of Trees &
      Shrubs of Japan

fr.   145

Ficus palmeri

Coyle, Jeanette
A Field Guide to the Common and
Interesting Plants of Baja California

hab.   p 69

Ficus pretoriae

Flowering Plants of Africa
vol 33   1959

fr.   Pl. 1319

Ficus pumila

Harrison, Richmond E.
Climbers and Trailers

hab.   p. 47   Pl. 94

Ficus pumila

Hay, Roy
The Color Dictionary of Flowers & Plants

hab.   p 67, Pl. 531

Ficus pumila

Hay, Roy
The Dictionary of House Plants

hab.   Pl. 239

Ficus pumila

Hersey, Jean
Woman's Day Book of House Plants

hab.   p. 61

Ficus pumila

Kiaer, Eigil
Indoor Plants in Colour

fr.   p. 58

Ficus pumila

Kitamura, Siro
Coloured Illustrations of Trees &
Shrubs of Japan

fr.   146

Fiscus pumila

Kroodijk, G.
200 House Plants in Colour

hab.   Pl. 94

Ficus pumila

Nicolaisen, Age
Pocket Encyclopedia of Indoor Plants

hab.   Pl. 50

Ficus pumila var.

Macoby, Stirling
What Flower is That

hab. P 133

Ficus radicans var.

Hay, Roy
The Color Dictionary of Flowers & Plants

hab.   p 67, Pl. 532

Ficus radicans var.

Hay, Roy
The Dictionary of House Plants

hab.   Pl. 240

Ficus radicans var.

Hersey, Jean
Woman's Day Book of House Plants

hab   p. 60

Ficus religiosa

Tsukamoto, Yotaro
Coloured Illustrations of Garden
Flowers Vol. 10

hab. Illus. 157   opp. p. 49

Ficus retusa

Flemer, William III
Shade and Ornamental Trees in Color

hab.   p. 93

Ficus rhodesiaca

Palgrave, K. C.
Trees of Central Africa

fr.   p. 281

Ficus rubiginosa var.

Oakman, Harry
Colorful Trees

hab.   p 111

Ficus schryveriana

Royal Hort. Soc.
Journal of the Royal Hort. Soc.
vol 90, No. 5   1965

hab.   p. 206   Pl. 94

Ficus soldanella

Flowering Plants of Africa
vol 31   1956

fr.   Pl. 1215

Ficus stipulata

Kitamura, Siro
Coloured Illustrations of Trees &
Shrubs of Japan

fr.   147

Ficus sycomorus

Wit, H. C. D. de
Plants of the World;
The Higher Plants, Vol I

fr.   p 103, Pl. 52

Ficus variegata

Everett, Thomas H.
Living Trees of the World

fr.   p. 148

Ficus wightiana

Kitamura, Siro
Coloured Illustrations of Trees & Shrubs of
Japan

fr.   143

Fiddleneck

see

Amsinckia douglasiana

Fiddleneck

see

Amsinckia intermedia

Fieldia australis

Morley, Brian D.
Wild Flowers of the World

fl.   Pl. 139

Fields, Gold

see

Baeria chrysostoma

Fig

see

Ficus

Fig, banjo

see

Ficus lyrata

Fig, Barbary

see

Opuntia ficus-indica

Fig, Cape

see

Ficus capensis

Fig, Climbing

see

Ficus pumila

Fig, Climbing

see

Ficus radicans

Fig, common
see
Ficus carica

Fig, Creeping
see
Ficus pumila

Fig, devil's
see
Argemone mexicana

Fig, fiddleleaf
see
Ficus lyrata

Fig, Hottentot
see
Carpobrotus edulis

Fig, Indian
see
Opuntia bergerana

Fig, Indian
see
Opuntia megacantha

Fig, Indian Laurel
see
Ficus retusa

Fig, lofty
see
Ficus altissima

Fig, Mistletoe
see
Ficus diversifolia

Fig, Port Jackson Variegated
see
Ficus rubiginosa var.

Fig, Red Hottentot
see
Carpobrotus acinaciformis

Fig, Rubber Variegated
see
Ficus elastica var.

Fig, sea
see
Carpobrotus chilensis

Fig, Sea
see
Mesembryanthemum chilense

Fig, Sour Giant
see
Carpobrotus quadrifidus

Fig, Strangler
see
Ficus aurea

Fig, weeping
see
Ficus benjamina

Fig, Wild
see
Ficus capensis

Fig, Wild
see
Ficus palmeri

Figwort
see
Scrophularia lanceolata

Figwort
see
Scrophularia nodosa var.

Figwort, Cape
see
Phygelius capensis

Figwort, Water
see
Scrophularia aquatica

Fig, Yellow Hottentot
see
Carpobrotus edulis

Filago apiculata
Martin, W. Keble
The Concise British Flora in Colour
fl. hab. Pl. 44

Filago gallica
Martin, W. Keble
The Concise British Flora in Colour
fl. hab. Pl. 44

Filago germanica
Ary, S.
The Oxford Book of Wildflowers
fl. p 32, Pl. 6

Filago germanica
Lindman, C. A. M.
Nordens Flora, Vol 9
fl. fr. hab. Pl. 585

Filago germanica
Martin, W. Keble
The Concise British Flora in Colour
fl. hab. Pl. 44

Filago minima
Ary, S.
The Oxford Book of Wildflowers
fl. p 32, Pl. 5

Filago minima
Lindman, C. A. M.
Nordens Flora, Vol 9
fl. fr. hab. Pl. 584

Filago minima
Martin, W. Keble
The Concise British Flora in Colour
fl. hab. Pl. 44

Filago spathulata

Martin, W. Keble
The Concise British Flora in Colour

fl. hab.   Pl. 44

Filago vulgaris

Polunin, Oleg
Flowers of Europe

fl. Pl. 143 #1375

Filaree, Redstem

see

Erodium cicutarium

Filbert

see

Corylus

Filbert, Purple-leafed

see

Corylus maxima

Filicium decipiens

Bruggeman, L.
Tropical Plants

hab.   Pl. 193

Filipendula hexapetala

Huxley, Anthony
Garden Perennials and Water Plants

fl.   Pl. 114

Filipendula hexapetala

Tsukamoto, Yotaro
Coloured Illustrations of Garden
    Flowers   Vol. 9

fl. opp. p. 60

Filipendula hexapetala var.

Hay, Roy
The Color Dictionary of Flowers &
    Plants

fl.   p. 142   Pl. 1135

Filipendula purpurea

Hay, Roy
The Color Dictionary of Flowers & Plants

fl.   p 142, Pl. 1136

Filipendula rubra

Courtenay, Booth
Wildflowers & Weeds

fl.   p 38

Filipendula rubra

Jennings, O. E.   vol II
Wild Flowers of Western Pennsylvania

fl.   Pl. 86

Filipendula rubra

Lemmon, Robert S.
Wildflowers of North America in
    Full Color

fl.   p. 183   Pl. 287

Filipendula ulmaria

Ary, S.
The Oxford Book of Wildflowers

fl.   p 80, Pl. 3

Filipendula ulmaria

Bianchini, Francesco
Health Plants of the World

fl.   p 141

Filipendula ulmaria

Kleijn, H.
Beauty of the Wild Plant

fl. opp. p. 17   Pl. 11

Filipendula ulmaria

Lindman, C. A. M.
Nordens Flora, Vol. 5

fl. fr.   Pl. 299

Filipendula ulmaria

Martin, W. Keble
The Concise British Flora in Colour

fl. hab.   Pl. 26

Filipendula ulmaria

Polunin, Oleg
Flowers of Europe

fl. Pl. 44 #421

Filipendula ulmaria var.

Royal Hort. Soc.
Journal of the Royal Hort. Soc.
    vol 100, No. 1   1975

hab.   p. 24   Pl. 4

Filipendula vulgaris

Felsko, Elsa
A Book of Wild Flowers
                2nd ser
fl.   p 123

Filipendula vulgaris

Lindman, C. A. M.
Nordens Flora, Vol 5

fl. hab.   Pl. 300

Filipendula vulgaris

Martin, W. Keble
The Concise British Flora in Colour

fl. hab.   Pl. 26

Filipendula vulgaris

Morley, Brian D.
Wild Flowers of the World

fl.   Pl. 9D

Filipendula vulgaris

Polunin, Oleg
Flowers of Europe

fl. Pl. 44 #422

Finger, Cut

see

Vinca major var.

Finger Flower

see

Cheiranthera linearis

Fingernail Plant

see

Neoregelia spectabilis

Fingers, Pink

see

Caladenia carnea

Fir

see

Abies

Fir, Alpine

see

Abies lasiocarpa

Fir, Balsam

see

Abies balsamea

Fir, Blue Noble

see

Abies procera var.

Fir, Blue Spanish

see

Abies pinsapo

Fir, Caucasian

see

Abies nordmanniana

Fir, China

see

Cunninghamia lanceolata

Fir, Colorado White

see

Abies concolor

Fir, Common Silver

see

Abies alba

Fir, Douglas

see

Pseudotsuga douglasii

Fir, Douglas

see

Pseudotsuga menziesii

Fir, Douglas

see

Pseudotsuga taxifolia

Fir, European Silver

see

Abies alba

Fir, Giant

see

Abies grandis

Fir, Giant Silver

see

Abies grandis

Fir, Grecian Silver

see

Abies cephalonica

Fir, Greek

see

Abies cephalonica

Fir, Korean silver

see

Abies koreana

Fir, Noble Silver

see

Abies procera

Fir, Oregon Douglas

see

Pseudotsuga menziesii

Fir, Pacific Coast

see

Pseudotsuga menziesii

Fir, Pacific Silver

see

Abies amabilis

Fir, Red

see

Abies magnifica

Fir, Spanish

see

Abies pinsapo

Fir, Veitch's Silver

see

Abies veitchii

Fir, white

see

Abies concolor

Fire Bush, Chilean

see

Embothrium coccineum

Firecracker, Brazilian

see

Manettia bicolor

Firecracker Flower

see

Brodiaea ida-maia

Fire dragon plant

see

Acalypha wilkesiana

Firethorn

see

Pyracantha

Firethorn, Laland

see

Pyracantha coccinea var.

Firethorn, orange

see

Pyracantha angustifolia

Firethorn, Santa Cruz

See

Pyracantha koidzumi

Firethorn, scarlet

see

Pyracantha coccinea var.

Firethorn, yellow

see

Pyracantha rodgersiana var.

Fire tree, Chilean

see

Embothrium coccineum var.

Fireweed

see

Epilobium angustifolium

Fireweed

see

Erechtites hieracifolia

Fireweed, dwarf

see

Epilobium latifolium

Fireweed, Mountain

see

Epilobium latifolium

Fire Wheel Tree

see

Stenocarpus sinuatus

Firewheels

see

Gaillardia pulchella

Firmiana colorata

Bruggeman, L.
Tropical Plants

fl.   Pl. 208

Firmiana colorata

Morley, Brian D.
Wild Flowers of the World

fl.   Pl. 116

Firmiana platanifolia

Kimura, Koiti
Japanese Medicinal Plants, Vol II

fl.   p 199

Firmiana platanifolia

Kitamura, Siro
Coloured Illustrations of Trees &
    Shrubs of Japan

fr.   338

Fittonia argyroneura

Addisonia
    vol 23   1954-59

fl.   p. 11   Pl. 742

Fittonia argyroneura

Bruggeman, L.
Tropical Plants

fl.   p 110

Fittonia argyroneura

Hay, Roy
The Dictionary of House Plants

hab.   Pl. 241

Fittonia argyroneura

Hersey, Jean
Woman's Day Book of House Plants

hab.   p 80

Fittonia argyroneura

Nicolaisen, Age
Pocket Encyclopedia of Indoor Plants

hab.   Pl. 127

Fittonia argyroneura

Tsukamoto, Yotaro
Coloured Illustrations of Garden
    Flowers   Vol. 9

hab.   opp. p. 47

Fittonia gigantea

Kiaer, Eigil
Indoor Plants in Colour

fl.   p. 60

Fittonia gigantea

Morley, Brian D.
Wild Flowers of the World

fl.   Pl. 184

Fittonia pearcei

Bruggeman, L.
Tropical Plants

fl.   Pl. 111

Fittonia pearcei

Massachusetts Hort. Soc.
Horticulture
    vol 32, No. 11   1954

hab.   p. 519

Fittonia sp.

Kroadijk, G.
200 House Plants in Colour

hab.   Pl. 95

Fittonia, tall

see

Fittonia gigantea

Fittonia verschaffeltii

Hay, Roy
The Color Dictionary of Flowers & Plants

hab.   p 67, Pl. 533

Fittonia verschaffeltii

Hay, Roy
The Dictionary of House Plants

hab.   Pl. 242

Fittonia verschaffeltii

Kiaer, Eigil
Indoor Plants in Colour

hab.   p. 60

Fittonia verschaffeltii

Macoby, Stirling
What Flower is That

hab.   p. 134

Fittonia verschaffelti

Massachusetts Hort. Soc.
Horticulture
    vol 35, No. 1   1957

hab.   p. 37 (see errata p. 108)

Fittonia verschaffeltii

Perry, Frances
Flowers of the World

fl.   p 13

Fittonia verschaffeltii var.

Kiaer, Eigil
Indoor Plants in Colour

hab.   p. 60

Fittonia verschaffeltii var.

Macoby, Stirling
What Flower is That

hab.   P. 134

Fittonia verschaffelti var.

Massachusetts Hort. Soc.
Horticulture
    vol 35, No. 1   1957

hab.   p. 37 (see errata p. 108)

Fittonia verschaffelti var.

Massachusetts Hort. Soc.
Horticulture
    vol 35, No. 11   1957

hab.   p. 556

Fittonia verschaffeltii var.

Massachusetts Hort. Soc.
Horticulture
    vol. 39, No. 12   1961

hab.   p. 599

Fivecorner fruit

see

Averrhoa carambola

Fivecorners, Pink

see

Styphelia triflora

Five-corners, red

see

Styphelia tubiflora

Five Finger

see

Potentilla canadensis

Five Finger

see

Pseudopanax arboreum

Fivefinger, Grayleaf

see

Potentilla glaucophylla

Fivefinger, Marsh

see

Potentilla palustris

Five fingers

see

Syngonium auritum

Five Leaves

see

Isotria verticillata

Fivespot

see

Nemophila maculata

Flacourtia cataphracta

Oakman, Harry
Colorful Trees

Fl.   P 64

Flacourtia indica

Palgrave, K. C.
Trees of Central Africa

fl., fr.   p. 191

Flacourtia inermis

Roxburgh, William
Icones Roxburghianae

fl., fr.   Pl. 8

Flag

See Iris

Flag, Blue

see

Iris tripetala

Flag, Carolina Blue

see

Iris darolina

Flag, Corn

see

Gladiolus

Flag, Long Purple

see

Patersonia longiscapa

Flag, Southern Blue

see

Iris virginica

Flag, Spanish

see

Quamoclit lobata

Flag, Spiral

see

Costus igneus

Flag, Spiral

see

Costus spicatus

Flag, Sweet

see

Acorus americanus

Flag, sweet

see

Acorus calamus

Flag, sweet

see

Acorus gramineus var.

Flag, yellow

see

Iris pseudacorus

Flagellarisaema urashima

Kimura, Koiti
Japanese Medicinal Plants, Vol II

hab.   p 131

Flambeau, glory

see

Brownea grandiceps

Flambeau tree

see

Spathodea campanulata

Flamboyant

see

Delonis regia

Flamboyant, Yellow

see

Peltophorum roxburghii

Flame Bush, Trinidad

see

Calliandra tweedii

Flame Flower

see

Macranthera flammea

Flame-of-the-forest

see

Butea frondosa

Flame of the Forest

see

Butea monosperma

Flame-of-the-forest

see

Delonix regia

Flame-of-the-Forest

see

Spathodea campanulata

Flame-of-the-Wood

see

Ixora macrothyrsa

Flame-Tree

see

Brachychiton acerifolium

Flame-Tree

see

Delonix regia

Flame Tree, Australian

see

Brachychiton acerifolium

Flame Tree, Chinese

see

Koelreuteria henryi

Flame Tree, Illawarra

see

Brachychiton acerifolium

Flame tree, pink

see

Brachychiton discolor

Flame Vine

see

Pyrostegia ignea

Flaming Sword

see

Vriesia splendens

Flamingo Flower

see

Anthurium andraeanum

Flamingo Flower

see

Anthurium scherzerianum

Flamingo Flower

see

Jacobinia suberecta

Flannel Bush

see

Fremontia californica

Flannel Bush

see

Fremontodendron californicum

Flannel Flower

see

Actinotus helianthi

Flannel-Plant

see

Verbascum thapsus

Flannel Plant, Big

see

Lachnostachys cliftonii

Flat-pea, Handsome

see

Platylobium formosum

Flaveria linearis
Fleming, Glenn
Wild Flowers of Florida

fl.   p 58

Flax

see

Linum

Flax, Alpine

see

Linum alpinum

Flax, Blue

see

Linum lewisii

Flax, Blue

see

Linum perenne var.

Flax, common

see

Linum usitatissimum

Flax, flowering

see

Linum grandiflorum

Flax, Mountain

see

Phormium colensoi

Flax, New Zealand

see

Phormium tenax

Flax, Purging

see

Linum catharticum

Flax, Slender Yellow

see

Linum virginianum

Flax, Small-leaved

see

Linum tenuifolium

Flax, Wild

see

Heliophila namaquana

Flax, Wild

see

Heliophila seselifolia

Flax, Wild

see

Linum perenne

Flax, wild blue

see

Linum lewisii

Flax, yellow

see

Reinwardtia trigyna

Fleabane

see

Erigeron

Fleabane

see

Pulicaria dysenterica

Fleabane, Alaskan

see

Erigeron caespitosus

Fleabane, Alaskan

see

Erigeron salsuginosus

Fleabane, Aleutian

see

Erigeron unalaschensis

Fleabane, Alpine

see

Erigeron alpinus

Fleabane, Annual

see

Erigeron annuus

Fleabane, Arctic

see

Erigeron hyperboreus

Fleabane, Beach

see

Senecio psuedo-arnica

Fleabane, bony tip

see

Erigeron karvinskianus

Fleabane, Boreal

see

Erigeron borealis

Fleabane, Common

see

Pulicaria dysenterica

Fleabane, Cutleaf

see

Erigeron compositus

Fleabane, Daisy

see

Erigeron philadelphicus

Fleabane, Golden

see

Erigeron aureus

Fleabane, llavender

see

Erigeron peregrinus

Fleabane, Marsh

see

Pluchea camphorata

Fleabane, Meadow

see

Erigeron speciosus

Fleabane, Philadelphia

see

Erigeron philadelphicus

Fleabane, Pink

see

Erigeron caespitosus

Fleabane, Triangular-Leafed

see

Senecio triangularis

Flindersia bennettiana

Blombery, Alec M.
What Wildflower is That

fl.   p 142, Pl. 377

Flixweed

see

Descurainia sophia

Floerkea proserpinacoides

Courtenay, Booth
Wildflowers & Weeds

fl.   p 67

Floppers

see

Bryophyllum pinnatum

Flor de Muerto

see

Lislanthius nigrescens

Floscopa glomerata

Moriarty, Audrey
Wild Flowers of Malawi

fl.   Pl. 20; 1

Floss flower

see

Ageratum houstonianum

Floss Silk Tree

see

Chorisia speciosa

Flower of a Day

see

Tradescantia virginiana

Flower-of-an-hour

see

Hibiscus trionum

Flower of Jupiter

see

Lychnis flos-jovis

Flower of the Ranges

see

Tetratheca ericifolia

Flower of Tigris

see

Tigridia pavonia

Fluellen, Round-leaved

see

Kickxia spuria

Flycatcher

see

Sarracenia

Fly-catcher Plant

see

Cephalotus follicularis

Fly-Poison

see

Amianthium muscaetoxicum

Foalswort

see

Tussilago farfara

Foamflower

see

Tiarella cordifolia

Foamflower

see

Tiarella trifoliata

Foamflower

see

Tiarella unifoliata

Fockea angustifolia

Flowering Plants of Africa
vol 43   1974-76

fl., fr.   Pl. 1711

Foeniculum officinale

Bianchini, F.
The Complete Book of Fruits & Vegetables

fr.   p 103

Foeniculum officinale

Perrot, Emile
Les Plantes Medicinales

fl. fr. hab.   p 96

Foeniculum officinale

Royal Hort. Soc.
Journal of the Royal Hort. Soc.
vol 96, No. 12   1971

hab.   p. 540   Pl. 229

Foeniculum vulgare

American Hort. Soc.
American Horticulturist
vol 53, No. 1   1974

hab., fr.   p. 37

Foeniculum vulgare

Ary, S.
The Oxford Book of Wildflowers

fl.   p 46, Pl. 5

Foeniculum vulgare

Bianchini, F.
The Complete Book of Fruits & Vegetables

hab.   p 103

Foeniculum vulgare

Bianchini, Francesco
Health Plants of the World

fl.   p 155

Foeniculum vulgare

Hay, Roy
The Color Dictionary of Flowers & Plants

hab.   p 143, Pl. 1137

Foeniculum vulgare

Hvass, Elsie
Plants That Feed and Serve Us

fl. fr.   p 67, Pl. 136

Foeniculum vulgare

Kariyone, Tatsuo
Atlas of Medicinal Plants

fl., fr.   Pl. 71

Foeniculum vulgare

Loewenfeld, Claire
The Complete Book of Herbs and Spices

fl. fr. hab.   p 128

Foeniculum vulgare

Martin, W. Keble
The Concise British Flora in Colour

fl. fr. hab.   Pl. 39

Foeniculum vulgare

Masefield, G. B.
The Oxford Book of Food Plants

fl. fr.   Pl. 139, 5

Foeniculum vulgare

Morley, Brian D.
Wild Flowers of the World

fl. fr.   Pl. 17C

Foeniculum vulgare

Rosengarten, Frederic, Jr.
The Book of Spices

fl. fr. p 240

Foeniculum vulgare

Takatori, Jisuke
Color Atlas of Medicinal Plants of Japan

fl.
fr.   Fig. 23B

Foeniculum vulgare var.

Masefield, G. B.
The Oxford Book of Food Plants

hab.   Pl. 149, 3

Fogfruit

see

Phyla lanceolata

Forbidden Fruit

see

Tabernaemontana dichotoma

Forestiera acuminata

Brown, Clair A.
Wildflowers of Louisiana and
Adjoining States

fl.　　　　　p 135

Forestiera ligustrina

Wharton, Mary E.
Trees & Shrubs of Kentucky

fl.　　　　p 111, Pl. 4.11
fr. hab.　　p 139, Pl. 2.32

Forget-me-not

see

Myosotis

Forget-me-not, Alpine

see

Eritrichium nanum

Forget-me-not, Alpine

see

Myosotis alpestris

Forget-me-not, Bur

see

Lappula diffusa

Forget-me-Not, Changing

see

Myosotis discolor

Forget-Me-Not, Chatham Islands

see

Myosotidium hortensia

Forget-me-not, Chinese

see

Cynoglossum amabile

Forget-me-not, Creeping

see

Omphalodes verna

Forget-me-not, dwarf

see

Eritrichium nanum

Forget-me-not, French

see

Malcolmia maritima

Forget-me-not, Giant

see

Myosotidium hortensia

Forget-me-not, Moss

see

Eritrichum elongatum

Forget-me-not, Mountain

see

Eritricium aretioides

Forget-me-not, summer

see

Anchusa capensis

Forget-me-not, summer white

see

Anchusa capensis var.

Forget-me-Not, Water

see

Myosotis palustris

Forget-me-not, Water

see

Myosotis scorpioides

Forget-me-not, white

see

Cryptantha nubigena

Forget-me-not, Woodland

see

Myosotis sylvatica

Forstera bellidifolia

Curtis, Winifred
The Endemic Flora of Tasmania
　　　　　　vol 3
fl.　　p 169

Forsythia europaea

Polunin, Oleg
Trees and Bushes of Europe

fl. hab.　　　p 164

Forsythia giraldiana

Curtis's Botanical Magazine
　Vol 164　　1943-48

fl.　　　　No. 9662

Forsythia hybrid

Crockett, James Underwood
Lawns & Ground Covers

fl.　p. 133

Forsythia hybrid

Curtis's Botanical Magazine
Vol 179　　　　1972

fl. fr.　　652

Forsythia hybrid.

Gault, S. Millar
The Color Dictionary of Shrubs

fl.　　　Pl. 196, 198

Forsythia hybrid

Harrison, R.E.
Trees and Shrubs

fl.　P 80

Forsythia x intermedia

Huxley, Anthony
Deciduous Garden Trees and Shrubs

fl.　　　　　Pl. 79

Forsythia x intermedia var.

Gault, S. Millar
The Color Dictionary of Shrubs

fl.　　　　　Pl. 197

Forsythia x intermedia var.

Hay, Roy
The Color Dictionary of Flowers &
　　Plants

fl.　　　p. 202　　Pl. 1611,1612

Forsythia x intermedia var.

Hellyer, A.G.L.
Shrubs in Colour

fl.   p. 54

Forsythia x intermedia var.

Menninger, Edwin A.
Flowering Vines of the World

fl.,hab.   Pl. 162

Forsythia x intermedia var.

Perry, Frances
Flowers of the World

fl.   p. 198

Forsythia x intermedia var.

Royal Hort. Soc.
Journal of the Royal Hort. Soc.
    vol 91, No. 5   1966

fl.   p. 206   Pl. 104

Forsythia koreana

Kitamura, Siro
Coloured Illustrations of Trees &
    Shrubs of Japan

fl.   417

Forsythia ovata

Curtis's Botanical Magazine
    Vol 159   1936

fl. fr.   No. 9437

Forsythia sp.

Everett, T. H., ed.
New Illustrated Encyclopedia of
    Gardening   Vol. 4

fl.   p. 695

Forsythia suspensa

Kariyone, Tatsuo
Atlas of Medicinal Plants

fl.   Pl. 64

Forsythia suspensa

Kimura, Koiti
Japanese Medicinal Plants, Vol I

fl.   p 72

Forsythia suspensa

Macoby, Stirling
What Flower is That

fl.   P 134

Forsythia suspensa

Takatori, Jisuke
Color Atlas of Medicinal Plants of Japan

fl.
fr.   Fig. 20

Forsythia suspensa var.

Gault, S. Millar
The Color Dictionary of Shrubs

fl.   Pl. 199

Fortunella japonica

Macoby, Stirling
What Flower is That

fr. P 135

Fortunella japonica

Nicolaisen, Age
Pocket Encyclopedia of Indoor
    Plants

fr.   Pl. 78

Fortunella sp.

Masefield, G. B.
The Oxford Book of Food Plants

fr.   Pl. 89, 2

Fosterella penduliflora

Wilson, Robert Gardner
Bromeliads in Cultivation

hab.   Pl. 114

Fothergilla gardenii

Hay, Roy
The Color Dictionary of Flowers & Plants

fl.   p 202, Pl. 1613

Fothergilla gardenii

Massachusetts Hort. Soc.
Horticulture
    vol 55, No. 12   1977

fl., hab.   p. 38

Fothergilla major

Bartels, Andreas
Das Grosse Buch der Gartengeholze

fl.   p 152

Fothergilla major

Gault, S. Millar
The Color Dictionary of Shrubs

fl.   Pl. 200, 201

Fothergilla major

Hellyer, A.G.L.
Shrubs in Colour

fl.   p. 54

Fothergilla major

Huxley, Anthony
Deciduous Garden Trees and Shrubs

fl.   Pl. 78

Fothergilla monticola

Bartels, Andreas
Das Grosse Buch der Gartengeholze

fl.   p 153

Fothergilla monticola

Hay, Roy
The Color Dictionary of Flowers & Plants

hab.   p 202, Pl. 1614

Fothergilla monticola

Massachusetts Hort. Soc.
Horticulture
    vol 41, No. 4   1963

fl.   cover

Fothergilla monticola

Perry, Frances
Flowers of the World

fl.   p 136

Fothergilla monticola

Royal Hort. Soc.
The Garden
    vol 101, No. 1 1976

fl.   p. 28

Fountain Grass

see

Pennisetum ruppelii

Fountain tree

See Spathodea campanulata

Fouquieria columnaris

American Hort. Soc.
American Horticulturist
    vol 56, No. 6   1977

hab.   p. 27

Fouquieria diguetii

Coyle, Jeanette
A Field Guide to the Common and
    Interesting Plants of Baja California

hab.   p 127

Fouquieria fasciculata

Cactus & Succulent Society of America
Journal
    vol 41, No. 4   1969

hab.   cover

Fouquieria formosa

O'Gorman, Helen
Mexican Flowering Trees and Plants

fl. hab.   p 27

Fouquieria splendens

Cactus & Succulent Society of America
Journal
vol 39, No. 3   1967

hab.   p. 101

Fouquieria splendens

Coyle, Jeanette
A Field Guide to the Common and
    Interesting Plants of Baja California

hab.   p 126

Fouquieria splendens

Lamb, Edgar
Colorful Cacti of the American Deserts

fl. hab.   Pl. 43,44, 55, 71, 96

Fouquieria splendens

Lemmon, Robert S.
Wildflowers of North America in
    Full Color

fl. hab.   pg. 76   Pl. 123

Fouquieria splendens

Mathias, Mildred E.
Color for the Landscape

fl.   p. 195

Fouquieria splendens

Milne, Lonus
Living Plants of the World

fl. hab.   p. 158

Fouquieria splendens

Morley, Brian D.
Wild Flowers of the World

fl.   Pl. 160

Fouquieria splendens

Orr, Robert T.
Wildflowers of Western America

fl.   Pl. 199

Fouquieria splendens

Royal Hort. Soc.
Journal of the Royal Hort. Soc.
    vol 86, No. 8   1961

fl.   p. 354   Pl. 109

Fouquieria splendens

Walcott, Mary Vaux
North American Wild Flowers
    vol. 5
fl.   Pl. 396

Fouquieria splendens

Wit, H. C. D. de
Plants of the World;
    The Higher Plants, Vol. I

fl.   p. 176   Pl. 106

Four-O'Clock

see

Mirabilis froebelii

Four-o'clock

see

Mirabilis jalapa

Four O'Clock, Giant

see

Mirabelis multiflora

Four-o'clock, Wild

see

Mirabilis nyctaginea

Four O'Clock, Wild

see

Oxybaphus nyctagineus

Fox and Cubs

see

Hieracium aurantiacum

Foxglove

see

Digitalis

Foxglove, annual

see

Digitalis purpurea

Foxglove, Bastard

see

Maurandia erubescens var.

Foxglove, Canary

see

Isoplexis canariensis

Foxglove, False

see

Agalinis fasciculata

Foxglove, False

see

Aureolaria

Foxglove, False

see

Gerardia flava

Foxglove, Fernleaf False

see

Aureolaria pedicularia

Foxglove, Mullein

see

Dasistoma macrophylla

Foxglove, Pink

see

Agalinis fasciculata

Foxglove, Purple

see

Digitalis purpurea

Foxglove, Red

see

Digitalis purpurea var.

Foxglove, Smooth False

see

Aureolaria laevigata

Foxglove, Sticky

see

Gerardia pectinata

Foxglove, Straw

see

Digitalis lutea

Foxglove Tree

see

Paulownia fortunei

Foxglove-Tree

see

Paulownia tomentosa

Foxglove, Wooly

see

Digitalis lanata

Foxglove, Woolly

see

Pittyrodia axillaris

Foxglove, Yellow

see

Digitalis grandiflora

Foxglove, Yellow

see

Digitalis lutea

Foxglove, Yellow False

see

Aureolaria flava

Fox Tail Grass

see

Alopecurus pratensis

Foxtail, green

see

Setaria viridis

Fragaria x Ananassa

Masefield, G. B.
The Oxford Book of Food Plants

fl. fr.   Pl. 75, 1

Fragaria chiloensis

American Hort. Soc.
American Horticultural Magazine
vol 49, No. 1   1970

fl., fr.   backcover

Fragaria chiloensis

Crockett, James Underwood
Lawns & Ground Covers

fl.   p. 133

Fragaria chiloensis

Heller, Christine
Wild Flowers of Alaska

fl.   Pl. 150

Fragaria chiloensis

Lemmon, Robert S.
Wildflowers of North America in
Full Color

fl.   pg.24   pl. 40

Fragaria chiloensis

Mathias, Mildred E.
Color for the Landscape

fl. hab.   p 140

Fragaria chiloensis

Munz, Philip A.
California Spring Wildflowers

fl.   p 33, Pl. 20

Fragaria chiloensis

Orr, Robert T.
Wildflowers of Western America

fl.   Pl. 63

Fragaria chiloensis var.

Macoby, Stirling
WhatFlower is That

fr.   P 135

Fragaria glauca

Porsild, A. E.
Rocky MOuntain Wild Flowers

fl. hab.   p 235

Fragaria glauca

Walcott, Mary Vaux
North American Wild Flowers
vol. 5

fl.   Pl. 362

Fragaria grandiflora

Riefel, Carlos von
A Folio of Fruit

fr.   Pl. 4

Fragaria indica

Kromdijk, G.
200 House Plants in Colour

fr.   Pl. 96

Fragaria indica

Macoby, Stirling
What Flower is That

fl.fr.   P 135

Fragaria vesca

Ary, S.
The Oxford Book of Wildflowers

fl. fr.   p 80, Pl. 4

Fragaria vesca

Barneby, T. P.
European Alpine Flowers in Colour

fl.,fr.   Pl. 36; 6

Fragaria vesca

Bianchini, F.
The Complete Book of Fruits & Vegetables

fl. fr.   p 159

Fragaria vesca

Felsko, Elsa
A Book of Wildflowers

fl., fr.   p. 149

Fragaria vesca

Furrer, D.
Blumen am Wege

fl. fr. hab.   p 35

Fragaria vesca

Hvass, Elsie
Plants That Feed and Serve Us

fl. fr. hab.   p 59, Pl. 112

Fragaria vesca

Kleijn, H.
Beauty of the Wild Plant

fl. opp. p. 45   Pl. 53

Fragaria vesca

Lindman, C. A. M.
Nordens Flora, Vol 5

fr.   Pl. 310A

Fragaria vesca

Martin, W. Keble
The Concise British Flora in Colour

fl. fr. hab.   Pl. 27

Fragaria vesca

Perrot, Emile
Les Plantes Medicinales

fl. fr. hab.   p 188

Fragaria vesca

Shaw, Richard J.
Field Guide to the Vascular Plants of
Grand Teton National Park

fl.   Pl. 9

Fragaria vesca

Wit, H. C. D. de
Plants of the World;
    The Higher Plants, Vol I

fl.   p 273, Pl. 147

Fragaria vesca var.

Crockett, James Underwood
Annuals

fl.fr.   P 119

Fragaria vesca var.

Masefield, G. B.
The Oxford Book of Food Plants

fl. fr.   Pl. 75, 2

Fragaria virginiana

Brown, Clair A.
Wildflowers of Louisiana and
    Adjoining States

fl. hab.   p 67

Fragaria virginiana

Campbell, Carlos C.
Great Smoky Mountain Wildflowers

fl.   p 27

Fragaria virginiana

Clark, Lewis J.
Wild Flowers of British Columbia

fl. fr.   p 242

Fragaria virginiana

Klimas, John E.
Wildflowers of Eastern America

fl.   Pl. 100

Fragaria virginiana

Lemmon, Robert S.
Wildflowers of North America in
    Full Color

fl.   pg. 182   pl. 285

Fragaria virginiana

Pond, Barbara
A Sampler of Wayside Herbs

fl. fr.   Pl. IV

Fragaria virginiana

Welsh, Stanley L.
Flowers of the Mountain Country

fl.   p. 13

Fragaria virginiana

Wharton, Mary E.
A Guide to the Wildflowers & Ferns
    of Kentucky

fl.   p 114   Pl. 1.41

Fragaria viridis

Lindman, C. A. M.
Nordens Flora, Vol 5

fl. fr. hab.   Pl. 310B

Fragrance, Golden

see

Pittosporum napaulense

Frailea asterioides

Kupper, Walter
Cacti

fl.,hab.   p. 65   Pl. 29

Frailea asterioides

Lamb, Edgar
The Pocket Encyclopedia of Cacti
    and Succulents in Color

fl. hab.   Pl. 55

Frailea castanea

Backeberg, Curt
Cactus Lexicon

hab.   p 609

Frailea matoana

Cactus & Succulent Society of America
Journal
    vol 43, No. 4   1971

fl.,, hab.   p. 139

Francoa appendiculata

Munoz Pizarro, Carlos
Flores Silvestres de Chile

fl.,, hab.   Pl. 34

Francoa sonchifolia

Kiaer, Eigil
Indoor Plants in Colour

hab.   p. 60

Frangipani

see

Plumeria

Frangipani Australian

see

Hymenosporum flavum

Frangipanni, native

see

Hymenosporum flavum

Frangipani, red

see

Plumeria rubra

Frangipani, White

see

Plumieria alba

Frangula alnus

Ary, S.
The Oxford Book of Wildflowers

fl. fr.   p 182, Pl. 2

Frangula alnus

Martin, W. Keble
The Concise British Flora in Colour

fl. fr. hab.   Pl. 20

Frangula alnus

Polunin, Oleg
Trees and Bushes of Europe

fl. fr. hab.   p 136

Frankenia grandifolia

Coyle, Jeanette
A Field Guide to the Common and
    Interesting Plants of Baja California

fl.   p 125

Frankenia laevis

Ary, S.
The Oxford Book of Wildflowers

fl.   p 112, Pl. 7

Frankenia laevis

Martin, W. Keble
The Concise British Flora in Colour

fl. hab.   Pl. 13

Frankenia palmeri

Coyle, Jeanette
A Field Guide to the Common and
    Interesting Plants of Baja California

fl.   p 125

Frankenia pauciflora

Blombery, Alec M.
What Wildflower is That

fl.   p 142, Pl. 378

Franklinia alatamaha

American Hort. Soc.
National Horticultural Magazine
    vol 34, No. 4   1955

fl.   cover

Franklinia alatamaha

American Hort. Soc.
American Horticulturist
vol 56, No. 4   1977

fl., fr., hab.   p. 39

Franklinia alatamaha

Massachusetts Hort. Soc.
Horticulture
vol 41, No. 9   1963

fl.   cover

Franklinia alatamaha

Massachusetts Hort. Soc.
Horticulture
vol 48, No. 9   1970

fl.   cover, backcover

Franklinia alatamaha

Massachusetts Hort. Soc.
Horticulture
vol 55, No. 5   1977

fl.   p. 36

Franklinia alatamaha

Pennsylvania Hort. Soc.
The Green Scene
vol 4, No. 4   1976

fl.   cover

Franklinia alatamaha

Walcott, Mary Vaux
North American Wild Flowers
vol. 4

fl.   Pl. 244

Franseria chamissonis var.

Clark, Lewis J.
Wild Flowers of British Columbia

fl.   p 551

Franseria chenopodifolia

Coyle, Jeanette
A Field Guide to the Common and
    Interesting Plants of Baja California

fr.   p 171

Franseria dumosa

Coyle, Jeanette
A Field Guide to the Common and
    Interesting Plants of Baja California

fr. hab.   p 173

Frasera speciosa

Lemmon, Robert S.
Wildflowers of North America in
    Full Color

fl.   pg. 144   pl. 230

Frasera speciosa

Pacific Hort. Foundation
Pacific Horticulture
vol 39, No. 2   1978

hab., fl.   p. 28

Fraxinus americana

Boom, B. K.
The glory of the tree

hab.   p. 110   pl. 188

Fraxinus americana

Edlin, Herbert
The Illustrated Encyclopedia
    of Trees

fr. hab.   p 205

Fraxinus americana

Flemer, William III
Shade and Ornamental Trees in Color

hab.   p 65

Fraxinus angustifolia

Polunin, Oleg
Trees and Bushes of Europe

fl. fr. hab.   p 161

Fraxinus angustifolia var.

Polunin, Oleg
Trees and Bushes of Europe

fr. hab.   p 162

Fraxinus excelsior

Boom, B.K.
The Glory of the Tree

fl., fr.   p. 110   Pl. 189-90

Fraxinus excelsior

Edlin, Herbert
The Illustrated Encyclopedia
    of Trees

fr. hab.   p 204-5

Fraxinus excelsior

Huxley, Anthony
Deciduous Garden Trees and Shrubs

fl. fr.   Pl. 80,80a,80b

Fraxinus excelsior

Hvass, Elsie
Plants That Feed and Serve Us

fr.   p 124, Pl. 273

Fraxinus excelsior

Lindman, C. A. M.
Nordens Flora, Vol 7

fr.   Pl. 477

Fraxinus excelsior

Macoby, Stirling
What Flower is That

fr.   P 135

Fraxinus excelsior

Martin, W. Keble
The Concise British Flora in Colour

fr.   Pl. 58

Fraxinus excelsior

Perrot, Emile
Les Plantes Medicinales

fl. fr. hab.   p 99

Fraxinus excelsior

Polunin, Oleg
Trees and Bushes of Europe

fl. fr. hab.   p 161

Fraxinus excelsior

Tosco, Uberto
The World of Mountain Flowers

hab.   p 27

Fraxinus excelsior

Vedel, H.
Arbres et Arbustes

fl. fr. hab.   p 104, 105, No. 110

Fraxinus excelsior var.

Edlin, Herbert
The Illustrated Encyclopedia
    of Trees

hab.   p 205

Fraxinus excelsior var.

Harrison, R. E.
Trees and Shrubs

hab.   P. 79

Fraxinus excelsior var.

Hay, Roy
The Color Dictionary of Flowers & Plants

hab.   p 202, Pl. 1615

Fraxinus excelsior var.

Oakman, Harry
Colorful Trees

Hab.   P 104

Fraxinus ornus

Bianchini, Francesco
Health Plants of the World

hab.   p 59

Fraxinus ornus

Boom, B. K.
The Glory of the Tree

fl. fr. hab.   Pl. 185, 186, 187

Fraxinus ornus

Edlin, Herbert
The Illustrated Encyclopedia
of Trees

fl. hab.   p 204-5

Fraxinus ornus

Huxley, Anthony
Deciduous Garden Trees and Shrubs

fl.   Pl. 81

Fraxinus ornus

Polunin, Oleg
Flowers of Europe

fl.   Pl. 94 #979

Fraxinus ornus

Polunin, Oleg
Trees and Bushes of Europe

fl. hab.   p 162, 163

Fraxinus oxycarpa var.

Oakman, Harry
Colorful Trees

Hab.   P. 105

Fraxinus pallisiae

Curtis's Botanical Magazine, v. 173
1960

fl.   Pl. 370

Fraxinus pallisiae

Polunin, Oleg
Trees and Bushes of Europe

fl. hab.   p 162

Fraxinus pennsylvanica

Flemer, William III
Shade and Ornamental Trees in Color

hab.   p. 65

Fraxinus pennsylvanica

Polunin, Oleg
Trees and Bushes of Europe

hab.   p 163

Fraxinus pennsylvanica var.

Flemer, William
Shade and Ornamental Trees in Color

hab.   p. 65

Fraxinus quadrangulata

Wharton, Mary E.
Trees & Shrubs of Kentucky

fl.   p 111, Pl. 4.10

Fraxinus sieboldiana var.

Kitamura, Siro
Coloured Illustrations of Trees &
Shrubs of Japan

fl.   418

Fraxinus trifoliata

Coyle, Jeanette
A Field Guide to the Common and
Interesting Plants of Baja California

fr.   p 149

Fraxinus uhdei var.

Crockett, James Underwood
Evergreens

hab.   p. 126

Fraxinus velutina var.

Flemer, William
Shade and Ornamental Trees in Color

hab.   p. 65

Freckle face

see

Hypoestes sanguinolenta

Freesia, Flame

see

Tritonia crocata

Freesia hybrid

Macoby, Stirling
What Flower is That

fl.   P. 136

Freesia hybrid

Miles, Bebe
Bulbs for the Home Gardener

fl.   p 165

Freesia hybrid

Perry, Frances
Flowers of the World

fl.   p. 149

Freesia x hybrida

Hay, Roy
The Dictionary of House Plants

fl.   Pl. 243

Freesia x hybrida

Huxley, Anthony
Garden Annuals and Bulbs

hab.,f.   Pl.159, 213

Freesia mendota

Macoby, Stirling
What Flower is That

fl.   P 136

Freesia, orange

see

Freesia mendota

Freesia refracta

Hay, Roy
The Color Dictionary of Flowers & Plants

fl.   p 92, Pl. 734

Freesia refracta

Macoby, Stirling
What Flower is That

fl.   P 136

Freesia refracta

Tsukamoto, Yotaro
Coloured illustrations of Garden
Flowers, v.9
fl.
opp.p.9

Fremontia californica

Harrison, R.E.
Trees and Shrubs

fl. P 79

Fremontia californica

Hay, Roy
The Color Dictionary of Flowers & Plants

fl.   p 248, Pl. 1977

Fremontia californica

Hellyer, A. G. L.
Shrubs in Color

fl.   p. 55

Fremontia californica

Lemmon, Robert S.
Wildflowers of North America in
Full Color

fl.   pg. 137   Pl. 217

Fremontia californica

Munz, Philip A.
California Desert Wildflowers

fl.   p 38, Pl. 34

Fremontia californica

Munz, Philip A.
California Spring Wildflowers

fl.   p 42, Pl. 46

Fremontia californica
Perry, Frances
Flowers of the World

fl.   p 288

Fremontia californica
Royal Hort. Soc.
Journal of the Royal Hort. Soc.
     vol 87, No. 4   1962

hab., fl.   p. 174   Pl. 52

Fremontia, Mexican
see
Fremontodendron mexicanum

Fremontodendron californicum
Mathias, Mildred E.
Color for the Landscape

fl. hab.   p 182

Fremontodendron californicum
Pacific Hort. Foundation
Pacific Horticulture
     vol 38, No. 4   1977

fl.   p. 40

Fremontodendron californicum var.
Gault, S. Millar
The Color Dictionary of Shrubs

fl.   Pl. 202

Fremontodendron hybrid
Mathias, Mildred E.
Color for the Landscape

fl.   p 168

Fremontodendron mexicanum
Walcott, Mary Vaux
North American Wild Flowers

fl. hab.   Pl. 206

Freycinetia arborea
Pacific Tropical Botanical Garden
Bulletin
     vol 2, No. 3   1972

fl., hab.   cover

Freylinia tropica
Flowering Plants of Africa
     vol 33   1959

fl.   Pl. 1320

Friar's Cowl
see
Arisarum vulgare

Fried-egg Plant
see
Oncoba spinosa

Friendly neighbor
see
Bryophyllum tubiflorum

Friendship Plant
see
Crassula arborescens

Fringe Cups
see
Tellima grandiflora

Fringe flower
see
Loropetalum chinense

Fringe Tree
see
Chionanthus virginica

Fringe tree
see
Garrya elliptica

Fringetree, Chinese
see
Chionanthus retusus

Frithia pulchra
American Hort. Soc.
American Horticulturist
     vol 53, No. 3   1974

fl.   p. 34

Frithia pulchra
Cactus & Succulent Society of America
Journal
     vol 36, No. 1   1964

fl.   cover (p. 1)

Frithia pulchra
Herre, H.
The Genera of the Mesembryanthemaceae

fl. fr.   p 161

Frithia pulchra
Subik, Rudolf
Decorative Cacti

fl. hab.   p 117

Fritillaria acmopetala
Hay, Roy
The Color Dictionary of Flowers & Plants

fl.   p 92, Pl. 736

Fritillaria acmopetala
Megaw, Elektra
Wild Flowers of Cyprus

fl., hab.   Pl 31

Fritillaria acmopetala
Polunin, Oleg
Flowers of the Mediterranean

fl.   No. 229

Fritillaria acmopetala
Royal Hort. Soc.
Journal of the Royal Hort. Soc.
     vol 91, No. 7   1966

fl.   p. 294   Pl. 156

Fritillaria alburyana
Mathew, Brian
Dwarf Bulbs

fl.   p. 128   Pl. 44

Fritillaria biflora
Orr, Robert T.
Wildflowers of Western America

fl.   Pl. 289

Fritillaria camschatcensis
Clark, Lewis J.
Wild Flowers of British Columbia

fl.   p 39

Fritillaria camschatcensis
Curtis's Botanical Magazine
Vol 166   1949

fl.   63

Fritillaria caucasica
Beck, Christabel
Fritillaries

fl.   p 35

Fritillaria cirrhosa
Curtis's Botanical Magazine
Vol 170   1954-55

fl.   255

Fritillaria cirrhosa

Royal Hort. Soc.
Journal of the Royal Hort. Soc.
vol 94, No. 5   1969

fl.   p. 214   Pl. 103

Fritillaria citrina

Curtis's Botanical Magazine
Vol 162   1939

fl.   No. 9560

Fritillaria citrina

Morley, Brian D.
Wild Flowers of the World

fl.   Pl. 41

Fritillaria crassifolia var.

Royal Hort. Soc.
Journal of the Royal Hort. Soc.
vol 89, No. 3   1964

fl.   p. 114   Pl. 40

Fritillaria davisii

Goulimis, Constantine N.
Wild Flowers of Greece

fl.   p 145

Fritillaria drenovskii

Curtis's Botanical Magazine
Vol 163   1940 - 42

fl.   No. 9625

Fritillaria drenovskii

Goulimis, Constantine N.
Wild Flowers of Greece

fl. hab.   p 147

Fritillaria ehrharti

Chrtis's Botanical Magazine
Vol 163   1940 - 42

fl. hab.   No. 9635

Fritillaria gibbosa

Mathew, Brian
Dwarf Bulbs

fl.   p. 112   Pl. 41

Fritillaria gibbosa

Royal Hort. Soc.
Journal of the Royal Hort. Soc.
vol 90, No. 11   1965

fl.   p. 470   Pl. 216

Fritillaria gibbosa

Wendelbo, Per
Tulips and Irises of Iran and
   Their Relatives

fl.   p 29, Pl. 26

Fritillaria glaucoviridis

Curtis's Botanical Magazine
Vol 159   1936

fl.   No. 9462

Fritillaria gracilis

Curtis's Botanical Magazine
Vol 160   1937

fl.   No. 9500

Fritillaria graeca

Goulimis, Constantine N.
Wild Flowers of Greece

fl. hab. p 145

Fritillaria graeca

Polunin, Oleg
Flowers of the Mediterranean

fl.   No. 228

Fritillaria hispanica

Taylor, A. W.
Wildflowers of Spain and Portugal

fl.   p. 10

Fritillaria imperialis

Huxley, Anthony
Garden Annuals and Bulbs

hab., fl.   Pl. 157, 218

Fritillaria imperialis

Massachusetts Hort. Soc.
Horticulture
vol 43, No. 5   1965

fl.   p. 32

Fritillaria imperialis

Massachusetts Hort. Soc.
Horticulture
vol 48, No. 11   1970

fl.   p. 25

Fritillaria imperialis

Massachusetts Hort. Soc.
Horticulture
vol 50, No. 4   1972

fl.   p. 33

Fritillaria imperialis

Miles, Bebe
Bulbs for the Home Gardener

fl.   p 73

Fritillaria imperialis

Perry, Frances
Flowers of the World

fl.   p 171

Fritillaria imperialis

Royal Hort. Soc.
Journal of the Royal Hort. Soc.
vol 88, No. 5   1963

fl.   p. 210   Pl. 65

Fritillaria imperialis

Wendelbo, Per
Tulips and Irises of Iran and
   Their Relatives

fl. hab.   p 27, Pl. 23

Fritillaria imperialis var.

Hay, Roy
The Color Dictionary of Flowers & Plants

fl.   p 93, Pl. 737, 738

Fritillaria involucrata

Barneby, T. P.
European Alpine Flowers in Colour

fl.   Pl. 2, 5

Fritillaria involucrata

Beck, Christabel
Fritillaries

fl.   p 51

Fritillaria kamchatcensis

Alaska-Yukon Wild Flowers Guide

fl.   p 9

Fritillaria kamchatcensis

Heller, Christine
Wild Flowers of Alaska

fl., hab.   Pl. 284-285

Fritillaria karelinii

Royal Hort. Soc.
Journal of the Royal Hort. Soc.
vol 90, No. 1   1965

fl.   p. 22   Pl. 11

Fritillaria kotschyana

Wendelbo, Per
Tulips and Irises of Iran and
   Their Relatives

fl.   p 31, Pl. 28

Fritillaria lanceolata

Clark, Lewis J.
Wild Flowers of British Columbia

fl.   p 39, 43

Fritillaria lanceolata

Ferguson, Mary
Wildflowers

fl.   p 22

Fritillaria lanceolata

Lemmon, Robert S.
Wildflowers of North America in
    Full Color

fl.   pg. 230   pl. 365

Fritillaria lanceolata

Orr, Robert T.
Wildflowers of Western America

fl.   Pl. 231

Fritillaria latifolia

Beck, Christabel
Fritillaries

fl.   p 41

Fritillaria libanotica

Megaw, Elektra
Wild Flowers of Cyprus

fl., hab.   Pl. 31

Fritillaria libanotica

Polunin, Oleg
Flowers of the Mediterranean

fl.   No. 231

Fritillaria liliacea

Curtis's Botanical Magazine
    Vol 161   1938

fl.   No. 9541

Fritillaria lutea

Beck, Christabel
Fritillaries

fl.   p 41

Fritillaria meleagris

Ary, S.
The Oxford Book of Wildflowers

fl.   p 162, Pl. 4

Fritillaria meleagris

Barneby, T. P.
European Alpine Flowers in Colour

fl.   Pl. 2, 4

Fritillaria meleagris

Felsko, Elsa
A Book of Wild Flowers, 2nd ser.

fl.   p 74

Fritillaria meleagris

Huxley, Anthony
Garden Annuals and Bulbs

hab., fl.   Pl. 158 219

Fritillaria meleagris

Kleijn, H.
Beauty of the Wild Plant

fl.   opp. p. 32   Pl. 34

Fritillaria meleagris

Lindman, C. A. M.
Nordens Flora, Vol 1

fl. fr.   Pl. 57

Fritillaria meleagris

Martin, W. Keble
The Concise British Flora in Colour

fl. hab.   Pl. 85

Fritillaria meleagris

Miles, Bebe
Bulbs for the Home Gardener

fl.   p 72

Fritillaria meleagris

Polunin, Oleg
Flowers of Europe

fl.   Pl. 167 #1622

Fritillaria meleagris

Royal Hort. Soc.
The Garden
    vol 100, No. 11   1975

fl.   p. 551

Fritillaria meleagris var.

Hay, Roy
The Color Dictionary of Flowers & Plants

fl.   p 93, Pl. 739, 740

Fritillaria meleagris var.

Perry, Frances
Flowers of the World

fl.   p. 171

Fritillaria messanensis

Curtis's Botanical Magazine
    Vol 164   1943-48

fl.   No. 9659

Fritillaria messanensis

Polunin, Oleg
Flowers of the Mediterranean

fl.   No. 227

Fritillaria michailovskyi

Mathew, Brian
Dwarf Bulbs

fl.   p. 128   Pl. 43

Fritillaria nigra

Morley, Brian D.
Wild Flowers of the World

fl., hab.   Pl. 24

Fritillaria nobilis

Beck, Christabel
Fritillaries

fl.   p 41

Fritillaria obliqua

Beck, Christabel
Fritillaries

fl. fr.   frontis.

Fritillaria olivieri

Royal Hort. Soc.
Journal of the Royal Hort. Soc.
    vol 88, No. 5   1963

fl.   p. 210   Pl. 70

Fritillaria pallidiflora

Hay, Roy
The Color Dictionary of Flowers & Plants

fl.   p 93, Pl. 741

Fritillaria pallidiflora

Morley, Brian D.
Wild Flowers of the World

fl., hab.   Pl. 24

Fritillaria pallidiflora

Royal Hort. Soc.
Journal of the Royal Hort. Soc.
    vol 85, No. 9   1960

fl.   p. 398   Pl. 127

Fritillaria pallidiflora

Royal Hort. Soc.
Journal of the Royal Hort. Soc.
    vol 92, No. 6   1967

fl.   p. 250   Pl. 126

Fritillaria persica

Morley, Brian D.
Flowers of the World

fl.   Pl. 51

Fritillaria persica

Wendelbo, Per
Tulips and Irises of Iran and
    Their Relatives

fl.   p 29, Pl. 24

Fritillaria pinardii

Curtis's Botanical Magazine
    Vol 170   1954-55

fl. hab.   227

Fritillaria pinetorum

Munz, Philip A.
California Mountain Wildflowers

fl.   p 29, Pl. 8

Fritillaria pontica var.

Curtis's Botanical Magazine
Vol. 165   1948

fl.   8

Fritillaria pudica

Addisonia, v. 22, 1943-46

fl.   opp. p. 37   Pl. 723

Fritillaria pudica

Clark, Lewis J.
Wild Flowers of British Columbia

fl.   p 39

Fritillaria pudica

Curtis's Botanical Magazine
Vol 163   1940-42

fl. hab.   No. 9617

Fritillaria pudica

Lemmon, Robert S.
Wildflowers of North America in
     full color

fl.   pg. 115   Pl. 183

Fritillaria pudica

Welsh, Stanley L.
Flowers of the Mountain Country

fl.   p. 52

Fritillaria purdyi

Mathew, Brian
Dwarf Bulbs

fl.   p. 128   Pl. 45

Fritillaria pyrenaica

Royal Hort. Soc.
Journal of the Royal Hort. Soc.
     vol 92, No. 6   1967

fl.   p. 250   Pl. 126

Fritillaria pyrenaica var.

Alpine Garden Society
Bulletin
     vol. 38, No. 2   1970

fl.   p. 139

Fritillaria pyrenaica var.

Royal Hort. Soc.
Journal of the Royal Hort. Soc.
     vol 93, No. 9   1968

fl.   p. 382   Pl. 195

Fritillaria raddeana

Mathew, Brian
Dwarf Bulbs

fl.   p. 112   Pl. 42

Fritillaria raddeana

Royal Hort. Soc.
Journal of the Royal Hort. Soc.
     vol 90, No. 11   1965

fl.   p. 470   Pl. 215

Fritillaria recurva

Lemmon, Robert S.
Wildflowers of North America in
     Full Color

fl.   pg. 115   Pl. 184

Fritillaria recurva

Massachusetts Hort. Soc.
Horticulture
     vol 48, No. 11   1970

fl.   p. 15

Fritillaria recurva

Miles, Bebe
Bulbs for the Home Gardener

fl.   p 72

Fritillaria recurva

Munz, Philip A.
California Mountain Wildflowers

fl.   p 29, Pl. 7

Fritillaria recurva

Royal Hort. Soc.
Journal of the Royal Hort. Soc.
     vol 92, No. 6   1967

fl.   p. 250   Pl. 125

Fritillaria reuteri

Curtis's Botanical Magazine
Vol 179   1972

fl. hab.   658

Fritillaria reuteri

Royal Hort. Soc.
Journal of the Royal Hort. Soc.
     vol 94, No. 3   1969

fl.   p. 118   Pl. 61

Fritillaria reuteri

Wendelbo, Per
Tulips and Irises of Iran and
     Their Relatives

fl.   p 29, Pl. 25

Fritillaria reuteri var.

Royal Hort. Soc.
Journal of the Royal Hort. Soc.
     vol 94, No. 3   1969

fl.   p. 118   Pl. 62

Fritillaria roylei

Morley, Brian D.
Wild Flowers of the World

fl.   Pl. 108

Fritillaria sewerzowii

Mathew, Brian
Dwarf Bulbs

fl.   p. 128   Pl. 46

Fritillaria sibthorpiana

Curtis's Botanical Magazine
Vol 167   1950

fl.   129

Fritillaria straussii

Wendelbo, Per
Tulips and Irises of Iran and
     Their Relatives

fl. hab.   p 31, Pl. 29

Fritillaria thunbergii

Kariyone, Tatsuo
Atlas of Medicinal Plants

fl., hab.   Pl. 19

Fritillaria thunbergii

Kimura, Koiti
Japanese Medicinal Plants, Vol I

fl.   p 10

Fritillaria tubiformis

Beck, Christabel
Fritillaries

fl.   p 35

Fritillaria tubiformis var.

Beck, Christabel
Fritillaries

fl.   p 35

Fritillaria verticillata var.

Takatori, Jisuke
Color Atlas of Medicinal Plants
     of Japan

fl.   Pl. 72B

Fritillaria zagrica

Wendelbo, Per
Tulips and Irises of Iran and
     Their Relatives

fl. hab.   p 31, Pl. 27

Fritillary, Scarlet

see

Fritillaria recurva

Fritillary, snake's head

see

Fritillaria meleagris

Fritillary, yellow

see

Fritillaria pudica

Froelichia floridana

Courtenay, Booth
Wildflowers & Weeds

fl.   p 69

Froelichia floridana

Duncan, Wilbur H.
Wildflowers of the Southeastern
     United States

fl.   p 31

Frog Belly

see

Sarracenia alata

Frogbit

see

Hydrocharis morsus-ranae

Frog Fruit

see

Lippia nodiflora

Frog's Britches

see

Sarracenia purpurea

Frogsmouth

see

Philydrum lanuginosum

Frost-Flower

see

Aster pilosus

Frostweed

see

Helianthemum canadense

Frostweed

see

Helianthemum corymbosum

Fruit Salad, New Zealand

see

Actinidia chinensis

Fruit salad plant

see

Monstera deliciosa

Fruit Salad Tree

see

Feijoa sellowiana

Frying Pans

see

Eschscholtzia lobbii

Fuchsia aborescens

Harrison, R.E.
Trees and Shrubs

fl.   P 81

Fuchsia, Australian

see

Epacris longiflora

Fuchsia, California

see

Ribes speciosum

Fuchsia, California

see

Zauschneria

Fuchsia, Cape

see

Phygelius capensis

Fuchsia cordifolia

Royal Hort. Soc.
Journal of the Royal Hort. Soc.
     vol 93, No. 4   1968

fl.   p. 164   Pl. 83

Fuchsia corymbiflora

Wit, H. C. D. de
Plants of the World:
     The Higher Plants, Vol II

fl.   p 51, Pl. 7

Fuchsia, Creeping

see

Heeria elegans

Fuchsia excorticata

Morley, Brian D.
Wildflowers of the World

fl.   Pl. 133

Fuchsia excorticata

Royal Hort. Soc.
The Garden
     vol 103, No. 10   1978

hab.   p. 404

Fuchsia fulgens

Morley, Brian D.
Wild Flowers of the World

fl.   Pl. 173

Fuchsia fulgens

Nicolaisen, Age
Pocket Encyclopedia of Indoor Plants

fl.   Pl. 100b

Fuchsia fulgens

O'Gorman, Helen
Mexican Flowering Trees and Plants

fl. hab.   p 61

Fuchsia fulgens var.

Huxley, Anthony
Garden Annuals and Bulbs

fl.   Pl. 46

Fuchsia gracilis var.

Royal Hort. Soc.
The Garden
     vol 100, No. 9   1975

fl., hab.   p. 422-23

Fuchsia, Hardy

see

Fuchsia magellanica

Fuchsia, Hardy

see

Fuchsia riccartonii

Fuchsia hybrid

Blunt, Wilfrid
Flora Superba

fl.   Pl. VIII

Fuchsia hybrid

Everett, T.H., ed.
New Illustrated Encyclopedia of
    Gardening   Vol IV

fl.   p. 695

Fuchsia hybrid.

Gault, S. Millar
The Color Dictionary of Shrubs

fl.   Pl. 203, 204, 205, 207, 208

Fuchsia hybrid

Harrison, R.E.
Trees and Shrubs

fl.   p 81

Fuchsia hybrid

Hay, Roy
The Color Dictionary of Flowers &
    Plants

fl.   p. 67-68, 202-03   Pl. 534-39,
                1616,1618

Fuchsia hybrid.

Hay, Roy
The Dictionary of House Plants

fl.   Pl. 244 - 248

Fuchsia hybrid

Hellyer, A.G.L.
Shrubs in Colour

fl.   p. 55

Fuchsia hybrid

Huxley, Anthony
Garden Annuals and Bulbs

fl.   Pl. 45, 47, 48, 50

Fuchsia hybrid

Kromdijk, G.
200 House Plants in Colour

fl.   Pl. 97

Fuchsia hybrid

Macoby, Stirling
What Flower is That

fl.   p. 136, 137

Fuchsia hybrid

Mathias, Mildred E.
Color for the Landscape

fl.   p 68

Fuchsia hybrid

Perry, Frances
Flowers of the World

fl.   p. 201-203

Fuchsia hybrid

Tsukamoto, Yotaro
Coloured Illustrations of Garden
    Flowers Vol 9

fl.   p. 47

Fuchsia x hybrida

Harrison, R. E.
Trees and Shrubs

fl.   p. 82

Fuchsia x hybrida

Nicolaisen, Age
Pocket Encyclopedia of Indoor Plants

fl.   Pl. 100

Fuchsia x hybrida var.

Kiaer, Eigil
Indoor Plants in Colour

fl.   p. 61

Fuchsia magellanica

American Hort. Soc.
American Horticulturist
    vol 57, No. 2   1978

fl.   p. 26

Fuchsia magellanica

Hay, Roy
The Color Dictionary of Flowers & Plants

fl. hab.   p 203, Pl. 1617

Fuchsia magellanica

Massachusetts Hort. Soc.
Horticulture
    vol 32, No. 11   1954

fl., hab.   p. 519

Fuchsia magellanica

Munoz Pizarro, Carlos
Flores Silvestres de Chile

fl.   Pl. 49

Fuchsia magellanica

Perry, Frances
Flowers of the World

fl.   p 201

Fuchsia magellanica

Polunin, Oleg
Trees and Bushes of Europe

fl. hab.   p 151

Fuchsia magellanica

Wit, H. C. D. de
Plants of the World;
    The Higher Plants, Vol II

fl.   p 52, Pl. 11

Fuchsia magellanica var.

Gault, S. Millar
The Color Dictionary of Shrubs

fl.   Pl. 206

Fuchsia magellanica var.

Hellyer, A.G.L.
Shrubs in Colour

fl.   p. 55

Fuchsia magellanica var.

Hersey, Jean
Woman's Day Book of House Plants

fl.   p. 64

Fuchsia magellanica var.

Macoby, Stirling
What Flower is That

fl.   p. 136

Fuchsia, native

see

Correa reflexa

Fuchsia, Native

see

Epacris longiflora

Fuchsia procumbens

Harrison, Richmond E.
Climbers and Trailers

fl. fr.   p 93   Pl. 231

Fuchsia procumbens

Morley, Brian D.
Wild Flowers of the World

fl. fr.   Pl. 133

Fuchsia procumbens

Perry, Frances
Flowers of the World

fl.   p 201

Fuchsia riccartonii

Huxley, Anthony
Deciduous Garden Trees and Shrubs

fl.   Pl. 82

Fuchsia simplicicaulis

Morley, Brian D.
Wild Flowers of the World

fl.   p. 173

Fuchsia sp.

Milne, Lorus
Living Plants of the World

fl.   p 170

Fuchsia splendens

Chickering, Carol Rogers
Flowers of Guatemala

fl.   p 65

Fuchsia splendens

Harrison, R.E.
Trees and Shrubs

fl. P 81

Fuchsia, Trailing

see

Fuchsia procumbens

Fuchsia, Tree

see

Schotia brachypetala

Fuchsia triphylla var.

Harrison, R. E.
Trees and Shrubs

fl.   p. 81

Fuchsia triphylla var.

Macoby, Stirling
What Flower is That

fl.   p. 137

Fuchsia triphylla var.

Tsukamoto, Yotaro
Coloured illustrations of Garden
     Flowers   Vol 9

fl.   p. 47

Fuirena squarrosa

Tarver, David P.
Aquatic and Wetland Plants
     of Florida

fl. hab.   p 83

Fumaria bastardii

Martin, W. Keble
The Concise British Flora in Colour

fl. fr. hab.   Pl. 6

Fumaria capreolata

Martin, W. Keble
The Concise British Flora in Colour

fl. fr. hab.   Pl. 6

Fumaria capreolata

Polunin, Oleg
Flowers of Europe

fl.   Pl. 30 #280

Fumaria capreolata

Polunin, Oleg
Flowers of the Mediterranean

fl.   No. 37

Fumaria densiflora

Martin, W. Keble
The Concise British Flora in Colour

fl. fr. hab.   Pl. 6

Fumaria martinii

Martin, W. Keble
The Concise British Flora in Colour

fl. fr. hab.   Pl. 6

Fumaria muralis var.

Martin, W. Keble
The Concise British Flora in Colour

fl. fr. hab.   Pl. 6

Fumaria occidentalis

Martin, W. Keble
The Concise British Flora in Colour

fl. fr. hab.   Pl. 6

Fumaria officinalis

Ary, S.
The Oxford Book of Wildflowers

fl.   p 136, Pl. 1

Fumaria officinalis

Bianchini, Francesco
Health Plants of the World

fl.   p 143

Fumaria officinalis

Courtenay, Booth
Wildflowers & Weeds

fl.   p 22

Fumaria officinalis

Felsko, Elsa
A Book of Wild Flowers, 2nd Ser.

fl.   p 83

Fumaria officinalis

Furrer, D.
Blumen am Wege

fl. fr. hab.   p 29

Fumaria officinalis

Lindman, C. A. M.
Nordens Flora, Vol 4

fl. fr.   Pl. 250

Fumaria officinalis

Martin, W. Keble
The Concise British Flora in Colour

fl. fr. hab.   Pl. 6

Fumaria officinalis

Perrot, Emile
Les Plantes Medicinales

fl. fr. hab.   p 100

Fumaria parviflora

Martin, W. Keble
The Concise British Flora in Colour

fl. fr. hab.   Pl. 6

Fumaria purpurea

Martin, W. Keble
The Concise British Flora in Colour

fl. fr. hab.   Pl. 6

Fumaria vaillantii

Martin, W. Keble
The Concise British Flora in Colour

fl. fr. hab.   Pl. 6

Fumeroot, Pink

see

Capnoides sempervirens

Fume-root, pink

see

Corydalis sempervirens

Fumewort, golden

see

Corydalis aurea

Fumitory

see

Fumaria officinalis

Fumitory, Hollow

see

Corydalis cava

Fumitory, Yellow

see

Corydalis lutea

Funkia

see

Hosta plantaginea

Fur Bush, Golden

see

Callicarpa loureiroi

Furcraea bedinghausii

Royal Hort. Soc.
Journal of the Royal Hort. Soc.
vol 93, No. 8   1968

hab.   p. 336   Pl. 171

Furcraea gigantea

Curtis's Botanical Magazine
vol 48   1821

fl.   Pl. 2250

Furcraea longaeva

Royal Hort. Soc.
The Garden
vol 101, No. 4   1976

fl., hab.   p.218

Furcraea macrophylla

Addisonia, v. 22, 1943-46

fl.   Pl. 714

Furze

see

Ulex europaeus

Furze, Dwarf

see

Ulex minor

Gagea arvensis

Felsko, Elsa
A Book of Wild Flowers
2nd ser.
fl. hab.   p 3

Gagea arvensis

Polunin, Oleg
Flowers of the Mediterranean

fl.   No. 236

Gagea confusa

Wendelbo, Per
Tulips and Irises of Iran and
Their Relatives

fl. hab.   P. 42   Pl. 42

Gagea fistulosa

Barneby, T. P.
European Alpine Flowers in Colour

fl. hab.   Pl. 2, 6

Gagea fistulosa

Kohlhaupt, Paula
Fleurs des Alpages
vol 2
fl.   P 8

Gagea fistulosa

Polunin, Oleg
Flowers of Europe

fl.   Pl. 164 #1601

Gagea graeca

Polunin, Oleg
Flowers of Europe

Gagea hispanica

Taylor, A. W.
Wildflowers of Spain and Portugal

fl.   p. 11

Gagea lutea

Lindman, C. A. M.
Nordens Flora, Vol 1

fl. hab.   Pl. 55A

Gagea lutea

Martin, W. Keble
The Concise British Flora in Colour

fl. hab.   Pl. 86

Gagea minima

Lindman, C. A. M.
Nordens Flora, Vol 1

fl. hab.   Pl. 55B

Gagea peduncularis

American Hort. Soc.
American Horticulturist
vol 53, No. 5 1974

fl.   p. 16

Gagea peduncularis

Megaw, Elektra
Wild Flowers of Cyprus

fl., hab.   Pl. 26

Gagea reticulata

Wendelbo, Per
Tulips and Irises of Iran and
Their Relatives

fl.   p 42, Pl. 41

Cahnia clarkei

Blombery, Alec M.
What Wildflower is That

fl.   p 143, Pl. 379

Cahnia graminifolia

Curtis, Winifred
The Endemic Flora of Tasmania, Vol 5

fl. fr. hab.   Pl. 160

Cahnia radula

Cochrane, G. R.
Flowers and Plants of Victoria

fl.   Pl. 14

Cahnia sieberiana

Cochrane, G. R.
Flowers and Plants of Victoria

fl. fr. hab.   Pl. 356, 357

Gaillardia aristata

Lemmon, Robert S.
Wildflowers of North America in
Full Color

fl.   p. 162, 211   Pl. 259, 334

Gaillardia aristata

Macoby, Stirling
What Flower is That

fl.   P 137

Gaillardia aristata

Neufeld, J.B.
Wild Flowers of the Prairies

fl.   p. 13

Gaillardia aristata

Porsild, A. E.
Rocky Mountain Wild Flowers

fl.   p 417

Gaillardia aristata

Tsukamoto, Yotaro
Coloured Illustrations of Garden
Flowers

fl.   Illus. 63   opp. p. 20

Gaillardia aristata

Walcott, Mary Vaux
North American Wild Flowers
vol. 5
fl. Pl. 354

Gaillardia aristata

Welsh, Stanley L.
Flowers of the Mountain Country

fl. p. 53

Gaillardia aristata var.

Huxley, Anthony
Garden Perennials and Water Plants

fl. Pl. 116

Gaillardia hybrid

Hay, Roy
The Color Dictionary of Flowers & Plants

fl. p 143, Pl. 1138, 1139

Gaillardia pinnatifida

Lemmon, Robert S.
Wildflowers of North America in
Full Color

fl. p. 211 Pl. 334

Gaillardia pinnatifida

Welsh, Stanley L.
Flowers of the Canyon Country

fl. p 39

Gaillardia pulchella

Batson, Wade T.
Wild Flowers in South Carolina

fl. p 123

Gaillardia pulchella

Brown, Clair A.
Wildflowers of Louisiana and
Adjoining States

fl. hab. p 202

Gaillardia pulchella

Bruggeman, L.
Tropical Plants

fl. Pl. 41 & 42

Gaillardia pulchella

Crockett, James Underwood
Annuals

fl. P 120

Gaillardia pulchella

Dean, Blanche E.
Wildflowers of Alabama and
Adjoining States

fl. p 201

Gaillardia pulchella

Duncan, Wilbur H.
Wildflowers of the Southeastern
United States

fl. p 223

Gaillardia pulchella

Morley, Brian D.
Wild Flowers of the World

fl. Pl. 165

Gaillardia pulchella

Tsukamoto, Yotaro
Coloured Illustrations of Garden
Flowers

fl. opp. p. 48

Gaillardia pulchella var.

Everett, T.H., ed.
New Illustrated Encyclopedia of
Gardening vol 5

fl. p. 742

Gaillardia pulchella var.

Hay, Roy
The Color Dictionary of Flowers &
Plants

fl. p. 37 Pl. 296

Gaillardia pulchella var.

Huxley, Anthony
Garden Annuals and Bulbs

fl. Pl. 51-2

Gaillardia, red

see

Gaillardia aristata

Gaillardia, whole-coloured

see

Gaillardia aristata

Gaimardia fitzgeraldii

Curtis, Winifred
The Endemic Flora of Tasmania
Vol VI

fl. fr. hab. Pl. 204

Galactia elliottii

Duncan, Wilbur H.
Wildflowers of the Southeastern
United States

fl. p 87

Galactia minor

Duncan, Wilbur H.
Wildflowers of the Southeastern
United States

fl. p 87

Galactia regularis

Fleming, Glenn
Wild Flowers of Florida

fl. p 64

Galactia volubis

Wharton, Mary E.
A Guide to the Wildflowers & Ferns
of Kentucky

fl. p 186 Pl. 1.16

Galactites tomentosa

Polunin, Oleg
Flowers of Europe

fl. Pl. 154 #1493

Galactites tomentosa

Polunin, Oleg
Flowers of the Mediterranean

fl. No. 208

Galactites tomentosa

Royal Hort. Soc.
The Garden
vol 103, No. 2 1978

fl. p. 55

Galanthus allenii

Royal Hort. Soc.
Journal of the Royal Hort. Soc.
vol 79, No. 10 1954

fl. p. 436 Pl. 117

Galanthus allenii

Stern, F. C.
Snowdrops and Snowflakes

fl. Pl. 5

Galanthus caucasicus

Stern, F. C.
Snowdrops and Snowflakes

fl. hab. Frontis.

Galanthus elwesii

Hay, Roy
The Color Dictionary of Flowers & Plants

fl. p 93, Pl. 742

Galanthus elwesii

Morley, Brian D.
Wild Flowers of the World

fl.,hab. Pl. 42

Galanthus elwesii

Pennsylvania Hort. Soc.
The Green Scene
vol 5, No. 4 1977

fl. p. 15 Pl. 6

Galanthus elwesii

Tsukamoto, Yotaro
Coloured Illustrations of Garden
    Flowers

fl.   opp. p. 9

Galanthus ikariae

Curtis's Botanical Magazine
    Vol 160   1937

fl. hab.   No. 9474

Galanthus ikariae

Hay, Roy
The Color Dictionary of Flowers & Plants

fl.   p 93, Pl. 743

Galanthus latifolius

Curtis's Botanical Magazine
    Vol 164   1943-48

fl.   No. 9669

Galanthus nivalis

American Hort. Soc.
American Horticulturist
    vol 52, No. 1   1973

fl.   cover

Galanthus nivalis

Everett, T.H., ed.
New Illustrated Encyclopedia of
    Gardening vol 5

fl.   p. 742

Galanthus nivalis

Felsko, Elsa
A Book of Wildflowers

fl.   P 125

Galanthus nivalis

Hay, Roy
The Color Dictionary of Flowers & Plants

fl.   p 93, Pl. 744

Galanthus nivalis

Huxley, Anthony
Garden Annuals and Bulbs

hab., fl.   Pl. 160 , 220

Galanthus nivalis

Martin, W. Keble
The Concise British Flora in Colour

fl. hab.   Pl. 83

Galanthus nivalis

Massachusetts Hort. Soc.
Horticulture
    vol 42, No. 3   1964

fl.   p. 15

Galanthus nivalis

Polunin, Oleg
Flowers of Europe

fl.   Pl. 173 #1663

Galanthus nivalis

Tosco, Uberto
The World of Mountain Flowers

fl. hab.   p 27

Galanthus nivalis

Vilmorin, Roger de
Plantes Alpines dans les Jardins

fl. hab.   Pl. IV

Galanthus nivalis

Wit, H. C. D. de
Plants of the World;
    The Higher Plants, Vol II

fl.   Pl. 135

Galanthus nivalis var.

Hay, Roy
The Color Dictionary of Flowers &
    Plants

fl.   p. 94   Pl. 745-46

Galanthus nivalis var.

Miles, Bebe
Bulbs for the Home Gardener

fl. hab.   p 74

Galanthus nivalis var.

Perry, Frances
Flowers of the World

fl.   p. 23   Pl. 2

Galanthus nivalis var.

Stern, F. C.
Snowdrops and Snowflakes

fl. hab.   Pl. 4

Galanthus plicatus

Hay, Roy
The Color Dictionary of Flowers & Plants

fl.   p 94, Pl. 746

Galanthus plicatus var.

Royal Hort. Soc.
Journal of the Royal Hort. Soc.
    vol 92, No. 6   1967

fl.   p. 250   Pl. 114

Galanthus sp.

Massachusetts Hort. Soc.
Horticulture
    vol 38, No. 10   1960

fl.   p. 530

Galanthus transcaucasicus

Wendelbo, Per
Tulips and Irises of Iran and
    Their Relatives

fl.   p 55, Pl. 57

Galax aphylla

Batson, Wade T.
Wild Flowers in South Carolina

fl.   p 89

Galax aphylla

Campbell, Carlos C.
Great Smoky Mountain Wildflowers

fl.   p 55

Galax aphylla

Crockett, James Underwood
Lawns & Ground Covers

fl. p. 133

Galax aphylla

Dean, Blanche E.
Wildflowers of Alabama and
    Adjoining States

fl.   p 127

Galax aphylla

Duncan, Wilbur H.
Wildflowers of the Southeastern
    United States

fl.   p 119

Galax aphylla

Klimas, John E.
Wildflowers of Eastern America

fl.   Pl. 31

Galax aphylla

Massachusetts Hort. Soc.
Horticulture
    vol 52, No. 10   1974

fl., hab.   p. 34

Galax aphylla

Wharton, Mary E.
A Guide to the Wildflowers & Ferns
    of Kentucky

fl.   p 154   Pl. 2.13

Gale, Sweet

    see

Myrica gale

Galeandra beyrichii

Luer, Carlyle A.
The Native Orchids of Florida

fl. hab.   Pl. 74; 5-8

Galeandra devoniana

Kramer, Jack
Orchids; Flowers of Romance
and Mystery

fl.   p 146

Galeandra devoniana

Massachusetts Hort. Soc.
Horticulture
    vol 45, No. 3   1967

fl.   p. 47

Galeandra devoniana

Pabst, G. F. J.
Orchidaceae Brasilienses,
    Vol 1

fl.   p 184

Galeandra dives

Pabst, G. F. J.
Orchidaceae Brasilienses,
    Vol 1

fl.   p 184

Galeandra lacustris

Pabst, G. F. J.
Orchidaceae Brasilienses,
    Vol 1

fl.   p 184

Galeandra leptoceras

Ospina, Mariano
Orquideas de las americas

fl.   Pl. 143

Galeandra montana

Pabst, G. F. J.
Orchidaceae Brasilienses,
    Vol 1

fl.   p 184

Galearis spectabilis

Luer, Carlyle A.
The Native Orchids of the
    United States and Canada

fl. hab.   Pl. 36

Galega officinalis

Bianchini, Francesco
Health Plants of the World

fl.   p 155

Galega officinalis

Perrot, Emile
Les Plantes Medicinales

fl. fr. hab.   p 199

Galega officinalis

Polunin, Oleg
Flowers of Europe

fl.   Pl. 53 #524

Galega officinalis var.

Hay, Roy
The Color Dictionary of Flowers &
    Plants

fl.   p. 143   Pl. 1140,1141

Galega officinalis var.

Huxley, Anthony
Garden Perennials and Water Plants

fl.   Pl. 117

Galeobdolon luteum

Ary, S.
The Oxford Book of Wildflowers

fl.   p 28, Pl. 3

Galeobdolon luteum

Felsko, Elsa
A Book of Wild Flowers
            2nd ser
fl. hab.   p 7

Galeobdolon luteum

Kleijn, H.
The Beauty of the Wild Plant

fl.   p. 49   Pl. 60

Galeobdolon luteum

Martin, W. Keble
The Concise British Flora in Colour

fl. hab.   Pl. 70

Galeobdolon luteum

Polunin, Oleg
Flowers of Europe

fl.   Pl. 112 #1132

Galeola cassythoides

Blombery, Alec M.
What Wildflower is That

fl.   p 143, Pl. 380

Galeola cassythoides

Cady, Leo
Australian Native Orchids in Colour

fl. hab.   Pl. 72

Galeopsis angustifolia

Ary, S.
The Oxford Book of Wildflowers

fl.   p 148, Pl. 1

Galeopsis angustifolia

Martin, W. Keble
The Concise British Flora in Colour

fl. hab.   Pl. 69

Galeopsis segetum

Martin, W. Keble
The Concise British Flora in Colour

fl. hab.   Pl. 69

Galeopsis speciosa

Kleijn, H.
The Beauty of the Wild Plant

fl.   p. 33   Pl. 37

Galeopsis speciosa

Lindman, C. A. M.
Nordens Flora, Vol 8

fl. fr.   Pl. 502

Galeopsis speciosa

Martin, W. Keble
The Concise British Flora in Colour

fl. hab.   Pl. 69

Galeopsis speciosa

Polunin, Oleg
Flowers of Europe

fl.   Pl. 111 #1126

Galeopsis tetrahit

Ary, S.
The Oxford Book of Wildflowers

fl.   p 148, Pl. 2

Galeopsis tetrahit

Courtenay, Booth
Wildflowers & Weeds

fl.   p 95

Galeopsis tetrahit

Martin, W. Keble
The Concise British Flora in Colour

fl. hab.   Pl. 69

Galeopsis tetrahit

Polunin, Oleg
Flowers of Europe

fl.   Pl. 111 #1126

Galeottia jprisiana

Dunsterville, G. C. K
Introduction to the World
    of Orchids

fl.   Pl. 10

Galinsoga ciliata

Courtenay, Booth
Wildflowers & Weeds

fl.   p 113

Galinsoga ciliata

Furrer, D.
Blumen am Wege

fl. hab.   p 121

Galinsoga ciliata

Weeds of the Southern United States

hab., fl.   p. 12

Galinsoga parviflora

Polunin, Oleg
Flowers of Europe

fl.   Pl. 147 #1407

Galinsoga quadriradiata

Felsko, Elsa
A Book of Wildflowers

fl.   P 146

Galium albescens

Curtis, Winifred
The Endemic Flora of Tasmania
vol 5

fl., fr., hab.   Pl. 193

Galium aparine

Ary, S.
The Oxford Book of Wildflowers

fl. fr.   p 92, Pl. 1

Galium aparine

Courtenay, Booth
Wildflowers & Weeds

fl.   p 72

Galium aparine

Lindman, C. A. M.
Nordens Flora, Vol 8

fl. fr.   Pl. 552

Galium aparine

Martin, W. Keble
The Concise British Flora in Colour

fl. fr. hab.   Pl. 42

Galium aparine

Perrot, Emile
Les Plantes Medicinales

fl. fr. hab.   p 102

Galium boreale

Alaska-Yukon Wild Flower Guide

fl.   p 156

Galium boreale

Clark, Lewis J.
Wild Flowers of British Columbia

fl.   p 494

Galium boreale

Courtenay, Booth
Wildflowers & Weeds

fl.   p 72

Galium boreale

Lemmon, Robert S.
Wildflowers of North America in
Full Color

fl.   pg. 203   Pl. 320

Galium boreale

Lindman, C. A. M.
Nordens Flora, Vol 8

fl. fr. hab.   Pl. 549

Galium boreale

Martin, W. Keble'
The Concise British Flora in Colour

fl. hab.   Pl. 42

Galium boreale

Moyle, John B.
Northland Wild Flowers

fl. hab.   p 150, Pl. 173

Galium boreale

Porsild, A. E.
Rocky Mountain Wild Flowers

fl.   p 365

Galium boreale

Walcott, Mary Vaux
North American Wild Flowers

fl.   Pl. 63

Galium concinnum

Courtenay, Booth
Wildflowers & Weeds

fl.   p 72

Galium concinnum

Wharton, Mary E.
A Guide to the Wildflowers & Ferns
of Kentucky

fl.   p 171   Pl. 2.50

Galium cruciata

Ary, S.
The Oxford Book of Wildflowers

fl.   p 30, Pl. 1

Galium cruciata

Perrot, Emile
Les Plantes Medicinales

fl. fr. hab.   p 102

Galium debile

Martin, W. Keble
The Concise British Flora in Colour

fl. fr. hab.   Pl. 42

Galium gaudichaudii

Cochrane, G. R.
Flowers and Plants of Victoria

fl.   Pl. 215

Galium helveticum

Barneby, T. P.
European Alpine Flowers in Colour

fl.   Pl. 79, 3

Galium hercynicum

Ary, S.
The Oxford Book of Wildflowers

fl.   p 92, Pl. 4

Galium mollugo

Ary, S.
The Oxford Book of Wildflowers

fl.   p 92, Pl. 2

Galium mollugo

Perrot, Emile
Les Plantes Medicinales

fl. fr. hab.   p 102

Galium mollugo var.

Martin, W. Keble
The Concise British Flora in Colour

fl. fr. hab.   Pl. 42

Galium odoratum

Lindman, C. A. M.
Nordens Flora, Vol 8

fl. fr. hab.   Pl. 550

Galium odoratum

Martin, W. Keble
The Concise British Flora in Colour

fl. fr. hab.   Pl. 42

Galium odoratum

Polunin, Oleg
Flowers of Europe

fl.   Pl. 99 #1027

Galium palustre

Martin, W. Keble
The Concise British Flora in Colour

fl. fr. hab.   Pl. 42

Galium palustre

Polunin, Oleg
Flowers of Europe

fl.   Pl. 99 #1022

Galium parisiense

Martin, W. Keble
The Concise British Flora in Colour

fl. hab.   Pl. 42

Galium pumilum

Barneby, T. P.
European Alpine Flowers in Colour

fl.   Pl. 79, 2

Galium pumilum

Martin, W. Keble
The Concise British Flora in Colour

fl. fr. hab.   Pl. 42

Galium pusilum

Tosco, Uberto
The World of Mountain Flowers

fl. hab.   p. 78

Galium saxatile

Martin, W. Keble
The Concise British Flora in Colour

fl. fr. hab.   Pl. 42

Galium sp.

Klimas, John E.
Wildflowers of Eastern America

fl.   Pl. 68

Galium stenophyllum

Moriarty, Audrey
Wild Flowers of Malawi

fl.   Pl. 48; 1

Galium sylvaticum

Felsko, Elsa
A Book of Wild Flowers
            2nd ser
fl.   p 136

Galium tricornutum

Martin, W. Keble
The Concise British Flora in Colour

fl. fr. hab.   Pl. 42

Galium triflorum

Courtenay, Booth
Wildflowers & Weeds

fl.   p 72

Galium uliginosum

Martin, W. Keble
The Concise British Flora in Colour

fl. fr. hab.   Pl. 42

Galium verum

Ary, S.
The Oxford Book of Wildflowers

fl.   p 30, Pl. 2

Galium verum

Courtenay, Booth
Wildflowers & Weeds

fl.   p 72

Galium verum

Lemmon, Robert S.
Wildflowers of North America in
    Full Color

fl.   pg. 202   Pl. 319

Galium verum

Lindman, C. A. M.
Nordens Flora, Vol 8

fl. fr.   Pl. 551

Galium verum

Martin, W. Keble
The Concise British Flora in Colour

fl. hab.   Pl. 42

Galium verum

Perrot, Emile
Les Plantes Medicinales

fl. fr. hab.   p 102

Galium verum

Pond, Barbara
A Sampler of Wayside Herbs

fl.   Pl. XIII

Galleta Grass

see

Hilaria rigida

Gallineta

see

Mascagnia macroptera

Calphimia

see

Thryallis glauca

Galphimia glauca

O'Gorman, Helen
Mexican Flowering Trees and Plants

fl. hab.   p 63

Galpinia transvaalica

Flowering Plants of Africa
    vol 40   1969-70

fl.   Pl. 1591

Galtonia candicans

Hay, Roy
The Color Dictionary of Flowers & Plants

fl.   p 94, Pl. 747

Galtonia candicans

Macoby, Stirling
What Flower is That

fl.   p. 138

Galtonia candicans

Miles, Bebe
Bulbs for the Home Gardener

fl.   p 75

Galtonia candicans

Tsukamoto, Yotaro
Coloured Illustrations of Garden
    Flowers   Vol. 9

fl.   opp. p. 9

Galtonia viridiflora

Flowering Plants of Africa
    vol 30   1954-55

fl.   Pl. 1188

Galvezia speciosa

Orr, Robert T.
Wildflowers of Western America

fl.   Pl. 191

Gamolepis chrysanthemoides

Macoby, Stirling
What Flower is That

fl.   P 138

Gamolepis tagetes

Crockett, James Underwood
Annuals

fl.   P 120

Garberia heterophylla

Fleming, Glenn
Wild Flowers of Florida

fl.   p 86

Garcinia cowa

Roxburgh, William
Icones Roxburghianae

fl, fr.   Pl. 13

Garcinia kydia

Roxburgh, William
Icones Roxburghianae

fl., fr.   Pl. 14

Garcinia lanceofoelia

Roxburgh, William
Icones Roxburghianae

fl., fr.   Pl. 15

Garcinia mangostana

Masefield, G. B.
The Oxford Book of Food Plants

fr.   Pl. 101, 3

Garcinia mangostana

Morley, Brian D.
Wild Flowers of the World

fl.fr.   Plate 118

Garcinia multiflora

Walden, Beryl M.
Wild Flowers of Hong Kong

fl.   Pl. 30(81)

Garcinia paniculata

Roxburgh, William
Icones Roxburghianae

fl., fr.   Pl. 16

Garcinia pedunculata

Roxburgh, William
Icones Roxburghianae

fl., fr.   Pl. 17, 18

Gardenia asperula

Palgrave, K. C.
Trees of Central Africa

fl., fr.   p. 387

Gardenia augusta

Bruggeman, L.
Tropical Plants

fl.   Pl. 240

Gardenia cornuta

Flowering Plants of South Africa
vol XXIII   1943

fr.   Pl. 918

Gardenia gjellerupii

Bruggeman, L.
Tropical Plants

fl.   Pl. 209

Gardenia hybrid

Harrison, R.E.
Trees and Shrubs

fl.   P 82

Gardenia jasminoides

Crockett, James Underwood
Evergreens

hab.,fl.   p. 127

Gardenia jasminoides

Everett, T.H., ed.
New Illustrated Encyclopedia of
    Gardening   vol 5

fl.   p. 775

Gardenia jasminoides

Hersey, Jean
Woman's Day Book of House Plants

Gardenia jasminoides

Kiaer, Eigil
Indoor Plants in Colour

fl.   p. 62

Gardenia jasminoides

Kromdijk, G.
200 House Plants in Colour

fl.   Pl. 98

Gardenia jasminoides

Nicolaisen, Age
Pocket Encyclopedia of Indoor Plants

fl.   Pl. 142

Gardenia jasminoides

Takatori, Jisuke
Color Atlas of Medicinal Plants of Japan

fl.
fr.   Fig. 9

Gardenia jasminoides

Walden, Beryl M.
Wild Flowers of Hong Kong

fl.   Pl. 42(116)

Gardenia jasminoides var.

Encke, Fritz
Zimmerpflanzen

fl.   p. 78

Gardenia jasminoides var.

Hay, Roy
The Dictionary of House Plants

fl.   Pl. 249

Gardenia jasminoides var.

Kariyone, Tatsuo
Atlas of Medicinal Plants

fl., fr.   Pl. 56

Gardenia jasminoides var.

Kitamura, Siro
Coloured Illustrations of Trees &
    Shrubs of Japan

fl.   433

Gardenia jasminoides var.

Macoby, Stirling
What Flower is That

fl. P 138

Gardenia jasminoides var.

Perry, Frances
Flowers of the World

fl.   p. 266

Gardenia jasminoides var.

Tsukamoto, Yotaro
Coloured Illustrations of Garden
    Flowers   vol 10

fl.   p. 49   Pl. 158

Gardenia neuberia

Flowering Plants of Africa
vol 31   1956

fl., fr.   Pl. 1229

Gardenia rothmannia

Flowering Plants of South Africa
vol XVII   1937

fl., fr.   Pl. 657

Gardenia, Scentless

    see

Tabernaemontana dichotoma

Gardenia spatulifolia

Codd, L.E.W.
Trees & Shrubs of the Kruger
    National Park

fl., fr.   p. 162   Pl. VI

Gardenia spatulifolia

Flowering Plants of Africa
vol 32   1957-58

fl, fr.   Pl. 1241

Gardenia spatulifolia

Morley, Brian D.
Wild Flowers of the World

fl.,fr.   Pl. 79

Gardenia spatulifolia

Palgrave, K. C.
Trees of Central Africa

fl., fr.   p. 391

Gardenia thunbergia

Macoby, Stirling
What Flower is That

fl.   P 138

Gardenia, Transvaal

see

Gardenia spatulifolia

Gardenia, tree

see

Rothmannia globosa

Gardinia subacaulis

Moriarty, Audrey
Wild Flowers of Malawi

fl.   Pl. 48; 3

Gardoquia coccinea

Addisonia
vol 21   1939-42

fl.   p. 45   Pl. 695

Garland Flower

see

Daphne cneorum

Garland Flower

see

Hedychium coronarium

Garland, mountain

see

Clarkia elegans

Garlic

see

Allium sativum

Garlic, Broad Leaved

see

Allium ursinum

Garlic, Crow

see

Allium vineale

Garlic, Elegant

see

Allium pulchellum

Garlic, False

see

Nothoscordum bivalve

Garlic, False

see

Nothoscordum bivalve

Garlic, Field

see

Allium vineale

Garlic, Meadow

see

Allium canadense

Garlic, Rose

see

Allium Roseum

Garlic, Society

see

Tulbaghia violacea

Garlic, Southern

see

Allium insubricum

Garlic, Stag's

see

Allium victorialis

Garlic, Sweet

see

Tulbaghia violacea

Garlic, Wild

see

Allium canadense

Garlic, wild

see

Allium vineale

Garrya elliptica

Curtis's Botanical Magazine
Vol 170   1954-55

fl.   220

Garrya elliptica

Gault, S. Millar
The Color Dictionary of Shrubs

fl.   Pl. 209

Garrya elliptica

Harrison, R.E.
Trees and Shrubs

fl.   P 82

Garrya elliptica

Hay, Roy
The Color Dictionary of Flowers & Plants

fr.   p 203, Pl. 1619

Garrya elliptica

Hellyer, A.G.L.
Shrubs in Colour

fl.   p. 55

Garrya elliptica

Macoby, Stirling
What Flower is That

fl.   P 139

Garrya elliptica

Morley, Brian D.
Wild Flowers of the World

fl. fr.   Pl. 155

Garrya elliptica

Perry, Frances
Flowers of the World

fl.   p 120

Garuleum album

Flowering Plants of Africa
vol 40   1969-70

fl.   Pl. 1593

Gas Plant

see

Dictamus albus

Gasteria armstrongii

Subik, Rudolf
Decorative Cacti

fl. hab.   p 119

Gasteria beckeri

Lamb, Edgar
The Pocket Encyclopedia of Cacti
and Succulents in Color

hab.   Pl. 257

Gasteria brevifolia

Hay, Roy
The Dictionary of House Plants

hab.   Pl. 250

Gasteria candicans

Morley, Brian D.
Wild Flowers of the World

fl.   Pl. 85

Gasteria excelsa

Flowering Plants of South Africa
vol XVII   1937

fl.   Pl. 665

Gasteria liliputana

Lamb, Edgar
The Pocket Encyclopedia of Cacti
and Succulents in Color

hab.   Pl. 259

Gasteria liliputana

Lamb, Edgar
Popular Exotic Cacti in Color

hab.   p 93

Gasteria maculata

Subik, Rudolf
Decorative Cacti

fl. hab.   p 121

Gasteria, ox-tongue

see

Gasteria verrucosa

Gasteria pulchra

Addisonia
vol 23   1954-59

fl.   p. 31   Pl. 752

Gasteria rawlinsonii

Flowering Plants of Africa
vol 43   1974-76

fl.   Pl. 1701

Gasteria transvaalensis

Flowering Plants of Africa
vol 38   1967

fl.   Pl. 1501

Gasteria verrucosa

Encke, Fritz
Zimmerpflanzen

hab.   p. 78

Gasteria verrucosa

Kiaer, Eigil
Indoor Plants in Colour

fl.   p. 62

Gastrochilus bellinus

Kamemoto, Haruyuki
Beautiful Thai Orchid Species

fl.   p 149

Gastrochilus dasypogon

Kamemoto, Haruyuki
Beautiful Thai Orchid Species

fl.   p 149

Gastrodia elata

Kariyone, Tatsuo
Atlas of Medicinal Plants

fl., fr., hab.   Pl. 3

Gastrodia elata

Kimura, Koiti
Japanese Medicinal Plants, Vol I

fl.   p 18

Gastrodia sesamoides

Cady, Leo
Australian Native Orchids in Colour

fl.   Pl. 73

Gastrodia sesamoides

Cochrane, G. R.
Flowers and Plants of Victoria

fl.   Pl. 437

Gastrodia sesamoides

Morcombe, M. K.
Australia's Western Wildflowers

fl.   p 102

Gastrolobium villosum

Blombery, Alec M.
What Wildflower is That

fl.   p 144, Pl. 381

Gaultheria appressa

Cochrane, G. R.
Flowers and Plants of Victoria

fl.   Pl. 392

Gaultheria codonantha

Curtis's Botanical Magazine
Vol 159   1936

fl. fr.   No. 9456

Gaultheria elliptica

Curtis's Botanical Magazine
vol 175   1964-65

fl.   Pl. 449

Gaultheria eriophylla

Curtis's Botanical Magazine
Vol 170   1954-55

fl. fr.   254

Gaultheria hispida

Blombery, Alec M.
What Wildflower is That

fl.   p 144, Pl. 382

Gaultheria hispida

Curtis, Winifred
The Endemic Flora of Tasmania
vol 3

fl. fr.   p 193

Gaultheria hispidula

Clark, Lewis J.
Wild Flowers of British Columbia

fr.   p 391

Gaultheria hispidula

Courtenay, Booth
Wildflowers & Weeds

fl.   p 77

Gaultheria miqueliana

Curtis's Botanical Magazine
Vol 163   1940-42

fl. fr.   No. 9629

Gaultheria ovatifolia

Clark, Lewis J.
Wild Flowers of British Columbia

fr.   p 391

Gaultheria procumbens

Bartels, Andreas
Das Grosse Buch der Gartengeholze

fr. hab.   p 153

Gaultheria procumbens

Campbell, Carlos C.
Great Smoky Mountain Wildflowers

fl.   p 81

Gaultheria procumbens

Courtenay, Booth
Wildflowers & Weeds

fl.   p 79

Gaultheria procumbens

Ferguson, Mary
Wildflowers

fr.   p 176

Gaultheria procumbens

Hay, Roy
The Color Dictionary of Flowers & Plants

fr.   p 203, Pl. 1620

Gaultheria procumbens

Huxley, Anthony
Evergreen Garden Trees and Shrubs

fr. hab.   Pl. 123

Gaultheria procumbens

Klimas, John E.
Wildflowers of Eastern America

fl.   Pl. 76

Gaultheria procumbens

Moyle, John B.
Northland Wild Flowers

fl. fr.   p 118, Pl. 118, 118a

Gaultheria procumbens

Perry, Frances
Flowers of the World

fr. p 108

Gaultheria procumbens

Royal Hort. Soc.
Journal of the Royal Hort. Soc.
vol 92, No. 4   1967

fr.   p. 160   Pl. 71

Gaultheria procumbens

Turner, Nancy J.
Wild Coffee and Tea Substitutes of
          Canada

fr.   p. 53

Gaultheria procumbens

Wharton, Mary E.
A Guide to the Wildflowers & Ferns
          of Kentucky

fl. p 173   Pl. 2,53

Gaultheria semi-infera

Curtis's Botanical Magazine
Vol 169   1952-53

fl. fr.   197

Gaultheria shallon

Clark, Lewis J.
Wild Flowers of British Columbia

fl. fr.   p 387

Gaultheria shallon

Fries, Mary A.
Wildflowers of Mount Ranier and
          the Cascades

fl.   P 56

Gaultheria shallon

Hellyer, A.G.L.
Shrubs in Colour

fl., fr.   p. 58

Gaultheria shallon

Massachusetts Hort. Soc.
Horticulture
vol 47, No. 2   1969

fl.   p. 32-33

Gaultheria shallon

Morley, Brian D.
Wild Flowers of the World

fl. fr.   Pl. 161

Gaultheria shallon

Orr, Robert T.
Wildflowers of Western America

fl.   Pl. 33

Gaultheria shallon

Taylor, Ronald J.
Mountain Wild Flowers

fl.   p 52

Gaultheria stapfiana

Curtis's Botanical Magazine
Vol 179   1972

fl.   651

Gaultheria tetramera

Curtis's Botanical Magazine
Vol 163   1940 - 42

fl. fr.   No. 9618

Gaultheria wardii

Curtis's Botanical Magazine
Vol 161   1938

fl. fr.   No. 9516

Gaura biennis

Courtenay, Booth
Wildflowers & Weeds

fl.   p 75

Gaura biennis

Dean, Blanche E.
Wildflowers of Alabama and
          Adjoining States

fl.   p 119

Gaura coccinea

Moyle, John B.
Northland Wild Flowers

fl.   p 98, Pl. 87

Gaura coccinea

Welsh, Stanley L.
Flowers of the Canyon Country

fl.   p 22

Gaura lindheimeri

Brown, Clair A.
Wildflowers of Louisiana and
          Adjoining States

fl.   p 121

Gayfeather

see

Liatris spicata

Gayfeather

see

Liatris squarrosa

Gayfeather, Round-Headed

see

Liatris scariosa

Gayfeather, spike

see

Liatris spicata

Gaywings

see

Polygala paucifolia

Gaylussacia baccata

Courtenay, Booth
Wildflowers & Weeds

fl.   p. 75

Gaylussacia baccata

Wharton, Mary E.
Trees & Shrubs of Kentucky

fl.   p 76, Pl. 2.14b
fr.   p 141, Pl. 2.37a

Gaylussacia brachycera

Walcott, Mary Vaux
North American Wild Flowers

fl. hab.   Pl. 229

Gaylussacia brachycera

Wharton, Mary E.
Trees & Shrubs of Kentucky

fl.   p 76, Pl. 2.14a
fr.   p 141, Pl. 2.37b

Gaylussacia frondosa

Brown, Clair A.
Wildflowers of Louisiana and
    Adjoining States

fl. hab.   p 127

Gazania hybrid

Macoby, Stirling
What Flower is That

fl.   P 139

Gazania hybrid

Mathias, Mildred E.
Color for the Landscape

fl. hab.   p. 145

Gazania hybrid

Perry, Frances
Flowers of the World

fl.   p. 87

Gazania krebsiana

Eliovson, Sima
Namaqualand in Flower

fl.   Pl. 8

Gazania krebsiana var.

Curtis's Botanical Magazine
    vol 174   1962-63

fl.   Pl. 418

Gazania lichtensteinii

Eliovson, Sima
Namaqualand in Flower

fl.   Pl. 8

Gazania longiscapa

Crockett, James Underwood
Annuals

fl.   P 120

Gazania longiscapa var.

Harrison, Richmond E.
Climbers and Trailers

fl.   p. 95   Pl. 237

Gazania munroi

Flowering Plants of South Africa
    vol XVII   1937

fl., hab.   Pl. 659

Gazania pavonia

Everett, T. H., ed.
New Illustrated Encyclopedia of
        Gardening Vol. 5

fl.   p. 775

Gazania pinnata

American Hort. Soc.
American Horticulturist
    vol 53, No. 3   1974

fl.   p. 34

Gazania pinnata

Royal Hort. Soc.
Journal of the Royal Hort. Soc.
    vol 90, No. 8   1965

fl.   p. 338   Pl. 142

Gazania rigens

Kiaer, Eigil
Indoor Plants in Colour

fl.   p. 62

Gazania rigens var.

Morley, Brian D.
Wild Flowers of the World

fl.   Pl. 83

Gazania splendens

Hay, Roy
The Color Dictionary of Flowers & Plants

fl.   p 38, Pl. 297

Gazania splendens

Huxley, Anthony
Garden Annuals and Bulbs

fl.   Pl. 54

Gazania splendens

Macoby, Stirling
What Flower is That

fl.   P 139

Gazania splendens

Nicolaisen, Age
Pocket Encyclopedia of Indoor Plants

fl.   Pl. 149

Gazania, Trailing

see

Gazania uniflora

Gazania uniflora

Crockett, James Underwood
Lawns & Ground Covers

fl.   p. 134

Gazania uniflora

Macoby, Stirling
What Flower is That

fl.   P 139

Gean, Double

see

Prunus avium var.

Geebung, Australian

see

Persoonia pinifolia

Geebung, Broad-leaf

see

Persoonia levis

Geebung, pineleaf

see

Persoonia pinifolia

Geebung, Prickly

see

Persoonia juniperina

Geiger

see

Cordia dodecandra

Geiger Tree

see

Cordia sebestena

Geigera parviflora

Blombery, Alec M.
What Wildflower is That

fl.   p 145, Pl. 386

Geissorhiza namaquensis

Flowering Plants of South Africa
    Vol. XVIII   1938

fl., hab.   Pl. 688

Geissorhiza rochensis

American Hort. Soc.
American Horticulturist
    vol 53, No. 3   1974

fl.   p. 35

Geissorrhiza rochensis

American Hort. Soc.
American Horticulturist
    vol 55, No. 2   1976

fl.   cover

Geissorhiza rochensis

Morley, Brian D.
Wild Flowers of the World

fl., hab.   Pl. 87

Geleznowia verrucosa

Macoby, Stirling
What Flower is That

fl.   P 140

Geleznowia verrucosa

Morcombe, M. K.
Australia's Western Wildflowers

fl.   p 92

Gelsemium rankinii

Lemmon, Robert S.
Wildflowers of North America in
    Full Color

fl.   Pg. 38   pl. 64

Gelsemium sempervirens

American Hort. Soc.
American Horticulturist
    vol 55, No. 6 1976

fl., hab.   p. 6

Gelsemium sempervirens

Batson, Wade T.
Wild Flowers in South Carolina

fl.   p 92

Gelsemium sempervirens

Brown, Clair A.
Wildflowers of Louisiana and
    Adjoining States

fl. hab.   p 136

Gelsemium sempervirens

Dean, Blanche E.
Wildflowers of Alabama and
    Adjoining States

fl.   p 131

Gelsemium sempervivens

Everett, T. H.
New Illustrated Encyclopedia of
    Gardening, V. 5

fl.   opp. p. 743

Gelsemium sempervirens

Fleming, Glenn
Wild Flowers of Florida

fl.   p 61

Gelsemium sempervirens

Harrison, Richmond E.
Climbers and Trailers

fl.   p. 47   Pl. 95

Gelsemium sempervirens

Macoby, Stirling
What Flower is That

fl.   P 140

Gelsemium sempervirens

Mathias, Mildred E.
Color for the Landscape

fl. hab.   p 109

Gelsemium sempervirens

Menninger, Edwin A.
Flowering Vines of the World

fl.   Pl. 112

Gelsemium sempervirens

Walcott, Mary Vaux
North American Wild Flowers

fl.   Pl. 220

Genepy

see

Artemisia genipii

Geniosporum paludosum

Moriarty, Audrey
Wild Flowers of Malawi

fl.   Pl. 53; 6

Genista acanthoclados

Polunin, Oleg
Flowers of the Mediterranean

fl.   No. 60

Genista aethnensis

Gault, S. Millar
The Color Dictionary of Shrubs

fl.   Pl. 210

Genista aethnensis

Hellyer, A.G.L.
Shrubs in Color
fl.
p. 58

Genista aethnensis

Pacific Hort. Foundation
Pacific Horticulture
    vol. 39, No. 2 1978

hab., fl.   p. 40

Genista aethnensis

Polunin, Oleg
Trees and Bushes of Europe

fl. fr.   p. 109

Genista anglica

Martin, W. Keble
The Concise British Flora in Colour

fl. fr. hab.   Pl. 21

Genista canariensis

Macoby, Stirling
What Flower is That

fl.   P 140

Genista cinerea

Gault, S. Millar
The Color Dictionary of Shrubs

fl.   Pl. 211

Genista cinerea

Hay, Roy
The Color Dictionary of Flowers & Plants

fl.   p 203, Pl. 1621

Genista cinerea

Polunin, Oleg
Flowers of the Mediterranean

fl.   No. 57

Genista dalmatica

Alpine Garden Society
Bulletin
    vol. 39, No. 3   1971

fl.   p. 225

Genista equisetiformis

Polunin, Oleg
Flowers of the Mediterranean

fl.   No. 58

Genista equisetiformis

Taylor, A. W.
Wild Flowers of Spain and Portugal

fl.   p. 58

Genista hirsuta

Polunin, Oleg
Flowers of the Mediterranean

fl.   No. 59

Genista hispanica

Bartels, Andreas
Das Grosse Buch der Gartengeholze

fl. hab.   p 153

Genista hispanica

Hay, Roy
The Color Dictionary of Flowers & Plants

fl.   p 203, Pl. 1622

Genista hispanica

Hellyer, A. G. L.
Shrub in Color

fl.   p. 58

Genista hispanica

Polunin, Oleg
Flowers of Europe

fl.   Pl. 52 #511

Genista hispanica var.

Vilmorin, Roger de
Plantes Alpines dans les Jardins

fl. hab.   Pl. III

Genista januensis

Curtis's Botanical Magazine
    Vol 162   1939

fl.   No. 9574

Genista juncea

Massachusetts Hort. Soc.
Horticulture
    vol 54, No. 6   1976

fl., fr.   p. 45

Genista lydia

Curtis's Botanical Magazine
Vol 171   1956-57

fl.   292

Genista lydia

Hay, Roy
The Color Dictionary of Flowers & Plants

fl. hab.   p 203, Pl. 1623

Genista lydia

Hellyer, A.G.L.
Shrubs in Color
fl.
p. 58

Genista lydia

Royal Hort. Soc.
The Garden
    vol 100, No. 10   1975

fl.   p. 474

Genista lydia

Royal Hort. Soc.
The Garden
    vol 102, No. 12 1977

fl., hab.   p. 499

Genista monosperma

Macoby, Stirling
What Flower is That

fl.   P 140

Genista pilosa

Barneby, T. P.
European Alpine Flowers in Colour

fl.   Pl. 39, 2

Genista pilosa

Martin, W. Keble
The Concise British Flora in Colour

fl. fr. hab.   Pl. 21

Genista sagittalis

Barneby, T. P.
European Alpine Flowers in Colour

fl.   Pl. 39, 4

Genista sagittalis

Curtis's Botanical Magazine, v.172
1958-59
fl.
pl. 332

Genista sagittalis

Gault, S. Millar
The Color Dictionary of Shrubs

fl. hab.   Pl. 212

Genista sp.

Everett, T. H.
New Illustrated Encyclopedia of
Gardening, Vol. 10

fl., hab.   opp. p. 1830

Genista sphacelata

Feinbrun-Dothan, Naomi
Wild Plants in the Land
of Israel

fl. fr.   p 48

Genista tinctoria

Ary, S.
The Oxford Book of Wildflowers

fl.   p 18, Pl. 1

Genista tinctoria

Barneby, T. P.
European Alpine Flowers in Colour

fl.   Pl. 39, 3

Genista tinctoria

Felsko, Elsa
A Book of Wildflowers

fl.   P 31

Genista tinctoria

Huxley, Anthony
Deciduous Garden Trees and Shrubs

fl.   Pl. 83

Genista tinctoria

Lindman, C. A. M.
Nordens Flora, Vol 5

fl. fr.   Pl. 335

Genista tinctoria

Martin, W. Keble
The Concise British Flora in Colour

fl. hab.   Pl. 21

Genista tinctoria var.

Hay, Roy
The Color Dictionary of Flowers & Plants

fl.   p 203, Pl. 1624

Genista tinctoria var.

Massachusetts Hort. Soc.
Horticulture
    vol. 40, No. 6   1962

fl.   inside cover

Genista virgata

Hellyer, A. G. L.
Shrubs in Color

fl.   p. 58

Genlisea hispidula
Moriarty, Audrey
Wild Flowers of Malawi
fl.   Pl. 38; 3

Gentian, Allegheny Glade
see
Gentiana saponaria var.

Gentian, Alpine
see
Gentiana newberryi

Gentian, Alpine
see
Gentiana nivalis

Gentian, Arctic
see
Gentiana algida

Gentian, Austral
see
Gentianella diemensis

Gentian, Bavarian
see
Gentiana bavarica

Gentian, Bladder
see
Gentiana utriculosa

Gentian, Bluegreen
see
Gentiana glauca

Gentian, Catchfly
see
Eustoma exaltatum

Gentian, Closed
see
Gentiana andrewsii

Gentian, Creeping
see
Crawfurdia fasciculata

Gentian, Desert
see
Eustoma exaltatum

Gentian, Downy
see
Gentiana puberula

Gentian, Eastern Fringed
see
Gentiana crinata

Gentian, Favrat's
see
Gentiana orbicularis

Gentian, Fringed
see
Gentiana ciliata

Gentian, Fringed
see
Gentiana crinita

Gentian, Fringed
see
Gentiana thermalis

Gentian, German
see
Gentianella germanica

Gentian, Horse
see
Triosteum aurantiacum

Gentian, Horse Yellow
see
Triosteum angustifolium

Gentian, Hungarian
see
Gentiana pannonica

Gentian, Ladder
see
Gentiana acuta

Gentian, Marsh
see
Gentiana pneumonanthe

Gentian, Meadow
see
Gentianella campestris

Gentian, Moss
see
Gentiana prostrata

Gentian, Pinebarren
see
Gentiana porphyrio

Gentian, Prairie
see
Eustoma grandiflorum

Gentian, Prairie
see
Gentiana affinis

Gentian, Purple
see
Gentiana purpurea

Gentian, Rainier pleated
see
Gentiana calycosa

Gentian, Riverbank
see
Gentiana affinis

Gentian, Rose

see

Sabatia

Gentian, Rose Saltmarsh

see

Sabbatia stellaris

Gentian, Ruff

see

Gentiana calycosa

Gentian, Schleicher's

see

Gentiana bavarica var.

Gentian, Sierran

see

Frasera speciosa

Gentian, Slender

see

Gentiana tenella

Gentian, Small Alpine

see

Gentiana nivalis

Gentian, Snow

see

Gentiana nivalis

Gentian, Soapwort

see

Gentiana saponaria

Gentian, Spotted

see

Gentiana punctata

Gentian, Spring

see

Gentiana verna

Gentian, star

see

Swertia perennis

Gentian, Stemless

see

Gentiana acaulis

Gentian, Stemless

see

Gentiana clusii

Gentian, Stiff

see

Gentiana quinquefolia

Gentian, Striped

see

Gentiana villosa

Gentian, Swallow-root

see

Gentiana asclepiadea

Gentian, Swamp

see

Gentiana douglasiana

Gentian, Trumpet

see

Gentiana clusii

Gentian, western fringed

see

Gentiana thermalis

Gentian, Whitish

see

Gentian algida

Gentian, Willow

see

Gentiana asclepiadea

Gentian, Yellow

see

Gentiana lutea

Gentian, Yellowish

see

Gentiana flavida

Gentiana acaulis

Bartlett, Mary
Gentians

fl.   p 52

Gentiana acaulis

Bianchini, Francesco
Health Plants of the World

fl.   p 25

Gentiana acaulis

Color Treasury of Herbs & Other
     Medicinal Plants

fl.   p 44   Pl. 59

Gentiana acaulis

Felsko, Elsa
A Book of Wildflowers

fl.   P 53

Gentiana acaulis

Hay, Roy
The Color Dictionary of Flowers & Plants

fl.   p 9, Pl. 71

Gentiana acaulis

Huxley, Anthony
Garden Perennials and Water Plants

fl.   Pl. 119

Gentiana acaulis

Macoby, Stirling
What Flower is That

fl.   P 141

Gentiana acaulis

Massachusetts Hort. Soc.
Horticulture
     vol 49, No. 10 1971

fl.   p. 26

Gentiana acaulis

Massachusetts Hort. Soc.
Horticulture
     vol 52, No. 2   1974

fl., hab.   p. 36

Gentiana acaulis

Royal Hort. Soc.
Journal of Royal Hort. Soc.
vol 83, No. 4   1958

fl.   p. 162   Pl. 41

Gentiana acaulis

Royal Hort. Soc.
Journal of the Royal Hort. Soc.
vol 92, No. 10   1967

fl.   p. 432   Pl. 213

Gentiana acaulis

Tosco, Uberto
The World of Mountain Flowers

fl. hab.   p 92, 116

Gentiana acuta

Walcott, Mary Vaux
North American Wild Flowers
vol. 4

fl.   Pl. 294

Gentiana affinis

Porsild, A. E.
Rocky Mountain Wild Flowers

fl.   p 329

Gentiana affinis

Walcott, Mary Vaux
North American Wild Flowers

fl.   Pl. 87

Gentiana algida

Alaska-Yukon Wild Flower Guide

fl. hab.   p 128, 129

Gentiana algida

Heller, Christine
Wild Flowers of Alaska

fl.   Pl. 186

Gentiana algida

Welsh, Stanley L.
Flowers of the Mountain Country

fl.   p. 14

Gentiana alpina

Bartlett, Mary
Gentians

fl.   p 33

Gentiana alpina

Royal Hort. Soc.
Journal of the Royal Hort. Soc.
vol 86, No. 7   1961

fl.   p. 310   Pl. 84

Gentiana alpina

Royal Hort. Soc.
Journal of the Royal Hort. Soc.
vol 87, No. 1   1962

fl.   p. 22   Pl. 8

Gentiana andrewsii

Bartlett, Mary
Gentians

fl.   p 70

Gentiana andrewsii

Courtenay, Booth
Wildflowers & Weeds

fl.   p 80

Gentiana andrewsii

Ferguson, Mary
Wildflowers

fl.   p 51

Gentiana andrewsii

Jennings, O. E.
Wild Flowers of Western Pennsylvania
v. 2
fl.   Pl. 116

Gentiana andrewsii

Klimas, John E.
Wildflowers of Eastern America

fl.   Pl. 294

Gentiana andrewsii

Lemmon, Robert S.
Wildflowers of North America in
    Full Color

fl.   pg. 268   Pl. 425

Gentiana andrewsii

Moyle, John B.
Northland Wild Flowers

fl.   p 122, Pl. 125

Gentiana arctophila

Porsild, A. E.
Rocky Mountain Wild Flowers

fl. hab.   p 329

Gentiana asclepiadea

Barneby, T. P.
European Alpine Flowers in Colour

fl.   Pl. 63, 4

Gentiana asclepiadea

Bianchini, Francesco
Health Plants of the World

fl. hab.   p 25

Gentiana asclepiadea

Curtis's Botanical Magazine
Vol 169   1952-53

fl.   191

Gentiana asclepiadea

Felsko, Elsa
A Book of Wild Flowers
2nd Ser.
fl.   p 64

Gentiana asclepiadea

Hay, Roy
The Color Dictionary of Flowers & Plants

fl.   p 143, Pl. 1142

Gentiana asclepiadea

Huxley, Anthony
Garden Perennials and Water Plants

fl.   Pl. 115

Gentiana asclepiadea

Kohlhaupt, Paula
Fleurs des Alpages, v.1

fl.   P 80

Gentiana asclepiadea

Perry, Frances
Flowers of the World

fl.   p 121

Gentiana asclepiadea

Polunin, Oleg
Flowers of Europe

fl.   Pl. 97 #994

Gentiana asclepiadea

Robert, Paul A
Alpine Flowers

fl., hab.   Pl. 8

Gentiana asclepiadea

Tosco, Uberto
The World of Mountain Flowers

fl.   p 113

Gentiana asclepiadea var.

Royal Hort. Soc.
Journal of the Royal Hort. Soc.
vol 98, No. 4   1973

fl.   p. 168   Pl. 104

Gentiana bavarica

Barneby, T. P.
European Alpine Flowers in Colour

fl.   Pl. 64, 2

Gentiana bavarica

Kohlhaupt, Paula
Fleurs des Alpages , v.1

fl. P. 74

Gentiana bavarica var.

Barneby, T. P.
European Alpine Flowers in Colour

fl.   Pl. 64, 3

Gentiana bellidifolia

Royal Hort. Soc.
The Garden
    vol 103, No. 10 1978
fl.   p. 404

Gentiana brachyphylla

Alpine Garden Society
Bulletin
    vol. 6, No. 4   1938

fl.   p. 332

Gentiana brachyphylla

Barneby, T. P.
European Alpine Flowers in Colour

fl. hab.   Pl. 64, 6

Gentiana bulgarica

Bartlett, Mary
Gentians

fl.   p 143

Gentiana cachemirica

Morley, Brian D.
Wild Flowers of the World

fl.   Plate 97

Gentiana calycosa

Fries, Mary A.
Wildflowers of Mount Ranier and
the Cascades

fl.   P 189

Gentiana calycosa

Lemmon, Robert S.
Wildflowers of North America in
    Full Color

fl.   pg. 145   Pl. 232

Gentiana calycosa

Walcott, Mary Vaux
North American Wild Flowers
    vol. 4

fl.   Pl. 318

Gentiana calycosa

Welsh, Stanley L.
Flowers of the Mountain Country

fl.   p. 68

Gentiana campestris

Kohlhaupt, Paula
Fleurs des Alpages , v.2

Fl.   P 83

Gentiana cephalantha

Curtis's Botanical Magazine
    Vol 159   1936

fl.   No. 9468

Gentiana cephalantha

Morley, Brian D.
Wild Flowers of the World

fl.   Plate 97

Gentiana ciliata

Barneby, T. P.
European Alpine Flowers in Colour

fl.   Pl. 64, 1

Gentiana ciliata

Felsko, Elsa
A Book of Wild Flowers , 2nd Ser.

fl.   p 65

Gentiana clusii

Kohlhaupt, Paula
Fleurs des Alpages
    vol 1
fl. P 75

Gentiana clusii

Polunin, Oleg
Flowers of Europe

fl.   Pl. 96 #992

Gentiana clusii

Tosco, Uberto
The World of Mountain Flowers

fl. hab.   p 65

Gentiana clusii var.

Barneby, T. P.
European Alpine Flowers in Colour

fl.   Pl. 63, 5

Gentiana concinna

Morley, Brian D.
Wild Flowers of the World

fl.   Pl. 4H

Gentiana crinita

Courtenay, Booth
Wildflowers & Weeds

fl.   p 80

Gentiana crinita

Ferguson, Mary
Wildflowers

fl.   p 54

Gentiana crinita

Jennings, O. E.
Wild Flowers of Western Pennsylvania
    v. 2
fl.   Pl. 115

Gentiana crinita

Klimas, John E.
Wildflowers of Eastern America

fl.   Pl. 295

Gentiana crinita

Klute, Jeannette
Woodland Portraits

fl.   Pl. 46

Gentiana crinita

Lemmon, Robert S.
Wildflowers of North America in
    Full Color

fl.   pg. 38   Pl. 65

Gentiana crinita

Massachusetts Hort. Soc.
Horticulture
    vol 37, No. 9 1959

fl.   p. 477

Gentiana crinita

Massachusetts Hort. Soc.
Horticulture
    Vol. 42, No. 5 1964

fl.   p. 14

Gentiana crinita

Milne, Lorus
Living Plants of the World

fl.   p 194

Gentiana crinita

Moyle, John B.
Northland Wild Flowers

fl.   p 123, Pl. 128

Gentiana crinita

Pennsylvania Hort. Soc.
The Green Scene
    vol 5, No. 1   1976

fl.   p. 33

Gentiana crinita

Walcott, Mary Vaux
North American Wild Flowers
    vol. 5

fl.   Pl. 336

Gentiana cruciata

Polunin, Oleg
Flowers of Europe

fl.   Pl. 96 #989

Gentiana decora

Dean, Blanche E.
Wildflowers of Alabama and
     Adjoining States

fl.   p 135

Gentiana dendrologii

Bartlett, Mary
Gentiana

fl.   p 108

Gentiana depressa

Curtis's Botanical Magazine
Vol 170   1954-55

fl.   230

Gentiana depressa

Morley, Brian D.
Wild Flowers of the World

fl.   Plate 97

Gentiana dinarica

Royal Hort. Soc.
Journal of the Royal Hort. Soc.
     vol 87, No. 1   1962

fl.   p. 22   Pl. 1

Gentiana douglasiana

Heller, Christine
Wild Flowers of Alaska

fl.   Pl. 187

Gentiana excisa

Morley, Brian D.
Wild Flowers of the World

fl.   Plate 97

Gentiana excisa

Perry, Frances
Flowers of the World

fl.   p 120

Gentiana fasta var.

Royal Hort. Soc.
Journal of the Royal Hort. Soc.
     vol 86, No. 7   1961

fl.   p. 310   Pl. 77

Gentiana flavida

Courtenay, Booth
Wildflowers & Weeds

fl.   p 80

Gentiana flavida

Moyle, John B.
Northland Wild Flowers

fl. hab.   p 122, Pl. 126

Gentiana gelida

Bartlett, Mary
Gentians

fl.   p 87

Gentiana germanica

Barneby, T. P.
European Alpine Flowers in Colour

fl. hab.   Pl. 62, 4

Gentiana germanica

Martin, W. Keble
The Concise British Flora in Colour

fl. hab.   Pl. 59

Gentiana glauca

Alaska-Yukon Wild Flower Guide

fl. hab.   p 131

Gentiana glauca

Heller, Christine
Wild Flowers of Alaska

fl.   Pl. 258

Gentiana glauca

Walcott, Mary Vaux
North American Wild Flowers

fl. hab.   Pl. 108

Gentiana gracilipes

Vilmorin, Roger de
Plantes Alpines dans les Jardins

fl. hab.   Pl. XXXVIII

Gentiana hybrid

Bartlett, Mary
Gentians

fl.   p 88-9, 124,
                 126

Gentiana hybrid

Hay, Roy
The Color Dictionary of Flowers
     & Plants

fl.   p. 9

Gentiana hybrid

Royal Hort. Soc.
Journal of the Royal Hort. Soc.
     vol 89, No. 2   1964

fl.   p. 84   Pl. 31

Gentiana incurva

Bartlett, Mary
Gentians

fl.   p 70

Gentiana juniperina

Royal Hort. Soc.
Journal of the Royal Hort. Soc.
     vol 88, No. 3   1963

fl.   p. 118   Pl. 45

Gentiana kochiana

Barneby, T. P.
European Alpine Flowers in Colour

fl.   Pl. 63, 6

Gentiana kochiana

Kohlhaupt, Paula
Fleurs des Alpages , v.1

fl. P 76

Gentiana kochiana

Polunin, Oleg
Flowers of Europe

fl.   Pl. 96 #992

Gentiana kochiana

Wit, H. C. D. de
Plants of the World;
     The Higher Plants, Vol II

fl.   p 106, Pl. 62

Gentiana lagodechiana

Bartlett, Mary
Gentians

fl. hab.   p 69

Gentiana lagodechiana

Hay, Roy
The Color Dictionary of Flowers & Plants

fl.   p 10, Pl. 73

Gentiana linearis

Addisonia, v. 22, 1943-46

fl. opp. p. 43   pl. 726

Gentiana linearis

Bartlett, Mary
Gentians

fl.   p 70

Gentiana linearis

Campbell, Carlos C.
Great Smoky Mountain Wildflowers

fl.   p 65

Gentiana lourelroi

Walden, Beryl M.
Wild Flowers of Hong Kong

fl.   Pl. 6(18)

Gentiana lutea

Barneby, T. P.
European Alpine Flowers in Colour

fl.   Pl. 62, 1

Gentiana lutea

Bianchini, Francesco
Health Plants of the World

hab.   p 25

Gentiana lutea

Felsko, Elsa
A Book of Wild Flowers
2nd Ser.
fl.   p 33

Gentiana lutea

Hvass, Elsie
Plants That Feed and Serve Us

fl. fr.   p 94, Pl. 199

Gentiana lutea

Kleijn, H.
The Beauty of the Wild Plant

fl.   p. 96   Pl. 133

Gentiana lutea

Kohlhaupt, Paula
Fleurs des Alpages , v.2

Fl.   P 81-82

Gentiana lutea

Perrot, Emile
Les Plantes Medicinales

fl. fr. hab.   p. 108

Gentiana lutea

Perry, Frances
Flowers of the World

fl.   p 122

Gentiana lutea

Polunin, Oleg
Flowers of Europe

fl.   Pl. 95 #996

Gentiana x macaulayi

Perry, Frances
Flowers of the World

fl.   p. 121

Gentiana x macaulayi

Royal Hort. Soc.
Journal of the Royal Hort. Soc.
vol 89, No. 2   1964

fl.   p. 84   Pl. 31

Gentiana x macaulayi var.

Royal Hort. Soc.
Journal of the Royal Hort. Soc.
vol. 86, No. 7   1961

fl.   p. 310   Pl. 78

Gentiana Macounii

Porsild, A. E.
Rocky Mountain Wild Flowers

fl. hab.   p 331

Gentiana makinoi

Bartlett, Mary
Gentians

fl.   p 51

Gentiana makinoi

Curtis's Botanical Magazine,
V. 173, 1960

fl.   Pl. 390

Gentiana newberryi

Munz, Philip A.
California Mountain Wildflowers

fl.   p 49, Pl. 69

Gentiana nivalis

Barneby, T. P.
European Alpine Flowers in Colour

fl. hab.   Pl. 63, 1

Gentiana nivalis

Lindman, C. A. M.
Nordens Flora, Vol 7

fl. fr. hab.   Pl. 473

Gentiana nivalis

Martin, W. Keble
The Concise British Flora in Colour

fl. hab.   Pl. 59

Gentiana nivalis

Royal Hort. Soc.
The Garden
vol 102, No. 5   1977

fl., hab.   p. 199

Gentiana olivieri

Curtis's Botanical Magazine
Vol 177   1969-70

fl.   533

Gentiana orbicularis

Barneby, T. P.
European Alpine Flowers in Colour

fl. hab.   Pl. 64, 5

Gentiana platypetala

Alaska-Yukon Wild Flower Guide

fl.   p 130

Gentiana platypetala

Heller, Christine
Wild Flowers of Alaska

fl.   Pl. 259

Gentiana pneumonanthe

Ary, S.
The Oxford Book of Wildflowers

fl.   p 176, Pl. 6

Gentiana pneumonanthe

Barneby, T. P.
European Alpine Flowers in Colour

fl.   Pl. 63, 3

Gentiana pneumonanthe

Kleijn, H.
The Beauty of the Wild Plant

fl.   p. 57   Pl. 74

Gentiana pneumonanthe

Lindman, C. A. M.
Nordens Flora, Vol 7

fl. hab.   Pl. 472

Gentiana pneumonanthe

Martin, W. Keble
The Concise British Flora in Colour

fl. hab.   Pl. 59

Gentiana pneumonanthe

Polunin, Oleg
Flowers of Europe

fl.   Pl. 96 #993

Gentiana porphyrio

Walcott, Mary Vaux
North American Wild Flowers

fl.   Pl. 8

Gentiana propinqua

Heller, Christine
Wild Flowers of Alaska

fl.   Pl. 261

Gentiana prostrata

Heller, Christine
Wild Flowers of Alaska

fl.   Pl. 260

Gentiana prostrata

Porsild, A. E.
Rocky Mountain Wild Flowers

fl. hab.   p 331

Gentiana prostrata

Walcott, Mary Vaux
North American Wild Flowers

fl. hab.   Pl. 178

Gentiana puberula

Courtenay, Booth
Wildflowers & Weeds

fl.   p 80

Gentiana puberula

Moyle, John B.
Northland Wild Flowers

fl. hab.   p 123, Pl. 127

Gentiana punctata

Barneby, T. P.
European Alpine Flowers in Colour

fl.   Pl. 62, 3

Gentiana punctata

Bartlett, Mary
Gentians

fl.   p 108

Gentiana punctata

Kohlhaupt, Paula
Fleurs des Alpages, V. 1

fl.   P. 77-78

Gentiana punctata

Polunin, Oleg
Flowers of Europe

fl.   Pl. 97 #992

Gentiana punctata

Tosco, Uberto
The World of Mountain Flowers

fl. hab.   p 9, 45, 97, 113

Gentiana purpurea

Barneby, T. P.
European Alpine Flowers in Colour

fl.   Pl. 62, 2

Gentiana purpurea

Felsko, Elsa
A Book of Wild Flowers , 2nd Ser.

fl.

Gentiana purpurea

Kohlhaupt, Paula
Fleurs des Alpages , v.1

fl. P 79

Gentiana purpurea

Polunin, Oleg
Flowers of Europe

fl.   Pl. 97 #998

Gentiana pyrenaica

Bartlett, Mary
Gentians

fl.   p 87

Gentiana pyrenaica

Royal Hort. Soc.
Journal of the Royal Hort. Soc.
vol 92, No. 10   1967

fl.   p. 432   Pl. 227

Gentiana quinquefolia

Courtenay, Booth
Wildflowers & Weeds

fl.   p 80

Gentiana quinquefolia

Duncan, Wilbur H.
Wildflowers of the Southeastern
United States

fl.   p 125

Gentiana saponaria

American Hort. Soc.
American Horticulturist
vol 53, No. 5 1974

fl.   cover

Gentiana saponaria

Brown, Clair A.
Wildflowers of Louisiana and
Adjoining States

fl.   p 137

Gentiana saponaria

Dean, Blanche E.
Wildflowers of Alabama and
Adjoining States

fl.   p 135

Gentiana saponaria

Duncan, Wilbur H.
Wildflowers of the Southeastern
United States

fl.   p 125

Gentiana saponaria

Walcott, Mary Vaux
North American Wild Flowers

fl. hab.   Pl. 161

Gentiana saponaria

Wharton, Mary E.
A Guide to the Wildflowers & Ferns
of Kentucky

fl.   p 128   Pl. 1.68

Gentiana saponaria .var.

Jennings, O. E.
Wild Flowers of Western Pennsylvania
v. 2
fl.   Pl. 117

Gentiana saxosa

Bartlett, Mary
Gentians

fl. hab.   p 90

Gentiana saxosa

Curtis's Botanical Magazine
Vol 171   1956-57

fl.   303

Gentiana saxosa

Hay, Roy
The Color Dictionary of Flowers & Plants

fl.   p 10, Pl. 74

Gentiana saxosa

Perry, Frances
Flowers of the World

fl.   p 122

Gentiana scabra var.

Kariyone, Tatsuo
Atlas of Medicinal Plants

fl., hab.   Pl. 62

Gentiana scabra var.

Kimura, Koiti
Japanese Medicinal Plants, Vol II

fl.   p 213

Gentiana scabra var.

Takatori, Jisuke
Color Atlas of Medicinal Plants of Japan

fl.   Fig. 19A

Gentiana sceptrum

Clark, Lewis J.
Wild Flowers of British Columbia

fl.   p 423

Gentiana septemfida

Bartlett, Mary
Gentians

fl.   p 34

Gentiana septemfida

Huxley, Anthony
Garden Perennials and Water Plants

fl.   Pl. 118

Gentiana septemfida

Perry, Frances
Flowers of the World

fl.   p 122

Gentiana sino-ornata

Alpine Garden Society
Bulletin
    vol. 2, No. 2   1933

fl., hab.   p. 38

Gentiana sino-ornata

Bartlett, Mary
Gentians

fl.   p 51

Gentiana sino-ornata

Hay, Roy
The Color Dictionary of Flowers & Plants

fl.   p 10, Pl. 75

Gentiana sino-ornata

Morley, Brian D.
Wild Flowers of the World

fl.   Plate 97

Gentiana sino-ornata

Perry, Frances
Flowers of the World

fl.   p 122

Gentiana sino-ornata

Royal Hort. Soc.
Journal of the Royal Hort. Soc.
    vol. 78, No. 12 1953

fl.   p. 450   Pl. 129

Gentiana sino-ornata

Royal Hort. Soc.
Journal of the Royal Hort. Soc.
    vol 98, No. 3   1973

fl.   p. 122   Pl. 55

Gentiana tenella

Barneby, T. P.
European Alpine Flowers in Colour

fl. hab.   Pl. 62, 5

Gentiana tenella

Morley, Brian D.
Wild Flowers of the World

fl.   Pl. 4C

Gentiana thermalis

Lemmon, Robert S.
Wildflowers of North America in
    Full Color

fl.   pg. 146   Pl. 233

Gentiana thermalis

Orr, Robert T.
Wildflowers of Western America

fl.   Pl. 225

Gentiana thibetica

Tosco, Uberto
The World of Mountain Flowers

fl.   p 92

Gentiana trichotoma

Curtis's Botanical Magazine
Vol 163   1940-42

fl.   No. 9638

Gentiana utriculosa

Barneby, T. P.
European Alpine Flowers in Colour

fl.   Pl. 63,2

Gentiana utriculosa

Polunin, Oleg
Flowers of Europe

fl.   Pl. 96 #990

Gentiana verna

Ary, S.
The Oxford Book of Wildflowers

fl.   p 176, Pl. 5

Gentiana verna

Barneby, T. P.
European Alpine Flowers in Colour

fl.   Pl. 64, 4

Gentiana verna

Bartlett, Mary
Gentians

fl. hab.   p 107

Gentiana verna

Color Treasury of Herbs & Other
    Medicinal Plants

fl. p 44   Pl. 60

Gentiana verna

Hay, Roy
The Color Dictionary of Flowers & Plants

fl.   p 10, Pl. 76, 77

Gentiana verna

Kleijn, H.
The Beauty of the Wild Plant

fl.   p. 93   Pl.129

Gentiana verna

Kohlhaupt, Paula
Fleurs des Alpages. v.1

fl. P 73

Gentiana verna

Martin, W. Keble
The Concise British Flora in Colour

fl. hab.   Pl. 59

Gentiana verna

Perry, Frances
Flowers of the World

fl.   p 121

Gentiana verna

Polunin, Oleg
Flowers of Europe

fl.   Pl. 96 #991

Gentiana verna

Royal Hort. Soc.
Journal of the Royal Hort. Soc.
    vol 78, No. 8   1953

fl.   p. 284   Pl. 92

Gentiana verna

Royal Hort Soc
The Garden
    vol 102, No. 5   1977

fl., hab.   p. 197

Gentiana verna var.

Royal Hort. Soc.
Journal of the Royal Hort. Soc.
    vol 92, No. 10   1967

fl.   p. 432   Pl. 214, 226

Gentiana villosa

Brown, Clair A.
Wildflowers of Louisiana and
    Adjoining States

fl. hab.   p 138

Gentiana villosa

Dean, Blanche E.
Wildflowers of Alabama and
    Adjoining States

fl.   p 135

Gentiana villosa

Duncan, Wilbur H.
Wildflowers of the Southeastern
United States

fl.   p 125

Gentiana villosa

Wharton, Mary E.
A Guide to the Wildflowers & Ferns
of Kentucky

fl.   p 144   Pl. 1.100

Gentianella amarella

Ary, S.
The Oxford Book of Wildflowers

fl.   p 138, Pl. 6

Gentianella amarella

Clark, Lewis J.
Wild Flowers of British Columbia

fl.   p 422

Gentianella amarella

Tosco, Uberto
The World of Mountain Flowers

fl.   p 117

Gentianella amarella var.

Martin, W. Keble
The Concise British Flora in Colour

fl. hab.   Pl. 59

Gentianella anglica

Martin, W. Keble
The Concise British Flora in Colour

fl. hab.   Pl. 59

Gentianella campestris

Ary, S.
The Oxford Book of Wildflowers

fl.   p 138, Pl. 7

Gentianella campestris

Barneby, T. P.
European Alpine Flowers in Colour

fl.   Pl. 62, 6

Gentianella campestris

Lindman, C. A. M.
Nordens Flora, Vol 7

fl. fr. hab.   Pl. 474

Gentianella campestris

Polunin, Oleg
Flowers of Europe

fl.   Pl. 96 #999

Gentianella campestris var.

Martin, W. Keble
The Concise British Flora in Colour

fl. hab.   Pl. 59

Gentianella ciliata

Tosco, Uberto
The World of Mountain Flowers

fl.   p 117

Gentianella diemensis

Cochrane, G. R.
Flowers and Plants of Victoria

fl.   Pl. 515

Gentianella diemensis

Mullins, Barbara
Australian Wildflowers in Colour

fl.   p. 85   Pl. 83

Gentianella germanica

Felsko, Elsa
A Book of Wild Flowers, 2nd Ser.

fl. hab.   p 66

Gentianella procera

Bartlett, Mary
Gentians

fl.   p 144

Geoblasta penicillata

Pabst, G. F. J.
Orchidaceae Brasilienses,
Vol 1

fl.   p 177

Geocaulon lividum

Heller, Christine
Wild Flowers of Alaska

fr.   Pl. 302

Geocaulon lividum

Porsild, A. E.
Rocky Mountain Wild Flowers

fr.   p 139

Geodorum densiflorum

Hu, Shiu-ying
The Genera of Orchidaceae in
Hong Kong

fl.   p 123

Geodorum pictum

Blombery, Alec M.
What Wildflower is That

fl.   p 146, Pl. 387

Geodorum pictum

Cady, Leo
Australian Native Orchids in Colour

fl. hab.   Pl. 83

Ceogenanthus undatus

Hay, Roy
The Dictionary of House Plants

hab.   Pl. 251

Geogenanthus undatus

Hersey, Jean
Woman's Day Book of House Plants

hab.   p. 94

Geogenanthus undatus

Massachusetts Hort. Soc.
Horticulture
vol 35, No. 1   1957

hab.   p. 37 (see errata p. 108)

Geranium

see

Pelargonium

Geranium antrorsum

Blombery, Alec M.
What Wildflower is That

fl. hab.   p 145, Pl. 385

Geranium argenteum

Alpine Garden Society
Bulletin
vol. 38, No. 2   1970

fl., hab.   p. 158

Geranium argenteum

Kohlhaupt, Paula
Fleurs des Alpages, V. 1

fl.   p. 63

Geranium argenteum var.

Barneby, T. P.
European Alpine Flowers in Colour

fl.   Pl. 47, 4

Geranium bohemicum

Lindman, C. A. M.
Nordens Flora, Vol 6

fl. fr.   Pl. 369

Geranium canariense

Bramwell, David
Wild Flowers of the Canary Islands

fl., hab.   Pl. 179

Geranium canariense

Royal Hort. Soc.
Journal of the Royal Hort. Soc.
vol 95, No. 9   1970

fl.   p. 398   Pl. 206 -07

Geranium carolinianum

Brown, Clair A.
Wildflowers of Louisiana and
Adjoining States

fl. hab.   p 97

Geranium carolinianum

Duncan, Wilbur H.
Wildflowers of the Southeastern
United States

fl.   p 89

Geranium carolinianum

Weeds of the Southern United States

hab., fl.   p. 23

Geranium cinereum var.

Hay, Roy
The Color Dictionary of Flowers &
Plants

fl.   p. 10   Pl. 78

Geranium , Climbing

see

Pelargonium peltatum

Geranium collinum

Royal Hort. Soc.
Journal of Royal Hort. Soc.
vol 81, No. 1   1956

fl.   p. 26   Pl. 13

Geranium columbinum

Martin, W. Keble
The Concise British Flora in Colour

fl. fr. hab.   Pl. 19

Geranium, common

see

Pelargonium hortorum

Geranium dalmaticum

Pennsylvania Hort. Soc.
The Green Scene
vol 6, No. 2   1977

fl., hab.   p. 32

Geranium dissectum

Ary, S.
The Oxford Book of Wildflowers

fl.   p 128, Pl. 4

Geranium dissectum

Crockett, James Underwood
Lawns & Ground Covers

fl.   p. 71

Geranium dissectum

Martin, W. Keble
The Concise British Flora in Colour

fl. fr. hab.   Pl. 19

Geranium endressii

Royal Hort. Soc.
Journal of the Royal Hort. Soc.
vol 92, No. 5   1967

fl.   p. 206   Pl. 96

Geranium erianthum

Alaska-Yukon Wild Flower Guide

fl.   p 93, 164

Geranium erianthum

Clark, Lewis J.
Wild Flowers of British Columbia

fl.   p 314

Geranium erianthum

Heller, Christine
Wild Flowers of Alaska

fl.   Pl. 254-255

Geranium grandiflorum var.

Everett, T. H.
New Illustrated Encyclopedia of
Gardening, Vol. 5

fl.   p. 774

Geranium hybrid

Macoby, Stirling
What Flower is That

fl. P 141

Geranium ibericum

Meikle, R. D.
Garden Flowers

fl.   p. 129   Pl. 3

Geranium ibericum var.

Hay, Roy
The Color Dictionary of Flowers &
Plants

fl.   p. 143,144   Pl. 1143-1145

Geranium incanum var.

Moriarty, Audrey
Wild Flowers of Malawi

fl.   Pl. 56; 3

Geranium, Ivy

see

Pelargonium hybrid

Geranium, Jungle

see

Ixora macrothyrsa

Geranium, Lady Washington

see

Pelargonium domesticum

Geranium lamberti

Hara, Hiroshi, comp.
Photo-Album of Plants of
Eastern Himalaya

fl.   Pl. 133

Geranium lanuginosum

Lindman, C. A. M.
Nordens Flora, Vol 6

fl. fr.   Pl. 368

Geranium lucidum

Ary, S.
The Oxford Book of Wildflowers

fl.   p 128, Pl. 2

Geranium lucidum

Lindman, C. A. M.
Nordens Flora, Vol 6

fl. fr. hab.   Pl. 371A

Geranium lucidum

Martin, W. Keble
The Concise British Flora in Colour

fl. fr. hab.   Pl. 19

Geranium lucidum

Polunin, Oleg
Flowers of Europe

fl.   Pl. 63 #651

Geranium macrorrhizum

Goulimis, Constantine N.
Wild Flowers of Greece

fl.   p 51

Geranium macrorrhizum

Huxley, Anthony
Garden Perennials and Water Plants

fl.   Pl. 120

Geranium macrorrhizum

Kohlhaupt, Paula
Fleurs des Alpages , v.1

fl.   P 64

Geranium macrorrhizum

Meikle, R.D.
Garden Flowers

fl.   p. 129   Pl. 3

Geranium macrorrhizum

Polunin, Oleg
Flowers of Europe

fl.   Pl. 64 #647

Geranium macrorrhizum

Royal Hort. Soc.
Journal of the Royal Hort. Soc.
   vol 92, No. 5   1967

fl.   p. 206   Pl. 97

Geranium macrorrhizum var.

Hay, Roy
The Color Dictionary of Flowers &
    Plants

fl.   p. 144   Pl. 1146

Geranium maculatum

American Hort. Soc.
American Horticulturist
   vol 53, No. 5 1974

fl.   p. 24

Geranium maculatum

Batson, Wade T.
Wild Flowers in South Carolina

fl.   p 68

Geranium maculatum

Campbell, Carlos C.
Great Smoky Mountain Wildflowers

fl.   p 57

Geranium maculatum

Courtenay, Booth
Wildflowers & Weeds

fl.   p 42

Geranium maculatum

Dean, Blanche E.
Wildflowers of Alabama and
    Adjoining States

fl.   p 93

Geranium maculatum

Duncan, Wilbur H.
Wildflowers of the Southeastern
    United States

fl.   p 89

Geranium maculatum

Ferguson, Mary
Wildflowers

fl.   p 164

Geranium maculatum

Jennings, O. E.
Wild Flowers of Western Pennsylvania
v. 2
fl.   Pl. 52

Geranium maculatum

Klimas, John E.
Wildflowers of Eastern America

fl.   Pl. 186

Geranium maculatum

Lemmon, Robert S.
Wildflowers of North America in
    Full Color

fl.   pg. 28   Pl. 50

Geranium maculatum

Moyle, John B.
Northland Wild Flowers

fl. hab.   p 86, Pl. 66

Geranium maculatum

Roberts, June Carver
Born in the Spring

fl. fr. hab.   p 93

Geranium maculatum

Wharton, Mary E.
A Guide to the Wildflowers & Ferns
    of Kentucky

fl. fr.   p 126   Pl. 1.63

Geranium maderense

Royal Hort. Soc.
Journal of the Royal Hort. Soc.
   vol 95, No. 9   1970

fl.   p. 398   Pl. 209 -10

Geranium x magnificum

Perry, Frances
Flowers of the World

fl.   p. 123

Geranium, Marsh

see

Geranium palustre

Geranium, Martha Washington

see

Pelargonium domesticum

Geranium, Miniature

see

Pelargonium hortorum var.

Geranium molle

Ary, S.
The Oxford Book of Wildflowers

fl.   p 128, Pl. 6

Geranium molle

Clark, Lewis J.
Wild Flowers of British Columbia

fl.   p 310

Geranium molle

Lindman, C. A. M.
Nordens Flora, Vol 6

fl. fr. hab.   Pl. 370A

Geranium molle

Martin, W. Keble
The Concise British Flora in Colour

fl. fr. hab.   Pl. 19

Geranium molle var.

Polunin, Oleg
Flowers of the Mediterranean

fl. fr.   No. 82

Geranium nepalense

Kariyone, Tatsuo
Atlas of Medicinal Plants

fl., hab.   Pl. 97

Geranium nepalense

Takatori, Jisuke
Color Atlas of Medicinal Plants of Japan

fl.
fr.   Fig..36

Geranium nepalense var.

Kimura, Koiti
Japanese Medicinal Plants, Vol I

fl.   p 40

Geranium nervosum

Lemmon, Robert S.
Wildflowers of North America in
    Full Color

fl.   pg. 135   Pl. 214

Geranium nodosum

Barneby, T. P.
European Alpine Flowers in Colour

fl.   Pl. 48, 1

Geranium nodosum

Furrer, D.
Blumen am Wege

fl. hab.   p 53

Geranium nodosum

Polunin, Oleg
Flowers of Europe

fl.   Pl. 63 #642

Geranium, Oak Leaf

see

Pelargonium quercifolium

Geranium palmatum

Royal Hort. Soc.
Journal of the Royal Hort. Soc.
    vol 91, No. 11   1966

fl.   p. 470   Pl. 244

Geranium palmatum

Royal Hort. Soc.
Journal of the Royal Hort. Soc.
    vol 95, No. 9   1970

fl.   p. 398   Pl. 208

Geranium palustre

Barneby, T. P.
European Alpine Flowers in Colour

fl.   Pl. 48, 2

Geranium palustre

Felsko, Elsa
A Book of Wildflowers

fl.fr.   P 103

Geranium, Peppermint

see

Pelargonium tomentosum

Geranium phaeum

Felsko, Elsa
A Book of Wildflowers

fl.   P 69

Geranium phaeum

Lindman, C. A. M.
Nordens Flora, Vol 6

fl. fr. hab.   Pl. 367

Geranium phaeum

Martin, W. Keble
The Concise British Flora in Colour

fl. fr. hab.   Pl. 19

Geranium phaeum

Morley, Brian D.
Wild Flowers of the World

fl. fr.   Pl. 14D

Geranium phaeum

Polunin, Oleg
Flowers of Europe

fl.   Pl. 64 #646

Geranium phaeum var.

Barneby, T. P.
European Alpine Flowers in Colour

fl.   Pl. 47, 5

Geranium pratense

Ary, S.
The Oxford Book of Wildflowers

fl.   p 168, Pl. 2

Geranium pratense

Kleijn, H.
The Beauty of the Wild Plant

fl.   p. 17   Pl. 14

Geranium pratense

Martin, W. Keble
The Concise British Flora in Colour

fl. fr. hab.   Pl. 19

Geranium pratense

Morley, Brian D.
Wild Flowers of the World

fl. fr.   Pl. 14B

Geranium pratense

Perry, Frances
Flowers of the World

fl.   p 123

Geranium pratense var.

Hay, Roy
The Color Dictionary of Flowers &
    Plants

fl.   p. 144   Pl. 1147

Geranium procurrens

Curtis's Botanical Magazine
Vol 179   1972

fl.   644

Geranium psilostemon

Perry, Frances
Flowers of the World

fl.   p 123

Geranium purpureum

Martin, W. Keble
The Concise British Flora in Colour

fl. fr. hab.   Pl. 19

Geranium pusillum

Lindman, C. A. M.
Nordens Flora, Vol 6

fl.   Pl. 370B

Geranium pusillum

Martin, W. Keble
The Concise British Flora in Colour

fl. fr. hab.   Pl. 19

Geranium pyrenaicum

Felsko, Elsa
A Book of Wild Flowers
2nd Ser.
fl. fr. hab.   p 43

Geranium pyrenaicum

Furrer, D.
Blumen am Wege

fl. fr. hab.   p 55

Geranium pyrenaicum

Martin, W. Keble
The Concise British Flora in Colour

fl. fr. hab.   Pl. 19

Geranium pyrenaicum

Polunin, Oleg
Flowers of Europe

fl.   Pl. 63 #642

Geranium richardsonii

Welsh, Stanley L.
Flowers of the Canyon Country

fl.   p 11

Geranium rivulare

Barneby, T. P.
European Alpine Flowers in Colour

fl.   Pl. 48, 3

Geranium rivulare

Kohlhaupt, Paula
Fleurs des Alpages, v.2

Fl.   P 67

Geranium robertianum

Ary, S.
The Oxford Book of Wildflowers

fl.   p 128, Pl. 1

Geranium robertianum

Bianchini, Francesco
Health Plants of the World

fl. hab.   p 165

Geranium robertianum

Clark, Lewis J.
Wild Flowers of British Columbia

fl.   p 311

Geranium robertianum

Courtenay, Booth
Wildflowers & Weeds

fl.   p 42

Geranium robertianum

Furrer, D.
Blumen am Wege

fl. fr. hab.   p 57

Geranium robertianum

Lindman, C. A. M.
Nordens Flora, Vol 6

fl. fr.   Pl. 371B

Geranium robertianum

Perrot, Emile
Les Plantes Medicinales

fl. fr. hab.   p 109

Geranium robertianum var.

Martin, W. Keble
The Concise British Flora in Colour

fl. fr. hab.   Pl. 19

Geranium, Rose
see
Pelargonium graveolens

Geranium rotundifolium

Martin, W. Keble
The Concise British Flora in Colour

fl. fr. hab.   Pl. 19

Geranium rubescens

Royal Hort. Soc.
Journal of the Royal Hort. Soc.
vol 95, No. 9   1970

fl.   p. 398   Pl. 205

Geranium sanguineum

Ary, S.
The Oxford Book of Wildflowers

fl.   p. 128   Pl. 3

Geranium sanguineum

Barneby, T. P.
European Alpine Flowers in Colour

fl.   Pl. 47, 6

Geranium sanguineum

Felsko, Elsa
A Book of Wild Flowers
2nd Ser.
fl.   p 84

Geranium sanguineum

Hay, Roy
The Color Dictionary of Flowers & Plants

fl.   p 144, Pl. 1148

Geranium sanguineum

Kleijn, H.
The Beauty of the Wild Plant

fl.   p. 88   Pl. 118

Geranium sanguineum

Lindman, C. A. M.
Nordens Flora, Vol 6

fl. fr. hab.   Pl. 365

Geranium sanguineum

Martin, W. Keble
The Concise British Flora in Colour

fl. fr. hab.   Pl. 19

Geranium sanguineum

Perrot, Emile
Les Plantes Medicinales

fl. fr. hab.   p 109

Geranium sanguineum

Polunin, Oleg
Flowers of Europe

fl.   Pl. 63 #641

Geranium sanguineum var.

Everett, T. H., ed.
New Illustrated Encyclopedia
of Gardening Vol. 5

fl.   P. 774

Geranium sanguineum var.

Royal Hort. Soc.
The Garden
vol 102, No. 8   1977

fl.   p. 334

Geranium sanguineum var.

Vilmorin, Roger de
Plantes Alpines dans les Jardins

fl. hab.   Pl. XXXIX

Geranium solanderi

Cochrane, G. R.
Flowers and Plants of Victoria

fl.   Pl. 59

Geranium, strawberry
see
Saxifraga sarmentosa

Geranium, Strawberry
see
Saxifraga stolonifera

Geranium subcaulescens

Everett, T.H., ed.
New Illustrated Encyclopedia of
Gardening vol. 5

fl.   p. 774

Geranium subcaulescens

Hay, Roy
The Color Dictionary of Flowers & Plants

fl.   p 10, Pl. 79

Geranium sylvaticum

Barneby, T. P.
European Alpine Flowers in Colour

fl.   Pl. 47, 2, 3

Geranium sylvaticum

Lindman, C. A. M.
Nordens Flora, Vol. 6

fl. fr. hab.   Pl. 366

Geranium sylvaticum

Martin, W. Keble
The Concise British Flora in Colour

fl. fr. hab.   Pl. 19

Geranium sylvaticum

Polunin, Oleg
Flowers of Europe

fl.   Pl. 63 #645

Geranium sylvaticum

Tosco, Uberto
The World of Mountain Flowers

fl.   p. 20

Geranium sylvaticum

Webster, Mary
Flora of Moray, Nairn & East
Inverness

fl.   p. 68   Pl. 6

Geranium sylvaticum

Wit, H. C. D. de
Plants of the World;
    The Higher Plants, Vol II

fl.   p 62, Pl. 36

Geranium sylvaticum var.

Hay, Roy
The Color Diction ary of Flowers & Plants

fl.   p 144, Pl. 1149

Geranium sylvaticum var.

Royal Hort. Soc.
The Garden
    vol 102, No. 8   1977

fl.   p. 334

Geranium Tree

        see

Cordia sebestena

Geranium tuberosum

Morley, Brian D.
Wild Flowers of the World

fl.,hab.   Pl. 32

Geranium tuberosum

Polunin, Oleg
Flowers of the Mediterranean

fl.   No. 83

Geranium vagans

Moriarty, Audrey
Wild Flowers of Malawi

fl.   Pl. 56; 2

Geranium versicolor

Martin, W. Keble
The Concise British Flora in Colour

fl. fr. hab.   Pl. 19

Geranium viscosissimum

Clark, Lewis J.
Wild Flowers of British Columbia

fl. fr.   p 311

Geramium viscosissimum

Porsild, A. E.
Rocky Mountain Wild Flowers

fl. fr.   p 271

Geranium viscosissimum

Walcott, Mary Vaux
North American Wild Flowers
        vol. 4

fl.   Pl. 307

Geranium wallichianum

Macoby, Stirling
What Flower is That

fl.   p. 141

Geranium wallichianum var.

Hay, Roy
The Color Dictionary of Flowers & Plants

fl.   p 144, Pl. 1150

Geranium, Wild

        see

Erodium texanum

Geranium, Wild

        see

Geranium maculatum

Geranium, wild

        see

Geranium nervosum

Geranium, zonal

        see

Pelargonium zonale

Gerardia acuta

Addisonia, v. 22, 1943-46

fl., fr.   opp. p. 33   Pl. 721

Gerardia aspera

Moyle, John B.
Northland Wild Flowers

fl.   p 147, Pl. 168

Gerardia pectinata

Batson, Wade T.
Wild Flowers in South Carolina

fl.   p 103

Gerardia, Purple

        see

Agalinis purpurea

Gerardia purpurea

Batson, Wade T.
Wild Flowers in South Carolina

fl.   p 104

Gerardia purpurea

Campbell, Carlos C.
Great Smoky Mountain Wildflowers

fl.   p 91

Gerardia purpurea

Courtenay, Booth
Wildflowers & Weeds

fl.   p 99

Gerardia purpurea

Klimas, John E.
Wildflowers of Eastern America

fl.   Pl. 275

Gerardia purpurea

Wharton, Mary E.
A Guide to the Wildflowers & Ferns
    of Kentucky

fl.   p 227   Pl. 3.34

Gerardia, Rough

        see

Gerardia aspera

Gerardia tenuifolia

Courtenay, Booth
Wildflowers & Weeds

fl.   p 99

Gerbera ambigua

Moriarty, Audrey
Wild Flowers of Malawi

fl.   Pl. 80; 1

Gerbera discolor

Flowering Plants of South Africa
    vol XXIV   1944

fl., hab.   Pl. 938

Gerbera jamesonii

Bruggeman, L.
Tropical Plants

fl.   Pl. 112

Gerbera jamesonii

Hay, Roy
The Dictionary of House Plants

fl.   Pl. 252

Gerbera jamesonii

Kiaer, Eigil
Indoor Plants in Colour

fl.   p. 63

Gerbera jamesonii

Macoby, Stirling
What Flower is That

fl.   p. 141

Gerbera jamesonii

Tsukamoto, Yotaro
Coloured Illustrations of
Garden Flowers, v.9
fl.
opp.p.48

Gerbera jamesonii var.

Hay, Roy
The Dictionary of House Plants

fl.   Pl. 253

Gerbera jamesonii var.

Macoby, Stirling
What Flower is That

fl.   p. 141

Gerbera viridifolia

Moriarty, Audrey
Wild Flowers of Malawi

fl.   Pl. 80; 3

Gerbera welwitschii

Moriarty, Audrey
Wild Flowers of Malawi

fl.   Pl. 80; 2

Germander
see
Teucrium canadense

Germander, American
see
Teucrium canadense var.

Germander, Chamaedrys
see
Teucrium chamaedrys

Germander, Mountain
see
Teucrium montanum

Germander, Wall
see
Teucrium chamaedrys

Gerrerdiina angolensis

Moriarty, Audrey
Wild Flowers of Malawi

fl.   Pl. 40;2

Gesneria cuneifolia var.

Elbert, Virginie F.
The Miracle Houseplants

fl. hab.   p 116

Gesneria humilis

American Gloxinia & Gesneriad Society
The Gloxinian
vol 21, No. 3   1971

fl.   cover (p. 1)

Gethyllis britteniana

Flowering Plants of Africa
vol 36   1963-64

fl., hab.   Pl. 1428

Gethyllis lanuginosa

Flowering Plants of Africa
vol 42   1972-73

fl., hab.   Pl. 1649

Gethyllis linearis

Flowering Plants of Africa
vol XXVI   1947

fl., hab.   Pl. 1018

Gethyllis villosa

Flowering Plants of Africa
vol 42   1972-73

fl., hab.   Pl. 1650

Geum aleppicum

Courtenay, Booth
Wildflowers & Weeds

fl.   p 40

Geum aleppicum

Klimas, John E.
Wildflowers of Eastern America

fl.   Pl. 145

Geum aleppicum var.

Moyle, John B.
Northland Wild Flowers

fl. fr. hab.   p 68, Pl. 40

Geum x borisii

Hay, Roy
The Color Dictionary of Flowers &
Plants

fl.   p. 144   Pl. 1151

Geum x borisii

Macoby, Stirling
What Flower is That

fl.   p. 142

Geum canadense

Courtenay, Booth
Wildflowers & Weeds

fl.   p 40

Geum canadense

Klimas, John E.
Wildflowers of Eastern America

fl.   Pl. 92

Geum canadense

Wharton, Mary E.
A Guide to the Wildflowers & Ferns
of Kentucky

fl. fr.   p 115   Pl. 1.43

Geum chiloense

Macoby, Stirling
What Flower is That

hab.   p. 142

Geum chiloense

Tsukamoto, Yotaro
Coloured Illustrations of
Garden Flowers vol 10

fl.   p. 21   Pl. 65

Geum chiloense var.

Hay, Roy
The Color Dictionary of Flowers & Plants

fl.   p 144, Pl. 1152

Geum coccineum

Curtis's Botanical Magazine
Vol 169   1952-53

fl.   212

Geum coccineum

Coulimis, Constantine N.
Wild Flowers of Greece

fl.   p 35

Geum coccineum

Morley, Brian D.
Wild Flowers of the World

fl.,hab.   Pl. 28

Geum coccineum

Royal Hort. Soc.
The Garden
vol 102, No. 12   1977

fl.   p. 498

Geum glaciale

Alaska-Yukon Wild Flowers Guide

fl. hab.   p 80

Geum glaciale

Heller, Christine
Wild Flowers of Alaska

fl.   Pl. 14-15

Geum glaciale

Massachusetts Hort. Soc.
Horticulture
vol 48, No. 7 1970

fl.   p. 29

Geum hybrid

Huxley, Anthony
Garden Perennials and Water Plants

fl.   Pl. 121

Geum hybrid

Massachusetts Hort. Soc.
Horticulture
    vol 42, No. 5   1964

fl.   cover, backcover

Geum hybrid

Perry, Frances
Flowers of the World

fl.   p. 263

Geum japonicum

Kimura, Koiti
Japanese Medicinal Plants, Vol II

fl.   p 181

Geum macrophyllum

Clark, Lewis J.
Wild Flowers of British Columbia

fl.   p 243

Geum macrophyllum

Heller, Christine
Wild Flowers of Alaska

fl.   Pl. 13

Geum montanum

Barneby, T. P.
European Alpine Flowers in Colour

fl. hab.   Pl. 38, 4

Geum montanum

Felsko, Elsa
A Book of Wildflowers

fl. hab.   P. 18

Geum montanum

Tosco, Uberto
The World of Mountain Flowers

fl.   p 71

Geum, mountain

        see

Geum montanum

Geum pentapetalum

Alaska-Yukon Wild Flower Guide

fl.   p 81

Geum radiatum

Campbell, Carlos C.
Great Smoky Mountain Wildflowers

fl.   p 81

Geum reptans

Barneby, T. P.
European Alpine Flowers in Colour

fl.   Pl. 38, 3

Geum reptans

Kohlhaupt, Paula
Fleurs des Alpages
            vol 1
fl. P 55-56

Geum reptans

Tosco, Uberto
The World of Mountain Flowers

fl. hab.   p 71

Geum rivale

Ary, S.
The Oxford Book of Wildflowers

fl. fr.   p 116, Pl. 5

Geum rivale

Barneby, T. P.
European Alpine Flowers in Colour

fl.   Pl. 38, 5

Geum rivale

Courtenay, Booth
Wildflowers & Weeds

fl.   p 40

Geum rivale

Felsko, Elsa
A Book of Wild Flowers
2nd Ser.
fl.   p 75

Geum rivale

Jennings, O. E.
Wild Flowers of Western Pennsylvania
2nd ed.

fl.   Pl. 85

Geum rivale

Klimas, John E.
Wildflowers of Eastern America

fl.   Pl. 270

Geum rivale

Lemmon, Robert S.
Wildflowers of North America in
    Full Color

fl.   pg. 183   Pl. 288

Geum rivale

Lindman, C. A. M.
Nordens Flora, Vol 5

fl. fr. hab.   Pl. 334

Geum rivale

Martin, W. Keble
The Concise British Flora in Colour

fl. fr. hab.   Pl. 29

Geum rivale

Morley, Brian D.
Wild Flowers of the World

fl. fr.   Pl. 16D

Geum rivale

Polunin, Oleg
Flowers of Europe

fl.   Pl. 45 #443

Geum rivale

Webster, Mary
Flora of Moray, Nairn & East
    Inverness

fl.   p. 164   Pl. 7

Geum rossii

Alaska-Yukon Wild Flower Guide

fl. hab.   p 79

Geum rossii

Heller, Christine
Wild Flowers of Alaska

fl.   Pl. 16

Geum talbotianum

Curtis, Winifred
The Endemic Flora of Tasmania, Vol. 5

fl. hab.   Pl. 159

Geum triflorum

Clark, Lewis J.
Wild Flowers of British Columbia

fl.   p 247

Geum triflorum

Courtenay, Booth
Wildflowers & Weeds

fl.   p 40

Geum triflorum

Ferguson, Mary
Wildflowers

fl.   p 103

Geum triflorum

Moyle, John B.
Northland Wild Flowers

fl. hab.   p 69, Pl. 41

Geum triflorum

Orr, Robert T.
Wildflowers of Western America

fl. fr.   Pl. 149

Geum triflorum

Porsild, A. E.
Rocky Mountain Wild Flowers

fl. hab.   p 237

Geum urbanum

Ary, S.
The Oxford Book of Wildflowers

fl.   p 16, Pl. 1

Geum urbanum

Lindman, C. A. M.
Nordens Flora, Vol 5

fl. fr.   Pl. 333

Geum urbanum

Martin, W. Keble
The Concise British Flora in Colour

fl. fr. hab.   Pl. 29

Geum urbanum

Perrot, Emile
Les Plantes Medicinales

fl. fr. hab.   p 30

Gherkin

see

Cucumis anguria

Ghost Flower

see

Mohavea confertiflora

Ghostflower

see

Monotropa uniflora

Ghost pipe

see

Orobanche uniflora

Ghostpipe

see

Thalesia uniflora

Gibasis oaxacana

Curtis's Botanical Magazine
Vol 179   1972

fl.   624

Gibasis schiedeana

Curtis's Botanical Magazine
Vol 179   1972

fl.   636

Gibbaeum gibbosum

Morley, Brian D.
Wild Flowers of the World

fl.,hab.   Pl. 73

Gibbaeum heathii

Cactus & Succulent Society of America
Journal
vol 46, No. 6   1974

hab.   p. 286

Gibbaeum perviride

Flowering Plants of Africa
vol 29   1952-53

fl.   Pl. 1153

Gibbaeum pubescens

Herre, H.
The Genera of the Mesembryanthemaceae

fl.   fr.   p 163

Gilia aggregata

Clark, Lewis J.
Wild Flowers of British Columbia

fl.   p 427

Gilia aggregata

Ferguson, Mary
Wildflowers

fl.   p 115

Gilia aggregata

Fries, Mary A.
Wildflowers of Mount Ranier and
the Cascades

fl. P 80

Gilia aggregata

Lemmon, Robert S.
Wildflowers of North America in
  Full Color

fl.   p. 147   Pl. 234

Gilia aggregata

Orr, Robert T.
Wildflowers of Wetsern America

fl.   Pl. 155

Gilia aggregata

Shaw, Richard J.
Field Guide to the Vascular Plants of
  Grand Teton National Park

fl.   Pl. 11

Gilia aggregata

Welsh, Stanley L.
Flowers of the Canyon Country

fl.   p. 27

Gilia aurea

Lemmon, Robert S.
Wildflowers of North America in
  Full Color

fl.   p. 103   Pl. 165

Gilia, Blue

see

Gilia rigidula

Gilia cana var.

Munz, Philip A.
California Desert Wildflowers

fl.   p 46, Pl. 58

Gilia capitata

Crockett, James Underwood
Annuals
fl.
p. 121

Gilia capitata

Munz, Philip A.
California Spring Wildflowers

fl.   p 87, Pl. 85

Gilia coronopifolia

Tsukamoto, Yotaro
Coloured Illustrations of
Garden Flowers   Vol. 10

fl. Illus. 64 opp. p. 21

Gilia dianthoides

Lemmon, Robert S.
Wildflowers of North America in
Full Color

fl.   p. 42   Pl. 70

Gilia, downy

see

Gilia floccosa

Gilia floccosa

Lemmon, Robert S.
Wildflowers of North America in
Full Color

fl.   p. 103   Pl. 165

Gilia, golden

see

Gilia aurea

Gilia hybrida var.

Hay, Roy
The Color Dictionary of Flowers &
Plants

fl.   p. 38   Pl. 298

Gilia latiflora

Munz, Philip A.
California Desert Wildflowers

fl.   p 45, Pl. 57

Gilia rigidula

Orr, Robert T.
Wildflowers of Western America

fl.   Pl. 273

Gilia, scarlet

see

Gilia aggregata

Gilia, Scarlet

see

Ipomopsis aggregata var.

Gilia, Skyrocket

see

Gilia aggregata

Gilia tricolor

Huxley, Anthony
Garden Annuals and Bulbs

fl.   Pl. 53

Gilia tricolor

Munz, Philip A.
California Spring Wildflowers

fl.   p 35, Pl. 25

Gill over the Ground

see

Glecoma hederacea

Gill-over-the-ground

see

Nepeta hederacea var.

Gillenia stipulata

Jennings, O. E.
Wild Flowers of Western Pennsylvania
v. 2
fl.   Pl. 81

Gillenia stipulata

Wharton, Mary E.
A Guide to the Wildflowers & Ferns
of Kentucky

fl.   p 115   Pl. 1.42

Gillenia trifoliata

Batson, Wade T.
Wild Flowers in South Carolina

fl.   p 54

Gillenia trifoliata

Courtenay, Booth
Wildflowers & Weeds

fl.   p 38

Gillenia trifoliata

Dean, Blanche E.
Wildflowers of Alabama and
Adjoining States

fl.   p 83

Gillenia trifoliata

Duncan, Wilbur H.
Wildflowers of the Southeastern
United States

fl.   p 63

Gillenia trifoliata

Klimas, John E.
Wildflowers of Eastern America

fl.   Pl. 97

Gillenia trifoliata

Perry, Frances
Flowers of the World

fl.   p 265

Gillyflower

see

Matthiola incana

Gin, Black

see

Kingia australis

Ginger

see

Zingiber officinale

Ginger, Acuminate Wild

see

Asarum canadense var.

Ginger, British Columbia Wild

see

Asarum caudatum

Ginger, Crape

see

Costus speciosus

Ginger, Dwarf

see

Globba schomburgkii

Ginger Flower, Wild

see

Alpinia chinensis

Ginger, Indian Head

see

Costus spicatus

Ginger, Kahili

see

Hedychium gardneranum

Ginger, Malay

see

Costus speciosus

Ginger, Malaysian

see

Tapeinochilos ananassae

Ginger, Olena

see

Curcuma domestica

Ginger, Orange

see

Costus igneus

Ginger, Ostrich Plume

see

Alpinia purpurata

Ginger, Pineapple

see

Tapeinochilos ananassae

Ginger, porcelain

see

Alpinia speciosa

Ginger, Red

see

Alpinia purpurata

Ginger, Shell

see

Alpinia nutans

Ginger, shell

see

Alpinia speciosa

Ginger, Shell

see

Alpinia zerumbet

Ginger Shell

see

Catimbium speciosum

Ginger, Spiral

see

Costus igneus

Ginger, Torch

see

Nicolaia elatior

Ginger, Torch

see

Phaeomeria magnifica

Ginger, Variegated

see

Alpinia sanderae

Ginger, White

see

Hedychium coronarium

Ginger, wild

see

Asarum canadense

Ginger, Wild

see

Asarum caudatum

Ginger, Wild

see

Asarum europaeum

Ginger, Wild

see

Hexastylis arifolia

Ginger, Wood

see

Anemone ranunculoides

Ginger, Yellow

see

Hedychium flavescens

Ginger, Yellow

see

Hedychium gardnerianum

Ginkgo biloba
American Hort. Soc.
American Horticulturist
vol 56, No. 5 1977
hab. p. 30-31

Ginkgo biloba
Bartels, Andreas
Das Grosse Buch der Gartengeholze
hab. p 252

Ginkgo biloba
Bianchini, F.
The Complete Book of Fruits & Vegetables
fr. p 189

Ginkgo biloba
Boom, B.K.
The Glory of the Tree
fr. hab. p. 12 Pl. 2,4

Ginko biloba
Edlin, Herbert
The Illustrated Encyclopedia
of Trees
fl. fr. hab. p 29, 86

Ginkgo biloba
Flemer, William
Shade and Ornamental Trees in Color
hab. p. 89

Ginkgo biloba
Harrison, R. E.
Trees and Shrubs
hab. Pl. 573

Ginkgo biloba
Hay, Roy
The Color Dictionary of Flowers & Plants
hab. p 252, Pl. 2014

Ginkgo biloba
Huxley, Anthony
Evergreen Garden Trees and Shrubs
fr. hab. Pl. 26

Ginkgo biloba

Kitamura, Siro
Coloured Illustrations of Trees &
    Shrubs of Japan

fr. hab.   3, 4

Ginkgo biloba

Oakman, Harry
Colorful Trees

hab.   p. 129

Ginkgo biloba

Pennsylvania Hort. Soc.
The Green Scene
    vol 5, No. 3   1977

hab.   p. 22-23, backcover

Ginkgo biloba

Royal Hort. Soc.
The Garden
    vol 101, No. 4   1976

hab.   p. 203

Ginkgo biloba

Wit, H. C. D. de
Plants of the World;
    The Higher Plants, Vol. I

fl.   p. 50   Pl. 3

Ginseng

    see

Panax quinquefolium

Ginseng, Dwarf

    see

Panax trifolius

Gipsywort

    see

Lycopus europaeus

x Gladanthera hybrid

Royal Hort. Soc.
The Garden
    vol 101, No. 8   1976

fl.   p. 427

Gladiolus alatus var.

Lewis, G. Joyce
Gladiolus; A Revision of The South
    African Species

fl., hab.   p. 160   Pl. 15

Gladiolus atropurpureus

Flowering Plants of Africa
    vol 44   1977

fl., hab.   Pl. 1760

Gladiolus atropurpureus

Moriarty, Audrey
Wild Flowers of Malawi

fl.   Pl. 8; 2

Gladiolus atroviolaceus

Wendelbo, Per
Tulips and Irises of Iran and
    Their Relatives

fl.   p 79, Pl. 84

Gladiolus aurantiacus

Lewis, G. Joyce
Gladiolus; A Revision of The South
    African Species

fl., hab.   p. 285   Pl. 32

Gladiolus aureus

Curtis's Botanical Magazine, v.175
1964-65
fl.
pl. 479

Gladiolus blandus var.

Curtis's Botanical Magazine
Vol 166   1949

fl.   81

Gladiolus brevifolius

Flowering Plants of Africa
    vol 43   1974-76

fl., hab.   Pl. 1683

Gladiolus brevifolius var.

Lewis, G. Joyce
Gladiolus; A Revision of The South
    African Species

fl., hab.   p. 269,284   Pl.30-31

Gladiolus bullatus

Lewis, G. Joyce
Gladiolus; A Revision of The South
    African Species

fl., hab.   p. 241   Pl. 22

Gladiolus byzantinus

Hay, Roy
The Color Dictionary of Flowers & Plants

fl.   p 94, Pl. 748

Gladiolus byzantinus

Miles, Bebe
Bulbs for the Home Gardener

fl.   p 74

Gladiolus byzantinus

Polunin, Oleg
Flowers of Europe

fl.   Pl. 176 #1695

Gladiolus callianthus

Moriarty, Audrey
Wild Flowers of Malawi

fl.   Pl. 9; 1

Gladiolus cardinalis

Flowering Plants of Africa
    vol 43   1974-76

fl., hab.   Pl. 1682

Gladiolus cardinalis

Lewis, G. Joyce
Gladiolus; A Revision of the South
    African Species

fl., hab.   p. 24   Pl. 1

Gladiolus cardinalis

Morley, Brian D.
Wild Flowers of the World

fl.   Pl. 87

Gladiolus carinatus var.

Lewis, G. Joyce
Gladiolus; A Revision of The South
    African Species

fl., hab.   p. 208   Pl. 17

Gladiolus carneus

Lewis, G. Joyce
Gladiolus; A Revision of The South
    African Species

fl., hab.   p. 96   Pl. 9

Gladiolus ceresianus

Curtis's Botanical Magazine
Vol 167   1950

fl.   104

Gladiolus coluillei var.

Tsukamoto, Yotaro
Coloured Illustrations of Garden
    Flowers, Vol. 9

fl.   opp. p. 11

Gladiolus communis

Polunin, Oleg
Flowers of the Mediterranean

fl.   No. 256

Gladiolus comptonii

Lewis, G. Joyce
Gladiolus; A Revision of The South
    African Species

fl., hab.   p. 225   Pl. 20

Gladiolus densiflorus

Flowering Plants of Africa
    vol 42   1972-73

fl.   Pl. 1651

Gladiolus ecklonii

Flowering Plants of Africa
vol 32   1957-58

fl., hab.   Pl. 1265

Gladiolus edulis

Flowering Plants of Africa
Vol. XXVI   1947

fl., hab.   Pl. 1017

Gladiolus elliotii

Lewis, G. Joyce
Gladiolus; A Revision of The South
African Species

fl., hab.   p. 31   Pl. 5

Gladiolus equitans

Eliovson, Sima
Namaqualand in Flower

fl.   Pl. 29, 2

Gladiolus erectiflorus

Moriarty, Audrey
Wild Flowers of Malawi

fl.   Pl. 8; 3

Gladiolus floribundus var.

Lewis G. Joyce
Gladiolus; A Revision of The South
African Species

fl., hab.   p. 97   Pl. 10

Gladiolus gracilis var.

Lewis, G. Joyce
Gladiolus; A Revision of The South
African Species

fl., hab.   p. 224   Pl. 19

Gladiolus guthriei

Lewis, G. Joyce
Gladiolus; A Revision of The South
African Species

fl., hab.   p. 260   Pl. 25

Gladiolus hollandii

Lewis, G. Joyce
Gladiolus; A Revision of The South
African Species

fl., hab.   p. 49   Pl. 8

Gladiolus hybrid

Hay, Roy
The Color Dictionary of Flowers & Plants

fl.   p 94 - 96, Pl. 749 thru 763

Gladiolus hybrid

Hay, Roy
The Dictionary of House Plants

fl.   Pl. 254, 255

Gladiolus hybrid

Huxley, Anthony
Garden Annuals and Bulbs

fl.   Pl. 221

Gladiolus hybrid

Macoby, Stirling
What Flower is That

fl.   p. 142

Gladiolus hybrid

Miles, Bebe
Bulbs for the Home Gardener

fl.   p 166 - 167

Gladiolus hybrid

Perry, Frances
Flowers of the World

fl.   p. 146-47

Gladiolus illyricus

Martin, W. Keble
The Concise British Flora in Colour

fl. hab.   Pl. 83

Gladiolus laxiflorus

Flowering Plants of Africa
vol 33   1959

fl., hab.   Pl. 1287

Gladiolus laxiflorus

Moriarty, Audrey
Wild Flowers of Malawi

fl.   Pl. 8; 4

Gladiolus lewisiae

Flowering Plants of Africa
vol 40   1969-70

fl., hab.   Pl. 1596

Gladiolus lewisiae

Lewis, G. Joyce
Gladiolus; A Revision of The South
African Species

fl., hab.   p. 112   Pl. 11

Gladiolus maculatus

Flowering Plants of Africa
vol 43   1974-76

fl., hab.   Pl. 1684

Gladiolus maculatus var.

Lewis, G. Joyce
Gladiolus; A Revision of The South
African Species

fl., hab.   p. 260   Pl. 27-28

Gladiolus melleri

Flowering Plants of South Africa
vol XX   1940

fl., hab.   Pl. 768

Gladiolus melleri

Moriarty, Audrey
Wild Flowers of Malawi

fl.   Pl. 8;1

Gladiolus monticola

Flowering Plants of Africa
vol 34   1960-61

fl., hab.   Pl. 1339

Gladiolus monticola

Lewis, G. Joyce
Gladiolus; A Revision of The South
African Species

fl., hab.   p. 260   Pl. 26

Gladiolus nanus

Tsukamoto, Yotaro
Coloured Illustrations of Garden
Flowers, V. 9

fl.   opp. p. 11

Gladiolus nanus var.

Hay, Roy
The Color Dictionary of Flowers &
Plants

fl.   p. 94   Pl. 749-51

Gladiolus nanus var.

Macoby, Stirling
What Flower is That

fl.   p. 142

Gladiolus natalensis

Moriarty, Audrey
Wild Flowers of Malawi

fl.   Pl. 8; 5, 6

Gladiolus natalensis var.

Lewis, G. Joyce
Gladiolus; A Revision of The South
African Species

fl., hab.   p. 32   Pl. 6

Gladiolus natalensis var.

Royal Hort. Soc.
The Garden
vol 101, No. 8   1976

fl.   p. 427

Gladiolus nerineoides

Flowering Plants of Africa
vol XXV   1945-46

fl., hab.   Pl. 994

Gladiolus nerineoides

Lewis, G. Joyce
Gladiolus; A Revision of The South
    African Species

fl., hab.   p. 253   Pl. 24

Gladiolus odoratus

Curtis's Botanical Magazine
Vol 170   1954-55

fl.   223

Gladiolus oppositiflorus var.

Lewis, G. Joyce
Gladiolus; A Revision of The South
    African Species

fl., hab.   p. 25   Pl. 4

Gladiolus orchidiflorus

Lewis, G. Joyce
Gladiolus; A Revision of The South
    African Species

fl., hab.   p. 144   Pl. 13

Gladiolus ornatus

Lewis, G. Joyce
Gladiolus; A Revision of The South
    African Species

fl., hab.   p. 209   Pl. 18

Gladiolus papilio

Lewis, G. Joyce
Gladiolus; A Revision of The South
    African Species

fl., hab.   p. 48   Pl. 7

Gladiolus papilio

Royal Hort. Soc.
Journal of the Royal Hort. Soc.
    vol 98, No. 4   1973

fl.   p. 168   Pl. 105

Gladiolus papilio

Royal Hort. Soc.
The Garden
    vol 101, No. 8   1976

fl.   p. 427

Gladiolus, Parrot

see

Gladiolus psittacinus

Gladiolus pillansii var.

Flowering Plants of Africa
    vol 29   1952-53

fl.,hab.   Pl. 1159

Gladiolus pole-evansii

Flowering Plants of Africa
    vol 35   1962

fl.   Pl. 1373

Gladiolus primulinus var.

Everett, T. H.
New Illustrated Encyclopedia of
    Gardening, V. 5

fl. opp. p. 791

Gladiolus prismatosiphon

Flowering Plants of Africa
    vol 29   1952-53

fl., hab.   Pl. 1160

Gladiolus psittacinus

Lind, E.M.
Some Common Flowering Plants of
    Uganda

fl.   Pl. 13a

Gladiolus psittacinus

Macoby, Stirling
What Flower is That

fl.   p. 142

Gladiolus rogersii

Flowering Plants of South Africa
    vol XXIII   1943

fl.   Pl. 919

Gladiolus rogersii var.

Lewis, G. Joyce
Gladiolus; A Revision of The South
    African Species

fl., hab.   p. 240   Pl. 21

Gladiolus saundersii

Lewis, G. Joyce
Gladiolus; A Revision of The South
    African Species

fl., hab.   p. 24   Pl. 3

Gladiolus scullyi

Eliovson, Sima
Namaqualand in Flower

fl.   Pl. 29, 1

Gladiolus scullyi

Lewis, G. Joyce
Gladiolus; A Revision of The South
    African Species

fl., hab.   p. 113   Pl. 12

Gladiolus segetum

Barneby, T. P.
European Alpine Flowers in Colour

fl.   Pl. 7, 2

Gladiolus segetum

Polunin, Oleg
Flowers of Europe

fl.   Pl. 176 #1695

Gladiolus segetum

Royal Hort. Soc.
Journal of the Royal Hort. Soc.
    vol 88, No. 10   1963

fl.   p. 434   Pl. 165

Gladiolus segetum

Taylor, A. W.
Wild Flowers of Spain and Portugal

fl.   p. 22

Gladiolus segetum

Wendelbo, Per
Tulips and Irises of Iran and
    Their Relatives

fl.   p 79, Pl. 83

Gladiolus sempervirens

Lewis, G. Joyce
Gladiolus; A Revision of The South
    African Species

fl., hab.   p. 24   Pl. 2

Gladiolus sp.

Huxley, Anthony
Garden Annuals and Bulbs

hab.   Pl. 161

Gladiolus stanfordiae

Curtis's Botanical Magazine
Vol 161   1938

fl.   No. 9522

Gladiolus stokoei

Flowering Plants of Africa
    vol XXVI   1947

fl., hab.   Pl. 1004

Gladiolus stokoei

Lewis, G. Joyce
Gladiolus; A Revision of The South
    African Species

fl., hab.   p. 252   Pl. 23

Gladiolus subcaeruleus

Flowering Plants of Africa
    vol 29   1952-53

fl., hab.   Pl. 1158

Gladiolus trichonemifolius var.

Curtis's Botanical Magazine
    vol 173   1960

fl.   Pl. 357

Gladiolus triphyllus

Megaw, Elektra
Wild Flowers of Cyprus

fl., hab.   p. 14   Pl. 26

Gladious tristis

Royal Hort. Soc.
The Garden
Vol. 101, No. 8   1976

fl.   p. 426

Gladiolus tristis var.

Hay, Roy
The Dictionary of House Plants

fl.   Pl. 256

Gladiolus tristis var.

Lewis, G. Joyce
Gladiolus; A Revision of The South
African Species

fl., hab.   p. 161   Pl. 16

Gladiolus vaginatus var.

Lewis, G. Joyce
Gladiolus; A Revision of The South
African Species

fl., hab.   p. 268   Pl. 29

Gladiolus varius var.

Flowering Plants of South Africa
vol XX   1940

fl., hab.   Pl. 791

Gladiolus venustus

Flowering Plants of South Africa
vol XXIII   1943

fl., hab.   Pl. 895

Gladiolus vinoso-maculatus

Flowering Plants of Africa
vol 29   1952-53

fl., hab.   Pl. 1123

Gladiolus virescens var.

Lewis, G. Joyce
Gladiolus; A Revision of The South
African Species

fl., hab.   p. 145   Pl. 14

Gladiolus woodii var.

Flowering Plants of Africa
vol 35   1962

fl., hab.   Pl. 1378

Gland Flower, Hairy

See

Adenanthos barbigera

Glasswort

see

Salicornia

Glasswort, Beaded

see

Salicornia quinqueflora

Glasswort, Desert

see

Pachycornia triandra

Glaucidium palmatum

Curtis's Botanical Magazine
Vol. 159   1936

fl.   No. 9432

Glaucidium palmatum

Royal Hort. Soc.
Journal of the Royal Hort. Soc.
vol 86, No. 7   1961

fl.   p. 310   Pl. 89

Glaucidium palmatum

Royal Hort. Soc.
The Garden
vol 101, No. 7   1976

fl., hab.   p. 361

Glaucidium palmatum

Royal Hort. Soc.
The Garden
vol 101, No. 8   1976

fl.   p. 409

Glaucidium palmatum var.

Hay, Roy
The Color Dictionary of Flowers &
Plants

fl.   p. 10   Pl. 80

Glaucium corniculatum

Polunin, Oleg
Flowers of Europe

fl.   Pl. 29 #273

Glaucium corniculatum

Polunin, Oleg
Flowers of the Mediterranean

fl.   No. 35

Glaucium flavum

Ary, S.
The Oxford Book of Wildflowers

fl.   p 6, Pl. 1

Glaucium flavum

Lindman, C. A. M.
Nordens Flora, Vol 4

fl. fr.   Pl. 247

Glaucium flavum

Martin, W. Keble
The Concise British Flora in Colour

fl. hab.   Pl. 5

Glaucium flavum

Polunin, Oleg
Flowers of Europe

fl.   Pl. 29 #272

Glaucium flavum var.

Goulimis, Constantine N.
Wild Flowers of Greece

fl.   p 27

Glaucium flavum var.

Hay, Roy
The Color Dictionary of Flowers &
Plants

fl.   p. 38   Pl. 299

Glaux maritima

Ary, S.
The Oxford Book of Wildflowers

fl.   p 124, Pl. 8

Glaux maritima

Felsko, Elsa
A Book of Wild Flowers
2nd Ser.

fl. hab.   p 85

Glaux maritima

Lindman, C. A. M.
Nordens Flora, Vol 7

fl. fr. hab.   Pl. 468

Glaux maritima

Polunin, Oleg
Flowers of Europe

fl.   Pl. 92 #968

Glechoma hederacea

Ary, S.
The Oxford Book of Wildflowers

fl.   p 144, Pl. 6

Glechoma hederacea

Clark, Lewis J.
Wild Flowers of British Columbia

fl.   p. 447

Glechoma hederacea

Courtenay, Booth
Wildflowers and Weeds

fl.   p. 95

Glechoma hederacea
Crockett, James Underwood
Lawns & Ground Covers
fl.   p. 71

Glechoma hederacea
Felsko, Elsa
A Book of Wildflowers
fl.   P 47

Glechoma hederacea
Furrer, D.
Blumen am Wege
fl. hab.   p. 87

Glechoma hederacea
Kimura, Koiti
Japanese Medicinal Plants, Vol I
fl.   p 80

Glechoma hederacea
Lindman, C. A. M.
Nordens Flora, Vol 8
fl. fr. hab.   Pl. 500

Glechoma hederacea
Martin, W. Keble
The Concise British Flora in Colour
fl. hab.   Pl. 68

Glechoma hederacea
Moyle, John B.
Northland Wild Flowers
fl. hab.   p 138, Pl. 153

Glechoma hederacea
Perrot, Emile
Les Plantes Medicinales
fl. fr. hab.   p 135

Glechoma hederacea
Polunin, Oleg
Flowers of Europe
fl.   Pl. 109 #1118

Glechoma hederacea
Wharton, Mary E.
A Guide to the Wildflowers & Ferns
of Kentucky
fl. p 206   Pl. 2.31

Glechoma hederacea var.
Kariyone, Tatsuo
Atlas of Medicinal Plants
fl., hab.   Pl. 46

Gleditsia japonica
Boom, B. K.
The glory of the tree
fr.   between p. 84 & 85   Pl. 139

Gleditsia japonica
Kariyone, Tatsuo
Atlas of Medicinal Plants
fl., fr.   Pl. 100

Gleditsia japonica
Kitamura, Siro
Coloured Illustrations of Trees &
Shrubs of Japan
fl. fr.   243, 244,

Gleditsia japonica
Takatori, Jisuke
Color Atlas of Medicinal Plants of
Japan
fl.   Pl. 39

Gleditsia triacanthos
Boom, B. K.
The glory of the tree
hab.   opp. p. 40   Pl. 52

Gleditsia triacanthos
Edlin, Herbert
The Illustrated Encyclopedia
of Trees
fl. fr. hab.   p. 183

Gleditsia triacanthos
Huxley, Anthony
Deciduous Garden Trees and Shrubs
hab.   Pl. 84

Gleditsia triacanthos
Polunin, Oleg
Trees and Bushes of Europe
fl. hab.   p 101

Gleditsia triacanthos var.
Everett, T. H.
New Illustrated Encyclopedia of
Gardening, V. 12
hab.   opp. flyleaf

Gleditsia triacanthos var.
Flemer, William
Shade and Ornamental Trees in Color
hab.   p. 67

Gleditsia triacanthos var.
Gault, S. Millar
The Color Dictionary of Shrubs
hab.   Pl. 213

Glehnia leiocarpa
Clark, Lewis J.
Wild Flowers of British Columbia
fr.   p 351

Glehnia littoralis
Kariyone, Tatsuo
Atlas of Medicinal Plants
fl., hab.   Pl. 72

Glehnia littoraris
Kimura, Koiti
Japanese Medicinal Plants, Vol I
fl. hab.   p 70

Gliricidia sepium
Chickering, Carol Rogers
Flowers of Guatemala
fl.   p 67

Gliricidia sepium
O'Gorman, Helen
Mexican Flowering Trees and Plants
fl. hab.   p 29

Gliricidia sepium
Pertchik, Bernard
Flowering Trees of the
Caribbean
fl.   p 71

Glischrocolla formosa
Flowering Plants of Africa
vol 40   1969-70
fl.   Pl. 1574

Globba atrosanguinea
Hay, Roy
The Dictionary of House Plants
fl.   Pl. 257

Globba schomburgkii
American Hort. Soc.
American Horticulturist
vol 53, No. 4 1974
fl.   p. 25

Globba winitii
Morley, Brian D.
Wild Flowers of the World
fl.   Plate 124

Globeflower
see
Trollius

Globe flower bush

see

Kerria japonica var.

Globeflower, White

see

Trollius albiflorus

Globularia alypum

Goulimis, Constantine N.
Wild Flowers of Greece

fl.   p. 97

Globularia alypum

Polunin, Oleg
Flowers of Europe

fl.   Pl. 129 #1262

Globularia alypum

Polunin, Oleg
Flowers of the Mediterranean

fl.   No. 179

Globularia ascanii

Bramwell, David
Wild Flowers of the Canary Islands

fl.   Pl. 269

Globularia cordifolia

Barneby, T. P.
European Alpine Flowers in Colour

fl. hab.   Pl. 77, 6

Globularia cordifolia

Polunin, Oleg
Flowers of Europe

fl.   Pl. 129 #1263

Globularia cordifolia

Robert, Paul A.
Alpine Flowers

Fl., hab.   Pl. 2

Globularia elongata

Barneby, T. P.
European Alpine Flowers in Colour

fl. hab.   Pl. 77, 5

Globularia nana

Kohlhaupt, Paula
Fleurs des Alpages, v.2

Fl.   P 97

Globularia nudicaulis

Barneby, T. P.
European Alpine Flowers in Colour

fl. hab.   Pl. 77, 4

Globularia nudicaulis

Kohlhaupt, Paula
Fleurs des Alpages, v.2

Fl.   P 98

Globularia vulgaris

Lindman, C. A. M.
Nordens Flora, Vol 8

fl. fr. hab.   Pl. 544

Globularia vulgaris

Polunin, Oleg
Flowers of Europe

fl.   Pl. 129 #1264

Globularia willkommii

Felsko, Elsa
A Book of Wild Flowers
2nd Ser.
fl.   p 44

Glochidion eriocarpum

Walden, Beryl M.
Wild Flowers of Hong Kong

fl.   Pl. 41(118)

Glochidion obovatum

Kitamura, Siro
Coloured Illustrations of Trees &
Shrubs of Japan

fr.   270

Gloriosa greeneae

Menninger, Edwin A.
Flowering Vines of the World

fl.   Pl. 114

Gloriosa, greenish-flowered

see

Gloriosa virescens

Gloriosa rothschildiana

Encke, Fritz
Zimmerpflanzen

fl.   p. 78

Gloriosa rothschildiana

Massachusetts Hort. Soc.
Horticulture
    vol 34, No. 8   1956
fl.   p. 427

Gloriosa rothschildiana

Massachusetts Hort. Soc.
Horticulture
    vol 42, No. 12   1964
fl.   p. 35

Gloriosa rothschildiana

Mathias, Mildred E.
Color for the Landscape

fl. hab.   p 112

Gloriosa rothschildiana

Menninger, Edwin A.
Flowering Vines of the World

fl.   Pl. 119

Gloriosa rothschildiana

Miles, Bebe
Bulbs for the Home Gardener

fl.   p 168

Gloriosa rothschildiana

Perry, Frances
Flowers of the World

fl.   p 176

Gloriosa rothschildiana

Wit, H. C. D. de
Plants of the World;
    The Higher Plants, Vol II

fl.   Pl. 120

Gloriosa simplex

Lind, E.M.
Some Common Flowering Plants of
    Uganda

fl.   Pl. 16

Gloriosa simplex

Morley, Brian D.
Wild Flowers of the World

fl.   Pl. 66

Gloriosa sp.

Kromdijk, G.
200 House Plants in Colour

fl.   Pl. 99

Gloriosa superba

American Hort. Soc.
American Horticulturist
    vol 53, No. 3   1974
fl.   p. 34

Gloriosa superba

Bruggeman, L.
Tropical Plants

fl.   Pl. 12

Gloriosa superba

Hall, Clarence E.
Flowers of the Islands in the Sun

fl.   p. 103   Pl. 22

Gloriosa superba

Hay, Roy
The Dictionary of House Plants

fl.   Pl. 258

Gloriosa superba

Kimura, Koiti
Japanese Medicinal Plants, Vol I

fl.   p 10

Gloriosa superba

Mathias, Mildred E.
Color for the Landscape

fl. hab.   p 112

Gloriosa superba

Milne, Lorus
Living Plants of the World

fl.   p 261

Gloriosa superba

Moriarty, Audrey
Wild Flowers of Malawi

fl.   Pl. 10; 2

Gloriosa superba var.

Royal Hort. Soc.
Journal of the Royal Hort. Soc.
   vol 94, No. 1   1969

fl.   p. 22   Pl. 1

Gloriosa virescens

Moriarty, Audrey
Wild Flowers of Malawi

fl.   Pl. 10; 1

Glory Bird, Morocco

     see

Convolvulus mauritanicus

Glory bower

     see

Clerodendrum speciosissimum

Glory bower

     see

Clerodendrum splendens

Glorybower, Fragrant

     see

Clerodendron fragrans

Glory-Bower, Harlequin

     see

Clerodendron Trichotomum

Glory Bush

     see

Tibouchina semidecandra

Glory Creeper, Golden

     see

Thunbergia gibsonii

Glory Flower, Chilean

     see

Eccremocarpus scaber

Glory of Texas

     see

Thelocactus bicolor

Glory of the Snows

     see

Chionodoxa luciliae

Glory of the sun

     see

Ipheion uniflorum

Glory of the Sun

     see

Leucocoryne ixioides

Glory Vine, Crimson

     see

Vitis coignetiae

Glory Vine, Golden

     see

Thunbergia alata

Glossodia major

Blombery, Alec M.
What Wildflower is That

fl.   p 146, Pl. 389

Glossodia major

Cady, Leo
Australian Native Orchids in Colour

fl.   Pl. 43

Glossodia major

Cochrane, G.R.
Flowers and Plnats of Victoria

fl.   Pl. 340

Glossodia major

Curtis's Botanical Magazine, v. 175
   1964-65

fl.   Pl. 441

Glossodia major

Perry, Frances
Flowers of the World

fl.   p 211

Glossodia minor

Cady, Leo
Australian Native Orchids in Colour

fl.   Pl. 44

Glossostelma carsonii

Moriarty, Audrey
Wild Flowers of Malawi

fl.   Pl. 46; 3

Glottidium vesicarium

Duncan, Wilbur H.
Wildflowers of the Southeastern
   United States

fl. hab.   p 79

Glottiphyllum arrectum

Lamb, Edgar
The Pocket Encyclopedia of Cacti
   and Succulents in Color

fl. hab.   Pl. 258

Glottiphyllum arrectum

Lamb, Edgar
Popular Exotic Cacti in Color

fl. hab.   p 94

Glottiphyllum fragrans

Morley, Brian D.
Wild Flowers of the World

fl.,hab.   Pl. 73

Glottiphyllum oligocarpum

Lamb, Edgar
The Pocket Encyclopedia of Cacti
and Succulents in Color

fl. hab.   Pl. 260

Glottiphyllum scalpratum

Herre, H.
The Genera of the Mesembryanthemaceae

fl.   fr.   p 165

x Gloxinera hybrid

American Gloxinia & Gesneriad Soc.
The Gloxinian
     vol 20, No. 6   1970

fl., hab.   cover

x Gloxinera hybrid

American Gloxinia & Gesneriad Society
The Gloxinian
     vol 23, No. 4   1973

fl.   cover

Gloxinia perennis

American Gloxinia & Gesneriad Society
The Gloxinian
     vol 27, No. 4   1977

fl., hab.   cover

Gloxinia perennis

Elbert, Virginie F.
The Miracle Houseplants

fl.   p 116

Gloxinia perennis

Wit, H. C. D. de
Plants of the World;
The Higher Plants, Vol II

fl.   Pl. 94

Gloxinia, hardy

     see

Incarvillea delavayi

Gloxinia

see also

Sinningia

Glyceria aquatica var.

Huxley, Anthony
Garden Perennials and Water Plants

hab.   Pl. 320

Glyceria declinata

Polunin, Oleg
Flowers of Europe

fr.   Pl. 180 #1756

Glyceria fluitans

Lindman, C. A. M.
Nordens Flora, Vol 2

fr. hab.   Pl. 71

Glyceria maxima

Lindman, C. A. M.
Nordens Flora, Vol 2

fr.   Pl. 72

Glyceria maxima

Polunin, Oleg
Flowers of Europe

fr.   Pl. 180 #1755

Glycine clandestina

Blombery, Alec M.
What Wildflower is That

fl.   p 147, Pl. 390

Glycine clandestina

Cochrane, G. R.
Flowers and Plants of Victoria

fl.   Pl. 457

Glycine max

Masefield, G. B.
The Oxford Book of Food Plants

fl. fr.   Pl. 25, 1

Glycine soja

Hvass, Elsie
Plants That Feed and Serve Us

fl. fr.   p 16, Pl. 23

Glycosmis citirifolia

Addisonia, v. 21, 1939-42

fl., fr.   opp. p. 29   Pl. 687

Glycyrrhiza echinata

Hvass, Elsie
Plants That Feed and Serve Us

fl., fr.   p. 93   Pl. 197

Glycyrrhiza glabra

Color Treasury of Herbs & Other
Medicinal Plants

fl.   p. 34   Pl. 40

Glycyrrhiza glabra

Kariyone, Tatsuo
Atlas of Medicinal Plants

fl.   Pl. 101

Glycyrrhiza glabra

Kimura, Koiti
Japanese Medicinal Plants, Vol II

fl.   p 189

Glycyrrhiza glabra

Masefield, G. B.
The Oxford Book of Food Plants

hab.   Pl. 135, 3

Glycyrrhiza glabra

Perrot, Emile
Les Plantes Medicinales

fl. hab.   p 199

Gmelina hystrix

Menninger, Edwin A.
Flowering Vines of the World

fl.   Pl. 194

Gmelina phillippensis

Hall, Clarence E.
Flowers of the Islands in the Sun

fl.   p. 135   Pl. 30

Gnaphalium luteo-album

Cochrane, G. R.
Flowers and Plants of Victoria

fl.   Pl. 300

Gnaphalium luteo-album

Walden, Beryl M.
Wild Flowers of Hong Kong

fl. fr.   Pl. 77 (235)

Gnaphalium norvegicum

Lindman, C. A. M.
Nordens Flora, Vol 9

fl. hab.   Pl. 589B

Gnaphalium norvegicum

Martin, W. Keble
The Concise British Flora in Colour

fl. hab.   Pl. 44

Gnaphalium norvegicum

Webster, Mary

Flora of Moray, Nairn & East
     Inverness

fl.   p. 292   Pl. 18

Gnaphalium obtusifolium

Courtenay, Booth
Wildflowers & Weeds

fl.   p 126

Gnaphalium obtusifolium

Duncan, Wilbur H.
Wildflowers of the Southeastern
United States

fl.   p 209

Gnaphalium obtusifolium

Jennings, O. E.
Wild Flowers of Western Pennsylvania
v. 2
fl. fr. hab.   Pl. 181

Gnaphalium obtusifolium

Moyle, John B.
Northland Wild Flowers

fl. hab.   p 184, Pl. 232

Gnaphalium obtusifolium

Wharton, Mary E.
A Guide to the Wildflowers & Ferns
of Kentucky

fl.   p 233   Pl. 1.2

Gnaphalium sp.

Weeds of the Southern United States

fl.   p. 12

Gnaphalium supinum

Barneby, T. P.
European Alpine Flowers in Colour

fl. hab.   Pl. 89, 4

Gnaphalium supinum

Lindman, C. A. M.
Nordens Flora, Vol 9

fl. hab.   Pl. 590B

Gnaphalium supinum

Martin, W. Keble
The Concise British Flora in Colour

fl. hab.   Pl. 44

Gnaphalium supinum

Raven, John
Mountain Flowers

hab.   p. 44   Pl. 3

Gnaphalium sylvaticum

Barneby, T. P.
European Alpine Flowers in Colour

fl.   Pl. 89, 3

Gnaphalium sylvaticum

Lindman, C. A. M.
Nordens Flora, Vol. 9

fl. hab.   Pl. 589A

Gnaphalium sylvaticum

Martin, W. Keble
The Concise British Flora in Colour

fl. hab.   Pl. 44

Gnaphalium uliginosum

Ary, S.
The Oxford Book of Wildflowers

fl.   p 32, Pl. 4

Gnaphalium uliginosum

Lindman, C. A. M.
Nordens Flora, Vol 9

fl. hab.   Pl. 590A

Gnaphalium uliginosum

Martin, W. Keble
The Concise British Flora in Colour

fl. hab.   Pl. 44

Gnaphalium uliginosum

Polunin, Oleg
Flowers of Europe

fl.   Pl. 144, #1380

Gnashacks

see

Arctostaphylos urva-ursi

Gnetum montanum

Menninger, Edwin A.
Flowering Vines of the World

fl.   Pl. 168

Gnidia anthylloides var.

Flowering Plants of South Africa
vol XVIII   1938

fl.   Pl. 718

Gnidia buchananii

Moriarty, Audrey
Wild Flowers of Malawi

fl.   Pl. 61; 2

Gnidia chrysantha

Moriarty, Audrey
Wild Flowers of Malawi

fl.   Pl. 61; 3

Gnidia kraussiana

Moriarty, Audrey
Wild Flowers of Malawi

fl.   Pl. 61; 4

Gnidia polystachia

Addisonia, v. 22, 1943-46

fl. opp. p. 29   Pl. 719

Gnome, Scarlet

see

Kalanchoe blossfeldiana

Goathorn Creeper

see

Strophanthus divaricatus

Goatnut

see

Simmondsia chinensis

Goat Root

see

Ononis natrix

Goat's-beard

see

Aruncus dioicus

Goat's Beard

see

Aruncus sylvester

Goat's beard

see

Rafinesquina neo-mexicana

Goat's-beard

see

Tragopogon pratensis

Goat's beard, Virginia

see

Krigia biflora

Goat's beard, yellow

see

Tragopogon pratensis

Goat's-rue

see

Tephrosia virginiana

Godetia amoena

Crockett, James Underwood
Annuals

fl.   P 121

Godetia amoena

Tsukamoto, Yotaro
Coloured Illustrations of
   Garden Flowers V. 10

fl. Illus. 66 opp. p. 21

Godetia grandiflora

Crockett, James Underwood
Annuals

fl.   P 121

Godetia grandiflora

Macoby, Stirling
What Flower is That

fl.   p. 143

Godetia grandiflora var.

Hay, Roy
The Color Dictionary of Flowers &
   Plants

fl.   p. 38   Pl.300,301

Godetia hybrid

Crockett, James Underwood
Annuals

fl.   p. 26-27

Godetia whitneyi var.

Huxley, Anthony
Garden Annuals and Bulbs

fl.   Pl. 55

Gold, Alpine

see

Hulsea algida

Gold Crest

see

Lophiola americana

Gold dust

see

Alyssum saxatile

Gold-dust plant

see

Aucuba japonica var.

Gold dust tree

see

Aucuba japonica var.

Gold flower

see

Hypericum

Goldstar grass

see

Hypexis hirsuta

Gold Tips, Broad

see

Leucadendron decorum

Gold Tips, Narrow

see

Leucadendron sabulosum

Gold Vine, Guinea

see

Hibbertia scandens

Goldenbush

see

Haplopappus linearifolius

Golden Chain

see

Cytisus laburnum

Golden chain

see

Laburnum

Golden Club

see

Orontium aquaticum

Golden flowers

see

Leucaena retusa

Goldenglow, Wild

see

Rudbeckia laciniata

Goldenhead

see

Acamptopappus shockleyi

Golden rain

see

Laburnum anagyroides

Golden Rain

see

Laburnum watereri

Golden-rain Tree

see

Koelreuteria paniculata

Goldenrod

see

Solidago

Goldenrod, Blue-stemmed

see

Solidago caesia

Goldenrod, Bog

see

Solidago uliginosa

Goldenrod, Broad-leaf

see

Solidago flexicaulis

Goldenrod, Canada

see

Solidago canadensis

Goldenrod, Early Plumed
see
Solidago juncea

Goldenrod, Elegant
see
Solidago lepida

Goldenrod, Elm-leaved
see
Solidago ulmifolia

Goldenrod, Erect
see
Solidago erecta

Goldenrod, False
see
Solidago sphacelata

Goldenrod, Golden
see
Solidago multiradiata

Goldenrod, Grass-leaved
see
Solidago graminifolia

Goldenrod, gray
see
Solidago nemorallis

Goldenrod, Hard-Leaved
see
Solidago rigida

Goldenrod, Northern
see
Solidago multiradiata

Goldenrod, Rayless
see
Bigelowia nudata

Goldenrod, Rough
see
Solidago rugosa var.

Goldenrod, Showy
see
Solidago speciosa

Goldenrod, Shrubby
see
Chrysoma pauciflosculosa

Goldenrod, Sidesaddle
see
Solidago ciliosa

Goldenrod, Slender
see
Solidago erecta

Goldenrod, Small-headed
see
Solidago microcephala

Goldenrod, Sweet
see
Solidago odora

Goldenrod, White
see
Solidago bicolor

Goldenrod, White-haired
see
Solidago albopilosa

Goldenrod, Wreath
see
Solidago caesia

Goldenrod, Zigzag
see
Solidago flexicaulis

Goldenseal
see
Hydrastis canadensis

Golden Tip
see
Goodia lotifolia

Golden Tuft
see
Alyssum saxatile

Golden Vine
see
Stigmaphyllon ciliatum

Goldenweed, Lyall's
see
Haplopappus lyallii

Goldilocks
see
Linosyris vulgaris

Goldilocks
see
Ranunculus auricomus

Golondrina
see
Euphorbia leucophylla

Gomesa alpina
Pabst, G. F. J.
Orchidaceae Brasilienses,
Vol 2
fl.   p 240

Gomesa divaricata
Pabst, G. F. J.
Orchidaceae Brasilienses,
Vol 2
fl.   p 241

Gomesa glaziovii
Pabst, G. F. J.
Orchidaceae Brasilienses,
Vol 2
fl.   p 241

Gomesa planifolia

Pabst, G. F. J.
Orchidaceae Brasilienses,
Vol 2

fl.   p 241

Gomesa recurva

Pabst, G. F. J.
Orchidaceae Brasilienses,
Vol 2

fl.   p 241

Gomesa verboonenii

Ospina, Mariano
Orquideas de las americas

fl.   Pl. 174

Gomphichis lancipetala

Ospina, Mariano
Orquideas de las americas

fl.   Pl. 27

Gompholobium ecostatum

Cochrane, G. R.
Flowers and Plants of Victoria

fl.   Pl. 15

Gompholobium grandiflorum

Macoby, Stirling
What Flower is That

fl.   p. 143

Gompholobium huegelii

Blombery, Alec M.
What Wildflower is That

fl.   p 147, Pl. 392

Gompholobium polymorphum

Morley, Brian D.
Wild Flowers of the World

fl.   Plate 129

Gompholobium species

Mullins, Barbara
Australian Wildflowers in Colour

fl.   P. 21   Pl. 11

Gompholobium uncinatum

Blombery, Alec M.
What Wildflower is That

fl.   p 147, Pl. 393

Gomphrena globosa

Bruggeman, L.
Tropical Plants

fl.   Pl. 43

Gomphrena globosa

Crockett, James Underwood
Annuals

fl.   P 121

Gomphrena globosa

Everett, T.H.
New Illustrated Encyclopedia of
Gardening, v.5
fl.
opp.p.790

Gomphrena globosa

Macoby, Stirling
What Flower is That

fl.   p. 144

Gomphrena globosa

Tsukamoto, Yotaro
Coloured Illustrations of Garden
Flowers Vol. 10

fl. Illus. 67 opp. p. 22

Gomphrena globosa var.

Hay, Roy
The Color Dictionary of Flowers &
Plants

fl.   p. 38   Pl. 302

Gonatopus boivinii

Moriarty, Audrey
Wild Flowers of Malawi

fl.   Pl. 1

Gongora armeniaca

Ebel, Friedrich
The Strange and Beautiful
World of Orchids

fl.   p 91

Gongora bufonia

Pabst, G. F. J.
Orchidaceae Brasilienses,
Vol 2

fl.   p 220

Gongora galeata

Ebel, Friedrich
The Strange and Beautiful
World of Orchids

fl.   p 71

Gongora galeata

Kupper, Walter
Orchid

fl. hab.   p 69

Gongora gratulabunda

Ospina, Mariano
Orquideas de las americas

fl.   Pl. 153

Gongora nigrita

Pabst, G. F. J.
Orchidaceae Brasilienses,
Vol 2

fl.   p 220

Gongora quinquenervis

American Hort. Soc.
American Horticulturist
vol 51, No. 4 1972

fl.   p. 29

Gongora quinquenervis

Dunsterville, C. C. K.
Introduction to the World
of Orchids

fl.   Pl. 11a,b

Gongora quinquenervis

Ebel, Friedrich
The Strange and Beautiful
World of Orchids

fl. hab.   p 125

Gongora quinquenervis

Pabst, G. F. J.
Orchidaceae Brasilienses,
Vol 2

fl.   p 221

Gongora scaphephorus

Kramer, Jack
Orchids; Flowers of Romance
and Mystery

fl.   p 179

Gonocaryum pyriforme

Bruggeman, L.
Tropical Plants

fl.,fr.   Pl. 194

Gonolobus caroliensis

Batson, Wade, T.
Wild Flowers in South Carolina

fl.   p 96

Gonolobus fulvidus

Curtis's Botanical Magazine
Vol 163   1940 - 42

fl.   No. 9611

Gonolobus shortii

Wharton, Mary E.
A Guide to the Wildflowers & Ferns
of Kentucky

fl. p 130 Pl. 1.71

Gonospermum canariense

Bramwell, David
Wild Flowers of the Canary Islands

fl.   Pl. 278

Good King Henry

see

Chenopodium bonus-henricus

Goodenia affinis

Cochrane, G. R.
Flowers and Plants of Victoria

fl.   Pl. 134

Goodenia barbata

Blombery, Alec M.
What Wildflower is That

fl.   p 147, Pl. 394

Goodenia decurrens

Blombery, Alec M.
What Wildflower is That

fl.   p 147, Pl. 395

Goodenia hederacea

Blombery, Alec M.
What Wildflower is That

fl.   p 148, Pl. 396

Goodenia hederacea var.

Cochrane, G. R.
Flowers and Plants of Victoria

fl. hab.   Pl. 523

Goodenia humilis

Cochrane, G. R.
Flowers and Plants of Victoria

fl. hab.   Pl. 61

Goodenia ovata

Blombery, Alec M.
What Wildflower is That

fl.   p 148, Pl. 397

Goodenia ovata

Cochrane, G. R.
Flowers and Plants of Victoria

fl.   Pl. 381

Goodenia scapigera

Blombery, Alec M.
What Wildflower is That

fl.   p 148, Pl. 398

Goodia lotifolia

Blombery, Alec M.
What Wildflower is That

fl.   p 148, Pl. 399

Goodia lotifolia

Cochrane, G. R.
Flowers and Plants of Victoria

fl.   Pl. 398

Goodyera hispida

Grubb, Roy
Selected Orchidaceous Plants
Vol II

fl.   p 19

Goodyera oblongifolia

Clark, Lewis J.
Wild Flowers of British Columbia

fl. hab.   p 82, 83

Goodyera oblongifolia

Luer, Carlyle A.
The Native Orchids of the
United States and Canada

fl. hab.   Pl. 34

Goodyera oblongifolia

Taylor, Ronald J.
Mountain Wild Flowers

fl. hab.   p. 49

Goodyera procera

Hu, Shiu-ying
The Genera of Orchidaceae in
Hong Kong

fl.   p 121

Goodyera pubescens

Campbell, Carlos C.
Great Smoky Mountain Wildflowers

fl. hab.   p 91

Goodyera pubescens

Courtenay, Booth
Wildflowers & Weeds

fl.   p 16

Goodyera pubescens

Dean, Blanche E.
Wildflowers of Alabama and
Adjoining States

fl. hab.   p 41

Goodyera pubescens

Duncan, Wilbur H.
Wildflowers of the Southeastern
United States

fl. hab.   p 277

Goodyera pubescens

Klimas, John E.
Wildflowers of Eastern America

fl.   Pl. 35

Goodyera pubescens

Lemmon, Robert S.
Wildflowers of North America in
Full Color

fl.   p. 242   Pl. 382

Goodyera pubescens

Luer, Carlyle A.
The Native Orchids of Florida

fl. hab.   Pl. 1; 1, 2

Goodyera pubescens

Luer, Carlyle A.
The Native Orchids of the
United States and Canada

fl. hab.   Pl. 33

Goodyera pubescens

Wharton, Mary E.
A Guide to the Wildflowers & Ferns
of Kentucky

fl. p 64 Pl. 2,3

Goodyera repens

Ary, S.
The Oxford Book of Wildflowers

fl.   p. 100   Pl. 1

Goodyera repens

Barneby, T. P.
European Alpine Flowers in Colour

fl.   Pl. 13, 3

Goodyera repens

Brooke, Jocelyn
The Wild Orchids of Britain

fl.   Pl. 11

Goodyera repens

Luer, Carlyle A.
The Native Orchids of the
United States and Canada

fl. hab.   Pl. 35; 5-8

Goodyera repens

Martin, W. Keble
The Concise British Flora in Colour

fl. hab.   Pl. 80

Goodyera repens

Moyle, John B.
Northland Wild Flowers

fl. hab.   p 225, Pl. 299

Goodyera repens

Polunin, Oleg
Flowers of Europe

fl.   Pl. 192 #1925

Goodyera rubicunda

Grubb, Roy
Selected Orchidaceous Plants
    Vol II

fl.   p 23

Goodyera striata

Ospina, Mariano
Orquideas de las americas

fl.   Pl. 24

Goodyera tesselata

Luer, Carlyle A.
The Native Orchids of the
    United States and Canada

fl. hab.   Pl. 35; 1-4

Goodyera vittata

Hara, Hiroshi, comp.
Photo-Album of Plants of
    Eastern Himalaya

fl.   Pl. 130

Goodyera youngsayei

Walden, Beryl M.
Wild Flowers of Hong Kong

fl.   Pl. 8(24)

Gooseberry

see

Ribes

Gooseberry, Barbados

see

Pereskia tampicana

Gooseberry, Cape

see

Physalis peruviana

Gooseberry, Chinese

see

Actinidia chinensis

Gooseberry, Chinese

see

Actinidia sinensis

Gooseberry, Fuschia-Flowered

see

Ribes speciosum

Gooseberry, Spanish

see

Pereskia tampicana

Goose Flower

see

Aristolochia grandiflora

Goosefoot

see

Chenopodium

Goosefoot, Stinking

see

Chenopodium vulvaria

Goosefoot, White

see

Chenopodium album

Goose Grass

see

Eleusine indica

Goose Grass

see

Galium aparine

Goose Grass

see

Potentilla anserina

Goose Tongue

see

Plantago maritima

Gordonia alatamaha

Everett, T. H.
New Illustrated Encyclopedia of
    Gardening, V. 5

fl.   p. 743

Gordonia alatamaha

Milne, Lorus
Living Plants of the World

fl.   p 151

Gordonia axillaris

Macoby, Stirling
What Flower is That

fl.   p. 144

Gordonia axillaris

Walden, Beryl M.
Wild Flowers of Hong Kong

fl. fr.   Pl. 69 (206)

Gordonia chrysandra

Curtis's Botanical Magazine
    Vol 171   1956-57

fl.   285

Gordonia lasianthus

Brown, Clair A.
Wildflowers of Louisiana and
    Adjoining States

fl. hab.   p 112

Gordonia lasianthus

Everett, T. H.
New Illustrated Encyclopedia of
    Gardening, V. 5

fl.   opp. p. 822

Gordonia lasianthus

Fleming, Glenn
Wild Flowers of Florida

fl. hab.   p 19

Gorgon's head

see

Euphorbia caput-medusae

Gorse

see

Ulex europaeus

Corteria diffusa var.

Eliovson, Sima
Namaqualand in Flower

fl.   Pl. 25, 3

Gossypium arboreum

Hvass, Elsie
Plants That Feed and Serve Us

fl.   p 101, Pl. 217

Gossypium australe

Blombery, Alec M.
What Wildflower is That

fl.   p. 148   Pl. 400

Gossypium barbadense

Bianchini, F.
The Complete Book of Fruits & Vegetables

fr.   p 239

Gossypium barbadense

Hvass, Elsie
Plants That Feed and Serve Us

fl.   p 101, Pl. 216

Gossypium barbadense

Massachusetts Hort. Soc.
Horticulture
   vol 54, No. 1   1976

fl., fr.   p. 40

Gossypium davidsonii

Coyle, Jeanette
A Field Guide to the Common and
   Interesting Plants of Baja California

fl.   p 123

Gossypium herbaceum

Perrot, Emile
Les Plantes Medicinales

fl. fr. hab.   p 79

Gossypium herbaceum

Polunin, Oleg
Flowers of Europe

fl.   Pl. 73 #750

Gossypium hirsutum

Hvass, Elsie
Plants That Feed and Serve Us

fl. fr.   p 101, Pl. 215

Gossypium hirsutum

Wit, H. C. D. de
Plants of the World;
   The Higher Plants, Vol I

fl.   p 219, Pl. 131

Gossypium sp.

Kariyone, Tatsuo
Atlas of Medicinal Plants

fl., fr.   Pl. 84

Gossypium species

Mullins, Barbara
Australian Wildflowers in Colour

fl.   P. 49   Pl. 44

Gossypium sturtianum

Blombery, Alec M.
What Wildflower is That

fl.   p 149, Pl. 401

Gourd
   see
Cucurbita pepo

Gourd, Bottle
   see
Lagenaria vulgaris

Gourd, Chito
   see
Cucumis melo

Gourd, Dipper
   see
Lagenaria siceraria

Gourd, Dishcloth
   see
Luffa sp.

Gourd, Miniature Bottle
   see
Cucurbita pepo var.

Gourd, Missouri
   see
Cucurbita foetidissima

Gourd, Nest Egg
   see
Cucurbita pepo var.

Gourd, Scallop
   see
Curcurbita pepo var.

Gourd, snake
   see
Trichosanthes cucumerina

Gourd, Sour Ethiopian
   see
Adansonia digitata

Gourd, Striped Pear
   see
Cucurbita pepo

Gourd, Turk's Turban
   see
Cucurbita maxima

Gourd, Warty Hardheads
   see
Cuburbita pepo

Gourd, Wild
   see
Cucurbita foetidissima

Goutweed
   see
Aegopodium podagraria

Goutweed, Silveredge
   see
Aegopodium podagraria var.

Govenia lilacea

Ospina, Mariano
Orquideas de las americas

fl.   Pl. 96

Covenia utriculata

Luer, Carlyle A.
The Native Orchids of Florida

fl. hab.   Pl. 74; 1-4

Govenia utriculata

Pabst, G. F. J.
Orchidaceae Brasilienses,
   Vol 1

fl.   p 241

Gowan, Lucky
   see
Trollius europaeus

Grain Tree
   see
Quercus coccifera

Gram, black

see

Phaseolus mungo

Gram, green

see

Phaseolus aureus

Grammangis ellisii

Stewart, Joyce
Orchids of Tropical Africa

fl.   Pl. 30

Grammatophyllum scriptum

Kramer, Jack
Orchids; Flowers of Romance
and Mystery

fl.   p 167

Grammatophyllum speciosum

Kamemoto, Haruyuki
Beautiful Thai Orchid Species

fl.   p 150

Granadilla

see

Passiflora edulis

Granadilla

see

Passiflora quadrangularis

Granadilla, giant

see

Passiflora quadrangularis

Granadilla, purple

see

Passiflora edulis

Grape

see

Vitis

Grapefruit

see

Citrus aurantium var.

Grapefruit

see

Citrus paradisi

Grape, Mt. Ida

see

Vaccinium vitis-idaea

Grape, Oregon

see

Berberis repens

Grape, Oregon

see

Mahonia aquifolium

Grape, Oregon Cascade

see

Berberis nervosa

Grape, Oregon Creeping

see

Mahonia repens

Grape, Oregon Tall

see

Berberis nervosa

Grape, Ornamental

see

Vitis coignetiae

Grape, sea

see

Coccoloba uvifera

Graphistemna pictum

Walden, Beryl M.
Wild Flowers of Hong Kong

fl.   Pl. 31(88)

Grapple Plant

see

Harpagophytum procumbens

Graptopetalum fruticosum

Cactus & Succulent Society of America
Journal
vol 40, No. 4   1968

fl.   p. 152

Graptopetalum pachyphyllum

Lamb, Edgar
The Pocket Encyclopedia of Cacti
and Succulents in Color

fl. hab.   Pl. 261

Graptopetalum pentandrum

Cactus & Succulent Society of America
Journal
vol 43, No. 6   1971

fl.   p. 255

Graptopetalum rusbyi

Lamb, Edgar
Colorful Cacti of the American Deserts

fl. hab.   Pl. 101

Graptophyllum pictum

Bruggeman, L.
Tropical Plants

hab.   Pl. 287-288

x Graptoveria hybrid

Cactus & Succulent Society of America
Journal
vol 50, No. 2   1978

hab.   p. 78

Grass of Parnassus

see

Parnassia

Grass-of-Parnassus, Fringed

see

Parnassia fimbriata

Grass of Parnassus, Northern

see

Parnassia palustris

Grass, Silk

see

Yucca filamentosa

Grass Tree

see

Kingia australis

Grass Tree

see

Xanthorrhoea

Grass Tree, Austral

see

Xanthorrhoea australis

Grass Tree, Giant

see

Richea pandanifolia

Gratiola aurea

Courtenay, Booth
Wildflowers & Weeds

fl.   p 99

Gratiola brevifolia

Brown, Clair A.
Wildflowers of Louisiana and
   Adjoining States

fl. hab.   p 168

Gratiola floridana

Duncan, Wilbur H.
Wildflowers of the Southeastern
   United States

fl.   p 171

Gratiola neglecta

Courtenay, Booth
Wildflowers & Weeds

fl.   p 99

Gratiola neglecta

Duncan, Wilbur H.
Wildflowers of the Southeastern
   United States

fl.   p 173

Gratiola officinalis

Perrot, Emile
Les Plantes Medicinales

fl, hab.   p. 211

Gratiola officinalis

Polunin, Oleg
Flowers of Europe

fl.   Pl. 124 #1218

Gratiola peruviana

Cochrane, C. R.
Flowers and Plants of Victoria

fl.   Pl. 302

Gratiola ramosa

Duncan, Wilbur H.
Wildflowers of the Southeastern
   United States

fl.   p 171

Graveyard flower

see

Plumeria acutifolia

Gravisia aquilega

Rauh, Werner
Bromeliads for Home, Garden and
   Greenhouse

fl.   Pl. 108

Gravisia brassicoides

Bromeliad Society
Bulletin
   vol 13, No. 3   1963

fl., hab.   p. 57

Gravisia fosteriana

Wilson, Robert Gardner
Bromeliads in Cultivation

fl.   Pl. 116

Grayia spinosa

Munz, Philip A.
California Desert Wildflowers

fl.   p 31, Pl. 14

Greasewood

see

Adenostoma fasciculatum

Green and Gold

see

Chrysogonum virginianum

Greenbriar

see

Smilax

Greenbriar, Common

see

Smilax rotundifolia

Greengage

see

Prunus italica

Greenhood, Bearded

see

Pterostylis barbata

Greenhood, King

see

Pterostylis baptistii

Greenhood, Nodding

see

Pterostylis nutans

Greenhood, Superb

see

Pterostylis grandiflora

Greenweed

see

Genista

Greenweed, Dyer's

see

Genista tinctoria

Greenovia aurea

Bramwell, David
Wild Flowers of the Canary Islands

fl.   Pl. 161

Greenovia aurea

Lamb, Edgar
The Pocket Encyclopedia of Cacti
   and Succulents in Color

fl. hab.   Pl. 322

Greenovia aurea

Lamb, Edgar
Popular Exotic Cacti in Color

hab.   p 95

Gregia spacelata

Munoz Pizarro, Carlos
Flores Silvestres de Chile

fl., hab   Pl. 31

Gregoria vitaliana

Kohlhaupt, Paula
Fleurs des Alpages, v.1

fl. P 41

Greigia van-hyningii

Wilson, Robert Gardner
Bromeliads in Cultivation

hab.　　　　Pl. 115

Grevillea acanthifolia

Blombery, Alec M.
What Wildflower is That

fl.　　　　p 150, Pl. 402

Grevillea alpina

Addisonia, V. 23, 1954-59

fl.　opp. p. 57　　Pl. 765

Grevillea alpina

Cochrane, G. R.
Flowers and Plants of Victoria

fl.　　　Pl. 311

Grevillea alpina

Morley, Brian D.
Wild Flowers of the World

fl.　　　Pl. 134

Grevillea annulifera

Morcombe, M. K.
Australia's Western Wildflowers

fl.　　　p 92

Grevillea aquifolium

Cochrane, G. R.
Flowers and Plants of Victoria

fl.　　　Pl. 102

Grevillea asplenifolia

Harrison, R.E.
Trees and Shrubs

fl.　　P 83

Grevillea banksii

Blombery, Alec M.
What Wildflower is That

fl.　　　p 150, Pl. 403

Grevillea banksii

Bruggeman, L.
Tropical Plants

fl.　　　Pl. 241

Grevillea banksii

Mathias, Mildred E.
Color for the Landscape

fl. hab.　　p 78

Grevillea banksii

Mullins, Barbara
Australian Wildflowers in Colour

fl.　　P. 31　　　Pl. 22

Grevillea banksii var.

Harrison, R. E.
Trees and Shrubs

fl.　　P. 83

Grevillea banksii var.

Macoby, Stirling
What Flower is That

fl.　　　p. 144

Grevillea banksii var.

Massachusetts Hort. Soc.
Horticulture
Vol. 44, No. 12　1966

fl.　　P. 20

Grevillea baueri

Blombery, Alec M.
What Wildflower is That

fl.　　　p 150, Pl. 404

Grevillea bipinnatifida

Blombery, Alec M.
What Wildflower is That

fl.　　　p 151, Pl. 405

Grevillea bipinnatifida

Menninger, Edwin A.
Flowering Vines of the World

fl.　　　Pl. 166

Grevillea bipinnatifida

Morcombe, M. K.
Australia's Western Wildflowers

fl.　　　p 11

Grevillea biternata

Macoby, Stirling
What Flower is That

fl.　　　p. 144

Grevillea buxifolia

Blombery, Alec M.
What Wildflower is That

fl.　　　p 153, Pl. 407

Grevillea buxifolia

Mullins, Barbara
Australian Wildflowers in Colour

fl.　　P. 29　　　Pl. 19

Grevillea candelabroides

Harrison, R.E.
Trees and Shrubs

fl.　　P 85

Grevillea candicans

Morcombe, M. K.
Australia's Western Wildflowers

fl.　　　p 34

Grevillea dimorpha

Cochrane, G. R.
Flowers and Plants of Victoria

fl.　　　Pl. 93

Grevillea drummondii

Morcombe, M. K.
Australia's Western Wildflowers

fl.　　　p 84

Grevillea eriostachya

Blombery, Alec M.
What Wildflower is That

fl.　　　p 153, Pl. 408

Grevillea eriostachya

Morcombe, M. K.
Australia's Western Wildflowers

fl.　　　p 32

Grevillea eriostachya

Mullins, Barbara
Australian Wildflowers in Colour

fl.　　P. 31　　　Pl. 21

Grevillea fasciculata

Menninger, Edwin A.
Flowering Vines of the World

fl.　　　Pl. 167

Grevillea, Fuchsia

see

Grevillea bipinnatifida

Grevillea glabrata

Blombery, Alec M.
What Wildflower is That

fl.　　　p 153, Pl. 409

Grevillea hilliana

Blombery, Alec M.
What Wildflower is That

fl.　　　p 153, Pl. 410

Grevillea hookeriana
Blombery, Alec M.
What Wildflower is That

fl.   p 153, Pl. 411

Grevillea hookeriana
Harrison, R.E.
Trees and Shrubs

fl. P 83

Grevillea hybrid
Blombery, Alec M.
What Wildflower is That

fl.   p. 152   Pl. 406

Grevillea hybrid
Harrison, R. E.
Trees and Shrubs

fl.   p. 83

Grevillea juncifolia
Blombery, Alec M.
What Wildflower is That

fl.   p 153, Pl. 412

Grevillea juniperina
Blombery, Alec M.
What Wildflower is That

fl.   p 153, Pl. 413

Grevillea juniperina
Harrison, R.E.
Trees and Shrubs

fl. P 84

Grevillea juniperina var.
Macoby, Stirling
What Flower is That

fl.   p. 145

Grevillea laurifolia
Blombery, Alec M.
What Wildflower is That

fl.   p 153, Pl. 414

Grevillea lavandulacea
Cochrane, G. R.
Flowers and Plants of Victoria

fl.   Pl. 98

Grevillea lavandulacea var.
Harrison, R.E.
Trees and Shrubs

fl. P 84

Grevillea longistyla
Harrison, R.E.
Trees and Shrubs

fl. P 82

Grevillea, Oak-leaved
    see
Grevillea quercifolia

Grevillea oleoides
Blombery, Alec M.
What Wildflower is That

fl.   p 155, Pl. 415

Grevillea oleoides
Mullins, Barbara
Australian Wildflowers in Colour

fl.   P. 31   Pl. 24

Grevillea oleoides var.
Harrison, R.E.
Trees and Shrubs

fl. P 84

Grevillea, olive
    see
Grevillea oleoides

Grevillea, Orange
    see
Grevillea eriostachya

Grevillea petrophiloides
Blombery, Alec M.
What Wildflower is That

fl. fr.   p 154, Pl. 416

Grevillea petrophiloides
Harrison, R.E.
Trees and Shrubs

fl. P 84

Grevillea petrophiloides
Morcombe, M. K.
Australia's Western Wildflowers

fl.   p 19

Grevillea pteridifolia
Mullins, Barbara
Australian Wildflowers in Colour

fl.   P. 29   Pl. 20

Grevillea pungens
Mullins, Barbara
Australian Wildflower in Colour

fl.   P. 31   Pl. 23

Grevillea punicea
Blombery, Alec M.
What Wildflower is That

fl.   p 155, Pl. 417

Grevillea punicea
Harrison, R.E.
Trees and Shrubs

fl.   p. 85

Grevillea punicea
Macoby, Stirling
What Flower is That

fl.   p. 145

Grevillea punicea
Morley, Brian D.
Wild Flowers of the World

fl. fr.   Pl. 134

Grevillea punicea
Mullins, Barbara
Australian Wildflowers in Colour

fl.   P. 29   Pl. 18

Grevillea quercifolia
Blombery, Alec M.
What Wildflower is That

fl.   p 155, Pl. 418

Grevillea robusta
American Hort. Soc.
American Horticulturist
   vol 52, No. 4   1973

fl.   p. 36, 40

Grevillea robusta
Blombery, Alec M.
What Wildflower is That

fl.   p 155, Pl. 419

Grevillea robusta
Harrison, R. E.
Trees and Shrubs

fl.   Pl. 85

Grevillea robusta
Hersey, Jean
Woman's Day Book of House Plants

hab.   p. 96

Grevillea robusta

Kiaer, Eigil
Indoor Plants in Colour

hab.   p. 63

Grevillea robusta

Kromdijk, G.
200 House Plants in Colour

hab.   Pl. 100

Grevillea robusta

Macoby, Stirling
What Flower is That

fl.   p. 145

Grevillea robusta

Mathias, Mildred E.
Color for the Landscape

fl. hab.   p 19

Grevillea robusta

Oakman, Harry
Colorful Trees

hab., fl.   p. 45

Grevillea, Rock

see

Grevillea petrophiloides

Grevillea rosmarinifolia

Cochrane, G. R.
Flowers and Plants of Victoria

fl.   Pl. 310

Grevillea rosmarinifolia

Harrison, R.E.
Trees and Shrubs

Grevillea rosmarinifolia

Hellyer, A.G.L.
Shrubs in Color
fl.
p. 59

Grevillea rosmarinifolia

Perry, Frances
Flowers of the World

fl.   p 245

Grevillea X semperflorens

Curtis's Botanical Magazine,
v. 173, 1960

fl.   Pl. 353

Grevillea sericea

Blombery, Alec M.
What Wildflower is That

fl.   p 156, Pl. 420

Grevillea, Smooth

see

Grevillea glabrata

Grevillea sp.

Everett, Thomas H.
Living Trees of the World

fl.   p. 148

Grevillea sulphurea

Hay, Roy
The Color Dictionary of Flowers & Plants

fl.   p 204, Pl. 1625

Grevillea, Toothbrush

see

Grevillea hookerana

Grevillea tridentifera

Blombery, Alec M.
What Wildflower is That

fl.   p 157, Pl. 421

Grevillea victoriae

Cochrane, G. R.
Flowers and Plants of Victoria

fl.   Pl. 506

Grevillea wickhamii

Blombery, Alec M.
What Wildflower is That

fl.   p 157, Pl. 422

Grevillea wilsonii

Blombery, Alec M.
What Wildflower is That

fl.   p. 157   Pl. 423

Grevillea wilsonii

Morcombe, M. K.
Australia's Western Wildflowers

fl.   p 45

Crewia biloba

Walden, Beryl M.
Wild Flowers of Hong Kong

fl.   Pl. 43(10?)

Grewia lasiocarpa

Flowering Plants of Africa
vol 44   1977

fl.   Pl. 1733

Greyia radlkoferi

Harrison, R.E.
Trees and Shrubs

fl. P 86

Greyia sutherlandii

Curtis's Botanical Magazine
v. 173, 1960

fl.   Pl. 374

Greyia sutherlandii

Flowering Plants of Africa
vol 44   1977

fl.   Pl. 1732

Greyia sutherlandii

Mathias, Mildred E.
Color for the Landscape

fl.   p 69

Greyia sutherlandii

Morley, Brian D.
Wild Flowers of the World

fl.   Pl. 77

Greyia sutherlandii

Palmer, Eve.
Trees of South Africa

fl.   p. 240   Pl. XIX

Greyia sutherlandii

Perry, Frances
Flowers of the World

fl.   p 133

Greyia sutherlandii

Royal Hort. Soc.
Journal of the Royal Hort. Soc.
Vol. 88, No. 8   1963

fl.   p. 342   Pl. 137

Greyia sutherlandii

Royal Hort. Soc.
Journal of the Royal Hort. Soc.
Vol. 91, No. 10   1966

fl.   p. 426   Pl. 228

Grielum humifusum

Eliovson, Sima
Namaqualand in Flower

fl.   Pl. 14, 1, 2
        Pl. 15, 1, 2

Grielum humifusum

Royal Hort. Soc.
Journal of the Royal Hort. Soc.
vol 90, No. 8   1965

fl.   p. 338   Pl. 141

Grielum tenuifolium

Royal Hort. Soc.
Journal of the Royal Hort. Soc.
vol 90, No. 8   1965

fl.   p. 338   Pl. 144

Griffinia hyacinthina

Everett, T. H.
New Illustrated Encyclopedia of
Gardening, V. 5

fl.   opp. p. 822

Grindelia chiloensis

Curtis's Botanical Magazine
Vol 160   1937

fl.   No. 9471

Grindelia integrifolia

Clark, Lewis J.
Wild Flowers of British Columbia

fl.   p 551

Grindelia integrifolia

Ferguson, Mary
Wildflowers

fl.   p 113

Grindelia integrifolia var.

Clark, Lewis J.
Wild Flowers of British Columbia

fl.   p 551

Grindelia oolepis

Addisonia, V. 21, 1939-42

fl., fr.   Pl. 704

Grindelia robusta

Bianchini, Francesco
Health Plants of the World

fl. hab.   p 97

Grindelia robusta

Lemmon, Robert S.
Wildflowers of North America in
Full Color

fl.   p. 53   Pl. 88

Grindelia squarrosa

Courtenay, Booth
Wildflowers & Weeds

fl.   p 118

Grindelia squarrosa

Klimas, John E.
Wildflowers of Eastern America

fl.   Pl. 172

Grindelia squarrosa

Moyle, John B.
Northland Wild Flowers

fl. hab.   p 169, Pl. 206

Grindelia squarrosa

Perrot, Emile
Les Plantes Medicinales

fl. fr. hab.   p 112

Grindelia squarrosa

Welsh, Stanley L.
Flowers of the Mountain Country

fl.   p. 54

Grindelia stricta

Orr, Robert T.
Wildflowers of Western America

fl.   Pl. 128

Grindelia stricta var.

Munz, Philip A.
California Spring Wildflowers

fl.   p 84, Pl. 76

Griselinia littoralis

Hay, Roy
The Color Dictionary of Flowers & Plants

hab.   p 204, Pl. 1626

Grobya amherstiae

Curtis's Botanical Magazine
Vol. 181   1976-77

fl. hab.   Pl. 720

Grobya amherstiae

Pabst, G. F. J.
Orchidaceae Brasilienses,
Vol. 1

fl.   p. 241

Grobya galeata

Pabst, G. F. J.
Orchidaceae Brasilienses,
Vol 1

fl.   p 241

Groenlandia densa

Martin, W. Keble
The Concise British Flora in Colour

fl. hab.   Pl. 89

Cromwell

see

Lithospermum incisum

Cromwell

see

Lithospermum officinale

Cromwell, blue

see

Lithospermum purpureo-caeruleum

Cromwell, Corn

see

Lithospermum arvense

Cromwell, False

see

Onosmodium molle

Ground Berry

see

Melastoma dodecandrum

Groundnut

see

Apios americana

Groundnut

see

Arachis hypogaea

Groundnut, Bambarra

see

Voandzeia subterranea

Groundnut, False

see

Desmodium heterocarpum

Groundsel

see

Baccharis halimifolia

Groundsel

see

Senecio

Groundsel, Arrowleaf

see

Senecio triangularis

Groundsel, bigleaf

see

Senecio grandifolius

Groundsel, Black-Tipped

see

Senecio lugens

Groundsel, golden

see

Senecio aureus

Groundsel, Mourning

see

Senecio lugens

Groundsel, peach-leaf

see

Senecio amygdalifolius

Groundsel, Rayless

see

Senecio pauciflorus

Groundsel, Variable

see

Senecio lautus

Groundsel, velvet

see

Senecio petasitis

Grouse Berry, Pink Fruited

see

Vaccinium scoparium

Grumolo

see

Cichorium endivia var.

Grusonia bradtiana

Backeberg, Curt
Cactus Lexicon

fl.   p 614

Guaiacum officinale

Hargreaves, Dorothy
Tropical Blossoms of the Caribbean
fl. hab.   p. 50

Guaiacum officinale

Hvass, Elsie
Plants That Feed and Serve Us

fl. fr.   p 129, Pl. 286

Guaiacum officinale

Morley, Brian D.
Wild Flowers of the World

fl. fr.   Pl. 178

Guaiacum officinale

Pertchik, Bernard
Flowering Trees of the
  Caribbean

fl.   p 107

Guaiacum sanctum

Chickering, Carol Rogers
Flowers of Guatemala

fl. fr.   p. 69

Guava

see

Psidium guajava

Guava, Pineapple

see

Feijoa sellowiana

Guava, pineapple

see

Psidium guajava

Guava, purple

see

Psidium cattleianum

Guava, strawberry

see

Psidium cattleianum

Guinea-flower, Erect

see

Hibbertia stricta

Guineagrass

see

Panicum maximum

Guitar Plant

see

Lomatia tinctoria

Guizotia abysinnica

Bianchini, F.
The Complete Book of Fruits & Vegetables

fl. fr.   p 239

Gum

see

Aphanopetalum resinosum

Gum Arabic

see

Acacia senegal

Gum Balata

see

Mimusops balata

Gum bell-fruited

see

Eucalyptus preissiana

Gum, Black

see

Nyssa sylvatica

Gum, Blue

see

Eucalyptus globulus

Gum, cider

see

Eucalyptus gunnii

Gum, Coral

see

Eucalyptus torquata

Gum, Cup

see

Eucalyptus cosmophylla

Gum, Fuchsia

see

Eucalyptus forrestiana

Gum, Ghost

see

Eucalyptus papuana

Gum, Gimlet

see

Eucalyptus salubris

Gum, Grampians

see

Eucalyptus alpina

Gum, Lemon-flowered

see

Eucalyptus woodwardii

Gum, Lemon Scented

see

Eucalyptus citriodora

Gum, Lindsay

see

Eucalyptus erythronema

Gum, Manna

see

Eucalyptus viminalis

Gum, Pink

see

Eucalyptus fasciculosa

Gum, pink-flowered blue

see

Eucalyptus leucoxylon

Gum Plant

see

Grindelia

Gum Plant, Salt Marsh

see

Grindelia stricta

Gum, Red Cap

see

Eucalyptus erythrocorys

Gum, red flowering

see

Eucalyptus ficifolia

Gum, Red Spotted

see

Eucalyptus mannifera var.

Gum, Ribbon

see

Eucalyptus viminalis

Gum, River Red

see

Eucalyptus camaldulensis

Gum, Rose

see

Eucalyptus grandis

Gum scarlet

see

Eucalyptus ficifolia

Gum, Scarlet Pear

see

Eucalyptus stoatei

Gum, Scribbly

see

Eucalyptus haemostoma

Gum, Scribbly

see

Eucalyptus rossii

Gum, silver

see

Eucalyptus cordata

Gum, Snow

see

Eucalyptus niphophila

Gum, Snow

see

Eucalyptus pauciflora

Gum, Sour

see

Nyssa sylvatica

Gum, South Australian Blue

see

Eucalyptus leucoxylon

Gum, Southern Blue

see

Eucalyptus globulus

Gum, Spinning

see

Eucalyptus perriniana

Gum, Spotted

see

Eucalyptus maculata

Gum, Strickland's

see

Eucalyptus stricklandii

Gum, Sugar

see

Eucalyptus cladocalyx

Gum, Sweet

see

Liquidambar styraciflua

Gum, Sydney Blue

see

Eucalyptus saligna

Gum, Tasmanian blue

see

Eucalyptus globulus

Gum, Tasmanian Snow

see

Eucalyptus coccifera

Gum Tragacanth

see

Astragalus gummifer

Gum, Urn

see

Eucalyptus urnigera

Gum, Varnished-Leaved

see

Eucalyptus vernicosa

Gum Vine

see

Aphanopetalum resinosum

Gum, Water

see

Tristania laurina

Gumweed

see

Grindelia integrifolia

Gumweed

see

Grindelia squarrosa

Gum, White

see

Eucalyptus mannifera var.

Gum, Yellow

see

Eucalyptus johnstonii

Gum, Yates

see

Eucalyptus lehmanii var.

Gum, Yellow

see

Eucalyptus leucoxylon

Gungurra

see

Eucalyptus caesia

Gunnera cordifolia

Curtis, Winifred
The Endemic Flora of Tasmania
Vol. 2

fl. fr.   Pl. 65

Gunnera magellanica

Morley, Brian D.
Wild Flowers of the World

fl.   Pl. 3G

Gutierrezia microphala

Welsh, Stanley L.
Flowers of the Canyon Country

fl.   p 41

Gutta-percha

see

Mimusops balata

Guzmania angustifolia

Rauh, Werner
Bromeliads for Home, Garden and
Greenhouse

fl.   Pl. 50

Guzmania berteroniana

Addisonia, v. 22, 1943-46

fl. opp. p. 23   Pl. 716

Guzmania berteroniana

Bromeliad Society
Bulletin
vol 13, No. 2 1963

fl.   p. 39

Guzmania berteroniana

Bromeliad Society
Journal
vol 23, No. 3 1973

fl.   p. 124

Guzmania berteroniana

Bromeliad Society
Journal
vol 26, No. 5 1976

fl.   cover (p. 177)

Guzmania berteroniana

Everett, T. H.
New Illustrated Encyclopedia
of Gardening, v. 5

fl.   opp. p. 775

Guzmania berteroniana

Padilla, Victoria
Bromeliads

fl.   p. 70   Pl. 8

Guzmania berteroniana

Padilla, Victoria, ed.
Bromeliads in Color and
Their Culture

fl.   p 31

Guzmania berteroniana

Rauh, Werner
Bromeliads for Home, Garden and
Greenhouse

fl.   Pl. 51

Guzmania cardinalis

Perry, Frances
Flowers of the World

fl.   p 55

Guzmania conifera

Bromeliad Society
Journal
vol 27, No. 2 1977

fl.   p. 96

Guzmania diffusa

Bromeliad Society
Journal
vol 27, No. 4   1977

fl., hab.   cover (p. 145)

Guzmania dissitiflora

**Bromeliad Society
Bulletin**
vol 13, No. 2   1963

fl.   p. 39

Guzmania dissitiflora

Bromeliad Society
Journal
vol 23, No. 3   1973

fl.   p. 124

Guzmania dissitiflora

Padilla, Victoria, ed.
Bromeliads in Color and
Their Culture

fl.   p 30

Guzmania donnell-smithii

Rauh, Werner
Bromeliads for Home, Garden and
Greenhouse

fl.   Pl. 52

Guzmania erythrolepis

Bromeliad Society
Bulletin
vol 19, No. 3   1969

fl.   p. 72

Guzmania gloriosa

Bromeliad Society
Journal
vol 21, No. 3   1971

hab.   p. 72

Guzmania gloriosa

Padilla, Victoria
Bromeliads

hab.   p. 67   Pl. 5

Guzmania graminifolia

Bromeliad Society
Journal
vol 22, No. 2   1972

fl.   cover (p. 26)

Guzmania hybrid.

Padilla, Victoria, ed.
Bromeliads in Color and
Their Culture

fl.   p 14

Guzmania lingulata

Hay, Roy
The Dictionary of House Plants

fl.   Pl. 259

Guzmania lingulata

Rauh, Werner
Bromeliads for Home, Garden and
Greenhouse

fl.   Pl. 6-7

Guzmania lingulata var.

**Bromeliad Society
Bulletin**
vol 13, No. 2   1963

hab., fl.   p. 34

Guzmania lingulata var.

Bromeliad Society
Bulletin
vol 17, No. 4   1967

fl.   cover (p. 73)

Guzmania lingulata var.

Bromeliad Society
Journal
vol 23, No. 3   1973

fl.   cover (p. 82)

Guzmania lingulata var.

Bromeliad Society
Journal
vol 26, No. 5   1976

fl., hab.   p. 198

Guzmania lingulata var.

Bromeliad Society
Journal
vol 27, No. 1   1977

hab., fl.   p. 34

Guzmania lingulata var.

Padilla, Victoria
Bromeliads

fl.   p. 67   Pl. 5

Guzmania lingulata var.

Padilla, Victoria, ed.
Bromeliads in Color and
Their Culture

fl.   p 30

Guzmania lychnis

Bromeliad Society
Journal
vol 26, No. 5   1976

fl., hab.   p. 198

Guzmania melinonis

Rauh, Werner
Bromeliads for Home, Garden and
Greenhouse

fl.   Pl. 55

Guzmania minor

Rauh, Werner
Bromeliads for Home, Garden and
Greenhouse

fl.   Pl. 54

Guzmania monostachia

Bromeliad Society
Bulletin
vol 17, No. 5   1967

fl.   cover (p. 97)

Guzmania monostachia

Luer, Carlyle A.
The Native Orchids of Florida

fl. hab.   Pl. 2; 3

Guzmania monostachia

Rauh, Werner
Bromeliads for Home, Garden and
Greenhouse

fl.   Pl. 9

Guzmania musaica

Addisonia, v. 22, 1934-36

fl.   opp. p. 17   Pl. 713

Guzmania musaica

**Bromeliad Society
Bulletin**
vol 18, No. 3   1968

fl., hab. cover (p. 49)

Guzmania musaica

Padilla, Victoria, ed.
Bromeliads in Color and
Their Culture

fl.   p 31

Guzmania musaica

Rauh, Werner
Bromeliads for Home, Garden and
Greenhouse

fl.   Pl. 56

Guzmania nicaraguensis

Bromeliad Society
Journal
vol 26, No. 5   1976

fl.   p. 199

Guzmania nicaraguensis

Rauh, Werner
Bromeliads for Home, Garden and
Greenhouse

fl.   Pl. 53

Guzmania quitense

Bromeliad Society
Bulletin
vol 18, No. 4   1968

fl.   p. 96

Guzmania sanguinea

Bromeliad Society
Bulletin
vol 18, No. 6   1968

hab., fl.   p. 142

Guzmania sanguinea

Bromeliad Society
Journal
    vol 25, No. 3   1975

hab.   cover (p. 81)

Guzmania sanguinea

**Bromeliad Society
Bulletin**
    vol 13, No. 2   1963

hab.   p. 34

Guzmania sanguinea

Bromeliad Society
Journal
    vol 26, No. 5   1976

hab., fl.   p. 199

Guzmania sanguinea

Padilla, Victoria
Bromeliads

hab.   p. 70   Pl. 8

Guzmania sanguinea

Padilla, Victoria, ed.
Bromeliads in Color and
    Their Culture

fl. hab.   p 30

Guzmania sanguinea

Rauh, Werner
Bromeliads for Home, Garden and
    Greenhouse

hab., fl.   Pl. 3

Guzmania sanguinea var.

Bromeliad Society
Journal
    vol 23, No. 4   1973

fl., hab.   p. 156

Guzmania scherzeriana

Bromeliad Society
Journal
    vol 28, No.2   1978

fl.   cover (p. 49)

Guzmania sibundoyorum

Bromeliad Society
Journal
    vol 23, No. 6   1973

fl.   cover (p. 205)

Guzmania sp.

Kromdijk, G.
200 House Plants in Colour

fl.   Pl. 101

Guzmania squarrosa

Bromeliad Society
Journal
    vol 26, No. 6   1976

fl., hab.   p. 237

Guzmania wittmackii

Bromeliad Society
Journal
    vol 27, No. 5   1977

fl.   p. 240

Guzmania zahnii

Bromeliad Society
Bulletin
    vol 19, No. 2   1969

fl., hab.   cover (p. 25)

Guzmania zahnii

Padilla, Victoria
Bromeliads

fl.   p. 67   Pl. 5

Guzmania zahnii

Padilla, Victoria, ed.
Bromeliads in Color and
    Their Culture

fl.   p 31

Guzmania zahnii

Rauh, Werner
Bromeliads for Home, Garden and
    Greenhouse

fl.   Pl. 57

Guzmania zahnii var.

Bromeliad Society
Journal
    vol 26, No. 5   1976

fl., hab.   p. 199

Gymnadenia conopsea

Ary, S.
The Oxford Book of Wildflowers

fl.   p 158, Pl. 6

Gymnadenia conopsea

Barneby, T. P.
European Alpine Flowers in Colour

fl.   Pl. 11, 2

Gymnadenia conopsea

Brooke, Jocelyn
The Wild Orchids of Britain

fl.   Pl. 18

Gymnadenia conopsea

Kohlhaupt, Paula
Fleurs des Alpages
v. 2

fl.   P 23

Gymnadenia conopsea

Lindman, C. A. M.
Nordens Flora, Vol 2

fl. hab.   Pl. 136A

Gymnadenia conopsea

Luer, Carlyle A.
The Native Orchids of the
    United States and Canada

fl.   Pl. 55; 8

Gymnadenia conopsea

Martin, W. Keble
The Concise British Flora in Colour

fl. hab.   Pl. 82

Gymnadenia conopsea

Polunin, Oleg
Flowers of Europe

fl.   Pl. 190 #1912

Gymnadenia conopsea

Webster, Mary
Flora of Moray, Nairn & East
    Inverness

fl.   p. 453   Pl. 20

Gymnadenia densiflora

Brooke, Jocelyn
The Wild Orchids of Britain

fl.   Pl. 19

Gymnadenia odoratissima

Barneby, T.P.
European Alpine Flowers in Colour

fl.   Pl. 11, 3

Gymnadenia odoratissima

Lindman, C. A. M.
Nordens Flora, Vol 2

fl.   Pl. 136B

Gymnocactus aguirreanus

Cactus & Succulent Society of America
Journal
    vol 48, No. 3   1976

hab., fl.   cover

Gymnocactus beguinii var.

Cactus & Succulent Soc. of America
Journal
    vol 44, No. 4   1972

fl.   cover (p. 137)

Gymnocactus subterraneus

Cactus & Succulent Soc. of America
Journal
    vol 50, No. 6   1978

fl., hab.   p. 281

Gymnocactus viereckii var.

Cactus & Succulent Soc. of America
Journal
    vol 50, No. 6   1978

fl., hab.   p. 284

Gymnocalycium baldianum

Hay, Roy
The Dictionary of House Plants

hab.   Pl. 260

Gymnocalycium baldianum

Kupper, Walter
Cacti

fl., hab.   p. 75   Pl. 34

Gymnocalycium baldianum

Subik, Rudolf
Decorative Cacti

fl. hab.   p 45

Gymnocalycium bruchii

Lamb, Edgar
The Pocket Encyclopedia of Cacti
   and Succulents in Color

fl. hab.   Pl. 57

Gymnocalycium castellanosii

Lamb, Edgar
The Pocket Encyclopedia of Cacti
   and Succulents in Color

fl. hab.   Pl. 56

Gymnocalycium denudatum

Lamb, Edgar
The Pocket Encyclopedia of Cacti
   and Succulents in Color

fl. hab.   Pl. 54

Gymnocalycium denudatum

Subik, Rudolf
Decorative Cacti

fl. hab.   p 47, 49

Gymnocalycium denudatum

Van Laren, A. J.
Cactus

fl.   P 55   Fig. 55

Gymnocalycium friedrichii

Subik, Rudolf
Decorative Cacti

fl. hab.   p 45

Gymnocalycium hammerschmidii

Backeberg, Curt
Cactus Lexicon

fl.   p 619

Gymnocalycium horstii

Cactus & Succulent Society of America
Journal
      vol 43, No. 6   1971

fl., hab.   p. 250

Gymnocalycium hossei

Lamb, Edgar
The Pocket Encyclopedia of Cacti
   and Succulents in Color

fl. hab.   Pl. 58

Gymnocalycium lafaldense

Kupper, Walter
Cacti

fl., hab.   p. 77   Pl. 35

Gymnocalycium mihanovichii

Lamb, Edgar
The Pocket Encyclopedia of Cacti
   and Succulents in Color

fl. hab.   Pl. 59

Gymnocalycium mihanovichii var.

Kupper, Walter
Cacti

fl., hab.   p. 77   Pl. 35

Gymnocalycium netrelianum

Lamb, Edgar
The Pocket Encyclopedia of Cacti
   and Succulents in Color

fl. hab.   Pl. 60

Gymnocalycium ochoterenai

Kupper, Walter
Cacti

fl., hab.   p. 79   Pl. 36

Gymnocalycium ourselianum

Kupper, Walter
Cacti

fl., hab.   p. 81   Pl. 37

Gymnocalycium platense

Kupper, Walter
Cacti

fl., hab. p. 41 Pl. 17

Gymnocalycium platense

Lamb, Edgar
The Pocket Encyclopedia of Cacti
   and Succulents in Color

fl. hab.   Pl. 61

Gymnocalycium platense

Van Laren, A. J.
Cactus

fl.   P 55   Fig. 58

Gymnocalycium prolifer

Lamb, Edgar
The Pocket Encyclopedia of Cacti
   and Succulents in Color

fl. hab.   Pl. 63

Gymnocalycium quehlianum

Van Laren, A. J.
Cactus

fl.   P 55   Fig. 59

Gymnocalycium saglione

Kupper, Walter
Cacti

fl., hab.   p. 83   Pl. 38

Gymnocalycium saglione

Lamb, Edgar
The Pocket Encyclopedia of Cacti
   and Succulents in Color

fl. hab.   Pl. 64

Gymnocalycium saglione

Van Laren, A. J.
Cactus

fl.   P 55   Fig. 56

Gymnocalycium schickendantzii

Lamb, Edgar
The Pocket Encyclopedia of Cacti
   and Succulents in Color

fl. hab.   Pl. 62

Gymnocalycium stuckertii

Lamb, Edgar
The Pocket Encyclopedia of Cacti
   and Succulents in Color

fl. hab.   Pl. 65

Gymnocarpos salsoloides

Bramwell, David
Wild Flowers of the Canary Islands

hab.   Pl. 130

Gymnostephium corymbosum

Flowering Plants of South Africa
      vol XXIV   1944

fl.   Pl. 952

Gynandriris sisyrinchium

Megaw, Elektra
Wild Flowers of Cyprus

fl., hab.   p. 13   Pl. 25

Gynandropsis speciosa

Bruggeman, L.
Tropical Plants

fl.   Pl. 44

Gynura aurantiaca

Hay, Roy
The Color Dictionary of Flowers & Plants

hab.   p 68, Pl. 540

Cynura aurantiaca

Hay, Roy
The Dictionary of House Plants

hab.   Pl. 261

Cynura aurantiaca

Hersey, Jean
Woman's Day Book of House Plants

hab.   p. 102

Cynura aurantiaca

Kromdijk, G.
200 House Plants in Colour

hab.   Pl.102

Cynura aurantiaca var.

Hay, Roy
The Color Dictionary of Flowers &
Plants

hab.   p. 68   Pl. 541

Cynura divaricata

Walden, Beryl M.
Wild Flowers of Hong Kong

fl.   Pl. 67 (202)

Cynura scandens

Encke, Fritz
Zimmerpflanzen

hab., fl.   p. 78

Cynura, Yellow

see

Cynura divaricata

Gypsophila cerastioides

Huxley, Anthony
Garden Perennials and Plants

fl.   Pl. 124

Gypsophila cerastioides

Vilmorin, Roger de
Plantes Alpines dans les Jardins

fl. hab.   Pl. XXXV

Gypsophila elegans

Crockett, James Underwood
Annuals

fl.   P 122

Gypsophila elegans

Huxley, Anthony
Garden Annuals and Bulbs

fl.   Pl. 56

Gypsophila elegans

Macoby, Stirling
What Flower is That

hab., fl.   p. 145

Gypsophila elegans

Tsukamoto, Yotaro
Coloured Illustrations of Garden
Flowers, v.10
fl.
ill. 68 opp.p.22

Gypsophila elegans var.

Hay, Roy
The Color Dictionary of Flowers &
Plants

fl.   p. 38   Pl. 303

Gypsophila fastigiata

Lindman, C. A. M.
Nordens Flora, Vol 4

fl. fr. hab.   Pl. 213A

Gypsophila oldhamiana

Curtis's Botanical Magazine
Vol 160   1937

fl.   No. 9484

Gypsophila paniculata

Huxley, Anthony
Garden Perennials and Water Plants

hab.   Pl. 11

Gypsophila paniculata var.

Hay, Roy
The Color Dictionary of Flowers &
Plants

fl.   p. 145   Pl. 1153,1154

Gypsophila paniculata var.

Huxley, Anthony
Garden Perennials and Water Plants

fl.   Pl. 123

Gypsophila repens

Barneby, T. P.
European Alpine Flowers in Color

fl.   Pl. 15, 6

Gypsophila repens

Felsko, Elsa
A Book of Wild Flowers
2nd Ser.
fl.   p 86

Gypsophila repens

Hay, Roy
The Color Dictionary of Flowers & Plants

fl.   p 11, Pl. 81

Gypsophila repens

Polunin, Oleg
Flowers of Europe

fl.   Pl. 16 #178

Gypsophila repens var.

Huxley, Anthony
Garden Perennials and Water Plants

fl.   Pl. 122

Gyrocarpus americanus

Palgrave, K. C.
Trees of Central Africa

fl., fr.   p. 194

Haageocereus acranthus

Lamb, Edgar
The Pocket Encyclopedia of Cacti
and Succulents in Color

hab.   Pl. 303

Haageocereus acranthus

Lamb, Edgar
Popular Exotic Cacti in Color

fl. hab.   p 100

Haageocereus albispinus

Backeberg, Curt
Cactus Lexicon

fl.   p 628

Haageocereus divaricatispinus

Lamb, Edgar
The Pocket Encyclopedia of Cacti
and Succulents in Color

hab.   Pl. 66

Haageocereus horrens

Backeberg, Curt
Cactus Lexicon

fl.   P. 629

Haageocereus olowinskianus

Lamb, Edgar
The Pocket Encyclopedia of Cacti
and Succulents in Color

hab.   Pl. 67

Haageocereus olowinskianus var.

Backeberg, Curt
Cactus Lexicon

fl.   P. 629

Haageocereus sp.

Lamb, Edgar
The Pocket Encyclopedia of Cacti
and Succulents in Color

hab.   Pl. 305

Habenaria aberrans

Flowering Plants of Africa
vol 29   1952-53

fl., hab.   Pl. 1133

Habenaria blephariglottis

Brown, Clair A.
Wildflowers of Louisiana and
Adjoining States

fl.   p 37

Habenaria blephariglottis

Dean, Blanche E.
Wildflowers of Alabama and
Adjoining States

fl.   p 45

Habenaria blephariglottis

Duncan, Wilbur H.
Wildflowers of the Southeastern
United States

fl.   p 273

Habenaria blephariglottis

Klimas, John E.
Wildflowers of Eastern America

fl.   Pl. 42

Habenaria blephariglottis var.

Dean, Blanche E.
Wildflowers of Alabama and
Adjoining States

fl.   p 45

Habenaria ciliaris

American Hort. Soc.
American Horticulturist
vol 53, No. 5 1974

fl., hab.   p. 26

Habenaria ciliaris

Batson, Wade T.
Wild Flowers in South Carolina

fl. hab.   p 39

Habenaria ciliaris

Brown, Clair A.
Wildflowers of Louisiana and
Adjoining States

fl.   p 37

Habenaria ciliaris

Campbell, Carlos C.
Great Smoky Mountain Wildflowers

fl.   p 59

Habenaria ciliaris

Courtenay, Booth
Wildflowers & Weeds

fl.   p 12

Habenaria ciliaris

Dean, Blanche E.
Wildflowers of Alabama and
Adjoining States

fl.   p 43

Habenaria ciliaris

Duncan, Wilbur H.
Wildflowers of the Southeastern
United States

fl.   p 273

Habenaria ciliaris

Klimas, John E.
Wildflowers of Eastern America

fl.   Pl. 134

Habenaria ciliaris

Lemmon, Robert S.
Wildflowers of North America in
Full Color

fl.   p. 12   Pl. 16

Habenaria ciliaris

Royal Hort. Soc.
Journal of the Royal Hort. Soc.
vol. 100, No. 1   1975

fl.   p. 15   Pl. 17

Habenaria ciliaris

Walcott, Mary Vaux
North American Wild Flowers
vol. 5

fl.   Pl. 340

Habenaria ciliaris

Wharton, Mary E.
A Guide to the Wildflowers & Ferns
of Kentucky

fl. p 72 Pl. 2.15

Habenaria clavellata

Courtenay, Booth
Wildflowers & Weeds

fl.   p 12

Habenaria clavellata

Duncan, Wilbur H.
Wildflowers of the Southeastern
United States

fl.   p 271

Habenaria columbae

Kamemoto, Haruyuki
Beautiful Thai Orchid Species

fl.   p 152

Habenaria cornuta

Flowering Plants of Africa
vol 36   1963-64

fl.   Pl. 1404

Habenaria cristata

Brown, Clair A.
Wildflowers of Louisiana and
Adjoining States

fl.   p 38

Habenaria dentata

Curtis's Botanical Magazine
Vol 164   1943-48

fl.   No. 9663

Habenaria dentata

Walden, Beryl M.
Wild Flowers of Hong Kong

fl.   Pl. 59 (170)

Habenaria dilatata

Clark, Lewis J.
Wild Flowers of British Columbia

fl.   p 87

Habenaria dilatata

Fries, Mary A.
Wildflowers of Mount Ranier and
the Cascades

fl.   p. 84

Habenaria dilatata

Lemmon, Robert S.
Wildflowers of North America in
Full Color

fl.   p. 174   Pl. 274

Habenaria dilatata

Orr, Robert T.
Wildflowers of Western America

fl.   Pl. 10

Habenaria dilatata

Taylor, Ronald J.
Mountain Wild Flowers

fl.   p 83

Habenaria dilatata

Welsh, Stanley L.
Flowers of the Mountain Country

fl.   p. 15

Habenaria dilatata var.

Clark, Lewis J.
Wild Flowers of British Columbia

fl.   p 98

Habenaria distans

Luer, Carlyle A.
The Native Orchids of Florida

fl. hab.   Pl. 46;6-8

Habenaria englerana

Curtis's Botanical Magazine
Vol 179   1972

fl.   631

Habenaria fastor

Pabst, G. F. J.
Orchidaceae Brasilienses,
   Vol 1

fl.   p 177

Habenaria ferdinandii

Cady, Leo
Australian Native Orchids in Colour

fl.   Pl. 3

Habenaria fimbriata

Klimas, John E.
Wildflowers of Eastern America

fl.   Pl. 284

Habenaria flava

Dean, Blanche E.
Wildflowers of Alabama and
   Adjoining States

fl.   p 43

Habenaria flava

Wharton, Mary E.
A Guide to the Wildflowers & Ferns
   of Kentucky

fl.   p 72   Pl. 2.14

Habenaria gonatosiphon

Moriarty, Audrey
Wild Flowers of Malawi

fl.   Pl. 26; 1

Habenaria grandiflora

Walcott, Mary Vaux
North American Wild Flowers
   vol. 4

fl.   Pl. 243

Habenaria hexaptera

Pabst, G. F. J.
Orchidaceae Brasilienses,
   Vol 1

fl.   p 177

Habenaria hookeri

Courtenay, Booth
Wildflowers & Weeds

fl.   p 11

Habenaria hyperborea

American Hort. Soc.
American Horticulturist
   vol 57, No. 3   1978

fl.   p. 38

Habenaria hyperborea

Porsild, A. E.
Rocky Mountain Wild Flowers

fl. hab.   p 117

Habenaria integra

Dean, Blanche E.
Wildflowers of Alabama and
   Adjoining States

fl.   p 43

Habenaria lacera

Dean, Blanche E.
Wildflowers of Alabama and
   Adjoining States

fl.   p 41

Habenaria lacera

Duncan, Wilbur H.
Wildflowers of the Southeastern
   United States

fl.   p 271

Habenaria lacera

Jennings, O. E.
Wild Flowers of Western Pennsylvania
v. 2
fl.   Pl. 39

Habenaria lacera

Walcott, Mary Vaux
North American Wild Flowers

fl. hab.   Pl. 215

Habenaria leucophaea

Courtenay, Booth
Wildflowers & Weeds

fl.   p 12

Habenaria linguella

Hu, Shiu-ying
The Genera of Orchidaceae in
   Hong Kong

fl.   p 120

Habenaria linguella

Walden, Beryl M.
Wild Flowers of Hong Kong

fl.   Pl. 58 (169)

Habenaria, Long Horned

see

Habenaria quinqueseta var.

Habenaria macrostele

Moriarty, Audrey
Wild Flowers of Malawi

fl.   Pl. 26; 4

Habenaria micrum

Williamson, Graham
The Orchids of South Central Africa

fl.   Pl. 44

Habenaria malacophylla

Flowering Plants of South Africa
   vol XIX   1939

fl., hab.   Pl. 725

Habenaria medioflexa

Kamemoto, Haruyuki
Beautiful Thai Orchid Species

fl.   p 152

Habenaria miersiana

Royal Hort. Soc.
Journal of the Royal Hort. Soc.
   vol 100, No. 1   1975
fl.   p. 25   Pl. 16

Habenaria monorhyza

Ospina, Mariano
Orquideas de las americas

fl.   Pl. 31

Habenaria nicholsonii

Williamson, Graham
The Orchids of South Central Africa

fl.   Pl. 45

Habenaria nivea

Brown, Clair A.
Wildflowers of Louisiana and
   Adjoining States

fl.   p 38

Habenaria nivea

Dean, Blanche E.
Wildflowers of Alabama and
   Adjoining States

fl.   p 43

Habenaria nivea

Duncan, Wilbur H.
Wildflowers of the Southeastern
   United States

fl.   p 273

Habenaria obtusata

Clark, Lewis J.
Wild Flowers of British Columbia

fl.   p 99

Habenaria obtusata

Porsild, A. E.
Rocky Mountain Wild Flowers

fl. hab.   p 119

Habenaria obtusata

Walcott, Mary Vaux
North American Wild Flowers

fl. hab.   Pl. 76

Habenaria odontopetala

Luer, Carlyle A.
The Native Orchids of Florida

fl., hab.   Pl. 4;4,
                    44

Habenaria orbiculata

Clark, Lewis J.
Wild Flowers of British Columbia

fl.   p 98

Habenaria orbiculata

Courtenay, Booth
Wildflowers & Weeds

fl.   p 11

Habenaria orbiculata

Lemmon, Robert S.
Wildflowers of North America in
    Full Color

fl.   p. 240   Pl. 378

Habenaria parvifolia

Williamson, Graham
The Orchids of South Central Africa

fl.   Pl. 40

Habenaria peramoena

Jennings, O. E.
Wild Flowers of Western Pennsylvania
    vol II
fl.   Pl. 40

Habenaria peramoena

Wharton, Mary E.
A Guide to the Wildflowers & Ferns
    of Kentucky

fl.   p 71   Pl. 2.12

Habenaria praestans

Williamson, Graham
The Orchids of South Central Africa

fl.   Pl. 39

Habenaria psycodes

Campbell, Carlos C.
Great Smoky Mountain Wildflowers

fl.   Cover

Habenaria psycodes

Courtenay, Booth
Wildflowers & Weeds

fl.   p 12

Habenaria psycodes

Duncan, Wilbur H.
Wildflowers of the Southeastern
    United States

fl.   p 271

Habenaria psycodes

Ferguson, Mary
Wildflowers

fl.   p 174

Habenaria psycodes

Jennings, O. E.
Wild Flowers of Western Pennsylvania
    vol II
fl.   Pl. 41

Habenaria psycodes

Moyle, John B.
Northland Wild Flowers

fl. hab.   p. 222   Pl. 295

Habenaria psycodes

Wharton, Mary E.
A Guide to the Wildflowers & Ferns
    of Kentucky

fl.   p 71   Pl. 2.13

Habenaria psycodes var.

Lemmon, Robert S.
Wildflowers of North America in
    Full Color

fl.   p. 241, 242, 379, 381

Habenaria pubipetala

Moriarty, Audrey
Wild Flowers of Malawi

fl.   Pl. 26; 5

Habenaria quinqueseta

Fleming, Glenn
Wild Flowers of Florida

fl. hab.   p 40

Habenaria quinqueseta

Luer, Carlyle A.
The Native Orchids of Florida

fl.   Pl. 45; 1,2,
                4,5

Habenaria quinqueseta var.

Luer, Carlyle A.
The Native Orchids of Florida

fl.   Pl. 45; 3

Habenaria repens

Brown, Clair A.
Wildflowers of Louisiana and
    Adjoining States

fl.   p 39

Habenaria repens

Luer, Carlyle A.
The Native Orchids of Florida

fl. hab.   Pl. 46; 1,2,3,
                    4,5

Habenaria rhodocheila

Kamemoto, Haruyuki
Beautiful Thai Orchid Species

fl.   p 151

Habenaria rhodocheila

Perry, Frances
Flowers of the World

fl.   p 212

Habenaria rhodocheila

Walden, Beryl M.
Wild Flowers of Hong Kong

fl.   Pl. 58 (171)

Habenaria rhopalostigma

Williamson, Graham
The Orchids of South Central Africa

fl.   Pl. 43

Habenaria sp.

Moriarty, Audrey
Wild Flowers of Malawi

fl.   Pl. 26; 6

Habenaria sp.

Williamson, Graham
The Orchids of South Central Africa

fl.   Pl. 38, 41

Habenaria stenorhynchos

Williamson, Graham
The Orchids of South Central Africa

fl.   Pl. 42

Habenaria tentaculigera

Moriarty, Audrey
Wild Flowers of Malawi

fl.   Pl. 26; 3

Habenaria unalascensis

Clark, Lewis J.
Wild Flowers of British Columbia

fl.   p 99

Habenaria viridis

Courtenay, Booth
Wildflowers & Weeds

fl.   p 13

Habenaria viridis var.

Clark, Lewis J.
Wild Flowers of British Columbia

fl.   p 99

Habenaria walleri

Moriarty, Audrey
Wild Flowers of Malawi

fl.   Pl. 26; 2

Haberlea rhodopensis

Goulimis, Constantine N.
Wild Flowers of Greece

fl. hab.   p 99

Haberlea rhodopensis

Hay, Roy
The Color Dictionary of Flowers & Plants

fl.   p 11, Pl. 82

Haberlea rhodopensis

Meikle, R. D.
Garden Flowers

fl.   opp. p. 352   Pl. 10

Haberlea rhodopensis var.

American Gloxinia & Gesneriad Society
The Gloxinian
    vol 28, No. 5   1978

fl.   p. 17

Habranthus andersonii

Addisonia, V. 22, 1943-46

fl., fr.   opp. p. 45   Pl. 727

Habranthus brachyandrus

Massachusetts Hort. Soc.
Horticulture
    vol 53, No. 3   1975

fl.   p. 82

Habranthus pratensis

Royal Hort. Soc.
The Garden
    vol 101, No. 7 1976

fl.   p. 361

Habranthus texanus

Brown, Clair A.
Wildflowers of Louisiana and
    Adjoining States

fl.   p 24

Hackelia deflexa

Lindman, C. A. M.
Nordens Flora, Vol 8

fl. fr. hab.   Pl. 490

Hackelia longituba

Orr, Robert T.
Wildflowers of Western America

fl.   Pl. 245

Hackelia virginiana

Courtenay, Booth
Wildflowers & Weeds

fl.   p 86

Hacquetia epipactis

Alpine Garden Society
Bulletin
    vol. 37, No. 4   1969

fl., hab.   p. 309

Hacquetia epipactis

Felsko, Elsa
A Book of Wild Flowers, 2nd Ser.

fl. hab.   p 6

Hacquetia epipactis

Hay, Roy
The Color Dictionary of Flowers & Plants

fl.   p 11, Pl. 83

Hacquetia epipactis

Perry, Frances
Flowers of the World

fl.   p 300

Hacquetia epipactis

Polunin, Oleg
Flowers of Europe

fl.   Pl. 83 #851

Hacquetia epipactis

Vilmorin, Roger de
Plantes Alpines dans les Jardins

fl. hab.   Pl. XXXVI

Haemanthus albiflos

Morley, Brian D.
Wild Flowers of the World

fl.,fr.,hab.   Pl. 86

Haemanthus albiflos

Royal Hort. Soc.
The Garden
    vol 103, No. 11   1978

fl.   p. 439

Haemanthus albiflos var.

Encke, Fritz
Zimmerpflanzen

fl.   p. 79

Haemanthus amaryllotdes

Curtis's Botanical Magazine, v. 174
    1962-63

fl.   Pl. 415

Haemanthus cinnabarinus

Royal Hort. Soc.
Journal of the Royal Hort. Soc.
    vol 92, No. 5   1967

fl.   p. 206   Pl. 112

Haemanthus cinnabarinus

Wit, H. C. D. de
Plants of the World;
    The Higher Plants, Vol II

fl.   Pl. 130

Haemanthus coccineus

Flowering Plants of Africa
    vol 31   1956

fl.   Pl. 1239

Haemanthus coccineus

Macoby, Stirling
What Flower is That

fl.   p. 146

Haemanthus coccineus

Perry, Frances
Flowers of the World

fl.   p 25

Haemanthus coccineus

Royal Hort. Soc.
The Garden
    vol 100, No. 12   1975

fl.   p. 582

Haemanthus hybrid

Miles, Bebe
Bulbs for the Home Gardener

fl.   p 171

Haemanthus katherinae

American Hort. Soc.
American Horticulturist
    vol 53, No. 3   1974

fl.   p. 34

Haemanthus katherinae

Everett, T.H.
New Illustrated Encyclopedia of Gardening
v.5
fl.
opp.p.790

Haemanthus katherinae

Hay, Roy
The Color Dictionary of Flowers & Plants

fl.   p 68, Pl. 542

Haemanthus katherinae

Hay, Roy
The Dictionary of House Plants

fl.   Pl. 262

Haemanthus katharinae

Kiaer, Eigil
Indoor Plants in Colour

fl.   p. 71

Haemanthus katherinae

Tsukamoto, Yotaro
Coloured Illustrations of Garden
    Flowers, V. 9

fl.   opp. p. 49

Haemanthus magnificus

Flowering Plants of Africa
    Vol. 43   1974-76

fl.   Pl. 1681

Haemanthus multiflorus

Bruggeman, L.
Tropical Plants

fl.   Pl. 113

Haemanthus multiflorus

Macoby, Stirling
What Flower is That

fl.   p. 146

Haemanthus multiflorus

Moriarty, Audrey
Wild Flowers of Malawi

fl.   Pl. 3

Haemanthus multiflorus

Royal Hort. Soc.
Journal of the Royal Hort. Soc.
    vol 94, No. 1   1969

fl., fr.   p. 22   Pl. 9

Haemanthus namaquensis

Flowering Plants of South Africa
    vol XX   1940

fl., hab.   Pl. 793

Haemanthus nelsonii

Flowering Plants of South Africa
    vol XVIII   1938

fl., hab.   Pl. 695

Haemanthus pole-evansii

Curtis's Botanical Magazine
Vol 178   1970-72

fl.   572

Haemanthus pole-evansii

Flowering Plants of Africa
    vol 37   1965-66

fl.   Pl. 1452

Haemanthus pole-evansii

Royal Hort. Soc.
Journal of the Royal Hort. Soc.
    vol 87, No. 7   1962

fl.   p. 310   Pl. 96

Haemanthus puniceus

Nicolaisen, Age
Pocket Encyclopedia of Indoor Plants

fl.   Pl. 35

Haemanthus puniceus

Royal Hort. Soc.
The Garden
    vol 103, No. 11   1978

fl.   p. 439

Haemanthus sp.

Royal Hort. Soc.
The Garden
    vol 102, No. 12   1977

fl.   p. 508

Haemanthus tigrinus

Flowering Plants of Africa
    vol 32   1957-58

fl.   Pl. 1257

Haemaria discolor

Grubb, Roy
Selected Orchidaceous Plants
    Vol II

hab.   p 31

Haemodorum distichophyllum

Curtis, Winifred
The Endemic Flora of Tasmania, Vol. 5

fl. fr. hab.   Pl. 197

Hairbrush Vine

    see

Combretum farinosum

Hakea bakerana

Blombery, Alec M.
What Wildflower is That

fl.   p 158, Pl. 424

Hakea, Bird Beak

    see

Hakea orthorrhyncha

Hakea, bottlebush

    see

Hakea bucculenta

Hakea bucculenta

Blombery, Alec M.
What Wildflower is That

fl.   p 158, Pl. 425

Hakea bucculenta

Morcombe, M. K.
Australia's Western Wildflowers

fl.   p. 33

Hakea bucculenta

Mullins, Barbara
Australian Wildflowers in Colour

fl.   P. 39   Pl. 32

Hakea corymbosa

Blombery, Alec M.
What Wildflower is That

fl.   p 158, Pl. 426

Hakea cucullata

Blombery, Alec M.
What Wildflower is That

fl.   p 159, Pl. 428

Hakea, dagger

    see

Hakea teretifolia

Hakea epiglottis

Curtis, Winifred
The Endemic Flora of Tasmania
    Vol VI

fl.fr.   Pl. 253

Hakea ferruginea

Blombery, Alec M.
What Wildflower is That

fl.   p 159, Pl. 429

Hakea invaginata

Harrison, R.E.
Trees and Shrubs

fl.   P 86

Hakea laurina

Blombery, Alec M.
What Wildflower is That

fl.   p 161, Pl. 430

Hakea laurina
Harrison, R.E.
Trees and Shrubs

fl. P 87

Hakea laurina

Macoby, Stirling
What Flower is That

fl.   p. 146

Hakea laurina

Morcombe, M. K.
Australia's Western Wildflowers

fl.   p 48

Hakea microcarpa

Hay, Roy
The Color Dictionary of Flowers & Plants

fl.   p 204, Pl. 1627

Hakea multilineata

Blombery, Alec M.
What Wildflower is That

fl.   p 160, Pl. 431

Hakea myrtoides

Harrison, R.E.
Trees and Shrubs

fl. P 87

Hakea nodosa

Cochrane, G. R.
Flowers and Plants of Victoria

fl.   Pl. 11

Hakea orthorrhyncha

Blombery, Alec M.
What Wildflower is That

fl.   p 160, Pl. 432

Hakea orthorrhyncha

Mullins, Barbara
Australian Wildflowers in Color

fl.   pl. 30   p. 39

Hakea, Pincushion
   see
Hakea laurina

Hakea, pincushion
   see
Hakea orthorrhyncha

Hakea purpurea

Blombery, Alec M.
What Wildflower is That

fl.   p 161, Pl. 433

Hakea purpurea

Harrison, R.E.
Trees and Shrubs

fl. P 87

Hakea sericea
Harrison, R.E.
Trees and Shrubs

fl. P 87

Hakea sp.

Milne, Lorus
Living Plants of the World

fl.   p 73

Hakea suberea

Blombery, Alec M.
What Wildflower is That

fl.   p 161, Pl. 434

Hakea tenuifolia

Curtis's Botanical Magazine
Vol 170   1954-55

fl. fr.   229

Hakea tenuifolia

Macoby, Stirling
What Flower is That

fl.   p. 146

Hakea teretifolia

Blombery, Alec M.
What Wildflower is That

fl.   p 161, Pl. 435

Hakea teretifolia

Cochrane, G. R.
Flowers and Plants of Victoria

fl.   Pl. 13

Hakea teretifolia

Mullins, Barbara
Australian Wildflowers in Colour

fl.   P. 39   Pl. 31

Hakea ulicina

Cochrane, G. R.
Flowers and Plants of Victoria

fl.   Pl. 12

Hakea victoriae

Blombery, Alec M.
What Wildflower is That

hab.   p 161, Pl. 436

Hakea victoriae

Harrison, R. E.
Trees and Shrubs

hab.   P. 86

Hakea vittata

Cochrane, G. R.
Flowers and Plants of Victoria

fl.   Pl. 150

Hakonechloa macra var.

Royal Hort. Soc.
Journal of the Royal Hort. Soc.
   vol 99, No. 11   1974

hab.   p. 490   Pl. 229

Halenbergia hypertrophicum

Herre, H.
The Genera of the Mesembryanthemaceae

fl.   fr.   p 167

Halenia deflexa

Courtenay, Booth
Wildflowers & Weeds

fl.   p 80

Halesia carolina

Batson, Wade T.
Wild Flowers in South Carolina

fl.   p 91

Halesia carolina

Campbell, Carlos C.
Great Smoky Mountain Wildflowers

fl.   p 49

Halesia carolina

Dean, Blanche E.
Wildflowers of Alabama and
   Adjoining States

fl.   p 131

Halesia carolina

Flemer, William
Shade and Ornamental Trees in Color

fl.   p. 78

Halesia carolina

Huxley, Anthony
Deciduous Garden Trees and Shrubs

fl. Pl. 85

Halesia carolina

Massachusetts Hort. Soc.
Horticulture
vol 56, No. 6   1978

fl., fr.   p. 36-39

Halesia carolina

Perry, Frances
Flowers of the World

fl.   p 290

Halesia carolina

Wharton, Mary E.
Trees & Shrubs of Kentucky

fl.   p 56, Pl. 1.18

Halesia diptera

Brown, Clair A.
Wildflowers of Louisiana and
   Adjoining States

fl. hab.   p 134

Halesia diptera var.

Royal Hort. Soc.
Journal of the Royal Hort. Soc.
   vol 94, No. 2   1969

fl.   p. 86   Pl. 41

Halesia monticola

Edlin, Herbert
The Illustrated Encyclopedia
   of Trees

fl. fr. hab.   p 200-1

Halesia monticola

Hay, Roy
The Color Dictionary of Flowers & Plants

fl.   p 204, Pl. 1628

Half-mens

   see

Pachypodium namaquanum

Halgania lavandulacea

Cochrane, G. R.
Flowers and Plants of Victoria

fl.   Pl. 158

Halgania littoralis

Blombery, Alec M.
What Wildflower is That

fl.   p 162, Pl. 43?

xHalimiocistus sahucii

Gault, S. Millar
The Color Dictionary of Shrubs

fl.   Pl. 214

xHalimiocistus wintonensis

Gault, S. Millar
The Color Dictionary of Shrubs

fl.   Pl. 215

Halimione pedunculata

Lindman, C. A. M.
Nordens Flora, Vol 3

fl. fr. hab.   Pl. 188

Halimione pedunculata

Martin, W. Keble
The Concise British Flora in Colour

hab.   Pl. 72

Halimione portulacoides

Ary, S.
The Oxford Book of Wildflowers

fl.   p 56, Pl. 2

Halimione portulacoides

Martin, W. Keble
The Concise British Flora in Colour

fl. hab.   Pl. 72

Halimium atriplicifolium

Pacific Hort. Foundation
Pacific Horticulture
   vol 39, No. 3   1978

fl.   p. 36

Halimium atriplicifolium

Polunin, Oleg
Flowers of the Mediterranean

fl.   No. 108

Halimium commutatum

Polunin, Oleg
Flowers of Europe

fl.   Pl. 78 #796

Halimium lasianthum

Hay, Roy
The Color Dictionary of Flowers & Plants

fl.   p 204, Pl. 1629

Halimium lasianthum

Hellyer, A. G. L.
Shrubs in Color

fl.   p. 59

Halimium lasianthum var.

Perry, Frances
Flowers of the World

fl.   p. 75

Halleria lucida

Flowering Plants of Africa
   vol XVI   1945-46

fl.   Pl. 961

Hamadryas argentea

Morley, Brian D.
Wild Flowers of the World

fl.   Pl. 2F

Hamamelis x intermedia var.

Gault, S. Millar
The Color Dictionary of Shrubs

fl.   Pl. 216

Hamamelis x intermedia var.

Perry, Frances
Flowers of the World

fl.   p. 136

Hamamelis x intermedia var.

Royal Hort. Soc.
Journal of the Royal Hort. Soc.
   vol 94, No. 2   1969

fl.   p. 86   Pl. 43

Hamamelis x intermedia var.

Royal Hort. Soc.
Journal of the Royal Hort. Soc.
   vol 99, No. 1   1974

fl.   p. 22   Pl. 5

Hamamelis japonica

American Hort. Soc.
American Horticulturist
   vol 56, No. 5 1977

fl.   p. 23

Hamamelis japonica

Huxley, Anthony
Deciduous Garden Trees and Shrubs

fl. fr.   Pl. 86

Hamamelis japonica

Kimura, Koiti
Japanese Medicinal Plants, Vol I

fl.   p 42

Hamamelis japonica

Kitamura, Siro
Coloured Illustrations of Trees &
   Shrubs of Japan

fl.   208

Hamamelis japonica

Milne, Lorus
Living Plants of the World

fl.   p 107

Hamamelis japonica
Royal Hort. Soc.
Journal of the Royal Hort. Soc.
vol 99, No. 1   1974

fl.   p. 22   Pl. 3

Hamamelis japonica
Wit, H. C. D. de
Plants of the World;
The Higher Plants, Vol I

fl.   p 100, Pl. 46

Hamamelis japonica var.
Curtis's Botanical Magazine, v.174,
1962-63
fl., fr.
pl.420

Hamamelis japonica var.
Hay, Roy
The Color Dictionary of Flowers &
Plants

fl.   p. 204, Pl. 1630-31

Hamamelis japonica var.
Hellyer, A.G.L.
Shrubs in Color
fl.
p. 59

Hamamelis japonica var.
Kitamura, Siro
Coloured Illustrations of Trees &
Shrubs of Japan

fl.   209

Hamamelis japonica var.
Royal Hort. Soc.
Journal of the Royal Hort. Soc.
vol 99, No. 1   1974

fl.   p. 22   Pl. 4

Hamamelis mollis
American Hort. Soc.
American Horticulturist
vol 56, No. 5 1977

fl.   p. 22

Hamamelis mollis
Bartels, Andreas
Das Grosse Buch der Gartengeholze

hab.   p 156

Hamamelis mollis
Harrison, R.E.
Trees and shrubs

fl. P 86

Hamamelis mollis
Hay, Roy
The Color Dictionary of Flowers &
Plants

fl.   p. 204   Pl. 1632

Hamamelis mollis
Hellyer, A. G. L.
Shrubs in Color

fl.   p. 59

Hamamelis mollis
Huxley, Anthony
Deciduous Garden Trees and Shrubs

hab.   Pl. 87

Hamamelis mollis
Macoby, Stirling
What Flower is That

fl.   p. 147

Hamamelis mollis
Massachusetts Hort. Soc.
Horticulture
vol 45, No. 1 1967

fl.   p. 29

Hamamelis mollis
Morley, Brian D.
Wild Flowers of the World

fl.   Plate 94

Hamamelis mollis
Perry, Frances
Flowers of the World

fl.   p 136

Hamamelis mollis
Royal Hort. Soc.
Journal of the Royal Hort. Soc.
vol 99, No. 1   1974

fl.   p. 22   Pl. 1

Hamamelis mollis var.
American Hort. Soc.
American Horticulturist
vol 55, No. 6   1976

fl.   p. 39

Hamamelis mollis var.
Gault, S. Millar
The Color Dictionary of Shrubs

fl.   Pl. 217, 218

Hamamelis mollis var.
Hay, Roy
The Color Dictionary of Flowers &
Plants

fl.   p. 205   Pl. 1633-34

Hamamelis mollis var.
Royal Hort. Soc.
Journal of the Royal Hort. Soc.
vol 94, No. 2   1969

fl.   p. 86   Pl. 42

Hamamelis mollis var.
Royal Hort. Soc.
Journal of the Royal Hort. Soc.
vol 99, No. 1   1974

fl.   p. 22   Pl. 2, 6

Hamamelis vernalis
Morley, Brian D.
Wild Flowers of the World

fl.   Pl. 155

Hamamelis vernalis var.
Royal Hort. Soc.
The Garden
vol 101, No. 2   1976

hab.   p. 103

Hamamelis virginiana
Bianchini, Francesco
Health Plants of the World

fl. hab.   p 81

Hamamelis virginiana
Campbell, Carlos C.
Great Smoky Mountain Wildflowers

fl.   p 103

Hamamelis virginiana
Jennings, O. E.
Wild Flowers of Western Pennsylvania
v. 2
fl. fr.   Pl. 79

Hamamelis virginiana
Lemmon, Robert S.
Wildflowers of North America in
Full Color

fl.   p. 258   Pl. 408

Hamamelis virginiana
Massachusetts Hort. Soc.
Horticulture
vol 50, No. 10 1972

fl.   p. 36

Hamamelis virginiana
Perrot, Emile
Les Plantes Medicinales

fl. fr. hab.   p 115

Hamamelis virginiana
Walcott, Mary Vaux
North American Wild Flowers
vol. 5

fl.   Pl. 323

Hamamelis virginiana
Wharton, Mary E.
Trees & Shrubs of Kentucky

fl.   p 44, Pl. 1.6

Hamamelis virginiana var.

Royal Hort. Soc.
Journal of the Royal Hort. Soc.
vol 94, No. 2   1969

fl.   p. 86   Pl. 44

Hamatocactus hamatacanthus

Lamb, Edgar
Colorful Cacti of the American Deserts

hab.   Pl. 48

Hamatocactus setispinus

Lamb, Edgar
Colorful Cacti of the American Deserts

hab.   Pl. 139

Hamatocactus setispinus

an Laren, A. J.
Cactus

fl.   P 45   Fig. 45

Hamatocactus uncinatus

Lamb, Edgar
Colorful Cacti of the American Deserts

fl. fr. hab.   Pl. 20 - 22

Hamelia erecta

O'Gorman, Helen
Mexican Flowering Trees and Plants

fl. hab.   p 65

Hammarbya paludosa

Lindman, C. A. M.
Nordens Flora, Vol 3

fl. hab.   Pl. 153A

Hammarbya paludosa

Martin, W. Keble
The Concise British Flora in Colour

fl. hab.   Pl. 80

Hand Flower Tree

see

Chiranthodendron pentadactylon

Handkerchief Tree

see

Davidia involucrata

Handsome Harry

see

Rhexia virginica

Haplopappus acaulis

Welsh, Stanley L.
Flowers of the Mountain Country

fl.   p. 54

Haplopappus linearifolius

Munz, Philip A.
California Desert Wildflowers

fl.   p 52, Pl. 77

Haplopappus Lyallii

Porsild, A. E.
Rocky Mountain Wild Flowers

fl. hab.   p 417

Haplopappus lyallii

Taylor, Ronald J.
Mountain Wild Flowers

fl.   p 104

Harbinger-of-Spring

see

Erigenia bulbosa

Hardhack

see

Spiraea tomentosa

Hard Heads

see

Eriocaulon decangulare

Hardenbergia comptoniana

Blombery, Alec M.
What Wildflower is That

fl.   p 162, Pl. 438

Hardenbergia comptoniana

Mathias, Mildred E.
Color for the Landscape

fl. hab.   p 114

Hardenbergia violacea

Blombery, Alec M.
What Wildflower is That

fl.   p 162, Pl. 439

Hardenbergia violacea

Cochrane, G. R.
Flowers and Plants of Victoria

fl. hab.   Pl. 352

Hardenbergia violacea

Harrison, Richmond E.
Climbers and Trailers

fl. p 47   Pl. 96

Hardenbergia violacea

Macoby, Stirling
What Flower is That

fl.   p. 147

Hardenbergia violacea

Morley, Brian D.
Wild Flowers of the World

fl.   Plate 130

Harebell

see

Campanula rotundifolia

Harebell

see

Scilla nonscripta

Harebell, Alpine

see

Campanula lasiocarpa

Harebell, Appalachian

see

Campanula divaricata

Harebell, Arctic

see

Campanula uniflora

Harebell, Carpathian

see

Campannla carpatica

Harebell, Mountain

see

Campanula lasiocarpa

Harebell, Southern

see

Campanula divaricata

HARE'S EAR

see

Bupleurum

Hare's Tail

see

Eriophorum spissum

Harlequin flower

see

Sparaxis tricolor

Harlequin, Yellow

see

Corydalis micrantha

Harpagophytum procumbens

Morley, Brian D.
Wild Flowers of the World

fl.,fr.   Pl. 79

Harpullia arborea

Mathias, Mildred E.
Color for the Landscape

fr. hab.   p 20

Harpullia pendula

Oakman, Harry
Colorful Trees

Hab. Fr.   P 51

Harrisella porrecta

Luer, Carlyle A.
The Native Orchids of Florida

fl. fr. hab.   Pl. 82

Harrisella porrecta

Ospina, Mariano
Orquideas de las americas

fl. hab.   Pl. 137

Harrisia bonplandii

Lamb, Edgar
The Pocket Encyclopedia of Cacti
   and Succulents in Color

fl. hab.   Pl. 69

Harrisia bonplandii

Massachusetts Hort. Soc.
Horticulture
   vol 31, No. 10   1953

fl.   p. 418

Harrisia fragrans

Lamb, Edgar
Popular Exotic Cacti in Color

fl. hab.   p 99

Harrisia jusbertii

Lamb, Edgar
The Pocket Encyclopedia of Cacti
   and Succulents in Color

fl. hab.   Pl. 68

Harrisia martinii

Lamb, Edgar
The Pocket Encyclopedia of Cacti
   and Succulents in Color

fr. hab.   Pl. 69

Harveya capensis

Morley, Brian D.
Wild Flowers of the World

fl.   Pl. 81

Harveya hyobanchoides

Flowering Plants of Africa
   vol 32   1957-58

fl.   Pl. 1261

Hatpins

see

Eriocaulon decangulare

Hat plant, Chinese

see

Holmskioldia sanguinea

Hatiora salicornioides

Lamb, Edgar
The Pocket Encyclopedia of Cacti
   and Succulents in Color

fl. hab.   Pl. 70

Hatiora salicornioides

Lamb, Edgar
Popular Exotic Cacti in Color

fl. hab.   p 108

Hatiora salicornioides

Van Laren, A. J.
Cactus

fl.   P 94   Fig. 110

Hau tree

see

Hibiscus tiliaceus

Haw, Black

see

Viburnum obovatum

Haw, Blue

see

Viburnum cassinoides

Haw, Harbison's

see

Crataegus harbisonii

Haw, May

see

Crataegus opaca

Haw, Possum

see

Ilex decidua

Haw, Possum

see

Viburnum nudum

Haw, Red

see

Crataegus mollis

Hawthorn

see

Crataegus

Hawthorn, Chinese

see

Photina serrulata

Hawthorn, downy

see

Crataegus mollis

Hawthorn, English

see

Crataegus oxycantha

Hawthorn, giant

see

Crataegus ellwangeriana

Hawthorn, Hong Kong

see

Raphiolepis indica

Hawthorn, Indian

see

Raphiolepis indica

Hawthorn, Mexican

see

Crataegus mexicana

Hawthorn, Midland

see

Crataegus oxyacantha

Hawthorn, Paul's scarlet

see

Crataegus oxyacantha var.

Hawthorn, Pink English

see

Crataegus oxycantha

Hawthorn, River

see

Crataegus douglasii

Hawthorn, Riverflat

see

Crataegus opaca

Hawthorn, Washington

see

Crataegus phaenopyrum

Hawthorn, Water

see

Aponogeton distachyos

Hawthorn, Yeddo

see

Raphiolepis umbellata var.

Hawaii Gold

see

Leucospermum hybrid

Hawkbit, Autumnal

see

Leontodon autumnalis

Hawkbit, Rough

see

Leontodon hispidus

Hawkweed

see

Hieracium

Hawkweed, Chicory Leaved

see

Hieracium intybaceum

Hawkweed, Endive-leaved

see

Hieracium intybaceum

Hawkweed, mouse-ear

see

Heiracium pilosella

Hawkweed, Orange

see

Hieracium aurantiacum

Hawkweed, Woolly

see

Hieraceium triste

Hawkweed, Woolly

see

Hieracium villosum

Hawk's beard

see

Crepis rubra

Hawksbeard, Alpine

see

Crepis nana

Hawk's-beard, Beaked

see

Crepis taraxacifolia

Hawk's Beard, Cushion

see

Crepis nana

Hawk's Beard, Golden

see

Crepis aurea

Hawk's-beard, Smooth

see

Crepis capillaris

Haworthia attenuata

Hay, Roy
The Dictionary of House Plants

hab.   Pl. 263

Haworthia attenuata var.

Cactus & Succulent Society of America
Journal
    vol 34, No. 4   1962

hab.   cover (p. 97)

Haworthia attenuata var.

Lamb, Edgar
Popular Exotic Cacti in Color

hab.   p 103

Haworthia blackburniae

Flowering Plants of South Africa
    vol XXII   1942

fl., hab.   Pl. 880

Haworthia bolusii

Cactus & Succulent Society of America
Journal
    vol 20, No. 8   1948

hab.   cover (p. 109)

Haworthia coarctata

Cactus & Succulent Society of America
Journal
vol 22, No. 6   1950

hab.   cover (p. 161)

Haworthia cuspidata

Addisonia
vol 23   1954-59

fl.   p. 9   Pl. 741

Haworthia fasciata

Lamb, Edgar
The Pocket Encyclopedia of Cacti
and Succulents in Color

hab.   Pl. 265

Haworthia fasciata

Subik, Rudolf
Decorative Cacti

fl. hab.   p 123

Haworthia granulata

Cactus & Succulent Society of America
Journal
vol 50, No. 2   1978

hab.   p. 74

Haworthia herrei var.

Lamb, Edgar
The Pocket Encyclopedia of Cacti
and Succulents in Color

hab.   Pl. 263

Haworthia hurlingii

Flowering Plants of Africa
vol 29   1952-53

fl., hab.   Pl. 1149

Haworthia koelmaniora

Flowering Plants of Africa
vol 38   1967

fl., hab.   Pl. 1502

Haworthia limifolia var.

Cactus & Succulent Society of America
Journal
vol 21, No. 5   1949

hab.   cover (p. 129)

Haworthia lockwoodii

Flowering Plants of South Africa
vol XX   1940

fl., hab.   Pl. 792

Haworthia longiana

Flowering Plants of South Africa
vol XXII   1942

fl., fr., hab.   Pl. 842

Haworthia margaritifera

Kiaer, Eigil
Indoor Plants in Colour

hab.   p. 63

Haworthia margaritifera var.

Cactus & Succulent Society of America
Journal
vol 20, No. 3   1948

hab.   cover (p. 29)

Haworthia margaritifera var.

Cactus & Succulent Society of America
Journal
vol 20, No. 4   1948

hab.   cover (p. 45)

Haworthia margaritifera var.

Cactus & Succulent Society of America
Journal
vol 20, No. 12   1948

hab.   cover (p. 175)

Haworthia maughanii

Cactus & Succulent Society of America
Journal
vol 27, No. 3   1955

hab.   p. 67

Haworthia pallida var.

Flowering Plants of Africa
vol XXV   1945-46

fl., hab.   Pl. 989

Haworthia reinwardtii

Flowering Plants of Africa
vol 29   1952-53

fl.   Pl. 1148

Haworthia reinwardtii

Lamb, Edgar
The Pocket Encyclopedia of Cacti
and Succulents in Color

hab.   Pl. 264

Haworthia tauteae

Flowering Plants of Africa
vol XXV   1945-46

fl., hab.   Pl. 992

Haworthia tessellata var.

Cactus & Succulent Society of America
Journal
vol 20, No. 2   1948

hab.   cover (p. 13)

Haworthia truncata

Cactus & Succulent Society of America
Journal
vol 27, No. 2   1955

hab.   cover (p. 33)

Haworthia truncata

Subik, Rudolf
Decorative Cacti

fl. hab.   p 125

Haworthia ubomboensis

Flowering Plants of South Africa
vol XXI   1941

fl., hab.   Pl. 818

Hayseed, Greek

see

Trigonella foenum-graecum

Hazel

see

Corylus

Hazel, Beaked

see

Corylus cornuta

Hazel, Chinese Witch

see

Hamamelis mollis

Hazel, Common

see

Corylus avellana

Hazel, Corkscrew

see

Corylus avellana var.

Hazelnut

see

Corylus maxima var.

Hazel, Winter

see

Corylopsis spicata

Hazel, Witch

see

Hamamelis

Hazel, Witch Japanese

see

Hamamelis japonica

Heal-all

see

Prunella vulgaris

Heart Berry

see

Aristotelia peduncularis

Heart, Floating

see

Nymphoides aquatica

Heart, Floating Yellow

see

Nymphoides peltata

Heart-flower

see

Talauma mexicana

Heartleaf

see

Hexastylis arifolia

Heart, Mother's

see

Capsella bursa-pastoris

Heart of flame

see

Neoregelia concentrica

Heart Seed

see

Cardiospermum haliacacabum

Heartwood

see

Cassia abbreviata

Hearts-A-Bustin

see

Euonymus americanus

Hearts, Broken

see

Clerodendron thomsonae

Hearts-Bursting-With-Love

see

Euonymus americanus

Heartsease

see

Viola tricolor var.

Hearts, Floating

see

Nymphoides aquatica

Hearts, White

see

Buchnera cruciata

Heath

see

Erica

Heath, Albertinia

see

Erica bauera

Heath, Alkali

see

Frankenia grandifolia

Heath, Alpine

see

Erica carnea

Heath,berry

see

Erica baccans

Heath, Biscay

see

Erica erigena

Heath, Blue

see

Phyllodoce caerulea

Heath,bridal

see

Erica bauera

Heath, Brush

see

Brachyloma ericoides

Heath, Candle

see

Richea continentis

Heath, Carpet

see

Pentachondra pumila

Heath Chilean

see

Fabiana imbricata

Heath, Common

see

Epacris impressa

Heath, Connemara

see

Daboecia cantabrica

Heath, Coral

see

Epacris microphylla

Heath, Cornish

see

Erica vagans

Heath, Corsican

see

Erica terminalis

Heath, Cranberry

see

Styphelia humifusa

Heath, Cross-leaved

see

Erica tetralix

Heath, Daphne

see

Brachyloma daphnoides

Heath, Darley

see

Erica darleyensis

Heath, Fine-leaved

see

Erica cinerea

Heath, Flame

see

Styphelia behrii

Heath, fuschia

see

Epacris longiflora

Heath, Gallipoli

see

Physostegia virginiana var.

Heath, Golden

see

Styphelia adscendens

Heath, Grampians

see

Epacris impressa var.

Heath, Irish

see

Daboecia cantabrica

Heath, lantern

see

Erica blenna

Heath, Mediterranean

see

Erica erigena

Heath, moss-leafed

see

Astroloma ciliata

Heath, Peach

see

Styphelia strigosa

Heath, Pine

see

Styphelia pinifolia

Heath, Pink Beard

see

Styphelia ericoides

Heath, pink shower

see

Erica quadrangularis

Heath, Pink Swamp

see

Sprengelia incarnata

Heath, Portugal

see

Erica lusitanica

Heath, Prince of Wales

see

Erica x wilmorei

Heath, Purple Mountain

see

Phyllodoce empetriformis

Heath, red hairy

see

Erica cerinthoides

Heath, red signal

see

Erica mammosa

Heath, Royal

see

Erica regia var.

Heath, St. Dabeoc's

see

Daboecia cantabrica

Heath, St. Dabeoc's

see

Daboecia x scotica

Heath, Sea

see

Frankenia laevis

Heath, Snow

see

Erica herbacea

Heath, Southern

see

Erica australis

Heath, Spanish

see

Erica australis

Heath, Spanish

see

Erica lusitanica

Heath, Spring

see

Erica carnea

Heath, spring

see

Erica herbacea

Heath, Swamp Pink

see

Sprengelia incarnata

Heath, Thyme

see

Epacris serpyllifolia

Heath, transluscent

see

Erica diaphana

Heath, Tree

see

Erica arborea

Heath, Urn

see

Styphelia urceolata

Heath, velvet bell

see

Erica peziza

Heath, Walker's

see

Erica walkeria

Heath, wax

see

Erica ventricosa

Heath, Winter

see

Erica herbacea

Heath, Woolly Style

see

Epacris lanuginosa

Heath, Yellow Mountain

see

Phyllodoce glandulifera

Heather, Alaska Moss

see

Cassiope stelleriana

Heather, Aleutian

see

Phyllodoce aleutica

Heather, Artic Bell

see

Cassiope tetragona

Heather, Bell

see

Erica cinerea

Heather, Christmas

see

Erica gracilis

Heather, Dorset

see

Erica ciliaris

Heather, False

see

Hudsonia tomentosa

Heather, Green

see

Erica scoparia

Heather, Mountain

see

Cassiope mertensiana

Heather, Mountain

see

Phyllodoce breweri

Heather, Pink Mountain

see

Phyllodoce empetriformis

Heather, red

see

Phyllodoce empetriformis

Heather, Red Mountain

see

Phyllodoce empetriformis

Heather, Scotch

see

Calluna vulgaris

Heather, Twisted

see

Erica cinerea

Heather, white

see

Cassiope mertensiana

Heather, White Mountain

see

Cassiope mertensiana

Heather, White Rocky Mountain

see

Cassiope tetragona

Heather, Yellow

see

Phyllodoce aleutica

Heather, Yellow Mountain

see

Phyllodoce glanduliflora

Heavenly Blue

see

Lithospermum diffusum

Hebe albicans

Gault, S. Millar
The Color Dictionary of Shrubs

fl. hab.   Pl. 219

Hebe x andersonii var.

Hellyer, A.G.L.
Shrubs in Color

fl.   p. 62

Hebe armstrongii

Harrison, R. E.
Trees and Shrubs

hab.   P. 88

Hebe armstrongii

Huxley, Anthony
Evergreen Garden Trees and Shrubs

hab.   Pl. 124

Hebe brachysiphon

Hay, Roy
The Color Dictionary of Flowers & Plants

fl.   p 205, Pl. 1636

Hebe x franciscana var.

Royal Hort. Soc.
The Garden
   vol 103, No. 7   1978

hab.   p. 269

Hebe hulkeana

Harrison, R.E.
Trees and Shrubs

fl.   P 88

Hebe hulkeana

Hay, Roy
The Color Dictionary of Flowers & Plants

fl.   p 205, Pl. 1637

Hebe hybrid.

Gault, S. Millar
The Color Dictionary of Shrubs

fl.   Pl. 220, 221, 222, 223, 224

Hebe hybrid

Harrison, Richmond E.
Climbers and Trailers

fl.   p. 93   Pl. 233

Hebe hybrid

Harrison, R. E.
Trees and Shrubs

fl.   P. 88

Hebe hybrid.

Hay, Roy
The Color Dictionary of Flowers & Plants

fl.   p 205, Pl. 1635, 1638

Hebe hybrid

Hellyer, A. G. L.
Shrubs in Colour

fl.   p. 62

Hebe lavaudiana

Royal Hort. Soc.
Journal of the Royal Hort. Soc.
   vol 97 No. 1 1972

fl.   p. 22   Pl. 20

Hebe macrantha

Curtis's Botanical Magazine
Vol 169   1952-53

fl.   177

Hebe macrocarpa var.

Curtis's Botanical Magazine, v.173, 1960
fl.
pl. 358

Hebe, showy

see

Hebe speciosa

Hebe speciosa

Crockett, James Underwood
Evergreens

hab. fl.   P. 127

Hebe speciosa

Macoby, Stirling
What Flower is That

fl.   p. 147

Hebe speciosa var.

Harrison, R.E.
Trees and Shrubs

fl. P 88

Hebe speciosa var.

Hay, Roy
The Color Dictionary of Flowers & Plants

fl.   p 205, 206; Pl. 1639, 1640, 1641

Hebe speciosa var.

Perry, Frances
Flowers of the World

fl.   p. 279

Hebe vernicosa var.

Curtis's Botanical Magazine
Vol 168   1951

fl.   136

Hebenstretia dentata

Moriarty, Audrey
Wild Flowers of Malawi

fl.   Pl. 62; 3

Hechtia epigyna

Rauh, Werner
Bromeliads for Home, Garden and
   Greenhouse

fl.   Pl. 126

Hechtia rosea

Rauh, Werner
Bromeliads for Home, Garden and
   Greenhouse

fl.   Pl. 127

Hedeoma ciliolata

Royal Hort. Soc.
Journal of the Royal Hort. Soc.
   vol 97, No. 10   1972

fl.   p. 442   Pl. 236

Hedeoma hispida

Courtenay, Booth
Wildflowers & Weeds

fl.   p. 95

Hedeoma pulegioides

Wharton, Mary E.
A Guide to the Wildflowers & Ferns
   of Kentucky

fl.   p 206   Pl. 2.29

Hedera arborescens

American Hort. Soc.
American Horticulturist
   vol 54, No. 3   1975

hab.   p. 6

Hedera canariensis

Crockett, James Underwood
Lawns & Ground Covers

hab.   p. 134

Hedera canariensis

Kiaer, Eigil
Indoor Plants in Colour

hab.   p. 64

Hedera canariensis

Nicolaisen, Age
Pocket Encyclopedia of Indoor Plants

hab.   Pl. 104

Hedera canariensis var.

American Hort. Soc.
American Horticulturist
    vol 54, No. 3 1975

hab.   p. 6

Hedera canariensis var.

Harrison, Richmond E.
Climbers and Trailers

hab.   p. 48   Pl. 97

Hedera canariensis var.

Hay, Roy
The Color Dictionary of Flowers & Plants

hab.   p 68, Pl. 543

Hedera canariensis var.

Hay, Roy
The Dictionary of House Plants

hab.   Pl. 264

Hedera canariensis var.

Perry, Frances
Flowers of the World

hab.   p. 36

Hedera colchica

Macoby, Stirling
What Flower is That

hab.   p. 148

Hedera colchica

Polunin, Oleg
Trees and Bushes of Europe

fr. hab.   p 155

Hedera colchica var.

Hellyer, A. G. L.
Shrubs in Color

hab.   p. 63

Hedera colchica var.

Huxley, Anthony
Evergreen Garden Trees and Shrubs

hab.   Pl. 125, 126

Hedera colchica var.

Kiaer, Eigil
Indoor Plants in Colour

hab.   p. 64

Hedera colchica var.

Royal Hort. Soc.
Journal of the Royal Hort. Soc.
vol 92, No. 4   1967

hab.   p. 160   Pl. 72

Hedera helix

Ary, S.
The Oxford Book of Wildflowers

fl.   p 64, Pl. 3

Hedera helix

Bianchini, Francesco
Health Plants of the World

fr. hab.   p 105

Hedera helix

Color Treasury of Herbs & Other
    Medicinal Plants

fl. fr.   p 40   Pl. 52

Hedera helix

Crockett, James Underwood
Lawns & Ground Covers

hab. P. 134

Hedera helix

Felsko, Elsa
A Book of Wild Flowers
    2nd Ser.

fl.   p. 140

Hedera helix

Huxley, Anthony
Evergreen Garden Trees and Shrubs

fl. hab.   Pl. 127

Hedera helix

Kiaer, Eigil
Indoor Plants in Colour

hab.   p. 64

Hedera helix

Lindman, C. A. M.
Nordens Flora, Vol 6

fl. fr.   Pl. 413

Hedera helix

Macoby, Stirling
What Flower is That

hab.   p. 148

Hedera helix

Martin, W. Keble
The Concise British Flora in Colour

fl. fr. hab.   Pl. 41

Hedera helix

Nicolaisen, Age
Pocket Encyclopedia of Indoor Plants

hab.   Pl. 105

Hedera helix

Polunin, Oleg
Flowers of Europe

fl.   Pl. 82 #848

Hedera helix

Polunin, Oleg
Trees and Bushes of Europe

fl. fr. hab.   p 154

Hedera helix

Vedel, H.
Arbres et Arbustes

fl.   p 102, No. 107

Hedera helix var.

Everett, T.H., ed.
The New Illustrated Encyclopedia of
    Gardening vol. 5

hab.   p. 823

Hedera helix var.

Gault, S. Millar
The Color Dictionary of Shrubs

hab.   Pl. 225

Hedera helix var.

Harrison, Richmond E.
Climbers and Trailers

hab.   p. 48, 49   Pl. 98-100

Hedera helix var.

Hay, Roy
The Color Dictionary of Flowers & Plants

hab.   p 68, 69; Pl. 544, 545, 546

Hedera helix var.

Hay, Roy
The Dictionary of House Plants

hab.   Pl. 265 - 267

Hedera helix var.

Hellyer, A.G.L.
Shrubs in Colour

hab.   p. 63

Hedera helix var.

Hersey, Jean
Woman's Day Book of House Plants

hab.   p. 71

Hedera helix var.

Kiaer, Eigil
Indoor Plants in Colour

hab.   p. 64

Hedera helix var.

Kromdijk, G.
200 House Plants in Colour

hab.   Pl. 103

Hedera helix var.

Macoby, Stirling
What Flower is That

hab.   p. 148

Hedera helix var.

Perry, Frances
Flowers of the World

fl.   p. 36

Hedera helix var.

Royal Hort. Soc.
Journal of the Royal Hort. Soc.
    vol 100, No. 2   1975

hab.   p. 78   Pl. 25

Hedera nepalensis

Hara, Hiroshi, comp.
Photo-Album of Plants of
    Eastern Himalaya

fr.   Pl. 68

Hedera rhombea

Kitamura, Siro
Coloured Illustrations of Trees &
    Shrubs of Japan

fl.   363

Hedgehog, Marine

see

Echinocereus maritimus

Hedycarya arborea

Harvey, Norman B.
New Zealand Botanical Paintings

fr.   p. 20   Pl. 7

Hedychium coccineum

Hay, Roy
The Dictionary of House Plants

fl.   Pl. 268

Hedychium coronarium

American Hort. Soc.
American Horticulturist
    vol 52, No. 4   1973

fl.   p. 11

Hedychium coronarium

American Hort. Soc.
American Horticulturist
    vol 53, No. 4   1974

fl.   p. 26

Hedychium coronarium

Bruggeman, L.
Tropical Plants

fl.   Pl. 114

Hedychium coronarium

Macoby, Stirling
What Flower is That

fl.   p. 148

Hedychium coronarium

Walden, Beryl M.
Wild Flowers of Hong Kong

fl.   Pl. 52

Hedychium densiflorum

Curtis's Botanical Magazine, v.172
1958-59
fl.
pl. 325

Hedychium densiflorum

Royal Hort. Soc.
Journal of the Royal Hort. Soc.
    vol 90, No. 1   1965

fl.   p. 22   Pl. 14

Hedychium flavescens

American Hort. Soc.
American Horticulturist
    vol 53, No. 4   1974

fl.   p. 27

Hedychium flavescens

Morley, Brian D.
Wild Flowers of the World

fl.   Plate 124

Hedychium gardneranum

American Hort. Soc.
American Horticulturist
    Vol. 53, No. 4   1974

fl.   p. 26

Hedychium gardneranum

Hay, Roy
The Dictionary of House Plants

fl.   Pl. 269

Hedychium gardneranum

Macoby, Stirling
What Flower is That

fl.   p. 148

Hedychium gardneranum

Perry, Frances
Flowers of the World

fl.   p 311

Hedychium gardneranum

Royal Hort. Soc.
Journal of the Royal Hort. Soc.
    Vol. 92, No. 11   1967

fl.   p. 478   Pl. 264

Hedychium gardneranum

Tsukamoto, Yotaro
Coloured Illustrations of Garden Flowers
    vol 9

fl.   p. 27

Hedychium sp.

Everett, T. H.
New Illustrated Encyclopedia of
    Gardening, V. 5

fl.   opp. p. 790

Hedychium sp.

Royal Hort. Soc.
Journal of the Royal Hort. Soc.
    vol 85, No. 9   1960

fl., hab.   p. 398   Pl. 117

Hedychium spicatum

Morley, Brian D.
Wild Flowers of the World

fr.   Plate 124

Hedyotis bracteosa

Walden, Beryl M.
Wild Flowers of Hong Kong

fl.   Pl. 45(127)

Hedyotis consangiunea

Walden, Beryl M.
Wild Flowers of Hong Kong

fl.   Pl. 45(128)

Hedyotis crassifolia

Duncan, Wilbur H.
Wildflowers of the Southeastern
    United States

fl.   p 187

Hedyotis procumbens

Fleming, Glenn
Wild Flowers of Florida

fl. hab.   p 20

Hedyotis purpurea

Duncan, Wilbur H.
Wildflowers of the Southeastern
    United States

fl.   p 187

Hedypnois rhagadioloides

Polunin, Oleg
Flowers of Europe

fl.   Pl. 158 #1519

Hedysarum alpinum var

Heller, Christine
Wild Flowers of Alaska

fl.   Pl. 101

Hedysarum boreale

Welsh, Stanley L.
Flowers of the Canyon Country

fl.   p 13

Hedysarum boreale var.

Alaska-Yukon Wild Flower Guide

fl. hab.   p 92

Hedysarum coronarium

Polunin, Oleg
Flowers of Europe

fl. Pl. 162 #634

Hedysarum coronarium

Polunin, Oleg
Flowers of the Mediterranean

fl.   No. 74

Hedysarum glomeratum

Polunin, Oleg
Flowers of Europe

fl. Pl. 62 #634

Hedysarum hedysaroides

Kohlhaupt, Paula
Fleurs des Alpages , v.2

Fl. P 64

Hedysarum hedysaroides

Polunin, Oleg
Flowers of Europe

fl. Pl. 61 #635

Hedysarum mackenzii

Heller, Christine
Wild Flowers of Alaska

fl.   Pl. 102

Hedysarum Mackenzii

Porsild, A.E.
Rocky Mountain Wild Flowers

fl. fr.   p. 261

Hedysarum mackenzii

Walcott, Mary Vaux
North American Wild Flowers

fl. hab.   Pl. 97

Hedysarum multifugum

Hellyer, A.G.L.
Shrubs in Colour

fl.   p. 62

Hedysarum obscurum

Barneby, T. P.
European Alpine Flowers in Colour

fl.   Pl. 45, 1

Hedysarum sulphurescens

Clark, Lewis J.
Wild Flowers of British Columbia

fl. fr.   p 283

Heeria elegans

Macoby, Stirling
What Flower is That

fl.   p. 149

Heeria reticulata

Palgrave, K. C.
Trees of Central Africa

fl., fr.   p. 3

Heimerliodendron brunonianum var.

Encke, Fritz
Zimmerpflanzen

hab.   p. 79

Helenium amarum

Brown, Clair A.
Wildflowers of Louisiana and
    Adjoining States

fl. hab.   p 207

Helenium amarum

Duncan, Wilbur H.
Wildflowers of the Southeastern
    United States

fl.   p 223

Helenium amarum

Weeds of the Southern United States

fl.   p. 13

Helenium autumnale

Brown, Clair A.
Wildflowers of Louisiana and
    Adjoining States

fl. hab.   p 207

Helenium autumnale

Courtenay, Booth
Wildflowers & Weeds

fl.   p 115

Helenium autumnale

Dean, Blanche E.
Wildflowers of Alabama and
    Adjoining States

fl.   p 201

Helenium autumnale

Huxley, Anthony
Garden Perennials and Water Plants

fl.   Pl. 125

Helenium autumnale

Jennings, O. E.
Wild Flowers of Western Pennsylvania
    v. 2
fl.   Pl. 189

Helenium autumnale

Macoby, Stirling
What Flower is That

fl.   p. 149

Helenium autumnale

Moyle, John B.
Northland Wild Flowers

fl. hab.   p 163, Pl. 196

Helenium autumnale

Tsukamoto, Yotaro
Coloured Illustrations of Garden Flowers, v.9
fl.
opp.p.49

Helenium autumnale

Wharton, Mary E.
A Guide to the Wildflowers & Ferns
    of Kentucky

fl. p 253 Pl. 3.19

Helenium autumnale var.

Hay, Roy
The Color Dictionary of Flowers & Plants

fl.   p 145, Pl. 1155 thru 1158

Helenium autumnale var.

Huxley, Anthony
Garden Perennials and Water Plants

fl.   pl. 125a, b
                    pl. 127a, b, c, d

Helenium flexuosum

Duncan, Wilbur H.
Wildflowers of the Southeastern
    United States

fl.   p 223

Helenium hoopesii

Huxley, Anthony
Garden Perennials and Water Plants

fl.   Pl. 126

Helenium hybrid

Perry, Frances
Flowers of the World

fl.   p. 83

Helenium nudiflorum

Addisonia, V. 20, 1937-38

fl. opp. p. 53   Pl. 667

Helenium nudiflorum

Batson, Wade T.
Wild Flowers in South Carolina

fl.   p 122

Helenium nudiflorum

Klimas, John E.
Wildflowers of Eastern America

fl.   Pl. 149

Helenium nudiflorum

Wharton, Mary E.
A Guide to the Wildflowers & Ferns
   of Kentucky

fl.   p 254   Pl. 3.20

Helenium pinnatifidum

Lemmon, Robert S.
Wildflowers of North America in
   Full Color

fl.   p. 54   Pl. 91

Helenium sp.

Huxley, Anthony
Garden Perennials and Water Plants

hab.   Pl. 12

Helenium vernale

Duncan, Wilbur H.
Wildflowers of the Southeastern
   United States

fl.   p 223

Helianthella uniflora

Tosco, Uberto
The World of Mountain Flowers

fl. hab.   p 82-3

Helianthemum alpestre

Barneby, T. P.
European Alpine Flowers in Colour

fl.   Pl. 50, 2

Helianthemum apenninum

Barneby, T. P.
European Alpine Flowers in Colour

fl.   Pl. 50, 3

Helianthemum apenninum

Goulimis, Constantine N.
Wild Flowers of Greece

fl.   p 65

Helianthemum apenninum

Martin, W. Keble
The Concise British Flora in Colour

fl. fr. hab.   Pl. 11

Helianthemum apenninum

Polunin, Oleg
Flowers of Europe

fl. fr. Pl. 78 #803

Helianthemum apenninum var.

Vilmorin, Roger de
Plantes Alpines dans les Jardins

fl., hab.   Pl. VIII

Helianthemum canadense

Courtenay, Booth
Wildflowers & Weeds

fl.   p 36

Helianthemum canadense

Klimas, John E.
Wildflowers of Eastern America

fl.   Pl. 119

Helianthemum canum

Martin, W. Keble
The Concise British Flora in Colour

fl. fr. hab.   Pl. 11

Helianthemum carolinianum

Dean, Blanche E.
Wildflowers of Alabama and
   Adjoining States

fl.   p 109

Helianthemum carolinianum

Duncan, Wilbur H.
Wildflowers of the Southeastern
   United States

fl.   p 101

Helianthemum chamaecistus

Ary, S.
The Oxford Book of Wildflowers

fl.   p 14, Pl. 6

Helianthemum chamaecistus

Felsko, Elsa
A Book of Wild Flowers
   2nd Ser.

fl.   p 11

Helianthemum chamaecistus

Harrison, Richmond E.
Climbers and Trailers

fl.   p 94, Pl. 236

Helianthemum chamaecistus

Martin, W. Keble
The Concise British Flora in Colour

fl. fr. hab.   Pl. 11

Helianthemum chamaecistis var.

Huxley, Anthony
Garden Perennials and Water Plants
fl.
pl. 128a-e

Helianthemum corymbosum

Fleming, Glenn
Wild Flowers of Florida

fl. hab.   p 56

Helianthemum guttatum

Kleijn, H.
Beauty of the Wild Plant

fl. opp. p. 117   Pl. 174

Helianthemum hybrid

Hay, Roy
The Color Dictionary of Flowers & Plants

fl.   p 11, Pl. 84, 85

Helianthemum hybrid

Perry, Frances
Flowers of the World

fl.   p. 74

Helianthemum lunulatum

Gault, S. Millar
The Color Dictionary of Shrubs

fl.   Pl. 226

Helianthemum nummularium

Barneby, T. P.
European Alpine Flowers in Colour

fl.   Pl. 50, 1

Helianthemum nummularium

Crockett, James Underwood
Lawns & Ground Covers

fl.   p. 135

Helianthemum nummularium

Lindman, C. A. M.
Nordens Flora, Vol 6

fl. fr. hab.   Pl. 392

Helianthemum nummularium

Polunin, Oleg
Flowers of Europe

fl.   Pl. 78 #802

Helianthemum nummularium var.

Gault, S. Millar
The Color Dictionary of Shrubs

fl.   Pl. 227, 228, 229

Helianthemum nummularium var.

Macoby, Stirling
What Flower is That

fl.   p. 149

Helianthemum scoparium

Munz, Philip A.
California Spring Wildflowers

fl.   p 90, Pl. 96

Helianthemum vulgare

Wit, H. C. D. de
Plants of the World;
   The Higher Plants, Vol II

fl.   Pl. 123

Helianthemum vulgare var.

Vilmorin, Roger de
Plantes Alpines dans les Jardins

fl. hab.   Pl. XXXIV

Helianthocereus hybrid

Backeberg, Curt.
Cactus Lexicon

fl.   p 637

Helianthocereus pecheretianus

Backeberg, Curt
Cactus Lexicon

fl. hab.   p 635

Helianthocereus poco var.

Backeberg, Curt
Cactus Lexicon

fl. hab.   p 636

Helianthocereus pseudocandicans var.

Backeberg, Curt
Cactus Lexicon

hab.   p 637

Helianthus angustifolius

Addisonia, V. 19, 1935-36

fl. fr. opp. p. 53   Pl. 635

Helianthus angustifolius

Brown, Clair A.
Wildflowers of Louisiana and
   Adjoining States

fl. hab.   p 203

Helianthus angustifolius

Bruggeman, L.
Tropical Plants

fl.   Pl. 115

Helianthus angustifolius

Dean, Blanche E.
Wildflowers of Alabama and
   Adjoining States

fl.   p 197

Helianthus angustifolius

Duncan, Wilbur H.
Wildflowers of the Southeastern
   United States

fl.   p 215

Helianthus angustifolius

Fleming, Glenn
Wild Flowers of Florida

fl.   p 46

Helianthus annuus

Bianchini, F.
The Complete Book of Fruits & Vegetables

fr.   p 239

Helianthus annuus

Brown, Clair A.
Wildflowers of Louisiana and
   Adjoining States

fl. hab.   p 203

Helianthus annuus

Bruggeman, L.
Tropical Plants

fl.   Pl. 45

Helianthus annuus

Clark, Lewis J.
Wild Flowers of British Columbia

fl.   p 554

Helianthus annuus

Color Treasury of Herbs & Other
   Medicinal Plants

fl.   p 61 Pl. 98

Helianthus annuus

Crockett, James Underwood
Annuals

fl.   P 123

Helianthus annuus

Fleming, Glenn
Wild Flowers of Florida

fl.   p 62

Helianthus annuus

Hay, Roy
The Color Dictionary of Flowers & Plants

fl.   p 38, Pl. 304

Helianthus annuus

Hvass, Elsie
Plants That Feed and Serve Us

fl. fr.   p 15, Pl. 21

Helianthus annuus

Jones, Paul
Flora Magnifica

fl.   Pl. 1

Helianthus annuus

Kimura, Koiti
Japanese Medicinal Plants, Vol II

fl.   p 239

Helianthus annuus

Klimas, John E.
Wildflowers of Eastern America

fl.   Pl. 146

Helianthus annuus

Lemmon, Robert S.
Wildflowers of North America in
   Full Color

fl.   p. 208   Pl. 330

Helianthus annuus

Macoby, Stirling
What Flower is That

fl.   p. 150

Helianthus annuus

Masefield, G. B.
The Oxford Book of Food Plants

fl. fr.   Pl. 25, 2

Helianthus annuus

Massachusetts Hort. Soc.
Horticulture
   vol 39, No. 9   1961

fl.   p. 470

Helianthus annuus
Massachusetts Hort. Soc.
Horticulture
    vol 54, No. 1   1976
fl.   p. 43

Helianthus annuus
Moyle, John B.
Northland Wild Flowers
fl. hab.   p 157, Pl. 183

Helianthus annuus
Perrot, Emile
Les Plantes Medicinales
fl. fr. hab.   p 235

Helianthus annuus
Welsh, Stanley L.
Flowers of the Canyon Country
fl.   p 30

Helianthus annuus var.
Huxley, Anthony
Garden Annuals and Bulbs
fl.   Pl. 57, 59

Helianthus annuus var.
Tsukamoto, Yotaro
Coloured Illustrations of Garden
    Flowers   Vol. 10
fl.   Illus. 69,70 opp. p. 22

Helianthus anomalus
Welsh, Stanley L.
Flowers of the Canyon Country
fl. hab.   p 31

Helianthus argophyllus
Bruggeman, L.
Tropical Plants
fl.   Pl. 46

Helianthus atrorubens
Duncan, Wilbur H.
Wildflowers of the Southeastern
    United States
fl. hab.   p 215

Helianthus cucumerifolius
Brown, Clair A.
Wildflowers of Louisiana and
    Adjoining States
fl. hab.   p 204

Helianthus cucumerifolius
Huxley, Anthony
Garden Annuals and Bulbs
fl.   Pl. 58

Helianthus debilis
Bruggeman, L.
Tropical Plants
fl.   Pl. 47

Helianthus decapetalus
Batson, Wade T.
Wild Flowers in South Carolina
fl.   p 120

Helianthus decapetalus
Huxley, Anthony
Garden Perennials and Water Plants
fl.
pl. 130

Helianthus decapetalus
Lemmon, Robert S.
Wildflowers of North America in
    Full Color
fl.   p. 210   Pl. 333

Helianthus decapetalus
Wharton, Mary E.
A Guide to the Wildflowers & Ferns
    of Kentucky
fl.   p 260   Pl. 3.34

Helianthus decapetalus var.
Macoby, Stirling
What Flower is That
fl.   p. 150

Helianthus divaricatus
Brown, Clair A.
Wildflowers of Louisiana and
    Adjoining States
fl. hab.   p 205

Helianthus divaricatus
Moyle, John B.
Northland Wild Flowers
fl. hab.   p 157, Pl. 184

Helianthus divaricatus
Wharton, Mary E.
A Guide to the Wildflowers & Ferns
    of Kentucky
fl.   p 259   Pl. 3.32

Helianthus giganteus
Batson, Wade T.
Wild Flowers in South Carolina
fl.   p. 120

Helianthus grosseserratus
Courtenay, Booth
Wildflowers & Weeds
fl.   p 118

Helianthus heterophyllus
Brown, Clair A.
Wildflowers of Louisiana and
    Adjoining States
fl.   p 205

Helianthus hirsutus
Duncan, Wilbur H.
Wildflowers of the Southeastern
    United States
fl.   p 215

Helianthus hybrid
Perry, Frances
Flowers of the World
fl.   p. 82

Helianthus laetiflorus
Courtenay, Booth
Wildflowers & Weeds
fl.   p. 118

Helianthus maximiliani
Moyle, John B.
Northland Wild Flowers
fl. hab.   p 158, Pl. 186

Helianthus microcephalus
Wharton, Mary E.
A Guide to the Wildflowers & Ferns
    of Kentucky
fl.   p 255   Pl. 3.23

Helianthus mollis
Addisonia, V. 20, 1937-38
fl. fr. opp. p. 5   Pl. 643

Helianthus mollis
Brown, Clair A.
Wildflowers of Louisiana and
    Adjoining States
fl. hab.   p 204

Helianthus nuttallii
Orr, Robert T.
Wildflowers of Western America
fl.   Pl. 96

Helianthus occidentalis
Courtenay, Booth
Wildflowers & Weeds
fl.   p 116

Helianthus petiolaris
Ferguson, Mary
Wildflowers
fl.   p 125

Helianthus rigidus

Huxley, Anthony
Garden Perennials and Water Plants

fl.   Pl. 129

Helianthus salicifolius

Huxley, Anthony
Garden Perennials and Water Plants

hab.   Pl. 131

Helianthus simulans

Brown, Clair A.
Wildflowers of Louisiana and
      Adjoining States

fl. hab.   p. 206

Helianthus strumosus

Brown, Clair A.
Wildflowers of Louisiana and
      Adjoining States

fl. hab.   p 206

Helianthus strumosus

Courtenay, Booth
Wildflowers & Weeds

fl.   p 118

Helianthus strumosus

Jennings, O. E.
Wild Flowers of Western Pennsylvania
v. 2
fl. hab.   Pl. 186

Helianthus strumosus

Wharton, Mary E.
A Guide to the Wildflowers & Ferns
      of Kentucky

fl.   p 259   Pl. 3.33

Helianthus subtuberosus

Neufeld, J.B.
Wild Flowers of the Prairies

fl.   p. 7

Helianthus tomentosus

Addisonia, V. 19, 1935-36

fl. fr.   opp. p. 35   Pl. 626

Helianthus tuberosus

Bianchini, F.
The Complete Book of Fruits & Vegetables

hab.   p 227

Helianthus tuberosus

Courtenay, Booth
Wildflowers & Weeds

fl.   p 116

Helianthus tuberosus

Hvass, Elsie
Plants That Feed and Serve Us

fl. hab.   p 25, Pl. 41

Helianthus tuberosus

Kimura, Koiti
Japanese Medicinal Plants, Vol I

fl.   p 102

Helianthus tuberosus

Masefield, G. B.
The Oxford Book of Food Plants

hab.   Pl. 179, 1

Helianthus tuberosus

Moyle, John B.
Northland Wild Flowers

fl. hab.   p 158, Pl. 185

Helianthus tuberosus

Szczawinski, Adam F.
Edible Garden Weeds of Canada

fl., hab.   p. 52

Helianthus tuberosus

Turner, Nancy J.
Wild Coffee and Tea Substitutes of
      Canada

fl. hab.   p. 40

Helianthus tuberosus

Wharton, Mary E.
A Guide to the Wildflowers & Ferns
      of Kentucky

fl.   p 260   Pl. 3.35

x Heliaporus mallisonii

Hay, Roy
The Color Dictionary of Flowers & Plants

fl.   p 69, Pl. 547

Helichrysum acuminatum

Cochrane, G. R.
Flowers and Plants of Victoria

fl.   Pl. 533

Helichrysum adenophorum var.

Cochrane, G. R.
Flowers and Plants of Victoria

fl.   Pl. 539

Helichrysum agyrophyllum

Harrison, Richmond E.
Climbers and Trailers

fl. p 93   Pl. 232

Helichrysum angustifolium

Hay, Roy
The Color Dictionary of Flowers & Plants

fl.   p 145, Pl. 1159

Helichrysum antennaria

Curtis, Winifred
The Endemic Flora of Tasmania Vol. 2

fl.   Pl. 46

Helichrysum apiculatum

Blombery, Alec M.
What Wildflower is That

fl.   p 162, Pl. 440

Helichrysum apiculatum

Cochrane, G. R.
Flowers and Plants of Victoria

fl.   Pl. 243

Helichrysum arenarium

Felsko, Elsa
A Book of Wildflowers

fl.   P 29

Helichrysum arenarium

Lindman, C. A. M.
Nordens Flora, Vol 9

fl. hab.   Pl. 591

Helichrysum augustifolium

Hay, Roy
The Color Dictionary of Flowers and Plants

fl.   p 145

Helichrysum backhousii var.

Curtis, Winifred
The Endemic Flora of Tasmania, Vol. 5

fl. fr.   Pl. 181, 182, 184

Helichrysum baxteri

Cochrane, G. R.
Flowers and Plants of Victoria

fl.   Pl. 122

Helichrysum bracteatum

Blombery, Alec M.
What Wildflower is That

fl.   p 163, Pl. 441

Helichrysum bracteatum

Crockett, James Underwood
Annuals

fl.   P 123

Helichrysum bracteatum

Hay, Roy
The Color Dictionary of Flowers & Plants

fl.   p 39, Pl. 305

Helichrysum bracteatum

Macoby, Stirling
What Flower is That

fl.   p. 150

Helichrysum bracteatum

Morley, Brian D.
Wild Flowers of the World

fl.   Pl. 140

Helichrysum bracteatum

Mullins, Barbara
Australian Wildflowers in Colour

fl.   P. 99   Pl. 98

Helichrysum bracteatum

Perry, Frances
Flowers of the World

fl.   p 86

Helichrysum bracteatum

Tsukamoto, Yotaro
Coloured Illustrations of Garden Flowers, v.10
fl.
ill. 71 opp.p.23

Helichrysum bracteatum var.

Huxley, Anthony
Garden Annuals and Bulbs

fl.   Pl. 60

Helichrysum bracteatum var.

Royal Hort. Soc.
The Garden
   vol 103, No. 12   1978
fl.   p. 490

Helichrysum brassii

Moriarty, Audrey
Wild Flowers of Malawi

fl.   Pl. 76; 3

Helichrysum cassinianum

Blombery, Alec M.
What Wildflower is That

fl. hab.   p 164, Pl. 442

Helichrysum costatifructum

Curtis, Winifred
The Endemic Flora of Tasmania
v. 3
fl. p 163

Helichrysum davenportii

Morcombe, M. K.
Australia's Western Wildflowers

fl.   p 13

Helichrysum diosmifolium

Blombery, Alec M.
What Wildflower is That

fl.   p 164, Pl. 443

Helichrysum diosmifolium

Harrison, R. E.
Trees and Shrubs

fl.   P. 89

Helichrysum doerfleri

Goulimis, Constantine N.
Wild Flowers of Greece

fl. hab. p 133

Helichrysum elatum

Blombery, Alec M.
What Wildflower is That

fl.   p 164, Pl. 444

Helichrysum ericeteum

Curtis, Winifred
The Endemic Flora of Tasmania Vol. 1

fl.   Pl. 23a, b

Helichrysum expansifolium

Curtis, Winifred
The Endemic Flora of Tasmania
vol 2

fl.   Pl. 45
          (see errata, vol 6)

Helichrysum flammeiceps

Moriarty, Audrey
Wild Flowers of Malawi

fl.   Pl. 76; 2

Helichrysum frigidum

Hay, Roy
The Color Dictionary of Flowers & Plants

fl.   p 11, Pl. 86

Helichrysum herbaceum

Moriarty, Audrey
Wild Flowers of Malawi

fl.   Pl. 75; 3

Helichrysum lanatum

Hay, Roy
The Color Dictionary of Flowers & Plants

fl.   p 206, Pl. 1642

Helichrysum lastii

Moriarty, Audrey
Wild Flowers of Malawi

fl.   Pl. 75; 4

Helichrysum ledifolium

Curtis, Winifred
The Endemic Flora of Tasmania
       Vol. 1

fl.   Pl. 24

Helichrysum lycopodioides

Curtis, Winifred
The Endemic Flora of Tasmania
       vol 2

fl.   Pl. 47
          (see errata, vol 6)

Helichrysum meyeri-johannis

Morley, Brian D.
Wild Flowers of the World

fl.,hab.   Pl. 65

Helichrysum milfordiae

Hay, Roy
The Color Dictionary of Flowers & Plants

fl.   p 11, Pl. 87

Helichrysum milliganii

Curtis, Winifred
The Endemic Flora of Tasmania Vol. 1

fl.   Pl. 22

Helichrysum monogynum

Bramwell, David
Wild Flowers of the Canary Islands

fl.   Pl. 276

Helichrysum nitens

Moriarty, Audrey
Wild Flowers of Malawi

fl.   Pl. 76; 4

Helichrysum obcordatum

Cochrane, G. R.
Flowers and Plants of Victoria

fl.   Pl. 316

Helichrysum obtusifolium

Cochrane, G. R.
Flowers and Plants of Victoria

fl.   Pl. 20

Helichrysum paralium

Cochrane, G. R.
Flowers and Plants of Victoria

fl.   Pl. 278

Helichrysum patulifolium

Moriarty, Audrey
Wild Flowers of Malawi

fl.   Pl. 75; 5

Helichrysum pumilum

Curtis, Winifred
The Endemic Flora of Tasmania   Vol. 1

fl.   Pl. 21

Helichrysum purpurascens

Curtis, Winifred
The Endemic Flora of Tasmania
Vol VI

fl.   Pl. 231

Helichrysum reticulatum

Curtis, Winifred
The Endemic Flora of Tasmania
v. 3

fl.   p 163

Helichrysum rosmarinifolium

Curtis, Winifred
The Endemic Flora of Tasmania
vol 1

fl., fr.   Pl. 20
          (see errata, vol 6)

Helichrysum rutidolepis

Cochrane, G. R.
Flowers and Plants of Victoria

fl.   Pl. 501

Helichrysum, sand

see

Helichrysum arenarium

Helichrysum sanguineum

Polunin, Oleg
Flowers of the Mediterranean

fl.   No. 194

Helichrysum scorpoides

Blombery, Alec M.
What Wildflower is That

fl.   p 165, Pl. 446

Helichrysum scutellifolium

Curtis, Winifred
The Endemic Flora of Tasmania
vol 2

fl.   Pl. 48

Helichrysum semipapposum

Alpine Flowers of the Kosciusko
State Park

fl.   Pl. 1

Helichrysum semipapposum

Blombery, Alec M.
What Wildflower is That

fl.   p 164, Pl. 445

Helichrysum semipapposum

Cochrane, G. R.
Flowers and Plants of Victoria

fl.   Pl. 216

Helichrysum serpyllifolium

Kiaer, Eigil
Indoor Plants in Colour

hab.   p. 65

Helichrysum setosum

Moriarty, Audrey
Wild Flowers of Malawi

fl.   Pl. 75; 2

Helichrysum splendidum

Gault, S. Millar
The Color Dictionary of Shrubs

fl.   Pl. 230

Helichrysum stirlingii

Cochrane, G. R.
Flowers and Plants of Victoria

fl.   Pl. 402

Helichrysum stoechas

Polunin, Oleg
Flowers of Europe

fl.   Pl. 144 #1385

Helichrysum stoechas

Polunin, Oleg
Flowers of the Mediterranean

fl.   No. 195

Helichrysum stoechas

Taylor, A. W.
Wild Flowers of Spain and Portugal

fl.   p. 74

Helichrysum virgineum

Royal Hort. Soc.
The Garden
vol 101, No. 11 1976

fl.   P. 555

Helichrysum whyteanum

Moriarty, Audrey
Wild Flowers of Malawi

fl.   Pl. 76; 1

Helicia cochinchinensis

Kitamura, Siro
Coloured Illustrations of Trees &
Shrubs of Japan

fr.   144

Heliconia acuminata

Mee, Margaret
Flowers of the Brazilian Forests

fl.   Pl. 21

Heliconia bihai

Chickering, Carol Rogers
Flowers of Guatemala

fl. fr.   p 71

Heliconia bihai

Milne, Lorus
Living Plants of the World

fl.   p 316

Heliconia bihai

Perry, Frances
Flowers of the World

fl.   p. 137

Heliconia bourgeana

Pacific Hort. Foundation
Pacific Horticulture
vol 38, No. 2   1977

fl.   p. 29

Heliconia collinsiana

Hargreaves, Dorothy
Tropical Blossoms of the Caribbean

fl.   p. 16

Heliconia collinsiana

Macoby, Stirling
What Flower is That

fl.   p. 151

Heliconia elongata

Hargreaves, Dorothy
Tropical Blossoms of the Caribbean

fl.   p. 15

Heliconia, Fishpole

see

Heliconia collinsiana

Heliconia, Hanging

see

Heliconia collinsiana

Heliconia humilis

American Hort. Soc.
American Horticulturist
vol 54, No. 4   1975

fl.   p. 24

Heliconia humilis

Hall, Clarence E.
Flowers of the Islands in the Sun

fl.   p. 95   Pl. 20

Heliconia humilis

Macoby, Stirling
What Flower is That

fl.   p. 151

Heliconia latispatha

O'Gorman, Helen
Mexican Flowering Trees and Plants

fl. hab.   p 159

Heliconia pendula

Royal Hort. Soc.
The Garden
vol 101, No. 1   1976

fl.   p. 19

Heliconia psittacorum

Morley, Brian D.
Wild Flowers of the World

fl. fr.   Pl. 189

Heliconia rostrata

American Hort. Soc.
American Horticulturist
vol 54, No. 4   1975

fl.   p. 26

Heliconia rostrata

Pacific Tropical Botanical Garden
Bulletin
vol 7, No. 2   1977

fl.   cover

Heliconia sp.

Hargreaves, Dorothy
Tropical Blossoms of the Caribbean

fl.   p. 17

Heliconia subulata

Mee, Margaret
Flowers of the Brazilian Forests

fl., hab.   Pl. 22

Helicteres angustifolia

Walden, Beryl M.
Wild Flowers of Hong Kong

fl.   Pl. 70 (212)

Helictotrichon sempervirens

Hay, Roy
The Color Dictionary of Flowers &
    Plants

fr.,hab.   p. 145   Pl. 1160

Helictotrichon sempervirens

Huxley, Anthony
Garden Perennials and Water Plants

fr.   Pl. 314

Helictotrichon sempervirens

Royal Hort. Soc.
Journal of the Royal Hort. Soc.
vol 93, No. 10   1968

hab.   p. 426   Pl. 231

Heliocereus speciosus

Hay, Roy
The Dictionary of House Plants

fl.   Pl. 270

Heliocereus speciosus

Van Laren, A. J.
Cactus

fl.   P 20   Fig. 16

Heliophila bulbostyla

Eliovson, Sima
Namaqualand in Flower

fl   Pl. 11, 1

Heliophila coronopifolia

Eliovson, Sima
Namaqualand in Flower

fl.   Pl. 11, 2

Heliophila cuneata

Flowering Plants of Africa
vol 43   1974-76

fl.   Pl. 1707

Heliophila longifolia

Hay, Roy
The Color Dictionary of Flowers & Plants

fl.   p 39, Pl. 306

Heliophila namaquana

Eliovson, Sima
Namaqualand in Flower

fl.   Pl. 10

Heliophila seselifolia

Eliovson, Sima
Namaqualand in Flower

fl.   Pl. 10

Heliopsis helianthoides

Courtenay, Booth
Wildflowers & Weeds

fl.   p 116

Heliopsis helianthoides

Jennings, O. E.
Wild Flowers of Western Pennsylvania
v. 2
fl. hab.   Pl. 184

Heliopsis helianthoides

Moyle, John B.
Northland Wild Flowers

fl. hab.   p 159, Pl. 187

Heliopsis helianthoides

Wharton, Mary E.
A Guide to the Wildflowers & Ferns
    of Kentucky

fl.   p 258   Pl. 3.31

Heliopsis scabra var.

Hay, Roy
The Color Dictionary of Flowers &
    Plants

fl.   p. 146   Pl. 1161, 1162

Heliopsis scabra var.

Huxley, Anthony
Garden Perennials and Water Plants

fl.   Pl. 132

Heliotrope, common

see

Heliotropium arborescens

Heliotrope, common

see

Heliotropium peruvianum

Heliotrope, Wild

see

Phacelia crenulata

Heliotrope, Wild

see

Phacelia sericea

Heliotropium arborescens

Harrison, R.E.
Trees and Shrubs

fl. P 89

Heliotropium arborescens

Kiaer, Eigil
Indoor Plants in Colour

fl.   p. 65

Heliotropium arborescens var.

Everett, T. H.
New Illustrated Encyclopedia of
        Gardening, Vol. 5

fl.   opp. p. 823

Heliotropium convolvulaceum

Welsh, Stanley L.
Flowers of the Canyon Country

fl.   p 4

Heliotropium corymbosum

Massachusetts Hort. Soc.
Horticulture
        Vol. 55, No. 1   1977

fl.   P. 36

Heliotropium curassavicum var.

Coyle, Jeanette
A Field Guide to the Common and
    Interesting Plants of Baja California

fl.   p 155

Heliotropium curassavicum var.

Lemmon, Robert S.
Wildflowers of North America in
        Full Color

fl.   p. 198   Pl. 310

Heliotropium europaeum

Polunin, Oleg
Flowers of Europe

fl.   Pl. 101 #1045

Heliotropium hybrid

Perry, Frances
Flowers of the World

fl.   p. 49

Heliotropium hybridum

Hay, Roy
The Dictionary of House Plants

fl.   Pl. 271

Heliotropium indicum

Dean, Blanche E.
Wildflowers of Alabama and
        Adjoining States

fl.   p 147

Heliotropium indicum

Duncan, Wilbur H.
Wildflowers of the Southeastern
        United States

fl.   p 149

Heliotropium peruvianum

Crockett, James Underwood
Annuals

fl.   P 124

Heliotropium peruvianum

Hvass, Elsie
Plants That Feed and Serve Us

fl.   p 113, Pl. 245

Heliotropium peruvianum

Macoby, Stirling
What Flower is That

fl.   p. 151

Heliotropium peruvianum var.

Huxley, Anthony
Garden Annuals and Bulbs

fl.   Pl. 61

Helipterum albicans

Blombery, Alec M.
What Wildflower is That

fl.   p 165, Pl. 447

Helipterum albicans

Cochrane, G. R.
Flowers and Plants of Victoria

fl.   Pl. 538

Helipterum albicans var.

Curtis, Winifred
The Endemic Flora of Tasmania   Vol. 2

fl.   Pl. 56

Helipterum albicans var.

Macoby, Stirling
What Flower is That

fl.   p. 152

Helipterum anethemoides

Blombery, Alec M.
What Wildflower is That

fl.   p 165, Pl. 448

Helipterum canescens

Morley, Brian D.
Wild Flowers of the World

fl.   Pl. 83

Helipterum corymbiflorum

Cochrane, G. R.
Flowers and Plants of Victoria

fl.   Pl. 168

Helipterum craspedioides

Morcombe, M. K.
Australia's Western Wildflowers

fl.   p 23, 24

Helipterum floribundum

Mullins, Barbara
Australian Wildflowers in Colour

fl.   P. 93   Pl. 91

Helipterum incanum

Royal Hort. Soc.
Journal of the Royal Hort. Soc.
        vol 89, No. 9   1964

fl.   p. 376   Pl. 151

Helipterum manglesii

Crockett, James Underwood
Annuals

fl.   P 124

Helipterum manglesii

Hay, Roy
The Color Dictionary of Flowers & Plants

fl.   p 39, Pl. 307

Helipterum roseum

Blombery, Alec M.
What Wildflower is That

fl. hab.   p 165, Pl. 449

Helipterum roseum

Huxley, Anthony
Garden Annuals and Bulbs

fl.   Pl. 63

Helipterum roseum

Tsukamoto, Yotar
Coloured Illustrations of
        Garden Flowers vol 10

fl.   Pl. 72

Helipterum rubellum

Morcombe, M. K.
Australia's Western Wildflowers

fl.   p 22

Helipterum splendidum

Morcombe, M. K.
Australia's Western Wildflowers

fl.   p 23

Helipterum variegatum

Flowering Plants of South Africa
        vol XXI   1941

fl.   Pl. 817

Helixyra simulans

Flowering Plants of South Africa
vol XVI   1936

fl., hab.   Pl. 623

Hellebore, American White
see
Veratrum eschscholtzii

Hellebore, False
see
Veratrum californicum

Hellebore, false
see
Veratrum viride

Hellebore, Green
see
Helleborus viridis

Hellebore, green
see
Veratrum viride

Hellebore, Green False
see
Veratrum viride

Hellebore, Stinking
see
Helleborus foetidus

Hellbore, White
see
Veretrum album

Hellebore, white
see
Veratrum californicum

Helleborine
see
Cephalanthera

Helleborine
see
Epipactis

Helleborine, Broad Leaved
see
Epipactis helleborine

Helleborine, Broad-Leaved
see
Epipactis latifolia

Helleborine, Dark Red
see
Epipactis atrorubens

Helleborine, Giant
see
Epipactis gigantea

Helleborine, Long-Leaved
see
Cephalanthera longifolia

Helleborine, Marsh
see
Epipactis palustris

Helleborine, White
see
Cephalanthera damasonium

Helleborus argutifolius

Curtis's Botanical Magazine
vol 178   1970-72

fl.   Pl. 578

Helleborus argutifolius

Perry, Frances
Flowers of the World

fl.   p 251

Helleborus atrorubens

Curtis's Botanical Magazine
vol 177   1969-70

fl.   Pl. 545

Helleborus atrorubens

Huxley, Anthony
Garden Perennials and Water Plants

fl.   Pl. 135

Helleborus corsicus

Hay, Roy
The Color Dictionary of Flowers & Plants

fl.   p 146, Pl. 1163

Helleborus corsicus

Royal Hort. Soc.
The Garden
vol 103, No. 3   1978

fl.   cover

Helleborus cyclophyllus

Polunin, Oleg
Flowers of Europe

fl.   Pl. 18   #200

Helleborus dumetorum var.

Morley, Brian D.
Wild Flowers of the World

fl.   Pl. 8B

Helleborus dumetorum var.

Royal Hort. Soc.
Journal of the Royal Hort. Soc.
vol 91, No. 8   1966

fl.   p. 338   Pl. 186

Helleborus foetidus

Barneby, T. P.
Europaen Alpine Flowers in Color

fl.   Pl. 21, 5

Helleborus foetidus

Felsko, Elsa
A Book of Wild Flowers
2nd Ser.

fl.   p. 70

Helleborus foetidus

Martin, W. Keble
The Concise British Flora in Colour

fl. hab.   Pl. 4

Helleborus foetidus

Perrot, Emile
Les Plantes Medicinales

fl. hab.   p 116

Helleborus foetidus

Perry, Frances
Flowers of the World

fl.   p 251

Helleborus foetidus

Polunin, Oleg
Flowers of Europe
fl.
pl. 19 no.199

Helleborus hybrid

Kromdijk, G.
200 House Plants in Colour

fl.   Pl. 104

Helleborus multifidus

Curtis's Botanical Magazine
    Vol. 180   1974

fl.   Pl. 698

Helleborus niger

American Hort. Soc.
American Horticulturist
    vol 56, No. 3   1977

fl.   p. 15

Helleborus niger

Bianchini, Francesco
Health Plants of the World

fl. hab.   p 53

Helleborus niger

Felsko, Elsa
A Book of Wildflowers

fl.   P 160

Helleborus niger

Hay, Roy
The Color Dictionary of Flowers & Plants

fl.   p 146, Pl. 1164

Helleborus niger

Huxley, Anthony
Garden Perennials and Warer Plants

fl.   Pl. 136

Helleborus niger

Kohlhaupt, Paula
Fleurs des Alpages, v.1

fl. P 12

Helleborus niger

Massachusetts Hort. Soc.
Horticulture
    vol. 31, No. 12   1953

fl.   p. 474

Helleborus niger

Massachusetts Hort. Soc.
Horticulture
    vol 36, No. 12   1958

fl.   inside cover

Helleborus niger

Massachusetts Hort. Soc.
Horticulture
    vol 43, No. 12   1965

fl.   cover, backcover, p. 37

Helleborus niger

Milne, Lorus
Living Plants of the World

fl.   p 54

Helleborus niger

Perrot, Emile
Les Plantes Medicinales

fl. fr. hab.   p 116

Helleborus niger

Perry, Frances
Flowers of the World

fl.   p. 250

Helleborus niger

Polunin, Oleg
Flowers of Europe
fl.
pl. 19 no.201

Helleborus niger

Royal Hort. Soc.
The Garden
    vol 103, No. 1   1978

fl.   cover

Helleborus niger

Tosco, Uberto
The World of Mountain Flowers

fl.   p 98

Helleborus niger

Tsukamoto, Yotaro
Coloured Illustrations of Garden Flowers, v.9
fl.
opp.p.49

Helleborus niger var.

Alpine Garden Society
Bulletin
    vol. 33, No. 2   1965

fl.   p. 117

Helleborus niger var.

Hay, Roy
The Color Dictionary of Flowers & Plants

fl.   p. 146   Pl. 1165

Helleborus niger var.

Royal Hort. Soc.
The Garden
    vol 102, No. 1   1977

fl.   p. 9

Helleborus x nigricors

Royal Hort. Soc.
The Garden
    vol 100, No. 7   1975

fl.   p. 318

Helleborus x nigricors var.

Royal Hort. Soc.
Journal of the Royal Hort. Soc.
    vol 97, No. 3   1972

fl.   p. 122   Pl. 50

Helleborus orientalis

Hay, Roy
The Color Dictionary of Flowers & Plants

fl.   p 146, Pl. 1166

Helleborus orientalis

Jones, Paul
Flora Magnifica

fl.   Pl. 5

Helleborus orientalis

Macoby, Stirling
What Flower is That

fl.   p. 152

Helleborus orientalis

Perry, Frances
Flowers of the World

fl.   p 250

Helleborus orientalis

Royal Hort. Soc.
Journal of the Royal Hort. Soc.
    vol 85, No. 9   1960

fl.   p. 398   Pl. 119

Helleborus orientalis

Tsukamoto, Yotaro
Coloured Illustrations of Garden Flowers, v.9
fl.
opp.p.49

Helleborus x sternii

Curtis's Botanical Magazine
    Vol. 171   1956-57

fl.   291

Helleborus vesicarius

Curtis's Botanical Magazine
    Vol 167   1950

fl. fr.   116

Helleborus vesicarius

Royal Hort. Soc.
Journal of the Royal Hort. Soc.
    vol 91, No. 8   1966

fr.   p. 338   Pl. 185

Helleborus viridis

Ary, S.
The Oxford Book of Wildflowers

fl.   p 50, Pl. 3

Helleborus viridis

Bianchini, Francesco
Health Plants of the World

fl. hab.   p 53

Helleborus viridis

Felsko, Elsa
A Book of Wild Flowers
2nd Ser.
fl.   p 112

Helleborus viridis

Perrot, Emile
Les Plantes Medicinales

hab.   p 116

Helleborus viridis

Tosco, Uberto
The World of Mountain Flowers

fl.   p 32

Helleborus viridis var.

Barneby, T. P.
European Alpine Flowers in Colour

fl.   Pl. 21, 6

Helleborus viridis var.

Martin, W. Keble
The Concise British Flora in Colour

fl. hab.   Pl. 4

Helleriella nicaraguensis

Ospina, Mariano
Orquideas de las americas

fl. hab.   Pl. 60

Helmet Flower

see

Scutellaria integrifolia

Helonias bullata

Duncan, Wilbur H.
Wildflowers of the Southeastern
    United States

fl.   p 247

Helonias bullata

Klimas, John E.
Wildflowers of Eastern America

fl.   Pl. 200

Helxine soleirolii

Crockett, James Underwood
Lawns & Ground Covers

hab. P. 135

Helxine soleirolii

Hersey, Jean
Woman's Day Book of House Plants

hab.   p. 32

Helxine soleirolii

Kiaer, Eigil
Indoor Plants in Colour

hab.   P. 65

Helxine soleirolii

Macoby, Stirling
What Flower is That

hab.   p. 152

Hemerocallis aurantiaca

Macoby, Stirling
What Flower is That

fl.   p. 153

Hemerocallis aurantiaca

Tsukamoto, Yotaro
Coloured Illustrations of Garden
    Flowers V. 9

fl.   opp. p. 50

Hemerocallis flava

Bruggeman, L.
Tropical Plants

fl.   Pl. 116

Hemerocallis flava

Huxley, Anthony
Garden Perennials and Water Plants

fl.   Pl. 133

Hemerocallis fulva

Courtenay, Booth
Wildflowers and weeds

fl. p. 4

Hemerocallis fulva

Duncan, Wilbur H.
Wildflowers of the Southeastern
    United States

fl. hab.   p 249

Hemerocallis fulva

Kimura, Koiti
Japanese Medicinal Plants, Vol II

fl.   p 137

Hemerocallis fulva

Klimas, John E.
Wildflowers of Eastern America

fl.   Pl. 125

Hemerocallis fulva

Moyle, John B.
Northland Wild Flowers

fl. hab.   p 206, Pl. 269

Hemerocallis fulva

Wharton, Mary E.
A Guide to the Wildflowers & Ferns
    of Kentucky

fl.   p 49 Pl. 1.4

Hemerocallis fulva var.

Crockett, James Underwood
Lawns & Ground Covers

fl.   P. 135

Hemerocallis fulva var.

Macoby, Stirling
What Flower is That

fl.   p. 153

Hemerocallis hybrid

Everett, T.H., e.d.
New Illustrated Encyclopedia of
    Gardening   Vol. 5

fl.   p. 838, 839

Hemerocallis hybrid

Hay, Roy
The Color Dictionary of Flowers & Plants

fl.   p 146, 147; Pl. 1167 thru 1171

Hemerocallis hybrid

Huxley, Anthony
Garden Perennials and Water Plants

fl.   Pl 134

Hemerocallis lilioasphodelus

Polunin, Oleg
Flowers of Europe

fl.   Pl. 164 #1597

Hemiandra pungens

Blombery, Alec M.
What Wildflower is That

fl.   p 166, Pl. 450

Hemiandra pungens

Morcombe, M. K.
Australia's Western Wildflowers

fl.   p 57

Hemigenia purpurea
Blombery, Alec M.
What Wildflower is That
fl.   p 167, Pl. 451

Hemigraphis colorata
Bruggeman, L.
Tropical Plants
fl.   Pl. 117

Hemigraphis colorata
Kiaer, Eigil
Indoor Plants in Colour
fl.   p. 65

Hemigraphis repanda
Bruggeman, L.
Tropical Plants
hab.   Pl. 118

Hemitomes congestum
Clark, Lewis J.
Wild Flowers of British Columbia
fl.   p 375

Hemizygia bracteosa
Moriarty, Audrey
Wild Flowers of Malawi
fl.   Pl. 55; 3

Hemlock, Canada
see
Tsuga canadensis

Hemlock, Carolina
see
Tsuga caroliana

Hemlock, Chinese
see
Tsuga chinensis

Hemlock, Eastern
see
Tsuga canadensis

Hemlock, Mountain
see
Tsuga mertensiana

Hemlock, Northern Japanese
see
Tsuga diversifolia

Hemlock, Poison
see
Conium maculatum

Hemlock, Sargent's weeping
see
Tsuga canadensis var.

Hemlock, Spotted
see
Conium maculatum

Hemlock, Striped
see
Molospermum peloponesciacum

Hemlock, Water
see
Cicuta maculata

Hemlock Water Dropwrot
see
Oenanthe crocata

Hemlock, Water Mackenzie
see
Cicuta mackenzieana

Hemlock, Water Poison
see
Cicuta mackenzieana

Hemlock, Water Spotted
see
Cicuta maculata

Hemlock, Water Western
see
Cicuta douglasii

Hemlock, Western
see
Tsuga heterophylla

Hemp
see
Cannabis sativa

Hemp, African
see
Sparmannia africana

Hemp, Bow-string
see
Sanseviera trifasciata

Hemp, Indian
see
Apocynum cannabinum

Hemp, Indian
see
Apocynum sibiricum

Hemp, Manila
see
Musa textilis

Hemp, Sida
see
Sida rhombifolia

Hempweed, Climbing
see
Mikania scandens

Hen and chickens
see
Echáveria

Hen and chickens
see
Sempervivum tectorum

Henbane

see

Hyoscyamus niger

Henbane, Yellow

see

Hyoscyamus aureus

Henbit

see

Lamium amplexicaule

Henbit Dead-Nettle

see

Lamium amplexicaule

Henna

see

Lawsonia inermis

Hepatica acutiloba

Campbell, Carlos C.
Great Smoky Mountain Wildflowers

fl.   p 11

Hepatica acutiloba

Courtenay, Booth
Wildflowers & Weeds

fl.   p 35

Hepatica acutiloba

Dean, Blanche E.
Wildflowers of Alabama and
      Adjoining States

fl.   p 65

Hepatica acutiloba

Ferguson, Mary
Wildflowers

fl. hab.   p 178

Hepatica acutiloba

Moyle, John B.
Northland Wild Flowers

fl. hab.   p 45, Pl. 1

Hepatica acutiloba

Roberts, June Carver
Born in the Spring

fl. fr. hab.   p 23

Hepatica acutiloba

Wharton, Mary E.
A Guide to the Wildflowers & Ferns
      of Kentucky

fl.   p 108 Pl. 1.29
fl.   p. 140 pl. 1.91

Hepatica americana

American Hort. Soc.
American Horticulturist
      vol 55, No. 6   1976

fl.   p. 15

Hepatica americana

Dean, Blanche E.
Wildflowers of Alabama and
      Adjoining States

fl.   p 65

Hepatica americana

Duncan, Wilbur H.
Wildflowers of the Southeastern
      United States

fl.   p 45

Hepatica americana

Jennings, O. E.
Wild Flowers of Western Pennsylvania
v. 2
fl.   Pl. 19

Hepatica americana

Klimas, John E.
Wildflowers of Eastern America

fl.   Pl. 183

Hepatica americana

Lemmon, Robert S.
Wildflowers of North America in
      Full Color

fl.   p. 250   Pl. 393

Hepatica americana

Massachusetts Hort. Soc.
Horticulture
      vol 40, No. 3   1962

fl.   inside backcover

Hepatica americana

Massachusetts Hort. Soc.
Horticulture
      vol 53, No. 2   1975

fl.   p. 46

Hepatica americana

Walcott, Mary Vaux
North American Wild Flowers

fl. hab.   Pl. 125

Hepatica americana var.

Klute, Jeannette
Woodland Portraits

fl. hab.   Pl. 3

Hepatica x ballardii

Hay, Roy
The Color Dictionary of Flowers &
      Plants

fl.   p. 11   Pl. 88

Hepatica, Blunt-Lobed

see

Hepatica americana

Hepatica hybrid

Alpine Garden Society
Bulletin
      vol. 37, No. 3   1969

fl.   p. 275

Hepatica nobilis

Barneby, T. P.
European Alpine Flowers in Colour

fl.   Pl. 23, 4

Hepatica nobilis

Perry, Frances
Flowers of the World

fl.   p 253

Hepatica nobilis

Polunin, Oleg
Flowers of Europe

fl.   Pl. 20 #210

Hepatica nobilis

Tosco, Uberto
The World of Mountain Flowers

fl.   p 26, 111

Hepatica, Round-Lobed

see

Hepatica americana

Hepatica, Sharp Lobed

see

Hepatica acutiloba

Hepatica transsilvanica

Huxley, Anthony
Garden Perennials and Water Plants

fl.   Pl. 137

Hepatica transsilvanica

Royal Hort. Soc.
The Garden
      vol 102, No. 11 1977

fl.   p. 470

Hepatica triloba

Hay, Roy
The Color Dictionary of Flowers & Plants

fl.   p 12, Pl. 89

Hepatica triloba

Perrot, Emile
Les Plantes Medicinales

fl. fr. hab.   p 86

Hepatica triloba var.

Huxley, Anthony
Garden Perennials and Water Plants

fl.   Pl. 138

Heracleum lanatum

Alaska-Yukon Wild Flower Guide

fl.   p 105

Heracleum lanatum

Clark, Lewis J.
Wild Flowers of British Columbia

fl.   p 354

Heracleum lanatum

Courtenay, Booth
Wildflowers & Weeds

fl.   p 58

Heracleum lanatum

Fries, Mary A.
Wildflowers of Mount Ranier and the
      Cascades

fl.   Pl. 108

Heracleum lanatum

Heller, Christine
Wild Flowers of Alaska

fl.   Pl. 207-208

Heracleum lanatum

Moyle, John B.
Northland Wild Flowers

fl. hab.   p 102, Pl. 95

Heracleum lanatum

Munz, Philip A.
California Mountain Wildflowers

fl.   p 43, Pl. 49

Heracleum lanatum

Orr, Robert T.
Wildflowers of Western America

fl.   Pl. 67

Heracleum lanatum

Taylor, Ronald J.
Mountain Wild Flowers

fl. hab.   p 92

Heracleum lanatum

Welsh, Stanley L.
Flowers of the Mountain Country

fl.   p. 15

Heracleum mantegazzianum

Polunin, Oleg
Flowers of Europe

fl.   Pl. 86 #902

Heracleum maximum

Jennings, O. E.
Wild Flowers of Western Pennsylvania
v. 2
fl.   Pl. 106

Heracleum maximum

Klimas, John E.
Wildflowers of Eastern America

fl.   Pl. 88

Heracleum moellendorffi

Kimura, Koiti
Japanese Medicinal Plants, Vol II

fl.   p 207

Heracleum sphondylium

Ary, S.
The Oxford Book of Wildflowers

fl.   p 86, Pl. 2

Heracleum sphondylium

Bianchini, Francesco
Health Plants of the World

fr.   p 117

Heracleum sphondylium

Color Treasury of Herbs & Other
    Medicinal Plants

fl.   p 41   Pl. 54

Heracleum sphondylium

Martin, W. Keble
The Concise British Flora in Colour

fl. fr. hab.   Pl. 40

Heracleum sphondylium var.

Lindman, C. A. M.
Nordens Flora, Vol 7

fl. fr.   Pl. 433

Herb Bennet

see

Herb Christopher

see

Actaea spicata

Herb Mercury

see

Mercurialis annua

Herb of Grace

see

Ruta graveolens

Herb Paris

see

Paris quadrifolia

Herb Robert

see

Geranium robertianum

Herbertia

see

Alophia drummondii

Herbertia drummondii

Addisonia, v. 20, 1937-38

fl., fr. opp. p. 3   Pl. 642

Hercules' Club

see

Aralia spinosa

Hereroa bergeriana

Herre, H.
The Genera of the Mesembryanthemaceae

fl., fr.   p. 169

Hereroa dyeri

Lamb, Edgar
The Pocket Encyclopedia of Cacti
   and Succulents in Color

fl.   Pl. 262

Hermannia althaeifolia

Flowering Plants of Africa
vol 41   1970-71

fl.   Pl. 1603

Hermannia angularis

Flowering Plants of Africa
vol 41   1970-71

fl.   Pl. 1604

Hermannia comosa

Flowering Plants of Africa
vol 41   1970-71

fl.   Pl. 1605

Hermannia concinnifolia

Flowering Plants of Africa
Vol. 43   1974-76

fl.   Pl. 1691

Hermannia confusa

Flowering Plants of Africa
vol 43   1974-76

fl.   Pl. 1718

Hermannia cristata

Flowering Plants of Africa
vol 30   1954-55

fl.   Pl. 1169

Hermannia cuneifolia

Eliovson, Sima
Namaqualand in Flower

fl.   Pl. 42

Hermannia diversistipula var.

Flowering Plants of Africa
vol 41   1970-71

fl.   Pl. 1620

Hermannia prismatocarpa

Flowering Plants of Africa
vol 41   1970-71

fl.   Pl. 1628

Hermannia stricta

Eliovson, Sima
Namaqualand in Flower

fl.   Pl. 41, 1/2

Hermannia stricta

Flowering Plants of Africa
vol 39   1968-69

fl.   Pl. 1539

Herminium monorchis

Ary, S.
The Oxford Book of Wildflowers

fl.   p 44, Pl. 3

Herminium monorchis

Brooke, Jocelyn
The Wild Orchids of Britain

fl., hab.   Pl. 16

Herminium monorchis

Lindman, C. A. M.
Nordens Flora, Vol 3

fl. hab.   Pl. 145

Herminium monorchis

Martin, W. Keble
The Concise British Flora in Colour

fl. hab.   Pl. 82

Herminium monorchis

Polunin, Oleg
Flowers of Europe

fl.   Pl. 190 #1909

Hermodactylus tuberosus

Goulimis, Constantine N.
Wild Flowers of Greece

fl. hab.   p. 181

Hermodactylus tuberosus

Hay, Roy
The Color Dictionary of Flowers & Plants

fl.   p 96, Pl. 764

Hermodactylus tuberosus

Macoby, Stirling
What Flower is That

fl.   p. 153

Hermodactylus tuberosus

Morley, Brian D.
Wild Flowers of the World

fl.,hab.   Pl. 42

Hermodactylus tuberosus

Perry, Frances
Flowers of the World

fl.   p 146

Hermodactylus tuberosus

Polunin, Oleg
Flowers of Europe

fl.   Pl. 174 #1683

Hermodactylus tuberosus

Polunin, Oleg
Flowers of the Mediterranean

fl.   No. 262

Herniaria glabra

Lindman, C. A. M.
Nordens Flora, Vol 3

fl. fr. hab.   Pl. 204

Herniaria glabra

Martin, W. Keble
The Concise British Flora in Colour

fl. hab.   Pl. 71

Herniaria glabra

Perrot, Emile
Les Plantes Medicinales
Vol. 1

fl., fr., hab.   p. 118

Heronbill

Erodium cicutarium

Herpetospermum pedunculosum

Curtis's Botanical Magazine
Vol 159   1936

fl.   No. 9463

Herpolirion novae-zelandiae

Cochrane, G. R.
Flowers and Plants of Victoria

fl.   Pl. 529

Herrea gydouwensis

Herre, H.
The Genera of the Mesembryanthemaceae

fl. fr.   p 171

Herreanthus meyeri

Herre, H.
The Genera of the Mesembryanthemaceae

fl. fr.   p 173

Herschelia charpentierana

American Hort. Soc.
American Horticulturist
Vol. 53, No. 3   1974

fl.   p. 35

Herschelia charpentierana

Flowering Plants of Africa
vol 42   1972-73

fl.   Pl. 1673

Herschelia charpentieriana

Morley, Brian D.
Wild Flowers of the World

fl.,hab.   Pl. 89

Herschelia graminifolia

Flowering Plants of Africa
    vol 30   1954-55

fl., hab.   Pl. 1172

Herschelia graminifolia

Morley, Brian D.
Wild Flowers of the World

fl.,hab.   Pl. 89

Herschelia hians

Flowering Plants of Africa
    vol 42   1972-73

fl., hab.   Pl. 1674

Hesperantha pauciflora

Flowering Plants of South Africa
    vol XVIII   1938

fl., hab.   Pl. 682

Hesperantha petitiana var.

Moriarty, Audrey
Wild Flowers of Malawi

fl.   Pl. 7; 4

Hesperis matronalis

Clark, Lewis J.
Wild Flowers of British Columbia

fl.   p 194

Hesperis matronalis

Courtenay, Booth
Wildflowers & Weeds

fl.   p 20

Hesperis matronalis

Lemmon, Robert S.
Wildflowers of North America in
    Full Color

fl.   p. 179   Pl. 282

Hesperis matronalis

Martin, W. Keble
The Concise British Flora in Colour

fl. hab.   Pl. 8

Hesperis matronalis

Moyle, John B.
Northland Wild Flowers

fl. hab.   p 61, Pl. 28

Hesperis matronalis

Polunin, Oleg
Flowers of Europe

fl.   Pl. 31 #296

Hesperis matronalis

Wharton, Mary E.
A Guide to the Wildflowers & Ferns
    of Kentucky

fl. p 124 Pl. 1.60

Hesperocallis undulata

Munz, Philip A.
California Desert Wildflowers

fl.   p 27, Pl. 1

Hesperocallis undulata

Orr, Robert T.
Wildflowers of Western America

fl.   Pl. 58

Hesperoyucca whipplei

Cactus & Succulent Society of America
Journal
    vol 7, No. 12   1936

fl.   p. 179

Hessea tenella

Flowering Plants of Africa
    vol 36   1963-64

fl., hab.   Pl. 1413

Hetaeria cristata

Walden, Beryl M.
Wild Flowers of Hong Kong

fl.   Pl. 61 (196)

Hetaeria nitida

Walden, Beryl M.
Wild Flowers of Hong Kong

fl. hab.   Pl. 25(67)

Heteranthera dubia

Stodola, Jiri
Encyclopedia of Water Plants

fl.   P. 262

Heteranthera limosa

Stodola, Jiri
Encyclopedia of Water Plants

fl.   P. 263

Heteranthera reniformis

Stodola, Jiri
Encyclopedia of Water Plants

fl.   P. 263

Heteranthera zosterifolia

Stodola, Jiri
Encyclopedia of Water Plants

hab.   P. 266

Heterocentron roseus

Harrison, Richmond E.
Climbers and Trailers

fl.   p 94   Pl. 235

Heteromeles arbutifolia

Coyle, Jeanette
A Field Guide to the Common and
    Interesting Plants of Baja California

fl.   p 83

Heteromeles arbutifolia

Flemer, William
Shade and Ornamental Trees in Color

hab.   p. 83

Heteromeles arbutifolia

Mathias, Mildred E.
Color for the Landscape

fr. hab.   p 185

Heterostemon ellipticus

Mee, Margaret
Flowers of the Brazilian Forests.

fl.   Pl. 6

Heterostemon ellipticus

Royal Hort. Soc.
Journal of the Royal Hort. Soc.
    vol 93, No. 7   1968

fl.   p. 290   Pl. 160

Heterotheca mariana

Dean, Blanche E.
Wildflowers of Alabama and
    Adjoining States

fl.   p 193

Heterotheca mariana

Duncan, Wilbur H.
Wildflowers of the Southeastern
    United States

fl.   p 199

Heterotheca pilosa

Dean, Blanche E.
Wildflowers of Alabama and
    Adjoining States

fl.   p 193

Heterotheca pinifolia

Duncan, Wilbur H.
Wildflowers of the Southeastern
    United States

fl.   p 201

Heterotheca scabrella

Fleming, Glenn
Wild Flowers of Florida

fl. hab.   p 47

Heterotheca subaxillaris

Batson, Wade T.
Wild Flowers in South Carolina

fl.   p 114

Heterotheca subaxillaris

Brown, Clair A.
Wildflowers of Louisiana and
   Adjoining States

fl.   p 208

Heterotheca subaxillaris

Duncan, Wilbur H.
Wildflowers of the Southeastern
   United States

fl.   p 201

Heterotheca subaxillaris

Lemmon, Robert S.
Wildflowers of North America in
   Full Color

fl.   p. 57   Pl. 95

Heterotheca subaxillaris

Weeds of the Southern United States

hab.   p. 13

Heuchera, Alpine

see

Heuchera glabra

Heuchera americana

Dean, Blanche E.
Wildflowers of Alabama and
   Adjoining States

fl. hab.   p 79

Heuchera americana

Wharton, Mary E.
A Guide to the Wildflowers & Ferns
   of Kentucky

fl.   p 166   Pl. 2.39

Heuchera x brizoides

Huxley, Anthony
Garden Perennials and Water Plants

fl.   Pl. 141

Heuchera cylindrica

Clark, Lewis J.
Wild Flowers of British Columbia

fl.   p 199

Heuchera glabra

Clark, Lewis J.
Wild Flowers of British Columbia

fl.   p 207

Heuchera glabra

Heller, Christine
Wild Flowers of Alaska

fl.   Pl. 179

Heuchera micrantha

Clark, Lewis J.
Wild Flowers of British Columbia

fl.   p 206

Heuchera ovalifolia

Porsild, A.E.
Rocky Mountain Wild Flowers

fl. hab.   p 209

Heuchera richardsonii

Courtenay, Booth
Wildflowers & Weeds

fl.   p 42

Heuchera richardsonii

Moyle, John B.
Northland Wild Flowers

fl. hab.   p 63, Pl. 31

Heuchera rubescens

Munz, Philip A.
California Mountain Wildflowers

fl.   p 31, Pl. 14

Heuchera sanguinea

Huxley, Anthony
Garden Perennials and Water Plants

fl.   Pl. 140

Heuchera sanguinea

Macoby, Stirling
What Flower is That

fl.   p. 153

Heuchera sanguinea var.

Hay, Roy
The Color Dictionary of Flowers &
   Plants

fl.   p. 147   Pl. 1172,1173

Heuchera sanguinea var.

Perry, Frances
Flowers of the World

fl.   p. 273

Heuchera sp.

Mathias, Mildred E.
Color for the Landscape

fl. hab.   p 144

Heuchera villosa

Duncan, Wilbur H.
**Wildflowers of the Southeastern
   United States**

fl.   p 61

Heuchera villosa

Wharton, Mary E.
A Guide to the Wildflowers & Ferns
   of Kentucky

fl.   p 167   Pl. 2.40

x Heucherella tiarelloides

Curtis's Botanical Magazine
Vol 165   1948

fl.   31

Hevea brasiliensis

Hvass, Elsie
Plants That Feed and Serve Us

fl.   p 106, Pl. 227

Hewittia sublobata

Moriarty, Audrey
Wild Flowers of Malawi

fl.   Pl. 59; 3

Hexadesmia Dunstervillei

Dunsterville, G. C. K.
Introduction to the World
   of Orchids

fl.   Pl. 12

Hexadesmia sessilis

Pabst, G. F. J.
Orchidaceae Brasilienses,
   Vol 1

fl.   p 225

Hexalectris grandiflora

Luer, Carlyle A.
The Native Orchids of the
   United States and Canada

fl. hab.   Pl. 75

Hexalectris nitida

Luer, Carlyle A.
The Native Orchids of the
   United States and Canada

fl. hab.   Pl. 77; 1-3

Hexalectris revoluta

Luer, Carlyle A.
The Native Orchids of the
   United States and Canada

fl. hab.   Pl. 77; 4-6

Hexalectris spicata

Brown, Clair A.
Wildflowers of Louisiana and
Adjoining States

fl.   p 39

Hexalectris spicata

Dean, Blanche E.
Wildflowers of Alabama and
Adjoining States

fl.   p 41

Hexalectris spicata

Luer, Carlyle A.
The Native Orchids of Florida

fl. hab.   Pl. 51

Hexalectris spicata

Luer, Carlyle A.
The Native Orchids of the
United States and Canada

fl. hab.   Pl. 74

Hexalectris spicata

Wharton, Mary E.
A Guide to the Wildflowers & Ferns
of Kentucky

fl.   p 92 Pl. 3,4

Hexalectris warnockii

Luer, Carlyle A.
The Native Orchids of the
United States and Canada

fl. hab.   Pl. 76

Hexalobus glabrescens

Flowering Plants of Africa
vol 30   1954-55

fl., fr.   Pl. 1195

Hexastylis arifolia

Batson, Wade T.
Wild Flowers in South Carolina

hab.   p 40

Hexastylis arifolia

Brown, Clair A.
Wildflowers of Louisiana and
Adjoining States

fl. hab.   p 46

Hexastylis arifolia

Dean, Blanche E.
Wildflowers of Alabama and
Adjoining States

fl. hab.   p 49

Hexastylis arifolia

Duncan, Wilbur H.
Wildflowers of the Southeastern
United States

fl. hab.   p 25

Hexastylis heterophylla

Duncan, Wilbur H.
Wildflowers of the Southeastern
United States

fl. hab.   p 25

Hexastylis minor

Dean, Blanche E.
Wildflowers of Alabama and
Adjoining States

fl.   p 51

Hexastylis shuttleworthii

Dean, Blanche E.
Wildflowers of Alabama and
Adjoining States

fl.   p 51

Hexastylis speciosa

Dean, Blanche E.
Wildflowers of Alabama and
Adjoining States

fl. hab.   p 49

Hexisea bidentata

Addisonia, v. 22, 1943-46

fl.   opp. p. 1   Pl. 705

Hexisea bidentata

Dunsterville, G. C. K.
Introduction to the World
of Orchids

fl.   Pl. 13

Hexisea bidentata

Ospina, Mariano
Orquideas de las americas

fl.   Pl. 61

Hexisea bidentata

Pabst, G. F. J.
Orchidaceae Brasilienses,
Vol 1

fl.   p 186

Hibbertia acerosa

Blombery, Alec M.
What Wildflower is That

fl.   p 167, Pl. 452

Hibbertia bracteata

Blombery, Alec M.
What Wildflower is That

fl.   p 167, Pl. 453

Hibbertia dentata

Menninger, Edwin A.
Flowering Vines of the World

fl.   Pl. 73

Hibbertia fasciculata

Cochrane, G. R.
Flowers and Plants of Victoria

fl.   Pl. 39

Hibbertia glomerosa

Blombery, Alec M.
What Wildflower is That

fl.   p 167, Pl. 454

Hibbertia hirsuta

Curtis, Winifred
The Endemic Flora of Tasmania
v. 3

fl.   p. 173

Hibbertia linearis var.

Cochrane, G. R.
Flowers and Plants of Victoria

fl.   Pl. 325

Hibbertia quadricolor

Morcombe, M. K.
Australia's Western Wildflowers

fl.   p 18

Hibbertia racemosa

Morcombe, M. K.
Australia's Western Wildflowers

fl.   p 107

Hibbertia scandens

Blombery, Alec M.
What Wildflower is That

fl.   p 168, Pl. 455

Hibbertia scandens

Macoby, Stirling
What Flower is That

fl.   p. 154

Hibbertia scandens

Mathias, Mildred E.
Color for the Landscape

fl. hab.   p 109

Hibbertia scandens

Morley, Brian D.
Wild Flowers of the World

fl.   Pl. 128

Hibbertia scandens

Perry, Frances
Flowers of the World

fl.   p 100

Hibbertia scandens

Wit, H. C. D. de
Plants of the World;
The Higher Plants, Vol I

fl.   p 224, Pl. 146

Hibbertia stellaris

Blombery, Alec M.
What Wildflower is That

fl. hab.   p 169, Pl. 456

Hibbertia stellaris

Harrison, Richmond E.
Climbers and Trailers

fl.   p 49   Pl. 101

Hibbertia stricta

Blombery, Alec M.
What Wildflower is That

fl.   p 169, Pl. 457

Hibbertia stricta

Cochrane, C. R.
Flowers and Plants of Victoria

fl.   Pl. 7

Hibiscus aculeatus

Brown, Clair A.
Wildflowers of Louisiana and
Adjoining States

fl.   p 109

Hibiscus aculeatus

Dean, Blanche E.
Wildflowers of Alabama and
Adjoining States

fl.   p 105

Hibiscus aculeatus

Duncan, Wilbur H.
Wildflowers of the Southeastern
United States

fl.   p 101

Hibiscus, Beach

see

Hibiscus tillaceus

Hibiscus, Blue

see

Hibiscus huegelii

Hibiscus burtt-daveyi

Royal Hort. Soc.
Journal of the Royal Hort. Soc.
vol 94, No. 1   1969

fl.   p. 22   Pl. 7

Hibiscus calyphyllus

Flowering Plants of Africa
vol 37   1965-66

fl.   Pl. 1470

Hibiscus cannabinus

Flowering Plants of Africa
vol 34   1960-61

fl., fr., hab.   Pl. 1335

Hibiscus, Chinese

see

Hibiscus rosa-sinensis

Hibiscus, Coral

see

Hibiscus schizopetalus

Hibiscus coccineus

Macoby, Stirling
What Flower is That

fl.   p. 154

Hibiscus elatus

Hargreaves, Dorothy
Tropical Blossoms of the Caribbean

fl.   p. 20

Hibiscus elatus

Oakman, Harry
Colorful Trees

Hab. Fl.   P 133

Hibiscus esculentus

Masefield, G. B.
The Oxford Book of Food Plants

fl. fr.   Pl. 163, 3

Hibiscus, Fringed

see

Hibiscus schizopetalus

Hibiscus, Fuschia

see

Hibiscus schizopetalus

Hibiscus fuscus

Moriarty, Audrey
Wild Flowers of Malawi

fl.   Pl. 35; 2

Hibiscus hamabo

Kitamura, Siro
Coloured Illustrations of Trees &
Shrubs of Japan

fl.   337

Hibiscus, Hawaiian Tree

see

Hibiscus tiliaceus

Hibiscus heterophyllus

Blombery, Alec M.
What Wildflower is That

fl.   p 170, Pl. 458

Hibiscus huegelii

Curtis's Botanical Magazine, v. 172
1958-59

fl.   Pl. 313

Hibiscus huegelii

Mathias, Mildred E.
Color for the Landscape

fl. hab.   p 70

Hibiscus huegelii

Morley, Brian D.
Wild Flowers of the World

fl.   Pl. 135

Hibiscus huegelii

Mullins, Barbara
Australian Wildflowers in Colour

fl.   P. 49   Pl. 43

Hibiscus hybrid

Bruggeman, L.
Tropical Plants

fl.   Pl. 242

Hibiscus hybrid

Harrison, R. E.
Trees and Shrubs

fl.   P. 89

Hibiscus hybrid

Macoby, Stirling
What Flower is That

fl.   P. 155

Hibiscus hybrid

Tsukamoto, Yotaro
Coloured Illustrations of Garden
Flowers, V. 10

fl.   Ill. 160   opp. p. 50

Hibiscus insularis

Morley, Brian D.
Wild Flowers of the World

fl.   Pl. 147

Hibiscus irritans

Flowering Plants of Africa
     vol 27   1948-49

fl.   Pl. 1050

Hibiscus lasiocarpos

Brown, Clair A.
Wildflowers of Louisiana and
     Adjoining States

fl. hab.   p 109

Hibiscus, lilac

     see

Hibiscus huegelii

Hibiscus manihot

Kariyone, Tatsuo
Atlas of Medicinal Plants

fl.   Pl. 85

Hibiscus manihot

Takatori, Jisuke
Color Atlas of Medicinal Plants of Japan

fl.   Fig. 28

Hibiscus meeusei

Flowering Plants of Africa
     vol 34   1960-61

fl., fr., hab.   Pl. 1359

Hibiscus militaris

Brown, Clair A.
Wildflowers of Louisiana and
     Adjoining States

fl. hab.   p 110

Hibiscus militaris

Courtenay, Booth
Wildflowers & Weeds

fl.   p 44

Hibiscus militaris

Duncan, Wilbur H.
Wildflowers of the Southeastern
     United States

fl.   p 99

Hibiscus militaris

Wharton, Mary E.
A Guide to the Wildflowers & Ferns
     of Kentucky

fl.   p 127   Pl. 1.65

Hibiscus moscheutos

Dean, Blanche E.
Wildflowers of Alabama and
     Adjoining States

fl.   p 105

Hibiscus moscheutos

Duncan, Wilbur H.
Wildflowers of the Southeastern
     United States

fl.   p 99

Hibiscus moscheutos

Everett, T. H.
New Illustrated Encyclopedia of
     Gardening, V. 5

fl.   opp. p. 871

Hibiscus moscheutos

Klimas, John E.
Wildflowers of Eastern America

fl.   Pl. 61

Hibiscus moscheutos

Massachusetts Hort. Soc.
Horticulture
     vol 32, No. 8   1954

fl.   cover

Hibiscus mutabilis

Bruggeman, L.
Tropical Plants

fl.   Pl. 243

Hibiscus mutabilis

Macoby, Stirling
What Flower is That

fl.   p. 154

Hibiscus mutabilis var.

Tsukamoto, Yotaro
Coloured Illustrations of Garden
     Flowers, V. 10

fl. Ill. 159   opp. p. 50

Hibiscus, Norfolk Island

     see

Lagunaria patersoni

Hibiscus palustris

Dean, Blanche E.
Wildflowers of Alabama and
     Adjoining States

fl.   p 105

Hibiscus palustris

Lemmon, Robert S.
Wildflowers of North America in
     Full Color

fl.   P. 31   Pl. 52

Hibiscus palustris

Massachusetts Hort. Soc.
Horticulture
     vol 40, No. 8   1962

fl.   cover

Hibiscus panduriformis

Blombery, Alec M.
What Wildflower is That

fl.   p 171, Pl. 459

Hibiscus pedunculatus

Curtis's Botanical Magazine, v.173
1960
fl.
pl. 369

Hibiscus pedunculatus

Flowering Plants of Africa
     vol 29   1952-53

fl.   Pl. 1157

Hibiscus, Pepper

     see

Malvaviscus arboreus

Hibiscus, Pineland

     see

Hibiscus aculeatus

Hibiscus rhodanthus

Moriarty, Audrey
Wild Flowers of Malawi

fl.   Pl. 35, 3

Hibiscus, Rhodesian Tree

     see

Azanza gardkeana

Hibiscus rosa-sinensis

Bruggeman, L.
Tropical Plants

fl.   Pl. 244

Hibiscus rosa-sinensis

Crockett, James Underwood
Evergreens

hab.,fl.   p. 128

Hibiscus rosa-sinensis

Encke, Fritz
Zimmerpflanzen

fl.   p. 79

Hibiscus rosa-sinensis

Hall, Clarence E.
Flowers of the Islands in the Sun

fl.   p. 75   Pl. 15

Hibiscus rosa-sinensis

Harrison, R. E.
Trees and Shrubs

fl.   P. 89

Hibiscus rosa-sinensis

Hay, Roy
The Color Dictionary of Flowers & Plants

fl.   p 69, Pl. 548

Hibiscus rosa-sinensis

Hay, Roy
The Dictionary of House Plants

fl.   Pl. 272

Hibiscus rosa-sinensis

Hersey, Jean
Woman's Day Book of House Plants
fl.
P. 70

Hibiscus rosa-sinensis

Kromdijk, G.
200 House Plants in Colour

fl.   Pl. 105

Hibiscus rosa-sinensis

Macoby, Stirling
What Flower is That

fl.   p. 154

Hibiscus rosa-sinensis

Massachusetts Hort. Soc.
Horticulture
    vol 40, No. 2   1962

fl.   cover

Hibiscus rosa-sinensis

Mathias, Mildred E.
Color for the Landscape

fl. hab.   p 70

Hibiscus rosa-sinensis

Nicolaisen, Age
Pocket Encyclopedia of Indoor Plants

fl.   Pl. 82

Hibiscus rosa-sinensis

Perry, Frances
Flowers of the World

fl.   p 183

Hibiscus rosa-sinensis

Wit, H. C. D. de
Plants of the World;
The Higher Plants, Vol I

fl.   p 220, Pl. 134

Hibiscus rosa-sinensis var.

Blunt, Wilfrid
Flora Superba

fl.   Pl. XI

Hibiscus rosa-sinensis var.

Everett, T. H.
New Illustrated Encyclopedia of
Gardening, v. 5

fl. opp. p. 822, 823

Hibiscus rosa-sinensis var.

Hall, Clarence E.
Flowers of the Islands in the Sun

fl.   p. 71, 79   Pl. 14, 16

Hibiscus rosa-sinensis var.

Hay, Roy
The Dictionary of House Plants

fl.   Pl. 273 - 275

Hibiscus rosa-sinensis var.

Kiaer, Eigil
Indoor Plants in Colour

fl.   p. 66

Hibiscus rosa-sinensis var.

Macoby, Stirling
What Flower is That

fl.   p. 155

Hibiscus rosa-sinensis var.

Tsukamoto, Yotaro
Coloured Illustrations of Garden
    Flowers   Vol. 10

fl.   opp. p. 50   Pl. 161,162

Hibiscus rostellatus

Menninger, Edwin A.
Flowering Vines of the World

fl.   Pl. 128

Hibiscus saintjohnianus

American Hort. Soc.
American Horticulturist
    vol 57, No. 4   1978

fl.   p. 10

Hibiscus saint johnianus

Pacific Tropical Botanical Garden
Bulletin
    vol 8, No. 1   1978

fl.   cover

Hibiscus schizopetalus

Hall, Clarence E.
Flowers of the Islands in the Sun

fl.   p. 67   Pl. 13

Hibiscus schizopetalus

Hargreaves, Dorothy
Tropical Blossoms of the Caribbean

fl.   p. 18

Hibiscus schizopetalus

Macoby, Stirling
What Flower is That

fl.   P 155

Hibiscus schizopetalus

Morley, Brian D.
Wild Flowers of the World

fl.   Pl. 57

Hibiscus schizopetalus

Perry, Frances
Flowers of the World

fl.   p 183

Hibiscus schizopetalus

Tsukamoto, Yotaro
Coloured Illustrations of Garden
    Flowers, v. 10

fl.   opp. p. 51   ill. 163

Hibiscus schizopetalus

Wit, H. C. D. de
Plants of the World;
The Higher Plants, Vol I

fl.   p 221, Pl. 137

Hibiscus sinosyriacus

Royal Hort. Soc.
Journal of the Royal Hort. Soc.
    vol 86, No. 3   1961

fl.   p. 118   Pl. 31

Hibiscus, skeleton

see

Hibiscus schizopetalus

Hibiscus, Sleeping

see

Malvaviscus arboreus

Hibiscus splendens

Blombery, Alec M.
What Wildflower is That

fl.   p 172, Pl. 460

Hibiscus syriacus

Hay, Roy
The Color Dictionary of Flowers &
  Plants

fl.   p. 206   Pl. 1643

Hibiscus syriacus

Huxley, Anthony
Deciduous Garden Trees and Shrubs

fl.   Pl. 88

Hibiscus syriacus

Macoby, Stirling
What Flower is That

fl.   P 155

Hibiscus syriacus

Polunin, Oleg
Trees and Bushes of Europe

fl.   p 142

Hibiscus syriacus var.

Bartels, Andreas
Das Grosse Buch der Gartengeholze

fl.   p 157

Hibiscus syriacus var.

Everett, T.H.
New Illustrated Encyclopedia
  of Gardening, v. 5

fl.   opp. p. 822

Hibiscus syriacus var.

Gault, S. Millar
The Color Dictionary of Shrubs

fl.   Pl. 231, 232, 233, 234, 235,
                236

Hibiscus syriacus var.

Harrison, R.E.
Trees and Shrubs

fl. P 89

Hibiscus syriacus var.

Hay, Roy
The Color Dictionary of Flowers &
  Plants

fl.   p. 206   Pl. 1644-46

Hibiscus syriacus var.

Hellyer, A.G.L.
Shrubs in Colour

fl.   p. 63

Hibiscus syriacus var.

Perry, Frances
Flowers of the World

fl.   p. 183

Hibiscus syriacus var.

Tsukamoto, Yotaro
Coloured Illustrations of Garden
  Flowers Vol. 10

fl.   p. 51   Pl. 164,165

Hibiscus tiliaceus

Macoby, Stirling
WhatFlower is That

fl.   P 155

Hibiscus tiliaceus

Oakman, Harry
Colorful Trees

Fl.   P 80

Hibiscus tiliaceus

Pacific Tropical Botanical Garden
Bulletin
    vol 4, No. 3   1974

fl.   cover

Hibiscus tiliaceus

Walden, Beryl M.
Wild Flowers of Hong Kong

fl. fr.   Pl. 65

Hibiscus, Tree

  see

Hibiscus elatus

Hibiscus trionum

Courtenay, Booth
Wildflowers & Weeds

fl.   p 44

Hibiscus trionum

Crockett, James Underwood
Annuals

fl.   P 124

Hibiscus trionum

Klimas, John E.
Wildflowers of Eastern America

fl.   Pl. 167

Hibiscus trionum

Polunin, Oleg
Flowers of Europe

fl. Pl. 73 #752

Hibiscus venustus

Bruggeman, L.
Tropical Plants

fl.   Pl. 245

Hibiscus vitifolius var.

Royal Hort. Soc.
Journal of the Royal Hort. Soc.
    vol 94, No. 1   1969

fl.   p. 22   Pl. 7

Hibiscus waimeae

Hay, Roy
The Dictionary of House Plants

fl.   Pl. 276

Hickenia microsperma

Van Laren, A. J.
Cactus

fl.   P 62   Fig. 74

Hickory, Shag-bark

  see

Carya ovata

Hieracium albiflorum

Clark, Lewis J
Wild Flowers of British Columbia

fl.   p. 554

Hieracium alpicola

Barneby, T. P.
European Alpine Flowers in Colour

fl.   Pl. 96, 5

Hieracium anglicum

Martin, W. Keble
The Concise British Flora in Colour

fl. hab.   Pl. 51

Hieracium aurantiacum

Ary, S.
The Oxford Book of Wildflowers

fl.   p 104, Pl. 5

Hieracium aurantiacum

Barneby, T. P.
European Alpine Flowers in Colour

fl.   Pl. 96, 2

Hieracium aurantiacum

Clark, Lewis J.
Wild Flowers of British Columbia

fl.   p 554

Hieracium aurantiacum

Courtenay, Booth
Wildflowers & Weeds

fl.   p 105

Hieracium aurantiacum

Crockett, James Underwood
Lawns & Ground Covers

fl.   p. 72

Hieracium aurantiacum

Felsko, Elsa
A Book of Wildflowers

fl.   P 98

Hieracium aurantiacum

Ferguson, Mary
Wildflowers

fl.   p 99

Hieracium aurantiacum

Jennings, O. E.
Wild Flowers of Western Pennsylvania
v. 2
fl.   Pl. 199

Hieracium aurantiacum

Klimas, John E.
Wildflowers of Eastern America

fl.   Pl. 133

Hieracium aurantiacum

Kohlhaupt, Paula
Fleurs des Alpages, v.2

Pl.   P 116

Hieracium aurantiacum

Lemmon, Robert S.
Wildflowers of North America in
Full Color

fl.   p. 217   Pl. 344

Hieracium aurantiacum

Lindman, C. A. M.
Nordens Flora, Vol 10

fl. hab.   Pl. 663

Hieracium aurantiacum

Moyle, John B.
Northland Wild Flowers

fl.   p 197, Pl. 257

Hieracium britannicum

Martin, W. Keble
The Concise British Flora in Colour

fl. fr. hab.   Pl. 51

Hieracium brunneocroceum

Morley, Brian D.
Wild Flowers of the World

fl.,hab.   Pl. 22

Hieracium canadense

Courtenay, Booth
Wildflowers & Weeds

fl.   p 105

Hieracium canadense

Moyle, John B.
Northland Wild Flowers

fl.   p 197, Pl. 256

Hieracium echioides

Barneby, T. P.
European Alpine Flowers in Colour

fl.   Pl. 96, 4

Hieracium exotericum var.

Martin, W. Keble
The Concise British Flora in Colour

fl. hab.   Pl. 51

Hieracium gracile

Porsild, A. E.
Rocky Mountain Wild Flowers

fl. hab.   p 419

Hieracium holosericeum

Martin, W. Keble
The Concise British Flora in Colour

fl. hab.   Pl. 51

Hieracium intybaceum

Barneby, T. P.
European Alpine Flowers in Colour

fl. hab.   Pl. 96, 3

Hieracium intybaceum

Felsko, Elsa
A Book of Wild Flowers
2nd Ser.
fl.   p 34

Hieracium lachenalii

Martin, W. Keble
The Concise British Flora in Colour

fl. hab.   Pl. 52

Hieracium latobrigorum

Martin, W. Keble
The Concise British Flora in Colour

fl. hab.   Pl. 52

Hieracium lingulatum

Martin, W. Keble
The Concise British Flora in Colour

fl. hab.   Pl. 51

Hieracium longipilum

Courtenay, Booth
Wildflowers & Weeds

fl.   p 105

Hieracium maculatum

Martin, W. Keble
The Concise British Flora in Colour

fl. hab.   Pl. 52

Hieracium, orange

see

Hieracium aurantiacum

Hieracium peleterianum

Martin, W. Keble
The Concise British Flora in Colour

fl. hab.   Pl. 52

Hieracium perpropinquum

Martin, W. Keble
The Concise British Flora in Colour

fl. hab.   Pl. 52

Hieracium pilosella

Barneby, T. P.
European Alpine Flowers in Colour

fl. hab.   Pl. 96, 6

Hieracium pilosella

Felsko, Elsa
A Book of Wildflowers

fl.   P 22

Hieracium pilosella

Lemmon, Robert S.
Wildflowers of North America in
Full Color

fl.   p. 51   Pl. 84

Hieracium pilosella

Lindman, C. A. M.
Nordens Flora, Vol 10

fl. fr. hab.   Pl. 662

Hieracium pilosella

Martin, W. Keble
The Concise British Flora in Colour

fl. hab.   Pl. 51

Hieracium pilosella

Morley, Brian D.
Wild Flowers of the World

fl.,hab.   Pl. 22

Hieracium pilosella

Polunin, Oleg
Flowers of Europe

fl.   Pl. 160 #1550

Hieracium pratense

Campbell, Carlos C.
Great Smoky Mountain Wildflowers

fl.   p 59

Hieracium pratense

Jennings, O. E.
Wild Flowers of Western Pennsylvania
                                   vol II
fl.   Pl. 199

Hieracium pratense

Klimas, John E.
Wildflowers of Eastern America

fl.   Pl. 174

Hieracium schmidtii

Martin, W. Keble
The Concise British Flora in Colour

fl. hab.   Pl. 51

Hieracium triangulare

Lindman, C. A. M.
Nordens Flora, Vol 10

fl. fr. hab.   Pl. 661

Hieracium trichocaulon

Martin, W. Keble
The Concise British Flora in Colour

fl. hab.   Pl. 52

Hieracium triste

Heller, Christine
Wild Flowers of Alaska

fl.   Pl. 48

Hieracium umbellatum

Lindman, C. A. M.
Nordens Flora, Vol 10

fl. fr.   Pl. 660

Hieracium umbellatum

Martin, W. Keble
The Concise British Flora in Colour

fl. hab.   Pl. 52

Hieracium venosum

Fleming, Glenn
Wild Flowers of Florida

fl.   p 48

Hieracium venosum

Jennings, O. E.
Wild Flowers of Western Pennsylvania

fl. hab.   Pl. 200

Hieracium venosum

Wharton, Mary E.
A Guide to the Wildflowers & Ferns
    of Kentucky

fl. p 242 Pl. 2,4

Hieracium villosum

Barneby, T. P.
European Alpine Flowers in Colour

fl.   Pl. 96, 1

Hieracium villosum

Hay, Roy
The Color Dictionary of Flowers & Plants

fl.   p 147, Pl. 1174

Hieracium vulgatum

Martin, W. Keble
The Concise British Flora in Colour

fl. hab.   Pl. 51

Hierochloe odorata

Lindman, C. A. M.
Nordens Flora, Vol 2

fr. hab.   Pl. 103

Hierochloe odorata

Porsild, A. E.
Rocky Mountain Wild Flowers

fr.   p. 55

Hilaria rigida

Coyle, Jeanette
A Field Guide to the Common and
Interesting Plants of Baja California

fl.   p 47

Himantoglossum hircinum

Barneby, T. P.
European Alpine Flowers in Colour

fl.   Pl. 10, 2

Himantoglossum hircinum

Brooke, Jocelyn
The Wild Orchids of Britain

fl.   Pl. 22

Himantoglossum hircinum

Martin, W. Keble
The Concise British Flora in Colour

fl.   Pl. 81

Himantoglossum hircinum

Polunin, Oleg
Flowers of Europe

fl.   Pl. 190 #1907

Himantoglossum hircinum

Royal Hort. Soc.
The Garden
       vol 102, No. 6 1977

fl.   p. 262

Himantoglossum hircinum var.

Couliais, Constantine N.
Wild Flowers of Greece

fl. hab. p 191

Himantoglossum longibracteatum

Polunin, Oleg
Flowers of Europe

fl.   Pl. 190 #1906

Himantoglossum longibracteatum

Polunin, Oleg
Flowers of the Mediterranean

fl.   No. 291

Hintonella mexicana

Ospina, Mariano
Orquideas de las americas

fl.   Pl. 131

Hippeastrum aulicum

Hay, Roy
The Color Dictionary of Flowers & Plants

fl.   p 69, Pl. 549

Hippeastrum aulicum

Hay, Roy
The Dictionary of House Plants

fl.   Pl. 280

Hippeastrum blumenavia

Curtis's Botanical Magazine
    Vol 160   1937

fl.   No. 9504

Hippeastrum calyptratum

Mee, Margaret
Flowers of the Brazilian Forests

fl., hab. Pl. 29

Hippeastrum equestre

Hargreaves, Dorothy
Tropical Blossoms of the Caribbean

fl.   p. 5

Hippeastrum hortorum

Nicolaisen, Age
Pocket Encyclopedia of
Indoor Plants

fl.   Pl. 30

Hippeastrum hybrid

Encke, Fritz
Zimmerpflanzen

fl.   p. 79

Hippeastrum hybrid

Hay, Roy
The Color Dictionary of Flowers & Plants

fl.   p 69, Pl.   551

Hippeastrum hybrid

Hay, Roy
The Dictionary of House Plants

fl.   Pl. 277-79

Hippeastrum hybrid

Kroadijk, G.
200 House Plants in Colour

fl.   Pl. 106

Hippeastrum hybrid

Macoby, Stirling
What Flower is That

fl.   P 155

Hippeastrum hybrid

Massachusetts Hort. Soc.
Horticulture
vol 56, No. 12   1978

fl.   p. 22

Hippeastrum hybrid

Miles, Bebe
Bulbs for the Home Gardener

fl.   p 170

Hippeastrum hybrid

Perry, Frances
Flowers of the World

fl.   p. 24

Hippeastrum hybrid

Tsukamoto, Yotaro
Coloured Illustrations of Garden
Flowers V. 9

fl.   opp. p. 2

Hippeastrum x hybridum

Bruggeman, L.
Tropical Plants

fl.   Pl. 119

Hippeastrum igneum

Munoz Pizarro, Carlos
Flores Silvestres de Chile

fl., hab.   Pl. 13

Hippeastrum x johnsonii

Curtis's Botanical Magazine
vol 167   1950

fl.   Pl. 122

Hippeastrum x johnsonii

Hay, Roy
The Color Dictionary of Flowers &
Plants

fl.   p. 69   Pl. 550

Hippeastrum pardinum

Pacific Hort. Foundation
Pacific Horticulture
vol 38, No. 3   1977

fl.   p. 36

Hippeastrum sp.

Hersey, Jean
Woman's Day Book of House Plants

fl.   p 29

Hippeastrum vittatum

Kiaer, Eigil
Indoor Plants in Colour

fl.   P. 67

Hippobroma longiflora

Morley, Brian D.
Wild Flowers of the World

fl. fr.   Pl. 172

Hippocrepis comosa

Ary, S.
The Oxford Book of Wildflowers

fl.   p 22, Pl. 2

Hippocrepis comosa

Huxley, Anthony
Garden Perennials and Water Plants

fl.   Pl. 139

Hippocrepis comosa

Martin, W. Keble
The Concise British Flora in Colour

fl. fr. hab.   Pl. 24

Hippocrepis comosa

Robert, Paul A.
Alpine Flowers

fl.   Pl. 2

Hippocrepis comosa

Tosco, Uberto
The World of Mountain Flowers

fl. hab.   p 70

Hippophae rhamnoides

Ary, S.
The Oxford Book of Wildflowers

fl. fr.   p 182, Pl. 4

Hippophae rhamnoides

Bartels, Andreas
Das Grosse Buch der Gartengeholze

fr.   p 157

Hippophae rhamnoides

Hansen, Richard
Baume und Straucher im Garten

fr.   p. 100

Hippophae rhamnoides

Bianchini, Francesco
Health Plants of the World

fr.   p. 49

Hippophae rhamnoides

Hay, Roy
The Color Dictionary of Flowers & Plants

fr.   p 206, Pl. 1647

Hippophae rhamnoides

Huxley, Anthony
Deciduous Garden Trees and Shrubs

fr.   Pl. 89

Hippophae rhamnoides

Lindman, C. A. M.
Nordens Flora, Vol 6

fl. fr.   Pl. 401

Hippophae rhamnoides

Martin, W. Keble
The Concise British Flora in Colour

fr. hab.   Pl. 75

Hippophae rhamnoides

Perry, Frances
Flowers of the World

fr.   p 104

Hippophae rhamnoides

Polunin, Oleg
Flowers of Europe

fr.   Pl. 74 #761

Hippophae rhamnoides

Polunin, Oleg
Trees and Bushes of Europe

fr. hab.   p 142, 143

Hippophae rhamnoides

Royal Hort. Soc.
Journal of the Royal Hort. Soc.
vol 98, No. 2   1973

fr. hab. p. 56   Pl. 36

Hippophae rhamnoides

Vedel, H.
Arbres et Arbustes

fl. fr.   p 101, No. 106

Hippuris vulgaris

Ary, S.
The Oxford Book of Wildflowers

fl.   p 52, Pl. 4

Hippurus vulgaris

Heller, Christine
Wild Flowers of Alaska

hab.   Pl. 294

Hippuris vulgaris

Lindman, C. A. M.
Nordens Flora, Vol. 6

fl. hab.   Pl. 410

Hippuris vulgaris

Martin, W. Keble
The Concise British Flora in Colour

fr. hab.   Pl. 34

Hippuris vulgaris

Stodola, Jiri
Encyclopedia of Water Plants

hab.   P. 270

Hippuris vulgaris

Webster, Mary
Flora of Moray, Nairn & East
        Inverness

hab.   p. 164   Pl. 10

Hiptage benghalensis

Walden, Beryl M.
Wild Flowers of Hong Kong

fl.   Pl. 16(46)

Hirpicium angustifolium

Curtis's Botanical Magazine
        Vol. 180   1974

fl. hab.   Pl. 676

Hissing Tree

        see

Parinari mobola

Hobblebush

        see

Viburnum alnifolium

Hobble, Dog

        see

Leucothoe axillaris

Hobble, Dog

        see

Leucothoe editorum

Hobble, Witch

        see

Viburnum alnifolium

Hodgsonia heteroclita

Morley, Brian D.
Wild Flowers of the World

fl.fr.   Plate 100

Hoehneella gehrtiana

Pabst, G. F. J.
Orchidaceae Brasiliensos,
        Vol 2

fl.   p 231

Hoffmanseggia drepanocarpa

Orr, Rober T.
Wildflowers of Western America

fl., fr.   Pl. 106

Hoffmanseggia microphylla

Coyle, Jeanette
A Field Guide to the Common and
        Interesting Plants of Baja California

fl. fr.   p 95

Hogweed

        see

Heracleum sphondyllium

Hohenbergia horrida

Smith, Lyman B.
The Bromeliads

fl.   Pl. 22

Hohenbergia stellata

Bromeliad Society
Bulletin
        vol 6, No. 2   1956

fl.   p. 53

Hohenbergia stellata

Padilla, Victoria
Bromeliads

fl.   p. 67   Pl. 5

Hohenbergia stellata

Padilla, Victoria, ed.
Bromeliads in Color and
        Their Culture

fl.   p 78

Hohenbergia stellata

Rauh, Werner
Bromeliads for Home, Garden and
        Greenhouse

fl.   Pl. 109

Hoheria lyallii

Morley, Brian D.
Wild Flowers of the World

fl.   Pl. 135

Hoheria populnea

Harvey, Norman B.
New Zealand Botanical Paintings

fl.   p. 38   Pl. 16

Hoheria populnea var.

Harrison, R. E.
Trees and Shrubs

hab.   P. 89

Hoheria sexstylosa

Hay, Roy
The Color Dictionary of Flowers & Plants

fl.   p 206, Pl. 1648

Hoheria sexstylosa

Perry, Frances
Flowers of the World

fl.   p 184

Holarrhena febrifuga

Palgrave, K. C.
Trees of Central Africa

fl., fr.   p. 27

Holarrhena pubescens

Flowering Plants of Africa
        vol 44   1977

fl.   Pl. 1758

Holboellia coriacea

Curtis's Botanical Magazine, v.175,
1964-65.
fl.
pl. 447.

Holboellia coriacea

Royal Hort. Soc.
The Garden
    vol 101, No. 10   1976

fr.   p. 524

Holcus lanatus

Lindman, C. A. M.
Nordens Flora, Vol 2

fr. hab.   Pl. 95

Holcus lanatus

Polunin, Oleg
Flowers of Europe

fr.   Pl. 181 #1770

Holcus mollis var.

Perry, Frances
Flowers of the World

fr.   p. 131

Holly

see

Ilex

Holly, American

see

Ilex opaca

Holly, Chinese

see

Osmanthus ilicifolius var.

Holly, Christmas

see

Heteromeles arbutifolia

Holly, English

see

Ilex aquifolium

Holly, False

see

Osmanthus ilicifolius var.

Hollygrape, Beal's

see

Mahonia bealei

Holly grape, Chinese

see

Mahonia lomariifolia

Hollygrape, Creeping

see

Berberis repens

Hollygrape, Leatherleaf

see

Mahonia bealei

Holly grape, Oregon

see

Mahonia aquifolium

Hollyhock

see

Althea rosea

Hollyhock, Australian

see

Lavatera plebeja

Hollyhock, Desert

see

Sphaeralcea ambigua

Hollyhock, mountain

see

Iliamna rivularis

Holly, Japanese

see

Ilex crenata

Holly, native

see

Oxylobrium ilicifolium

Holly, Rhodesian

see

Psorospermum febrifugum var.

Holly, Sea

see

Eryngium amethystinum

Holly, Sea

see

Eryngium maritimum

Holly, sea

see

Eryngium planum

Holly, Sea Alpine

see

Eryngium alpinum

Holly, Singapore

see

Malpighia coccigera

Holly, Swamp

see

Ilex decidua

Holly, variegated

see

Ilex aquifolium var.

Holly, Winterberry

see

Ilex verticillata

Holly, yellow

see

Ilex aquifolium var.

Holmskioldia sanguinea

Bruggeman, L.
Tropical Plants

fl.   Pl. 246

Holmskioldia sanguinea

Hall, Clarence E.
Flowers of the Islands in the Sun

fl.   p. 43   Pl. 7

Holmskioldia sanguinea

Harrison, Richmond E.
Climbers and Trailers

fl.   p 49   Pl. 102

Holmskioldia sanguinea

Macoby, Stirling
What Flower is That

fl.   P 156

Holmskioldia tettensis

Morley, Brian D.
Wild Flowers of the World

fl.   Pl. 80

Holodiscus discolor

Clark, Lewis J.
Wild Flowers of British Columbia

fl. hab.   p 243, 246

Holodiscus discolor

Munz, Philip A.
California Spring Wildflowers

fl.   p 27, Pl. 1

Holodiscus discolor

Perry, Frances
Flowers of the World

fl.   p 261

Holodiscus discolor

Taylor, Ronald J.
Mountain Wild Flowers

fl.   p 35

Holodiscus discolor var.

Hansen, Richard
Baume und Straucher im Garten

fl.   p. 154

Holosetum umbellatum

Martin, W. Keble
The Concise British Flora in Colour

fl. fr. hab.   Pl. 14

Holothrix longiflora

Williamson, Graham
The Orchids of South Central Africa

fl.   Pl. 23

Holothrix orthoceras

Flowering Plants of Africa
vol 37   1965-66

fl., hab.   Pl. 1469

Holothrix papillosa

Williamson, Graham
The Orchids of South Central Africa

fl.   Pl. 22

Holothrix puberula

Williamson, Graham
The Orchids of South Central Africa

fl.   Pl. 24

Holubia saccata

Flowering Plants of South Africa
vol XVII   1937

fl., fr.   Pl. 661

Holy Tree

see

Melia azedarach

Homalocephala texensis

Backeberg, Curt
Cactus Lexicon

fr.   p 639

Homalocephala texensis

Lamb, Edgar
Colorful Cacti of the American Deserts

fl. hab.   Pl. 2, 3

Homalocephala texensis

Lamb, Edgar
The Pocket Encyclopedia of Cacti
   and Succulents in Color

fl. hab.   Pl. 71

Homalocephala texensis

Van Laren, A. J.
Cactus

fl.   P 56   Fig. 63

Homeria collina

Curtis's Botanical Magazine
Vol 160   1937

fl.   No. 9487

Homeria miniata

Eliovson, Sima
Namaqualand in Flower

fl.   Pl. 33, 1

Homeria pillansii

Flowering Plants of South Africa
vol XVIII   1938

fl.   Pl. 684

x Homoglad hybrid

Royal Hort. Soc.
Journal of the Royal Hort. Soc.
   vol 92, No. 7   1967

fl.   p. 296   Pl. 164

Homoglossum merianella

Curtis's Botanical Magazine
Vol 160   1937

fl.   No. 9510

Homogyne alpina

Barneby, T. P.
European Alpine Flowers in Colour

fl.   Pl. 86, 6

Homogyne alpina

Morley, Brian D.
Wild Flowers of the World

fl.,hab.   Pl. 22

Homogyne alpina

Polunin, Oleg
Flowers of Europe

fl.   Pl. 150 #1443

Honewort

see

Cryptotaenia canadensis

Honesty

see

Lunaria annua

Honesty

see

Lunaria biennis

Honey Cup

see

Zenobia pulverulenta

Honey flower

see

Lambertia formosa

Honey flower, Cape

see

Protea

Honeyplant, beautiful

see

Hoya bella

Honeysuckle

see

Aquilegia canadensis

Honeysuckle

see

Lambertia formosa

Honeysuckle

see

Lonicera

Honeysuckle, Alpine

see

Lonicera alpigena

Honeysuckle, Australian

see

Banksia sp.

Honeysuckle, Bearberry

see

Lonicera involucrata

Honeysuckle, Black

see

Lonicera nigra

Honeysuckle, Blue

see

Lonicera coerulea

Honeysuckle, Burmese

see

Lonicera hildebrandiana

Honeysuckle, Bush

see

Diervilla sessilifolia

Honeysuckle, bush

see

Lonicera tatarica

Honeysuckle, California

see

Lonicera ledebourii

Honeysuckle, Cape

see

Protea mellifera

Honeysuckle, Cape

see

Tecomaria capensis

Honeysuckle, Coral

see

Lonicera sempervirens

Honeysuckle, Douglas

see

Lonicera glaucescens

Honeysuckle, Fly

see

Lonicera canadensis

Honeysuckle, Fly

see

Lonicera xylosteum

Honeysuckle, giant

see

Lonicera hildebrandiana

Honeysuckle, hairpin

see

Banksia collina

Honeysuckle, Himalaya

see

Leycesteria formosa

Honeysuckle, Japanese

see

Lonicera japonica

Honeysuckle, Kaffir

see

Tecomaria capensis

Honeysuckle, late Dutch

see

Lonicera periclymenum var.

Honeysuckle, Limber

see

Lonicera dioica var.

Honeysuckle, Perfoliate

see

Lonicera caprifolium

Honeysuckle, Pink

see

Rhododendron nudiflorum

Honeysuckle, Red

see

Banksia serrata

Honeysuckle, Trumpet

see

Lonicera sempervirens

Honeysuckle, Utah

see

Lonicera utahensis

Honeysuckle, White

see

Banksia integrifolia

Honeysuckle, winter

See

Lonicera fragrantissima

Honeywort

see

Cerinthe aspera

Honkenya ficifolia

Bruggeman, L.
Tropical Plants

fl.   Pl. 247

Honkenya peploides

Ary, S.
The Oxford Book of Wildflowers

fl.   p 76, Pl. 2

Honkenya peploides

Heller, Christine
Wild Flowers of Alaska

fl.   Pl. 158

Honkenya peploides

Martin, W. Keble
The Concise British Flora in Colour

fl. hab.   Pl. 16

Honkenya peploides

Polunin, Oleg
Flowers of Europe

fl.   Pl. 11 #133

Honkenya peploides var.

Clark, Lewis J.
Wild Flowers of British Columbia

fl.   p 139

Hoodia bainii

Lamb, Edgar
The Pocket Encyclopedia of Cacti
and Succulents in Color

fl. hab.   Pl. 269

Hoodia bainii

Morley, Brian D.
Wild Flowers of the World

fl.,hab.   Pl. 78

Hoodia currorii

Flowering Plants of Africa
vol 40   1969-70

fl.   Pl. 1568

Hoodia currorii var.

Flowering Plants of Africa
vol 37   1965-66

fl.   Pl. 1474

Hoodia dregei

Flowering Plants of South Africa
vol XVII   1937

fl.   Pl. 670

Hoodia gordonii

Lamb, Edgar
The Pocket Encyclopedia of Cacti
and Succulents in Color

fl. hab.   Pl. 323

Hoodia lugardi

Flowering Plants of South Africa
vol XVI   1936

fl.   Pl. 617

Hoodia lugardi

Flowering Plants of Africa
vol XXV   1945-46

fl., hab.   Pl. 977

Hoodia rosea

Flowering Plants of South Africa
vol XVI   1936

fl.   Pl. 615

Hook Thorn Tree

see

Acacia polyacantha var.

Hookera pauciflora

Walcott, Mary Vaux
North American Wild Flowers
vol. 5

fl.   Pl. 389

Hooks, Nine

see

Alchemilla vulgaris

Hoop petticoats

see

Narcissus bulbicodium

Hop

see

Humulus

Hop Bush

see

Dodonaea pinnata

Hop Bush

see

Dodonaea triquetra

Hopbush

see

Dodonaea viscosa

Hop, Bush

see

Hymenoclea salsola

Hop, Golden common

see

Humulus lupulus var.

Hop, native

see

Dodonaea ericifolia

Hop Tree

see

Ptelea trifoliata

Hop vine, Common

see

Humulus lupulus

Hop vine, Japanese

see

Humulus scandens

Hopea odorata

Roxburgh, William
Icones Roxburghianae

fl., fr.   Pl. 24

Hordeum distichum

Hvass, Elsie
Plants That Feed and Serve Us

fr.   p 5, Pl. 2a

Hordeum distichum

Masefield, G. B.
The Oxford Book of Food Plants

fr.   Pl. 5, 3

Hordeum jubatum

Crockett, James Underwood
Annuals

fr.   P 125

Hordeum jubatum

Hay, Roy
The Color Dictionary of Flowers &
    Plants

fr.   p. 39   Pl. 308

Hordeum jubatum

Huxley, Anthony
Garden Annuals and Bulbs

fr.   Pl. 62

Hordeum jubatum

Royal Hort. Soc.
The Garden
    vol 101, No. 9   1976

hab., fr.   p 448

Hordeum murinum

Polunin, Oleg
Flowers of Europe

fr. Pl. 178 #1735

Hordeum pusillum

Weeds of the Southern United States

hab., fr.   p. 25

Hordeum vulgare

Hvass, Elsie
Plants That Feed and Serve Us

fr.   p 5, Pl. 2

Hordeum vulgare

Masefield, G. B.
The Oxford Book of Food Plants

fr.   Pl. 5, 4

Hordeum vulgare

Perrot, Emile
Les Plantes Medicinales

fr. hab.   p 56

Hordeum vulgare var.

Bianchini, F.
The Complete Book of Fruits & Vegetables

fr.   p 21

Horehound, False
  see
Eupatorium rotundifolium

Horehound, Water
  see
Lycopus uniflorus

Horehound, White
  see
Marrubium vulgare

Horizontal
  see
Anodopetalum biglandulosum

Horminum pyrenaicum

Barneby, T. P.
European Alpine Flowers in Color

fl. 6   Pl. 69, 1

Horminum pyrenaicum

Morley, Brian D.
Wild Flowers of the World

fl.,fr.   Pl. 37

Horminum pyrenaicum

Polunin, Oleg
Flowers of Europe

fl. Pl. 114 #1151

Hormuzakia aggregata

Feinbrun-Dothan, Naomi
Wild Plants in the Land
   of Israel

fl. hab.   p 130

Hornbeam
  see
Carpinus

Hornbeam, Common
  see
Carpinus betulus

Hornbeam, European Hop
  see
Ostrya carpinifolia

Hornbeam, Japanese
  see
Carpinus japonica

Horn, Huntsman's
  see
Sarracenia flava

Horn, Little
  see
Diascia thunbergiana

Horn-wort
  see
Ceratophyllum demersum

Horn, Yellow
  see
Xanthoceras sorbifolium

Hornungia petraea

Martin, W. Keble
The Concise British Flora in Colour

fl. fr. hab.   Pl. 10

Horridocactus tuberisulcatus

Backeberg, Curt
Cactus Lexicon

fl.   p 639, 640

Horridocactus tuberisulcatus var.

Backeberg, Curt
Cactus Lexicon

fl.   p. 639

Horse Crippler
  see
Homalocephala texensis

Horseradish
  see
Armoracia lapathifolia

Horseradish
  see
Armoracia rusticana

Horseradish

see

Cochlearia armoracia

Horseradish

see

Nasturtium armoracia

Horsetail tree

see

Casuarina stricta

Horseweed

see

Conyza canadensis

Horseweed

see

Erigeron canadensis

Horsfieldia iryaghedi

Bruggeman, L.
Tropical Plants

fl.   Pl. 210

Hortensia

see

Hydrangea macrophylla

Hosta crispula

Hay, Roy
The Color Dictionary of Flowers &
Plants

hab.   p. 147   Pl. 1175

Hosta crispula

Royal Hort. Soc.
Journal of the Royal Hort. Soc.
vol 93, No. 9   1968

hab.   p. 382   Pl. 187

Hosta fortunei

Huxley, Anthony
Garden Perennials and Water Plants

fl.   Pl. 143

Hosta fortunei var.

Hay, Roy
The Color Dictionary of Flowers & Plants

hab.   p 147, Pl. 1176

Hosta fortunei var.

Macoby, Stirling
What Flower is That

hab.   P. 156

Hosta fortunei var.

Royal Hort. Soc.
Journal of the Royal Hort. Soc.
vol 92, No. 4   1967

fl., hab.   p. 160   Pl. 65

Hosta lancifolia

Meikle, R. D.
Garden Flowers

fl., hab.   p. 417   Pl. 15

Hosta lancifolia var.

American Hort. Soc.
American Horticulturist
vol 56, No. 4   1977

hab.   p. 6

Hosta lancifolia var.

Huxley, Anthony
Garden Perennials and Water Plants

fl.   Pl. 142

Hosta plantaginea

American Hort. Soc.
American Horticulturist
vol 56, No. 4   1977

fl., hab.   p. 7

Hosta plantaginea

Macoby, Stirling
What Flower is That

hab.   p. 156

Hosta plantaginea var.

Hay, Roy
The Color Dictionary of Flowers & Plants

fl.   p 148, Pl. 1177

Hosta plantaginea var.

Royal Hort. Soc.
Journal of the Royal Hort. Soc.
vol 98, No. 7   1973

fl.   P. 306   Pl. 163

Hosta rectifolia

Curtis's Botanical Magazine
Vol 168   1951

fl.   138

Hosta sieboldiana

Hay, Roy
The Color Dictionary of Flowers & Plants

hab.   p 148, Pl. 1178

Hosta sieboldiana var.

Royal Hort. Soc.
Journal of the Royal Hort. Soc.
vol 92, No. 4   1967

fl., hab.   p. 160   Pl. 65

Hosta sieboldiana var.

Royal Hort. Soc.
Journal of the Royal Hort. Soc.
vol 99, No. 11   1974

hab.   p. 490   Pl. 230

Hosta tardiflora

Curtis's Botanical Magazine
Vol 169   1952-53

fl.   204

Hosta tardiflora

Royal Hort. Soc.
Journal of the Royal Hort. Soc.
vol 93, No. 9   1968

hab.   p. 382   Pl. 186

Hosta tokudama

Royal Hort. Soc.
Journal of the Royal Hort. Soc.
vol 93, No. 9   1968

hab.   p. 382   Pl. 184

Hosta undulata

Hay, Roy
The Color Dictionary of Flowers &
Plants

hab.   p. 148   Pl. 1179

Hosta ventricosa var.

Royal Hort. Soc.
Journal of the Royal Hort. Soc.
vol 93, No. 9   1968

hab.   p. 382   Pl. 185

Hosta venusta

Curtis's Botanical Magazine
vol 176   1966-68

fl.   Pl. 499

Hot Water Plant

see

Achimenes longiflora

Hottonia palustris

Ary, S.
The Oxford Book of Wildflowers

fl.   p 108, Pl. 3

Hottonia palustris

Huxley, Anthony
Garden Perennials and Water Plants

fl.   Pl. 289

Hottonia palustris

Lindman, C. A. M.
Nordens Flora, Vol 7

fl. fr. hab.   Pl. 462

Hottonia palustris

Martin, W. Keble
The Concise British Flora in Colour

fl. hab.   Pl. 57

Hottonia palustris

Polunin, Oleg
Flowers of Europe

fl. Pl. 92 #957

Hottonia palustris

Stodola, Jiri
Encyclopedia of Water Plants

fl.   P. 267

Hottonia, Water

see

Hottonia palustris

Houlletia brocklehurstiana

Pabst, G. F. J.
Orchidaceae Brasilienses,
Vol 2

fl.   p 215

Houlletia juruensis

Pabst, G. F. J.
Orchidaceae Brasilienses,
Vol 2

fl.   p 215

Houlletia wallisii

Ospina, Mariano
Orquideas de las americas

fl.   Pl. 154

Hound's Tongue

see

Cynoglossum

Hound's Tongue

see

Trilissa paniculata

Hounds-Tongue, Sweet

see

Cynoglossum suaveolens

Hound's tongue, western

see

Cynoglossum grande

Houseleek

see

Sempervivum

Houseleek, Mountain

see

Sempervivum montanum

Houstonia caerulea

Batson, Wade T.
Wild Flowers in South Carolina

fl.   p 107

Houstonia caerulea

Brown, Clair A.
Wildflowers of Louisiana and
Adjoining States

fl. hab.   p 177

Houstonia caerulea

Courtenay, Booth
Wildflowers & Weeds

fl.   p 70

Houstonia caerulea

Jennings, O. E.
Wild Flowers of Western Pennsylvania
v. 2
fl.   Pl. 104

Houstonia caerulea

Lemmon, Robert S.
Wildflowers of North America in
Full Color

fl.   p. 158   Pl. 253

Houstonia caerulea

Macoby, Stirling
What Flower is That

fl. P 156

Houstonia caerulea

Roberts, June Carver
Born in the Spring

fl. fr. hab.   p 69

Houstonia caerulea

Walcott, Mary Vaux
North American Wild Flowers

fl. hab.   Pl. 59

Houstonia caerulea

Wharton, Mary E.
A Guide to the Wildflowers & Ferns
of Kentucky

fl. p 137   Pl. 1.86

Houstonia lanceolata

Wharton, Mary E.
A Guide to the Wildflowers & Ferns
of Kentucky

fl. p 172   Pl. 2.51

Houstonia longifolia

Courtenay, Booth
Wildflowers & Weeds

fl.   p 70

Houstonia patens

Wharton, Mary E.
A Guide to the Wildflowers & Ferns
of Kentucky

fl. p 177   Pl. 2.64

Houstonia purpurea

Campbell, Carlos C.
Great Smoky Mountain Wildflowers

fl.   p 75

Houstonia purpurea

Dean, Blanche E.
Wildflowers of Alabama and
Adjoining States

fl.   p 175

Houstonia pusilla

Dean, Blanche E.
Wildflowers of Alabama and
Adjoining States

fl.   p 175

Houstonia pygmaea

Dean, Blanche E.
Wildflowers of Alabama and
Adjoining States

fl.   p 175

Houstonia serpyllifolia

Campbell, Carlos C.
Great Smoky Mountain Wildflowers

fl.   p 75

Houstonia sp.

Klimas, John E.
Wildflowers of Eastern America

fl.   Pl. 247

Houttuynia cordata

Kariyone, Tatsuo
Atlas of Medicinal Plants

fl.   Pl. 119

Houttuynia cordata

Kimura, Koiti
Japanese Medicinal Plants, Vol I

fl.   p. 20

Houttuynia cordata

Perry, Frances
Flowers of the World

fl.   p 271

Houttuynia cordata

Takatori, Jisuke
Color Atlas of Medicinal Plants of Japan

fl.   Fig. 64

Houttuynia cordata

Wit, H. C. D. de
Plants of the World;
    The Higher Plants, Vol I

fl.   p 101, Pl. 48

Hovea chorizemifolia

Morcombe, M. K.
Australia's Western Wildflowers

fl.   p 108

Hovea elliptica

Blombery, Alec M.
What Wildflower is That

fl.   p 173, Pl. 461

Hovea elliptica

Harrison, R.E.
Trees and Shrubs

fl.   p. 89

Hovea heterophylla

Cochrane, C. R.
Flowers and Plants of Victoria

fl.   Pl. 369

Hovea lanceolata

Blombery, Alec M.
What Wildflower is That

fl.   p 173, Pl. 463

Hovea linearis

Blombery, Alec M.
What Wildflower is That

fl.   p 173, Pl. 462

Hovea longifolia

Macoby, Stirling
What Flower is That

fl.   P 157

Hovea longifolia var.

Cochrane, C. R.
Flowers and Plants of Victoria

fl.   Pl. 492

Hovea longifolia var.

Curtis's Botanical Magazine
Vol 171   1956-57

fl.   305

Hovea longifolia var.

Morley, Brian D.
Wild Flowers of the World

fl. fr.   Pl. 129

Hovea pungens

Blombery, Alec M.
What Wildflower is That

fl.   p 173, Pl. 464

Hovea rosmarinifolia

Cochrane, C. R.
Flowers and Plants of Victoria

fl.   Pl. 396

Hovea trisperma

Blombery, Alec M.
What Wildflower is That

fl.   p 173, Pl. 465

Hovenia dulcis

Kitamura, Siro
Coloured Illustrations of Trees &
    Shrubs of Japan

hab.   329, 330

Howeia belmoreana

Kiaer, Eigil
Indoor Plants in Colour

hab.   p. 68

Howeia belmoreana

Nicolaisen, Age
Pocket Encyclopedia of Indoor Plants

hab.   Pl. 6

Howeia forsteriana

Hersey, Jean
Woman's Day Book of House Plants

hab.   p. 76

Howittia trilocularis

Blombery, Alec M.
What Wildflower is That

fl.   p. 174   Pl. 466

Howittia trilocularis

Cochrane, C. R.
Flowers and Plants of Victoria

fl.   Pl. 469

Hoya australis

Blombery, Alec M.
What Wildflower is That

fl.   p 174, Pl. 467

Hoya australis

Harrison, Richmond E.
Climbers and Trailers

fl.   p 50   Pl. 105

Hoya australis

Lamb, Edgar
Popular Exotic Cacti in Color

fl. hab.   p 105

Hoya australis

Royal Hort. Soc.
Journal of the Royal Hort. Soc.
    vol 98, No. 5   1973

fl.   p. 216   Pl. 116

Hoya bella

Hay, Roy
The Dictionary of House Plants

fl.   Pl. 282

Hoya bella

Kiaer, Eigil
Indoor Plants in Colour

fl.   p. 67

Hoya bella

Kromdijk, G.
200 House Plants in Colour

fl.   Pl. 107

Hoya bella

Macoby, Stirling
What Flower is That

fl.   P 157

Hoya bella

Massachusetts Hort. Soc.
Horticulture
    vol 34, No. 1   1956

fl.   p. 20

Hoya bella

Menninger Edwin A.
Flowering Vines of the World

fl.   Pl. 25

Hoya bella

Nicolaisen, Age
Pocket Encyclopedia of Indoor
    Plants

fl., hab.   Pl. 137

Hoya bella

Royal Hort. Soc.
Journal of the Royal Hort. Soc.
    vol 98, No. 5   1973

fl.   p. 216   Pl. 115

Hoya bella

Wit, H. C. D. de
Plants of the World;
    The Higher Plants, Vol II

fl.   p 107, Pl. 66

Hoya carnosa

Encke, Fritz
Zimmerpflanzen

fl.   p. 97

Hoya carnosa

Harrison, Richmond E.
Climbers and Trailers

fl.   p 51   Pl. 106

Hoya carnosa

Hay, Roy
The Dictionary of House Plants

fl.   Pl. 283

Hoya carnosa

Kiaer, Eigil
Indoor Plants in Colour

fl.   p. 67

Hoya carnosa

Kromdijk, G.
200 House Plants in Colour

fl.   Pl. 108

Hoya carnosa

Lamb, Edgar
The Pocket Encyclopedia of Cacti
    and Succulents in Color

fl.   Pl. 270

Hoya carnosa

Macoby, Stirling
What Flower is That

fl.   P 157

Hoya carnosa

Massachusetts Hort. Soc.
Horticulture
    vol 41, No. 1   1963

fl.   inside cover

Hoya carnosa

Massachusetts Hort. Soc.
Horticulture
    vol 45, No. 6   1967

fl.   p. 25

Hoya carnosa

Nicolaisen, Age
Pocket Encyclopedia of Indoor
    Plants

fl., hab.   Pl. 138

Hoya carnosa

Tsukamoto, Yotaro
Coloured Illustrations of Garden Flowers,
    v.9
fl.
opp.p.50

Hoya carnosa

Walden, Beryl M.
Wild Flowers of Hong Kong

fl.   Pl. 34(89)

Hoya carnosa var.

Hersey, Jean
Woman's Day Book of House Plants
fl.
p. 106

Hoya carnosa var.

Menninger, Edwin A.
Flowering Vines of the World

fl.   Pl. 22

Hoya carnosa var.

Perry, Frances
Flowers of the World

fl.   p 38

Hoya coriacea

Morley, Brian D.
Wild Flowers of the World

fl.   Pl. 120

Hoya hybrid

Menninger, Edwin A.
Flowering Vines of the World

fl.   Pl. 24

Hoya imperialis

Harrison, Richmond E.
Climbers and Trailers

fl.   p 51   Pl. 107

Hoya imperialis

Royal Hort. Soc.
Journal of the Royal Hort. Soc.
    vol 98, No. 5   1973

fl.   p. 216   Pl. 117

Hoya purpurea-fusca

Morley, Brian D.
Wild Flowers of the World

fl.   Pl. 120

Hoya purpurea-fusca var.

Harrison, Richmond E.
Climbers and Trailers

fl.   p. 51   Pl. 108

Hoya sussuela

Morley, Brian D.
Wild Flowers of the World

fl.   Pl. 120

Huckleberry

see

Gaylussacia baccata

Huckleberry

see

Vaccinium deliciosum

Huckleberry, Cascade

see

Vaccinium deliciosum

Huckleberry, Fool's

see

Menziesia ferruginea

Huckleberry, garden

see

Solanum intrusum

Huckleberry, Mountain

see

Vaccinium membranaceum

Huckleberry, Oval-leaf

see

Vaccinium ovalifolium

Huckleberry, red

see

Vaccinium parvifolium

Huckleberry, Squaw

see

Polycodium stamineum

Huckleberry, Tree

see

Vaccinium arboreum

Huckleberry, Winter

see

Vaccinium arboreum

Hudsonia tomentosa

Courtenay, Booth
Wildflowers & Weeds

fl.   p 36

Huernia aspera

Flowering Plants of Africa
vol XXV   1945-46

fl., hab.   Pl. 984

Huernia aspera

Lamb, Edgar
Stapeliads in Cultivation

fl.   p 71

Huernia campanulata

Lamb, Edgar
Stapeliads in Cultivation

fl.   p 72

Huernia clavigera

Lamb, Edgar
Stapeliads in Cultivation

fl.   p 71

Huernia concinna

Flowering Plants of Africa
vol XXV   1945-46

fl., hab.   Pl. 999

Huernia erinacea

Flowering Plants of Africa
vol 31   1956

fl.   Pl. 1206

Huernia hislopii

Flowering Plants of South Africa
vol XIX   1939

fl., hab.   Pl. 758

Huernia hystrix

Flowering Plants of South Africa
vol XIX   1939

fl.   Pl. 757

Huernia hystrix

Morley, Brian D.
Wild Flowers of the World

fl.   Pl. 78

Huernia keniensis

Lamb, Edgar
The Pocket Encyclopedia of Cacti
and Succulents in Color

fl. hab.   Pl. 267

Huernia keniensis var.

Flowering Plants of Africa
vol 37   1965-66

fl.   Pl. 1472

Huernia keniensis var.

Flowering Plants of Africa
vol 38   1967

fl.   Pl. 1511

Huernia kirkii

Lamb, Edgar
Stapeliads in Cultivation

fl.   p 71

Huernia leachii

Flowering Plants of Africa
vol 37   1965-66

fl.   Pl. 1473

Huernia levyi

Flowering Plants of South Africa
vol XVI   1936

fl., hab.   Pl. 616

Huernia longituba

Lamb, Edgar
Stapeliads in Cultivation

fl. hab.   p 54

Huernia macrocarpa

Lamb, Edgar
The Pocket Encyclopedia of Cacti
and Succulents in Color

fl. hab.   Pl. 268

Huernia namaquensis

Lamb, Edgar
Stapeliads in Cultivation

fl.   p 71

Huernia oculata

Flowering Plants of South Africa
vol XVI   1936

fl., hab.   Pl. 619

Huernia pendula

Flowering Plants of Africa
vol 28   1950-51

fl., hab.   Pl. 1108

Huernia pillansii

Flowering Plants of South Africa
vol XXII   1942

fl., hab.   Pl. 843

Huernia reticulata

Lamb, Edgar
Stapeliads in Cultivation

fl.   p 89

Huernia similis

Flowering Plants of Africa
vol 38   1967

fl., hab.   Pl. 1519a

Huernia somalica

Flowering Plants of Africa
vol 31   1956

fl.   Pl. 1238

Huernia transmutata

Cactus & Succulent Society of America
Journal
vol 41, No. 6   1969

fl.   p. 276

Huernia transmutata

Lamb, Edgar
Popular Exotic Cacti in Color

fl. hab.   p 107

Huernia urceolata

Flowering Plants of Africa
vol 39   1968-69

fl., hab.   Pl. 1550

Huernia volkartii var.

Flowering Plants of Africa
vol 38   1967

fl., hab.   Pl. 1519b

Huernia whitesloaneana

Flowering Plants of South Africa
vol XVI   1936

fl., hab.   Pl. 632

Huernia zebrina var.

Flowering Plants of South Africa
vol XVI   1936

fl., hab.   Pl. 613

Hug me tight

see

Doxanthus unguis-cati

Hugeria erythrocarpa

Addisonia, v. 21, 1939-42

fl., fr.   opp. p. 3   Pl. 674

Huisache

see

Acacia farnesiana

Hulsea algida

Lemmon, Robert S.
Wildflowers of North America in
Full Color

fl.   p. 163   Pl. 260

Hulsea algida

Munz, Philip A.
California Mountain Wildflowers

fl.   p 57, Pl. 92

Hulsea algida

Orr, Robert T.
Wildflowers of Western America

fl.   Pl. 69

Hulthemia berberidifolia

Royal Hort. Soc.
Journal of the Royal Hort. Soc.
vol 88, No. 4   1963

fl.   p. 164   Pl. 63

Hulthemia berberidifolia

Royal Hort. Soc.
Journal of the Royal Hort. Soc.
vol 90, No. 2   1965

fl.   p. 84   Pl. 31

Hulthemia persica

Morley, Brian D.
Wild Flowers of the World

fl.   Pl. 45

x Hulthemosa hardii

Royal Hort. Soc.
Journal of the Royal Hort. Soc.
vol 91, No. 8   1966

fl.   p. 338   Pl. 163

Humea elegans

Blombery, Alec M.
What Wildflower is That

fl.   p 174, Pl. 468

Humea elegans

Macoby, Stirling
What Flower is That

hab P 157

Humilis

see

Rosmarinus officinalis var.

Hummingbird

see

Beloperone californica

Hummingbird Flower

see

Erythrina flabelliformis

Humulus lupulus

Ary, S.
The Oxford Book of Wildflowers

fl.   p 64, Pl. 5

Humulus lupulus

Bianchini, F.
The Complete Book of Fruits & Vegetables

fr.   p 215

Humulus lupulus

Bianchini, Francesco
Health Plants of the World

fr.   p 107

Humulus lupulus

Courtenay, Booth
Wildflowers & Weeds

fl.   p 69

Humulus lupulus

Crockett, James Underwood
Annuals

fr.   p. 125

Humulus lupulus

Hvass, Elsie
Plants That Feed and Serve Us

fl.   fr.   p 79, Pl. 163

Humulus lupulus

Kariyone, Tatsuo
Atlas of Medicinal Plants

fr.   Pl. 140

Humulus Lupulus

Kimura, Koiti
Japanese Medicinal Plants, Vol I

fl.   p 20

Humulus lupulus

Lindman, C. A. M.
Nordens Flora, Vol 3

fr.   Pl. 173

Humulus lupulus

Martin, W. Keble
The Concise British Flora in Colour

fl. hab.   Pl. 76

Humulus lupulus

Masefield, G. B.
The Oxford Book of Food Plants

fl. fr.   Pl. 137, 1

Humulus lupulus

Perrot, Emile
Les Plantes Medicinales

fl. fr. hab.   p 120

Humulus lupulus

Polunin, Oleg
Flowers of Europe

fr.   Pl. 6 #63

Humulus lupulus

Takatori, Jisuke
Color Atlas of Medicinal Plants of Japan

fl.
fr.   Fig. 62

Humulus lupulus var.

Harrison, Richmond E.
Climbers and Trailers

fl.   p 50   Pl. 103

Humulus scandens

Crockett, James Underwood
Annuals

hab.   p. 125

Humulus scandens var.

Hay, Roy
The Color Dictionary of Flowers &
Plants

hab.   p. 39   Pl. 309

Hunnemannia fumariifolia

Crockett, James Underwood
Annuals

fl.   p. 126

Hunnemannia fumariifolia

Royal Hort. Soc.
Journal of the Royal Hort. Soc.
     vol 95, No. 8   1970

fl.   p. 352   Pl. 182

Hunter's Horn

          see

     Sarracenia flava

Huntleya burtii

Grubb, Roy
Selected Orchidaceous Plants
     Vol I

fl.   p 127

Huntleya burtii

Ospina, Mariano
Orquideas de las americas

fl.   Pl. 105

Huntleya burtii

Royal Hort. Soc.
Journal of the Royal Hort. Soc.
     vol 94, No. 7   1969

fl.   p. 302   Pl. 164

Huntleya lucida

Dunsterville, G. C. K.
Introduction to the World
     of Orchids

fl. hab.   Pl. 14

Huntleya lucida

Grubb, Roy
Selected Orchidaceous Plants
     Vol II

fl. hab.   p 75

Huntleya meleagris

Pabst, G. F. J.
Orchidaceae Brasilienses,
     Vol 2

fl.   p 231

Huntsman's Cap

          see

     Sarracenia purpurea

Hutchinsia alpina

Barneby, T. P.
European Alpine Flowers in Colour

fl. hab.   Pl. 30, 2

Hutchinsia alpina

Felsko, Elsa
A Book of Wild Flowers
2nd Ser.
fl. hab.   p 124

Hutchinsia alpina

Huxley, Anthony
Garden Perennials and Water Plants

fl.   Pl. 144

Hutchinsia alpina

Kohlhaupt, Paula
Fleurs des Alpages
     vol 2

fl.   Pl. 76

Huttonaea pulchra

Morley, Brian D.
Wild Flowers of the World

fl.   Pl. 89

Hyacinth

     see

Hyacinthus

Hyacinth bean

     see

     Dolichos lablab

Hyacinth, Desert

     see

Brodiaea pulchella var.

Hyacinth, Dutch

     see

Hyacinthus orientalis var.

Hyacinth, French

     see

     Hyacinthus orientalis var.

Hyacinth, Grape

     see

Muscari

Hyacinth, Pine

     see

     Clematis baldwinii

Hyacinth, Pine

     see

Clematis crispa

Hyacinth, Roman

     see

Hyacinthus orientalis var.

Hyacinth Shrub

     see

Xanthoceras sorbifolium

Hyacinth, summer

     see

Galtonia candicans

Hyacinth, Tassel

     see

Muscari comosum

Hyacinth, water

     see

Eichhornia crassipes

Hyacinth, Wild

     see

Brodiaea pulchella

Hyacinth, wild

     see

Camassia scilloides

Hyacinth, Wild

     see

Endymion non-scriptus

Hyacinth, wild

     see

Scilla nonscripta

Hyacinth, wood

     see

     Scilla nonscripta

Hyacinthella azurea

Vilmorin, Roger de
Plantes Alpines dans les Jardins

fl. hab.   Pl. II

Hyacinthus amethystinus

Hay, Roy
The Color Dictionary of Flowers & Plants

fl.   p 96, Pl. 765

Hyacinthus amethystinus

Miles, Bebe
Bulbs for the Home Gardener

fl.   p. 77

Hyacinthus azureus

Miles, Bebe
Bulbs for the Home Gardener

fl.   p 77

Hyacinthus candicans

Huxley, Anthony
Garden Annuals and Bulbs

fl.   Pl. 225

Hyacinthus hybrid

Hay, Roy
The Color Dictionary of Flowers & Plants

fl.   p 96, 97, Pl. 766 thru 772

Hyacinthus hybrid

Hay, Roy
The Dictionary of House Plants

fl.   Pl. 284 - 286

Hyacinthus hybrid

Kromdijk, G.
200 House Plants in Colour

fl.   Pl. 109

Hyacinthus hybrid

Macoby, Stirling
What Flower is That

fl.   p. 158

Hyacinthus litwinowii

Wendelbo, Per
Tulips and Irises of Iran and
    Their Relatives

fl. hab.   p 44, Pl. 45

Hyacinthus orientalis

Huxley, Anthony
Garden Annuals and Bulbs

hab.   Pl. 162

Hyacinthus orientalis

Polunin, Oleg
Flowers of Europe

fl.   Pl. 171 #1643

Hyacinthus orientalis

Tsukamoto, Yotaro
Coloured Illustrations of Garden
    Flowers, v. 9

fl.   opp. p. 12

Hyacinthus orientalis var.

Huxley, Anthony
Garden Annuals and Bulbs

fl.   Pl. 222-4

Hyacinthus orientalis var.

Kiaer, Eigil
Indoor Plants in Colour

fl.   p. 69, 70

Hyacinthus orientalis var.

Macoby, Stirling
What Flower is That

fl.   p 158

Hyacinthus orientalis var.

Nicolaisen, Age
Pocket Encyclopedia of Indoor Plants

fl.   p 23

Hyacinthus transcaspicus

Wendelbo, Per
Tulips and Irises of Iran and
    Their Relatives

fl. hab.   p 45, Pl. 46

Hybanthus concolor

Dean, Blanche E.
Wildflowers of Alabama and
    Adjoining States

fl.   p 109

Hybanthus concolor

Klimas, John E.
Wildflowers of Eastern America

fl.   Pl. 8

Hybanthus enneaspermus var.

Moriarty, Audrey
Wild Flowers of Malawi

fl.   Pl. 62; 5

Hybanthus floribundus

Cochrane, G. R.
Flowers and Plants of Victoria

fl.   Pl. 181

Hybanthus monopetalus

Morley, Brian D.
Wild Flowers of the World

fl.   Pl. 135

Hybanthus vernonii

Blombery, Alec M.
What Wildflower is That

fl.   p 174, Pl. 469

Hybanthus vernonii

Cochrane, G. R.
Flowers and Plants of Victoria

fl.   Pl. 466

Hydatella filamentosa

Curtis, Winifred
The Endemic Flora of Tasmania
Vol VI

fl. fr. hab.   Pl. 203

Hydnocarpus alpina

Kariyone, Tatsuo
Atlas of Medicinal Plants

fl., fr.   Pl. 82

Hydnora africana

Morley, Brian D.
Wild Flowers of the World

fl.   Pl. 53

Hydrangea anomela var.

American Hort. Soc.
American Horticulturist
    vol 52, No. 1 1973

fl., hab.   p. 29

Hydrangea arborescens

Batson, Wade T.
Wild Flowers in South Carolina

fl.   p 53

Hydrangea arborescens

Campbell, Carlos C.
Great Smoky Mountain Wildflowers

fl.   p 71

Hydrangea arborescens

Wharton, Mary E.
Trees & Shrubs of Kentucky

fl.   p 88, Pl. 2.27

Hydrangea aspera

Gault, S. Millar
The Color Dictionary of Shrubs

fl.   Pl. 237

Hydrangea aspera var.

Morley, Brian D.
Wild Flowers of the World

fl.   Plate 99

Hydrangea, bigleaf
see
Hydrangea macrophylla

Hydrangea, Chinese
see
Hydrangea macrophylla

Hydrangea, Climbing
see
Decumaria barbara

Hydrangea, Climbing
see
Hydrangea petiolaris

Hydrangea, florists'
see
Hydrangea macrophylla

Hydrangea, French
see
Hydrangea macrophylla

Hydrangea grandiflora

Everett, T.H.
New Illustrated Encyclopedia of
    Gardening, v. 5

hab., fl.   p. 886

Hydrangea heteromalla

Gault, S. Millar
The Color Dictionary of Shrubs

fl.   Pl. 238

Hydrangea, hills of snow
see
Hydrangea grandiflora

Hydrangea hirta

Kitamura, Siro
Coloured Illustrations of Trees &
    Shrubs of Japan

fl.   199

Hydrangea, House
see
Hydrangea macrophylla

Hydrangea hybrid

Harrison, R.E.
Trees and Shrubs

fl.   P 91

Hydrangea hybrid.

Hay, Roy
The Color Dictionary of Flowers & Plants

fl.   p 207, Pl. 1649 thru 1651

Hydrangea hybrid.

Hay, Roy
The Dictionary of House Plants

fl.   Pl. 287 - 289

Hydrangea hybrid

Kromdijk, G.
200 House Plants in Colour

fl.   Pl. 110

Hydrangea hybrid

Perry, Frances
Flowers of the World

fl.   p. 139

Hydrangea integerrima

Curtis's Botanical Magazine
Vol 168   1951

fl.   153

Hydrangea involucrata

Kitamura, Siro
Coloured Illustrations of Trees &
    Shrubs of Japan

fl.   197

Hydrangea, Lacecap
see
Hydrangea macrophylla

Hydrangea macrophylla

Encke, Fritz
Zimmerpflanzen

fl.   p. 97

Hydrangea macrophylla

Harrison, R.E.
Trees and Shrubs

fl. P 91

Hydrangea macrophylla

Huxley, Anthony
Deciduous Garden Trees and Shrubs

fl.   Pl. 90

Hydrangea macrophylla

Kiaer, Eigil
Indoor Plants in Colour

fl.   p. 68

Hydrangea macrophylla

Macoby, Stirling
What Flower is That

fl.   P 158

Hydrangea macrophylla var.

Bartels, Andreas
Das Grosse Buch der Gartengeholze

fl.   p 160

Hydrangea macrophylla var.

Everett, T.H., ed.
New Illustrated Encyclopedia of
    Gardening vol. 5

fl.   p. 887

Hydrangea macrophylla var.

Gault, S. Millar
The Color Dictionary of Shrubs

fl.   Pl. 239-46

Hydrangea macrophylla var.

Hellyer, A. G. L.
Shrubs in Colour

fl.   p. 66,67

Hydrangea macrophylla var.

Huxley, Anthony
Deciduous Garden Trees and Shrubs

fl.   Pl. 91

Hydrangea macrophylla var.

Macoby, Stirling
What Flower is That

fl P 158

Hydrangea macrophylla var.

Massachusetts Hort. Soc.
Horticulture
    vol 39, No. 2   1961

fl.   p. 86

Hydrangea macrophylla var.

Perry, Frances
Flowers of the World

fl.   p. 140

Hydrangea macrophylla var.

Tsukamoto, Yotaro
Coloured Illustrations of Garden Flowers,
v.10
fl.
ill. 166 opp.p. 52

Hydrangea, Oak-leaved

see

Hydrangea quercifolia

Hydrangea paniculata

Curtis's Botanical Magazine
Vol 171   1956-57

fl.   301

Hydrangea paniculata

Huxley, Anthony
Deciduous Garden Trees and Shrubs

fl.   Pl. 92

Hydrangea paniculata

Kitamura, Siro
Coloured Illustrations of Trees &
Shrubs of Japan

fl.   196

Hydrangea paniculata

Macoby, Stirling
What Flower is That

fl.   P 158

Hydrangea paniculata var.

Everett, T.H.
New Illustrated Encyclopedia of Gardening,
v.5
hab., fl.
opp.p.886

Hydrangea paniculata var.

Gault, S. Millar
The Color Dictionary of Shrubs

fl.   Pl. 250, 251, 252

Hydrangea paniculata var.

Harrison, R. E.
Trees and Shrubs

fl., hab.   p. 91

Hydrangea paniculata var.

Hay, Roy
The Color Dictionary of Flowers & Plants

fl.   p 207, Pl. 1652

Hydrangea paniculata var.

Hellyer, A.G.L.
Shrubs in Color
fl.
p. 66

Hydrangea paniculata var.

Perry, Frances
Flowers of the World

fl.   p. 140

Hydrangea paniculata var.

Tsukamoto, Yotaro
Coloured Illustrations of Garden Flowers,v.10
fl.
ill. 167 opp.p.52

Hydrangea, peegee

see

Hydrangea paniculata var.

Hydrangea petiolaris

Hellyer, A.G.L.
Shrubs in Colour

fl.   p. 67

Hydrangea petiolaris

Huxley, Anthony
Deciduous Garden Trees and Shrubs

fl.   Pl. 290

Hydrangea petiolaris

Kitamura, Siro
Coloured Illustrations of Trees &
Shrubs of Japan

fl.   195

Hydrangea petiolaris

Massachusetts Hort. Soc.
Horticulture
vol 47, No. 9 1969

fl.   p. 37

Hydrangea, plumed

see

Hydrangea paniculata

Hydrangea quercifolia

American Hort. Soc.
American Horticulturist
Vol. 54, No. 5 1975

hab.   P. 15

Hydrangea quercifolia

Dean, Blanche E.
Wildflowers of Alabama and
Adjoining States

fl.   p 79

Hydrangea quercifolia

Harrison, R.E.
Trees and Shrubs

fl.   P 91

Hydrangea sargentiana

Royal Hort. Soc.
The Garden
vol 100, No. 12 1975

hab., fl.   p. 599

Hydrangea serrata

Hellyer, A.G.L.
Shrubs in Colour

fl.   p. 66

Hydrangea serrata

Kitamura, Siro
Coloured Illustrations of Trees &
Shrubs of Japan

fl.   198

Hydrangea serrata var.

Gault, S. Millar
The Color Dictionary of Shrubs

fl.   Pl. 247, 248, 249

Hydrangea serrata var.

Hay, Roy
The Color Dictionary of Flowers & Plants

fl.   p 207, Pl. 1653

Hydrangea serrata var.

Hellyer, A.G.L.
Shrubs in Color

fl.   p. 67

Hydrangea serrata var.

Kariyone, Tatsuo
Atlas of Medicinal Plants

fl.   Pl. 110

Hydrangea serrata var.

Royal Hort. Soc.
Journal of the Royal Hort. Soc.
vol 86, No. 3 1961

fl.   p. 118   Pl. 33

Hydrangea serrata var.

Takatori, Jisuke
Color Atlas of Medicinal Plants
of Japan

fl.   Pl. 43A

Hydrangea serrata var.

Wit, H. C. D. de
Plants of the World:
The Higher Plants, Vol I

fl.   p 279, Pl. 161

Hydrangea villosa

Hay, Roy
The Color Dictionary of Flowers & Plants

fl.   p 207, Pl. 1654

Hydrangea villosa

Hellyer, A. G. L.
Shrubs in Colour

fl.   p. 67

Hydrangea, Wild

see

Hydrangea arborescens

Hydrastis canadensis

Courtenay, Booth
Wildflowers & Weeds

fl.   p 30

Hydrastis canadensis

Duncan, Wilbur H.
Wildflowers of the Southeastern
     United States

fl.   p 41

Hydrastis canadensis

Hvass, Elsie
Plants That Feed and Serve Us

fl.   fr.   p 97, Pl. 207

Hydrastis canadensis

Jennings, O. E.
Wild Flowers of Western Pennsylvania
v.2
fl.   Pl. 93

Hydrastis canadensis

Klimas, John E.
Wildflowers of Eastern America

fl.   Pl. 9

Hydrastis canadensis

Perrot, Emile
Les Plantes Medicinales

fl. fr. hab.   p 122

Hydrastis canadensis

Wharton, Mary E.
A Guide to the Wildflowers & Ferns
     of Kentucky

fl.   p 274   Pl. 1,3

Hydrilla verticillata

Stodola, Jiri
Encyclopedia of Water Plants

hab.   P. 270

Hydrilla verticillata

Tarver, David P.
Aquatic and Wetland Plants
    of Florida

fl. hab.   p 50-1

Hydrocharis morsus-ranae

Ary, S.
The Oxford Book of Wildflowers

fl.   p 102, Pl. 2

Hydrocharis morsus-ranae

Huxley, Anthony
Garden Perennials and Water Plants

fl.   Pl. 290

Hydrocharis morsus-ranae

Lindman, C. A. M.
Nordens Flora, Vol 1

fl. fr. hab.   Pl. 45

Hydrocharis morsus-ranae

Martin, W. Keble
The Concise British Flora in Colour

fl. hab.   Pl. 79

Hydrocharis morsus-ranae

Morley, Brian D.
Wild Flowers of the World

fl.,hab.   Pl. 23

Hydrocharis morsus-ranae

Polunin, Oleg
Flowers of Europe

fl.   Pl. 161 #1566

Hydrocharis morsus-ranae

Stodola, Jiri
Encyclopedia of Water Plants

fl.   P. 50

Hydrochloa caroliniensis

Tarver, David P.
Aquatic and Wetland Plants
    of Florida

hab.   p 88

Hydrocleys nymphoides

Stodola, Jiri
Encyclopedia of Water Plants

fl.   p. 219

Hydrocleys nymphoides

Wit, H. C. D. de
Plants of the World;
    The Higher Plants, Vol. II
fl.   Pl. 121

Hydrocotyle bonariensis

Brown, Clair A.
Wildflowers of Louisiana and
    Adjoining States

fl. hab.   p 124

Hydrocotyle lexiflora

Cochrane, G. R.
Flowers and Plants of Victoria

fl.   Pl. 350

Hydrocotyle rotundifolia

Crockett, James Underwood
Lawns & Ground Covers

fl.   p. 73

Hydrocotyle umbellata

Brown, Clair A.
Wildflowers of Louisiana and
    Adjoining States

fl. hab.   p 125

Hydrocotyle umbellata

Burgis, D.S.
Florida Weeds

hab.   p. 10

Hydrocotyle umbellata

Dean, Blanche E.
Wildflowers of Alabama and
    Adjoining States

fr.   p 121

Hydrocotyle umbellata

Tarver, David P.
Aquatic and Wetland Plants
    of Florida

fl. hab.   p 42

Hydrocotyle vulgaris

Ary, S.
The Oxford Book of Wildflowers

fl.   p 46, Pl. 4

Hydrocotyle vulgaris

Lindman, C. A. M.
Nordens Flora, Vol 6

fl. fr. hab.   Pl. 414

Hydrocotyle vulgaris

Martin, W. Keble
The Concise British Flora in Colour

fl. fr. hab.   Pl. 36

Hydrocotyle vulgaris

Stodola, Jiri
Encyclopedia of Water Plants

fl.   P. 235

Hydrodea sarcocalycantha

Herre, H.
The Genera of the Mesembryanthemaceae

fl.   fr.   p 175

Hydrolea corymbosa

Duncan, Wilbur H.
Wildflowers of the Southeastern
United States

fl.   p 147

Hydrolea corymbosa

Fleming, Glenn
Wild Flowers of Florida

fl.,hab.   p. 90

Hydrolea ovata

Brown, Clair A.
Wildflowers of Louisiana and
Adjoining States

fl.   p 152

Hydrolea quadrivalvis

Dean, Blanche E.
Wildflowers of Alabama and
Adjoining States

fl.   p 147

Hydrolea quadrivalvis

Duncan, Wilbur H.
Wildflowers of the Southeastern
United States

fl.   p 147

Hydrophyllum appendiculatum

Courtenay, Booth
Wildflowers & Weeds

fl.   p 88

Hydrophyllum appendiculatum

Jennings, O. E.
Wild Flowers of Western Pennsylvania
Vol. 2

fl.   Pl. 126

Hydrophyllum appendiculatum

Lemmon, Robert S.
Wildflowers of North America in
Full Color

fl.   p. 268   Pl. 427

Hydrophyllum appendiculatum

Wharton, Mary E.
A Guide to the Wildflowers & Ferns
of Kentucky

fl.   p 136   Pl. 1.83

Hydrophyllum canadense

Duncan, Wilbur H.
Wildflowers of the Southeastern
United States

fl.   p 145

Hydrophyllum canadense

Wharton, Mary E.
A Guide to the Wildflowers & Ferns
of Kentucky

fl.   p 119   Pl. 1.51

Hydrophyllum capitatum var.

Clark, Lewis J.
Wild Flowers of British Columbia

fl.   p 438

Hydrophyllum fendleri

Taylor, Ronald J.
Mountain Wild Flowers

fl. hab.   p 147

Hydrophyllum macrophyllum

Wharton, Mary E.
A Guide to the Wildflowers & Ferns
of Kentucky

fl.   p 119   Pl. 1.50

Hydrophyllum tenuipes

Clark, Lewis J.
Wild Flowers of British Columbia

fl.   p 434

Hydrophyllum virginianum

Courtenay, Booth
Wildflowers & Weeds

fl.   p 88

Hydrophyllum virginianum

Jennings, O. E.
Wild Flowers of Western Pennsylvania
v. 2
fl.   Pl. 125

Hydrophyllum virginianum

Klimas, John E.
Wildflowers of Eastern America

fl.   Pl. 73

Hydrophyllum virginianum

Moyle, John B.
Northland Wild Flowers

fl. hab.   p 132, Pl. 143

Hygrophila angustifolia

Stodola, Jiri
Encyclopedia of Water Plants

hab.   p. 275

Hygrophila polysperma

Stodola, Jiri
Encyclopedia of Water Plants

hab.   P. 271

Hylocereus cubensis

Lamb, Edgar
Popular Exotic Cacti in Color

fl. hab.   p 109

Hylocereus undatus

Everett, T.H.
New Illustrated Encyclopedia of
Gardening, v. 5

fl.   opp. p. 823

Hylocereus undatus

Hargreaves, Dorothy
Tropical Blossoms of the Caribbean

fl.   p. 10

Hylocereus undatus

Jones, Paul
Flora Magnifica

fl.   Pl. 4

Hylocereus undatus

Macoby, Stirling
What Flower is That

fl.   p. 159

Hylomecon japonicum

Hay, Roy
The Color Dictionary of Flowers & Plants

fl.   p 148, Pl. 1180

Hymenanthera crassifolia

Curtis's Botanical Magazine
Vol 159   1936

fl.   No. 9426

Hymenanthera dentata

Cochrane, G. R.
Flowers and Plants of Victoria

fl.   Pl. 232

Hymenocallis americana

Macoby, Stirling
What Flower is That

fl.   P 159

Hymenocallis calathina

Everett, T.H.
New Illustrated Encyclopedia of
Gardening, v. 5

fl.   opp. p. 887

Hymenocallis calathina

Hersey, Jean
Woman's Day Book of House Plants
fl.
p. 87

Hymenocallis calathina

Massachusetts Hort. Soc.
Horticulture
vol 38, No. 4   1960

fl.   p. 215

Hymenocallis calathina

Miles, Bebe
Bulbs for the Home Gardener

fl.   p 172

Hymenocallis caribaea

Hay, Roy
The Dictionary of House Plants

fl.   Pl. 290

Hymenocallis caroliniana

Duncan, Wilbur H.
Wildflowers of the Southeastern
    United States

fl.   p 263

Hymenocallis coronaria

Dean, Blanche E.
Wildflowers of Alabama and
    Adjoining States

fl. hab.   p 27

Hymenocallis coronaria

Lemmon, Robert S.
Wildflowers of North America in
    Full Color

fl.   p. 9   Pl. 11

Hymenocallis hybrid

Perry, Frances
Flowers of the World

fl.   p. 26

Hymenocallis littoralis

Bruggeman, L.
Tropical Plants

fl.   Pl. 120

Hymenocallis littoralis

Hay, Roy
The Dictionary of House Plants

fl.   Pl. 291

Hymenocallis occidentalis

Brown, Clair A.
Wildflowers of Louisiana and
    Adjoining States

fl.   p 26

Hymenocallis occidentalis

Dean, Blanche E.
Wildflowers of Alabama and
    Adjoining States

fl.   p 27

Hymenocallis occidentalis

Miles, Bebe
Bulbs for the Home Gardener

fl.,hab.   p.78, 144-45

Hymenocallis occidentalis

Wharton, Mary E.
A Guide to the Wildflowers & Ferns
    of Kentucky

fl.   p 56   Pl. 1.18

Hymenocallis rotata

Walcott, Mary Vaux
North American Wild Flowers

fl.   Pl. 154

Hymenocallis speciosa

Kiaer, Eigil
Indoor Plants in Colour

fl.   p. 71

Hymenocallis speciosa

Tsukamoto, Yotaro
Coloured Illustrations of Garden
Flowers, v.9
fl.
opp.p.12

Hymenoclea salsola

Coyle, Jeanette
A Field Guide to the Common and
    Interesting Plants of Baja California

hab.   p 173

Hymenoclea salsola

Munz, Philip A.
California Desert Wildflowers

fl.   p 54, Pl. 83

Hymenogyne glabra

Herre, H.
The Genera of the Mesembryanthemaceae

fl.   fr.   p 177

Hymenopappus scabiosaeus

Brown, Clair A.
Wildflowers of Louisiana and
    Adjoining States

fl.   p 208

Hymenopogon parasiticus

Royal Hort. Soc.
The Garden
    vol 103, No. 7   1978

hab.   p. 262

Hymenosporum flavum

Blombery, Alec M.
What Wildflower is That

fl.   p 175, Pl. 470

Hymenosporum flavum

Harrison, R.E.
Trees and Shrubs

fl.   P 92

Hymenosporum flavum

Macoby, Stirling
What Flower is That

fl.   P 159

Hymenosporum flavum

Mathias, Mildred E.
Color for the Landscape

fl. hab.   p 21

Hymenosporum flavum

Oakman, Harry
Colorful Trees

Hab. Fl.   P 41

Hymenoxys acaulis

Munz, Philip A.
California Desert Wildflowers

fl.   p 55, Pl. 86

Hymenoxys argentea

Lemmon, Robert S.
Wildflowers of North America in
    Full Color

fl.   p. 107   Pl. 173

Hymenoxys grandiflora

Orr, Robert T.
Wildflowers of Western America

fl.   Pl. 68

Hymenoxys subintegra

Welsh, Stanley L.
Flowers of the Canyon Country

fl.   p 35

Hyobanche sanguinea

Eliovson, Sima
Namaqualand in Flower

fl.   Pl. 26, 3

Hyoscyamus albus

Polunin, Oleg
Flowers of Europe

fl.   Pl. 117 #1177

Hyoscyamus albus

Polunin, Oleg
Flowers of the Mediterranean

fl.   No. 165

Hyoscyamus aureus

Morley, Brian D.
Wild Flowers of the World

fl.,fr.   Pl. 38

Hyoscyamus aureus

Polunin, Oleg
Flowers of Europe

fl.   Pl. 117 #1177

Hyoscyamus aureus

Polunin, Oleg
Flowers of the Mediterranean

fl.   No. 166

Hyoscyamus aureus

Royal Hort. Soc.
Journal of the Royal Hort. Soc.
     vol 88, No. 9   1963

fl.   p. 388   Pl. 149

Hyoscyamus niger

Ary, S.
The Oxford Book of Wildflowers

fl.   p 28, Pl. 6

Hyoscyamus niger

Bianchini, Francesco
Health Plants of the World

fl.   p 119

Hyoscyamus niger

Color Treasury of Herbs & Other
     Medicinal Plants

fl.   p 48   Pl. 70

Hyoscyamus niger

Felsko, Elsa
A Book of Wildflowers

fl.   p. 68

Hyoscyamus niger

Hvass, Elsie
Plants That Feed and Serve Us

fl.   p 87, Pl. 183

Hyoscyamus niger

Kimura, Koiti
Japanese Medicinal Plants, Vol II

fl.   p 223

Hyoscyamus niger

Lemmon, Robert S.
Wildflowers of North America in
     Full Color

fl.   p. 201   Pl. 314

Hyoscyamus niger

Lindman, C. A. M.
Nordens Flora, Vol 8

fl. fr.   Pl. 517

Hyoscyamus niger

Loewenfeld, Claire
The Complete Book of Herbs and Spices

fl. hab.   p 128

Hyoscyamus niger

Martin, W. Keble
The Concise British Flora in Colour

fl. hab.   Pl. 61

Hyoscyamus niger

Perrot, Emile
Les Plantes Medicinales

fl. fr. hab.   p 128

Hyoscyamus niger

Polunin, Oleg
Flowers of Europe

fl.   Pl. 117 #1176

Hyoscyamus niger

Webster, Mary
Flora of Moray, Nairn & East
     Inverness

fl.   p. 292   Pl. 16

Hyparrhenia hirta

Feinbrun-Dothan, Naomi
Wild Plants in the Land
     of Israel

fr. hab.   p 68

Hyparrhenia hirta

Polunin, Oleg
Flowers of Europe

fr.   Pl. 182 #1809

Hypecoum imberbe

Polunin, Oleg
Flowers of Europe

fl.   Pl. 30 #276

Hypecoum procumbens

Polunin, Oleg
Flowers of the Mediterranean

fl.   No. 36

Hypericophyllum scabridum

Moriarty, Audrey
Wild Flowers of Malawi

fl.   Pl. 78; 2

Hypericum acmosepalum

Royal Hort. Soc.
Journal of the Royal Hort. Soc.
     vol 95, No. 11   1970

fl.   p. 488   Pl. 238

Hypericum anagalloides

Clark, Lewis J.
Wild Flowers of British Columbia

fl.   p 318

Hypericum androsaemum

Martin, W. Keble
The Concise British Flora in Colour

fl. fr. hab.   Pl. 17

Hypericum androsaemum

Polunin, Oleg
Flowers of Europe

fl. fr.   Pl. 74 #763

Hypericum ascyron

Kimura, Koiti
Japanese Medicinal Plants, Vol II

fl. fr.   p 201

Hypericum beanii

Huxley, Anthony
Deciduous Garden Trees and Shrubs

fl.   Pl. 93

Hypericum beanii

Royal Hort. Soc.
Journal of the Royal Hort. Soc.
     vol 95, No. 11   1970

fl.   p. 488   Pl. 235

Hypericum beanii var.

Royal Hort. Soc.
Journal of the Royal Hort. Soc.
     vol 95, No. 11   1970

fl.   p. 488   Pl. 236

Hypericum bellum

Royal Hort. Soc.
Journal of the Royal Hort. Soc.
     vol 95, No. 11   1970

fl.   p. 488   Pl. 239

Hypericum calycinum

Crockett, James Underwood
Lawns & Ground Covers

fl.   p. 136

Hypericum calycinum

Hay, Roy
The Color Dictionary of Flowers & Plants

fl.   p 207, Pl. 1655

Hypericum calycinum

Huxley, Anthony
Evergreen Garden Trees and Shrubs

fl. fr.   Pl. 128

Hypericum calycinum

Martin, W. Keble
The Concise British Flora in Colour

fl. hab.   Pl. 17

Hypericum calycinum

Morley, Brian D.
Wild Flowers of the World

fl.,hab.   Pl. 45

Hypericum calycinum

Perry, Frances
Flowers of the World

fl.   p 134

Hypericum calycinum

Polunin, Oleg
Flowers of Europe

fl. Pl. 75 #763

Hypericum canariense

Bramwell, David
Wild Flowers of the Canary Islands

fl.   Pl. 198

Hypericum cernuum

Roxburgh, William
Icones Roxburghianae

fl.   Pl. 12

Hypericum chinense

Macoby, Stirling
What Flower is That

fl.   P 160

Hypericum chinense

Royal Hort. Soc.
Journal of the Royal Hort. Soc.
  vol 96, No. 1   1971

fl.   p. 22   Pl. 19

Hypericum chinense

Tsukamoto, Yotaro
Coloured Illustrations of Garden
  Flowers V. 10

fl. Illus. 168 opp. p. 52

Hypericum cistifolium

Brown, Clair A.
Wildflowers of Louisiana and
  Adjoining States

fl.   p 113

Hypericum cistifolium

Fleming, Glenn
Wild Flowers of Florida

fl. hab.   p 57

Hypericum cuneatum

Royal Hort. Soc.
Journal of the Royal Hort. Soc.
  vol 98, No. 6   1973

fl.   p. 260   Pl. 139

Hypericum densiflorum

Brown, Clair A.
Wildflowers of Louisiana and
  Adjoining States

fl. hab.   p 113

Hypericum denticulatum

Batson, Wade T.
Wild Flowers in South Carolina

fl.   p 76

Hypericum dolabriforme

Wharton, Mary E.
A Guide to the Wildflowers & Ferns
  of Kentucky

fl. p 100   Pl. 1.12

Hypericum elatum

Curtis's Botanical Magazine, v. 173
  1960

fl., fr.   Pl. 378

Hypericum elatum var.

Hay, Roy
The Color Dictionary of Flowers & Plants

fl. fr.   p 207, Pl. 1656

Hypericum elodes

Kleijn, H.
Beauty of the Wild Plant

fl. opp. p. 56   Pl. 67

Hypericum elodes

Martin, W. Keble
The Concise British Flora in Colour

fl. fr. hab.   Pl. 17

Hypericum elodes

Polunin, Oleg
Flowers of Europe

fl.   Pl. 75 #766

Hypericum empetrifolium

Pacific Hort. Foundation
Pacific Horticulture
  vol 38, No. 1   1977

fl.   P. 18

Hypericum erectum

Kimura, Koiti
Japanese Medicinal Plants, Vol I

fl.   p 74

Hypericum ericoides

Curtis's Botanical Magazine
Vol 165   1948

fl.   36

Hypericum floribundum

Mathias, Mildred E.
Color for the Landscape

fl. hab.   p 71

Hypericum formosum

Clark, Lewis J.
Wild Flowers of British Columbia

fl.   p 314

Hypericum formosum

Orr, Robert T.
Wildflowers of Western America

fl.   Pl. 84

Hypericum frondosum

Batson, Wade T.
Wild Flowers in South Carolina

fl.   p 75

Hypericum galioides

Dean, Blanche E.
Wildflowers of Alabama and
  Adjoining States

fl.   p 107

Hypericum gentianoides

Courtenay, Booth
Wildflowers & Weeds

fl.   p 27

Hypericum gentianoides

Dean, Blanche E.
Wildflowers of Alabama and
  Adjoining States

fl.   p 107

Hypericum gentianoides

Duncan, Wilbur H.
Wildflowers of the Southeastern
  United States

fl.   p 101

Hypericum glandulosum

Bramwell, David
Wild Flowers of the Canary Islands

fl.   Pl. 199

Hypericum graveolens

Campbell, Carlos C.
Great Smoky Mountain Wildflowers

fl.   p 83

Hypericum hirsutum

Ary, S.
The Oxford Book of Wildflowers

fl.   p 12, Pl. 2

Hypericum hirsutum

Martin, W. Keble
The Concise British Flora in Colour

fl. hab. Pl. 17

Hypericum hirsutum

Polunin, Oleg
Flowers of Europe

fl. Pl. 75 #765

Hypericum humifusum

Ary, S.
The Oxford Book of Wildflowers

fl.   p 12, Pl. 3

Hypericum humifusum

Martin, W. Keble
The Concise British Flora in Colour

fl. fr. hab. Pl. 17

Hypericum hybrid

Crockett, James Underwood
Evergreens

fl. p. 68, 128

Hypericum hybrid.

Gault, S. Millar
The Color Dictionary of Shrubs

fl. Pl. 254

Hypericum hybrid

Hay, Roy
The Color Dictionary of Flowers &
 Plants

fl. p. 208 Pl. 1658

Hypericum hybrid

Perry, Frances
Flowers of the World

fl. p. 135

Hypericum hypericoides

Dean, Blanche E.
Wildflowers of Alabama and
 Adjoining States

fl. p 109

Hypericum indorum var.

Gault, S. Millar
The Color Dictionary of Shrubs

fl. fr. Pl. 253

Hypericum japonicum

Cochrane, G. R.
Flowers and Plants of Victoria

fl. Pl. 363

Hypericum kalmianum

Courtenay, Booth
Wildflowers & Weeds

fl. p. 26

Hypericum kalmianum

Ferguson, Mary
Wildflowers

fl. p 129

Hypericum kelleri

Pacific Hort. Foundation
Pacific Horticulture
 vol 38, No. 1 1977

fl., hab. p. 19

Hypericum kouytchense

Royal Hort. Soc.
Journal of the Royal Hort. Soc.
 vol 95, No. 11 1970

fl. p. 488 Pl. 240

Hypericum leschenaultii

Harrison, R.E.
Trees and Shrubs

fl. P 92

Hypericum leschenaultii

Morley, Brian D.
Wild Flowers of the World

fl. Plate 121

Hypericum leucoptychodes

Flowering Plants of South Africa
 vol XX 1940

fl. Pl. 787

Hypericum linarifolium

Martin, W. Keble
The Concise British Flora in Colour

fl. hab. Pl. 17

Hypericum maculatum

Lindman, C. A. M.
Nordens Flora, Vol 6

fl. Pl. 389B

Hypericum maculatum

Martin, W. Keble
The Concise British Flora in Colour

fl. hab. Pl. 17

Hypericum montanum

Martin, W. Keble
The Concise British Flora in Colour

fl. hab. Pl. 17

Hypericum montanum

Polunin, Oleg
Flowers of Europe

fl. Pl. 75 #764

Hypericum x moseranum

Gault, S. Millar
The Color Dictionary of Shrubs

fl. Pl. 255

Hypericum x moseranum

Hellyer, A.G.L.
Shrubs in Colour

fl. p. 70

Hypericum x moseranum var.

Gault, S. Millar
The Color Dictionary of Shrubs

fl.,hab. Pl. 256

Hypericum x moseranum var.

Harrison, R.E.
Trees and Shrubs

hab. p. 92

Hypericum x moseranum var.

Perry, Frances
Flowers of the World

fl. p. 134

Hypericum myrtifolium

Dean, Blanche E.
Wildflowers of Alabama and
 Adjoining States

fl. p 109

Hypericum nummularium

Perrot, Emile
Les Plantes Medicinales

fl. fr. hab. p 151

Hypericum olympicum

Goulimis, Constantine N.
Wild Flowers of Greece

fl. p 33

Hypericum olympicum

Vilmorin, Roger de
Plantes Alpines dans les Jardins

fl. hab. Pl. XXII

Hypericum patulum

Bartels, Andreas
Das Grosse Buch der Gartengeholze

fl.   p 160

Hypericum patulum var.

Everett, T.H.
New Illustrated Encyclopedia of
Gardening, v.5
fl.
opp.p.887

Hypericum patulum var.

Hansen, Richard
Baume und Straucher im Garten

fl.   p. 172

Hypericum patulum var.

Harrison, R.E.
Trees and Shrubs

fl. P 92

Hypericum patulum var.

Hay, Roy
The Color Dictionary of Flowers & Plants

fl.   p 208, Pl. 1657

Hypericum patulum var.

Hellyer, A.G.L.
Shrubs in Color
fl.
p. 70

Hypericum patulum var.

Royal Hort. Soc.
Journal of the Royal Hort. Soc.
    vol 80, No. 12   1955

    fl.   p. 548   Pl. 129

Hypericum patulum var.

Royal Hort. Soc.
The Garden
    vol 103, No. 12   1978

fl., hab.   p. 465

Hypericum peplidifolium

Moriarty, Audrey
Wild Flowers of Malawi

fl.   Pl. 73: 2

Hypericum perforatum

Ary, S.
The Oxford Book of Wildflowers

fl.   p 12, Pl. 1

Hypericum perforatum

Bianchini, Francesco
Health Plants of the World

fl.   p 171

Hypericum perforatum

Clark, Lewis J.
Wild Flowers of British Columbia

fl.   p 330

Hypericum perforatum

Courtenay, Booth
Wildflowers & Weeds

fl.   p 26

Hypericum perforatum

Felsko, Elsa
A Book of Wildflowers

fl.   P 30

Hypericum perforatum

Klimas, John E.
Wildflowers of Eastern America

fl.   Pl. 164

Hypericum perforatum

Lindman, C. A. M.
Nordens Flora, Vol 6

fl.   Pl. 389A

Hypericum perforatum

Martin, W. Keble
The Concise British Flora in Colour

fl. hab.   Pl. 17

Hypericum perforatum

Moyle, John B.
Northland Wild Flowers

fl. hab.   p 89, Pl. 72

Hypericum perforatum

Perrot, Emile
Les Plantes Medicinales

fl. fr. hab.   p 151

Hypericum perforatum

Polunin, Oleg
Flowers of Europe

fl.   Pl. 75 #768

Hypericum perforatum

Pond, Barbara
A Sampler of Wayside Herbs

fl.   Pl. X

Hypericum perforatum

Wharton, Mary E.
A Guide to the Wildflowers & Ferns
    of Kentucky

fl.   p 99 Pl. 1.11

Hypericum polyphyllum

Hay, Roy
The Color Dictionary of Flowers & Plants

fl.   p 12, Pl. 90

Hypericum polyphyllum

Pacific Hort. Foundation
Pacific Horticulture
    vol 38, No. 1   1977

fl.   p. 18

Hypericum pulchrum

Lindman, C. A. M.
Nordens Flora, Vol 6

fl. hab.   Pl. 388

Hypericum pulchrum

Martin, W. Keble
The Concise British Flora in Colour

fl. hab.   Pl. 17

Hypericum punctatum

Courtenay, Booth
Wildflowers & Weeds

fl.   p 27

Hypericum punctatum

Duncan, Wilbur H.
Wildflowers of the Southeastern
    United States

fl.   p 101

Hypericum pyramidatum

Courtenay, Booth
Wildflowers & Weeds

fl.   p 26

Hypericum pyramidatum

Moyle, John B.
Northland Wild Flowers

fl. hab.   p 89, Pl. 73

Hypericum revolutum

Moriarty, Audrey
Wild Flowers of Malawi

fl.   Pl. 73: 1

Hypericum rhodopaeum

Hay, Roy
The Color Dictionary of Flowers & Plants

fl.   p 12, Pl. 91

Hypericum scouleri

Lemmon, Robert S.
Wildflowers of North America in
    Full Color

fl.   p. 136   Pl. 216

Hypericum Scouleri

Porsild, A. E.
Rocky Mountain Wild Flowers

fl. hab.   p 277

Hypericum sonderi

Flowering Plants of South Africa
        vol XXIII   1943

fl.   Pl. 897

Hypericum spathulatum

Wharton, Mary E.
A Guide to the Wildflowers & Ferns
of Kentucky

fl.   p 98 Pl. 1.9

Hypericum spathulatum

Wharton, Mary E.
Trees & Shrubs of Kentucky

fl.   p 54, Pl. 1.15

Hypericum sp.

Milne, Lorus
Living Plants of the World

fl.   p 152

Hypericum sphaerocarpum

Wharton, Mary E.
A Guide to the Wildflowers & Ferns
of Kentucky

fl.   p 100 Pl. 1.13

Hypericum stans

Dean, Blanche E.
Wildflowers of Alabama and
Adjoining States

fl.   p 107

Hypericum stellatum

Royal Hort. Soc.
Journal of the Royal Hort. Soc.
vol 95, No. 11   1970

fl.   p. 488   Pl. 237

Hypericum tetrapterum

Martin, W. Keble
The Concise British Flora in Colour

fl. fr. hab.   Pl. 17

Hypericum trichocaulon

Royal Hort. Soc.
Journal of the Royal Hort. Soc.
vol 85, No. 9   1960

fl.   p. 398   Pl. 126

Hypericum undulatum

Martin, W. Keble
The Concise British Flora in Colour

fl. fr. hab.   Pl. 17

Hyphaene thebaica

Edlin, Herbert
The Illustrated Encyclopedia
of Trees

hab.   p 232

Hypocalymma angustifolium

Blombery, Alec M.
What Wildflower is That

fl.   p 175, Pl. 471

Hypocalymma cordifolium

Blombery, Alec M.
What Wildflower is That

fl.   p 175, Pl. 472

Hypocalymma robustum

Blombery, Alec M.
What Wildflower is That

fl.   p 175, Pl. 473

Hypocalymma robustum

Harrison, R.E.
Trees and Shrubs

fl.   P 93

Hypocalymma robustum

Morcombe, M. K.
Australia's Western Wildflowers

fl.   p 1

Hypochoeris glabra

Martin, W. Keble
The Concise British Flora in Colour

fl. fr. hab.   Pl. 50

Hypochoeris maculata

Lindman, C. A. M.
Nordens Flora, Vol 10

fl. hab.   Pl. 638

Hypochoeris maculata

Martin, W. Keble
The Concise British Flora in Colour

fl. hab.   Pl. 50

Hypochoeris radicata

Ary, S.
The Oxford Book of Wildflowers

fl.   p 36, Pl. 5

Hypochoeris radicata

Lindman, C. A. M.
Nordens Flora, Vol 10

fl. fr. hab.   Pl. 639

Hypochoeris radicata

Martin, W. Keble
The Concise British Flora in Colour

fl. fr. hab.   Pl. 50

Hypochoeris uniflora

Barneby, T. P.
European Alpine Flowers in Colour

fl. hab.   Pl. 94, 2

Hypochoeris uniflora

Kohlhaupt, Paula
Fleurs des Alpages, v.1

fl.   P 121

Hypochoeris uniflora

Polunin, Oleg
Flowers of Europe

fl.   Pl. 158 #1521

Hypocyrta glabra

Encke, Fritz
Zimmerpflanzen

fl., hab.   p. 97

Hypocyrta glabra

Hay, Roy
The Dictionary of House Plants

fl.   Pl. 295

Hypocyrta hybrid

American Gloxinia & Gesneriad Society
The Gloxinian
vol 19, No. 6   1969

fl.   cover

Hypocyrta nummularia

Curtis's Botanical Magazine
vol 176   1966-68

fl.   Pl. 497

Hypocyrta nummularia

Morley, Brian D.
Wild Flowers of the World

fl.   Pl. 185

Hypocyrta radicans

Nicolaisen, Age
Pocket Encyclopedia of Indoor Plants

fl.   Pl. 120

Hypocyrta strigillosa

Kromdijk, G.
200 House Plants in Colour

fl.   Pl. 111

Hypoestes sanguinolenta

Kromdijk, G.
200 House Plants in Colour

hab.   Pl. 112

Hypoestes sanguinolenta

Macoby, Stirling
What Flower is That

hab P 160

Hypoestes taeniata

Curtis's Botanical Magazine, v.175
1964-65
fl.
pl. 477

Hypoestes taeniata

Hay, Roy
The Dictionary of House Plants

fl.   Pl. 292

Hypoestes taeniata

Morley, Brian D.
Wild Flowers of the World

fl.   Pl. 64

Hypopitys americana

Walcott, Mary Vaux
North American Wild Flowers

fl. hab.   Pl. 157

Hypopitys lanuginosa

Walcott, Mary Vaux
North American Wild Flowers

fl. hab.   Pl. 213

Hypopitys latisquama

Heller, Christine
Wild Flowers of Alaska

hab.   Pl. 276

Hypopitys monotropa

Clark, Lewis J.
Wild Flowers of British Columbia

fl.   p 378

Hypopitys monotropa

Fries, Mary A.
Wildflowers of Mount Ranier and
the Cascades

fl. P 61

Hypopitys monotropa

Taylor, Ronald J.
Mountain Wild Flowers

fl. hab.   p 44

Hypoxis dregei

Moriarty, Audrey
Wild Flowers of Malawi

fl.   Pl. 4; 4

Hypoxis hirsuta

Batson, Wade T.
Wild Flowers in South Carolina

fl. hab.   p 36

Hypoxis hirsuta

Brown, Clair A.
Wildflowers of Louisiana and
Adjoining States

fl. hab.   p 26

Hypoxis hirsuta

Campbell, Carlos C.
Great Smoky Mountain Wildflowers

fl. hab.   p 31

Hypoxis hirsuta

Courtenay, Booth
Wildflowers & Weeds

fl.   p 9

Hypoxis hirsuta

Dean, Blanche E.
Wildflowers of Alabama and
Adjoining States

fl.   p 29

Hypoxis hirsuta

Duncan, Wilbur H.
Wildflowers of the Southeastern
United States

fl. hab.   p 265

Hypoxis hirsuta

Jennings, O. E.
Wild Flowers of Western Pennsylvania
v. 2

fl. hab.   Pl. 32

Hypoxis hirsuta

Klimas, John E.
Wildflowers of Eastern America

fl.   Pl. 142

Hypoxis hirsuta

Lemmon, Robert S.
Wildflowers of North America in
Full Color

fl.   p. 168   Pl. 265

Hypoxis hirsuta

Miles, Bebe
Bulbs for the Home Gardener

fl.   p 78

Hypoxis hirsuta

Moyle, John B.
Northland Wild Flowers

fl. hab.   p 215, Pl. 285

Hypoxis hirsuta

Wharton, Mary E.
A Guide to the Wildflowers & Ferns
of Kentucky

fl. hab. p 51 Pl. 1.8

Hypoxis hygrometrica

Blombery, Alec M.
What Wildflower is That

fl.   p 175, Pl. 474

Hypoxis juncea

Fleming, Glenn
Wild Flowers of Florida

fl. hab.   p 54

Hypoxis longifolia

Macoby, Stirling
What Flower is That

fl. P 160

Hypoxis nitida

Flowering Plants of Africa
vol 27   1948-49

fl.   Pl. 1058

Hypoxis obtusa

Moriarty, Audrey
Wild Flowers of Malawi

fl.   Pl. 4; 5

Hypoxis rigidula

Flowering Plants of Africa
vol XXVI   1947

fl.   Pl. 1021

Hypoxis sp.

Morley, Brian D.
Wild Flowers of the World

fl., hab.   Pl. 66

Hypsela reniformis

Curtis's Botanical Magazine
vol 176   1966-68

fl.   Pl. 395B

Hypsela reniformis

Morley, Brian D.
Wild Flowers of the World

fl.   Pl. 172

Hyptis alata

Fleming, Glenn
Wild Flowers of Florida

fl.   p 27

Hyptis emoryi

Coyle, Jeanette
A Field Guide to the Common and
Interesting Plants of Baja California

fl.   p 155

Hyssop
see
Hyssopus

Hyssop, Fragrant Giant
see
Agastache foeniculum

Hyssop, Giant
see
Agastache nepetoides

Hyssop, Giant
see
Agastache scrophulariaefolia

Hyssop, Hedge
see
Gratiola brevifolia

Hyssop, Hedge
see
Gratiola neglecta

Hyssopus aristatus

Hay, Roy
The Color Dictionary of Flowers & Plants

fl.   p 148, Pl. 1181

Hyssopus officinalis

Bianchini, Francesco
Health Plants of the World

fl. hab.   p 43

Hyssopus officinalis

Massachusetts Hort. Soc.
Horticulture
   vol 54, No. 3   1976

fl.   p. 52

Hyssopus officinalis

Perrot, Emile
Les Plantes Medicinales

fl. fr. hab.   p 123

Hyssopus officinalis

Polunin, Oleg
Flowers of Europe

fl.   Pl. 115 #1159

Hyssopus officinalis var.

Royal Hort. Soc.
Journal of the Royal Hort. Soc.
   vol 96, No. 12   1971

fl.   p. 540   Pl. 231

Hystrix patula

Wharton, Mary E.
A Guide to the Wildflowers & Ferns
   of Kentucky

fl. p 88 Pl. 2.11

Iberis amara

Macoby, Stirling
What Flower is That

fl.   p 161

Iberis amara

Martin, W. Keble
The Concise British Flora in Colour

fl. fr. hab.   Pl. 10

Iberis amara

Polunin, Oleg
Flowers of Europe

fl.   Pl. 36 #348

Iberis amara

Tsukamoto, Yotaro
Coloured Illustrations of Garden
   Flowers, v. 10

fl. ill. 73 opp. p. 23

Iberis gibraltarica

Royal Hort. Soc.
The Garden
   vol 101, No. 4   1976

fl., hab.   p. 187

Iberis saxatilis

Barneby, T. P.
European Alpine Flowers in Colour

fl. hab.   Pl. 31, 4

Iberis saxatilis

Hay, Roy
The Color Dictionary of Flowers & Plants

fl.   p 12, Pl. 92

Iberis sempervirens

Crockett, James Underwood
Lawns & Ground Covers

fl. p. 136

Iberis sempervirens var.

Hay, Roy
The Color Dictionary of Flowers & Plants

fl.   p 148, Pl. 1182

Iberis sempervirens var.

Huxley, Anthony
Garden Perennials and Water Plants

fl.   Pl. 145

Iberis sempervirens var.

Vilmorin, Roger de
Plantes Alpines dans les Jardins

fl.   Pl. XXXIII

Iberis umbellata

Crockett, James Underwood
Annuals

fl.   P 126

Iberis umbellata

Huxley, Anthony
Garden Annuals and Bulbs

fl.   Pl. 64

Iberis umbellata var.

Hay, Roy
The Color Dictionary of Flowers &
   Plants

fl.   p. 39   Pl. 310

Iberis welwitschii

Curtis's Botanical Magazine
   Vol 162   1939

fl.   No. 9573

Ibidium cernuum

Walcott, Mary Vaux
North American Wild Flowers

fl.   Pl. 4

Ibidium gracile

Walcott, Mary Vaux
North American Wild Flowers

fl.   Pl. 4

Ibidium strictum

Walcott, Mary Vaux
North American Wild Flowers
   vol. 5

fl.   Pl. 356

Flowering Plant Index of Illustration and Information

Iboza riparia
Flowering Plants of South Africa vol XX 1940
fl. Pl. 767

Iboza riparia
Macoby, Stirling
What Flower is That
fl. P. 161

Ice plant
see
Dorotheanthus bellidiformis

Ice Plant
see
Lampranthus

Ice Plant
see
Mesembryanthemum crystallinum

Ice-plant
see
Sedum spectabile

Ice-plant, orange trailing
see
Lampranthus aurantiacus

Ice plant, pink
see
Lampranthus spectabilis

Ice Plant, Rosea
see
Drosanthemum hispidum

Ice Plant, Trailing
see
Lampranthus spectabilis

Idesia polycarpa
Curtis's Botanical Magazine
Vol 179 1972
fl. fr. 649

Idesia polycarpa
Harrison, R. E.
Trees and Shrubs
fr. P. 93

Idesia polycarpa
Kitamura, Siro
Coloured Illustrations of Trees & Shrubs of Japan
fl. fr. 351, 352

Idria columnaris
Coyle, Jeanette
A Field Guide to the Common and Interesting Plants of Baja California
hab. p 129

Idria columnaris
Lamb, Edgar
The Pocket Encyclopedia of Cacti and Succulents in Color
hab. Pl. 324

Idria columnaris
Wit, H. C. D. de
Plants of the World; The Higher Plants, Vol I
hab. p 209, Pl. 108

Ilex x altaclarensis
Gault, S. Millar
The Color Dictionary of Shrubs
hab. Pl. 257

Ilex aquifolium
American Hort. Soc.
American Horticultural Magazine vol 49, No. 4 1970
fr. cover

Ilex aquifolium
American Hort. Soc.
American Horticulturist vol 54, No. 6 1975
fr. p. 22

Ilex aquifolium
Ary, S.
The Oxford Book of Wildflowers
fl. fr. p 182, Pl. 3

Ilex aquifolium
Bartels, Andreas
Das Grosse Buch der Gartengeholze
fl. p 162

Ilex aquifolium
Bianchini, Francesco
Health Plants of the World
fr. hab. p 17

Ilex aquifolium
Crockett, James Underwood
Evergreens
hab., fr. p. 129

Ilex aquifolium
Edlin, Herbert
The Illustrated Encyclopedia of Trees
fl. fr. hab. p 186-7

Ilex aquifolium
Flemer, William
Shade and Ornamental Trees in Color
hab. p. 83

Ilex aquifolium
Hay, Roy
The Color Dictionary of Flowers & Plants
fr. p. 208 Pl. 1659

Ilex aquifolium
Huxley, Anthony
Evergreen Garden Trees and Shrubs
fr. Pl. 129

Ilex aquifolium
Lindman, C. A. M.
Nordens Flora, Vol 6
fl. fr. Pl. 378

Ilex aquifolium
Martin, W. Keble
The Concise British Flora in Colour
fl. fr. hab. Pl. 20

Ilex aquifolium
Massachusetts Hort. Soc.
Horticulture vol 32, No. 5 1954
hab. fr. p. 250

Ilex aquifolium
Massachusetts Hort. Soc.
Horticulture vol 36, No. 12 1958
fr. backcover

Ilex aquifolium
Massachusetts Hort. Soc.
Horticulture vol 38, No. 12 1960
fr. backcover

Ilex aquifolium
Massachusetts Hort. Soc.
Horticulture vol 44, No. 12 1966
fr. inside backcover

722

Ilex aquifolium

Massachusetts Hort. Soc.
Horticulture
vol 51, No. 12 1973

fr.   p. 30

Ilex aquifolium

Meikle, R.D.
British Trees and Shrubs

fl. fr.   Pl. 1

Ilex aquifolium

Polunin, Oleg
Flowers of Europe

fl.   Pl. 71 #717

Ilex aquifolium

Polunin, Oleg
Trees and Bushes of Europe

fl. fr. hab.   p 129

Ilex aquifolium

Tosco, Uberto
The World of Mountain Flowers

fr.   p 28

Ilex aquifolium

Vedel, H.
Arbres et Arbustes

fr. hab.   p. 90, No. 93

Ilex aquifolium var.

American Hort. Soc.
American Horticulturist
vol. 54, No. 6 1975

hab.,fr.   p. 19, 22

Ilex aquifolium var.

Harrison, R. E.
Trees and Shrubs

fr.   P. 94

Ilex aquifolium var.

Hay, Roy
The Color Dictionary of Flowers & Plants

fr. hab.   p 208, Pl. 1659 thru 1662

Ilex aquifolium var.

Huxley, Anthony
Evergreen Garden Trees and Shrubs

hab.   Pl. 130

Ilex aquifolium var.

Macoby, Stirling
What Flower is That

fr.   P 162

Ilex aquifolium var.

Massachusetts Hort. Soc.
Horticulture
vol 32, No. 5   1954

hab., fr.   p. 250

Ilex aquifolium var.

Massachusetts Hort. Soc.
Horticulture
vol 33, No. 11 1955

fr.   p. 525

Ilex aquifolium var.

Massachusetts Hort. Soc.
Horticulture
vol 43, No. 12   1965

fr.   p. 22

Ilex asprella

Walden, Beryl M.
Wild Flowers of Hong Kong

fl.   Pl. 13(39)

Ilex canariensis

Bramwell, David
Wild Flowers of the Canary Islands

fl.   Pl. 194

Ilex cassine

Fleming, Glenn
Wild Flowers of Florida

fr. hab.   p 75

Ilex cornuta

Massachusetts Hort. Soc.
Horticulture
vol 32, No. 5   1954

hab.   p. 250

Ilex cornuta var.

Crockett, James Underwood
Evergreens

hab.,fr.   p. 129

Ilex cornuta var.

Massachusetts Hort. Soc.
Horticulture
vol 32, No. 5   1954

fr.   p. 250

Ilex crenata

American Hort. Soc.
American Horticulturist
vol 54, No. 6 1975

fr.   p. 23

Ilex crenata

Crockett, James Underwood
Evergreens

hab. fr.   p. 130

Ilex crenata

Kitamura, Siro
Coloured Illustrations of Trees &
Shrubs of Japan

fr.   288

Ilex crenata

Massachusetts Hort. Soc.
Horticulture
vol 32, No. 5   1954

hab.   p. 250

Ilex crenata var.

Crockett, James Underwood
Evergreens

hab., fr.   p. 130

Ilex crenata var.

Kitamura, Siro
Coloured Illustrations of Trees &
Shrubs of Japan

           289

Ilex crenata var.

Massachusetts Hort. Soc.
Horticulture
vol 32, No. 5   1954

hab.   p. 250

Ilex decidua

Brown, Clair A.
Wildflowers of Louisiana and
Adjoining States

fr.   p 105

Ilex decidua

Wharton, Mary E.
Trees & Shrubs of Kentucky

fl.   p 96, Pl. 2.31b
fr.   p 124, Pl. 2.10b

Ilex fargesii

Curtis's Botanical Magazine
Vol 164   1943-48

fl. fr.   No. 9670

Ilex glabra

Massachusetts Hort. Soc.
Horticulture
vol 32, No. 5   1954

hab.   p. 250

Ilex hybrid

Gault, S. Millar
The Color Dictionary of Shrubs

hab.   Pl. 257

Ilex hybrid

Royal Hort. Soc.
Journal of the Royal Hort. Soc.
vol 97, No. 12   1972

fr.   cover

Ilex integra

Kitamura, Siro
Coloured Illustrations of Trees &
Shrubs of Japan

fr.   293

Ilex laevigata

Batson, Wade T.
Wild Flowers in South Carolina

fr.   p 72

Ilex latifolia

Kitamura, Siro
Coloured Illustrations of Trees &
Shrubs of Japan

fr.   291

Ilex macrocarpa

Curtis's Botanical Magazine
Vol 166   1949

fl. fr.   72

Ilex macropoda

Kitamura, Siro
Coloured Illustrations of Trees &
Shrubs of Japan

fr.   284

Ilex melanotricha

Curtis's Botanical Magazine
Vol 166   1949

fl. fr.   84

Ilex montana

Wharton, Mary E.
Trees & Shrubs of Kentucky

fl.   p 96, Pl. 2.31d
fr.   p 125, Pl. 2.10d

Ilex opaca

American Hort. Soc.
American Horticulturist
    vol 54, No. 6 1975

fr.   p. 18

Ilex opaca

Brown, Clair A.
Wildflowers of Louisiana and
    Adjoining States

fr.   p 105

Ilex opaca

Campbell, Carlos C.
Great Smoky Mountain Wildflowers

fr.   p 105

Ilex opaca

Crockett, James Underwood
Evergreens

hab.,fr.   p. 130

Ilex opaca

Everett, T.H.
New Illustrated Encyclopedia of
    Gardening, v. 11

hab.,fr.   opp. p. 2039

Ilex opaca

Flemer, William III
Shade and Ornamental Trees in Color

fr., hab.   p. 94, 100

Ilex opaca

Massachusetts Hort. Soc.
Horticulture
    vol 31, No. 12   1953

fr.   p. 474

Ilex opaca

Massachusetts Hort. Soc.
Horticulture
    vol 40, No. 12   1962

fr.   p. 622

Ilex opaca

Massachusetts Hort. Soc.
Horticulture
    vol 47, No. 12   1969

fr.   p. 18

Ilex opaca

Massachusetts Hort. Soc.
Horticulture
    vol 49, No. 12   1971

fl.   p. 23

Ilex opaca

Walcott, Mary Vaux
North American Wild Flowers
    vol. 4

fr.   Pl. 266

Ilex opaca

Wharton, Mary E.
Trees & Shrubs of Kentucky

fl.   p 95, Pl. 2.31a
fr.   p 124, Pl. 2.10a

Ilex opaca var.

Massachusetts Hort. Soc.
Horticulture
    vol 32, No. 5   1954

fr.   p. 250

Ilex paraguariensis

Hvass, Elsie
Plants That Feed and Serve Us

fl. fr.   p 78, Pl. 162

Ilex pedunculosa

American Hort. Soc.
American Horticulturist
    vol 54, No. 6 1975

fr.   p. 19

Ilex pedunculosa

Kitamura, Siro
Coloured Illustrations of Trees &
Shrubs of Japan

fr.   290

Ilex pernyi

Huxley, Anthony
Evergreen Garden Trees and Shrubs

hab.   Pl. 131

Ilex rotunda

Kitamura, Siro
Coloured Illustrations of Trees &
Shrubs of Japan

fr.   292

Ilex serrata

Kitamura, Siro
Coloured Illustrations of Trees &
Shrubs of Japan

fr.   285

Ilex sugeroki var.

Kitamura, Siro
Coloured Illustrations of Trees &
Shrubs of Japan

fr.   287

Ilex verticillata

American Hort. Soc.
American Horticulturist
    vol 53, No. 4 1974

fl., fr.   p. 34-37

Ilex verticillata

American Hort. Soc.
American Horticulturist
    vol 54, No. 6 1975

fr.   p. 23

Ilex verticillata

American Hort. Soc.
American Horticulturist
    vol 56, No. 3 1977

fr.   p. 34

Ilex verticillata

Everett, T.H.
New Illustrated Encyclopedia of
Gardening, v. 5

fr.   opp. p. 887

Ilex verticillata

Ferguson, Mary
Wildflowers

fr.   p 23

Ilex verticillata

Massachusetts Hort. Soc.
Horticulture
    vol 32, No. 5   1954

fr.   p. 250

Ilex verticillata

Massachusetts Hort. Soc.
Horticulture
vol 33, No. 10   1955

fr.   p. 486

Ilex verticillata

Massachusetts Hort. Soc.
Horticulture
vol 38, No. 12   1960

fr.   cover

Ilex verticillata

Massachusetts Hort. Soc.
Horticulture
vol 50, No. 11   1972

hab., fr.   p. 31

Ilex verticillata

Pennsylvania Hort. Soc.
The Green Scene
vol 4, No. 2   1975

fr.   cover

Ilex verticillata

Royal Hort. Soc.
Journal of the Royal Hort. Soc.
vol 94, No. 2   1969

fr.   p. 86   Pl. 31

Ilex verticillata

Walcott, Mary Vaux
North American Wild Flowers

fr.   Pl. 54

Ilex verticillata

Wharton, Mary E.
Trees & Shrubs of Kentucky

fl.   p 96, Pl. 2.31c
fr.   p 125, Pl. 2.10c

Ilex vomitoria

American Hort. Soc.
American Horticulturist
vol 54, No. 6   1975

fr.   p. 19

Ilex vomitoria

Brown, Clair A.
Wildflowers of Louisiana and
Adjoining States

fr.   p 106

Ilex vomitoria

Crockett, James Underwood
Evergreens

hab., fr.   p. 131

Ilex vomitoria

Fleming, Glenn
Wild Flowers of Florida

fr. hab.   p 77

Ilex vomitoria

Walcott, Mary Vaux
North American Wild Flowers

fr. hab.   Pl. 226

Ilex vomitoria var.

American Hort. Soc.
American Horticulturist
vol 54, No. 6   1975

fr.   p. 19

Iliamna rivularis

Clark, Lewis J.
Wild Flowers of British Columbia

fl.   p 319

Iliamna rivularis

Lemmon, Robert S.
Wildflowers of North America in
Full Color

fl.   p. 136   Pl. 215

Iliamna rivularis

Shaw, Richard J.
Field Guide to the Vascular Plants of
Grand Teton National Park

fl.   Pl. 4

Illecebrum verticillatum

Martin, W. Keble
The Concise British Flora in Colour

fl. hab.   Pl. 71

Illecebrum verticillatum

Polunin, Oleg
Flowers of Europe

fl.   Pl. 13 #152

Illicium anisatum

Bianchini, F.
The Complete Book of Fruits & Vegetables

fr.   p 103

Illicium anisatum

Perry, Frances
Flowers of the World

fl.   p 142

Illicium anisatum

Royal Hort. Soc.
Journal of the Royal Hort. Soc.
vol 97, No. 12   1972

fl.   p. 528   Pl. 259

Illicium anisatum

Takatori, Jisuke
Color Atlas of Medicinal Plants of Japan

fl.
fr.   Fig. 49

Illicium dunnianum

Walden, Beryl E.
Wild Flowers of Hong Kong

fl.   Pl. 9(29)

Illicium floridanum

Brown, Clair A.
Wildflowers of Louisiana and
Adjoining States

fl. hab.   p 55

Illicium floridanum

Dean, Blanche E.
Wildflowers of Alabama and
Adjoining States

fl.   p 67

Illicium floridanum

Morley, Brian D.
Wild Flowers of the World

fl.   Pl. 150

Illicium floridanum

Royal Hort. Soc.
Journal of the Royal Hort. Soc.
vol 97, No. 12   1972

fl.   p. 528   Pl. 261-62

Illicium henryi

Royal Hort. Soc.
Journal of the Royal Hort. Soc.
vol 97, No. 12   1972

fl.   p. 528   Pl. 263

Illicium religiosum

Kitamura, Siro
Coloured Illustrations of Trees &
Shrubs of Japan

fl. fr.   174

Illicium verum

Bianchini, Francesco
Health Plants of the World

fr.   p 155

Illicium verum

Perrot, Emile
Les Plantes Medicinales

fl. fr. hab.   p 25

Illicium verum

Rosengarten, Frederic, Jr.
The Book of Spices

fl. fr.   p 105

Illicium verum

Wit, H. C. D. de
Plants of the World;
The Higher Plants, Vol I

fl.   p 57, Pl. 19

Imitaria muirii

Lamb, Edgar
The Pocket Encyclopedia of Cacti
   and Succulents in Color

fl. hab.   Pl. 266

Immortelle, common

      see

Xeranthemum annuum

Immortelle, Mountain

      see

Erythrina poeppigiana

Immortelle, Orange

      see

Waitzia acuminata

Immortelle, Swamp

      see

Erythrina glauca

Impatiens arguta

Hara, Hiroshi, comp.
Photo-Album of Plants of
   Eastern Himalaya

fl.   Pl. 87

Impatiens balsamina

Bruggeman, L.
Tropical Plants

fl.   Pl. 48

Impatiens balsamina

Crockett, James Underwood
Annuals

fl. P 127

Impatiens balsamina

Hay, Roy
The Dictionary of House Plants

fl.   Pl. 293

Impatiens balsamina

Huxley, Anthony
Garden Annuals and Bulbs

fl.   Pl. 65

Impatiens balsamina

Tsukamoto, Yotaro
Coloured Illustrations of Garden Flowers,
v.10
fl.
ill. 74 opp.p.23

Impatiens balsamina var.

Hay, Roy
The Color Dictionary of Flowers &
   Plants

fl.   p. 39   Pl. 311

Impatiens balsamina var.

Macoby, Stirling
What Flower is That

fl. P 163

Impatiens bicornuta

Hara, Hiroshi, comp.
Photo-Album of Plants of
   Eastern Himalaya

fl.   Pl. 86

Impatiens biflora

Batson, Wade T.
Wild Flowers in South Carolina

fl.   p 73

Impatiens biflora

Courtenay, Booth
Wildflowers & Weeds

fl.   p 57

Impatiens biflora

Lemmon, Robert S.
Wildflowers of North America in
   Full Color

fl.   p. 188   Pl. 295

Impatiens biflora

Moyle, John B.
Northland Wild Flowers

fl. hab.   p 88, Pl. 71

Impatiens capensis

Ary, S.
The Oxford Book of Wildflowers

fl.   p 12, Pl. 6

Impatiens capensis

Dean, Blanche E.
Wildflowers of Alabama and
   Adjoining States

fl.   p 103

Impatiens capensis

Duncan, Wilbur H.
Wildflowers of the Southeastern
   United States

fl.   p 99

Impatiens capensis

Ferguson, Mary
Wildflowers

fl.   p 37

Impatiens capensis

Klimas, John E.
Wildflowers of Eastern America

fl.   Pl. 136

Impatiens capensis

Martin, W. Keble
The Concise British Flora in Colour

fl. fr. hab.   Pl. 20

Impatiens capensis

Polunin, Oleg
Flowers of Europe

fl.   Pl. 70 #714

Impatiens capensis

Pond, Barbara
A Sampler of Wayside Herbs

fl.   Pl. XIV

Impatiens capensis

Wharton, Mary E.
A Guide to the Wildflowers & Ferns
   of Kentucky

fl.   p 212   Pl. 3.3

Impatiens chinensis

Walden, Beryl M.
Wild Flowers of Hong Kong

fl. fr.   Pl. 63 (184)

Impatiens congolensis var.

Perry, Frances
Flowers of the World

fl.   p. 39

Impatiens elegantissima

Royal Hort. Soc.
Journal of the Royal Hort. Soc.
   vol 95, No. 11   1970

fl.   p. 488   Pl. 246

Impatiens epiphytica

Curtis's Botanical Magazine
   Vol. 181 1976-77

fl.   Pl. 723

Impatiens glandulifera

Ary, S.
The Oxford Book of Wildflowers

fl.   p 108, Pl. 1

Impatiens glandulifera

Martin, W. Keble
The Concise British Flora in Colour

fl. hab.   Pl. 20

Impatiens glandulifera

Polunin, Oleg
Flowers of Europe

fl.   Pl. 70 #716

Impatiens gomphophylla

Moriarty, Audrey
Wild Flowers of Malawi

fl.   Pl. 47; 1

Impatiens graciliflora

Hara, Hiroshi, comp.
Photo-Album of Plants of
   Eastern Himalaya

fl.   Pl. 88

Impatiens herzogii

American Hort. Soc.
American Horticulturist
   vol 52, No. 3 1973

fl.   p. 18

Impatiens hochstetteri

Moriarty, Audrey
Wild Flowers of Malawi

fl.   Pl. 47; 4

Impatiens holstii

Bruggeman, L.
Tropical Plants

fl.   Pl. 49

Impatiens holstii

Kiaer, Eigil
Indoor Plants in Colour

fl.   p. 72

Impatiens hybrid

American Hort. Soc.
American Horticulturist
   vol 52, No. 3 1973

fl.   p. 17-22

Impatiens hybrid

American Hort. Soc.
American Horticulturist
   vol 53, No. 1 1974

fl.   p. 14-18

Impatiens hybrid

Kromdijk, G.
200 House Plants in Colour

fl.   P. 113

Impatiens hybrid

Perry, Frances
Flowers of the World

fl.   p. 39

Impatiens insignis

Hara, Hiroshi, comp.
Photo-Album of Plants of
   Eastern Himalaya

fl.   Pl. 82

Impatiens Jurpia

Hara, Hiroshi, comp.
Photo-Album of Plants of
   Eastern Himalaya

fl.   Pl. 84

Impatiens linearifolia

American Hort. Soc.
American Horticulturist
   vol 52, No. 3 1973

fl.   p. 19

Impatiens noli-tangere

Alaska-Yukon Wild Flower Guide

fl.   p 94

Impatiens noli-tangere

Heller, Christine
Wild Flowers of Alaska

fl.   Pl. 20

Impatiens noli-tangere

Kleijn, H.
Beauty of the Wild Plant

fl.   opp. p. 49   Pl. 58

Impatiens noli-tangere

Lindman, C. A. M.
Nordens Flora, Vol 6

fl.   Pl. 382

Impatiens noli-tangere

Martin, W. Keble
The Concise British Flora in Colour

fl.   hab.   Pl. 20

Impatiens noli-tangere

Polunin, Oleg
Flowers of Europe

fl.   Pl. 70 #714

Impatiens oliveri

Macoby, Stirling
What Flower is That

fl. P 163

Impatiens pallida

Campbell, Carlos C.
Great Smoky Mountain Wildflowers

fl.   p 55

Impatiens pallida

Wharton, Mary E.
A Guide to the Wildflowers & Ferns
   of Kentucky

fl.   p 212   Pl. 3.4

Impatiens parviflora

Felsko, Elsa
A Book of Wildflowers

fl.fr.   P 38

Impatiens parviflora

Polunin, Oleg
Flowers of Europe

fl.   Pl. 70 #715

Impatiens petersiana

Hay, Roy
The Color Dictionary of Flowers & Plants

fl.   p 69, Pl. 552

Impatiens puberula

Hara, Hiroshi, comp.
Photo-Album of Plants of
   Eastern Himalaya

fl.   Pl. 85

Impatiens richardsiae

Moriarty, Audrey
Wild Flowers of Malawi

fl.   Pl. 47; 3

Impatiens schlechteri

Curtis's Botanical Magazine
   Vol. 181   1976-77

fl.   Pl. 710

Impatiens schulziana

Moriarty, Audrey
Wild Flowers of Malawi

fl.   Pl. 47; 2

Impatiens sodenii

Curtis's Botanical Magazine
   Vol. 181   1976-77

fl.   Pl. 744

Impatiens stenantha

Hara, Hiroshi, comp.
Photo-Album of Plants of
   Eastern Himalaya

fl.   Pl. 81

Impatiens sultanii

Bruggeman, L.
Tropical Plants

fl.   Pl. 50

Impatiens sultanii

Hersey, Jean
Woman's Day Book of House Plants

fl.   p 86

Impatiens sultanii

Kiaer, Eigil
Indoor Plants in Colour

fl.   p. 72

Impatiens sultanii

Macoby, Stirling
What Flower is That

fl.   P. 163

Impatiens sultanii

Tsukamoto, Yotaro
Coloured Illustrations of Garden
    Flowers, v. 10

fl.   opp. p. 27   ill. 85

Impatiens sultanii var.

Everett, T.H.
New Illustrated Encyclopedia of
    Gardening, v. 6

fl.   opp. p. 934

Impatiens tricolor var.

Hay, Roy
Color Dictionary of Flowers and
    Plants

fl.   p. 39

Impatiens tripetala

Hara, Hiroshi, comp.
Photo-Album of Plants of
    Eastern Himalaya

fl.   Pl. 83

Impatiens wallerana

Crockett, James Underwood
Annuals

fl.   P 127

Impatiens wallerana

Encke, Fritz
Zimmerpflanzen

fl.   p. 97

Impatiens wallerana

Nicolaisen, Age
Pocket Encyclopedia of Indoor Plants

fl.   Pl. 77

Impatiens wallerana

Perry, Frances
Flowers of the World

fl.   p 39

Impatiens wallerana

Wit, H. C. D. de
Plants of the World;
    The Higher Plants, Vol. II

fl.   p. 61   Pl. 34

Impatiens wallerana var.

Hay, Roy
The Dictionary of House Plants

fl.   Pl. 294, 296

Imperata cylindrica var.

Kimura, Koiti
Japanese Medicinal Plants, Vol I

fr.   p 2

Incarvillea delavayi

Hay, Roy
The Color Dictionary of Flowers & Plants

fl.   p 148, Pl. 1183

Incarvillea delavayi

Huxley, Anthony
Garden Perennials and Water Plants

fl.   Pl. 146

Incarvillea delavayi

Macoby, Stirling
What Flower is That

fl.   P 163

Incarvillea delavayi

Perry, Frances
Flowers of the World

fl.   p 46

Incarvillea grandiflora

Hay, Roy
The Color Dictionary of Flowers & Plants

fl.   p 148, Pl. 1184

Incarvillea grandiflora var.

Royal Hort. Soc.
Journal of the Royal Hort. Soc.
    vol 86, No. 7   1961

fl.   p. 310   Pl. 91

Incarvillea lutea var.

Royal Hort. Soc.
Journal of the Royal Hort. Soc.
    vol 89, No. 4   1964

fl.   p. 158   Pl. 73

Incarvillea younghusbandii

Morley, Brian D.
Wild Flowers of the World

fl.   Plate 103

Incense Plant

see

Calomeria amaranthoides

Incense Plant

see

Humea elegans

Inch Plant

see

Tradescantia albiflora

Inch plant

see

Tradescantia fluminensis var

Indian Cup

see

Silphium perfoliatum

Indian Shot

see

Canna indica

Indigo

see

Dalea schottii

Indigo

see

Indigofera gerardiana

Indigo

see

Indigofera tinctoria

Indigo, Austral

see

Indigofera australis

Indigo, Blue false

see

Baptisia australis

Indigo, Cream Wild

see

Baptisia leucophaea

Indigo, False

see

Amorpha fruticosa

Indigo, False

see

Baptisia alba

Indigo, False

see

Baptisia bracteata

Indigo, Nodding

see

Baptisia leucophaea

Indigo, Nuttall

see

Baptisia nuttaliana

Indigo, Prairie Wild

see

Baptisia leucophaea

Indigo, White

see

Baptisia leucantha

Indigo, Wild

see

Baptisia alba

Indigo, Wild

see

Baptisia leucantha

Indigo, Wild

see

Baptisia tinctoria

Indigo, Yellow Wild

see

Baptisia tinctoria

Indigofera atriceps var.

Moriarty, Audrey
Wild Flowers of Malawi

fl. fr.   Pl. 66; 1

Indigofera australis

Blombery, Alec M.
What Wildflower is That

fl.   p 176, Pl. 475

Indigofera australis

Cochrane, G. R.
Flowers and Plants of Victoria

fl.   Pl. 384

Indigofera burkeana

Flowering Plants of South Africa
vol XXIV   1944

fl.   Pl. 939

Indigofera decora

Harrison, R. E.
Trees and Shrubs

fl.   P. 93

Indigofera decora

Kitamura, Siro
Coloured Illustrations of Trees &
Shrubs of Japan

fl.   256

Indigofera decora

Macoby, Stirling
What Flower is That

fl P 163

Indigofera dendroides

Moriarty, Audrey
Wild Flowers of Malawi

fl,   Pl. 66; 2

Indigofera gerardiana

Hellyer, A.G.L.
Shrubs in Colour
fl.
p. 70

Indigofera gerardiana

Huxley, Anthony
Deciduous Garden Trees and Shrubs

fl.   Pl. 94

Indigofera hilaris

Flowering Plants of Africa
vol 32   1957-58

fl., hab.   Pl. 1247

Indigofera pseudo-tinctoria

Royal Hort. Soc.
Journal of the Royal Hort. Soc.
vol 90, No. 9   1965

fl.   p. 382   Pl. 180

Indigofera suffruticosa

Brown, Clair A.
Wildflowers of Louisiana and
Adjoining States

fl. hab.   p 79

Indigofera tinctoria

Hvass, Elsie
Plants That Feed and Serve Us

fl.   p 110, Pl. 236

Innocence

see

Collinsia heterophylla

Innocence

see

Hedyotis procumbens

Innocence

see

Houstonia caerulea

Inside-Out Flower

see

Vancouveria hexandra

Inula britannica

Polunin, Oleg
Flowers of Europe

fl.   Pl. 145 #1391

Inula conyza

Ary, S.
The Oxford Book of Wildflowers

fl.   p 34, Pl. 5

Inula conyza

Martin, W. Keble
The Concise British Flora in Colour

fl. hab.   Pl. 45

Inula conyza

Perrot, Emile
Les Plantes Medicinales

fr.   p. 24

Inula conyza

Polunin, Oleg
Flowers of Europe

fl.   Pl. 144 #1387

Inula crithmoides

Ary, S.
The Oxford Book of Wildflowers

fl.   p 38, Pl. 2

Inula crithmoides

Martin, W. Keble
The Concise British Flora in Colour

fl. hab.   Pl. 45

Inula crithmoides

Polunin, Oleg
Flowers of Europe

fl.   Pl. 145 #1390

Inula ensifolia

Alpine Garden Society
Bulletin
    vol. 38, No. 4   1970

fl., hab.   p. 391

Inula ensifolia

Huxley, Anthony
Garden Perennials and Water Plants

fl.   Pl. 147

Inula ensifolia var.

Hay, Roy
The Color Dictionary of Flowers &
    Plants

fl.   p. 149   Pl. 1185

Inula glomerata

Moriarty, Audrey
Wild Flowers of Malawi

fl.   Pl. 78; 4

Inula helenium

Bianchini, Francesco
Health Plants of the World

fl.   p 93

Inula helenium

Color Treasury of Herbs & Other
    Medicinal Plants

fl.   p 60   Pl. 96

Inula helenium

Courtenay, Booth
Wildflowers & Weeds

fl.   p 118

Inula helenium

Klimas, John E.
Wildflowers of Eastern America

fl.   Pl. 151

Inula helenium

Lindman, C. A. M.
Nordens Flora, Vol 9

fl.   Pl. 592

Inula helenium

Martin, W. Keble
The Concise British Flora in Colour

fl. hab.   Pl. 45

Inula helenium

Perrot, Emile
Les Plantes Medicinales

fl. fr. hab.   p 24

Inula helenium

Polunin, Oleg
Flowers of Europe

fl.   Pl. 144 #1388

Inula hookeri

Hay, Roy
The Color Dictionary of Flowers & Plants

fl.   p 149, Pl. 1186

Inula hookeri

Royal Hort. Soc.
Journal of the Royal Hort. Soc.
    vol 98, No. 8   1973

fl.   p. 354   Pl. 184

Inula hybrid

Lindman, C. A. M.
Nordens Flora, Vol 9

fl. hab.   Pl. 594

Inula orientalis

Royal Hort. Soc.
Journal of the Royal Hort. Soc.
    vol 87, No. 1   1962

fl.   p. 22   Pl. 3

Inula salicina

Lindman, C. A. M.
Nordens Flora, Vol 9

fl. hab.   Pl. 593A

Inula viscosa

Feinbrun-Dothan, Naomi
Wild Plants in the Land
    of Israel

fl.   p 172

Iochroma cyaneum

Harrison, R.E.
Trees and Shrubs

fl. P 93

Iochroma cyaneum

Herklots, Geoffrey
Flowering Tropical Climbers

fl.   p. 82   Pl. 8

Iochroma cyaneum

Macoby, Stirling
What Flower is That

fl.   P 164

Iodanthus pinnatifidus

Jennings, O. E.
Wild Flowers of Western Pennsylvania
v. 2

fl.   Pl. 70

Iodanthus pinnatifidus

Wharton, Mary E.
A Guide to the Wildflowers & Ferns
    of Kentucky

fl.   p 142   Pl. 1.96

Ionopsis, Delicate

see

Ionopsis utricularioides

Ionopsis utricularioides

American Hort. Soc.
American Horticulturist
    vol 51, No. 4   1972

fl.   p. 29

Ionopsis utricularioides

Lemmon, Robert S.
Wildflowers of North America in
    Full Color

fl.   p. 12   Pl. 17

Ionopsis utricularioides

Luer, Carlyle A.
The Native Orchids of Florida

fl. hab.   Pl. 81

Ionopsis utricularioides

Ospina, Mariano
Orquideas de las americas

fl.   Pl. 175

Ionopsis utricularioides

Pabst, G. F. J.
Orchidaceae Brasilienses,
Vol 2

fl.   p 261

Ipe

see

Tabebuia impetiginosa

Ipecac, American

see

Gillenia stipulata

Ipecacuanha

see

Cephaelis ipecacuanha

Ipheion uniflorum

Batson, Wade T.
Wild Flowers in South Carolina

fl.   p 35

Ipheion uniflorum

Curtis's Botanical Magazine
Vol 169   1952-53

fl.   185

Ipheion uniflorum

Hay, Roy
The Color Dictionary of Flowers & Plants

fl.   p 97, Pl. 773

Ipheion uniflorum

Macoby, Stirling
What Flower is That

fl.   P 164

Ipheion uniflorum

Miles, Bebe
Bulbs for the Home Gardener

fl.   p 79

Ipheion uniflorum

Pennsylvania Hort. Soc.
The Green Scene
      Vol. 6, No. 1   1977

fl.   P. 15

Ipheion uniflorum

Perry, Frances
Flowers of the World

fl.   p. 173

Ipheion uniflorum

Royal Hort. Soc.
Journal of the Royal Hort. Soc.
   vol 95, No. 1   1970

fl.   p. 22   Pl. 22

Ipheion uniflorum

Royal Hort Soc.
The Garden
      vol 100, No. 10   1975

fl.   p. 463

Ipomoea acuminata

Menninger, Edwin A.
Flowering Vines of the World

fl.   Pl. 67

Ipomoea alpina

Moriarty, Audrey
Wild Flowers of Malawi

fl.   Pl. 59; 2

Ipomoea aquatica

Menninger, Edwin A.
Flowering Vines of the World

fl.   Pl. 69

Ipomoea arachnosperma

Flowering Plants of Africa
      vol 31   1956

fl., fr.   Pl. 1203

Ipomoea arborescens

O'Gorman, Helen
Mexican Flowering Trees and Plants

fl. hab.   p 31

Ipomoea batatas

Bianchini, F.
The Complete Book of Fruits &
   Vegetables

hab.   p. 227

Ipomoea batatas

Hvass, Elsie
Plants That Feed and Serve Us

fl.   fr.   p 10, Pl. 10

Ipomoea batatas

Masefield, G. B.
The Oxford Book of Food Plants

fl.   Pl. 183, 1

Ipomoea batatas

Massachusetts Hort. Soc.
Horticulture
   vol 56, No. 11   1978

hab.   p. 44

Ipomoea bisavium

Flowering Plants of Africa
   vol 34   1960-61

fl., fr.   Pl. 1360

Ipomoea bracteata

Menninger, Edwin A.
Flowering Vines of the World

fl.   Pl. 65

Ipomoea cairica

Harrison, Richmond E.
Climbers and Trailers

fl.   p 52   Pl. 109

Ipomoea carnea

Chickering, Carol Rogers
Flowers of Guatemala

fl.   p 73

Ipomoea clavata

Herklots, Geoffrey
Flowering Tropical Climbers

fl.   p. 82   Pl. 8

Ipomoea coccinea

Duncan, Wilbur H.
Wildflowers of the Southeastern
   United States

fl.   p 139

Ipomoea coptica

Flowering Plants of Africa
      vol 31   1956

fl., fr., hab.   Pl. 1217a

Ipomoea crassipes

Morley, Brian D.
Wild Flowers of the World

fl.   Pl. 79

Ipomoea gracilisepala

Flowering Plants of Africa
      vol 31   1956

fl., fr.   Pl. 1217b

Ipomoea hederacea

Brown, Clair A.
Wildflowers of Louisiana and
   Adjoining States

fl. hab.   p 150

Ipomoea hederacea

Duncan, Wilbur H.
Wildflowers of the Southeastern
   United States

fl.   p 139

Ipomoea hederacea

Klimas, John E.
Wildflowers of Eastern America

fl.   Pl. 266

Ipomoea hederacea

Polunin, Oleg
Flowers of Europe

fl.   Pl. 100 #1033

Ipomoea hederacea

Wharton, Mary E.
A Guide to the Wildflowers & Ferns
    of Kentucky

fl.   p 131   Pl. 1.74

Ipomoea hederifolia

Fleming, Glenn
Wild Flowers of Florida

fl.   p 78

Ipomoea hederifolia

Perry, Frances
Flowers of the World

fl.   p 91

Ipomoea hochstetteri

Flowering Plants of Africa
    vol 30   1954-55

fl., fr.   Pl. 1189

Ipomoea horsfalliae

Bruggeman, L.
Tropical Plants

fl.   Pl. 13

Ipomoea horsfalliae

Harrison, Richmond E.
Climbers and Trailers

fl.   p 52   Pl. 110

Ipomoea horsfalliae

Menninger, Edwin A.
Flowering Vines of the World

fl.   Pl. 66

Ipomoea hybrid

Kromdijk, G.
200 House Plants in Colour

fl.   Pl. 114

Ipomoea hybrid

Perry, Frances
Flowers of the World

fl.   p. 90

Ipomoea lacunosa

Brown, Clair A.
Wildflowers of Louisiana and
    Adjoining States

fl. hab.   p 147

Ipomoea lapathifolia

Flowering Plants of Africa
    vol 31   1956

fl., fr.   Pl. 1209

Ipomoea learii

Bruggeman, L.
Tropical Plants

fl.   Pl. 14

Ipomoea learii

Harrison, Richmond E.
Climbers and Trailers

fl.   p 52   Pl. 111

Ipomoea learii

Macoby, Stirling
What Flower is That

fl.   P 165

Ipomoea learii

Morley, Brian D.
Wild Flowers of the World

fl.   Pl. 180

Ipomoea learii

Perry, Frances
Flowers of the World

fl.   p 91

Ipomoea lobata

Harrison, Richmond E.
Climbers and Trailers

fl.   p 53   Pl. 112

Ipomoea magnusiana var.

Flowering Plants of Africa
    vol 31   1956

fl., fr.   Pl. 1201

Ipomoea obscura var.

Flowering Plants of Africa
    vol 31   1956

fl., fr.   Pl. 1222

Ipomoea ochracea

Flowering Plants of Africa
    vol 31   1956

fl.   Pl. 1221

Ipomoea ochracea

Herklots, Geoffrey
Flowering Tropical Climbers

fl.   p. 98   Pl. 9

Ipomoea palmata

Macoby, Stirling
What Flower is That

fl.   P 165

Ipomoea palmata

Menninger, Edwin A.
Flowering Vines of the World

fl.   Pl. 70

Ipomoea pandurata

Batson, Wade T.
Wild Flowers in South Carolina

fl.   p 96

Ipomoea pandurata

Brown, Clair A.
Wildflowers of Louisiana and
    Adjoining States

fl. hab.   p 147

Ipomoea pandurata

Courtenay, Booth
Wildflowers & Weeds

fl.   p. 91

Ipomoea pandurata

Dean, Blanche E.
Wildflowers of Alabama and
    Adjoining States

fl.   p 141

Ipomoea pandurata

Duncan, Wilbur H.
Wildflowers of the Southeastern
    United States

fl.   p 139

Ipomoea pandurata

Weeds of the Southern United States

fl., hab.   p. 17

Ipomoea pandurata

Wharton, Mary E.
A Guide to the Wildflowers & Ferns
    of Kentucky

fl.   p. 118   Pl. 1.49

Ipomoea pes-caprae

Brown, Clair A.
Wildflowers of Louisiana and
    Adjoining States

fl. hab.   p. 148

Ipomoea pes-caprae

Dean, Blanche E.
Wildflowers of Alabama and
Adjoining States

fl. p 141

Ipomoea pes-caprae

Moriarty, Audrey
Wild Flowers of Malawi

fl. Pl. 59; 4

Ipomoea plummerae

Orr, Robert T.
Wildflowers of Western America

fl. Pl. 233

Ipomoea purga

Perrot, Emile
Les Plantes Medicinales

fl. fr. hab. p 126

Ipomoea purpurea

Batson, Wade T.
Wild Flowers in South Carolina

fl. p 97

Ipomoea purpurea

Crockett, James Underwood
Annuals

fl. P 127

Ipomoea purpurea

Dean, Blanche E.
Wildflowers of Alabama and
Adjoining States

fl. p 141

Ipomoea purpurea

Duncan, Wilbur H.
Wildflowers of the Southeastern
United States

fl. p 137

Ipomoea purpurea

Huxley, Anthony
Garden Annuals and Bulbs

fl. Pl. 66

Ipomoea purpurea

Tsukamoto, Yotaro
Coloured Illustrations of Garden
Flowers, v.10
fl.
ill. 90 opp.p.28

Ipomoea purpurea

Weeds of the Southern United States

fl., hab. p. 17

Ipomoea purpurea

Wharton, Mary E.
A Guide to the Wildflowers & Ferns
of Kentucky

fl. p 131 Pl. 1.73

Ipomoea purpurea var.

Hay, Roy
The Dictionary of House Plants

fl. Pl. 297

Ipomoea quamoclit

Brown, Clair A.
Wildflowers of Louisiana and
Adjoining States

fl. hab. p 148

Ipomoea quamoclit

Dean, Blanche E.
Wildflowers of Alabama and
Adjoining States

fl. p 141

Ipomoea quamoclit

Weeds of the Southern United States

fl., hab. p. 17

Ipomoea sagittata

Brown, Clair A.
Wildflowers of Louisiana and
Adjoining States

fl. p 149

Ipomoea sagittata

Dean, Blanche E.
Wildflowers of Alabama and
Adjoining States

fl. p 143

Ipomoea stolonifera

Brown, Clair A.
Wildflowers of Louisiana and
Adjoining States

fl. hab. p 149

Ipomoea stolonifera

Dean, Blanche E.
Wildflowers of Alabama and
Adjoining States

fl. p 143

Ipomoea stolonifera

Duncan, Wilbur H.
Wildflowers of the Southeastern
United States

fl. p 139

Ipomoea stolonifera

Goulimis, Constantine N.
Wild Flowers of Greece

fl. p 89

Ipomoea tenuirostris

Lind, E.M.
Some Common Flowering Plants of
Uganda

fl. Pl. 8

Ipomoea trichocarpa

Duncan, Wilbur H.
Wildflowers of the Southeastern
United States

fl. p 141

Ipomoea tricolor

Hay, Roy
The Dictionary of House Plants

fl. Pl. 298

Ipomoea tricolor

Herklots, Geoffrey
Flowering Tropical Climbers

fl. p. 82 Pl. 8

Ipomoea tricolor

O'Gorman, Helen
Mexican Flowering Trees and Plants

fl. hab. p 101

Ipomoea tricolor

Wit, H. C. D. de
Plants of the World;
The Higher Plants, Vol II

fl. Pl. 75

Ipomoea tricolor var.

Hay, Roy
The Color Dictionary of Flowers &
Plants

fl. p. 39 Pl. 312

Ipomoea tricolor var.

Kiaer, Eigil
Indoor Plants in Colour

fl. p. 72

Ipomoea trifida

Fleming, Glenn
Wild Flowers of Florida

fl. p 65

Ipomoea tuberosa

Hargreaves, Dorothy
Tropical Blossoms of the Caribbean

fl.,fr. p. 21

Ipomopsis aggregata

Wit, H. C. D. de
Plants of the World;
The Higher Plants, Vol II

fl. p 112, Pl. 77

Ipomopsis aggregata var.

Munz, Philip A.
California Desert Wildflowers

fl.   p. 44   Pl. 52,53

Ipomopsis rubra

Brown, Clair A.
Wildflowers of Louisiana and
   Adjoining States

fl.   p 151

Ipomopsis rubra

Dean, Blanche E.
Wildflowers of Alabama and
   Adjoining States

fl.   p 143

Iresine herbstii

Bruggeman, L.
Tropical Plants

hab.   Pl. 121

Iresine herbstii

Crockett, James Underwood
Annuals

hab.   P 128

Iresine herbstii

Hersey, Jean
Woman's Day Book of House Plants

hab.   p. 37

Iresine herbstii

Kromdijk, G.
200 House Plants in Colour

hab.   Pl. 115

Iresine herbstii

Macoty, Stirling
What Flower is That

hab P 165

Iris acutiloba

Hay, Roy
The Color Dictionary of Flowers & Plants

fl.   p 99, Pl. 791

Iris acutiloba

Royal Hort. Soc.
Journal of the Royal Hort. Soc.
   vol 88, No. 4   1963

fl.   p. 164   Pl. 56

Iris acutiloba

Royal Hort. Soc.
Journal of the Royal Hort. Soc.
   vol 90, No. 1   1965

fl.   p. 52   Pl. 27

Iris acutiloba

Royal Hort. Soc.
Journal of the Royal Hort. Soc.
   vol 93, No. 9   1968

fl.   p. 382   Pl. 204

Iris afghanica

Curtis's Botanical Magazine
   Vol. 180   1974

fl. hab.   Pl. 668

Iris, Algerian

see

Iris stylosa

Iris, Algerian

see

Iris unguicularis

Iris attica

Morley, Brian D.
Wild Flowers of the World

fl.,hab.   Pl. 42

Iris aucheri

Hay, Roy
The Color Dictionary of Flowers & Plants

fl.   p 97, Pl. 774

Iris aucheri

Wendelbo, Per
Tulips and Irises of Iran and
   Their Relatives

fl. hab.   p 77, Pl. 81

Iris bakeriana

Hay, Roy
The Color Dictionary of Flowers & Plants

fl.   p 97, Pl. 775

Iris bakeriana

Morley, Brian D.
Wild Flowers of the World

fl.,hab.   Pl. 51

Iris baldschuanica

Mathew, Brian
Dwarf Bulbs

fl.   p. 152   Pl. 52

Iris barnumae

Wendelbo, Per
Tulips and Irises of Iran and
   Their Relatives

fl.   p 73, Pl. 75

Iris barnumae var.

Wendelbo, Per
Tulips and Irises of Iran and
   Their Relatives

fl.   p 73, Pl. 75

Iris, Blueflag

see

Iris versicolor

Iris bracteata

Lemmon, Robert S.
Wildflowers of North America in
   Full Color

fl.   p. 235   Pl. 371

Iris bracteata

Royal Hort. Soc.
Journal of the Royal Hort. Soc.
   vol 91, No. 4   1966

fl.   p. 160   Pl. 92

Iris brevicaulis

Brown, Clair A.
Wildflowers of Louisiana and
   Adjoining States

fl. hab.   p 30

Iris bucharica

Hay, Roy
The Color Dictionary of Flowers & Plants

fl.   p 97, Pl. 776

Iris bucharica

Wit, H. C. D. de
Plants of the World;
   The Higher Plants, Vol II

fl.   Pl. 143

Iris, bush

see

Patersonia sericea

Iris caucasica

Curtis's Botanical Magazine, v.174
1962-63
fl.
pl. 405

Iris caucasica

Mathew, Brian
Dwarf Bulbs

fl.   p. 160   Pl. 61

Iris caucasica var.

Royal Hort. Soc.
Journal of the Royal Hort. Soc.
   vol 90, No. 1   1965

fl.   p. 22   Pl. 10

Iris chamaeiris

Kleijn, H.
The Beauty of the Wild Plant

fl.   p. 112   Pl. 160

Iris chamaeiris

Morley, Brian D.
Wild Flowers of the World

fl.,hab.   Pl. 42

Iris chamaeiris

Polunin, Oleg
Flowers of the Mediterranean

fl.   No. 265, 266

Iris chamaeiris

Vilmorin, Roger de
Plantes Alpines dans les Jardins

fl. hab.   Pl. XXVIII

Iris chamaeiris var.

Kleijn, H.
Beauty of the Wild Plant

fl.   opp. p. 112   Pl. 160

Iris chrysaeola

American Hort. Soc.
National Horticultural Magazine
     vol 12, No. 1   1933

fl.   opp. p. 77

Iris clarkei

Royal Hort. Soc.
Journal of the Royal Hort. Soc.
     vol 88, No. 6   1963

fl.   p. 254   Pl. 92

Iris, Copper Colored

see

Iris fulva

Iris cretensis

Royal Hort. Soc.
The Garden
     vol 101, No. 6   1976

fl.   p. 336

Iris cretica

Polunin, Oleg
Flowers of the Mediterranean

fl.   No. 264

Iris cristata

Batson, Wade T.
Wild Flowers in South Carolina

fl.   p 37

Iris cristata

Campbell, Carlos C.
Great Smoky Mountain Wildflowers

fl.   p 37

Iris cristata

Dean, Blanche E.
Wildflowers of Alabama and
     Adjoining States

fl.   p 33

Iris cristata

Jennings, O. E.
Wild Flowers of Western Pennsylvania
v. 2
fl.   Pl. 34

Iris cristata

Pennsylvania Hort. Soc.
The Green Scene
     vol 3, No. 3   1975

fl.   backcover

Iris cristata

Roberts, June Carver
Born in the Spring

fl.   p. 74

Iris cristata

Royal Hort. Soc.
Journal of the Royal Hort. Soc.
     vol 88, No. 7   1963

fl.   p. 298   Pl. 115

Iris cristata

Walcott, Mary Vaux
North American Wild Flowers

fl. hab.   Pl. 33

Iris cristata

Wharton, Mary E.
A Guide to the Wildflowers & Ferns
     of Kentucky

fl.   p 60 Pl. 1,26

Iris cycloglossa

Curtis's Botanical Magazine
     Vol. 181   1976-77

fl.   Pl. 708

Iris danfordiae

Hay, Roy
The Color Dictionary of Flowers & Plants

fl.   p 98, Pl. 777

Iris danfordiae

Huxley, Anthony
Garden Annuals and Bulbs

hab.   Pl. 164

Iris danfordiae

Massachusetts Hort. Soc.
Horticulture
     vol 44, No. 9   1966

fl.   p. 16

Iris danfordiae

Miles, Bebe
Bulbs for the Home Gardener

fl.   p 80

Iris danfordiae

Pennsylvania Hort. Soc.
The Green Scene
     vol 6, No. 1   1977

fl.   p. 14

Iris danfordiae

Perry, Frances
Flowers of the World

fl.   p 142

Iris decora

Royal Hort. Soc.
Journal of the Royal Hort. Soc.
     vol 94, No. 5   1969

fl.   p. 214   Pl. 100

Iris delavayi

Hay, Roy
The Color Dictionary of Flowers & Plants

fl.   p 149, Pl. 1187

Iris delavayi

Royal Hort. Soc.
Journal of the Royal Hort. Soc.
     vol 88, No. 6   1963

fl.   p. 254   Pl. 91

Iris demawendica

Curtis's Botanical Magazine, v.175
1964-65
fl.
pl. 448

Iris demawendica

Wendelbo, Per
Tulips and Irises of Iran and
     Their Relatives

fl.   p 73, Pl. 76

Iris doabensis

Curtis's Botanical Magazine
Vol 179   1972

fl. hab.   620

Iris douglasiana

Hay, Roy
The Color Dictionary of Flowers & Plants

fl.   p 12, Pl. 93

Iris douglasiana

Lemmon, Robert S.
Wildflowers of North America in
Full Color

fl.   P. 119   Pl. 190

Iris douglasiana

Orr, Robert T.
Wildflowers of Western America

fl.   Pl. 276

Iris drepanophylla

Mathew, Brian
Dwarf Bulbs

fl.   p. 160 Pl. 60

Iris drepanophylla

Royal Hort. Soc.
Journal of the Royal Hort. Soc.
vol 93, No. 9 1968

fl.   p. 382   Pl. 206

Iris drepanophylla

Royal Hort. Soc.
Journal of the Royal Hort. Soc.
vol 99, No. 4 1974

fl.   p. 156 Pl. 85

Iris, Dutch

see

Iris xiphioides var.

Iris, Dwarf

see

Iris verna

Iris, dwarf lake

see

Iris lacustris

Iris eantanesii

Tsukamoto, Yotaro
Coloured Illustrations of Garden
Flowers, v.9
fl.
opp.p.13

Iris ensata var.

Royal Hort. Soc.
Journal of the Royal Hort. Soc.
vol 88, No. 7 1963

fl.   p. 298   Pl. 112

Iris ensata var.

Tsukamoto, Yotaro
Coloured Illustrations of Garden
Flowers, v. 9

fl.   opp. p. 52

Iris ewbankiana

Royal Hort. Soc.
Journal of the Royal Hort. Soc.
vol 93, No. 9 1968

fl.   p. 382   Pl. 204

Iris, Fan

see

Neomarica northiana

Iris fernaldii

Royal Hort. Soc.
Journal of the Royal Hort. Soc.
vol 91, No. 4 1966

fl.   p. 160   Pl. 89

Iris filifolia

Royal Hort. Soc.
Journal of the Royal Hort. Soc.
vol 89, No. 1 1964

fl.   p. 22   Pl. 8

Iris florentina

Bianchini, Francesco
Health Plants of the World

fl. hab.   p 45

Iris florentina

Kimura, Koiti
Japanese Medicinal Plants, Vol I

fl.   p 16

Iris florentina

Massachusetts Hort. Soc.
Horticulture
vol 54, No. 5 1976

fl.   p. 36

Iris florentina

Perrot, Emile
Les Plantes Medicinales

fl.   p 125

Iris florentina

Polunin, Oleg
Flowers of the Mediterranean

fl.   No. 259

Iris foetidissima

Ary, S.
The Oxford Book of Wildflowers

fl.   p 162, Pl. 1

Iris foetidissima

Hay, Roy
The Color Dictionary of Flowers & Plants

fr.   p 149, Pl. 1188

Iris foetidissima

Martin, W. Keble
The Concise British Flora in Colour

fl. fr. hab.   Pl. 83

Iris foetidissima

Perry, Frances
Flowers of the World

fr.   p 144

Iris foliosa

Hay, Roy
The Color Dictionary of Flowers & Plants

fl.   p 149, Pl. 1189

Iris forrestii

Royal Hort. Soc.
Journal of the Royal Hort. Soc.
vol 98, No. 3 1973

fl.   p. 122   Pl. 65

Iris fosterana

Mathew, Brian
Dwarf Bulbs

fl.   p. 152   Pl. 53

Iris fosterana

Royal Hort. Soc.
Journal of the Royal Hort. Soc.
Vol. 90, No. 11 1965

fl.   P. 470   Pl. 211

Iris fosterana

Wendelbo, Per
Tulips and Irises of Iran and
Their Relatives

fl. hab.   p 75, Pl. 78

Iris fulva

Brown, Clair A.
Wildflowers of Louisiana and
Adjoining States

fl. hab.   p 30

Iris fulva

Lemmon, Robert S.
Wildflowers of North America in
Full Color

fl.   P. 171   Pl. 269

Iris fulva

Morley, Brian D.
Wild Flowers of the World

fl.   Pl. 166

Iris fulva

Royal Hort. Soc.
Journal of the Royal Hort. Soc.
vol 88, No. 7 1963

fl.   p. 298   Pl. 113

Iris x germanica

Brown, Clair A.
Wildflowers of Louisiana and
    Adjoining States

fl.   p. 31

Iris x germanica

Huxley, Anthony
Garden Perennials and Water Plants

hab.   Pl. 13

Iris x germanica

Kimura, Koiti
Japanese Medicinal Plants, Vol. II

fl.   p. 145

Iris x germanica

Perrot, Emile
Les Plantes Medicinales

fl.   p. 125

Iris x germanica

Polunin, Oleg
Flowers of Europe

fl.   Pl.175 #1693

Iris x germanica

Royal Hort. Soc.
The Garden
    vol. 102, No. 6   1977

fl.   p. 260

Iris x germanica

Takatori, Jisuke
Color Atlas of Medicinal Plants of
    Japan

fl.   fig. 67A

Iris x germanica var.

Huxley, Anthony
Garden Perennials and Water Plants

fl.   Pl.149,150

Iris x germanica var.

Perry, Frances
Flowers of the World

fl.   p. 143

Iris x germanica var.

Tsukamoto, Yotaro
Coloured Illustrations of Garden
    Flowers   Vol. 9

fl.   opp. p. 51

Iris, Giant Blue

    see

Iris giganticaerulea

Iris giganticaerulea

Brown, Clair A.
Wildflowers of Louisiana and
    Adjoining States

fl.   p 31

Iris gracilipes

Tsukamoto, Yotaro
Coloured Illustrations of Garden
Flowers, v.9
fl.
opp.p.51

Iris graeberiana

Curtis's Botanical Magazine
Vol 167   1950

fl.   126

Iris graeberiana

Huxley, Anthony
Garden Annuals and Bulbs

fl., hab.   Pl. 229, 163

Iris graminea

Addisonia, v. 24, 1960-64

fl. fr.   opp. p. 39   Pl. 788

Iris graminea

Polunin, Oleg
Flowers of Europe

fl.   Pl. 175 #1686

Iris, Ground

    see

Iris macrosiphon

Iris hallandica

Tsukamoto, Yotaro
Coloured Illustrations of Garden
Flowers, v.9
fl.
opp.p.13

Iris hartwegii

Munz, Philip A.
California Mountain Wildflowers

fl.   p 52, Pl. 78

Iris heweri

Royal Hort. Soc.
Journal of the Royal Hort. Soc.
    vol 99, No. 4   1974

fl.   p. 156   Pl. 87

Iris hexagona var.

Fleming, Glenn
Wild Flowers of Florida

fl. hab.   p 87

Iris histrio

Morley, Brian D.
Wild Flowers of the World

fl.,hab.   Pl. 51

Iris histrio

Royal Hort. Soc.
Journal of the Royal Hort. Soc.
    vol 91, No. 8   1966

fl.   p. 338   Pl. 187

Iris histrio var.

Mathew, Brian
Dwarf Bulbs

fl.   p. 144   Pl. 51

Iris histrio var.

Royal Hort. Soc.
Journal of the Royal Hort. Soc.
    vol 91, No. 7   1966

fl.   p. 294   Pl. 158

Iris histrioides

Perry, Frances
Flowers of the World

fl.   p 142

Iris histrioides

Polunin, Oleg
Flowers of the Mediterranean

fl.   Pl. 267

Iris histrioides

Royal Hort. Soc.
Journal of Royal Hort. Soc.
    vol 81, No. 7   1956

fl.   p. 310   Pl. 93

Iris histrioides

Royal Hort. Soc.
Journal of the Royal Hort. Soc.
    vol 92, No. 6   1967

fl.   p. 250   Pl. 115

Iris histrioides

Royal Hort. Soc.
Journal of the Royal Hort. Soc.
    vol 97, No. 4   1972

fl.   p. 170   Pl. 69

Iris histrioides var.

Hay, Roy
The Color Dictionary of Flowers & Plants

fl.   p 98, Pl. 778

Iris histrioides var.

Royal Hort. Soc.
Journal of the Royal Hort. Soc.
    vol 88, No. 7   1963

fl.   p. 298   Pl. 102

Iris, Hong Kong

see

Iris speculatrix

Iris hookeriana

Royal Hort. Soc.
Journal of the Royal Hort. Soc.
vol 93, No. 6   1968

fl.   p. 246   Pl. 128

Iris hybrid

Hay, Roy
The Color Dictionary of Flowers &
Plants

fl.   p. 99,151-153 Pl. 788-792,
1206 thru 1221

Iris hybrid

Jones, Paul
Flora Magnifica

fl.   Pl. 14

Iris hybrid

Macoby, Stirling
What Flower is That

fl.   p. 166

Iris hymenospatha

Wendelbo, Per
Tulips and Irises of Iran and
Their Relatives

fl.   p 76, Pl. 80

Iris iberica

Royal Hort. Soc.
Journal of the Royal Hort. Soc.
vol 86, No. 6   1961

fl.   p. 266   Pl. 64

Iris imbricata

Royal Hort. Soc.
Journal of the Royal Hort. Soc.
vol 90, No. 1   1965

fl.   p. 52   Pl. 28

Iris imbricata

Wendelbo, Per
Tulips and Irises of Iran and
Their Relatives

fl. hab.   p 67, Pl. 70

Iris innominata

Curtis's Botanical Magazine
Vol 163   1940-42
fl. hab.   No. 9628

Iris innominata

Hay, Roy
The Color Dictionary of Flowers & Plants

fl.   p 149, Pl. 1190

Iris innominata

Lemmon, Robert S.
Wildflowers of North America in
Full Color

fl.   p. 119   Pl. 189

Iris innominata

Royal Hort. Soc.
Journal of the Royal Hort. Soc.
vol 91, No. 4   1966

fl.   p. 160   Pl. 94

Iris innominata

Royal Hort. Soc.
Journal of the Royal Hort. Soc.
vol 93, No. 8   1968

fl.   cover

Iris, Japanese

see

Iris kaempferi

Iris, Japanese

see

Iris laevigate

Iris japonica var.

Hay, Roy
The Color Dictionary of Flowers & Plants

fl.   p 149, Pl. 1191

Iris japonica var.

Royal Hort. Soc.
Journal of the Royal Hort. Soc.
vol 88, No. 7   1963

fl.   p. 298   Pl. 114

Iris kaempferi

Macoby, Stirling
What Flower is That

fl.   P 166

Iris kaempferi var.

American Hort. Soc.
American Horticulturist
vol 57, No. 2   1978

fl.   p. 6-7

Iris kaempferi var.

Hay, Roy
The Color Dictionary of Flowers & Plants

fl.   p 149, 150; Pl. 1192, 1193

Iris kaempferi var.

Huxley, Anthony
Garden Perennials and Water Plants

fl.   Pl. 151

Iris kaempferi var.

Perry, Frances
Flowers of the World

fl.   p. 145

Iris kamaonensis

Hara, Hiroshi, comp.
Photo-Album of Plants of
Eastern Himalaya

fl. hab.   Pl. 204-206

Iris kopetdagensis

Mathew, Brian
Dwarf Bulbs

fl.   p. 160   Pl. 62

Iris kopetdagensis

Royal Hort. Soc.
Journal of the Royal Hort. Soc.
Vol. 93, No. 9   1968

fl.   P. 382   Pl. 205

Iris kopetdagensis

Wendelbo, Per
Tulips and Irises of Iran and
Their Relatives

fl. hab.   p 79, Pl. 82

Iris korolkowii

Royal Hort. Soc.
Journal of the Royal Hort. Soc.
vol 93, No. 2   1968

fl.   p. 94   Pl. 36

Iris kumaonensis

Morley, Brian D.
Wild Flowers of the World

fl.   Plate 107

Iris kumaonensis

Royal Hort. Soc.
The Garden
Vol. 102, No. 11   1977

fl. hab.   p. 452

Iris kuschkensis

Royal Hort. Soc.
Journal of the Royal Hort. Soc.
vol 99, No. 4   1974

fl.   p. 156   Pl. 88

Iris lacustris

Courtenay, Booth
Wildflowers & Weeds

fl.   p 10

Iris lacustris

Hay, Roy
The Color Dictionary of Flowers & Plants

fl.   p 12, Pl. 94

Flowering Plant Index of Illustration and Information

Iris laevigata

Perry, Frances
Flowers of the World

fl.   p 145

Iris laevigata

Tsukamoto, Yotaro
Coloured Illustrations of Garden
  Flowers, v. 9

fl.   opp. p. 52

Iris laevigata var.

Hay, Roy
The Color Dictionary of Flowers & Plants

fl.   p 150, Pl. 1194, 1195

Iris laevigata var.

Huxley, Anthony
Garden Perennials and Water Plants

fl.   Pl. 152

Iris laevigata var.

Royal Hort. Soc.
Journal of the Royal Hort. Soc.
  vol 88, No. 6   1963

fl.   p. 254   Pl. 93

Iris lazica

Royal Hort. Soc.
Journal of the Royal Hort. Soc.
  vol 90, No. 1   1965

fl.   p. 22   Pl. 12

Iris lineolata

Wendelbo, Per
Tulips and Irises of Iran and
  Their Relatives

fl. hab.   p 71, Pl. 74

Iris linifolia

Royal Hort. Soc.
Journal of the Royal Hort. Soc.
  vol 93, No. 1   1968

fl.   p. 24   Pl. 5

Iris longipetala

Royal Hort. Soc.
Journal of the Royal Hort. Soc.
  vol 88, No. 7   1963

fl.   p. 298   Pl. 111

Iris lortetii

Morley, Brian D.
Wild Flowers of the World

fl.   Pl. 51

Iris lycotis

Curtis's Botanical Magazine
Vol 178   1970-72

fl.   580

Iris lycotis

Wendelbo, Per
Tulips and Irises of Iran and
  Their Relatives

fl.   p 71, Pl. 73

Iris macrosiphon

Orr, Robert T.
Wildflowers of Western America

fl.   Pl. 12, 76

Iris macrosiphon

Royal Hort. Soc.
Journal of the Royal Hort. Soc.
  vol 91, No. 4   1966

fl.   p. 160   Pl. 88

Iris meda

Royal Hort. Soc.
Journal of the Royal Hort. Soc.
  vol 88, No. 4   1963

fl.   p. 164   Pl. 58

Iris meda

Wendelbo, Per
Tulips and Irises of Iran and
  Their Relatives

fl.   p 75, Pl. 77

Iris mellori

Royal Hort. Soc.
Journal of the Royal Hort. Soc.
  vol 88, No. 7   1963

fl.   p. 298   Pl. 101

Iris microglossa

Mathew, Brian
Dwarf Bulbs

fl.   p. 152   Pl. 55

Iris milesii

Pacific Hort. Foundation
Pacific Horticulture
  vol 38, No. 4   1977

fl.   p. 37

Iris, Morning

see

Orthrosanthus laxus

Iris, Mountain

see

Iris douglasiana

Iris munzii var.

Pacific Hort. Foundation
Pacific Horticulture
  vol 37, No. 2   1976

fl.   p. 52

Iris, Native

see

Patersonia

Iris nepalensis

Massachusetts Hort. Soc.
Horticulture
  vol 54, No. 5   1976

fl.   p. 36

Iris ochroleuca

Hay, Roy
The Color Dictionary of Flowers & Plants

fl.   p 150, Pl. 1196

Iris ochroleuca

Macoby, Stirling
What Flower is That

fl.   P 111

Iris orchioides

Mathew, Brian
Dwarf Bulbs

fl.   p. 160   Pl. 63

Iris orchioides

Royal Hort. Soc.
Journal of the Royal Hort. Soc.
  vol 91, No. 1   1966

fl.   p. 22   Pl. 13

Iris orientalis

Goulimis, Constantine N.
Wild Flowers of Greece

fl.   p 183

Iris orientalis

Perry, Frances
Flowers of the World

fl.   p 145

Iris pallasii

Kimura, Koiti
Japanese Medicinal Plants, Vol II

fl.   p 147

Iris pallida

Alpine Garden Society
Bulletin
  vol. 9, No. 3   1941

fl., hab.   p. 230

739

Iris pallida
Perrot, Emile
Les Plantes Medicinales

fl. fr. hab.   p 125

Iris pallida var.
Hay, Roy
The Color Dictionary of Flowers & Plants

fl.   p 150, Pl. 1197

Iris pamphylica
Curtis's Botanical Magazine
Vol 179   1972

fl.   648

Iris pamphylica
Mathew, Brian
Dwarf Bulbs

fl.   p. 144   Pl. 50

Iris pamphylica
Royal Hort. Soc.
Journal of the Royal Hort. Soc.
vol 96, No. 8   1971

fl.   p. 354   Pl. 151

Iris paradoxa
Morley, Brian D.
Wild Flowers of the World

fl.,hab.   Pl. 51

Iris paradoxa var.
Royal Hort. Soc.
Journal of the Royal Hort. Soc.
vol 90, No. 1   1965

fl.   p. 52   Pl. 21

Iris , peacock
see
Moraea pavonia

Iris persica var.
Royal Hort. Soc.
Journal of the Royal Hort. Soc.
vol 90, No. 1   1965

fl.   p. 22   Pl. 13

Iris planifolia
Hay, Roy
The Color Dictionary of Flowers & Plants

fl.   p 98, Pl. 779

Iris planifolia
Taylor, A. W.
Wildflowers of Spain and Portugal

fl.   p. 18

Iris, Prairie
see
Iris hexagona var.

Iris prismatica
Jennings, O. E.
Wild Flowers of Western Pennsylvania
v. 2

fl.   Pl. 33

Iris pseudacorus
Ary, S.
The Oxford Book of Wildflowers

fl.   p 28, Pl. 1

Iris pseudacorus
Brown, Clair A.
Wildflowers of Louisiana and
Adjoining States

fl.   p 32

Iris pseudacorus
Clark, Lewis J.
Wild Flowers of British Columbia

fl.   p 67

Iris pseudacorus
Courtenay, Booth
Wildflowers and weeds

fl. p. 10

Iris pseudacorus
Felsko, Elsa
A Book of Wildflowers

fl.   P. 17

Iris pseudacorus
Hay, Roy
The Color Dictionary of Flowers & Plants

fl.   p 150, Pl. 1198

Iris pseudacorus
Huxley, Anthony
Garden Perennials and Water Plants

fl.   Pl. 291

Iris pseudacorus
Kleijn, H.
The Beauty of the Wild Plant

fl.   p. 22   Pl. 16

Iris pseudacorus
Klimas, John E.
Wildflowers of Eastern America

fl.   Pl. 155

Iris pseudacorus
Lindman, C. A. M.
Nordens Flora, Vol 1

fl. fr.   Pl. 62

Iris pseudacorus
Martin, W. Keble
The Concise British Flora in Colour

fl. hab.   Pl. 83

Iris pseudacorus
Moyle, John B.
Northland Wild Flowers

fl. hab.   p 217, Pl. 287

Iris pseudacorus
Polunin, Oleg
Flowers of Europe

fl.   Pl. 174 #1690

Iris pseudacorus
Pond, Barbara
A Sampler of Wayside Herbs

fl.   Pl. V

Iris pseudocaucasica
Wendelbo, Per
Tulips and Irises of Iran and
Their Relatives

fl. hab.   p 75, Pl. 79

Iris pumila
Morley, Brian D.
Wild Flowers of the World

fl.,hab.   Pl. 42

Iris pumila
Polunin, Oleg
Flowers of Europe

fl. Pl. 176 #1691

Iris pumila var.
Huxley, Anthony
Garden Perennials and Water Plants

fl.   Pl. 148

Iris pumila var.
Polunin, Oleg
Flowers of the Mediterranean

fl.   No. 260, 261

Iris pumila var.
Royal Hort. Soc.
Journal of the Royal Hort. Soc.
vol 88, No. 10   1963

fl.   p. 434   Pl. 166-67

Iris, Red

see

Iris fulva

Iris reticulata

Huxley, Anthony
Garden Annuals and Bulbs

fl., hab.   Pl. 226, 166

Iris reticulata

Kromdijk, G.
200 House Plants in Colour

fl.   Pl. 116

Iris reticulata

Miles, Bebe
Bulbs for the Home Gardener

fl.   p. 80, 191

Iris reticulata

Pennsylvania Hort. Soc.
The Green Scene
    vol 6, No. 1   1977

fl.   p. 15

Iris reticulata

Royal Hort. Soc.
Journal of the Royal Hort. Soc.
    vol 92, No. 6   1967

fl.   p. 250   Pl. 113

Iris reticulata

Royal Hort. Soc.
Journal of the Royal Hort. Soc.
    vol 98, No. 7   1973

fl.   p. 306   Pl. 149

Iris reticulata

Wendelbo, Per
Tulips and Irises of Iran and
    Their Relatives

fl. hab.   p 67, Pl. 71

Iris reticulata var.

Hay, Roy
The Color Dictionary of Flowers & Plants

fl.   p 98, Pl. 781, 782, 783

Iris reticulata var.

Miles, Bebe
Bulbs for the Home Gardener

fl.   p 80

Iris reticulata var.

Perry, Frances
Flowers of the World

fl.   p. 142

Iris reticulata var.

Royal Hort. Soc.
Journal of the Royal Hort. Soc.
    vol 94, No. 12   1969

fl.   p. 514   Pl. 266

Iris rosenbachiana

Curtis's Botanical Magazine, v. 175
fl.   1964-65
pl. 483

Iris rosenbachiana

Royal Hort. Soc.
Journal of the Royal Hort. Soc.
    vol 91, No. 1   1966

fl.   p. 22   Pl. 10

Iris sari

Alpine Garden Society
Bulletin
    vol. 45, No. 2   1977

fl., hab.   p. 113

Iris serotina

Curtis's Botanical Magazine
    Vol. 181   1976-77

fl.   Pl. 733

Iris setosa

Alaska-Yukon Wild Flowers Guide

fl. hab.   p 12

Iris setosa

Heller, Christine
Wild Flowers of Alaska

fl.   Pl. 111-112

Iris setosa var.

Clark, Lewis J.
Wild Flowers of British Columbia

fl.   P. 78

Iris shrevei

Courtenay, Booth
Wildflowers & Weeds

fl.   p 10

Iris sibirica

Felsko, Elsa
A Book of Wildflowers

fl.   P 51

Iris sibirica

Huxley, Anthony
Garden Perennials and Water Plants

fl.   Pl. 153

Iris sibirica

Perry, Frances
Flowers of the World

fl.   p 144

Iris sibirica var.

Hay, Roy
The Color Dictionary of Flowers & Plants

fl.   p 150, Pl. 1199, 1200

Iris sibirica var.

Huxley, Anthony
Garden Perennials and Water Plants

fl.   Pl. 154

Iris x sindpers

Curtis's Botanical Magazine, v. 174
    1962-63

fl.   Pl. 419

Iris sintenisii

Goulimis, Constantine N.
Wild Flowers of Greece

fl. hab.   p 185

Iris sisyrinchium

Polunin, Oleg
Flowers of Europe

fl.   Pl. 175 #1684

Iris sisyrinchium

Polunin, Oleg
Flowers of the Mediterranean

fl.   No. 269

Iris sisyrinchium

Taylor, A. W.
Wild Flowers of Spain and Portugal

fl.   p. 23

Iris sisyrinchium

Wendelbo, Per
Tulips and Irises of Iran and
    Their Relatives

fl. hab.   p 69, Pl. 72

Iris, Snakes' Head

see

Hermodactylis tuberosus

Iris sofarana

Royal Hort. Soc.
Journal of the Royal Hort. Soc.
    vol 89, No. 3   1964

fl.   p. 114   Pl. 39

Iris songarica

Wendelbo, Per
Tulips and Irises of Iran and
   Their Relatives

fl.   p 67, Pl. 69

Iris, Spanish

see

Iris xiphium

Iris sp.

Color Treasury of Herbs & Other
   Medicinal Plants

fl.   p 24   Pl. 17

Iris speculatrix

Walden, Beryl M.
Wild Flowers of Hong Kong

fl.   Pl. 35(78)

Iris spuria

Martin, W. Keble
The Concise British Flora in Colour

fl.   Pl. 83

Iris spuria

Wendelbo, Per
Tulips and Irises of Iran and
   Their Relatives

fl.   p 65, Pl. 68

Iris spuria var.

Hay, Roy
The Color Dictionary of Flowers & Plants

fl.   p 151, Pl. 1201

Iris stylosa

Macoby, Stirling
What Flower is That

fl.   P 166

Iris susiana

Massachusetts Hort. Soc.
Horticulture
      vol 54, No. 5 1976

fl.   p. 36

Iris susiana var.

Curtis's Botanical Magazine
Vol. 177   1969-70

fl.   550

Iris tectorum

Hay, Roy
The Color Dictionary of Flowers & Plants

fl.   p 151, Pl. 1202

Iris tectorum

Kimura, Koiti
Japanese Medicinal Plants, Vol II

fl.   p 147

Iris tectorum

Royal Hort. Soc.
Journal of the Royal Hort. Soc.
      vol 88, No. 7   1963

fl.   p. 298   Pl. 116

Iris tenax

Lemmon, Robert S.
Wildflowers of North America in
   Full Color

fl.   p. 235   Pl. 372

Iris tenax

Morley, Brian D.
Wild Flowers of the World

fl. fr.   Pl. 166

Iris tenax

Royal Hort. Soc.
Journal of the Royal Hort. Soc.
      vol 91, No. 4   1966

fl.   p. 160   Pl. 95

Iris tenax var.

Royal Hort. Soc.
Journal of the Royal Hort. Soc.
      vol 91, No. 4   1966

fl.   p. 160   Pl. 91

Iris tenuissima

Royal Hort. Soc.
Journal of the Royal Hort. Soc.
      vol 91, No. 4   1966

fl.   p. 160   Pl. 90

Iris tingitana

Hay, Roy
The Color Dictionary of Flowers & Plants

fl.   p 98, Pl. 784

Iris, tough-leaved

see

Iris tenax

Iris tridentata

Massachusetts Hort. Soc.
Horticulture
      vol 54, No. 5 1976

fl.   p. 36

Iris unguicularis

Hay, Roy
The Color Dictionary of Flowers & Plants

fl.   p 151, Pl. 1203

Iris unguicularis

Meikle, R. D.
Garden Flowers

hab.   opp. p. 385   Pl. 13

Iris unguicularis

Perry, Frances
Flowers of the World

fl.   p 146

Iris unguicularis var.

Hay, Roy
The Color Dictionary of Flowers &
   Plants

fl.   p. 151   Pl. 1204

Iris unguicularis var.

Royal Hort. Soc.
Journal of the Royal Hort. Soc.
      vol 96, No. 2   1971

fl. p 56   Pl. 22, 30

Iris verna

Dean, Blanche E.
Wildflowers of Alabama and
   Adjoining States

fl.   p 33

Iris verna

Duncan, Wilbur H.
Wildflowers of the Southeastern
   United States

fl.   p 265

Iris verna

Hay, Roy
The Color Dictionary of Flowers & Plants

fl.   p 12, Pl. 95

Iris verna

Jennings, O. E.
Wild Flowers of Western Pennsylvania
   v.2
fl. hab.   Pl. 33

Iris verna

Massachusetts Hort. Soc.
Horticulture
      vol 54, No. 5 1976

fl.   p. 36

Iris verna

Walcott, Mary Vaux
North American Wild Flowers

fl. hab.   Pl. 13

Iris verna

Wharton, Mary E.
A Guide to the Wildflowers & Ferns
   of Kentucky

fl. p 60 Pl. 1.27

Iris versicolor

Ferguson, Mary
Wildflowers

fl.   p 25

Iris versicolor

Jennings, O. E.
Wild Flowers of Western Pennsylvania
v.2
fl.   Pl. 33

Iris versicolor

Klimas, John E.
Wildflowers of Eastern America

fl.   Pl. 233

Iris versicolor

Lemmon, Robert S.
Wildflowers of North America in
       Full Color

fl.   p. 170   P. 268

Iris versicolor

Moyle, John B.
Northland Wild Flowers

fl. hab.   p 216, Pl. 286

Iris versicolor

Pond, Barbara
A Sampler of Wayside Herbs

fl.   Pl. V

Iris versicolor

Walcott, Mary Vaux
North American Wild Flowers
       vol. 5

fl.   Pl. 332

Iris versicolor var.

Hay, Roy
The Color Dictionary of Flowers & Plants

fl.   p 151, Pl. 1205

Iris virginica

Batson, Wade T.
Wild Flowers in South Carolina

fl.   p 37

Iris virginica

Brown, Clair A.
Wildflowers of Louisiana and
       Adjoining States

fl.   p 32

Iris virginica

Duncan, Wilbur H.
Wildflowers of the Southeastern
       United States

fl.   p 265

Iris virginica

Lemmon, Robert S.
Wildflowers of North America in
       Full Color

fl.   p. 10   Pl. 14

Iris virginica var.

Wharton, Mary E.
A Guide to the Wildflowers & Ferns
       of Kentucky

fl.   p. 61   Pl. 1.28

Iris warleyensis

Mathew, Brian
Dwarf Bulbs

fl.   p. 152   Pl. 54

Iris warleyensis

Royal Hort. Soc.
Journal of the Royal Hort. Soc.
       vol 91, No. 7   1966

fl.   p. 294   Pl. 159

Iris wattii

Curtis's Botanical Magazine
       Vol 162   1939

fl.   No. 9590

Iris wattii

Morley, Brian D.
Wild Flowers of the World

fl.   Plate 107

Iris, Widow

       see

Hermodactylis tuberosus

Iris, wild

       see

Dietes grandiflora

Iris, wild

       see

Iris shrevei

Iris, wild

       see

Patersonia glauca

Iris winogradowii

Alpine Garden Society
Bulletin
       vol. 35, No. 3   196?

fl.   p. 225

Iris winogradowii

Royal Hort. Soc.
Journal of the Royal Hort. Soc.
       vol 87, No. 8   1962

fl.   p. 358   Pl. 114

Iris, winter

       see

Iris stylosa

Iris xanthochlora

Royal Hort. Soc.
Journal of the Royal Hort. Soc.
       vol 99, No. 4   1974

fl.   p. 156   Pl. 86

Iris xiphioides

Kohlhaupt, Paula
Fleurs des Alpages , v.2

fl.   P 13

Iris xiphioides

Perry, Frances
Flowers of the World

fl.   p 143

Iris xiphioides

Polunin, Oleg
Flowers of Europe

fl.   Pl. 175 #1685

Iris xiphioides

Tsukamoto, Yotaro
Coloured Illustrations of Garden
       Flowers vol 9

fl.   p. 13

Iris xiphioides var.

Macoby, Stirling
What Flower is That

fl.   P 167

Iris xiphium

Everett, T.H.
New Illustrated Encyclopedia of
       Gardening, v. 6

fl.   opp. p. 934

Iris xiphium

Hay, Roy
The Color Dictionary of Flowers & Plants

fl.   p 99, Pl. 785

Iris xiphium

Huxley, Anthony
Garden Annuals and Bulbs

hab.   Pl. 165

Iris xiphium

Polunin, Oleg
Flowers of Europe

fl. Pl. 175 #1685

Iris xiphium

Polunin, Oleg
Flowers of the Mediterranean

fl.   No. 263

Iris xiphium var.

Hay, Roy
The Color Dictionary of Flowers &
    Plants

fl.   p. 99   Pl.786-87

Iris xiphium var.

Huxley, Anthony
Garden Annuals and Bulbs

fl.   Pl. 228a, b

Iris, Yellow

see

Iris pseudacorus

Iris, Zig Zag

see

Iris brevicaulis

Ironbark, narrowleaf

see

Eucalyptus racemosa

Ironbark, pink-flowered

see

Eucalyptus sideroxylon

Iron Bark, Red

see

Eucalyptus sideroxylon

Ironbark, Red Broadleaved

see

Eucalyptus siderophloia

Ironbark, White

see

Eucalyptus leucoxylon

Ironheart tree

see

Swartzia madagascariensis

Ironweed

see

Sida acuta

Ironweed

see

Vernonia

Ironweed, New York

see

Veronica noveboracensis

Ironweed, Tall

see

Vernonia altissima

Ironweed, Yellow

see

Actinomeris alternifolia

Ironwood

see

Casuarina stricta

Ironwood

see

Olneya tesota

Ironwood, Catalina

see

Lyonothamnus floribundus var.

Ironwood, Lemon

see

Backhousia citriodora

Ironwood, Rhodesian

see

Colophospermum mopane

Ironwood, Santa Cruz

see

Lyonothamnus asplenifolius

Isabelia virginalis

Pabst, G.F.J.
Orchidaceae Brasilienses
    vol 1

hab., fl.   p. 55 , 220

Isabelia virginalis

Pabst, G.F.J.
Orchidaceae Brasilienses
    vol 2

hab.   p. 174

x Isanitella pabstii

Pabst, G. F. J.
Orchidaceae Brasilienses,
    Vol 2

fl.   p 213

Isatis tinctoria

Ary, S.
The Oxford Book of Wildflowers

fl. fr.   p 10, Pl. 1

Isatis tinctoria

Lindman, C. A. M.
Nordens Flora, Vol 4

fl. fr.   Pl. 256

Isatis tinctoria

Martin, W. Keble
The Concise British Flora in Colour

fl. fr. hab.   Pl. 10

Isatis tinctoria

Perrot, Emile
Les Plantes Medicinales

fl. fr. hab.   p 172

Isatis tinctoria

Polunin, Oleg
Flowers of Europe

fl. Pl. 32 #292

Islaya krainziana

Backeberg, Curt
Cactus Lexicon

fl.   p 641

Islaya solitaria

Lamb, Edgar
Popular Exotic Cacti in Color

fl. hab.   p 116

Ismene

see

Hymenocallis americana

Ismene

see

Hymenocallis calathina

Ismene

see

Hymenocallis speciosa

Isoberlinia globiflora

Palgrave, K. C.
Trees of Central Africa

fl., fr.   p. 109

Isochilus linearis

Ospina, Mariano
Orquideas de las americas

fl.   Pl. 45

Isochilus linearis

Pabst, G. F. J.
Orchidaceae Brasilienses,
   Vol 1

fr.   p 225

Isochilus linearis

Pabst, G.F.J.
Orchidaceae Brasilienses
   vol 2

hab.   p. 175

Isoglossa grandiflora

Moriarty, Audrey
Wild Flowers of Malawi

fl.   Pl. 42; 4

Isoloma hirsutum

Massachusetts Hort. Soc.
Horticulture
   vol 34, No. 1   1956

fl.   p. 20

Isomeris arborea

Lemmon, Robert S.
Wildflowers of North America in
   Full Color

fl. fr.   p. 20   Pl. 31

Isomeris arborea

Munz, Philip A.
California Desert Wildflowers

fl.   p 33, Pl. 21

Isophysis tasmanica

Curtis, Winifred
The Endemic Flora of Tasmania
   Vol. 2

fl.   Pl. 41

Isoplexis canariensis

Bramwell, David
Wild Flowers of the Canary Islands

fl.   Pl. 260, 261

Isoplexis canariensis

Curtis's Botanical Magazine
Vol 177   1969-70

fl.   559

Isoplexis canariensis

Morley, Brian D.
Wild Flowers of the World

fl.   Pl. 43

Isoplexis isabelliana

Bramwell, David
Wild Flowers of the Canary Islands

fl.   Pl. 259

Isopogon anemonifolius

Blombery, Alec M.
What Wildflower is That

fl.   p 176, Pl. 476

Isopogon anemonifolius

Mullins, Barbara
Australian Wildflowers in Colour

fl.   P. 41   Pl. 33

Isopogon anethifolius

Blombery, Alec M.
What Wildflower is That

fl.   p 177, Pl. 477

Isopogon anethifolius

Harrison, R.E.
Trees and Shrubs

fl.   P 95

Isopogon ceratophyllus

Cochrane, G. R.
Flowers and Plants of Victoria

fl.   Pl. 2

Isopogon cuneatus

Harrison, R.E.
Trees andShrubs

fl.   P 95

Isopogon dubius

Blombery, Alec M.
What Wildflower is That

fl.   p 177, Pl. 478

Isopogon latifolius

Harrison, R.E.
Trees and Shrubs

fl.   P 95

Isopogon latifolius

Morcombe, M. K.
Australia's Western Wildflowers

fl.   p 5, 99

Isopogon villosus

Curtis's Botanical Magazine
   Vol. 180   1974

fl.   Pl. 694

Isopyrum biternatum

Courtenay, Booth
Wildflowers & Weeds

fl.   p 33

Isopyrum biternatum

Dean, Blanche E.
Wildflowers of Alabama and
   Adjoining States

fl.   p 63

Isopyrum biternatum

Moyle, John B.
Northland Wild Flowers

fl. hab.   p 49, Pl. 7

Isopyrum biternatum

Wharton, Mary E.
A Guide to the Wildflowers & Ferns
   of Kentucky

fl. hb.   p 109 Pl. 1.31

Isopyrum thalictroides

Felsko, Elsa
A Book of Wild Flowers
2nd Ser.
fl. hab.   p 117

Isotoma axillaris

Blombery, Alec M.
What Wildflower is That

fl.   p 177, Pl. 479

Isotoma axillaris

Cochrane, G. R.
Flowers and Plants of Victoria

fl. hab.   Pl. 201, 202

Isotoma fluviatilis

Cochrane, G. R.
Flowers and Plants of Victoria

fl. hab.   Pl. 254

Isotrema griffithii

Curtis's Botanical Magazine
Vol 178   1970-72

fl.   576

Isotria medeoloides

Luer, Carlyle A.
The Native Orchids of the
    United States and Canada

fl. fr. hab.   Pl. 68

Isotria verticillata

Campbell, Carlos C.
Great Smoky Mountain Wildflowers

fl.   p 17

Isotria verticillata

Courtenay, Booth
Wildflowers & Weeds

fl.   p 14

Isotria verticillata

Dean, Blanche E.
Wildflowers of Alabama and
    Adjoining States

fl.   p 37

Isotria verticillata

Klimas, John E.
Wildflowers of Eastern America

fl.   Pl. 109

Isotria verticillata

Luer, Carlyle A.
The Native Orchids of Florida

fl. hab.   Pl. 9; 1 - 5

Isotria verticillata

Luer, Carlyle A.
The Native Orchids of the
    United States and Canada

fl. fr. hab.   Pl. 67

Isotria verticillata

Wharton, Mary E..
A Guide to the Wildflowers & Ferns
    of Kentucky

fl.   p. 69   Pl. 2.9

Isotropis cuneifolia

Blombery, Alec M.
What Wildflower is That

fl.   p 178, Pl. 480

Itea ilicifolia

Gault, S. Millar
The Color Dictionary of Shrubs

fl.   Pl. 258

Itea ilicifolia

Hay, Roy
The Color Dictionary of Flowers & Plants

fl.   p 208, Pl. 1663

Itea japonica

Kitamura, Siro
Coloured Illustrations of Trees &
    Shrubs of Japan

fl.   204

Itea virginica

Batson, Wade T.
Wild Flowers in South Carolina

fl.   p 54

Itea virginica

Wharton, Mary E.
Trees & Shrubs of Kentucky

fl.   p 72, Pl. 2.7

Iva xanthifolia

Courtenay, Booth
Wildflowers & Weeds

fl.   p 113

Ivesia gordonii

Welsh, Stanley L.
Flowers of the Mountain Country

fl.   p. 55

Ivory Nut Palm

see

Phytelephas macrocarpa

Ivy

See Hedera

Ivy

see

Kalmia latifolia

Ivy

see

Rhododendron minus

Ivy, Algerian

see

Hedera canariensis

Ivy, birdsfoot

see

Hedera helix var.

Ivy, Boston

see

Parthenocissus tricuspidata

Ivy, Canary Island

see

Hedera canariensis

Ivy, Cape

see

Senecio macroglossus

Ivy, devil's

see

Scindapsus aureus

Ivy, English

see

Hedera helix

Ivy, Feather

see

Hedera helix var.

Ivy, German

see

Senecio mikanioides

Ivy, Glacier

see

Hedera helix var.

Ivy, grape

see

Cissus rhombifolia

Ivy, Ground

see

Glechoma hederacea

Ivy, ground

see

Nepeta hederacea var.

Ivy, Irish

see

Hedera helix var.

Ivy, Japanese

see

Fatshedera lizei

Ivy, Japanese

see

Parthenocissus tricuspidata

Ivy, Kenilworth

see

Cymbalaria muralis

Ivy, Parlor

see

Senecio mikanioides

Ivy, Persian

see

Hedera colchica

Ivy, poison

see

Rhus radicans

Ivy, Prickly

see

Smilax aspera

Ivy, thinleaf

see

Hedera colchica var.

Ivy, tree

see

Fatshedera lizei

Ivy, Tree

see

Hedera arborescens

Ivy, Violet

see

Cobaea scandens

Ivy, Water

see

Senecio mikanioides

Ixeria denticulata

Walden, Beryl M.
Wild Flowers of Hong Kong

fl.   Pl. 70 (200)

Ixia, blue

see

Ixia viridiflora

Ixia cochlearis

Flowering Plants of Africa
        vol XXV   1945-46

fl.   Pl. 969

Ixia, green

see

Ixia viridiflora

Ixia hybrid.

Hay, Roy
The Dictionary of House Plants

fl.   Pl. 299

Ixia hybrid

Huxley, Anthony
Garden Annuals and Bulbs

fl.   Pl. 227

Ixia maculata

Macoby, Stirling
What Flower is That

fl.   P 168

Ixia maculata

Tsukamoto, Yotaro
Coloured Illustrations of Garden
   Flowers, v. 9

fl.   opp. p. 14

Ixia polystachya

Flowering Plants of Africa
      vol XXV   1945-46

fl.   Pl. 968

Ixia scillaris

Eliovson, Sima
Namaqualand in Flower

fl.   Pl. 28, 1

Ixia viridiflora

American Hort. Soc.
American Horticulturist
   vol 53, No. 3 1974

fl.   p. 35

Ixia viridiflora

Macoby, Stirling
What Flower is That

fl.   P 168

Ixia viridiflora

Morley, Brian D.
Wild Flowers of the World

fl.,hab.   Pl. 87

Ixianthes retzioides

Morley, Brian D.
Wild Flowers of the World

fl.   Pl. 81

Ixiolirion pallasi

Miles, Bebe
Bulbs for the Home Gardener

fl.   p 81

Ixiolirion tataricum

Wendelbo, Per
Tulips and Irises of Iran and
   Their Relatives

fl. hab.   p 55, Pl. 56

Ixodia achilleoides

Cochrane, G. R.
Flowers and Plants of Victoria

fl.   Pl. 17

Ixodia achlaena

Curtis, Winifred
The Endemic Flora of Tasmania
   Vol VI

fl.   Pl. 202

Ixodia angusta

Curtis, Winifred
The Endemic Flora of Tasmania
V. 3

fl.   p. 197

Ixora borbonica

Royal Hort. Soc.
Journal of the Royal Hort. Soc.
vol 90, No. 11   1965

hab.   p. 470   Pl. 232

Ixora chinensis var.

Macoby, Stirling
What Flower is That

fl.   P 168

Ixora coccinea

Crockett, James Underwood
Evergreens

hab. fl.   p. 131

Ixora coccinea

Macoby, Stirling
What Flower is That

fl.   P 168

Ixora coccinea

Nicolaisen, Age
Pocket Encyclopedia of Indoor Plants

fl.   Pl. 143

Ixora coccinea

Pennsylvania Hort. Soc.
The Green Scene
vol 2, No. 3   1974

fl.   cover

Ixora, Flame

see

Ixora macrothyrsa

Ixora griffithiana

de Wit, H. C. D.
Plants of the World; The Higher Plants Vol II

fl.   p 108, Pl. 67, 68

Ixora hybrid

Kiaer, Eigil
Indoor Plants in Colour

fl.   P. 73

Ixora javanica

Bruggeman, L.
Tropical Plants

fl.   Pl. 248

Ixora macrothyrsa

Everett, T. H.
New Illustrated Encyclopedia of
    Gardening, V. 6

fl.   opp. p. 934

Ixora macrothyrsa

Hargreaves, Dorothy
Tropical Blossoms of the Caribbean

fl.,hab.   p. 22

Ixora species

Hall, Clarence E.
Flowers of the Islands in the Sun

fl.   p. 83   Pl. 17

Ixora sp.

Kromdijk, G.
200 House Plants in Colour

fl.   Pl. 117

Ixora stricta

Bruggeman, L.
Tropical Plants

fl.   Pl. 249

Ixora williamsii

Curtis's Botanical Magazine, v. 173
1960

fl.   Pl. 352

Jaborandi

see

Pilocarpus pennatifolius

Jacamilla

see

Adenoropium berlandieri

Jacaranda acutifolia

Everett, Thomas H.
Living Trees of the World

hab.,fl.   p. 296

Jacaranda acutifolia

Flemer, William
Shade and Ornamental Trees in Color

fl.   p. 94

Jacaranda acutifolia

Hargreaves, Dorothy
Tropical Blossoms of the Caribbean

fl.   p. 49

Jacaranda acutifolia

Oakman, Harry
Colorful Trees

Hab. Fl.   P 44

Jacaranda acutifolia

O'Gorman, Helen
Mexican Flowering Trees and Plants

fl. fr. hab.   p 33

Jacaranda acutifolia

Royal Hort. Soc.
Journal of the Royal Hort. Soc.
vol 98, No. 6   1973

hab., fl.   p. 260   Pl. 148

Jacaranda cuspidifolia

Hvass, Elsie
Plants That Feed and Serve Us

fl.   p 128, Pl. 284

Jacaranda filicifolia

Bruggeman, L.
Tropical Plants

fl.   Pl. 211

Jacaranda filicifolia

Pertchik, Bernard
Flowering Trees of the
    Caribbean

fl.   p 111

Jacaranda mimosifolia

Edlin, Herbert
The Illustrated Encyclopedia
    of Trees

fl.   p 225

Jacaranda mimosifolia

Macoby, Stirling
What Flower is That

fl.   p. 169

Jacaranda mimosifolia

Mathias, Mildred E.
Color for the Landscape

fl. hab.   p 22

Jacaranda mimosifolia

Perry, Frances
Flowers of the World

fl.   p 44

Jacaranda, Rhodesian

see

Stereospermum kunthianum

Jack-by-the-hedge

see

Alliaria petiolata

Jack-go-to-bed-at-noon

see

Tragopogon pratensis

Jack-in-the-pulpit

see

Arisaema atrorubens

Jack-in-the-pulpit

see

Arisaema triphyllum

Jack-in-the-pulpit, Stewardson's

see

Arisaema stewardsonii

Jack-in-the-Pulpit, Zebra-Striped

see

Arisaema atrorubens var.

Jackscrew Root

see

Shortia galacifolia

Jacksonia scoparia

Blombery, Alec M.
What Wildflower is That

fl.   p 179, Pl. 481

Jacobinia aurea

American Hort. Soc.
American Horticulturist
   vol 54, No. 4 1975

fl.   p. 29

Jacobinia carnea

Everett, T.H.
New Illustrated Encyclopedia of
   Gardening, v. 6

fl.   opp. p. 966

Jacobinia carnea

Harrison, R. E.
Trees and Shrubs

fl.   P. 96

Jacobinia carnea

Hay, Roy
The Dictionary of House Plants

fl.   Pl. 300

Jacobinia carnea

Macoby, Stirling
What Flower is That

fl.   p. 169

Jacobinia carnea

Mathias, Mildred E.
Color for the Landscape

fl. hab.   p 72

Jacobinia carnea

Tsukamoto, Yotaro
Coloured Illustrations of Garden
   Flowers, v. 10

fl.   opp. p. 53   ill. 169

Jacobinia carnea

Wit, H. C. D. de
Plants of the World;
   The Higher Plants, Vol II

fl.   Pl. 98, 99

Jacobinia coccinea

Hay, Roy
The Dictionary of House Plants

fl.   Pl. 301

Jacobinia pauciflora

Harrison, R. E.
Trees and Shrubs

hab.   P. 96

Jacobinia pauciflora

Hay, Roy
The Dictionary of House Plants

fl.   Pl. 302

Jacobinia pauciflora

Kiaer, Eigil
Indoor Plants in Colour

fl.   p. 73

Jacobinia spicigera

Addisonia, v. 20, 1937-38

fl.   opp. p. 51   Pl. 666

Jacobinia suberecta

Hay, Roy
The Dictionary of House Plants

fl.   Pl. 303

Jacobinia suberecta

Hersey, Jean
Woman's Day Book of House Plants
fl.
p. 63

Jacobinia umbrosa

Chickering, Carol Rogers
Flowers of Guatemala

fl.   p 75

Jacobinia umbrosa

O'Gorman, Helen
Mexican Flowering Trees and Plants

Jacob's Ladder

see

Polemonium

Jacob's Ladder, Blue

see

Polemonium caeruleum

Jacob's Ladder, Common

see

Polemonium reptans

Jacob's-ladder, Elegant

see

Polemonium elegans

Jacob's Ladder, Marsh

see

Polemonium van bruntiae

Jacob's-ladder, Showy

see

Polemonium pulcherrimum

Jacob's Ladder, Tall

see

Polemonium acutiflorum

Jacobsenia kolbei

Herre, H.
The Genera of the Mesembryanthemaceae

fl.   fr.   p 181

Jacquemontia corymbulosa

Curtis's Botanical Magazine
von 167   1950

fl.   Pl. 102

Jacquemontia pentantha

Menninger, Edwin A.
Flowering Vines of the World

fl.   Pl. 63

Jacquemontia tamnifolia

Brown, Clair A.
Wildflowers of Louisiana and
   Adjoining States

fl. hab.   p 150

Jacquemontia tamnifolia

Duncan, Wilbur H.
Wildflowers of the Southeastern
   United States

fl.   p 137

Jacquemontia tamnifolia

Weeds of the Southern United States

fl.   p. 18

Jacquinia aurantica

O'Gorman, Helen
Mexican Flowering Trees and Plants

fl. hab.   p 69

Jacquinia, orange-flowered

see

Jacquinia aurantiaca

Jade Plant

see

Crassula arborescens

Jade Plant

see

Crassula portulacea

Jade plant

see

Portulacaria afra

Jade Vine

see

Strongylodon macrobotrys

Jakfruit

see

Artocarpus

Jamberry

see

Physalis ixocarpa

Jankaea heldreichii

Hay, Roy
The Color Dictionary of Flowers & Plants

fl.   p 12, Pl. 96

Jankaea heldreichii

Royal Hort. Soc.
Journal of the Royal Hort. Soc.
vol 89, No. 1   1964

fl.   p. 22   Pl. 16

Japonica, white

see

Chaenomeles lagenaria var.

Jasione montana

Ary, S.
The Oxford Book of Wildflowers

fl.   p 178, Pl. 3

Jasione montana

Kleijn, H.
Beauty of the Wild Plant

fl.   opp. p. 56   Pl. 68

Jasione montana

Lindman, C. A. M.
Nordens Flora, Vol 9

fl. fr.   Pl. 572

Jasione montana

Martin, W. Keble
The Concise British Flora in Colour

fl. hab.   Pl. 54

Jasione montana

Polunin, Oleg
Flowers of Europe

fl.   Pl. 141 #1355

Jasmine

see

Jasminum

Jasmine, Cape

see

Gardenia jasminoides

Jasmine, Carolina

see

Gelsemium sempervirens

Jasmine, Chilean

see

Mandevilla laxa

Jasmine, Chilean

see

Mandevilla suaveolens

Jasmine, Chinese

see

Jasminum mesnyi

Jasmine, crape

see

Ervatamia coronaria

Jasmine, False

see

Clerodendrum inerme

Jasmine, king

see

Jasminum rex

Jasmine, Large-flowered

see

Jasminum grandiflorum

Jasmine, Madagascar

see

Stephanotis floribunda

Jasmine, primrose

see

Jasminum mesnyi

Jasmine, Red

see

Plumeria rubra

Jasmine, Rock

see

Androsace chamaejasme

Jasmine, Rock

see

Androsace sarmentosa

Jasmine, Star

see

Trachelospermum jasminoides

Jasmine tree

see

Holarrhena febrifuga

Jasmine, Winter

see

Jasminum nudiflorum

Jasminum azoricum

Harrison, Richmond E.
Climbers and Trailers

fl.   p 53   Pl. 113

Jasminum azoricum

Macoby, Stirling
What Flower is That

fl.   P 170

Jasminum dispermum

Curtis's Botanical Magazine
Vol 162   1939

fl.   No. 9567

Jasminum floridum

Huxley, Anthony
Evergreen Garden Trees and Shrubs

fl.   Pl. 185

Jasminum fruticans

Polunin, Oleg
Flowers of Europe

fl.   Pl. 95 #983

Jasminum fruticans

Polunin, Oleg
Trees and Bushes of Europe

fl.   p 163

Jasminum grandiflorum

Hvass, Elsie
Plants That Feed and Serve Us

fl.   p 116, Pl. 253

Jasminum grandiflorum

Perrot, Emile
Les Plantes Medicinales

fl. fr. hab.   p 127

Jasminum humile var.

Hellyer, A.G.L.
Shrubs in Colour

fl.   p. 70

Jasminum kedahense

Curtis's Botanical Magazine
Vol 177   1969-70

fl.   547

Jasminum mesnyi

Crockett, James Underwood
Evergreens

fl.   p. 132

Jasminum mesnyi

Harrison, Richmond E.
Climbers and Trailers

fl.   p 53   Pl. 114

Jasminum mesnyi

Harrison, R.E.
Trees and Shrubs

fl. P 95

Jasminum mesnyi

Hay, Roy
The Dictionary of House Plants

fl.   Pl. 304

Jasminum mesnyi

Kiaer, Eigil
Indoor Plants in Colour

fl.   p. 73

Jasminum mesnyi

Macoby, Stirling
What Flower is That

fl.   p. 170

Jasminum meyeri-johannis

Moriarty, Audrey
Wild Flowers of Malawi

fl.   Pl. 73; 4

Jasminum multipartitum

Flowering Plants of Africa
vol 32   1957-58

fl., fr.   Pl. 1272

Jasminum nitidum

Harrison, Richmond E.
Climbers and Trailers

fl.   p. 54   Pl. 115-116

Jasminum nudiflorum

Bartels, Andreas
Das Grosse Buch der Gartengeholze

fl.   p 162

Jasminum nudiflorum

Hay, Roy
The Color Dictionary of Flowers & Plants

fl.   p 208, Pl. 1664

Jasminum nudiflorum

Huxley, Anthony
Deciduous Garden Trees and Shrubs

fl.   Pl. 291

Jasminum nudiflorum

Macoby, Stirling
What Flower is That

fl.   P 170

Jasminum nudiflorum

Massachusetts Hort. Soc.
Horticulture
vol 55, No. 1   1977

fl.   p. 36

Jasminum nudiflorum

Perry, Frances
Flowers of the World

fl.   p 199

Jasminum nudiflorum

Polunin, Oleg
Trees and Bushes of Europe

fl.   p 164

Jasminum officinale

Perry, Frances
Flowers of the World

fl.   p 199

Jasminum officinale

Polunin, Oleg
Flowers of Europe

fl.   Pl. 95 #984

Jasminum officinale

Polunin, Oleg
Trees and Bushes of Europe

fl. hab.   p 164

Jasminum polyanthum

Curtis's Botanical Magazine
    vol 161   1938-39

fl.   Pl. 9545

Jasminum polyanthum

Harrison, Richmond E.
Climbers and Trailers

fl.   p 54   Pl. 117

Jasminum polyanthum

Hay, Roy
The Color Dictionary of Flowers & Plants

fl.   p 248, Pl. 1978

Jasminum polyanthum

Macoby, Stirling
What Flower is That

fl.   P 170

Jasminum polyanthum

Morley, Brian D.
Wild Flowers of the World

fl.fr.   Plate 95

Jasminum polyanthum

Nicolaisen, Age
Pocket Encyclopedia of Indoor Plants

fl.   Pl. 133

Jasminum polyanthum

Royal Hort. Soc.
Journal of the Royal Hort. Soc.
vol 91, No.10   1966

fl.   p. 426   Pl. 208

Jasminum revolutum

Hay, Roy
The Color Dictionary of Flowers & Plants

fl.   p 248, Pl. 1979

Jasminum rex

Macoby, Stirling
What Flower is That

fl. P 170

Jasminum rex

Morley, Brian D.
Wild Flowers of the World

fl.   Plate 121

Jasminum rex

Perry, Frances
Flowers of the World

fl.   p 199

Jasminum sp.

Kromdijk, G.
200 House Plants in Colour

fl.   Pl. 118

Jasminum x stephanense

Gault, S. Millar
The Color Dictionary of Shrubs

fl.   Pl. 259

Jasminum unnanensis

Tsukamoto, Yotaro
Coloured Illustrations of Garden
Flowers, v.10
fl.
ill. 170 opp.p.53

Jasminum urophyllum var.

Curtis's Botanical Magazine
Vol 168   1951

fl.   148

Jasminum vanprukii

Bruggeman, L.
Tropical Plants

fl.   Pl. 16

Jatropha cinerea

Coyle, Jeanette
A Field Guide to the Common and
    Interesting Plants of Baja California

fl. fr.   p 111

Jatropha cuneata

Coyle, Jeanette
A Field Guide to the Common and
    Interesting Plants of Baja California

fr.   p 111

Jatropha integerrima

Morley, Brian D.
Wild Flowers of the World

fl.   Pl. 178

Jatropha multifida

Bruggeman, L.
Tropical Plants

fl.   Pl. 250

Jatropha pandurifolia

Bruggeman, L.
Tropical Plants

fl.   Pl. 251

Jatropha podagrica

Cactus & Succulent Society of America
Journal
    vol 48, No. 6   1976

hab.   cover

Jatropha podagrica

Perry, Frances
Flowers of the World

fl.   p 116

Jeffersonia, Barton's

    see

Jeffersonia diphylla

Jeffersonia diphylla

Courtenay, Booth
Wildflowers and weeds

fl. p. 20

Jeffersonia diphylla

Dean, Blanche E.
Wildflowers of Alabama and
    Adjoining States

fl.   p 69

Jeffersonia diphylla

Duncan, Wilbur H.
Wildflowers of the Southeastern
    United States

fl. hab.   p 49

Jeffersonia diphylla

Jennings, O. E.
Wild Flowers of Western Pennsylvania
    V. 2

fl. fr. hab.   Pl. 66

Jeffersonia diphylla

Massachusetts Hort. Soc.
Horticulture
    vol 54, No. 6   1976

fl., fr., hab.   cover

Jeffersonia diphylla

Walcott, Mary Vaux
North American Wild Flowers

fl. hab.   Pl. 72

Jeffersonia diphylla

Wharton, Mary E.
A Guide to the Wildflowers & Ferns
    of Kentucky

fl.   p 112 Pl. 1.36

Jeffersonia dubia

Curtis's Botanical Magazine
Vol 164   1943-48

fl. hab.   No. 9681

Jeffersonia dubia

Hay, Roy
The Color Dictionary of Flowers & Plants

fl.   p 13, Pl. 97

Jeffersonia dubia

Massachusetts Hort. Soc.
Horticulture
vol 54, No. 6   1976

fl., hab.   backcover

Jeffersonia dubia

Perry, Frances
Flowers of the World

fl.   p 233

Jelly beans
see
Sedum rubrotinctum

Jenny, Creeping
see
Lysimachia nummularia

Jessamine, carolina yellow
see
Gelsemium sempervivens

Jessamine, night
see
Cestrum nocturnum

Jessamine, Night-Blooming
see
Cestrum parqui

Jessamine, orange
see
Murraya exotica

Jessamine, Willow-Leaved
see
Cestrum parqui

Jessamine, yellow
see
Gelsemium rankinii

Jetbead
see
Rhodotypos kerrioides

Jew, Wandering
see
Cyanotis veldthoutiana

Jew, Wandering
see
Tradescantia

Jew, Wandering
see
Zebrina

Jewel Flower, Mexican
see
Tacitus bellus

Jewelweed
see
Impatiens

Jewelweed, Pale
see
Impatiens pallida

Jewelweed, spotted
see
Impatiens biflora

Jewelweed, Spotted
see
Impatiens capensis

Jimson weed
see
Datura discolor

Jimson Weed
see
Datura meteloides

Jimson Weed
see
Datura stramonium

Job's tears
see
Coix lacryma-jobi

Joe-Pye-Weed
see
Eupatorium

Joe-Pye Weed, Hollow
see
Eupatorium fistulosum

Joe-Pye Weed, Purple-Stemmed
see
Eupatorium fistulosum

Johnny jump-up
see
Viola tricolor

Johnny tuck
see
Orthocarpus tenuifolius

Johnsongrass
see
Sorghum halepense

Johnsonia lupina

Blombery, Alec M.
What Wildflower is That

fl.   p 179, Pl. 482

Jointweed
see
Polygonella articulata

Jointweed

see

Polygonella polygama

Jonquil

see

Narcissus jonquilla

Joseph's Coat

see

Amaranthus gangeticus var.

Joseph's Coat

see

Amaranthus tricolor

Joshua tree

see

Yucca brevifolia

Jovellana violacea

Hay, Roy
The Dictionary of House Plants

fl.   Pl. 305

Juanulloa aurantiaca

Perry, Frances
Flowers of the World

fl.   p 287

Juanulloa mexicana

Morley, Brian D.
Wild Flowers of the World

fl.   Pl. 183

Jubaea spectabilis

Huxley, Anthony
Evergreen Garden Trees and Shrubs

hab.   Pl. 193

Jubaea spectabilis

Polunin, Oleg
Trees and Bushes of Europe

hab.   p. 184

Judas Tree

see

Cercis canadensis

Judas tree

see

Cercis siliquastrum

Jug Plant

see

Hexastylis arifolia

Juglans ailantifolia var.

Bartels, Andreas
Das Grosse Buch der Gartengeholze

fl.   p 163

Juglans cinerea

Edlin, Herbert
The Illustrated Encyclopedia
of Trees

fr.   p 137

Juglans cinerea

Masefield, G. B.
The Oxford Book of Food Plants

fr.   Pl. 29, 3

Juglans cinerea

Massachusetts Hort. Soc.
Horticulture
    vol 55, No. 12   1977

fl., hab.   p. 32

Juglans cinerea

Wharton, Mary E.
Trees & Shrubs of Kentucky

fl.   p 100, Pl. 3.1

Juglans cordiformis

Wit, H. C. D. de
Plants of the World;
    The Higher Plants, Vol I

fl.   p 109, Pl. 65

Juglans mandschurica

Kitamura, Siro
Coloured Illustrations of Trees &
    Shrubs of Japan

fl. fr.   89, 90

Juglans nigra

Boom, B. K.
The glory of the tree

fl.   p.57, 60   Pl. 86, 88

Juglans nigra

Edlin, Herbert
The Illustrated Encyclopedia
of Trees

fr. hab.   p 137

Juglans nigra

Masefield, G. B.
The Oxford Book of Food Plants

fr.   Pl. 29, 2

Juglans nigra

Polunin, Oleg
Trees and Bushes of Europe

fr. hab.   p 42

Juglans regia

Bianchini, F.
The Complete Book of Fruits & Vegetables

fl. fr.   p 193

Juglans regia

Bianchini, Francesco
Health Plants of the World

fr. hab.   p 157

Juglans regia

Boom, B. K.
The glory of the tree

fr.   opp. p. 60   Pl. 90

Juglans regia

Edlin, Herbert
The Illustrated Encyclopedia
of Trees

fl. fr. hab.   p 136-7

Juglans regia

Huxley, Anthony
Deciduous Garden Trees and Shrubs

fr.   Pl. 95, 95a

Juglans regia

Hvass, Elsie
Plants That Feed and Serve Us

fl. fr.   p 20, Pl. 30

Juglans regia

Masefield, G. B.
The Oxford Book of Food Plants

fl. fr.   Pl. 29, 1

Juglans regia

Perrot, Emile
Les Plantes Medicinales

fl. fr. hab.   p 162

Juglans regia

Polunin, Oleg
Flowers of Europe

fl.   Pl. 4 #35

Juglans regia

Polunin, Oleg
Trees and Bushes of Europe

fl. fr. hab.   p 41, 42

Jujube

see

Ziziphus jujuba

Jujube, Christ Thorn

see

Zizyphus spina-Christi

Jumellea comorensis

Stewart, Joyce
Orchids of Tropical Africa

fl. hab.   Pl. 31

Jump Seed

see

Polygonum virginianum

Juncus acutiflorus

Martin, W. Keble
The Concise British Flora in Colour

fr. hab.   Pl. 87

Juncus acutus

Feinbrun-Dothan, Naomi
Wild Plants in the Land
of Israel

fr.   p 172

Juncus acutus

Martin, W. Keble
The Concise British Flora in Colour

fr. hab.   Pl. 87

Juncus acutus

Polunin, Oleg
Flowers of Europe

fr.   Pl. 177 #1705

Juncus albescens

Porsild, A. E.
Rocky Mountain Wild Flowers

fl. hab.   p 87

Juncus antarcticus

Morley, Brian D.
Wild Flowers of the World

fr.   Pl. 7C

Juncus arcticus

Lindman, C. A. M.
Nordens Flora, Vol 1

fl. fr. hab.   Pl. 63E

Juncus articulatus

Martin, W. Keble
The Concise British Flora in Colour

fr. hab.   Pl. 87

Juncus articulatus

Polunin, Oleg
Flowers of Europe

fl. Pl. 177 #1707

Juncus balticus

Martin, W. Keble
The Concise British Flora in Colour

fr. hab.   Pl. 86

Juncus bufonius

Martin, W. Keble
The Concise British Flora in Colour

fr. hab.   Pl. 86

Juncus bulbosus

Martin, W. Keble
The Concise British Flora in Colour

fr. hab.   Pl. 87

Juncus capitatus

Martin, W. Keble
The Concise British Flora in Colour

fr. hab.   Pl. 87

Juncus castaneus

Martin, W. Keble
The Concise British Flora in Colour

fr. hab.   Pl. 87

Juncus castaneus

Porsild, A. E.
Rocky Mountain Wild Flowers

fr. hab.   p 87

Juncus compressus

Martin, W. Keble
The Concise British Flora in Colour

fr. hab.   Pl. 86

Juncus conglomeratus

Lindman, C. A. M.
Nordens Flora, Vol 1

fl.   Pl. 63B

Juncus conglomeratus

Martin, W. Keble
The Concise British Flora in Colour

fl. hab.   Pl. 86

Juncus drummondii

Porsild, A. E.
Rocky Mountain Wild Flowers

fl. hab.   p 89

Juncus effusus

Lindman, C. A. M.
Nordens Flora, Vol 1

fl. fr. hab.   Pl. 63 A

Juncus effusus

Martin, W. Keble
The Concise British Flora in Colour

fr. hab.   Pl. 86

Juncus effusus

Polunin, Oleg
Flowers of Europe

fr.   Pl. 177 #1699

Juncus effusus

Tarver, David P.
Aquatic and Wetland Plants
of Florida

fl. hab.   p 99

Juncus effusus

Wharton, Mary E.
A Guide to the Wildflowers & Ferns
of Kentucky

fl.   p 87 Pl. 2.9

Juncus filiformis

Lindman, C. A. M.
Nordens Flora, Vol 1

fl. hab.   Pl. 63D

Juncus filiformis

Martin, W. Keble
The Concise British Flora in Colour

fr. hab.   Pl. 86

Juncus gerardii

Martin, W. Keble
The Concise British Flora in Colour

fr. hab.   Pl. 86

Juncus inflexus

Lindman, C. A. M.
Nordens Flora, Vol 1

fl.   Pl. 63C

Juncus inflexus

Martin, W. Keble
The Concise British Flora in Colour

fr.   Pl. 86

Juncus inflexus

Polunin, Oleg
Flowers of Europe

fl.   Pl. 177 #1698

Juncus jacquinii

Polunin, Oleg
Flowers of Europe

fr. Pl. 177 #1701

Juncus kochii

Martin, W. Keble
The Concise British Flora in Colour

fr. hab.   Pl. 87

Juncus maritimus

Martin, W. Keble
The Concise British Flora in Colour

fr. hab.   Pl. 87

Juncus Mertensianus

**Porsild, A. E.
Rocky Mountain Wild Flowers**

fl. hab.   p 91

Juncus mutabilis

Martin, W. Keble
The Concise British Flora in Colour

fr. hab.   Pl. 87

Juncus roemerianus

**Duncan, Wilbur H.
Wildflowers of the Southeastern
United States**

fl.   p 245

Juncus roemerianus

**Tarver, David P.
Aquatic and Wetland Plants
of Florida**

fl. hab.   p 100

Juncus sp.

Coyle, Jeanette
A Field Guide to the Common and
Interesting Plants of Baja California

hab.   p 53

Juncus squarrosus

Martin, W. Keble
The Concise British Flora in Colour

fr. hab.   Pl. 86

Juncus squarrosus

Polunin, Oleg
Flowers of Europe

fl.   Pl. 177 #1702

Juncus subnodulosus

Martin, W. Keble
The Concise British Flora in Colour

fr. hab.   Pl. 87

Juncus subuliflorus

Polunin, Oleg
Flowers of Europe

fr.   Pl. 177 #1700

Juncus tenuis

Martin, W. Keble
The Concise British Flora in Colour

fr. hab.   Pl. 86

Juncus trifidus

Martin, W. Keble
The Concise British Flora in Colour

fr. hab.   Pl. 86

Juncus triglumis

Martin, W. Keble
The Concise British Flora in Colour

fr. hab.   Pl. 87

Juneberry

see

Amelanchier alnifolia

Juneberry

see

Amelanchier canadensis

Juneberry, Common

see

Amelanchier arborea

Juniper

see

Juniperus

Juniper, Chinese

see

Juniperus chinensis

Juniper, Common

see

Juniperus communis

Juniper, creeping

see

Juniperis horizontalis

Juniper, Creeping

see

Juniperus media var.

Juniper, Greek

see

Juniperus excelsa

Juniper, Irish

see

Juniperus communis var.

Juniper, Mountain

see

Juniperus sibirica

Juniper, Phoenician

see

Juniperus phoenicea

Juniper, Prickly

see

Juniperus oxycedrus

Juniper, prostrate

see

Juniperus horizontalis

Juniper, Savin

see

Juniperus sabina

Juniper, Spanish

see

Juniperus sabina var.

Juniper, Swedish

see

Juniperus communis var.

Juniper, Syrian

see

Juniperus drupacea

Juniper, tamarix

see

Juniperus sabina **var.**

Juniper tree

see

Juniperus oxycedrus

Juniper, Utah

see

Juniperus osteosperma

Juniper, Wilton Carpet

see

Juniperus horizontalis **var.**

Juniperus cedrus

Bramwell, David
Wild Flowers of the Canary Islands

fr.   Pl. 124

Juniperus chinensis

Edlin, Herbert
The Illustrated Encyclopedia
of Trees

fr. hab.   p 93

Juniperus chinensis

Huxley, Anthony
Evergreen Garden Trees and Shrubs

hab.   Pl. 27

Juniperus chinensis

Kitamura, Siro
Coloured Illustrations of Trees &
Shrubs of Japan

hab.   71

Juniperus chinensis

Wit, H. C. D. de
Plants of the World;
The Higher Plants, Vol I

fr.   p 55, Pl. 15

Juniperus chinensis var.
Crockett, James Underwood
Evergreens

hab.   p. 96, 97

Juniperus chinensis var.

Everett, T. H.
New Illustrated Encyclopedia of
Gardening, v. 6

hab.   opp. p. 935

Juniperus chinensis var.

Gault, S. Millar
The Color Dictionary of Shrubs

hab.   Pl. 265

Juniperus chinensis var.

Hansen, Richard
Baume und Straucher im Garten

hab.   p. 190

Juniperus chinensis var.

Harrison, R. E.
Trees and Shrubs

hab.   Pl. 574, 577

Juniperus communis

Ary, S.
The Oxford Book of Wildflowers

fl. fr.   p 184, Pl. 4

Juniperus communis

Bianchini, F.
The Complete Book of Fruits & Vegetables

fr.   p 207

Juniperus communis

Bianchini, Francesco
Health Plants of the World

fr.   p 137

Juniperus communis

Edlin, Herbert
The Illustrated Encyclopedia
of Trees

fr. hab.   p. 92

Juniperus communis

Heller, Christine
Wild Flowers of Alaska

fr.   Pl. 280

Juniperus communis

Lindman, C. A. M.
Nordens Flora, Vol. 1

fr.   Pl. 26

Juniperus communis

Loewenfeld, Claire
The Complete Book of Herbs and Spices

fr. hab.   p 209

Juniperus communis

Masefield, G. B.
The Oxford Book of Food Plants

fr., hab.   Pl. 137, 2

Juniperus communis

Perrot, Emile
Les Plantes Medicinales

fl. fr. hab.   p 106

Juniperus communis

Polunin, Oleg
Flowers of Europe

fr.   Pl. 2 #12

Juniperus communis

Polunin, Oleg
Flowers of the Mediterranean

fr.   No. 1

Juniperus communis

Polunin, Oleg
Trees and Bushes of Europe

fr. hab.   p 26, 27

Juniperus communis

Vedel, H.
Arbres et Arbustes

fr. hab.   p 35, No. 23

Juniperus communis var.

Everett, T.H.
New Illustrated Encyclopedia of
Gardening, v.6
hab.
opp.p.935

Juniperus communis var.

Gault, S. Millar
The Color Dictionary of Shrubs

hab.   Pl. 260, 261

Juniperus communis var.

Hansen, Richard
Baume und Straucher im Garten

hab.   p. 190

Juniperus communis var.

Harrison, Richmond E.
Climbers and Trailers

hab. p 95   Pl. 238

Juniperus communis var.

Harrison, R. E.
Trees and Shrubs

hab.   Pl. 575

Juniperus communis var.

Hay, Roy
The Color Dictionary of Flowers & Plants

hab.   p 252, 253, Pl. 2015 thru 2017

Juniperus communis var.

Huxley, Anthony
Evergreen Garden Trees and Shrubs

fr. hab.   Pl. 31, 32, 33, 34

Juniperus communis var.

Kitamura, Siro
Coloured Illustrations of Trees &
   Shrubs of Japan

hab.   70

Juniperus communis var.

Porsild, A. E.
Rocky Mountain Wild Flowers

fr.   p 33

Juniperus communis var.

Tosco, Uberto
The World of Mountain Flowers

hab.   p 44, 57

Juniperus conferta

Gault, S. Millar
The Color Dictionary of Shrubs

hab.   Pl. 262

Juniperus conferta

Kitamura, Siro
Coloured Illustrations of Trees &
   Shrubs of Japan

fr.   69

Juniperus coxii var.

Harrison, R. E.
Trees and Shrubs

hab.   Pl. 576

Juniperus drupacea

Polunin, Oleg
Trees and Bushes of Europe

fr. hab.   p 27

Juniperus excelsa

Polunin, Oleg
Trees and Bushes of Europe

fr. hab.   p 29

Juniperus excelsa var.

Huxley, Anthony
Evergreen Garden Trees and Shrubs

hab.   Pl. 42

Juniperus foetidissima

Polunin, Oleg
Trees and Bushes of Europe

fr. hab.   p 28, 29

Juniperus horizontalis

Macoby, Stirling
What Flower is That

hab P 171

Juniperus horizontalis

Porsild, A. E.
Rocky Mountain Wild Flowers

fr.   p 35

Juniperus horizontalis

Walcott, Mary Vaux
North American Wild Flowers
vol. 5

fr.   Pl. 379

Juniperus horizontalis var.

American Hort. Soc.
American Horticulturist
vol 56, No. 5 1977

hab.   P. 27

Juniperus horizontalis var.

Crockett, James Underwood
Evergreens

hab.   p. 98

Juniperus horizontalis var.

Crockett, James Underwood
Lawns & Ground Covers

hab p. 137

Juniperus horizontalis var.

Gault, S. Millar
The Color Dictionary of Shrubs

hab.   Pl. 264

Juniperus horizontalis var.

Harrison, Richmond E.
Climbers and Trailers

hab.   P. 95-96   Pl. 239-240

Juniperus horizontalis var.

Huxley, Anthony
Evergreen Garden Trees and Shrubs

hab.   Pl. 43

Juniperus hybrid

Gault, S. Millar
The Color Dictionary of Shrubs

hab.   Pl. 263

Juniperus hybrid

Huxley, Anthony
Evergreen Garden Trees and Shrubs

hab.   Pl. 41

Juniperus x media var.

Hay, Roy
The Color Dictionary of Flowers &
   Plants

hab.   p. 253   Pl. 2019

Juniperus x media var.

Huxley, Anthony
Evergreen Garden Trees and Shrubs

hab.   Pl.28,29,30

Juniperus osteosperma

Royal Hort. Soc.
Journal of the Royal Hort. Soc.
vol 95, No. 5   1970

fr.   p. 214   Pl. 124

Juniperus oxycedrus

Color Treasury of Herbs & Other
   Medicinal Plants

fr.   p 21   Pl. 9

Juniperus oxycedrus

Perrot, Emile
Les Plantes Medicinales

fl. fr. hab.   p 107

Juniperus oxycedrus

Polunin, Oleg
Flowers of Europe

fr.   Pl. 2 #13

Juniperus oxycedrus

Polunin, Oleg
Trees and Bushes of Europe

fr. hab.   p 27

Juniperus phoenicea

Braswell, David
Wild Flowers of the Canary Islands

hab.   Pl. 125

Juniperis phoenicea

Polunin, Oleg
Flowers of Europe

fr. Pl. 2 #14

Juniperus phoenicea

Polunin, Oleg
Flowers of the Mediterranean

fr.   No. 2

Juniperus phoenicea

Polunin, Oleg
Trees and Bushes of Europe

fr. hab.   p 28

Juniperus procumbens

Harrison, Richmond E.
Climbers and Trailers

hab.   p 96   Pl. 241

Juniperus procumbens var.

American Hort. Soc.
American Horticulturist
    vol 56, No. 5 1977

hab.   p. 27

Juniperus recurva var.

Hay, Roy
The Color Dictionary of Flowers & Plants

hab.   p 253, Pl. 2020

Juniperus rigida

Kitamura, Siro
Coloured Illustrations of Trees &
    Shrubs of Japan

fr.   68

Juniperus sabina

Bianchini, Francesco
Health Plants of the World

fr.   p 151

Juniperus sabina

Huxley, Anthony
Evergreen Garden Trees and Shrubs

hab.   Pl. 35

Juniperus sabina

Perrot, Emile
Les Plantes Medicinales

fl. fr. hab.   p 209

Juniperus sabina

Polunin, Oleg
Trees and Bushes of Europe

fr.   p 28

Juniperus sabina var.

American Hort. Soc.
American Horticulturist
    vol 56, No. 5 1977

hab.   p. 27

Juniperus sabina var.

Crockett, James Underwood
Evergreens

hab.   p. 98

Juniperus sabina var.

Huxley, Anthony
Evergreen Garden Trees and Shrubs

hab.   Pl. 36, 37

Juniperus scopulorum

Flemer, William
Shade and Ornamental Trees in Color
hab.
p. 85 .

Juniperus sibirica

Walcott, Mary Vaux
North American Wild Flowers

fr. hab.   Pl. 86

Juniperus squamata var.

Harrison, R.E.
Trees and Shrubs

hab.   Pl. 578

Juniperus squamata var.

Hay, Roy
The Color Dictionary of Flowers & Plants

hab.   p 253, Pl. 2021

Juniperus squamata var.

Huxley, Anthony
Evergreen Garden Trees and Shrubs

hab.   Pl. 38

Juniperus virginiana

Boom, B. K.
The glory of the tree

fr.   between p. 40 & 41   Pl. 57

Juniperus virginiana

Edlin, Herbert
The Illustrated Encyclopedia
    of Trees

fr. hab.   p 92

Juniperus virginiana

Hvass, Elsie
Plants That Feed and Serve Us

fr.   p 120, Pl. 264

Juniperus virginiana

Polunin, Oleg
Trees and Bushes of Europe

hab.   p 29

Juniperus virginiana

Wharton, Mary E.
Trees & Shrubs of Kentucky

fr. hab.   p 145, Pl. 2.40

Juniperus virginiana var.

Crockett, James Underwood
Evergreens

hab. fr.   p. 99

Juniperus virginiana var.

Gault, S. Millar
The Color Dictionary of Shrubs

hab.   Pl. 267

Juniperus virginiana var.

Hansen, Richard
Baume und Straucher im Garten

hab.   p. 190

Juniperus virginiana var.

Huxley, Anthony
Evergreen Garden Trees and Shrubs

fr.,hab.   Pl.39, 40

Jupiter's Beard

see

Centranthus ruber

Jupiter's beard

see

Kentranthus ruber

Jupiter's Distaff

see

Salvia glutinosa

Jussiaea diffusa

Addisonia, v. 21, 1939-42

fl. fr.   opp. p. 35   Pl. 690

Jussiaea peruviana

Lemmon, Robert S.
Wildflowers of North America in
    Full Color

fl.   p. 263   Pl. 418

Jussiaea repens

Wharton, Mary E.
A Guide to the Wildflowers & Ferns
    of Kentucky

fl.   p 102 Pl. 1.17

Justicia americana

Courtenay, Booth
Wildflowers & Weeds

fl.   p 92

Justicia americana

Dean, Blanche E.
Wildflowers of Alabama and
    Adjoining States

fl.   p 173

Justicia americana

Duncan, Wilbur H.
Wildflowers of the Southeastern
    United States.

fl.   p. 187

Justicia americana

Jennings, O. E.
Wild Flowers of Western Pennsylvania
v. 2
fl. hab.   Pl. 148

Justicia americana

Klimas, John E.
Wildflowers of Eastern America

fl.   Pl. 259

Justicia americana

Wharton, Mary E.
A Guide to the Wildflowers & Ferns
    of Kentucky

fl.   p 210   Pl. 2.38

Justicia betonica

Hargreaves, Dorothy
Tropical Blossoms of the
    Caribbean

fl.   p. 34

Justicia coccinea

Macoby, Stirling
What Flower is That

fl.   P 171

Justicia incerta

Flowering Plants of Africa
    vol 31   1956

fl.   Pl. 1237

Justicia pedunculosa

Curtis's Botanical Magazine
    vol 50

fl.   Pl. 2367

Justicia peruviana

Macoby, Stirling
What Flower is That

fl.   P 171

Justicia procumbens

Walden, Beryl M.
Wild Flowers of Hong Kong

fl.   Pl. 55 (164)

Justicia, red

see

Justicia coccinea

Justicia sp.

Lind, E.M.
Some Common Flowering Plants of
    Uganda

fl.   Pl. 12a

Justicia striata

Moriarty, Audrey
Wild Flowers of Malawi

fl.   Pl. 42; 6

Jute

see

Corchorus capsularis

Juttadinteria elizae

Herre, H.
The Genera of the Mesembryanthemaceae

fl.   fr.   p 185

Juttadinteria simpsonii

Flowering Plants of Africa
    vol 32   1957-58

fl., hab.   Pl. 1273